Synthesis: Working Backwards

Granting that we know the chemistry of the individual steps, how do we go about planning a route to more complicated compounds—alcohols, say? In almost every organic synthesis it is best to begin with the molecule we want—the *target* molecule—and *work backwards* from it. There are relatively few ways to make a complicated alcohol, for example; there are relatively few ways to make a Grignard reagent or an aldehyde or ketone; and so on back to our ultimate starting materials. On the other hand, our starting materials can undergo so many different reactions that, if we go at the problem the other way around, we find a bewildering number of paths, few of which take us where we want to go.

We try to limit a synthesis to as few steps as possible, but nevertheless do not sacrifice purity for time. To avoid a rearrangement in the preparation of an alkene, for example, we take two steps via the halide rather than the single step of dehydration.

Table 29.1 WOODWARD-HOFFMAN RULES FOR ELECTROCYCLIC REACTIONS

Number of π electrons	Reaction	Motion
$4n$	thermal	conrotatory
$4n$	photochemical	disrotatory
$4n + 2$	thermal	disrotatory
$4n + 2$	photochemical	conrotatory

Table 29.2 WOODWARD-HOFFMAN RULES FOR $[i + j]$ CYCLOADDITIONS

$i + j$	Thermal	Photochemical
$4n$	supra-antara antara-supra	supra-supra antara-antara
$4n + 2$	supra-supra antara-antara	supra-antara antara-supra

Study Guide

to

ORGANIC CHEMISTRY

Third Edition

by

R. T. Morrison
R. N. Boyd

Allyn and Bacon, Inc.

Boston · London · Sydney · Toronto

Printing number and year (last digits):
10   —   80

Acknowledgments

Our thanks to Sadtler Research Laboratories for the infrared spectra labeled "Sadtler" and to the Infrared Data Committee of Japan for those labeled "IRDC," and to Dr. David Kritchevsky of the Wistar Institute for permission to quote the words of his song, "Farnesol."

To the Student

Right now, confronted with the array of unfamiliar material in your textbook, you must be wondering: *what am I expected to get out of all this?*

The best way to find out is to work problems: first, to see if you understand the facts and principles you have been reading about; second, and more important, to learn how to *use* this chemistry in the same practical ways that an organic chemist does.

Give yourself a fair chance to work each problem. Don't give up too easily. Re-read the pertinent part of the text. *Think* about it. Use paper and pencil and really work at it.

Only after you have done all that, check your answer against the one in this Study Guide. If you were on the right track, fine. If you went off the track, try to see *where*. Follow through the explanation carefully to see how you should approach this kind of problem next time

You must *learn* how to use your brand-new organic chemistry, and to do this you must push yourself. You must try to work difficult problems, and you will not always succeed. But you can learn from your failures as well as your successes.

With some answers, we have given references to the chemical literature. It is not necessary for you to read all of these papers—or even any of them. But if some topic catches your fancy, follow it up. And, in any case, read one—or two or three—of these papers, so that you can see in down-to-earth detail the kind of experimental work that underlies any science.

Robert Thornton Morrison
Robert Neilson Boyd

Note

Reference to a page in this Study Guide will be given as "page 000 of this Study Guide." Any other reference is understood to be to a page in *Organic Chemistry, Third Edition*, by R. T. Morrison and R. N. Boyd.

Chapter 1 | Structure and Properties

1.1 Ionic: a, e, f.

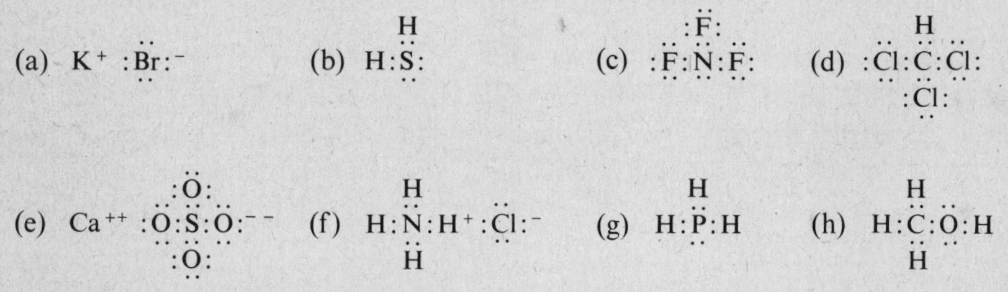

(a) K^+ $:\ddot{Br}:^-$

(b) $H:\ddot{S}:$ (with H above)
$\quad\quad H$
$H:\ddot{S}:$

(c) $:\ddot{F}:\ddot{N}:\ddot{F}:$ (with :F: above N)
$\quad\quad :\ddot{F}:$
$:\ddot{F}:\ddot{N}:\ddot{F}:$

(d) $:\ddot{Cl}:\ddot{C}:\ddot{Cl}:$ (with H above, Cl below)
$\quad\quad H$
$:\ddot{Cl}:\ddot{C}:\ddot{Cl}:$
$\quad\quad :\ddot{Cl}:$

(e) Ca^{++}
$\quad\quad :\ddot{O}:$
$:\ddot{O}:\ddot{S}:\ddot{O}:^{--}$
$\quad\quad :\ddot{O}:$

(f)
$\quad\quad H$
$H:\ddot{N}:H^+ :\ddot{Cl}:^-$
$\quad\quad H$

(g)
$\quad\quad H$
$H:\ddot{P}:H$
$\quad\quad H$

(h)
$\quad\quad H$
$H:\ddot{C}:\ddot{O}:H$
$\quad\quad H$

1.2 (a) $H:\ddot{O}:\ddot{O}:H$ (b) $:N:::N:$ (c) $H:\ddot{O}:\ddot{N}::\ddot{O}:$

$\quad\quad\quad\quad\quad\quad\quad\quad\quad\quad\quad\quad\quad\quad :\ddot{O}:$
(c) $H:\ddot{O}:\ddot{N}::\ddot{O}:$

$\quad\quad\quad\quad\quad\quad :\ddot{O}:$
(d) $:\ddot{O}:\ddot{N}::\ddot{O}:^-$

(e) $H:C:::N:$ (f) $:\ddot{O}::C::\ddot{O}:$

(g)
$H:\ddot{O}:C:\ddot{O}:H$
$\quad\quad :\ddot{O}:$

(h)
$\quad\quad H\ H$
$H:\ddot{C}:\ddot{C}:H$
$\quad\quad H\ H$

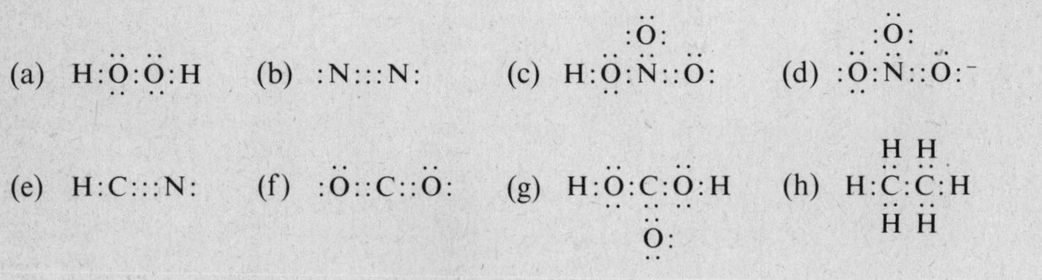

1.3 (a)

Na	2	8	1		
Mg	2	8	2		
Al	2	8	2	1	
Si	2	8	2	1	1
P	2	8	2	1 1 1	
S	2	8	2	2 1 1	
Cl	2	8	2	2 2 1	
Ar	2	8	2	2 2 2	

(b) Elements of same family have same electronic configuration for highest energy level.

(c) Metallic elements on left lose electrons to give 2,8 configuration; non-metallic elements on right gain electrons to give 2,8,8 configuration.

1.4 All are tetrahedral (sp^3) like CH_4, NH_3, or H_2O.

(a) Tetrahedral about N; (b) about O, with one 2e lobe; (c) about C and O, with two 2e lobes; (d) about C and N, with one 2e lobe.

1.5 Structure (a), not (b), since in (a) the dipoles would balance each other out.

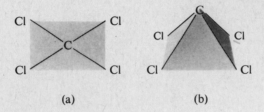

(a)   (b)

1.6 Linear (sp hybridization).

1.7 (a) We would expect zero, but in fact it is 1.46 D. Therefore molecule is not planar, and not sp^2-hybridized.

(b) If p-hybridized, molecule would be pyramidal, with unshared pair in $2s$ orbital symmetrical about N. We would expect NF_3 to have much larger dipole than NH_3, since there would be no opposing dipole from unshared pair. In fact, NF_3 has much smaller dipole than NH_3.

1.8 Associated: a, e, f.

$$CH_3-O\cdots H-O \qquad CH_3-\overset{\overset{\displaystyle H}{|}}{N}\cdots H-\overset{\overset{\displaystyle H}{|}}{N} \qquad CH_3-\overset{\overset{\displaystyle H}{|}}{N}\cdots H-\overset{\overset{\displaystyle CH_3}{|}}{N}$$
$$\underset{\displaystyle H}{|} \qquad \underset{\displaystyle CH_3}{|} \qquad\qquad \underset{\displaystyle H}{|} \qquad \underset{\displaystyle CH_3}{|} \qquad\qquad \underset{\displaystyle CH_3}{|} \qquad \underset{\displaystyle CH_3}{|}$$

1.9 (a) $CH_3OH > CH_3NH_2$ (b) $CH_3SH > CH_3OH$ (c) $H_3O^+ > NH_4^+$

1.10 (a) H_3O^+ (b) NH_4^+ (c) H_2S (d) H_2O (e) positive charge $\longrightarrow$ increased acidity.

1.11 (a) $CH_3^- > NH_2^- > OH^- > F^-$ (b) $NH_3 > H_2O > HF$ (c) $SH^- > Cl^-$

(d) $F^- > Cl^- > Br^- > I^-$ (e) $OH^- > SH^- > SeH^-$

1.12 $CH_3NH_2 > CH_3OH > CH_3F$

1.13 (a) $OH^- > H_2O > H_3O^+$ (b) $NH_2^- > NH_3$ (c) $S^{--} > HS^- > H_2S$
(d) negative charge $\longrightarrow$ increased basicity

1.14 $NH_3 > NF_3$. F withdraws electrons from N, and makes them less available for sharing.

1. Ionic: a, d, e, g.

(a) Mg^{++} $2\,:\!\ddot{C}l\!:^-$ (b) $:\!\ddot{C}l\!:\!C\!:\!\ddot{C}l\!:$ (with H above and below C)   (c) $:\!\ddot{I}\!:\!\ddot{C}l\!:$ (d) Na^+ $:\!\ddot{O}\!:\!\ddot{C}l\!:^-$

(e) K^+ $:\!\ddot{O}\!:\!\ddot{C}l\!:\!\ddot{O}\!:^-$ (with $:\!\ddot{O}\!:$ above and below Cl)   (f) $:\!\ddot{C}l\!:\!Si\!:\!\ddot{C}l\!:$ (with $:\!\ddot{C}l\!:$ above and below Si)   (g) Ba^{++} $:\!\ddot{O}\!:\!S\!:\!\ddot{O}\!:^{--}$ (with $:\!\ddot{O}\!:$ above and below S)   (h) $H\!:\!\ddot{C}\!:\!\ddot{N}\!:\!H$ (with H below C and H below N)

2. (a) $H\!:\!\ddot{N}\!:\!\ddot{N}\!:\!H$ (with H below each N)   (b) $H\!:\!\ddot{O}\!:\!S\!:\!\ddot{O}\!:\!H$ (with $:\!\ddot{O}\!:$ above and below S)   (c) $H\!:\!\ddot{O}\!:\!S\!:\!\ddot{O}\!:^-$ (with $:\!\ddot{O}\!:$ above and below S)   (d) $:\!\ddot{C}l\!:\!C\!:\!:\!\ddot{O}\!:$ (with $:\!\ddot{C}l\!:$ above C)

(e) $H\!:\!\ddot{O}\!:\!\ddot{N}\!:\!:\!\ddot{O}\!:$ (f) $:\!\ddot{O}\!:\!\ddot{N}\!:\!:\!\ddot{O}\!:^-$ (g) $:\!\ddot{O}\!:\!C\!:\!:\!\ddot{O}\!:^{--}$ (with $:\!\ddot{O}\!:$ above C)   (h) $H\!:\!\ddot{C}\!:\!:\!\ddot{C}\!:\!H$ (with H above each C)

(i) $H\!:\!C\!:\!:\!:\!C\!:\!H$ (j) $H\!:\!\ddot{C}\!:\!:\!\ddot{O}\!:$ (with H above C)   (k) $H\!:\!\ddot{C}\!:\!\ddot{O}\!:\!H$ (with $:\!\ddot{O}\!:$ below C)   (l) $H\!:\!\ddot{C}\!:\!\ddot{C}\!:\!\ddot{C}\!:\!H$ (with H above and below each C)

3. a, c: trigonal, like BF_3. Others are tetrahedral like CH_4, NH_3, or H_2O: b, h with one 2e lobe; d, e, f with two 2e lobes.

4. Octahedral

5. (a) Toward Br; (b) toward Cl; (c) non-polar; (d) away from H atoms, bisecting angle between Cl atoms; (e) $180°$ away from C—H bond; (f) similar to water molecule; (g) similar to water molecule; (h) toward unshared electron pair on N; (i) away from Cl atoms, bisecting angle between F atoms.

6. (a) F is more electronegative than Cl; although d is smaller, e is bigger.

(b) Less toward D than toward H; that is, D attracts electrons less than H does.

7. Li compound: ionic, salt-like. Be compound: non-ionic, covalent.

8. Boiling point is raised by H-bonding between (like) molecules of a compound; solubility is increased by H-bonding between solute molecules and solvent molecules. Alcohol molecules form H-bonds to each other and to water molecules. Ether molecules can form H-bonds only to water, which furnishes H attached to O.

9. (a)
$$H_3O^+ \; + \; HCO_3^- \; \longrightarrow \; H_2CO_3 \; + \; H_2O$$
Stronger acid   Stronger base   Weaker acid   Weaker base

(b)
$$OH^- \; + \; HCO_3^- \; \longrightarrow \; CO_3^{--} \; + \; H_2O$$
Stronger base   Stronger acid   Weaker base   Weaker acid

(c)
$$NH_3 \; + \; H_3O^+ \; \longrightarrow \; NH_4^+ \; + \; H_2O$$
Stronger base   Stronger acid   Weaker acid   Weaker base

(d)
$$CN^- \; + \; H_2O \; \longleftarrow \; HCN \; + \; OH^-$$
Weaker base   Weaker acid   Stronger acid   Stronger base

(e)
$$H^- \; + \; H_2O \; \longrightarrow \; H_2 \; + \; OH^-$$
Stronger base   Stronger acid   Weaker acid   Weaker base

(f)
$$C_2^{--} \; + \; H_2O \; \longrightarrow \; 2\,OH^- \; + \; C_2H_2$$
Stronger base   Stronger acid   Weaker base   Weaker acid

10. (a) H_3O^+; (b) HCl; (c) HCl in benzene. In aqueous solution, HCl reacts to yield the weaker acid, H_3O^+.

11. Reversible protonation of an unshared electron pair on an oxygen atom (compare page 33).

$$-\ddot{O}{:} + H_2SO_4 \; \rightleftharpoons \; -\ddot{O}{:}H^+ + HSO_4^-$$

12. (a)

```
    H  H  H              H  H  H
    |  |  |              |  |  |
H—C—C—N—H          H—C—N—C—H
    |  |                 |     |
    H  H                 H     H
```

(b)

```
    H  H  H
    |  |  |
H—C—C—C—H
    |  |  |
    H  H  H
```

(c)

```
    H  H  H  H                    H                H  H  H
    |  |  |  |                    |                |  |  |
H—C—C—C—C—H              H—C—H          H—C—C—C—H
    |  |  |  |                |   |   |            |  |  |
    H  H  H  H           H—C—C—C—H            H  H  H
                             |   |   |
                             H  H  H
```

(d)

```
    H  H  H                    H  H  H
    |  |  |                    |  |  |
H—C—C—C—Cl              H—C—C—C—H
    |  |  |                    |  |  |
    H  H  H                    H  Cl H
```

(e)

$$H-\underset{\underset{H}{|}}{\overset{\overset{H}{|}}{C}}-\underset{\underset{H}{|}}{\overset{\overset{H}{|}}{C}}-\underset{\underset{H}{|}}{\overset{\overset{H}{|}}{C}}-O-H \qquad H-\underset{\underset{H}{|}}{\overset{\overset{H}{|}}{C}}-\underset{\underset{O-H}{|}}{\overset{\overset{H}{|}}{C}}-\underset{\underset{H}{|}}{\overset{\overset{H}{|}}{C}}-H \qquad H-\underset{\underset{H}{|}}{\overset{\overset{H}{|}}{C}}-O-\underset{\underset{H}{|}}{\overset{\overset{H}{|}}{C}}-\underset{\underset{H}{|}}{\overset{\overset{H}{|}}{C}}-H$$

(f)

$$H-\underset{\underset{H}{|}}{\overset{\overset{H}{|}}{C}}-C=O \qquad H-\underset{}{\overset{\overset{H}{|}}{C}}=\underset{}{\overset{\overset{H}{|}}{C}}-O-H \qquad H-\underset{}{\overset{\overset{H}{|}}{C}}-\underset{}{\overset{\overset{H}{|}}{C}}-H$$

13. To minimize decomposition of an unstable compound.

Chapter 2 | Methane

Energy of Activation. Transition State

2.1 (a) $CH_3-H + Br-Br \longrightarrow CH_3-Br + H-Br$

$\underset{\underset{150}{}}{104 \qquad 46}$ $\qquad$ $\underset{\underset{158}{}}{70 \qquad 88}$ $\qquad$ $\Delta H = -8$ kcal

(b) $CH_3-H + I-I \longrightarrow CH_3-I + H-I$

$\underset{\underset{140}{}}{104 \qquad 36}$ $\qquad$ $\underset{\underset{127}{}}{56 \qquad 71}$ $\qquad$ $\Delta H = +13$ kcal

(c) $CH_3-H + F-F \longrightarrow CH_3-F + H-F$

$\underset{\underset{142}{}}{104 \qquad 38}$ $\qquad$ $\underset{\underset{244}{}}{108 \qquad 136}$ $\qquad$ $\Delta H = -102$ kcal

2.2 (a) $\underset{(46)}{Br-Br} \longrightarrow 2Br\cdot$ $\qquad$ $\Delta H = +46$ kcal

$\underset{(104)}{Br\cdot + CH_3-H} \longrightarrow CH_3\cdot + \underset{(88)}{H-Br}$ $\qquad$ $\Delta H = +16$ kcal

$\underset{(46)}{CH_3\cdot + Br-Br} \longrightarrow \underset{(70)}{CH_3-Br} + Br\cdot$ $\qquad$ $\Delta H = -24$ kcal

(b) $\underset{(36)}{I-I} \longrightarrow 2I\cdot$ $\qquad$ $\Delta H = +36$ kcal

$\underset{(104)}{I\cdot + CH_3-H} \longrightarrow CH_3\cdot + \underset{(71)}{H-I}$ $\qquad$ $\Delta H = +33$ kcal

$\underset{(36)}{CH_3\cdot + I-I} \longrightarrow \underset{(56)}{CH_3-I} + I\cdot$ $\qquad$ $\Delta H = -20$ kcal

(c) $F\!-\!F \longrightarrow 2F\cdot$ $\Delta H = +38$ kcal
           (38)

$F\cdot + CH_3\!-\!H \longrightarrow CH_3\cdot + H\!-\!F$ $\Delta H = -32$ kcal
           (104)   (136)

$CH_3\cdot + F\!-\!F \longrightarrow CH_3\!-\!F + F\cdot$ $\Delta H = -70$ kcal
           (38)   (108)

2.3 CH_3^{+} is sp^2, with the p orbital empty; flat, trigonal, $120°$ angles. $CH_3:^-$ is sp^3, with the fourth sp^3 orbital occupied by an unshared pair of electrons; pyramidal.

2.4 (a) Forms insoluble silver halide in presence of nitric acid.

(b) Boiling removes volatile HCN and H_2S which otherwise would interfere (AgCN and Ag_2S) with halide test.

2.5 (a) $(\%C + \%H) < 100\%$ (b) 34.8% O $(=100.0\% - (52.1\% \, C + 13.1\% \, H))$

2.6 (a) $\text{wt. Cl} = 20.68 \times \dfrac{Cl}{AgCl} = 20.68 + \dfrac{35.46}{143.34}$ mg

$\%Cl = \text{wt. Cl/wt. sample} \times 100 = 20.68 \times \dfrac{35.46}{143.34} \times \dfrac{1}{7.36} \times 100$

$\%Cl = 69.6\%$

(b) $\%Cl = \dfrac{Cl}{CH_3Cl} \times 100 = \dfrac{35.46}{50.46} \times 100 = 70.4\%$

(c) $\text{wt. AgCl} = \text{wt. sample} \times \dfrac{2AgCl}{CH_2Cl_2} = 7.36 \times \dfrac{286.68}{84.92} = 24.85$ mg

(d) $\text{wt. AgCl} = \text{wt. sample} \times \dfrac{3AgCl}{CHCl_3} = 7.36 \times \dfrac{430.02}{119.38} = 26.49$ mg

(e) $\text{wt. AgCl} = \text{wt. sample} \times \dfrac{4AgCl}{CCl_4} = 7.36 + \dfrac{573.36}{153.84} = 27.44$ mg

2.7 (a) $\text{wt. C} = 8.86 \times \dfrac{C}{CO_2} = 8.86 \times \dfrac{12.01}{44.01}$ mg

$\%C = \text{wt. C/wt. sample} \times 100 = 8.86 \times \dfrac{12.01}{44.01} \times \dfrac{1}{3.02} \times 100$

$\%C = 80.0\%$

$\text{wt. H} = 5.43 \times \dfrac{2H}{H_2O} = 5.43 \times \dfrac{2.016}{18.02}$ mg

$\%H = \text{wt. H/wt. sample} \times 100 = 5.43 \times \dfrac{2.016}{18.02} \times \dfrac{1}{3.02} \times 100$

$\%H = 20.2\%$

C: $\dfrac{80.0}{12.01} = 6.65$ gram-atoms   H: $\dfrac{20.2}{1.008} = 20.0$ gram-atoms

C: $6.65/6.65 = 1.0$ H: $20.0/6.65 = 3.0$

Empirical formula $= CH_3$

(b) As in (a) for %C and %H, using proper weights for sample, CO_2, and H_2O, we get

$$C: 2.67 \text{ gram-atoms} \quad H: 5.38 \text{ gram-atoms}$$

$$\text{wt. Cl} = 13.49 \times \frac{Cl}{AgCl} = 13.49 \times \frac{35.46}{143.34} \text{ mg}$$

$$\%Cl = \text{wt. Cl/wt. sample} \times 100 = 13.49 \times \frac{35.46}{143.34} \times \frac{1}{5.32} \times 100$$

$$\%Cl = 62.8\%$$

$$Cl: \frac{62.8}{35.46} = 1.77 \text{ gram-atoms}$$

$$C: \frac{2.67}{1.77} = 1.51 \quad H: \frac{5.38}{1.77} = 3.04 \quad Cl: \frac{1.77}{1.77} = 1$$

But we cannot accept $C_{1.5}H_3Cl$, so we multiply by 2 to get a whole number of each kind of atom; this gives us an empirical formula of $C_3H_6Cl_2$.

2.8 $CH = 13$; $78/13 = 6(CH)$ units; molecular formula is C_6H_6.

2.9
$$\text{wt. C} = 10.32 \times \frac{C}{CO_2} = 10.32 \times \frac{12.01}{44.01} \text{ mg}$$

$$\%C = \text{wt. C/wt. sample} \times 100 = 10.32 \times \frac{12.01}{44.01} \times \frac{1}{5.17} \times 100$$

$$\%C = 54.5$$

$$\text{wt. H} = 4.23 \times \frac{2H}{H_2O} = 4.23 \times \frac{2.016}{18.02} \text{ mg}$$

$$\%H = \text{wt. H/wt. sample} \times 100 = 4.23 \times \frac{2.016}{18.02} \times \frac{1}{5.17} \times 100$$

$$\%H = 9.1$$

The deficiency of 36.4% $(100 - (54.5 + 9.1))$ is due to oxygen.

$$C: \frac{54.5}{12.01} = 4.53 \text{ gram-atoms} \quad H: \frac{9.1}{1.0} = 9.1 \text{ gram-atoms} \quad O: \frac{36.4}{16.0} = 2.28 \text{ gram-atoms}$$

$$C: \frac{4.53}{2.26} = 2.0 \qquad H: \frac{9.1}{2.28} = 4.0 \qquad O: \frac{2.28}{2.28} = 1$$

Empirical formula $= C_2H_4O$; each unit has weight of 44.

$88/44 = 2\ C_2H_4O$ units; thus molecular formula $= C_4H_8O_2$.

1. Follow procedures of Problems 2.5, 2.6, and 2.7.

A: 93.9% C, 6.3% H.　　B: 64.0% C, 4.5% H, 31.4% Cl.　　C: 62.0% C, 10.3% H, 27.7% O.

For compound C, 27.7% O calculated from $100 - (62.0\%\ C + 10.3\%\ H)$.

2. (a) $$C_3H_7Cl = 78.55 \text{ m.w.}$$

$$\%C = \frac{3C}{C_3H_7Cl} \times 100 = \frac{3 \times 12.01}{78.55} \times 100 = 45.9\% \text{ C}$$

$$\%H = \frac{7H}{C_3H_7Cl} \times 100 = \frac{7 \times 1.008}{78.55} \times 100 = 8.9\% \text{ H}$$

$$\%Cl = \frac{Cl}{C_3H_7Cl} \times 100 = \frac{35.46}{78.55} \times 100 = 45.2\% \text{ Cl}$$

Use of this procedure leads to the following:

(b) 52.1% C   (c) 54.5% C   (d) 41.8% C   (e) 20.0% C   (f) 55.6% C
     13.1% H   9.1% H   4.7% H   6.7% H   6.2% H
     34.8% O   36.3% Cl   18.6% O   26.6% O   10.8% O
                                          16.3% N   46.7% N   27.4% Cl
                                          18.6% S

3. Follow procedures of Problem 2.7. In (c), (d), and (f), oxygen is determined by difference.

(a) CH_2 (b) CH (c) CH_2O (d) C_2H_5OCl (e) $C_3H_{10}N_2$ (f) $C_3H_4O_2Cl_2$

The formula in (f) can be calculated as follows:

$$
\begin{array}{l}
25.2\% \text{ C} \\
2.8\% \text{ H} \\
\underline{49.6\% \text{ Cl}} \\
77.6\%\text{: make up to 100\% by 22.4\% O}
\end{array}
$$

$$C: \frac{25.2}{12.0} = 2.1 \quad H: \frac{2.8}{1.0} = 2.8 \quad Cl: \frac{49.6}{35.5} = 1.4 \quad O: \frac{22.4}{16.0} = 1.4$$

$$C: \frac{2.1}{1.4} = 1.5 \quad H: \frac{2.8}{1.4} = 2 \quad Cl: \frac{1.4}{1.4} = 1 \quad O: \frac{1.4}{1.4} = 1$$

But we cannot accept $C_{1.5}H_2ClO$, so we multiply by 2 to get whole number of each kind of atom; this gives us an empirical formula of $C_3H_4Cl_2O_2$.

4. 18.9% O calculated by difference.

$$C: \frac{70.8}{12.0} = 5.9 \quad H: \frac{6.2}{1.0} = 6.2 \quad O: \frac{19.9}{16.0} = 1.2 \quad N: \frac{4.1}{14.0} = 0.3$$

$$C: 5 \cdot 9/0.3 = 20 \quad H: 6.2/0.3 = 21 \quad O: 1.2/0.3 = 4 \quad N: 0.3/0.3 = 1$$

Empirical formula of papaverine = $C_{20}H_{21}O_4N$.

5. In similar fashion (14.7% O calculated by difference):

$$C = 4.3 \quad H = 4.3 \quad O = 0.9 \quad N = 0.9 \quad S = 0.3 \quad Na = 0.3$$

$$C = \frac{4.3}{0.3} = 14 \quad H = \frac{4.3}{0.3} = 14 \quad O = \frac{0.9}{0.3} = 3 \quad N = \frac{0.9}{0.3} = 3 \quad S = \frac{0.3}{0.3} = 1 \quad Na = \frac{0.3}{0.3} = 1$$

Empirical formula of methyl orange = $C_{14}H_{14}O_3N_3SNa$.

6. (a) As in Problem 2.7: 85.8% C, 14.3% H.

(b) $C: \dfrac{85.8}{12.0} = 7.1 \quad H: \dfrac{14.3}{1.0} = 14.3$

Empirical formula = CH_2; unit weight = 14

(c) $84/14 = 6 \, CH_2$ units.   Molecular formula = C_6H_{12}

7. 53.3% O calculated by difference.

C: 40.0% of m.w. of 60 $\longrightarrow$ 24 g C/mole $\longrightarrow$ 2 C per molecule

H: 6.7% of m.w. of 60 $\longrightarrow$ 4.0 g H/mole $\longrightarrow$ 4 H per molecule

O: 53.3% of m.w. of 60 $\longrightarrow$ 32 g O/mole $\longrightarrow$ 2 O per molecule

Molecular formula = $C_2H_4O_2$

8. Half as many atoms as in Problem 7 = CH_2O.

9. As in Problem 7:

12.2% O calculated by difference.

C: $0.733 \times 262 = 192$ g C/mole $\longrightarrow$ 16 C per molecule

H: $0.038 \times 262 = 10.0$ g H/mole $\longrightarrow$ 10 H per molecule

O: $0.122 \times 262 = 32.0$ g O/mole $\longrightarrow$ 2 O per molecule

N: $0.107 \times 262 = 28.0$ g N/mole $\longrightarrow$ 2 N per molecule

Molecular formula of indigo = $C_{16}H_{10}O_2N_2$

10. (a) $\dfrac{S}{\text{mol. wt.}} \times 100 = \% \, S = 3.4 = \dfrac{32}{\text{mol. wt.}}$

mol. wt. = 942 for minimum unit (one S per molecule)

(b) $5734/942 =$ approx. 6 of the minimum salts $\longrightarrow$ 6 S per molecule

11. As in Problem 2.1:

(a) −130 kcal (b) −44 kcal (c) −26 kcal (d) −2 kcal (e) −13 kcal (f) −8 kcal
(g) −1 kcal

(h) As in Problem 2.2:

1st step: Br—Br $\longrightarrow$ 2Br. $\Delta H = +46$ kcal

	C_2H_6	$C_6H_5CH_3$	$H_2C{=}CHCH_3$
2nd steps: $\Delta H =$	+10 kcal	−3 kcal	0 kcal
3rd steps: $\Delta H =$	−23 kcal	−5 kcal	−1 kcal

12. (a) (1) $\qquad$ Cl—Cl $\longrightarrow$ 2Cl· $\qquad$ $\Delta H = +58$ kcal
$\qquad\qquad\qquad$ (58)

$\qquad$ (2) Cl· + CH$_3$—H $\longrightarrow$ CH$_3$—Cl + H· $\qquad$ $\left. \begin{array}{l} \Delta H = +20 \text{ kcal} \\ (E_{\text{act}} \geq 20 \text{ kcal}) \\ \Delta H = -45 \text{ kcal} \end{array} \right\}$ **Chain-carrying steps**
$\qquad\qquad$ (104) $\qquad\qquad$ (84)

$\qquad$ (3) $\quad$ H· + Cl—Cl $\longrightarrow$ H—Cl + Cl·
$\qquad\qquad\qquad$ (58) $\qquad\qquad$ (103)

(b) E_{act} of a chain-carrying step (step 2) $\geq$ 20 kcal, which is too high.

13. (a) E_{act} of reaction (ii) $\geq$ 16 kcal; E_{act} of reaction (i) *could* be zero (is actually 13 kcal).

(b) Highly improbable, since E_{act} for reaction with Cl$_2$ is much smaller.

14. (a) CH$_3$· can react not only with Br$_2$ but also with HBr (reverse of step 2, page 60).

(b) Reaction of CH$_3$· with HBr (E_{act} 2 kcal) can compete with (easy) reaction with X$_2$; reaction of CH$_3$· with HCl (E_{act} 3 kcal) cannot compete as readily.

(c) Halogenation increasingly reversed by HBr that accumulates as reaction product.

15. (a) $\qquad\qquad\qquad\qquad$ Cl—Cl $\xrightarrow{\text{light}}$ 2Cl·

$\qquad\qquad\qquad$ Cl· + H—H $\longrightarrow$ HCl + H·

$\qquad\qquad\qquad$ H· + Cl—Cl $\longrightarrow$ HCl + Cl· etc.

(b) E_{act} of a chain-carrying step $\geq$ 33 kcal.

$\qquad\qquad$ I· + H—H $\longrightarrow$ H—I + H· $\quad$ $\Delta H = +33$ kcal
$\qquad\quad$ (104) $\qquad\qquad$ (71) $\qquad\qquad$ ($E_{\text{act}} \geq 33$ kcal)

16. (a) $\qquad$ (CH$_3$)$_4$Pb $\xrightarrow{\text{heat}}$ Pb + 4CH$_3$· $\xrightarrow[\text{Old mirror}]{\text{Pb}}$ (CH$_3$)$_4$Pb
$\qquad\qquad\qquad\qquad\qquad$ New $\qquad\qquad\qquad\qquad\qquad\qquad$ Effluent
$\qquad\qquad\qquad\qquad\qquad$ mirror

(b) More chance for

$\qquad\qquad\qquad$ CH$_3$· + ·CH$_3$ $\longrightarrow$ CH$_3$—CH$_3$

and hence fewer CH$_3$· radicals are available for removal of mirror.

$\qquad$ (This work is summarized in G. W. Wheland, *Adv. Org. Chem.*, pp. 733–737, and, in more detail, in E. W. R. Steacie, *Atom and Free Radical Reactions*, 2nd ed., vol. 1, Reinhold, New York, 1954, pp. 37–53. For a fascinating outgrowth of this chemistry, see F. O. Rice, "The Chemistry of Jupiter," *Sci. American*, June 1956.)

17. $\qquad\qquad\qquad\qquad$ (C$_2$H$_5$)$_4$Pb $\xrightarrow{140°}$ Pb + 4C$_2$H$_5$·

We expect C$_2$H$_5$· formed in this way to react with Cl$_2$ in the same manner as CH$_3$· to give ethyl chloride and Cl·, via

$\qquad\qquad\qquad$ C$_2$H$_5$· + Cl—Cl $\longrightarrow$ C$_2$H$_5$Cl + Cl·

Then Cl· generated in this way can start a halogenation chain involving methane.

Chapter 3

Alkanes
Free-Radical Substitution

3.1 (a) 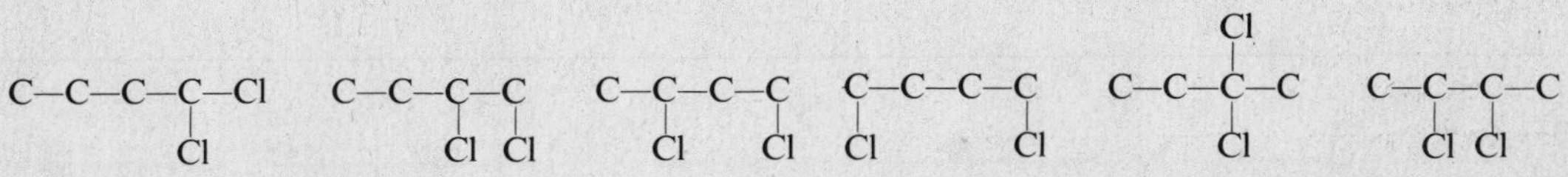

(b) 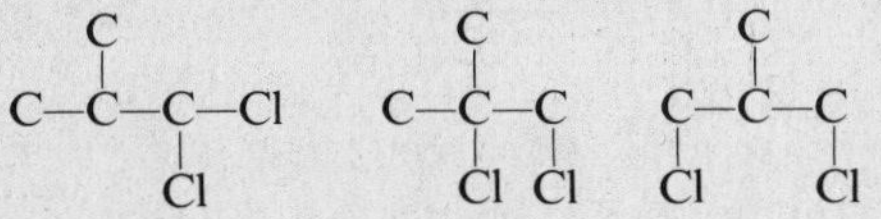

3.2 No. Each can give rise to two monochloro compounds.

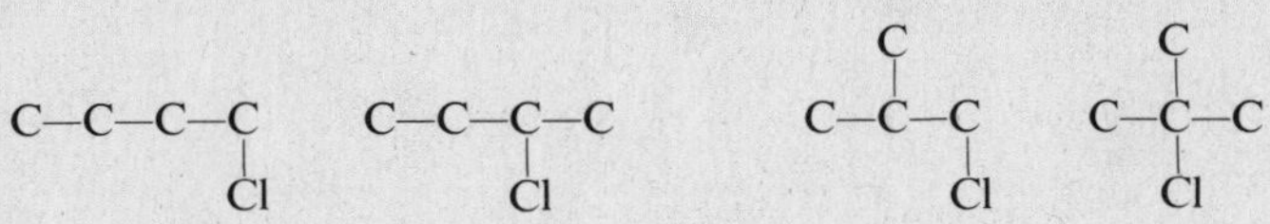

3.3 Van der Waals repulsion between "large" methyls.

3.4 (a)

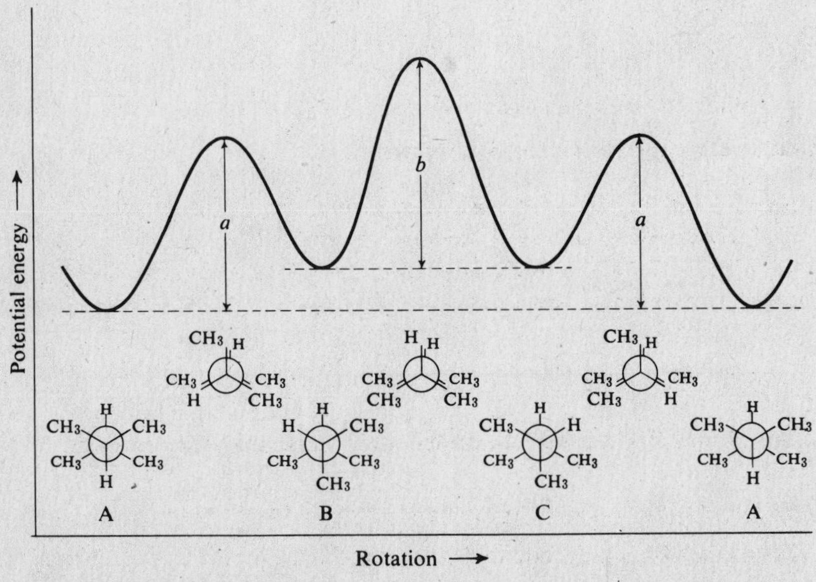

(b)

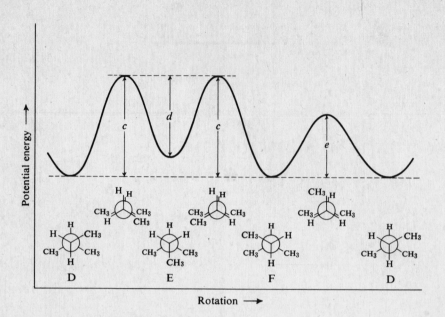

(c)

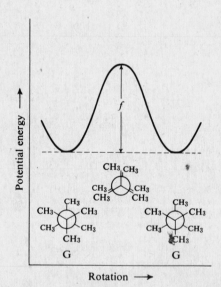

(d) On the assumption of 0.8 kcal per methyl–methyl gauche interaction, and of 3.0 kcal torsional energy plus 0.4 kcal for two methyl–hydrogen eclipsings and 2.2–3.9 kcal per methyl–methyl eclipsing (from Fig. 3.4), one arrives at the following tentative predictions:

$$b > 4.4\text{--}6.1 > a > 3.4; \qquad c > 4.4\text{--}6.1 > d > 3.4 > e;$$

size of f depends on value of methyl–methyl eclipsing.

3.5 (a)

```
C–C–C–C–C–C–C      C–C–C–C–C–C      C–C–C–C–C–C      C–C–C–C–C
                          |                |                |
                          C                C                C
                                                            |
                                                            C
```

```
C–C–C–C–C      C–C–C–C–C      C–C–C–C–C      C–C–C–C–C      C–C–C–C–C
    |   |          |   |          |              |              |   |
    C   C          C   C          C              C              C   C
                                  |              |
                                  C              C
```

(b)

```
C–C–C–C–C      C–C–C–C–C      C–C–C–C–C      C–C–C–C
        |              |              |              |  |
        Cl             Cl             Cl             C  Cl
                                                     |
                                                     C
```

```
    C              C              C              C
    |              |              |              |
C–C–C–C      C–C–C–C      C–C–C–C      C–C–C
    |              |          |              |  |
    Cl             Cl         Cl             C  Cl
```

(c)

```
C–C–C–C–Br      C–C–C–C      C–C–C–C      C–C–C–C      C–C–C–C
        |          |  |          |  |          |  |          |
        Br         Br Br         Br Br         Br Br         Br
                                                              |
                                                              Br
```

```
C–C–C–C      C–C–C–Br      C–C–C      C–C–C
    |  |        |  |          |  |        |  |
    Br Br       C  Br         C  |        C  |
                              |  Br Br     |  Br Br
```

3.6 (a) *n*-hexane
    2-methylpentane
    3-methylpentane
    2,2-dimethylbutane
    2,3-dimethylbutane

(b) Order of isomers identical to Problem 3.5(a):

n-heptane	2,4-dimethylpentane
2-methylhexane	3,3-dimethylpentane
3-methylhexane	3-ethylpentane
2,2-dimethylpentane	2,2,3-trimethylbutane
2,3-dimethylpentane	

3.7 (a) Order of isomers identical to Problem 3.5(b):
    1-chloropentane
    2-chloropentane
    3-chloropentane
    1-chloro-2-methylbutane
    2-chloro-2-methylbutane
    3-chloro-2-methylbutane
    1-chloro-3-methylbutane
    1-chloro-2,2-dimethylpropane

(b) Order of isomers identical to Problem 3.5(c):
    1,1-dibromobutane
    1,2-dibromobutane
    1,3-dibromobutane
    1,4-dibromobutane
    2,2-dibromobutane
    2,3-dibromobutane
    1,1-dibromo-2-methylpropane
    1,2-dibromo-2-methylpropane
    1,3-dibromo-2-methylpropane

3.8 All three graphs show a rate of increase falling off with increasing carbon number.

3.9 Hydrogen (H or D) becomes attached to the same carbon that held Mg.

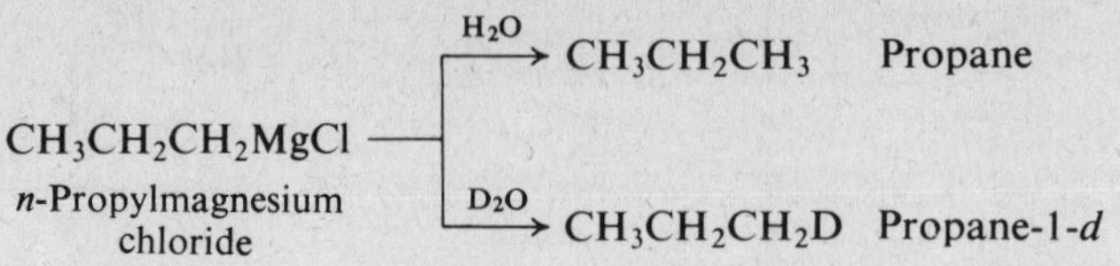

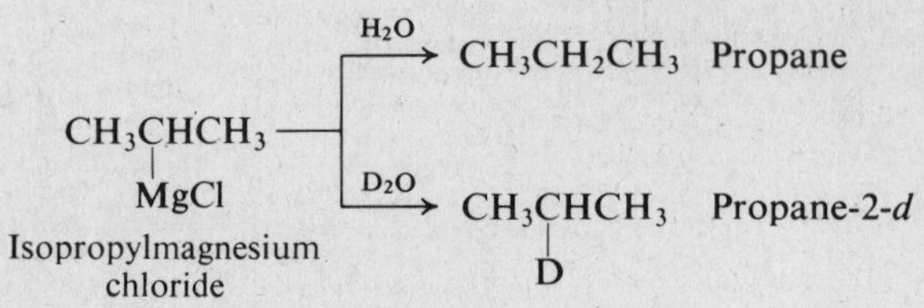

$$CH_3CH_2CH_2MgCl \xrightarrow{\begin{array}{c}H_2O\\D_2O\end{array}} \begin{array}{l} CH_3CH_2CH_3 \quad \text{Propane}\\ CH_3CH_2CH_2D \quad \text{Propane-1-}d\end{array}$$

n-Propylmagnesium chloride

$$\underset{\underset{\text{Isopropylmagnesium chloride}}{MgCl}}{CH_3CHCH_3} \xrightarrow{\begin{array}{c}H_2O\\D_2O\end{array}} \begin{array}{l} CH_3CH_2CH_3 \quad \text{Propane}\\ CH_3\underset{D}{CH}CH_3 \quad \text{Propane-2-}d\end{array}$$

3.10 (a) All that have carbon skeleton of *n*-pentane:

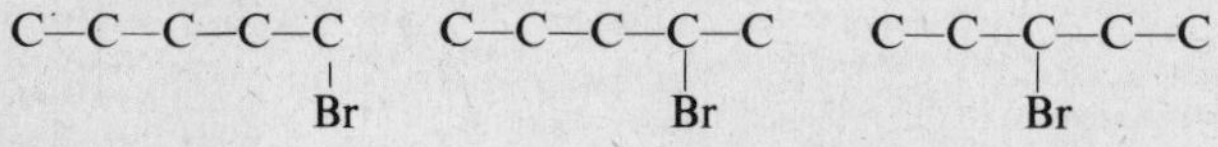

$$\underset{Br}{C-C-C-C-C} \qquad \underset{Br}{C-C-C-C-C} \qquad \underset{Br}{C-C-C-C-C}$$

(b) All that have carbon skeleton of 2-methylbutane (isopentane):

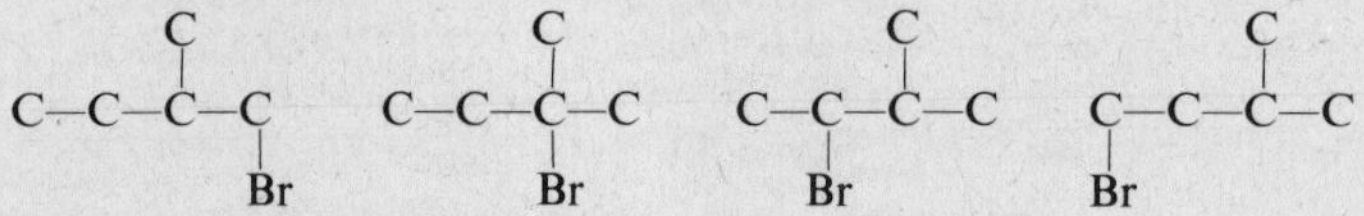

$$\overset{C}{\underset{Br}{C-C-C-C}} \qquad \overset{C}{\underset{Br}{C-C-C-C}} \qquad \overset{C}{\underset{Br}{C-C-C-C}} \qquad \overset{C}{\underset{Br}{C-C-C-C}}$$

(c) All that have carbon skeleton of 2,3-dimethylbutane:

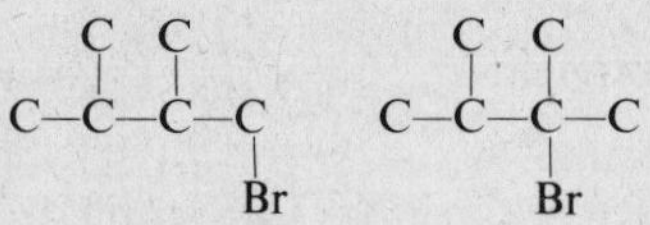

$$\overset{C\ \ C}{\underset{Br}{C-C-C-C}} \qquad \overset{C\ \ C}{\underset{Br}{C-C-C-C}}$$

(d) Only neopentyl bromide has proper carbon skeleton:

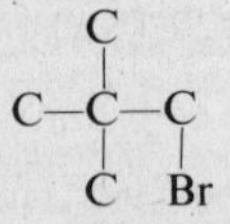

$$\overset{C}{\underset{C\ \ Br}{C-C-C}}$$

3.11 (a)

$$\underset{\text{CH}_3}{CH_3-CH-\!\{CH_2-CH_2-CH_3\}} \longleftarrow \begin{array}{l} (CH_3-\overset{CH_3}{CH})_2CuLi + X-CH_2-CH_2-CH_3\\[2em] CH_3-\overset{CH_3}{CH}-X + LiCu(CH_2-CH_2-CH_3)_2\end{array}$$

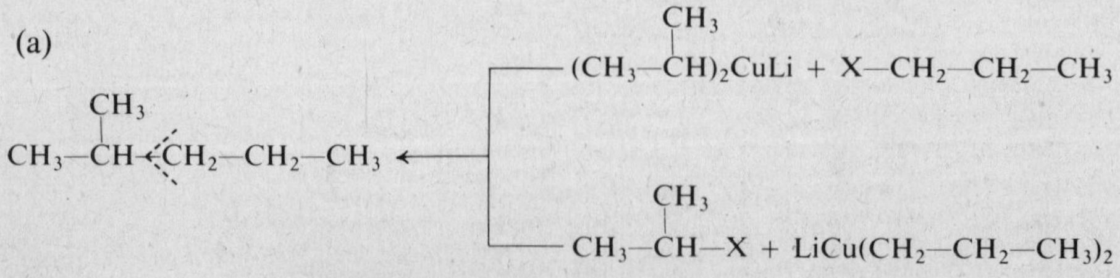

(b) Use the first, since in this synthesis R'X is primary.

(E. J. Corey and G. H. Posner, "Carbon–Carbon Bond Formation by Selective Coupling of *n*-Alkylcopper Reagents with Organic Halides," J. Am. Chem. Soc., **90**, 5615 (1968); H. O. House *et al.*, "Reaction of Lithium Dialkyl- and Diarylcuprates with Organic Halides," J. Am. Chem. Soc., **91**, 4871 (1969).)

3.12 (a)

$$C-C-C-C-C \atop \qquad\qquad\quad| \atop \qquad\qquad\; Cl \qquad\qquad C-C-C-C-C \atop \qquad\qquad\qquad| \atop \qquad\qquad\quad Cl \qquad\qquad C-C-C-C-C \atop \qquad\qquad| \atop \qquad\quad Cl$$

(b)

$$\underset{Cl}{\overset{C}{C-\underset{|}{\overset{|}{C}}-C-C}} \qquad \underset{Cl}{\overset{C}{C-\underset{|}{\overset{|}{C}}-C-C}} \qquad \underset{Cl}{\overset{C}{C-\underset{|}{\overset{|}{C}}-C-C}} \qquad \underset{Cl}{\overset{C}{C-\underset{|}{\overset{|}{C}}-C-C}}$$

3.13 $(CH_3)_3CCH_2X$ is the only possible substitution product.

3.14

(a)	(b)	(c)	(d)
44% 1-Cl	64% 1°	55% 1°	21% 1-Cl
56% 2-Cl	36% 3°	45% 3°	53% 2-Cl
			26% 3-Cl

(e) 28% 1-Cl-2-Me
    23% 2-Cl-2-Me
    35% 3-Cl-2-Me
    14% 1-Cl-3-Me

(f) 45% 1-Cl-2,2,3-triMe
    25% 3-Cl-2,2,3-triMe
    30% 1-Cl-2,3,3-triMe

(g) 33% 1-Cl-2,2,4-triMe
    28% 3-Cl-2,2,4-triMe
    18% 4-Cl-2,2,4-triMe
    22% 1-Cl-2,4,4-triMe

These predicted proportions of products can be calculated by the method shown on page 100. For (a), this would be:

$$\frac{n\text{-PrCl}}{i\text{-PrCl}} = \frac{\text{no. of } 1° \text{ H}}{\text{no. of } 2° \text{ H}} \times \frac{\text{reactivity of } 1° \text{ H}}{\text{reactivity of } 2° \text{ H}} = \frac{6}{2} \times \frac{1.0}{3.8} = \frac{6.0}{7.6}$$

Then,

$$\% \ 1° = \frac{6.0}{6.0 + 7.6} \times 100 = 44\%$$

$$\% \ 2° = \frac{7.6}{6.0 + 7.6} \times 100 = 56\%$$

For (d):
    1-Cl: 6H, each with reactivity 1.0 $\longrightarrow$ 6.0
    2-Cl: 4H, each with reactivity 3.8 $\longrightarrow$ 15.2
    3-Cl: 2H, each with reactivity 3.8 $\longrightarrow$ 7.6
             number × reactivity total = 28.8

$$\% \ 1\text{-Cl} = \frac{\text{reactivity 1-Cl}}{\text{total reactivity of molecule}} \times 100 = \frac{6.0}{28.8} \times 100 = 21\%$$

$$\% \ 2\text{-Cl} = \frac{15.2}{28.8} \times 100 = 53\%$$

$$\% \ 3\text{-Cl} = \frac{7.6}{28.8} \times 100 = 26\%$$

3.15 (a) 4% 1-Br   (b) 0.6% 1°   (c) 0.3% 1°   (d) 1% 1-Br

           96% 2-Br   99.4% 3°   99.7% 3°   66% 2-Br

                                                                               33% 3-Br

(e) 0.3% 1-Br-2-Me   (f) 0.6% 1-Br-2,2,3-triMe   (g) 0.5% 1-Br-2,2,4-triMe

     90% 2-Br-2-Me   99% 3-Br-2,2,3-triMe   9% 3-Br-2,2,4-triMe

     9% 3-Br-2-Me   0.4% 1-Br-2,3,3-triMe   90% 4-Br-2,2,4-triMe

     0.2% 1-Br-3-Me   0.3% 1-Br-2,4,4-triMe

These predicted proportions are calculated by using the relative activities of 1600:82:1 given on page 100.

3.16 Instead of 400/1, the ratio of products would be $400/(1 \times 10) = 40:1$, a ratio that would be much easier to measure accurately.

3.17 $(2.3/2) = 1.15$ times more reactive.

3.18 (a) DCl:HCl ratio would have been less than t-BuCl:i-BuCl ratio.

(b) Same as (a).

     (H. C. Brown and G. A. Russell, "Photochlorination of 2-Methylpropane-2-d and α-d_1-Toluene; the Question of Free Radical Rearrangement or Exchange in Substitution Reactions," J. Am. Chem. Soc., **74**, 3995 (1952).)

3.19 t-Bu—D $\xleftarrow{\text{D}_2\text{O}}$ t-BuMgCl $\xleftarrow[\text{Et}_2\text{O}]{\text{Mg}}$ t-BuCl

3.20 Add DBr, and see if unconsumed methane contains deuterium.

3.21 See if $^{35}\text{Cl}^{36}\text{Cl}$ (mass 71) and/or $^{36}\text{Cl}^{37}\text{Cl}$ (mass 73) shows up in the mass spectrum.

$$^{35}\text{Cl}\cdot + {}^{36}\text{Cl}\text{—}^{36}\text{Cl} \longrightarrow {}^{35}\text{Cl}\text{—}^{36}\text{Cl} + {}^{36}\text{Cl}\cdot$$
$$\text{Mass} = 71$$

$$^{37}\text{Cl}\cdot + {}^{36}\text{Cl}\text{—}^{36}\text{Cl} \longrightarrow {}^{37}\text{Cl}\text{—}^{36}\text{Cl} + {}^{36}\text{Cl}\cdot$$
$$\text{Mass} = 73$$

Other (normal) mass values would be 70 (^{35}Cl—^{35}Cl), 72 (^{35}Cl—^{37}Cl), and 74 (^{37}Cl—^{37}Cl).

3.22 n-BuBr + LiCu(Bu-t)$_2$ $\longrightarrow$ n-Bu—Bu-t (2,2-Dimethylhexane)

1. (a)
```
        C  C
        |  |
   C—C—C—C—C
        |  |
        C  C
```
(b)
```
   C—C—C—C
      |  |
      C  C
```
(c)
```
          C
          |
   C—C—C—C—C—C—C
      |     |
      C     C
```
(d)
```
              C
              |
   C—C—C—C—C—C—C—C
          |  |
          C  C
          |
          C
```

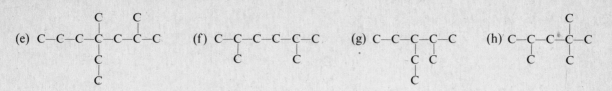

2. (a) 2-methylpentane
   (c) 3-methyl-3-ethylpentane
   (e) 3-methyl-5-ethyloctane
   (g) 2-methyl-5-ethylheptane
   (i) 2-methyl-3,3-diethylhexane

   (b) 3,3-dimethylpentane
   (d) 2,3,4-trimethylhexane
   (f) 2,2,4,4-tetramethylpentane
   (h) 2,3,5-trimethyl-5-ethylheptane
   (j) 3-methyl-5-isopropyloctane

3. (a) 1a, 2b, 2c, 2f
   (c) 1b, 1c, 1f, 1g, 2e, 2g, 2h
   (e) 1a, 1h, 2d, 2f

   (b) 1d, 1e, 1h, 2a, 2i
   (d) 1b
   (f) 2c

4. (a) 1e, 1g, 1h, 2a, 2d, 2g, 2h, 2i, 2j
   (d) 1f
   (g) 1a, 1h
   (j) 1h

   (b) 1b, 1f
   (e) 1d, 2d, 2e, 2j
   (h) 2f
   (k) 1d, 2e, 2j

   (c) 1e, 1h, 2a, 2g
   (f) 1c
   (i) 2d, 2j

5. (a) 2,3-dimethylbutane
   (c) 3-methylpentane
   (e) 6

   (b) *n*-hexane, 2,2-dimethylbutane
   (d) 2-methylpentane

6. One monochloro, three dichloro, four trichloro. (This neglects stereoisomerism, Chapter 4.)

7. c, b, e, a, d.

8. (a) isobutylmagnesium bromide
   (c) isobutane
   (e) $(CH_3)_2CHCH_2D$
   (g) 3-methylpentane

   (b) *tert*-butylmagnesium bromide
   (d) isobutane
   (f) (*sec*-Bu)$_2$CuLi

9. (a) $CH_3CH_2CH_2CH_2Br \xrightarrow{Mg} CH_3CH_2CH_2CH_2MgBr \xrightarrow{H_2O} CH_3CH_2CH_2CH_3$

   (b) $CH_3CH_2\overset{\overset{\displaystyle CH_3}{|}}{C}HBr \xrightarrow{Mg} CH_3CH_2\overset{\overset{\displaystyle CH_3}{|}}{C}HMgBr \xrightarrow{H_2O} CH_3CH_2CH_2CH_3$

   (c) $CH_3CH_2Cl \xrightarrow{Li} CH_3CH_2Li \xrightarrow{CuCl} (CH_3CH_2)_2CuCl \xrightarrow{CH_3CH_2Cl} CH_3CH_2CH_2CH_3$

   (d) $CH_3CH_2CH{=}CH_2 \xrightarrow{H_2,\ Ni} CH_3CH_2CH_2CH_3$

   (e) $CH_3CH{=}CHCH_3 \xrightarrow{H_2,\ Ni} CH_3CH_2CH_2CH_3$

10. (a)

C—C—C—C—C—C   C—C—C—C—C—C   C—C—C—C—C—C
             | | |
             Cl   Cl   Cl

(b)

$$\underset{\text{Cl}}{\underset{|}{C}}-\underset{\overset{|}{C}}{\overset{|}{C}}-C-C-C \qquad \underset{\text{Cl}}{\underset{|}{C}}-\underset{\overset{|}{C}}{\overset{|}{C}}-C-C-C \qquad C-\underset{\overset{|}{C}}{\overset{|}{C}}-\underset{\text{Cl}}{\underset{|}{C}}-C-C$$

(showing five isomers with a methyl branch on C-2 and Cl substituents)

C—C—C—C—C (with C above second carbon, Cl below end) ; C—C—C—C—C ; C—C—C—C—C ; C—C—C—C—C ; C—C—C—C—C

(c)

(four isomers, each with two methyl branches and one Cl)

(d)

(three isomers, each with two methyl branches and one Cl)

11. Order of isomers as in Problem 10:

  (a) 16, 42, 42%   (b) 21, 17, 26, 26, 10%
  (c) 33, 28, 18, 22%   (d) 46, 39, 15%

Calculations performed as in Problem 3.14, page 100.

12. (a) Water rapidly destroys a Grignard reagent.

(b) The —OH group is acidic enough to destroy a Grignard reagent, if indeed one could even be formed.

13. Allyl, benzyl > 3° > 2° > 1° > methyl, vinyl

14. Allylic, benzylic > 3° > 2° > 1° > CH_4, vinylic

15. Rearrangement, *by migration of Br*, of initially formed 1° radical into more stable 2° or 3° radical:

$$\text{Cl·} \begin{cases} \xrightarrow{CH_3CH_2CH_2Br} CH_3\dot{C}HCH_2Br \\ \xrightarrow{CH_3CHBrCH_3} CH_3CHBr\dot{C}H_2 \xrightarrow{\text{rearr.}} CH_3\dot{C}HCH_2Br \end{cases} \xrightarrow{Cl_2} CH_3CHClCH_2Br$$

$$\text{Cl·} \begin{cases} \xrightarrow{(CH_3)_2CHCH_2Br} (CH_3)_2\dot{C}CH_2Br \\ \xrightarrow{(CH_3)_3CBr} (CH_3)_2CBr\dot{C}H_2 \xrightarrow{\text{rearr.}} (CH_3)_2\dot{C}CH_2Br \end{cases} \xrightarrow{Cl_2} (CH_3)_2CClCH_2Br$$

Although alkyl free radicals seldom if ever rearrange by migration of hydrogen or alkyl, it seems clear that they can rearrange by migration of halogen.

16. (a)
$$n\text{-}C_{14}H_{30} + \frac{43}{2}\,O_2 \longrightarrow 14CO_2 + 15H_2O$$

From Table 3.3, $n\text{-}C_{14}H_{30}$ (m.w. 198) has a density of 0.764 g/ml. One liter of kerosine thus weighs 764 g, and is $\dfrac{764 \text{ g}}{198 \text{ g/mole}}$ moles. This amount of kerosine requires $\dfrac{764}{198} \times \dfrac{43}{2}$ moles of oxygen, which weighs $\dfrac{764}{198} \times \dfrac{43}{2} \times 32$ g, or 2650 g.

(b)

$$CH_3(CH_2)_{12}CH_3: \quad 12 \;-\!CH_2\!- \cdot \longrightarrow \quad 12 \times 157 \text{ kcal} = 1884 \text{ kcal}$$
$$2 \quad CH_3\!- \longrightarrow \quad 2 \times 186 \text{ kcal} = \underline{327 \text{ kcal}}$$
$$2256 \text{ kcal/mole}$$

One liter of kerosine:

$$\frac{764}{198} \text{ mole} \times 2256 \text{ kcal/mole} \longrightarrow 8710 \text{ kcal}$$

(c)
$$H\cdot + \cdot H = H_2 \quad \Delta H = -104 \text{ kcal/mole}$$

To produce 8710 kcal requires

$$\frac{8710 \text{ kcal}}{104 \text{ kcal/mole}} = 84 \text{ moles } H_2 = 168 \text{ g}$$

17. Carius: mono, 45.3% Cl; di, 62.8% Cl.
Mol. wt. detn.: mono, 78.5; di, 113.

18. Try to synthesize it by Corey-House method from isopentyl bromide.

$$(CH_3\overset{\overset{\displaystyle CH_3}{|}}{C}HCH_2CH_2)_2CuLi \xleftarrow{\text{CuI}} \xleftarrow{\text{Li}} CH_3\overset{\overset{\displaystyle CH_3}{|}}{C}HCH_2CH_2Br$$

$$CH_3\overset{\overset{\displaystyle CH_3}{|}}{C}HCH_2CH_2 \!\! \mid \!\! CH_2CH_2\overset{\overset{\displaystyle CH_3}{|}}{C}HCH_3 \longleftarrow$$

$$BrCH_2CH_2\overset{\overset{\displaystyle CH_3}{|}}{C}HCH_3$$

19. (a) Methane, formed by $CH_3OH + CH_3MgI \longrightarrow CH_4 + CH_3OMgI$.

$$1.04 \text{ cc gas is } \frac{1.04 \text{ cc}}{22.4 \text{ cc/mmole}} \text{ or } \frac{1.04}{22.4} \text{ mmole } CH_4,$$

evolved from the same number of millimoles of CH_3OH, m.w. 32.

$$\text{wt. } CH_3OH = \frac{1.04}{22.4} \text{ mmole} \times 32 \text{ mg/mmole} = 1.49 \text{ mg } CH_3OH$$
$$\text{(no. of moles)} \times \text{(wt. per mole)}$$

(b)   no. mmole alcohol $= \dfrac{1.57 \text{ cc}}{22.4 \text{ cc/mmole}}$

m.w. alc. (mg/mmole) $= \dfrac{\text{wt. alc. (mg)}}{\text{no. mmoles alc}} = \dfrac{4.12 \text{ mg}}{1.57/22.4} = 59$

$59 = C_nH_{2n+1}OH = 12n + (2n+1)(1) + 16 + 1 = 14n + 19$

$n = 3 \longrightarrow C_3H_7OH$, which is n-PrOH or iso-PrOH.

(c)   mmoles of alcohol $= \dfrac{1.79 \text{ mg}}{90 \text{ mg/mmole}} =$ approx. 0.02 mmole alcohol

0.02 mmole alcohol $\longrightarrow \dfrac{1.34 \text{ cc}}{22.4 \text{ cc/mmole}} = 0.06$ mmole H_2

each mole alcohol $\longrightarrow$ 3 moles H_2

alcohol contains 3 —OH groups

m.w. 90 $= C_nH_{2n-1}(OH)_3 = 12n + (2n-1)(1) + (3 \times 16) + (3 \times 1)$

$90 = 14n + 50$

$n = 3$ gives $\underset{\underset{OH}{|}}{CH_2} - \underset{\underset{OH}{|}}{CH} - \underset{\underset{OH}{|}}{CH_2},$ glycerol (Chap. 15).

20. (a) (1) $(CH_3)_3CO-OC(CH_3)_3 \xrightarrow{130°} 2(CH_3)_3CO\cdot$

(2) $(CH_3)_3CO\cdot + (CH_3)_3CH \longrightarrow (CH_3)_3COH + (CH_3)_3C\cdot$

(3) $(CH_3)_3C\cdot + CCl_4 \longrightarrow (CH_3)_3CCl + Cl_3C\cdot$

(4) $Cl_3C\cdot + (CH_3)_3CH \longrightarrow Cl_3CH + (CH_3)_3C\cdot$

then (3), (4), (3), (4), etc.

(b) (1) $(CH_3)_3C-O-Cl \xrightarrow{\text{light}} (CH_3)_3CO\cdot + Cl\cdot$

(2) $(CH_3)_3CO\cdot + RH \longrightarrow (CH_3)_3COH + R\cdot$

(3) $R\cdot + (CH_3)_3C-O-Cl \longrightarrow RCl + (CH_3)_3CO\cdot$

then (2), (3), (2), (3), etc.

Chapter 4

Stereochemistry I. Stereoisomers

4.1 If methane were a pyramid with a rectangular base:

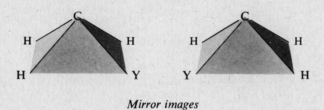

Mirror images

4.2 (a) If methane were rectangular:

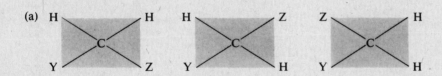

(b) If methane were square:

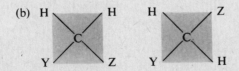

(c) If methane were a pyramid with a square base:

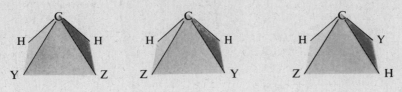

Mirror images

(d) If methane were tetrahedral:

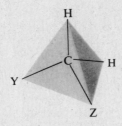

4.3 (a) $$[\alpha] = \frac{\alpha}{l \times d} = \frac{-1.2°}{0.5 \text{ dm} \times \frac{6.15}{100}} = -39.0°$$

(b) $-39° = \dfrac{\alpha}{1.0 \times \dfrac{6.15}{100}}$ A tube twice as long $\longrightarrow$ doubled observed rotation

$$\alpha = -2.4°$$

(c) $-39.0° = \dfrac{\alpha}{0.5 \times \dfrac{6.15}{200}}$ Halved conc. $\longrightarrow$ halved no. of molecules $\longrightarrow$ halved rotn.

$$\alpha = -0.6°$$

4.4 Use a shorter or longer tube, and measure rotation again. For example, if α_{obs} were actually $+45°$, a 2.5-cm tube $\longrightarrow$ α_{obs} of $+11.25°$. Other values of α_{obs} would give α_{obs} as follows:

$$-315° \longrightarrow -78.75°; \quad +405° \longrightarrow +101.25°; \quad +765° \longrightarrow +191.25°.$$

4.5 Chiral: b, d, f, g, h.

(b) C—C—C—$\overset{*}{\text{C}}$H—C
          |
          Cl

(d) C—C—C—$\underset{*}{\text{C}}$H—CH$_2$Cl

(f) C—C—$\overset{*}{\text{C}}$H—C—C
          | |
          Cl   C

(g) C—$\overset{*}{\text{C}}$H—C—C—C
          | |
          Cl   C

(h) C—C—$\overset{*}{\text{C}}$H—CH$_2$Cl
              |
              Br

4.6 (a) $CH_3CH_2\overset{*}{C}HDCl$ $CH_3\overset{*}{C}HDCH_2Cl$ $DCH_2CH_2CH_2Cl$ $CH_3\overset{*}{C}HClCH_2D$ $(CH_3)_2CDCl$

(b) Chiral centers marked by $\overset{*}{C}$.

4.7 (a)–(c) Work with models; (d) mirror images: a, b.

4.8 (a)

Cl |
I—C—SO₃H
H
R

Cl |
HO₃S—C—I
H
S

(b)

D |
CH₃—C—Br
H
R

D |
Br—C—CH₃
H
S

4.9 3°, 2°, 1°, CH₃.

4.10 (a)

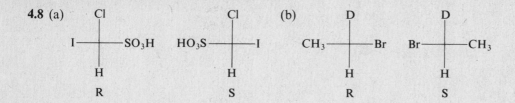

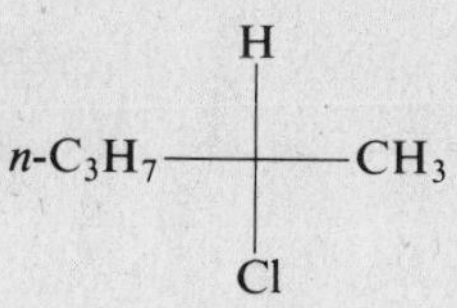

H |
n-C₃H₇—C—CH₃
Cl
R

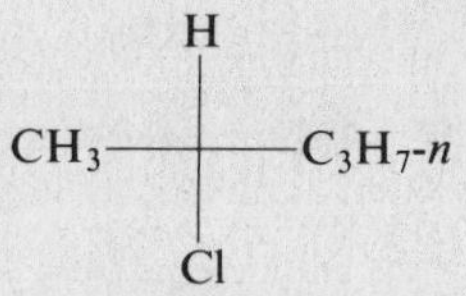

H |
CH₃—C—C₃H₇-n
Cl
S

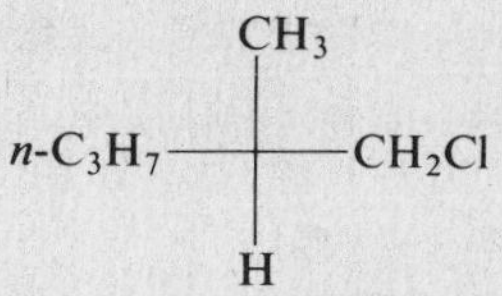

CH₃ |
n-C₃H₇—C—CH₂Cl
H
R

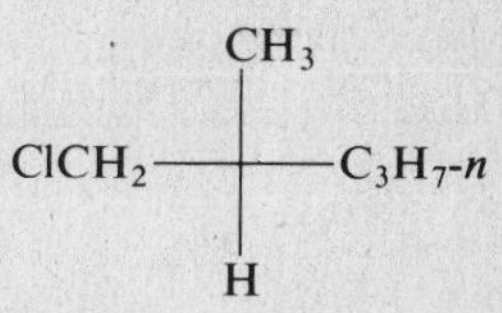

CH₃ |
ClCH₂—C—C₃H₇-n
H
S

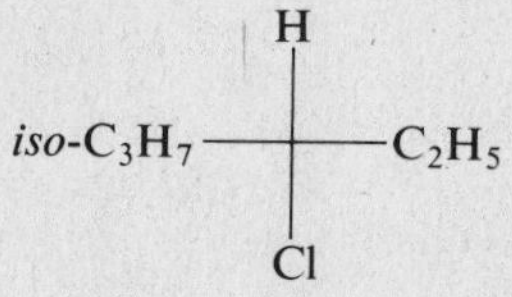

H |
iso-C₃H₇—C—C₂H₅
Cl
R

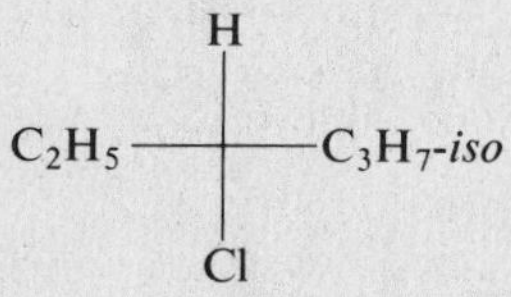

H |
C₂H₅—C—C₃H₇-iso
Cl
S

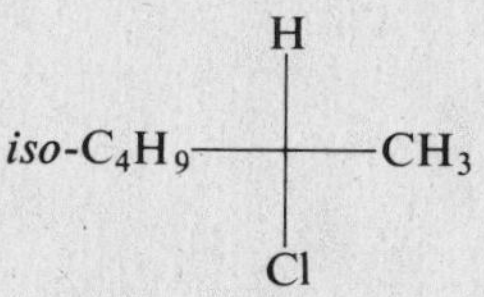

H |
iso-C₄H₉—C—CH₃
Cl
R

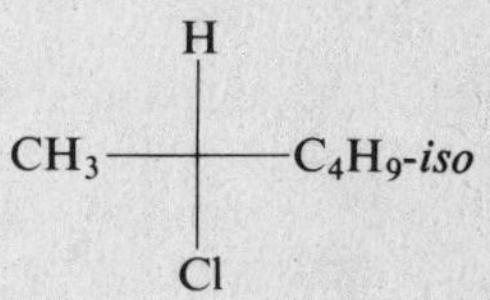

H |
CH₃—C—C₄H₉-iso
Cl
S

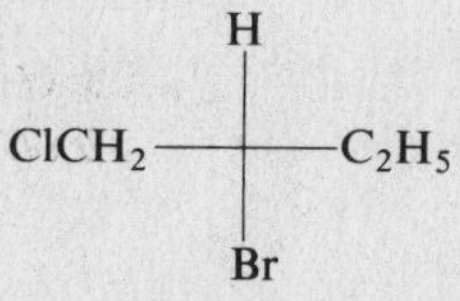

H |
ClCH₂—C—C₂H₅
Br
R

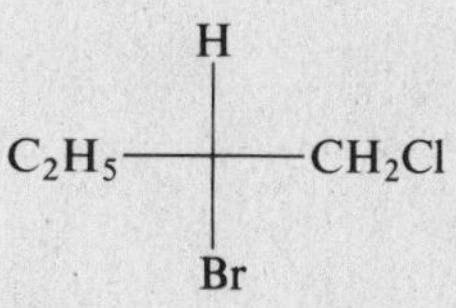

H |
C₂H₅—C—CH₂Cl
Br
S

(b)

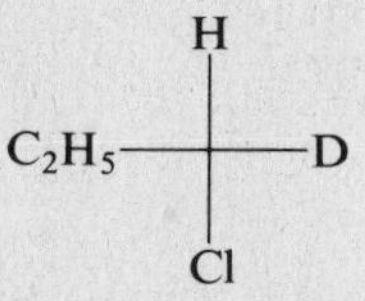

H |
C₂H₅—C—D
Cl
R

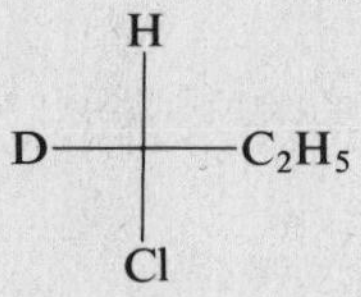

H |
D—C—C₂H₅
Cl
S

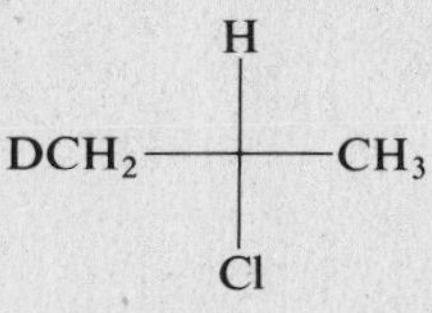

H |
DCH₂—C—CH₃
Cl
R

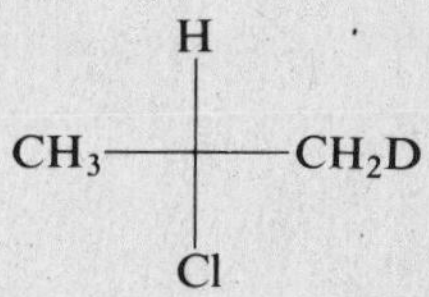

H |
CH₃—C—CH₂D
Cl
S

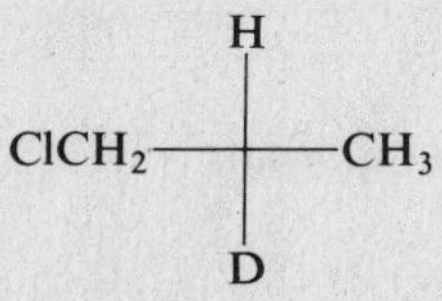

H |
ClCH₂—C—CH₃
D
R

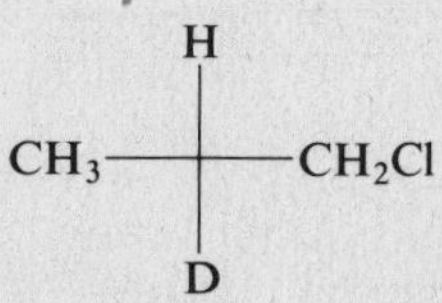

H |
CH₃—C—CH₂Cl
D
S

25

4.11 (a)

```
        Cl                              Cl
         |                               |
C2H5 ——————— CH=CH2        CH2=CH ——————— C2H5
         |                               |
         H                               H
         R                               S
```

(b)

```
            Cl                                Cl
             |                                 |
(CH3)2CH ——————— CH=CH2        CH2=CH ——————— CH(CH3)2
             |                                 |
             H                                 H
             R                                 S
```

(c)

```
         H                               H
         |                               |
HOOC ——————— CH2COOH        HOOCCH2 ——————— COOH
         |                               |
         OH                              OH
         R                               S
```

(d)

```
         H                               H
         |                               |
C6H5 ——————— CH3             CH3 ——————— C6H5
         |                               |
         NH2                             NH2
         R                               S
```

(e)

```
            CH3                              CH3
             |                                |
(CH3)2CH ——————— C3H7-n        n-C3H7 ——————— CH(CH3)2
             |                                |
             C2H5                             C2H5
             R                                S
```

(f)

```
         H                               H
         |                               |
HOOC ——————— C6H5           C6H5 ——————— COOH
         |                               |
         OH                              OH
         R                               S
```

(g)

```
         H                               H
         |                               |
HOOC ——————— CH3            CH3 ——————— COOH
         |                               |
         NH2                             NH2
         R                               S
```

4.12 (a), (g): 2 enantiomers, both active
(b), (c), (e): 2 enantiomers, both active; 1 inactive *meso* compound.
(f), (h): 2 pairs of enantiomers, all active
(d): 4 pairs of enantiomers, all active.

4.13 The *meso* compounds are:

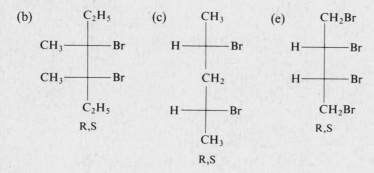

4.14 Refer to curves on pages 13–14 of Study Guide.

(a) 3 conformers: A, B, C; B and C are enantiomers
(b) 3 conformers: D, E, F; D and F are enantiomers
(c) 1 conformer: G

In (a), B and C less abundant (more Me–Me *gauche* interactions).

In (b), D and F more abundant (fewer Me–Me *gauche* interactions).

4.15 (a)

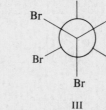

The two forms are made up of I, and II plus III.

(b) Neither form is active: I is achiral, II plus III is a racemic modification.

1. (a) Secs. 4.3 and 4.4   (b) Sec. 4.4   (c) Sec. 4.4   (d) Sec. 4.5
(e) Sec. 4.9   (f) Sec. 4.9   (g) Sec. 4.10   (h) Sec. 4.2
(i) Secs. 4.7 and 4.11   (j) Secs. 4.7 and 4.17 (k) Sec. 4.18   (l) Sec. 4.12
(m) Sec. 4.14   (n) Sec. 3.3   (o) Sec. 4.15   (p) Sec. 4.15
(q) Sec. 4.4   (r) Sec. 4.4   (s) Sec. 4.20   (t) Sec. 4.20

2. (a) Chirality. (b) Chirality. (c) Usually an excess of one enantiomer that persists long enough to permit measurement. (d) Mirror image not superimposable on original. (e) Restrictions on planar formulas are discussed in Sec. 4.10. In models, no bond to a chiral carbon should be broken. (f) Draw a picture or build a model of the molecule; then follow the steps in Sec. 4.15 and Sequence Rules in Sec. 4.16.

3. Equal but opposite specific rotations; opposite R/S specifications; all other properties are the same.

4. (a) Screw, scissors, spool of thread.
　　(b) Glove, shoe, coat sweater, tied scarf.
　　(c) Helix, double helix.
　　(d) Football (laced), tennis racket (looped trim), golf club, rifle barrel.
　　(e) Hand, foot, ear, nose, yourself.

5. (a) Sawing. (b) Opening milk bottle. (c) Throwing a ball.

6. (a)

$$
\begin{array}{c}
\text{H} \\
| \\
n\text{-}C_3H_7 \text{——} C_2H_5 \\
| \\
\text{Br} \\
\text{R}
\end{array}
\qquad
\begin{array}{c}
\text{H} \\
| \\
C_2H_5 \text{——} C_3H_7\text{-}n \\
| \\
\text{Br} \\
\text{S}
\end{array}
$$

　　(b) Achiral.

　　(c)

$$
\begin{array}{c}
\text{CH}_3 \\
| \\
\text{BrCH}_2 \text{——} C_2H_5 \\
| \\
\text{Br} \\
\text{R}
\end{array}
\qquad
\begin{array}{c}
\text{CH}_3 \\
| \\
C_2H_5 \text{——} \text{CH}_2\text{Br} \\
| \\
\text{Br} \\
\text{S}
\end{array}
$$

　　(d)

$$
\begin{array}{c}
\text{H} \\
| \\
\text{ClCH}_2\text{CH}_2 \text{——} \text{CH}_2\text{CH}_3 \\
| \\
\text{Cl} \\
\text{R}
\end{array}
\qquad
\begin{array}{c}
\text{H} \\
| \\
\text{CH}_3\text{CH}_2 \text{——} \text{CH}_2\text{CH}_2\text{Cl} \\
| \\
\text{Cl} \\
\text{S}
\end{array}
$$

　　(e)

$$
\begin{array}{c}
\text{H} \\
| \\
t\text{-}C_4H_9 \text{——} C_4H_9\text{-}iso \\
| \\
\text{Cl} \\
\text{R}
\end{array}
\qquad
\begin{array}{c}
\text{H} \\
| \\
iso\text{-}C_4H_9 \text{——} C_4H_9\text{-}t \\
| \\
\text{Cl} \\
\text{S}
\end{array}
$$

(f)

$$
\begin{array}{c}
\text{H} \\
| \\
n\text{-}C_3H_7\!\!-\!\!C\!\!-\!\!D \\
| \\
\text{Cl} \\
R
\end{array}
\qquad\qquad
\begin{array}{c}
\text{H} \\
| \\
D\!\!-\!\!C\!\!-\!\!C_3H_7\text{-}n \\
| \\
\text{Cl} \\
S
\end{array}
$$

7. (a) and (b) 3-Methylhexane and 2,3-dimethylpentane.

8. 2 pairs of enantiomers: a, b, e, k.　　　1 pair of enantiomers, 2 *meso*: g.
　　1 pair of enantiomers, 1 *meso*: c, d, h.　　2 diastereomers: i.
　　4 pairs of enantiomers: f.　　　　　　　1 pair of enantiomers: j.

9. A, $CH_3CCl_2CH_3$　　B, $ClCH_2CH_2CH_2Cl$　　　　C, $CH_3CHClCH_2Cl$, chiral
　　D, $CH_3CH_2CHCl_2$　　E, $CH_3CHClCHCl_2$, chiral

　　Inactive trichloro products from active C were: $CH_3CCl_2CH_2Cl$ and $ClCH_2CHClCH_2Cl$.

10. (a) None. (b) One pair of configurational enantiomers. (c) F and G are conformational
　　enantiomers. I, J, and K are conformers of one configurational enantiomer; L, M, and N
　　are conformers of the other.

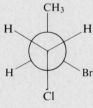

11. Attractive dipole–dipole interaction between —Cl and —CH$_3$. The groups can get far enough apart to avoid van der Waals repulsion.

12. (a) Zero. (b) The *gauche* conformer must be present. (c) 12% *gauche* (as non-resolvable racemic modification), 88% *anti*.

$$\mu^2 = N_x\mu_x^2 + N_y\mu_y^2$$

$$(1.12)^2 = N_{anti} \times 0 + N_{gauche}(3.2)^2$$

$$N_{gauche} = 0.12 \quad N_{anti} = 1 - N_{gauche} = 0.88$$

Chapter 5

Alkenes I. Structure and Preparation
Elimination

5.1 (a) 1-Pentene
(Z)-2-pentene
(E)-2-pentene
2-methyl-1-butene
3-methyl-1-butene
2-methyl-2-butene

(b) (Z)-1-chloropropene
(E)-1-chloropropene
2-chloropropene
3-chloropropene

(c) (Z)-1-chloro-1-butene
(E)-1-chloro-1-butene
2-chloro-1-butene
3-chloro-1-butene
4-chloro-1-butene
(Z)-1-chloro-2-butene
(E)-1-chloro-2-butene
(Z)-2-chloro-2-butene
(E)-2-chloro-2-butene
1-chloro-2-methylpropene
3-chloro-2-methylpropene

Z isomers are:

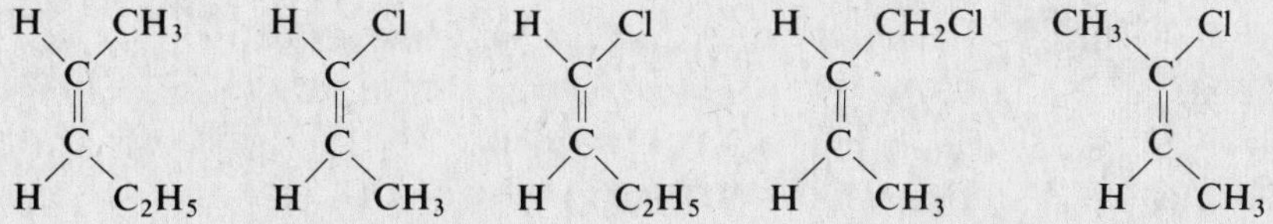

5.2 (a) $(CH_3)_2C{=}C(CH_3)_2$ (c)

(b) $H_2C{=}C(CH_3)CH_2Br$

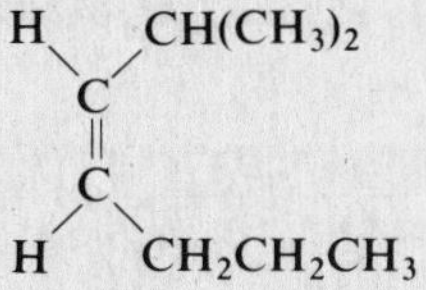

(d)

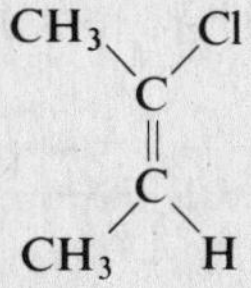

5.3 Names given in answer to Problem 5.1.

5.4 (a) *trans*-1,2-Dichloroethene is non-polar; in 1,1- and *cis*-1,2-dichloroethene, dipole lies in plane of molecule along bisector of angle between Cl atoms.

$\mu = 0$ net dipole

(b) The C_4 compound has the larger dipole moment because of electron release by the two —CH_3 groups in the same direction as the C—Cl dipole.

(c) Net dipole in direction of Cl atoms, but smaller because C—Br dipoles oppose the C—Cl dipoles.

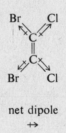

net dipole

5.5 (a) 1-Pentene   (b) 1-pentene   (c) (*Z*)- and (*E*)-2-pentene
                             (*Z*)- and (*E*)-2-pentene

(d) 2-methyl-2-butene   (e) 2-methyl-2-butene   (f) 2,3-dimethyl-2-butene   (g) none
    2-methyl-1-butene   3-methyl-1-butene   2,3-dimethyl-1-butene

5.6 (a) *tert*-butyl or isobutyl   (b) *n*-pentyl   (c) 3-pentyl
    (d) $CH_3CH_2CH(CH_3)CH_2X$ (e) none   (f) isopentyl

5.7 Where a choice arises, the more substituted alkene—the more stable alkene—is expected to predominate.

(a) 1-pentene   (b) 2-pentene (predominantly *E*)   (c) 2-pentene (predominantly *E*)
(d) 2-methyl-2-butene   (e) 2-methyl-2-butene   (f) 2,3-dimethyl-2-butene
(g) none

5.8 See equations on page 193.

5.9 The more highly substituted alkene—the more stable alkene—is expected to be the preferred product.

(a) 2-methyl-2-butene (b) 2-methyl-2-butene (c) 2,3-dimethyl-2-butene

1. (a) $CH_3CH_2CH(CH_3)CH_2CH_2CH(CH_3)CH=CH_2$ (b) $ClCH_2CH=CH_2$ (c) $(CH_3)_3CCH=C(CH_3)_2$

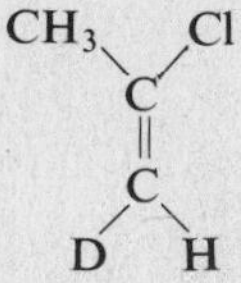

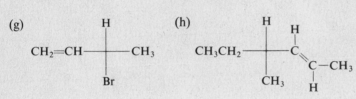

(g)

$CH_2=CH$—$\overset{\overset{\displaystyle H}{|}}{\underset{\underset{\displaystyle Br}{|}}{C}}$—$CH_3$

(h)

CH_3CH_2—$\overset{\overset{\displaystyle H}{|}}{\underset{\underset{\displaystyle CH_3}{|}}{C}}$—$\overset{\overset{\displaystyle H}{}}{\underset{\underset{\displaystyle H}{}}{C=C}}$—$CH_3$

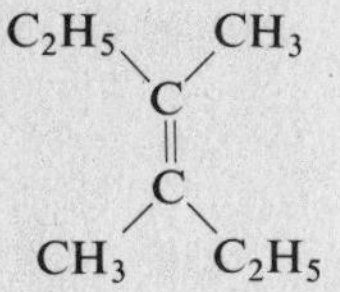

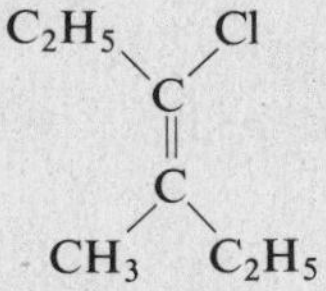

2. (a) 2-methylpropene   (b) (Z)-3-hexene   (c) 3,3-dimethyl-1-butene
   (d) (E)-2,5-dimethyl-3-hexene (e) 2,5-dimethyl-2-hexene   (f) 2-ethyl-1-butene

3. b, d, g, h, i, k (3 isomers: Z,Z-, E,E-, Z,E-).

4. (a) 1-hexene✓   3-methyl-1-pentene✓
      2-hexene✓   3-methyl-2-pentene✓
      3-hexene✓   3,3-dimethyl-1-butene
      2-methyl-1-pentene✓ 2,3-dimethyl-1-butene
      2-methyl-2-pentene✓ 2,3-dimethyl-2-butene
   ✓4-methyl-1-pentene   2-ethyl-1-butene
      4-methyl-2-pentene✓

   (b) (Z)- and (E)-2-hexene   (Z)- and (E)-3-hexene
       (Z)- and (E)-4-methyl-2-pentene (Z)- and (E)-3-methyl-2-pentene

   (c) (R)- and (S)-3-methyl-1-pentene

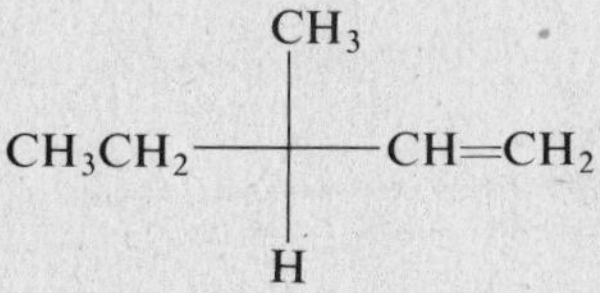

$$CH_3CH_2-\overset{\overset{\displaystyle CH_3}{|}}{\underset{\underset{\displaystyle H}{|}}{C}}-CH=CH_2 \text{ is } R$$

5. Differ in all except (h); dipole moment would tell.

6. (a) $CH_3CH_2CH_2OH \xrightarrow{H^+, \text{ heat}} CH_3CH=CH_2 + H_2O$

   (b) $CH_3CHOHCH_3 \xrightarrow{H^+, \text{ heat}} CH_3CH=CH_2 + H_2O$

   (c) $CH_3CHClCH_3 + KOH \longrightarrow CH_3CH=CH_2 + KCl + H_2O$

   (d) $CH_3C\equiv CH + H_2 \xrightarrow{\text{Pd or Ni-B}} CH_3CH=CH_2$ (See Sec. 8.9)

   (e) $CH_3CHBrCH_2Br + Zn \longrightarrow CH_3CH=CH_2 + ZnBr_2$

7. (a) 1-hexene   (b) 1-hexene   (c) 2-methyl-1-pentene
                                   (Z)-2-hexene
                                   (E)-2-hexene

   (d) 2-methyl-2-pentene   (e) 2-methyl-2-pentene   (f) 4-methyl-1-pentene
       2-methyl-1-pentene   (Z)-4-methyl-2-pentene   (Z)-4-methyl-2-pentene
                                   (E)-4-methyl-2-pentene   (E)-4-methyl-2-pentene

   (g) 4-methyl-1-pentene   (h) 2,3-dimethyl-2-pentene
                                   (Z)-3,4-dimethyl-2-pentene
                                   (E)-3,4-dimethyl-2-pentene
                                   2-ethyl-3-methyl-1-butene

8. The more highly substituted alkene is usually the major product.

   (b) 2-hexene (predominantly E)   (d) 2-methyl-2-pentene
   (e) 2-methyl-2-pentene   (f) 4-methyl-2-pentene (predominantly E)
   (h) 2,3-dimethyl-2-pentene

9. The order of reactivity is generally $3° > 2° > 1°$.

   (a) $CH_3CH_2CH_2CHOHCH_3$ (b) $(CH_3)_2C(OH)CH_2CH_3$ (c) $(CH_3)_2CHC(OH)(CH_3)_2$

10. (a) Br_2, heat, light; KOH(alc), heat.

11. n-Propyl cation from n-PrF rearranges (by hydride shift) into the more stable (secondary) isopropyl cation, the species formed directly from i-PrF.

$$n\text{-PrF} + SbF_5 \longrightarrow n\text{-Pr}^+SbF_6^- \longrightarrow i\text{-Pr}^+SbF_6^- \longleftarrow i\text{-PrF} + SbF_5$$

   (For an early paper on carbonium ions, and a much later one, see F. C. Whitmore, "The Common Basis of Intramolecular Rearrangements," J. Am Chem. Soc., **54**, 3274 (1932), and G. A. Olah *et al.*, "Stable Carbonium Ions. V. Alkylcarbonium Hexafluoroantimonates," J. Am. Chem. Soc., **86**, 1360 (1964).)

12. (a) $(CH_3)_2C=CHCH_3$ (major product) and $CH_2=C(CH_3)C_2H_5$, both via rearrangement.

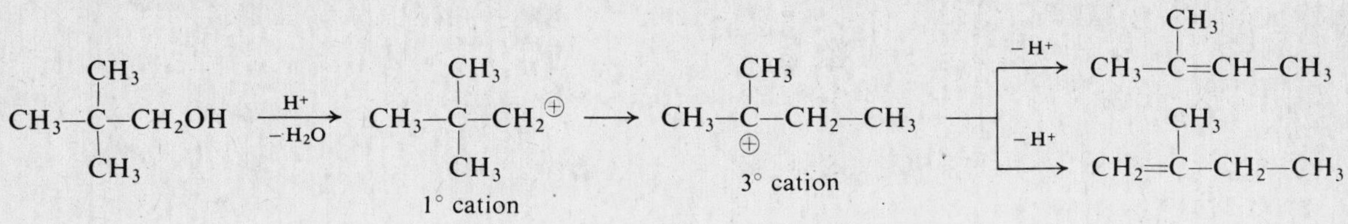

$$
\begin{array}{c}
\overset{\displaystyle CH_3}{\underset{\displaystyle CH_3}{CH_3-\overset{|}{\underset{|}{C}}-CH_2OH}} \xrightarrow[-H_2O]{H^+} \overset{\displaystyle CH_3}{\underset{\displaystyle \underset{1°\ cation}{CH_3}}{CH_3-\overset{|}{\underset{|}{C}}-CH_2^{\oplus}}} \longrightarrow \underset{3°\ cation}{CH_3-\overset{\displaystyle CH_3}{\overset{|}{\underset{\oplus}{C}}}-CH_2-CH_3}
\end{array}
$$

$$
\xrightarrow{-H^+}\ CH_3-\overset{\displaystyle CH_3}{\overset{|}{C}}=CH-CH_3
$$

$$
\xrightarrow{-H^+}\ CH_2=\overset{\displaystyle CH_3}{\overset{|}{C}}-CH_2-CH_3
$$

The major product is the more highly substituted alkene, as usual.

(b) No. Slow ionization of halide to yield neopentyl cation, then the same alkenes as in (a).
Acid catalysis is not required because Br^- is a better leaving group than OH^-.

13. Rearrangement of a 2° cation into a 3° cation appears likely. See page 197 for equations.

Chapter 6 | Alkenes II. Reactions of the Carbon–Carbon Double Bond
Electrophilic and Free-Radical Addition

6.1 (a) $C_4H_8 + 6O_2 \longrightarrow 4CO_2 + 4H_2O$ (b) same as (a).

(c) 1-butene   649.8   (d) 1-pentene   806.9
    cis-2-butene   648.1   *cis*-2-pentene   805.3
    trans-2-butene 647.1   *trans*-2-pentene 804.3

The most highly branched alkene is usually the most stable (lowest energy content); a *trans*-isomer is usually more stable (lower energy content) than a *cis*-isomer.

6.2 (a) H_3O^+; HBr.   (b) HBr.   (c) HBr.

In aqueous solution, HBr reacts to yield the weaker acid, H_3O^+. (Compare Problem 10, Chapter 1.)

6.3 If carbonium ions were formed reversibly and rapidly, some of them would lose H^+ rather than the newly acquired D^+, to form $(CH_3)_2C{=}CDCH_3$, which would be found in the unconsumed alkene at the end of the reaction.

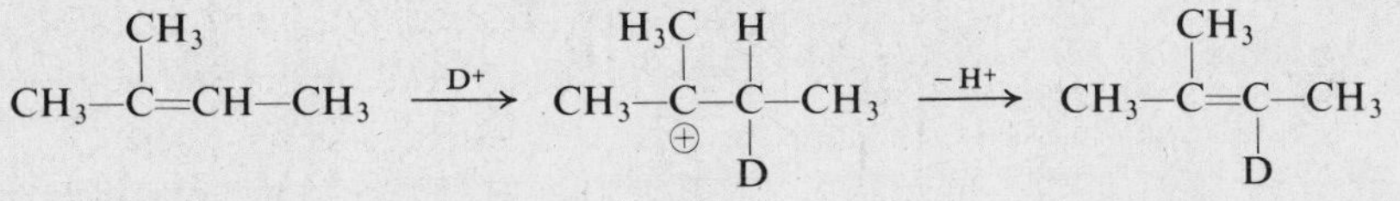

6.4 3-Chloro- and 2-chloro-2-methylbutane; the second product results from attachment of Cl^- to a 3° cation formed by a hydride shift.

$$CH_3-\underset{\underset{H}{|}}{\overset{\overset{CH_3}{|}}{C}}-CH=CH_2 \xrightarrow{H^+} CH_3-\underset{\underset{H}{|}}{\overset{\overset{CH_3}{|}}{C}}-\overset{\oplus}{C}H-CH_3 \xrightarrow{rearr.} CH_3-\underset{\oplus}{\overset{\overset{CH_3}{|}}{C}}-CH_2-CH_3$$

$$\downarrow Cl^- \qquad\qquad\qquad\qquad \downarrow Cl^-$$

$$CH_3-\overset{\overset{CH_3}{|}}{C}-\underset{\underset{Cl}{|}}{\overset{\overset{|}{}}{C}H}-CH_3 \qquad\qquad CH_3-\underset{\underset{Cl}{|}}{\overset{\overset{CH_3}{|}}{C}}-CH_2-CH_3$$

6.5 The skeletal rearrangement is identical to that in the addition reaction on page 197, and strongly suggests that the same carbonium ions are intermediates. The secondary cation is formed in a familiar way (Sec. 5.20),

$$CH_3-\underset{\underset{OH}{|}}{\overset{\overset{CH_3}{|}}{C}}\underset{H_3C}{}-CH-CH_3 \rightleftharpoons CH_3-\underset{H_3C\ \ OH_2^+}{\overset{\overset{CH_3}{|}}{C}}-CH-CH_3 \xrightarrow{-H_2O} CH_3-\underset{\underset{CH_3}{|}}{\overset{\overset{CH_3}{|}}{C}}-\overset{\oplus}{C}H-CH_3$$

2° cation

and rearranges to the tertiary cation, which combines with chloride ion.

$$CH_3-\underset{CH_3}{\overset{\overset{CH_3}{|}}{C}}-\overset{\oplus}{C}H-CH_3 \xrightarrow{rearr.} CH_3-\underset{\underset{CH_3}{|}}{\overset{\overset{CH_3}{|}}{\underset{\oplus}{C}}}-CH-CH_3 \xrightarrow{Cl^-} CH_3-\underset{\underset{Cl\ \ CH_3}{}}{\overset{\overset{CH_3}{|}}{C}}-CH-CH_3$$

2° cation   3° cation

The net result is *substitution*, via a combination of steps we have encountered before: the first, in elimination, and the last, in addition.

6.6 In view of chemistry we have learned, combination of isobutane with ethylene would seem to require generation of a *tert*-butyl cation,

$$CH_2{=}CH_2 \xrightarrow{H^+} CH_3CH_2^{\oplus} \xrightarrow{t\text{-}BuH} CH_3CH_3 + CH_3-\underset{\underset{CH_3}{|}}{\overset{\overset{CH_3}{|}}{\overset{\oplus}{C}}}$$

followed by its addition to ethylene.

$$CH_3-\underset{\underset{CH_3}{|}}{\overset{\overset{CH_3}{|}}{\overset{\oplus}{C}}} + CH_2{=}CH_2 \longrightarrow CH_3-\underset{\underset{CH_3}{|}}{\overset{\overset{CH_3}{|}}{C}}-CH_2-CH_2^{\oplus}$$

The skeleton of the alkylation product requires rearrangement of this 1° cation: by first a hydride shift to give a 2° cation, and then an alkyl shift to give a 3° cation, which has the required carbon skeleton.

$$CH_3-\underset{\underset{H_3C\ \ H}{}}{\overset{\overset{CH_3}{|}}{C}}-CH-CH_2^{\oplus} \longrightarrow CH_3-\underset{CH_3}{\overset{\overset{CH_3}{|}}{C}}-\overset{\oplus}{C}H-CH_3 \longrightarrow CH_3-\underset{\underset{CH_3}{|}}{\overset{\overset{CH_3}{|}}{\underset{\oplus}{C}}}-CH-CH_3$$

1° cation   2° cation   3° cation

Finally, the 3° cation abstracts hydride from isobutane to give the alkylation product and a new *tert*-butyl cation, which continues the chain.

$$
\begin{array}{ccc}
& CH_3 & CH_3 & CH_3 & CH_3 \\
CH_3-\overset{\oplus}{C}-CH-CH_3 + H-\overset{|}{C}-CH_3 & \longrightarrow & CH_3-\overset{|}{C}-CH-CH_3 + \overset{\oplus}{C}-CH_3 \\
\underset{CH_3}{|} & \underset{CH_3}{|} & H\ CH_3 & CH_3
\end{array}
$$

6.7 Same mechanism as on page 205, with radicals in steps (2) and (4) abstracting: (a) H, (b) Br, (c) Br, (d) H from S, (e) H. For example, in (e):

(1)   peroxide $\longrightarrow$ Rad·

(2)   Rad· + R—C=O $\longrightarrow$ Rad:H + R—C=O
                                           $\ddot{H}$ ·

(3)   R—C=O + *n*-C$_6$H$_{13}$CH=CH$_2$ $\longrightarrow$ *n*-C$_6$H$_{13}$ĊH—CH$_2$—C—R
                       ·   ‖
                                                                                          O

(4)   *n*-C$_6$H$_{13}$ĊH—CH$_2$—C—R + R—C=O $\longrightarrow$ *n*-C$_6$H$_{13}$CH$_2$—CH$_2$—C—R + R—Ċ=O
                                 ‖ $\ddot{H}$ ‖
                                 O   O

   then (3), (4), (3), (4), etc.

(Key steps in development of the theory of free-radical addition are described in: M. S. Kharasch and F. R. Mayo, J. Am. Chem. Soc., **55**, 2468 (1933); M. S. Kharasch, H. Engelmann, and F. R. Mayo, J. Org. Chem., **2**, 288 (1937); M. S. Kharasch, E. V. Jensen, and W. H. Urry, Science, **102**, 128 (1945). For a personal account of the discovery of the peroxide effect, see F. R. Mayo in W. A. Waters (ed.), *Vistas in Free Radical Chemistry*, Pergamon Press, New York, 1959, pp. 139–142.)

6.8 The radical produced in step (3) of the CCl$_4$ sequence on page 205 adds to RCH=CH$_2$, and the new radical so formed then attacks CCl$_4$ in the manner of step (4).

   RĊH—CH$_2$—CCl$_3$ + RCH=CH$_2$ $\longrightarrow$ RCH—CH$_2$—ĊH—CH$_2$—CCl$_3$
                                                              |
                                                              R

   RCH—CH$_2$—ĊH—CH$_2$—CCl$_3$ + Cl:CCl$_3$ $\longrightarrow$ RCH—CH$_2$—CH—CH$_2$—CCl$_3$ + ·CCl$_3$
                  | | |
                  R   Cl   R

There is thus *competition* between two familiar reactions of free radicals (in this case, of the RCHCH$_2$CCl$_3$ radical): addition to a double bond, or abstraction of an atom. This competition exists in all reactions of this kind; the relative importance of the two paths depends on how reactive the alkene is, how reactive CX$_4$ is, and how selective RCHCH$_2$CCl$_3$ is. (See Problem 7, page 1050.)

6.9 (a) No free radicals are formed in the dark, and the ionic reaction is too slow with de-activated tetrachloroethylene.

(b) (1) $\qquad\qquad$ $Cl_2 \xrightarrow{\text{light}} 2Cl\cdot$

(2) $\quad Cl\cdot + Cl_2C{=}CCl_2 \longrightarrow Cl_3C{-}\overset{\cdot}{C}Cl_2$

(3) $\quad Cl_3C{-}\overset{\cdot}{C}Cl_2 + Cl_2 \longrightarrow Cl_3C{-}CCl_3 + Cl\cdot$

then (2), (3), (2), (3), etc.

Oxygen stops the chain by reacting with the free radical or $Cl\cdot$.

6.10 (a) Orlon, $CH_2{=}CH{-}CN$, acrylonitrile
(b) Saran, $CH_2{=}CCl_2$, 1,1-dichloroethene (vinylidene chloride)
(c) Teflon, $CF_2{=}CF_2$, tetrafluoroethylene

6.11 If the more stable radical is formed faster in each step—a 2° radical, say, rather than a 1° radical—orientation will always be the same.

6.12 React with HCl (minimum E_{act} 15 kcal).

6.13

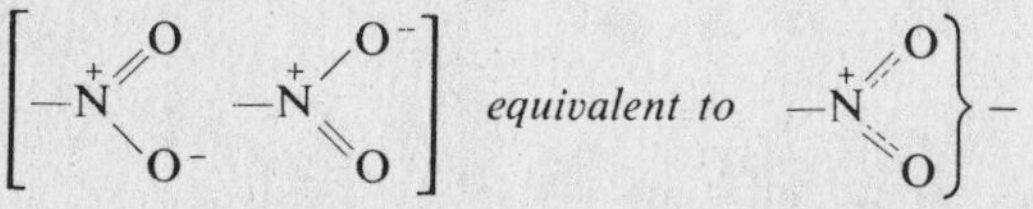

$$\left[-\overset{+}{N}\underset{O^-}{\overset{O}{\Big\langle}}\quad -\overset{+}{N}\underset{O}{\overset{O^-}{\Big\langle}} \right] \quad equivalent\ to\quad \left. -\overset{+}{N}\underset{O}{\overset{O}{\Big\langle}} \right\} -$$

6.14

$$\left[O{=}C\underset{O^-}{\overset{O^-}{\Big\langle}}\quad {}^-O{-}C\underset{O^-}{\overset{O}{\Big\langle}}\quad {}^-O{-}C\underset{O}{\overset{O^-}{\Big\langle}} \right] \quad equivalent\ to\quad \left. O{=\!=}C\underset{O}{\overset{O}{\Big\langle}} \right\}{-}{-}$$

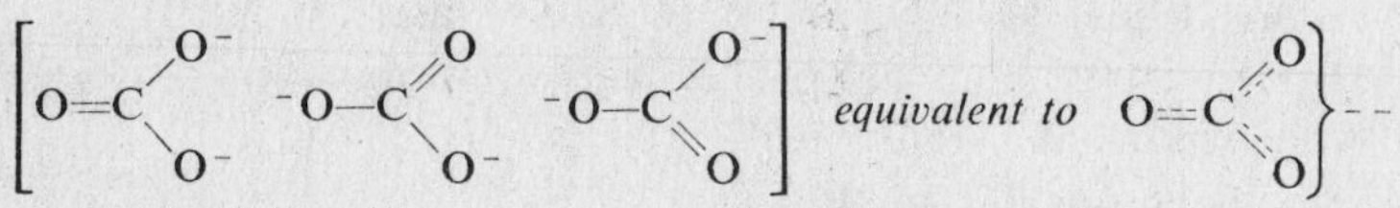

6.15 (a) 188 + 88 (Table 1.2, page 21) = 276 kcal. It is 24 kcal more stable than ethyl cation (300 kcal, Fig. 5.9, page 165), about as stable as isopropyl cation (277 kcal).

(b) $\left[\underset{+}{C}H_2{-}CH{=}CH_2 \quad CH_2{=}CH{-}\underset{+}{C}H_2 \right]$ *equivalent to* $\quad \underbrace{CH_2{=\!=}CH{=\!=}CH_2}_{+}$

Orbital picture as in Fig. 6.15 (p. 214), except only two π electrons instead of three.

6.16 (a) No. The measured heat of hydrogenation (49.8 kcal/mole) is 36.0 kcal/mole less than that calculated for 3 double bonds (3 × 28.6 = 85.8 kcal/mole).

(b) A better representation is:

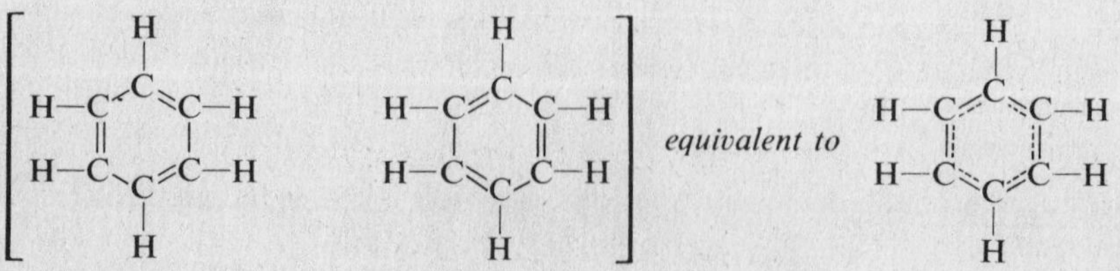

6.17 Hyperconjugation involving structures like I–IV on page 217 with $+$ replacing odd electron; in orbital picture on page 217, p orbital contains "no e" instead of "1 e".

6.18 (a) $(CH_3)_3CCH_2C{=}O$ and $O{=}CH_2$; $(CH_3)_3C{-}C{=}O$ and $O{=}C(CH_3)_2 \cdot$

      | |
      CH_3 H

(b) $(CH_3)_3CCH_2C{=}O$ and CO_2; $(CH_3)_3C{-}COOH$ and $O{=}C(CH_3)_2 \cdot$

      |
      CH_3

Ozonolysis or cleavage by $KMnO_4$ effectively replaces a double bond to carbon by a double bond to oxygen. $KMnO_4$ oxidizes aldehydes further to carboxylic acids,

$$R{-}C{=}O \longrightarrow R{-}C{=}O$$
$$\ \ \ \ |\ |$$
$$\ \ \ \ H\ \ \ \ \ \ \ \ \ \ \ \ \ \ \ \ \ \ OH$$

Aldehydes   Carboxylic
         acids

and specifically formaldehyde is oxidized to carbon dioxide.

$$H{-}C{=}O \longrightarrow H{-}C{=}O \longrightarrow HO{-}C{=}O \xrightarrow{-H_2O} O{=}C{=}O$$
$$\ \ \ \ |\ \ \ \ \ \ \ \ \ \ \ \ \ \ \ \ \ \ |\ \ \ \ \ \ \ \ \ \ \ \ \ \ \ \ \ |$$
$$\ \ \ \ H\ \ \ \ \ \ \ \ \ \ \ \ \ \ \ OH\ \ \ \ \ \ \ \ \ \ OH$$

Formaldehyde Formic   Carbonic   Carbon
       acid   acid   dioxide
             Unstable

6.19 (a) Br_2/CCl_4 or $KMnO_4$, or conc. H_2SO_4; (b) Br_2/CCl_4 or $KMnO_4$, or conc. H_2SO_4, or halogen test; (c) Br_2/CCl_4, or CrO_3/H_2SO_4; (d) only alkene gives positive Br_2/CCl_4 test; only alkyl halide gives halogen test; only 2° alcohol gives positive CrO_3/H_2SO_4 test.

6.20 A, alkane B, 2° alcohol C, alkyl halide D, alkene E, 3° alcohol

1. (a) 1,2-dibromoethane
 (c) bromoethene (vinyl bromide)
 (e) 1,2-propanediol
 (g) bromoethene (bromoethylene)

 (b) bromoethane
 (d) 1,2-ethanediol
 (f) 1-bromo-2-propanol
 (h) 3-chloro-1-propene

2. (a) isobutane
 (c) 1,2-dibromo-2-methylpropane
 (e) *tert*-butyl bromide
 (g) *tert*-butyl iodide
 (i) *tert*-butyl hydrogen sulfate
 (k) 1-bromo-2-methyl-2-propanol
 (m) 2,4,4-trimethyl-1-pentene and 2,4,4-trimethyl-2-pentene
 (o) 2-methyl-1,2-propanediol (isobutylene glycol)
 (q) same as (o)

 (b) 1,2-dichloro-2-methylpropane
 (d) no reaction
 (f) isobutyl bromide
 (h) *tert*-butyl iodide
 (j) *tert*-butyl alcohol
 (l) 1-bromo-2-chloro-2-methylpropane and products (c) and (k)
 (n) 2,2,4-trimethylpentane
 (p) acetone and carbon dioxide
 (r) acetone and formaldehyde.

3. (a) propylene   (b) ethylene   (c) 2-butene   (d) isobutylene
   (e) vinyl chloride (f) 2-methyl-1-butene (g) ethylene   (h) propylene

Relative reactivities follow the sequence given on page 195. The electronegative oxygens of —COOH make this group electron withdrawing, like the halogens (see page 196).

4. (a) 2-iodobutane   (b) 3-iodopentane and 2-iodopentane
   (c) *tert*-pentyl iodide   (d) *tert*-pentyl iodide
   (e) 3-iodo-2-methylbutane and   (f) 1-bromo-1-iodoethane
       2-iodo-2-methylbutane
   (g) 2-iodo-2,3-dimethylbutane   (h) 4-iodo-2,2,4-trimethylpentane

In each case orientation follows Markovnikov's rule (Secs. 6.6 and 6.11). In (b), the rule indicates no preference—a 2° cation is formed either way—and both products are obtained. In (e), some of the intermediate 2° cation rearranges to the more stable 3° cation.

5.

$$\overset{\overset{\displaystyle H_3C}{|}}{CH_3-C}-\overset{\overset{\displaystyle H}{|}}{\underset{\underset{\displaystyle H}{|}}{C}}-\overset{\overset{\displaystyle H}{|}}{\underset{\underset{\displaystyle H}{|}}{C}}-\overset{\overset{\displaystyle H}{|}}{C}=\overset{\overset{\displaystyle H}{|}}{C}-CH_3$$

$$\underset{H\ H\ H}{}$$

    5   3   4   1   6   6   2   *order of reactivity*

H that is both 2° and allylic is more reactive than H that is 1° and allylic.

6. The 3° radical is more stable than the 2° radical, and forms faster.

7. Methyl alcohol (rather than water) combines with the carbonium ion; subsequent loss of a proton leads to an ether.

$$\underset{Br}{CH_2}-\overset{\oplus}{CH_2} + CH_3OH \longrightarrow \underset{Br}{CH_2}-\underset{\overset{\oplus}{O}-CH_3}{CH_2} \xrightarrow{-H^+} \underset{Br}{CH_2}-\underset{OCH_3}{CH_2}$$

8. If alkene were intermediate, it should undergo reaction with D^+ as well as with H^+, and the product (alcohol) should contain D attached to carbon.

$$\underset{H}{-C}-\overset{\oplus}{C}- \xrightarrow{-H^+} -C=C- \xrightarrow{D^+} \underset{D}{-C}-\overset{\oplus}{C}- \xrightarrow{water} \text{C-deuteriated alcohol}$$

9. (a) Less likely: formation of C—Br bond less exothermic than formation of C—H bond; (4a) would have minimum E_{act} of 19 kcal for addition to ethylene, even bigger for other alkenes.

(b) Intermediate radical (step 3, page 203) must be

$$\underset{Br}{-C}-C-$$

since otherwise (step 3a, page 222) it would be

$$-\overset{\displaystyle |}{C}-\overset{\displaystyle |}{\underset{\displaystyle H}{C}}-$$

or

$$-\overset{\displaystyle |}{C}-\overset{\displaystyle |}{\underset{\displaystyle D}{C}}-$$

depending on whether HBr or DBr was used.

10. (a) See page 203.

    (b) As in (a), with Cl for Br.

    (c) Using 68 kcal for π bond and 95 kcal for 2° R—H, one calculates:

        for HBr, step (3) -1 kcal, step (4), -7 kcal;
        for HCl, step (3) -14 kcal, step (4), $+8$ kcal.

    (d) Steps (2) and (4) are too difficult with HCl, and cannot compete with ionic addition.

11. (a) $Cl^{14}CH_2C(CH_3){=}CH_2$

    (b) Ionic. No light or peroxides to generate free radicals; if free-radical chains *were* started (by, say, concerted homolysis, page 206), oxygen would break many chains and thus lower the yield of methallyl chloride—contrary to fact.

    (c) $(CH_3)_2C{=}CH_2 + Cl_2 \longrightarrow (CH_3)_2\overset{\displaystyle \oplus}{C}{-}CH_2Cl + Cl^-$

$$(CH_3)_2\overset{\displaystyle \oplus}{C}{-}CH_2Cl \longrightarrow CH_2{=}\overset{\displaystyle \overset{\textstyle CH_3}{|}}{C}{-}CH_2Cl + H^+$$

    (d) More stable 3° cation less reactive toward Cl^-, and more prone to lose one of six hydrogens to form branched alkene.

    (e) $(CH_3)_3CCH{=}CH_2 + Cl_2 \longrightarrow (CH_3)_3CCH{-}\overset{\displaystyle \oplus}{}CH_2Cl + Cl^-$

$$(CH_3)_3CCH{-}\overset{\displaystyle \oplus}{}CH_2Cl \xrightarrow{\text{rearr.}} (CH_3)_2\overset{\displaystyle \oplus}{C}{-}CH(CH_3){-}CH_2Cl$$

$$(CH_3)_2\overset{\displaystyle \oplus}{C}{-}CH(CH_3){-}CH_2Cl \longrightarrow CH_2{=}\overset{\displaystyle \overset{\textstyle CH_3}{|}}{C}{-}CH(CH_3){-}CH_2Cl + H^+$$

12. See pp. 597–598.

13. (a) Bromine can become attached to either C–1 or C–3 of intermediate radical,

$$C_5H_{11}\underbrace{CH{=\!\cdot\!=}CH{=\!\cdot\!=}CH_2}$$

(b) ^{14}C will be distributed equally ("scrambled") between C–1 and C–3.

$$^{14}CH_2{=}CH{-}CH_3 \xrightarrow{\;Br\cdot\;} \left[\begin{array}{l} ^{14}CH_2{=}CH{-}CH_2\cdot \\ ^{14}\dot{C}H_2{-}CH{=}CH_2 \end{array} \right] \textit{equivalent to } \; ^{14}CH_2{=}\!\!=\!CH\!=\!\!=CH_2$$

$$^{14}CH_2{=}\!\!=\!CH\!=\!\!=CH_2$$

$$\Big\downarrow Br_2$$

$$^{14}CH_2{=}CH{-}CH_2Br \qquad\qquad Br^{14}CH_2{-}CH{=}CH_2$$

Equal amounts
(^{14}C label is "scrambled")

14. (a) n-Pr$-\overset{\overset{\displaystyle H}{|}}{C}{=}O + O{=}\overset{\overset{\displaystyle H}{|}}{C}H \xleftarrow{\;O_3\;} n$-Pr$-\overset{\overset{\displaystyle H}{|}}{C}{=}\overset{\overset{\displaystyle H}{|}}{C}H$

                                    1-Pentene

(b) i-Pr$-\overset{\overset{\displaystyle H}{|}}{C}{=}O + O{=}\overset{\overset{\displaystyle H}{|}}{C}{-}CH_3 \xleftarrow{\;O_3\;} CH_3{-}\overset{\overset{\displaystyle H}{|}}{\underset{\underset{\displaystyle CH_3}{|}}{C}}{-}\overset{\overset{\displaystyle H}{|}}{C}{=}\overset{\overset{\displaystyle H}{|}}{C}{-}CH_3$

                                    4-Methyl-2-pentene

(c) $CH_3{-}\overset{\overset{\displaystyle CH_3}{|}}{C}{=}O + O{=}\overset{\overset{\displaystyle CH_3}{|}}{C}{-}CH_3 \xleftarrow{\;O_3\;} CH_3{-}\overset{\overset{\displaystyle H_3C}{|}}{C}{=}\overset{\overset{\displaystyle CH_3}{|}}{C}{-}CH_3$

                                    2,3-Dimethyl-2-butene

(d) $CH_3{-}\overset{\overset{\displaystyle H}{|}}{C}{=}O + O{=}\overset{\overset{\displaystyle H}{|}}{C}{-}CH_2{-}\overset{\overset{\displaystyle H}{|}}{C}{=}O + O{=}\overset{\overset{\displaystyle H}{|}}{C}H \xleftarrow{\;O_3\;} CH_3{-}\overset{\overset{\displaystyle H}{|}}{C}{=}\overset{\overset{\displaystyle H}{|}}{C}{-}CH_2{-}\overset{\overset{\displaystyle H}{|}}{C}{=}\overset{\overset{\displaystyle H}{|}}{C}H$

                                                      1,4-Hexadiene

(e)   Carboxylic acids (RCOOH) instead of aldehydes (RCHO); CO_2 instead of formaldehyde (HCHO); ketones (RCOR′) in (c).

15. (a) $AgNO_3$ (b) Br_2/CCl_4 or $KMnO_4$, or $AgNO_3$, or
                                                                conc. H_2SO_4

(c) Br_2/CCl_4 or $KMnO_4$, or conc. H_2SO_4 (d) $AgNO_3$

(e) CrO_3/H_2SO_4, or conc. H_2SO_4 (actually,   (f) Br_2/CCl_4, or CrO_3/H_2SO_4
      alcohol is H_2O-sol.)

(g) conc. H_2SO_4 (actually, alcohol is   (h) Br_2/CCl_4
      H_2O-sol.)

16. 3-Hexene.   The carbon skeleton must be the same as that of n-hexane; there is one double bond, which must be located in the center of the chain (three carbons on each side).

17. (a) $CH_2{=}CH_2$ $\xleftarrow{\text{KOH(alc)}}$ CH_3CH_2Br $\xleftarrow[\text{light}]{Br_2}$ CH_3CH_3

(b) $CH_3CH{=}CH_2$ $\xleftarrow{\text{KOH(alc)}}$ $n\text{-PrBr} + i\text{-PrBr}$ $\xleftarrow[\text{light}]{Br_2}$ $CH_3CH_2CH_3$

(c) CH_3CH_2I $\xleftarrow{\text{HI}}$ $CH_2{=}CH_2$ $\longleftarrow$ (a)

(d) $CH_3CHBrCH_3$ $\xleftarrow{\text{HBr}}$ $CH_3CH{=}CH_2$ $\longleftarrow$ (b)

(e) $CH_3CHBrCH_2Br$ $\xleftarrow{\text{Br}_2/\text{CCl}_4}$ $CH_3CH{=}CH_2$ $\longleftarrow$ (b)

(f) $CH_3CH_2CHBrCH_2Br$ $\xleftarrow{\text{Br}_2/\text{CCl}_4}$ $CH_3CH_2CH{=}CH_2$ $\xleftarrow{\text{KOH(alc)}}$ $n\text{-BuBr}$

(g) $CH_3CH_2CHICH_3$ $\xleftarrow{\text{HI}}$ $CH_3CH_2CH{=}CH_2$ $\xleftarrow{\text{KOH(alc)}}$ $n\text{-BuCl}$

(h) $CH_3\underset{\underset{CH_3}{|}}{C}H{-}C_3H_7\text{-}n$ $\longleftarrow$
$\qquad (i\text{-Pr})_2CuLi$ $\xleftarrow{\text{CuX}}$ $i\text{-PrLi}$ $\xleftarrow{\text{Li}}$ $i\text{-PrBr}$ $\longleftarrow$ (d)
$\qquad n\text{-PrBr}$ $\xleftarrow{\text{HBr, peroxides}}$ $CH_3CH{=}CH_2$

(i) $CH_3CH_2\underset{\underset{CH_3}{|}}{C}H{-}Bu\text{-}n$ $\longleftarrow$
$\qquad (sec\text{-Bu})_2CuLi$ $\xleftarrow{\text{CuX}}$ $sec\text{-BuLi}$ $\xleftarrow{\text{Li}}$ $sec\text{-BuBr}$
$\qquad n\text{-BuBr}$ $\xrightarrow{\text{KOH(alc)}}$ $CH_3CH_2CH{=}CH_2$ $\xuparrow{\text{HBr}}$

(j) $(CH_3)_2CBrCH_2Br$ $\xleftarrow{Br_2}$ $(CH_3)_2C{=}CH_2$ $\xleftarrow{\text{KOH}}$
$\qquad\quad i\text{-BuBr}$ $\longleftarrow$
$\qquad\quad t\text{-BuBr}$ $\longleftarrow$ $\xleftarrow[\text{light}]{Br_2}$ $(CH_3)_3CH$

(k) $CH_3CH_2CHICH_3$ $\xleftarrow{\text{HI}}$ $\underset{Chiefly}{CH_3CH{=}CHCH_3}$ $\xleftarrow[\text{heat}]{H_2SO_4}$ $n\text{-BuOH}$

(l) $CH_3CH_2CH_2Br$ $\xleftarrow{\text{HBr, peroxides}}$ $CH_3CH{=}CH_2$ $\xleftarrow{\text{KOH(alc)}}$ $i\text{-PrBr}$

(m) $CH_3CHOHCH_2Cl$ $\xleftarrow{\text{Cl}_2,\ \text{H}_2\text{O}}$ $CH_3CH{=}CH_2$ $\xleftarrow{\text{KOH(alc)}}$ $n\text{-PrI}$

(n) $(CH_3)_2CHCH_2CH_2CH_3$ $\xleftarrow{\text{H}_2,\ \text{Ni}}$ $(CH_3)_2C{=}CHCH_2CH_3$ $\xleftarrow[\text{warm}]{H_2SO_4}$ $(CH_3)_2C(OH)CH_2CH_2CH_3$

(o) $(CH_3)_3CCH_2CH_3$ $\xleftarrow{\text{H}_2\text{O}}$ $(CH_3)_3CCH(MgCl)CH_3$ $\xleftarrow{\text{Mg, Et}_2\text{O}}$ $(CH_3)_3CCHClCH_3$

Chapter 7

Stereochemistry II. Preparation and Reactions of Stereoisomers

7.1 (a) Abstraction of either 2° hydrogen is equally likely, giving equal amounts of enantiomeric pyramidal radicals, each of which reacts equally readily with chlorine, thus giving equal amounts of enantiomeric *sec*-BuCl molecules (i.e., the racemic modification); in any case, a pyramidal radical would probably undergo rapid inversion and lose configuration.

(b) Displacement of either 2° hydrogen is equally likely, giving equal amounts of enantiomeric *sec*-BuCl molecules (racemic modification).

7.2 (a) 4 fractions. (b) 1-Cl-3-Me, 3-Cl-2-Me (2 enantiomers), 2-Cl-2-Me, 1-Cl-2-Me (2 enantiomers).

(c) No fraction is optically active: each fraction is either a single achiral compound or a racemic modification. (Inactive reactants give inactive products.)

(d) Replacement of either H on —CH$_2$— is equally likely, and replacement of any H on either —CH$_3$ is equally likely, in each case giving rise to the racemic modification.

7.3 (a)

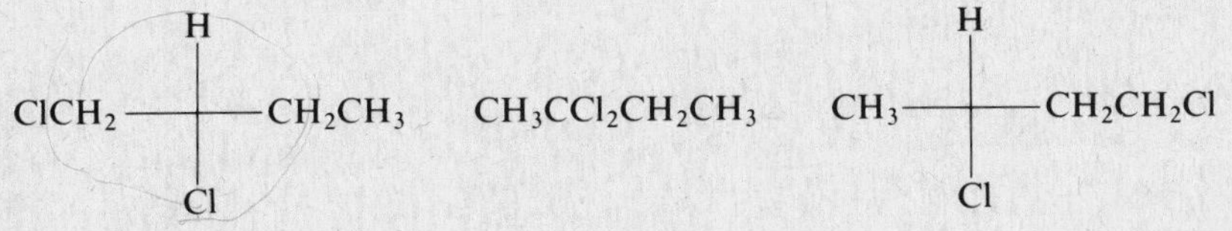

$$ClCH_2 \overset{\displaystyle H}{\underset{\displaystyle Cl}{\rule{0pt}{1pt}|\!\!-\!\!\!-\!\!|}} CH_2CH_3 \qquad CH_3CCl_2CH_2CH_3 \qquad CH_3 \overset{\displaystyle H}{\underset{\displaystyle Cl}{\rule{0pt}{1pt}|\!\!-\!\!\!-\!\!|}} CH_2CH_2Cl$$

$$\text{R} \qquad\qquad\qquad\qquad \textit{Achiral} \qquad\qquad\qquad \text{S}$$

(b) 1,2- and 1,3-dichlorobutane will be optically active; achiral 2,2-dichlorobutane will be optically inactive.

7.4 c, d, e, g. In others, a bond to chiral carbon is broken.

7.5 (a) C—O bond is broken; acid catalysis suggests that this happens by separation of H_2O from protonated alcohol. Clearly, C—O bond is re-formed; most probably this happens by attachment of H_2O to carbon: either *after* (carbonium ion mechanism) or *simultaneously with* (Sec. 14.9) loss of H_2O. Whatever the mechanism, the new C—O bond (nearly always) involves, not the original oxygen, but oxygen from the vastly more abundant solvent.

(b) Same as in (a), with C—I bond and I^- involved.

(c) Try the experiments with $H_2{}^{18}O$ and radioactive I^-: expect to find ^{18}O and radioactive I in products.

7.6
$$(+3.12°/+5.756°) \times (-1.64°) = -0.89°.$$
Purity of product same as purity of starting material

7.7 (a) V is mirror image of III; VI is mirror image of IV.
(b) V and VI are diastereomers.   (c) Ratio is the same.
(d) V $\longrightarrow$ *meso*; VI $\longrightarrow$ (R,R)-isomer.   (e) *meso* product.
(f) R,R:*meso* = 29:71.

7.8 (a) 5 diastereomeric fractions: two inactive, three active.

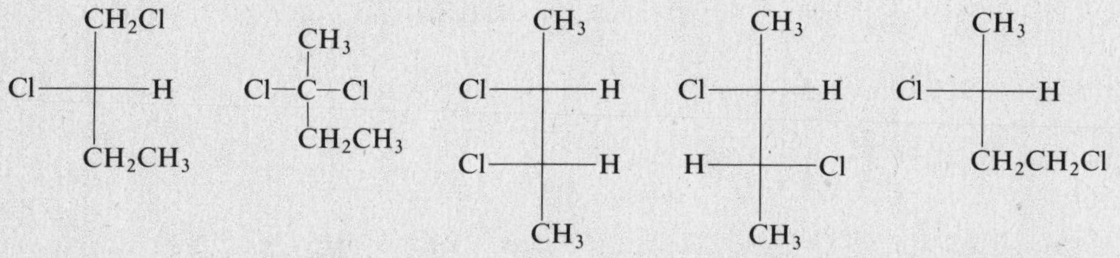

S	Achiral	R,S	R,R	R
Active	*Inactive*	*Meso* *Achiral* *Inactive*	*Active*	*Active*

(b) 5 diastereomeric fractions: all inactive. Same as in (a), except that each chiral structure is now accompanied by its enantiomer.

(c) 6 diastereomeric fractions: all inactive.

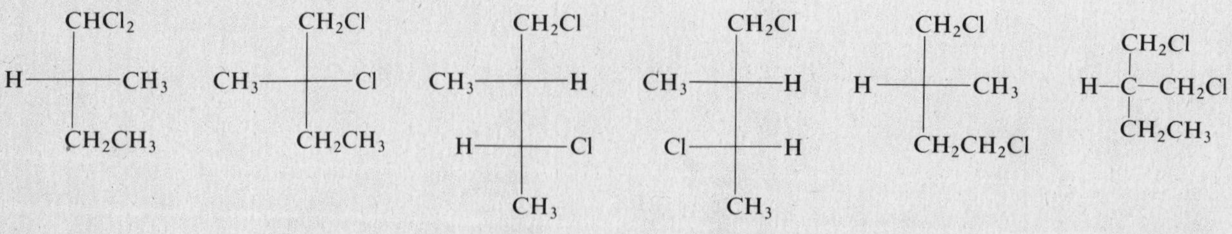

R	R	2R,3R	2R,3S	R	Achiral
+enantiomer (S) *Inactive*	+enantiomer (S) *Inactive*	+enantiomer (2S,3S) *Inactive*	+enantiomer (2S,3R) *Inactive*	+enantiomer (S) *Inactive*	*Inactive*

(d) 2 diastereomeric fractions: both active.

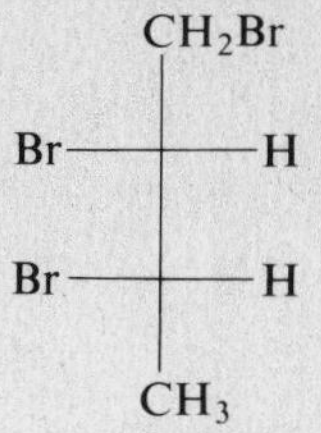

$$
\begin{array}{cc}
CH_2Br & CH_2Br \\
Br{-}|{-}H & H{-}|{-}Br \\
Br{-}|{-}H & Br{-}|{-}H \\
CH_3 & CH_3 \\
2R,3S & 2S,3S \\
\textit{Active} & \textit{Active}
\end{array}
$$

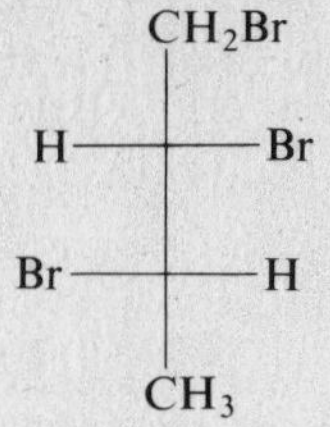

7.9 A pyramidal radical that undergoes inversion (page 18) faster than it reacts with chlorine, leading to equal amounts of the enantiomers.

7.10 As always, note that: (1) unless a bond to the original chiral center is broken, configuration is maintained at that center; and (2) if a new chiral center is generated, *both* possible configurations about the new center result.

In each active fraction there is only one compound, which is chiral.

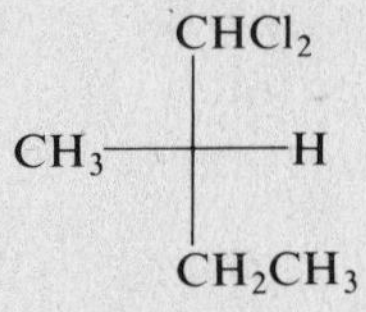

$$
\begin{array}{cccc}
CHCl_2 & CH_2Cl & CH_2Cl & CH_2Cl \\
CH_3{-}|{-}H & CH_3{-}|{-}H & CH_3{-}|{-}H & CH_3{-}|{-}H \\
CH_2CH_3 & H{-}|{-}Cl & Cl{-}|{-}H & CH_2CH_2Cl \\
 & CH_3 & CH_3 & \\
S & 2R,3R & 2R,3S & S
\end{array}
$$

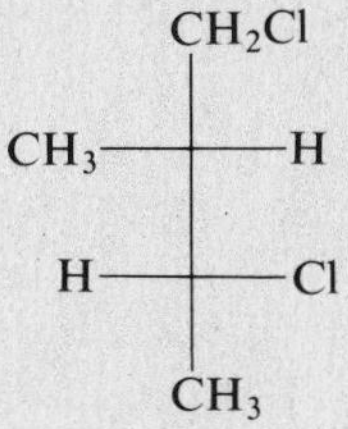

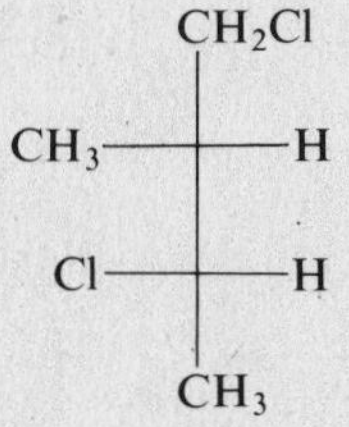

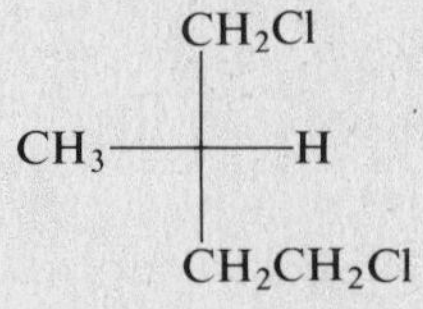

In one of the inactive fractions there is only one compound; it is achiral (only 3 different groups on C–2).

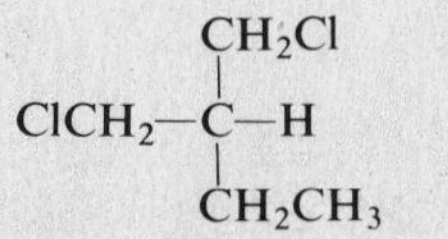

$$
\begin{array}{c}
CH_2Cl \\
ClCH_2{-}C{-}H \\
CH_2CH_3 \\
\textit{Achiral}
\end{array}
$$

The other inactive fraction is a racemic modification.

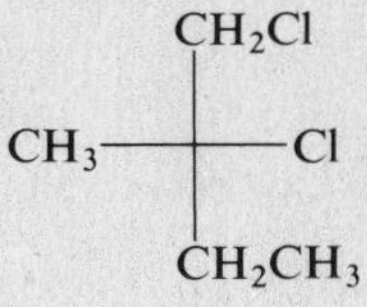

$$
\begin{array}{cc}
CH_2Cl & CH_2Cl \\
CH_3{-}|{-}Cl & Cl{-}|{-}CH_3 \\
CH_2CH_3 & CH_2CH_3 \\
R & S
\end{array}
$$

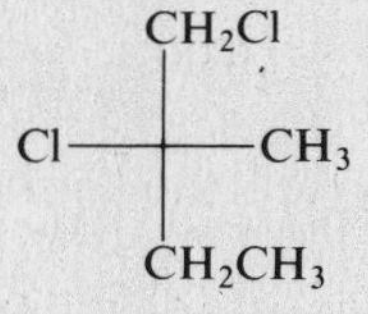

Enantiomers

C–2 is attacked, and the resultant planar free radical gives equal numbers of enantiomeric product molecules.

(H. C. Brown, M. S. Kharasch, and T. H. Chao, J. Am. Chem. Soc., **62**, 3435 (1940).)

7.11 (a) The glycol of m.p. 19°, inactive but resolvable, is a racemic modification that consists of equal amounts of (2R,3R)- and (2S,3S)-2,3-butanediol. The non-resolvable glycol of m.p. 34° is *meso*-2,3-butanediol.

(b) Permanganate gives *syn*-hydroxylation. Top-side and bottom-side attachments equally likely.

syn-Addition

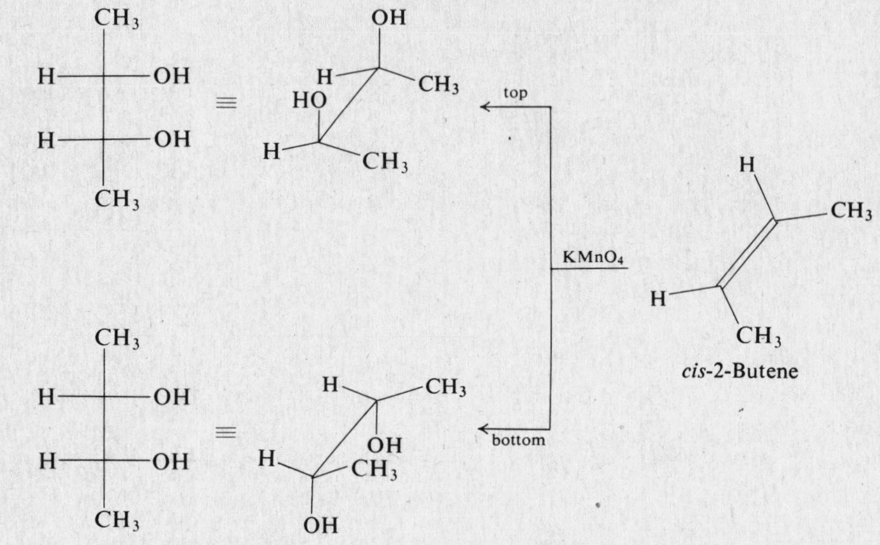

CH₃
H——OH
H——OH
CH₃
≡

OH
H——CH₃
HO
H——CH₃

top

KMnO₄

H
——CH₃

H——
CH₃
cis-2-Butene

CH₃
H——OH
H——OH
CH₃
≡

H——CH₃
OH
H——CH₃
OH

bottom

meso-2,3-Butanediol
M.p. 34°

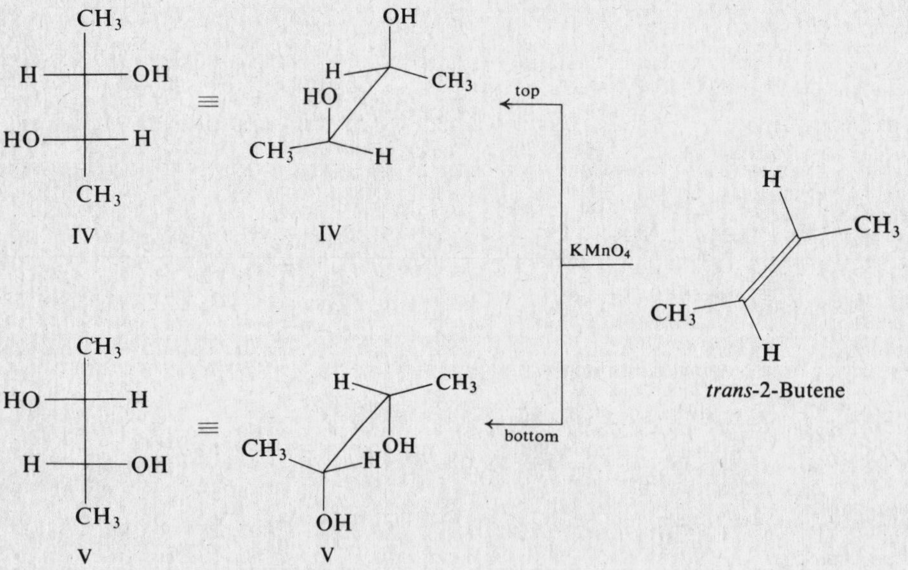

CH₃
H——OH
HO——H
CH₃
IV
≡

OH
H——CH₃
HO
CH₃——H
IV

top

KMnO₄

H
——CH₃

CH₃——
H
trans-2-Butene

CH₃
HO——H
H——OH
CH₃
V
≡

H——CH₃
CH₃——H
OH
OH
V

bottom

IV *and* V *are enantiomers*
Racemic 2,3-butanediol
M.p. 19°

(c) Peroxy acids give *anti*-hydroxylation. Attachments as in *a* and *b* (or *c* and *d*) equally likely.

***anti*-Addition**

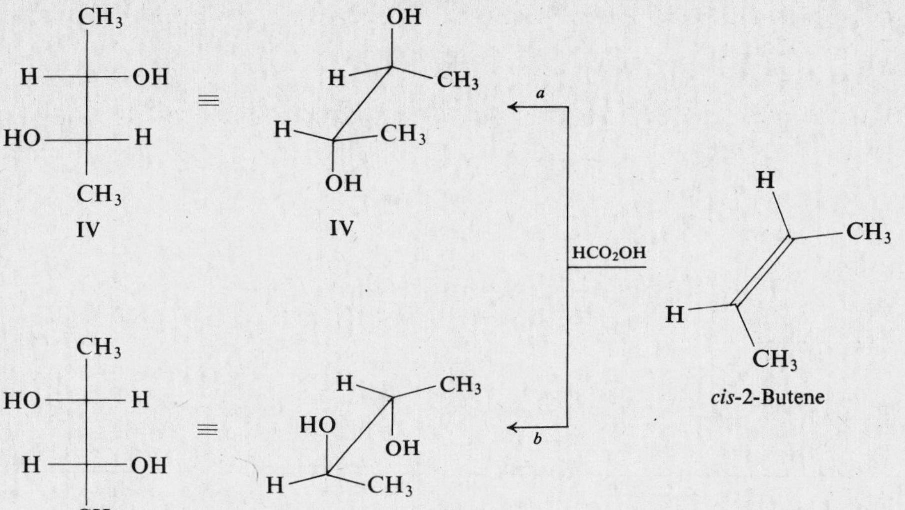

IV IV

V V

cis-2-Butene

IV *and* V *are enantiomers*

Racemic 2,3-butanediol

M.p. 19°

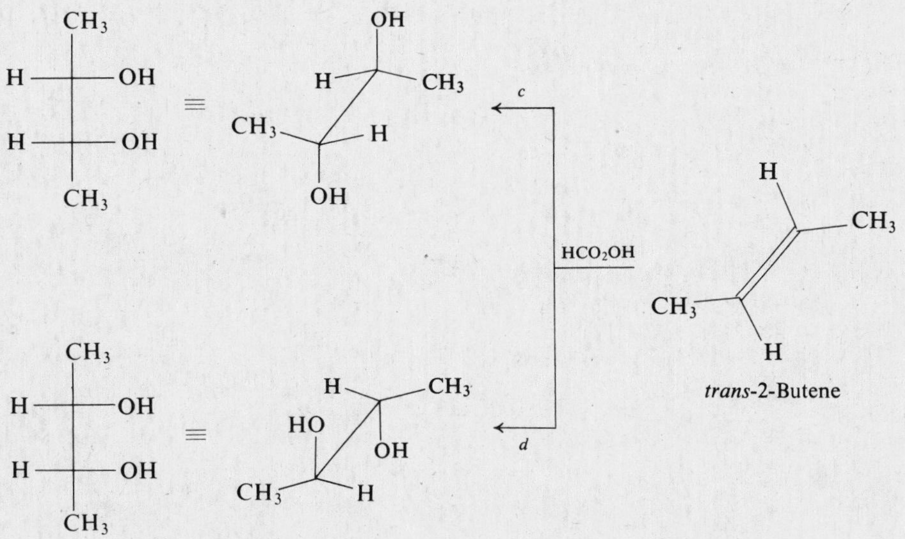

trans-2-Butene

meso-2,3-Butanediol

M.p. 34°

7.12 (a) Enantiomers, formed in equal amounts.

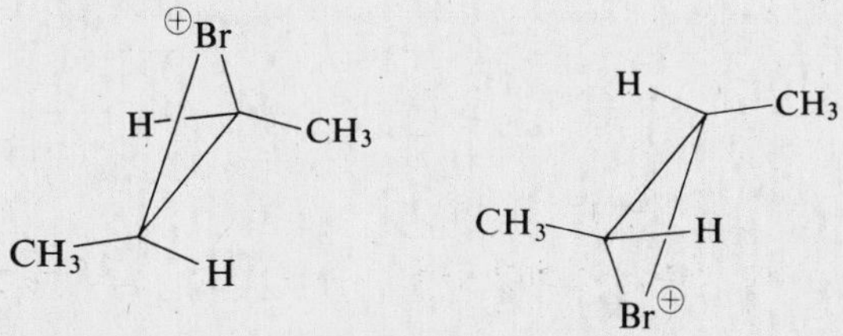

(b) Identical, achiral.

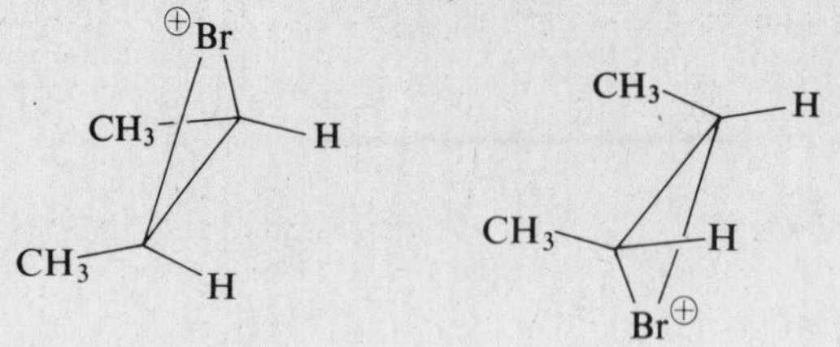

(c) Enantiomers, formed in equal amounts.

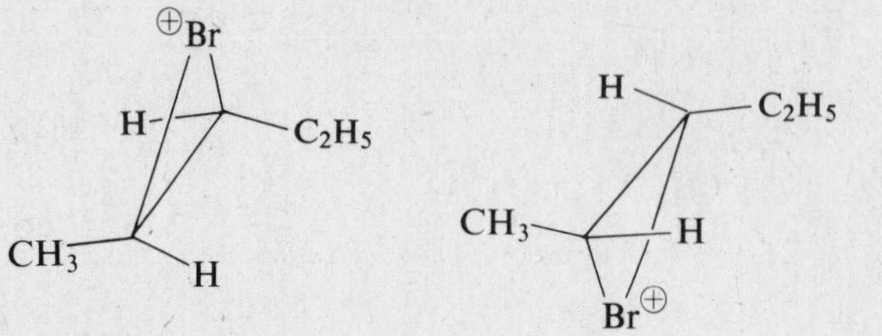

(d) Enantiomers, formed in equal amounts.

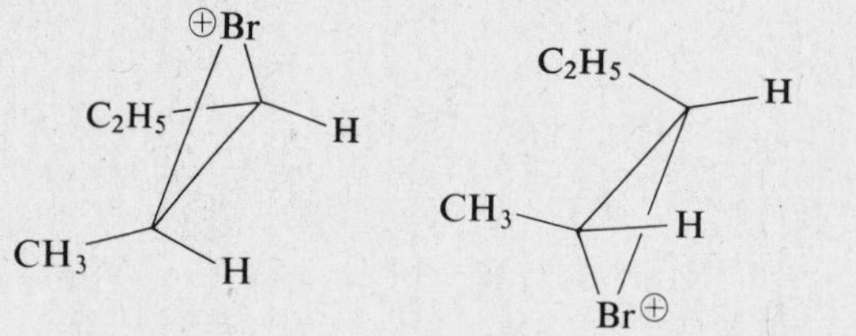

(I. Roberts and G. E. Kimball, "The Halogenation of Ethylenes," J. Am. Chem. Soc., **59**, 947 (1937).)

7.13 (a) Even though attacks by the two paths—at the methyl end and at the ethyl end—are not equally likely, the produce is racemic: (2R,3S)- and (2S,3R)-2,3-dibromopentane. There are equal amounts of the enantiomeric cyclic bromonium ions (see Problem 7.12c) undergoing attack. The product from one bromonium ion undoubtedly consists of unequal amounts of the two possible enantiomeric dibromides; if, say, attack at the methyl end were preferred, then R,S > S,R.

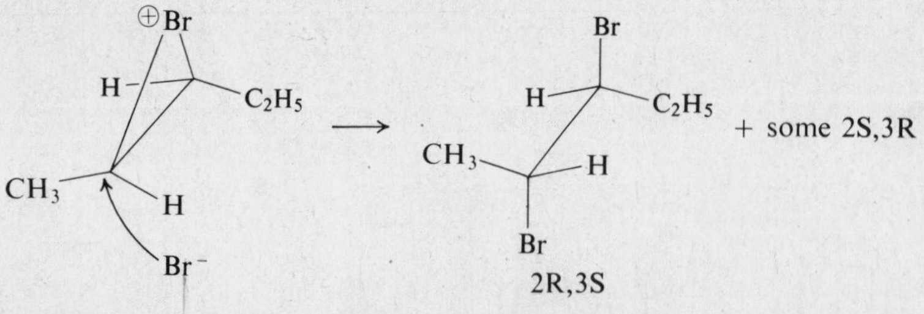

$+$ some 2S,3R

2R,3S

But this would be exactly balanced by the same preference for attack at the methyl end of the other (enantiomeric) bromonium ion, to give S,R > R,S.

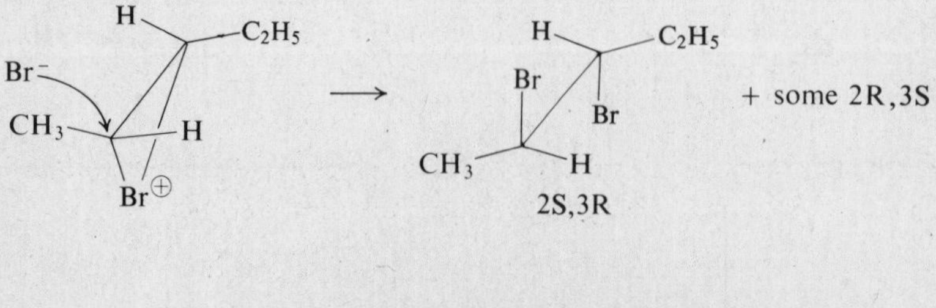

$+$ some 2R,3S

2S,3R

(b) Similar to (a), with the enantiomeric products (2R,3R)- and (2S,3S)-2,3-dibromo-pentane. Here, if R,R > S,S from one bromonium ion, it would be balanced by S,S > R,R from the other.

7.14 Highly electronegative F is less able than Cl, Br, or I to share electrons to form 3-membered ring.

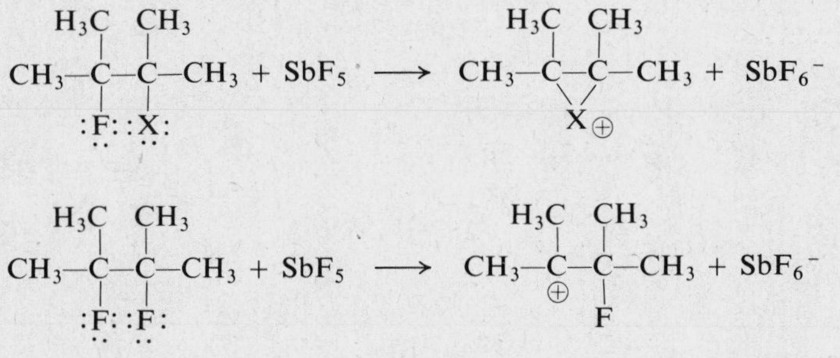

$X = Cl, Br, I$

1. Read general instructions for Problem 7.10.

(a) 3 fractions: all inactive (achiral or racemic).

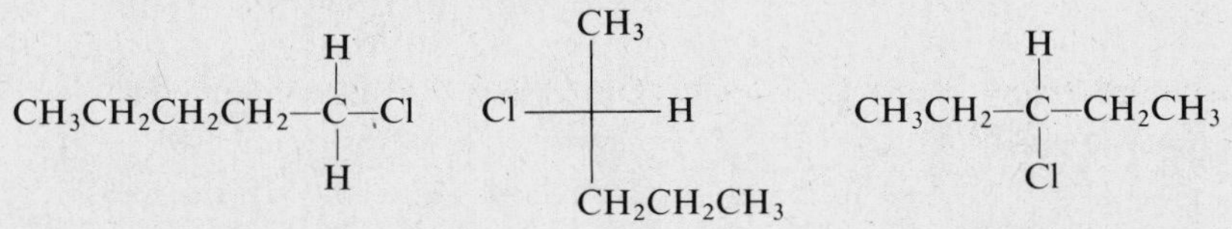

Achiral   R   *Achiral*

+enantiomer (S)

(b) 5 fractions: all inactive (achiral or racemic).

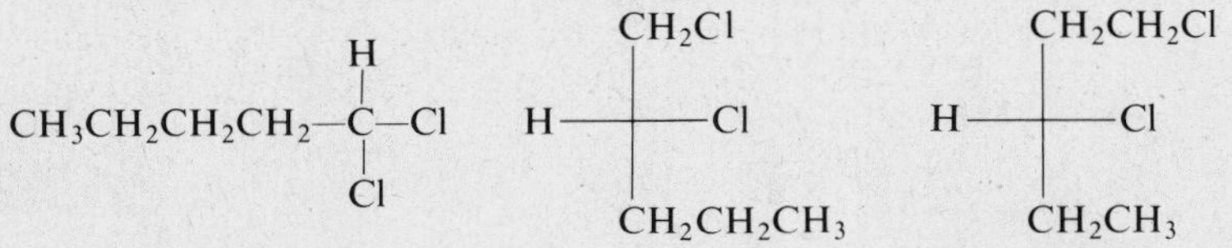

Achiral   R   R

+enantiomer (S)   +enantiomer (S)

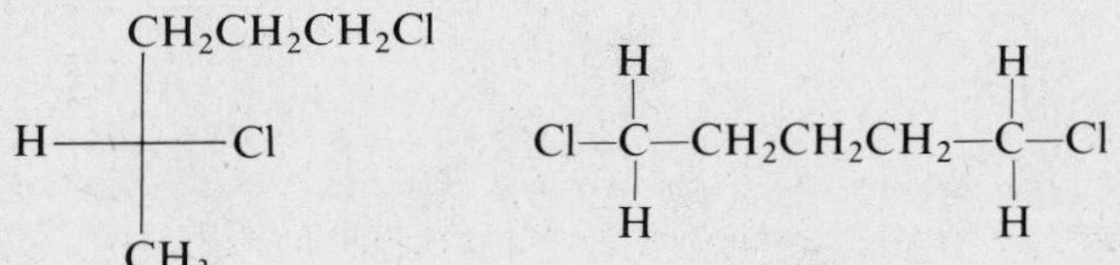

R   *Achiral*

+enantiomer (S)

(c) 7 fractions: five active, two inactive (both achiral).

Note that *configuration* about the original chiral center (the lower one in these formulas) is maintained—although the *specification* may change (as it does in the last formula).

$$
\begin{array}{ccccc}
& \text{CH}_3 & \text{CH}_3 & & \\
& \overset{|}{\underset{}{\text{Cl}-\!\!\!-\!\!\!-\text{H}}} & \text{H}-\!\!\!-\!\!\!-\text{Cl} & \text{CH}_2\text{CH}_3 & \text{CH}_2\text{CH}_3 \\
\text{CH}_2\text{CH}_2\text{CH}_2\text{Cl} & \text{CH}_2 & \text{CH}_2 & \text{Cl}-\!\!\!-\text{H} & \text{H}-\!\!\!-\text{Cl} \\
\text{Cl}-\!\!\!-\text{H} & \text{Cl}-\!\!\!-\text{H} \quad \text{Cl}-\!\!\!-\text{H} & & \text{Cl}-\!\!\!-\text{H} & \text{Cl}-\!\!\!-\text{H} \\
\text{CH}_3 & \text{CH}_3 & \text{CH}_3 & \text{CH}_3 & \text{CH}_3 \\
\textit{S} & \textit{S,R} & \textit{S,S} & \textit{2S,3R} & \textit{2S,3S} \\
\textit{Active} & \textit{Achiral (meso)} & \textit{Active} & \textit{Active} & \textit{Active} \\
& \textit{Inactive} & & &
\end{array}
$$

$$
\begin{array}{cc}
& \text{CH}_2\text{CH}_2\text{CH}_3 \\
\text{CH}_2\text{CH}_2\text{CH}_3 & \\
\text{Cl}-\text{C}-\text{Cl} & \text{Cl}-\!\!\!-\text{H} \\
\text{CH}_3 & \text{CH}_2\text{Cl} \\
\textit{Achiral} & \textit{R} \\
\textit{Inactive} & \textit{Active}
\end{array}
$$

(d) 7 fractions: six active, one inactive (racemic).

Here again, note that configuration about the original chiral center (C–3) is maintained— except, of course, when a bond to it is broken. In this particular case, bond-breaking gives an intermediate free radical, which loses configuration, and a racemic modification is formed.

$$
\begin{array}{ccccc}
\text{CH}_2\text{Cl} & \text{CH}_3 & \text{CH}_2\text{Cl} & \text{CH}_3 & \text{CH}_3 \\
\text{Cl}-\overset{2}{\underset{}{-}}\text{CH}_3 & \text{Cl}-\overset{2}{\underset{}{-}}\text{CH}_2\text{Cl} \equiv & \text{CH}_3-\overset{2}{\underset{}{-}}\text{Cl} & \text{CH}_3-\text{C}-\text{Cl} & \text{CH}_3-\text{C}-\text{Cl} \\
\text{H}-\overset{3}{\underset{}{-}}\text{CH}_3 & \text{H}-\overset{3}{\underset{}{-}}\text{CH}_3 & \text{H}-\overset{3}{\underset{}{-}}\text{CH}_3 & \text{Cl}-\text{CH}_3 & \text{CH}_3-\text{Cl} \\
\text{CH}_2\text{CH}_3 & \text{CH}_2\text{CH}_3 & \text{CH}_2\text{CH}_3 & \text{CH}_2\text{CH}_3 & \text{CH}_2\text{CH}_3 \\
\textit{2S,3R} & \textit{2R,3R} & & \textit{S} & \textit{R} \\
\textit{Active} & \textit{Active} & & \multicolumn{2}{c}{\text{Racemic modification}} \\
& & & \multicolumn{2}{c}{\textit{Inactive fraction}}
\end{array}
$$

$$
\begin{array}{cccc}
\text{CH}_3 & \text{CH}_3 & \text{CH}_3 & \text{CH}_3 \\
\text{CH}_3-\text{C}-\text{Cl} & \text{CH}_3-\text{C}-\text{Cl} & \text{CH}_3-\text{C}-\text{Cl} & \text{CH}_3-\text{C}-\text{Cl} \\
\text{H}-\text{CH}_2\text{Cl} & \text{H}-\overset{3}{\underset{}{-}}\text{CH}_3 & \text{H}-\overset{3}{\underset{}{-}}\text{CH}_3 & \text{H}-\text{CH}_3 \\
\text{CH}_2\text{CH}_3 & \text{H}-\overset{4}{\underset{}{-}}\text{Cl} & \text{Cl}-\overset{4}{\underset{}{-}}\text{H} & \text{CH}_2\text{CH}_2\text{Cl} \\
& \text{CH}_3 & \text{CH}_3 & \\
\textit{S} & \textit{3R,4R} & \textit{3R,4S} & \textit{R} \\
\textit{Active} & \textit{Active} & \textit{Active} & \textit{Active}
\end{array}
$$

(e) 1 fraction: inactive (racemic).   (Oxidation of either —CH$_2$OH.)

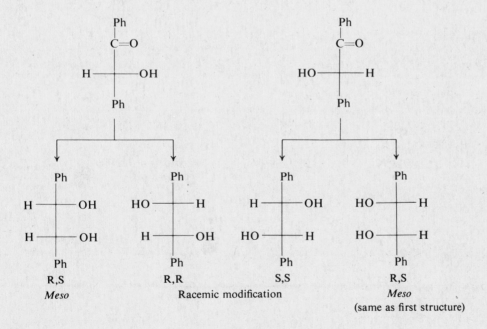

$$
\begin{array}{ccccc}
\text{COOH} & & \text{CH}_2\text{OH} & & \text{COOH} \\
\text{H}\!-\!\!\underset{2}{\big|}\!\!-\!\text{OH} & & \text{H}\!-\!\!\underset{3}{\big|}\!\!-\!\text{OH} & & \text{HO}\!-\!\!\underset{2}{\big|}\!\!-\!\text{H} \\
\text{H}\!-\!\!\underset{3}{\big|}\!\!-\!\text{OH} & \equiv & \text{H}\!-\!\!\underset{2}{\big|}\!\!-\!\text{OH} & & \text{HO}\!-\!\!\underset{3}{\big|}\!\!-\!\text{H} \\
\text{CH}_2\text{OH} & & \text{COOH} & & \text{CH}_2\text{OH} \\
2R,3R & & & & 2S,3S
\end{array}
$$

Racemic modification

(f) 3 fractions: all inactive.   Products are *trans*-2-butene, *cis*-2-butene, and 1-butene, all achiral.

(g) 2 fractions: one active, one inactive (*meso* compound).

$$
\begin{array}{ccc}
\text{CH}_3 & & \text{CH}_3 \\
\text{H}\!-\!\!\big|\!\!-\!\text{Cl} & & \text{Cl}\!-\!\!\big|\!\!-\!\text{H} \\
\text{Cl}\!-\!\!\big|\!\!-\!\text{H} & & \text{Cl}\!-\!\!\big|\!\!-\!\text{H} \\
\text{CH}_3 & & \text{CH}_3 \\
S,S & & S,R
\end{array}
$$

Active / Achiral (*meso*) Inactive

(h) 2 fractions: both inactive (one racemic, one *meso*).

$$
\begin{array}{ccccc}
\text{Ph} & & & \text{Ph} \\
\text{C}\!=\!\text{O} & & & \text{C}\!=\!\text{O} \\
\text{H}\!-\!\!\big|\!\!-\!\text{OH} & & & \text{HO}\!-\!\!\big|\!\!-\!\text{H} \\
\text{Ph} & & & \text{Ph}
\end{array}
$$

$$
\begin{array}{cccc}
\text{Ph} & \text{Ph} & \text{Ph} & \text{Ph} \\
\text{H}\!-\!\!\big|\!\!-\!\text{OH} & \text{HO}\!-\!\!\big|\!\!-\!\text{H} & \text{H}\!-\!\!\big|\!\!-\!\text{OH} & \text{HO}\!-\!\!\big|\!\!-\!\text{H} \\
\text{H}\!-\!\!\big|\!\!-\!\text{OH} & \text{H}\!-\!\!\big|\!\!-\!\text{OH} & \text{HO}\!-\!\!\big|\!\!-\!\text{H} & \text{HO}\!-\!\!\big|\!\!-\!\text{H} \\
\text{Ph} & \text{Ph} & \text{Ph} & \text{Ph} \\
R,S & R,R & S,S & R,S \\
Meso & \text{Racemic modification} & & Meso
\end{array}
$$

(same as first structure)

2. See Secs. 17.10 and 17.12 for mechanism.   In Figures 7.6 and 7.7 (pages 244 and 245), replace Br in structures III and VI by oxygen, and in IV, V, VII, and VIII replace —Br by —OH.

3. A, (S,S)   B, (R,S) (*meso*)   C, (S,S)   D, (S,S)   E, (2R,3S)-4-Br-1,2,3-butanetriol   F, (R,R)
G, (R,S) (*meso*).

4. The enantiomeric acids react at different rates with the optically active reagent, the alcohol, to form diastereomeric esters. Unequal amounts of the acids are consumed, to give unequal amounts of the esters, and to leave unconsumed unequal amounts of the enantiomeric acids. The three products are, then: the two diastereomeric esters, and unconsumed acid that now contains an excess of one enantiomer and hence is optically active.

5. Chlorohydrin formation is *anti*. Like addition of chlorine, it proceeds via an intermediate

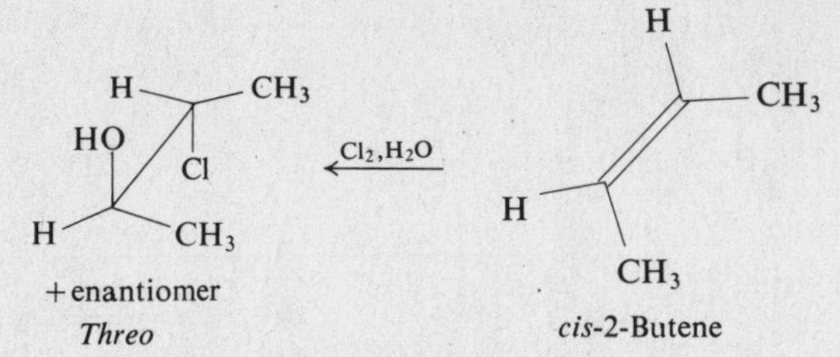

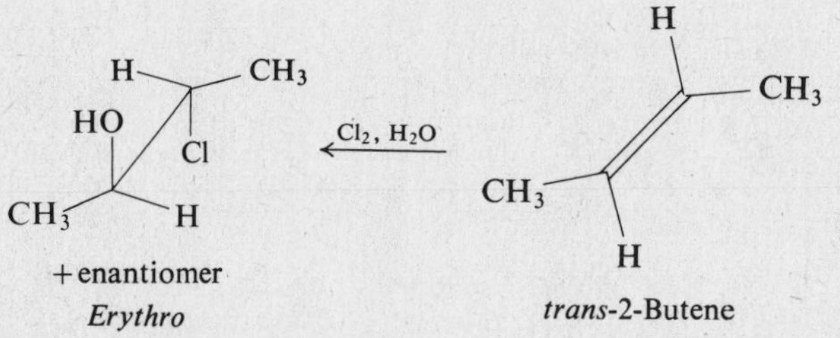

chloronium ion. Attack on the intermediate by chlorine ion gives the dichloro product; attack by water gives the chlorohydrin.

6. (a) Reaction proceeds (partly, at least) via a bridged bromonium ion formed from (or perhaps instead of) the open carbonium ion resulting from addition of a proton. Subsequent attack by water cleaves either C—Br bond.

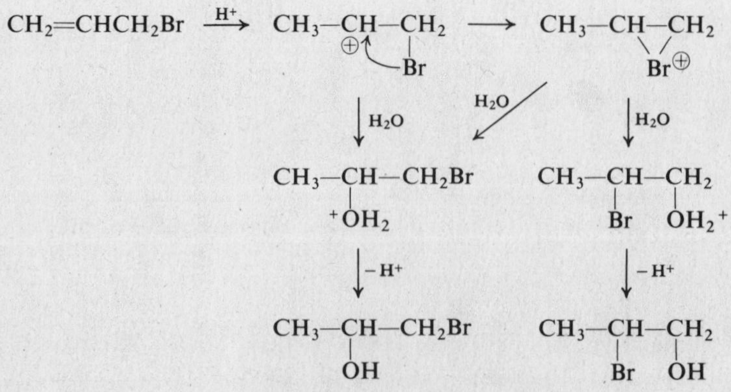

(b) Bridged chloronium ion is less easily formed, due to reluctance of more electronegative Cl to share its electrons; open carbonium ion is relatively more stable.

7. (a) Intermediate iodonium ion undergoes opposite-side attack by N_3^- ion.

(b) Polar solvents favor ionic reaction like that of IN_3.   Light or peroxides cause free-radical addition:

$$Rad\cdot + BrN_3 \longrightarrow RadBr + N_3\cdot$$

$$N_3\cdot + RCH{=}CH_2 \longrightarrow R\overset{.}{C}HCH_2N_3$$

Oxygen breaks chain by combining with organic free radicals.   In non-polar solvent, slow spontaneous free radical formation

$$BrN_3 \longrightarrow Br\cdot + N_3\cdot$$

makes itself evident; conceivably, there is concerted homolysis,

$$BrN_3 + RCH{=}CH_2 \longrightarrow R\overset{.}{C}HCH_2Br + N_3\cdot$$

(Alfred Hassner, "Regiospecific and Stereospecific Introduction of Azide Function into Organic Molecules," Accounts Chem. Res., **4**, 9 (1971).)

Chapter

8 | Alkynes and Dienes

8.1 CO_2: $O=C=O$, linear; C uses two *sp* orbitals, and two *p* orbitals at right angles (compare allene, Problem 15, page 280). H_2O: sp^3, tetrahedral, with unshared pairs in two lobes.

8.2 (a) $CH_3CBr=CBrCH_3$, $CH_3CBr_2CBr_2CH_3$; (b) decrease (compare vinyl bromide); (c) the dibromo compound is made a poorer competitor; (d) excess butyne; (e) drip Br_2 into a solution of 2-butyne.

8.3

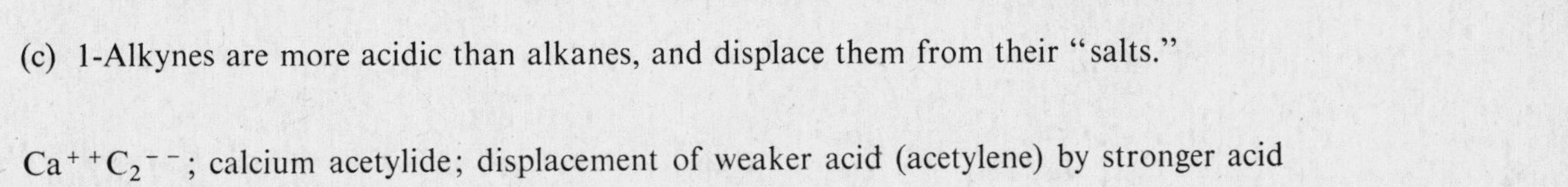

$$\underset{\textit{cis}\text{-2-Pentene}}{\overset{\displaystyle CH_3 \qquad\qquad C_2H_5}{\underset{\displaystyle H \qquad\qquad H}{C=C}}} \xleftarrow[\text{Pd or Ni-B(P-2)}]{H_2} CH_3C\equiv CC_2H_5 \xleftarrow{NaNH_2} \xleftarrow{KOH(alc)} \underset{\displaystyle Br \quad Br}{CH_3CH-CHC_2H_5} \xleftarrow{Br_2} \underset{\textit{cis-trans}\text{ mixture}}{\text{2-Pentene}}$$

8.4 (a) Propane.

(b) $n\text{-}C_4H_9C\equiv CH + CH_3CH_2CH_2MgBr \longrightarrow n\text{-}C_4H_9C\equiv CMgBr + CH_3CH_2CH_3$

(c) 1-Alkynes are more acidic than alkanes, and displace them from their "salts."

8.5 $Ca^{++}C_2^{--}$; calcium acetylide; displacement of weaker acid (acetylene) by stronger acid (water).

8.6 H goes to terminal C.

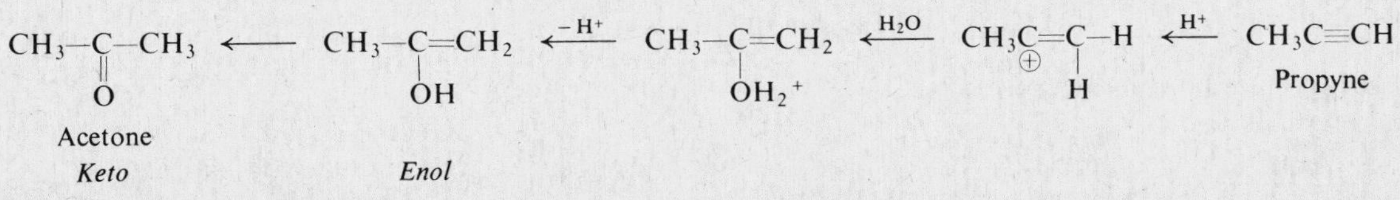

$$\underset{\substack{\displaystyle O \\ Acetone \\ Keto}}{CH_3-\overset{\displaystyle \|}{C}-CH_3} \longleftarrow \underset{\substack{\displaystyle OH \\ Enol}}{CH_3-\overset{\displaystyle |}{C}=CH_2} \xleftarrow{-H^+} \underset{\displaystyle OH_2{}^+}{CH_3-\overset{\displaystyle |}{C}=CH_2} \xleftarrow{H_2O} \underset{\displaystyle H}{CH_3\overset{\displaystyle \oplus}{C}=C-H} \xleftarrow{H^+} \underset{Propyne}{CH_3C\equiv CH}$$

8.7 1,3-Hexadiene, the more stable conjugated diene.

8.8 (a) Assuming a value of 28–30 kcal for each double bond:

$$CH_2=C=CH_2 + 2H_2 \longrightarrow CH_3CH_2CH_3 \quad \Delta H = -56 \text{ to } -60 \text{ kcal}$$
$$(2 \times 28) \ (2 \times 30)$$

(b) Cumulated double bonds are unstable relative to conjugated or isolated double bonds.

8.9 Hyperconjugation involving structures like I–IV on page 217 with + replacing odd electron; in orbital picture on page 217, p orbital contains "no e" instead of "1 e".

8.10 (a) Initial attachment of proton must be to C–1, since only in this way can the observed products be formed, by 1,2- and 1,4-addition.

Attachment of the proton to either end of the conjugated system gives an allylic cation. But attachment to the C–1 end is favored, since this allylic cation is also a 3° cation. That is to say, part of the positive charge develops on the carbon (C–2) to which the methyl group is attached; electron release by methyl helps to stabilize the incipient carbonium ion.

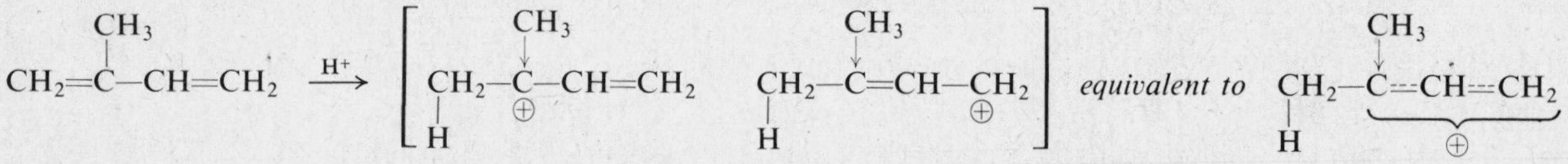

(b) Similar to (a), with initial attachment of Br^+ to C–1 end of conjugated system.

8.11 (a) Yes, a doubly allylic cation.

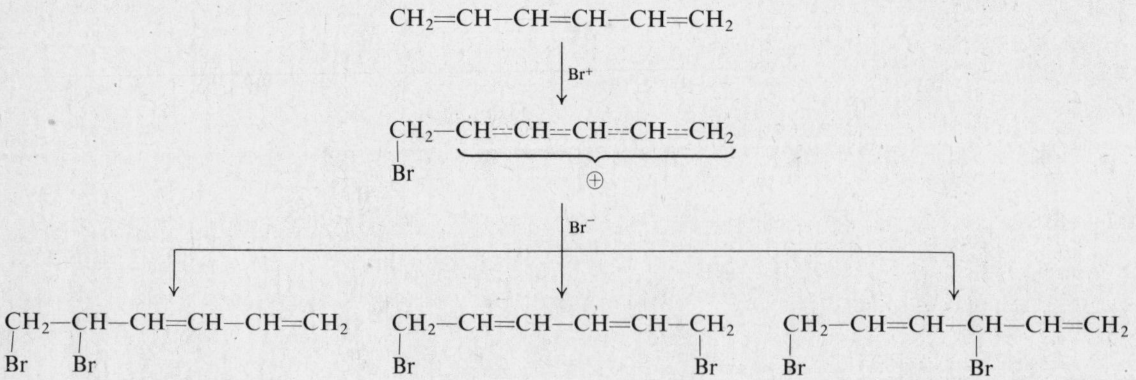

CH$_2$—CH—CH=CH—CH=CH$_2$ CH$_2$—CH=CH—CH=CH—CH$_2$ CH$_2$—CH=CH—CH—CH=CH$_2$
| | | | | |
Br Br   Br   Br   Br   Br

5,6-Dibromo-1,3-hexadiene   1,6-Dibromo-2,4-hexadiene   3,6-Dibromo-1,4-hexadiene
Sole products   *Not formed*

(b) 3,6-Dibromo-1,4-hexadiene.

(c) Equilibrium control, which favors the more stable products, the two conjugated dienes.

8.12 (a)

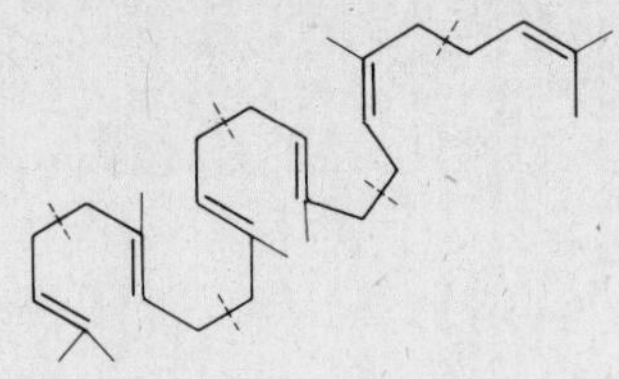

Squalene

(b) At the center; molecule is probably made by head-to-head combination of two identical C_{15} units.

(c) Four ring closures, and a methyl migration. Loss of three methyl groups.

8.13 (a) $n\text{-PrC}{\equiv}\text{CH} \longrightarrow n\text{-PrCOOH} + CO_2$

(b) $\text{EtC}{\equiv}\text{CCH}_3 \longrightarrow \text{EtCOOH} + \text{HOOCCH}_3$

(c) $i\text{-PrC}{\equiv}\text{CH} \longrightarrow i\text{-PrCOOH} + CO_2$

(d) $CH_3CH{=}CH{-}CH{=}CH_2 \longrightarrow CH_3CHO + OHC{-}CHO + HCHO$

(e) $CH_2{=}CH{-}CH_2{-}CH{=}CH_2 \longrightarrow HCHO + OHC{-}CH_2{-}CHO + HCHO$

(f) $CH_2{=}\underset{\underset{CH_3}{|}}{C}{-}CH{=}CH_2 \longrightarrow HCHO + O{=}\underset{\underset{CH_3}{|}}{C}{-}CHO + HCHO$

8.14 (a) $n\ CH_2{=}\underset{\underset{CH{=}CH_2}{|}}{CH} \xrightarrow{1,2} (-CH_2{-}\underset{\underset{CH{=}CH_2}{|}}{CH}{-})_n \xrightarrow{O_3} (-CH_2{-}\underset{\underset{CHO}{|}}{CH}{-})_n + n\ HCHO$

(b) $n\ CH_2{=}CH{-}CH{=}CH_2 \xrightarrow{1,4} (-CH_2{-}CH{=}CH{-}CH_2{-}CH_2{-}CH{=}CH{-}CH_2{-})_{n/2}$

$\Big\downarrow O_3$

$n\ O{=}\overset{\overset{H}{|}}{C}{-}CH_2{-}CH_2{-}\overset{\overset{H}{|}}{C}{=}O$

8.15 In natural rubber, the isoprene units are mostly combined as though by head-to-tail 1,4-addition.

$O{=}CH{-}CH_2{-}CH_2{-}\overset{\overset{CH_3}{|}}{C}{=}O$

$\Big\uparrow O_3$

$\sim\sim CH_2{-}\overset{\overset{CH_3}{|}}{C}{=}CH{-}CH_2{-}CH_2{-}\overset{\overset{CH_3}{|}}{C}{=}CH{-}CH_2{-}CH_2{-}\overset{\overset{CH_3}{|}}{C}{=}CH{-}CH_2\sim\sim$

Isoprene unit   Isoprene unit   Isoprene unit

1.   C—C—C—C—C≡CH   C—C—C—C≡C—C   C—C—C≡C—C—C
           1-Hexyne   2-Hexyne   3-Hexyne
     (*n*-Butylacetylene)   (*n*-Propylmethylacetylene)   (Diethylacetylene)

                        Forms salts

   Gives: *n*-BuCOOH + CO_2 *n*-PrCOOH + HOOCCH$_3$ 2EtCOOH

                    C   C
                    | |
     C—C—C—C≡CH   C—C—C≡C—C
        4-Methyl-1-pentyne   4-Methyl-2-pentyne
        (Isobutylacetylene)   (Isopropylmethylacetylene)

              Forms salts

   Gives: iso-BuCOOH + CO_2 *i*-PrCOOH + HOOCCH$_3$

                    C   C
                    | |
     C—C—C—C≡CH   C—C—C≡CH
                                                         |
        3-Methyl-1-pentyne   C
        (*sec*-Butylacetylene)   3,3-Dimethyl-2-butyne
                                                (*tert*-Butylacetylene)

              Forms salts   *Forms salts*

   Gives: *sec*-BuCOOH + CO_2 *t*-BuCOOH + CO_2

2.   C—C—C=C—C=C   C—C=C—C—C=C
                  1,3-Hexadiene   1,4-Hexadiene

                   Conjugated

                   Geom. isom.   *Geom. isom.*

   Gives: C—C—CHO + OHC—CHO + HCHO   C—CHO + OHC—C—CHO + HCHO

                C=C—C—C—C=C   C—C=C—C=C—C
                  1,5-Hexadiene   2,4-Hexadiene

                                                  Conjugated

                                                  Geom. isom.

   Gives: HCHO + OHC—C—C—CHO + HCHO   C—CHO + OHC—CHO + OHC—C

                C—C=C—C=C   C=C—C—C=C
                    | |
                    C   C
          2-Methyl-1,3-pentadiene   2-Methyl-1,4-pentadiene

                Conjugated

                Geom. isom.

   Gives: C—CHO + OHC—C=O + HCHO   HCHO + OHC—C—C=O + HCHO
                          | |
                          C   C

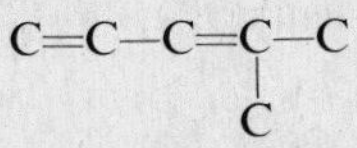

C=C—C=C—C
       |
       C

4-Methyl-1,3-pentadiene

Conjugated

Gives: HCHO + OHC—CHO + O=C—C
                                        |
                                        C

C—C=C—C=C
     |
     C

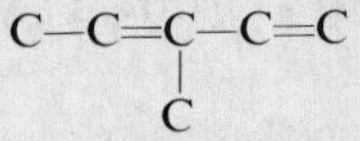

3-Methyl-1,3-pentadiene

Conjugated

Geom. isom.

C—CHO + O=C—CHO + HCHO
                    |
                    C

C=C—C—C=C
       |
       C

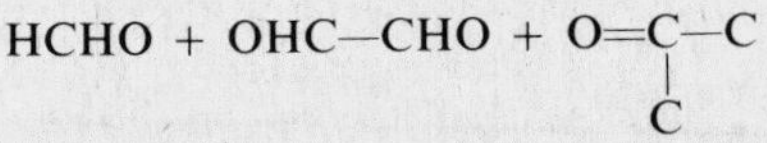

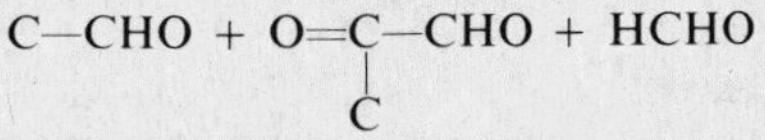

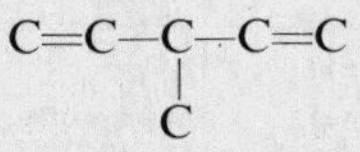

3-Methyl-1,4-pentadiene

Gives: HCHO + OHC—C—CHO + HCHO
                         |
                         C

C=C—C=C
   | |
   C   C

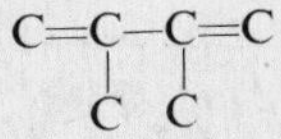

2,3-Dimethyl-1,3-butadiene

Conjugated

HCHO + O=C—C=O + HCHO
          | |
          C   C

C=C—C=C
     |
     C
     |
     C

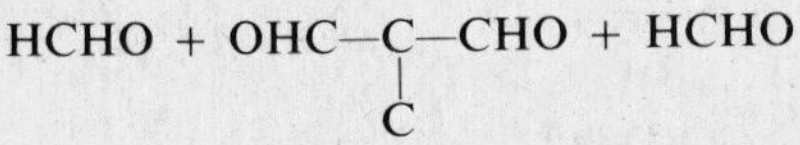

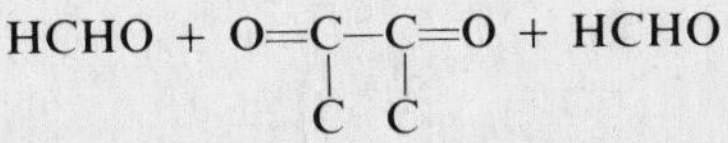

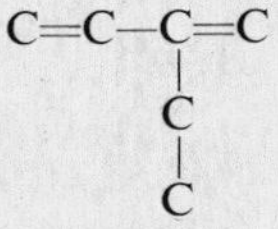

2-Ethyl-1,3-butadiene

Conjugated

Gives: HCHO + OHC—C=O + HCHO
                         |
                         C
                         |
                         C

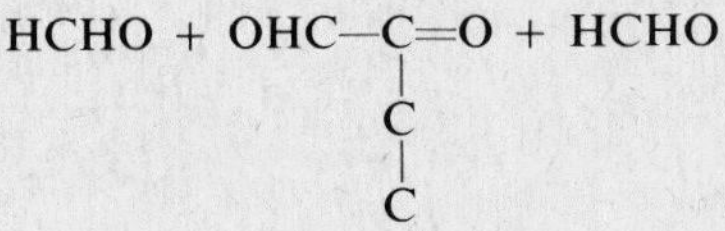

3. $CaCO_3$ + heat $\longrightarrow$ CaO + CO_2

                    coal + heat $\longrightarrow$ C(coke)

                    CaO + 3C + heat $\longrightarrow$ CaC_2 + CO

                    CaC_2 + $2H_2O$ $\longrightarrow$ HC≡CH + $Ca(OH)_2$

4. (a) Remove 2HBr: KOH(alc), heat   (b) Br$_2$/CCl$_4$; KOH(alc), heat

    (c) KOH(alc), heat; then as in (b)   (d) Br$_2$, heat, light; mixture of isomers,
                                                                KOH(alc), heat; then as in (b)

    (e) Acid, heat; then as in (b)   (f) Remove 2HCl: KOH(alc), heat; or
                                                                  KOH(alc), heat, followed by NaNH$_2$

    (g) Attach methyl group

$$CH_3C{\equiv}CH \xleftarrow{CH_3I} NaC{\equiv}CH \xleftarrow{NaNH_2} HC{\equiv}CH$$

    (h) Remove all Br's: Zn, heat

5. (a) 1H$_2$, cat.   (b) As in (a); then excess H$_2$, Ni, heat,
                                                                   pressure

    (c) $CH_3{-}CHBr_2 \xleftarrow{HBr} CH_2{=}CHBr \xleftarrow{HBr} HC{\equiv}CH$ (d) HCl, CuCl

    (e) $CH_2Cl{-}CH_2Cl \xleftarrow{Cl_2} CH_2{=}CH_2 \longleftarrow$ (a)   (f) H$_2$O, H$^+$, Hg^{++}

    (g) $CH_3{-}C{\equiv}CH \xleftarrow{CH_3I} NaC{\equiv}CH \xleftarrow{NaNH_2} HC{\equiv}CH$

    (h) $C_2H_5{-}C{\equiv}CH \xleftarrow{C_2H_5Br} NaC{\equiv}CH \xleftarrow{NaNH_2} HC{\equiv}CH$

    (i) $CH_3{-}C{\equiv}C{-}CH_3 \xleftarrow{CH_3I} NaC{\equiv}C{-}CH_3 \xleftarrow{NaNH_2} HC{\equiv}C{-}CH_3 \longleftarrow$ (g)

    (j) $cis\text{-}CH_3CH{=}CHCH_3 \xleftarrow[\text{Pd or Ni-B (P-2)}]{H_2} CH_3C{\equiv}CCH_3 \longleftarrow$ (i)

    (k) $trans\text{-}CH_3CH{=}CHCH_3 \xleftarrow{Na, NH_3} CH_3C{\equiv}CCH_3 \longleftarrow$ (i)

    (l) $n\text{-}Pr{-}C{\equiv}CH \xleftarrow{n\text{-}PrBr} NaC{\equiv}CH \xleftarrow{NaNH_2} HC{\equiv}CH$

    (m) $CH_3{-}C{\equiv}C{-}Et \xleftarrow{CH_3I} NaC{\equiv}C{-}Et \xleftarrow{NaNH_2} HC{\equiv}C{-}Et \longleftarrow$ (h)

    (n) $Et{-}C{\equiv}C{-}Et \xleftarrow{EtBr} NaC{\equiv}C{-}Et \xleftarrow{NaNH_2} HC{\equiv}C{-}Et \longleftarrow$ (h)

6. (a) 1-butene   (b) n-butane   (c) 1,2-dibromo-1-butene
    (d) 1,1,2,2-tetrabromobutane   (e) 2-chloro-1-butene   (f) 2,2-dichlorobutane
    (g) methyl ethyl ketone   (h) $AgC{\equiv}CC_2H_5$ (i) 1-butyne
           (2-butanone)   (j) $NaC{\equiv}CC_2H_5$ (k) 3-hexyne
    (l) 1-butyne and isobutylene (m) $C_2H_5C{\equiv}CMgBr$ and   (n) 1-butyne
    (o) $CH_3CH_2COOH +$ ethane   (p) $CH_3CH_2COOH + CO_2$
        HCOOH

7. (a) 1- and 2-butene
   (c) 1,4-dibromo-2-butene and
       3,4-dibromo-1-butene
   (f) 1,3- and 2,3-dichlorobutane
   (p) $4CO_2$

   (b) *n*-butane
   (d) 1,2,3,4-tetrabromobutane
   (e) 1-chloro-2-butene and 3-chloro-1-butene
   (o) $HCHO + OHC—CHO + HCHO$

No reaction in (g) through (n).

8. (a) 1-pentene
   (c) 4,5-dibromo-1-pentene
   (e) 4-chloro-1-pentene
   (o) $HCHO + OHCCH_2CHO + HCHO$

   (b) *n*-pentane
   (d) 1,2,4,5-tetrabromopentane
   (f) 2,4-dichloropentane
   (p) $CO_2 + HOOCCH_2COOH + CO_2$

No reaction in (g) through (n).

9. The first-named alkene of each set is the major product. We expect this to be the more stable product: the more highly branched alkene, or a conjugated diene.

   (a) 1-butene; *cis*- and *trans*-2-butene, 1-butene.
   (b) 1-butene; 1,3-butadiene.
   (c) 2-methyl-2-butene, 2-methyl-1-butene; 2-methyl-2-butene, 3-methyl-1-butene.
   (d) 2-methyl-1-butene; 3-methyl-1-butene.
   (e) 2,3-dimethyl-1-butene; 2,3-dimethyl-2-butene, 2,3-dimethyl-1-butene.
   (f) 1,3-butadiene; 1,4-pentadiene.

10. We predict the one that gives the more stable alkene or diene to be the more reactive.

    (a) 2-chlorobutane
    (c) 2-bromo-2-methylbutane
    (e) 2-chloro-2,3-dimethylbutane

    (b) 4-chloro-1-butene
    (d) 1-bromo-2-methylbutane
    (f) 4-chloro-1-butene

11. The initial attack is by H^+.

$$1,3\text{-Butadiene} \longrightarrow \underset{\underset{H}{|}}{CH_2}-CH=CH-\underset{\underset{Cl}{|}}{CH_2} \quad \text{and} \quad \underset{\underset{H}{|}}{CH_2}-\underset{\underset{Cl}{|}}{CH}-CH=CH_2$$

The other products are:

(a) $\underset{\underset{H}{|}}{CH_2}-\underset{\underset{Cl}{|}}{CH}-CH_2CH_3$

(b) $\underset{\underset{H}{|}}{CH_2}-\underset{\underset{Cl}{|}}{CH}-CH_2-CH=CH_2$

(c) $\underset{\underset{H}{|}}{CH_2}-\overset{\overset{CH_3}{|}}{\underset{\underset{Cl}{|}}{C}}-CH=CH_2 \quad \text{and} \quad \underset{\underset{H}{|}}{CH_2}-\overset{\overset{CH_3}{|}}{C}=CH-\underset{\underset{Cl}{|}}{CH_2}$

(d) $\underset{\underset{H}{|}}{CH_2}CH=CH-\underset{\underset{Cl}{|}}{CH}-CH_3$

12. The initial attack is by $\cdot CCl_3$.

$$1,3\text{-Butadiene} \longrightarrow \underset{\underset{Cl_3C}{|}}{CH_2}-CH=CH-\underset{\underset{Br}{|}}{CH_2} \quad \text{and} \quad \underset{\underset{Cl_3C}{|}}{CH_2}-\underset{\underset{Br}{|}}{CH}-CH=CH_2$$

The other products are:

(a) $\underset{\underset{Cl_3C}{|}}{CH_2}-\underset{\underset{Br}{|}}{CH}-CH_2-CH_3$

(b) $\underset{\underset{Cl_3C}{|}}{CH_2}-\underset{\underset{Br}{|}}{CH}-CH_2-CH=CH_2$

(c) $\underset{\underset{Cl_3C}{|}}{CH_2}-\overset{\overset{CH_3}{|}}{C}-CH=CH_2$ and $\underset{\underset{Cl_3C}{|}}{CH_2}-\overset{\overset{CH_3}{|}}{C}=CH-\underset{\underset{Br}{|}}{CH}$

(d) $\underset{\underset{Cl_3C}{|}}{CH_2}-CH=CH-\underset{\underset{Br}{|}}{CH}-CH_3$

13. The more reactive compound is the one that gives the more stable carbonium ion: diene > alkene; conjugated diene > unconjugated; branched diene > unbranched.

(a) 1,3-butadiene
(b) 1,3-butadiene
(c) 2-methyl-1,3-butadiene
(d) 1,3-pentadiene

14. (a) $HC\equiv CH + 2H_2 \longrightarrow C_2H_6 \qquad \Delta H = -75$ kcal

$-(H_2C=CH_2 + H_2 \longrightarrow C_2H_6 \qquad \Delta H = -32.8$ kcal$)$

$\overline{\qquad HC\equiv CH + H_2 \longrightarrow H_2C=CH_2 \quad \Delta H = -42.2 \text{ kcal}}$

(b) Alkyne less stable relative to alkene than alkene relative to alkane.

(c) Expect the less stable (higher energy) compound, acetylene, to react faster.

(d) $CH_3-CH=CH\cdot$ less stable than $CH_3-CH_2-CH_2\cdot$ (vinylic < 1°); expect slower addition to acetylene.

(e) Apparently here, as in many other instances, product stability is more important than reactant stability. That is, factors stabilizing the transition state (through its product character) are stronger than factors stabilizing the reactant.

15. (a) CH_2 planes are perpendicular to each other.

(b) Yes; even though no chiral carbon is present, the molecules are chiral.

(c) Central carbon sp, linear; terminal carbons sp^2, trigonal; planes of the two double bonds perpendicular. Leads to same shape as in (a) and (b).

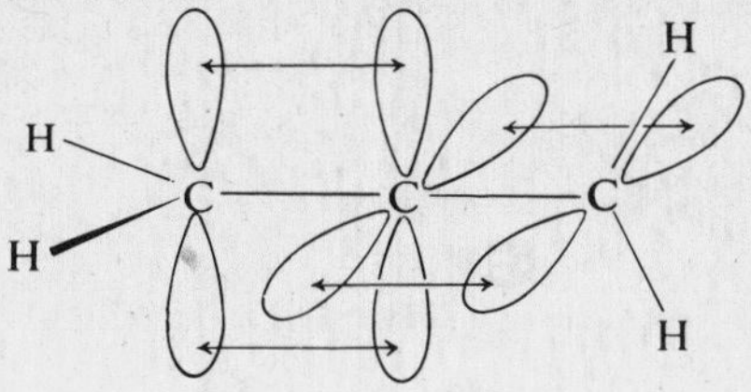

Allene

(See E. L. Eliel, *Stereochemistry of Carbon Compounds*, pp. 307–311.)

16. (a) *n*-Hexane, 1,2-dibromohexane, 1,2,5,6-tetrabromohexane.

(b) Heat middle fraction with Zn (or NaI in acetone), collect 1-hexene (b.p. 63.5°).

(c) Add Br_2; fractionate mixture; then heat bromo compound(s) with Zn (or NaI in acetone).

17. (a) *anti*-Reduction, then *anti*-addition of bromine.

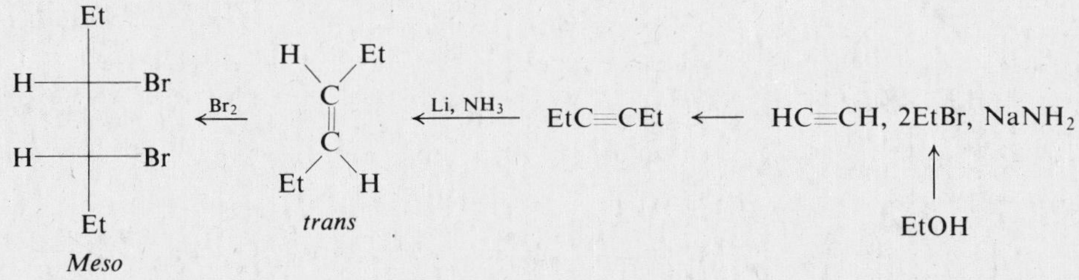

(b) Either *anti*-reduction, *syn*-hydroxylation; or *syn*-reduction, *anti*-hydroxylation.

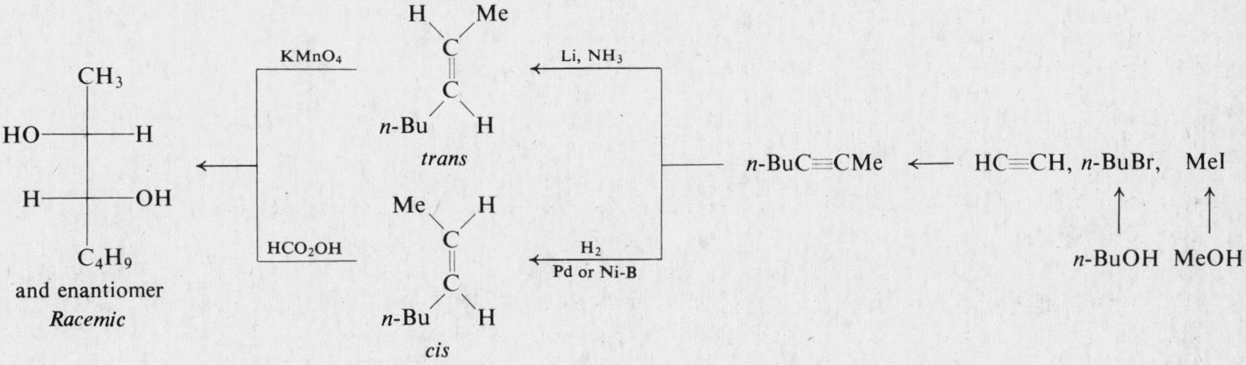

18.

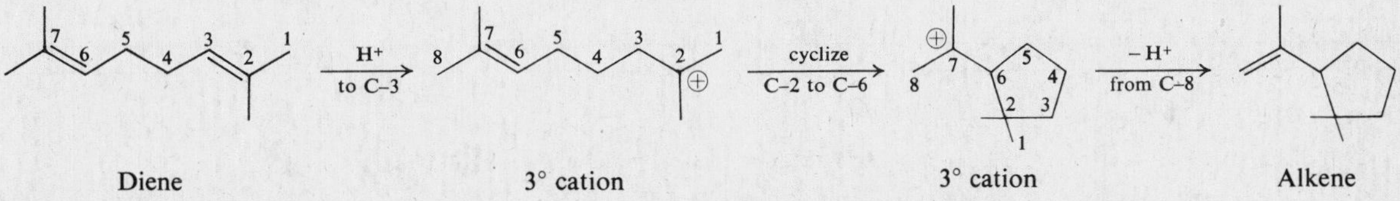

(Location of double bond in product is not definite, but infrared spectrum supports location in side chain.)

19. Gutta percha is *trans*-isomer.

20. (a) $KMnO_4$ or Br_2/CCl_4 (b) $Ag(NH_3)_2OH$
     (c) $Ag(NH_3)_2OH$ (d) $KMnO_4$ or Br_2/CCl_4
     (e) $Ag(NH_3)_2OH$ (f) Br_2/CCl_4, or CrO_3/H_2SO_4
     (g) $AgNO_3$

21. (a) Quantitative hydrogenation; (b) same as (a); (c) ozonolysis, followed by identification of products; (d) same as (c).

22. Only *n*-pentane and methylene chloride give negative $KMnO_4$ test; distinguish by elemental analysis. Of the unsaturated compounds, only 1-chloropropene gives a positive halide test, and only 1-pentyne gives a precipitate with $Ag(NH_3)_2OH$. Distinguish the others by ozonolysis and identification of the cleavage products; or by infrared and nmr spectra.

23.

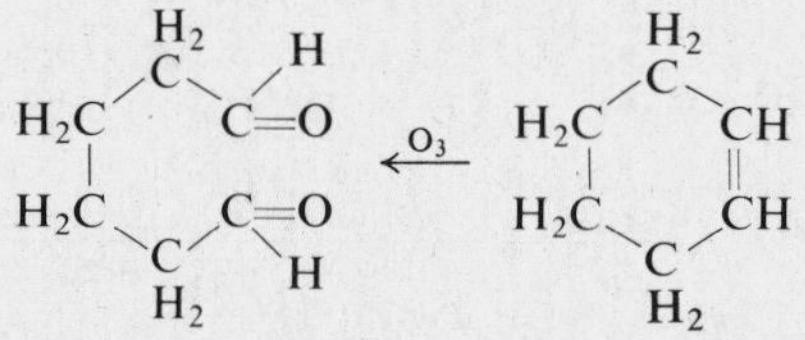

24. We can attack the problem as follows:

$$\text{mmoles } H_2 \text{ taken up} = \frac{8.40 \text{ cc}}{22.4 \text{ cc/mmole}}$$

$$\text{mmoles compound} \sim \frac{10.02 \text{ mg}}{80 \text{ mg/mmole}} \quad \text{(uncertainty in m.w.)}$$

$$\frac{\text{mmoles } H_2 \text{ taken up}}{\text{mmole compound}} = \frac{8.40/22.4}{10.02/80} = 3$$

The hydrocarbon therefore contains three double bonds, or one double bond and one triple bond. The ozonolysis results show only three carbons, but a m.w. in the range of 80–85 indicates six carbon atoms ($6 \times 12 = 72$). Evidently two moles of each ozonolysis product are obtained per mole of hydrocarbon. Since HCHO can only come from a terminal unsaturation, and the dialdehyde from an inner segment of a chain, there is only one way of putting the pieces together: the hydrocarbon is 1,3,5-hexatriene.

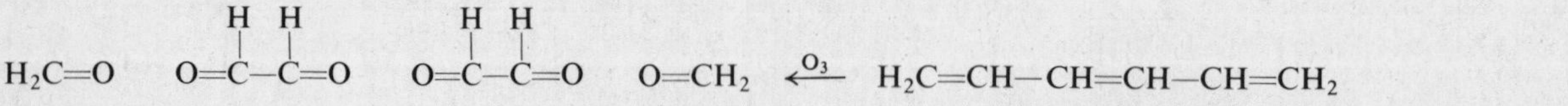

On checking, we find that the proposed triene structure accommodates all the analytical data. Any alkyne or cyclic structure will be found unsuitable.

25. Myrcene contains ten carbons, but the ozonolysis products show only nine; the missing carbon must be found in a second molecule of HCHO. This gives four molecules of ozonolysis products, which agrees with the three double bonds shown by hydrogenation. (Cleavage at three places gives four products.)

(a) The fragments can be put back together in three ways:

$$
\begin{array}{cccc}
\overset{\displaystyle C}{C-C=O} &
\overset{\displaystyle H}{O=C}-C-\overset{\displaystyle H}{C}-\overset{\displaystyle H}{C}-C=O &
O=CH_2 \leftarrow & C-\overset{\displaystyle C}{C}=C-C-C-C-C=C \\
& \overset{\displaystyle |}{O} & & \overset{\displaystyle |}{C} \\
& O=CH_2 & &
\end{array}
$$

$$
\begin{array}{cccc}
\overset{\displaystyle C}{C-C=O} &
\overset{\displaystyle H}{O=C}-C-C-\overset{\displaystyle H}{C}-C=O &
O=CH_2 \leftarrow & C-\overset{\displaystyle C}{C}=C-C-C-C-C=C \\
& \overset{\displaystyle |}{O} & & \overset{\displaystyle |}{C} \\
& O=CH_2 & &
\end{array}
$$

$$
\begin{array}{cccc}
\overset{\displaystyle C}{C-C=O} &
\overset{\displaystyle H}{O=C}-C-C-C=O &
O=CH_2 \leftarrow & C-\overset{\displaystyle C}{C}=C-C-C-C=C \\
& HC=O \quad O=CH_2 & & C=C
\end{array}
$$

(b) Myrcene is made up of two head-to-tail isoprene units.

$$
\underset{\text{Myrcene}}{CH_3-\overset{\displaystyle \overset{CH_3}{|}}{C}=CH-CH_2 \vdots CH_2-\overset{\displaystyle \underset{CH_2}{|}}{C}-CH=CH_2}
$$

26. Starting with the carbon skeleton of myrcene (Prob. 25, above), we find that the ten carbons of the oxidation fragments must be fitted together as follows:

$$
\begin{array}{cccc}
\overset{\displaystyle C}{C-C=O} &
O=C-C-C-C=O &
O=C-C \leftarrow & C-\overset{\displaystyle C}{C}=C-C-C-C=C-C \\
& \underset{HO}{|} \quad \underset{C}{|} & \underset{HO}{|} & \underset{C}{|} \\
& & & \text{Dihydromyrcene}
\end{array}
$$

Dihydromyrcene is, therefore, the result of 1,4-addition of hydrogen to the conjugated system of myrcene.

$$
\underset{\text{Myrcene}}{CH_3\overset{\displaystyle \overset{CH_3}{|}}{C}=CH-CH_2-CH_2-\overset{\displaystyle \overset{CH_2}{\|}}{C}-CH=CH_2} \xrightarrow[\text{1,4-addn.}]{H_2} \underset{\text{Dihydromyrcene}}{CH_3-\overset{\displaystyle \overset{CH_3}{|}}{C}=CH-CH_2-CH_2-\overset{\displaystyle \underset{CH_3}{|}}{C}=CH-CH_3}
$$

27. (a) Loss of OPP$^-$ to form dimethylallyl cation, which attacks terminal unsaturated carbon of isopentenyl pyrophosphate to form a 3° cation, which in turn loses a proton to yield geranyl pyrophosphate.

$$\underset{\text{Dimethylallyl pyrophosphate}}{CH_3-\overset{\overset{\displaystyle CH_3}{|}}{C}=CH-CH_2-OPP} \longrightarrow CH_3-\overset{\overset{\displaystyle CH_3}{|}}{\underset{\oplus}{\underbrace{C\cdots CH\cdots CH_2}}} + OPP^-$$

$$CH_3-\overset{\overset{\displaystyle CH_3}{|}}{\underset{\oplus}{\underbrace{C\cdots CH\cdots CH_2}}} + \underset{\text{Isopentenyl pyrophosphate}}{CH_2=\overset{\overset{\displaystyle CH_3}{|}}{C}-CH_2-CH_2-OPP} \longrightarrow CH_3-\overset{\overset{\displaystyle CH_3}{|}}{C}=CH-CH_2-CH_2-\overset{\overset{\displaystyle CH_3}{|}}{\underset{\oplus}{C}}-CH_2-CH_2-OPP$$

$$\Big\downarrow \scriptstyle{-H^+}$$

$$\underset{\text{Geranyl pyrophosphate}}{CH_3-\overset{\overset{\displaystyle CH_3}{|}}{C}=CH-CH_2-CH_2-\overset{\overset{\displaystyle CH_3}{|}}{C}=CH-CH_2-OPP}$$

(b) $(CH_3)_2C=CHCH_2CH_2C(CH_3)=CHCH_2CH_2C(CH_3)=CHCH_2OPP$
                                  Farnesyl pyrophosphate

(c) Two farnesyl units, head-to-head, form the squalene skeleton.

(d) Continuation of the sequence started in (a) and (b).

    We shall catch other glimpses of the organic chemistry underlying the biogenesis of these complex and important compounds. It begins with two-carbon acetate units, which combine to form four-carbon acetoacetate (Sec. 37.6), and then six-carbon mevalonate; this loses CO_2 (Problem 21, Chapter 26, page 864) to give five-carbon isopentenyl pyrophosphate with the characteristic isoprene skeleton.

    (R. B. Clayton, "Biosynthesis of Sterols, Steroids, and Terpinoids. Part I," Quart. Revs. (London), **19**, 168 (1965), especially pp. 169–170.)

FARNESOL

(*To the tune of:* "Jingle Bells")

Take an acetate,
Condense it with a mate,
Pretty soon you have
Acetoacetate.
Let 'em have a ball,
You get geraniol.
Add another isoprene
And you've got farnesol.

      Farnesol, farnesol, good old farnesol,
      First it goes to squalene, then you get cholesterol.
      Farnesol, farnesol, good old farnesol.
      First it goes to squalene, then you get cholesterol.

Now squalene makes a roll,
Becomes lanosterol.
The extra methyls do
Come off as CO_2.
Then comes zymosterol,
And then desmosterol,
If you don't take Triparanol,
You get cholesterol.

      Farnesol, farnesol, *etc.*

DAVID KRITCHEVSKY
The Wistar Institute
Philadelphia, Pennsylvania

Chapter 9 | Alicyclic Hydrocarbons

9.1

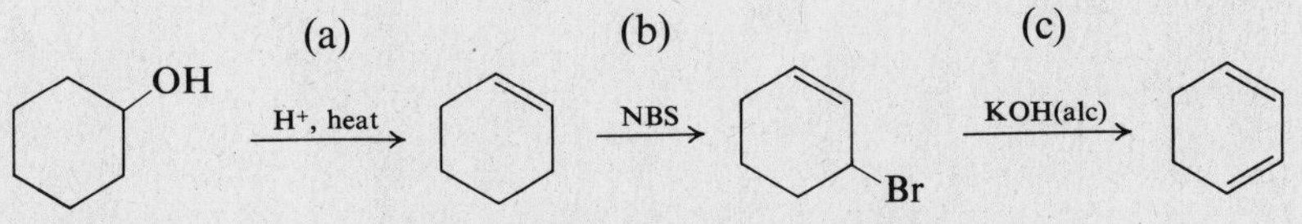

NBS = N-bromosuccinimide (p. 209)

9.2 Mg, anhyd. Et_2O; H_2O.

9.3 Using a model of the axial conformation, consider rotation about the bond between the alkyl group and C–1 of the ring. Ethyl and isopropyl can be rotated so that a hydrogen, —CH_2CH_3 or —$CH(CH_3)_2$, is nearest the axial hydrogens on C–3 and C–5; but no matter how *tert*-butyl is rotated, a large methyl group interferes with the axial hydrogens.

9.4 Only the *trans*-glycol is resolvable into enantiomers; the *cis*-glycol is a non-resolvable *meso* compound. (Compare Problem 7.11, p. 242.)

9.5 (a) H^+, heat; then $KMnO_4$ to give *syn*-hydroxylation.

(b) H^+, heat; then HCO_2OH to give *anti*-hydroxylation.

9.6 All the *cis* compounds, and *trans*-1,3-cyclobutanedicarboxylic acid.

9.7 (a), (d), (e): axial–equatorial = equatorial–axial.

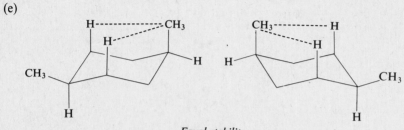

(a)   (d)

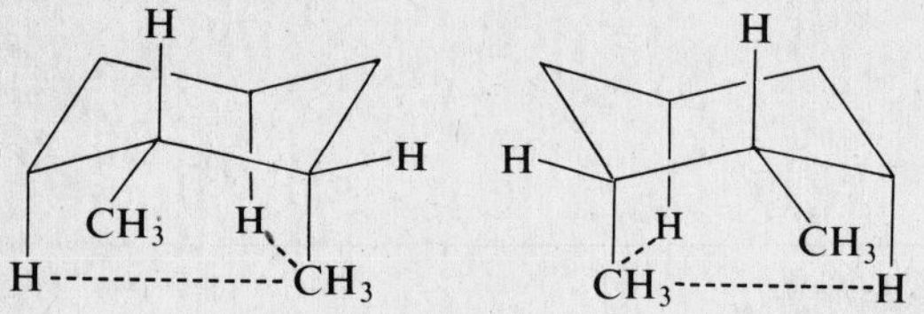

Equal stability
cis-1,2-Dimethylcyclohexane

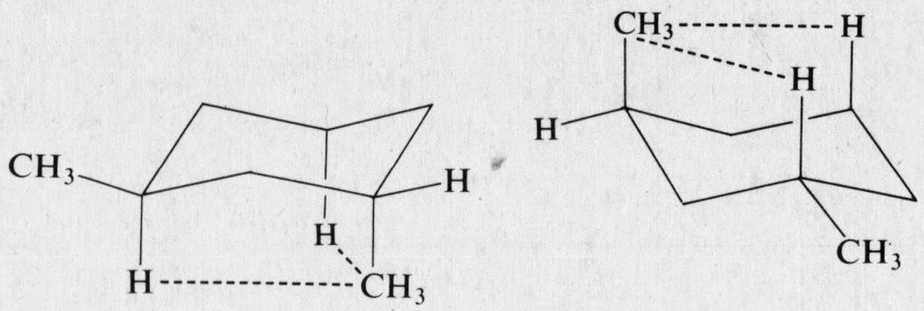

Equal stability
trans-1,3-Dimethylcyclohexane

(e)

Equal stability
cis-1,4-Dimethylcyclohexane

(b), (c), (f): equatorial–equatorial more stable than axial–axial.

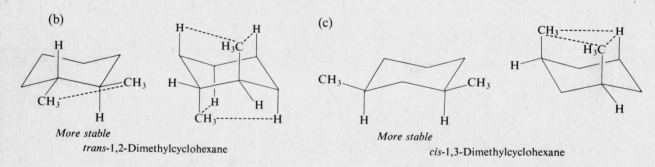

(b)   (c)

More stable
trans-1,2-Dimethylcyclohexane

More stable
cis-1,3-Dimethylcyclohexane

(f)

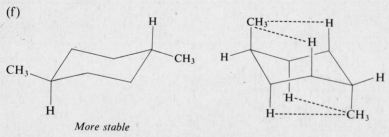

More stable
trans-1,4-Dimethylcyclohexane

(g) 0 kcal for (a), (d), (e): same number (2) of methyl–hydrogen interactions in each.
2.7 kcal for (b): one butane *gauche* (0.9 kcal) vs. four methyl–hydrogen interactions
(4 × 0.9 kcal).
5.4 kcal for (c): no interactions vs. two methyl–hydrogen interactions (2 × 0.9 kcal) +
methyl–methyl interaction (3.6 kcal, see Problem 9.8).
3.6 kcal for (f): no interactions vs. four methyl–hydrogen interactions (4 × 0.9 kcal).

9.8 (a) 1,3-Diaxial interaction of two —CH_3 groups (see 9.8(c), above).

(b) 3.6 kcal, the difference between total of 5.4 kcal and two methyl–hydrogen interactions (2 × 0.9 kcal).

(c) The *trans*-isomer exists as either of two equivalent chair conformations, with two methyl–hydrogen interactions and one methyl–methyl interaction. The *cis*-isomer exists (almost) exclusively in the chair conformation with only one axial methyl group and hence two methyl–hydrogen interactions.

trans:

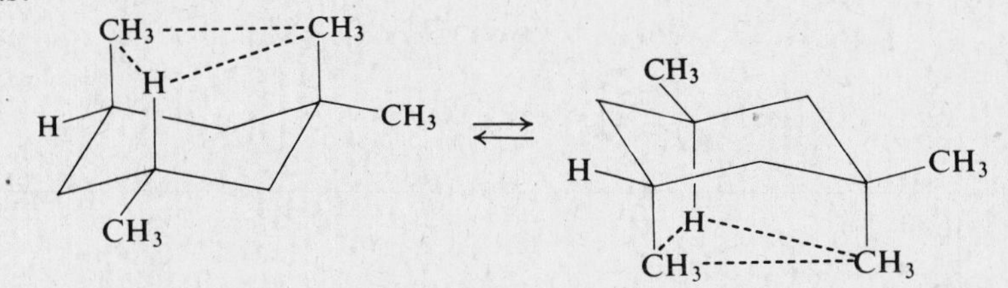

Equivalent
One methyl–methyl and two methyl–hydrogen interactions in each

cis:

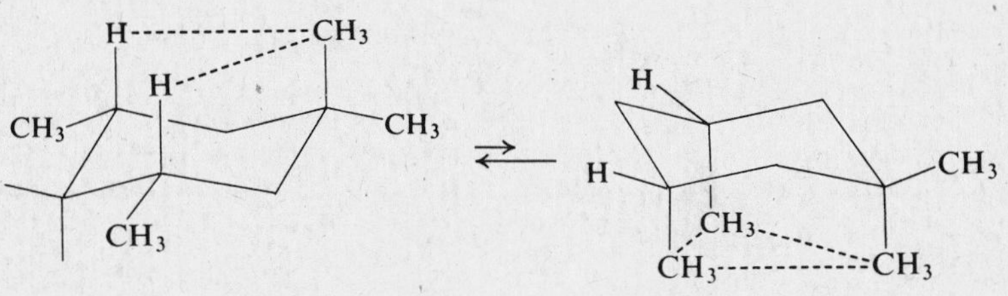

Greatly favored
Only two methyl–hydrogen
interactions

Can be neglected in equilibrium
Three methyl–methyl interactions

Thus the difference in stability between the two isomers (3.7 kcal) is due to one methyl–methyl interaction, in excellent agreement with Pitzer's calculations.

(N. L. Allinger and M. A. Miller, "The 1,3-Diaxial Methyl–Methyl Interaction," J. Am. Chem. Soc., **83**, 2145 (1961).)

9.9 (a) *cis*-Isomer more stable than *trans*.

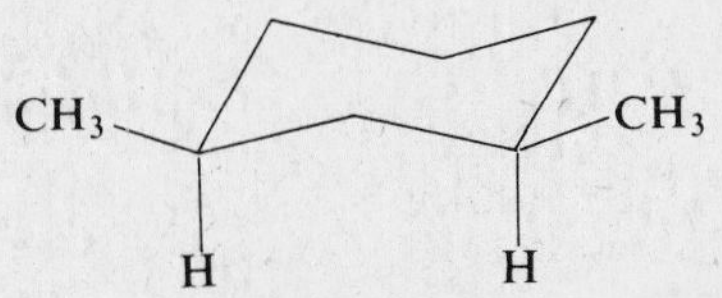

More stable
cis
No interactions
0 kcal

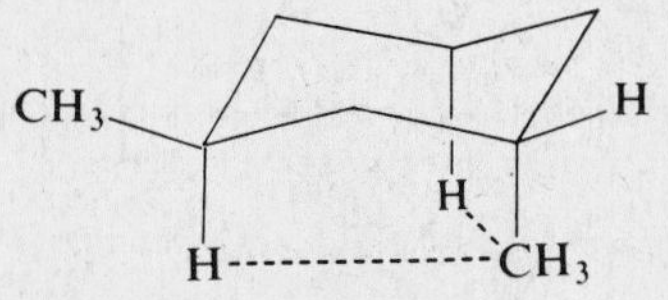

trans
Two methyl–hydrogen interactions
1.8 kcal

(b) *trans*-Isomer more stable than *cis*.

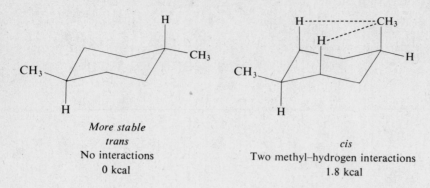

More stable
trans
No interactions
0 kcal

cis
Two methyl–hydrogen interactions
1.8 kcal

(c) Difference of 1.8 kcal/mole in each case.

9.10

(a)

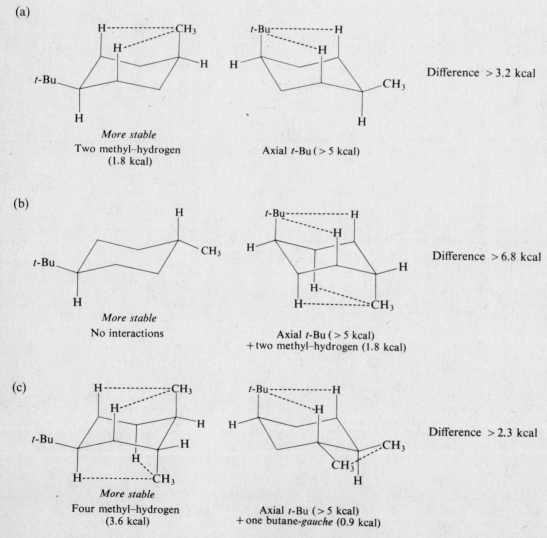

More stable
Two methyl–hydrogen
(1.8 kcal)

Axial *t*-Bu (> 5 kcal)

Difference > 3.2 kcal

(b)

More stable
No interactions

Axial *t*-Bu (> 5 kcal)
+ two methyl–hydrogen (1.8 kcal)

Difference > 6.8 kcal

(c)

More stable
Four methyl–hydrogen
(3.6 kcal)

Axial *t*-Bu (> 5 kcal)
+ one butane-*gauche* (0.9 kcal)

Difference > 2.3 kcal

9.11 Resolvable: b, d. *Meso:* c. (Neither e nor f contains chiral carbons.)

9.12 *cis*-1,2-Cyclohexanediol: a pair of conformational enantiomers.

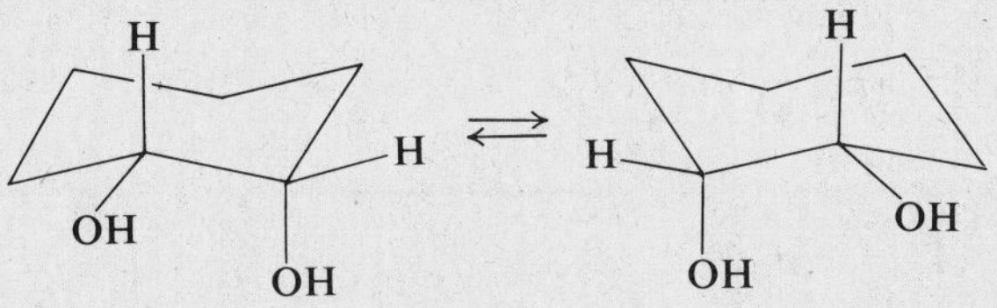

trans-1,2-Cyclohexanediol: a pair of configurational enantiomers, each of which exists as a pair of conformational diastereomers.

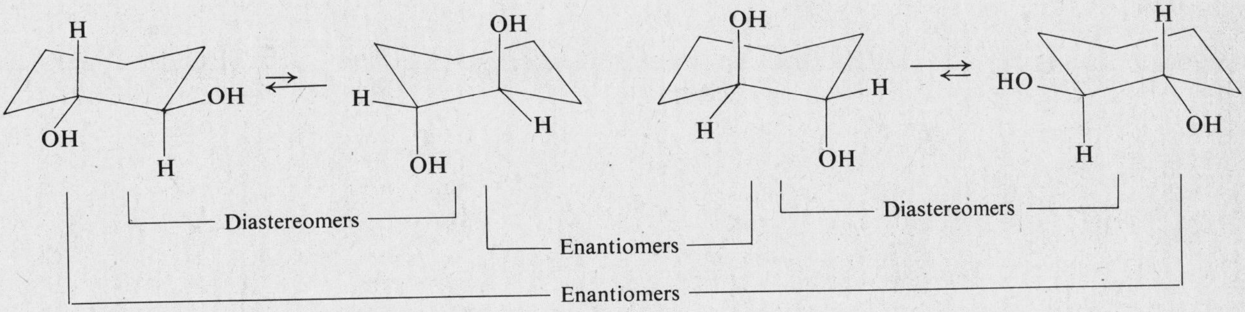

cis-1,3-Cyclohexanediol: a pair of conformational diastereomers.

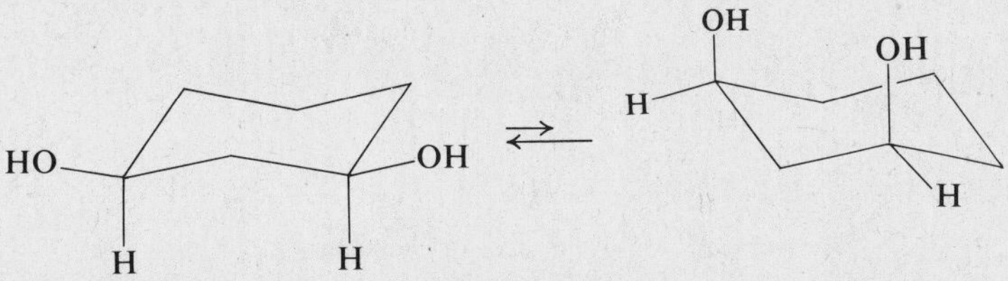

trans-1,3-Cyclohexanediol: a pair of configurational enantiomers, each of which exists as a single conformation. (Use models to convince yourself that certain structures are identical.)

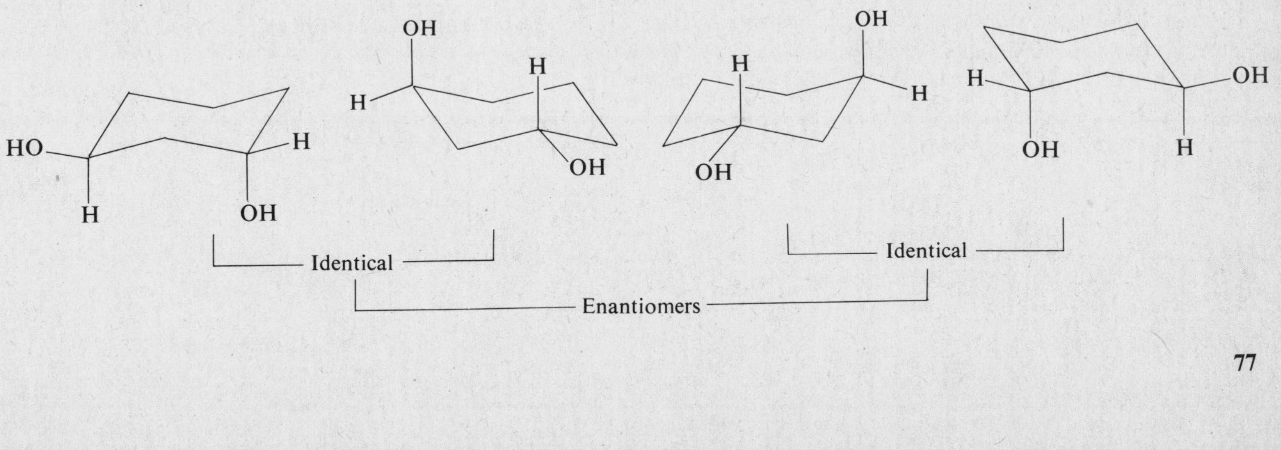

cis-1,4-Cyclohexanediol: exists as a single conformation.

Identical

trans-1,4-Cyclohexanediol: a pair of conformational diastereomers.

9.13. Pairs of enantiomers: a, b, c, d.
Achiral: e, f.
No *meso* compounds.
None are non-resolvable racemic modifications.

9.14. Oxygen reacts with triplet methylene, or with diradical intermediate, leaves only addition of singlet methylene. (Compare action of oxygen as inhibitor of free-radical chlorination, Sec. 2.14.)

9.15 (a)   (b)

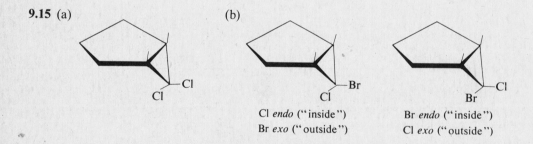

Cl *endo* ("inside")   Br *endo* ("inside")
Br *exo* ("outside")   Cl *exo* ("outside")

9.16 (a) $CHCl_3$ has no β-carbon. (b) Electron withdrawal by chlorines stabilizes anion, speeds up its formation and/or shifts equilibrium (1) (on page 311) to the right.

(Robert A. Moss, "Carbene Chemistry," Chem. Eng. News, June 16, 1969, pp. 60–68; June 30, 1969, pp. 50–58.)

9.17 (a) C_6H_{14}, C_6H_{12} (b) C_5H_{12}, C_5H_{10} (c) C_6H_{12}, C_6H_{10} (d) $C_{12}H_{26}$, $C_{12}H_{24}$, $C_{12}H_{22}$
(e) For the same degree of unsaturation, there are two fewer hydrogens for each ring that is present.

9.18 All are C_6H_{12}; no information about ring size.

9.19 α-Carotene:

$$C_{40}H_{78} \text{ sat'd}$$
$$- C_{40}H_{56} \text{ polyene}$$
$$\overline{\phantom{-C_{40}H_{56}}}$$
$$22H \text{ taken up to saturate 11 double bonds}$$

$$C_{40}H_{82} \text{ alkane}$$
$$- C_{40}H_{78}$$
$$\overline{\phantom{-C_{40}H_{78}}}$$
$$4H \text{ still missing}$$

The missing 4H means *two rings*.

β-Carotene: similarly, 11 double bonds and *two rings*

γ-Carotene:

$$C_{40}H_{80} \text{ sat'd}$$
$$- C_{40}H_{56} \text{ polyene}$$
$$\overline{\phantom{-C_{40}H_{56}}}$$
$$24H \text{ taken up to saturate 12 double bonds}$$

$$C_{40}H_{82} \text{ alkane}$$
$$- C_{40}H_{80}$$
$$\overline{\phantom{-C_{40}H_{80}}}$$
$$2H \text{ still missing}$$

The missing 2H means *one ring*.

Lycopene:

$$C_{40}H_{82} \text{ sat'd}$$
$$- C_{40}H_{56} \text{ polyene}$$
$$\overline{\phantom{-C_{40}H_{56}}}$$
$$26H \text{ taken up to saturate 13 double bonds}$$

$$C_{40}H_{82} \text{ alkane}$$
$$- C_{40}H_{82}$$
$$\overline{\phantom{-C_{40}H_{82}}}$$
$$\text{no H missing}$$

No missing H means *no ring*.

9.20 (a)

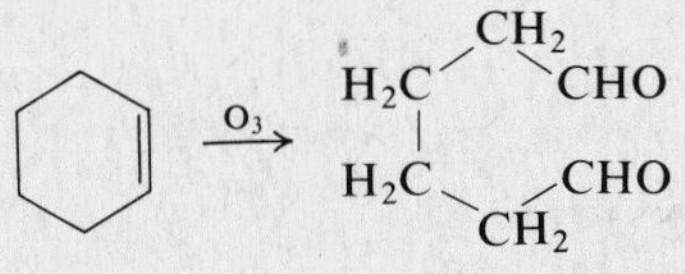

(b)

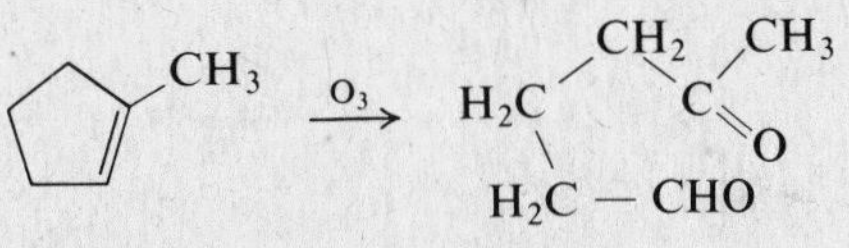

(c)

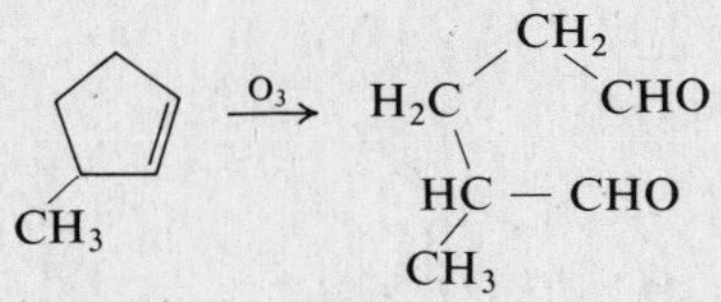

(d)

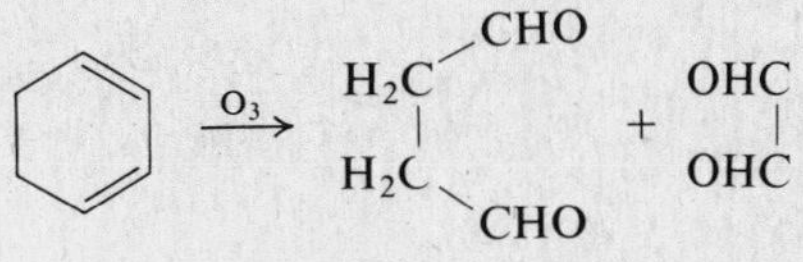

(e)

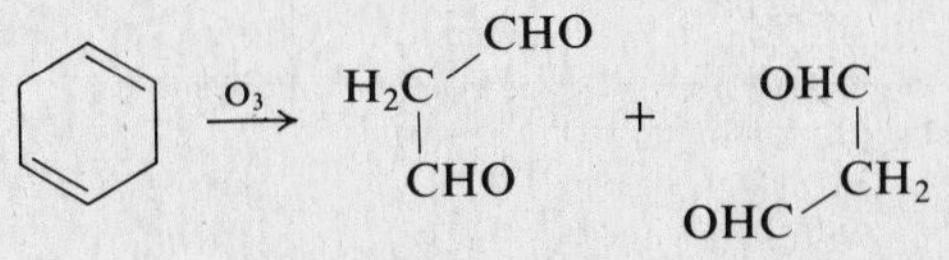

9.21 Diene takes up two moles H_2, cyclohexene only one. Diene yields HCHO, cyclohexene does not.

Stereochemical Formulas of Cyclic Compounds

For convenience, organic chemists use a variety of ways to show the stereochemistry of cyclic compounds.

A solid line indicates a bond coming *out of* the plane of the paper; a dashed line indicates a bond going *behind* the plane.

Alternatively, a round dot represents a hydrogen atom coming out of the plane of the paper; the other bond to that carbon is then understood to be going behind the plane. Where a dot is absent, hydrogen lies behind the plane, and the bond shown is understood to be coming out of the plane.

Thus, we may encounter *trans*-1,2-dibromocyclopentane represented as

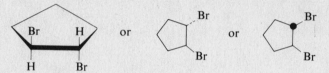

and the *cis*-isomer represented as

1. (a)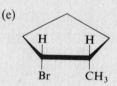

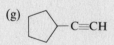

2. (a) ClCH$_2$CH$_2$CH$_2$Cl
     (*ring opens*)

(b)

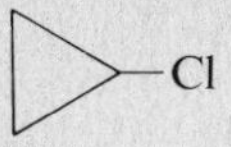

(c) CH$_3$CH$_2$CH$_2$OSO$_3$H
     (*ring opens*)

(d) no reaction

(e)

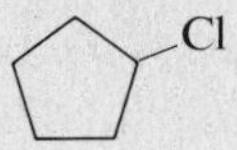

(f) no reaction

(g) *anti*-Addition

(h) Allylic substitution

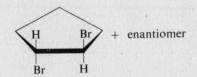 + enantiomer

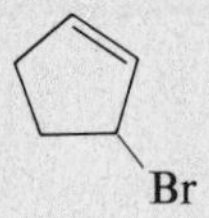

(j) *anti*-Addition via cyclic bromonium ion

(i)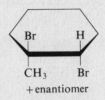

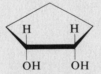
+enantiomer

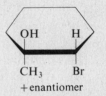
+enantiomer

(k) Free-radical addition of HBr is often stereospecific (*anti*) as it is here.

(l) Allylic substitution

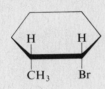

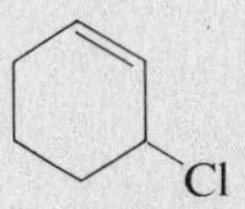

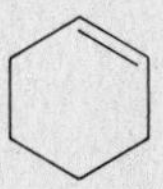

(o) *syn*-Hydroxylation

(p) *anti*-Hydroxylation

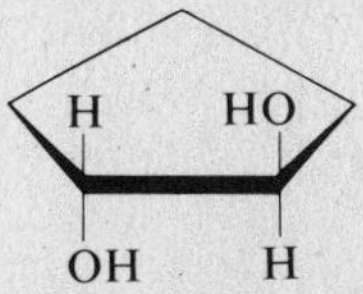

(m)

(n)

+enantiomer

(q) HOOC(CH$_2$)$_3$COOH

(r)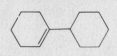

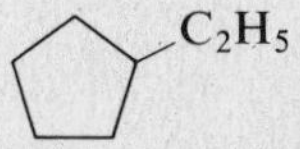

(s)

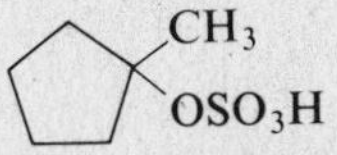

(t)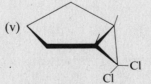

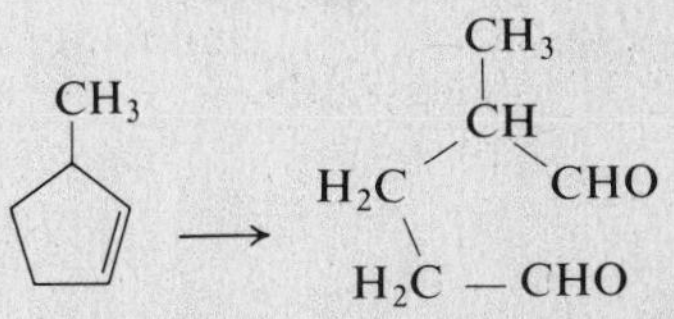

(u) Dimerization of alkene

(v)

Cl
Cl
6,6-Dichlorobicyclo[3.1.0]hexane

(w)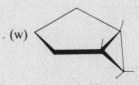

Bicyclo[3.1.0]hexane

3.

(a) H_2SO_4, heat

(b) Product (a), cat. H_2

(c) 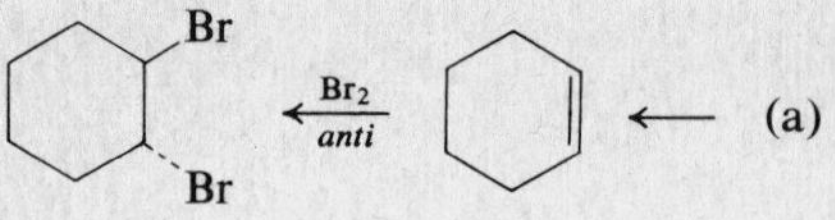 $\xleftarrow[anti]{Br_2}$ ← (a)

(d) 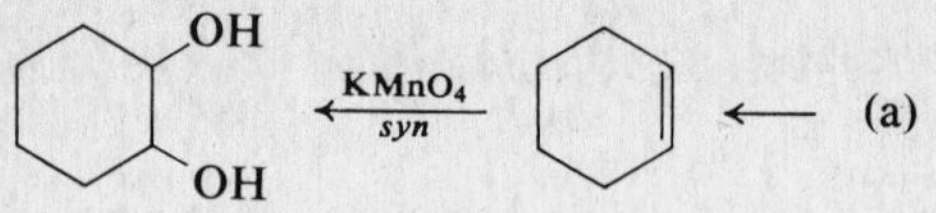 $\xleftarrow[syn]{KMnO_4}$ ← (a)

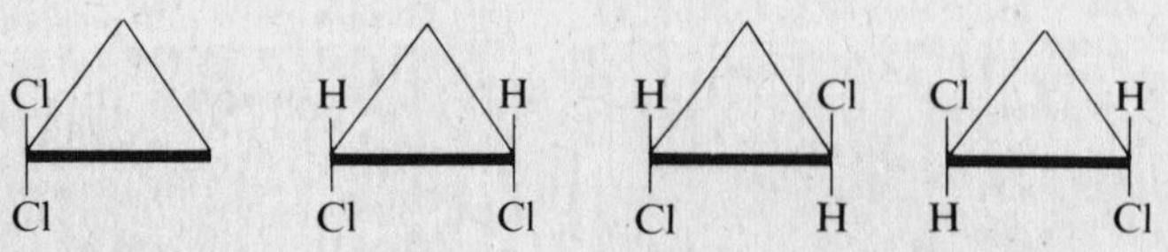

(e) $\xleftarrow[anti]{HCO_2OH}$ ← (a)

(f) $\xleftarrow[]{H_2O, Zn}$ $\xleftarrow[]{O_3}$ ← (a)

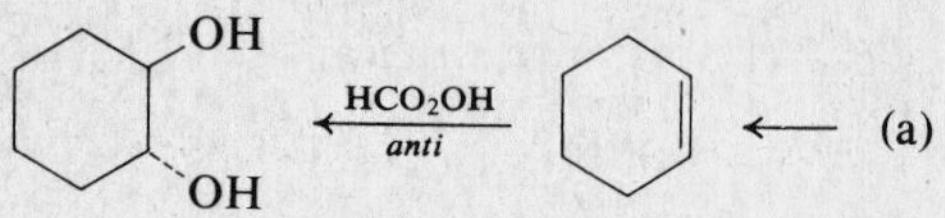

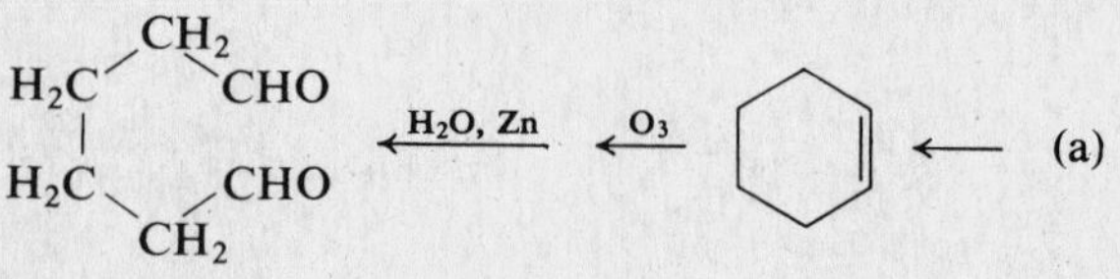

(g) $\xleftarrow[]{KMnO_4 \ (hot)}$ ← (a)

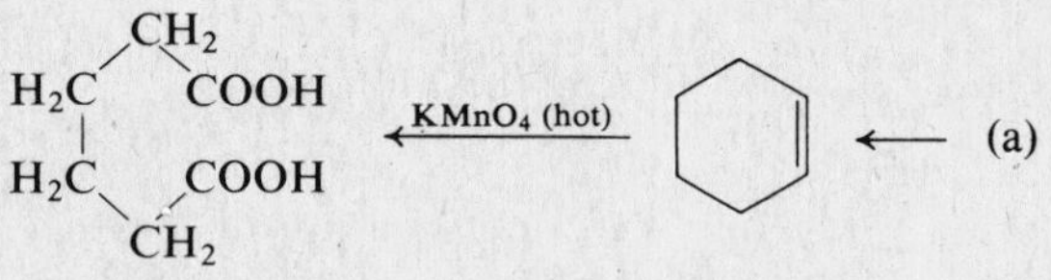

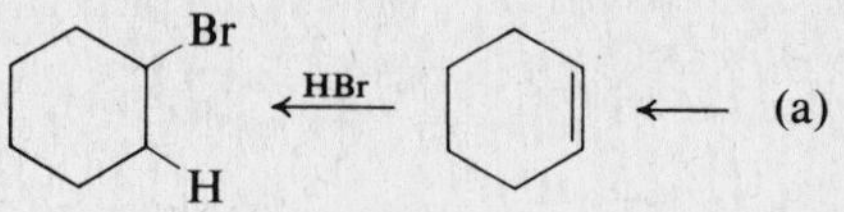

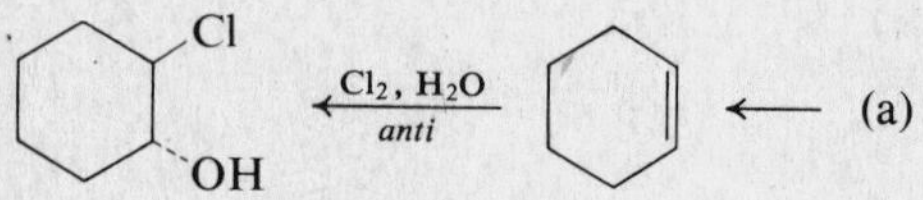

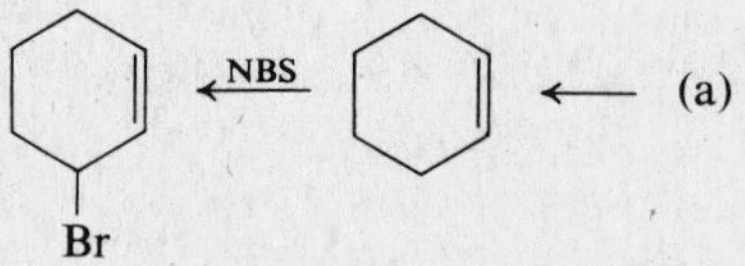

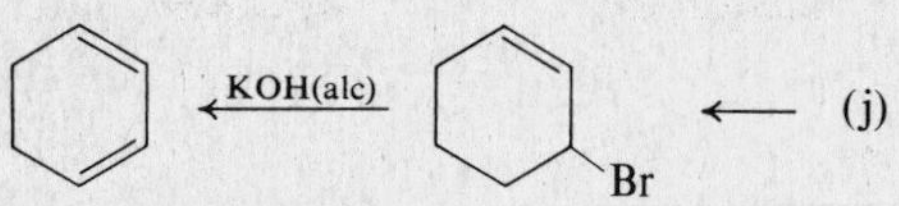

(h) $\xleftarrow[]{HBr}$ ← (a)

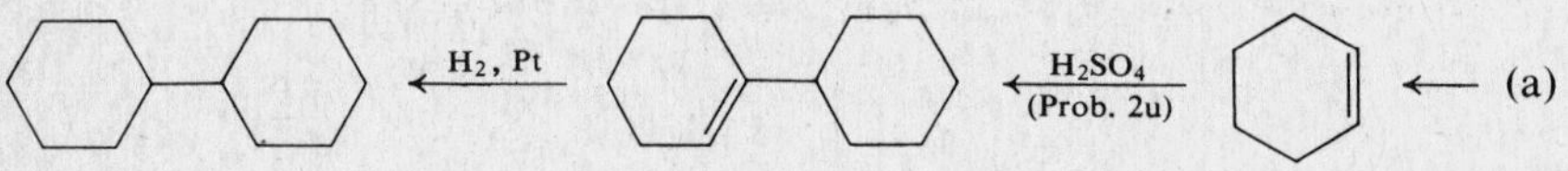

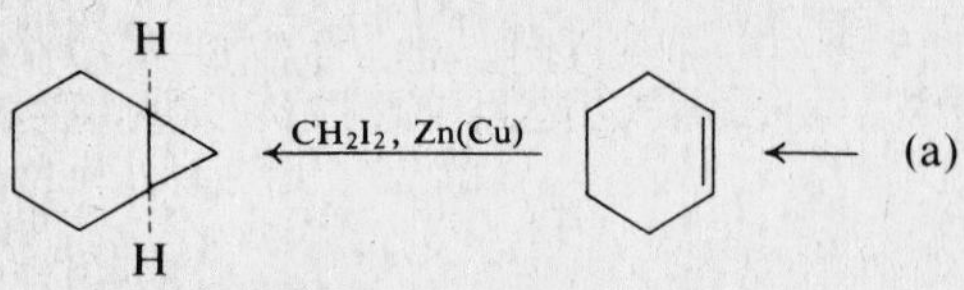

(i) $\xleftarrow[anti]{Cl_2, H_2O}$ ← (a)

(j) $\xleftarrow[]{NBS}$ ← (a)

(k) $\xleftarrow[]{KOH(alc)}$ ← (j)

(l) $\xleftarrow[]{H_2, Pt}$ $\xleftarrow[(Prob.\ 2u)]{H_2SO_4}$ ← (a)

(m) $\xleftarrow[]{CH_2I_2, Zn(Cu)}$ ← (a)

4. (a)

82

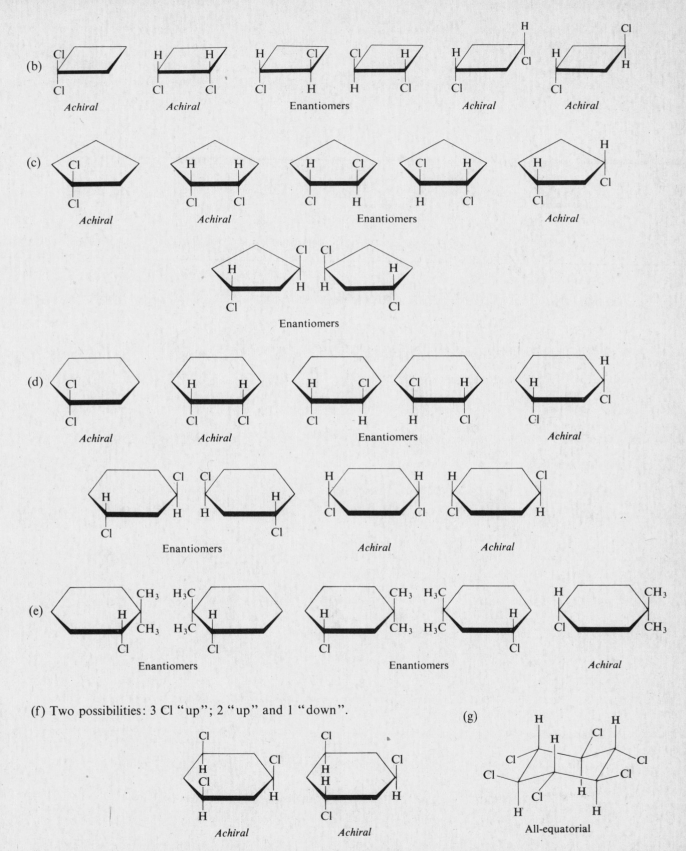

(f) Two possibilities: 3 Cl "up"; 2 "up" and 1 "down".

5. (a)

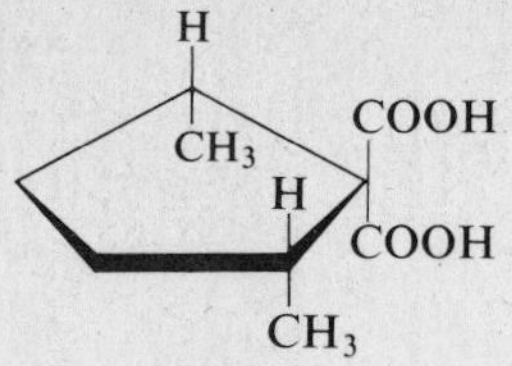

Methyls *cis* 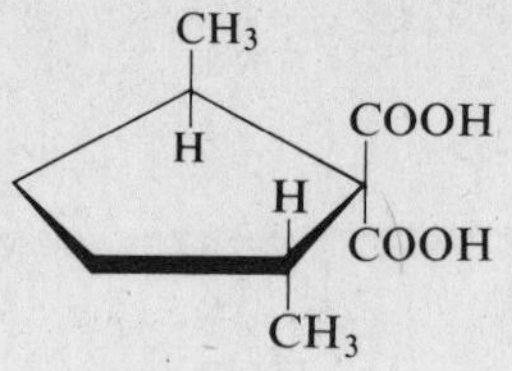   Methyls *trans*

(b)

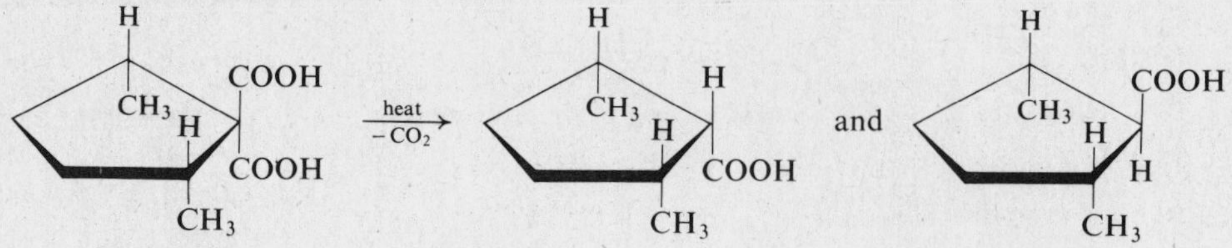

Methyls *cis*
Therefore, this must be A.   *Diastereomers*

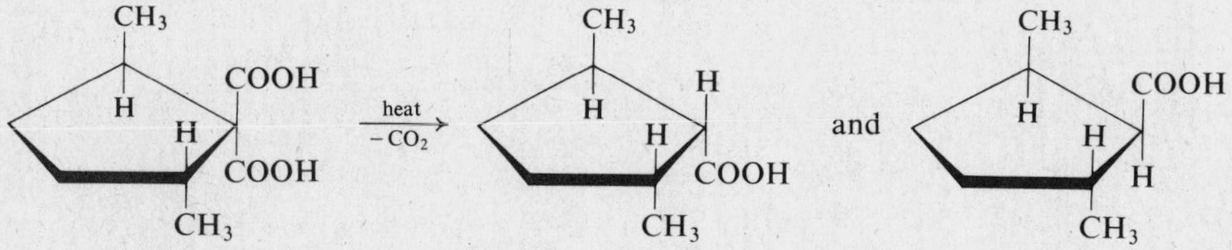

Methyls *trans*
Therefore, this must be B.   *Identical*

6. (a) No chiral carbons; molecules as whole are chiral.

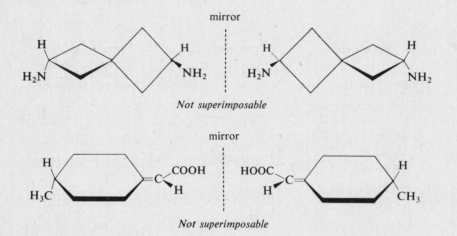

mirror

Not superimposable

mirror

Not superimposable

(b) *anti*-Addition takes place in two different ways,

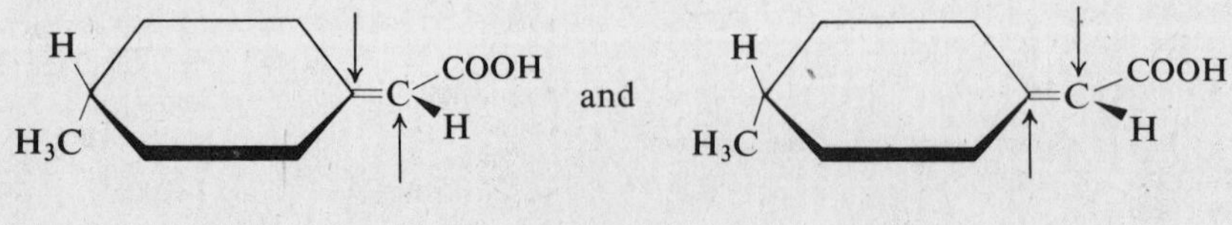

to give two diastereomeric products.

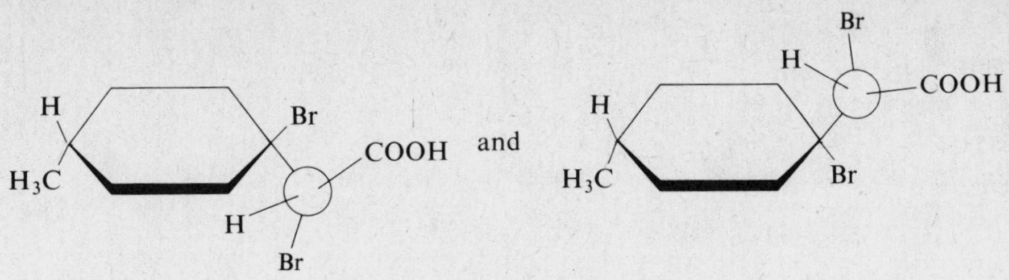

(See reference given in the answer to Problem 15, Chapter 8, page 280.)

7. (a) In the diequatorial conformation of the dibromo or dichloro compound, there is repulsion between the powerful dipoles of the C—X bonds (Sec. 9.10); this dipole–dipole repulsion is relieved in the diaxial conformation. As usual, however, steric (1,3) interactions are

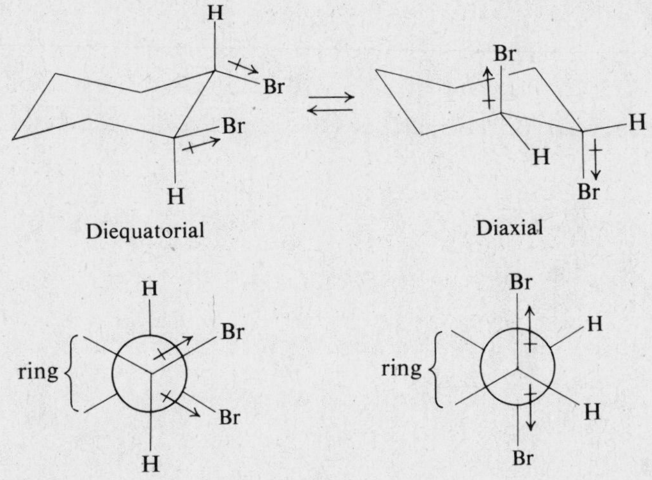

relieved in the diequatorial conformation. The two factors are evidently just about balanced.

The C—CH$_3$ dipoles are much weaker, and for the dimethyl compound steric interactions are the controlling factor.

(b) In the first structure there are four 1,3-diaxial interactions between Br's and H's; but there is relief of dipole–dipole repulsion (see part a). In the second structure there are no

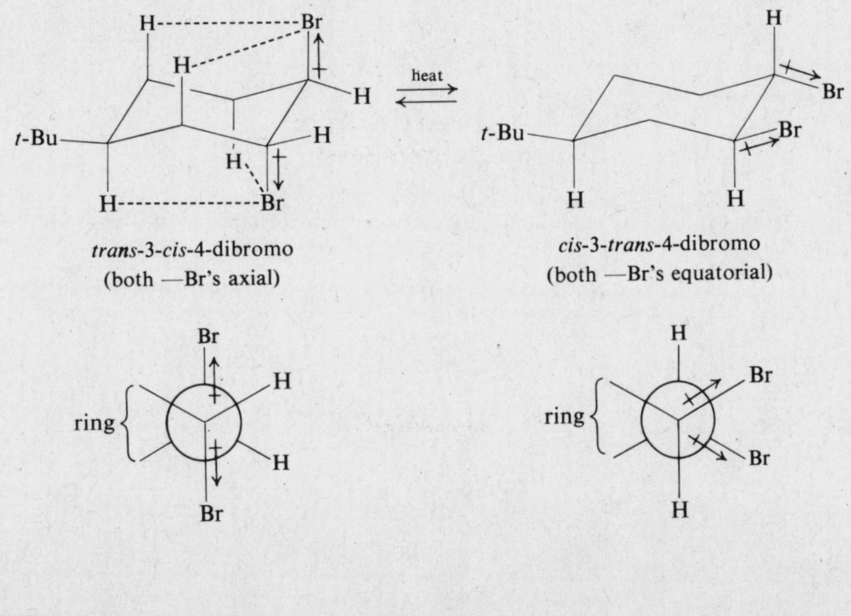

trans-3-*cis*-4-dibromo
(both —Br's axial)

cis-3-*trans*-4-dibromo
(both —Br's equatorial)

85

1,3-diaxial interactions between Br's and H's; but there is increased dipole–dipole repulsion. Dipole–dipole repulsions and 1,3-diaxial interactions just about balance each other, and the two diastereomers are of about equal stability.

(c) In each of the other possible diastereomers there are both 1,3-diaxial interactions and unrelieved dipole–dipole repulsion.

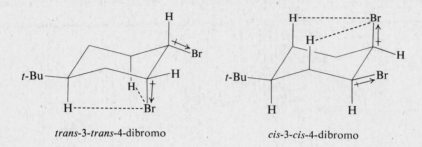

trans-3-*trans*-4-dibromo   *cis*-3-*cis*-4-dibromo

8. (a)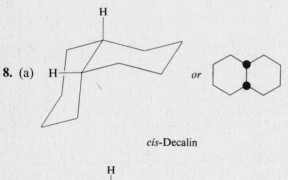

cis-Decalin

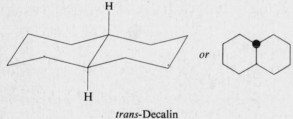

trans-Decalin

(b) *cis*:

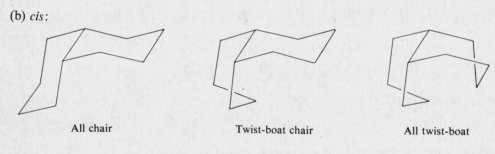

All chair   Twist-boat chair   All twist-boat

trans:

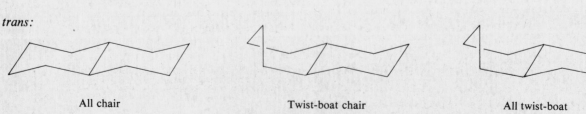

All chair   Twist-boat chair   All twist-boat

(c) The all-chair conformation is the most stable in each case.

(d) In the *trans*-decalin, both large substituents (the other ring) are equatorial.

In the *cis*-isomer, on the other hand, one of the two large substituents is axial.

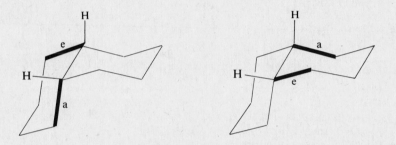

(e) The ease of interconversion does not depend on the energy difference between forms, but on the height of the energy barrier. There is a high-energy barrier between the decalins, since a carbon–carbon bond must be broken in the process of interconversion. There is a low-energy barrier between the chair and twist-boat forms of cyclohexane, since there inter- conversion requires only rotation about single bonds (with some increase in angle and tor- sional strain in the transition state).

9. The *cis*-isomer exists in a chair conformation with both *tert*-butyl groups equatorial. The *trans*-isomer exists in a twist-boat conformation that accommodates both *tert*-butyl groups in *quasi*-equatorial positions. The difference in energy (5.9 kcal) is thus due essentially to the interactions normally present in the twist conformation.

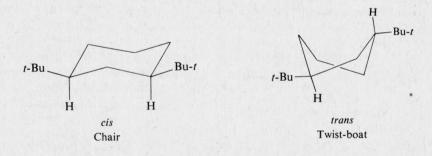

(N. L. Allinger and L. A. Freiberg, "The Energy of the Boat Form of the Cyclohexane Ring," J. Am. Chem. Soc., **82**, 2393 (1960); or E. L. Eliel, N. L. Allinger, S. J. Angyal, and G. A. Morrison, *Conformational Analysis*, pp. 38–39.)

10. (a) This is essentially an equatorial —CH_3.

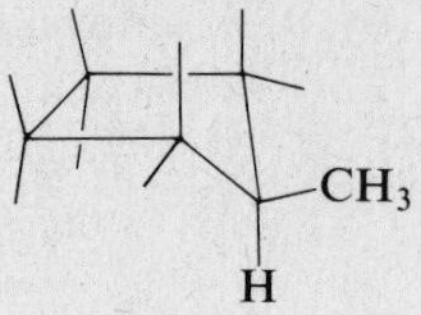

Methylcyclopentane

(b) The situation is very much as for the corresponding cyclohexane derivatives with regard to equatorial–axial relationships and diaxial interactions.

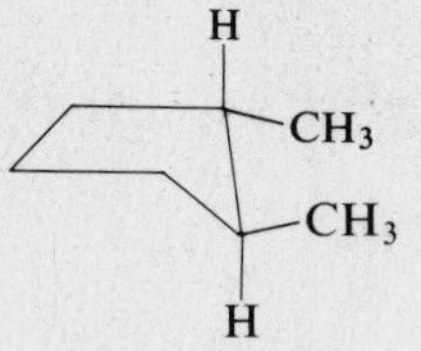

More stable
trans-1,2
Butane *gauche*
interactions

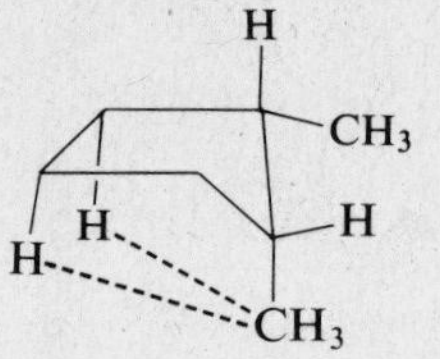

cis-1,2
Butane *gauche* + methyl–
hydrogen interactions

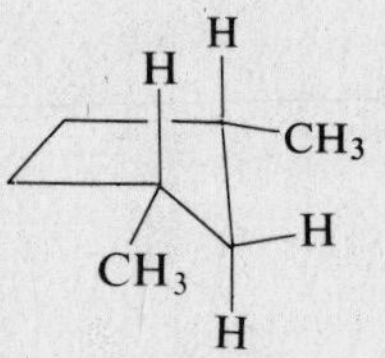

More stable
cis-1,3
No interactions

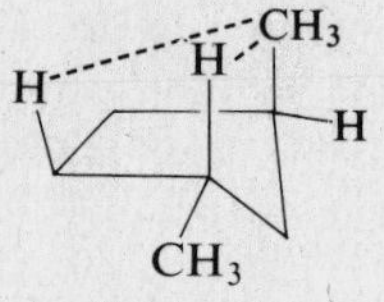

trans-1,3
Methyl–hydrogen
interactions

(c) The "folded" ring puts both substituents in the *cis*-isomer in *quasi*-equatorial positions.

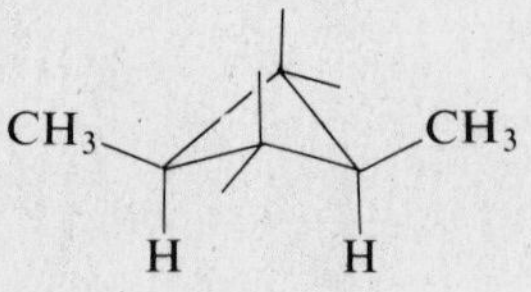

More stable
cis

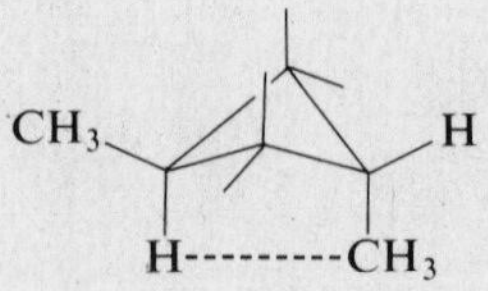

trans
Methyl–hydrogen interaction

(E. L. Eliel *et al.*, reference to Problem 9, above, pp. 201–202.)

11. (a) *syn*-Hydroxylation. Two fractions: one active, one inactive.

KMnO4
and

Meso *Inactive* — *Chiral* *Active*

(b) *anti*-Hydroxylation. Two fractions: one active, one inactive.

HCO2OH
and

Meso *Inactive* — *Chiral* *Active*

(c) Free radical substitution. Two fractions: one active, one inactive.

Cl2
300°
or
and
or

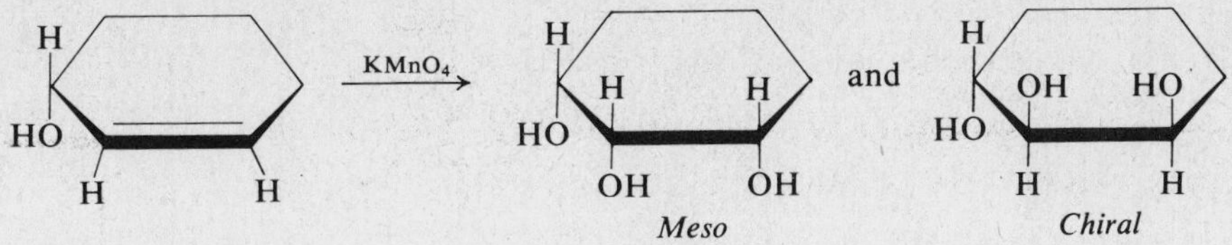

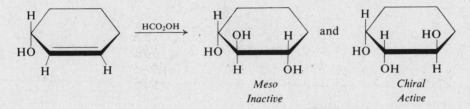

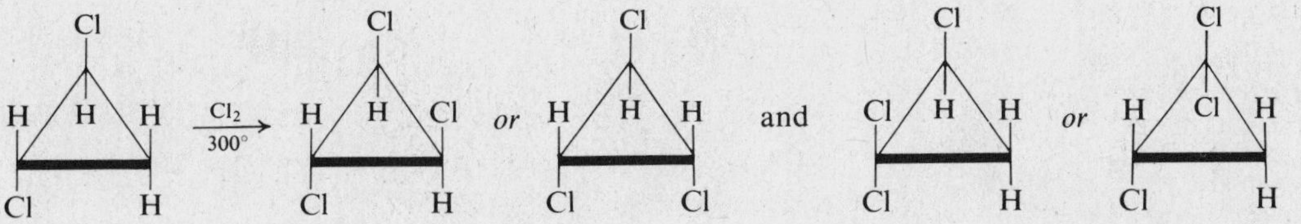

Identical *Meso* *Inactive* — Identical *Chiral* *Active*

(d) *anti*-Addition via cyclic bromonium ion. Two fractions: both inactive (racemic).

Br2
and

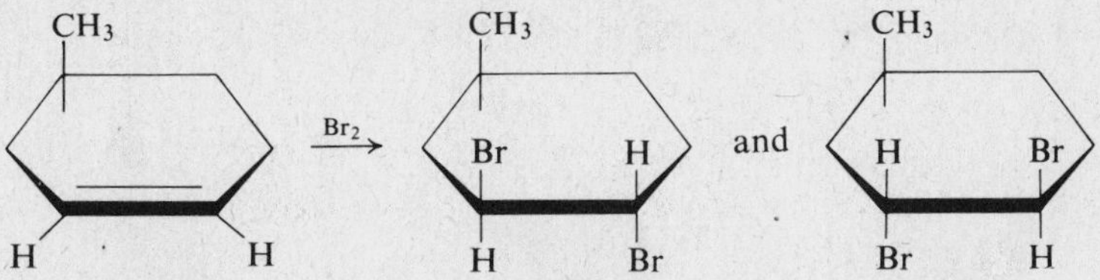

+enantiomer — +enantiomer Racemic *Inactive* — +enantiomer Racemic *Inactive*

12. (a)

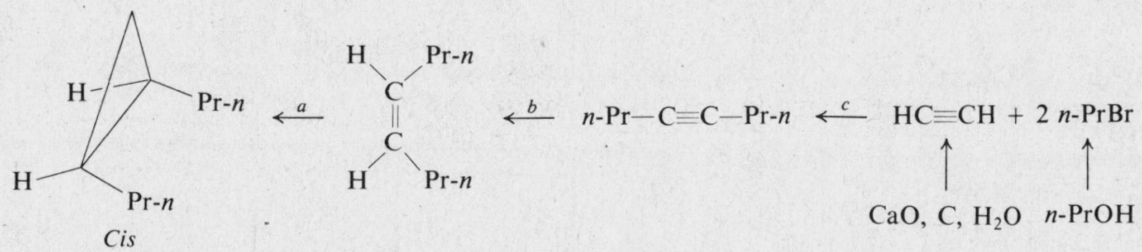

a
b
c
n-Pr—C≡C—Pr-n
HC≡CH + 2 n-PrBr
CaO, C, H2O
n-PrOH

Cis

a: CH_2I_2 (from reduction of CHI_3 obtained through action of I_2, OH^- on EtOH), Zn. (*syn*-Addition of methylene.)

b: H_2, Lindlar's or Brown's catalyst. (*syn*-Hydrogenation.)

c: 2 stages, each involving $NaNH_2$ and then *n*-PrBr (from *n*-PrOH).

(b)

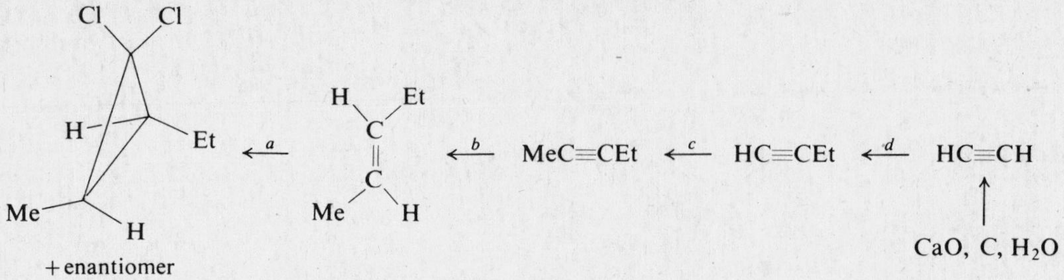

+ enantiomer
Racemic

a: $CHCl_3$ (from EtOH, Cl_2, OH^-), *t*-BuO^- (from *t*-BuOH). (*syn*-Addition of CCl_2.)

b: Li, NH_3. (*anti*-Hydrogenation.)

c: $NaNH_2$, then MeI (from MeOH).

d: $NaNH_2$, then EtBr (from EtOH).

13. (a) conc. H_2SO_4 (b) $KMnO_4$, or Br_2/CCl_4 (c) conc. H_2SO_4
      (d) $KMnO_4$, or Br_2/CCl_4 (e) $KMnO_4$, or Br_2/CCl_4 (f) $KMnO_4$, or Br_2/CCl_4
      (g) CrO_3/H_2SO_4 (h) Br_2/CCl_4, or CrO_3/H_2SO_4
      (i) Br_2/CCl_4 detects cyclohexene. CrO_3/H_2SO_4 detects cyclohexanol. Sodium fusion, then
          $AgNO_3$ detects bromocyclohexane. Cyclohexane is unaffected by any of these tests.

14. Compare Problem 9.19, page 313.

    (a) $C_{10}H_{22}$ alkane   (b) $C_{27}H_{56}$ alkane   (c) $C_{10}H_{22}$ alkane
        $-C_{10}H_{18}$ camphane $-C_{27}H_{48}$ cholestane $-C_{10}H_{16}Br_4$ (Br equiv. to H)

           4H missing   8H missing   2H missing
           Two rings   *Four rings*   *One ring*

    (d) $C_{28}H_{58}$ alkane   (e) $C_{28}H_{52}O$ sat'd derivative
        $-C_{28}H_{52}O$ (O replaces no H's) $-C_{28}H_{44}O$ ergocalciferol

           6H missing   8H missing
          Three rings   *Four double bonds*

    15. (a) C_6H_{14} alkane   (b) $C_{10}H_{22}$ alkane   (c) C_7H_{16} alkane
        $-C_6H_{12}$ $-C_{10}H_{18}$ $-C_7H_{14}$

           2H *One ring*   4H *Two rings*   2H *One ring*

(d) $C_{14}H_{30}$ alkane
$-C_{14}H_{24}$
6H *Three rings*

(e) $C_{14}H_{30}$ alkane
$-C_{14}H_{24}$
6H *Three rings*

(f) $C_{20}H_{42}$ alkane
$-C_{20}H_{32}$
10H *Five rings*

(g) $C_{18}H_{38}$ alkane
$-C_{18}H_{30}$
8H *Four rings*

16. (a) $C_{10}H_{22}$ alkane
$-C_{10}H_{18}$
4H *Two rings*

(b)

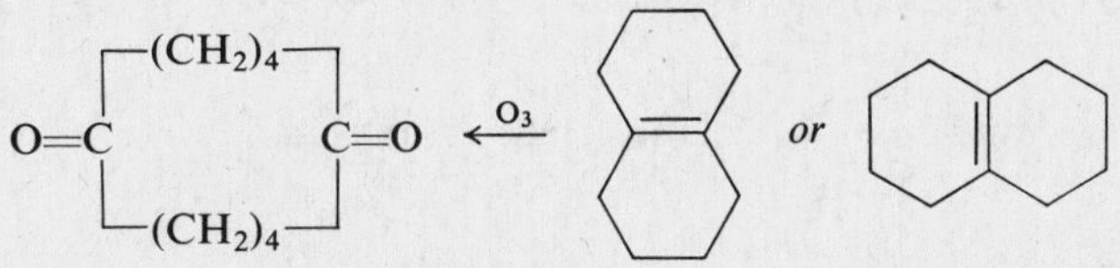

III

17. (a) $C_{10}H_{20}$ *p*-menthane
$-C_{10}H_{16}$ limonene
4H *Two double bonds* in limonene

$C_{10}H_{22}$ alkane
$-C_{10}H_{20}$ *p*-menthane
2H *One ring* in limonene

(b) One of the original 10 carbons is missing from the oxidation product IV; presumably it was lost as CO_2. On this assumption, and knowing that there are two double bonds and one ring, we arrive at the following possible structures for limonene:

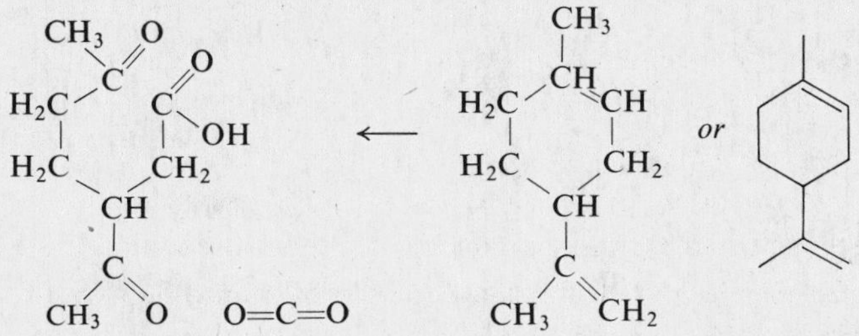

IV + CO_2

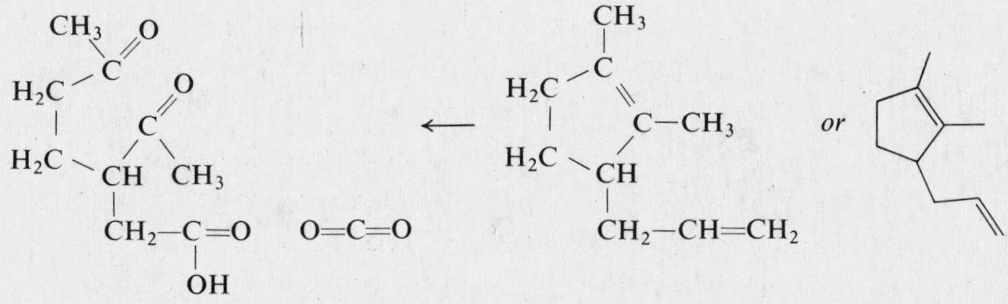

IV + CO_2

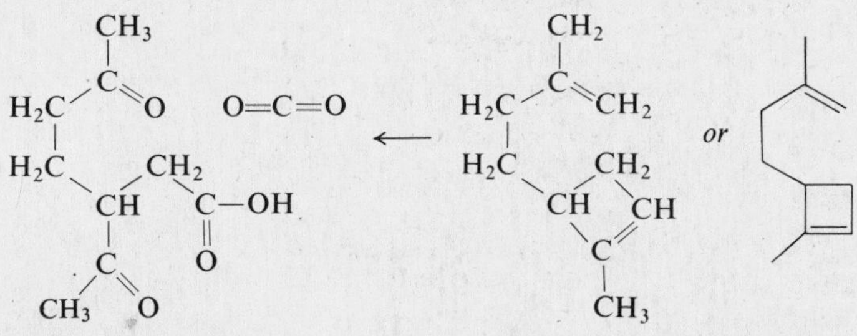

IV + CO_2

(c) The most likely structure is

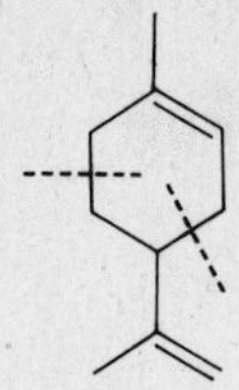

Limonene
Two isoprene units

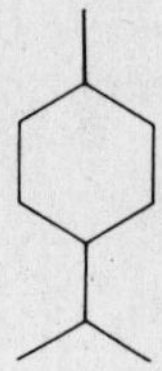

p-Menthane

(d) The alcohol could be either of the tertiary alcohols shown; actually, it is α-terpineol.

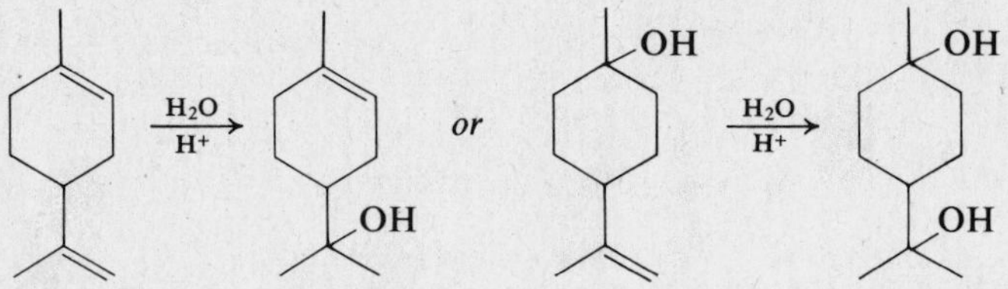

Limonene   α-Terpineol   Terpin hydrate

(e) Terpin hydrate is the di-*tert*-alcohol.

18. (a)   C$_{10}$H$_{20}$ *p*-menthane   C$_{10}$H$_{22}$ alkane
         $-$ C$_{10}$H$_{16}$ α-terpinene $-$ C$_{10}$H$_{20}$ *p*-menthane
         ───────────────   ───────────────
         4H *Two double bonds* in α-terpinene   2H *One ring* in α-terpinene

(b) Using the skeleton of the reduction product, *p*-menthane (Problem 17, above), as a clue, we can account for the eight-carbon oxidation product V and the missing two carbons as follows:

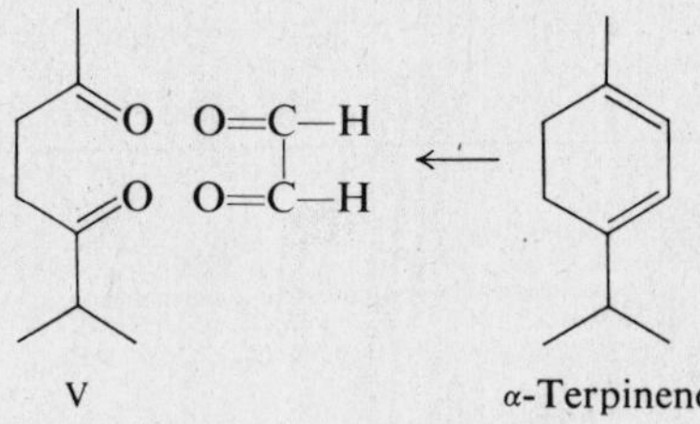

V   α-Terpinene

(c) VI is most likely formed as follows:

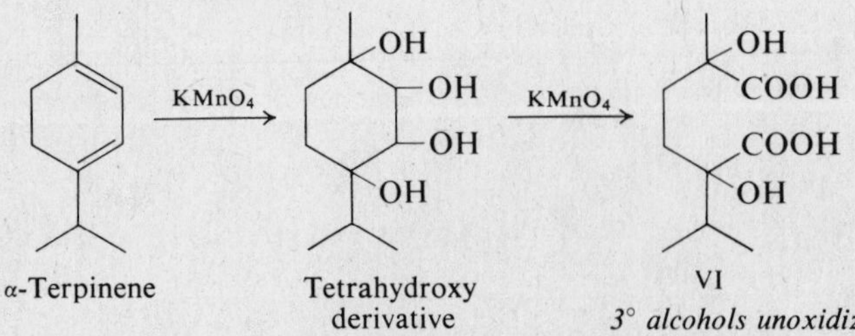

α-Terpinene   Tetrahydroxy   VI
                     derivative   *3° alcohols unoxidized*

92

19. The key step, ring-closure, involves the familiar addition of a carbonium ion to an alkene (Sec. 6.15).

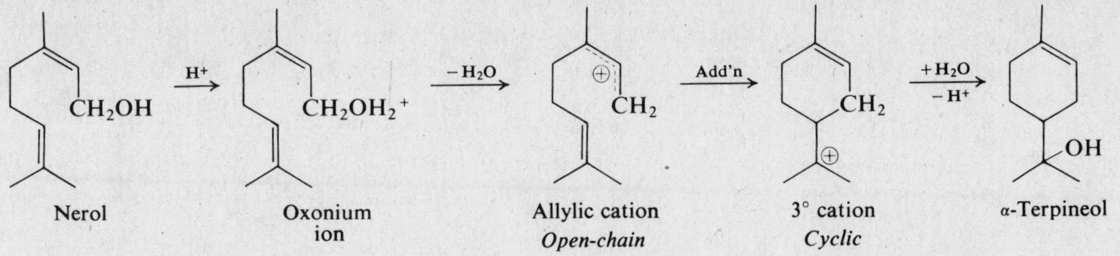

CH₂OH
H⁺
CH₂OH₂⁺
−H₂O
CH₂
Add'n
CH₂
+H₂O
−H⁺
OH
Nerol
Oxonium ion
Allylic cation
Open-chain
3° cation
Cyclic
α-Terpineol

Chapter 10 | Benzene

Aromatic Character

10.1 (a)

$$
\begin{aligned}
\text{benzene} + 3H_2 &= \text{cyclohexane} & \Delta H & \quad -49.8 \text{ kcal} \\
-(\text{cyclohexadiene} + 2H_2 &= \text{cyclohexane} & \Delta H & \quad -55.4 \text{ kcal}) \\
\hline
\text{benzene} + H_2 &= \text{cyclohexadiene} & \Delta H & \quad +5.6 \text{ kcal}
\end{aligned}
$$

(b)

$$
\begin{aligned}
\text{cyclohexadiene} + 2H_2 &= \text{cyclohexane} & \Delta H & \quad -55.4 \text{ kcal} \\
-(\text{cyclohexene} + H_2 &= \text{cyclohexane} & \Delta H & \quad -28.6 \text{ kcal}) \\
\hline
\text{cyclohexadiene} + H_2 &= \text{cyclohexene} & \Delta H & \quad -26.8 \text{ kcal}
\end{aligned}
$$

10.2 (a)

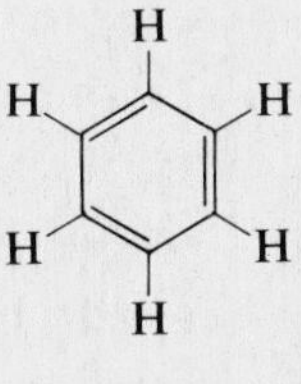

$$
\begin{aligned}
6 \text{ C—H bonds} &= 6 \times 54.0 = 324.0 \text{ kcal} \\
3 \text{ C==C bonds} &= 3 \times 117.4 = 352.2 \\
3 \text{ C—C bonds} &= 3 \times 49.3 = \underline{147.9} \\
\text{calcd.} &= 824.1 \text{ kcal}
\end{aligned}
$$

(b) $\underset{\text{(calc)}}{824.1} - \underset{\text{(obs)}}{789.1} = 35.0$ kcal greater than observed value.

10.3 The sp^2–s character of the C—H bond in benzene. It should be (and is) shorter, and thus stronger, than the sp^3–s C—H bond in cyclohexane.

10.4 (a) No; (b) see Sec. 31.6.

10.5 (a)

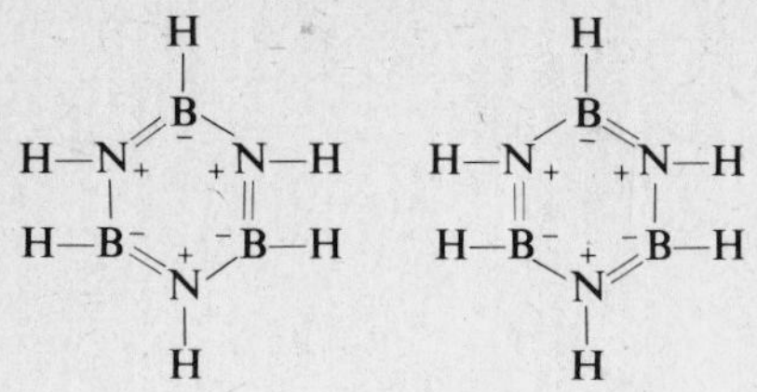

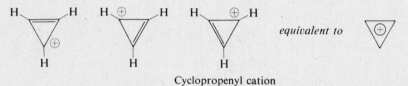

(b) Similar to Figs. 10.2 and 10.3.

(c) Six, two "from" each N.

10.6 (a) Compound I is the salt, cyclopropenyl hexachloroantimonate:

$$\text{(triangle with } \oplus \text{)} \quad SbCl_6^-$$

The same cation is formed by the AgBF$_4$ treatment. It is symmetrical, and all three protons are equivalent.

H—(triangle)—H ... equivalent to (triangle with ⊕)

Cyclopropenyl cation

The cyclopropenyl cation is even more stable than the allyl cation: 20 kcal more stable, relative to the parent chloride. The unusual stability suggests not just resonance stabilization, but aromaticity.

(b) The cyclopropenyl cation contains two π electrons, which fits the Hückel $4n + 2$ rule for $n = 0$.

(R. Breslow and J. T. Groves, "Cyclopropenyl Cation. Synthesis and Characterization," J. Am. Chem. Soc., **92**, 984 (1970).)

10.7 (a) Cyclic structure with alternating double and single bonds.

(b) The number of π electrons is eight, which is not a Hückel number.

(c) The number of π electrons in the anion is 10, fitting the Hückel $4n + 2$ rule for $n = 2$. Evidently stabilization due to aromaticity is enough to outweigh the double negative charge and angle strain (ring must be flat for π overlap, and hence must have C—C—C angles of 135°).

(d) Cyclooctatetraene: puckered rings maintaining geometry of carbon–carbon double bonds (x-ray diffraction shows it to be a "tub"). $C_8H_8^{--}$: flat, regular octagon.

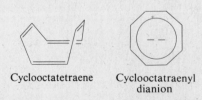

Cyclooctatetraene   Cyclooctatraenyl dianion

10.8

Six dibromonitrobenzenes

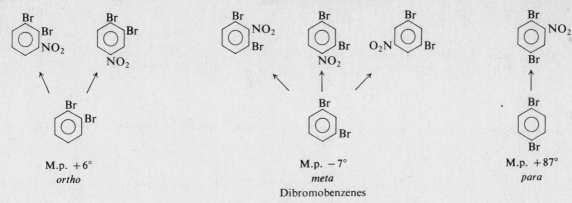

M.p. +6°   M.p. −7°   M.p. +87°
ortho   *meta*   *para*

Dibromobenzenes

10.9 KOH takes out CO_2 and H_2O; the other liquids would not do this.

10.10

partial pressure N_2 = total pressure − partial pressure H_2O
  (740 mm)   (746 mm)   (6 mm)

$$\text{vol. } N_2 \text{ at S.T.P.} = 1.31 \times \frac{273}{293} \times \frac{740}{760} \text{ cc}$$

$$\text{mmoles } N_2 = \frac{\text{cc gas (at S.T.P.)}}{22.4 \text{ cc/mmole}}$$

$$\text{wt. N} = \text{mmoles } N_2 \times 28.0 \text{ mg/mmole}$$

$$\%N = \frac{\text{wt. N}}{\text{wt. sample}} \times 100 = 1.31 \times \frac{273}{293} \times \frac{740}{760} \times \frac{1}{22.4} \times 28.0 \times \frac{1}{5.72} \times 100$$

$$\%N = 26.0\% \text{ N.}$$

10.11

$$\text{mmoles } NH_3 = \text{ml HCl} \times \text{conc. HCl} = 5.73 \text{ ml} \times 0.0110 \text{ mmole/ml}$$

$$\text{wt. N} = \text{mmoles } NH_3 \times 17.0 \text{ mg/mmole} \times \frac{N}{NH_3} = 5.73 \times 0.0110 \times 17.0 \times \frac{14.0}{17.0}$$

$$\%N = \frac{\text{wt. N}}{\text{wt. sample}} \times 100 = 5.73 \times 0.0110 \times 14.0 \times \frac{1}{3.88} \times 100$$

$$\%N = 22.8\% \text{ N.}$$

10.12

$$\text{wt. S} = \text{wt. } BaSO_4 \times \frac{S}{BaSO_4} = 6.48 \times \frac{32.0}{233.4} \text{ mg}$$

$$\%S = \frac{\text{wt. S}}{\text{wt. sample}} \times 100 = 6.48 \times \frac{32.0}{233.4} \times \frac{1}{4.81} \times 100 = 18.5\% \text{ S.}$$

10.13

$p\text{-}H_2NC_6H_4NH_2$ $\dfrac{2N}{C_6H_8N_2} \times 100 = \dfrac{28.0}{108} \times 100 = 25.9\%$ N (calcd.)

$HOCH_2CH_2NH_2$ $\dfrac{N}{C_2H_7ON} \times 100 = \dfrac{14.0}{61.0} \times 100 = 22.9\%$ N (calcd.)

$p\text{-}CH_3C_6H_4SO_3H$ $\dfrac{S}{C_7H_8O_3S} \times 100 = \dfrac{32.0}{172} \times 100 = 18.6\%$ S (calcd.)

1. (h) —SO$_3$H on C–1   (i) —CH$_3$ on C–1   (j) —COOH on C–1   (l) —OH on C–1

2. (a) 3 (o, m, p)   (b) 3 (o, m, p)

(c) 3:

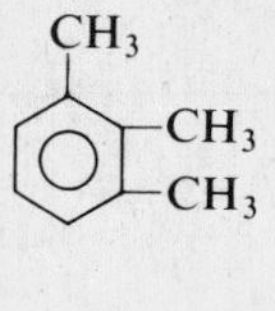

1,2,3 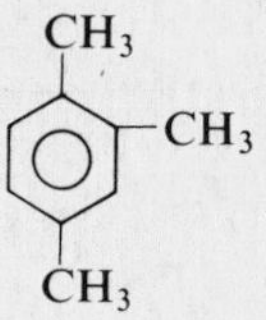   1,2,4 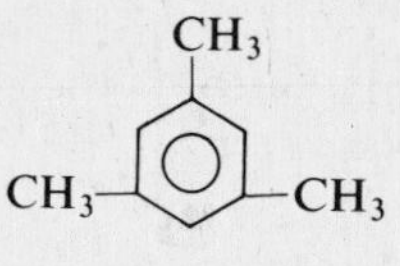   1,3,5

(d) 6: see answer to Problem 10.8

(e) 10:

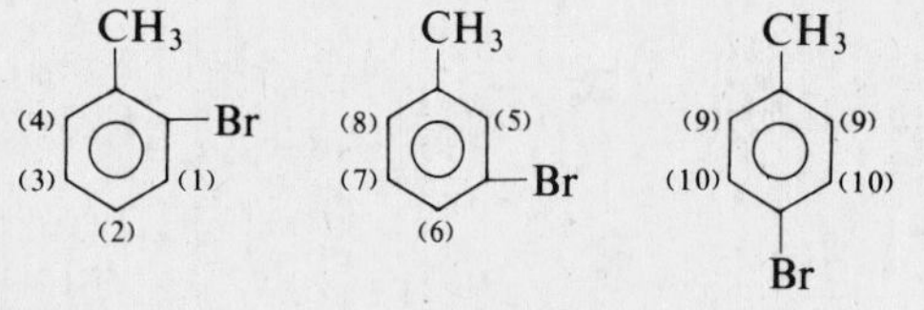

(*Each number in parentheses indicates a different isomer.*)

(f) 6: 2,3,4-, 2,3,5-, 2,3,6-, 2,4,5-, 2,4,6-, 3,4,5-

3. (a) 2, 3, 3, 1, 2.
   (b) 5, 5, 5, 2, 4 (neglecting stereoisomers).
   (c) None.

4. (a)

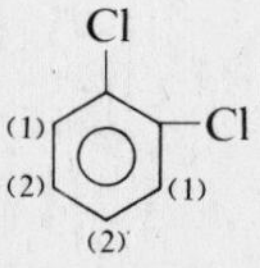

(b)

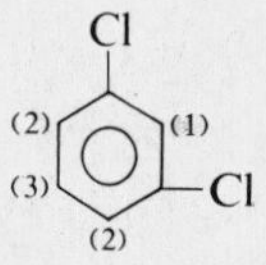

(c)

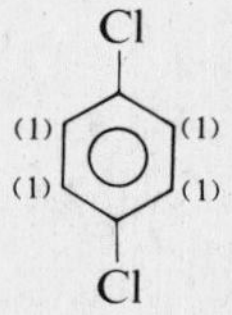

(d)

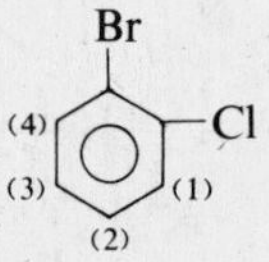

(e)

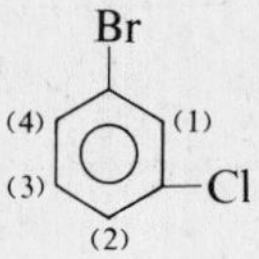

(f)

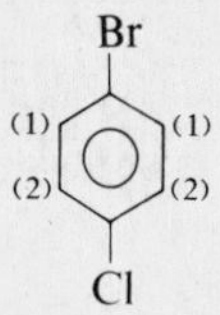

(g)

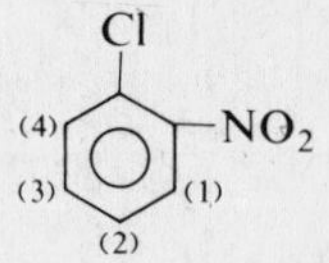

(h)

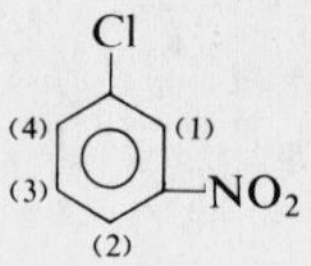

(i)

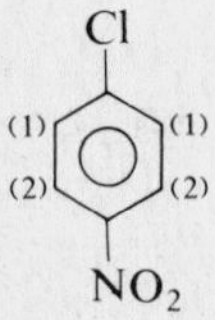

(j)

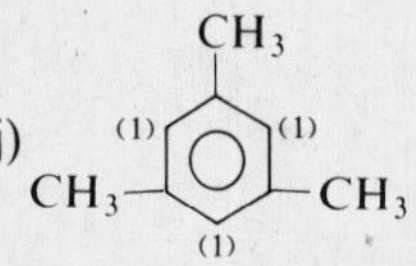

(k)

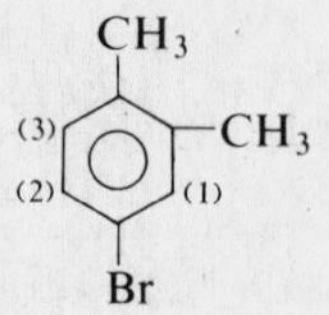

(l)

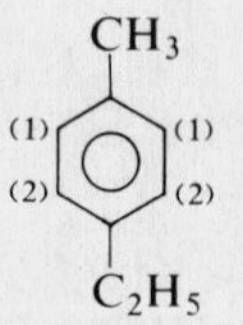

5. C$_8$H$_{10}$. Two carbons attached to ring: one Et, or two Me. The possible isomeric hydrocarbons—with the *different* substitution sites marked by numbers—are:

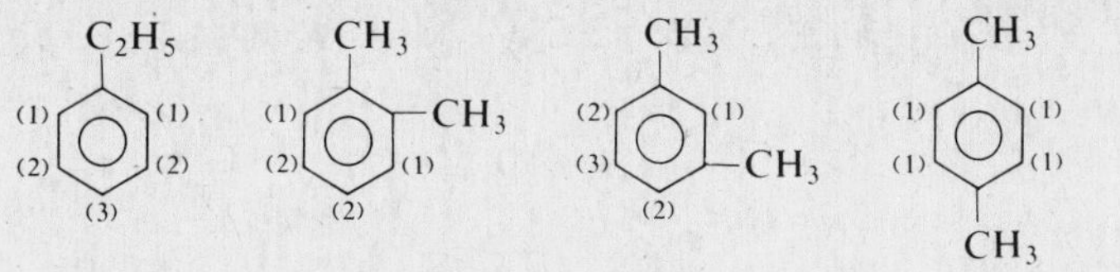

Three
(c)

Two
(b)

Three
(c)

One
(a)

C_9H_{12}. Three carbons attached to ring: one *n*-Pr or *i*-Pr; one Et and one Me; or three Me. The possible isomeric hydrocarbons—with the *different* substitution sites marked by numbers—are:

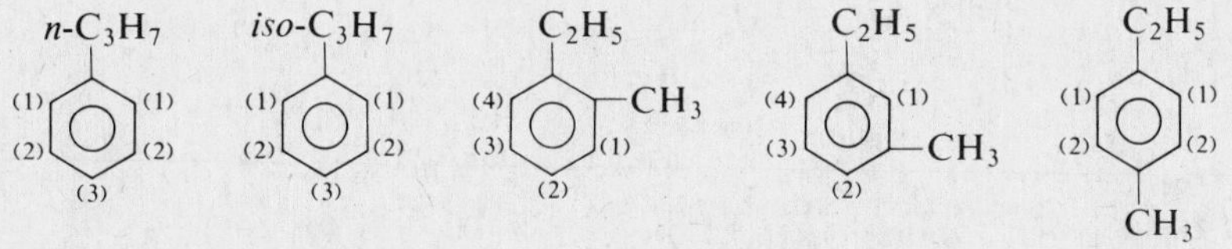

Three
(f)

Three
(f)

Four
(g)

Four
(g)

Two
(e)

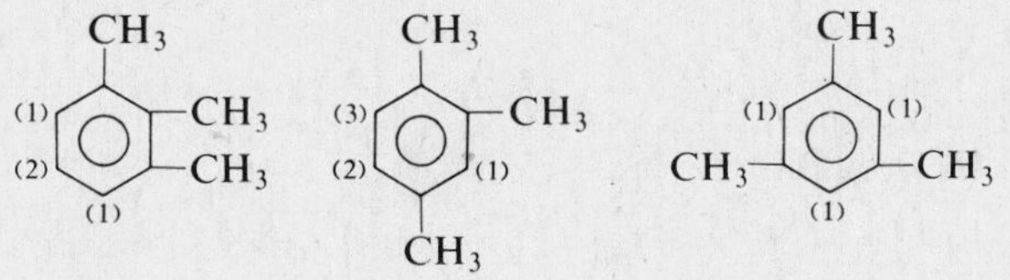

Two
(e)

Three
(f)

One
(d)

6. Yes. Each isomer has a different number of mononitro compounds that can be related to (derived from or convertible into) it.

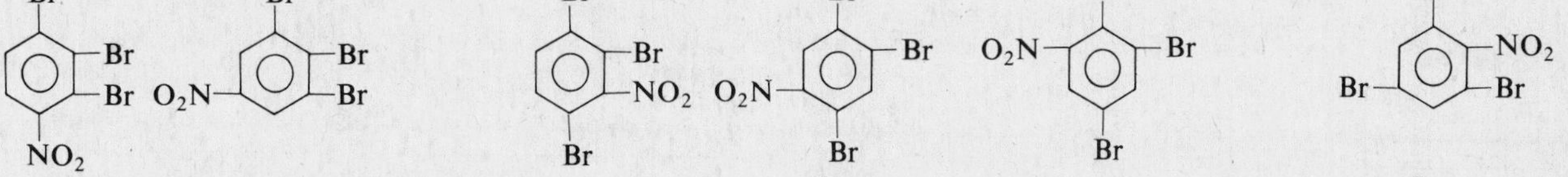

1,2,3-

1,2,4-

1,3,5-

Two

Three

One

7. (a) Two substituents can be attached to VI in three ways: (i) one above the other on any vertical edge; (ii) at opposite corners of any square; (iii) at two corners of one triangle.

(b)

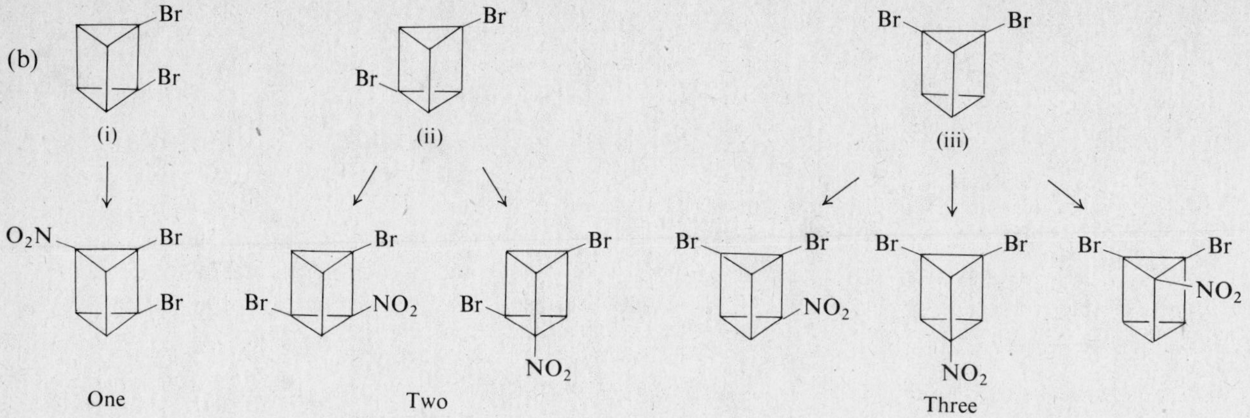

One   Two   Three

Therefore, (i) is *para*, (ii) is *ortho*, and (iii) is *meta*.

(c) No, the *ortho* isomer is chiral; enantiomeric structures are possible:

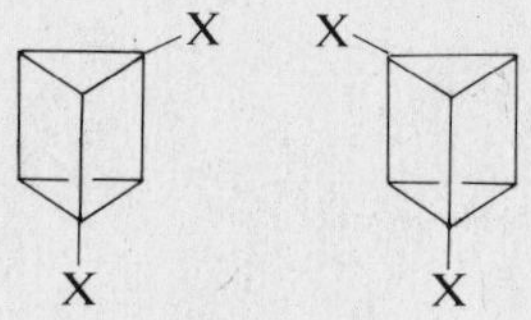

Enantiomers

8. The six possible diaminobenzoic acids and the structures of the diamines that could be formed from them are:

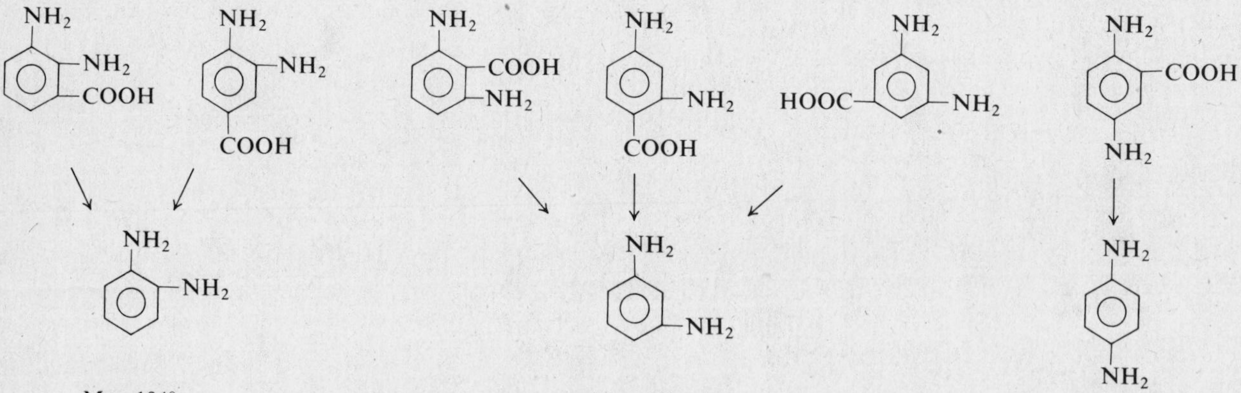

M.p. 104°   M.p. 63°   M.p. 142°

Ortho   *Meta*   *Para*

Diaminobenzenes

9. (a) To be aromatic, annulenes should follow the Hückel $4n + 2$ rule for the π electrons: 6 for $n = 1$, 10 for $n = 2$, 14 for $n = 3$, 18 for $n = 4$. The annulenes (with the number of π electrons indicated within the brackets in each name) that are expected to be aromatic are: [6]annulene (benzene); [10]annulene, [14]annulene, [18]annulene. (Actually, for [10] and [14]annulenes, the geometry is unfavorable: crowding of hydrogens inside ring prevents planarity and hence interferes with π overlap.)

(b)

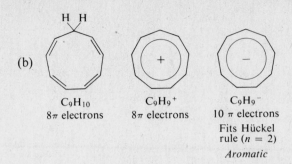

C_9H_{10}
8π electrons

$C_9H_9{}^+$
8π electrons

$C_9H_9{}^-$
$10\ \pi$ electrons
Fits Hückel
rule ($n = 2$)
Aromatic

10. Six π electrons (see Sec. 31.2), four from carbons, two from N.

11. (a)

$$\text{unit wt. CHCl} = 12 + 1 + 35.5 = 48.5$$

$$\frac{\text{wt./molecule}}{\text{wt./unit}} = \frac{291}{48.5} = 6 \text{ CHCl units per molecule}$$

$$\text{molecular formula} = C_6H_6Cl_6$$

(b) A possible compound is 1,2,3,4,5,6-hexachlorocyclohexane.

(c) Formed by an addition reaction:

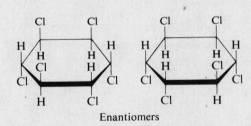

1,2,3,4,5,6-
Hexachlorocyclohexane

(d) Not aromatic: no π electrons available.

(e)–(f) These are stereoisomers, with various combinations of *cis* and *trans* Cl's. Actually, a total of nine such stereoisomers is possible. Two are a pair of enantiomers;

Enantiomers

the other seven are achiral.

12. The argument is an extension of that on pages 258–259. In the phenyl anion, $C_6H_5{}^-$, the unshared pair occupies an sp^2 orbital, which is intermediate in *s* character between the sp^3 orbital of the pentyl anion and the sp orbital of the acetylide anion. Electrons in the sp^2 orbital are held more tightly than those in an sp^3 orbital but not so tightly as those in an sp orbital; in basicity, therefore, phenyl anion lies between pentyl anion and acetylide anion.

Electrophilic Aromatic Substitution

11.1 $HONO_2 + HONO_2 \rightleftharpoons H_2\overset{+}{O}NO_2 + ONO_2^-$

$H_2\overset{+}{O}NO_2 \rightleftharpoons H_2O + {}^+NO_2$

$H_2O + HONO_2 \rightleftharpoons H_3O^+ + ONO_2^-$

$3HONO_2 \rightleftharpoons {}^+NO_2 + 2NO_3^- + H_3O^+$

The first step is a Lowry-Brønsted acid–base equilibrium in which one molecule of nitric acid serves as acid, and another as base.

11.2
$$\underset{O}{\overset{O}{HOSOH}} + \underset{O}{\overset{O}{HOSOH}} \rightleftharpoons \underset{O}{\overset{O}{HOSOH_2^+}} + \underset{O}{\overset{O}{OSOH^-}}$$

$$\underset{O}{\overset{O}{HOSOH_2^+}} \rightleftharpoons \underset{O}{\overset{O}{HOS^+}} + H_2O$$

$$\underset{O}{\overset{O}{HOS^+}} + \underset{O}{\overset{O}{OSOH^-}} \rightleftharpoons \underset{O\;O}{\overset{O\;O}{HOSOSOH}}$$

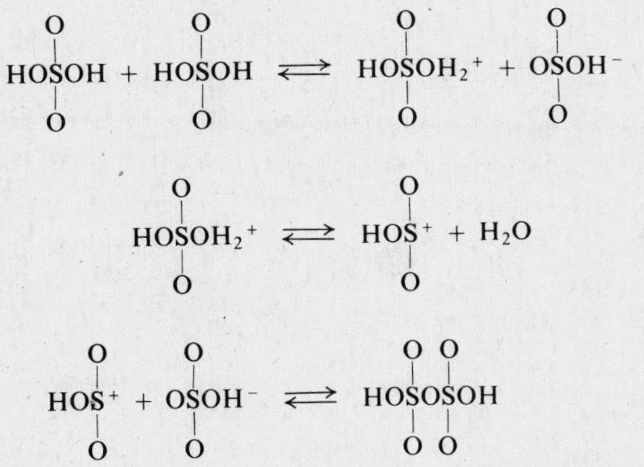

11.3 (a) $n\text{-Pr}^+ \longrightarrow iso\text{-Pr}^+$

(b) $iso\text{-Bu}^+ \longrightarrow t\text{-Bu}^+$

(c) $(CH_3)_3CCH_2^+ \longrightarrow tert\text{-pentyl cation}$

(d) carbonium ion mechanism

11.4 (a) $t\text{-BuOH} \xrightarrow{\text{H}^+} t\text{-BuOH}_2{}^+ \xrightarrow{-\text{H}_2\text{O}} t\text{-Bu}^{\oplus} \xrightarrow{\text{PhH}} \text{Ph—Bu-}t + \text{H}^+$

(b) $\text{CH}_3\text{CH}{=}\text{CH}_2 \xrightarrow{\text{H}^+} \underset{\oplus}{\text{CH}_3\text{CHCH}_3} \xrightarrow{\text{PhH}} \text{Ph—CH(CH}_3)_2 + \text{H}^+$

11.5 The chlorinating agent is $H_2\overset{+}{O}{-}Cl$ or possibly Cl^+, generated in the following manner:

$$\text{HOCl} + \text{H}^+ \rightleftharpoons H_2\overset{+}{O}{-}Cl \rightleftharpoons H_2\overset{\cdot}{O} + Cl^+$$

11.6 An acid–base complex is formed, which, like II on page 349, transfers halogen without its electrons. Because of its bulk, this reagent attacks a substituted aromatic ring most readily at the least hindered position: *para* to a substituent.

$$(\text{AcO})_3\text{Tl} + \text{Br}_2 \rightleftharpoons (\text{AcO})_3\overset{\ominus}{\text{Tl}}{-}\overset{\oplus}{\text{Br}}{-}\text{Br}$$

11.7 (a)

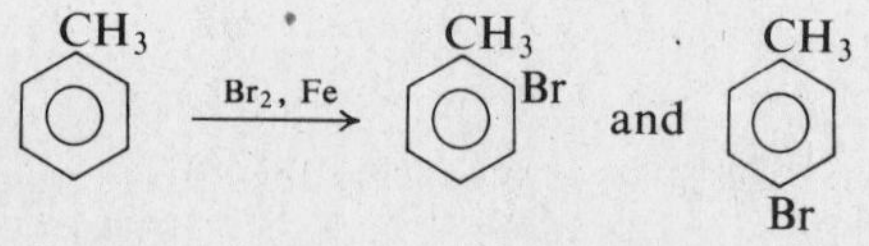

and (*Ortho* and *para*)

(b)

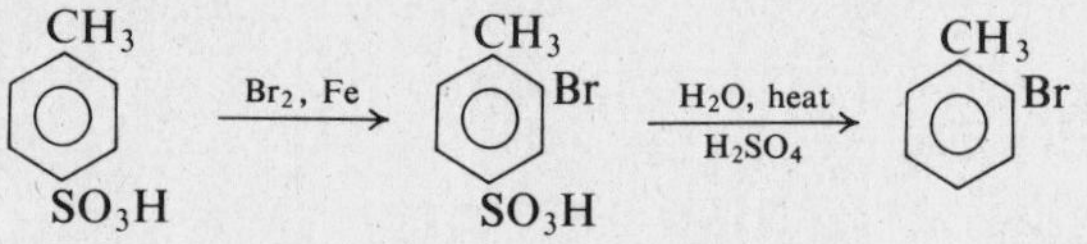

(*Ortho* only)

(c)

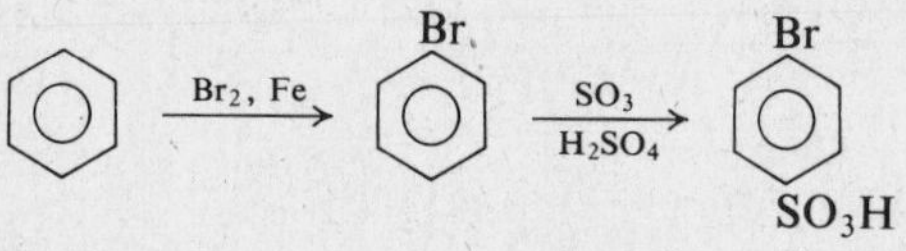

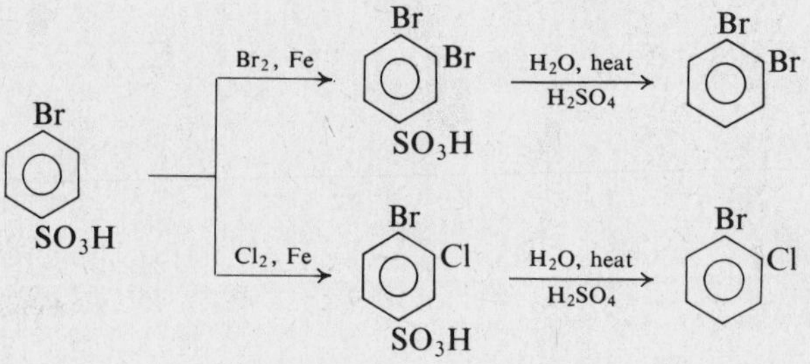

11.8 (a) $\underset{\overset{|}{\text{Cl}}}{\text{R—C}{=}\overset{..}{\text{O}}:} + \text{AlCl}_3 \longrightarrow (\text{R—C}{\equiv}\text{O}:)^+ \text{AlCl}_4{}^-$

(b) $(\text{Ar}'{-}\text{N}{\equiv}\text{N}:)^+\text{Cl}^-$

(c) $\text{H—}\overset{..}{\text{O}}{-}\overset{..}{\text{N}}{=}\overset{..}{\text{O}}: + \text{H}^+ \longrightarrow \underset{\overset{|}{\text{H}}}{\text{H—}\overset{..}{\text{O}}{-}\overset{+}{\text{N}}{=}\overset{..}{\text{O}}:} \longrightarrow (:\text{N}{\equiv}\text{O}:)^+ + \text{H}_2\text{O}$

In each case, the cation plays the role of Y^+ in a typical electrophilic aromatic substitution.

$$ArH + (R—C≡O:)^+ \longrightarrow Ar\overset{⊕}{\underset{\underset{R}{\overset{|}{C}=\ddot{O}:}}{\overset{H}{\diagup}}} \xrightarrow{-H^+} Ar—\underset{\underset{R}{|}}{C}=\ddot{O}: \qquad \text{Friedel-Crafts acylation}$$

Acylium ion   Aryl ketone

$$ArH + (Ar'—N≡N:)^+ \longrightarrow Ar\overset{⊕}{\underset{\underset{\cdot\cdot}{N=N—Ar'}}{\overset{H}{\diagup}}} \xrightarrow{-H^+} Ar—\ddot{N}=\ddot{N}—Ar' \qquad \text{Coupling}$$

Diazonium ion   Azo compound

$$ArH + (:N≡O:)^+ \longrightarrow Ar\overset{⊕}{\underset{\underset{\cdot\cdot}{N=O:}}{\overset{H}{\diagup}}} \xrightarrow{-H^+} Ar—\ddot{N}=\ddot{O}: \qquad \text{Nitrosation}$$

Nitrosonium   Nitroso
ion   compound

11.9 (a) Intermediate carbonium ion is $(ArHD)^+$, which can lose either H or D.

(b) Hydrogen ion is displaced from ring by another hydrogen (deuterium) ion in a typical electrophilic aromatic substitution (*deuterodeprotonation*) that is fast with (activated) phenol (PhOH), slow with benzene (PhH), and negligible with (deactivated) benzenesulfonic acid ($PhSO_3H$).

11.10 (a) $k^H/k^D = 2.05$, calculated as follows:

$$k^H/k^D = \text{rate per H/rate per D}$$

$$k^H/k^D = \frac{\text{moles HCl formed/H atoms available}}{\text{moles DCl formed/D atoms available}} = \frac{0.0868/2}{0.0212/1} = 2.05$$

(b)
$$k^H/k^D = 2.05 = \frac{\text{moles HCl}/1}{\text{moles DCl}/2}$$

$$\text{moles HCl/moles DCl} = 2.05/2 = 1.02$$

11.11 Shows that $Ar\overset{H}{\underset{H}{\diagup}}^+$ and presumably $Ar\overset{H}{\underset{Y}{\diagup}}^+$ intermediates can actually exist.

11.12 (a) $k^H/k^D = 6.77$, calculated as follows:

$$k^H/k^D = \text{rate per H/rate per D}$$

$$k^H/k^D = \frac{\text{moles H-cpd reacted/H atoms available}}{\text{moles D-cpd reacted/D atoms available}} = \frac{1.76/6}{0.26/6} = 6.77$$

(b) Yes, it is consistent with the one-step mechanism in which breaking of a C—H or C—D bond must necessarily take place in the rate-determining (and only) step.

(c) No, it is not consistent with breaking of a C—H or C—D bond in the fast step that is not rate-determining, since in this case the overall rates would be equally fast whichever compound was used.

(d) Yes, it is consistent with the breaking of a C—H or C—D bond in the slow, rate-determining step. (As we shall see in Sec. 14.20, there is other evidence to show that elimination generally follows mechanism (b), not mechanism (d).)

11.13 (a) $CH_2=CHCl + H^+ \longrightarrow CH_3-\overset{\overset{\textstyle H}{|}}{\underset{\oplus}{C}}-Cl \overset{I^-}{\longrightarrow} CH_3-\overset{\overset{\textstyle H}{|}}{\underset{\underset{\textstyle I}{|}}{C}}-Cl$

(b) $CH_2=CH_2 + H^+ \longrightarrow CH_3-\overset{\overset{\textstyle H}{|}}{\underset{\oplus}{C}}-H \overset{I^-}{\longrightarrow} CH_3-\overset{\overset{\textstyle H}{|}}{\underset{\underset{\textstyle I}{|}}{C}}-H$

(c) If, as usual, the more stable carbonium ion is formed faster, then $CH_3CH_2^+$ must be more stable than CH_3CHCl^+.

(d) Electron-withdrawing inductive effect of Cl tends to intensify positive charge and thus destabilizes cation.

(e) $CH_2=CHCl + H^+ \longrightarrow {}^{\oplus}CH_2-CH_2Cl \overset{I^-}{\longrightarrow} ICH_2CH_2Cl$ *Does not happen*

(f) Evidently CH_3CHCl^+ is more stable than $ClCH_2CH_2^+$ even though the former has the positive charge on carbon carrying electron-withdrawing chlorine.

(g) Through resonance, chlorine helps to accommodate the positive charge and stabilizes cation. *Such resonance is possible only for* CH_3CHCl^+.

$$\left[CH_3-\overset{\overset{\textstyle H}{|}}{\underset{\oplus}{C}}-\overset{\cdot\cdot}{\underset{\cdot\cdot}{Cl}}: \qquad CH_3-\overset{\overset{\textstyle H}{|}}{C}=\overset{\cdot\cdot}{\underset{\oplus}{Cl}}: \right]$$

(h) The inductive effect controls reactivity: formation of either cation from vinyl chloride is slower than formation of ethyl cation.

(i) The resonance effect controls orientation: formation of CH_3CHCl^+ is faster than formation of $^+CH_2CHCl$.

As with electrophilic substitution in chlorobenzene, the inductive effect is stronger, but the resonance effect is more selective.

1. Faster (activated): a, c, d, g, h, k. Slower (deactivated): b, e, f, i, j.

(a) [structure: benzene ring with NHCOCH₃ at top, Br at bottom] (b) [benzene ring with I at top, Br at bottom] + [benzene ring with I at top, Br on side] (c) [benzene ring with Bu-*sec* at top, Br at bottom] [benzene ring with Bu-*sec* at top, Br on side]

(very little *o*-isomer)

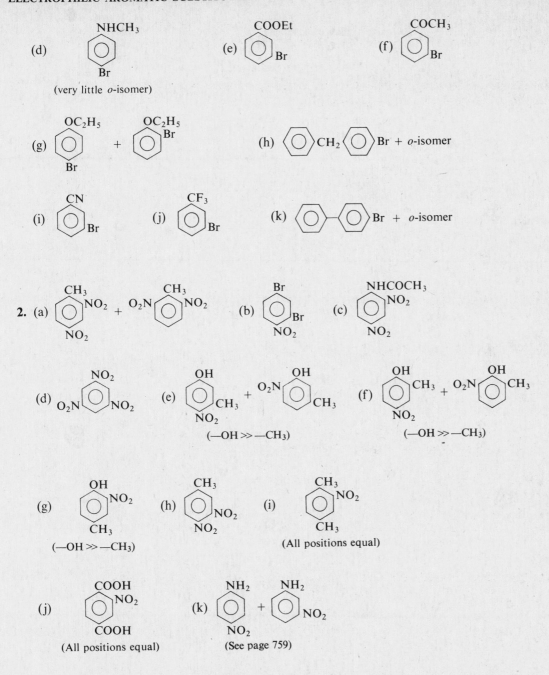

(d) NHCH₃ / Br (very little *o*-isomer)

(e) COOEt / Br

(f) COCH₃ / Br

(g) OC₂H₅ / Br + OC₂H₅ / Br

(h) ⟨⟩CH₂⟨⟩Br + *o*-isomer

(i) CN / Br

(j) CF₃ / Br

(k) ⟨⟩—⟨⟩Br + *o*-isomer

2. (a) CH₃ / NO₂ / NO₂ + O₂N / CH₃ / NO₂

(b) Br / Br / NO₂

(c) NHCOCH₃ / NO₂ / NO₂

(d) O₂N / NO₂ / NO₂

(e) OH / CH₃ / NO₂ + O₂N / OH / CH₃ (—OH ≫ —CH₃)

(f) OH / CH₃ / NO₂ + O₂N / OH / CH₃ (—OH ≫ —CH₃)

(g) OH / NO₂ / CH₃ (—OH ≫ —CH₃)

(h) CH₃ / NO₂ / NO₂

(i) CH₃ / NO₂ / CH₃ (All positions equal)

(j) COOH / NO₂ / COOH (All positions equal)

(k) NH₂ / NO₂ + NH₂ / NO₂ (See page 759)

3. When an *o,p*-directing group is present (as is true in every case except b and d), sulfonation *para* (and *ortho*) to that group predominates. All products are named as sulfonic acids, with —SO₃H considered to be on C–1; (d) is *m*-benzenedisulfonic acid.

4. *m*-Directing groups are deactivating; the more of them there are present, the lower the reactivity. Most *o,p*-directing groups (except halogen) are activating; the more activating groups there are present, and the more of them that activate the same position, the more reactive the compound. For example, in (a): the three —CH₃ groups in mesitylene (triply) activate the same positions, the two —CH$_3$ groups in *m*-xylene (doubly) activate the same positions, but the two —CH$_3$ groups in *p*-xylene activate different positions.

(a) mesitylene > *m*-xylene > *p*-xylene > toluene > benzene

(b) toluene > benzene > bromobenzene > nitrobenzene

(c) aniline > acetanilide > benzene > acetophenone

(d) *p*-xylene > toluene > *p*-toluic acid > terephthalic acid

(e) C$_6$H$_5$Cl > *p*-O$_2$NC$_6$H$_4$Cl > 2,4-(O$_2$N)$_2$C$_6$H$_3$Cl

(f) 2,4-dinitrophenol > 2,4-dinitrochlorobenzene

(g) 2,4-dinitrotoluene > *m*-dinitrobenzene

5. Toluene is more easily nitrated—and *poly*nitrated—than benzene.

6. (a) Substitution is faster in the ring that is not deactivated by —NO$_2$. Orientation is *o,p* to the other ring (see Table 11.3, page 342).

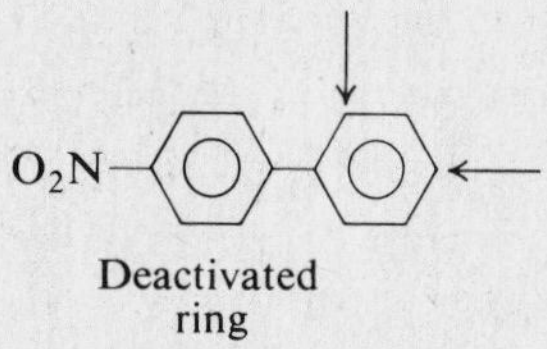

Deactivated
ring

(b) Substitution is faster in the ring that is not deactivated by —NO$_2$. Orientation is *o,p* to the substituent, —CH$_2$Ar.

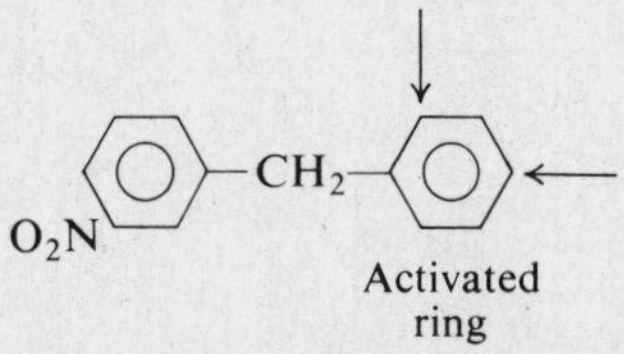

Activated
ring

(c) Substitution is faster in the ring that is activated by phenolic oxygen (similar to —OR). Other ring is actually deactivated by —COOR.

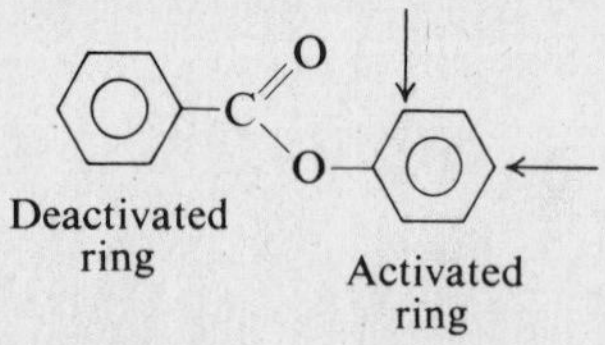

Deactivated
ring
Activated
ring

7. The farther away a deactivating —NR$_3$$^+$ or —NO$_2$ group is from the ring, the less effective it is. In (c), the more electron-withdrawing groups there are, the greater the deactivation.

(a) Ph(CH$_2$)$_3$NMe$_3$$^+$ > Ph(CH$_2$)$_2$NMe$_3$$^+$ > PhCH$_2$NMe$_3$$^+$ > PhNMe$_3$$^+$

(b) Ph(CH$_2$)$_2$NO$_2$ > PhCH$_2$NO$_2$ > PhNO$_2$

(c) PhCH$_3$ > PhCH$_2$COOEt > PhCH(COOEt)$_2$ > PhC(COOEt)$_3$

The most active compound in each set gives the lowest percentage of *m*-isomer, the least active the highest percentage.

8. The positive charge of the intermediate can be dispersed (through three extra structures) by the second phenyl group when attack is *ortho* or when it is *para*, but not when it is *meta*. For example:

9. (a) (Lewis) acidic thallium complexes with the basic oxygen of the alcohol group,

$$C_6H_5CH_2CH_2\overset{\oplus}{\underset{\underset{\ominus}{Tl(OOCCF_3)_3}}{\ddot{O}}}-H$$

and then reacts intramolecularly at the nearby *ortho* position (see page 351); finally thallium is displaced by iodide, still at the *ortho* position.

(b) As in (a), except that at elevated temperatures thallation is reversible, substitution becomes equilibrium-controlled, and the final product is the most stable *meta*-isomer.

(c) Thallium complexes with the more basic oxygen of the C=O group,

$$C_6H_5CH_2CH_2-O-\underset{\underset{\ominus \ddot{Tl}(OOCCF_3)_3}{\overset{\oplus}{\underset{O}{\parallel}}}}{C}-CH_3$$

and is held too far from the ring for intramolecular reaction: it evidently must leave oxygen and attack the ring *inter*molecularly, at the roomier *para* position.   (See page 351.)

(E. C. Taylor and A. McKillop, "Thallium in Organic Synthesis," *Accounts Chem. Res.*, **3**, 338 (1970).)

10. $HONO_2 + H_2SO_4 \rightleftarrows H_2\overset{+}{O}NO_2 + HSO_4^-$

$H_2\overset{+}{O}NO_2 \longrightarrow H_2O + {}^+NO_2$ *Slow*

$H_2O + H_2SO_4 \rightleftarrows H_3O^+ + HSO_4^-$

Protonated nitric acid forms NO_2^+ through loss of the good leaving group H_2O; unprotonated nitric acid would have to lose the more strongly basic OH^- ion. (See Sec. 5.20.)

11. The reactions are electrophilic substitution, with activation as usual by —NH$_2$ and —OH. The attacking electrophiles are the familiar Br$^+$ (or its equivalent) and NO$_2^+$. In each case, however, the displaced group is not the usual H^+ (deprotonation), but SO_3, as in desulfonation. Thus,

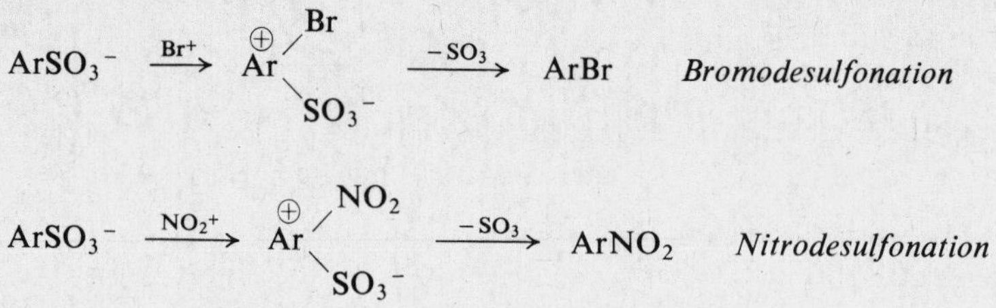

$$ArSO_3^- \xrightarrow{Br^+} \overset{Br}{\underset{SO_3^-}{Ar^{\oplus}}} \xrightarrow{-SO_3} ArBr \qquad \textit{Bromodesulfonation}$$

$$ArSO_3^- \xrightarrow{NO_2^+} \overset{NO_2}{\underset{SO_3^-}{Ar^{\oplus}}} \xrightarrow{-SO_3} ArNO_2 \qquad \textit{Nitrodesulfonation}$$

as compared with, say,

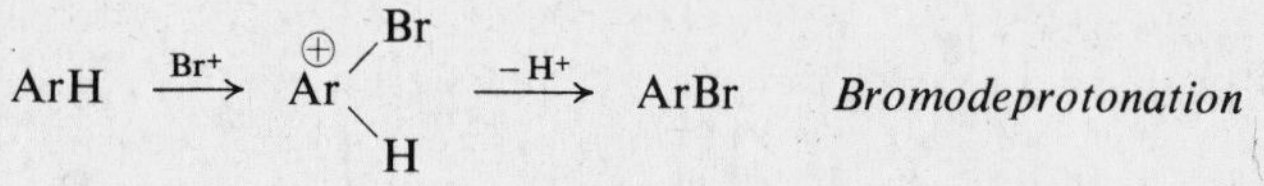

$$ArH \xrightarrow{Br^+} \overset{Br}{\underset{H}{Ar^{\oplus}}} \xrightarrow{-H^+} ArBr \qquad \textit{Bromodeprotonation}$$

or

$$ArSO_3^- \xrightarrow{H^+} \overset{H}{\underset{SO_3^-}{Ar^{\oplus}}} \xrightarrow{-SO_3} ArH \qquad \textit{Protodesulfonation}$$

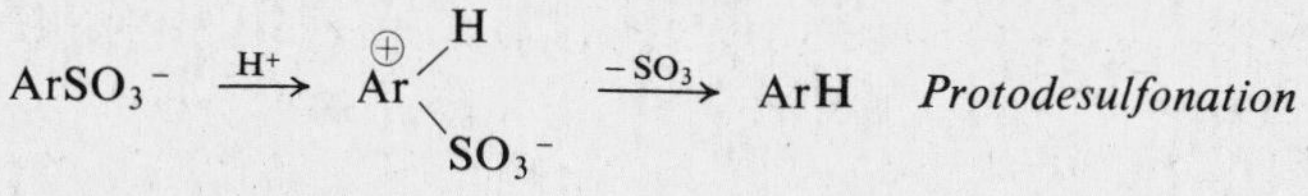

12. This is an example of *bromodealkylation*.

$$Br_2 + AlCl_3 \rightleftharpoons Cl_3\overset{\ominus}{Al}-\overset{\oplus}{Br}-Br$$

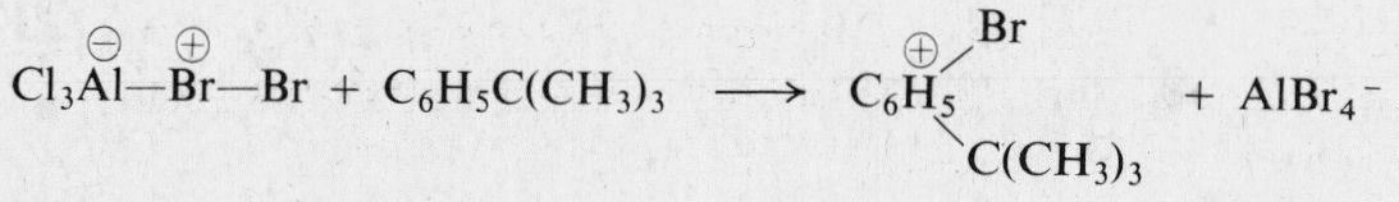

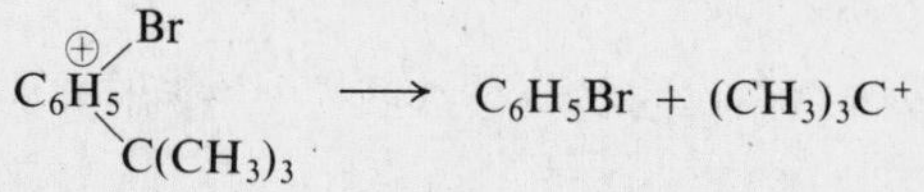

$$Cl_3\overset{\ominus}{Al}-\overset{\oplus}{Br}-Br + C_6H_5C(CH_3)_3 \longrightarrow \underset{C(CH_3)_3}{C_6\overset{\oplus}{H_5}}{}^{Br} + AlBr_4^-$$

$$\underset{C(CH_3)_3}{C_6\overset{\oplus}{H_5}}{}^{Br} \longrightarrow C_6H_5Br + (CH_3)_3C^+$$

$$(CH_3)_3C^+ + AlBr_4^- \longrightarrow (CH_3)_2C{=}CH_2 + HBr + AlBr_3$$

13. The substituent in each case, $-N(CH_3)_3{}^+$ or $-CF_3$, is powerfully electron-withdrawing and (in the absence of a stronger, opposing resonance effect) favors formation of the cation with charge on the more remote carbon.

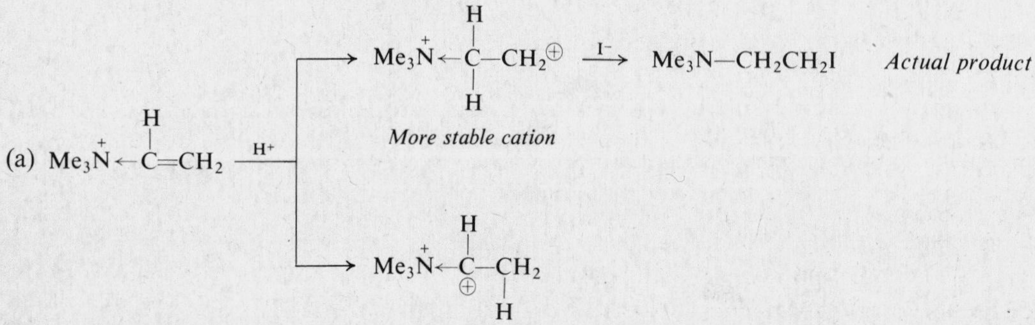

$$\text{(a)} \quad Me_3\overset{+}{N}{\leftarrow}\underset{H}{\overset{H}{C}}{=}CH_2 \xrightarrow{H^+}
\begin{cases}
Me_3\overset{+}{N}{\leftarrow}\underset{H}{\overset{H}{C}}{-}CH_2^{\oplus} \xrightarrow{I^-} Me_3N{-}CH_2CH_2I \quad \textit{Actual product} \\[2pt]
\textit{More stable cation} \\[6pt]
Me_3\overset{+}{N}{\leftarrow}\underset{H}{\overset{H}{\underset{\oplus}{C}}}{-}CH_2
\end{cases}$$

(b) $CH_2=C \to CF_3$ $\xrightarrow{H^+}$

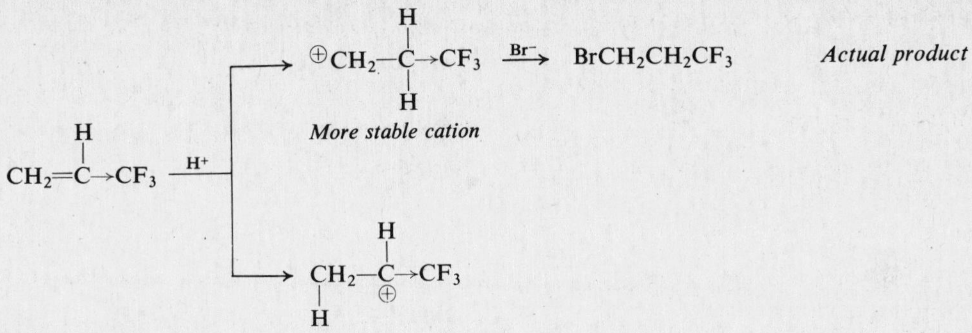

(c) $AlBr_3$ is used to provide the stronger acid ($HAlBr_4$) that is needed for attack on the highly deactivated alkene.

14. (a) Measure relative rates of reaction; expect lower rate with C_6D_6.

(b) Measure relative amounts of C_6H_5Y and C_6D_5Y by mass spectrometry; expect more C_6H_5Y.

(c) Carry out a substitution reaction with ordinary anisole,

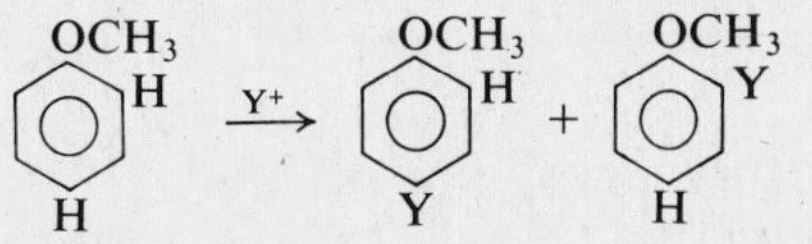

and determine (by gas chromatography, say) the ratio of o-product to p-product (o/p ratio). Now carry out the same reaction under the same conditions, but start with anisole-4-d.

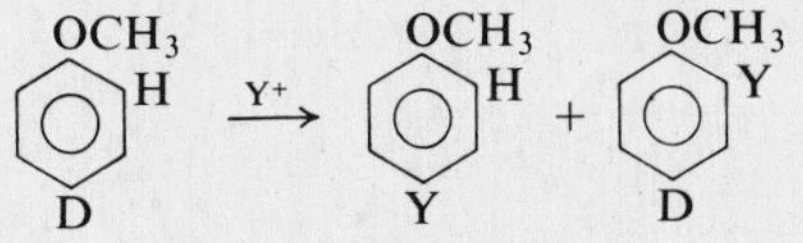

If there is an isotope effect, less of the p-product will be formed (C—D bond has to be broken) and the o/p ratio will be larger than before.

(d) Carry out a substitution reaction, and analyze the product for D/H ratio. If there is an isotope effect, fewer C—D bonds will be broken than C—H bonds, $C_6H_2D_3Y$ will exceed $C_6H_3D_2Y$, and the D/H ratio in the product will be higher than in the reactant.

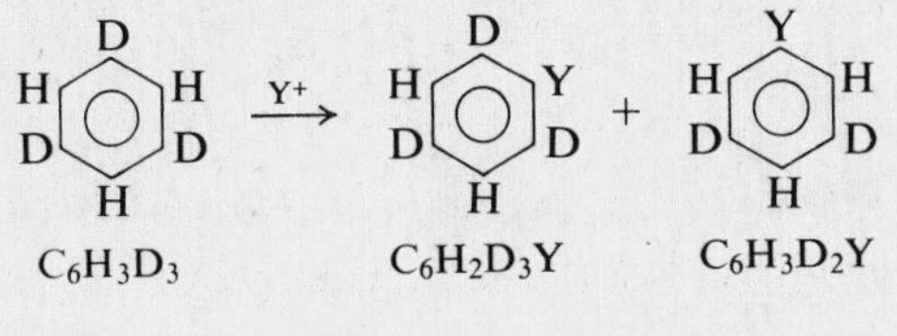

15. (a) $p\text{-}CH_3C_6H_4NO_2$ $\xleftarrow{HNO_3,\ H_2SO_4}$ $C_6H_5CH_3$

(b) $p\text{-}BrC_6H_4NO_2$ $\xleftarrow{HNO_3,\ H_2SO_4}$ BrC_6H_5 $\xleftarrow{Br_2,\ Fe}$ C_6H_6

(c) $p\text{-}ClC_6H_4Cl$ $\xleftarrow{Cl_2,\ Fe}$ ClC_6H_5 $\xleftarrow{Cl_2,\ Fe}$ C_6H_6

(d) $m\text{-}BrC_6H_4SO_3H$ $\xleftarrow{Br_2,\,Fe}$ $C_6H_5SO_3H$ $\xleftarrow{SO_3,\,H_2SO_4}$ C_6H_6

(e) $p\text{-}BrC_6H_4SO_3H$ $\xleftarrow{SO_3,\,H_2SO_4}$ BrC_6H_5 $\xleftarrow{Br_2,\,Fe}$ C_6H_6

(f) $p\text{-}BrC_6H_4COOH$ $\xleftarrow{KMnO_4}$ $p\text{-}BrC_6H_4CH_3$ $\xleftarrow{Br_2,\,Fe}$ $C_6H_5CH_3$

(g) $m\text{-}BrC_6H_4COOH$ $\xleftarrow{Br_2,\,Fe}$ C_6H_5COOH $\xleftarrow{KMnO_4}$ $C_6H_5CH_3$

(h)

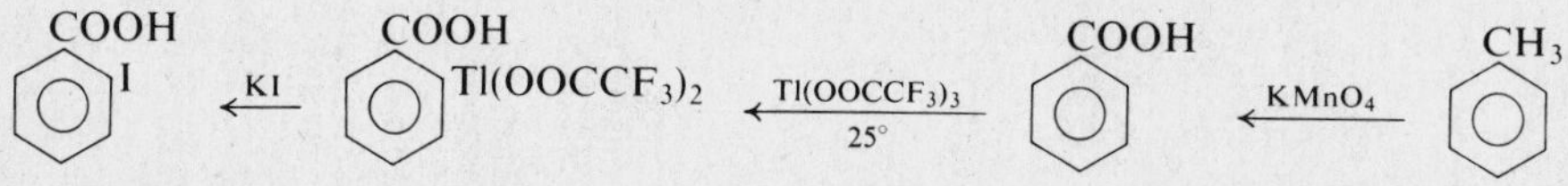

(i)

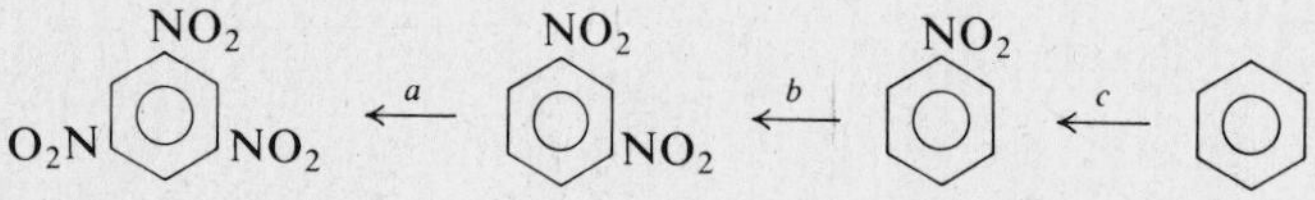

a: fuming HNO_3, fuming H_2SO_4, 100–110°, 5 days, 45% yield.

b: fuming HNO_3, conc. H_2SO_4, 95°, 84% yield.

c: conc. HNO_3, conc. H_2SO_4, 55–60°, 99% yield.

(j)

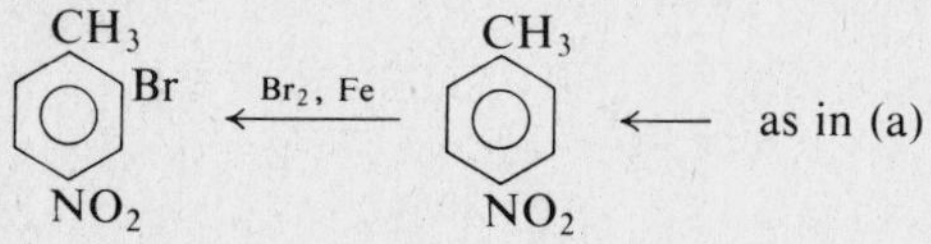

(k)

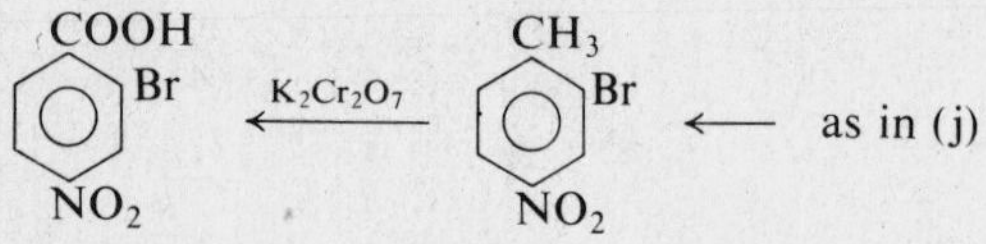

(l)

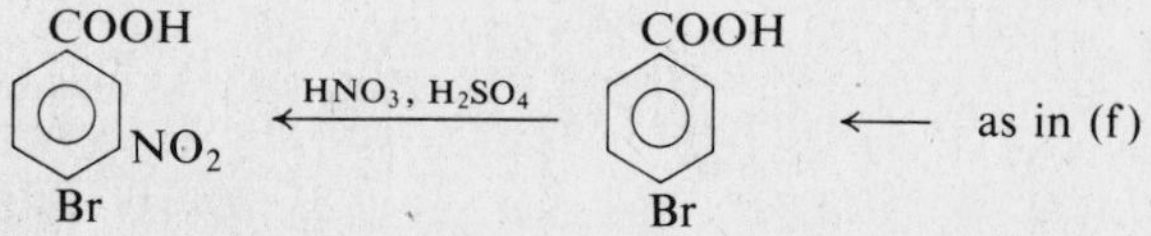

(m)

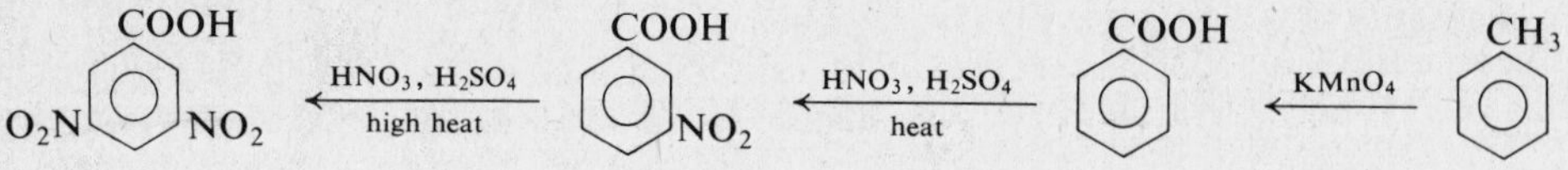

(n)

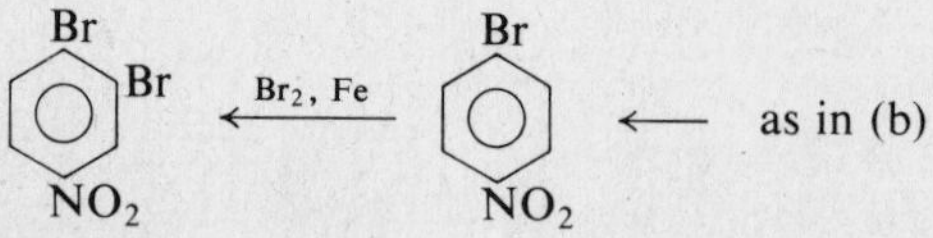

(o)

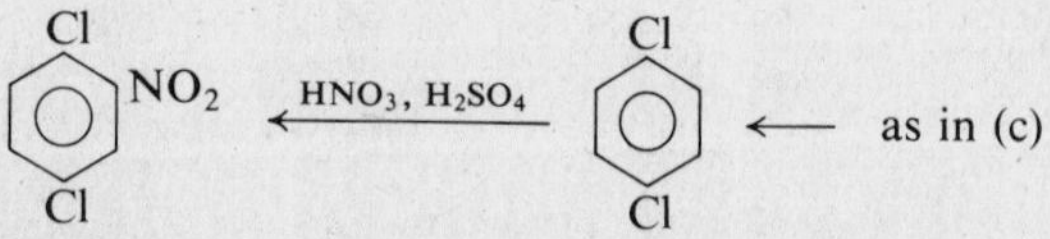

(p)

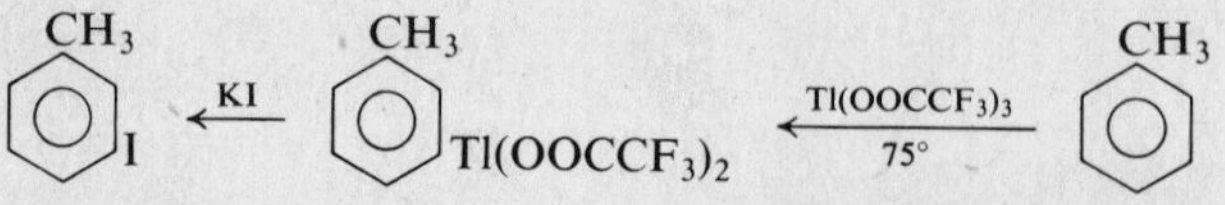

16. (a) 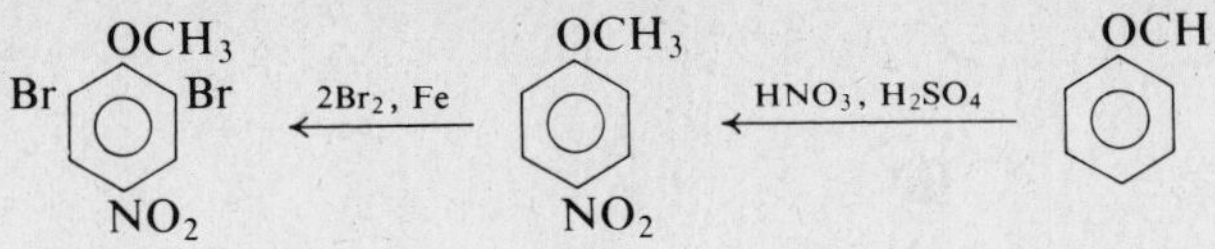 2,6-dibromo-4-nitroanisole $\xleftarrow{2Br_2, Fe}$ p-nitroanisole $\xleftarrow{HNO_3, H_2SO_4}$ anisole (OCH_3)

(b) 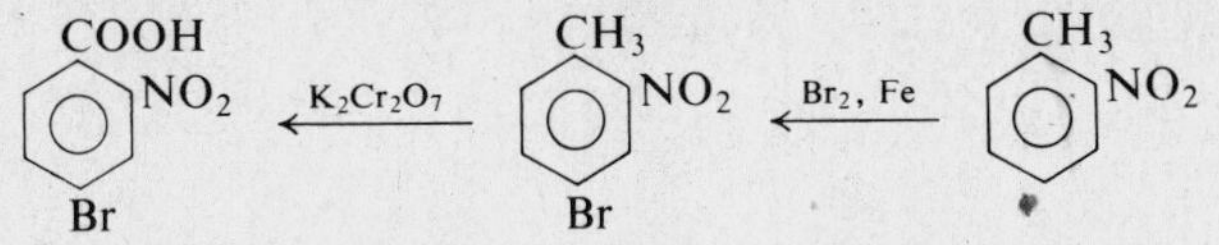 4-bromo-2-nitrobenzoic acid $\xleftarrow{K_2Cr_2O_7}$ 4-bromo-2-nitrotoluene $\xleftarrow{Br_2, Fe}$ o-nitrotoluene

(c) 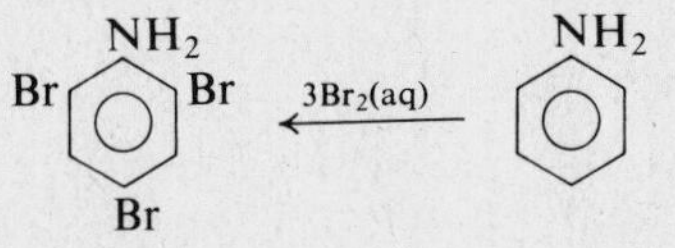 2,4,6-tribromoaniline $\xleftarrow{3Br_2(aq)}$ aniline (NH_2)

(d) 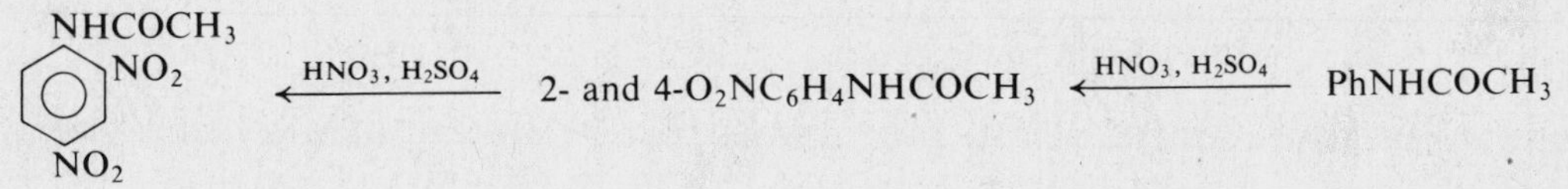 2,4-dinitroacetanilide ($NHCOCH_3$) $\xleftarrow{HNO_3, H_2SO_4}$ 2- and 4-$O_2NC_6H_4NHCOCH_3$ $\xleftarrow{HNO_3, H_2SO_4}$ PhNHCOCH$_3$

(e) 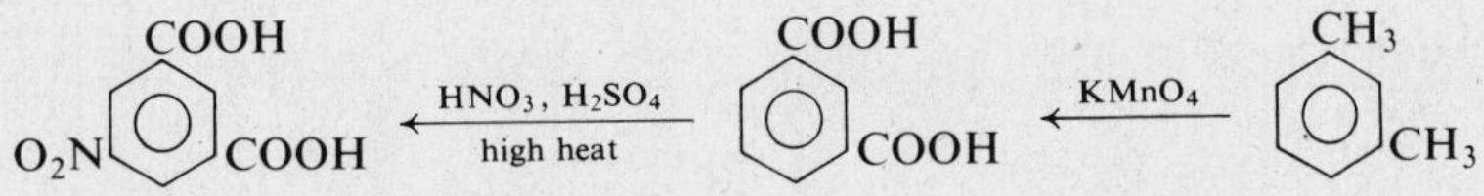 4-nitrophthalic acid (O_2N, COOH, COOH) $\xleftarrow[\text{high heat}]{HNO_3, H_2SO_4}$ phthalic acid (COOH, COOH) $\xleftarrow{KMnO_4}$ o-xylene (CH_3, CH_3)

(f) 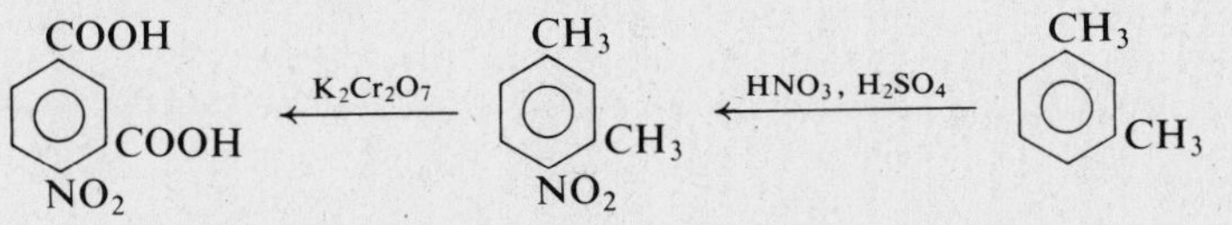 4-nitrophthalic... (COOH, COOH, NO$_2$) $\xleftarrow{K_2Cr_2O_7}$ 4-nitro-o-xylene (CH_3, CH_3, NO_2) $\xleftarrow{HNO_3, H_2SO_4}$ o-xylene (CH_3, CH_3)

(g) 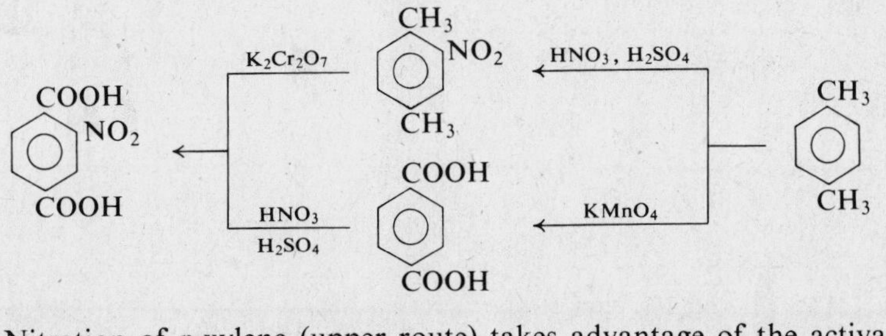 (COOH, NO$_2$, COOH) $\xleftarrow{K_2Cr_2O_7}$ nitro-p-xylene (CH_3, NO_2, CH_3) $\xleftarrow{HNO_3, H_2SO_4}$ p-xylene (CH_3, CH_3)

$\xleftarrow[\text{H}_2\text{SO}_4]{HNO_3}$ terephthalic acid (COOH, COOH) $\xleftarrow{KMnO_4}$ p-xylene (CH_3, CH_3)

(h) Nitration of p-xylene (upper route) takes advantage of the activating effect of the two —CH$_3$ groups. Nitration of a ring deactivated by two —COOH groups would be very difficult.

Chapter 12 | Arenes

12.1 (a) H_2SO_4, heat (b) $Zn(Hg)$, HCl
(c) H_2, Pt, heat, pressure (d) H_2SO_4, heat; then as in (c)
(e) KOH(alc); then as in (c). Or: Mg, anhyd. Et_2O; H_2O

12.2 PhCH—CHPh $\xleftarrow{\text{Na}}$ PhCHBr $\xleftarrow{\text{PBr}_3}$ PhCHOH
         | | | |
        CH_3 CH_3 CH_3 CH_3

12.3 *tert*-Pentylbenzene results in each case from attack on benzene by the *tert*-pentyl cation:

$$CH_3CH_2-\underset{\underset{CH_3}{|}}{\overset{\overset{CH_3}{|}}{C}}\oplus + C_6H_6 \longrightarrow C_6H_5\underset{\underset{CH_3}{|}}{\overset{\overset{CH_3}{|}}{C}}-CH_2CH_3 + H^+$$

This cation is formed as shown in Figure 12.3, page 116 of this Study Guide.

12.4 Because of its positive charge, the substituent in I should be strongly deactivating, comparable to $-NR_3^+$.

12.5 The situation reminds us of the competition between 1,2- and 1,4-addition (Sec. 8.22), and we make the following hypothesis.

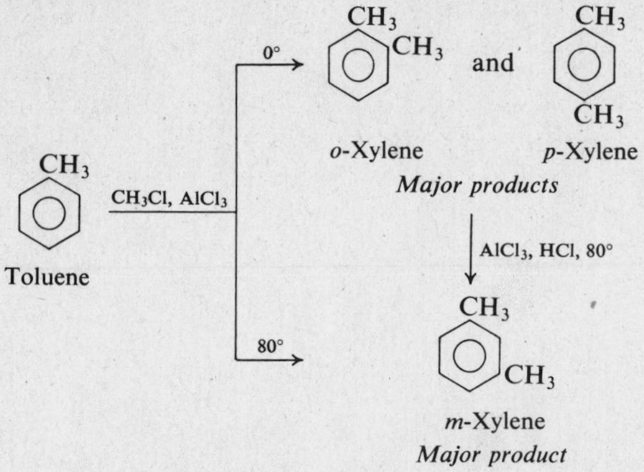

At 0°, we are observing rate-control: *o*- and *p*-xylenes are formed faster. At 80°, we are observing equilibrium-control: *m*-xylene is the more stable product.

Methyl activates most strongly at the *ortho* and *para* positions. This favors alkylation at these positions, but it also favors *dealkylation*—via electrophilic attack by a proton—at these same positions. The *ortho* and *para* isomers are formed more rapidly,

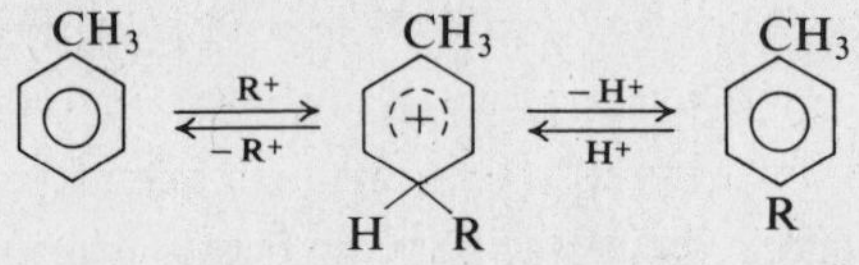

but are also dealkylated more rapidly; the *meta* isomer is formed more slowly, but, once formed, tends to persist.

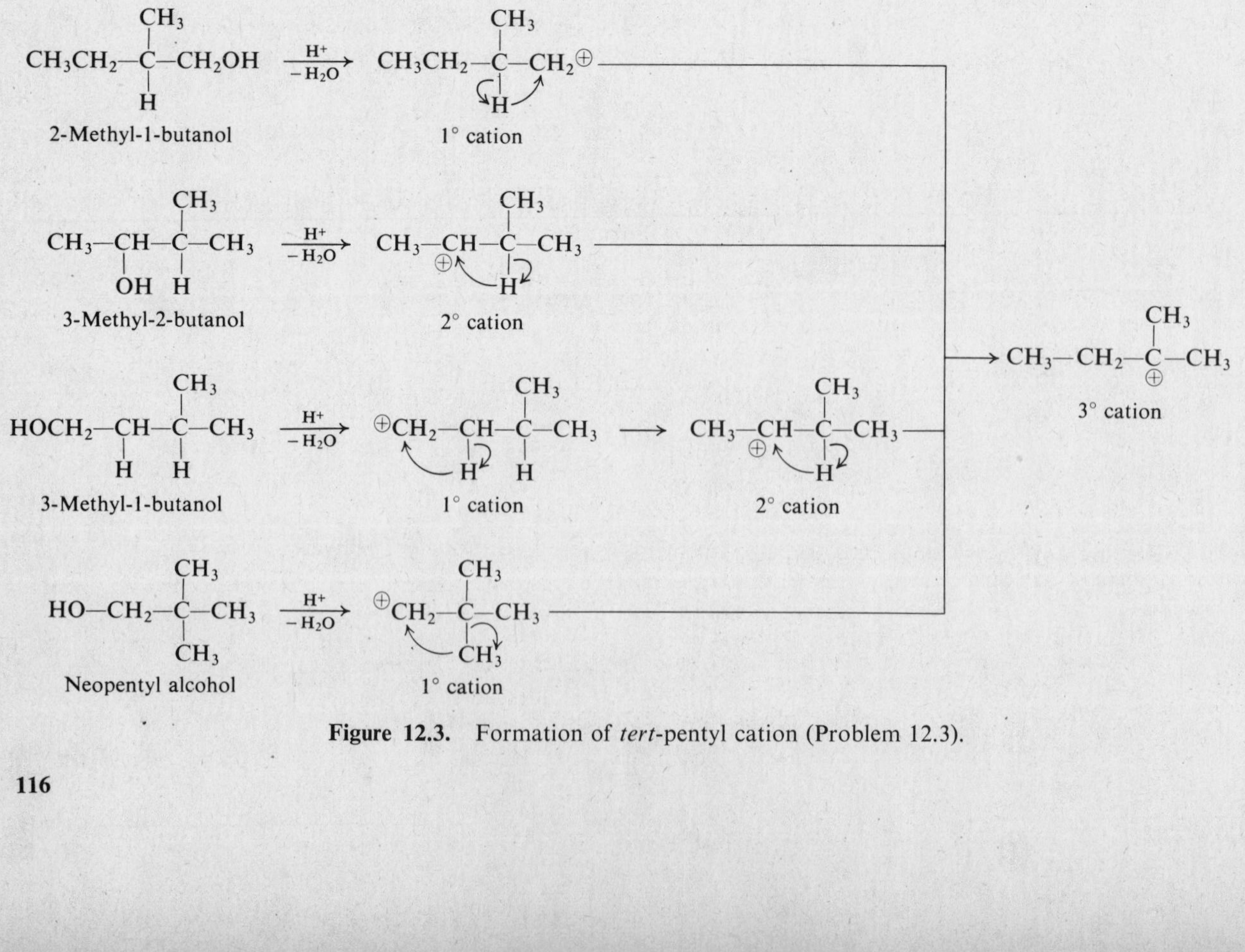

Figure 12.3. Formation of *tert*-pentyl cation (Problem 12.3).

Hydrogen chloride provides H^+ for reversal of alkylation:

$$HCl + AlCl_3 \rightleftharpoons H^+AlCl_4^-$$

Experiment has shown that the above hypothesis is correct in broad outline, but needs modification. Let us consider conversion of *p*-xylene as a rearrangement involving migration of an alkyl group. As we have pictured it, the alkyl group leaves (1) the *para* position—as the cation—and then attaches itself (2) to a *meta* position, most likely in another molecule of toluene. Reaction would thus be *intermolecular* (between molecules).

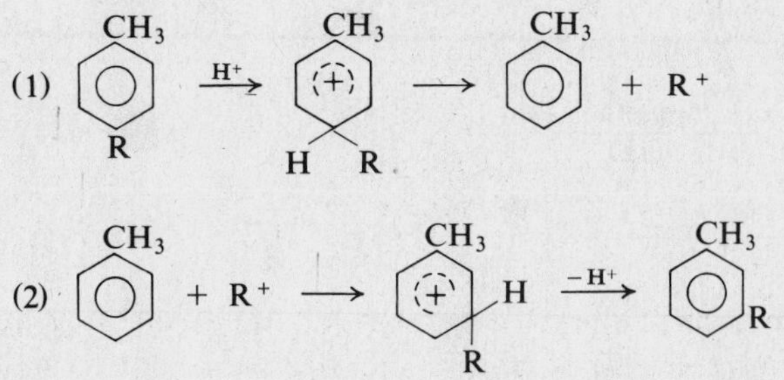

With alkyl groups bigger than methyl, this does seem to happen. But with methyl, there is evidence that rearrangement is *intramolecular* (within a molecule). A free methyl cation does not separate. Instead, in a 1,2-shift of exactly the kind we have already encountered (Sec. 5.22), methyl migrates from one position to the next in the intermediate carbonium ion.

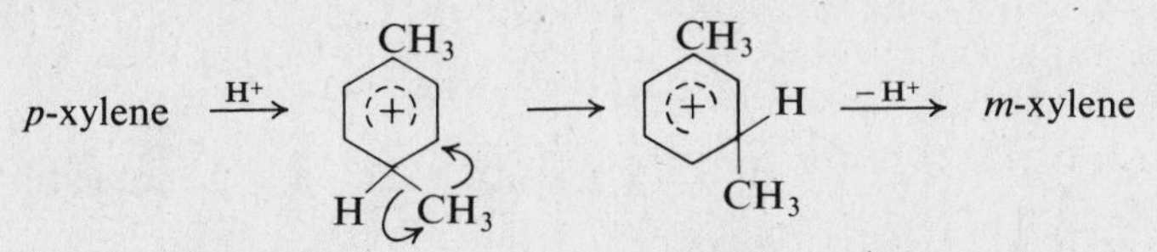

(For a general discussion, see R. M. Roberts, "Friedel-Crafts Chemistry," *Chem. Eng. News*, Jan. 25, 1965, p. 96, or R. O. C. Norman and R. Taylor, *Electrophilic Substitution in Benzenoid Compounds*, Elsevier, New York, 1965, pp. 160–168.)

12.6 The newly introduced alkyl group activates the aromatic ring toward further substitution; in the other reactions, the newly introduced group deactivates the ring.

12.7 To permit overlap of π cloud and p orbital. (See Figure 12.2, page 390.)

12.8 Hyperconjugation stabilizes *o*- or p-$CH_3C_6H_4CH_2\cdot$, through contribution from structures like these:

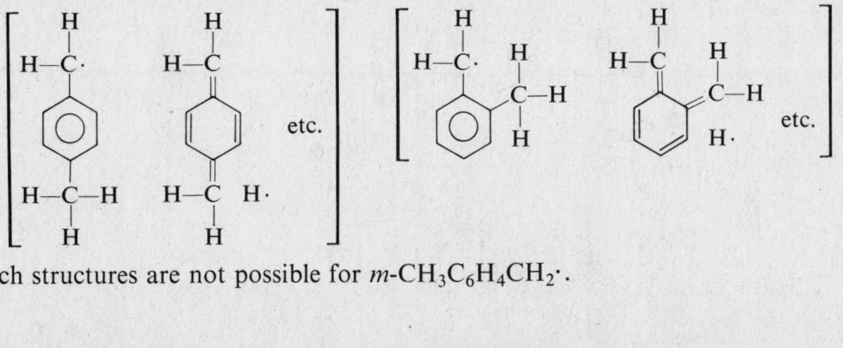

Such structures are not possible for m-$CH_3C_6H_4CH_2\cdot$.

12.9 (a) Similar to Figure 2.3, page 53, with E_{act} = 19 kcal, and ΔH = +11 kcal.

(b) 8 kcal (difference between E_{act} of 19 kcal and ΔH of +11 kcal).

(c) Steric hindrance to combination of the radicals.

12.10 The freezing point of a 1-molal solution (one mole of solute in 1000 g of solvent, or one millimole per gram of solvent) in benzene is 5° lower than that of pure benzene. If the freezing point of a solution is depressed by only one-tenth (0.5°/5.0°) of that amount, its concentration is only one-tenth as great, or 0.1 molal.

$$\text{molality} = \frac{\text{mmoles solute}}{\text{g solvent}} = \frac{\text{mg solute/m.w.}}{\text{g solvent}}$$

$$0.1 = \frac{1500 \text{ mg/m.w.}}{50} \qquad \text{m.w.} = 300$$

An apparent m.w. of 300 compared with an expected m.w. of 542 for $C_{42}H_{38}$ indicates considerable dissociation into free radicals.

12.11 (a) 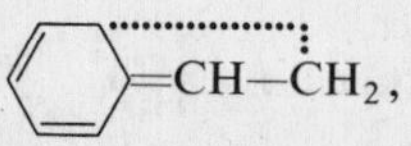 =CH—$\overset{..}{C}H_2$, etc.

(b) Same argument as for conjugated dienes (Sec. 8.19).

12.12 (a) $Ph_3C:\overset{..}{\underset{..}{C}}l: \longrightarrow Ph_3C^{\oplus} + :\overset{..}{\underset{..}{C}}l:^{-}$

Ph_3C^+ ion stabilized by dispersal of charge over 3 rings, and ionization is promoted by polar solvents.

(b) The same ion is formed from triphenylcarbinol.

$$Ph_3C—OH \xrightarrow{H^+} Ph_3C—\overset{\oplus}{\underset{H}{O}}H \longrightarrow Ph_3C^{\oplus} + H_2O$$
$$\qquad\qquad\qquad\qquad\qquad\qquad\qquad \textit{Yellow}$$

12.13 CCl_4

$\downarrow AlCl_3$

$^{\oplus}CCl_3 \xrightarrow{PhH} PhCCl_3$

$+$ $\qquad\qquad\quad \downarrow AlCl_3$

$AlCl_4^-$ $\qquad PhC\underset{\oplus}{C}l_2 \xrightarrow{PhH} Ph_2CCl_2$

$\qquad\qquad\quad + \qquad\qquad\quad \downarrow AlCl_3$

$\qquad\qquad AlCl_4^- \qquad Ph_2\underset{\oplus}{C}Cl \xrightarrow{PhH} Ph_3CCl$

$\qquad\qquad\qquad\qquad\quad + \qquad\qquad\quad \downarrow AlCl_3$

$\qquad\qquad\qquad\quad AlCl_4 \qquad Ph_3C^{\oplus} \xrightarrow{PhH} \text{no reaction}$

$\qquad\qquad\qquad\qquad\qquad\qquad + $

$\qquad\qquad\qquad\qquad\qquad AlCl_4^-$

Ph_3C^+ is too stable to react with C_6H_6 in final stage.

12.14 In Figure 8.9 (page 275), replace diene by alkenylbenzene, allyl free radical by benzylic cation, and alkyl free radical by alkyl cation.

12.15 Because of conjugation of both rings with the double bond, and perhaps between rings through the double bond, the reactant may be stabilized more than the transition state.

12.16 (a) $PhC\equiv CH \xleftarrow{NaNH_2} \xleftarrow{KOH(alc)} PhCHBrCH_2Br \xleftarrow{Br_2} PhCH=CH_2 \xleftarrow[-H_2]{cat., heat} PhCH_2CH_3$

(b)

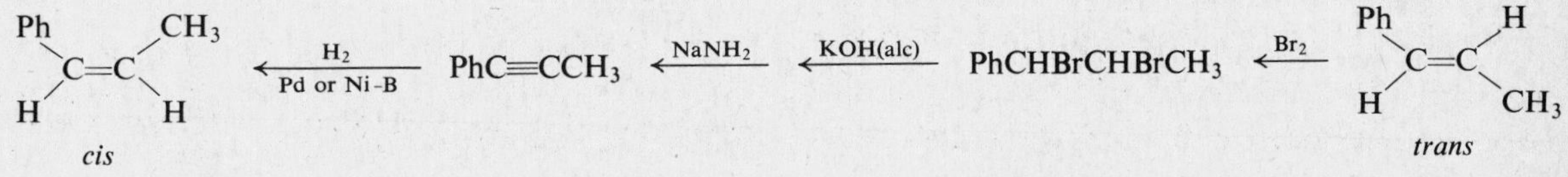

12.17 (a) elemental analysis   (b) $KMnO_4$ (c) $KMnO_4$
      (d) fuming sulfuric acid   (e) Br_2/CCl_4, or $KMnO_4$ (f) CrO_3/H_2SO_4

12.18 Upon oxidation by $KMnO_4$, n-butylbenzene gives C_6H_5COOH (m.p. 122°), and m-diethylbenzene gives m-$C_6H_4(COOH)_2$ (m.p. 348°).

12.19 (a) soluble (or polymerizes)   (b) discharge of color
      (c) discharge of color, brown MnO_2 (d) orange-red color
      (e) negative test (no color change)

12.20 (a) Br_2/CCl_4, or $KMnO_4$ (b) $Ag(NH_3)_2OH$ or $Cu(NH_3)_2OH$
      (c) oxidation of allylbenzene gives $PhCOOH$ (m.p. 122°); 1-nonene gives $C_7H_{15}COOH$
          (m.p. 16°, b.p. 239°)
      (d) CrO_3/H_2SO_4

1. (e) phenyl group in equatorial position   (h) (E)-$PhCH=CHPh$
   (i) *cis-trans* isomerism possible

2. (a) ethylene, HF   (b) cat. H_2
   (c) $2H_2$, Pt   (d) H_2SO_4, heat; H_2, Pt
   (e) H_2SO_4, heat; H_2, Pt. Or: PBr_3; Mg, Et_2O; H_2O
   (f) KOH(alc); H_2, Pt   (g) KOH(alc); H_2, Pt. Or: Mg, Et_2O; H_2O
   (h) Mg, Et_2O; H_2O (i) Zn(Hg), HCl

3. No reaction: a, c, f, g, m, q.

(b) *n*-propylcyclohexane
(d) benzoic acid (or salt)
(e) benzoic acid
(h) *o*- and *p*-*n*-$PrC_6H_4NO_2$
(i) *o*- and *p*-*n*-$PrC_6H_4SO_3H$
(j) *p*-*n*-$PrC_6H_4Tl(OOCCF_3)_2$
(k) *o*- and *p*-*n*-PrC_6H_4Cl
(l) *o*- and *p*-*n*-PrC_6H_4Br
(n) $PhCHBrCH_2CH_3$
(o) *o*- and *p*-*n*-$PrC_6H_4CH_3$
(p) *o*- and *p*-*n*-$PrC_6H_4CH_2C_6H_5$
(r) *p*-*t*-BuC_6H_4Pr-*n*
(s) *p*-*t*-BuC_6H_4Pr-*n*
(t) *p*-cyclohexyl-*n*-propylbenzene

4. (a) *n*-PrPh
(b) *n*-propylcyclohexane (alkyl group equatorial)
(c) $PhCHBrCHBrCH_3$ (for stereochemistry, wait for Problem 21, this chapter)
(d) *p*-$BrC_6H_4CHBrCHBrCH_3$
(e) $PhCHClCH_2CH_3$
(f) $PhCHBrCH_2CH_3$
(g) $PhCH_2CHBrCH_3$
(h) $PhCH(OSO_3H)CH_2CH_3$
(i) $PhCH(OH)—CH(Br)CH_3$
(j) *syn*-hydroxylation

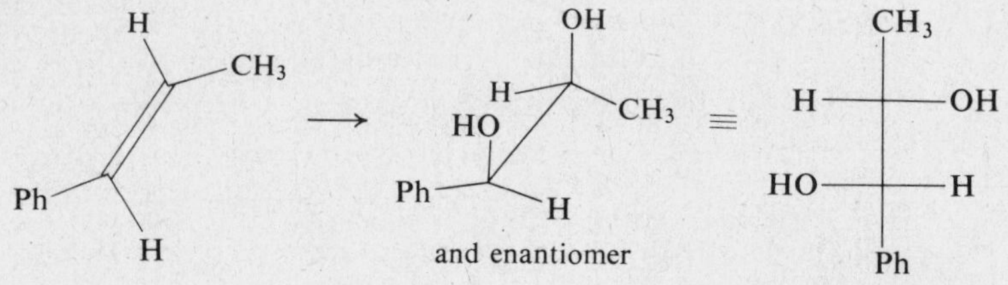

and enantiomer

and enantiomer
Threo
(1S,2S) and (1R,2R)

(k) $PhCOOH + CH_3COOH$
(l) *anti*-hydroxylation

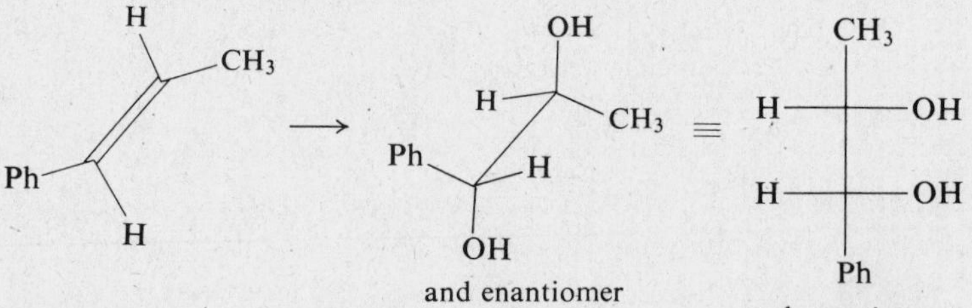

and enantiomer

and enantiomer
Erythro
(1R,2S) and (1S,2R)

(m) $PhCHO + OHCCH_3$
(n) $PhCH{=}CHCH_2Br$ (allylic substitution)
(o) *syn*-addition

$$CHBr_3 \xrightarrow{t\text{-BuOK}} :CBr_2$$

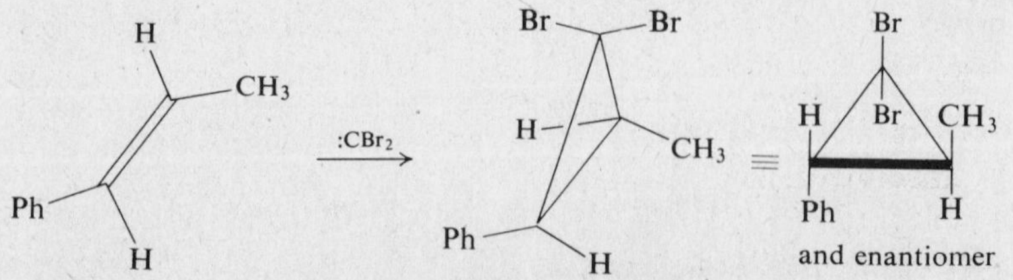

and enantiomer

and enantiomer

(p) $PhC{\equiv}CCH_3$

5. (a) cyclohexylbenzene   (b) PhC≡CAg

(c) m-$O_2NC_6H_4COOH$ (side-chain oxidation)

(d) $PhCH_2CHClCH_3$ (e) p-ClC_6H_4COOH

(f) isomerization to more stable, conjugated alkene (see Sec. 12.16)

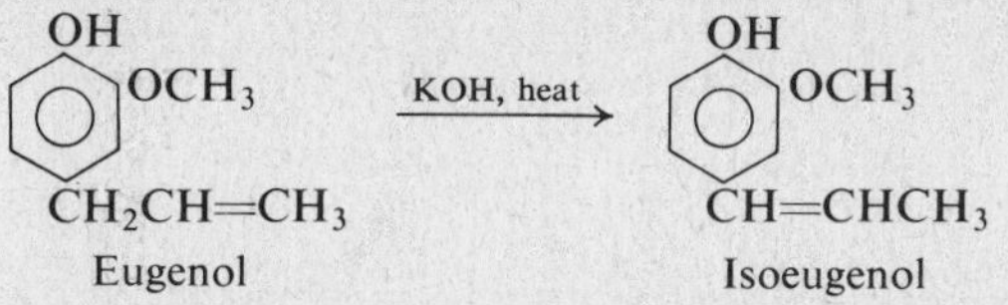

OH OCH₃ CH₂CH=CH₃	OH OCH₃ CH=CHCH₃
Eugenol	Isoeugenol

$$\xrightarrow{\text{KOH, heat}}$$

(g) $PhCH_2MgCl$ (h) $PhCH_3$ (i) 2-bromo-1,4-dimethylbenzene

(j) PhCH=CH—CH—CH₂   PhCH—CH=CH—CH₂   PhCH—CH—CH=CH₂
          | | | | |
          H   H   H   H   H

Only product reported   *Not formed*

Actually, only the most stable of the three possible products is obtained—the one in which the double bond is conjugated with the ring—suggesting that reaction is equilibrium-controlled. (Compare Problem 8.11.)

(k) PhCHO + PhCHO

(l) *syn*-hydrogenation ⟶ (Z)-PhCH=CHCH₂Ph

(m) *anti*-hydrogenation ⟶ (E)-PhCH=CHCH₂Ph

(n)

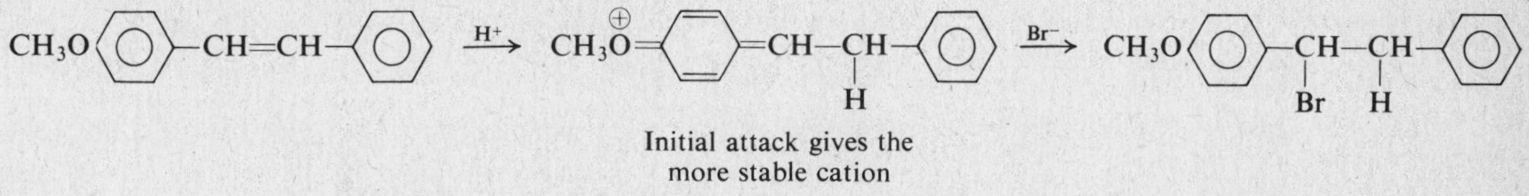

Initial attack gives the
more stable cation

6.

Benzyl cation

H⁺ | −H₂O

many repetitions

Polymer

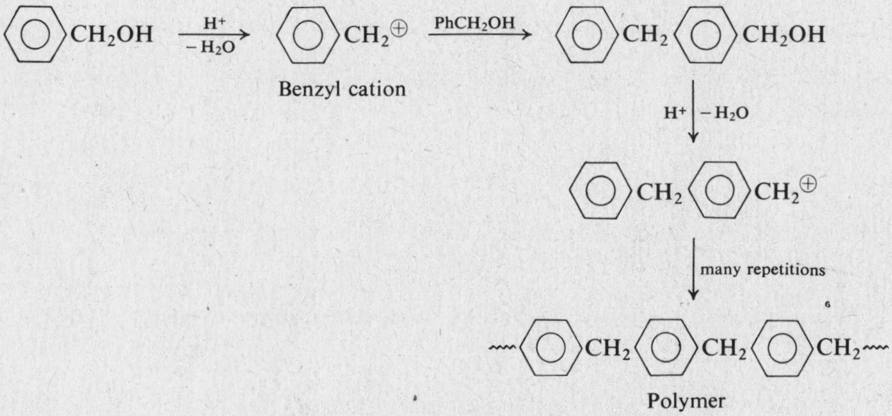

7. (a) The most reactive hydrogen is both benzylic and allylic.

```
      H  H  H  H  H
Ph—C—C=C—C—C—CH₃
      |        |  |
      H        H  H
      1  5  5  2  3  4
```

(b) The most reactive hydrogen is doubly benzylic.

$$CH_3-C_6H_4-CH_2-C_6H_4-CH_2-CH_2-CH_3$$
$$3 \qquad\quad 1 \qquad\qquad 2 \quad 4 \quad 5$$

(c) Removal of hydrogen from C–1 gives a benzylic free radical stabilized by hyperconjugation (see Problem 12.8, page 390).

$$1\text{-}CH_3 > 2\text{-}CH_3, 4\text{-}CH_3$$

(d) Removal of allylic hydrogen leads to two products via delocalized allylic radical.

(1) $PhCHCH{=}CHCH_2CH_2CH_3$ and $PhCH{=}CHCHCH_2CH_2CH_3$
   $|$ $|$
   Br   Br

(2) $PhCH_2CH{=}CHCHCH_2CH_3$ and $PhCH_2CHCH{=}CHCH_2CH_3$
         $|$ $|$
         Br   Br

(3) $PhCH_2CH{=}CHCH_2CHCH_3$
            $|$
            Br

(4) $PhCH_2CH{=}CHCH_2CH_2CH_2$
               $|$
               Br

(5) $PhCH_2C{=}CHCH_2CH_2CH_3$ or $PhCH_2CH{=}CCH_2CH_2CH_3$
   $|$ $|$
   Br   Br

8. We expect the preferred product (i) to have the double bond conjugated with the ring, and (ii) to be the less crowded of a pair of geometric isomers.

(a) $PhCH{=}CHCH_2CH_3$ (chiefly E) (b) $PhC{=}CH_2$ (c) major, $PhC{=}CHCH_3$ (chiefly E); $PhC{=}CH_2$
                      $|$ $|$ $|$
                     Et CH_3 Et

(d) major, $PhCH{=}CHCH_2CH_3$ (chiefly E); $PhCH_2CH{=}CHCH_3$ (chiefly E)

(e) major, $PhC{=}CHCH_3$ (chiefly E); $PhCHCH{=}CH_2$
     $|$ $|$
     CH_3 CH_3

9. We expect the more highly branched alkenes to predominate; these may be formed through rearrangement of the initially formed carbonium ions into more stable ones.

(a) major, $PhCH{=}CHCH_2CH_3$ (chiefly E) (b) major, $PhC{=}CHCH_3$ (chiefly E); $PhC{=}CH_2$
                                    $|$ $|$
                                   CH_3 Et

(c) as in (b) (d) major, $PhCH{=}CHCH_2CH_3$ (chiefly E) (e) as in (b)

10. We expect the alcohol that forms the most stable carbonium ion to be dehydrated fastest.

(a) $c > a > e, d > b$ (b) $Ph_2C(OH)CH_3 > PhCHOHCH_3 > PhCH_2CH_2OH$

11. We expect the alkene that forms the most stable carbonium ion to undergo addition fastest.

(a) $p\text{-}CH_3C_6H_4CH{=}CH_2 > C_6H_5CH{=}CH_2 > p\text{-}ClC_6H_4CH{=}CH_2$

(b) $p\text{-}H_2NC_6H_4CHOHCH_3 > C_6H_5CHOHCH_3 > p\text{-}O_2NC_6H_4CHOHCH_3$

12. (a) $PhCHCH=CHCH_2$ $PhCH-CHCH=CH_2$ $PhCH=CHCH-CH_2$
      | | | | |
      Br   Br   Br   Br   Br

(b) First and third structures, formed via initial addition to C–4.

(c) Most stable product (third structure) is formed, suggesting equilibrium control. (Compare Problem 8.11.)

13. (a) The *trans*-isomer is more stable by 5.7 kcal (26.3 − 20.6).

(b)

          cis-Stilbene   *trans*-Stilbene

The agent is Br·, formed in either of two familiar ways. Br· adds to *cis*-stilbene to give a free radical that rotates to another conformation (actually a more stable one, with Ph's farther apart). Before the second step of addition (reaction with *scarce* HBr or Br_2) can occur, Br· is lost, to yield *trans*-stilbene. (c) Equilibrium favors more stable stereoisomer.

14. Ionization of Ph_3COH in H_2SO_4 (to give Ph_3C^+) produces twice as many ions per mole as does MeOH (which gives $MeOH_2^+$), hence twice the lowering of the freezing point.

$$MeOH + H_2SO_4 \rightleftharpoons MeOH_2^+ + HSO_4^- \quad \textit{two ions}$$

$$Ph_3COH + H_2SO_4 \rightleftharpoons Ph_3COH_2^+ + HSO_4^-$$

$$Ph_3COH_2^+ \rightleftharpoons Ph_3C^+ + H_2O$$

$$\underline{H_2O + H_2SO_4 \rightleftharpoons H_3O^+ + HSO_4^-}$$

$$Ph_3COH + 2H_2SO_4 \rightleftharpoons Ph_3C^+ + H_3O^+ + 2HSO_4^- \quad \textit{four ions}$$

15. Resonance stabilization of the anion, with dispersal of negative charge, is greatest for Ph_3C^-, least for $C_5H_{11}^-$.

16. (a) (1) $CBrCl_3 \xrightarrow{\text{light}} Br\cdot + \cdot CCl_3$

(2) $\cdot CCl_3 + PhCH_3 \longrightarrow PhCH_2\cdot + HCCl_3$

(3) $PhCH_2\cdot + CBrCl_3 \longrightarrow PhCH_2Br + \cdot CCl_3$

        then (2), (3), (2), (3), etc.

(b) (4) $\cdot Br + PhCH_3 \longrightarrow PhCH_2\cdot + HBr$

    (5) $Cl_3C\cdot + \cdot CCl_3 \longrightarrow C_2Cl_6$

C_2Cl_6 is formed in the chain-terminating step (5).

Every time (1) occurs, there is formed not only $\cdot CCl_3$ but also $Br\cdot$. Like $\cdot CCl_3$, $Br\cdot$ abstracts hydrogen from toluene (4) to give $PhCH_2\cdot$ and thus starts a chain: (4), (3), (2), (3), (2), etc. For every (1), there are therefore two similar parallel chains, one started by the sequence (1), (2), and the other by the sequence (1), (4).

$CHCl_3$ is formed in each (2), (3) combination. HBr is formed only in (4), and is thus a measure of how many times (1) occurs (one HBr for each $Br\cdot$, and one $Br\cdot$ for each photon of light absorbed). The 20:1 ratio shows that the average chain length (*two* chains, remember) is 10.

17. 2-, 3-, 4-, 5-, and 6-phenyldodecanes, from rearrangement of initial secondary cation having charge on C–2 to other secondary cations with charge on C–3, C–4, C–5, or C–6.

18. The tricyclopropylmethyl free radical is stabilized—much as triphenylmethyl is—through delocalization of the odd electron over the rings. This is believed to involve overlap of the

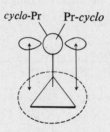

cyclo-Pr   Pr-cyclo

Tricyclopropylmethyl free radical

p orbital with the C—C bonds of the cyclopropane rings, which (Sec. 9.9) have considerable π character. (We shall encounter evidence of this kind of overlap in Problem 16, page 545.)

19. (a) The transition state is reached either earlier or later in chlorination than in bromination.

(b) Since $Cl\cdot$ is the more reactive reagent, it seems likely (Sec. 2.23) that the transition state is reached earlier in chlorination, and before bond breaking equals bond making.

20. (a) *m*-Xylene. Both —CH_3 groups in *m*-xylene activate the same positions toward electrophilic aromatic substitution. *m*-Xylene is thus preferentially sulfonated; the resulting sulfonic acid dissolves in the sulfuric acid, while unreacted *o*- and *p*-xylene remain insoluble.

(b) *m*-Xylene. Desulfonation is electrophilic aromatic substitution, and as in (a), the *m*-isomer is the most reactive. Preferential desulfonation thus frees insoluble (and volatile) *m*-xylene. The non-volatile *o*- and *p*-xylenesulfonic acids remain in the aqueous acid.

(c) *m*-Xylene.   Reaction with $H^+BF_4^-$ involves the first step of electrophilic aromatic substitution, formation of the benzenonium ion, or *sigma complex* (compare Problem 11.11).

$$CH_3 \langle \oplus \rangle \times {}^H_H \quad BF_4^-$$
$$CH_3$$

For the reasons given in (a), the sigma complex from *m*-xylene is the most stable, and hence the one favored by equilibrium.   The ionic sigma complex dissolves in the polar solvent BF_3/HF; *o*- and *p*-xylenes remain insoluble.

(d) *m*-Xylene.   A is m-$CH_3C_6H_4CH_2^-Na^+$.

We are dealing with equilibria involving *carbanions*: here, stability is decreased by electron-releasing groups, which tend to intensify the negative charge.   Equilibrium favors anions from the xylenes, rather than the one from isopropylbenzene with the two methyls attached to the negative site.

$$\begin{array}{c} CH_3 \\ \downarrow \\ C_6H_5C:^{\ominus}Na^+ + CH_3C_6H_4CH_3 \ \rightleftarrows\ C_6H_5Pr\text{-}iso + CH_3C_6H_4CH_2:^{\ominus}Na^+ \\ \uparrow \\ CH_3 \end{array}$$

More stable

Of the xylenes, *m*-xylene gives the most stable anion, m-$CH_3C_6H_4CH_2^-$.   Like all benzylic anions, it is stabilized by dispersal of the negative charge over the ring, particularly to the position *ortho* and *para* to the $-CH_2^-$.   When we draw the contributing structures for m-$CH_3C_6H_4CH_2^-$ we find that in none of them is the negative charge located on the carbon

to which the (destabilizing) methyl group is attached.   This is in contrast to what we find for *o*- or *p*-$CH_3C_6H_4CH_2^-$.   (*Draw these contributing structures.*)   Methyl thus destabilizes the *meta* isomer *least*.

Separation depends upon the relative non-volatility of organosodium compounds.
(See Chem. and Eng. News, June 14, 1971, pp. 30–32.)

21. Addition is predominantly, but not exclusively, *anti*.   This lack of (complete) stereospecificity indicates that much of the reaction proceeds via the open benzylic cation, which is subject to attack at either face, either before or after rotation.   For example, from a *trans*-alkene, as shown in Figure 12.4, page 127 of this Study Guide.

Through resonance, the electron-deficient carbon of the open benzylic cation gets electrons from the ring, and has less need of sharing an extra pair from bromine; bridging is weak, and easily broken.

The electron-releasing $-OCH_3$ group helps further to stabilize the benzylic cation, and hence increases the importance of the open cation in the reaction mechanism.

(R. C. Fahey and H.-J. Schneider, J. Am. Chem. Soc., **90**, 4429 (1968).)

22. (a) $PhCH_2CH_3 \xleftarrow{\text{HF}} CH_2=CH_2 + C_6H_6$

(b) $PhCH=CH_2 \xleftarrow{\text{KOH(alc)}} PhCHBrCH_3 \xleftarrow[\text{heat}]{\text{Br}_2} PhCH_2CH_3 \longleftarrow$ (a)

(c) $PhC\equiv CH$ $\xleftarrow{NaNH_2}$ $\xleftarrow{KOH(alc)}$ $\underset{\overset{|}{Br}\ \ \underset{}{Br}}{PhCH-CH_2}$ $\xleftarrow{Br_2,\ CCl_4}$ $PhCH=CH_2$ ⟵ (b)

(d) $PhCH(CH_3)_2$ $\xleftarrow{HF}$ $CH_3CH=CH_2 + C_6H_6$

(e) $Ph-\overset{\overset{\displaystyle CH_3}{|}}{C}=CH_2$ $\xleftarrow{KOH(alc)}$ $Ph-\underset{\overset{|}{Br}}{\overset{\overset{\displaystyle CH_3}{|}}{C}}-CH_3$ $\xleftarrow[heat]{Br_2}$ $PhCH(CH_3)_2$ ⟵ (d)

(f) $PhCH_2CH=CH_2$ $\xleftarrow{HF}$ $ClCH_2CH=CH_2 + C_6H_6$

(g) $PhC\equiv CCH_3$ ⟵

$\xrightarrow{CH_3I}$ $PhC\equiv CNa$ $\xleftarrow{NaNH_2}$ $PhC\equiv CH$ ⟵ (c)

$\xrightarrow{NaNH_2}$ $\xleftarrow{KOH(alc)}$ $\underset{\overset{|}{Br}\ \ \underset{}{Br}}{PhCH-CHPh}$ $\xleftarrow{Br_2}{CCl_4}$ $PhCH=CHCH_3$ $\xleftarrow[isom.]{KOH,\ heat}$ $PhCH_2CH=CH_2$

(f) ↑

(h) $\underset{Ph}{\overset{H}{}}\!\!\diagup\!\!\overset{}{C}=C\!\!\diagdown\!\!\underset{H}{\overset{CH_3}{}}$ $\xleftarrow{Li,\ NH_3}$ $PhC\equiv CCH_3$ ⟵ (g)

(i) $\underset{Ph}{\overset{H}{}}\!\!\diagup\!\!\overset{}{C}=C\!\!\diagdown\!\!\underset{CH_3}{\overset{H}{}}$ $\xleftarrow[Pd\ or\ Ni\text{-}B(P\text{-}2)]{H_2}$ $PhC\equiv CCH_3$ ⟵ (g)

(j) $p\text{-}(CH_3)_3CC_6H_4CH_3$ $\xleftarrow{HF}$ $(CH_3)_2C=CH_2 + C_6H_5CH_3$

(k) $O_2N\langle\bigcirc\rangle CH=CH_2$ $\xleftarrow{KOH}$ $O_2N\langle\bigcirc\rangle\underset{\overset{|}{Br}}{CHCH_3}$ $\xleftarrow[heat]{Br_2}$ $O_2N\langle\bigcirc\rangle CH_2CH_3$ $\xleftarrow[H_2SO_4]{HNO_3}$ $PhEt$ ⟵ (a)

(l) $Br\langle\bigcirc\rangle CH_2Br$ $\xleftarrow[heat]{Br_2}$ $Br\langle\bigcirc\rangle CH_3$ $\xleftarrow{Br_2,\ Fe}$ $C_6H_5CH_3$

(m) $O_2N\langle\bigcirc\rangle CHBr_2$ $\xleftarrow[heat]{2Br_2}$ $O_2N\langle\bigcirc\rangle CH_3$ $\xleftarrow{HNO_3,\ H_2SO_4}$ $C_6H_5CH_3$

(n) $p\text{-}BrC_6H_4COOH$ $\xleftarrow{KMnO_4}$ $p\text{-}BrC_6H_4CH_3$ $\xleftarrow{Br_2,\ Fe}$ $C_6H_5CH_3$

(o) $m\text{-}BrC_6H_4COOH$ $\xleftarrow{Br_2,\ Fe}$ C_6H_5COOH $\xleftarrow{KMnO_4}$ $C_6H_5CH_3$

(p)

$(PhCH_2)_2CuLi$ $\xleftarrow{CuX}$ $PhCH_2Li$ $\xleftarrow{Li}$

$PhCH_2-CH_2Ph$ ⟵

$PhCH_2Br$ $\xleftarrow[heat]{Br_2}$ $C_6H_5CH_3$

(q) $O_2N\langle\bigcirc\rangle CH_2\langle\bigcirc\rangle$ $\xleftarrow[AlCl_3]{C_6H_6}$ $O_2N\langle\bigcirc\rangle CH_2Cl$ $\xleftarrow[heat]{Cl_2}$ $O_2N\langle\bigcirc\rangle CH_3$ $\xleftarrow[H_2SO_4]{HNO_3}$ $C_6H_5CH_3$

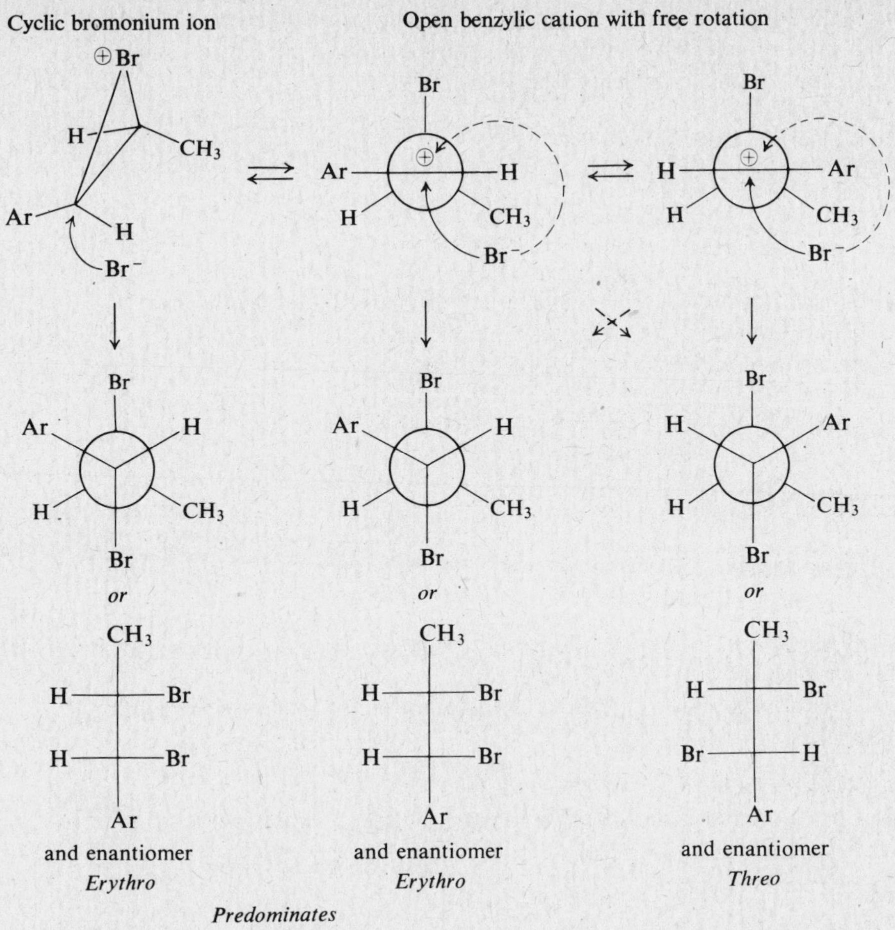

Figure 12.4. Addition of Br_2 to *trans*-$ArCH{=}CHCH_3$ (Problem 21).

23. (a) fuming sulfuric acid, or $CHCl_3/AlCl_3$ (b) Br_2/CCl_4, or $KMnO_4$
    (c) Br_2/CCl_4, or $KMnO_4$ (d) Br_2/CCl_4
    (e) CrO_3/H_2SO_4 (f) Br_2/CCl_4
    (g) $AgNO_3$ after sodium fusion   (h) nitrogen or bromine test after sodium
                                                    fusion

24. (a) Ozonolysis and identification of products.

(b) Oxidation and determination of m.p.'s of the resulting acids for the isomeric trimethyl-benzenes and ethyltoluenes; side-chain chlorination followed by dehydrohalogenation and then ozonolysis of the resulting alkenes will distinguish between *n*- and isopropylbenzene.

(c), (d), (e): Oxidation and determination of m.p.'s of the resulting acids.

25. Bromobenzene is the only one that will give a Br test.
    The three that give a Cl test can be distinguished from each other by oxidation to the acids and determination of m.p.'s.
    The two unsaturated compounds (positive $KMnO_4$ test) can be distinguished by ozonolysis and identification of the products.

The five arenes can be oxidized to carboxylic acids, which can be distinguished by their m.p.'s. The very high-melting acids from mesitylene and *m*-ethyltoluene can be further distinguished by their neutralization equivalents (Sec. 18.21).

26.

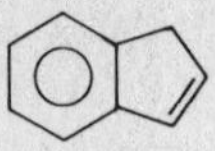

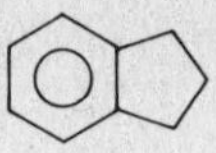

Indene   Indane

Absorption of $1H_2$ indicates one reactive double bond. Absorption of additional $3H_2$ indicates three more "double bonds"—probably an aromatic ring.

The presence of one benzene ring is confirmed by oxidation to

Determination of number of rings shows 2:

$$\begin{array}{l} C_9H_{20} \text{ alkane} \\ - C_9H_{16} \text{ sat'd product} \\ \hline \end{array}$$

missing 4H *means* 2 rings in indene and indane

This leads to the structures shown above. (Structures containing a four-membered ring are less likely to occur in coal tar.)

27. The empirical formulas of X, Y, and Z indicate that one phenyl group is present:

$$\begin{array}{l} C_8H_9 \\ - C_6H_5 \\ \hline C_2H_4 \end{array} \qquad\qquad \begin{array}{l} C_9H_{11} \\ - C_6H_5 \\ \hline C_3H_6 \end{array}$$

which suggests   *which suggests*

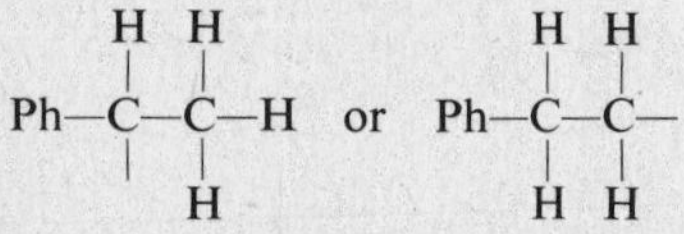

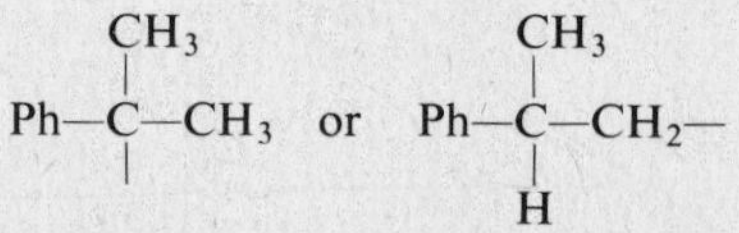

The aliphatic residues, C_2H_4— and C_3H_6—, are impossible for monosubstituted units. This predicament is eliminated by the m.w. determinations which show doubled empirical formulas, enabling us to write:

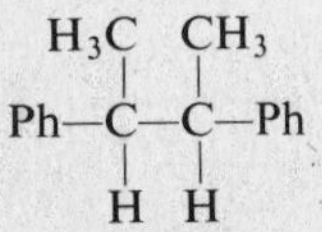

I   II
2,3-Diphenylbutane   1,4-Diphenylbutane

and

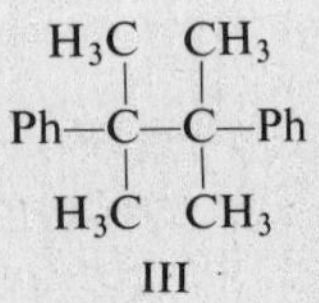

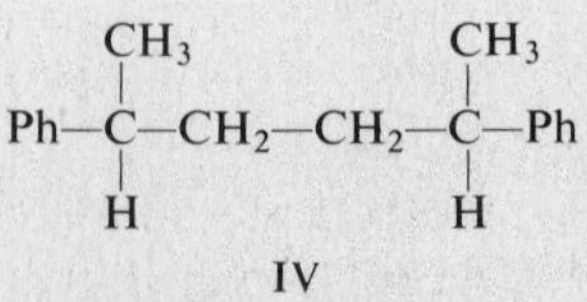

III   IV
2,3-Diphenyl-2,3- dimethylbutane   3,4-Diphenylhexane

We now can tackle the chemical reactions leading to X, Y, and Z. Free-radical attack seems likely under the conditions (see Problem 20, page 114). Knowing that benzylic hydrogen is most easily abstracted, we are led to write the following equations. For ethylbenzene:

$$t\text{-Bu}-\text{O}:\text{O}-\text{Bu-}t \longrightarrow 2t\text{-Bu}-\text{O}\cdot$$

$$\underset{\displaystyle \overset{|}{\text{CH}_3}}{2\text{Ph}-\text{CH}_2} + 2t\text{-Bu}-\text{O}\cdot \longrightarrow \underset{\displaystyle \overset{|}{\text{CH}_3}}{2\text{Ph}-\overset{\cdot}{\text{CH}}} + 2t\text{-BuOH}$$

$$\text{Ph}-\underset{\displaystyle \overset{}{\underset{\displaystyle \text{H}}{|}}}{\overset{\displaystyle \overset{\text{CH}_3}{|}}{\text{C}}}\cdot + \cdot\underset{\displaystyle \overset{}{\underset{\displaystyle \text{H}}{|}}}{\overset{\displaystyle \overset{\text{CH}_3}{|}}{\text{C}}}-\text{Ph} \longrightarrow \text{Ph}-\underset{\displaystyle \overset{}{\underset{\displaystyle \text{H}}{|}}}{\overset{\displaystyle \overset{\text{H}_3\text{C}}{|}}{\text{C}}}-\underset{\displaystyle \overset{}{\underset{\displaystyle \text{H}}{|}}}{\overset{\displaystyle \overset{\text{CH}_3}{|}}{\text{C}}}-\text{Ph}$$

The over-all reaction, then, is:

$$\underset{\displaystyle \overset{|}{\text{CH}_3}}{2\text{Ph}-\text{CH}_2} + (t\text{-Bu})_2\text{O}_2 \longrightarrow \text{Ph}-\underset{\displaystyle \overset{\text{H}_3\text{C}}{|}}{\text{CH}}-\underset{\displaystyle \overset{\text{CH}_3}{|}}{\text{CH}}-\text{Ph} + 2t\text{-BuOH}$$

$$\text{X and Y}$$

The observed yields fit this equation: there were obtained 0.02 mole of *tert*-butyl alcohol, and $(1 + 1)/210$ or about 0.01 mole of X plus Y.

For isopropylbenzene, in a similar manner:

$$\text{2Ph}-\underset{\displaystyle \overset{}{\underset{\displaystyle \text{CH}_3}{|}}}{\overset{\displaystyle \overset{\text{CH}_3}{|}}{\text{CH}}} + (t\text{-Bu})_2\text{O}_2 \longrightarrow \text{Ph}-\underset{\displaystyle \overset{}{\underset{\displaystyle \text{H}_3\text{C}}{|}}}{\overset{\displaystyle \overset{\text{H}_3\text{C}}{|}}{\text{C}}}-\underset{\displaystyle \overset{}{\underset{\displaystyle \text{CH}_3}{|}}}{\overset{\displaystyle \overset{\text{CH}_3}{|}}{\text{C}}}-\text{Ph} + 2t\text{-BuOH}$$

$$\text{Z}$$

with the observed yield again fitting the equation.

Now, why are there two products, X and Y? We might at first consider that they are I and II. But this would require formation of equal numbers of the benzylic free radicals $\text{Ph(CH}_3)\text{CH}\cdot$ and the primary free radicals $\text{PhCH}_2\text{CH}_2\cdot$, which seems highly unlikely. Besides, in that case we would expect a *third* isomer, formed by combination of unlike free radicals,

$$\underset{\displaystyle \overset{|}{\text{CH}_3}}{\text{Ph}-\text{CH}}-\text{CH}_2\text{CH}_2\text{Ph}$$

and only two isomers were actually obtained.

Examination of the formula for I shows that three stereoisomers are possible: a pair of enantiomers and a *meso* structure. It seems most likely, then, that X and Y are racemic and *meso*-2,3-diphenylbutane.

Enantiomers
One fraction

Meso

This conclusion is supported by the evidence from the isopropylbenzene experiment. The structure of III does not permit stereoisomerism; only one product is predicted, and only one was obtained.

Chapter 13 Spectroscopy and Structure

13.1 (a) $(CH_3)_3C^+$ $CH_2{=}CH{-}CH_2^+$ $CH_3CH_2^+$ $CH_2{=}CH^+$

(b) $C_5H_{12}^{\cdot +} \longrightarrow C_4H_9^+ + CH_3\cdot$

13.2

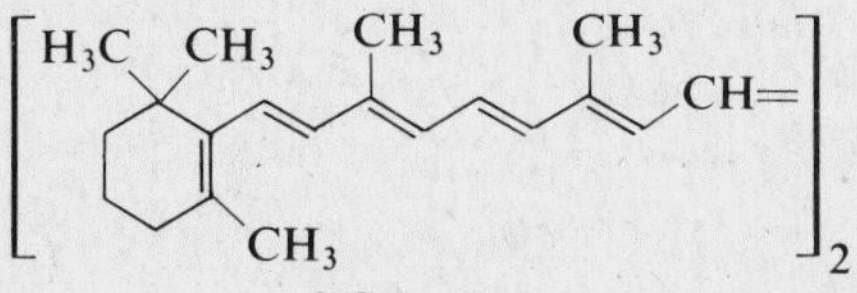

β-Carotene

13.3 (a) A, 1,4-pentadiene; B and C, (*Z*)- and (*E*)-1,3-pentadiene. (b) Heats of hydrogenation, infrared spectra.

13.4 (a) $\underset{a\ \ b}{CH_3CHCl_2}$, 2 signals; $\underset{a\ \ \ a}{CH_2ClCH_2Cl}$, 1 signal.

(b) $\underset{a\ \ \ \ a}{CH_3CBr_2CH_3}$, 1 signal; $\underset{b\ \ \ a\ \ \ b}{CH_2BrCH_2CH_2Br}$, 2 signals; $\underset{a\ \ \ b\ \ \ c}{CH_3CH_2CHBr_2}$, 3 signals;

$$\underset{a\qquad d}{CH_3CHBr}{-}\overset{\overset{b \text{ (or } c)}{H}}{\underset{\underset{c \text{ (or } b)}{H}}{C}}{-}Br$$, 4 signals.

(c) $\underset{c\ \ \ b\ \ \ a}{C_6H_5CH_2CH_3}$, 3 signals; $p\text{-}\underset{a\ \ \ \ b\ \ \ \ a}{CH_3C_6H_4CH_3}$, 2 signals.

(d) 1,3,5-$C_6H_3(CH_3)_3$, 2 signals; p-$CH_3C_6H_4CH_2CH_3$, 4 signals; $C_6H_5CH(CH_3)_2$, 3 signals.
            b a b d c a c b a

(e) CH_3CH_2OH, 3 signals; CH_3OCH_3, 1 signal.
       a b c a a

(f) $CH_3CH_2OCH_2CH_3$, 2 signals; $CH_3OCH_2CH_2CH_3$, 4 signals; $CH_3OCH(CH_3)_2$, 3 signals;
       a b b a c d b a b c a

     $CH_3CH_2CH_2CH_2OH$, 5 signals.
          a b c d e

(g) H_2C-CH_2 , 2 signals; CH_3-HC-C , 4 signals.
    (cyclopropane-type structures with labels a, b, c, d and b (or c), c (or b))

(h) CH_3CH_2CHO, 3 signals; CH_3COCH_3, 1 signal; (vinyl structure) , 5 signals.
       a b c a a

13.5

1,1-Dimethylcyclopropane — Two signals

trans-1,2-Dimethylcyclopropane — Three signals

cis-1,2-Dimethylcyclopropane — Four signals

13.6 One, because of rapid interconversion of equatorial and axial protons.

13.7 Electron release by methyl groups lowers deshielding of the ring protons.

13.8 The relative positions of protons are indicated by the sequence of letters in the answer to Problem 13.4; that is, a is farthest upfield, b is next, and so on. (Shift of —OH varies, Sec. 16.13.)

(a) a 3H, b 1H; a.
(b) a; a 2H, b 4H; a 3H, b 2H, c 1H; a 3H, b 1H, c 1H, d 1H.
(c) a 3H, b 2H, c 5H; a 6H, b 4H.
(d) a 9H, b 3H; a 3H, b 3H, c 2H, d 4H; a 6H, b 1H, c 5H.
(e) a 3H, b 2H, c 1H; a.
(f) a 6H, b 4H; a 3H, b 2H, c 3H, d 2H; a 6H, b 3H, c 1H; a 3H, b 2H, c 2H, d 2H, e 1H.
(g) a 2H, b 4H; a 3H, b 1H, c 1H, d 1H.
(h) a 3H, b 2H, c 1H; a; a 2H, b 1H, c, d, e 1H each.

Analyzing Spectra

Squeeze as much information as you can from the molecular formula: use chemical arithmetic, deciding where you can how many rings and/or double bonds are present. Combine this with characteristic infrared bands, δ values, proton counts, and splitting of various nmr signals to give you structural units. If the spectrum (or combination of spectra) is unambiguous, you should have only one possible structure left; go back and check this against all the information you have.

For problems on spectra, answers are presented in two stages: names of the unknown compounds are given in their proper sequence along with the other answers; then, at the end of the Study Guide, spectra are reproduced with infrared bands identified and nmr signals assigned. We suggest that you check each of your answers in two stages, too. First, check the name; if your answer is wrong, or if you have not been able to work the problem at all, return to the spectrum in the textbook and, knowing the correct structure, have another go at it: see if you can now identify bands, assign signals, and analyze spin–spin splittings. Then, finally, turn to the back of the Study Guide and check your answer against the analyzed spectrum.

13.9 See the labeled spectra on page 643 of this Study Guide.

(a) The relative peak heights are: a, 9H; b, 2H; c, 5H. Downfield peak c is clearly due to 5 aromatic protons: C_6H_5—. Peak a is clearly nine equivalent 1° aliphatic protons: $3CH_3$—. We do a little chemical arithmetic at this point,

$$
\begin{array}{ccc}
C_{11}H_{16} & C_5H_{11} & C_2H_2 \\
-\,C_6H_5- & -\,3CH_3- & -\,\overset{|}{\underset{|}{C}}- \quad \text{(to hold the 3Me's)} \\
\hline
C_5H_{11} & C_2H_2 & \\
& & \overline{-CH_2-}
\end{array}
$$

and end up with a residue of $-CH_2-$. This corresponds to peak b, which has a δ value of about that of benzylic H. There is only one way to put the pieces together:

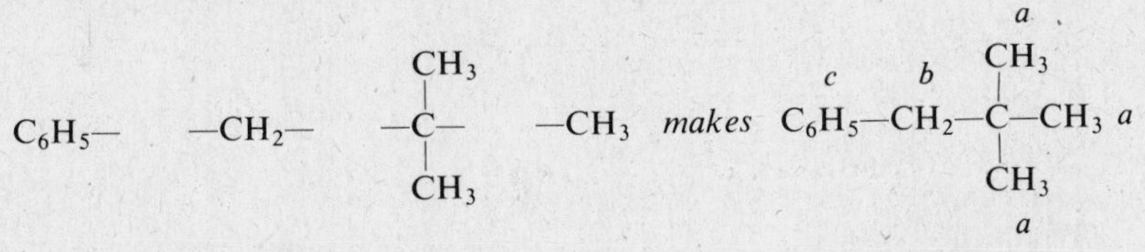

$$
C_6H_5- \qquad -CH_2- \qquad -\overset{CH_3}{\underset{CH_3}{\overset{|}{\underset{|}{C}}}}- \qquad -CH_3 \quad \textit{makes} \quad C_6H_5\overset{c}{-}CH_2\overset{b}{-}\overset{\overset{a}{CH_3}}{\underset{\underset{a}{CH_3}}{\overset{|}{\underset{|}{C}}}}-CH_3 \; a
$$

Neopentylbenzene

(b) The relative peak heights are $a:b = 3:1$. In view of the molecular formula, this means: a 6H; b, 2H. The molecule is saturated, open-chain ($C_4H_8Br_2$ corresponds to C_4H_{10}), and must have the carbon skeleton of either *n*-butane or isobutane. The six protons of signal a are equivalent, and are probably in two —CH_3 groups, shifted downfield

$$
\begin{array}{cc}
C_4H_8Br_2 & C_2H_2Br_2 \\
-\,2CH_3 & -\,CH_2 \\
\hline
C_2H_2Br_2 & CBr_2
\end{array}
$$

by —Br. Signal b is due, then, to —CH_2Br. On this assumption (supported by absence of any splitting in the signals due to protons on adjacent carbons, Sec. 13.10), we arrive at the following structure:

$$CH_3- \quad \overset{\overset{\displaystyle CH_3}{|}}{\underset{\underset{\displaystyle Br}{|}}{-C-}} \quad -CH_2- \quad -Br \quad makes \quad CH_3-\overset{\overset{\displaystyle \overset{a}{CH_3}}{|}}{\underset{\underset{\displaystyle Br}{|}}{C}}-CH_2-Br$$

$$a \qquad\qquad b$$

Isobutylene bromide

(c) The relative peak heights are: a, 1H; b, 2H; c, 5H. The broad signal a indicates acidic hydrogen; in view of the molecular formula, it must be attached to oxygen: —OH. Signal c clearly indicates aromatic protons: C_6H_5—. Some more chemical arithmetic,

$$\frac{\begin{array}{c} C_7H_8O \\ -C_6H_5- \end{array}}{CH_3O} \qquad \frac{\begin{array}{c} CH_3O \\ --OH \end{array}}{-CH_2-}$$

leaves us with a residue of —CH_2—. The pieces go together in only one way:

$$C_6H_5- \quad -CH_2- \quad -OH \quad makes \quad \overset{c}{C_6H_5}-\overset{b}{CH_2}-\overset{a}{OH}$$

Benzyl alcohol

13.10 Order of compounds same as in answer to Problem 13.4:

(a) a, doublet, 3H   (b) a, singlet
    b, quartet, 1H

    a, singlet a, quintet, 2H
                                 b, triplet, 4H

                                 a, triplet, 3H
                                 b, multiplet, 2H
                                 c, triplet, 1H

                                 a, doublet, 3H
                                 b, pair of doublets, 1H
                                 c, pair of doublets, 1H
                                 d, complex, 1H

(c) a, triplet, 3H   (d) a, singlet, 9H
    b, quartet, 2H b, singlet, 3H
    c, complex, 5H

    a, singlet, 6H a, triplet, 3H
    b, singlet, 4H b, singlet, 3H
                            c, quartet, 2H
                            d, complex, 4H

                            a, doublet, 6H
                            b, heptet, 1H
                            c, complex, 5H

(e) a, triplet, 3H   (f) a, triplet, 6H   (g) a, quintet, 2H   (h) a, triplet, 3H
    b, quartet, 2H b, quartet, 4H b, triplet, 4H b, multiplet, 2H
    c, singlet, 1H c, triplet, 1H

    a, singlet a, triplet, 3H a, doublet, 3H
                            b, multiplet, 2H b, pair of a, singlet
                            c, singlet, 3H   doublets, 1H
                            d, triplet, 2H c, pair of a, multiplet, 2H
                                                        doublets, 1H b, singlet, 1H
                            a, doublet, 6H d, complex, 1H c, d, and e, multi-
                            b, singlet, 3H   plets, 1H each
                            c, heptet, 1H

                            a, triplet, 3H
                            b, multiplet, 2H
                            c, multiplet, 2H
                            d, triplet, 2H
                            e, singlet, 1H

13.11 No; same compounds as in answer to Problem 13.9.

13.12 See the labeled spectra on page 644 of this Study Guide.

(a) The quartet-triplet combination of signals *a* and *b* is characteristic of a CH_3CH_2- group, as in Figure 13.8c: upfield triplet, splitting of $-CH_3$ by $-CH_2-$; downfield quartet, splitting of $-CH_2-$ by $-CH_3$; *J* values identical. Subtracting C_2H_5- from the molecular formula leaves C_6H_5-,

$$\begin{array}{r} C_8H_{10} \\ -C_2H_5- \\ \hline C_6H_5- \end{array}$$

which gives signal *c* in aromatic region. The compound is:

$$\overset{c}{}\overset{b}{}\overset{a}{}$$
$$C_6H_5CH_2CH_3$$
Ethylbenzene

The relative peak heights corroborate this: *a*, 3H; *b*, 2H; *c*, 5H.

(b) From its formula, it is clearly open-chain, saturated, and one of four possible isomeric dibromopropanes. The triplet-quintet combination indicates 4H's split by $-CH_2-$ (signal *b*), and 2H's split by four protons (signal *a*). Only one of the possible isomers fits this pattern,

$$\overset{b}{}\quad\overset{a}{}\quad\overset{b}{}$$
$$-CH_2-CH_2-CH_2-$$

and it takes no great imagination to attach 2Br atoms to give:

$$\overset{b}{}\quad\overset{a}{}\quad\overset{b}{}$$
$$BrCH_2-CH_2-CH_2Br$$
1,3-Dibromopropane

in which the signal for the terminal protons will be shifted far downfield by $-Br$.

(c) The formula requires a simple choice between *n*-propyl bromide and isopropyl bromide:

$$\overset{a}{}\quad\overset{b}{}\quad\overset{c}{}\qquad\qquad\overset{a}{}\quad\overset{b}{}\quad\overset{a}{}$$
$$CH_3-CH_2-CH_2-Br \qquad CH_3-CH-CH_3$$
$$\;|$$
$$\;Br$$

Expect three signals	*Expect two signals*
Triplet-multiplet-triplet	Septet-doublet
3H 2H 3H	6H 1H

Obviously, the compound is *n*-propyl bromide,

$$\overset{a}{}\quad\overset{b}{}\quad\overset{c}{}$$
$$CH_3-CH_2-CH_2-Br.$$
n-Propyl bromide

(See note on analyzing spectra, page 133 of this Study Guide.)

13.13 At room temperature, interconversion of the three possible conformers is so rapid that a single average signal is given; at $-120°$, interconversion is so slow that separate signals are given by the achiral structure I and (racemic) chiral structures II and III.

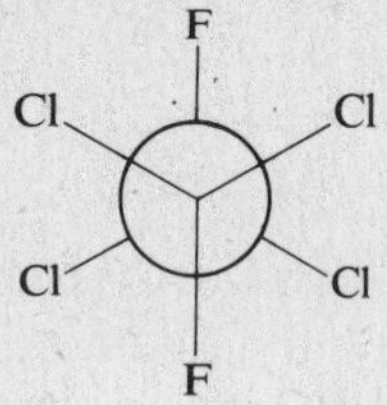

I 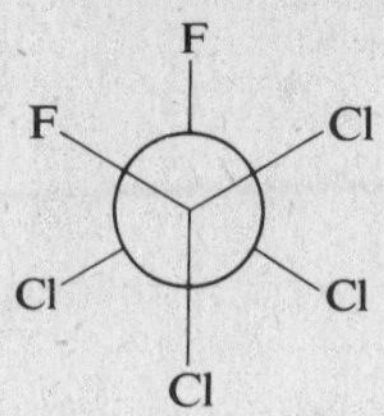   II 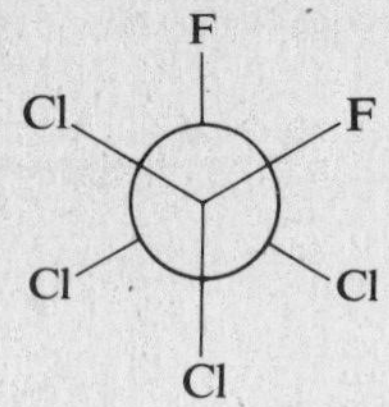   III

Unequal peak areas indicate different amounts of the two components; there is no splitting of signals because in any conformation the fluorines are equivalent.

13.14 At room temperature, interconversion of the three possible conformers is so rapid that a single average signal is given; at $-98°$, interconversion is so slow that separate signals are given by the achiral structure IV and (racemic) chiral structures V and VI.

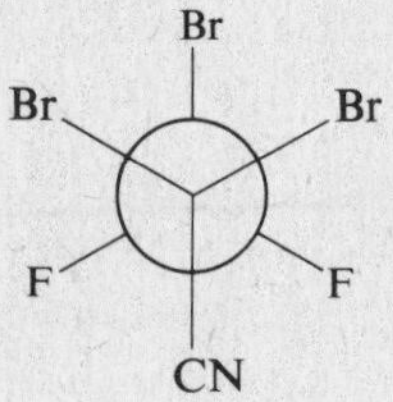

IV 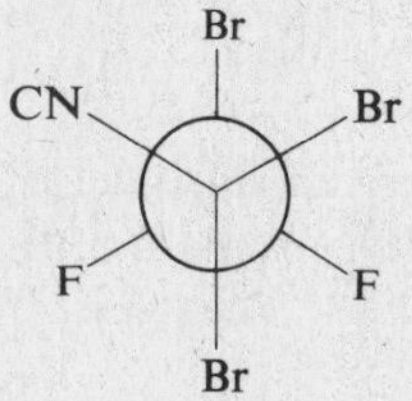   V 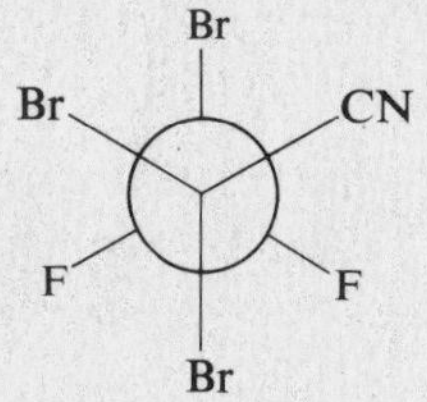   VI

The pair of doublets is given by V and VI: in each, the fluorines are not equivalent—one is "between" two —Br's, the other "between" —Br and —CN. Relative peak areas indicate that V and VI are favored: less crowding of Br atoms than in IV.

13.15 (a) Impossible to reverse spin of only one of a pair: violation of Pauli principle.

(b) If both spins of a pair are reversed, no net change in energy, no signal.

13.16 (a) $CH_3\cdot$ (b) $CH_3\overset{.}{C}HCH_3$ $CH_3CH_2\overset{.}{C}HCH_3$ (c) $Ph_3C\cdot$

13.17 No unsaturation apparent (no absorption in 1650 cm^{-1} range), so there must be one ring ($C_6H_{14} - C_6H_{12} = 2H$ missing, hence 1 ring). No indication of —CH$_3$ (2960 cm^{-1} and 2870 cm^{-1}), so the ring must have no side chain. The one compound that fits the data is cyclohexane.

1. (a) Formula shows open-chain, saturated structure ($C_3H_3Cl_5$ is equivalent to C_3H_8). One hydrogen split by two hydrogens (to give triplet) and two hydrogens split by one hydrogen (to give doublet) leave few choices for the structure:

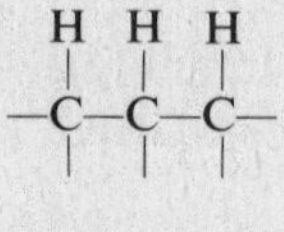

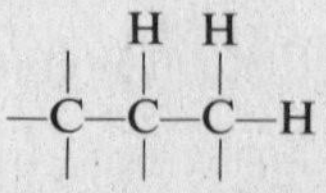

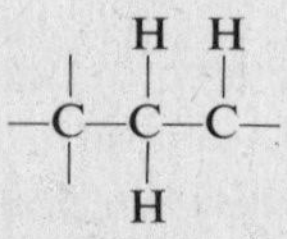

$$\underset{i}{\overset{\begin{matrix}H & H & H\end{matrix}}{Cl-\overset{Cl}{\underset{Cl}{C}}-\overset{}{\underset{Cl}{C}}-\overset{}{\underset{Cl}{C}}-Cl}} \quad or \quad \underset{ii}{\overset{\begin{matrix}Cl & H & H\end{matrix}}{Cl-\overset{}{\underset{Cl}{C}}-\overset{}{\underset{Cl}{C}}-\overset{}{\underset{Cl}{C}}-H}} \quad or \quad \underset{iii}{\overset{\begin{matrix}Cl & H & H\end{matrix}}{Cl-\overset{}{\underset{Cl}{C}}-\overset{}{\underset{H}{C}}-\overset{}{\underset{Cl}{C}}-Cl}}$$

Since the 2H signal is farther downfield, *ii* can be ruled out on the basis that 1° H (in —CH_2Cl) would be expected upfield from 2° H, and *iii* can be ruled out on the basis that —CH_2— would be expected upfield from —$CHCl_2$. This leaves only *i*:

$$\overset{b \qquad a \qquad b}{CHCl_2CHClCHCl_2}$$

1,1,2,3,3-pentachloropropane

1° H is shifted farther downfield by two —Cl's than is 2° H by one —Cl.

(b) Formula shows open-chain saturated structure. Unsplit singlet for 3H indicates —CH_3 attached to carbon carrying no hydrogen: C—C—CH_3. This leaves us with no other choice but to write:

$$\overset{b \qquad\qquad a}{CH_2Cl-CCl_2-CH_3}$$

1,2,2-Trichloropropane

(c) From the formula, it can be seen that we are dealing with one of four isomeric bromo-butanes:

$$CH_3CH_2CH_2CH_2Br \qquad CH_3CH_2CHBrCH_3 \qquad \underset{CH_3}{\overset{CH_3}{>}}CHCH_2Br \qquad CH_3-\overset{CH_3}{\underset{CH_3}{\overset{|}{\underset{|}{C}}}}-Br$$

$$\quad n\text{-BuBr} \qquad\qquad sec\text{-BuBr} \qquad\qquad\qquad iso\text{-BuBr} \qquad\qquad tert\text{-BuBr}$$

Only one of these, isobutyl bromide, can give rise to a 6H signal split into a doublet. Therefore, the compound must be:

$$\overset{a}{CH_3}\underset{\underset{a}{CH_3}}{\overset{}{>}}\overset{b \quad c}{CH-CH_2-Br}$$

Isobutyl bromide

The —CH_2— signal, split into a doublet by the 3° H, is shifted downfield by —Br.

(d) The formula, in conjunction with the 5H signal *b* at δ 7.28, indicates immediately the presence of a C_6H_5— group, and consequently ($C_{10}H_{14} - C_6H_5 = C_4H_9$) 4C and 9H attached to it. A single, unsplit 9H signal at δ 1.30 clearly indicates 3CH_3— groups, forcing us to conclude that C_4H_9 is *tert*-Bu, and that the compound is:

$$b\left\{ \hexagon\!\!-\overset{CH_3}{\underset{CH_3}{\overset{|}{\underset{|}{C}}}}-CH_3 \right\} a$$

tert-Butylbenzene

(e) This is a bit more formidable than (d), but a C_6H_5 group is present, and we need to work out only the C_4H_9 side chain, for which there are only 3 possibilities (the fourth was used in (d), above).

$$Ph—CH_2—CH_2—CH_2—CH_3 \qquad Ph—\overset{\overset{\displaystyle CH_3}{|}}{CH}—CH_2—CH_3 \qquad Ph—CH_2—\overset{\overset{\displaystyle CH_3}{\diagup}}{\underset{\diagdown CH_3}{CH}}$$

$$\textit{n}\text{-butyl} \qquad\qquad\qquad \textit{sec}\text{-butyl} \qquad\qquad\qquad \text{isobutyl}$$

Again (compare c, above), the only structure that can give a 6H signal split into a doublet is the isobutyl side chain, and the compound is isobutylbenzene:

The $—CH_2—$ doublet (signal c) is shifted downfield by the phenyl group.

(f) The presence of a disubstituted $(—C_6H_4—)$ benzene ring is shown by the downfield signal c (of 4H). This leaves $(C_9H_{10} - C_6H_4 = C_3H_6)$ 3C and 6H to be accounted for. The quintet-triplet combination of 2H (split into a quintet by 4H's) and 4H (split into a triplet by 2H's) leads to $—CH_2—CH_2—CH_2—$
$$\qquad\qquad\qquad\qquad\qquad\qquad\qquad\qquad\qquad b \qquad a \qquad b$$

This can be fitted to $C_6H_4\diagup$ in only one way:

Indane
(Problem 26, Chap. 12)

As benzylic H, signal b is shifted downfield to δ 2.91.

(g) Once again, a phenyl group $(C_6H_5—)$ shows up (signal c). Subtraction $(C_{10}H_{13}Cl - C_6H_5)$ leaves C_4H_8Cl, which indicates a single (only one group can be attached to $C_6H_5—$) saturated C_4 side chain. Signal a shows two unsplit $CH_3—$ groups; signal b shows an unsplit $—CH_2—$. This leaves one carbon and one —Cl. The pieces can be put together in two ways (of which the first is the actual compound).

$$\begin{array}{ccccc} & CH_3— & & & \\ C_6H_5— & & —CH_2— \quad —Cl & —\overset{|}{\underset{|}{C}}— & \textit{makes} \\ & CH_3— & & & \end{array}$$

2-Chloro-2-methyl-1-phenylpropane   or   1-Chloro-2-methyl-2-phenylpropane

(h) The two signals, a and b, are unusually far upfield, and indicate the presence of a cyclopropane ring (Table 13.4, page 421, or inside front cover) carrying 2 groups of 2H each. Signal c shows an unsplit $CH_3—$, and far downfield signal d shows a $C_6H_5—$ group. This accounts for all 10C and 12H, and the only reasonable combination of the fragments gives:

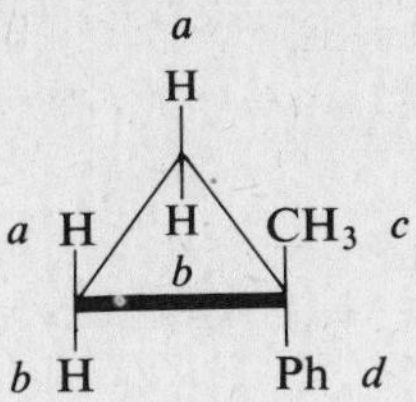

a
H

a H   H CH_3 c *a and b may be reversed*
     b

b H   Ph d

1-Methyl-1-phenylcyclopropane

(i) The presence of a C_6H_5— group (signal d, 5H) is easily seen, leaving a side chain of —C_3H_6Br to be figured out. The absence of any signal for 3H means the absence of any CH_3— group, and since any branched side chain,

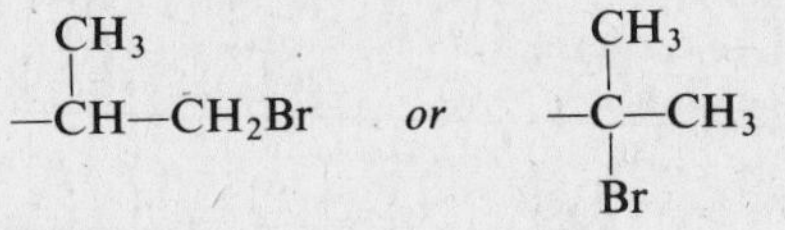

$$CH_3 \qquad\qquad CH_3$$
$$—CH—CH_2Br \quad or \quad —C—CH_3$$
$$\qquad\qquad\qquad\qquad Br$$

would contain a CH_3— group, we can eliminate those isomers. Of those remaining,

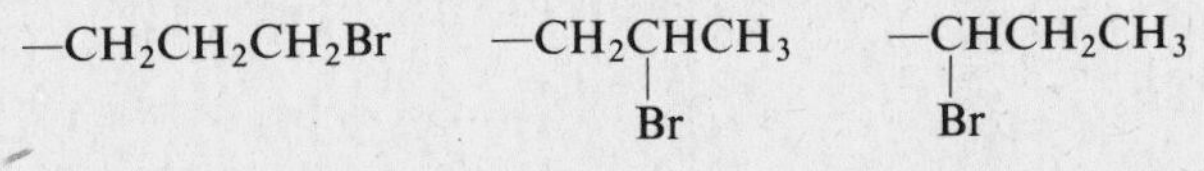

$$—CH_2CH_2CH_2Br \quad —CH_2CHCH_3 \quad —CHCH_2CH_3$$
$$\qquad\qquad\qquad\qquad Br \qquad\qquad Br$$

only the first has no CH_3— group. The compound must be, then:

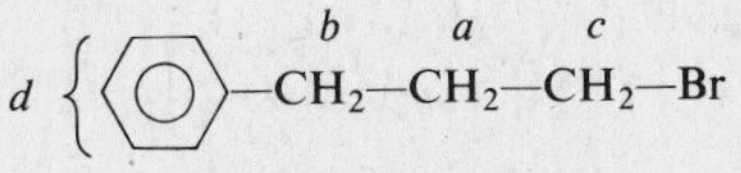

$$b \quad a \quad c$$
$$d \left\{ \bigcirc —CH_2—CH_2—CH_2—Br \right.$$

3-Bromo-1-phenylpropane

(j) Signal a indicates that a CH_3— is present in this saturated, open-chain molecule ($C_3H_5ClF_2$ equivalent to C_3H_8). That signal b is a triplet and not a quartet shows us that an ethyl group (triplet-quartet is characteristic) cannot be present. Hence, the only possible distribution of C's and H's is:

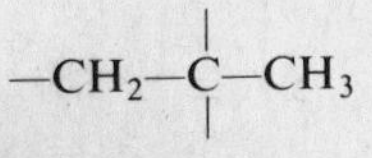

$$—CH_2—\overset{|}{\underset{|}{C}}—CH_3$$

The question still remains: what splits signal a into a triplet, and signal b also into a triplet? The answer to this is, of course, that the two ^{19}F atoms are doing it (Sec. 13.10, and Problems 13.13 and 13.14). The compound, therefore, must be:

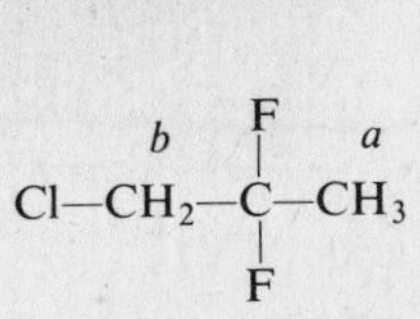

$$b \quad \overset{F}{\overset{|}{}} \quad a$$
$$Cl—CH_2—\overset{|}{\underset{|}{C}}—CH_3$$
$$\underset{F}{}$$

1-Chloro-2,2-difluoropropane

(Remember, *absorption* by fluorine does not appear in this pmr spectrum—only the *splitting* by fluorine.)

2. The two possible structures (both achiral) are:

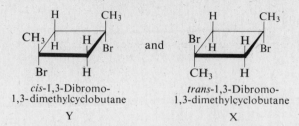

cis-1,3-Dibromo-   trans-1,3-Dibromo-
1,3-dimethylcyclobutane   1,3-dimethylcyclobutane
          Y   X

The ring protons in the *trans*-isomer are all equivalent (each is *cis* to a —Br and to a —CH₃) and will give rise to one unsplit 4H signal; the other singlet (6H) is of course due to the two —CH₃'s. Obviously, X is the *trans*-isomer. The ring protons in the *cis*-isomer fall into two groups: two (equivalent to each other) are *cis* to —CH₃ groups, two (equivalent to each other) are *cis* to —Br's. Thus there will be two signals for ring protons, each split into a doublet by H's on the opposite face of the ring.

3.

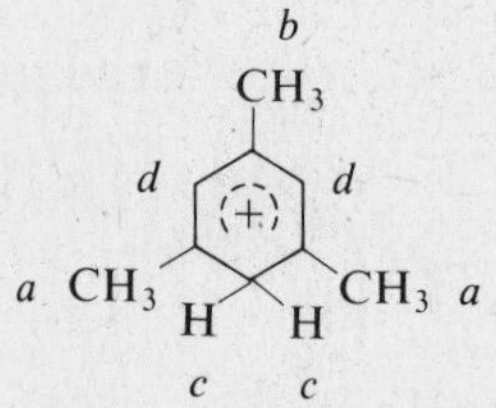

$$SbF_6^-$$

Shows that $Ar\langle\substack{H \\ H}$ ⊕ and presumably $Ar\langle\substack{H \\ Y}$ intermediates can exist.

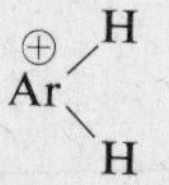

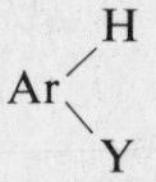

(See also Problem 11.11, page 358.)

4. (a)

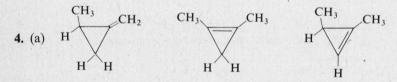

(b) The one with =CH₂ group.

(c) A, 1,2-dimethylcyclopropene, the only structure with but two sets of protons.

(d) The aromatic (two π electrons) cyclopropenyl cation formed by loss of a ring hydrogen from A (compare Problem 10.6, page 330);

(e) $CH_3C\equiv CCH_3 + CH_2N_2 +$ light.

5. See page 935.

6. (a) Two rings in C. (b) Two rings, two double bonds in B.

(c) B is bicyclo[2.2.0]hexa-2,5-diene ("Dewar benzene," pages 319 and 321). C is bicyclo-[2.2.0]hexane.

B,   C,

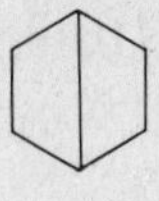

(d) The 3° H's are split by all the vinylic protons; in turn, these are split by both 3° H's.

7. (a)–(c)

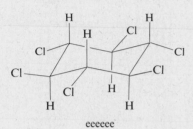

eeeeee

All H's equivalent (all axial)

One signal

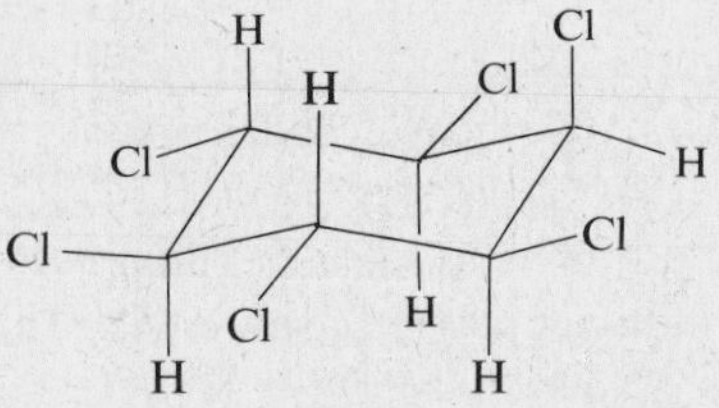

eeeeea

5 axial H's, 1 equatorial H

Two signals (5:1)

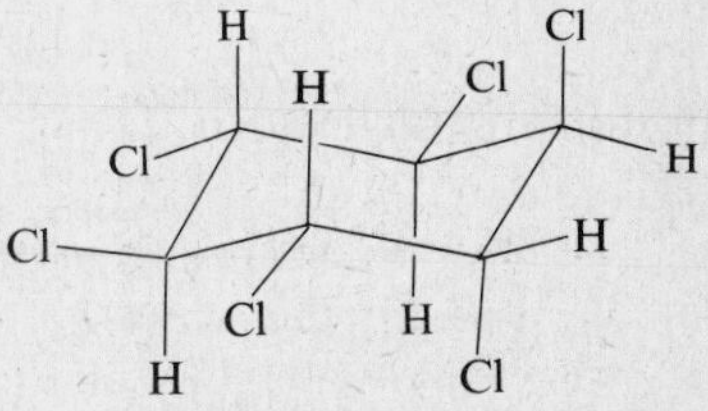

eeeeaa

4 axial H's, 2 equatorial H's

Two signals (4:2)

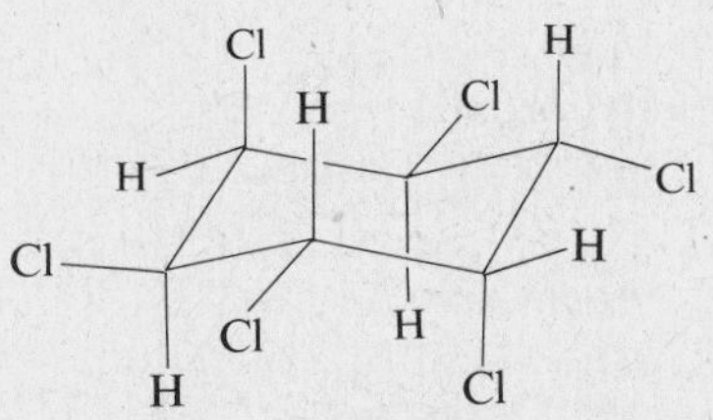

eeaeea

4 axial H's, 2 equatorial H's

Two signals (4:2)

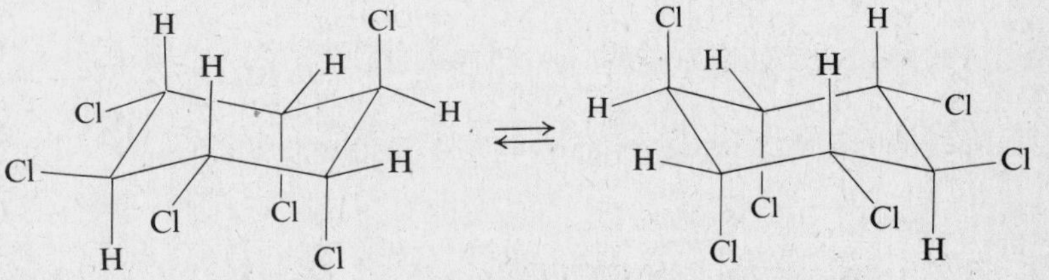

eeeaaa ⇌ aaaeee

Conformations of equal stability

All H's equivalent by rapid flip-flop between
axial and equatorial positions

One signal

(d) eeeeee: no change.

eeeaaa: split into two signals of equal area (one for 3 axial H's, one for 3 equatorial H's).

8. (a)

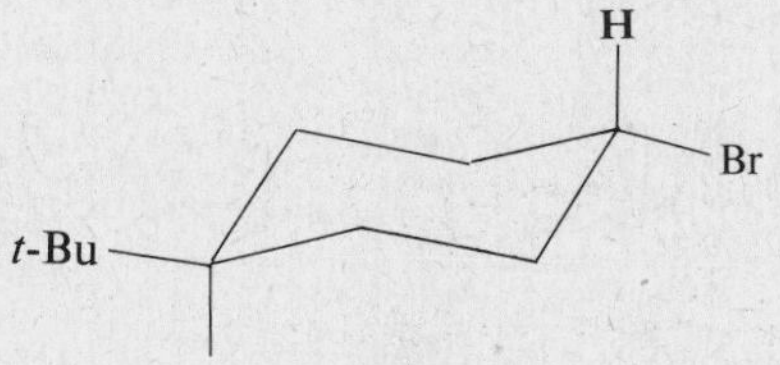

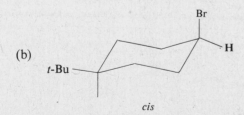

t-Bu

trans

Signal for axial —H on C-1 shifted
downfield (clear of other proton
signals) by adjacent —Br.

(b)

t-Bu

cis

Signal for equatorial —H on
C-1 even farther downfield than
signal for axial —H.

Generally, a signal for an axial —H will appear upfield from a signal for an equatorial —H.

9. At −75°, interconversion of chair conformers becomes so slow that signals for both axial and equatorial —H are seen. Conformation with equatorial —Br (axial —H, upfield signal) predominates, accounting for

$$\frac{4.6}{4.6 + 1.0} \times 100 = 82\%$$

of the molecules.

10. (a) isopropylbenzene   (b) isobutylene   (c) phenylacetylene

(Try to fit these answers to the spectra, following the general approach outlined on page 133 of this Study Guide.   Then, see the labeled spectra on page 645 of this Study Guide.)

11. (a) isobutylbenzene   (b) *tert*-butylbenzene   (c) *p*-cymene (*p*-isopropyltoluene)

(See the labeled spectra on page 646 of this Study Guide.)

12. (a) α-phenylethyl bromide, $C_6H_5CHBrCH_3$ (b) *tert*-pentylbenzene   (c) *sec*-butyl bromide

(See the labeled spectra on page 647 of this Study Guide.)

13. D, α-methylstyrene, $C_6H_5C(CH_3)=CH_2$

(See the labeled spectra on page 648 of this Study Guide.)

Chapter 14

Alkyl Halides
Nucleophilic Aliphatic Substitution. Elimination

14.1 (a)

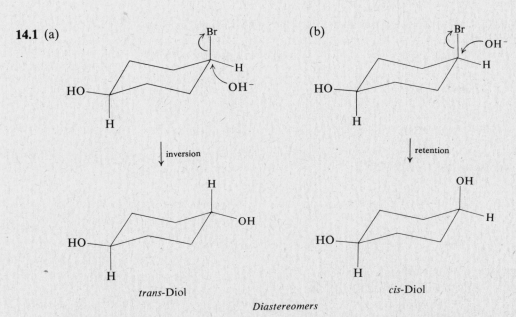

(b)

inversion

retention

trans-Diol

cis-Diol

Diastereomers

(c) Not if experiments can be designed to produce diastereomers, which can be differentiated by m.p., etc.

14.2 (a) CH₃—CH₂Br   CH₃CH₂—CH₂Br $\underset{\text{CH}_3}{\overset{\text{CH}_3}{\diagup}}$CH—CH₂Br $\text{CH}_3-\underset{\text{CH}_3}{\overset{\text{CH}_3}{\overset{|}{\underset{|}{\text{C}}}}}-\text{CH}_2\text{Br}$

G = CH₃—   CH₃CH₂— $\underset{\text{CH}_3}{\overset{\text{CH}_3}{\diagup}}$CH— $\text{CH}_3-\underset{\text{CH}_3}{\overset{\text{CH}_3}{\overset{|}{\underset{|}{\text{C}}}}}-$

143

(b) The larger G is, the lower the rate. The effect of *size* parallels the effect of *number* of groups.

14.3 The slow step—the one being measured in both reactions—is the dissociation of the dimer into free radicals. Subsequent steps are fast.

14.4 The rate is made up of two parts, S_N2 and S_N1:

$$\text{rate} = \overbrace{4.7 \times 10^{-5}[RX][OH^-]}^{S_N2} + \overbrace{0.24 \times 10^{-5}[RX]}^{S_N1}$$

From this we can establish the percentage due to S_N2:

$$\%S_N2 = \frac{S_N2}{S_N2 + S_N1} \times 100 = \frac{4.7 \times 10^{-5}[RX][OH^-]}{4.7 \times 10^{-5}[RX][OH^-] + 0.24 \times 10^{-5}[RX]} \times 100$$

$$\%S_N2 = \frac{4.7[OH^-]}{4.7[OH^-] + 0.24} \times 100$$

Finally, we can calculate the percentage due to S_N2 at various $[OH^-]$ values:

(a) When $[OH^-] = 0.001$, $\%S_N2 = \dfrac{4.7 \times 0.001}{4.7 \times 0.001 + 0.24} \times 100 = 1.9\%$

Similarly:

(b) $[OH^-] = 0.01$, $\%S_N2 = 16.4$ (c) $[OH^-] = 0.1$, $\%S_N2 = 66.2$

(d) $[OH^-] = 1.0$, $\%S_N2 = 95.1$ (e) $[OH^-] = 5.0$, $\%S_N2 = 99.0$

14.5 (a) If optically pure 2-bromooctane has a specific rotation of $-34.6°$, our reactant is

$$\frac{20.8}{34.6} \times 100 = 60\% \text{ optically pure}$$

and of the same configuration as shown on page 462.
    If optically pure 2-octanol has a specific rotation of $-9.9°$, our product is

$$\frac{3.96}{9.9} \times 100 = 40\% \text{ optically pure}$$

and of opposite configuration to that shown on page 462.

(b) The optical purity of the products is $\frac{2}{3}$ that of the reactant. Reaction proceeds with $\frac{2}{3}$ inversion and $\frac{1}{3}$ racemization.

(c) For each molecule that undergoes back-side attack, a racemic pair is formed, and the optical activity of *two* molecules is thus lost. In this reaction, then, $\frac{1}{6}$ of the molecules undergo front-side attack, and $\frac{5}{6}$ undergo back-side attack. The $\frac{1}{6}$ cancel the activity of another $\frac{1}{6}$, giving $\frac{1}{3}$ racemic and leaving the optical purity $\frac{2}{3}$ of the original value.

14.6 The S_N1 reaction is slow because the neopentyl cation is primary, and therefore slow to form. The S_N2 reaction is slow because of the steric factor: although there is only *one* substituent attached to —CH_2X, it is a very large one, *t*-Bu. (Compare Problem 14.2, page 465.)

14.7 From Sec. 14.8, Problem 14.4, and the present problem, we collect the following:

MeBr: rate = 0.0214[MeBr][OH$^-$] S_N2 *only*

EtBr: rate = 0.0017[EtBr][OH$^-$] S_N2 *only*

i-PrBr: rate = 4.7×10^{-5}[*i*-PrBr][OH$^-$] + 0.24×10^{-5}[*i*-PrBr]   *Mixed* $S_N2 + S_N1$

t-BuBr: rate = 0.010[*t*-BuBr] S_N1 *only*

For [OH$^-$] = 0.1, these reduce to:

MeBr: rate = 2.14×10^{-3}[MeBr]

EtBr: rate = 1.7×10^{-4}[EtBr]

i-PrBr: rate = 4.7×10^{-6}[*i*-PrBr] + 2.4×10^{-6}[*i*-PrBr] = 7.1×10^{-6}[*i*-PrBr]

t-BuBr: rate = 1.0×10^{-2}[*t*-BuBr]

Assuming identical concentrations for the halides, we get the following rates relative to *i*-PrBr (= 1):

$$\frac{\text{rate}_{\text{MeBr}}}{\text{rate}_{i\text{-PrBr}}} = \frac{2.14 \times 10^{-3}}{7.1 \times 10^{-6}} = 300$$

$$\frac{\text{rate}_{\text{EtBr}}}{\text{rate}_{i\text{-PrBr}}} = \frac{1.7 \times 10^{-4}}{7.1 \times 10^{-6}} = 24$$

$$\frac{\text{rate}_{i\text{-PrBr}}}{\text{rate}_{i\text{-PrBr}}} = 1$$

$$\frac{\text{rate}_{t\text{-BuBr}}}{\text{rate}_{i\text{-PrBr}}} = \frac{1.0 \times 10^{-2}}{7.1 \times 10^{-6}} = 1410$$

14.8 (a) Formic acid is more polar than dry alcohol and favors the S_N1 reaction of a neutral molecule, which proceeds via a transition state that is more polar than the reactants (see the upper equation on page 472). The less polar solvent, alcohol, favors the S_N2 reaction of a neutral molecule with a charged ion, which proceeds via a transition state that is less polar than the reactants (see the lower equation on page 472).

(b) Reinforcing the effects in (a), we have the following. The stronger nucleophile, OEt^-, is more likely to react by S_N2 than is the weak nucleophile, water.

(c) Electron release by $-CH_3$ speeds up S_N1, but has little effect on S_N2.

14.9 With the strong nucleophile, OEt^-, reaction is S_N2, the EtO^- displacing $-Cl$ from the 3-position. With the weak nucleophile, EtOH, reaction is S_N1, and proceeds via the hybrid allylic cation,

$$CH_3-CH\!=\!\!=\!CH\!=\!\!=\!CH_2 \underset{\oplus}{\underbrace{}} \xrightarrow{\quad} \begin{cases} \xrightarrow{a} CH_3CHCH\!=\!CH_2 \\ \qquad\qquad \overset{|}{Cl} \\ \xrightarrow{b} CH_3CH\!=\!CHCH_2Cl \end{cases}$$

which yields two products (compare pages 270–271).

14.10 (a) Large increase in rate.

$$H_3N + RX \longrightarrow \left[\overset{\delta+}{H_3N}\cdots R\cdots \overset{\delta-}{X} \right] \longrightarrow H_3\overset{+}{N}-R + X^-$$

Reactants   Transition state   Products

More polar than reactants: stabilized more by solvation

(b) Small decrease in rate.

$$R\!-\!\overset{+}{S}(CH_3)_2 \longrightarrow \left[\overset{\delta+}{R}\cdots \overset{\delta+}{S}(CH_3)_2 \right] \longrightarrow R^+ + S(CH_3)_2$$

Reactant   Transition state   Products

Concentrated charge: Dispersed charge

stabilized more than transition state by solvation

14.11 (a) Removal of hydrogen occurs from the carbon carrying the fewer hydrogens. ("From him that hath not, it shall be taken.")

(b) $PhCH\!=\!CHCH_2CH_3$ and $PhCH_2CH\!=\!CHCH_3$; no preference indicated by original rule.

(c) $PhCH\!=\!CHCH_2CH_3$, double bond conjugated with ring, more stable product.

14.12 In 2-phenylethyl bromide, the electron-withdrawing phenyl group is attached to the carbon from which H is lost; by dispersing a partial negative charge, phenyl stabilizes a carbanion-like transition state and speeds up reaction.

$$\begin{array}{c} \overset{\displaystyle Br}{\vdots} \\ CH_2\!=\!\!=\!\overset{\delta-}{CH}\!\longrightarrow\!Ph \\ \vdots \\ H\cdots B \end{array}$$

(This effect is *in addition to* the one exerted in both 2- and 1-phenylethyl bromides: through conjugation, phenyl stabilizes the incipient double bond of the transition state.)

14.13

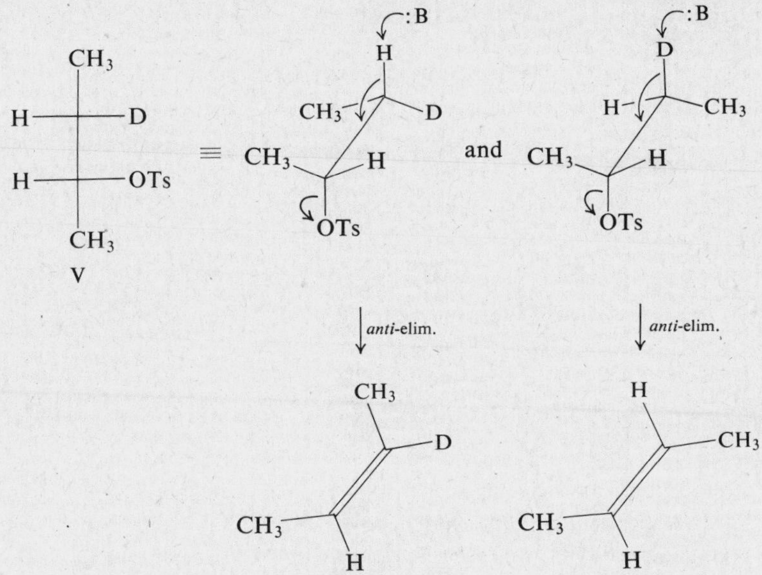

CH$_3$
H———D
H———OTs
CH$_3$
V

$\equiv$

:B
H
CH$_3$———D
CH$_3$ H
OTs

and

:B
D
H—   CH$_3$
CH$_3$ H
OTs

↓ *anti*-elim.

↓ *anti*-elim.

CH$_3$
D
CH$_3$
H
cis-2-Butene-2-*d*

H
CH$_3$
CH$_3$
H
trans-2-Butene

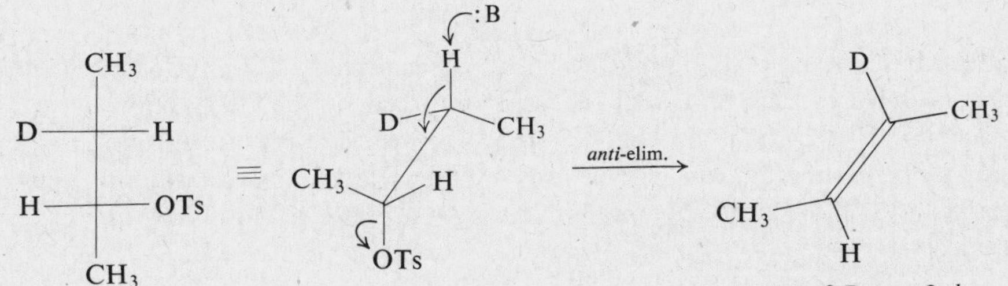

CH$_3$
D———H
H———OTs
CH$_3$
VI

$\equiv$

:B
H
D   CH$_3$
CH$_3$ H
OTs

$\xrightarrow{\text{\textit{anti}-elim.}}$

D
CH$_3$
CH$_3$
H
trans-2-Butene-2-*d*

The reactions must proceed as shown above, via *anti*-elimination of —OTs and —H (or —D). *syn*-Elimination from V, say, to give *cis*-2-butene would have to result in loss of deuterium, contrary to fact.

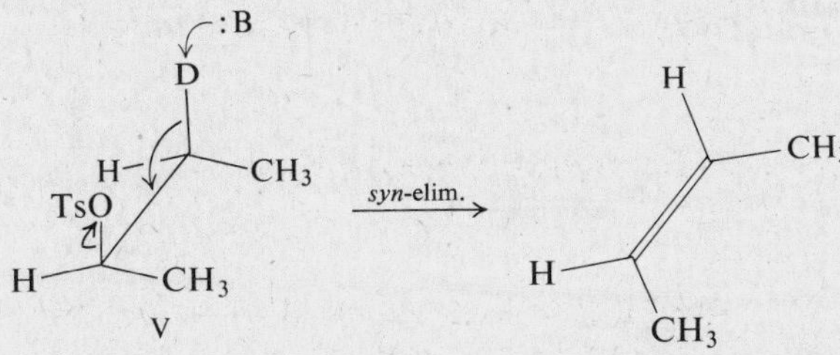

:B
D
H—   CH$_3$
TsO
H—   CH$_3$
V

$\xrightarrow{\text{\textit{syn}-elim.}}$

H
CH$_3$
H
CH$_3$
cis-2-Butene

Does not occur

14.14 The low rate of elimination raises the suspicion that in that one isomer it is impossible for —H and —Cl to get into an *anti*-periplanar relationship. Only one isomer,

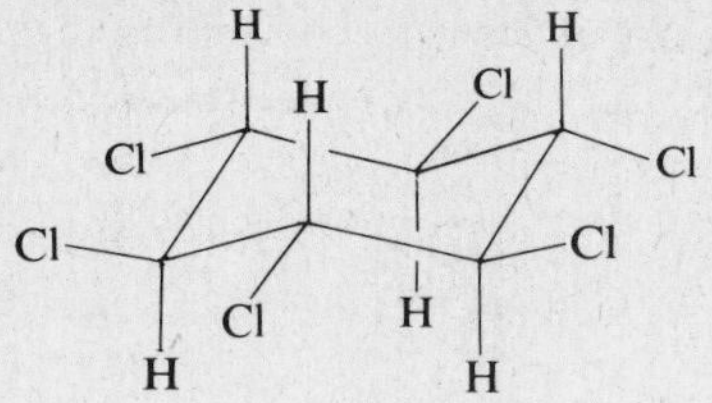

All Cl atoms equatorial;
no —Cl *anti* to —H.

fills the bill. (Draw some of the other structures to check the uniqueness of this isomer.)

14.15 *Crowding of —CH$_3$ groups*

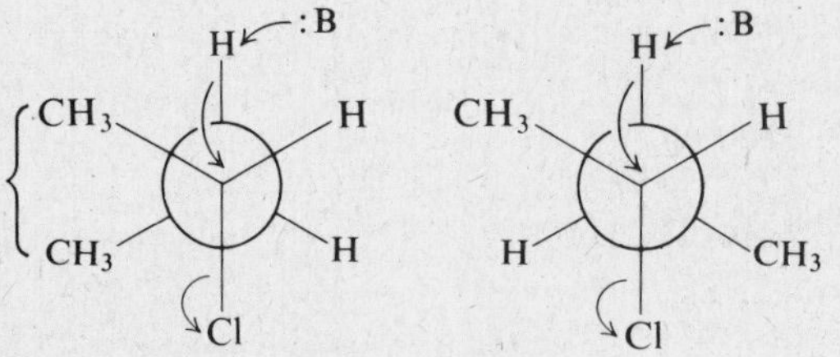

The transition state leading to the *cis*-alkene has crowding between the methyl groups, and is less stable. This interpretation is based on the fact that dehydrohalogenation is a relatively difficult reaction (E_{act} 20–25 kcal), and hence much slower than interconversion of conformations of reactant (see page 235). However, we would expect the same general results if product ratio were determined by conformational populations, since the conformation with the —CH$_3$ groups *anti* to each other would be more abundant.

14.16 In the absence of added base the E2 reaction of menthyl chloride, already slow (Problem 15, page 488), is slowed down to such an extent that it is outpaced by the E1 reaction. There are no stereochemical restrictions on this reaction, since the leaving groups are lost in different steps. The carbonium ion loses a proton from either of two positions to give both alkenes, with the more stable one predominating.

14.17 (a) Orientation of elimination is powerfully controlled by the phenyl group, which stabilizes the transition state in two ways: through conjugation with the incipient double bond, and through dispersal of the partial negative charge (see the answer to Problem 14.12).

With the nature of the product thus ordained, each tosylate reacts by the only periplanar elimination open to it: *anti* from the *cis*-isomer, and *syn* from the *trans*-isomer.

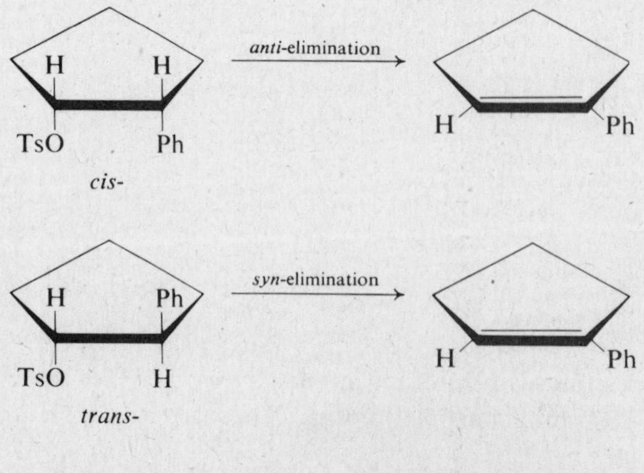

Although the *anti*-elimination requires some twisting of the ring, it is still faster than the *syn*-elimination. The significant thing is that *syn* elimination occurs as fast as it does; it requires eclipsing of groups, but in cyclopentane compounds, groups are already badly eclipsed. (The corresponding cyclohexane compounds give analogous results, but there the *cis* isomer reacts over 10,000 times as fast as the *trans*.)

(b) VII must react by *anti* elimination, and VIII by *syn* elimination. Here, *syn* elimination is actually faster. In this rigid bicyclic system, the twisting of VII required for *anti*-periplanar elimination is more difficult than in a simple cyclopentane derivative, whereas in VIII the leaving groups are already held *syn*-periplanar. (In addition, in VII both Cl's are tucked in a fold of the molecule, making solvation of a leaving Cl⁻ difficult.)

14.18 (a) EtBr. E2 elimination is significant for $PhCH_2CH_2Br$, which yields the more stable (conjugated) alkene. (See also the answer to Problem 14.12.)

(b) $PhCH_2CH_2Br$. The steric effect of Ph slows down the S_N2 reaction of $PhCHBrCH_3$. (This outweighs the somewhat faster elimination from $PhCH_2CH_2Br$.)

(c) *n*-BuBr. *iso*-BuBr undergoes elimination faster (yields a more stable alkene) and S_N2 more slowly (the steric effect of branching; Problem 14.2, page 465).

(d) *iso*-BuBr. This is a case of 1° vs. 3° RBr, as discussed on page 485.

14.19 (a) *t*-BuO⁻ in DMSO is not solvated via hydrogen bonding, and hence is a much stronger base than highly solvated OH⁻ in alcohol.

(b) F⁻ in DMSO is not solvated via hydrogen bonding, and is a strong base.

14.20 (a) CrO_3/H_2SO_4 (b) Br_2/CCl_4, or $KMnO_4$
(c) $AgNO_3$(alc), or fuming sulfuric acid (d) solubility in conc. H_2SO_4

1. (a) Br_2, heat or light. (b) HBr(g). (c) conc. HBr (method used in laboratory).

2. (a) b or c (b) none (c) c (d) c
(e) b or c (f) a or c (g) c (h) a or c

3. (a) *i*-PrBr $\xleftarrow{HBr}$ *i*-PrOH

(b) $CH_2=CH-CH_2Cl \xleftarrow[600°]{Cl_2} CH_2=CH-CH_3 \xleftarrow[heat]{acid}$ *i*-PrOH

(c) $CH_3CH-CH_2 \xleftarrow[H_2O]{Cl_2} CH_3CH=CH_2 \xleftarrow[heat]{acid}$ *i*-PrOH
    OH Cl

(d) CH_3CH-CH_2 $\xleftarrow{Br_2, CCl_4}$ $CH_3CH=CH_2$ $\xleftarrow[\text{heat}]{\text{acid}}$ $i\text{-PrOH}$
    $\quad\;\;|\quad\;\;|$
    $\quad\;\,Br\;\;\;Br$

(e) $CH_3-\overset{\displaystyle Br}{\underset{\displaystyle Br}{\overset{|}{\underset{|}{C}}}}-CH_3$ $\xleftarrow{2HBr}$ $CH_3C\equiv CH$ $\xleftarrow{NaNH_2}$ $\xleftarrow{KOH(alc)}$ CH_3CH-CH_2 $\longleftarrow$ (d)
    $\qquad\qquad\qquad\qquad\qquad\qquad\qquad\qquad\qquad\qquad\quad\;|\quad\;\;|$
    $\qquad\qquad\qquad\qquad\qquad\qquad\qquad\qquad\qquad\qquad\;Br\;\;\;Br$

(f) $CH_3C=CH_2$ $\xleftarrow{KOH(alc)}$ $CH_3-\overset{\displaystyle Br}{\underset{\displaystyle Br}{\overset{|}{\underset{|}{C}}}}-CH_3$ $\longleftarrow$ (e)
    $\quad\;\;|$
    $\quad\;Br$

(g) $CH_3CH=CH$ $\xleftarrow{KOH(alc)}$ CH_3CH-CH_2 $\longleftarrow$ (d)
    $\qquad\qquad|$ $\qquad\qquad\quad|\quad\;\;|$
    $\qquad\qquad Br$ $\qquad\qquad\;Br\;\;\;Br$

(h) $CH_2-CH-CH_2$ $\xleftarrow{Cl_2 / H_2O}$ $CH_2=CH-CH_2Cl$ $\longleftarrow$ (b)
    $\;\;|\quad\;|\quad\;\;|$
    $\;Cl\;\;OH\;\;Cl$

(i) $CH_2-CH-CH_2$ $\xleftarrow{Br_2, CCl_4}$ $CH_2=CH-CH_2$ $\xleftarrow[\text{H}_2\text{O}]{\text{OH}^-}$ $CH_2=CH-CH_2$ $\longleftarrow$ (b)
    $\;\;|\quad\;|\quad\;\;|$ $\qquad\qquad\quad|$ $\qquad\qquad\quad|$
    $\;Br\;\;Br\;\;OH$ $\qquad\qquad OH$ $\qquad\qquad\;Cl$

(j) $\xleftarrow{CHCl_3, \; t\text{-BuOK}}$ $CH_3CH=CH_2$ $\xleftarrow[\text{heat}]{\text{acid}}$ $i\text{-PrOH}$

4. (a) $\xleftarrow{HBr}$

(b) $\xleftarrow{HI \text{ or } P, I_2}$

(c) $\xleftarrow{Br_2, CCl_4}$ $\xleftarrow{H_2SO_4, \text{ heat}}$

(d) $\xleftarrow{NBS}$ $\xleftarrow{H_2SO_4, \text{ heat}}$

(e) $\xleftarrow{Cl_2, H_2O}$ $\xleftarrow{H_2SO_4, \text{ heat}}$

(f) $\xleftarrow{Na, MeOH}$ $\xleftarrow[t\text{-BuOK}]{CHCl_3}$ $\xleftarrow[\text{heat}]{H_2SO_4}$

5. (a)

$$\text{CH}_2\text{Cl–C}_6\text{H}_4\text{–Br} \xleftarrow{\text{Cl}_2, \text{ heat}} \text{CH}_3\text{–C}_6\text{H}_4\text{–Br} \xleftarrow{\text{Br}_2, \text{ Fe}} \text{CH}_3\text{–C}_6\text{H}_5$$

(b) $\text{Ph}_3\text{CCl} \xleftarrow{\text{AlCl}_3} 3\text{PhH} + \text{CCl}_4$ (Compare Problem 12.15, p. 399.)

(c) $\text{CH}_2{=}\text{CHCH}_2\text{I} \xleftarrow{\text{HI}} \text{CH}_2{=}\text{CHCH}_2\text{OH}$

(d) $\text{PhCHBr}_2 \xleftarrow{2\text{Br}_2, \text{ heat}} \text{PhCH}_3$

(e)

$$\text{CCl}_3\text{–C}_6\text{H}_3\text{–NO}_2 \xleftarrow{\text{HNO}_3, \text{ H}_2\text{SO}_4} \text{CCl}_3\text{–C}_6\text{H}_5 \xleftarrow{3\text{Cl}_2, \text{ heat}} \text{CH}_3\text{–C}_6\text{H}_5$$

(f)

$$\underset{\underset{\text{Cl}\quad\text{Cl}}{|\quad\;|}}{\text{PhCH–CH}_2} \xleftarrow{\text{Cl}_2} \text{PhCH}{=}\text{CH}_2 \xleftarrow[(-\text{H}_2)]{\text{cat., heat}} \text{PhCH}_2\text{CH}_3 \xleftarrow{\underset{\text{HF}}{\text{C}_2\text{H}_4}} \text{PhH}$$

(g)

$$\text{PhC}{\equiv}\text{CH} \xleftarrow{\text{NaNH}_2, \text{ NH}_3} \xleftarrow{\text{KOH(alc)}} \underset{\underset{\text{Cl}\quad\text{Cl}}{|\quad\;|}}{\text{PhCH–CH}_2} \xleftarrow{} \text{(f)}$$

(h)

$$\overset{\text{Ph}}{\triangle} \xleftarrow[\text{MeOH}]{\text{Na}} \overset{\text{Ph}}{\underset{\underset{\text{Cl}\;\;\text{Cl}}{}}{\triangle}} \xleftarrow[t\text{-BuOK}]{\text{CHCl}_3} \text{PhCH}{=}\text{CH}_2 \xleftarrow{} \text{as in (f)}$$

6. (a) 1-butanol   (b) 1-butene
   (c) no reaction   (d) *n*-butane
   (e) *n*-hexane   (f) *n*-butylmagnesium bromide
   (g) $\text{CH}_3\text{CH}_2\text{CH}_2\text{CH}_2\text{D}$, butane-1-*d*   (h) *n*-butane
   (i) no reaction   (j) 1-iodobutane
   (k) *sec*-butylbenzene (major) and *n*-butylbenzene
   (l) 2-heptyne   (m) 1-fluorobutane
   (n) no reaction

7. (a) *n*-BuNH$_2$ (b) $\underset{\underset{\text{Ph}}{|}}{n\text{-BuNH}}$ (c) *n*-BuCN

   (d) *n*-BuOEt   (e) $n\text{-BuO–}\underset{\underset{\text{O}}{\|}}{\text{C}}\text{–CH}_3$ (f) *n*-BuSCH$_3$

8. Competition is between substitution and elimination.

   (a) 1-butene   (b) 1-butene
   (c) 1-butanol and *n*-BuOMe (solvent is MeOH/H$_2$O)   (d) 1-butene

   With *tert*-BuBr, (a), (b), and (d) are more important, and (c) is less important.

9. Steric factors are controlling. The usual S_N2 order is $1° > 2° > 3°$, and slow for large G in $G-CH_2X$.

(a) $CH_3CH_2CH_2CH_2-\underset{\underset{\displaystyle 1°}{|}}{\underset{Br}{CH_2}} > CH_3CH_2CH_2-\underset{\underset{\displaystyle 2°}{|}}{\underset{Br}{CH}}-CH_3 > CH_3CH_2-\overset{\overset{\displaystyle CH_3}{|}}{\underset{\underset{\displaystyle 3°}{|}}{\underset{Br}{C}}}-CH_3$

(b) $\overset{\overset{\displaystyle CH_3}{|}}{CH_3CH}CH_2-\underset{\underset{\displaystyle 1°}{|}}{\underset{Br}{CH_2}} > \overset{\overset{\displaystyle CH_3}{|}}{CH_3CH}-\underset{\underset{\displaystyle 2°}{|}}{\underset{Br}{CH}}-CH_3 > CH_3-\overset{\overset{\displaystyle CH_3}{|}}{\underset{\underset{\displaystyle 3°}{|}}{\underset{Br}{C}}}-CH_2CH_3$

(c) $CH_3CH_2CH_2-CH_2Br > \overset{\overset{\displaystyle CH_3}{|}}{CH_3CH}CH_2-CH_2Br > CH_3CH_2\overset{\overset{\displaystyle CH_3}{|}}{CH}-CH_2Br > CH_3-\overset{\overset{\displaystyle CH_3}{|}}{\underset{\underset{\displaystyle CH_3}{|}}{C}}-CH_2Br$

$G = \qquad n\text{-Pr}- \qquad\qquad\qquad iso\text{-Bu}- \qquad\qquad\qquad sec\text{-Bu}- \qquad\qquad\qquad tert\text{-Bu}-$

10. Electronic factors are controlling: the more stable the cation being formed in the initial ionization, the faster the reaction. The usual order is

$$\text{benzylic} > 3° > 2° > 1°$$

(a) $CH_3CH_2-\overset{\overset{\displaystyle CH_3}{|}}{\underset{\underset{\displaystyle 3°}{|}}{\underset{Br}{C}}}-CH_3 > CH_3CH_2CH_2-\underset{\underset{\displaystyle 2°}{|}}{\underset{Br}{CH}}-CH_3 > CH_3CH_2CH_2CH_2-\underset{\underset{\displaystyle 1°}{|}}{\underset{Br}{CH_2}}$

(b) $CH_3-\overset{\overset{\displaystyle CH_3}{|}}{\underset{\underset{\displaystyle 3°}{|}}{\underset{Br}{C}}}-CH_2CH_3 > \overset{\overset{\displaystyle CH_3}{|}}{CH_3CH}-\underset{\underset{\displaystyle 2°}{|}}{\underset{Br}{CH}}-CH_3 > \overset{\overset{\displaystyle CH_3}{|}}{CH_3CH}CH_2-\underset{\underset{\displaystyle 1°}{|}}{\underset{Br}{CH_2}}$

(c) Electron-releasing substituents stabilize the incipient benzylic cation, and increase the rate; electron-withdrawing substituents destabilize the incipient benzylic cation, and decrease the rate.

MeO⟨○⟩CH_2Cl > Me⟨○⟩CH_2Cl > ⟨○⟩CH_2Cl > Cl⟨○⟩CH_2Cl > O_2N⟨○⟩CH_2Cl

(d) $Ph\underset{\underset{\displaystyle Br}{|}}{CH}CH_3 > PhCH_2Br > PhCH_2CH_2Br$

$\quad$ 2° benzylic $\qquad$ benzylic $\qquad\qquad$ 1°

11. Dehydrohalogenation follows Saytzeff orientation: stereochemistry permitting, the more stable alkene is formed faster.

(a) The more highly branched alkene is more stable: same order as in Problem 10(a).

(b) The more highly branched alkene is more stable: same order as in Problem 10(b).

(c) The ring-conjugated alkene is more stable.

$$PhCH_2\underset{\underset{Br}{|}}{C}HCH_3 > PhCH_2CH_2CH_2Br$$

(d) Stability of products:

benzene > conjugated diene > alkene.

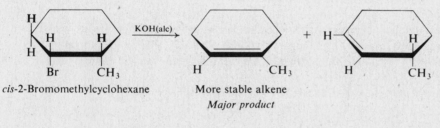

(e) *Anti*-periplanar elimination to the more stable branched alkene is possible from the *cis*-isomer, but not possible from the *trans*-isomer. (Compare reactions of neomenthyl and menthyl chlorides, pages 482–483.)

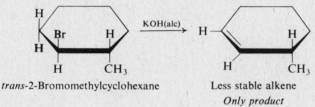

12.

	S$_N$2	S$_N$1
(a) Stereochemistry	Inversion	Racemization
(b) Kinetic order	2nd order	1st order
(c) Rearrangements	No	Yes
(d) Relative rates	Me > Et > *i*-Pr > *t*-Bu	*t*-Bu > *i*-Pr > Et > Me
(e) Relative rates	RI > RBr > RCl	RI > RBr > RCl
(f) Temperature inc.	Faster	Faster
(g) Doubling [RX]	Rate doubled	Rate doubled
(h) Doubling [OH⁻]	Rate doubled	Rate unaffected
(i) Increase H$_2$O	Little effect	Faster in more polar solvent
(j) Increase EtOH	Little effect	Slower

13. When *one* molecule of halide suffers exchange and inversion, the optical activity of *two* molecules is lost, since the optical rotation of the inverted molecule cancels the optical rotation of one unreacted molecule.

14. The protonated alcohol ionizes to a planar carbonium ion, hydration of which can occur in two ways to give racemic alcohol; or, the protonated alcohol suffers S_N2 attack by water, with inversion and loss of optical activity (see answer to Problem 13). (Actually, studies with $H_2{}^{18}O$ gave results similar to those in Problem 13.)

15. In the transition state for *anti* elimination from menthyl chloride, for —Cl to be in the required axial position, the bulky methyl and isopropyl groups must also be axial; in either transition state for *anti* elimination from neomenthyl chloride, the alkyl groups are equatorial.

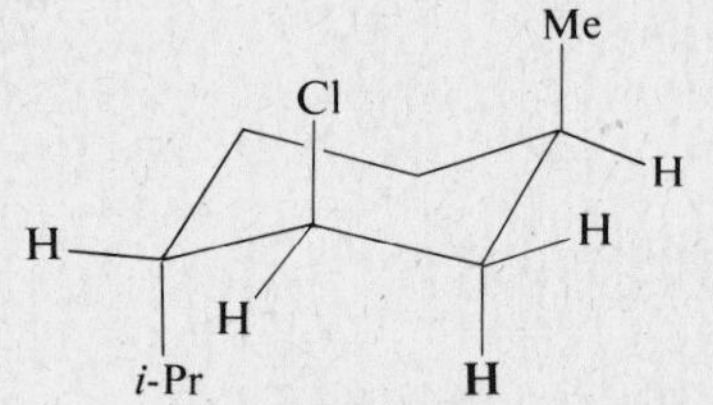

Menthyl chloride 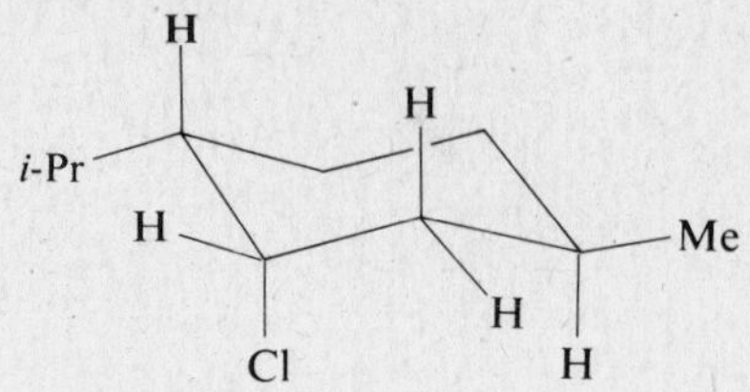   Neomenthyl chloride

16. (a) A and B are diastereomeric forms of $PhCHClCHClCH_3$. Their configurations were assigned by converting each by dehydrohalogenation into $PhC(Cl)\!=\!CHCH_3$, and determining the configuration of the alkene thus formed. *On the assumption that* this elimination is *anti*, then:

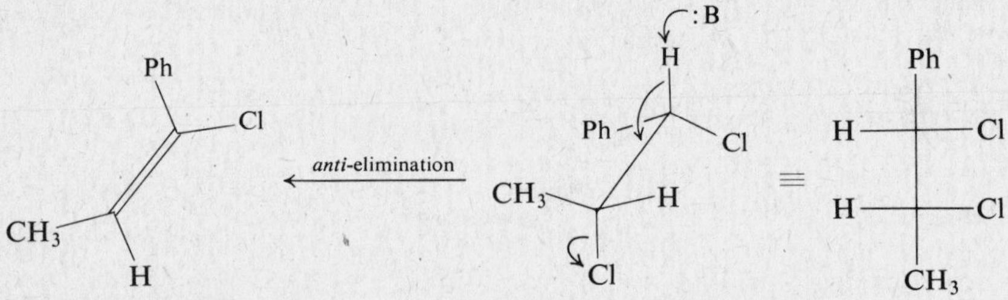

anti-elimination

(E)-1-Chloro-1-phenylpropene

and enantiomer

Erythro

A

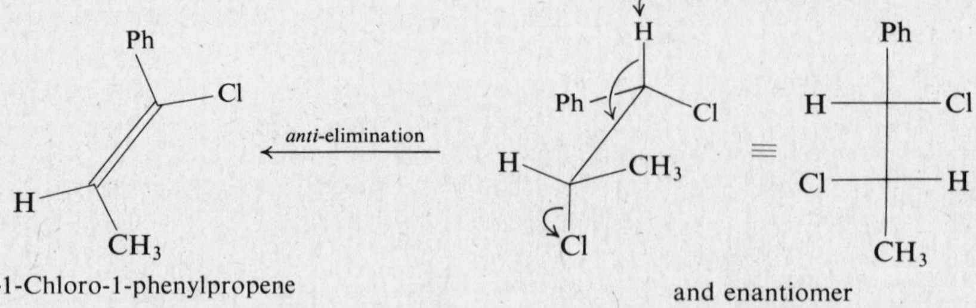

anti-elimination

(Z)-1-Chloro-1-phenylpropene

and enantiomer

Threo

B

(b) Once the configurations of A and B had been determined, the point of the study was to determine the stereochemistry of addition of chlorine to 1-phenylpropene, $PhCH\!=\!CHCH_3$. If addition were stereospecific, *anti*, taking place via bridged chloronium ions (Sec. 7.12), the *trans*-alkene would have given *only* A,

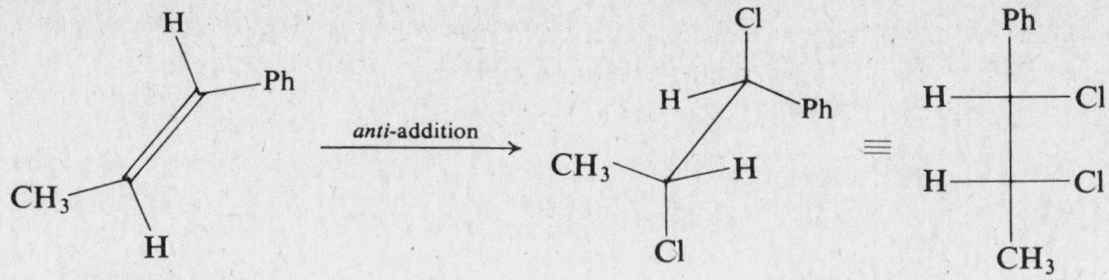

trans-1-Phenylpropene

anti-addition

and enantiomer

Erythro

A

and the *cis*-alkene would have given only B.

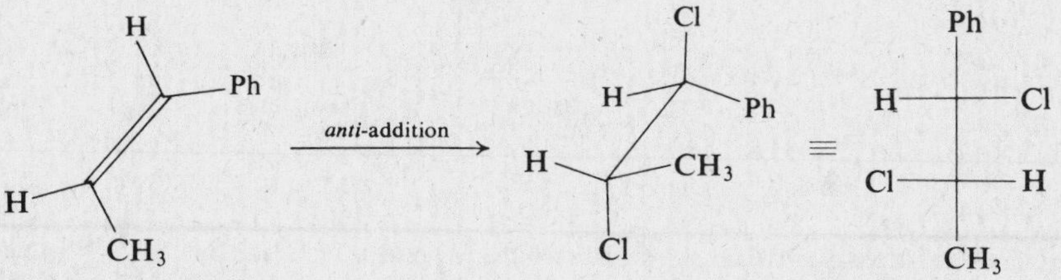

cis-1-Phenylpropene

anti-addition

and enantiomer

Threo

B

Since each geometric isomer actually gave *both* A and B, addition of chlorine to this alkene is *not* stereospecific. It was concluded that reaction here involves, not bridged chloronium ions, but open, resonance-stabilized benzylic cations. (See the answer to Problem 21, Chapter 12, which deals with addition of bromine to the same alkenes.)

(R. C. Fahey and C. Schubert, J. Am. Chem. Soc., **87**, 5172 (1965).)

17. The *cis*-isomer reacts via E2 *anti* elimination, with —H and —OTs in axial positions. In the *trans*-isomer, however, the leaving groups can be axial only if the bulky *tert*-butyl group is also put in an axial position; this is so difficult that the molecule reacts via a carbonium ion (E1) mechanism instead.

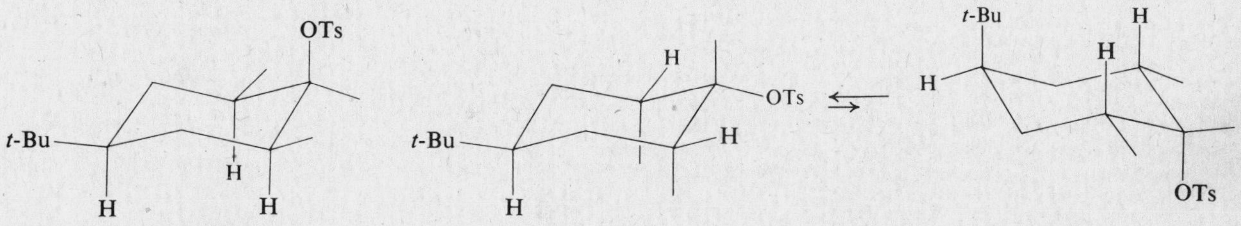

cis-Isomer

t-Bu equatorial
—OTs and —H axial

Anti E2 *not possible*

t-Bu equatorial
—OTs equatorial

Anti E2 *not possible*

trans-Isomer

t-Bu axial
Highly unstable

18. (a) (i) *Anti*-elimination, with I⁻ as base.

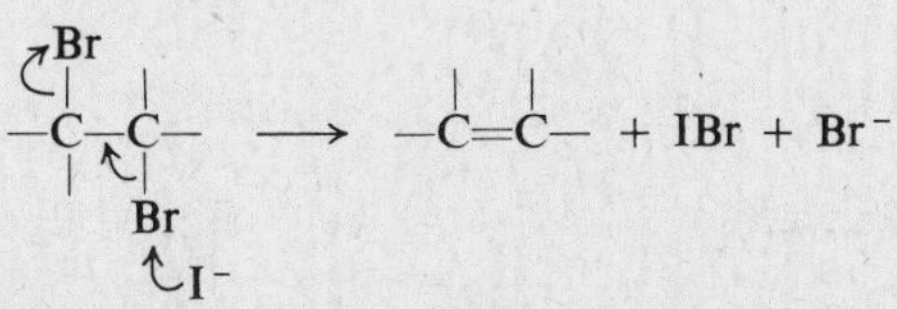

$$-C-C- \longrightarrow -C=C- + IBr + Br^-$$

155

(ii) Intermolecular nucleophilic attack, with inversion; followed by intramolecular nucleophilic attack, also with inversion; and finally elimination, which is necessarily *syn*. Net result: over-all *syn*-elimination.

(b) Only net *syn*-elimination (mechanism ii) can give the *cis*-CHD=CHD actually obtained from *meso*-CHDBr—CHDBr.

meso   *cis*

Only net *anti*-elimination (mechanism i) can give the observed results for the dibromobutanes.

meso   *trans*

and enantiomer
Racemic   *cis*

Both nucleophilic attacks of mechanism (ii) are more difficult at the 2° carbons of the dibromobutanes than at the 1° carbon of the dibromoethane. At the same time, the methyl groups of the dibromobutanes help to stabilize the incipient double bond of mechanism (i). The result is a shift in mechanism from (ii) for the dibromoethane to (i) for the dibromobutanes.

(c) Elimination (like addition) is *anti*, and removes the same (radioactive) Br's that were added.

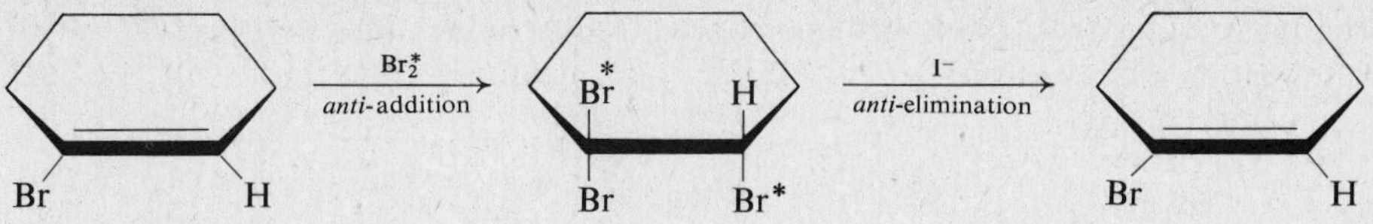

19. (a) The racemic dibromide undergoes the usual *anti* dehydrohalogenation, with pyridine attacking H.

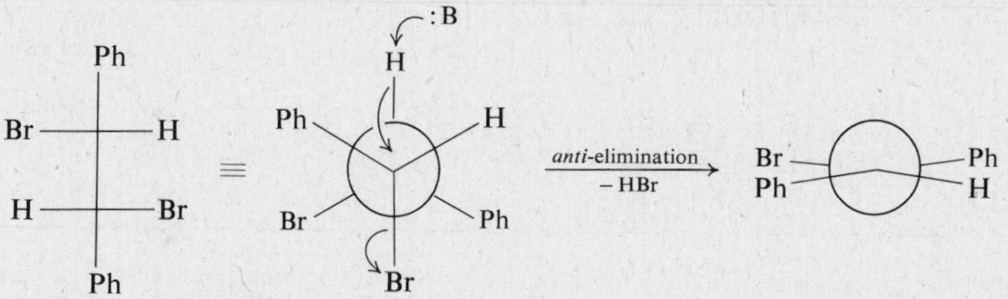

and enantiomer
Racemic

trans-1-Bromo-
1,2-diphenylethene

The *meso* dibromide undergoes *anti* elimination by mechanism (i) of preceding problem, · with pyridine (instead of iodide ion) attacking Br.

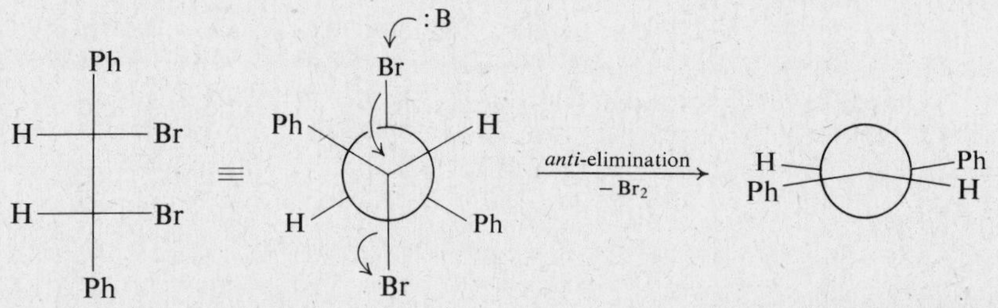

Meso

trans-1,2-Diphenylethene

(b) Each reaction proceeds via the transition state in which the leaving groups are *anti*-periplanar and the bulky phenyl groups are as far apart as possible.

20. The fact that the hydrocarbon does not decolorize permanganate, yet has the formula C_5H_{10}, shows that it contains one ring. Solubility in concentrated H_2SO_4 suggests that it is a cyclopropane; this is confirmed by the far upfield nmr peak at δ 0.20, characteristic of cyclopropane ring protons. The molecular formula requires it to be ethylcyclopropane or a dimethylcyclopropane. Of these, only one is consistent with the nmr spectrum: 1,1-dimethylcyclopropane. (See the answer to Problem 13.5.)

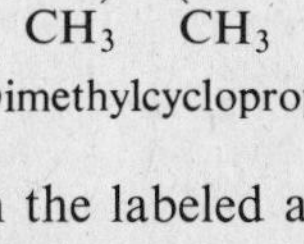

1,1-Dimethylcyclopropane

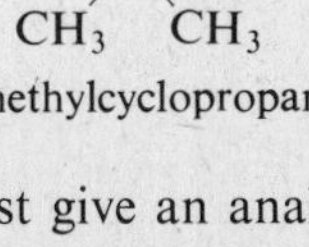

1,1-Dimethylcyclopropane-2-*d*

The reaction with the labeled alkyl halide must give an analogous hydrocarbon. The molecular weight of 71 shows that it contains *only one* deuterium atom per molecule, and is C_5H_9D.

These hydrocarbons are evidently formed through intramolecular insertion of a methylene generated by α-elimination (Sec. 9.16).

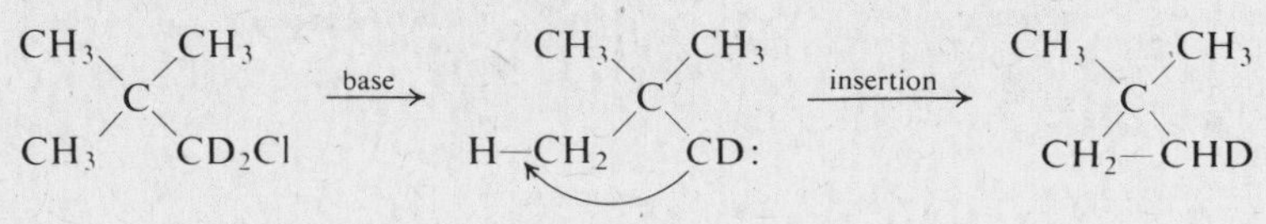

$$\underset{CH_3}{\overset{CH_3}{>}}C\underset{CD_2Cl}{\overset{CH_3}{<}} \xrightarrow{\text{base}} \underset{H-CH_2}{\overset{CH_3\quad CH_3}{>C<}}CD: \xrightarrow{\text{insertion}} \underset{CH_2-CHD}{\overset{CH_3\quad CH_3}{>C<}}$$

The labeling experiment rules out an alternative mechanism (involving "γ-elimination") proposed in 1942 by Frank Whitmore—before methylenes were conceived of, even by D. Duck. This mechanism would require *two* deuteriums per molecule of product, contrary to fact:

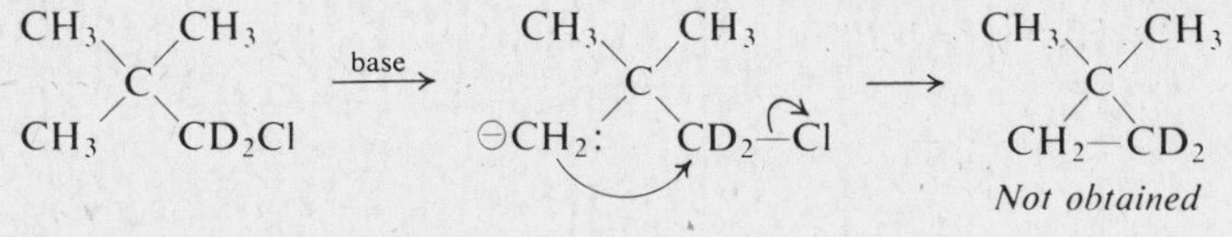

$$\underset{CH_3}{\overset{CH_3}{>}}C\underset{CD_2Cl}{\overset{CH_3}{<}} \xrightarrow{\text{base}} \underset{\ominus CH_2:}{\overset{CH_3\quad CH_3}{>C<}}CD_2-Cl \longrightarrow \underset{CH_2-CD_2}{\overset{CH_3\quad CH_3}{>C<}}$$
Not obtained

(L. Friedman and J. Berger, J. Am. Chem. Soc., **83**, 500 (1961).)

21. (a) *t*-BuF: doublet at δ 1.30 is the —CH$_3$ signal split by —F ($J = 20$).
   i-PrF: two doublets centered on δ 1.23 are the —CH$_3$ signal split by —H ($J = 4$) and split again by —F ($J = 23$);
   two multiplets centered on δ 4.64 are the —H signal split by —CH$_3$ ($J = 4$) and split again by —F ($J = 48$).

   (b) *t*-BuF + SbF$_5$ $\longrightarrow$ *t*-Bu$^+$SbF$_6^-$

   i-PrF + SbF$_5$ $\longrightarrow$ *i*-Pr$^+$SbF$_6^-$

Removed from the molecules, —F no longer splits the proton signals. At the same time, deshielding by the positive charge of the cation causes a very strong downfield shift. (See the spectrum in Figure 5.7, page 161. This was the first *direct observation* of simple alkyl cations, as discussed in Sec. 5.15.)

22. (a) 1-Methylcyclopropene, formed by intramolecular insertion of a methylene produced by α-elimination.

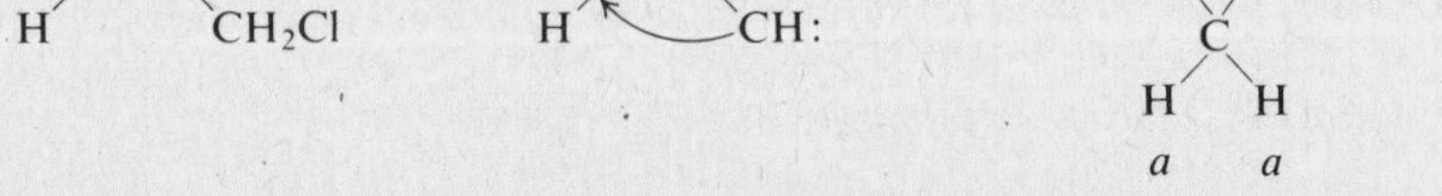

$$\underset{H}{\overset{H}{>}}C=C\underset{CH_2Cl}{\overset{CH_3}{<}} \xrightarrow{\text{base}} \underset{H}{\overset{H}{>}}C=C\underset{CH:}{\overset{CH_3}{<}} \xrightarrow{\text{insertion}} \underset{\underset{a\quad a}{H\quad H}}{\overset{\overset{c\qquad b}{H\quad CH_3}}{C=C}}C$$
1-Methylcyclopropene

(F. Fisher and D. E. Applequist, J. Org. Chem., **30**, 2089 (1965).)

(b) We would expect an analogous reaction to convert allyl chloride into cyclopropene.

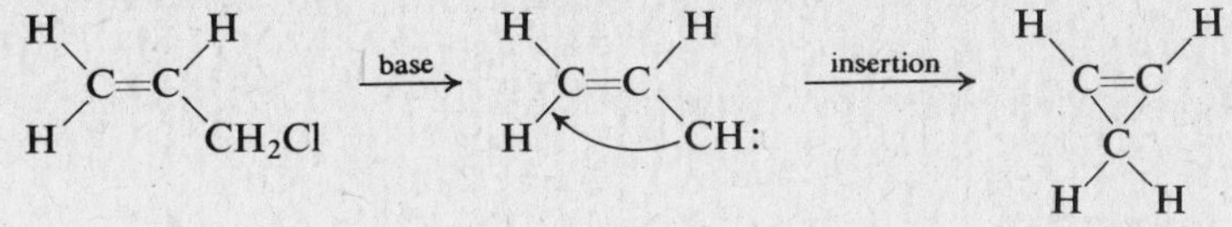

$$H_2C=CH-CH_2Cl \xrightarrow{base} H_2C=CH-CH: \xrightarrow{insertion} \text{Cyclopropene}$$

(A low yield of cyclopropene has actually been obtained in this manner by G. L. Closs and K. D. Krantz, J. Org. Chem., **31**, 638 (1966).)

23. (a) Hydrocarbon C is C_4H_6, and has the structure:

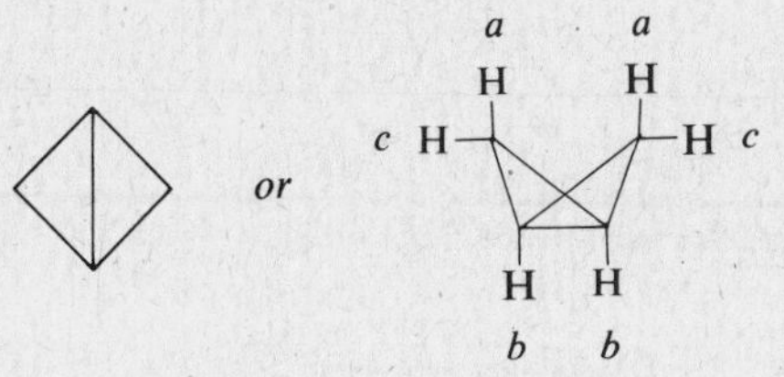

Bicyclo[1.1.0]butane

We can see the three sets of protons; nmr signals have been assigned as shown.

(b) Synthesis (i) is an intramolecular Wurtz reaction (page 93), and involves nucleophilic displacement of halide by carbanionoid carbon:

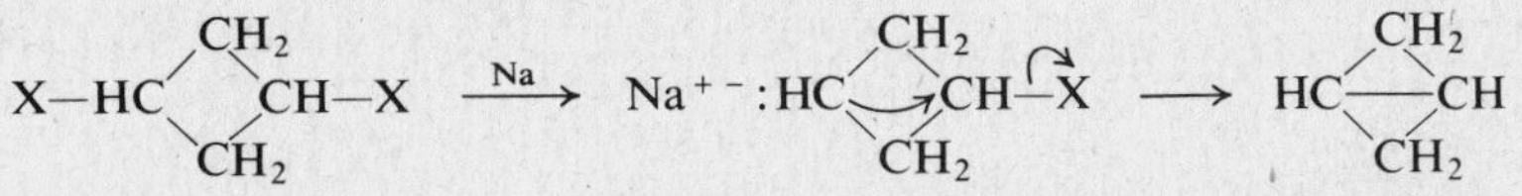

Bicyclobutane

Synthesis (ii) involves intramolecular addition of a methylene:

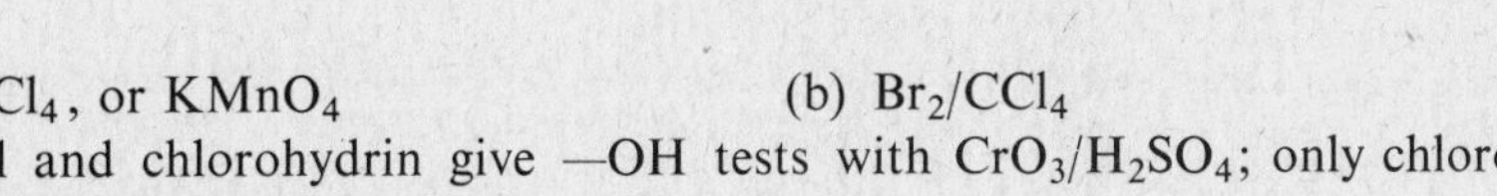

Bicyclobutane

We have now seen, in Problems 20 and 21 and the present problem, *intramolecular* (within-a-molecule) examples of the two principal reactions of methylenes: insertion and addition.

24. (a) Br_2/CCl_4, or $KMnO_4$ (b) Br_2/CCl_4
(c) Glycol and chlorohydrin give —OH tests with CrO_3/H_2SO_4; only chlorohydrin gives test for Cl
(d) Alkene detected by Br_2/CCl_4; alcohol detected by CrO_3/H_2SO_4
(e) *t*-BuCl gives test for Cl; alkene detected by Br_2/CCl_4
(f) $AgNO_3$ test positive for $PhCH_2Cl$, negative for aryl halide

25. (a) CH_3I, CH_2Cl_2, and possibly C_2H_5Br.   (b) CH_3I (violet color due to I_2).
(c) $Br^- \longrightarrow Br_2$, which is red-brown in CCl_4; Cl^- is unaffected.

26. (a) Bromo and chloro compounds by elemental analysis.

   n-Decane and *p*-cymene by insolubility in cold conc. H_2SO_4, but only *p*-cymene is soluble in fuming sulfuric acid.

   Limonene gives Br_2/CCl_4 test for unsaturation; the alcohol does not.

   (b) Chloro and iodo compounds by elemental analysis.

   Only propenylbenzene gives Br_2/CCl_4 test for unsaturation.

   Only 2-octanol gives iodoform when treated with I_2/OH^-.

   (c) *n*-Octyl chloride by elemental analysis.

   Only the alcohol is oxidized by CrO_3/H_2SO_4.

   Of the hydrocarbons, *trans*-decalin is insoluble in fuming sulfuric acid.

   Oxidation to acids and determination of m.p.'s of these identifies the arenes.

Chapter 15 | Alcohols I. Preparation and Physical Properties

15.1 Intramolecular hydrogen bond in *cis*-isomer (see Sec. 24.2).

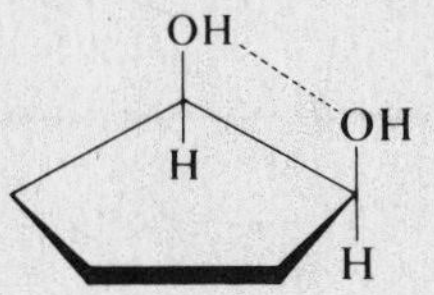

15.2 Carbon which (a) carries three —Cl's or (b) is *sp*-hybridized (Sec. 8.10) is, in effect, a more electronegative element.

15.3 Many —OH groups capable of hydrogen bonding with water molecules (see page 1119).

15.4 See answer to Problem 8, Chapter 1.

15.5 (a)

$$(CH_3)_2CHCH_2\overset{*}{C}H(NH_3{}^+)COO^- \longrightarrow (CH_3)_2CHCH_2CH_2OH$$

Leucine   Isopentyl alcohol
Chiral   *Achiral*

$$CH_3CH_2\overset{*}{C}H(CH_3)\overset{*}{C}H(NH_3{}^+)COO^- \longrightarrow CH_3CH_2\overset{*}{C}H(CH_3)CH_2OH$$

Isoleucine   Active amyl alcohol
Chiral   *Chiral*

(b) Chirality of one of the two original chiral carbons of isoleucine is retained.

15.6 The first material to distill is the ternary azeotrope; 100 g of it will distill, carrying over 7.5 g of water, 18.5 g of alcohol, and 74 g of benzene. This leaves 124 g of pure anhydrous alcohol. In actual practice, a slight excess of benzene is added; this is removed, after the distillation of the ternary mixture, as a binary azeotrope with alcohol (b.p. 68.3°).

15.7 If we take the raw data, and do a bit of "chemical arithmetic," we get:

$C_3H_6 \longrightarrow C_3H_5Cl$ Replace —H by —Cl
            A

$C_3H_5Cl \longrightarrow C_3H_6O$ Replace Cl by H and O (obviously —OH)
   A   B

$C_3H_6O \longrightarrow C_3H_7O_2Cl$ Add H, O, Cl (obviously HOCl, or —OH and —Cl)
   B   C

$C_3H_7O_2Cl \longrightarrow C_3H_8O_3$ Replace Cl by H and O (obviously —OH)
   C   D

Now, we apply our tentative conclusions to structural formulas, recalling that allylic substitution occurs at high temperatures (Sec. 6.21):

$$CH_3CH{=}CH_2 \xrightarrow[600°]{Cl_2} \underset{\underset{Cl}{|}}{CH_2}{-}CH{=}CH_2 \xrightarrow[OH^-]{H_2O} \underset{\underset{OH}{|}}{CH_2}{-}CH{=}CH_2$$

                                 A   B
                       Allyl chloride   Allyl alcohol

$$\underset{\underset{OH}{|}}{CH_2}{-}CH{=}CH_2 \xrightarrow[H_2O]{Cl_2} \underset{\underset{OH}{|}}{CH_2}{-}\underset{\underset{Cl}{|}}{CH}{-}\underset{\underset{OH}{|}}{CH_2} \xrightarrow[OH^-]{H_2O} \underset{\underset{OH}{|}}{CH_2}{-}\underset{\underset{OH}{|}}{CH}{-}\underset{\underset{OH}{|}}{CH_2}$$

          B   C   D
                                               Glycerol

15.8 Attack on benzylic carbon can take advantage of benzylic cation character in the transition state.

$$Ph{-}CH{=}CH_2 \xrightarrow{Hg^{++}} \underset{\underset{Hg_{++}}{|}}{Ph{-}CH{-}CH_2} \xrightarrow{CH_3OH} \left[\underset{\underset{Hg_{++}}{|}}{\overset{\overset{HOCH_3}{|}}{Ph{-}CH{-}CH_2}} \right] \xrightarrow{-H^+} \overset{\overset{OCH_3}{|}}{Ph{-}CH{-}CH_2Hg^+}$$

$$\downarrow NaBH_4$$

$$\overset{\overset{OCH_3}{|}}{Ph{-}CH{-}CH_3}$$

(H. C. Brown and P. Geoghegan, Jr., "The Oxymercuration-Demercuration of Representative Olefins. A Convenient, Mild Procedure for the Markovnikov Hydration of the Carbon–Carbon Double Bond," J. Am. Chem. Soc., **89**, 1522 (1967).)

(G. A. Olah and P. R. Clifford, "Organometallic Chemistry. I. The Ethylene- and Norbornylenemercurinium Ions," J. Am. Chem. Soc., **93**, 1261 (1971); "Organometallic Chemistry. II. Direct Mercuration of Olefins to Stable Mercurinium Ions," J. Am. Chem. Soc., **93**, 2320 (1971).)

15.9 *syn*-Hydration, with *anti*-Markovnikov orientation in each case.

(a)

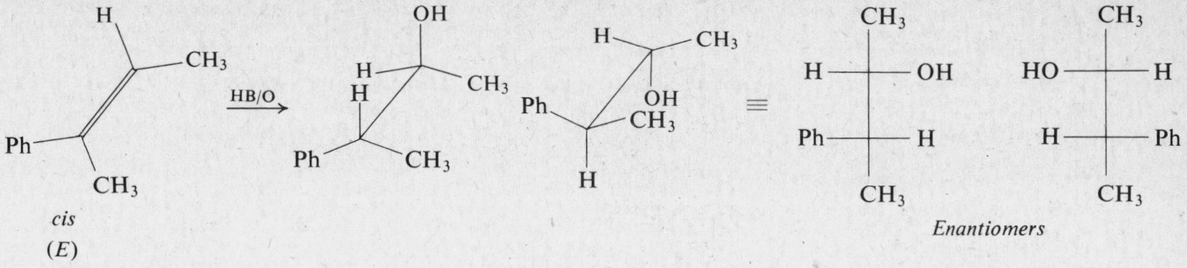

$\xrightarrow{\text{HB/O}}$

Ph   CH$_3$ cis (E)

OH   CH$_3$ Ph   CH$_3$ H   H   CH$_3$ Ph   OH   CH$_3$ H

$\equiv$

CH$_3$ H——OH   Ph——H   CH$_3$

CH$_3$ HO——H   H——Ph   CH$_3$

Enantiomers

(b)

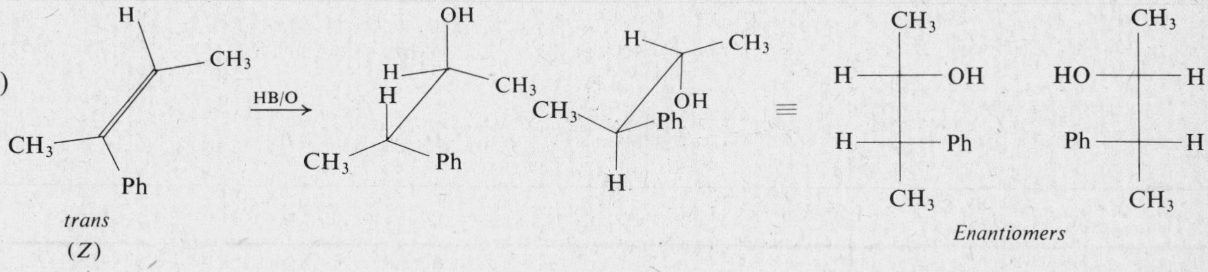

$\xrightarrow{\text{HB/O}}$

CH$_3$ Ph   *trans* (Z)

OH   CH$_3$ CH$_3$ Ph   H   H   CH$_3$ CH$_3$ Ph   OH   H

$\equiv$

CH$_3$ H——OH   H——Ph   CH$_3$

CH$_3$ HO——H   Ph——H   CH$_3$

Enantiomers

(c)

$\xrightarrow{\text{HB/O}}$

CH$_3$ H

CH$_3$ H   H   OH   CH$_3$ H   HO   CH$_3$ H

Enantiomers

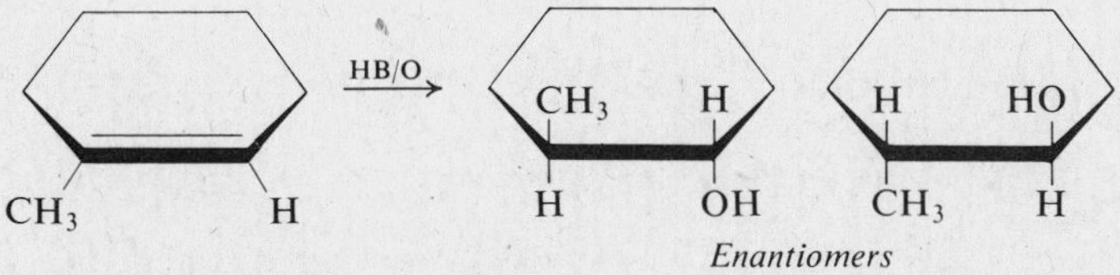

(H. C. Brown, *Hydroboration*, W. A. Benjamin, New York, 1962, especially Chapters 6–8; H. C. Brown, *Boranes in Organic Chemistry*, Cornell University Press, Ithaca, N. Y., 1972, pp. 255–280. This latter book contains a fascinating personal account not only of the discovery of organoboranes and their chemistry, but of the entire career—to date— of one of the most productive of all organic chemists.)

15.10 *syn*-Addition, with retention; or *anti*-addition, with inversion.

15.11

	(a)	(b)	(c)	(d)
Acid:	H_2O	Et_3B	$(BH_3)_2$	$(BH_3)_2$
Base:	H^-	NH_3	Me_3N	H^-

15.12 *syn*-Addition, with retention. (As we shall see in Problem 11, page 921, the mechanism of oxidation is the kind we would expect to give retention.)

1. (a)–(c)   CH$_3$CH$_2$CH$_2$CH$_2$CH$_2$OH   CH$_3$CH$_2$CH$_2$CHOHCH$_3$ CH$_3$CH$_2$CHOHCH$_2$CH$_3$

1-Pentanol   2-Pentanol   3-Pentanol

n-Butylcarbinol   Methyl-*n*-propylcarbinol   Diethylcarbinol

1°   2°   2°

$CH_3CH_2CH(CH_3)CH_2OH$ $CH_3CH_2CH(OH)(CH_3)_2$ $CH_3CHOHCH(CH_3)_2$

2-Methyl-1-butanol   2-Methyl-2-butanol   3-Methyl-2-butanol

sec-Butylcarbinol   Dimethylethylcarbinol   Methylisopropylcarbinol

1°   3°   2°

$(CH_3)_2CHCH_2CH_2OH$ $(CH_3)_3CCH_2OH$

3-Methyl-1-butanol   2,2-Dimethylpropanol

Isobutylcarbinol   *tert*-Butylcarbinol

1°   1°

(d) $(CH_3)_2CHCH_2CH_2OH$ $CH_3CH_2CH_2CH_2CH_2OH$ $CH_3CH_2C(OH)(CH_3)_2$

      Isopentyl alcohol   *n*-Pentyl alcohol   *tert*-Pentyl alcohol

(e) Examples are:

     n-$C_5H_{11}CH_2OH$ *n*-$C_4H_9CHOHCH_3$ *n*-$C_3H_7C(OH)(CH_3)_2$

         1°   2°   3°

(f) Examples are:

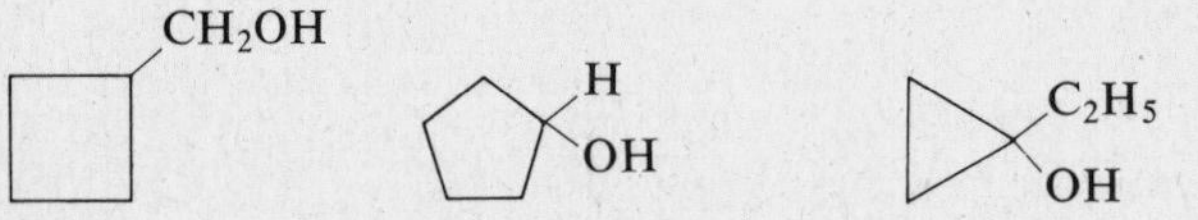

1°   2°   3°

2. d (highest b.p.), e, a, c, b.

3. (a) alcohols (Chapters 15 and 16)

       acids (Chapter 18)

       amides (in Chapter 20)

       amines (Chapters 22 and 23)

       phenols (Chapter 24)

       carbohydrates (Chapters 34 and 35)

       heterocyclic compounds with —N—H (Chapter 31)

       amino acids (Chapter 36)

       proteins (Chapter 36)

 (b) All, except hydrocarbons and halides.

4. (a) *p*-cresol   (b) propionic acid   (c) propionic acid

5. (a)–(c)

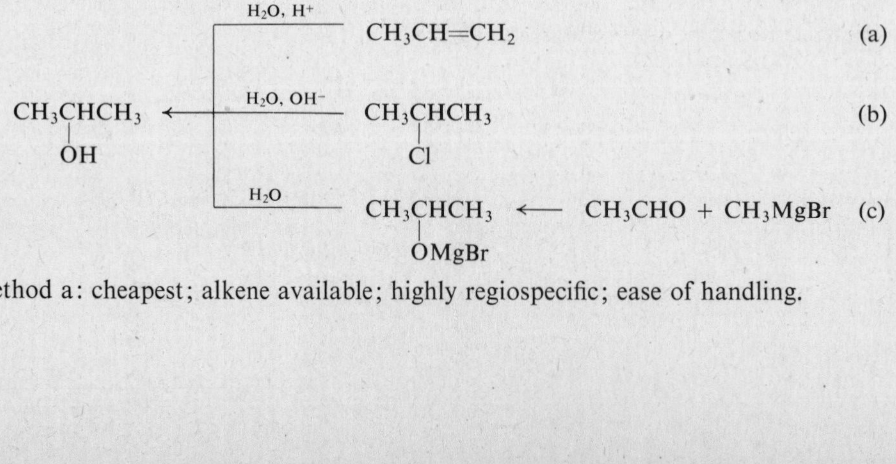

(d) Method a: cheapest; alkene available; highly regiospecific; ease of handling.

6. Hydrolysis of the intermediate Mg salt is not shown, but is part of each synthesis.

(a) $n\text{-Bu}\!\mid\!CH_2OH$ $\longleftarrow$ $n\text{-BuMgBr} + HCHO$

(b) $n\text{-Pr}\!\overset{a}{\mid}\!CHOH\!\overset{}{\mid}\!CH_3$ $\longleftarrow$
  $\overset{a}{}$ $n\text{-PrMgBr} + OHCCH_3$
  $\underset{b}{}$ $n\text{-PrCHO} + CH_3MgBr$

(c) $EtCHOH\!\mid\!Et$ $\longleftarrow$ $EtCHO + EtMgBr$

(d) $sec\text{-Bu}\!\mid\!CH_2OH$ $\longleftarrow$ $sec\text{-BuMgBr} + HCHO$

(e) $CH_3CH_2\!\mid\!\overset{a\ OH}{\underset{CH_3\,^b}{C}}\!\!\cdot\!CH_3$ $\longleftarrow$
  $\overset{a}{}$ $CH_3CH_2MgBr + O{=}C(CH_3)_2$
  $\underset{b}{}$ $CH_3CH_2\underset{CH_3}{C}{=}O + CH_3MgBr$

(f) $i\text{-Pr}\!\overset{a}{\mid}\!CHOH\!\overset{}{\mid}\!CH_3$ $\longleftarrow$
  $\overset{a}{}$ $i\text{-PrMgBr} + OHC{-}CH_3$
  $\underset{b}{}$ $i\text{-Pr}{-}CHO + CH_3MgBr$

(g) $iso\text{-Bu}\!\mid\!CH_2OH$ $\longleftarrow$ $iso\text{-BuMgBr} + HCHO$

(h) $t\text{-Bu}\!\mid\!CH_2OH$ $\longleftarrow$ $t\text{-BuMgBr} + HCHO$

(i) $Ph\!\mid\!CHOH\!\mid\!CH_2CH_3$ $\longleftarrow$
  $\overset{a}{}$ $PhMgBr + OHCCH_2CH_3$
  $\underset{b}{}$ $PhCHO + CH_3CH_2MgBr$

(j) $Ph\!\mid\!\overset{a\ OH}{\underset{CH_3}{C}}\!\!\cdot\!CH_3$ $\longleftarrow$
  $\overset{a}{}$ $PhMgBr + O{=}C(CH_3)_2$
  $\underset{b}{}$ $Ph\underset{CH_3}{C}{=}O + CH_3MgBr$

(k) $PhCH_2CHOH\!\mid\!CH_3$ $\longleftarrow$ $PhCH_2CHO + CH_3MgBr$

  (PhCH$_2$MgX reacts abnormally with CH$_3$CHO and many
  other aldehydes to give *ortho* substitution.)

(l) $PhCH_2CH_2\!\mid\!CH_2OH$ $\longleftarrow$ $PhCH_2CH_2MgBr + HCHO$

(m)

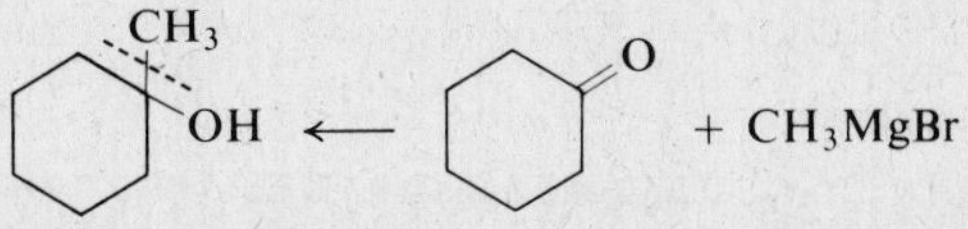

$\longleftarrow$ $+ CH_3MgBr$

(n)

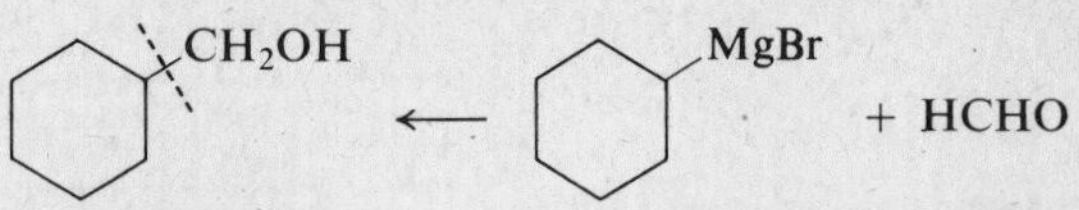

(o)

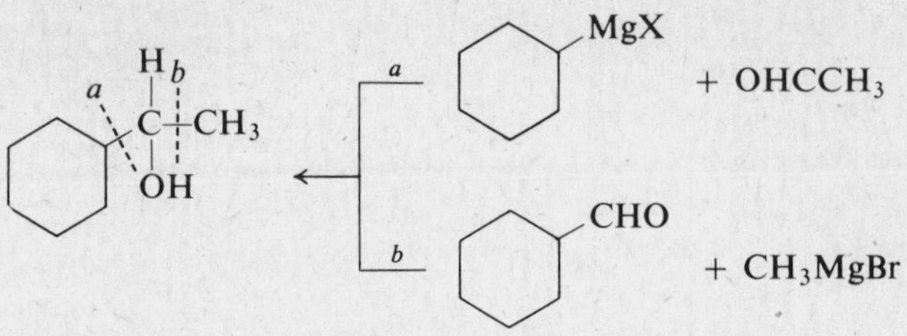

(p) $i\text{-Pr}\text{---}CHOH\text{---}Pr\text{-}i$ ⟵ $i\text{-PrCHO} + i\text{-PrMgBr}$

(q) $p\text{-}CH_3C_6H_4\text{---}CHOH\text{---}CH_3$ ⟵

a $p\text{-}CH_3C_6H_4MgBr + OHCCH_3$

b $p\text{-}CH_3C_6H_4CHO + CH_3MgBr$

(r)

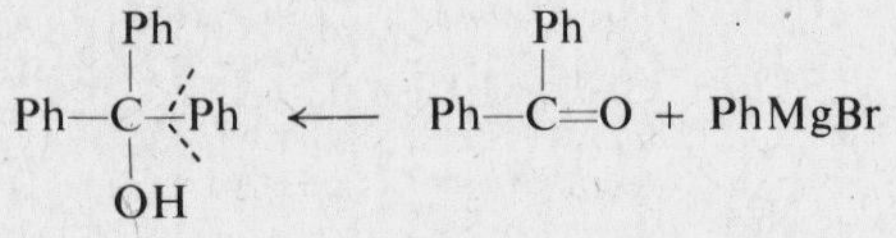

$$Ph\text{---}\underset{OH}{\overset{Ph}{\underset{|}{\overset{|}{C}}}}\text{---}Ph \quad \longleftarrow \quad Ph\text{---}\overset{Ph}{\overset{|}{C}}\text{=}O + PhMgBr$$

7. Intramolecular H-bond between —OH and —G stabilizes *gauche* conformation.

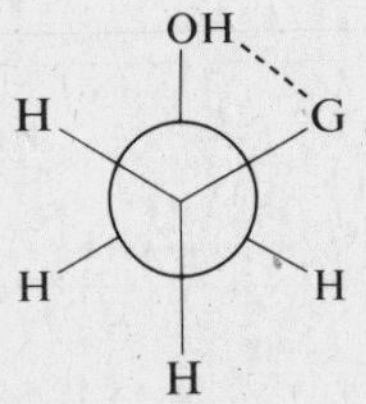

8. (*a*) *syn*-Hydration can take place in either of two ways: from "beneath" to yield cholestane-$3\beta,6\alpha$-diol or from "above" to yield the stereoisomeric coprostane-$3\beta,6\beta$-diol. Attack from "beneath" is less hindered than from "above," because of the substituents projecting "upward," particularly —CH_3 at C–10.

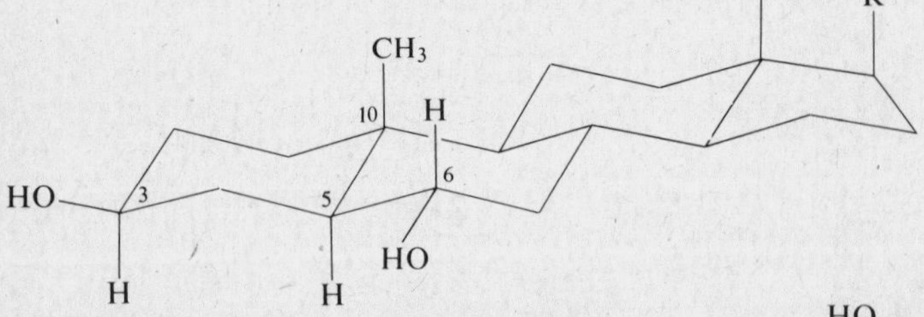

Cholestane-$3\beta,6\alpha$-diol
Greatly predominates

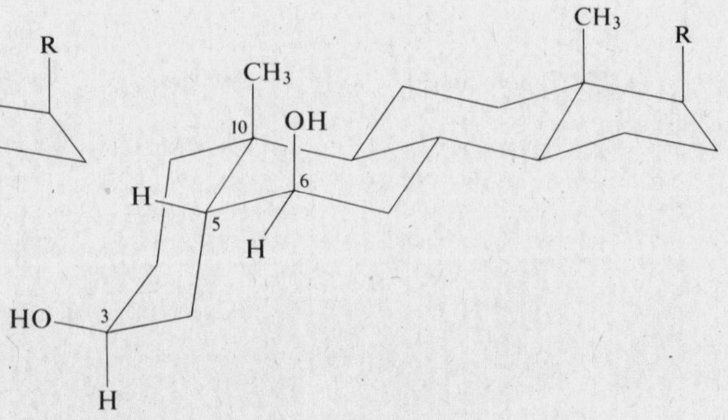

Coprostane-$3\beta,6\beta$-diol

(b) *syn*-Hydration from beneath gives α-OH at C–11, and α-H at C–9.

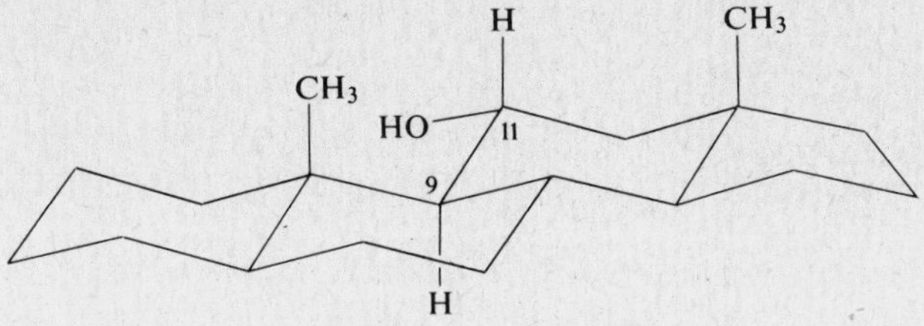

H   CH₃

CH₃

HO   11

9

H

9. (a)   HO   OH   OH

OH

H   H   H

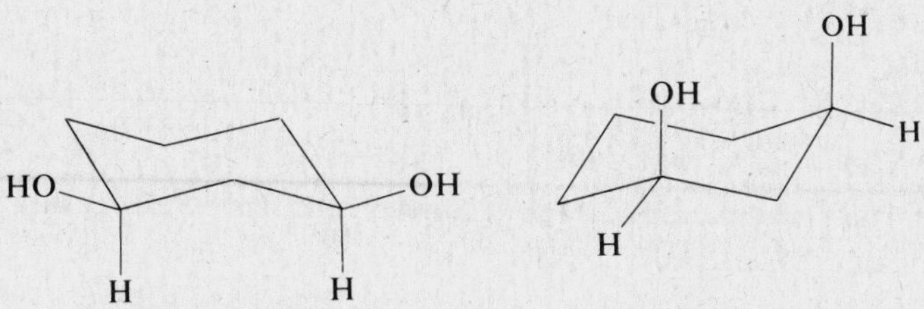

Both OH's equatorial   Both OH's axial

(b) On the basis of 1,3-interaction, one would predict the diequatorial conformation to be more stable.

(c) Intramolecular hydrogen bonding, which is a stabilizing factor, is possible only when both OH's are axial.

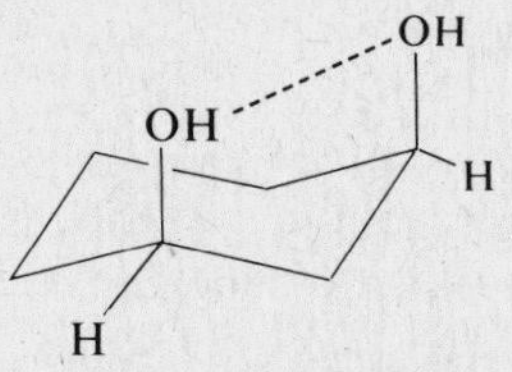

OH

OH

H

H

Diaxial OH's stabilized by
intramolecular H-bond

10. Twist-boat conformation: both *tert*-Bu groups are "equatorial," and both —OH groups are "flagpole" with intramolecular H-bonding.

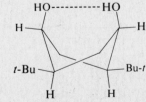

HO----------HO

H   H

t-Bu   Bu-*t*

H   H

2,5-Di-*tert*-butyl-1,4-cyclohexanediol
Twist-boat conformation

167

11. (a)

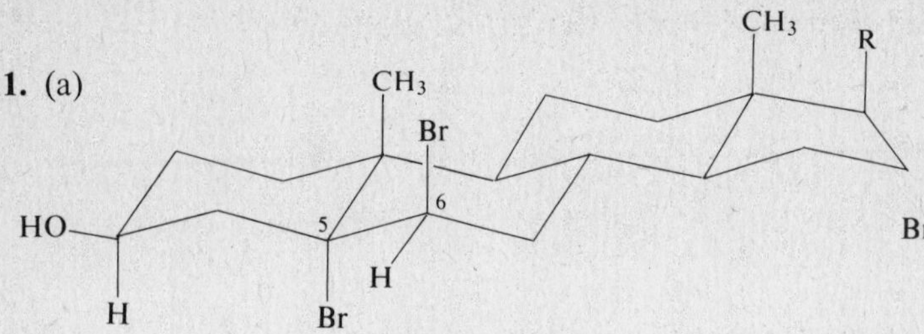

5α,6β-dibromo compound
Greatly predominates

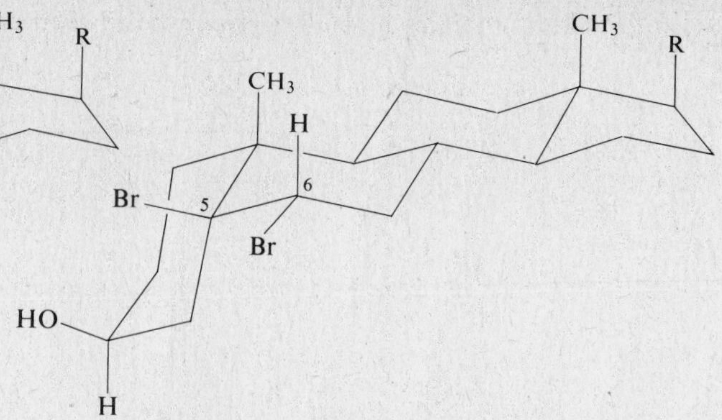

5β,6α-dibromo compound

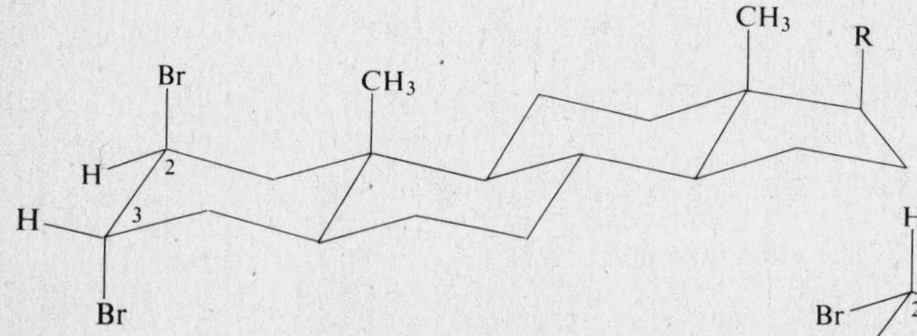

2β,3α-dibromo compound
Greatly predominates

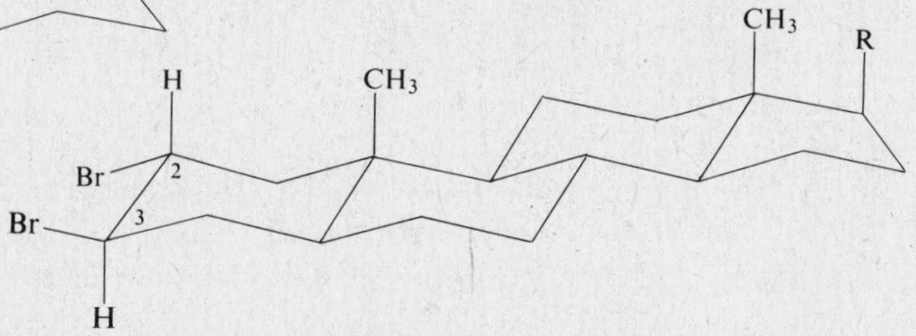

2α,3β-dibromo compound

(b) There is preferred formation of the bromonium ion by the less hindered attack from beneath. The bromonium ion is then opened via an *anti* transition state to yield a diaxial dibromide.

12. Two stereochemical combinations could give the observed results: (i) *anti*-addition of bromine, followed by *anti*-elimination; (ii) *syn*-addition followed by *syn*-elimination. Since (stereo-specific) addition of bromine is known to be *anti* (Sec. 7.11), elimination must be *anti*, too.

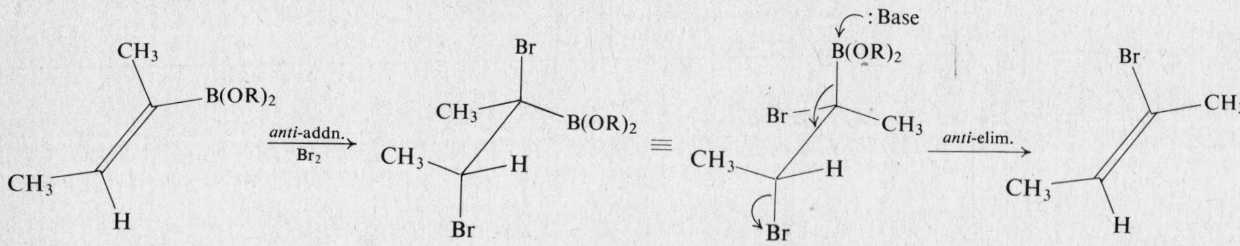

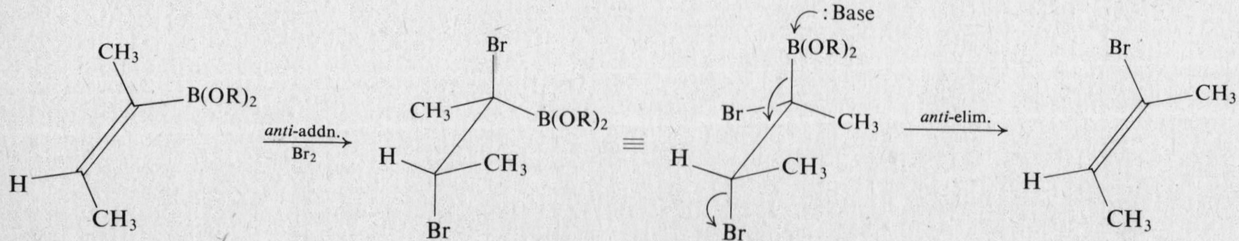

Water or acetylide ion (:Base in the above equations) attacks electron-poor (Lewis

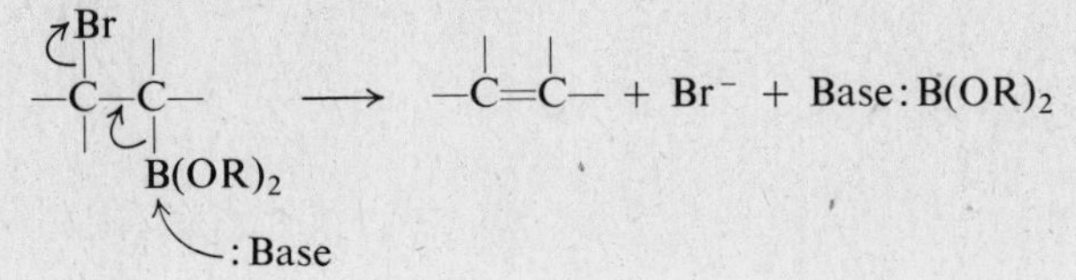

$$\underset{\underset{\displaystyle \,: Base}{B(OR)_2}}{\overset{Br}{\underset{|}{-C}} \underset{|}{\overset{|}{-C-}}} \longrightarrow -\overset{|}{C}=\overset{|}{C}- + Br^- + Base\!:\!B(OR)_2$$

acidic) boron, much as OH^- attacks —H in dehydrohalogenation, and brings about E2 elimination.

$$(RO)_2BCH_2CH_2Br + 3H_2O \longrightarrow CH_2{=}CH_2 + HBr + 2ROH + H_3BO_3$$

$$(RO)_2BCH_2CH_2Br + HC{\equiv}C^-Na^+ \longrightarrow CH_2{=}CH_2 + NaBr + HC{\equiv}CB(OR)_2$$

Alcohols II.   Reactions

16.1 Zinc chloride increases the acidity of the medium, possibly by

$$2HCl + ZnCl_2 \longrightarrow 2H^+ + ZnCl_4^{--}$$

16.2 Free radical chlorination of neopentane.   Simple alkyl free radicals show little tendency to rearrange.

16.3 Reaction, by the S_N1 mechanism, is favored by electron release.

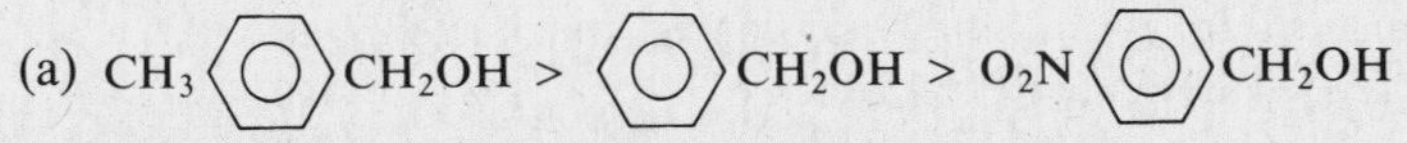

(a) CH_3⟨O⟩CH_2OH > ⟨O⟩CH_2OH > O_2N⟨O⟩CH_2OH

(Compare answer to Problem 10c, Chapter 14.)

(b) $PhCHOHCH_3$ > $PhCH_2OH$ > $PhCH_2CH_2OH$
    2° benzylic   benzylic   1°

16.4 (a) The allylic alcohol gives a hybrid allylic cation, which yields two products:

$$CH_3CHOHCH=CH_2 \xrightarrow[-H_2O]{H^+} CH_3\underset{\oplus}{\underline{CH=CH=CH_2}}$$

→ $CH_3\overset{|}{C}HCH=CH_2$ with Br

→ $CH_3CH=CHCH_2Br$

(b) This alcohol gives the same allylic cation, and hence the same products, as in (a):

$$CH_3CH=CHCH_2OH \xrightarrow[-H_2O]{H^+} CH_3-\underset{\oplus}{\underline{CH=CH=CH_2}} \longrightarrow \text{same products as in (a)}$$

(c) This is not really a rearrangement at all. There is no migration of any atom or group; in accordance with its hybrid nature, the intermediate cation is simply attacked by bromide at either of two positions, one of which is different from the position of attachment of the original —OH. (Compare the answer to Problem 14.9.)

16.5 (a) $ROH + HCl \rightleftharpoons ROH_2^+ + Cl^-$

$ROH_2^+ \rightleftharpoons R^+ + H_2O$ *Slow*

$R^+ + Cl^- \longrightarrow RCl$

(b) Formation of R^+. (c) ROH_2^+. (d) ROH and HCl. (e) On acid concentration also. Evidently, not always.

16.6 Electron-withdrawing substituents ($—Cl$, $—NO_2$, $—OH$) increase acid strength by stabilizing the anions.

(a) $ClCH_2CH_2OH > CH_3CH_2OH$ (b) $p\text{-}O_2NC_6H_4CH_2OH > C_6H_5CH_2OH$
(c) $HOCH_2CHOHCH_2OH > CH_3CH_2CH_2OH$

16.7 (a) $t\text{-BuOH} + Na \longrightarrow \frac{1}{2}H_2 + t\text{-BuO}^-Na^+$

$t\text{-BuO}^-Na^+ + EtBr \longrightarrow t\text{-BuOEt} + NaBr$

$EtOH + Na \longrightarrow \frac{1}{2}H_2 + EtO^-Na^+$

$EtO^-Na^+ + t\text{-BuBr} \longrightarrow (CH_3)_2C{=}CH_2 + EtOH + NaBr$

(b) In the first case, nucleophilic substitution, with $t\text{-BuO}^-$ acting as the nucleophile. In the second case, elimination, with EtO^- acting as the base.

(c) As usual (Sec. 14.23) substitution predominates with the 1° halide, and elimination predominates with the 3° halide.

16.8 $p\text{-CH}_3C_6H_4SO_2OBu\text{-}sec \xleftarrow[\text{base}]{sec\text{-BuOH}} p\text{-CH}_3C_6H_4SO_2Cl \xleftarrow{PCl_5} p\text{-CH}_3C_6H_4SO_3H \xleftarrow[SO_3]{H_2SO_4} C_6H_5CH_3$

Alternatively, a sulfonyl chloride is conveniently prepared by treatment of the aromatic substrate with excess *chlorosulfonic acid*, $ClSO_3H$.

$ArH + ClSO_3H \longrightarrow ArSO_3H + HCl$

$ArSO_3H + ClSO_3H \longrightarrow ArSO_2Cl + H_2SO_4$

16.9 There is clearly overall inversion of configuration of the alcohol. There are only two steps in the reaction sequence. The first, preparation of the *sec*-butyl tosylate, must involve cleavage of the O—H bond of the alcohol (see page 538), and hence cannot change the configuration about the chiral carbon. Inversion must, therefore, have occurred in the only other step, hydrolysis of the tosylate.

16.10

	conc. $H_2SO_4{}^a$	cold $KMnO_4{}^b$	$Br_2{}^b$	$CrO_3{}^c$	fum. $H_2SO_4{}^a$	$CHCl_3$, $AlCl_3{}^c$	Na^d
Alkanes	−	−	−		−	−	−
Alkenes	+	+	+	−	+	−	−
Alkynes	+	+	+	−	+	−	−^{e}
Alkyl halides	−	−	−	−	−	−	−
Alkylbenzenes	−	−^{f}	−	−	+	+	−
1° alcohols	+	−	−	+	+	−	+
2° alcohols	+	−	−	+	+	−	+
3° alcohols	+	−	−	−	+	−	+

^{a} Dissolves. ^{b} Decolorizes. ^{c} Changes color. ^{d} Hydrogen bubbles. ^{e} 1-Alkynes give test.
^{f} Decolorizes hot $KMnO_4$.

16.11 At site of cleavage:
(1) Replace bond to carbon by bond to —OH. Each such cleavage requires one HIO_4.
(2) If a resulting fragment is now a *gem*-diol (unstable),

$$ \overset{|}{\underset{OH}{-C-OH}} \longrightarrow H_2O + \overset{|}{-C=O} $$

A *gem*-diol   A carbonyl
Unstable   compound

remove H_2O to give stable carbonyl compound. For example:

$$ \underset{HO\ \ OH\ \ OH}{H_2C \vdots CH \vdots CH_2} \xrightarrow{2HIO_4} \underset{OH}{H_2C-OH} + \underset{OH}{HO-\overset{H}{C}-OH} + \underset{OH}{HO-CH_2} $$

A glycol

$$ \downarrow_{-H_2O} \qquad \downarrow_{-H_2O} \qquad \downarrow_{-H_2O} $$

$$ H_2C=O \qquad \underset{OH}{H-C\overset{O}{<}} \qquad O=CH_2 $$

Cleavage products

(a) $CH_3CHOH \vdots CH_2OH \xrightarrow{1HIO_4} CH_3CHO + HCHO$

(b) $CH_3CHOH \vdots CHO \xrightarrow{1HIO_4} CH_3CHO + HCOOH$

(c) $CH_2OH \vdots CHOHCH_2OCH_3 \xrightarrow{1HIO_4} HCHO + OHCCH_2OCH_3$

(d) $\underset{OCH_3}{CH_2OHCHCH_2OH} \longrightarrow$ no reaction with HIO_4

(e) 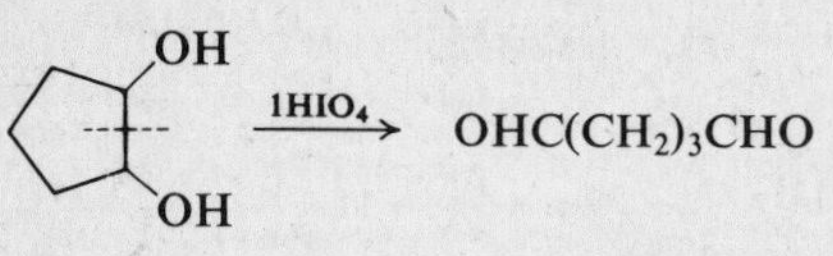 $\xrightarrow{\text{1HIO}_4}$ OHC(CH$_2$)$_3$CHO

(f) CH$_2$OH$\vdash$CHOH$\vdash$CHOH$\vdash$CHOH$\vdash$CHO $\xrightarrow{\text{4HIO}_4}$ HCHO + HCOOH + HCOOH + HCOOH + HCOOH

(g) CH$_2$OH$\vdash$CHOH$\vdash$CHOH$\vdash$CHOH$\vdash$CH$_2$OH $\xrightarrow{\text{4HIO}_4}$ HCHO + HCOOH + HCOOH + HCOOH + HCHO

16.12 To reconstruct (mentally) the compound from the cleavage fragments, reverse the procedure of the preceding problem: (1) Replace a carbonyl group in a cleavage product by two —OH's (unstable *gem*-diol). (2) Now delete —OH's in pairs, one from each of two fragments, and join the carbons together. For example:

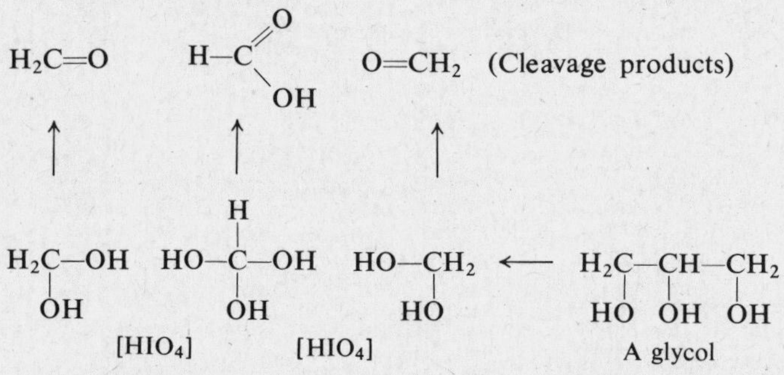

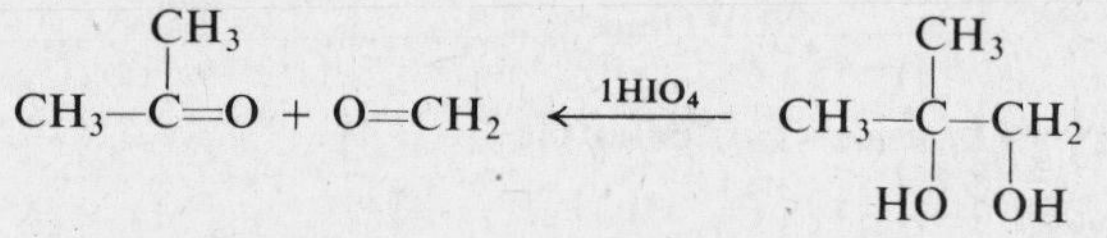

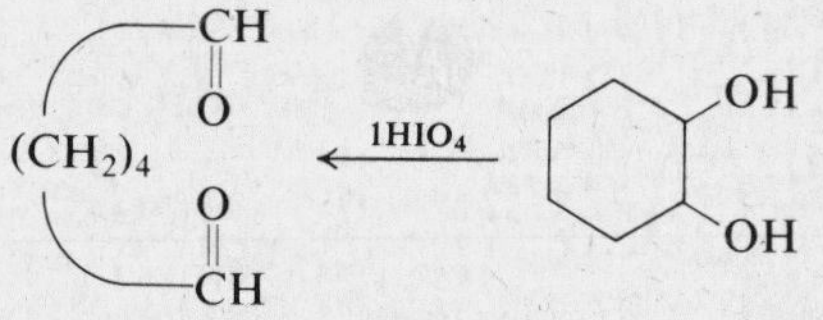

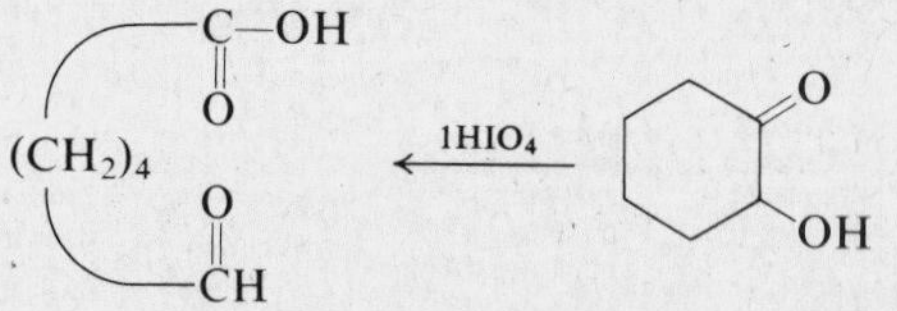

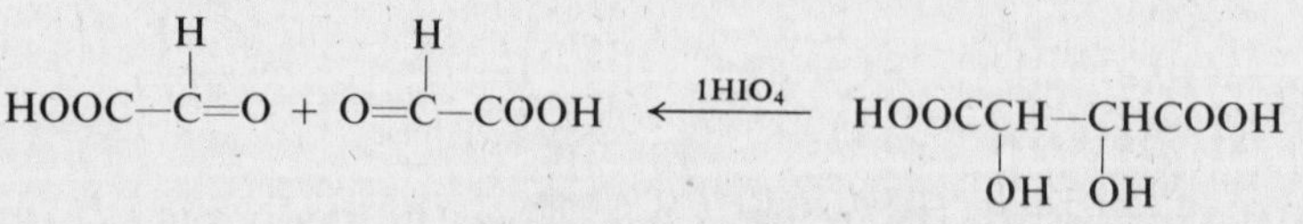

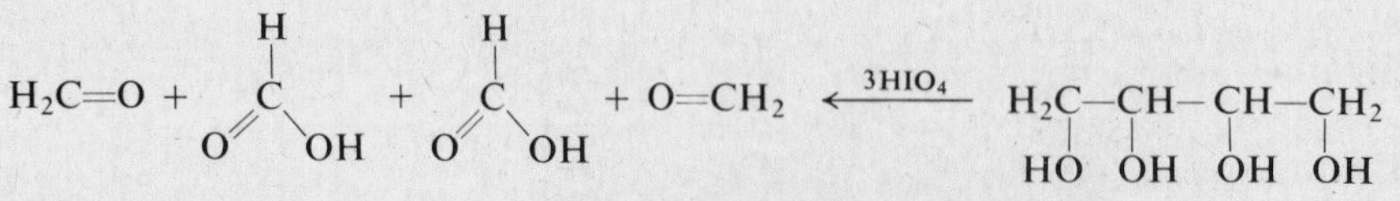

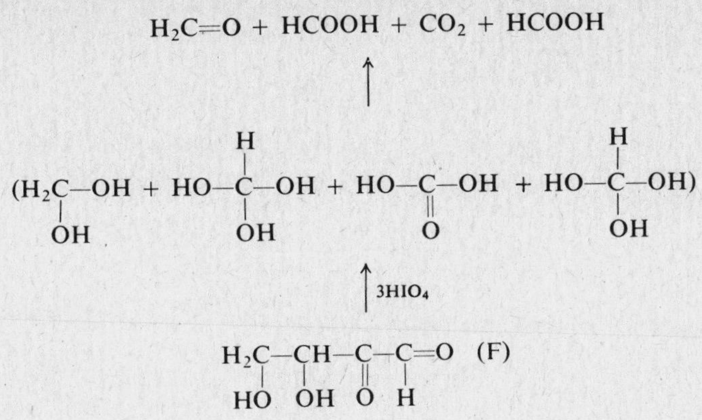

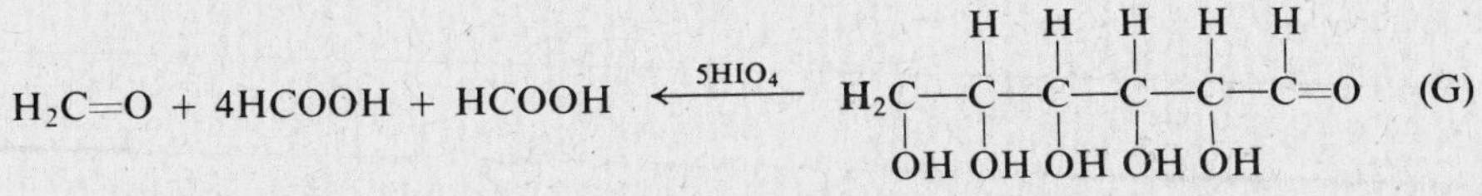

$$H_2C=O + HCOOH + CO_2 + HCOOH$$

$$\uparrow$$

$$(H_2C-OH + HO-\overset{H}{\underset{OH}{C}}-OH + HO-\overset{}{\underset{O}{C}}-OH + HO-\overset{H}{\underset{OH}{C}}-OH)$$

$$\overset{}{\underset{OH}{}}$$

$$\uparrow 3HIO_4$$

$$H_2C-CH-\overset{}{C}-C=O \quad (F)$$
$$\underset{HO}{}\ \underset{OH}{}\ \underset{O}{}\ \underset{H}{}$$

$$H_2C=O + 4HCOOH + HCOOH \xleftarrow{5HIO_4} H_2\overset{H}{\underset{OH}{C}}-\overset{H}{\underset{OH}{C}}-\overset{H}{\underset{OH}{C}}-\overset{H}{\underset{OH}{C}}-\overset{H}{\underset{OH}{C}}-C=O \quad (G)$$

16.13 Change the concentration.

16.14 (a) 1°, $-\overset{H}{\underset{H}{C}}-OH$ gives triplet (splitting by 2H on adjacent C);

2°, $-\overset{}{\underset{H}{C}}-OH$ gives doublet (splitting by 1H on adjacent C);

3°, $-\overset{}{\underset{}{C}}-OH$ gives singlet (no splitting, no H on adjacent C).

(b) Acid: $R*OH* + H:B \rightleftarrows \left[R*\overset{}{\underset{H}{O}}H*\right]^+ + :B^-$

$\left[R*\overset{}{\underset{H}{O}}H*\right]^+ + ROH \rightleftarrows R*OH + \left[\overset{}{\underset{H*}{R}OH}\right]^+$

$\left[\overset{}{\underset{H*}{R}OH}\right]^+ + :B^- \rightleftarrows ROH* + H:B$

Base: $R*OH* + :B \rightleftarrows R*O^- + [H*:B]^+$

$R*O^- + ROH \rightleftarrows R*OH + RO^-$

$RO^- + [H*:B]^+ \rightleftarrows ROH* + :B$

Answers to Synthesis Problems

To save space in this Study Guide, we have sometimes had to omit certain parts of a complete synthesis scheme. When we say "from RCH_2OH" or "from R_2CHOH" or "from ROH," for example, this is understood to mean

or

$$RCH_2OH \xrightarrow{K_2Cr_2O_7,\ H^+} RCHO$$

or

$$R_2CHOH \xrightarrow{CrO_3} R_2C{=}O$$

or

$$ROH \xrightarrow{PBr_3} RBr$$

$$ROH \xrightarrow{HBr} RBr$$

or whatever specific process is the correct one. We omit the treatment with water that is an essential step in the work-up of a Grignard reaction mixture:

$$RMgBr + {\scriptstyle\diagdown}C{=}O \longrightarrow {-}\underset{R}{\overset{|}{C}}{-}O^-MgBr^+ \xrightarrow{H_2O} {-}\underset{R}{\overset{|}{C}}{-}OH$$

Unless the ground rules for your particular course permit otherwise, you should show all these omitted steps in *your* synthesis schemes.

---

1. (a) 2-Pentanol and 3-methyl-2-butanol give iodoform; all others negative.

(b) 2-Methyl-2-butanol (3°) reacts rapidly with Lucas reagent; 2-pentanol, 3-pentanol, and 3-methyl-2-butanol (all 2°) react slowly; the others (1°) do not respond.

(c) 2-Methyl-2-butanol (3°) gives a negative test to CrO_3/H_2SO_4; all the others (1°, 2°) give a positive test.

(d) $n\text{-Bu}{\vdots}CH_2OH \longleftarrow n\text{-BuMgBr (from } n\text{-BuOH)} + HCHO \text{ (from MeOH)}$

$n\text{-Pr}\overset{a}{\vdots}CH\vdots CH_3 \longleftarrow$ ⎡ $\overset{a}{\quad}$ $n\text{-PrMgBr (from } n\text{-PrOH)} + OHCCH_3 \text{ (from EtOH)}$
$\phantom{n\text{-Pr}}\underset{OH\,b}{|} \quad$ ⎣ $\overset{b}{\quad}$ $n\text{-PrCHO (from } n\text{-BuOH)} + CH_3MgBr \text{ (from MeOH)}$

$Et{-}CHOH{\vdots}Et \longleftarrow EtCHO \text{ (from } n\text{-PrOH)} + EtMgBr \text{ (from EtOH)}$

$sec\text{-Bu}{\vdots}CH_2OH \longleftarrow sec\text{-BuMgBr (from } sec\text{-BuOH)} + HCHO \text{ (from MeOH)}$

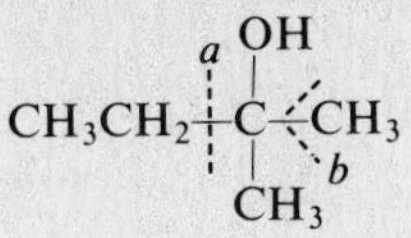

$$\text{CH}_3\text{CH}_2 \overset{a}{\underset{b}{|}} \overset{\text{OH}}{\underset{\text{CH}_3}{\overset{|}{\text{C}}}} \text{CH}_3 \quad \longleftarrow$$

- *a* — EtMgBr (from EtOH) + O=C(CH₃)₂ (from *i*-PrOH)

- *b* — Et—C=O (from *sec*-BuOH) + CH₃MgBr (from MeOH)
      $\quad\quad\;\;|$
      $\quad\quad\text{CH}_3$

$$i\text{-Pr} \overset{a}{\underset{b}{|}} \overset{\text{H}}{\underset{\text{OH}}{\overset{|}{\text{C}}}} \text{CH}_3 \quad \longleftarrow$$

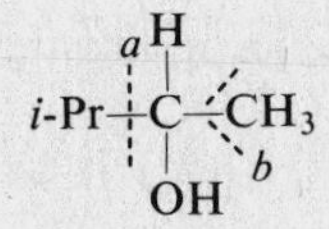

- *a* — *i*-PrMgBr (from *i*-PrOH) + OHCCH₃ (from EtOH)

- *b* — *i*-PrCHO (from *i*-BuOH) + CH₃MgBr (from MeOH)

iso-Bu┼CH₂OH ⟵ *iso*-BuMgBr (from *iso*-BuOH) + HCHO (from MeOH)

t-Bu┼CH₂OH ⟵ *t*-BuMgBr (from *t*-BuOH) + HCHO (from MeOH)

2. In the following, R = cyclohexyl, ⬡—

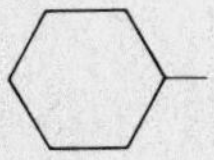

(a) ROSO₃H

(b) 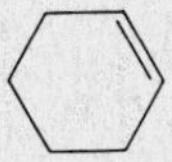 (cyclohexene)

(c) no reaction

(d) 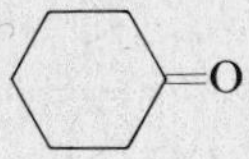 (cyclohexanone) =O

(e) no reaction

(f) RBr

(g) RI

(h) RO⁻ Na⁺

(i) R—O—C—CH₃
          ‖
          O

(j) no reaction

(k) ROMgBr + CH₄

(l) no reaction

(m) RMgBr

(n) 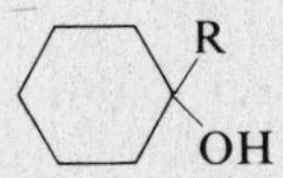 (cyclohexane with R and OH)

(o) (cyclohexane)—OH (*cis*)
          OH

(p) (cyclohexane)—Br (*trans*)
          Br

(q) (biphenyl-cyclohexyl)

(r) (cyclohexane)

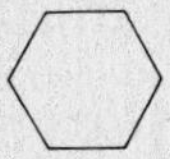

(s) *o*- and *p*-O₂NC₆H₄—R

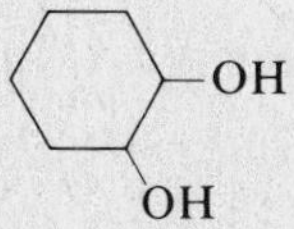

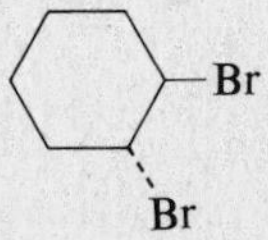

(t) 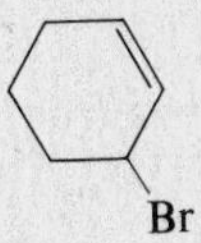 (cyclohexene)
          Br

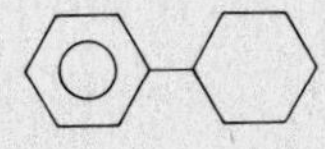

(u) 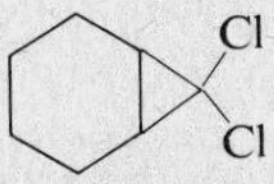 (bicyclic with) Cl, Cl
      7,7-Dichloronorcarane
      (7,7-Dichloro[4.1.0.]bicyclo-
      heptane)

(v) 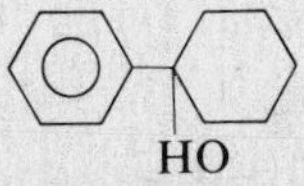 (fused ring with HO)

(w) 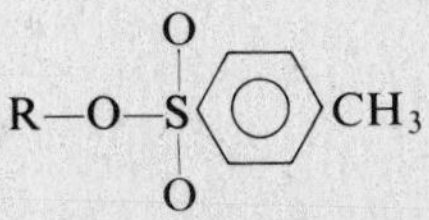 R—O—S—(⬡)—CH₃
              ‖
              O

(x) *t*-Bu—O—R

Synthesis: Working Backwards

Granting that we know the chemistry of the individual steps, how do we go about planning a route to more complicated compounds—alcohols, say? In almost every organic synthesis it is best to begin with the molecule we want—the *target* molecule—and *work backwards* from it. There are relatively few ways to make a complicated alcohol, for example; there are relatively few ways to make a Grignard reagent or an aldehyde or ketone; and so on back to our ultimate starting materials. On the other hand, our starting materials can undergo so many different reactions that, if we go at the problem the other way around, we find a bewildering number of paths, few of which take us where we want to go.

We try to limit a synthesis to as few steps as possible, but nevertheless do not sacrifice purity for time. To avoid a rearrangement in the preparation of an alkene, for example, we take two steps via the halide rather than the single step of dehydration.

---

3. (a) *n*-BuOH, HBr, heat   (b) Product (a), KOH(alc), heat

   (c) *n*-BuOH, cold, conc. H_2SO_4 (d) *n*-BuOH, K(metal)

   (e) *n*-BuOH, $K_2Cr_2O_7$ (f) *n*-BuOH, $KMnO_4$

   (g) Product (b), cat. H_2

   (h) $CH_3CH_2CH-CH_2$ $\xleftarrow{Br_2,\ CCl_4}$ $CH_3CH_2CH=CH_2$ $\longleftarrow$ (b)
            | |
          Br   Br

   (i) $CH_3CH_2CH-CH_2$ $\xleftarrow{Cl_2,\ H_2O}$ $CH_3CH_2CH=CH_2$ $\longleftarrow$ (b)
           | |
        OH   Cl

   (j) $CH_3CH_2C\equiv CH$ $\xleftarrow{NaNH_2}$ $\xleftarrow{KOH(alc)}$ $CH_3CH_2CH-CH_2$ $\longleftarrow$ (h)
                                 | |
                                Br   Br

   (k) Et—CH—CH$_2$ $\xleftarrow{Na \atop MeOH}$ Et—CH—CCl$_2$ $\longleftarrow$ Et—CH=CH$_2$ + :CCl$_2$ $\xleftarrow{t\text{-BuOK}}$ HCCl$_3$
              \| \|
            CH$_2$ CH$_2$

                                  ↑
                                (b)

   (l) $CH_3CH_2CH-CH_2$ $\xleftarrow{HCO_2OH\ or \atop cold\ KMnO_4}$ $CH_3CH_2CH=CH_2$ $\longleftarrow$ (b)
           | |
       OH   OH

                             ┌─────── *n*-BuBr $\longleftarrow$ (a)

   (m) *n*-C_4H_9┊C_4H_9-*n* $\longleftarrow$ ↓
                              └─── (*n*-Bu)$_2$CuLi $\xleftarrow{CuX}$ *n*-BuLi

(n) 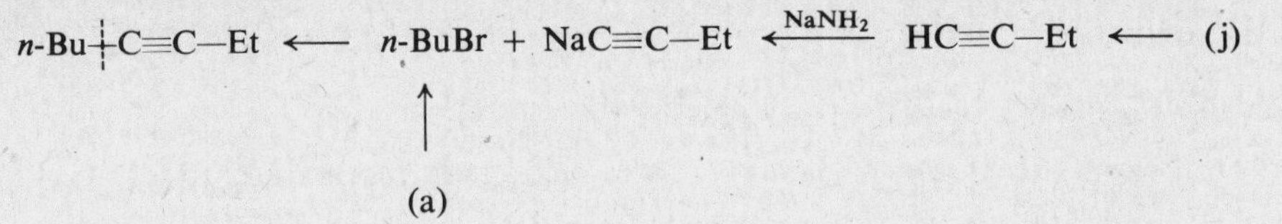 $n\text{-Bu} \vdots C \equiv C - Et \longleftarrow n\text{-BuBr} + NaC \equiv C - Et \xleftarrow{NaNH_2} HC \equiv C - Et \longleftarrow$ (j)

(a)

(o) 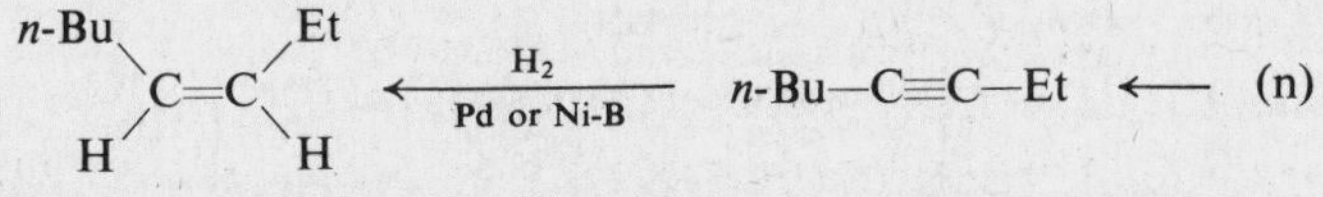 $\underset{H}{\overset{n\text{-Bu}}{>}} C = C \underset{H}{\overset{Et}{<}} \xleftarrow{\overset{H_2}{Pd \text{ or } Ni\text{-}B}} n\text{-Bu} - C \equiv C - Et \longleftarrow$ (n)

(p) 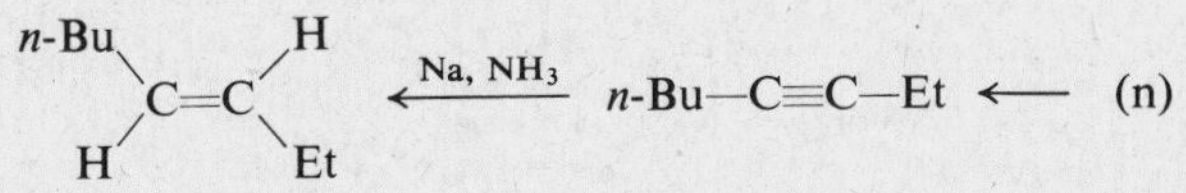 $\underset{H}{\overset{n\text{-Bu}}{>}} C = C \underset{Et}{\overset{H}{<}} \xleftarrow{Na, NH_3} n\text{-Bu} - C \equiv C - Et \longleftarrow$ (n)

(q) 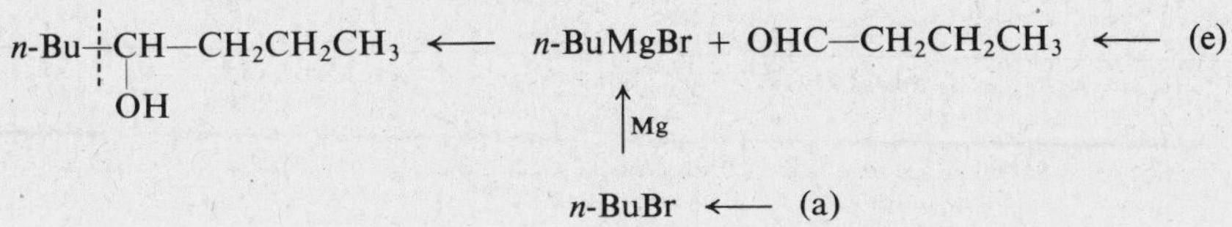 $n\text{-Bu} \vdots \underset{OH}{CH} - CH_2CH_2CH_3 \longleftarrow n\text{-BuMgBr} + OHC - CH_2CH_2CH_3 \longleftarrow$ (e)

$\uparrow Mg$

$n\text{-BuBr} \longleftarrow$ (a)

(r) 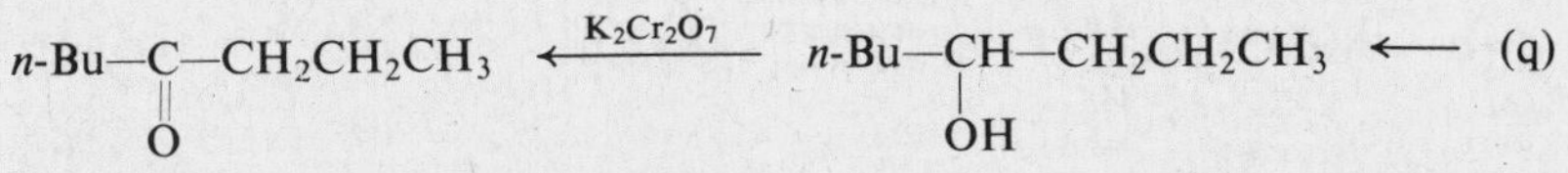 $n\text{-Bu} - \underset{O}{\overset{\|}{C}} - CH_2CH_2CH_3 \xleftarrow{K_2Cr_2O_7} n\text{-Bu} - \underset{OH}{CH} - CH_2CH_2CH_3 \longleftarrow$ (q)

(s) 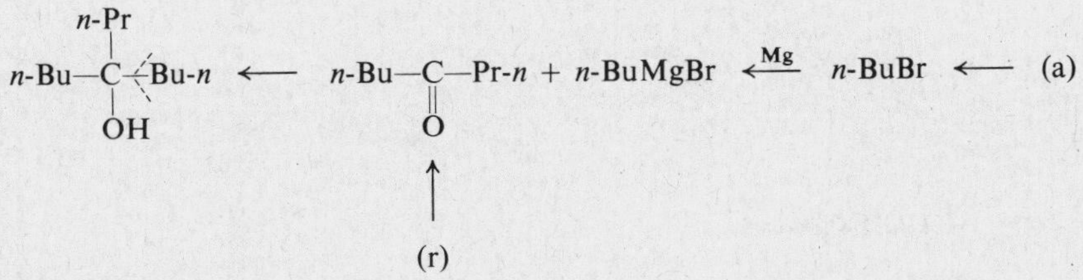 $n\text{-Bu} - \underset{OH}{\overset{n\text{-Pr}}{\underset{|}{C}}} \cdot Bu\text{-}n \longleftarrow n\text{-Bu} - \underset{O}{\overset{\|}{C}} - Pr\text{-}n + n\text{-BuMgBr} \xleftarrow{Mg} n\text{-BuBr} \longleftarrow$ (a)

$\uparrow$

(r)

(t) 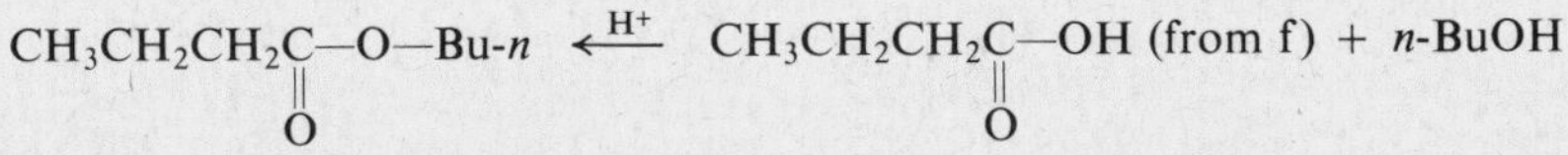 $CH_3CH_2CH_2\underset{O}{\overset{\|}{C}} - O - Bu\text{-}n \xleftarrow{H^+} CH_3CH_2CH_2\underset{O}{\overset{\|}{C}} - OH \text{ (from f)} + n\text{-BuOH}$

4. (a) $(PhCH_2O)_2Mg$ (b) $PhCOOCH_2CH(CH_3)_2$, isobutyl benzoate

(c) $\underset{OH \quad OH}{CH_2 - CH_2}$, ethylene glycol   (d) no reaction

(e) $CH_3CH - CH - CH_2$
       $\underset{OH \quad Br \quad OH}{}$ (f) $C_2H_6 + Mg(OCH_3)Br$

(g) $p\text{-BrC}_6H_4CH_2OH$ (h) $PhC(CH_3)_3$, *tert*-butylbenzene

5. Yellow dichromate is reduced to green Cr(III) by ethyl alcohol; the motorist is reduced to tears or rage and, often, to walking.

6. In general, the order of reactivity is

$$\text{allylic, benzylic} > 3° > 2° > 1° < \text{MeOH}$$

Electronic factors are important except for 1° alcohols, for which steric factors are controlling.

(a) $CH_3CH_2\underset{\underset{\text{OH}}{|}}{\overset{\overset{CH_3}{|}}{C}}CH_3$ > $CH_3\underset{\underset{\text{OH}}{|}}{CH}\overset{\overset{CH_3}{|}}{-}CH-CH_3$, $CH_3CH_2\underset{\underset{\text{OH}}{|}}{-}CH-CH_2CH_3$, $CH_3CH_2CH_2\underset{\underset{\text{OH}}{|}}{-}CH-CH_3$ >

        3°   2°   2°   2°

$CH_3CH_2CH_2CH_2\underset{\underset{\text{OH}}{|}}{-}CH_2$ > $CH_3\overset{\overset{CH_3}{|}}{C}HCH_2\underset{\underset{\text{OH}}{|}}{-}CH_2$ > $CH_3CH_2\overset{\overset{CH_3}{|}}{C}H\underset{\underset{\text{OH}}{|}}{-}CH_2$ > $CH_3\underset{\underset{CH_3}{|}}{\overset{\overset{CH_3}{|}}{C}}CH_2OH$

                  1° (rate falling off with increasing size of G in G—CH_2)

                                                           OH

(b) $Ph\underset{\underset{\text{OH}}{|}}{-}CH-CH_2CH_3$ $Ph\underset{\underset{\text{OH}}{|}}{-}CH_2-CH-CH_3$ $Ph\underset{\underset{\text{OH}}{|}}{-}CH_2CH_2-CH_2$

        benzylic   2°   1°

(c) $HO-\langle\bigcirc\rangle-CH_2OH$ > $\langle\bigcirc\rangle-CH_2OH$ > $N\equiv C-\langle\bigcirc\rangle-CH_2OH$

        (Compare answer to Problem 16.3a above)

(d) $CH_3CH=CHCH_2OH$ > $CH_2=CHCH_2CH_2OH$

        allylic   1°

(e) (cyclopentane with CH_3 and OH)   (cyclopentane with OH and CH_3)   (cyclopentane with CH_2OH)

          3°   2°   1°

(f) Ph_3COH > Ph_2CHOH > $PhCH_2OH$ > CH_3OH

    3° benzylic   2° benzylic   1° benzylic   MeOH

7. (a) $CH_3-\underset{\underset{H}{|}}{\overset{\overset{CH_3}{|}}{C}}-CH=CH_2$ $\xrightarrow{H^+}$ $CH_3-\underset{\underset{H}{|}}{\overset{\overset{CH_3}{|}}{C}}-\overset{\oplus}{CH}-\underset{\underset{H}{|}}{CH_2}$ $\xrightarrow[\text{shift}]{\text{hydride}}$ $CH_3-\underset{\underset{H}{|}}{\overset{\overset{CH_3}{|}}{C}}-\overset{\oplus}{C}H_2CH_3$

                                                  $\downarrow Cl^-$ $\downarrow Cl^-$

                                        $CH_3-\underset{\underset{Cl}{|}}{CH}-\overset{\overset{CH_3}{|}}{C}HCH_3$ $CH_3-\underset{\underset{Cl}{|}}{\overset{\overset{CH_3}{|}}{C}}-CH_2CH_3$

(b) 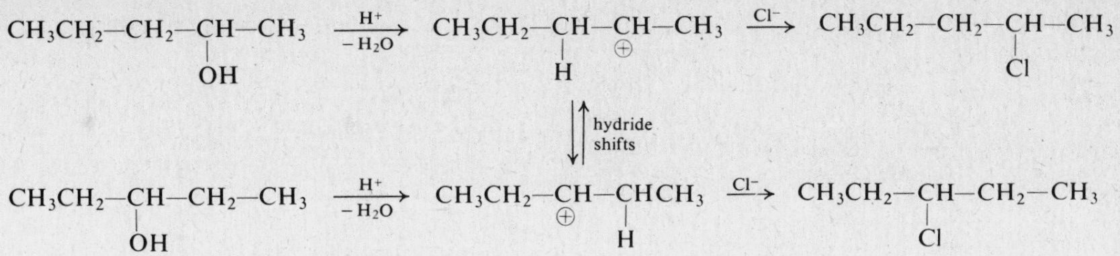 $CH_3CH_2-CH_2-\underset{\underset{\displaystyle OH}{|}}{CH}-CH_3 \xrightarrow[-H_2O]{H^+} CH_3CH_2-\underset{\underset{\displaystyle H}{|}}{CH}-\overset{\oplus}{CH}-CH_3 \xrightarrow{Cl^-} CH_3CH_2-CH_2-\underset{\underset{\displaystyle Cl}{|}}{CH}-CH_3$

$\Updownarrow$ hydride shifts

$CH_3CH_2-\underset{\underset{\displaystyle OH}{|}}{CH}-CH_2-CH_3 \xrightarrow[-H_2O]{H^+} CH_3CH_2-\underset{\oplus}{CH}-\underset{\underset{\displaystyle H}{|}}{CH}CH_3 \xrightarrow{Cl^-} CH_3CH_2-\underset{\underset{\displaystyle Cl}{|}}{CH}-CH_2-CH_3$

(c) The first step is the formation of a carbonium ion,

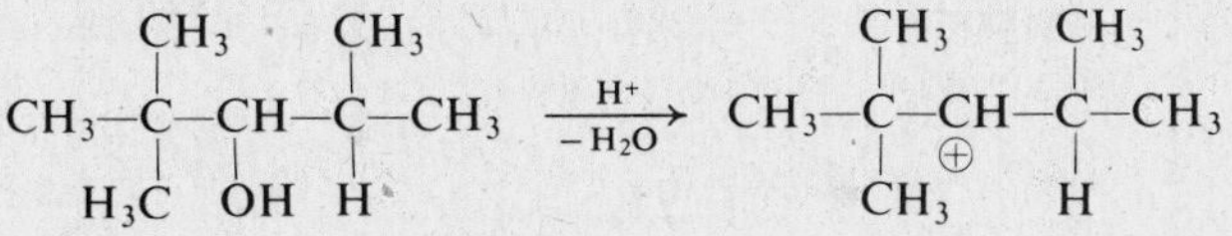

$$CH_3-\underset{\underset{\displaystyle H_3C}{|}}{\overset{\overset{\displaystyle CH_3}{|}}{C}}-\underset{\underset{\displaystyle OH}{|}}{CH}-\underset{\underset{\displaystyle H}{|}}{\overset{\overset{\displaystyle CH_3}{|}}{C}}-CH_3 \xrightarrow[-H_2O]{H^+} CH_3-\underset{\underset{\displaystyle CH_3}{|}}{\overset{\oplus}{C}}-CH-\underset{\underset{\displaystyle H}{|}}{\overset{\overset{\displaystyle CH_3}{|}}{C}}-CH_3$$

from which all the other products are derived:

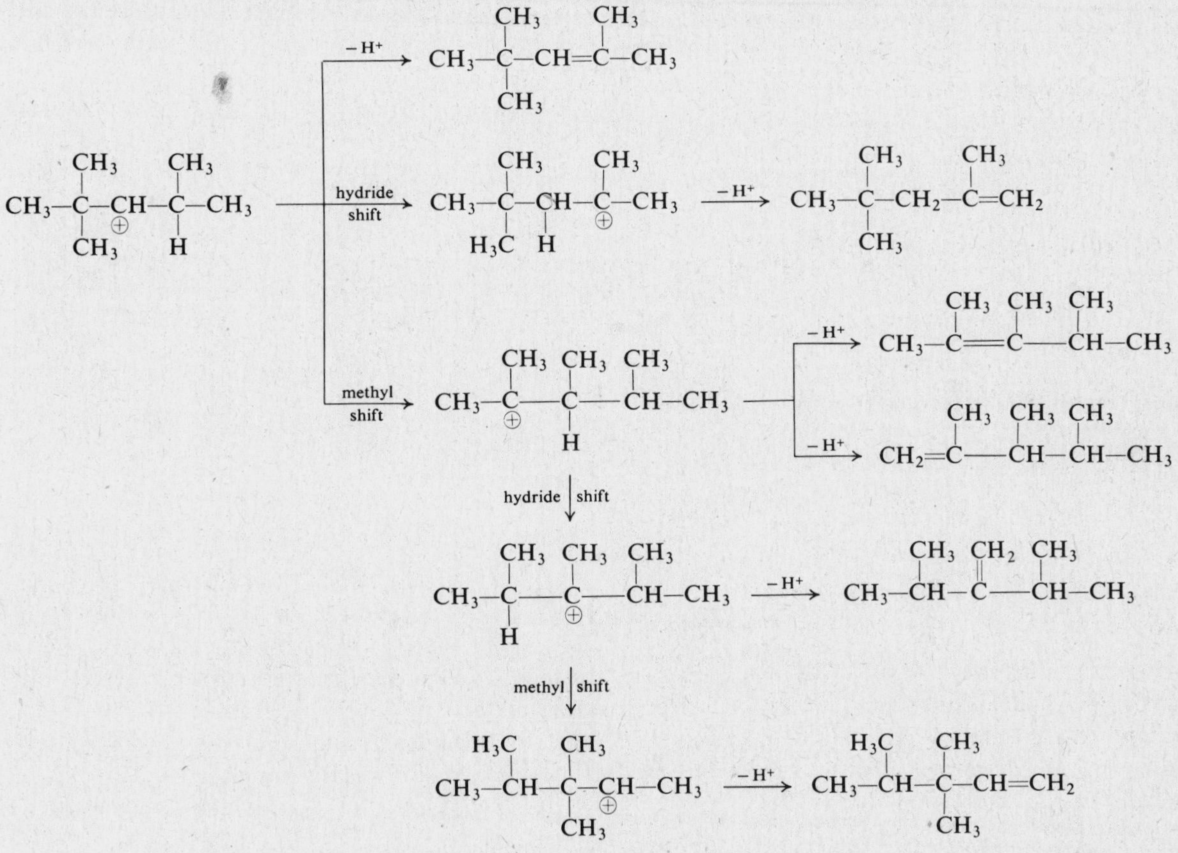

(d) First, a carbonium ion is formed, a 2° cation:

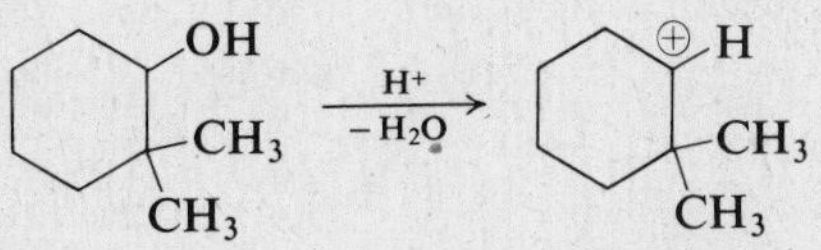

2° cation

181

This can rearrange by migration of a methyl group to give a 3° cation and eventually a highly branched alkene:

2° cation   3° cation   1,2-Dimethylcyclohexene
Highly branched

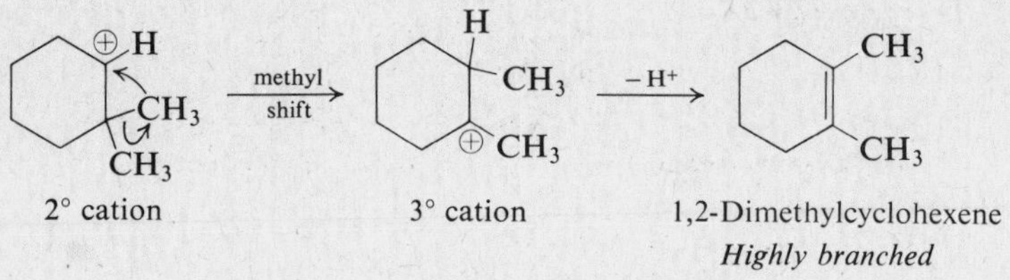

But we must not lose sight of the fact that carbons and hydrogens of the ring also make up alkyl groups, which are also capable of migration. One such migration gives a 3° cation and eventually an alkene. All this is accompanied by *ring contraction*:

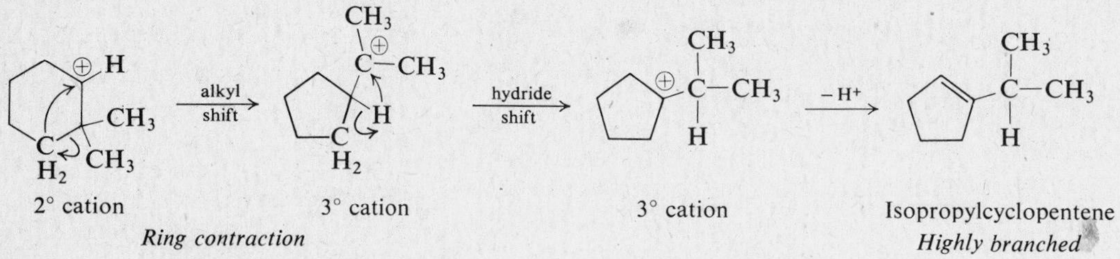

2° cation   3° cation   3° cation   Isopropylcyclopentene
Ring contraction   *Highly branched*

(e) A 3° cation is formed:

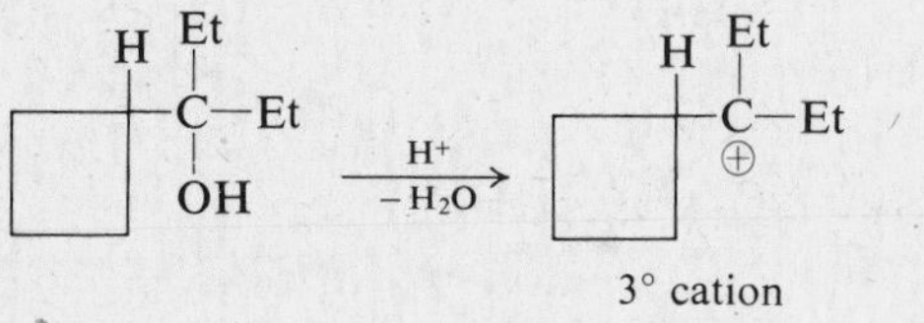

3° cation

Here, too, migration involves a carbon of the ring,

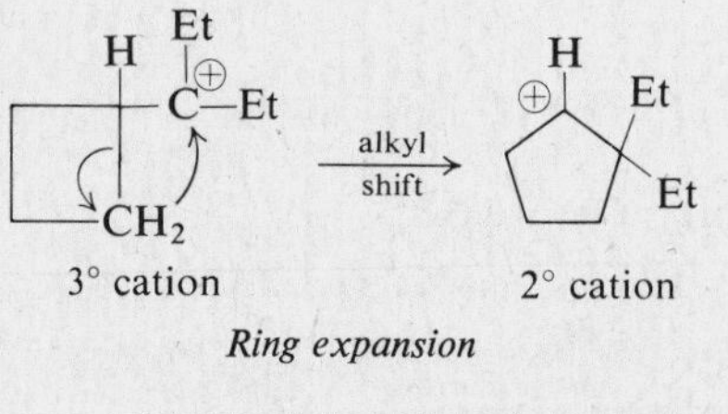

3° cation   2° cation
Ring expansion

and again the ring size changes. This time there is *ring expansion*. We notice that rearrangement converts a 3° cation into a 2° cation. The driving force here is the change from a strained four-membered ring into a relatively strain-free five-membered ring—plus the subsequent rearrangement to a 3° cation and eventual formation of a highly branched alkene:

2° cation   3° cation   1,2-Diethylcyclopentene
Highly branched

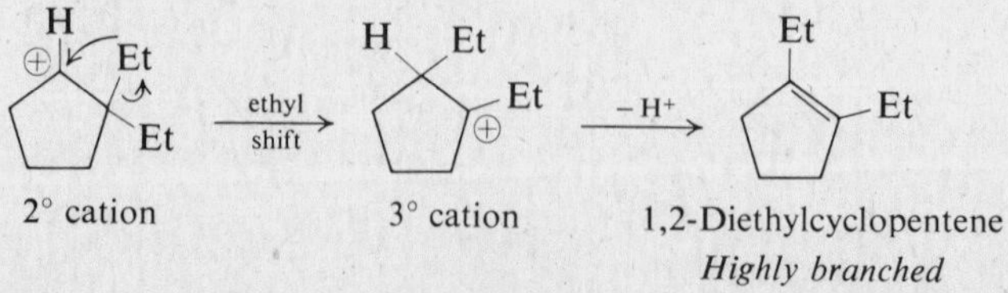

(f) Again ring expansion, helped along by formation of a 3° cation and eventually a highly branched alkene:

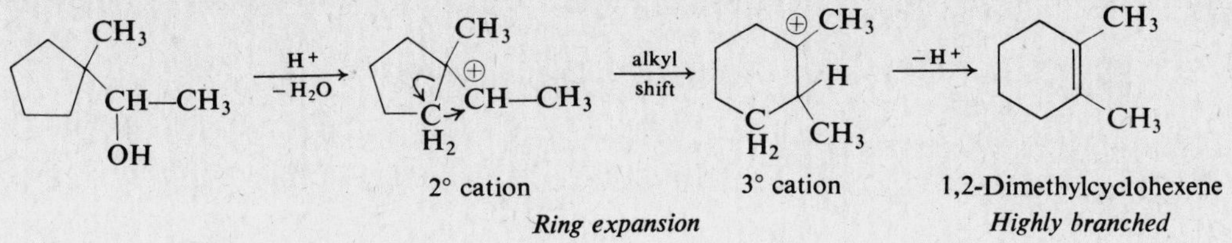

CH3
CH–CH3
OH
H+
–H2O
CH3
CH–CH3
H2
alkyl
shift
CH3
H
H2
CH3
–H+
CH3
CH3
2° cation
3° cation
1,2-Dimethylcyclohexene
Ring expansion
Highly branched

8. (a)

O
K2Cr2O7, H+
cyclohexanol

(b)

Br
PBr3
cyclohexanol

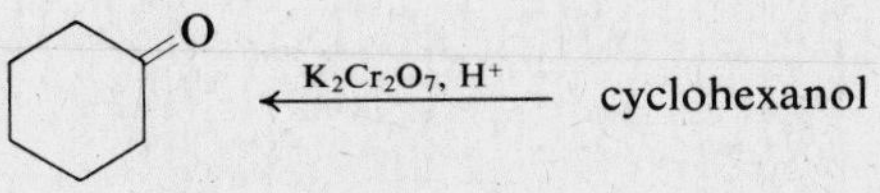

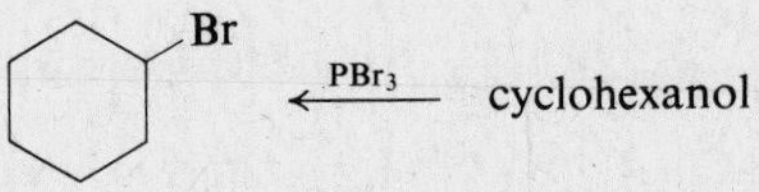

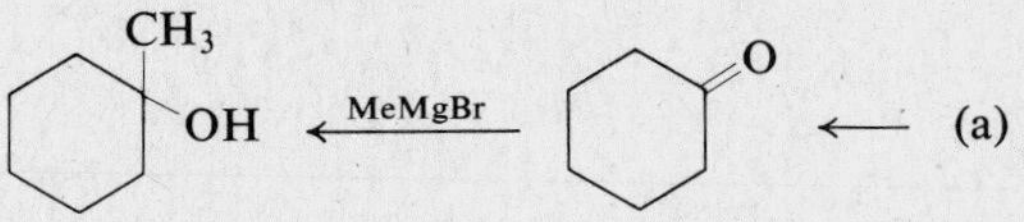

(c)

CH3
OH
MeMgBr
O
(a)

(d)

CH3
H+, heat
CH3
OH
(c)

(e)

CH3
OH
HB/O
CH3
(d) (HB/O = hydroboration-oxidation)

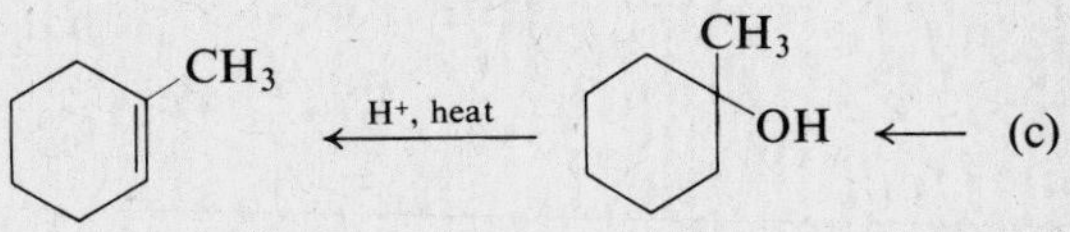

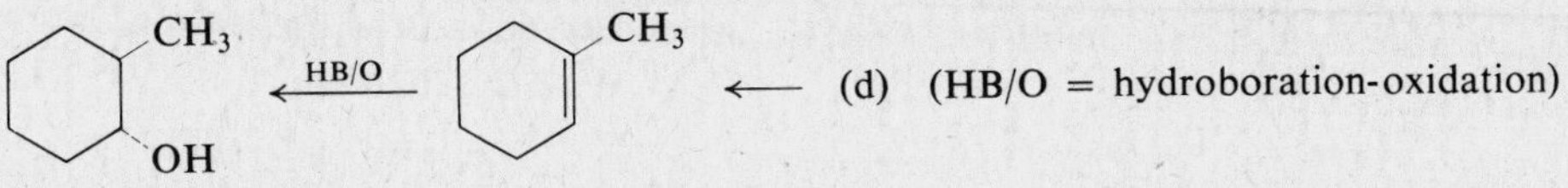

(f)

CH3CH–
OH
CH3CHO
BrMg–
Mg
Br–
(b)

(g)

Br
Br
Br2, CCl4
H+, heat
cyclohexanol

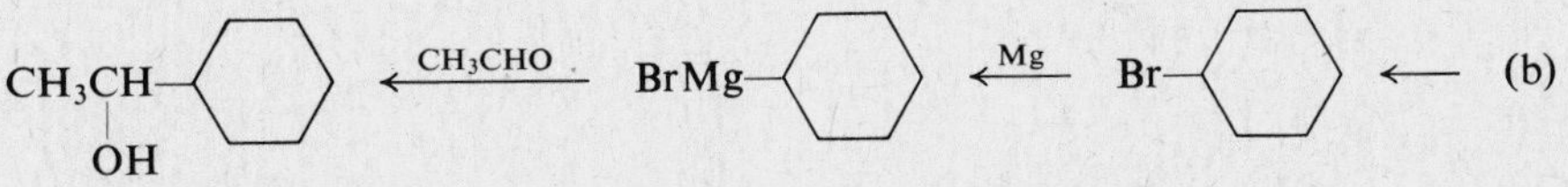

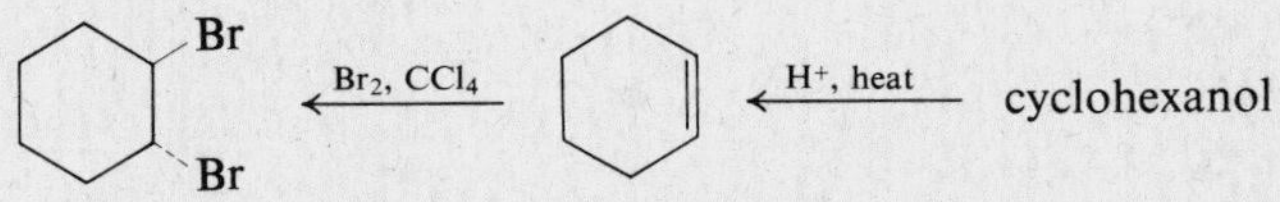

(h)

CH2OH
MgBr (as in f) + HCHO

(i)

Ph
Br
HBr
Ph
OH
PhMgBr
O
(a)

(j)

COOH
KMnO4
CH2OH
(h)
CO2
MgBr (as in f)

(k)

COOH
COOH
HNO3
heat
as in (g)

(l)

H
H
Na
MeOH
H
Cl
Cl
H
CCl4, t-BuOK
(as in g)

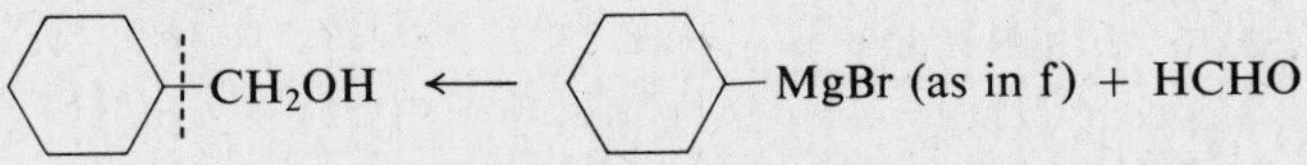

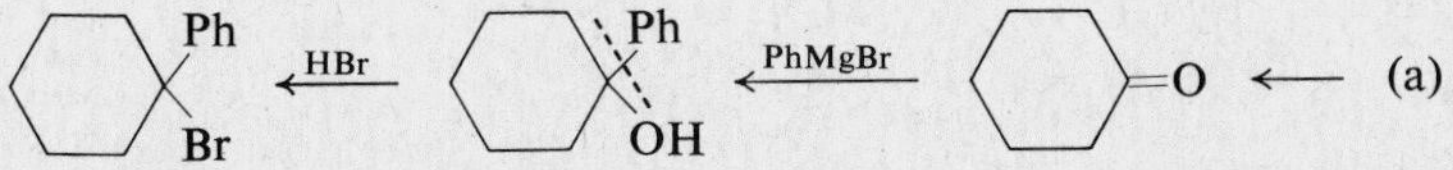

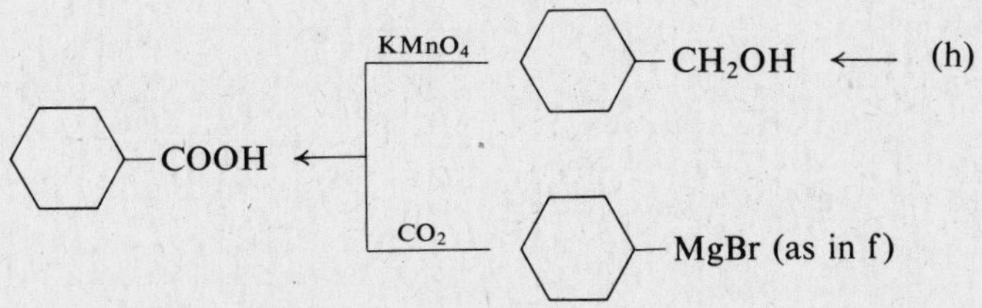

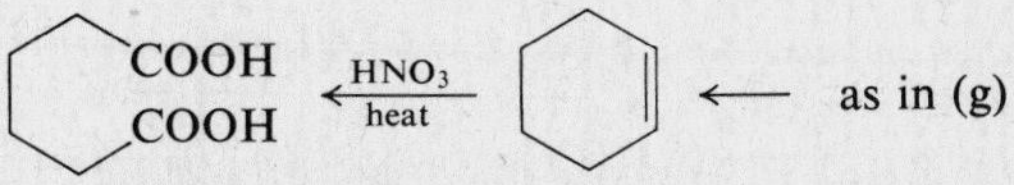

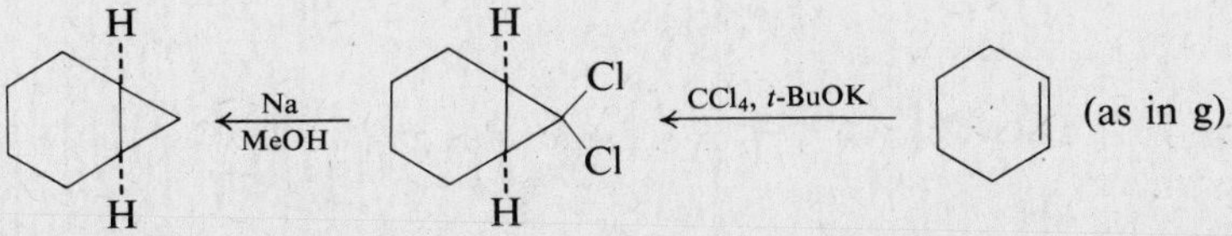

9. (a) $\underset{\underset{OH}{|}}{\overset{\overset{H_3C\quad CH_3}{|\quad\quad|}}{CH_3CH+C-CH_3}}$ ⟵ $\underset{\overset{CH_3}{|}}{CH_3CHMgBr}$ + $\underset{\overset{CH_3}{|}}{O=C-CH_3}$ (both from *iso*-PrOH)

(b) $\underset{\underset{OH}{|}}{\overset{\overset{Ph}{|}}{CH_3-C-CH_3}}$ ⟵ PhMgBr (from C_6H_6) + $\underset{\overset{||}{O}}{\overset{\overset{CH_3}{|}}{CH_3-C-CH_3}}$ (from *iso*-PrOH)

(c) $\underset{}{\overset{\overset{Ph}{|}}{CH_3-C=CH_2}}$ $\overset{H^+,\ heat}{\longleftarrow}$ $\underset{\underset{OH}{|}}{\overset{\overset{Ph}{|}}{CH_3-C-CH_3}}$ ⟵ (b)

(d) $\underset{}{\overset{\overset{CH_3}{|}}{CH_3CH_2-C=CH_2}}$ $\overset{KOH}{\underset{alc.}{\longleftarrow}}$ $\overset{PBr_3}{\longleftarrow}$ $\underset{}{\overset{\overset{H_3C}{|}}{CH_3CH_2C+CH_2OH}}$ ⟵ *sec*-BuMgBr + HCHO (from MeOH)

(e) $\underset{}{\overset{\overset{CH_3}{|}}{CH_3CH_2CHCH_3}}$ $\overset{H_2,\ Pt}{\longleftarrow}$ $\underset{}{\overset{\overset{CH_3}{|}}{CH_3CH_2C=CH_2}}$ ⟵ (d)

(f) $\underset{\underset{Br\ \ Br}{|\quad|}}{\overset{\overset{CH_3}{|}}{CH_3CH_2C-CH_2}}$ $\overset{Br_2,\ CCl_4}{\longleftarrow}$ $\underset{}{\overset{\overset{CH_3}{|}}{CH_3CH_2C=CH_2}}$ ⟵ (d)

(g) $\underset{\underset{OH}{|}^b}{n\text{-}Pr+CH+Et}^a$ ⟵ { ^{a} *n*-PrMgBr + OHC—Et (both from *n*-PrOH)
^{b} *n*-PrCHO (from *n*-BuOH) + EtMgBr (from EtOH) }

(h) $\underset{\underset{O}{||}}{n\text{-}Pr-C-Et}$ $\overset{K_2Cr_2O_7,\ H^+}{\longleftarrow}$ $\underset{\underset{OH}{|}}{n\text{-}Pr-CH-Et}$ ⟵ (g)

(i) $\underset{\underset{OH}{|}}{\overset{\overset{Et}{|}}{n\text{-}Pr-C-n\text{-}Pr}}$ ⟵ $\underset{\underset{O}{||}}{\overset{\overset{Et}{|}}{n\text{-}Pr-C}}$ (from h) + *n*-PrMgBr (from *n*-PrOH)

(j) $\underset{\underset{Br}{|}}{\overset{\overset{CH_3}{|}}{n\text{-}Bu-C-CH_3}}$ $\overset{HBr}{\longleftarrow}$ $\underset{\underset{OH}{|}}{\overset{\overset{CH_3}{|}}{n\text{-}Bu+C-CH_3}}$ ⟵ *n*-BuMgBr (from *n*-BuOH) + $\underset{\overset{CH_3}{|}}{O=C-CH_3}$ (from *i*-PrOH)

(k) Me—C≡CH $\overset{NaNH_2}{\longleftarrow}$ $\overset{KOH(alc)}{\longleftarrow}$ $\underset{\underset{Br\ \ Br}{|\quad|}}{Me-CH-CH_2}$ $\overset{Br_2,\ CCl_4}{\longleftarrow}$ MeCH=CH$_2$ $\overset{H^+,\ heat}{\longleftarrow}$ *iso*-PrOH

(l)

product (k) $\xrightarrow{\text{NaNH}_2}$ MeC≡CNa

(m) PhCHCH$_3$ $\xleftarrow{\text{HCl}}$ Ph$\vdash$CHCH$_3$ ⟵ PhMgBr + OHCCH$_3$ (from EtOH)
 | |
 Cl   OH

(n) C$_6$H$_5$$\vdash$CHCH$_2CH_3$ $\xleftarrow{\text{AlCl}_3}$ C$_6$H$_6$ + sec- BuCl (from sec-BuOH)
        |
        CH$_3$

(o) CH$_3$CH—C—CH$_3$ $\xleftarrow{\text{CrO}_3}$ CH$_3$CH$\vdash$CH—CH$_3$ ⟵ i-PrMgBr + OHCCH$_3$ (from EtOH)
      | ‖ | |
      CH$_3$ O   CH$_3$ OH

(p)
$\quad^a$ ⟶ (i-Pr)$_2$CuLi (from i-PrBr) + n-BuBr (from n-BuOH)

CH$_3$CH$_2$CH$_2$CH$_2$$\vdash$CHCH$_3$ ⟵
                    ^{a}|
                      CH$_3$

$\downarrow$ H$_2$, Pt

CH$_3$CH$_2$CH$_2$CH=C—CH$_3$ $\xleftarrow{\text{H}^+,\ \text{heat}}$ CH$_3$CH$_2$CH$_2$CH$_2$$\vdash$C—CH$_3$
              | ^{b}|
              CH$_3$ CH$_3$
                                                       |
                                                      OH

$\uparrow$ b

CH$_3$
|
O=C—CH$_3$ + n-BuMgBr
(from i-PrOH) (from n-BuOH)

(q) PhCH$_2$CCH$_3$ $\xleftarrow{\text{K}_2\text{Cr}_2\text{O}_7}$ PhCH$_2$CH$\vdash$CH$_3$ ⟵ CH$_3$MgBr + PhCH$_2$CHO
        ‖ |
        O   OH

PhCH$_2$CHO $\xleftarrow{\text{K}_2\text{Cr}_2\text{O}_7}$ Ph$\vdash$CH$_2$CH$_2$OH ⟵ PhMgBr + H$_2$C—CH$_2$ (Sec. 17.10)
                                                              _O_/

(See comment about reaction of PhCH$_2$MgX with aldehydes in answer to Problem 6k, Chapter 15.)

(r) CH$_3$CH$_2$CH$_2$CH$_2$$\vdash$C—CH$_3$ ⟵ (t-Bu)$_2$CuLi + n-BuBr
                  | (from t-BuOH) (from n-BuOH)
                CH$_3$
                 |
                CH$_3$

(s) PhCH$_2$CHCH$_3$ $\xleftarrow{\text{PBr}_3}$ PhCH$_2$CHCH$_3$ (made in q)
       | |
      Br   OH

(t) *n*-Pr┆C≡C—Et $\xleftarrow{\hspace{1cm}}$

                        n-PrBr (from *n*-PrOH)

                     NaC≡C—Et $\xleftarrow{\text{NaNH}_2}$ HC≡C—Et $\xleftarrow{\text{NaNH}_2}$ $\xleftarrow[\text{(alc)}]{\text{KOH}}$ CH$_2$—CH—Et
                                                               | |
                                                               Br   Br

                                                                Br$_2$ | CCl$_4$

                                                            CH$_2$=CH—Et
                                                           (from *n*-BuBr)

(u) CH$_3$CH$_2$C—OEt $\xleftarrow[\text{H}^+]{\text{EtOH}}$ CH$_3$CH$_2$COOH $\xleftarrow[\text{heat}]{\text{KMnO}_4}$ CH$_3$CH$_2$CH$_2$OH
            ‖
            O

10. (a) CH$_3$—CH—^{14}CH$_2$OH $\xleftarrow{i\text{-PrMgBr}}$ H^{14}CHO $\xleftarrow{\text{Cu, heat}}$ ^{14}CH$_3$OH
         |
       CH$_3$

(b) CH$_3$—^{14}CH—CH$_2$OH $\xleftarrow{\text{HCHO}}$ CH$_3$—^{14}CHMgBr $\xleftarrow{\text{Mg}}$ $\xleftarrow{\text{PBr}_3}$ CH$_3$—^{14}CHOH
        | |
     CH$_3$ CH$_3$

CH$_3$—^{14}CHOH $\xleftarrow{\text{CH}_3\text{MgBr}}$ CH$_3$—^{14}CHO $\xleftarrow{\text{K}_2\text{Cr}_2\text{O}_7}$ CH$_3$—^{14}CH$_2$OH
    |
 CH$_3$

CH$_3$—^{14}CH$_2$OH $\xleftarrow{\text{CH}_3\text{MgBr}}$ H^{14}CHO $\xleftarrow{\text{Cu, heat}}$ ^{14}CH$_3$OH

(c) ^{14}CH$_3$—CH—CH$_2$OH $\xleftarrow{\text{HCHO}}$ ^{14}CH$_3$—CHMgBr $\xleftarrow{\text{Mg}}$ $\xleftarrow{\text{PBr}_3}$ ^{14}CH$_3$—CHOH
         | |
      CH$_3$ CH$_3$

^{14}CH$_3$—CHOH $\xleftarrow{\text{CH}_3\text{CHO}}$ ^{14}CH$_3$MgBr $\xleftarrow{\text{Mg}}$ $\xleftarrow{\text{HBr}}$ ^{14}CH$_3$OH
      |
   CH$_3$

(d) CH$_3$—CH=^{14}CH$_2$ $\xleftarrow[\text{alc.}]{\text{KOH}}$ CH$_3$—CH$_2$—^{14}CH$_2$Br $\xleftarrow{\text{HBr}}$ CH$_3$CH$_2$—^{14}CH$_2$OH

CH$_3$CH$_2$—^{14}CH$_2$OH $\xleftarrow{\text{CH}_3\text{CH}_2\text{MgBr}}$ H^{14}CHO $\xleftarrow{\text{Cu, heat}}$ ^{14}CH$_3$OH

(e) CH$_3$—^{14}CH=CH$_2$ $\xleftarrow{\text{H}^+, \text{heat}}$ CH$_3$—^{14}CH—CH$_3$ (made in b)
                                             |
                                             OH

(f) $^{14}CH_3-CH=CH_2$ $\xleftarrow{KOH(alc)}$ $^{14}CH_3CH_2CH_2Br$ $\xleftarrow{PBr_3}$ $^{14}CH_3CH_2CH_2OH$

$^{14}CH_3CH_2CH_2OH$ $\xleftarrow{\underset{O}{H_2C-CH_2}}$ $^{14}CH_3MgBr$ (made in c)

(g) PhD $\xleftarrow{D_2O}$ PhMgBr $\xleftarrow{Mg}$ PhBr $\xleftarrow{Br_2, Fe}$ benzene

(h) PhCH$_2$D $\xleftarrow{D_2O}$ PhCH$_2$MgCl $\xleftarrow{Mg}$ PhCH$_2$Cl $\xleftarrow{Cl_2, heat}$ toluene

(i) p-CH$_3$C$_6$H$_4$D $\xleftarrow{D_2O}$ p-CH$_3$C$_6$H$_4$MgBr $\xleftarrow{Mg}$ p-CH$_3$C$_6$H$_4$Br $\xleftarrow{\frac{Br_2}{Fe}}$ toluene

(j) $\underset{D}{CH_3CH_2CH^{14}CH_3}$ $\xleftarrow{D_2O}$ $\underset{MgBr}{CH_3CH_2CH^{14}CH_3}$ $\xleftarrow{Mg}$ $\underset{Br}{CH_3CH_2CH^{14}CH_3}$

$\underset{Br}{CH_3CH_2CH^{14}CH_3}$ $\xleftarrow{PBr_3}$ $\underset{OH}{CH_3CH_2CH^{14}CH_3}$ $\xleftarrow{CH_3CH_2CHO}$ $^{14}CH_3MgBr$ (made in c)

11. (a) Elimination is evidently *anti*, which can take place only with the orientation observed:

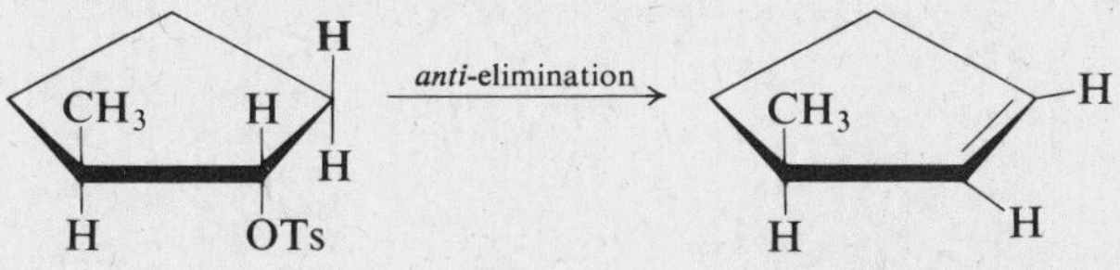

3-Methylcyclopentene

If it were *syn*, elimination could take place with either orientation, and almost certainly would give chiefly the more stable alkene, 1-methylcyclopentene:

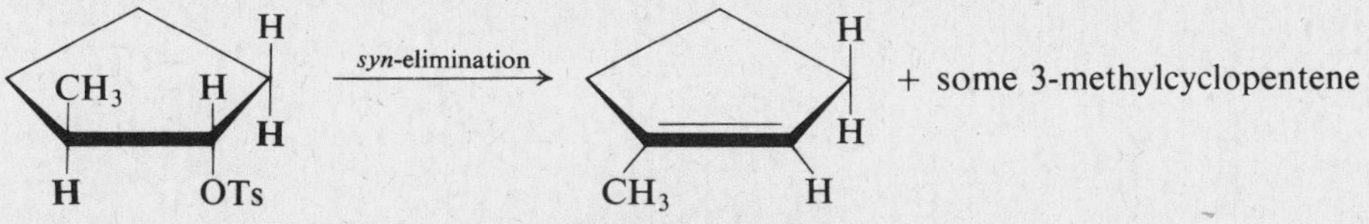

1-Methylcyclopentene
Not obtained

(Note the contrast to the behavior of *trans*-2-phenylcyclopentyl tosylate, Problem 14.17, page 484. There, the more powerful effect of the phenyl group forced Saytzeff orientation even though it had to be *syn*.)

(b)

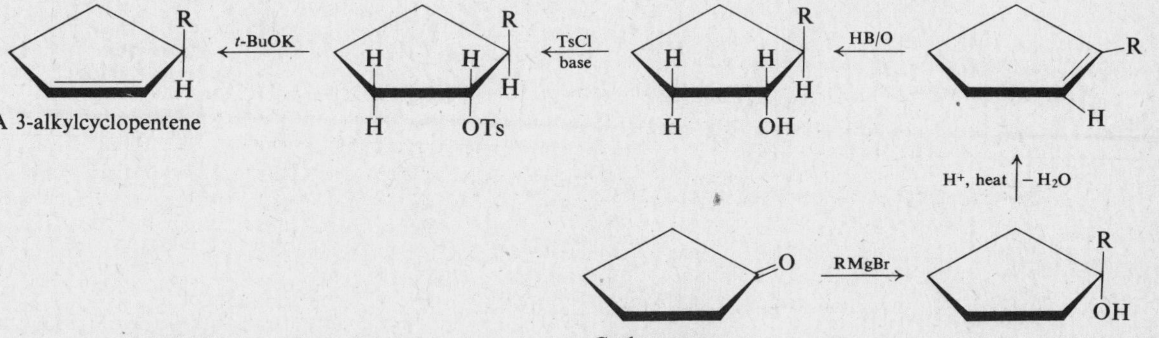

A 3-alkylcyclopentene

Cyclopentanone

(c) Inversion at C–2 during replacement of —OH by —X would give the *cis*-2-halo-1-alkyl-cyclopentane, which would undergo *anti*-elimination to yield chiefly 1-alkylcyclopentene, the more stable alkene. But this would simply bring us back to the same alkene that we had subjected to hydroboration-oxidation (HB/O).

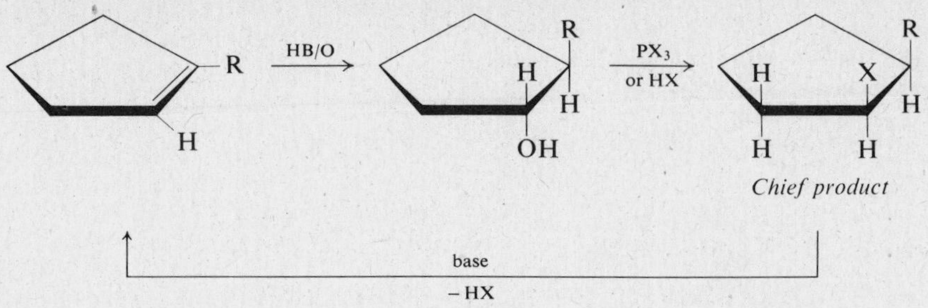

Chief product

base
– HX

12. These are not terribly difficult syntheses, if we realize that most of the molecule is simply going along for the ride.

(a) The key step is anti-Markovnikov introduction of —OH.

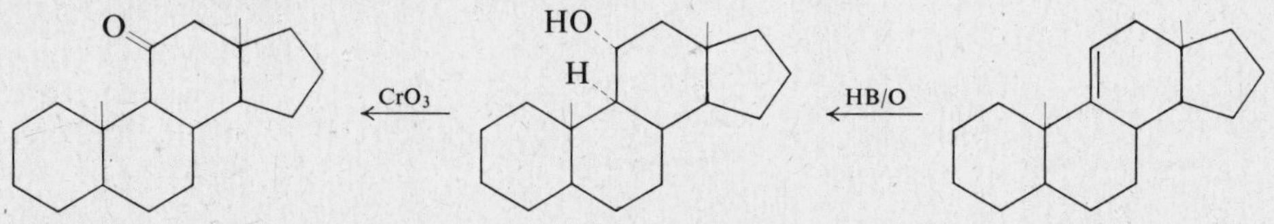

Androstan-11-one   Androstan-11α-ol   Androst-9(11)-ene

(b) Again, anti-Markovnikov introduction of —OH is needed.

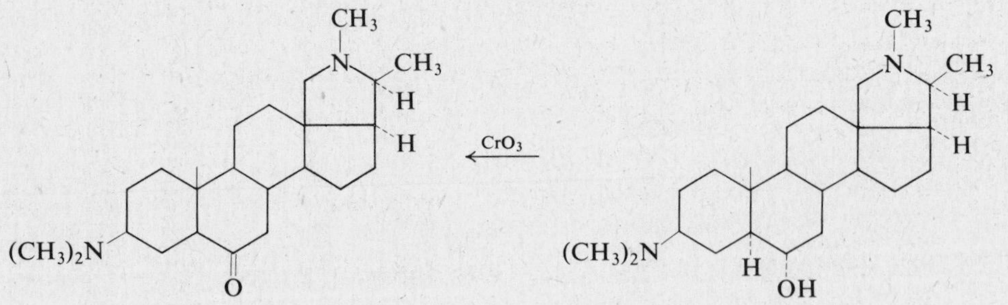

3β-Dimethylaminoconanin-6-one   3β-Dimethylaminoconanin-6α-ol

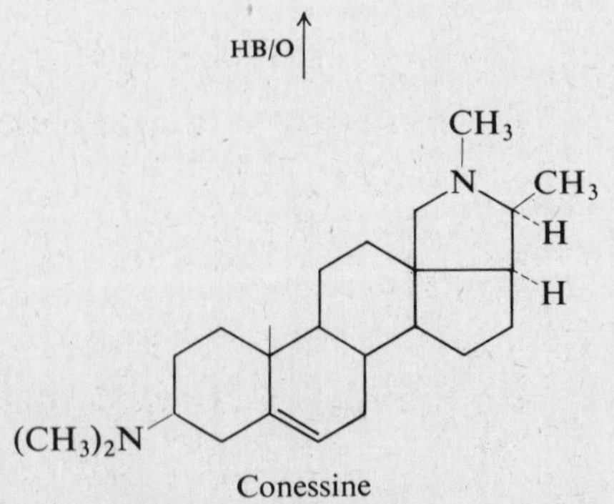

HB/O

Conessine
(3β-Dimethylaminocon-5-enine)

(c) The transformation at C–3 is done last:

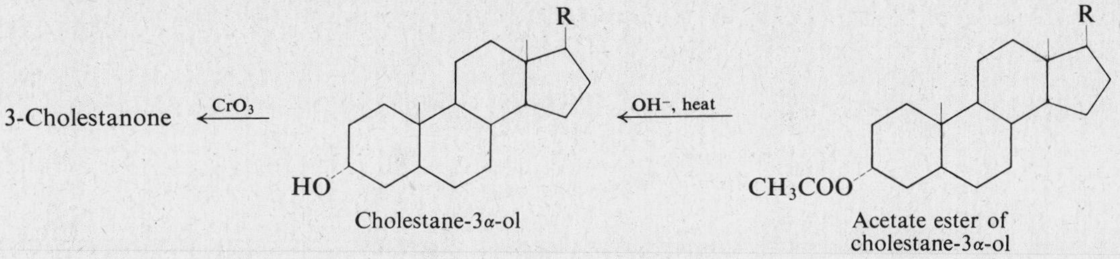

3-Cholestanone $\xleftarrow{\text{CrO}_3}$ $\xleftarrow{\text{OH}^-,\ \text{heat}}$

Cholestane-3α-ol   Acetate ester of
                                                                 cholestane-3α-ol

First, to get the acetate of cholestane-3α-ol, we build the saturated eight-carbon chain R:

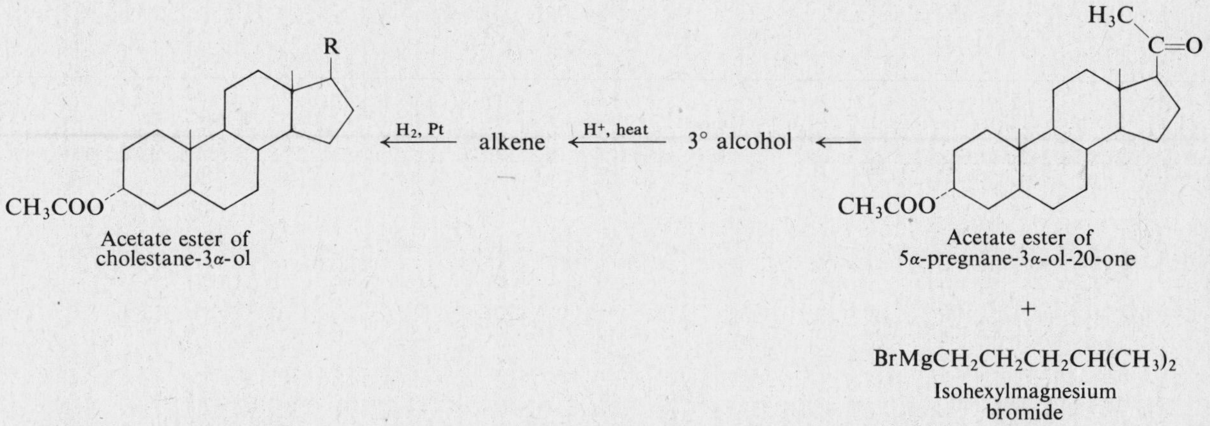

$\xleftarrow{\text{H}_2,\ \text{Pt}}$ alkene $\xleftarrow{\text{H}^+,\ \text{heat}}$ 3° alcohol $\longleftarrow$

Acetate ester of   Acetate ester of
cholestane-3α-ol   5α-pregnane-3α-ol-20-one

+

$BrMgCH_2CH_2CH_2CH(CH_3)_2$
Isohexylmagnesium
bromide

Alternative sequences would not work. We cannot introduce the C–3 keto group before we build the chain, since that keto group would react with the Grignard reagent, too. A free —OH at C–3 would decompose the Grignard reagent—or, if things got that far, undergo dehydration along with the 3° alcohol. These problems are solved by *protecting* the —OH as an ester. Esters do react with Grignard reagents, but not so rapidly as ketones. Once the side-chain has been built, the ester can conveniently be converted into the alcohol.

13. (a) $H_2C{=}CH_2 \xrightarrow[\text{H}_2\text{O}]{\text{Cl}_2}$ $\underset{\underset{\text{A}}{\overset{|}{\text{Cl}}\ \overset{|}{\text{OH}}}}{H_2C{-}CH_2}$ $\xrightarrow{\text{H}_2\text{O, base}}$ $\underset{\underset{\text{B}}{\overset{|}{\text{HO}}\ \overset{|}{\text{OH}}}}{H_2C{-}CH_2}$

$\downarrow{\scriptsize \text{oxidn.}}\ {\scriptsize \text{HNO}_3}$

(b) $\underset{\underset{\text{C}}{\overset{|}{\text{Cl}}}}{H_2C{-}COOH}$ $\xrightarrow{\text{H}_2\text{O}}$ $\underset{\underset{\text{D}}{\overset{|}{\text{OH}}}}{H_2C{-}COOH}$

(c) 6HCOOH ← $\underset{OH}{\overset{H}{HO-\overset{|}{\underset{|}{C}}-OH}}$ $\underset{OH}{\overset{H}{HO-\overset{|}{\underset{|}{C}}-OH}}$ $\underset{OH}{\overset{H}{HO-\overset{|}{\underset{|}{C}}-OH}}$ $\underset{OH}{\overset{H}{HO-\overset{|}{\underset{|}{C}}-OH}}$ $\underset{OH}{\overset{H}{HO-\overset{|}{\underset{|}{C}}-OH}}$ $\underset{OH}{\overset{H}{HO-\overset{|}{\underset{|}{C}}-OH}}$

$\uparrow$ 5HIO$_4$

$HO-\overset{H}{\underset{OH}{C}}+\overset{H}{\underset{OH}{C}}+\overset{H}{\underset{OH}{C}}+\overset{H}{\underset{OH}{C}}+\overset{H}{\underset{OH}{C}}+\overset{H}{\underset{OH}{C}}-OH$ ←HIO$_4$ $\begin{matrix} & CHOH & \\ HOCH & & CHOH \\ HOCH & & CHOH \\ & CHOH & \end{matrix}$

E

(d) $CH_3(CH_2)_7\overset{H}{C}=O + O=\overset{H}{C}(CH_2)_7COOH$ ←HIO$_4$ $CH_3(CH_2)_7-\underset{OH}{CH}+\underset{OH}{CH}-(CH_2)_7COOH$

G

$\uparrow$ HCO$_2$OH

$CH_3(CH_2)_7-CH=CH-(CH_2)_7COOH$

F

(e) $CH_2=CHCH_2OH$ →Br$_2$ $\underset{Br\ Br}{CH_2CHCH_2OH}$ $\xrightarrow[\text{oxidn.}]{HNO_3}$ $\underset{Br\ Br}{CH_2CHCOOH}$ →Zn $CH_2=CHCOOH$

H   I   J

(f) $\underset{Br\ Br\ Br}{H_2C-\overset{H}{C}-CH_2}$ →KOH $\underset{Br\quad Br}{H_2CCH=CH}$ $\xrightarrow{H_2O,\ OH^-}$ $\underset{OH\quad Br}{H_2CCH=CH}$ →KOH $\underset{OH}{H_2C-C\equiv CH}$

K   L   M

(Note that the allylic Br of K is displaced, not the vinylic Br.)

(g) $\underset{Cl}{\overset{Cl}{CH_3-\overset{|}{\underset{|}{C}}-CH_3}}$ →OH$^-$ $\left[\underset{OH}{\overset{OH}{CH_3-\overset{|}{\underset{|}{C}}-CH_3}}\right]$ $\xrightarrow{-H_2O}$ $\underset{O}{\overset{}{CH_3-\overset{||}{C}-CH_3}}$

N   O

Unstable gem-diol

(h) $CH_3C\equiv CH$ $\xrightarrow{Cl_2,\ H_2O}$ $\left[\underset{HO\ Cl}{\overset{HO\ Cl}{CH_3-\overset{|}{\underset{|}{C}}-\overset{|}{\underset{|}{CH}}}}\right]$ $\xrightarrow{-H_2O}$ $\underset{O\ Cl}{\overset{Cl}{CH_3-\overset{||}{C}-\overset{|}{CH}}}$ →Cl$_2$ $\underset{O}{CH_3-\overset{||}{C}-CCl_3}$

P   Q   R

Unstable gem-diol

190

That the orientation of addition to give P, Q, and R is as shown is proved by cleavage of R through the haloform reaction (Sec. 16.11):

$$CH_3-\underset{\underset{R}{\overset{\parallel}{O}}}{C}-CCl_3 \xrightarrow{NaOH} HCCl_3 + CH_3COO^-Na^+$$

Chloroform   S

(i)

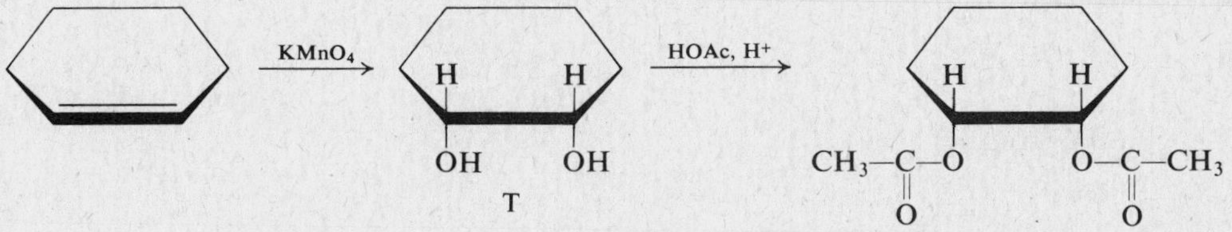

T   U

(j)
$$
\begin{array}{cc}
W & C_9H_{14}O_6 \\
V & -C_3H_8O_3 \\
\hline
 & C_6H_6O_3
\end{array}
\qquad
\begin{array}{c}
R-O-\overset{\overset{\displaystyle O}{\parallel}}{C}-CH_3 \\
-R-O-H \\
\hline
C_2H_2O
\end{array}
\qquad
\frac{C_6H_6O_3}{C_2H_2O} = 3
$$

For every —OH esterified by acetic acid, the formula is increased by C_2H_2O. Therefore V has 3 —OH groups. Since V is stable, it is not a *gem*-diol, and each —OH group must be on a different carbon. V is glycerol; W is its triacetate.

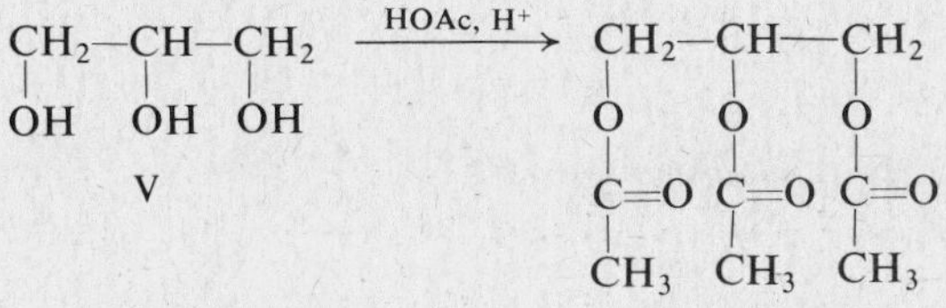

$$
\underset{V}{\underset{\underset{OH\ \ \ OH\ \ OH}{|\ \ \ \ |\ \ \ |}}{CH_2-CH-CH_2}} \xrightarrow{HOAc,\ H^+} \underset{W}{CH_2-CH-CH_2}
$$

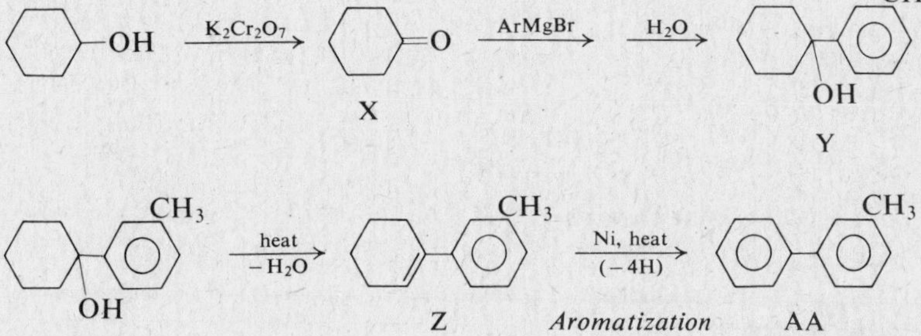

(k)

X   Y

Y   Z   *Aromatization*   AA

3-Methylbiphenyl

(l) We note application of our general principles of page 49 of this Study Guide: (1) since no bond is broken to the original chiral center, configuration is maintained about that center; (2) each time a new chiral center is generated, *both* possible configurations about that center result, and mixtures of diastereomers are obtained.

191

$$
\begin{array}{c}
\text{CH}_2\text{Br} \\
\text{H} \overset{|}{-\!\!-\!\!-} \text{CH}_3 \\
\text{i-Bu} \\
\text{(R)-}
\end{array}
\xrightarrow{\text{Mg}}
\begin{array}{c}
\text{CH}_2\text{MgBr} \\
\text{H} \overset{|}{-\!\!-\!\!-} \text{CH}_3 \\
\text{i-Bu} \\
\text{BB}
\end{array}
\xrightarrow{\text{i-BuCHO}}
\begin{array}{cc}
\begin{array}{c}
\text{i-Bu} \\
\text{H} \overset{|}{-\!\!-\!\!-} \text{OH} \\
\text{CH}_2 \\
\text{H} \overset{|}{-\!\!-\!\!-} \text{CH}_3 \\
\text{i-Bu}
\end{array}
&
\begin{array}{c}
\text{i-Bu} \\
\text{HO} \overset{|}{-\!\!-\!\!-} \text{H} \\
\text{CH}_2 \\
\text{H} \overset{|}{-\!\!-\!\!-} \text{CH}_3 \\
\text{i-Bu}
\end{array}
\end{array}
$$

Mixture of diastereomers
CC

$$
\text{CC} \xrightarrow{\text{CrO}_3}
\begin{array}{c}
\text{i-Bu} \\
\text{C}\!=\!\text{O} \\
\text{CH}_2 \\
\text{H} \overset{|}{-\!\!-\!\!-} \text{CH}_3 \\
\text{i-Bu} \\
\text{DD}
\end{array}
\xrightarrow[\text{}]{\text{CH}_3\text{MgBr}} \xrightarrow{\text{H}_2\text{O}}
\begin{array}{cc}
\begin{array}{c}
\text{i-Bu} \\
\text{CH}_3 \overset{|}{-\!\!-\!\!-} \text{OH} \\
\text{CH}_2 \\
\text{H} \overset{|}{-\!\!-\!\!-} \text{CH}_3 \\
\text{i-Bu}
\end{array}
&
\begin{array}{c}
\text{i-Bu} \\
\text{HO} \overset{|}{-\!\!-\!\!-} \text{CH}_3 \\
\text{CH}_2 \\
\text{H} \overset{|}{-\!\!-\!\!-} \text{CH}_3 \\
\text{i-Bu}
\end{array}
\end{array}
$$

Mixture of diastereomers
EE

$$
\text{EE} \xrightarrow[-\text{H}_2\text{O}]{\text{I}_2,\ \text{heat}}
\begin{array}{cc}
\begin{array}{c}
\text{i-Bu}\!-\!\text{C}\!-\!\text{CH}_3 \\
\parallel \\
\text{CH} \\
\text{H} \overset{|}{-\!\!-\!\!-} \text{CH}_3 \\
\text{i-Bu} \\
\text{Mixture of diastereomers}
\end{array}
&
\begin{array}{c}
\text{i-Pr}\!-\!\text{C}\!-\!\text{H} \\
\parallel \\
\text{C}\text{CH}_3 \\
\text{CH}_2 \\
\text{H} \overset{|}{-\!\!-\!\!-} \text{CH}_3 \\
\text{i-Bu} \\
\text{Mixture of diastereomers}
\end{array}
\end{array}
$$

FF

$$
\text{FF} \xrightarrow{\text{H}_2,\ \text{Ni}}
\begin{array}{cc}
\begin{array}{c}
\text{i-Bu} \\
\text{CH}_3 \overset{|}{-\!\!-\!\!-} \text{H} \\
\text{CH}_2 \\
\text{H} \overset{|}{-\!\!-\!\!-} \text{CH}_3 \\
\text{i-Bu} \\
\textit{Optically active} \\
\text{GG} \\
\text{(4R,6R)-}
\end{array}
&
\begin{array}{c}
\text{i-Bu} \\
\text{H} \overset{|}{-\!\!-\!\!-} \text{CH}_3 \\
\text{CH}_2 \\
\text{H} \overset{|}{-\!\!-\!\!-} \text{CH}_3 \\
\text{i-Bu} \\
\textit{Optically inactive (meso)} \\
\text{HH} \\
\text{(4R,6S)- or (4S,6R)-}
\end{array}
\end{array}
$$

2,4,6,8-Tetramethylnonane

14. (a)–(c) See Sec. 28.10.

(S. Winstein and H. J. Lucas, "Retention of Configuration in the Reaction of the 3-Bromo-2-butanols with Hydrogen Bromide," J. Am. Chem. Soc., **61**, 1576 (1939).)

(d) Read the first three paragraphs of Sec. 28.1.

The two bromohydrins yield the same product because they are both converted into the same intermediate bromonium ion. In the *trans* reactant, —Br is located in position for an S_N2-like displacement of —OH_2^+. For the *cis* reactant, only an S_N1-like reaction is possible: —Br waits until —OH_2^+ has left, and then forms the bridge.

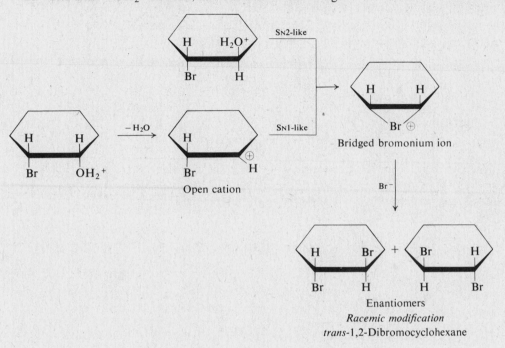

Bridged bromonium ion

Open cation

Enantiomers
Racemic modification
trans-1,2-Dibromocyclohexane

15. (a) The spectra are those of the protonated alcohols:

$$\underset{a\quad b}{CH_3OH_2^+} \qquad \underset{b}{\overset{\overset{a}{\overset{\displaystyle CH_3}{|}}}{\underset{|}{\overset{a}{CH_3}-\overset{|}{\underset{H}{C}}-\underset{c\quad d}{CH_2-OH_2^+}}}}$$

(b)–(c) Protonated isobutyl alcohol is slowly transformed (with rearrangement) into the *tert*-butyl cation,

$$\overset{a}{(CH_3)_3C^+}$$

the same cation that is given immediately, even at $-60°$, by protonated *tert*-butyl alcohol.

(G. A. Olah, J. Sommer, and E. Namanworth, "Stable Carbonium Ions. XLI. Protonated Aliphatic Alcohols and Their Cleavage to Carbonium Ions," J. Am. Chem. Soc., **89**, 3576 (1967).)

16. (a) R_3C^+ is formed in acid.

(i) $R_3COH + 2H_2SO_4 \longrightarrow R_3C^+ + H_3O^+ + 2HSO_4^-$

(ii) Conjugation of positive charge with cyclopropyl rings.

(iii) Downfield shift due to deshielding by positive charge.

The tricyclopropylcarbinyl cation, R_3C^+, is stabilized—much as the triphenylmethyl cation is (Problem 12.12, page 398)—through dispersal of the positive charge over the rings. This is believed to involve overlap of the empty p orbital with the C—C bonds of the cyclopropane rings, which (Sec. 9.9) have considerable π character. (See the answer to Problem 18, Chapter 12.)

(b)

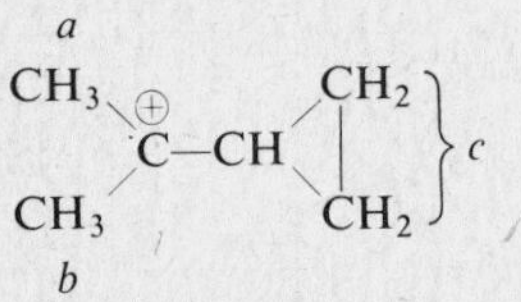

The two methyls are clearly not equivalent, since they give different nmr signals, and must therefore be located unsymmetrically. The plane of the methyls and trigonal carbon is perpendicular to, and bisects, the ring; this geometry permits the overlap described in (a):

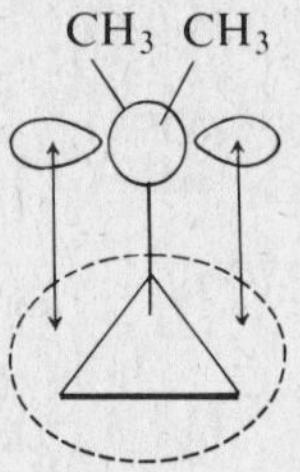

2-Cyclopropyl-2-propyl cation

The ring is *cis* to one methyl, *trans* to the other:

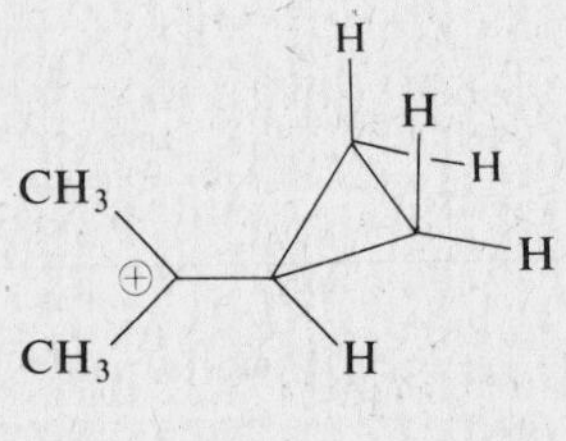

Either methyl can, of course, be *cis* or *trans* to the ring; when —CD$_3$, which gives no nmr signal, is present, only half as many molecules as before have —CH$_3$ *cis* and half as many have —CH$_3$ *trans*.

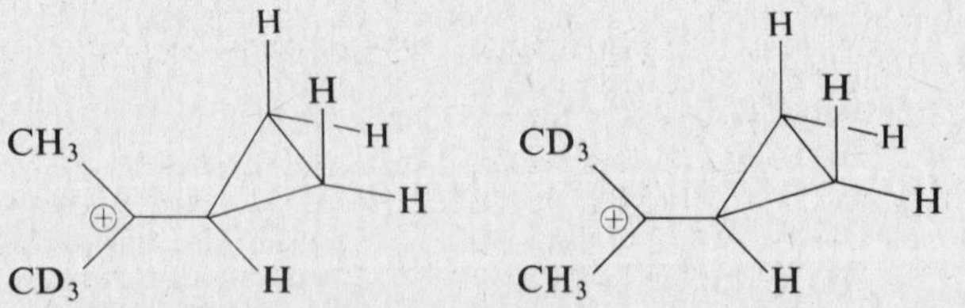

Each nmr signal is reduced to half its previous area.

(N. C. Deno *et al.*, "Carbonium Ions. XIX. The Intense Conjugation in Cyclopropyl Carbonium Ions," J. Am. Chem. Soc., **87**, 4533 (1965); C. U. Pittman, Jr., and G. A. Olah,

"Stable Carbonium Ions. XVII. Cyclopropyl Carbonium Ions and Protonated Cyclo-propyl Ketones," J. Am. Chem. Soc., **87**, 5123 (1965).)

17. II, 3-pentanol.
 JJ, 2-methyl-1-butanol or 3-methyl-1-butanol. Distinguish by α-naphthylurethane.
 KK, ethylene chlorohydrin or 4-methyl-2-pentanol. Distinguish by elemental analysis or by 3,5-dinitrobenzoate.
 LL, n-butyl alcohol or 2-pentanol. Distinguish by iodoform test.
 MM, 3-pentanol or n-butyl alcohol. Distinguish by 3,5-dinitrobenzoate.

18. (a) CrO_3/H_2SO_4 (b) Br_2/CCl_4, or CrO_3/H_2SO_4
 (c) CrO_3/H_2SO_4 (d) Br_2/CCl_4
 (e) Lucas reagent gives immediate reaction with allylic alcohol
 (f) Lucas reagent   (g) iodoform test
 (h) Br_2/CCl_4
 (i) HIO_4; then $AgNO_3$; 1,2-diol gives ppt. of $AgNO_3$
 (j) Lucas reagent, or CrO_3/H_2SO_4 (k) elemental analysis for Br
 (l) iodoform test

19. Electron-withdrawing chlorine decreases the stability of the carbonium ion, and thus lowers the rate of its formation.

20. (a) NN, $C_6H_5CH_2CHOHCH_3$ (b) OO, $C_6H_5CH(CH_3)CH_2OH$

 (1) acidic H: probably —OH   (1) same as NN

 (2) ester formed: —OH present   (2) same as NN

 (3) 1° or 2° alcohol   (3) same as NN

 (4) one side chain (3-carbon) on   (4) 1° alcohol $\longrightarrow$ —COOH
     phenyl ring

 (5) saturated side chain   (5) same as NN

 (6) $-\overset{|}{C}-\overset{|}{\underset{OH}{C}H}-CH_3$ present   (6) $-\overset{|}{C}-\overset{|}{\underset{OH}{C}H}-CH_3$ not present

 (7) chiral C present   (7) chiral C requires branched chain

21. PP, $\overset{d}{\underset{d}{Ph}}-\overset{}{\underset{H}{C}}-\overset{c}{\underset{OH}{C}}H-\overset{e}{Ph}$ QQ, $\overset{c}{Ph}-\overset{b}{C}H_2-\overset{c}{\underset{OH}{C}}-\overset{c}{Ph}$
          $\underset{b}{}\,\underset{a}{}$ $\underset{a}{}$

Distinguish 2° and 3° alcohols by CrO_3/H_2SO_4 test.

22. (a) CH₃CH₂CHCH₃   (b) CH₃—C—CH₂OH   (c) CH₃CH₂OCH₂CH₃

$$CH_3CH_2\underset{\overset{|}{OH}}{C}HCH_3 \quad CH_3-\underset{\overset{CH_3}{|}}{\overset{}{\underset{H}{C}}}-CH_2OH \quad CH_3CH_2OCH_2CH_3$$

     sec-Butyl alcohol   Isobutyl alcohol   Ethyl ether

(See the labeled spectra on page 649 of this Study Guide.)

23. (a) PhCHOHCH₃   (b) PhCH₂CH₂OH   (c) PhCH₂OCH₃

      α-Phenylethyl alcohol   β-Phenylethyl alcohol   Benzyl methyl ether

(See the labeled spectra on page 650 of this Study Guide.)

24.

$$H_2C=\underset{\overset{CH_3}{|}}{C}-CH_2OH \qquad CH_3-\underset{\overset{CH_3}{|}}{\overset{}{\underset{H}{C}}}-CH_2OH$$

  2-Methyl-2-propen-1-ol   Isobutyl alcohol,
    (Methallyl alcohol)   SS
               RR

(See the labeled spectra on page 651 of this Study Guide.)

25.

$$CH_3-\underset{\overset{}{\underset{H_3C}{|}}}{\overset{\overset{CH_3}{|}}{C}}-\underset{\overset{}{\underset{OH}{|}}}{C}H-CH_3$$

  3,3-Dimethyl-2-butanol
           TT

(See the labeled spectra on pages 651 and 652 of this Study Guide.)

26. (a) Aliphatic; 1° alcohol; C=C; yes.

(b) 9 (methyl, allylic?), 4 (methylene, allylic?), 1 (—OH), 2 (—CH₂OH), 2 (vinylic).

(c) Easily exchanged hydrogen, probably —OH.

(d) 3CH₃—, 2 —CH₂—, 2 olefinic H, one —CH₂OH.

(e) and (f)   (CH₃)₂C=CHCH₂CH₂C(CH₃)=CHCH₂OH
                        Geraniol
                Fits isoprene rule

(See the labeled spectra on page 652 of this Study Guide.)

27. Oxidation to RCHO and RCOOH without loss of carbon shows RCH_2OH.

(a)

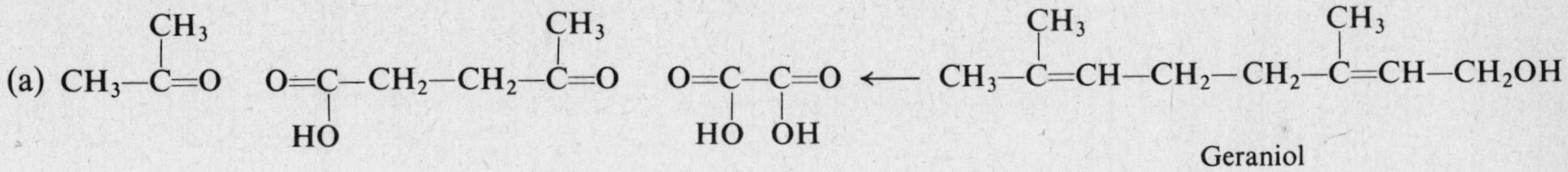

$$CH_3-\underset{\underset{HO}{|}}{\overset{\overset{CH_3}{|}}{C}}=O \qquad O=\underset{\underset{HO}{|}}{C}-CH_2-CH_2-\overset{\overset{CH_3}{|}}{C}=O \qquad O=\underset{\underset{HO}{|}}{C}-\underset{\underset{OH}{|}}{C}=O \quad \longleftarrow \quad CH_3-\overset{\overset{CH_3}{|}}{C}=CH-CH_2-CH_2-\overset{\overset{CH_3}{|}}{C}=CH-CH_2OH$$

Geraniol
(see Problem 26 above)

(b) Geometric isomers.

(c) Nerol must be the Z isomer, with —CH_2OH in the more favorable position for ring closure.

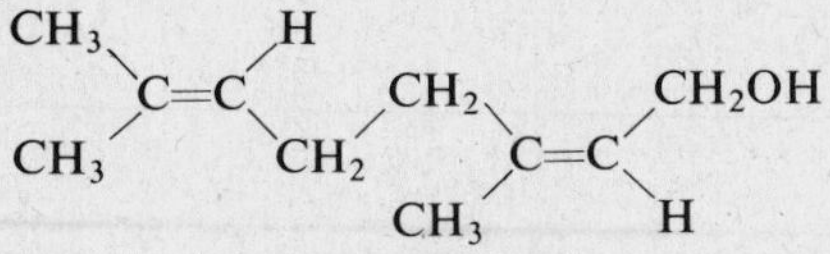

Nerol
Z

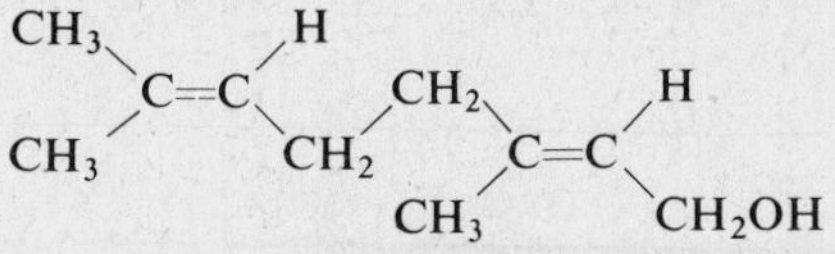

Geraniol
E

28. Both alcohols give the same allylic cation, and hence the same bromide.

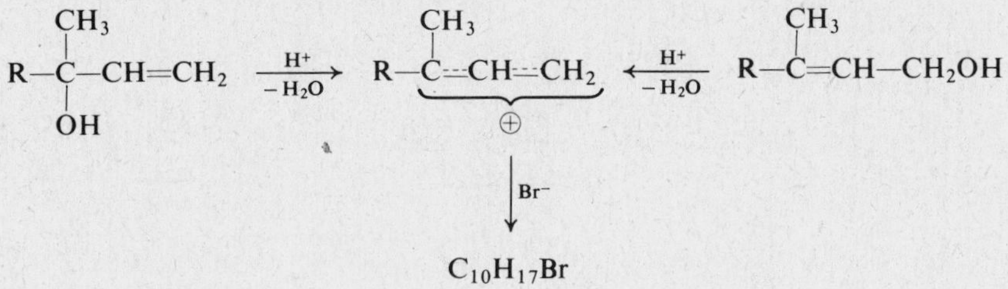

$$R-\underset{\underset{OH}{|}}{\overset{\overset{CH_3}{|}}{C}}-CH=CH_2 \quad \xrightarrow[-H_2O]{H^+} \quad R-\underset{\oplus}{\underbrace{\overset{\overset{CH_3}{|}}{C}{=}CH{=}CH_2}} \quad \xleftarrow[-H_2O]{H^+} \quad R-\overset{\overset{CH_3}{|}}{C}=CH-CH_2OH$$

$$\downarrow Br^-$$

$$C_{10}H_{17}Br$$

Chapter 17 | Ethers and Epoxides

17.1 (a) and (b)

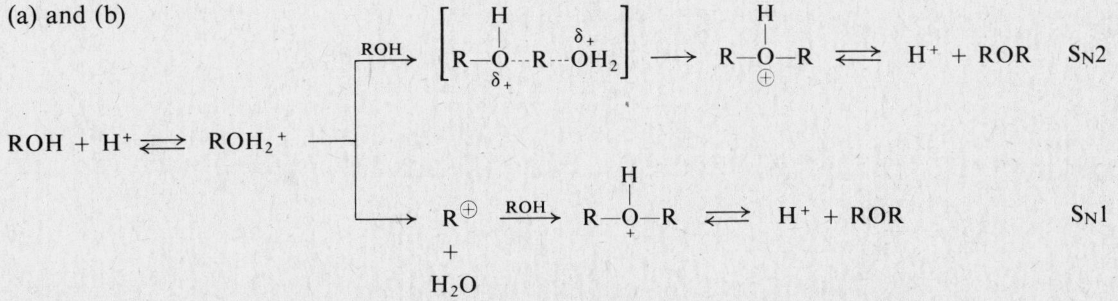

$$ROH + H^+ \rightleftharpoons ROH_2{}^+$$

$$\xrightarrow{ROH} \left[\begin{array}{c} H \\ | \\ R-\overset{\delta+}{O}\cdots R \cdots \overset{\delta+}{O}H_2 \end{array} \right] \longrightarrow R-\underset{\oplus}{\overset{H}{O}}-R \rightleftharpoons H^+ + ROR \qquad S_N2$$

$$\longrightarrow R^{\oplus} \xrightarrow{ROH} R-\underset{+}{\overset{H}{O}}-R \rightleftharpoons H^+ + ROR \qquad S_N1$$
$$+ $$
$$H_2O$$

(c) Since n-BuOH gives n-Bu$_2$O without rearrangement, presumably the reaction does not involve intermediate carbonium ions; hence it follows the S_N2 path. This is to be expected: 1° alcohols are least able to form carbonium ions but are most prone to back-side attack.

17.2 Alcohol $\longrightarrow$ alkene, most important for 3° alcohols.

17.3 (a) Ethyl ether, ethyl n-propyl ether, n-propyl ether.

(b) tert-Butyl ethyl ether in good yield, because t-BuOH gives a carbonium ion so much faster than EtOH.

$$t\text{-BuOH} \xrightarrow[-H_2O]{H^+} t\text{-Bu}^{\oplus} \xrightarrow{EtOH} t\text{-Bu}-\underset{\oplus}{\overset{H}{O}}-Et \xrightarrow{-H^+} t\text{-Bu}-O-Et$$

17.4 (a) The anions, $CH_3OSO_3{}^-$ and $SO_4{}^{--}$, are very weakly basic and hence are very good leaving groups.

(b) Sulfonic acid esters, such as the tosylates (Secs. 14.6 and 16.7).

17.5 (a) $t\text{-Bu}{-}O{+}Et$ ⟵

- $t\text{-BuO}^-\text{Na}^+$ $\xleftarrow{\text{Na}}$ $t\text{-BuOH}$
- EtBr $\xleftarrow{\text{PBr}_3}$ EtOH

(b) $n\text{-Pr}{+}O{-}Ph$ ⟵

- $n\text{-PrBr}$ $\xleftarrow{\text{PBr}_3}$ $n\text{-PrOH}$
- $\text{Na}^+\,{}^-\text{OPh}$ $\xleftarrow{\text{NaOH}}$ PhOH

(c) $sec\text{-Bu}{-}O{+}Bu\text{-}iso$ ⟵

- $sec\text{-BuO}^-\text{Na}^+$ $\xleftarrow{\text{Na}}$ $sec\text{-BuOH}$
- $iso\text{-BuBr}$ $\xleftarrow{\text{PBr}_3}$ $iso\text{-BuOH}$

(d) ⬡$-O{+}CH_3$ ⟵

- ⬡$-O^-\text{Na}^+$ $\xleftarrow{\text{Na}}$ ⬡$-OH$
- CH_3Br $\xleftarrow{\text{PBr}_3}$ CH_3OH

17.6

$PhOCH_2\langle\!\langle \bigcirc \rangle\!\rangle NO_2$ ⟵ $PhONa + BrCH_2\langle\!\langle \bigcirc \rangle\!\rangle NO_2$ $\xleftarrow[\text{heat}]{\text{Br}_2}$ $CH_3\langle\!\langle \bigcirc \rangle\!\rangle NO_2$ $\xleftarrow[\text{H}_2\text{SO}_4]{\text{HNO}_3}$ toluene

$\Big\uparrow$ NaOH

phenol

17.7 (a)–(b) No bond to the chiral carbon of 2-octanol is broken in the reaction sequence.

$$n\text{-Hex}{-}\overset{\displaystyle CH_3}{\underset{\displaystyle |}{CH}}{-}O{+}H \xrightarrow{\text{Na}} n\text{-Hex}{-}\overset{\displaystyle CH_3}{\underset{\displaystyle |}{CH}}{-}O^-\text{Na}^+ \xrightarrow{\text{EtBr}} n\text{-Hex}{-}\overset{\displaystyle CH_3}{\underset{\displaystyle |}{CH}}{-}O{-}Et$$

(−)-2-Octanol   (−)-2-Ethoxyoctane
Optical purity 83%   *Optical purity 83%*

As a result, the (−)-ether has the same configuration as the (−)-alcohol, and is of the same optical purity.

The starting alcohol is

$$\frac{8.24}{9.9} \times 100 = 83\% \text{ optically pure}$$

The product ether (−14.6°) is also 83% optically pure, which means that its rotation of −14.6° is 83% of the maximum rotation (for the pure ether). That is,

$$-14.6° = 0.83[\alpha]$$

and

$$[\alpha] = -17.6°$$

17.8 (a) The starting bromide is

$$\frac{30.3°}{34.6°} \times 100 = 87.6\% \text{ optically pure}$$

The product ether is

$$\frac{15.3°}{17.6°} \times 100 = 86.9\% \text{ optically pure}$$

and of opposite configuration. Since configuration is changed without significant loss of optical purity, the reaction proceeds with (practically) complete inversion.

(−)-2-Bromooctane (+)-2-Ethoxyoctane

Optical purity 87.6% *Optical purity 86.9%*

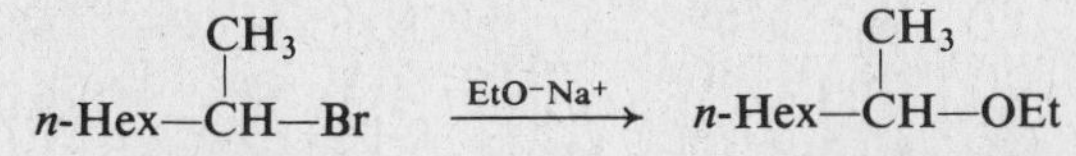

(b), (c) S_N2, as might be expected in a solvent of low polarity, and with a reactant of high basicity.

(d) S_N2 displacement of —Br from EtBr.

(e) In one there is attack (with inversion) on the chiral carbon; in the other there is retention of configuration since there is no attack on chiral carbon.

17.9

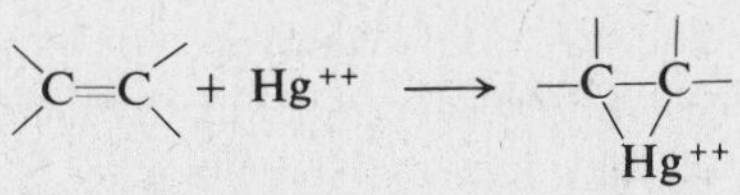

Cyclic mercurinium ion
(Sec. 15.8)

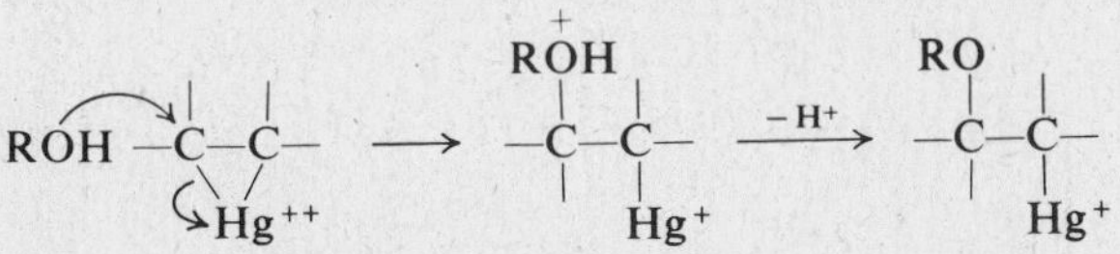

17.10 Acetate ion (OAc^-) competes with the alcohol (ROH) in attack on the cyclic mercurinium ion,

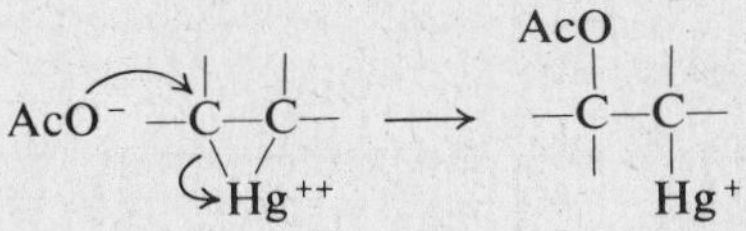

particularly when ROH is 2° or 3° and hence bulky. The less basic, less nucleophilic trifluoroacetate ion cannot compete successfully.

(H. C. Brown and M.-H. Rei, "The Solvomercuration-Demercuration of Representative Olefins in the Presence of Alcohols. Convenient Procedures for the Synthesis of Ethers," J. Am. Chem. Soc., **91**, 5646 (1969).)

17.11 (a) n-Hex—O—$\overset{\displaystyle CH_3}{\underset{}{CHCH_3}}$ ←

n-HexBr $\xleftarrow{PBr_3}$ n-HexOH

$\overset{\displaystyle CH_3}{\underset{}{CH_3CHONa}}$ $\xleftarrow{Na}$ $\overset{\displaystyle CH_3}{\underset{}{CH_3CHOH}}$

(b) n-Bu$\overset{\displaystyle CH_3}{\underset{}{CH}}$—O—$\overset{\displaystyle CH_3}{\underset{}{CHCH_3}}$ $\xleftarrow{NaBH_4}$ $\xleftarrow{Hg(OOCCF_3)_2}$

$\overset{\displaystyle CH_3}{\underset{}{CH_3CHOH}}$

n-BuCH=CH$_2$ $\xleftarrow{KOH(alc)}$ n-HexBr

n-HexBr $\xuparrow{}$ $\overset{PBr_3}{}$ n-HexOH

(c) $\langle \text{cyclohexyl} \rangle$—O—$\overset{\displaystyle CH_3}{\underset{\displaystyle CH_3}{C}}$—CH$_3$ $\xleftarrow{NaBH_4}$ $\xleftarrow{Hg(OOCCF_3)_2}$

CH_3—$\overset{\displaystyle CH_3}{\underset{\displaystyle CH_3}{C}}$—OH

$\langle \text{cyclohexene} \rangle$ $\xleftarrow{H^+,\ heat}$ $\langle \text{cyclohexyl} \rangle$—OH

(d) $\langle \text{cyclohexyl} \rangle$—O—$\langle \text{cyclohexyl} \rangle$ $\xleftarrow[-H_2O]{H_2SO_4}$ 2 $\langle \text{cyclohexyl} \rangle$—OH

17.12 Reaction of the protonated ether with bromide ion would be expected to take either of two courses: (a) if S$_N$2, by attack at the least hindered group, methyl, to yield MeBr and sec-BuOH, or (b) if S$_N$1, by formation of the more stable carbocation, sec-Bu$^+$, to yield sec-BuBr and MeOH. The evidence indicates that reaction is actually S$_N$2. (Since the sec-Bu—O bond is not broken, we observe retention of configuration and no loss of optical purity.)

17.13 (a) —OCH$_3$ > —CH$_3$

$\langle \text{benzene with } OCH_3, CH_3 \rangle$ → $\langle \text{benzene with } OCH_3, Br, CH_3 \rangle$

(b) —OCH$_3$ > —NO$_2$

$\langle \text{benzene with } OCH_3, NO_2 \rangle$ → $\langle \text{benzene with } OCH_3, NO_2, NO_2 \rangle$ + $\langle \text{benzene with } O_2N, OCH_3, NO_2 \rangle$

(c) —OR > —R

$\langle \text{benzene with } OCH_2Ph \rangle$ → $\langle \text{benzene with } OCH_2Ph, NO_2 \rangle$ + $\langle \text{benzene with } OCH_2Ph, NO_2 \rangle$

17.14

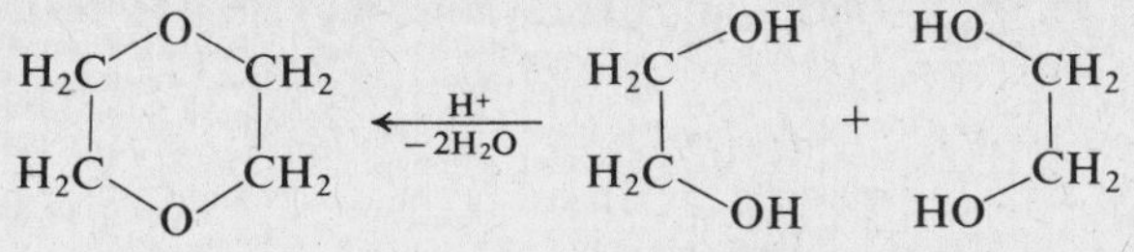

1,4-Dioxane $\xleftarrow[-2H_2O]{H^+}$ Ethylene glycol *Two moles*

17.15 Furan $\xrightarrow{H_2, Ni}$ Tetrahydrofuran $\xrightarrow[heat]{HCl}$ 4-Chloro-1-butanol $\xrightarrow[heat]{HCl}$ 1,4-Dichlorobutane

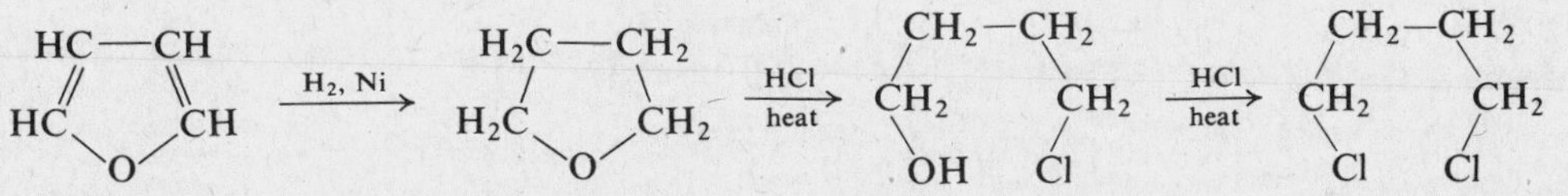

17.16 (a) $Et-O-CH_2CH_2-O \vdots CH_2CH_2-OH \xleftarrow[H^+]{} Et-O \vdots CH_2CH_2-OH \xleftarrow[H^+]{} EtOH$

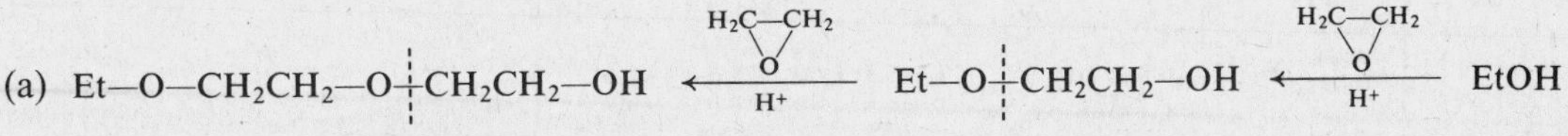

     (b) As in (a), replacing EtOH by PhOH.

     (c) As in (a), replacing EtOH by HOH.

     (d) $HO-CH_2CH_2-O-CH_2CH_2-O \vdots CH_2CH_2-OH \xleftarrow[H^+]{} HO-CH_2CH_2-O-CH_2CH_2-OH$
                                                           (From c)

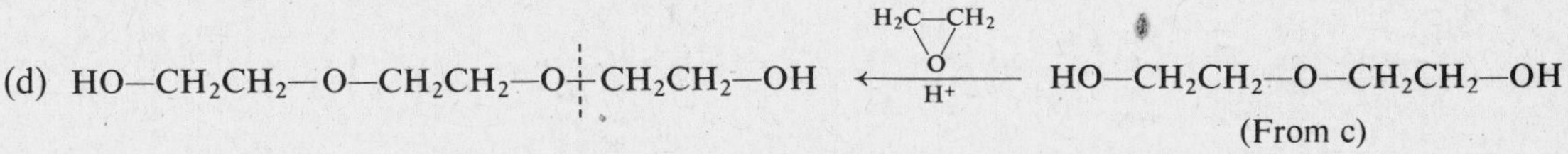

17.17 S_N1 $H^+ + CH_2-CH_2 \rightleftarrows CH_2-CH_2$

                         $CH_2-CH_2 \rightarrow CH_2-CH_2 \oplus$ *Slow*

            $H_2O + CH_2-CH_2 \oplus \rightarrow CH_2-CH_2 \xrightarrow{-H^+} CH_2-CH_2$

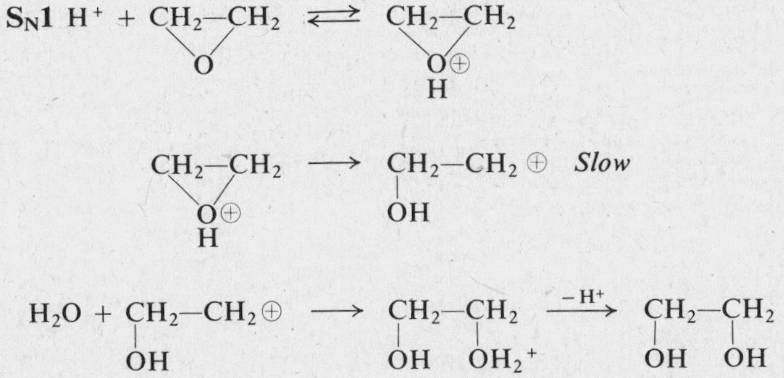

S_N2 as shown on page 569.

17.18   (i) Attachment of oxygen can occur with equal ease to either face of the alkene. In (a), the two (epoxide) structures are identical and achiral; the same is true in (b). In (c), (d), and (e), equal amounts of enantiomeric epoxides are formed. (Compare answer to Problem 7.12, page 245.)

     (ii) Back-side attack on the epoxide ring can occur with equal ease (is equally likely) at either carbon in (a), in (b), and in (c), giving in (a) and (b) equal amounts of enantiomers, in (c) the same inactive, *meso* compound, whichever enantiomer or whichever carbon is attacked. (See Figure 7.6 on page 244, and Figure 7.7 on page 245.)

     (iii) Even though attacks by the two paths—at the methyl end and at the ethyl end— are not equally likely in (d) and (e), the product is racemic in each case. There are equal amounts of enantiomeric epoxides undergoing attack. The product from one epoxide undoubtedly consists of unequal amounts of the two possible enantiomeric glycols; attack at the methyl end may, for example, be preferred. But any excess of, say, (S,S)-glycol from one epoxide stereoisomer would be exactly balanced by an excess of (R,R)-glycol from the enantiomeric epoxide. (Compare the answer to Problem 7.13, page 245.)

(a)

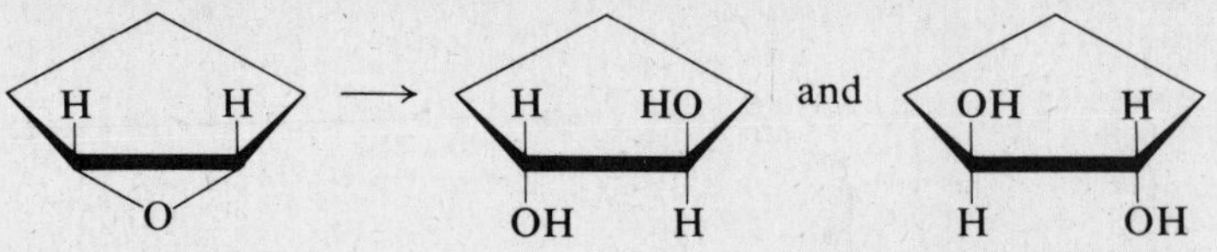

Achiral   trans-1,2-Cyclopentanediol
                                           Racemic modification

(b)

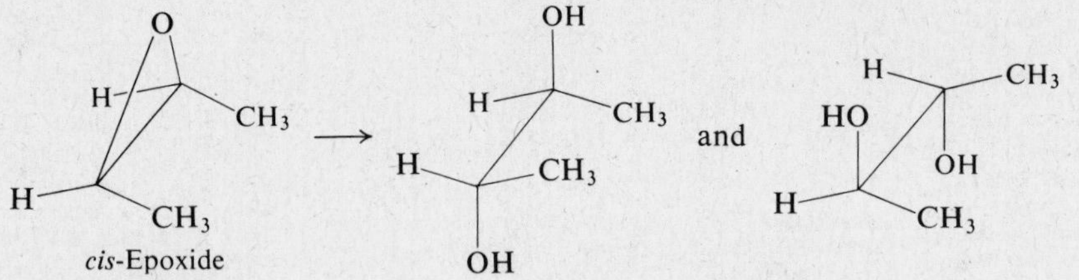

cis-Epoxide   2,3-Butanediol
Achiral   Racemic

(c)

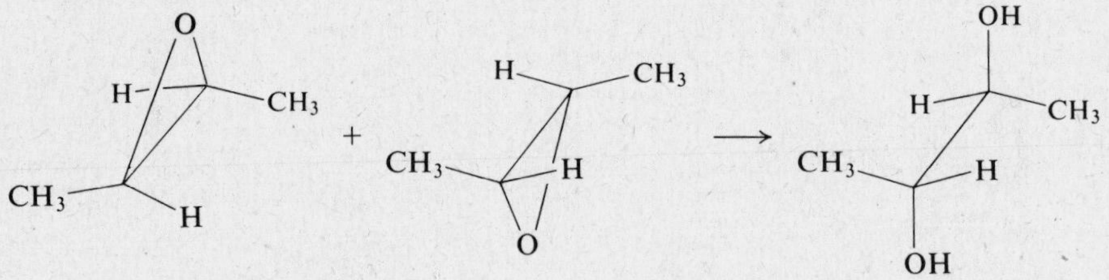

trans-Epoxide   2,3-Butanediol
Enantiomers   Meso

(d)

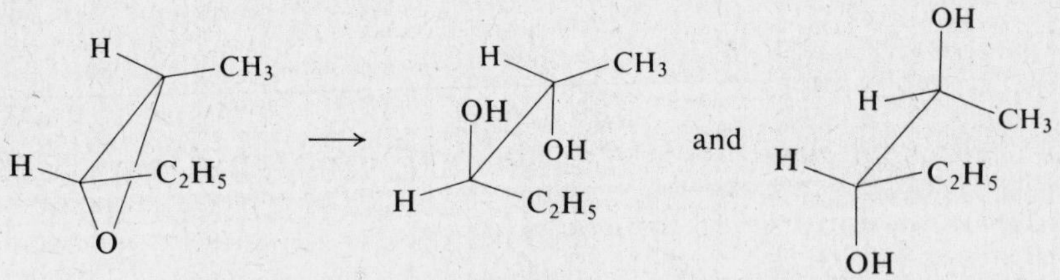

                        R,R-   S,S-

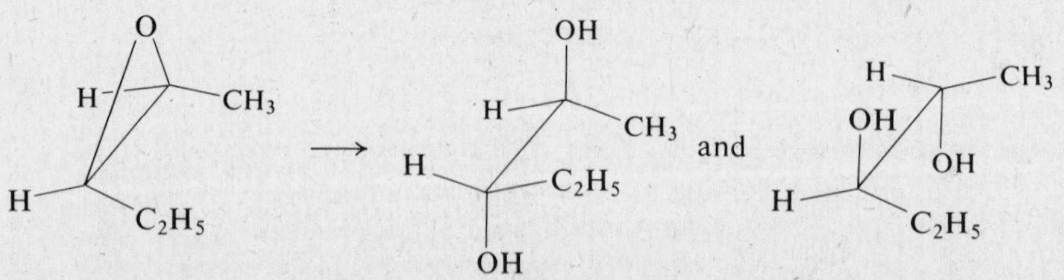

cis-Epoxide   S,S-   R,R-
Enantiomers   2,3-Pentanediol
                            Racemic modification

(e)

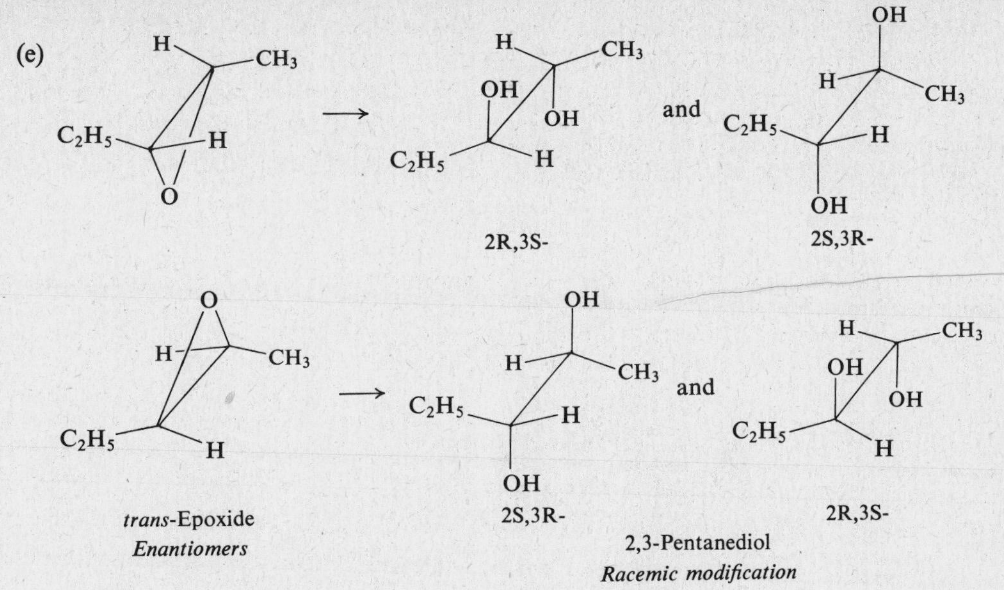

H   CH₃
C₂H₅   H
O

→

H   CH₃
OH
OH
C₂H₅   H

2R,3S-

and

OH
H
C₂H₅   CH₃
H
OH

2S,3R-

O
H   CH₃
C₂H₅   H

trans-Epoxide
Enantiomers

→

OH
H   CH₃
C₂H₅   H
OH

2S,3R-

and

H   CH₃
OH
C₂H₅   OH
H

2R,3S-

2,3-Pentanediol
Racemic modification

(f) None of the products, as obtained, would be optically active.

17.19 (a) $CH_3\overset{\frown}{O}H + H_2C\overset{\frown}{-}CH_2 \longrightarrow CH_3\overset{\underset{\oplus}{|}}{O}CH_2CH_2OH \xrightarrow{-H^+} CH_3OCH_2CH_2OH$

(with H on O⊕ and H above)

(b) $CH_3\overset{\frown}{O^-} + H_2C\overset{\frown}{-}CH_2 \longrightarrow CH_3OCH_2CH_2O^- \xrightarrow{+H^+} CH_3OCH_2CH_2OH$

(c) $Ph\overset{\frown}{N}H_2 + H_2C\overset{\frown}{-}CH_2 \longrightarrow Ph\overset{\underset{|}{\oplus}}{N}CH_2CH_2O^- \longrightarrow PhNCH_2CH_2OH$

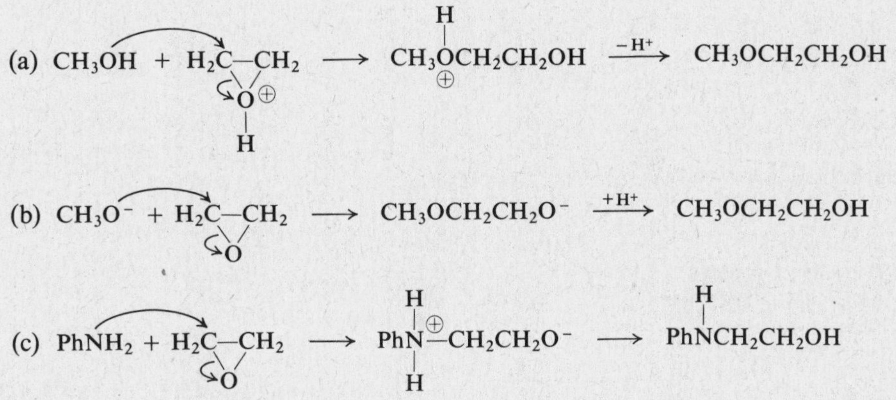

17.20 Phenoxide ion (ArO⁻) is more basic than ethylene oxide, and would gain the proton.

17.21 $CH_3CHOHCH_2OH + OH^- \rightleftarrows CH_3CHOHCH_2O^- + H_2O$

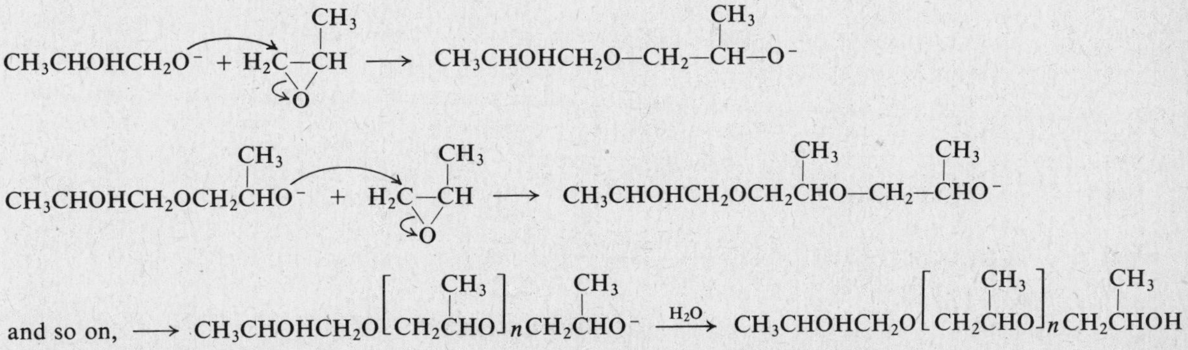

$CH_3CHOHCH_2O^- + H_2\overset{\frown}{C}\overset{\underset{\frown}{|}}{-}CH \longrightarrow CH_3CHOHCH_2O-CH_2-\overset{CH_3}{\underset{|}{CH}}-O^-$

(with CH₃ on epoxide CH and O)

$CH_3CHOHCH_2OCH_2\overset{CH_3}{\underset{|}{CH}}O^- + H_2C\overset{\underset{\frown}{|}}{-}CH \longrightarrow CH_3CHOHCH_2OCH_2\overset{CH_3}{\underset{|}{CH}}O-CH_2-\overset{CH_3}{\underset{|}{CH}}O^-$

and so on, $\longrightarrow CH_3CHOHCH_2O\left[CH_2\overset{CH_3}{\underset{|}{CH}}O\right]_n CH_2\overset{CH_3}{\underset{|}{CH}}O^- \xrightarrow{H_2O} CH_3CHOHCH_2O\left[CH_2\overset{CH_3}{\underset{|}{CH}}O\right]_n CH_2\overset{CH_3}{\underset{|}{CH}}OH$

205

17.22 (a) $Cl^- + PhHC{-}CH_2 \longrightarrow PhCH{-}CH_2$
                          O⁺   Cl   OH
                          H

(b) $PhHC{-}CH_2 + OCH_3^- \longrightarrow PhCH{-}CH_2 \xrightarrow{CH_3OH} PhCH{-}CH_2$
         O   O⁻   OCH₃   OH   OCH₃

(c) $CH_3HC{-}CH_2 + H_2NPh \longrightarrow CH_3CH{-}CH_2 \longrightarrow PhCH{-}CH_2$
         O   O⁻ NH₂Ph⁺   OH   NHPh

(d) $Cl^- + (CH_3)_2C{-}CHCH_3 \longrightarrow$ CH₃
                          O⁺ |
                          H   CH₃{-}C{-}CH{-}CH₃
                                                  | |
                                                  Cl OH

17.23 Formation of HBr as a product of a substitution reaction.

17.24 Add this row to the table.

	conc. H_2SO_4	cold $KMnO_4$	Br_2	CrO_3	fum. H_2SO_4	$CHCl_3$ $AlCl_3$	Na
Ethers	+	−	−	−	+	−	−

17.25 (a) solubility of ether in cold conc. H_2SO_4 (b) Br_2/CCl_4 or $KMnO_4$
(c) same as (b)   (d) same as (a)
(e) CrO_3/H_2SO_4 (f) warm acid followed by $KMnO_4$
(g) Br_2, accompanied by test for HBr

17.26

$$\text{mmoles AgI} = \text{mmoles } CH_3I$$

$$11.62 \text{ mg}/235 \text{ mg per mmole} = 11.62/235 \text{ mmoles } CH_3I$$

$$\text{papaverine } C_{20}H_{21}O_4N = \text{m.w. } 339$$

$$\frac{\text{mmoles } CH_3I}{\text{mmoles cpd.}} = \text{no. } OCH_3 \text{ groups per molecule} = \frac{11.62/235}{4.24/339} = 4$$

1. (a) CH_3OCH_3

(b) $(CH_3)_2CHOCH(CH_3)_2$

(c) $n\text{-}C_4H_9OCH_3$

(d) $(CH_3)_2CHCH_2OC(CH_3)_3$

(e) $n\text{-}C_3H_7CH(OCH_3)C_2H_5$

(f) $H_2C{=}CHOCH{=}CH_2$

(g) $CH_2{=}CHCH_2OCH_2CH{=}CH_2$

(h) $ClCH_2CH_2OCH_2CH_2Cl$

(i) $C_6H_5OCH_3$

(j) $C_6H_5OCH_2CH_3$

(k) $C_6H_5OC_6H_5$

(l)

(m) n-PrOCH$_2$⟨○⟩NO$_2$

(n) $CH_3CH_2CH_2HC\!-\!CH_2$
                    $\underset{O}{\diagdown\diagup}$

2. (a) isobutyl ether
   (c) ethyl *tert*-butyl ether
   (e) *p*-bromophenetole
   (g) 2,4-dibromoanisole

(b) methyl isopropyl ether
(d) 4-methoxyheptane
(f) *o*-nitrobenzyl phenyl ether

3. Each alkoxide, phenoxide, halide, or alkene is made from the corresponding alcohol or phenol.

(a) t-Bu—O$\,\vdots\,$Me ⟵ t-BuO$^-$Na$^+$ + MeBr

(b) Ph—O$\,\vdots\,$Et ⟵ PhO$^-$Na$^+$ + EtBr

(c) n-Bu$\,\vdots\,$O—⟨◯⟩ ⟵ n-BuBr + Na$^+$ $^-$O—⟨◯⟩

(d) CH$_3$⟨○⟩O$\,\vdots\,$CH$_2$⟨◯⟩ ⟵ CH$_3$⟨◯⟩O$^-$Na$^+$ + ClCH$_2$⟨○⟩

(e) i-Pr—O$\,\vdots\,$Bu-i ⟵ i-PrO$^-$Na$^+$ + i-BuBr

(f) $\underset{}{CH_3}$
      $\overset{\displaystyle CH_3 \quad\quad CH_3}{CH_3\!-\!CH\!-\!O\!-\!C\!-\!CH_3}$ $\xleftarrow{\text{NaBH}_4}$ $\xleftarrow{\text{Hg(OOCCF}_3)_2}$ $\overset{\displaystyle CH_3 \quad\quad\quad CH_3}{CH_3CHOH + CH_2\!=\!C\!-\!CH_3}$
               CH_3

(g) $\overset{\displaystyle OMe}{⟨○⟩}\!OMe$ $\xleftarrow[\text{or 2MeI}]{\text{2Me}_2\text{SO}_4}$ $\overset{\displaystyle O^-Na^+}{⟨○⟩}\!O^-Na^+$ $\xleftarrow{\text{2NaOH}}$ $\overset{\displaystyle OH}{⟨○⟩}\!OH$
                                          Resorcinol

4. Activating effect: —OH > —OMe > —CH$_3$ > —H > —Cl > —NO$_2$

(a) PhOH > PhOCH$_3$ > C$_6$H$_6$ > PhCl > PhNO$_2$

(b) m-HOC$_6$H$_4$OMe > m-MeC$_6$H$_4$OMe > o-MeC$_6$H$_4$OMe > C$_6$H$_5$OMe

(c) p-C$_6$H$_4$(OH)$_2$ > p-MeOC$_6$H$_4$OH > p-C$_6$H$_4$(OMe)$_2$

5. The products are:

(a) t-BuOEt
(c) Et$_2$O
(e) MeI and EtI
(g) (Et$_2$OH)$^+$HSO$_4$$^-$ (see page 33)
(i) PhOH and EtBr
(k) p-K$^+$ $^-$OOCC$_6$H$_4$OMe

(b) isobutylene
(d) no reaction
(f) no reaction
(h) 2EtOSO$_3$H
(j) p- and o-O$_2$NC$_6$H$_4$OEt
(l) p- and o-BrC$_6$H$_4$OCH$_2$C$_6$H$_5$

6. The ether is cleaved to isobutylene, which undergoes *acid-catalyzed* polymerization, an extension of dimerization (Sec. 6.15).

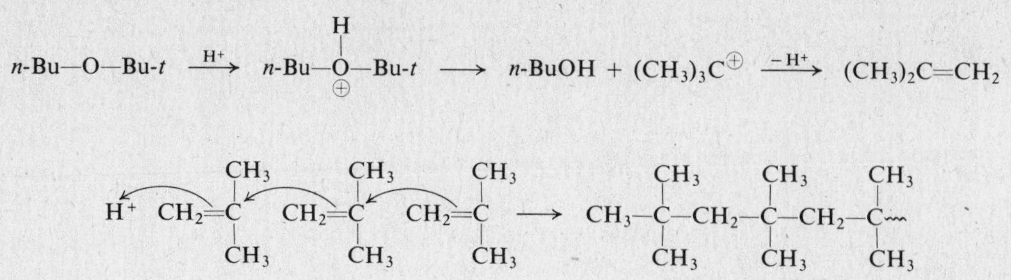

$$n\text{-Bu}-\text{O}-\text{Bu-}t \xrightarrow{\text{H}^+} n\text{-Bu}-\overset{\overset{\text{H}}{|}}{\underset{\oplus}{\text{O}}}-\text{Bu-}t \longrightarrow n\text{-BuOH} + (\text{CH}_3)_3\text{C}^{\oplus} \xrightarrow{-\text{H}^+} (\text{CH}_3)_2\text{C}{=}\text{CH}_2$$

Polyisobutylene

7. (a) CrO_3/H_2SO_4 (b) $AgNO_3$ (alc)   (c) Br_2/CCl_4 or $KMnO_4$
    (d) Br_2/CCl_4 or $KMnO_4$ (e) cold conc. H_2SO_4 (f) Br_2/CCl_4 or $KMnO_4$
    (g) cold conc. H_2SO_4

8. (a) Ether soluble in conc. H_2SO_4.

(b) Allylic ether oxidized by $KMnO_4$.

(c) Oxidation to acid and detn. of m.p.

(d) MeOH reacts with CrO_3/H_2SO_4; 1-hexene reacts with $KMnO_4$.

(e) Alcohol and ether are soluble in cold conc. H_2SO_4; alcohol reacts with CrO_3/H_2SO_4; determine others by elemental analysis.

(f) Isoprene reacts with Br_2/CCl_4 or with $KMnO_4$; pentane is insoluble in H_2SO_4.

(g) Oxidize methyl *o*-tolyl ether to acid; cleave phenetole to phenol or test for ease of substitution by bromine.

9. From the formula, A, B, and C are aromatic. A and C are aryl bromides, B has bromine on a side chain. None of them is a phenol or alcohol or aldehyde, so the oxygen is probably present in each of them in an ether linkage. There are many possibilities at this point. For example:

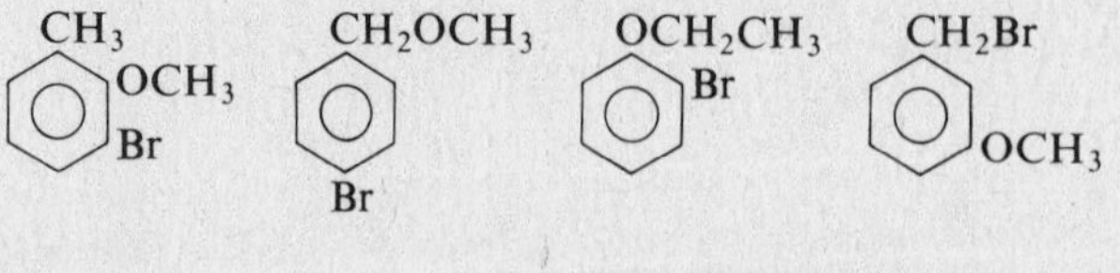

Any —CH_3 group attached to the ring could be detected by

$$\text{ArCH}_3 \xrightarrow[(-2\text{H}, +2\text{O})]{\text{KMnO}_4} \text{ArCOOH}$$

without loss of carbon.

$$C_8H_9OBr \xrightarrow[(-2H, +2O)]{KMnO_4} C_8H_7O_3Br \quad \text{Therefore A carries } -CH_3$$
$$\quad\;\; A \qquad\qquad\qquad\qquad\quad D$$

$$C_8H_9OBr \xrightarrow[(-H, -Br, +2O)]{KMnO_4} C_8H_8O_3 \quad \text{Therefore B carries } -CH_2Br$$
$$\quad\;\; B \qquad\qquad\qquad\qquad\quad\;\; E$$

$$C_8H_9OBr \xrightarrow{KMnO_4} \text{no reaction} \quad \text{Therefore C carries no } -CH_3$$
$$\quad\;\; C$$

Any $-OCH_3$ or $-OCH_2CH_3$ group can be detected by

$$ArOCH_3 \xrightarrow[(-CH_2)]{HBr} ArOH \quad \text{or} \quad ArOCH_2CH_3 \xrightarrow[(-C_2H_4)]{HBr} ArOH$$

Thus:
$$C_8H_9OBr \xrightarrow[(-CH_2)]{HBr} C_7H_7OBr \quad \text{Therefore A carries } -OCH_3$$
$$\qquad\;\; A \qquad\qquad\qquad\;\; F$$

$$C_8H_9OBr \xrightarrow[(-CH_2)]{HBr} C_7H_7OBr \quad \text{Therefore B carries } -OCH_3$$
$$\qquad\;\; B \qquad\qquad\qquad\;\; G$$

$$C_8H_9OBr \xrightarrow[(-C_2H_4)]{HBr} C_6H_5OBr \quad \text{Therefore C carries } -OCH_2CH_3$$
$$\qquad\;\; C \qquad\qquad\qquad\;\; H$$

At this point, C could be o-, m-, or p-$BrC_6H_4OC_2H_5$. Since H is identified as o-bromo-phenol, C must be o-$BrC_6H_4OC_2H_5$.

$$C_2H_5Br + \quad \underset{H}{\overset{OH}{\bigcirc}Br} \quad \xleftarrow{HBr} \quad \underset{C}{\overset{OC_2H_5}{\bigcirc}Br}$$

o-Bromophenol o-Bromophenetole

At this point, B could be o-, m-, or p-$CH_3OC_6H_4CH_2Br$. Therefore, E could be o-, m-, or p-$CH_3OC_6H_4COOH$. Cleavage of the ether group of E by HBr gives I, o-HOC_6H_4COOH, known as salicylic acid; thus B must be o-$CH_3OC_6H_4CH_2Br$.

$$CH_3Br + \underset{I}{\overset{OH}{\bigcirc}COOH} \xleftarrow{HBr} \underset{E}{\overset{OCH_3}{\bigcirc}COOH} \xleftarrow{KMnO_4} \underset{B}{\overset{OCH_3}{\bigcirc}CH_2Br} \xrightarrow{HBr} \underset{G}{\overset{OH}{\bigcirc}CH_2Br}$$

Salicylic acid o-Methoxybenzyl bromide o-Hydroxybenzyl bromide
(o-Hydroxybenzoic acid)

There are ten possibilities for A,

$$\begin{array}{c} CH_3 \\ | \\ C_6H_3-Br \\ | \\ OCH_3 \end{array}$$

a formidable list. But it is found that p-hydroxybenzoic acid can be converted into J and thence into D, showing that A is 2-bromo-4-methylanisole.

OH
(diagram, benzene ring with OH top and COOH bottom) →(CH₃)₂SO₄ / OH⁻→ →H⁺→ OCH₃ (benzene ring with OCH₃ top and COOH bottom) **J** →Br₂, Fe→ OCH₃ (benzene ring with OCH₃ top, Br right, COOH bottom) **D** ←KMnO₄← OCH₃ (benzene ring with OCH₃ top, Br right, CH₃ bottom) **A** →HBr→ OH (benzene ring with OH top, Br right, CH₃ bottom) **F**

p-Hydroxybenzoic acid

A — 2-Bromo-4-methylanisole

F — 2-Bromo-4-methylphenol

10.

a
CH₃ (benzene ring with CH₃ top labeled a, Br labeled, OCH₃ bottom labeled b, the bracket labeled c) **A**
a, singlet, 3H
b, singlet, 3H
c, multiplet, 3H

b
CH₂Br (benzene ring with CH₂Br top labeled b, OCH₃ labeled a, bracket labeled c) **B**
a, singlet, 3H
b, singlet, 2H
c, multiplet, 4H

b a
OCH₂CH₃ (benzene ring with OCH₂CH₃ top labeled b a, Br labeled, bracket labeled c) **C**
a, triplet, 3H
b, quartet, 2H
c, multiplet, 4H

11. (a) $HOCH_2CH_2OH$, ethylene glycol

(b) same as (a)

(c) $HOCH_2CH_2OEt$, 2-ethoxyethanol

(d) $HOCH_2CH_2OCH_2CH_2OEt$, ethyl ether of diethylene glycol (see e)

(e) $HOCH_2CH_2OCH_2CH_2OH$, diethylene glycol

(f) $HO(CH_2CH_2O)_3H$, triethylene glycol

(g) $BrCH_2CH_2OH$, 2-bromoethanol

(h) $N{\equiv}CCH_2CH_2OH$, 2-cyanoethanol

(i) $HCOOCH_2CH_2OH$, ethylene glycol monoformate

(j) $PhCH_2CH_2OH$, 2-phenylethanol (β-phenylethyl alcohol)

(k) $H_2NCH_2CH_2OH$, 2-aminoethanol (ethanolamine)

(l) $Et_2NCH_2CH_2OH$, 2-(N,N-diethylamino)ethanol

(m) $PhOCH_2CH_2OH$, 2-phenoxyethanol

(n) same as (m)

(o) $HOCH_2CH_2C{\equiv}CH$, 3-butyn-1-ol

12. Alkaline hydrolysis takes place with retention of configuration because attack by OH⁻ occurs at 1° carbon; no bond to the chiral carbon is broken. Acidic hydrolysis takes place with inversion of configuration because attack by H_2O on the protonated epoxide occurs at the 2° (in this case, chiral) carbon atom. (See Figure 17.7, page 211 of this Study Guide.)

13. (a) Nucleophilic substitution in which both nucleophile and substrate are parts of the same molecule.

(b)

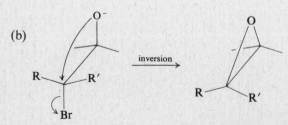

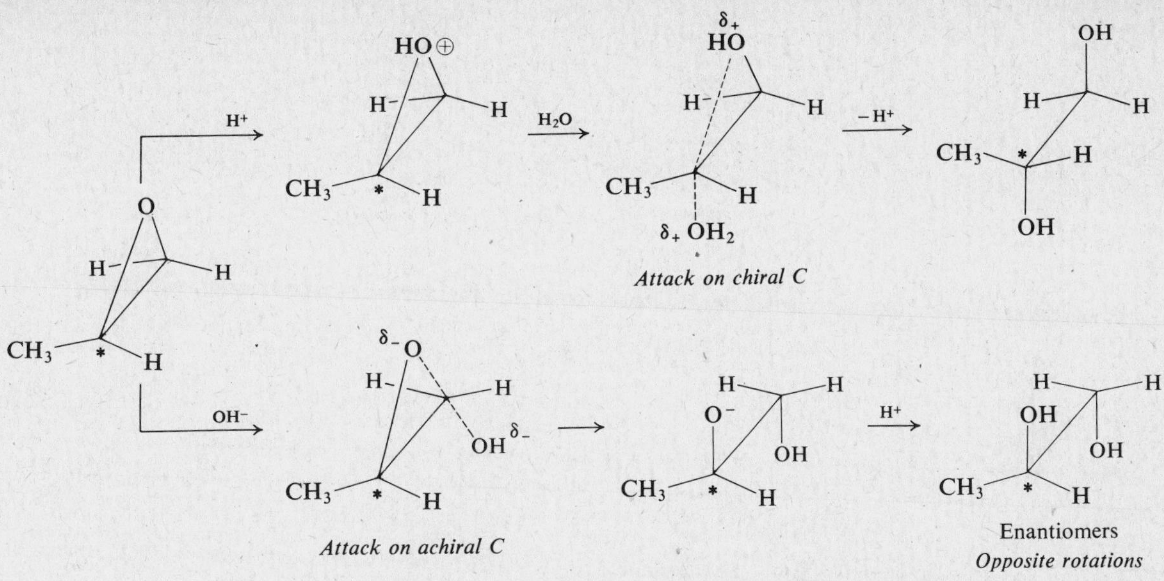

Attack on chiral C

Attack on achiral C

Enantiomers
Opposite rotations

Figure 17.7 Hydrolysis of optically active propylene oxide (Problem 12).

(c) In the *trans*-isomer, it is possible for both —O⁻ and —Cl to be axial and thus permit back-side (*anti*-periplanar) displacement.

trans-Isomer   *anti*-periplanar
displacement

cis-Epoxide

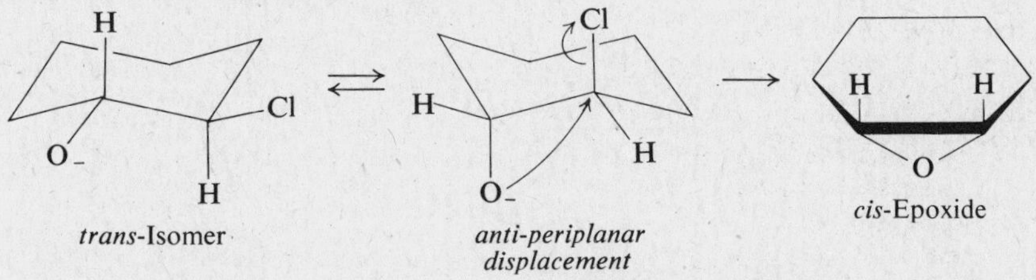

This is not possible for the *cis*-isomer, and reaction must follow another course.

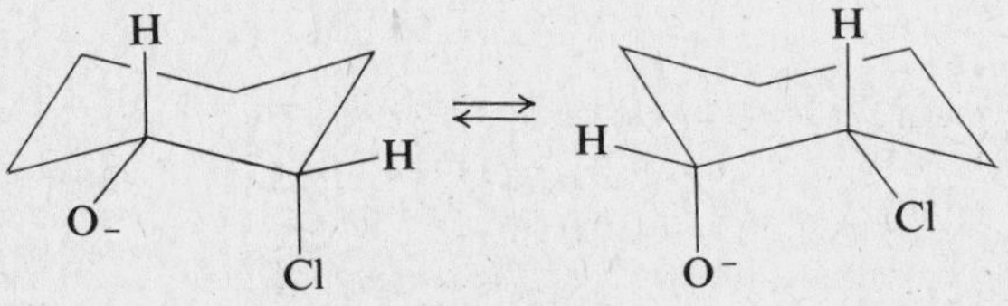

Anti-periplanar (back-side) displacement
of —Cl by —O⁻ not possible

cis-Isomer

(d) See Figure 17.8, page 212 of this Study Guide.

14. (a)

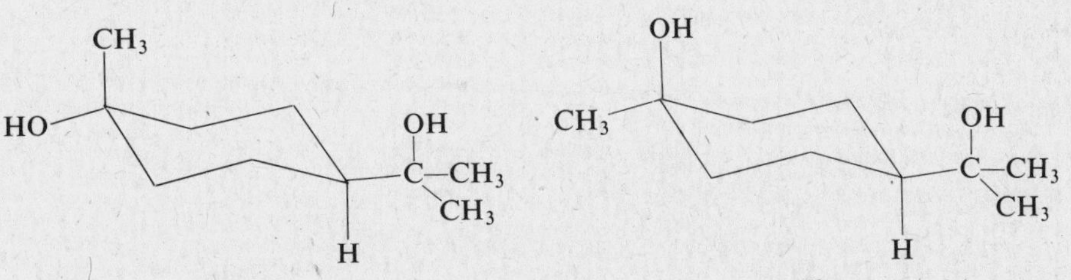

I*a*
Achiral

I*b*
Achiral

Diastereomers

211

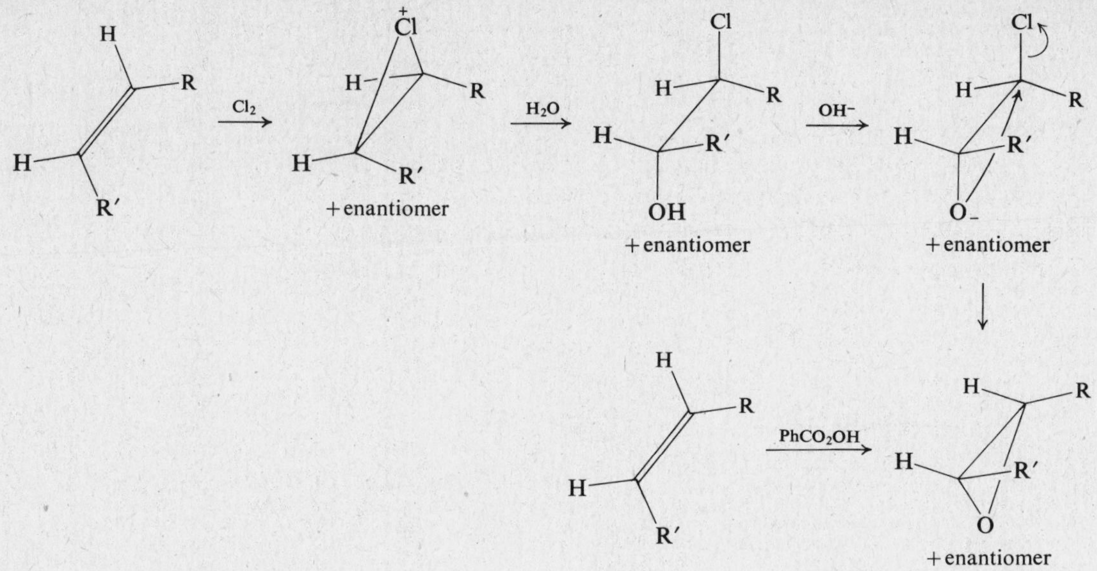

Figure 17.8. Conversion of oleic acid into the epoxide (Problem 13d).

$$R = -(CH_2)_7COOH, \quad R' = CH_3(CH_2)_7-$$

(b) Both are achiral, hence optically inactive.

(c) The ether is an internal cyclic ether. $(C_{10}H_{20}O_2 - H_2O \longrightarrow C_{10}H_{18}O)$. This can form only when the two —OH groups are on the same face of the molecule, as they are in Ib.

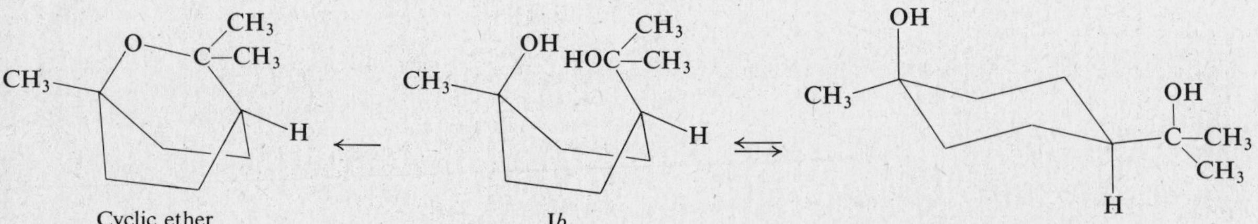

Cyclic ether
Bridged boat

Ib
Boat conformation

Ib
Chair conformation

15. (a) $CH_2{=}CH_2 \xrightarrow[\text{(+Cl, +O, +H)}]{Cl_2, H_2O}$ $\underset{\substack{| \quad | \\ Cl \quad OH \\ K}}{CH_2{-}CH_2}$

The transformation of K (2 carbons) into L (4 carbons) obviously requires 2 molecules of K, and an atom count shows the loss of H_2O in the reaction. An ether is formed.

$$ClCH_2CH_2OH + HOCH_2CH_2Cl \xrightarrow{-H_2O} ClCH_2CH_2{-}O{-}CH_2CH_2Cl \xrightarrow[-2HCl]{KOH} CH_2{=}CH{-}O{-}CH{=}CH_2$$

K (Two moles)   L   M

Vinyl ether

(b) 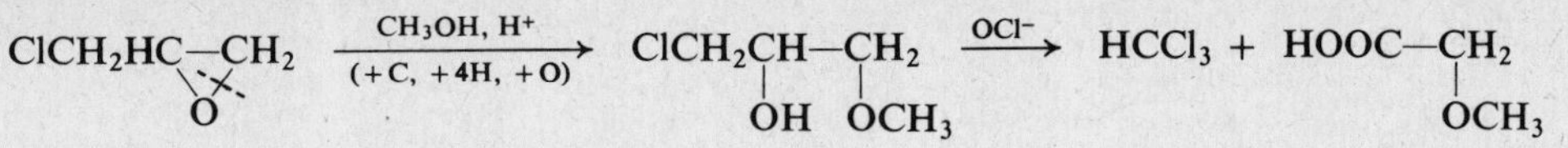 $ClCH_2HC-CH_2$ $\xrightarrow[(+C, +4H, +O)]{CH_3OH, H^+}$ $ClCH_2CH-CH_2$ $\xrightarrow{OCl^-}$ $HCCl_3$ + $HOOC-CH_2$

with O (epoxide), then OH OCH₃ (N), then OCH₃ (O)

The orientation of the initial methanolysis is shown by the fact that the haloform reaction (Secs. 16.11 and 19.9) takes place on compound N to give $HCCl_3$ and compound O. The alternative orientation would have produced a compound unresponsive to the haloform reaction.

$ClCH_2HC-CH_2$ $\longrightarrow$ $ClCH_2CH-CH_2$ *Not formed*

with O (epoxide), then CH₃O OH

No haloform reaction
(Sec. 16.11)

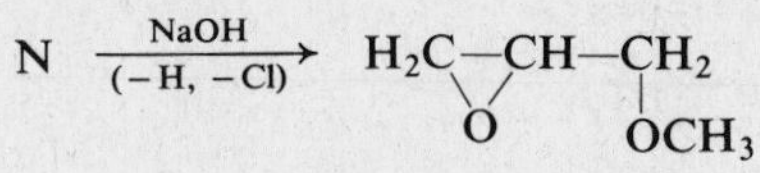

N $\xrightarrow[(-H, -Cl)]{NaOH}$ $H_2C-CH-CH_2$ *Intramolecular Williamson synthesis*

with O (epoxide) and OCH₃

P

(c) Intramolecular Williamson synthesis.

$CH_2CH_2CH_2$ $\xrightarrow[(-H, -Cl)]{KOH}$ H_2C-CH_2

with Cl and OH, then O-CH₂

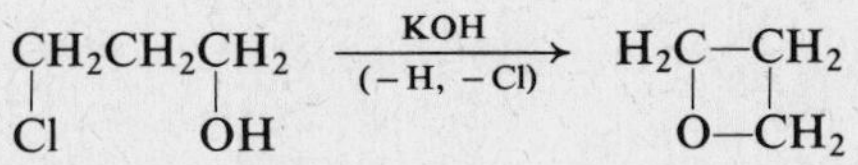

Q

(d) Friedel-Crafts alkylation.

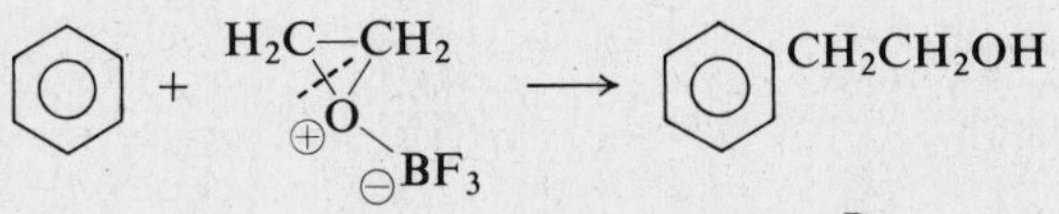

benzene + H_2C-CH_2 (epoxide O⁺, BF₃⁻) $\longrightarrow$ $C_6H_5CH_2CH_2OH$

R
β-Phenylethyl alcohol

(e) Intramolecular alkoxymercuration-demercuration $\longrightarrow$ cyclic ether.

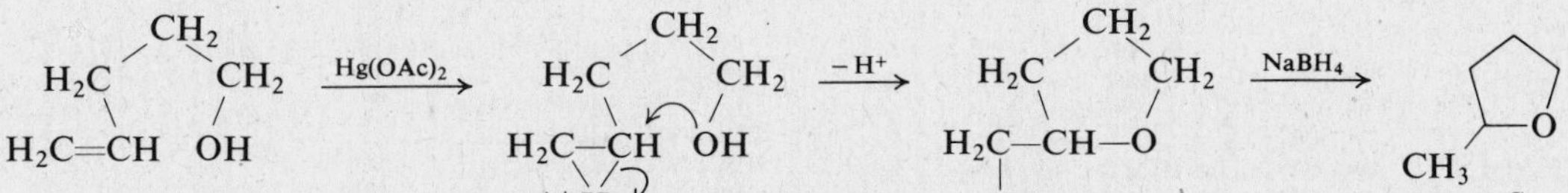

$\xrightarrow{Hg(OAc)_2}$ ... $\xrightarrow{-H^+}$ ... $\xrightarrow{NaBH_4}$

S

(f) As we shall see (Problem 15, page 649), enol ethers are readily cleaved by acids.

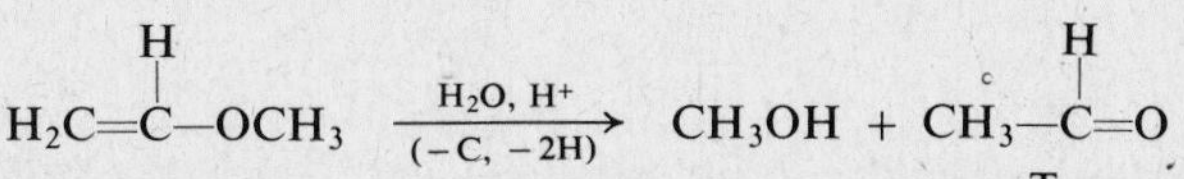

$H_2C=C-OCH_3$ $\xrightarrow[(-C, -2H)]{H_2O, H^+}$ CH_3OH + $CH_3-C=O$

with H

T

(g)

$\xrightarrow[(+H, +Cl)]{HCl}$

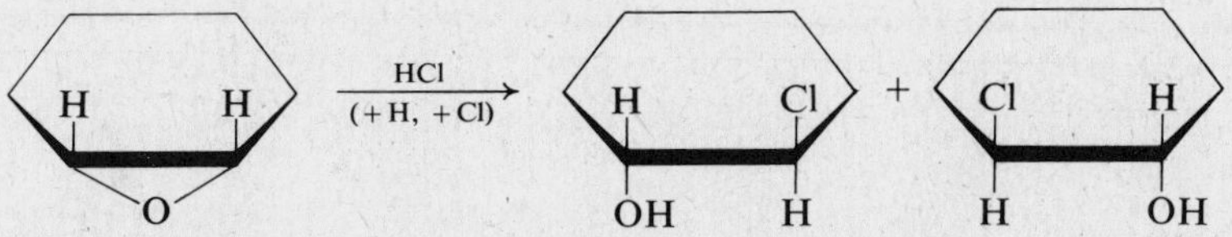

U
trans-2-Chlorocyclohexanol
Racemic

(h)

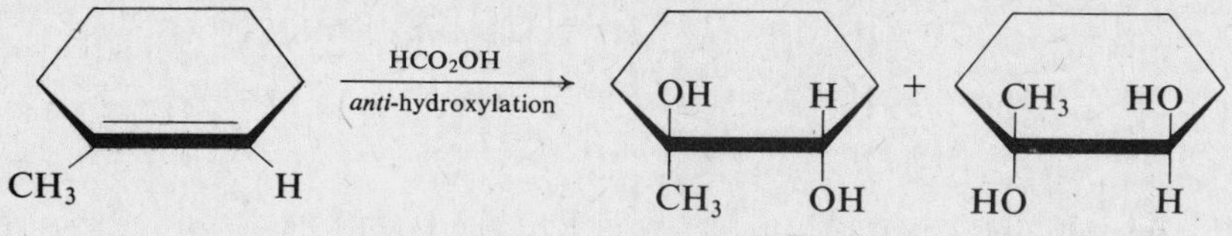

$$\text{(structure)} \xrightarrow[\text{anti-hydroxylation}]{\text{HCO}_2\text{OH}} \text{(structure with OH, H, CH}_3\text{, OH)} + \text{(structure with CH}_3\text{, HO, HO, H)}$$

V
Racemic

(i) *syn*-Hydroxylation of the double bond, producing one new chiral carbon (in both con-figurations), followed by cleavage of the epoxide.

Enantiomers

$$\xrightarrow{\text{KMnO}_4}$$

Enantiomers

$$\downarrow \text{H}_2\text{O, H}^+$$

Meso

Racemic

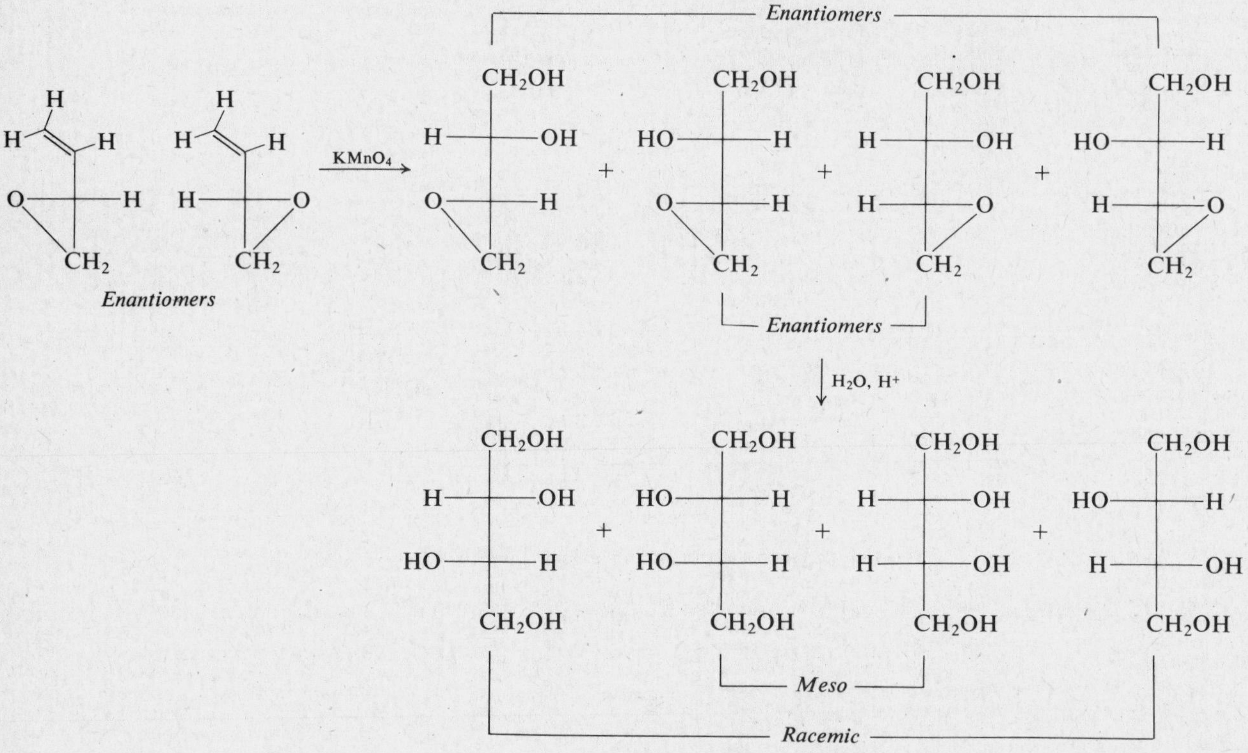

W

Whatever the mechanism of cleavage, a mixture (W) of *meso* and racemic products will be obtained.

(j) Chlorohydrin formation via a bridged chloronium ion, followed by an intramolecular Williamson synthesis to give an epoxide, and finally cleavage of the epoxide to a glycol. (See Figure 17.10, page 215 of this Study Guide.)

(k) Similar to (j), but starting with *trans*-2-butene, and ending up with *meso*-2,3-butanediol. (See Figure 17.11, page 216 of this Study Guide.)

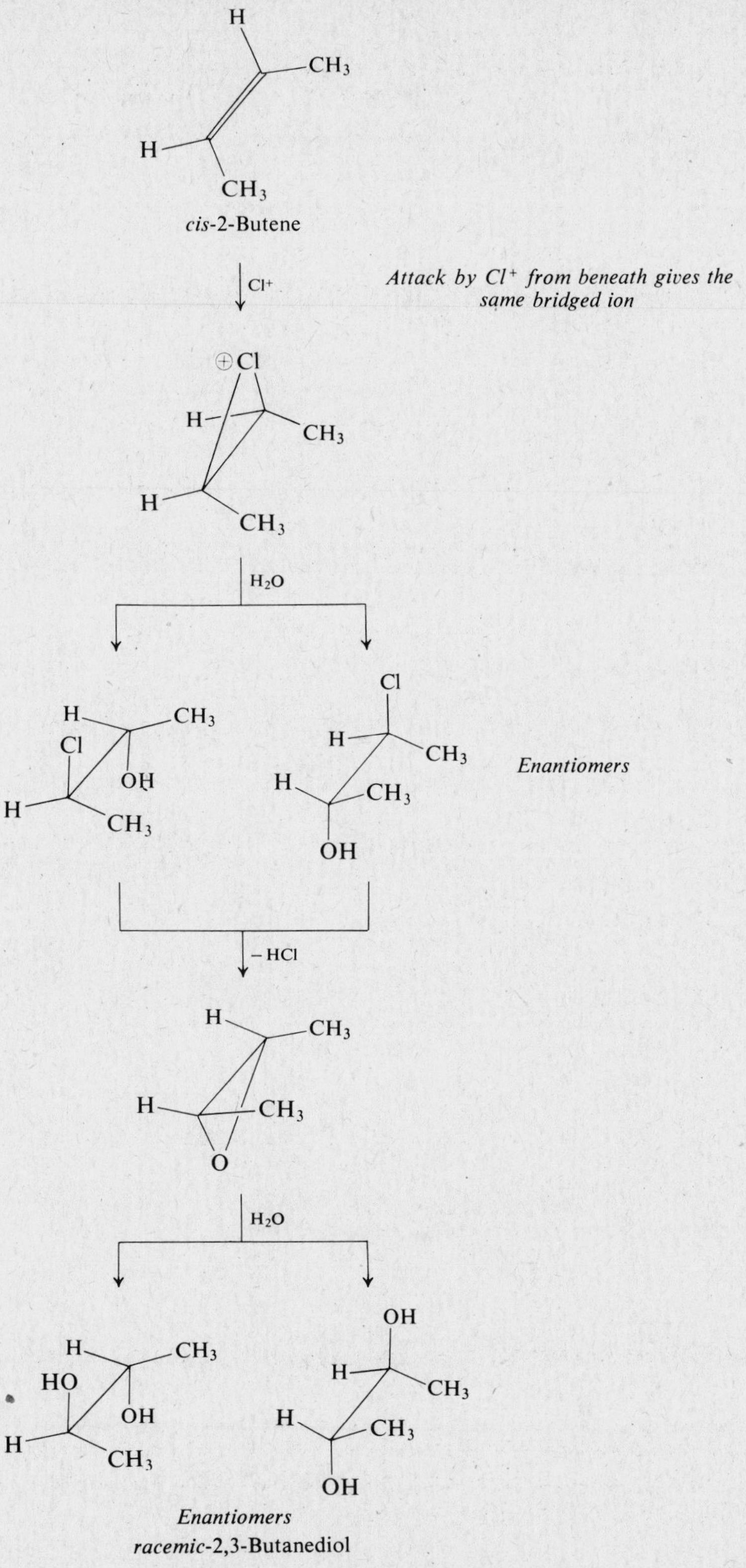

Figure 17.10. Reaction of *cis*-2-butene in Problem 15(j).

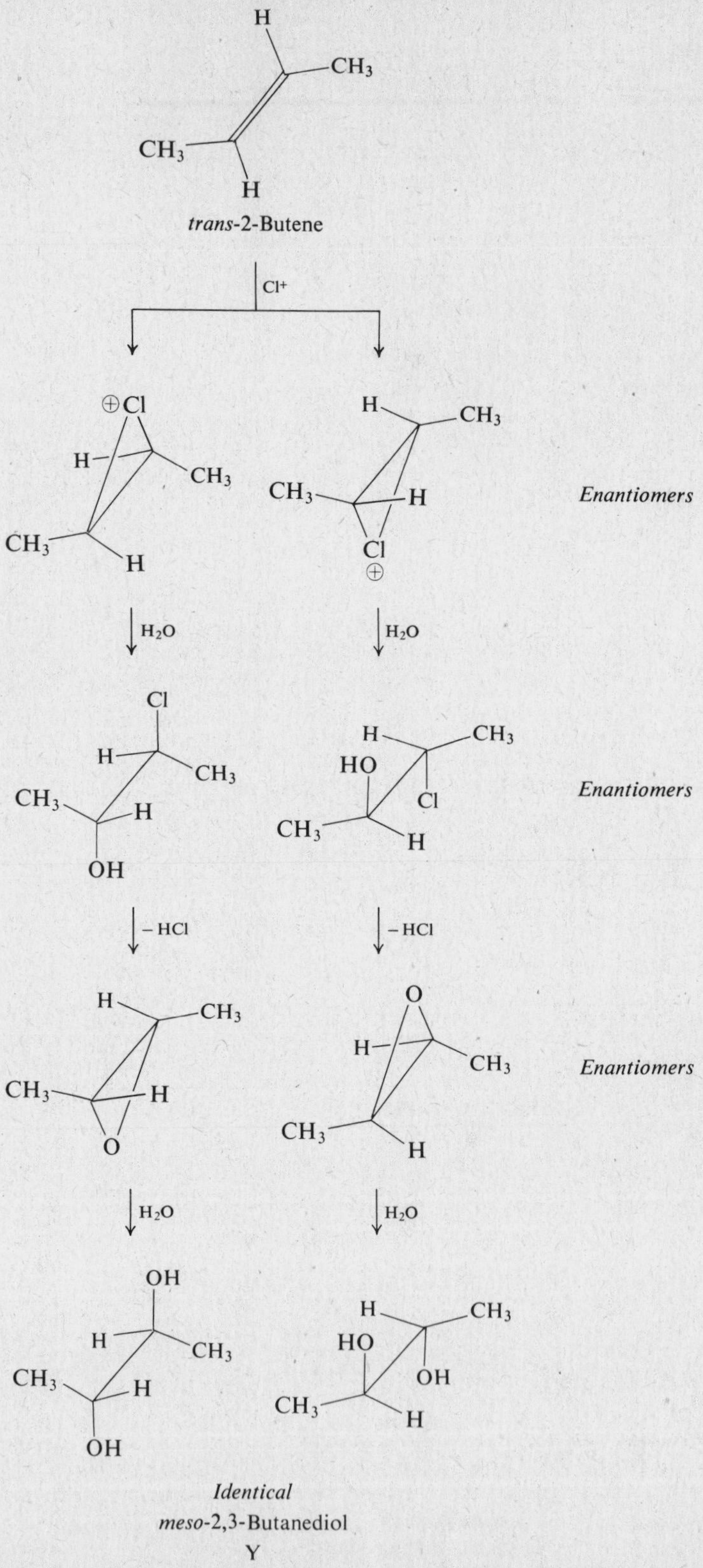

Figure 17.11. Reaction of *trans*-2-butene in Problem 15(k).

16. *m*-Methylanisole, *m*-$CH_3C_6H_4OCH_3$.

      (See the labeled spectrum on page 653 of this Study Guide.)

17. Z, anisyl alcohol, *p*-$CH_3OC_6H_4CH_2OH$.

      (See the labeled spectra on page 653 of this Study Guide.)

18. (a) $(CH_3)_3C-O-CH_2CH_3$ (b) $(CH_3CH_2CH_2)_2O$ (c) $(CH_3)_2CH-O-CH(CH_3)_2$
        tert-Butyl ethyl ether   *n*-Propyl ether   Isopropyl ether

      (See the labeled spectra on page 654 of this Study Guide.)

19. *p*-$CH_3C_6H_4OCH_2CH_3$ $C_6H_5CH_2OCH_2CH_3$ $C_6H_5CH_2CH_2CH_2OH$
      p-Methylphenetole   Benzyl ethyl ether   3-Phenyl-1-propanol
              AA   BB   CC

      (See the labeled spectra on pages 655 and 656 of this Study Guide.)

Chapter 18 | Carboxylic Acids

18.1 At 110°: $\dfrac{0.11 \text{ g}}{\text{m.w. (g/mole)}} = \dfrac{\text{vol. at STP}}{22400 \text{ cc/mole}} = \dfrac{63.7 \times \dfrac{273}{383} \times \dfrac{454}{760}}{22400}$

$\text{m.w.} = \dfrac{0.11 \times 22400}{63.7} \times \dfrac{383}{273} \times \dfrac{760}{454} = 91 \text{ (at } 110°)$

At 156°: $\dfrac{0.081}{\text{m.w.}} = \dfrac{66.4 \times \dfrac{273}{429} \times \dfrac{458}{760}}{22400}$

$\text{m.w.} = \dfrac{0.081 \times 22400}{66.4} \times \dfrac{429}{273} \times \dfrac{760}{458} = 71 \text{ (at } 156°)$

The monomer, CH_3COOH, would have m.w. 60. Association occurs even in the vapor phase, decreasing as the temperature increases.

18.2 (a) (1) $\qquad\qquad\qquad\qquad\qquad$ peroxide $\longrightarrow$ Rad·

(2) $\qquad\qquad$ Rad· + $CH_3CH_2CH_2COOH$ $\longrightarrow$ RadH + $CH_3CH_2\overset{.}{C}HCOOH$

(3) $\qquad$ n-BuCH=CH$_2$ + $CH_3CH_2\overset{.}{C}HCOOH$ $\longrightarrow$ n-BuCH—CH$_2$—CH—COOH
$\qquad\qquad\qquad\qquad\qquad\qquad\qquad\qquad\qquad\qquad\qquad\qquad$ $\overset{|}{C_2H_5}$

(4) n-Bu$\overset{.}{C}$HCH$_2$CHCOOH + $CH_3CH_2CH_2COOH$ $\longrightarrow$ n-BuCH$_2$CH$_2$CHCOOH + $CH_3CH_2\overset{.}{C}HCOOH$
$\qquad\qquad$ $\overset{|}{C_2H_5}$ $\qquad\qquad\qquad\qquad\qquad\qquad\qquad\qquad\qquad\qquad\qquad$ $\overset{|}{C_2H_5}$

then (3), (4), (3), (4), etc.

Free radical attack favors abstraction of an *alpha*-hydrogen, presumably because of resonance stabilization of the resulting free radical:

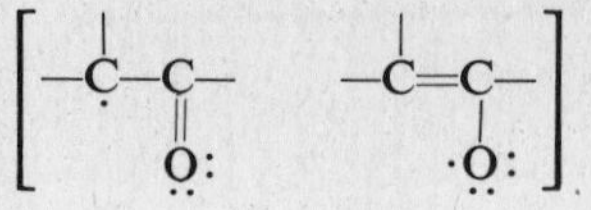

$$\left[\begin{array}{c} -\overset{|}{C}-\overset{|}{\underset{\displaystyle \overset{\cdot}{\underset{\cdot\cdot}{O}}:}{C}}- \end{array} \qquad \begin{array}{c} -\overset{|}{C}=\overset{|}{\underset{\displaystyle \cdot\underset{\cdot\cdot}{O}:}{C}}- \end{array} \right]$$

(b) 2-Methyldecanoic acid, $CH_3(CH_2)_7\underset{\displaystyle \overset{|}{CH_3}}{\overset{|}{C}}HCOOH$

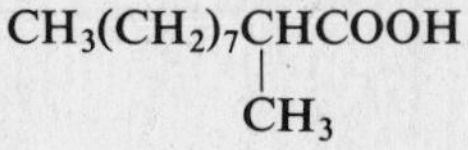

(c) 2,2-Dimethyldodecanoic acid, $CH_3(CH_2)_9\overset{\displaystyle CH_3}{\underset{\displaystyle CH_3}{\overset{|}{\underset{|}{C}}}}COOH$

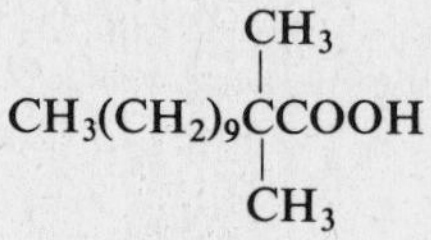

(d) Ethyl *n*-octylmalonate, $CH_3(CH_2)_7\overset{\displaystyle COOEt}{\underset{\displaystyle COOEt}{\overset{|}{\underset{|}{C}}}}H$

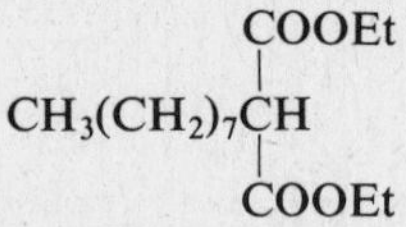

(A. D. Petrow, G. I. Nikischin, and J. N. Ogibin, "Freiradikale Anlagerung von Säuren und Alkoholen zu den α-Olefinen," Third International Congress on Surface-Active Agents, Cologne, September 1960, vol. 1, Sec. A, pp. 78–83.)

18.3 (a) Clearly, the *tert*-pentyl cation is an intermediate:

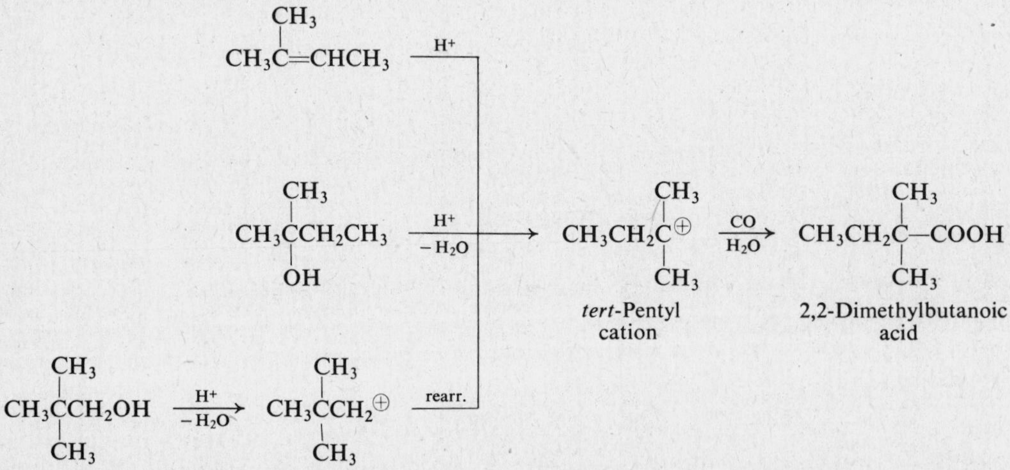

The most likely mechanism of its reaction with carbon monoxide involves the formation and hydration of an *acylium ion*, a cation that owes its considerable stability to the fact

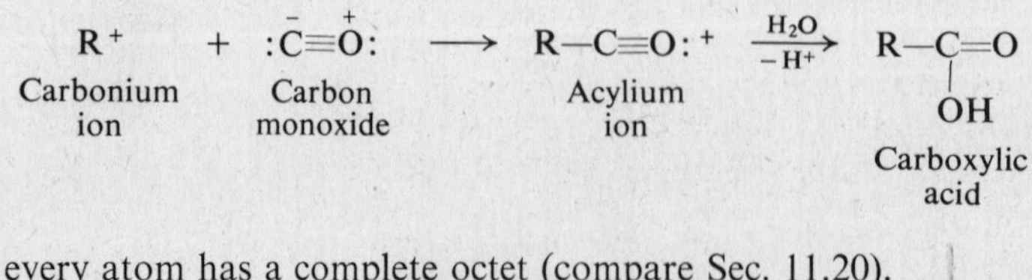

$$R^+ \quad + \quad :\overset{-}{C}\equiv\overset{+}{O}: \quad \longrightarrow \quad R-C\equiv O:^+ \quad \xrightarrow[-H^+]{H_2O} \quad R-\overset{\displaystyle C=O}{\underset{\displaystyle OH}{|}}$$

Carbonium   Carbon   Acylium   Carboxylic
ion   monoxide   ion   acid

that in it every atom has a complete octet (compare Sec. 11.20).

(b)

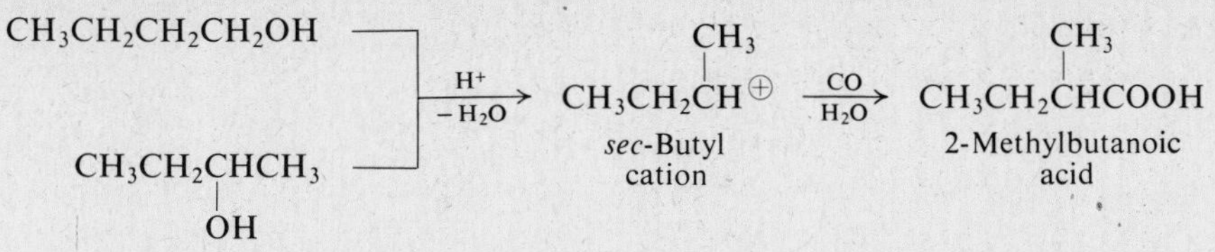

$CH_3CH_2CH_2CH_2OH$

$CH_3CH_2CHCH_3$ | OH

$\xrightarrow[-H_2O]{H^+}$ $CH_3CH_2\overset{CH_3}{\underset{|}{CH}}^{\oplus}$ $\xrightarrow[H_2O]{CO}$ $CH_3CH_2\overset{CH_3}{\underset{|}{CH}}COOH$

sec-Butyl cation   2-Methylbutanoic acid

(H. Koch and W. Haaf, "Über die Synthese verzweigter Carbonsäuren nach der Ameisensäure-Methode," Annalen der Chemie, **618**, 251 (1958).)

18.4 (a)

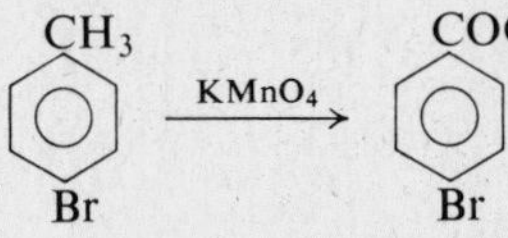

p-Bromobenzoic acid

(b)

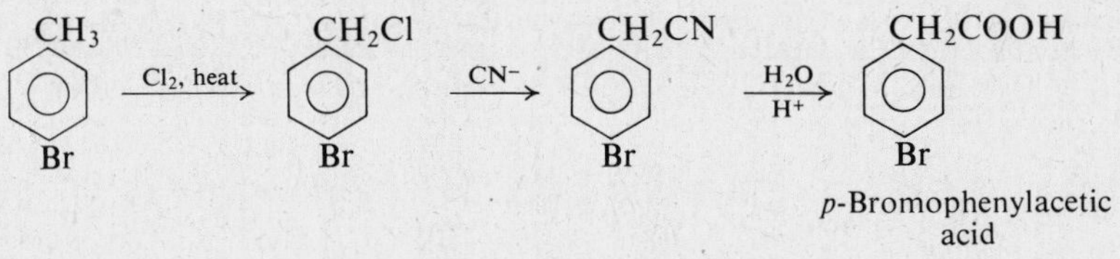

p-Bromophenylacetic acid

18.5 The carbonate ion is a hybrid of three structures of equal stability. Each C—O bond is double in one structure and single in two; it has less double bond character, and hence is longer, than the bonds in formate ion.

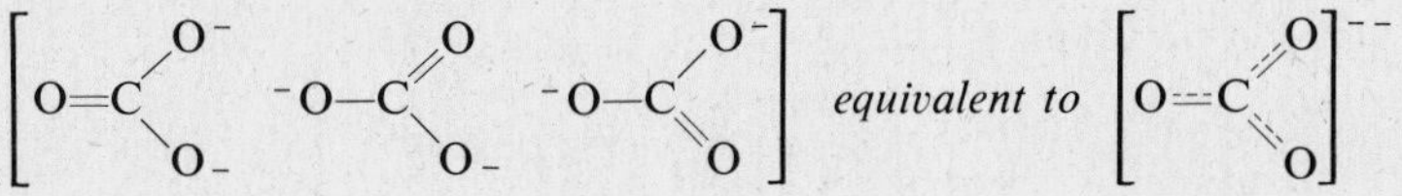

18.6 See Sec. 21.1.

18.7 (a) F > Cl > Br > I.

(b) Electron-withdrawing.

18.8

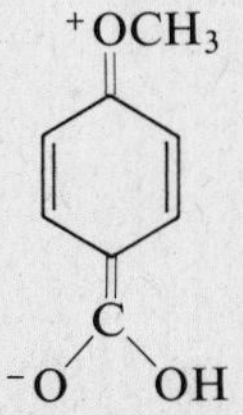

For the anion, such a structure would involve separation of positive and *double* negative charges.

18.9 (a) The C—OH bond is broken:

$$R-\overset{O}{\underset{OH}{\overset{\|}{C}}} \longrightarrow R-\overset{O}{\underset{Cl}{\overset{\|}{C}}}$$

(b) The C—OH bond of the acid and the CO—H bond of the alcohol are broken:

$$Ph-\overset{O}{\underset{OH}{\overset{\|}{C}}} + H\!-\!^{18}O\!-\!CH_3 \longrightarrow Ph-\overset{O}{\underset{^{18}OCH_3}{\overset{\|}{C}}} + HOH$$

18.10 (a) $n\text{-}C_{11}H_{23}CH_2Br \xleftarrow{\text{PBr}_3} n\text{-}C_{11}H_{23}CH_2OH \xleftarrow{\text{LiAlH}_4} n\text{-}C_{11}H_{23}COOH$

(b) $n\text{-}C_{11}H_{23}CH_2COOH \xleftarrow{\text{H}^+} \xleftarrow{\text{CO}_2} n\text{-}C_{11}H_{23}CH_2MgBr \xleftarrow{\text{Mg}} n\text{-}C_{11}H_{23}CH_2Br \longleftarrow$ (a)

   Or via nitrile: $RBr \longrightarrow RCN \longrightarrow RCOOH$

(c) $n\text{-}C_{11}H_{23}CH_2CH_2CH_2OH \xleftarrow{\overset{H_2C-CH_2}{\underset{O}{\diagdown\!\diagup}}} n\text{-}C_{11}H_{23}CH_2MgBr \longleftarrow$ (b)

(d) $n\text{-}C_{10}H_{21}CH\!=\!CH_2 \xleftarrow{\text{KOH(alc)}} n\text{-}C_{11}H_{23}CH_2Br \longleftarrow$ (a)

(e) $n\text{-}C_{10}H_{21}CH_2CH_3 \xleftarrow{\text{H}_2,\ \text{Pt}} n\text{-}C_{10}H_{21}CH\!=\!CH_2 \longleftarrow$ (d)

(f) $n\text{-}C_{10}H_{21}C\!\equiv\!CH \xleftarrow{\text{NaNH}_2} \xleftarrow{\text{KOH}} n\text{-}C_{10}H_{21}\underset{Br}{\overset{}{CH}}\!-\!\underset{Br}{\overset{}{CH_2}} \xleftarrow{\text{Br}_2,\ \text{CCl}_4} n\text{-}C_{10}H_{21}CH\!=\!CH_2 \longleftarrow$ (d)

(g) $n\text{-}C_{10}H_{21}\overset{}{\underset{O}{\overset{\|}{C}}}\!-\!CH_3 \xleftarrow[\text{(See Sec. 8.13)}]{\text{H}_2\text{O, H}_2\text{SO}_4,\ \text{Hg}^{++}} n\text{-}C_{10}H_{21}C\!\equiv\!CH \longleftarrow$ (f)

(h) $n\text{-}C_{10}H_{21}CHOH\!-\!CH_3 \xleftarrow{\text{H}_2,\ \text{Pt}} n\text{-}C_{10}H_{21}\overset{}{\underset{O}{\overset{\|}{C}}}\!-\!CH_3 \longleftarrow$ (g)

(i) $n\text{-}C_{10}H_{21}COOH \xleftarrow{\text{H}^+} \xleftarrow{\text{OI}^-} n\text{-}C_{10}H_{21}\overset{}{\underset{O}{\overset{\|}{C}}}\!-\!CH_3 \longleftarrow$ (g) (Compare Sec. 16.11)

(j) $n\text{-}C_{12}H_{25}CHOH\!-\!CH_3 \xleftarrow{\text{H}^+} \xleftarrow{\text{CH}_3\text{CHO}} n\text{-}C_{12}H_{25}MgBr \xleftarrow{\text{Mg}} n\text{-}C_{12}H_{25}Br \longleftarrow$ (a)

(k) $n\text{-}C_{12}H_{25}\underset{OH}{\overset{CH_3}{\overset{|}{\underset{|}{C}}}}\!-\!CH_3 \xleftarrow{\text{H}^+} \xleftarrow{\text{CH}_3\text{COCH}_3} n\text{-}C_{12}H_{25}MgBr \xleftarrow{\text{Mg}} n\text{-}C_{12}H_{25}Br \longleftarrow$ (a)

18.11 (a) CH_3CH_2COOH.

(b)

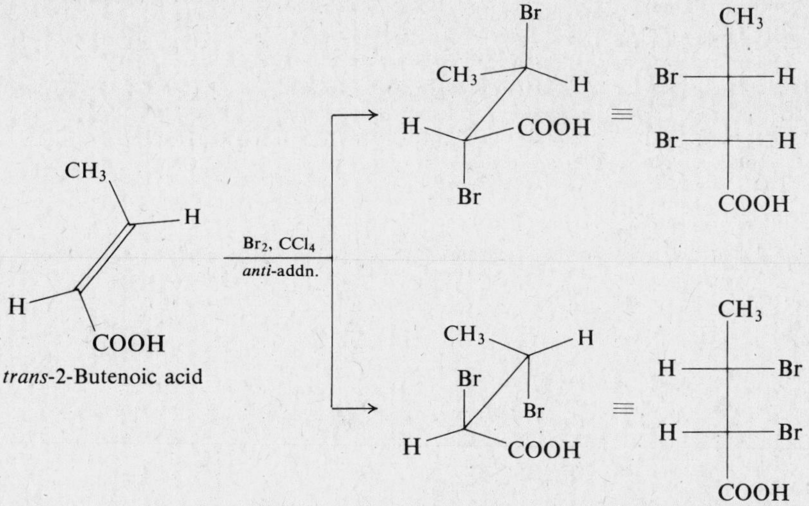

trans-2-Butenoic acid
$\xrightarrow[\text{anti-addn.}]{Br_2,\ CCl_4}$

Enantiomers
racemic-erythro- 2,3-Dibromobutanoic acid

(c) $PhCHOHCH_2COOH \xrightarrow[-H_2O]{H^+}$ (chiefly *trans*)

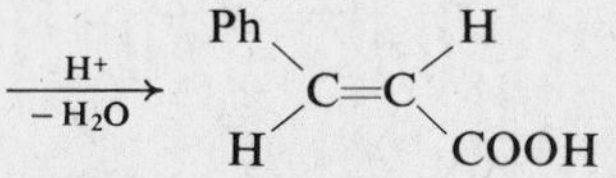

(d) The only way H_2O can be lost ($C_8H_8O_3 - H_2O = C_8H_6O_2$) within a single molecule (no increase in carbon number) is through intramolecular esterification to give a cyclic ester (a *lactone*, Sec. 20.15).

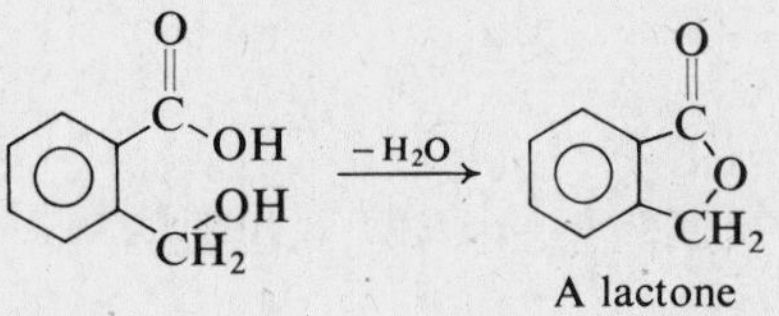

A lactone
A cyclic ester

18.12 To prevent generation of HCN by the action of the acid on sodium cyanide.

18.13 (a) We need to add two carbons:

$HOOC(CH_2)_3COOH \xleftarrow[\text{heat}]{H_2O,\ H^+} N\equiv C(CH_2)_3C\equiv N \xleftarrow{2CN^-} Br(CH_2)_3Br \xleftarrow[\text{excess}]{2HBr} HOCH_2CH_2CH_2OH$

(b) We need to break a C_{18}-acid into a C_9-fragment:

$CH_3(CH_2)_7CH=CH(CH_2)_7COOH \xrightarrow{O_3} \xrightarrow{H_2O,\ H^+} HOOC(CH_2)_7COOH$

(c) $HOCH_2C\equiv CCH_2OH \xrightarrow{H_2,\ Pt} HOCH_2CH_2CH_2CH_2OH \xrightarrow{KMnO_4} HOOCCH_2CH_2COOH$

18.14 (a) Diester, EtOOC(CH$_2$)$_4$COOEt.

(b) Monoester, EtOOC(CH$_2$)$_4$COOH.

(c) Diol, HOCH$_2$CH$_2$CH$_2$CH$_2$OH.

(d) Monobromo acid, HOOCCH$_2$CH$_2$CHBrCOOH.

(e) Diacyl chloride, p-C$_6$H$_4$(COCl)$_2$.

(f)

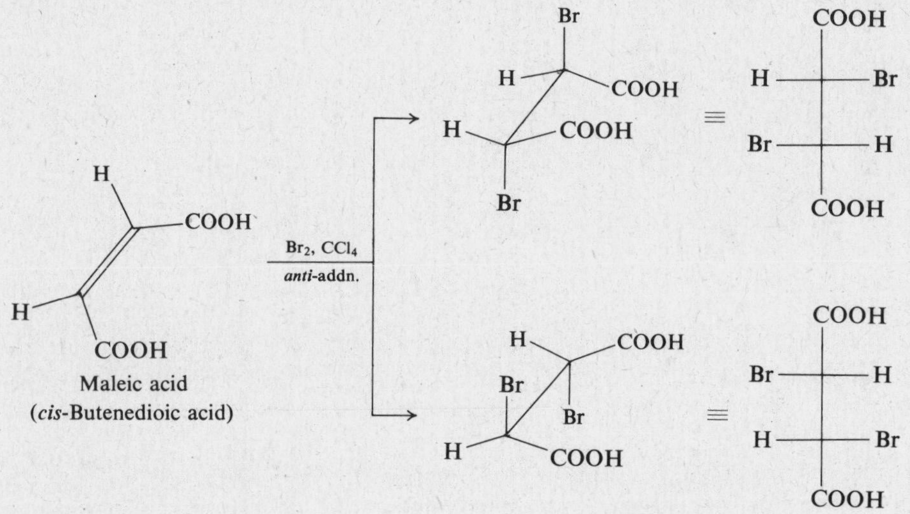

Maleic acid
(*cis*-Butenedioic acid)

Br$_2$, CCl$_4$
anti-addn.

Enantiomers
racemic-2,3-Dibromobutanedioic acid

18.15 The —COOH group is electron-withdrawing, acid-strengthening compared to —H or —CH$_3$.

HOOC—COOH   H—COOH   HOOC—CH$_2$—COOH   H—CH$_2$—COOH
K_a: 5400 × 10^{-5} > 17.7 × 10^{-5} 140 × 10^{-5} >   1.75 × 10^{-5}

18.16 As the distance between —COOH groups is increased, the inductive effect of one on the other weakens.

HOOC—COOH > HOOC—CH$_2$—COOH > HOOC—CH$_2$CH$_2$—COOH > HOOC—CH$_2$CH$_2$CH$_2$—COOH
K_a: 5400 × 10^{-5} 140 × 10^{-5} 6.4 × 10^{-5} 4.5 × 10^{-5}

18.17. (a)   C$_2$H$_2$O$_4$ oxalic acid
$\underline{+C_2H_6O_2}$ ethylene glycol
C$_4$H$_8$O$_6$
$\underline{-C_4H_4O_4}$ product
2H$_2$O   lost: there must be two esterifications

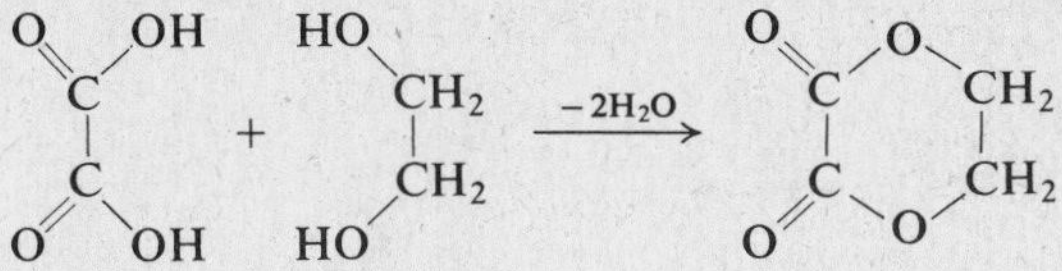

Oxalic acid   Ethylene glycol   *A cyclic diester*

(b) $C_4H_6O_4$ succinic acid
   $- C_4H_4O_3$ product
     H_2O lost: there must be intramolecular anhydride formation

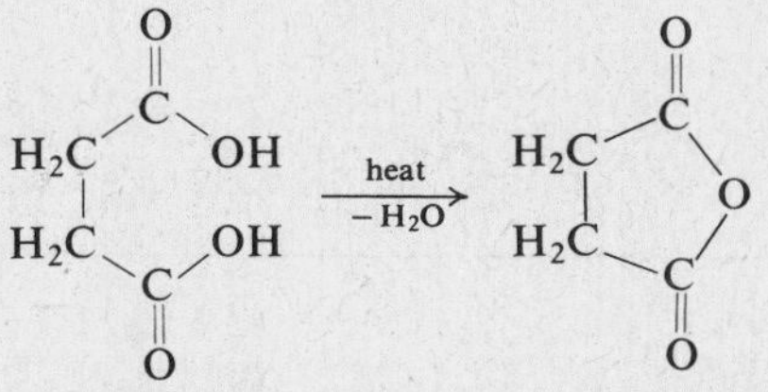

Succinic acid   Succinic anhydride
            A cyclic anhydride

(c) $C_8H_6O_4$ terephthalic acid
   $+ C_2H_6O_2$ ethylene glycol
   $\overline{C_{10}H_{12}O_6}$
   $- C_{10}H_8O_4$ Dacron unit
     $2H_2O$:   double esterification

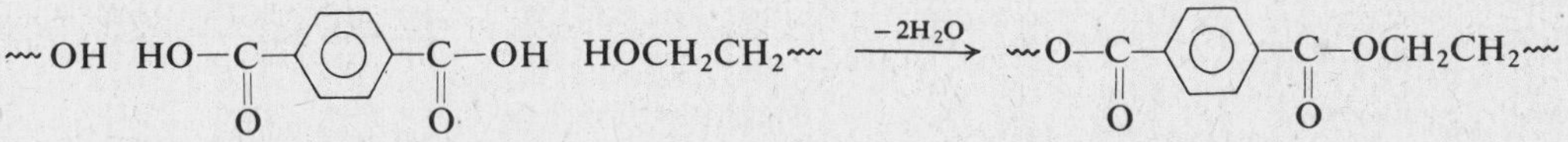

Dacron
A polyester

18.18

	conc. $H_2SO_4{}^a$	cold $KMnO_4{}^b$	$Br_2{}^b$	$CrO_3{}^c$	fum. $H_2SO_4{}^a$	$CHCl_3$, $AlCl_3{}^c$	Na^d	$NaOH^a$	$NaHCO_3{}^e$
Alkanes	−	−	−	−	−	−	−	−	−
Alkenes	+	+	+	−	+	−	−	−	−
Alkynes	+	+	+	−	+	−	$-^f$	−	−
Alkyl halides	−	−	−	−	−	−	−	−	−
Alkylbenzenes	−	$-^g$	−	−	+	+	−	−	−
1° alcohols	+	−	−	+	+	−	+	−	−
2° alcohols	+	−	−	+	+	−	+	−	−
3° alcohols	+	−	−	−	+	−	+	−	−
Ethers	+	−	−	−	+	−	−	−	−
Carboxylic acids	+	−	−	−	+	−	$+^h$	+	+
Phenols	+	−	$+^i$	−	+	−	+	+	$-^j$

^{a} Dissolve. ^{b} Decolorize. ^{c} Change color. ^{d} Hydrogen bubbles. ^{e} Dissolve, with CO_2 bubbles. ^{f} 1-Alkynes give test. ^{g} Decolorize hot $KMnO_4$. ^{h} Explosive reaction. ^{i} Evolve HBr. ^{j} Phenols with several electron-withdrawing substituents give test.

18.19   *o*-Chlorobenzoic acid, $C_7H_5O_2Cl$, m.w. (and equiv. weight) 156.5. (2,6-$Cl_2C_6H_3COOH$ has m.w. 191.)

18.20 (a)

$$\frac{18.7}{1000} \times 0.0972 \text{ equiv. of acid} = 0.187 \text{ g}$$

$$1 \text{ equiv. of acid} = 0.187 \times \frac{1000}{18.7} \times \frac{1}{0.0972} = 103 \text{ g}$$

(b) Ethoxyacetic acid, $C_2H_5OCH_2COOH$, m.w. (and equiv. weight) 104. The others are not possible: n-caproic acid, n-$C_5H_{11}COOH$, m.w. (and equiv. wt.) 116; methoxyacetic acid, CH_3OCH_2COOH, m.w. (and equiv. wt.) 90.

18.21 (a) Two; equiv. wt. = m.w./2 = 166/2 = 83.

(b) N.E. = mol.wt./number acidic H per molecule.

(c) 70 (m.w./3), 57 (m.w./6).

18.22 Sodium carbonate, Na_2CO_3.

1. C–1: formic, methanoic   C–2: acetic, ethanoic
   C–3: propionic, propanoic   C–4: n-butyric, butanoic
   C–5: n-valeric, pentanoic   C–6: n-caproic, hexanoic
   C–8: n-caprylic, octanoic   C–10: n-capric, decanoic
   C–12: lauric, dodecanoic   C–16: palmitic, hexadecanoic
   C–18: stearic, octadecanoic

2. (a) $(CH_3)_2CHCH_2COOH$
        3-Methylbutanoic acid

(b) $(CH_3)_3CCOOH$
      2,2-Dimethylpropanoic acid

(c) $CH_3CH_2CH_2CH(CH_3)CH(CH_3)COOH$
       2,3-Dimethylhexanoic acid

(d) n-$BuCH(Et)CH_2CH(CH_3)COOH$
     α-Methyl-γ-ethylcaprylic acid

(e) $PhCH_2COOH$
     Phenylethanoic acid

(f) $PhCH_2CH_2CH_2COOH$
     4-Phenylbutanoic acid

(g) $HOOCCH_2CH_2CH_2CH_2COOH$
     Hexanedioic acid

(h) p-$CH_3C_6H_4COOH$
     4-Methylbenzoic acid

(i) o-$C_6H_4(COOH)_2$
     1,2-Benzenedicarboxylic acid

(j) m-$C_6H_4(COOH)_2$
     1,3-Benzenedicarboxylic acid

(k) p-$C_6H_4(COOH)_2$
     1,4-Benzenedicarboxylic acid

(l) p-HOC_6H_4COOH
     4-Hydroxybenzoic acid

(m) $CH_3CH_2CH(CH_3)COOK$
     Potassium 2-methylbutanoate

(n) $(CH_3CHClCOO)_2Mg$
     Magnesium α-chloropropionate

(o) (Z)-$HOOC{-}CH{=}CH{-}COOH$
     (Z)-Butenedioic acid

(p) $HOOCCHBrCHBrCOOH$
     2,3-Dibromobutanedioic acid

(q) $(CH_3)_2CHC{\equiv}N$
    2-Methylethanenitrile

(r)   2,4-$(O_2N)_2C_6H_3C{\equiv}N$
     2,4-Dinitrobenzonitrile

3. (a) Same number of carbons:

$$PhCH_3 \xrightarrow[\text{heat}]{KMnO_4,\ OH^-} PhCOOH$$

(b) Add one carbon:

$$PhBr \xrightarrow{Mg} PhMgBr \xrightarrow{CO_2} PhCOOMgBr \xrightarrow{H^+} PhCOOH$$

(c) Same number of carbons:

$$PhC\equiv N \xrightarrow[H^+]{H_2O} PhCOOH + NH_4^+$$

(d) Same number of carbons:

$$PhCH_2OH \xrightarrow[\text{heat}]{KMnO_4,\ H^+} PhCOOH$$

(e) Same number of carbons: hydrolyze trihalide.

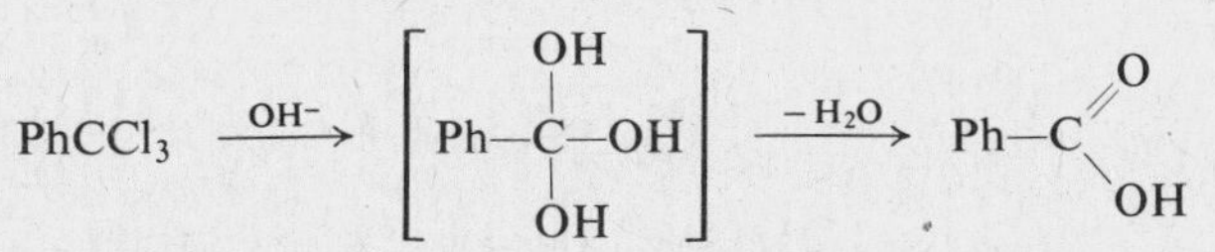

$$PhCCl_3 \xrightarrow{OH^-} \left[\ Ph-\overset{\displaystyle OH}{\underset{\displaystyle OH}{C}}-OH\ \right] \xrightarrow{-H_2O} Ph-C\overset{\displaystyle O}{\underset{\displaystyle OH}{\big\langle}}$$

(f) Lose one carbon (via haloform reaction, Sec. 16.11):

$$Ph-\underset{O}{\overset{}{C}}\!\!+\!CH_3 \xrightarrow{OI^-} \underset{\text{Iodoform}}{HCI_3} + PhCOO^- \xrightarrow{H^+} PhCOOH$$

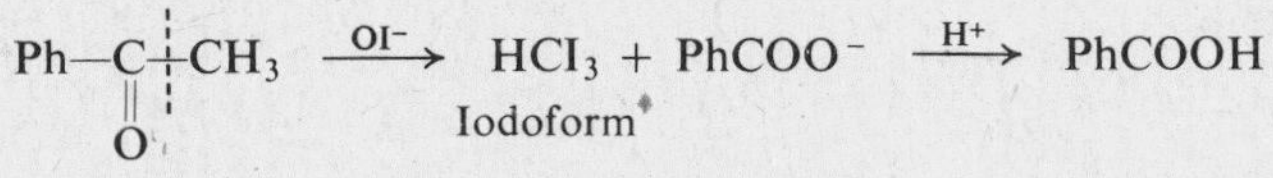

4. (a) Same number of carbons: 1° alcohol $\xrightarrow{\text{oxidn.}}$ acid.

$$n\text{-}PrCH_2OH \xrightarrow[\text{heat}]{KMnO_4,\ H^+} n\text{-}PrCOOH$$

(b) Add one carbon: via cyanide ion.

$$n\text{-}PrCOOH \xleftarrow[H^+]{H_2O} n\text{-}PrC\equiv N \xleftarrow{CN^-} n\text{-}PrBr \xleftarrow{PBr_3} n\text{-}PrOH$$

Cannot be used to prepare $(CH_3)_3$—COOH from t-BuBr. (See elimination vs. substitution, Secs. 8.12 and 14.23.)

(c) Add one carbon: via Grignard and CO_2.

$$n\text{-}PrCOOH \xleftarrow{H^+} \xleftarrow{CO_2} n\text{-}PrMgBr \xleftarrow{Mg} n\text{-}PrBr \xleftarrow{PBr_3} n\text{-}PrOH$$

(d) Remove one carbon: via haloform reaction, Sec. 16.11.

$$n\text{-}Pr-\underset{O}{\overset{}{C}}\!\!+\!CH_3 \xrightarrow{OI^-} \underset{\text{Iodoform}}{HCI_3} + n\text{-}PrCOO^- \xrightarrow{H^+} n\text{-}PrCOOH$$

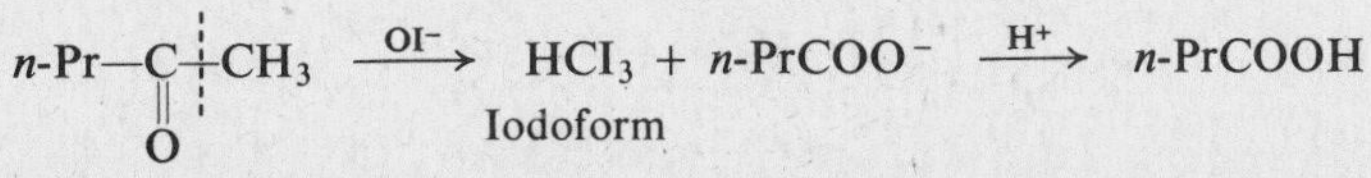

5. In every case, the ether has to be cleaved (ring opened) to yield the halide and/or alcohol.

(a) Same number of carbons:

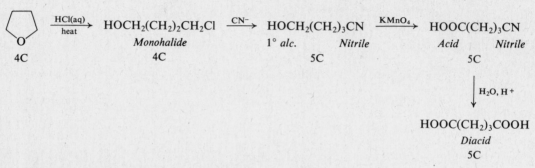

(b) We need to add one extra carbon, probably via CN^-:

$$
\underset{\substack{Ether\\4C}}{\text{(oxolane)}} \xrightarrow[\text{heat}]{\text{HCl(aq)}} \underset{\substack{Monohalide\\4C}}{\text{HOCH}_2\text{(CH}_2)_2\text{CH}_2\text{Cl}} \xrightarrow{CN^-} \underset{\substack{1°\ alc.\quad Nitrile\\5C}}{\text{HOCH}_2\text{(CH}_2)_3\text{CN}} \xrightarrow{\text{KMnO}_4} \underset{\substack{Acid\quad Nitrile\\5C}}{\text{HOOC(CH}_2)_3\text{CN}}
$$

$$
\xrightarrow[]{\text{H}_2\text{O, H}^+} \underset{\substack{Diacid\\5C}}{\text{HOOC(CH}_2)_3\text{COOH}}
$$

(c) We need to add two extra carbons, probably via CN^-:

$$
\underset{4C}{\text{(oxolane)}} \xrightarrow[\text{heat}]{\text{HCl(aq)}} \underset{\substack{Dihalide\\4C}}{\text{Cl(CH}_2)_4\text{Cl}} \xrightarrow{2CN^-} \underset{\substack{Dinitrile\\6C}}{\text{NC(CH}_2)_4\text{CN}} \xrightarrow[\text{H}^+]{\text{H}_2\text{O}} \underset{\substack{Diacid\\6C}}{\text{HOOC(CH}_2)_4\text{COOH}}
$$

6. (a) $PhCOOK$　　　　　　(b) $(PhCOO)_3Al$　　　　　(c) $(PhCOO)_2Ca$
　　(d) $PhCOONa$　　　　　(e) $PhCOONH_4$　　　　　(g) $PhCH_2OH$
　　(i) $PhCOCl$　　　　　　(j) $PhCOCl$　　　　　　　(k) $PhCOCl$

(l) ⬡ COOH, Br　　　　(n) ⬡ COOH, NO$_2$　　　　(o) ⬡ COOH, SO$_3$H

(q) ⬡ COOH, $Tl(OOCCF_3)_2$　　　　(r) $PhCOO\text{-}n\text{-}Pr$

No reaction: f, h, m, p.

7. (a) C_4H_9COOK　　　　　(b) $(C_4H_9COO)_3Al$　　　(c) $(C_4H_9COO)_2Ca$
　　(d) C_4H_9COONa　　　　(e) $C_4H_9COONH_4$　　　(g) $C_4H_9CH_2OH$
　　(i) C_4H_9COCl　　　　　(j) C_4H_9COCl　　　　　(k) C_4H_9COCl
　　(m) $C_3H_7\underset{\underset{\text{Br}}{|}}{\text{CH}}\text{COOH}$　　(r) $C_4H_9COOC_3H_7\text{-}n$

No reaction: f, h, l, n, o, p, q.

8. (a) $i\text{-Pr}-\overset{\displaystyle O}{\underset{\displaystyle OEt}{C}}$ $\xleftarrow{\text{H}^+}$ $i\text{-PrCOOH} + \text{HOEt}$

Ester

(b) $i\text{-Pr}\overset{\displaystyle O}{\underset{\displaystyle Cl}{C}}$ $\xleftarrow{\text{SOCl}_2}$ $i\text{-PrCOOH}$

Acid chloride

(c) $i\text{-Pr}-\overset{\displaystyle O}{\underset{\displaystyle NH_2}{C}}$ $\xleftarrow{\text{NH}_3}$ $i\text{-Pr}-\overset{\displaystyle O}{\underset{\displaystyle Cl}{C}}$ $\longleftarrow$ (b)

Amide

(d) $\left[i\text{-Pr}-\overset{\displaystyle O}{\underset{\displaystyle O}{C}} \right]_2^{-} \text{Mg}^{++}$ $\xleftarrow{\text{Mg}}$ $i\text{-Pr}-\overset{\displaystyle O}{\underset{\displaystyle OH}{C}}$

Salt

(e) $i\text{-Pr}-\text{CH}_2\text{OH}$ $\xleftarrow{\text{H}^+}$ $\xleftarrow{\text{LiAlH}_4}$ $i\text{-Pr}-\overset{\displaystyle O}{\underset{\displaystyle OH}{C}}$

1° alcohol

9. (a) $\text{PhCOO}^-\text{Na}^+$ $\xleftarrow{\text{NaOH}}$ PhCOOH

(b) $\text{PhC}\overset{\displaystyle O}{\underset{\displaystyle Cl}{}}$ $\xleftarrow{\text{PCl}_5}$ PhCOOH

(c) $\text{Ph}-\overset{\displaystyle O}{\underset{\displaystyle NH_2}{C}}$ $\xleftarrow{\text{NH}_3}$ $\text{Ph}-\overset{\displaystyle O}{\underset{\displaystyle Cl}{C}}$ $\longleftarrow$ (b)

(d) Get rid of —COOH. An old method is strong heating with soda lime (a combination of NaOH and CaO). Compare Problem 27, below.

$$\text{PhCOOH} \xrightarrow{\text{NaOH}} \text{PhCOONa} \xrightarrow[\text{strong heat}]{\text{soda lime}} \text{PhH}$$

(e) $\text{Ph}-\overset{\displaystyle O}{\underset{\displaystyle O-Pr\text{-}n}{C}}$ $\longleftarrow$ $\text{Ph}-\overset{\displaystyle O}{\underset{\displaystyle Cl}{C}}$ (from b) $+ n\text{-PrOH}$

(f) $\text{Ph}-\overset{\displaystyle O}{\underset{\displaystyle O-\text{⬡}CH_3}{C}}$ $\longleftarrow$ $\text{PhC}\overset{\displaystyle O}{\underset{\displaystyle Cl}{}}$ (from b) $+ \text{HO-⬡-}CH_3$

p-Cresol

(g) $\text{Ph}-\overset{\displaystyle O}{\underset{\displaystyle O-\text{⬡}\atop Br}{C}}$ $\xleftarrow{\text{OH}^-}$ $\text{Ph}-\overset{\displaystyle O}{\underset{\displaystyle Cl}{C}}$ (from b) $+ \text{HO-⬡}\atop Br$

m-Bromophenol

(h) PhCH_2OH $\xleftarrow{\text{H}^+}$ $\xleftarrow{\text{LiAlH}_4}$ PhCOOH

10. (a) Neutralize acid with NaOH(aq).

(b) Esterify acid with EtOH.

(c) Heat acid with $SOCl_2$.

(d) Allow acid chloride (from c) to react with NH_3.

(e) Brominate ring with Br_2, Fe.

(f) Nitrate ring with HNO_3/H_2SO_4.

(g) Reduce —COOH to —CH_2OH with $LiAlH_4$.

(h) Brominate side chain with Br_2/P (Hell-Volhard-Zelinsky reaction).

(i) Treat bromoacid (from h) with NH_3.

(j) Hydrolyze bromoacid (from h) with NaOH(aq), warm.

(k) Add an extra carbon: via CN^-, probably.

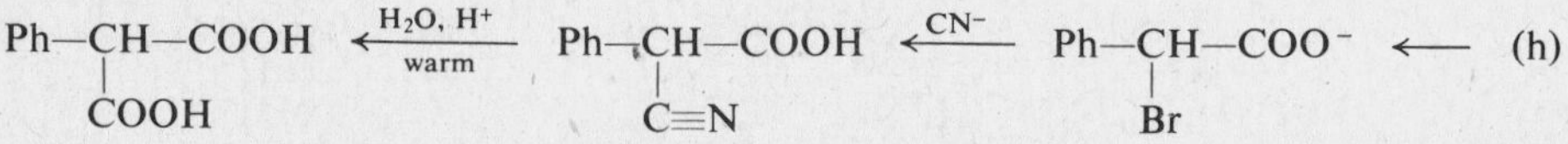

$$\underset{\text{COOH}}{\text{Ph—CH—COOH}} \xleftarrow[\text{warm}]{\text{H}_2\text{O, H}^+} \underset{\text{C}\equiv\text{N}}{\text{Ph—CH—COOH}} \xleftarrow{\text{CN}^-} \underset{\text{Br}}{\text{Ph—CH—COO}^-} \longleftarrow \text{(h)}$$

An alternative way will be found in Chapter 21: a crossed Claisen condensation between ethyl phenylacetate and ethyl carbonate,

$$\underset{\text{O}}{\overset{\displaystyle\parallel}{\text{EtO—C—OEt}}}$$

11. (a) PhCOOK

(b)

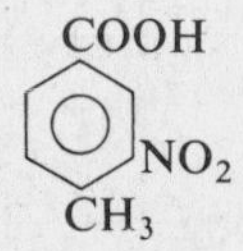

(c) $HOCH_2CH_2CH_2CH_2OH$

(d) $PhCOOCH_2Ph$, benzyl benzoate

(e) 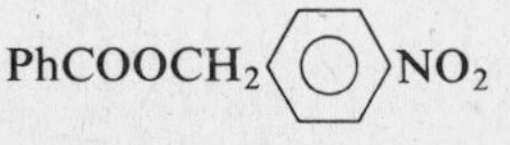 $PhCOOCH_2$⟨benzene ring⟩NO_2 (and *o*-isomer)

(f)

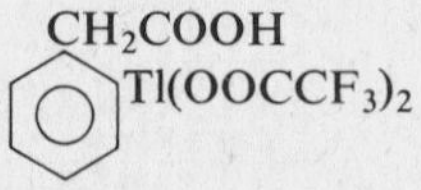

(g) 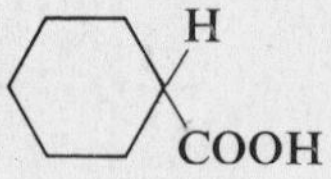
Cyclohexanecarboxylic acid

(h)

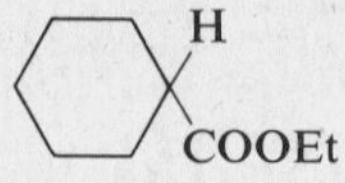

(i)

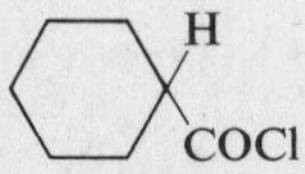

(j)

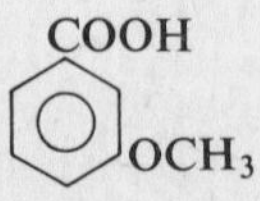

(k) $1,3,5\text{-}C_6H_3(COOH)_3$

(l) $(CH_3)_2CHCOOCH_2CH(CH_3)_2$, isobutyl isobutyrate

(m) Br—COOH OH (and COOH OH—Br)

(n) O_2N—CH_2O—$\overset{\displaystyle O}{\underset{\displaystyle \|}{C}}$—$CH_3$

p-Nitrobenzyl acetate

(o) n-$C_{17}H_{35}COOH$, stearic acid

(p) n-$C_8H_{17}COOH$ + $HOOC(CH_2)_7COOH$

(q) $CH_3(CH_2)_4\overset{\displaystyle H}{C}{=}O$ + $O{=}\overset{\displaystyle H}{C}CH_2\overset{\displaystyle H}{C}{=}O$ + $O{=}\overset{\displaystyle H}{C}(CH_2)_7COOH$

(r) same as product g

(s) $PhCOOCH_2CH_2OOCPh$, ethylene glycol dibenzoate

(t) o-$C_6H_4(COOEt)_2$, ethyl phthalate

(u) racemic *threo*-9,10-dibromooctadecanoic acid (*anti*-addition) (for *threo*, see page 247)

(v) $CH_3(CH_2)_7C{\equiv}C(CH_2)_7COOH$, 9-octadecynoic acid

(w) racemic *threo*-9,10-dihydroxyoctadecanoic acid (*anti*-hydroxylation) (for *threo*, see page 247)

12. (a) $CH_3CH_2CH_2{}^{14}COOH$ $\longleftarrow$ $CH_3CH_2CH_2MgBr$ + ${}^{14}CO_2$

(b) $CH_3CH_2{}^{14}CH_2COOH$ $\xleftarrow{CO_2}$ $Et{}^{14}CH_2MgBr$ $\xleftarrow{Mg}$ $\xleftarrow{PBr_3}$ $Et{}^{14}CH_2OH$

$Et{}^{14}CH_2OH$ $\xleftarrow{LiAlH_4}$ $Et{}^{14}COOH$ $\longleftarrow$ $EtMgBr$ + ${}^{14}CO_2$

(c) $CH_3{}^{14}CH_2CH_2COOH$ $\xleftarrow{KMnO_4}$ $CH_3{}^{14}CH_2CH_2CH_2OH$ $\longleftarrow$ $CH_3{}^{14}CH_2MgBr$ + $H_2C{-}CH_2$ (O)

$CH_3{}^{14}CH_2MgBr$ $\xleftarrow{Mg}$ $\xleftarrow{PBr_3}$ $CH_3{}^{14}CH_2OH$ $\xleftarrow{LiAlH_4}$ $CH_3{}^{14}COOH$ $\longleftarrow$ CH_3MgBr + ${}^{14}CO_2$

(d) ${}^{14}CH_3CH_2CH_2COOH$ $\xleftarrow{KMnO_4}$ ${}^{14}CH_3CH_2CH_2CH_2OH$ $\longleftarrow$ ${}^{14}CH_3CH_2MgBr$ + $H_2C{-}CH_2$ (O)

${}^{14}CH_3CH_2MgBr$ $\xleftarrow{Mg}$ $\xleftarrow{PBr_3}$ ${}^{14}CH_3CH_2OH$ $\xleftarrow{HCHO}$ ${}^{14}CH_3MgBr$ $\xleftarrow{Mg}$ $\xleftarrow{HBr}$ ${}^{14}CH_3OH$

13. (a) $PhCOOH$ $\xleftarrow{KMnO_4}$ $PhCH_3$

(b) $PhCH_2COOH$ $\xleftarrow[H^+]{H_2O}$ $PhCH_2C{\equiv}N$ $\xleftarrow{CN^-}$ $PhCH_2Cl$ $\xleftarrow[heat]{Cl_2}$ $PhCH_3$

(c) CH_3—COOH $\xleftarrow{CO_2}$ CH_3—MgBr $\xleftarrow{Mg}$ CH_3—Br $\xleftarrow{Br_2, Fe}$ $PhCH_3$

(d) COOH—Cl $\xleftarrow{Cl_2, Fe}$ COOH $\longleftarrow$ (a)

(e) $\underset{Cl}{\overset{COOH}{\bigcirc}}$ $\xleftarrow{KMnO_4}$ $\underset{Cl}{\overset{CH_3}{\bigcirc}}$ $\xleftarrow{Cl_2,\ Fe}$ $PhCH_3$

(f) Similar to (b), except brominate toluene first in the *para*-position.

(g) $\underset{Br}{PhCHCOOH}$ $\xleftarrow{Br_2,\ P}$ $PhCH_2COOH$ $\longleftarrow$ (b)

14. (a) $\underset{CH_3}{CH_3CH_2CHCOOEt}$ $\xleftarrow{EtOH,\ H^+}$ $\underset{CH_3}{CH_3CH_2CH-COOH}$ $\xleftarrow{CO_2}$ *sec*-BuMgBr
(from *sec*-BuOH)

(b) $\underset{NO_2}{\overset{COCl}{\underset{O_2N}{\bigcirc}}}$ $\xleftarrow{PCl_5}$ $\underset{NO_2}{\overset{COOH}{\underset{O_2N}{\bigcirc}}}$ $\xleftarrow[\text{twice}]{HNO_3,\ H_2SO_4}$ $\overset{COOH}{\bigcirc}$ $\xleftarrow{KMnO_4}$ $PhCH_3$

(c) $Br\underset{NH_2}{\bigcirc CHCOOH}$ $\xleftarrow{NH_3}$ $Br\underset{Br}{\bigcirc CHCOOH}$ $\xleftarrow{Br_2,\ P}$ $Br\bigcirc CH_2COOH$ $\longleftarrow$ 13(f)

(d) $\underset{OH}{CH_3CHCOOH}$ $\xleftarrow{OH^-}$ $\underset{Br}{CH_3CHCOOH}$ $\xleftarrow{\frac{Br_2}{P}}$ CH_3CH_2COOH $\xleftarrow{KMnO_4}$ $CH_3CH_2CH_2OH$

(e) $\underset{SO_3H}{\overset{COOH}{\bigcirc}}$ $\xleftarrow{KMnO_4}$ $\underset{SO_3H}{\overset{CH_3}{\bigcirc}}$ $\xleftarrow{H_2SO_4,\ SO_3}$ $PhCH_3$

(f) $CH_3CH_2CH=CHCOOH$ $\xleftarrow{KOH}$ $\underset{Br}{CH_3CH_2CH_2CHCOOH}$ $\xleftarrow{\frac{Br_2}{P}}$ $CH_3CH_2CH_2CH_2COOH$

$n\text{-BuCOOH}$ $\xleftarrow{CO_2}$ $n\text{-BuMgBr}$ $\xleftarrow{Mg}$ $\xleftarrow{PBr_3}$ $n\text{-BuOH}$

(g) $CH_3\underset{NH_2}{\bigcirc\overset{O}{C}}$ $\xleftarrow{NH_3}$ $CH_3\underset{Cl}{\bigcirc\overset{O}{C}}$ $\xleftarrow{SOCl_2}$ $CH_3\bigcirc COOH$ $\longleftarrow$ 13(c)

(h) $n\text{-BuCH}_2CH_2O\underset{O}{-C}-Ph$ $\xleftarrow{H^+}$ $n\text{-BuCH}_2CH_2OH + PhCOOH$ (from $PhCH_3$)
n-Hexyl alcohol

$n\text{-Bu}\vdots CH_2CH_2OH$ $\longleftarrow$ $n\text{-BuMgBr}$ (from *n*-BuOH) + $\underset{O}{H_2C-CH_2}$ (Sec. 17.10)

(i) 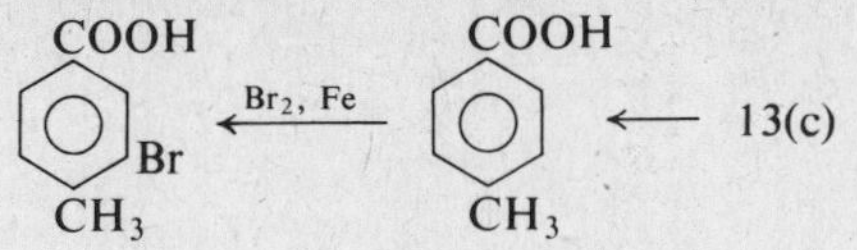
$\xleftarrow{\text{Br}_2, \text{Fe}}$
$\longleftarrow$ 13(c)

(j) PhCH⊢COOH $\xleftarrow{\text{CO}_2}$ PhCHMgCl $\xleftarrow{\text{Mg}}$ $\xleftarrow{\text{HCl}}$ Ph⊢CHOH $\longleftarrow$ PhMgBr (from C_6H_6) + CH_3CHO
      | | | (from EtOH)
      CH_3 CH_3 CH_3

(k) 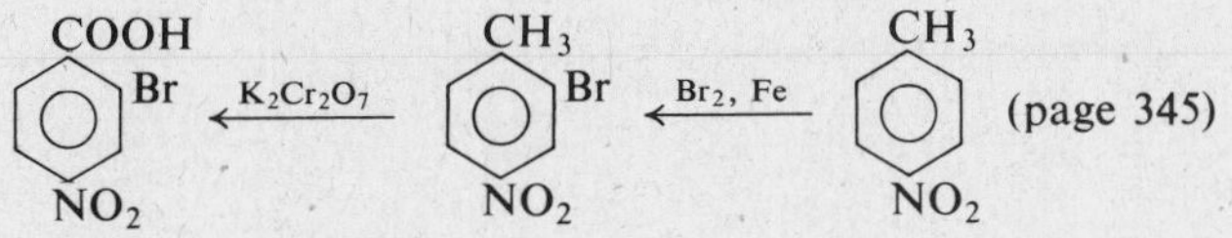
$\xleftarrow{\text{K}_2\text{Cr}_2\text{O}_7}$
$\xleftarrow{\text{Br}_2, \text{Fe}}$
(page 345)

(l) 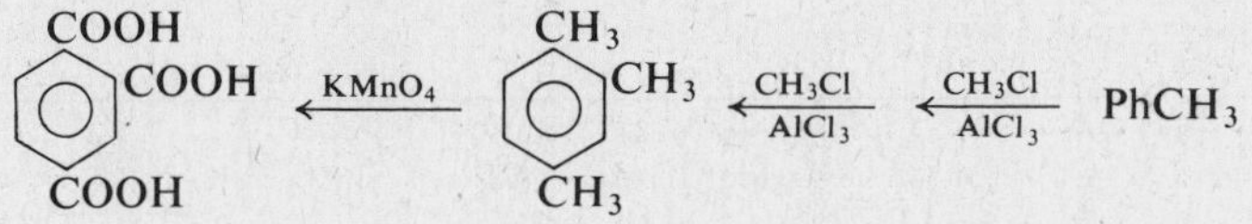
$\xleftarrow{\text{KMnO}_4}$
$\xleftarrow[\text{AlCl}_3]{\text{CH}_3\text{Cl}}$
$\xleftarrow[\text{AlCl}_3]{\text{CH}_3\text{Cl}}$
$PhCH_3$

15. Electron-withdrawing substituents increase acidity; electron-releasing substituents decrease acidity. The more substituents there are, the greater the effect. In saturated compounds, the closer the substituent is to —COOH, the greater its effect.

(a) $CH_3CH_2CHCOOH$ > CH_3CHCH_2COOH > $CH_2CH_2CH_2COOH$ > $CH_3CH_2CH_2COOH$
                | | |
               Br   Br   Br

(b) tri-Cl > di-Cl > mono-Cl > unsubstituted acid

(c)  O_2N⟨O⟩COOH > ⟨O⟩COOH > CH_3⟨O⟩COOH

(d) 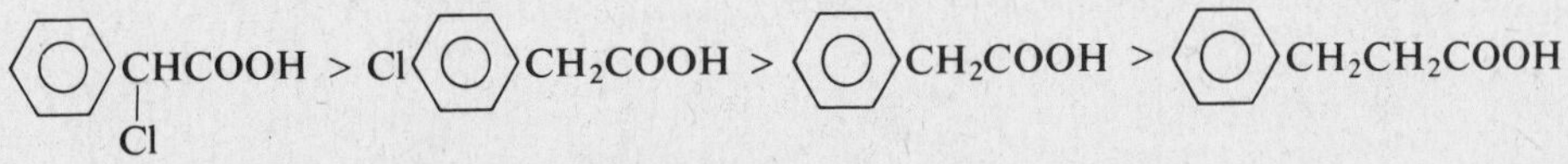 ⟨O⟩CHCOOH > Cl⟨O⟩CH_2COOH > ⟨O⟩CH_2COOH > ⟨O⟩CH_2CH_2COOH
       |
       Cl

⟨O⟩ is electron-withdrawing relative to —H

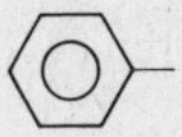

(e) 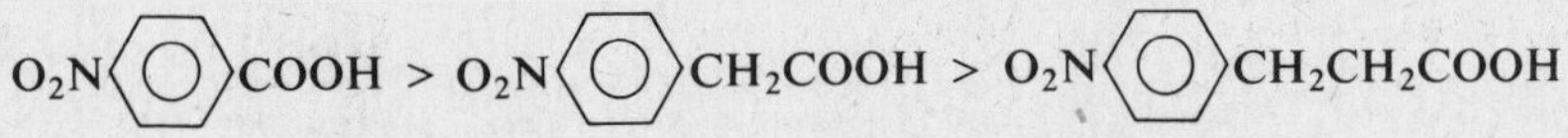 O_2N⟨O⟩COOH > O_2N⟨O⟩CH_2COOH > O_2N⟨O⟩CH_2CH_2COOH

(f) H_2SO_4 > CH_3COOH > H_2O > EtOH > HC≡CH > NH_3 > C_2H_6

(g) $HOOCCH_2COOH$ > $HOOCCH_2CH_2COOH$ > CH_3COOH

16. The weaker the acid, the more powerfully basic will be its anion.

NaC_2H_5 > $NaNH_2$ > NaC≡CH > NaOEt > NaOH > $NaOOCCH_3$ > $NaHSO_4$

233

17. The separation here is exactly as described in Sec. 18.4, with sodium formate taking the place of sodium bicarbonate. Sodium formate is chosen because formic acid is stronger than benzoic acid and weaker than *o*-chlorobenzoic acid.

$$o\text{-ClC}_6\text{H}_4\text{COOH} + \text{HCOO}^- \rightleftarrows o\text{-ClC}_6\text{H}_4\text{COO}^- + \text{HCOOH}$$

Stronger acid Stronger base Weaker base Weaker acid

K_a 120 × 10^{-5} K_a 17.7 × 10^{-5}

$$\text{C}_6\text{H}_5\text{COOH} + \text{HCOO}^- \rightleftarrows \text{C}_6\text{H}_5\text{COO}^- + \text{HCOOH}$$

Weaker acid Weaker base Stronger base Stronger acid

K_a 6.3 × 10^{-5} K_a 17.7 × 10^{-5}

18. Steric factors are controlling in esterification (or in its reverse, hydrolysis): the more hindered the acid or alcohol, the slower the reaction. In particular, the usual order for alcohols is MeOH > 1° > 2° > 3°.

(a) MeOH > *n*-PrOH > *sec*-BuOH > *tert*-pentyl alcohol

(b) $\text{C}_6\text{H}_5\text{COOH} > o\text{-CH}_3\text{C}_6\text{H}_4\text{COOH} > 2,6\text{-(CH}_3)_2\text{C}_6\text{H}_3\text{COOH}$

(c) $\text{HCOOH} > \text{CH}_3\text{COOH} > \text{CH}_3\text{CH}_2\text{COOH} > (\text{CH}_3)_2\text{CHCOOH} > (\text{CH}_3)_3\text{CCOOH}$

19. (a) See Figure 18.6, page 235 of this Study Guide.

(b) See Figure 18.7, page 235 of this Study Guide.

(c) $\text{C}_7\text{H}_8\text{Br}_2 - \text{C}_6\text{H}_8 = \text{CBr}_2$, showing that the methylene CBr_2 adds to the diene. The *cis*-structure of D, E, and F follows from the stereochemistry (*syn*) of the addition of CBr_2.

20. $\text{HC}{\equiv}\text{CH} + \text{CH}_3\text{MgBr} \longrightarrow \text{CH}_4 + \text{HC}{\equiv}\text{CMgBr}$ (See Secs. 3.16 and 8.10, and Problem 6m, page 280)

Stronger acid Weaker acid G

$$\text{HC}{\equiv}\text{CMgBr} + \text{CO}_2 \longrightarrow \text{HC}{\equiv}\text{CCOOMgBr} \xrightarrow{\text{H}^+} \text{HC}{\equiv}\text{CCOOH}$$

G   H   I

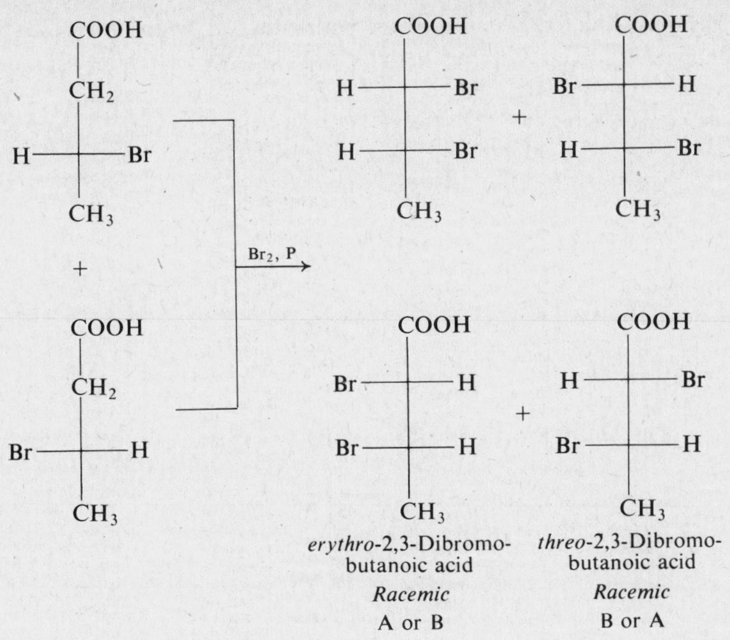

Figure 18.6. α-Bromination of β-bromobutyric acid (Problem 19a).

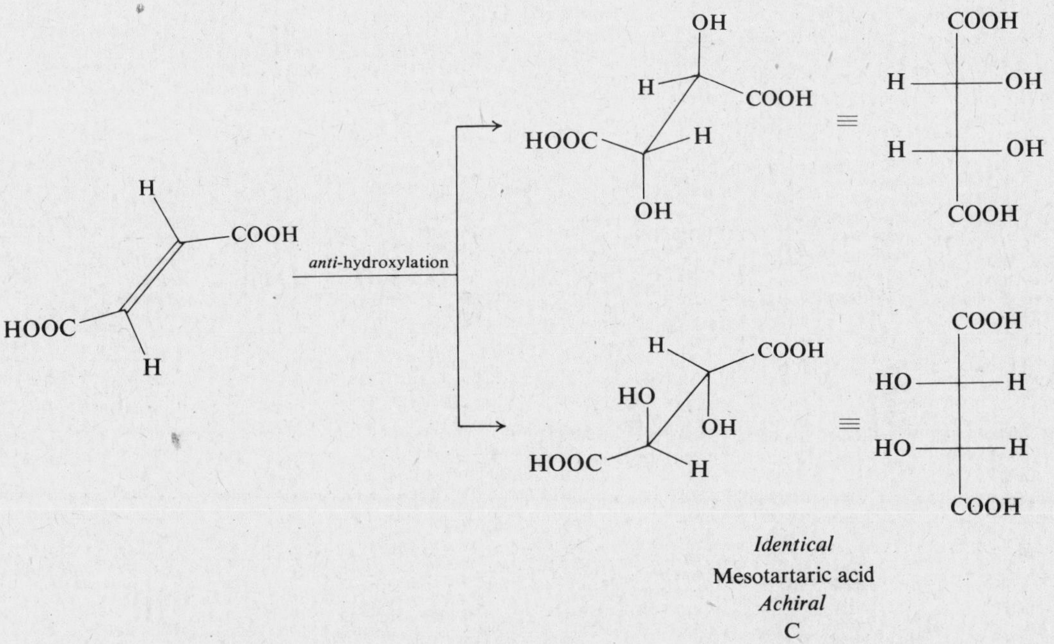

Figure 18.7. Hydroxylation of fumaric acid with HCO_2OH (Problem 19b).

Compound **I** undergoes hydration, which could take place with either of two orientations, *a* or *b*:

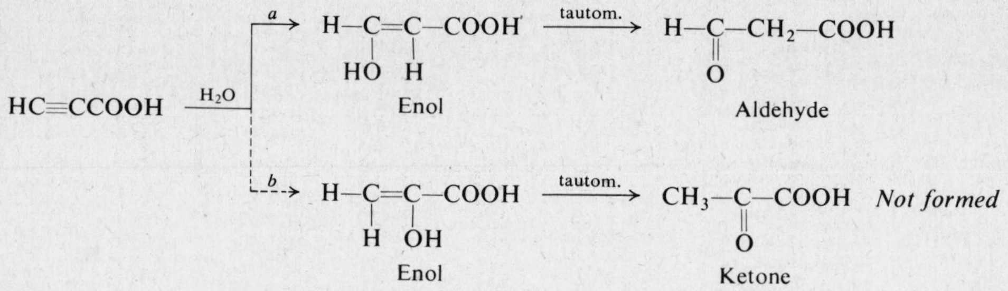

$$HC\!\equiv\!CCOOH \xrightarrow{\ H_2O\ }$$

a → H—C=C—COOH $\xrightarrow{\text{tautom.}}$ H—C—CH₂—COOH

                HO H   O

                 Enol   Aldehyde

b → H—C=C—COOH $\xrightarrow{\text{tautom.}}$ CH₃—C—COOH   *Not formed*

                 H OH   O

                 Enol   Ketone

The fact that **J** is oxidized to an acid without loss of carbon gives us its structure, and shows that orientation is as in *a*.

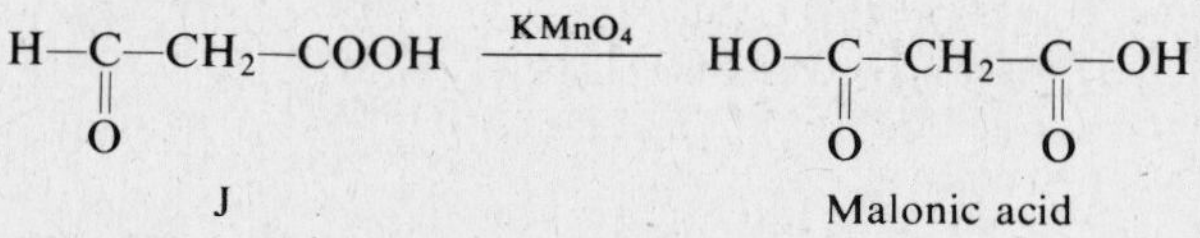

H—C—CH₂—COOH $\xrightarrow{\text{KMnO}_4}$ HO—C—CH₂—C—OH

     O   O   O

      J   Malonic acid

21. (a) $NaHCO_3$ (aq)   (b) $NaHCO_3$ (aq)   (c) $NaHCO_3$ (aq)

      (d) $AgNO_3$ (e) $NaHCO_3$ (aq)   (f) $NaHCO_3$ (aq)

Ppt. of AgCl in (d); evolution of CO_2 in all the others.

22.

	Benzoic acid	Sodium benzoate
(a)	low b.p.	high b.p. (*dec*)
(b)	low m.p.	high m.p. (*dec*)
(c)	insol. H_2O	sol. H_2O
(d)	sol. Et_2O	insol. Et_2O
(e)	partial ionization	complete ionization
(f)	stronger acid, weaker base	weaker acid, stronger base

This comparison generally holds for carboxylic acids and their salts. (Generally, only alkali metal salts are soluble in water.)

23. (a)

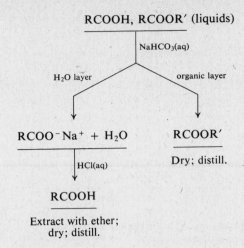

RCOOH, RCOOR' (liquids)

NaHCO₃(aq)

H₂O layer   organic layer

RCOO⁻Na⁺ + H₂O   RCOOR'

      Dry; distill.

HCl(aq)

RCOOH

Extract with ether;
dry; distill.

(b) *n*-Butyric acid is too soluble in water for us to use Procedure (a).

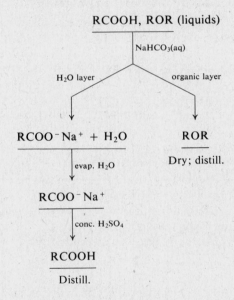

RCOOH, ROR (liquids)

NaHCO₃(aq)

H₂O layer   organic layer

RCOO⁻Na⁺ + H₂O   ROR

evap. H₂O   Dry; distill.

RCOO⁻Na⁺

conc. H₂SO₄

RCOOH

Distill.

(c) As in (b), with alcohol being in organic layer.

(d)

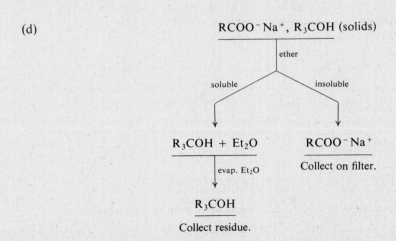

RCOO⁻Na⁺, R₃COH (solids)

ether

soluble   insoluble

R₃COH + Et₂O   RCOO⁻Na⁺

evap. Et₂O   Collect on filter.

R₃COH

Collect residue.

24. (a) $KMnO_4$.

(b) CrO_3/H_2SO_4.

(c) Mesotartaric acid reacts with hot $KMnO_4$; elemental analysis for other three acids.

(d) Valeric acid gives negative halogen test; use neutralization equivalents for others.

(e) Neutralization equivalents; or 3-nitrophthalic acid gives an anhydride (m.p. 162°, neutral compound) when heated.

(f) The cinnamic acid reacts with $KMnO_4$; elemental analysis for the others.

(g) Try the easiest tests first. If a test is negative, go on to next.

   (1) Only 4-methylpentanoic acid will react with $NaHCO_3$(aq) with evolution of CO_2, and go into solution.

   (2) Only α-phenylethyl chloride will react with $AgNO_3$ to ppt. AgCl.

   (3) Only o-toluidine is basic enough to dissolve in dilute HCl(aq).

   (4) Linalool is the only alcohol and the only compound that will respond quickly to CrO_3/H_2SO_4 reagent (orange-red $\longrightarrow$ green).

   (5) β-Chlorostyrene is the only alkene; try cold $KMnO_4$ (red-violet) $\longrightarrow$ MnO_2 (brown).

   (6) Try to dissolve the unknown in warm conc. H_2SO_4. Of the possibilities remaining at this point, only cis-decalin and 2,4-dichlorotoluene are insoluble in warm conc. H_2SO_4; and of these, only 2,4-dichlorotoluene will give a Cl test in an elemental analysis when subjected to sodium fusion. (Isodurene is easily sulfonated by conc. H_2SO_4; the others form soluble oxonium salts, $R_2OH^+HSO_4^-$.)
       If the unknown dissolves in warm conc. H_2SO_4, go on to (7). (Three possibilities left.)

   (7) Subject the unknown to sodium fusion and elemental analysis for halogen. If a positive test for Cl is observed, the unknown is o-chloroanisole.

   (8) If test (7) is negative, the unknown (now down to two possibilities) must be an ether or a hydrocarbon, p-cresyl ethyl ether or isodurene. Apply the Zeisel test (page 570) for ethers; if the test is negative, the unknown must be isodurene.

25. K, $m\text{-}ClC_6H_4COOH$; confirm by amide (m.p. 134°). ($m\text{-}BrC_6H_4COOH$ has N.E. 201.)

   L, adipic acid or $2,4,6\text{-}Me_3C_6H_2COOH$; distinguish by neutralization equivalent (73 vs. 164) or by p-nitrobenzyl ester (106° vs. 188°).

   M, $2,5\text{-}Cl_2C_6H_3COOH$ or $p\text{-}ClC_6H_4OCH_2COOH$; distinguish by amide (155° vs. 133°).

   N, trans-$CH_3CH{=}CHCOOH$ or $PhCH_2COOH$; distinguish by action of $KMnO_4$; or by oxidation of $PhCH_2COOH$ to benzoic acid (m.p. 122°).

O, $CH_3CH_2CHOHCOOH$ or ICH_2CH_2COOH; distinguish by elemental analysis.

P, $HOCH_2COOH$; confirm by amide (120°) or p-nitrobenzyl ester (107°). (α-Hydroxyiso-butyric acid is a 3° alcohol, negative to CrO_3/H_2SO_4.)

26.
$$\frac{12.4}{1000} \times 0.098 \text{ equiv. acid} = 0.201 \text{ g}$$

$$1 \text{ equiv. acid} = 0.201 \times \frac{1000}{12.4} \times \frac{1}{0.098} = 165$$

The acid is $o\text{-}O_2NC_6H_4COOH$, calcd. N.E. = 167. (Anthranilic acid, $o\text{-}H_2NC_6H_4COOH$, has N.E. 137.)

27. The best starting point for working out this problem is T, m.p. 121–2, N.E. 123 ± 2. Some chemical arithmetic,

$$
\begin{array}{rl}
123 \pm 2 & T \\
- \ 45 & -COOH \\
\hline
78 \pm 2 & \text{gives } C_6H_5\text{—, m.w. 77}
\end{array}
$$

and a look at Table 18.1, page 582, show this to be benzoic acid. Therefore, S must have had only one side chain. (More side chains would, of course, have led to an acid with more than one —COOH group.)

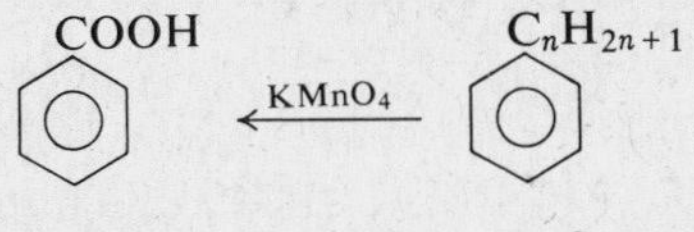

Table 12.1, page 375, shows ethylbenzene (b.p. 136°) to be the only possibility for S. Hence Q is one of three isomers, o-, m-, or p-ethylbenzoic acid.

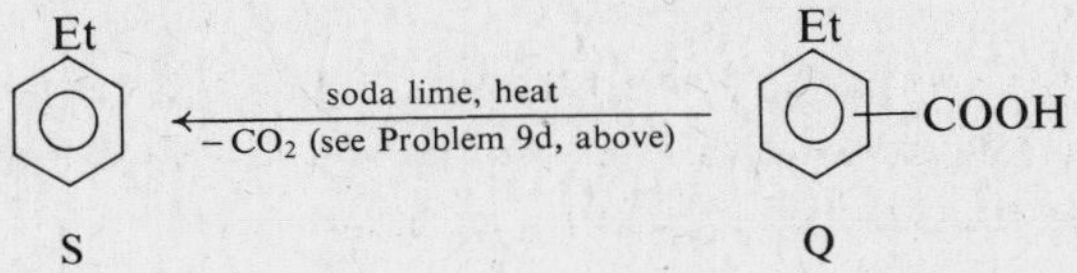

But oxidation of Q gave R, which must be isophthalic acid; since this is the *meta* isomer, Q must be the *meta* isomer, too.

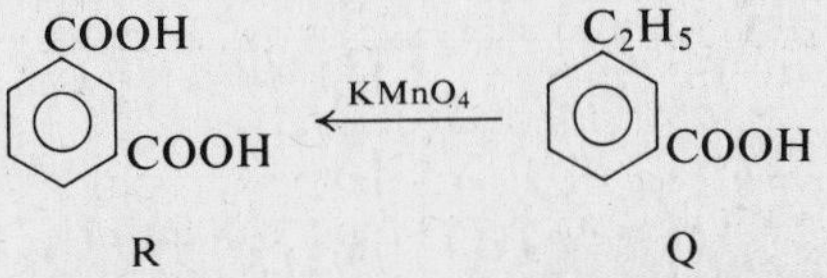

R
Isophthalic acid
M.p. 348°
N.E. = m.w./2 = 83

Q
m-Ethylbenzoic acid
N.E. calcd. 150

Oxidation of U gives an acid V, of very high m.p., that does not appear in either Table 18.1 or Table 18.3. Chemical arithmetic,

Try: N.E. = m.w./1   Try: N.E. = m.w./2   Try: N.E. = m.w./3

m.w. =	70			m.w. =	140			m.w. =	210
—COOH	− 45			2 —COOH	− 90			3 —COOH	−135
	25				50				75

not enough for   not enough for $C_6H_3 = 75$
benzene ring   benzene ring

shows that V must be a tricarboxylic acid, $C_6H_3(COOH)_3$. U, then, must be a dimethyl-benzoic acid, $(CH_3)_2C_6H_3COOH$.

More extensive tables of physical constants will show V to be $1,3,5\text{-}C_6H_3(COOH)_3$, giving the orientation of groups in U

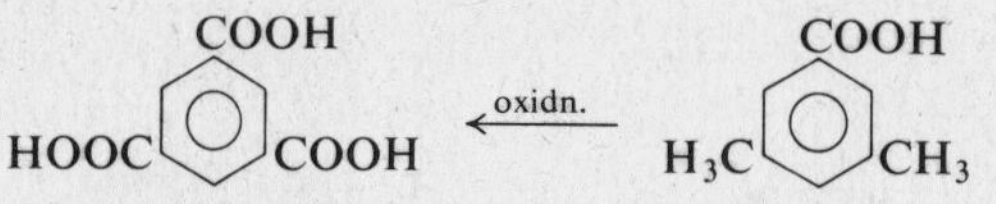

V
Trimesic acid
M.p. 380°
N.E. calcd. 70

U
3,5-Dimethylbenzoic acid
N.E. calcd. 150

28. (a) Tropic acid contains —COOH, and a 1° or 2° alcohol group; and because tropic acid yields benzoic acid, all this must be on only one side chain on a benzene ring.

$C_9H_{10}O_3$	C_8H_9O	C_8H_8
− COOH	− OH	− C_6H_5
C_8H_9O	C_8H_8	C_2H_3 left for side chain

Hydratropic acid has the same carbon skeleton; but, since —OH has been replaced by —H, the side chain carries no functional group other than —COOH.

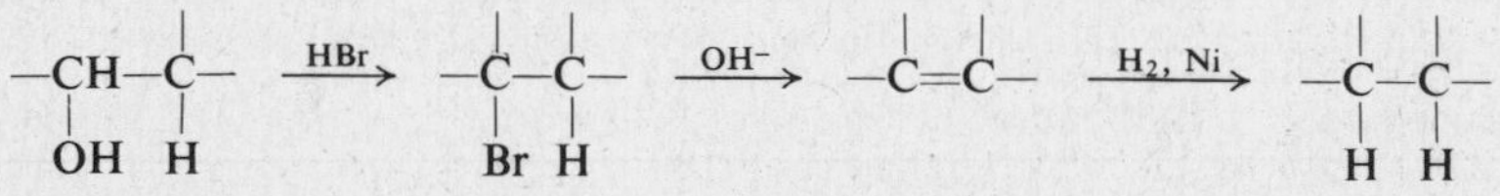

In tropic acid   *In atropic acid*   *In hydratropic acid*

At this point, the acids could be:

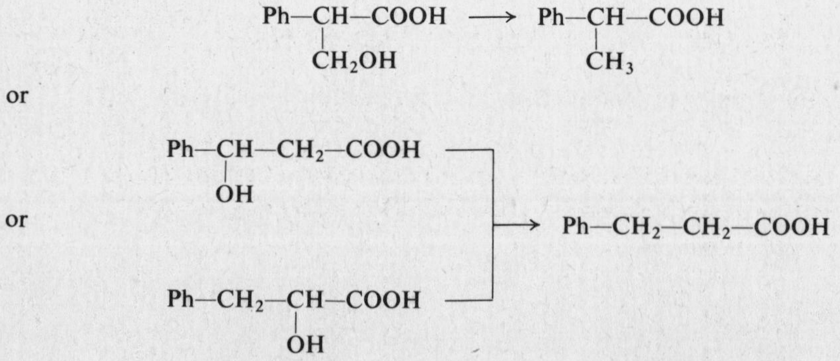

or

or

Tropic acid:
possible structures   Hydratropic acid:
                                             possible structures

(b) On the basis of the following synthesis,

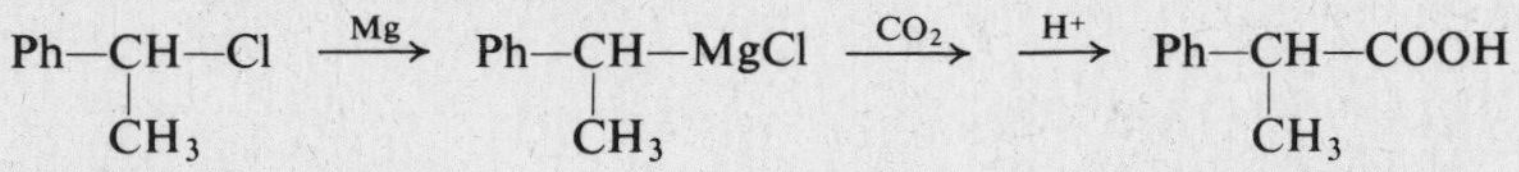

$$Ph-\underset{\underset{CH_3}{|}}{CH}-Cl \xrightarrow{Mg} Ph-\underset{\underset{CH_3}{|}}{CH}-MgCl \xrightarrow{CO_2} \xrightarrow{H^+} Ph-\underset{\underset{CH_3}{|}}{CH}-COOH$$

α-Phenylethyl chloride   Hydratropic acid

we can assign the following structures:

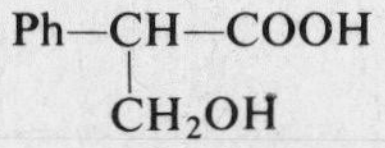

$$Ph-\underset{\underset{CH_2OH}{|}}{CH}-COOH \qquad Ph-\underset{\underset{CH_2}{\|}}{C}-COOH$$

Tropic acid 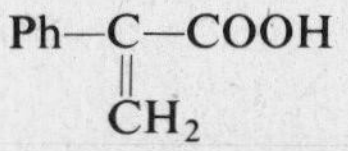   Atropic acid

29. Following the general procedure described in the box on page 133 of this Study Guide, we do some chemical arithmetic, draw tentative structures, eliminate the impossible isomers, and try to decide among the possible isomers.

(a) $C_3H_5ClO_2$

$$\frac{-\;COOH}{C_2H_4Cl} \qquad \begin{array}{l}\text{(tentative for } \delta\ 11.22 \text{ signal)}\\[4pt]\text{means a saturated group equivalent to } C_2H_5\end{array}$$

Two structures can be drawn:

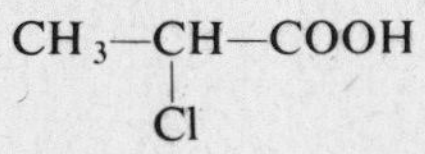

$$CH_3-\underset{\underset{Cl}{|}}{CH}-COOH \quad \text{and} \quad \underset{\underset{Cl}{|}}{CH_2}-CH_2-COOH$$

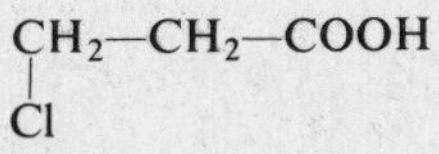

Of these, only the first can give a 3H signal.   The compound is

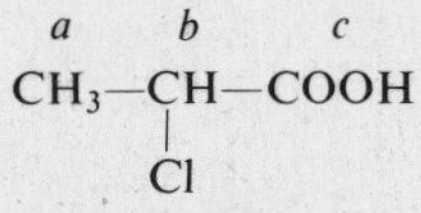

$$\overset{a}{CH_3}-\underset{\underset{Cl}{|}}{\overset{b}{CH}}-\overset{c}{COOH}$$

α-Chloropropionic acid

(b) The absence of any signal for acidic H leads us to consider esters (two oxygens), RCOOR′, for this compound.

$$\frac{-\;COO}{C_2H_5Cl, \text{ to be divided between R and R}'} \begin{array}{l}C_3H_5ClO_2\\[2pt]\text{from RCOOR}'\end{array}$$

Without looking further at the nmr data, we see four possibilities:

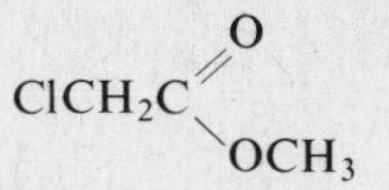

$$\underset{i}{\overset{\overset{\displaystyle O}{\|}}{ClCH_2C}\diagdown_{OCH_3}} \qquad \underset{ii}{\overset{\overset{\displaystyle O}{\|}}{CH_3C}\diagdown_{OCH_2Cl}} \qquad \underset{\underset{\underset{Cl}{|}}{iii}}{\overset{\overset{\displaystyle O}{\|}}{HC}\diagdown_{OCHCH_3}} \qquad \underset{iv}{\overset{\overset{\displaystyle O}{\|}}{HC}\diagdown_{OCH_2CH_2Cl}}$$

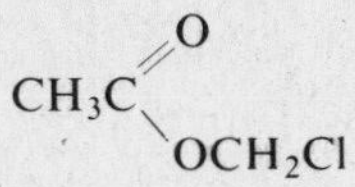

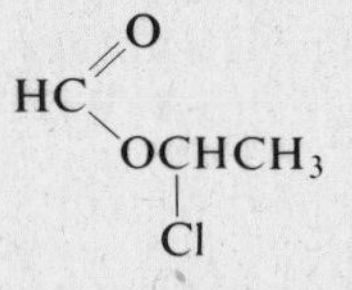

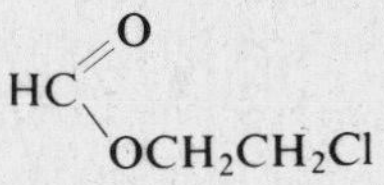

The appearance of two unsplit 3H and 2H signals in the nmr spectrum rules out all except *i* and *ii*. From *ii* we would expect 3H at about δ 2.0, and 2H considerably farther downfield than δ 4.08. The compound must be

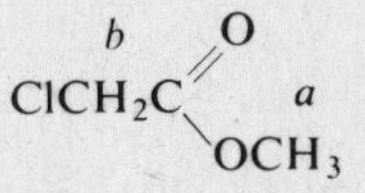

$$\underset{\text{Methyl chloroacetate}}{\overset{b}{ClCH_2}\overset{\overset{\displaystyle O}{\|}}{C}\diagdown_{OCH_3}\overset{a}{}}$$

(c) As in (b), an ester rather than an acid seems likely.

$$\begin{array}{l} C_4H_7BrO_2 \\ \underline{- \ COO} \quad \text{from } RCOOR' \\ C_3H_7Br, \text{ to be divided between R and } R' \end{array}$$

The 3H + 2H triplet-quartet points to a —CH_2CH_3, as usual,

$$\begin{array}{l} C_3H_7Br \\ \underline{- \ C_2H_5} \\ CH_2Br \end{array}$$

leaving only —CH_2Br to be fitted in. The 2H singlet at δ 3.77 is —CH_2— shifted downfield by —Br and —COOR' (compare signal b from compound b, above). The compound has to be

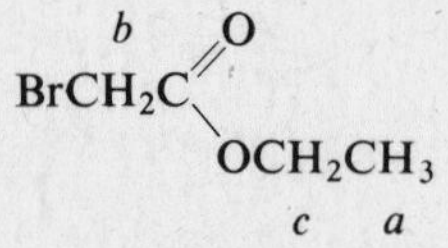

$$\underset{\underset{c \quad a}{OCH_2CH_3}}{\overset{\overset{b \quad O}{\parallel}}{BrCH_2C}}$$

Ethyl bromoacetate

(d) This looks like an acid (signal d at δ 10.97).

$$\begin{array}{l} C_4H_7BrO_2 \\ \underline{- \ COOH} \\ C_3H_6Br \end{array}$$

Signal a is —CH_3 split by two protons; signal b, a quintet, is —CH_2— split by —CH_3 and by

—CH—; signal c is —CH split by —CH_2— and shifted far downfield. The pieces

thus are:

$$CH_3\text{—} \qquad \text{—}CH_2\text{—} \qquad \text{—}CH\text{—} \qquad \text{—}Br \qquad \text{—}COOH$$

On the basis of the splittings of the signals and their δ values, only one structure is possible:

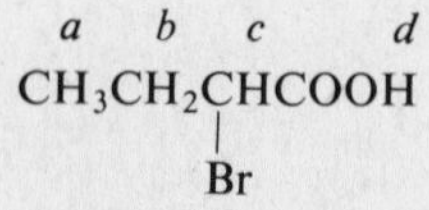

$$\overset{a \quad b \quad c \quad d}{CH_3CH_2CHCOOH}$$
$$\underset{Br}{|}$$

α-Bromobutyric acid

(e) The singlet at δ 10.95 indicates —COOH; the triplet-quartet system indicates —CH_2CH_3.

$$\begin{array}{lll} C_4H_8O_3 & C_3H_7O & CH_2O \\ \underline{- \ COOH} \ \ \text{signal } d & \underline{- \ C_2H_5} \ \ \text{signals } a \text{ and } b & \underline{- \ CH_2} \ \ \text{signal } c \\ C_3H_7O & CH_2O & O \qquad \text{oxygen} \end{array}$$

This leaves only oxygen.   In the absence of a signal for —OH, it must be in an ether linkage, and furthermore it must separate Et— from —CH$_2$— since signal b is unaffected by the protons of the —CH$_2$— group.   The compound can only be:

$$\overset{a}{\text{CH}_2}\overset{b}{\text{CH}_2}\text{O}\overset{c}{\text{CH}_2}\overset{d}{\text{COOH}}$$

Ethoxyacetic acid

30. (a) CH$_3$CH=CHCOOH   (b) C$_6$H$_5$CHOHCOOH   (c) p-O$_2$NC$_6$H$_4$COOH

    Crotonic acid   Mandelic acid p-Nitrobenzoic acid

(See the labeled spectra on page 657 of this Study Guide.)

Chapter 19

Aldehydes and Ketones
Nucleophilic Addition

19.1 A *gem*-diol, $RCH(OH)_2$.

19.2 See the answer to Problem 14, page 488. Carry out hydrolysis at reflux (reduced pressure if a temperature below 100° is desired) under a distilling column so that the *sec*-BuOH–water azeotrope distills out as rapidly as the alcohol is formed; use a non-volatile acid (H_2SO_4, $HClO_4$) to catalyze the hydrolysis. After drying the distillate over anhydrous K_2CO_3, measure the specific rotation of the alcohol layer in a polarimeter, and compare the value obtained with the specific rotation of the alcohol used to form the ester in the first place.

19.3 No, because one has to make the cadmium compound from the Grignard reagent, and the nitro group would interfere with any attempts to form that reagent.

19.4 (a)

$$CH_3CH_2CH_2 \overset{b}{\underset{|}{\mid}} \overset{a}{\underset{|}{\mid}} C \overset{}{\mid} CH_2CH_3 \xrightarrow{\text{oxid.}} \begin{cases} \overset{a}{\longrightarrow} CH_3CH_2CH_2COOH + HOOCCH_3 \\ \overset{b}{\longrightarrow} CH_3CH_2COOH + HOOCCH_2CH_3 \end{cases}$$

(b) 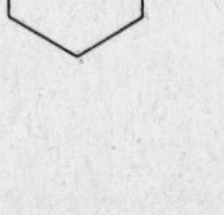 cyclohexanone $\xrightarrow{\text{oxid.}}$ $HO-\underset{O}{\overset{}{C}}(CH_2)_4\underset{O}{\overset{}{C}}-OH$, adipic acid

19.5

(a) 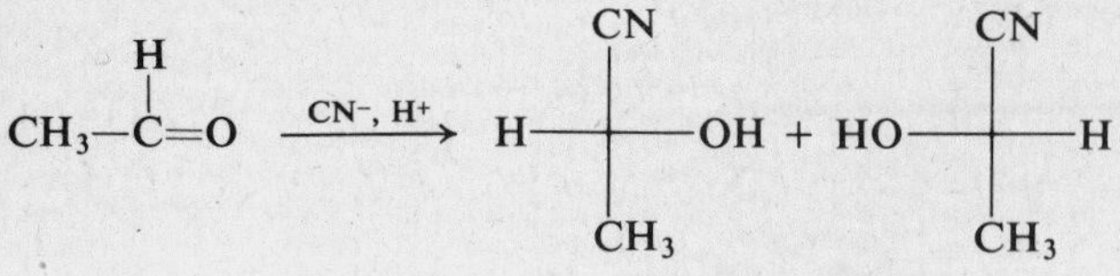 $CH_3-\overset{\overset{\displaystyle H}{|}}{C}=O$ $\xrightarrow{CN^-, \ H^+}$

Racemic modification
One fraction
Inactive, but resolvable

(b) 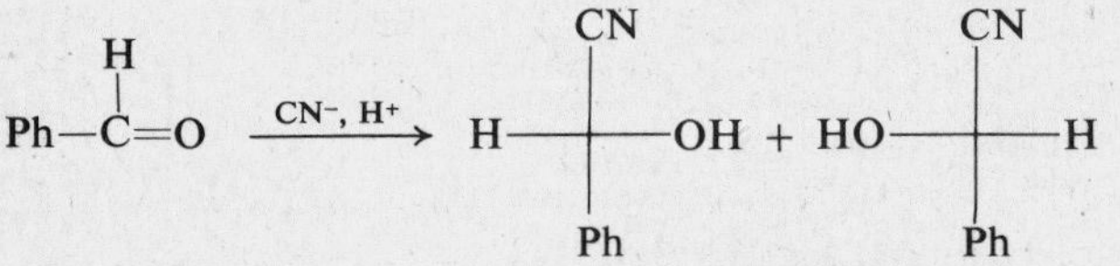 $Ph-\overset{\overset{\displaystyle H}{|}}{C}=O$ $\xrightarrow{CN^-, \ H^+}$

Racemic modification
One fraction
Inactive, but resolvable

(c) 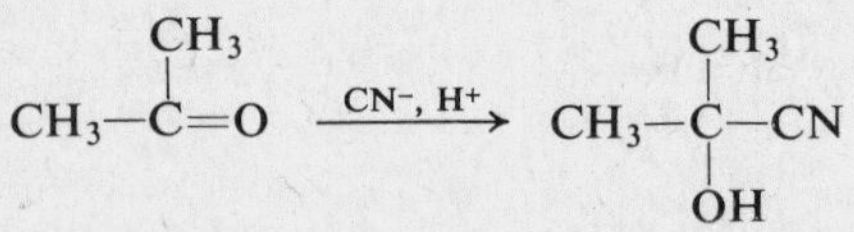 $CH_3-\overset{\overset{\displaystyle CH_3}{|}}{C}=O$ $\xrightarrow{CN^-, \ H^+}$ $CH_3-\overset{\overset{\displaystyle CH_3}{|}}{\underset{\underset{\displaystyle OH}{|}}{C}}-CN$

Single compound
One fraction
Inactive; not resolvable

(d)

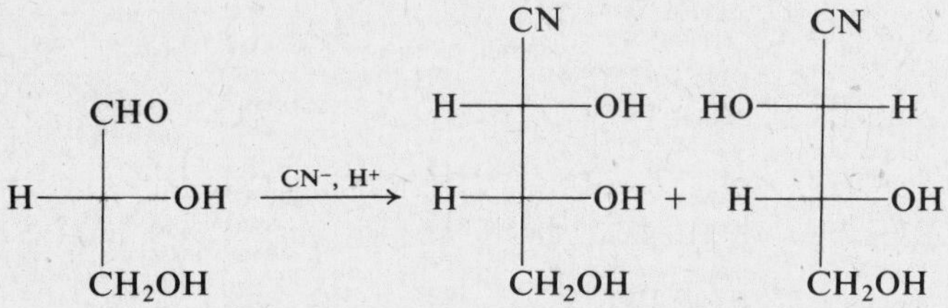

Diastereomers
Two fractions
Active   Active

(e)

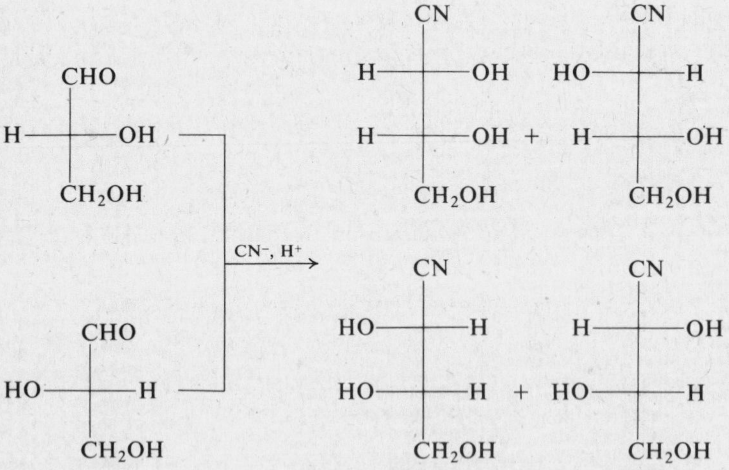

$\xrightarrow{CN^-, \ H^+}$

Racemic

Racemic   Racemic
Diastereomeric
Two fractions
Each inactive, but resolvable

(f) No change, since no bond to chiral carbon is broken.

19.6 (a) Oxidation of a 1° alcohol to an aldehyde: distill out the aldehyde, the alcohol being higher boiling (associated by H-bonding); or isolate the aldehyde as the crystalline bisulfite addition product.

(b) Reduction of an aldehyde to a 1° alcohol: shake the mixture with $NaHSO_3$(aq); separate the alcohol from the aqueous layer which contains the water-soluble bisulfite addition product of the aldehyde.

19.7 Semicarbazide formation is reversible. Cyclohexanone reacts more rapidly, but benzaldehyde gives the more stable product. Initially, one isolates the product of rate control; later, after equilibrium is established. one isolates the product of equilibrium control.

19.8 High alcohol, low water concentrations shift the hemiacetal–acetal and aldehyde–hemiacetal equilibria (pages 641–642) in direction of acetal. Low alcohol, high water concentrations shift equilibria in direction of aldehyde.

19.9 (a) Williamson synthesis of ethers.   (b) Acetals (cyclic).

(c) Treatment with acid gives HCHO and catechol, $o\text{-}C_6H_4(OH)_2$; there is no reaction with base, typical of acetals (and other ethers).

19.10 Convert PhCHO into the water-soluble bisulfite addition product, leaving water-insoluble acetal untouched.

19.11 An attempt to hydroxylate acrolein itself would result in oxidation of the —CHO group. The acetal grouping, however, is not affected by permanganate or the alkaline reaction medium, and is converted into —CHO after hydroxylation is complete.

$$CH_2{=}CH{-}CH(OEt)_2 \xrightarrow{KMnO_4} \underset{\underset{OH}{|}\ \underset{OH}{|}}{CH_2{-}CH{-}CH(OEt)_2} \xrightarrow{H_2O,\ H^+} \underset{\underset{OH}{|}\ \underset{OH}{|}}{CH_2{-}CH{-}CHO}$$

Acrolein diethyl acetal   Glyceraldehyde

19.12 The rate-determining step is formation of a cation analogous to I (page 642); the more stable the cation, the faster it is formed.

(a) $R{-}C{\underset{OR'}{\overset{OR'}{\big\langle}}} \Big\} \oplus \ >\ R{-}C{\underset{H}{\overset{^+OR'}{\big\langle}}}\ >\ R{-}CH_2{}^+$

From: ortho ester   acetal   ether

(b) $R_2C{\overset{+}{=}}OR'\ >\ RCH{\overset{+}{=}}OR'\ >\ H_2C{\overset{+}{=}}OR'$

From: ketal   acetal   formal

In (b), R stabilizes the cation (relative to the reactant) more than H does because (*i*) R releases electrons, and (*ii*) R is bigger and hence favors change from the tetrahedral (sp^3) reactant to the trigonal (sp^2) product.

19.13

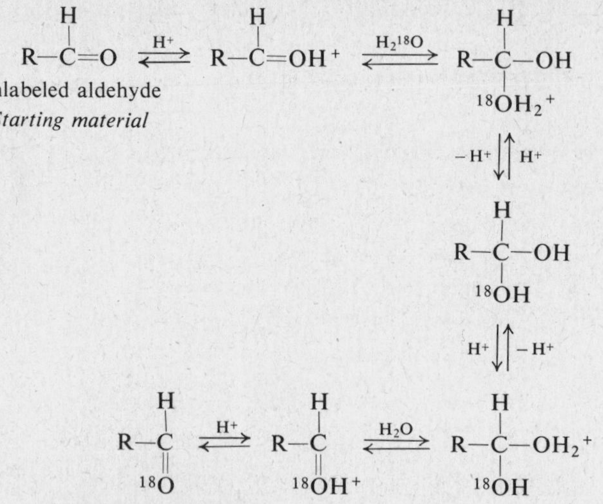

Unlabeled aldehyde
Starting material

Labeled aldehyde
Exchange product

19.14 (a)

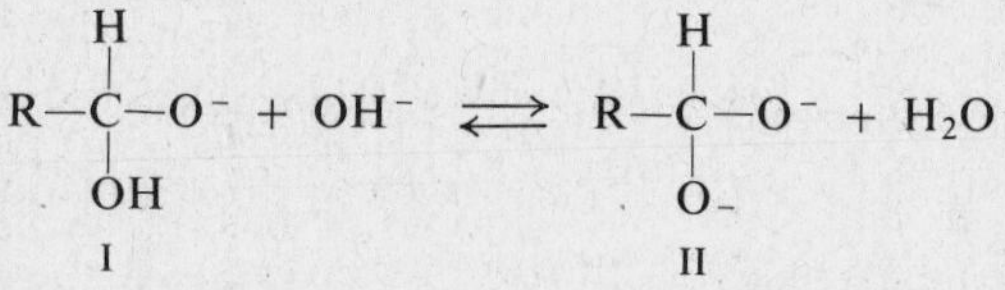

(b) Loss of hydride ion yields resonance-stabilized $RCOO^-$ (III) directly.

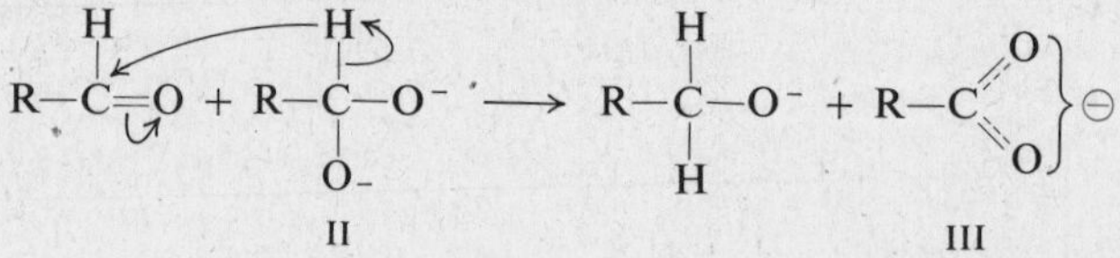

19.15 Use R—C(D)=O, and see if a second D shows up in the alcohol, R—CD$_2$—OH.

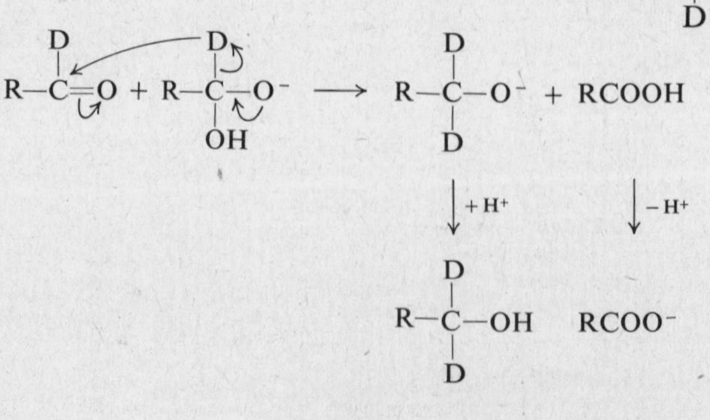

19.16 On both electronic and steric grounds (Sec. 19.8), one would expect reaction (1), page 644, to be *faster* for HCHO than for another aldehyde, and the position of *equilibrium* to lie farther to the right. If reaction (1) is rate-determining, HCHO is the chief hydride donor because it forms I faster than the other aldehyde does. If reaction (2) is rate-determining, HCHO is the chief hydride donor because equilibrium (1) provides more I derived from HCHO than from the other aldehyde. (Kinetics studies indicate that reaction (2) is rate-determining.)

19.17 Internal crossed Cannizzaro reaction:

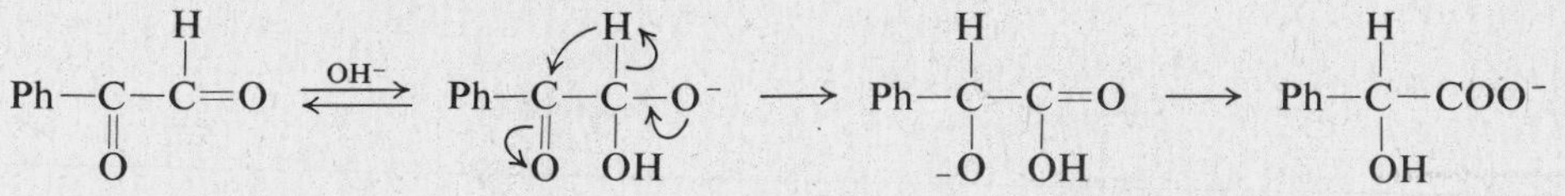

19.18

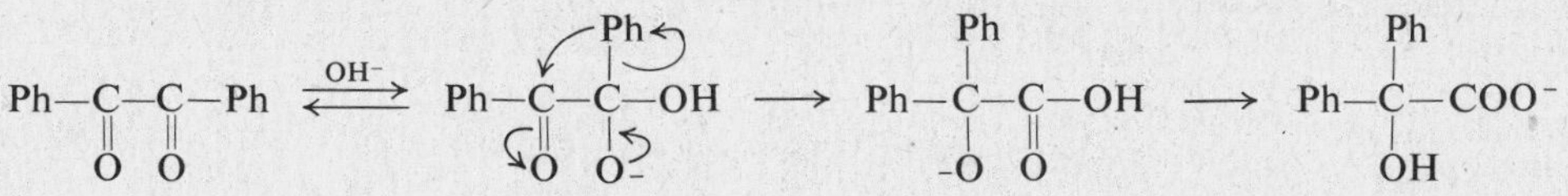

Use of OCH_3^- instead of OH^- gives the ester by the same route.

19.19

	cold KMnO4[a]	hot KMnO4[a]	CrO3[b]	I2/NaOH[c]	HIO4[d]	Tollens'[e]
Alkanes	−	−	−	−	−	−
Alkenes	+	+	−	−	−	−
Alkynes	+	+	−	−	−	−
Alkylbenzenes	−	+	−	−	−	−
1° alcohols	+	+	+	−[f]	−[g]	−
2° alcohols	+	+	+	−[h]	−[g]	−
RCH(OH)CH3	+	+	+	+	−[g]	−
RCH(OH)CH(OH)R[i]	+	+	+	−	+	−
3° alcohols	−	−	−	−	−	−
Ethers	−	−	−	−	−	−
Carboxylic acids	−	−	−	−	−	−
Aldehydes	+	+	+	−[j]	−[g]	+
Ketones	−	−	−	−[g]	−[g]	−
RCOCH3[i]	−	−	−	+	−[g]	−
RCOCH(OH)R[i]	+	+	+	−	+	−
RCOCOR[i]	−	−	−	−	+	−

[a] Decolorizes. [b] Color changes. [c] Yellow CHI3 precipitates. [d] AgNO3 gives white ppt. of AgClO3. [e] Silver mirror. [f] EtOH gives test. [g] Exceptions below. [h] RCH(OH)CH3 gives test. [i] R may be H. [j] CH3CHO gives test.

19.20 Oxime formation consumes the base, hydroxylamine; the change in pH is shown by the change in color of the indicator.

19.21 Add the following to the Table in the answers to Problems 16.10, 17.24, and 18.18. For oxidizing agents, see answer to Problem 19.19, above.

	conc. $H_2SO_4{}^a$	cold $KMnO_4{}^b$	$Br_2{}^b$	$CrO_3{}^c$	fum. $H_2SO_4{}^a$	$CHCl_3$, $AlCl_3{}^c$	Na^d
Aldehydes	+	+	$-{}^e$	+	+	−	−
Ketones	+	−	$-{}^e$	−	+	−	−

^{a} Dissolves. ^{b} Decolorizes. ^{c} Color changes. ^{d} Hydrogen bubbles. ^{e} Slow, liberates HBr.

1. (a) $CH_3CH_2CH_2CH_2CHO$ $CH_3CH_2CH_2COCH_3$ $CH_3CH_2COCH_2CH_3$
        n-Valeraldehyde   Methyl *n*-propyl ketone   Ethyl ketone
        Pentanal   2-Pentanone   3-Pentanone

        $CH_3CH_2CH(CH_3)CHO$ $(CH_3)_2CHCH_2CHO$ $(CH_3)_2CHCOCH_3$
        α-Methylbutyraldehyde   Isovaleraldehyde   Methyl isopropyl ketone
        2-Methylbutanal   3-Methylbutanal   3-Methyl-2-butanone

        $(CH_3)_3CCHO$
        Trimethylacetaldehyde
        2,2-Dimethylpropanal

   (b) $PhCH_2CHO$ $PhCOCH_3$ $o\text{-}CH_3C_6H_4CHO$
        Phenylacetaldehyde   Acetophenone   *o*-Tolualdehyde
        Phenylethanal

        $m\text{-}CH_3C_6H_4CHO$ $p\text{-}CH_3C_6H_4CHO$
        m-Tolualdehyde   *p*-Tolualdehyde

2. (a) CH_3COCH_3 (b) $PhCHO$

   (c) $CH_3COCH_2CH(CH_3)_2$ (d) $(CH_3)_3CCHO$

   (e) $PhCOCH_3$ (f) $PhCH=CHCHO$

   (g) $(CH_3)_2CHCH_2CH_2CHO$ (h) $PhCH_2CHO$

   (i) $PhCOPh$ (j) $C_2H_5CHCH_2CHCHO$
                                                CH_3 CH_3

   (k) $C_2H_5CH(CH_3)COCH_3$ (l) $CH_3CH=CHCHO$

   (m) $(CH_3)_2C=CHCOCH_3$ (n) $PhCH=CHCOPh$

   (o) $CH_3CH_2CHCH_2CHO$ (p) $PhCH_2COPh$
                   OH

   (q) $o\text{-}HOC_6H_4CHO$ (r) $(p\text{-}HOC_6H_4)_2C=O$

   (s) $m\text{-}CH_3C_6H_4CHO$

3. (a) $PhCH_2COO^- + Ag$ (b) $PhCH_2COOH$

(c) $PhCH_2COOH$ (d) $PhCOOH$

(e) $PhCH_2CH_2OH$ (f) $PhCH_2CH_2OH$

(g) $PhCH_2CH_2OH$ (h) $PhCH_2CHOHPh$

(i) $PhCH_2CHOHCH(\dot{C}H_3)_2$ (j) $PhCH_2CHOHSO_3Na$

(k) $PhCH_2CHOHCN$ (l) $PhCH_2CH{=}NOH$

(m) $PhCH_2CH{=}NNHPh$ (n) $PhCH_2CH{=}NNHAr$

(o) $PhCH_2CH{=}NNHCONH_2$ (p) $PhCH_2CH(OEt)_2$

4. (a) no reaction   (b) no reaction   (c) no reaction

(d) $HOOC(CH_2)_4COOH$ (e) cyclohexanol   (f) cyclohexanol

(g) cyclohexanol

(h)  [cyclohexane ring with Ph and OH substituents]

(i) 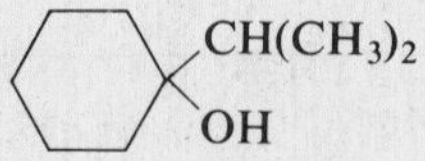 [cyclohexane ring with $CH(CH_3)_2$ and OH substituents]

(j) 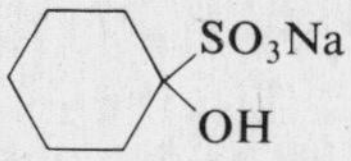 [cyclohexane ring with SO_3Na and OH substituents]

(k) 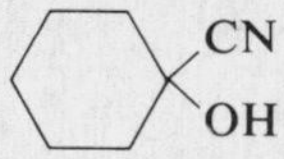 [cyclohexane ring with CN and OH substituents]

(l) 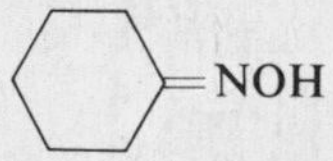 [cyclohexane ring with =NOH]

(m) 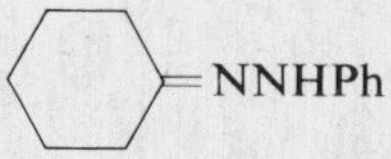 [cyclohexane ring with =NNHPh]

(n) 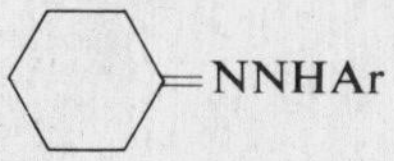 [cyclohexane ring with =NNHAr]

(o) 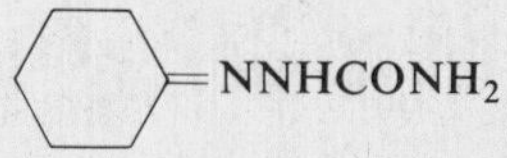 [cyclohexane ring with =NNHCONH$_2$]

(p) Might predict no reaction; actually get the ketal [cyclohexane ring with two OEt substituents]

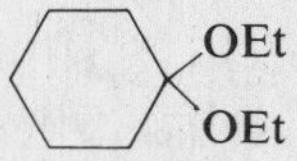

5. (a) Cannizzaro reaction:

$$2PhCHO \xrightarrow{OH^-} PhCH_2OH + PhCOO^-Na^+$$

(b) Crossed Cannizzaro reaction:

$$PhCHO + HCHO \xrightarrow{OH^-} PhCH_2OH + HCOO^-Na^+$$

(c) $PhCHOHCN$, mandelonitrile   (d) $PhCHOHCOOH$, mandelic acid

(e) $PhCHOHCH_3$, α-phenylethyl alcohol   (f) $PhCH{=}CH_2$, styrene

(g) $PhCHOH^{14}CH(CH_3)_2$ (h) $PhCH^{18}O$

6. (a) $CH_3CH_2CH_2OH \xleftarrow[\text{or } H_2,\text{ Ni}]{LiAlH_4} CH_3CH_2CHO$

(b) $CH_3CH_2COOH \xleftarrow{KMnO_4} CH_3CH_2CHO$

(c) $CH_3CH_2\underset{OH}{C}HCOOH \xleftarrow{H_2O,\ H^+} CH_3CH_2\underset{OH}{C}HC{\equiv}N \xleftarrow{CN^-,\ H^+} CH_3CH_2CHO$

(d) $CH_3CH_2CH\!\dashv\!CH_3$ ⟵ $CH_3CH_2CHO + CH_3MgBr$
$\quad\quad\quad\;\;|$
$\quad\quad\quad OH$

(e) $CH_3CH_2CH\!\dashv\!Ph$ ⟵ $CH_3CH_2CHO + PhMgBr$
$\quad\quad\quad\;\;|$
$\quad\quad\quad OH$

(f) $CH_3CH_2CCH_3 \xleftarrow{CrO_3} CH_3CH_2CHCH_3$ ⟵ (d)
$\quad\quad\quad\;||\quad\quad\quad\quad\quad\quad\quad\;|$
$\quad\quad\quad\;O\quad\quad\quad\quad\quad\quad\quad OH$

(g) $CH_3CH_2C\overset{O}{\overset{||}{}}\!\!\!\diagdown$
$\quad\quad\quad\quad\quad OCH_2CH_2CH_3$ $\xleftarrow{H^+}$ CH_3CH_2COOH (from b) $+ CH_3CH_2CH_2OH$ (from a)

(h) $CH_3CH_2CH\!\dashv\!CH(CH_3)_2$ ⟵ $CH_3CH_2CHO + BrMgCH(CH_3)_2$
$\quad\quad\quad\;\;|$
$\quad\quad\quad OH$

7. (a) $PhCH_2CH_3 \xleftarrow{Zn(Hg),\ H^+} PhCOCH_3$

(b) $PhCOOH \xleftarrow{H^+} \xleftarrow{NaOI} PhCOCH_3$

(c) $PhCHCH_3 \xleftarrow[\text{or }H_2,\ Ni]{LiAlH_4} PhCCH_3$
$\quad\quad|\quad\quad\quad\quad\quad\quad\quad\quad||$
$\quad\quad OH\quad\quad\quad\quad\quad\quad\;O$

(d) $Ph-\overset{CH_3}{\underset{OH}{\overset{|}{\underset{|}{C}}}}\!\!\dashv\!CH_2CH_3$ ⟵ $PhCCH_3 + C_2H_5MgBr$
$\quad\quad\quad\quad\quad\quad\quad\quad\quad\quad\quad||$
$\quad\quad\quad\quad\quad\quad\quad\quad\quad\quad\quad O$

(e) $Ph-\overset{CH_3}{\underset{OH}{\overset{|}{\underset{|}{C}}}}\!\!\dashv\!Ph$ ⟵ $PhCCH_3 + PhMgBr$
$\quad\quad\quad\quad\quad\quad\quad\quad\quad||$
$\quad\quad\quad\quad\quad\quad\quad\quad\quad O$

(f) $Ph-\overset{CH_3}{\underset{OH}{\overset{|}{\underset{|}{C}}}}-COOH \xleftarrow{H_2O,\ H^+} Ph-\overset{CH_3}{\underset{OH}{\overset{|}{\underset{|}{C}}}}-C\equiv N \xleftarrow{CN^-,\ H^+} PhCCH_3$
$\quad\quad\quad\quad\quad\quad\quad\quad\quad\quad\quad\quad\quad\quad\quad\quad\quad\quad\quad\quad\quad\quad\quad\quad\quad||$
$\quad\quad\quad\quad\quad\quad\quad\quad\quad\quad\quad\quad\quad\quad\quad\quad\quad\quad\quad\quad\quad\quad\quad\quad\quad O$

8. (a) $(CH_3)_2CHCHO \xleftarrow{K_2Cr_2O_7} iso\text{-}BuOH$

(b) $PhCH_2CHO \xleftarrow{K_2Cr_2O_7} Ph\!\dashv\!CH_2CH_2OH \xleftarrow{\quad}$ ⎰ PhMgBr (from C_6H_6)
$\quad\quad\quad\quad\quad\quad\quad\quad\quad\quad\quad\quad\quad\quad\quad\quad\quad\quad\quad\quad$ ⎱ H_2C-CH_2 (Sec. 17.10)
$\quad\quad\quad\quad\quad\quad\quad\quad\quad\quad\quad\quad\quad\quad\quad\quad\quad\quad\quad\quad\quad\quad\quad\quad\;\;\diagdown\!O\!\diagup$

(c) $Br\langle\bigcirc\rangle CHO \xleftarrow[H^+]{H_2O} \xleftarrow[Ac_2O]{CrO_3} Br\langle\bigcirc\rangle CH_3 \xleftarrow[Fe]{Br_2} \langle\bigcirc\rangle CH_3$

(d) $CH_3CH_2COCH_3 \xleftarrow{CrO_3} sec\text{-}BuOH$

(e) $\overset{CHO}{\underset{NO_2}{\langle\bigcirc\rangle}}NO_2 \xleftarrow{H_2O} \overset{CHCl_2}{\underset{NO_2}{\langle\bigcirc\rangle}}NO_2 \xleftarrow[heat]{2Cl_2} \overset{CH_3}{\underset{NO_2}{\langle\bigcirc\rangle}}NO_2 \xleftarrow[\text{in stages}]{\text{nitration}} \overset{CH_3}{\langle\bigcirc\rangle}$

(f) $O_2N\langle\bigcirc\rangle\underset{O}{\overset{}{C}}Ph \xleftarrow[AlCl_3]{C_6H_6} O_2N\langle\bigcirc\rangle\underset{O}{\overset{}{C}}Cl \xleftarrow{SOCl_2} O_2N\langle\bigcirc\rangle\underset{O}{\overset{}{C}}OH$ (page 345)

(g) $Et\overset{}{\underset{O}{-}}C\text{--}CH(CH_3)_2 \longleftarrow$
       Et$_2$Cd (from EtMgBr)

       $ClC\text{--}CH(CH_3)_2 \xleftarrow{SOCl_2} HOC\text{--}CH(CH_3)_2 \xleftarrow{KMnO_4} iso\text{-}BuOH$
       $\underset{O}{}$

(h) $PhCH_2\underset{O}{\overset{}{C}}CH_3 \xleftarrow{CrO_3} PhCH_2\underset{OH}{\overset{}{C}}H\text{--}CH_3 \longleftarrow PhCH_2CHO$ (from b) $+ CH_3MgBr$

(i) $\underset{O_2N}{\langle\bigcirc\rangle}\underset{O}{\overset{}{C}}Ph \xleftarrow[AlCl_3]{C_6H_6} \underset{O_2N}{\langle\bigcirc\rangle}\underset{O}{\overset{}{C}}Cl \xleftarrow{SOCl_2} \underset{O_2N}{\langle\bigcirc\rangle}\underset{O}{\overset{}{C}}OH$ (page 345)

(j) $n\text{-}PrC\langle\bigcirc\rangle CH_3 \xleftarrow[AlCl_3]{PhCH_3} n\text{-}Pr\underset{O}{\overset{}{C}}Cl \xleftarrow{SOCl_2} n\text{-}C_3H_7\underset{O}{\overset{}{C}}OH \xleftarrow{KMnO_4} n\text{-}BuOH$

(k) $CH_3CH_2\underset{CH_3}{\overset{}{C}}HCHO \xleftarrow{K_2Cr_2O_7} CH_3CH_2\underset{CH_3}{\overset{}{C}}H\text{--}CH_2OH \longleftarrow$
       $HCHO \xleftarrow{Cu,\ heat} MeOH$

       $sec\text{-}BuMgBr$

(l) $i\text{-}Bu\text{--}\underset{O}{\overset{}{C}}\text{--}Bu\text{-}n \longleftarrow$
       $(n\text{-}Bu)_2Cd$ (from $n\text{-}BuMgBr$)

       $i\text{-}Bu\text{--}\underset{O}{\overset{}{C}}\text{--}Cl \xleftarrow{SOCl_2} i\text{-}Bu\text{--}\underset{O}{\overset{}{C}}\text{--}OH \xleftarrow[]{H^+} \xleftarrow{CO_2} i\text{-}BuMgBr$

(m) $O_2N\langle\bigcirc\rangle\underset{O}{\overset{}{C}}\text{--}CH_3 \longleftarrow$
       Me_2Cd (from $MeMgBr$)

       $O_2N\langle\bigcirc\rangle\underset{O}{\overset{}{C}}Cl \xleftarrow{SOCl_2} O_2N\langle\bigcirc\rangle\underset{O}{\overset{}{C}}OH$ (page 345)

(n) CH_3⟨O⟩$\overset{|}{\underset{O}{C}}$⟨O⟩$NO_2$ $\xleftarrow[\text{AlCl}_3]{\text{PhCH}_3}$ $Cl\overset{}{\underset{O}{C}}$⟨O⟩$NO_2$ $\xleftarrow{\text{SOCl}_2}$ $HO\overset{}{\underset{O}{C}}$⟨O⟩$NO_2$ (page 345)

(o) O_2N⟨O⟩$\overset{|}{\underset{O}{C}}$Et ⟵ ┌ Et_2Cd (from EtMgBr)

└ O_2N⟨O⟩$\overset{}{\underset{O}{C}}Cl$ $\xleftarrow{\text{SOCl}_2}$ O_2N⟨O⟩$\overset{}{\underset{O}{C}}OH$ (page 345)

9. (a) $PhCH_2CH_2CH_2CH_3$ $\xleftarrow[\text{HCl}]{\text{Zn(Hg)}}$ $Ph-\overset{}{\underset{O}{C}}CH_2CH_2CH_3$ $\xleftarrow[\text{AlCl}_3]{\text{C}_6\text{H}_6}$ $Cl\overset{}{\underset{O}{C}}CH_2CH_2CH_3$

(b) $n\text{-PrCHCOOH}$ $\xleftarrow[\text{H}^+]{\text{H}_2\text{O}}$ $n\text{-PrCHC}\equiv N$ $\xleftarrow[\text{H}^+]{\text{CN}^-}$ $n\text{-PrCHO}$ $\xleftarrow{\text{K}_2\text{Cr}_2\text{O}_7}$ $n\text{-PrCH}_2\text{OH}$
      $\underset{OH}{|}$ $\underset{OH}{|}$

(c) $CH_3CH_2CH_2CH_2\overset{|}{+}CH_2CHCH_3$ $\xleftarrow[\text{HCl}]{\text{Zn(Hg)}}$ $CH_3CH_2CH_2CH_2\overset{CH_3}{\underset{O}{+}\underset{|}{C}CHCH_3}$ ⟵ ┌ $(CH_3CH_2CH_2CH_2)_2Cd$
              $\underset{CH_3}{|}$ (from n-BuOH)

└ $\underset{CH_3CHCOCl}{\overset{CH_3}{|}}$
 (from i-BuOH as in 8g)

(d) $\underset{CH_3}{\overset{CH_3}{|}}CH_3CHCH_2\overset{|}{-}\overset{CH_3}{\underset{HO}{|}}C-\overset{}{\underset{CH_3}{|}}CHCH_3$ ⟵ ┌ CH_3MgBr
                                                                                              (from MeOH)

└ $CH_3CHCH_2\overset{CH_3}{+}\overset{}{\underset{O}{C}}-\overset{}{\underset{CH_3}{|}}CHCH_3$ ⟵ ┌ $(i\text{-Bu})_2Cd$
    $\underset{CH_3}{|}$ (from i-BuOH)

└ $\underset{CH_3}{\overset{}{CH_3CHCOCl}}$
    (as in 8g)

(e) O_2N⟨O⟩$\overset{}{\underset{OH}{CH}}COOH$ $\xleftarrow[\text{H}^+]{\text{H}_2\text{O}}$ O_2N⟨O⟩$\overset{}{\underset{OH}{CH}}C\equiv N$ $\xleftarrow[\text{H}^+]{\text{CN}^-}$ O_2N⟨O⟩CHO (page 622)

(f) $PhCH_2\overset{Ph}{+}\overset{}{\underset{OH}{C}}-CH_3$ ⟵ $PhCH_2MgCl + PhCOCH_3$ (page 623)

(g) Br⟨O⟩$\overset{Ph}{\underset{OH}{C}}Et$ ⟵ ┌ $EtMgBr$

└ Br⟨O⟩$\overset{}{\underset{O}{C}}Ph$ $\xleftarrow{\text{AlCl}_3}$ ┌ $PhBr$ $\xleftarrow{\text{Br}_2, \text{Fe}}$ C_6H_6

└ $PhCCl$ $\xleftarrow{\text{PCl}_5}$ $PhCOH$
    $\underset{O}{|}$ $\underset{O}{|}$

(h) 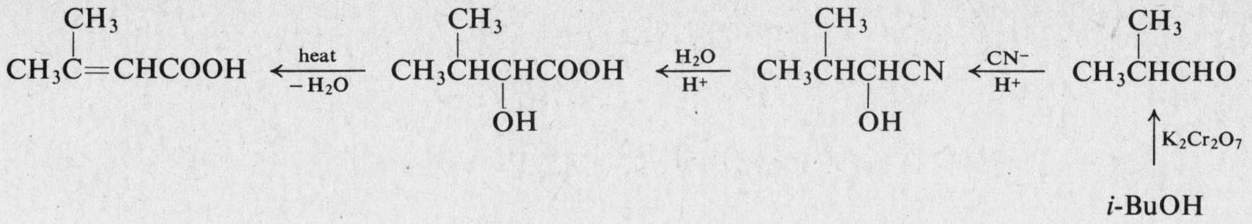 $CH_3\overset{\displaystyle CH_3}{\underset{}{C}}=CHCOOH \xleftarrow[-H_2O]{heat} CH_3\overset{\displaystyle CH_3}{\underset{\displaystyle OH}{CHCH}}CHCOOH \xleftarrow[H^+]{H_2O} CH_3\overset{\displaystyle CH_3}{\underset{\displaystyle OH}{CHCH}}CHCN \xleftarrow[H^+]{CN^-} CH_3\overset{\displaystyle CH_3}{\underset{}{CHCHO}}$

$\uparrow K_2Cr_2O_7$

i-BuOH

10. 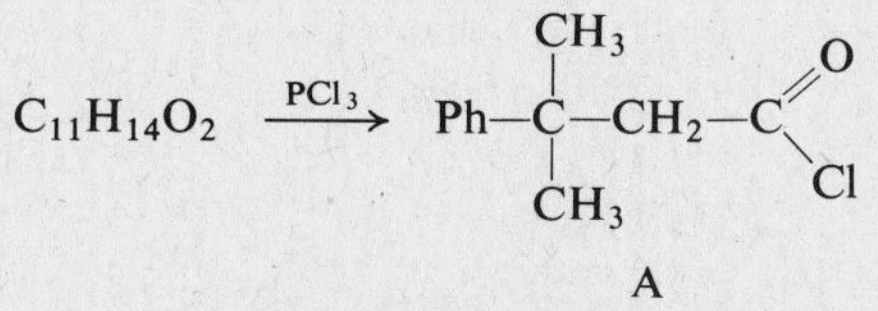 $C_{11}H_{14}O_2 \xrightarrow{PCl_3}$ $Ph-\overset{\displaystyle CH_3}{\underset{\displaystyle CH_3}{C}}-CH_2-\overset{\displaystyle O}{\underset{\displaystyle Cl}{C}}$

A

Loss of HCl but no change in carbon number in going from A to B indicates that the Friedel-Crafts acylation is intramolecular:

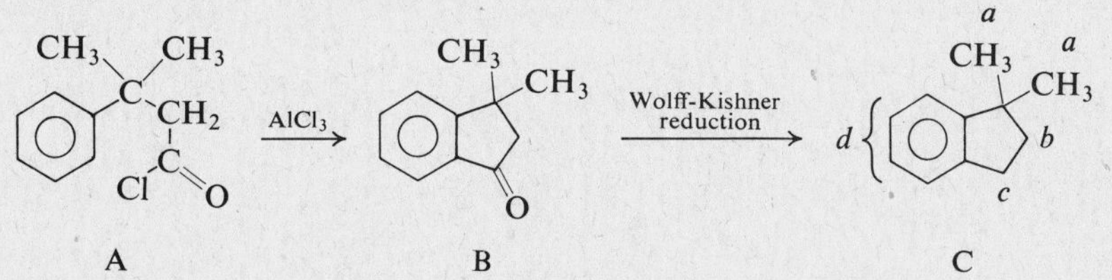

       A   B   C

(b)   C was also produced by Friedel-Crafts alkylation.

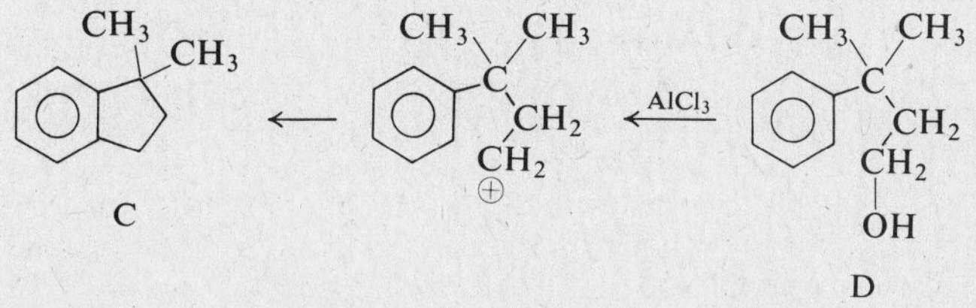

     C   D

11.   Once formed, the aldehyde RCHO reacts with unconsumed alcohol to form the hemiacetal $RCH(OH)OCH_2R$. This, like the hydrate $RCH(OH)_2$, forms a chromate ester (VII) analogous to the one on page 634. Elimination from this gives the ester:

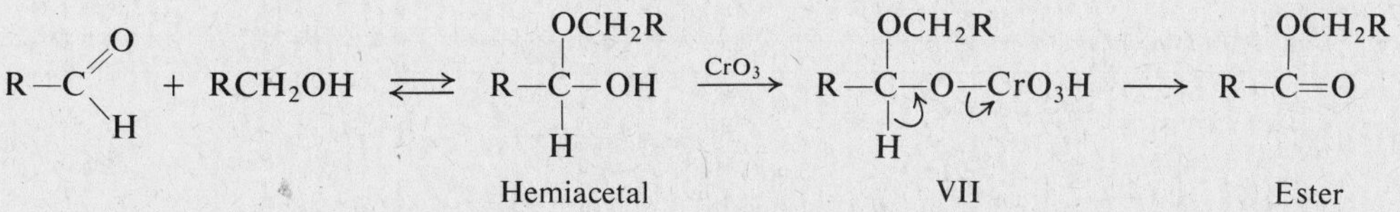

                              Hemiacetal   VII   Ester

Again, what we are dealing with is oxidation of a special kind of secondary alcohol (page 529).

12. Compare Sec. 34.14.

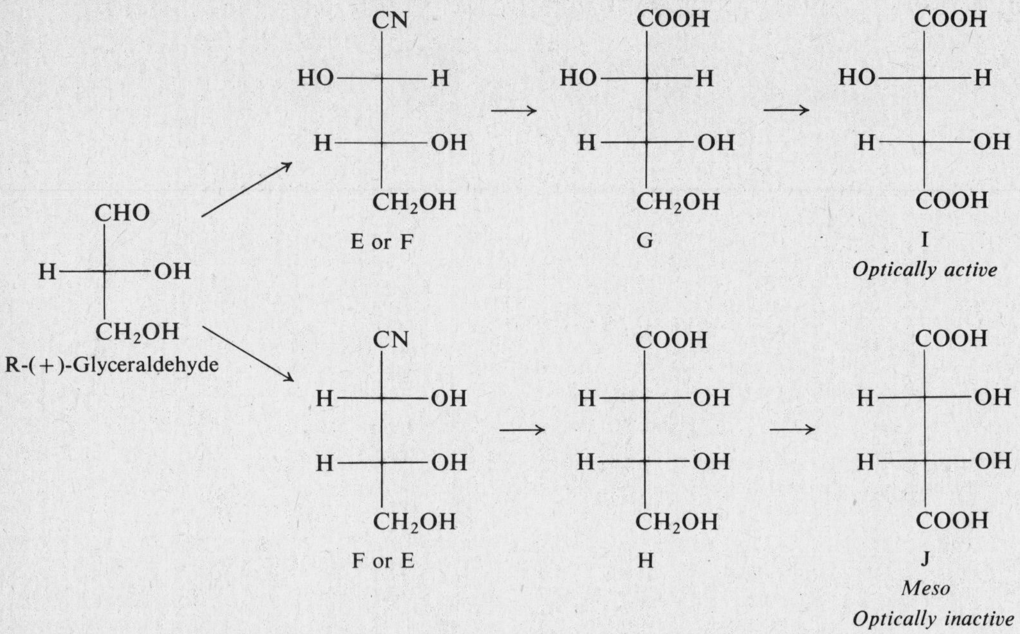

E or F   G   I
                       Optically active

F or E   H   J
                       Meso
                       Optically inactive

R-(+)-Glyceraldehyde

13. (a)

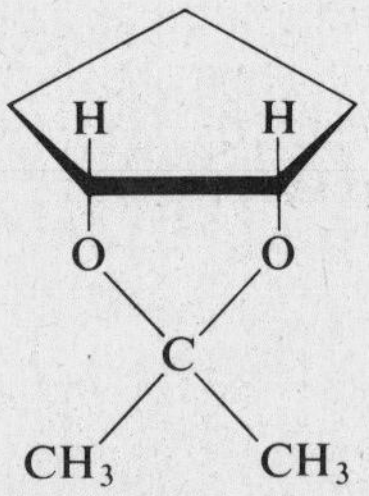

K
A cyclic ketal

(b) The —OH groups in the *trans*-isomer are too far apart for the cyclic structure to form.

14.

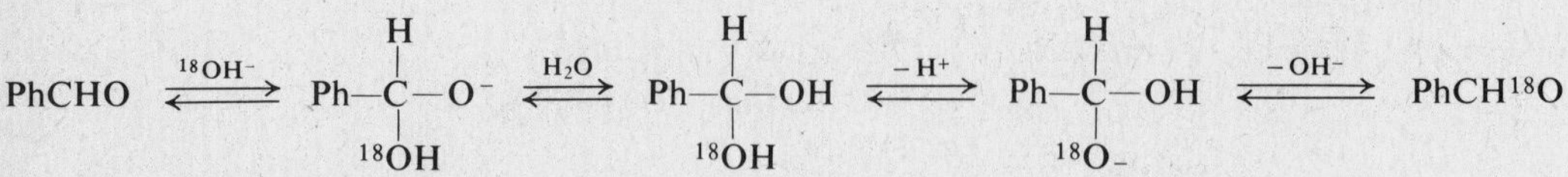

$$PhCHO \xrightleftharpoons{^{18}OH^-} Ph-\underset{^{18}OH}{\overset{H}{C}}-O^- \xrightleftharpoons{H_2O} Ph-\underset{^{18}OH}{\overset{H}{C}}-OH \xrightleftharpoons{-H^+} Ph-\underset{^{18}O_-}{\overset{H}{C}}-OH \xrightleftharpoons{-OH^-} PhCH^{18}O$$

15. A proton adds very rapidly to the double bond since this addition gives the comparatively stable oxonium ion VIII. Completion of addition then yields the hemiacetal, which is, of course, rapidly hydrolyzed.

$$R-\overset{\overset{H}{|}}{C}=\overset{\overset{H}{|}}{C}-OR' \underset{}{\overset{H^+}{\rightleftharpoons}} R-\overset{\overset{H}{|}}{\underset{\underset{H}{|}}{C}}-\overset{H}{C}=\overset{\oplus}{O}R' \overset{H_2{}^{18}O}{\rightleftharpoons} RCH_2-\overset{\overset{H}{|}}{\underset{\underset{\oplus}{{}^{18}OH_2}}{C}}-OR' \overset{-H^+}{\rightleftharpoons} RCH_2-\overset{\overset{H}{|}}{\underset{\underset{}{{}^{18}OH}}{C}}-OR'$$

VIII

Hemiacetal

$$RCH_2-\overset{\overset{H}{|}}{\underset{\underset{}{{}^{18}OH}}{C}}-OR' \overset{H^+}{\rightleftharpoons} RCH_2-\overset{\overset{H}{|}\;\overset{H}{|}}{\underset{\underset{\oplus}{{}^{18}OH}}{C}}-OR' \overset{-R'OH}{\rightleftharpoons} RCH_2-\overset{\overset{H}{|}}{C}={}^{18}\overset{\oplus}{OH} \overset{-H^+}{\rightleftharpoons} RCH_2CH^{18}O$$

16. (a) A double inversion would give net retention, as is observed.

$$H_2O + ROBs \xrightarrow{\text{inversion}} H_2\overset{+}{O}R \longrightarrow OBs^-$$

$$\longrightarrow H^+ + ROH$$

Inversion

$+ ROBs \xrightarrow{\text{inversion}}$ $\overset{+}{O}R + OBs^-$

$$\xrightarrow[\text{inversion}]{H_2O}$$

$+ R\overset{+}{O}H_2 \longrightarrow H^+ + ROH$

Net retention

(b) The mixed ketal IX is formed (1) and, in the presence of the H$^+$ formed along with it, undergoes methanolysis (2) to yield 2-octanol (and ketal X). When the base pyridine is

$$(1)\quad CH_3-\overset{\overset{CH_3}{|}}{C}=O + ROBs \longrightarrow CH_3-\overset{\overset{CH_3}{|}}{\underset{\underset{+}{}}{C}}=OR + OBs^-$$

$$\downarrow CH_3OH$$

$$CH_3-\overset{\overset{CH_3}{|}}{\underset{\underset{\underset{H}{|}}{CH_3\overset{+}{O}}}{C}}-OR \longrightarrow CH_3-\overset{\overset{CH_3}{|}}{\underset{\underset{}{CH_3O}}{C}}-OR + H^+$$

IX

Mixed ketal

$$(2)\quad IX + H^+ \longrightarrow CH_3-\overset{\overset{CH_3}{|}}{\underset{\underset{}{CH_3O\;\;H}}{C}}-\overset{+}{O}R \longrightarrow CH_3-\overset{\overset{CH_3}{|}}{\underset{\underset{}{CH_3\overset{+}{O}}}{C}} + ROH$$

2-Octanol

$$\downarrow CH_3OH$$

$$CH_3-\overset{\overset{CH_3}{|}}{\underset{\underset{\underset{H}{|}}{CH_3\overset{+}{O}}}{C}}-OCH_3 \longrightarrow CH_3-\overset{\overset{CH_3}{|}}{\underset{\underset{}{CH_3O}}{C}}-OCH_3 + H^+$$

X

Ketal

(3) IX + H₂O + H⁺ ⟶ (CH₃)₂C=O + ROH + CH₃OH

               └⟶ 2,4-dinitrophenylhydrazone

present to consume H⁺, the mixed ketal IX persists and is the substance isolated in impure form; it contains no carbonyl group but, in the acidic 2,4-dinitrophenylhydrazine solution, is hydrolyzed (3) to acetone, which then gives a 2,4-dinitrophenylhydrazone.

    (H. Weiner and R. A. Sneen, "Substitution at a Saturated Carbon. IV. Clarification of the Mechanism of Solvolyses of 2-Octyl Sulfonates. Stereochemical Considerations," J. Am. Chem. Soc., **87**, 287 (1965).)

17. This is an example of electrophilic aromatic substitution, with the leaving group XII a protonated aldehyde. The intermediate *sigma* complex, XI, is an oxonium ion with most of

(1) CH₃O⟨◯⟩CHOH⟨◯⟩G + Br₂ ⇌ CH₃O⁺=⟨◯⟩(Br)CHOH⟨◯⟩G + Br⁻

                                XI

(2) CH₃O⁺=⟨◯⟩(Br)CHOH⟨◯⟩G ⟶ CH₃O⟨◯⟩Br + HO⁺=C(H)⟨◯⟩G

                                        XII

                        ↓ −H⁺

                      OHC⟨◯⟩G

the charge located on the *p*-methoxy group. Formation (1) of XI is reversible; addition of Br⁻ speeds up reversal of (1) and raises the fraction of XI that reverts to starting material. (That is, $k_{-1}[Br^-]$ is increased relative to k_2.) Electron-release by G stabilizes the leaving group XII and the transition state leading to its formation; k_2 is thus increased relative to $k_{-1}[Br^-]$, and the rate of formation of products is speeded up.

18. The initial product of the Grignard reaction is, of course, the magnesium salt of the alcohol, benzhydrol. This is oxidized to the ketone benzophenone, presumably by a hydride transfer—to the excess benzaldehyde—which is aided by the negative charge on alkoxide oxygen.

PhMgBr + PhCHO ⟶ Ph−C(H)(Ph)−O⁻(MgBr)⁺

Ph−C(H)(Ph)−O⁻ + Ph−C(H)=O ⟶ Ph−C(Ph)=O + Ph−C(H)(H)−O⁻ —H₂O→ PhCH₂OH

                                                    Benzyl alcohol

Anion of benzhydrol   Benzaldehyde *Excess*   Benzophenone

Although an unwanted side-reaction here, hydride transfer from an alkoxide to a carbonyl group is the basis of the *Oppenauer oxidation* (C. Djerassi, O. R. VI-5),

$$R_2CHOH + CH_3COCH_3 \xrightleftharpoons{t\text{-BuOK}} R_2C{=}O + CH_3CHOHCH_3$$

and of its reverse, the *Meerwein-Pondorf-Verley* reduction (A. L. Wilds, O. R. II-5),

$$R_2C{=}O + CH_3CHOHCH_3 \xrightleftharpoons{Al(OPr\text{-}i)_3} R_2CHOH + CH_3COCH_3\uparrow$$

both very mild, selective synthetic reactions.

19. (a) Eugenol and isoeugenol are isomers ($C_{10}H_{12}O_2$), and eugenol is an allylbenzene (page 395): about the only thing that can happen is a shift in position of the double bond, into conjugation with the ring.

Eugenol $C_{10}H_{12}O_2$ — Isoeugenol $C_{10}H_{12}O_2$ — L $C_{12}H_{14}O_3$

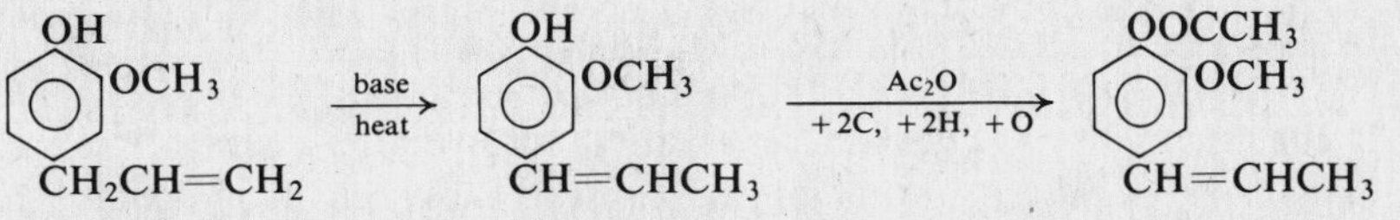

This isomerization is confirmed by the loss of two carbons in the subsequent oxidation.

L $C_{12}H_{14}O_3$ — M $C_{10}H_{10}O_4$ — Vanillin $C_8H_8O_3$

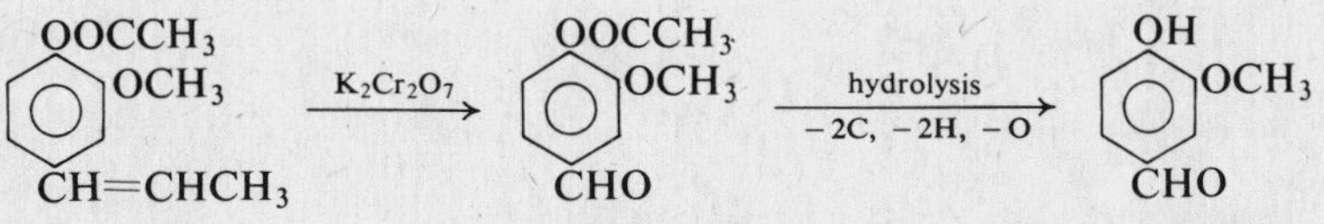

(b) Since the reagent is a base, its most likely function is to abstract a proton. Abstraction from the carbon next to the ring gives a resonance-stabilized allylic-benzylic anion,

$$HO^- + ArCH_2CH{=}CH_2 \rightleftharpoons \underbrace{ArCH{=\!=}CH{=\!=}CH_2}_{\ominus} + H_2O \rightleftharpoons \underset{\text{More stable}}{ArCH{=}CHCH_3} + OH^-$$

which can take the proton back at either of two positions. Equilibrium favors the more stable alkene in which the double bond is conjugated with the ring.

(c)

Safrole — Isosafrole — Piperonal

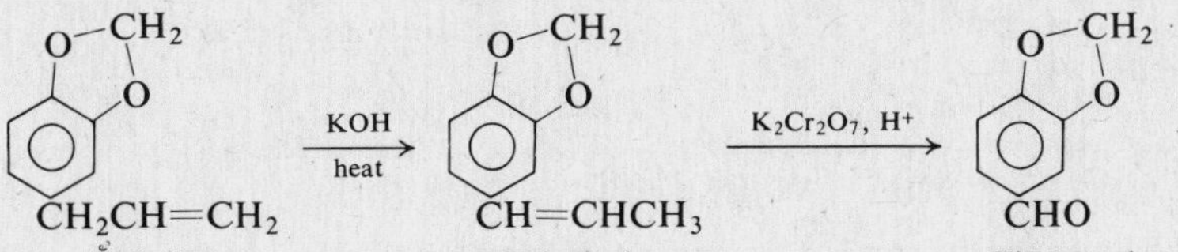

20. The protonated aldehyde is an electrophile, the double bond is a nucleophile; reaction between the two functions closes the ring; hydration of the resulting carbonium ion produces a diol.

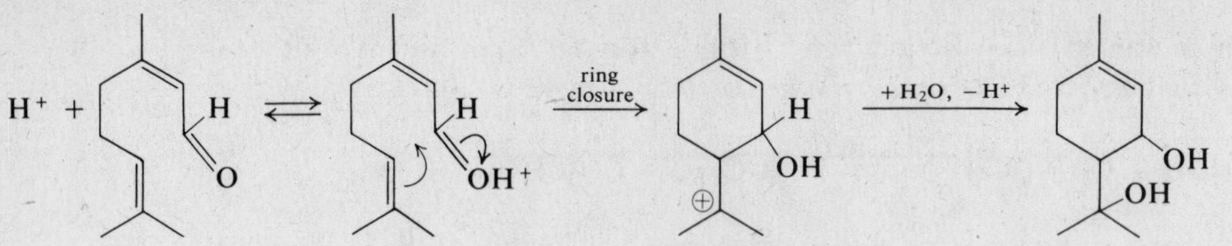

Protonated aldehyde   3° cation   Diol

A six-membered ring

21. N and O are stereoisomers. For each the preferred conformation is the one with a majority of bulky —CCl_3 groups in equatorial positions. In N, all H's are equivalent and axial (δ 4.28). In O, there are two axial H's (δ 4.63) and one equatorial H (further downfield at δ 5.50).

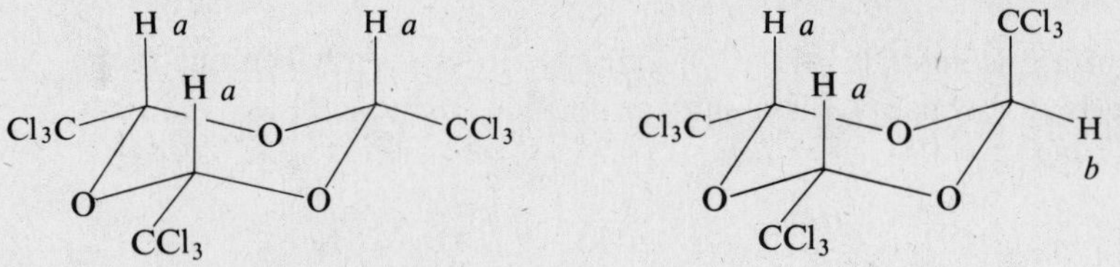

L   M

This interpretation is consistent with the general observation (Problem 8, Chapter 13, page 447) that, other things being equal, equatorial protons absorb downfield from axial protons.

22.

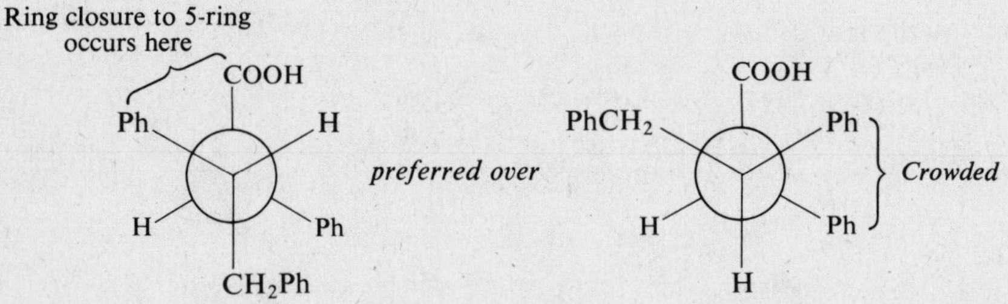

Ring closure to 5-ring occurs here

IV

Ring closure to 6-ring occurs here

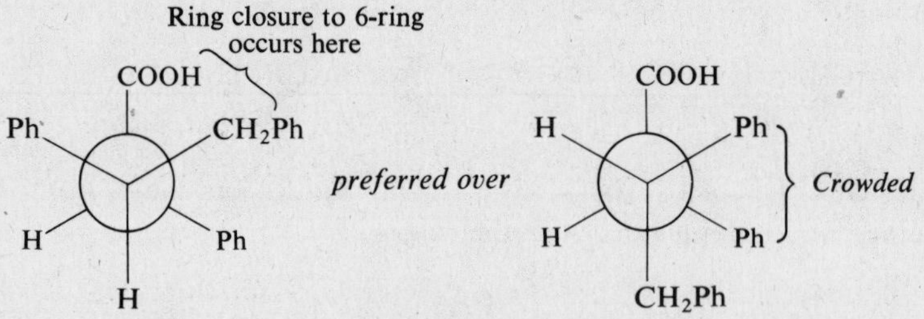

V

23. (a)

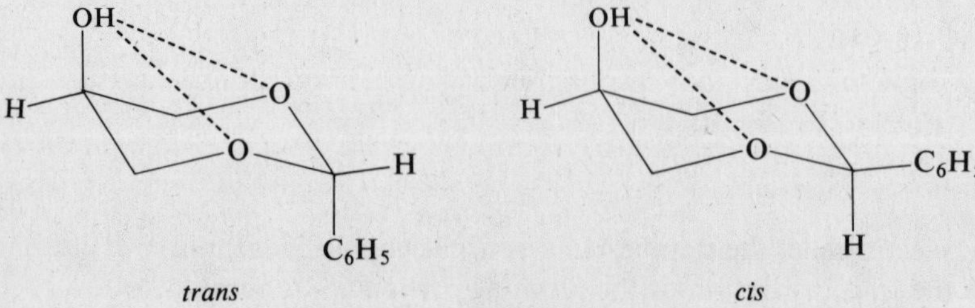

trans   *cis*

VI

(b) The *trans*-isomer has axial phenyl; hydrogen bonding involving —OH and the ring oxygens helps to stabilize this conformation.

24. (a) Tollens' reagent   (b) Tollens' reagent
    (c) 2,4-dinitrophenylhydrazine   (d) iodoform test
    (e) Schiff or Tollens' test   (f) Tollens' test
    (g) acid, then Schiff or Tollens' test   (h) iodoform test
    (i) 2,4-dinitrophenylhydrazine, or CrO_3/H_2SO_4 (j) acid, then Schiff or Tollens' test
    (k) acid and heat, then Schiff or Tollens' test

25. (a) $PhCOCH_3$ gives a positive iodoform test.
    $PhCH_2CHO$ gives positive Tollens' test.
    Oxidize the isomeric tolualdehydes to the toluic or phthalic acids and identify by m.p.

    (b) The acid gives CO_2 with $NaHCO_3$(aq).
    The methyl ketone gives a positive iodoform test.
    The hydrocarbon is insoluble in cold conc. H_2SO_4.
    The ester dissolves slowly when heated with NaOH(aq).
    The alcohol can be oxidized to PhCOOH (m.p. 122°).

    (c) The unsaturated ketone decolorizes Br_2/CCl_4.
    The alkane is insoluble in cold conc. H_2SO_4.
    The esters dissolve slowly when heated with NaOH(aq), but of these only ethyl benzoate yields a ppt. (of PhCOOH) when the resulting alkaline solution is acidified.
    The alcohol immediately gives a green color with CrO_3/H_2SO_4.

    (d) p-$ClC_6H_4CH_2Cl$ gives AgCl with $AgNO_3$.
    p-$ClC_6H_4COCH_3$ gives a positive iodoform test.
    The ester slowly dissolves in hot NaOH(aq).
    The nitro compound gives a positive test for N after sodium fusion.

26. The Tollens' test shows citral to be an aldehyde; cleavage into 3 fragments by oxidation indicates 2 double bonds. Putting the fragments together according to the isoprene rule,

we arrive at:

Citral *a*, like geraniol, has H and CH_3 *trans*; citral *b*, like nerol, has H and CH_3 *cis*.

27. The compound is evidently unsaturated ($KMnO_4$ positive) and a ketone (NH_2OH positive, Tollens' negative). The first hydrogenation gives a saturated ketone ($KMnO_4$ negative, NH_2OH positive), the second leads to an alcohol ($KMnO_4$ negative, NH_2OH negative, CrO_3/H_2SO_4 positive). The formula of the alcohol shows the presence of one ring.

$$\begin{array}{ll} C_{10}H_{22}O & \text{satd. open-chain alcohol} \\ - \; C_{10}H_{20}O & \text{carvomenthol} \\ \hline & \text{2H missing means 1 ring} \end{array}$$

Fitting together the oxidation fragments, with an eye on the isoprene rule,

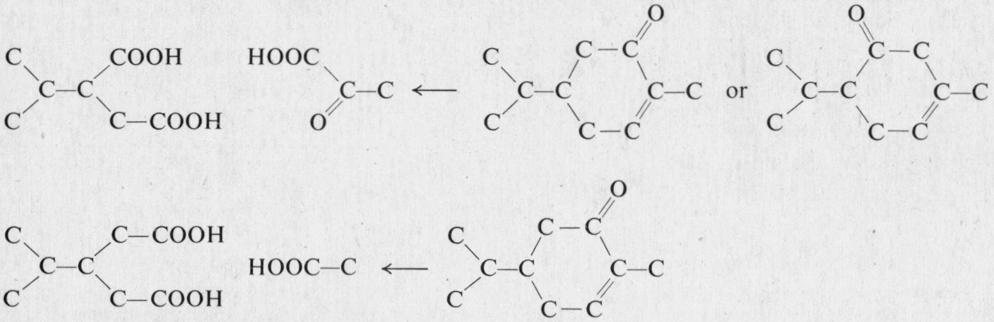

we conclude that carvotanacetone is

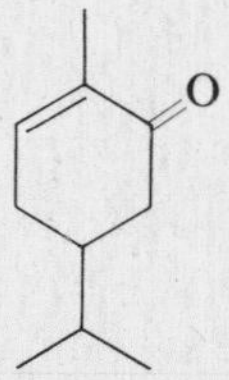

Carvotanacetone

28. (a) $CH_3CH_2CCH_3$ (with O double bond below C)　(b) $(CH_3)_2CH-CHO$　(c) $CH_2=CH-CH-CH_3$ (with OH below)

　　　2-Butanone　　　　　　Isobutyraldehyde　　　　　3-Buten-2-ol

(See the labeled spectra on page 658 of this Study Guide.)

29. (a) $CH_3CH_2CH_2CCH_3$ (with O double bond below)　(b) $CH_3-CH-C-CH_3$ (with CH_3 above, O below)　(c) $CH_3CH_2CCH_3$ (with O below)

　　　2-Pentanone　　　　Methyl isopropyl ketone　　　Methyl ethyl ketone

(See the labeled spectra on page 659 of this Study Guide.)

30.　$p\text{-}CH_3OC_6H_4CHO$　　　$p\text{-}CH_3OC_6H_4CCH_3$ (with O below)　　$C_6H_5-C-CH(CH_3)_2$ (with O below)

　p-Methoxybenzaldehyde　　p-Methoxyacetophenone　　Isobutyrophenone
　　　　P　　　　　　　　　　　Q　　　　　　　(Isopropyl phenyl ketone)

　　　　　　　　　　　　　　　　　　　　　　　　　　　　　　R

(See the labeled spectra on pages 660 and 661 of this Study Guide.)

Chapter 20

Functional Derivatives of Carboxylic Acids
Nucleophilic Acyl Substitution

20.1 (a)

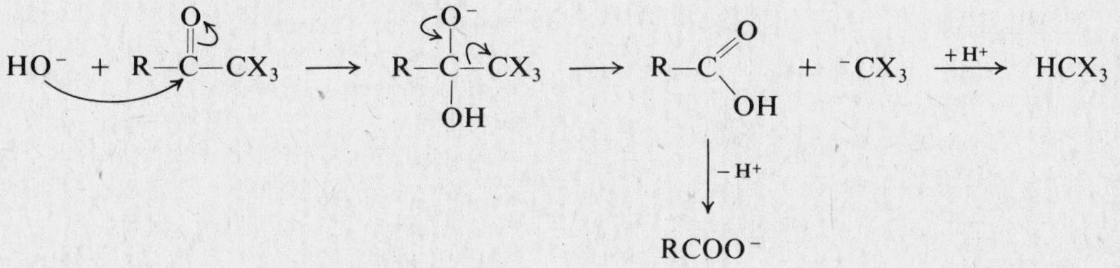

$$HO^- + R\overset{O}{\underset{}{C}}-CX_3 \longrightarrow R\overset{O^-}{\underset{OH}{C}}CX_3 \longrightarrow R-C\overset{O}{\underset{OH}{}} + {}^-CX_3 \xrightarrow{+H^+} HCX_3$$

$$\downarrow {}_{-H^+}$$

$$RCOO^-$$

This is an exception to our generalization of Sec. 20.4; here a ketone undergoes nucleophilic *substitution*—but only because the three halogen atoms make $CX_3{}^-$ a weaker base and hence better leaving group than most R^-.

(b)

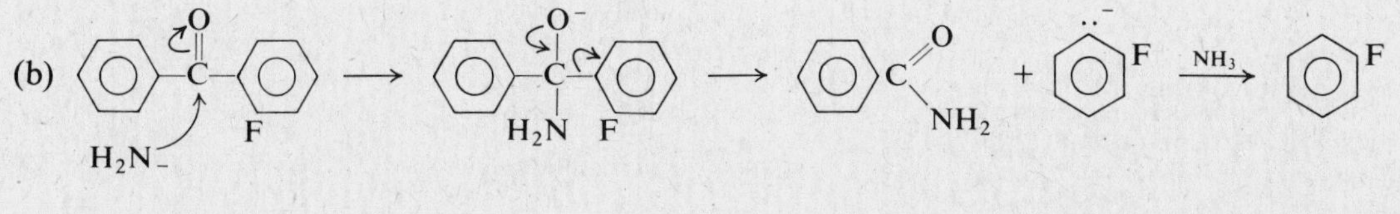

Here is another exception, but again one we can understand: the powerfully electron-withdrawing *o*-F stabilizes the leaving anion, making it a weaker base and better leaving group than most Ar^-.

20.2 The —COOH groups are (very nearly) the same distance apart in the *cis*- and *trans*-1,2-cyclohexanedicarboxylic acids, and the anhydride bridge can form easily from either. See the answer to Problem 20.8 (b).

In *trans*-1,2-cyclopentanedicarboxylic acid, formation of the anhydride bridge would place prohibitive strain on the molecule.

20.3 Maleic acid is *cis*- and fumaric acid is *trans*-butenedioic acid. Only maleic acid can form the anhydride without rotation about the carbon–carbon bond.

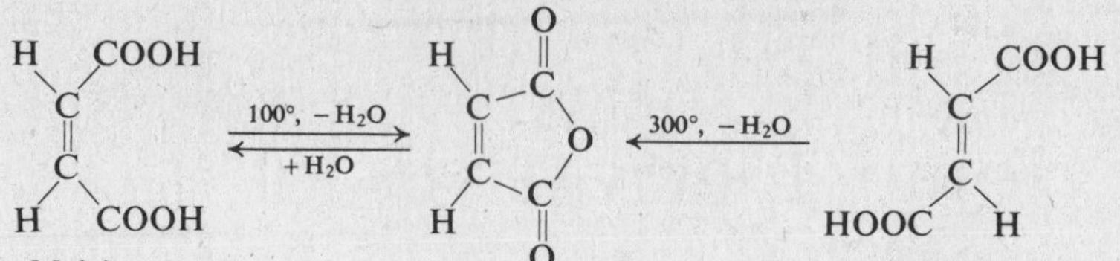

Maleic acid *cis* — Maleic anhydride — Fumaric acid *trans*

20.4 See Figure 30.2, page 987. A, β-benzoylpropionic acid. B, γ-phenylbutyric acid. C, acid chloride. D, α-tetralone. E, alcohol. F, alkene. G, naphthalene.

20.5 (a)

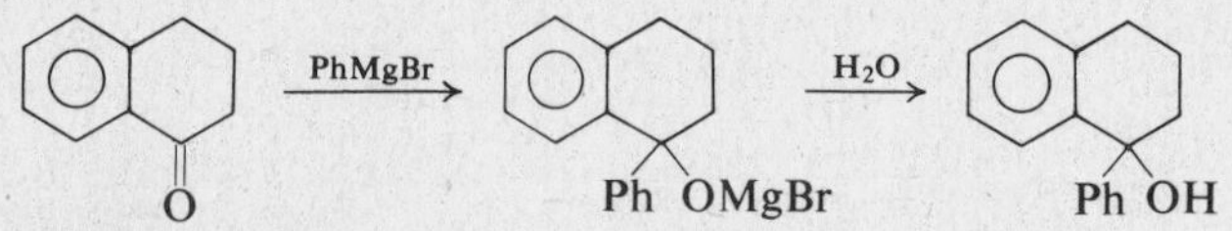

D (Problem 20.4)

(b)

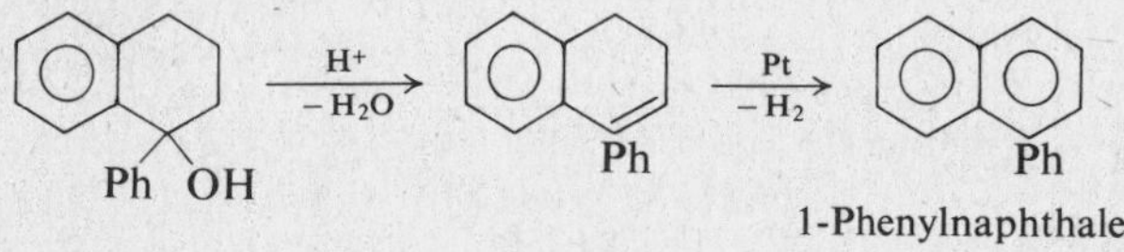

1-Phenylnaphthalene

20.6 Friedel-Crafts acylation.

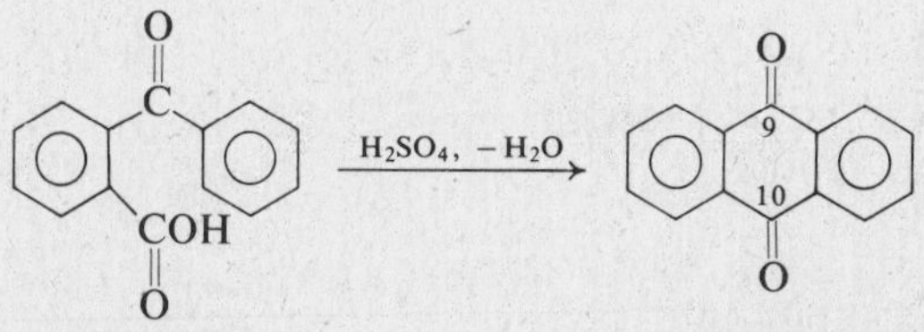

$C_{14}H_{10}O_3$ *o*-Benzoylbenzoic acid — $C_{14}H_8O_2$ 9,10-Anthraquinone (See Sec. 30.18)

20.7 See page 993.

20.8 (a) The *cis*-acid is the only one capable of forming a cyclic anhydride.

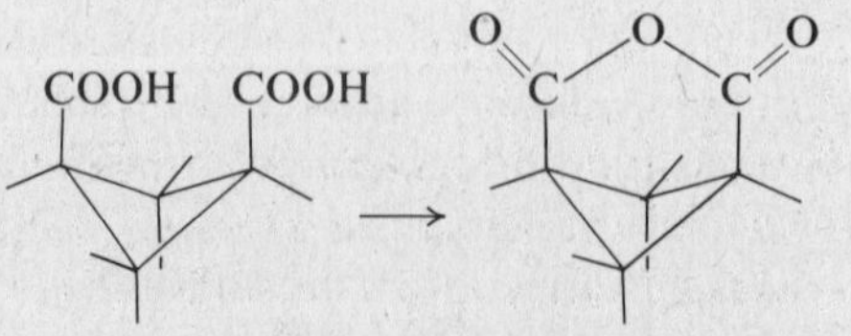

cis-Acid — Cyclic anhydride

(b) In both acids, the two —COOH groups are the same distance apart.

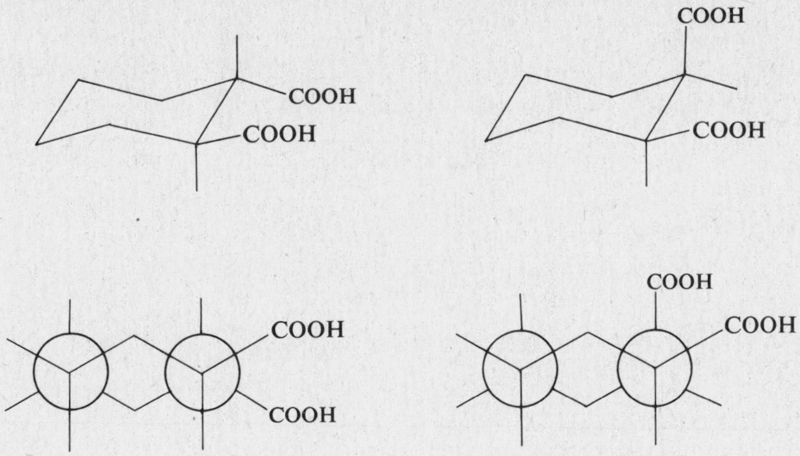

trans-1,2-Diacid   cis-1,2-Diacid

(c) Yes; only the *cis*-acid can form the cyclic (diaxial) anhydride.

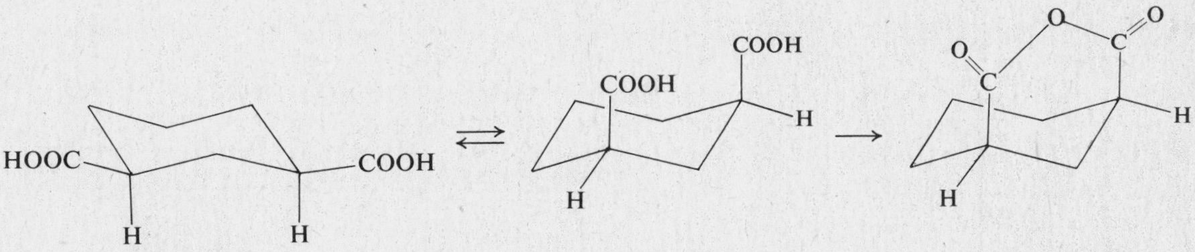

cis-1,3-Diacid   cis-1,3-Diacid   Cyclic anhydride
Diequatorial conformation   Diaxial conformation

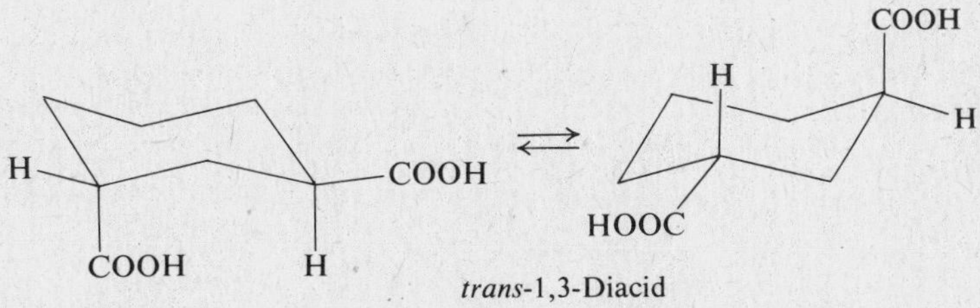

No cyclic anhydride possible

trans-1,3-Diacid

20.9 Reaction of phthalic anhydride with a racemic alcohol yields an alkyl hydrogen phthalate, which has an acidic "handle." This handle allows reaction with an optically active base to form diastereomeric salts, which can be separated by fractional crystallization. Once separated, each ester can be hydrolyzed to an optically active alcohol. Since hydrolysis of a carboxylic ester does not usually involve cleavage of the alkyl–oxygen bond (Sec. 20.17), there is no loss of optical activity in the hydrolysis step.

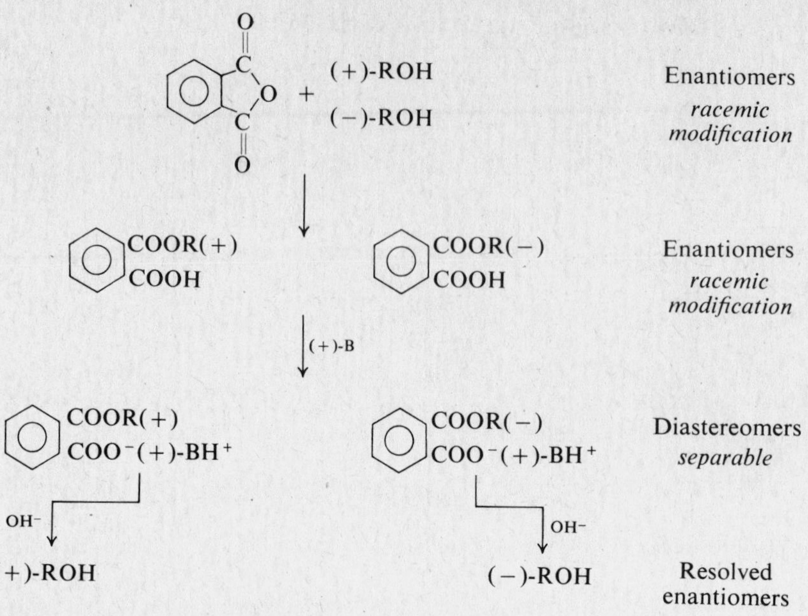

20.10

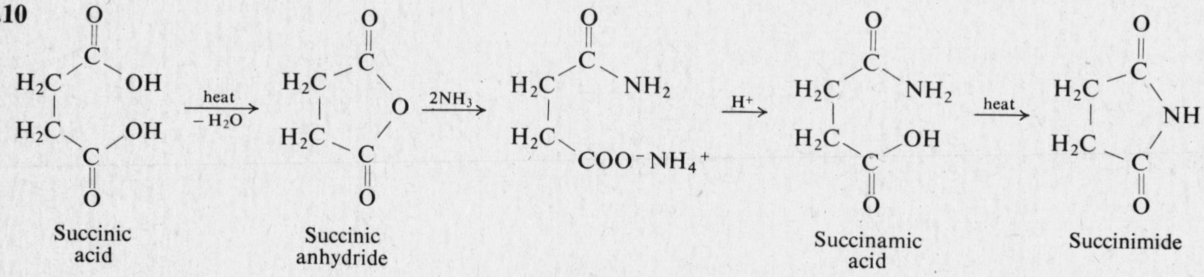

20.11 In the anion from ammonia, the negative charge is localized on nitrogen. In the anion from benzamide, it is accommodated by nitrogen and one oxygen, and in the anion from phthalimide by nitrogen and *two* oxygens.

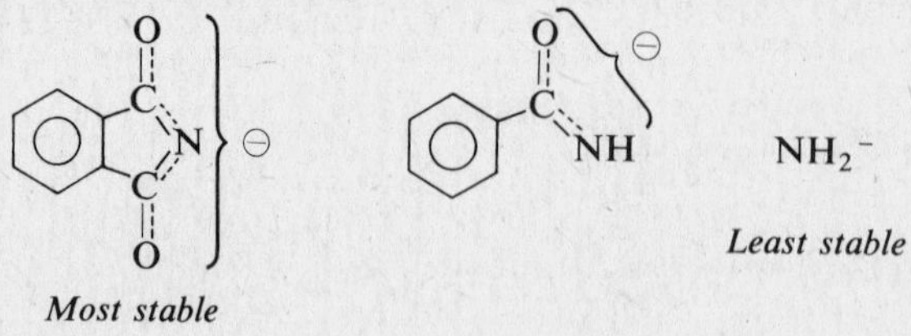

Most stable

Least stable

20.12 The mixture is shaken with benzene, which extracts the ester and benzoic acid. Treatment of the benzene extract with Na_2CO_3(aq) removes benzoic acid as the water-soluble sodium salt. Distillation of the wet benzene layer removes water and benzene first, and the dry ester is finally collected at 200°.

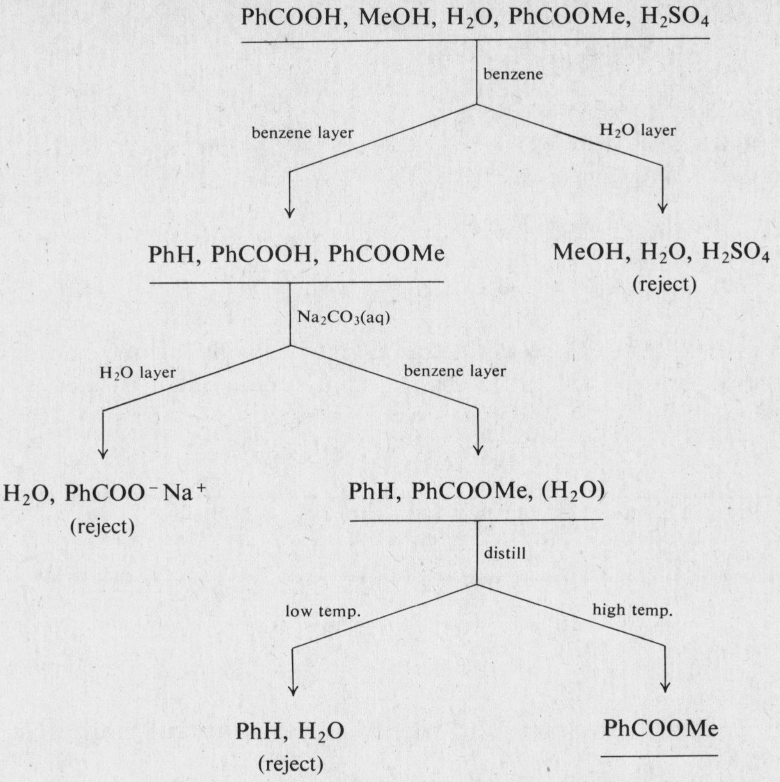

PhCOOH, MeOH, H_2O, PhCOOMe, H_2SO_4
_____

| benzene

benzene layer H_2O layer

PhH, PhCOOH, PhCOOMe   MeOH, H_2O, H_2SO_4
                                              (reject)

| Na_2CO_3(aq)

H_2O layer   benzene layer

H_2O, $PhCOO^- Na^+$ PhH, PhCOOMe, (H_2O)
(reject)

| distill

low temp.   high temp.

PhH, H_2O PhCOOMe
(reject)

20.13 (a) A six-carbon product requires combination of two lactic acid residues:

$$
\begin{array}{ll}
C_6H_{12}O_6 & \text{2 lactic acid} \\
- \underline{C_6H_8O_4} & \text{product} \\
4H + 2O & \text{means } 2H_2O \text{ removed}
\end{array}
$$

Water is lost through a double esterification involving opposite ends of two reactant molecules.

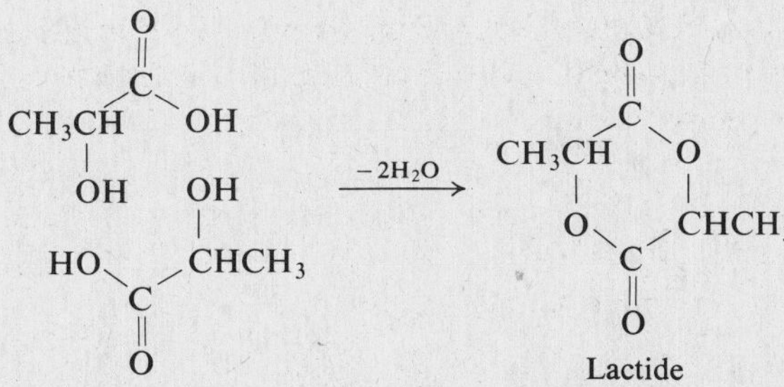

Lactic acid
Two moles

Lactide
A cyclic double ester

(b) Linkage of polyfunctional molecules through esterification of —OH on one molecule by —COOH from another to give a linear polymer. (Such a process is called *step-reaction polymerization*, page 1029.)

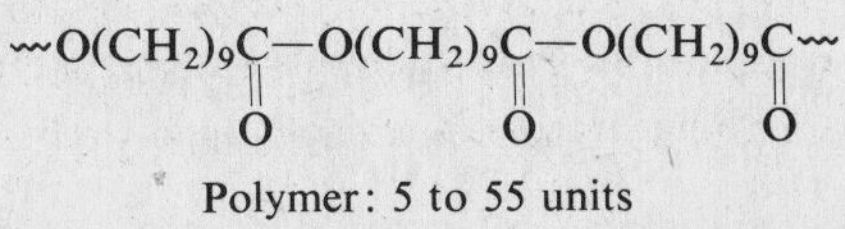

$$\sim\!\!O(CH_2)_9C\!-\!O(CH_2)_9C\!-\!O(CH_2)_9C\!\!\sim$$
$$\quad\quad\;\; \parallel \quad\quad\quad\; \parallel \quad\quad\quad\; \parallel$$
$$\quad\quad\;\; O \quad\quad\quad\; O \quad\quad\quad\; O$$

Polymer: 5 to 55 units

267

20.14 (a) Electron-withdrawal by G speeds up reaction by speeding up the first step: it helps to disperse the negative charge developing on oxygen.

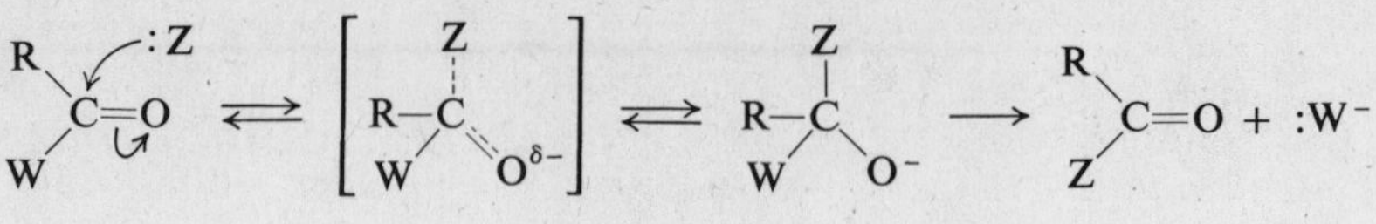

$$R = p\text{-}GC_6H_4$$

(b) Br, activating; NH_2, deactivating; t-Bu, deactivating.

(c) $NO_2 > H > CH_3 > NH_2$. Here, electron-withdrawal speeds up not only the first step as in (a), but also the second step: it lowers the basicity of W^-, and makes it a better leaving group.

20.15 (a) Steric effect, and electron release by alkyl groups (compare page 629)

(b)   formate > acetate > propionate > isobutyrate > trimethylacetate

$HCOOMe > CH_3COOMe > CH_3CH_2COOMe > (CH_3)_2CHCOOMe > (CH_3)_3CCOOMe$

20.16 The basicity of the leaving group is one factor: the weaker bases leave the intermediate faster.

Basicity: $Cl^- < RCOO^- < OR^- < NH_2^-$

20.17

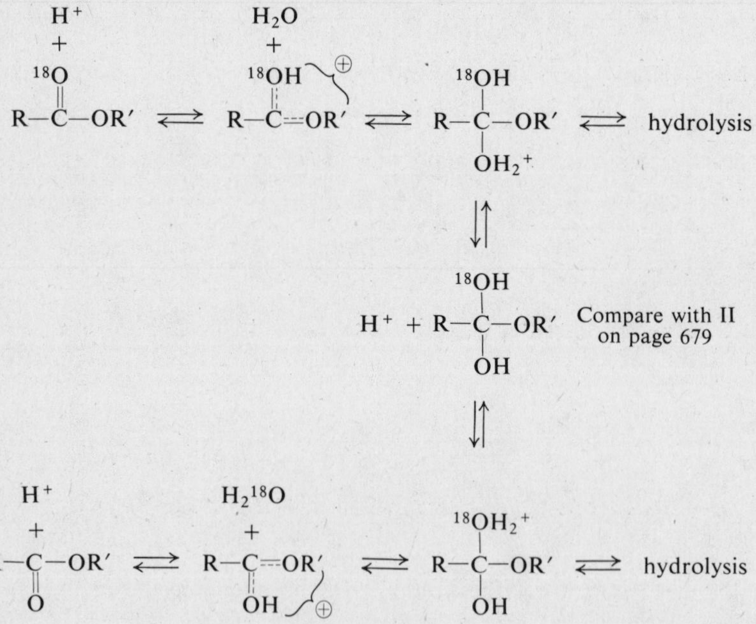

Compare with II on page 679

20.18 Crowding in the tetrahedral intermediates by bulky groups makes them less stable, and the corresponding transition states more difficult to achieve, thus decreasing reaction rates.

20.19 (a) Hydrolysis of these esters of 3° alcohols evidently proceeds via the 3° cations, in an S_N1 reaction of the protonated esters.

$$CH_3-\overset{\overset{\displaystyle O}{\|}}{C}-O-R + H^+ \rightleftharpoons \quad \begin{matrix} CH_3-\overset{\overset{\displaystyle OH}{|}}{C}\!=\!\overset{\oplus}{O}-R \\ \text{and/or} \\ CH_3-\overset{\overset{\displaystyle O}{\|}}{C}-\overset{\oplus}{\underset{\underset{\displaystyle H}{|}}{O}}-R \end{matrix} \quad \longrightarrow \quad CH_3COOH + R^{\oplus}$$

$$\xrightarrow{H_2O} ROH_2{}^+ \xrightarrow{-H^+} ROH$$

(b) Here, as in S_N2 vs. S_N1, we have competition between a bimolecular reaction and a unimolecular reaction. But in ester hydrolysis we have many possible reactions, and competition is between the *easiest* of these. The easiest unimolecular reaction is S_N1. The easiest bimolecular reaction is attack by water, not at alkyl carbon, but at acyl carbon (see Sec. 20.4). The change in mechanism results in, among other things, a change in point of cleavage.

tert-Alkyl esters undergo S_N1 fastest for electronic reasons, and undergo acyl substitutions slowest for steric reasons. If there is to be a shift in mechanism of ester hydrolysis, this is where we would certainly expect to observe it.

20.20

$$
\begin{array}{c}
\underset{\overset{|}{\underset{\displaystyle \overset{|}{OH}}{n\text{-Bu}-C-Et}}}{\overset{\displaystyle Et}{}} \\
\text{3-Ethyl-3-heptanol}
\end{array}
\quad
\begin{matrix}
\xleftarrow{a} & \xleftarrow{\text{EtMgBr}} & n\text{-Bu}-\overset{\overset{\displaystyle O}{\|}}{C}-Et & \xleftarrow{\text{Et}_2\text{Cd}} & n\text{-BuCOCl} \\
 & & & & \uparrow \text{SOCl}_2 \\
\xleftarrow{b} & \xleftarrow{\text{2EtMgBr}} & n\text{-BuCOOMe} & \xleftarrow{\text{MeOH, H}^+} & n\text{-BuCOOH}
\end{matrix}
$$

20.21 (a) Esters of formic acid.

$$
\underset{\text{A 2° alcohol}}{\overset{\overset{\displaystyle R}{|}}{\underset{\underset{\displaystyle OH}{|}}{H-C-R}}} \quad \longleftarrow \quad \underset{\text{A formate}}{H-\overset{\overset{\displaystyle O}{\|}}{C}-OR'} \quad + \ 2RMgX
$$

(b) $\underset{\underset{\displaystyle OH}{|}}{\overset{\overset{\displaystyle H}{|}}{n\text{-Pr}-C-\text{Pr-}n}}$ $\longleftarrow$ $2n$-PrMgBr (from n-PrOH) + HCOOEt (from MeOH and EtOH)

4-Heptanol

20.22 1-Octadecanol, $CH_3(CH_2)_{16}CH_2OH$, and 1-butanol. The conditions of hydrogenolysis lead also to hydrogenation of the double bond.

20.23 (a)

$$\underset{\substack{\text{Amide-ester}}}{H_2N-\underset{O}{\overset{}{C}}-O-\overset{CH_3}{\overset{|}{C}HCH_2CH_2CH_2}} \xleftarrow{NH_3} \underset{\substack{\text{Ester-acid chloride}}}{Cl-\underset{O}{\overset{}{C}}-OPe-2} \longleftarrow \underset{\substack{\text{Phosgene}}}{Cl-\underset{O}{\overset{}{C}}-Cl} + 2\text{-PeOH}$$

(b)

$$\underset{\substack{\text{Carbobenzoxychloride} \\ \text{Ester-acid chloride}}}{PhCH_2O-\underset{O}{\overset{}{C}}-Cl} \longleftarrow \underset{\substack{\text{Phosgene}}}{Cl-\underset{O}{\overset{}{C}}-Cl} + \underset{\substack{\text{Benzyl alcohol} \\ \text{One mole}}}{PhCH_2OH}$$

20.24 Great resonance stabilization of cation, $[C(NH_2)_3]^+$, with contribution from three equivalent structures, and accommodation of positive charge by three nitrogens.

$$\left[\underset{\overset{\oplus}{N}H_2}{\underset{C}{\overset{H_2N\qquad NH_2}{\diagdown\diagup}}} \quad \underset{NH_2}{\underset{C}{\overset{H_2\overset{\oplus}{N}\quad NH_2}{\diagdown\diagup}}} \quad \underset{NH_2}{\underset{C}{\overset{H_2N\quad \overset{\oplus}{N}H_2}{\diagdown\diagup}}} \right] \; \textit{equivalent to} \; \left\{ \underset{NH_2}{\underset{C}{\overset{H_2N\qquad NH_2}{\diagdown\diagup}}} \right\}^+$$

20.25

$$\left[^{--}:\ddot{N}-C\equiv N: \quad :N\equiv C-\ddot{N}:^{--} \quad {}^{-}:\ddot{N}=C=\ddot{N}:^{-} \right]$$

Linear: *sp* hybridization of carbon. Bonds of equal length (longer than triple, shorter than single); charge equally distributed between nitrogens.

20.26 Hydrolysis of nitrile and amide (or of di-imide). Products are urea (intermediate), calcium carbonate, ammonia.

20.27 (a) Carbon–nitrogen triple bond (nitrile group). For example:

$$\underset{:NH_3}{H_2N-C\equiv N} \longrightarrow \underset{+NH_3}{H_2N-C=N^-} \longrightarrow \underset{\substack{NH_2 \\ \text{Guanidine}}}{H_2N-C=NH}$$

(b) Nucleophilic addition (followed by proton shifts).

(c) H_2O, $MeOH$, H_2S, NH_3 are nucleophilic reagents that add to the carbon–nitrogen triple bond of cyanamide. In acid, protonation of $\equiv N$ occurs first; in aqueous base, OH^- is nucleophile.

20.28 (a) RCOCl

(b) $RCOO^-NH_4^+$, $RCONH_2$, RCN, amides of low mol. wt. amines

(c) $RCOO^-NH_4^+$ (d) $(RCO)_2O$ (e) RCOOR′

20.29 (a)

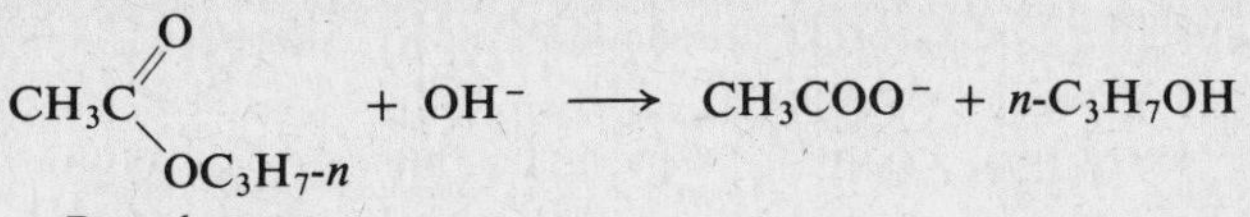

$$CH_3C \begin{smallmatrix} O \\ \\ OC_3H_7\text{-}n \end{smallmatrix} + OH^- \longrightarrow CH_3COO^- + n\text{-}C_3H_7OH$$

n-Propyl acetate

One mole of the ester requires one mole of OH^- for hydrolysis (saponification): the saponification equivalent (S.E.) = m.w. of the ester = 102 ($C_5H_{10}O_2$).

(b) All eight have the same formula, $C_5H_{10}O_2$, with four carbons divided between acid and alcohol. In RCOOR′, therefore, R can be H, Me, Et, *n*-Pr, *iso*-Pr; R′ can be Me, Et, *n*-Pr, *iso*-Pr, *n*-Bu, *iso*-Bu, *sec*-Bu, *tert*-Bu (but not H). The nine allowable combinations are:

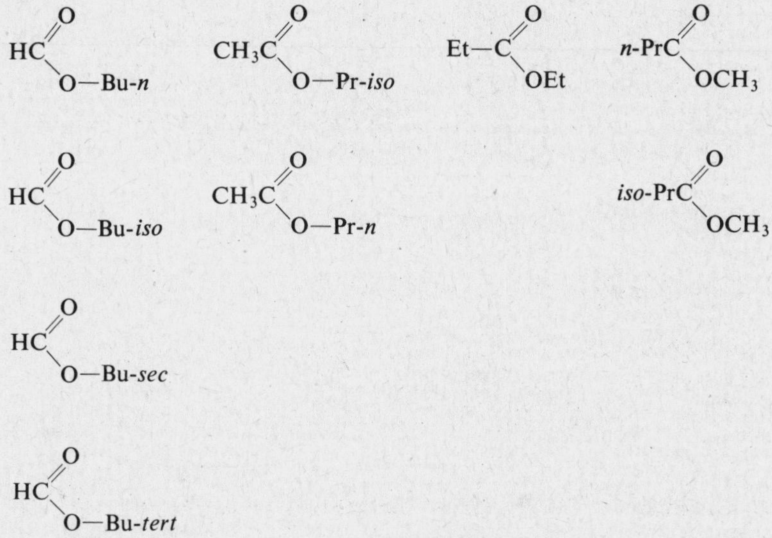

(c) In RCOOH, $C_5H_{10}O_2$, R must be a four-carbon group: *n*-Bu, *iso*-Bu, *sec*-Bu, *tert*-Bu are possible, giving rise to only four acids of equiv. wt. 102.

(d) Less useful, because of the great number of possible combinations of R and R′ in RCOOR′.

20.30 (a)

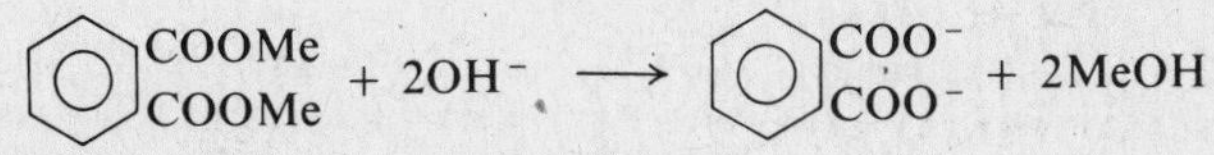

Methyl phthalate
M.w. 194

Uses 2 moles OH^-; S.E. = m.w./2 = 194/2 = 97.

(b) S.E. = m.w./number ester groups per molecule.

(c) S.E. = m.w./3 = $C_{57}H_{110}O_6$/3 = 890/3 = 297.

1. (a) In the order in which their structures appear in the answer to Problem 20.29(b) above:

n-butyl formate	isopropyl acetate	ethyl propionate	methyl *n*-butyrate
isobutyl formate	*n*-propyl acetate		methyl isobutyrate
sec-butyl formate			
tert-butyl formate			

(b) These obviously contain an aromatic ring. For example, $C_8H_8O_2 - COOCH_3 = C_6H_5$.

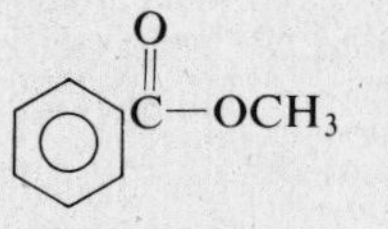

Methyl benzoate 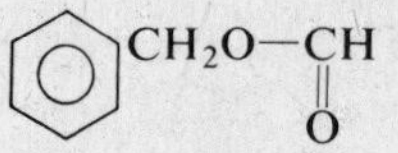   Benzyl formate   Phenyl acetate

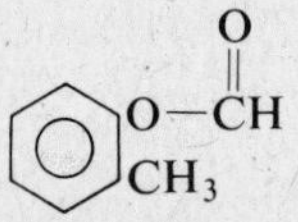

o-Cresyl formate 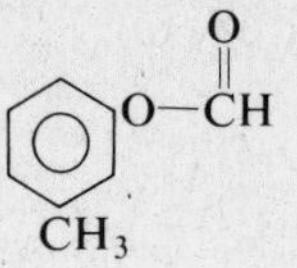   *m*-Cresyl formate 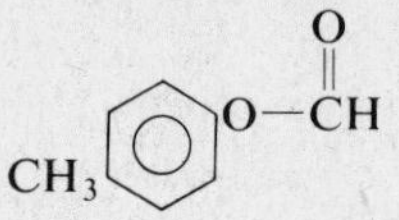   *p*-Cresyl formate

(c) These are diesters (accounting for the four oxygens):

$$C_7H_{12}O_4$$
$$\underline{-\ 2\ COOCH_3}$$

C_3H_6 means satd. open-chain: $-C-C-C-$, $-C-C-$, $-\overset{\displaystyle C}{\underset{\displaystyle C}{C}}-$

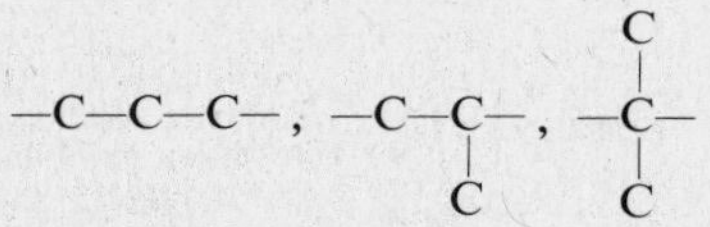

The three possibilities are:

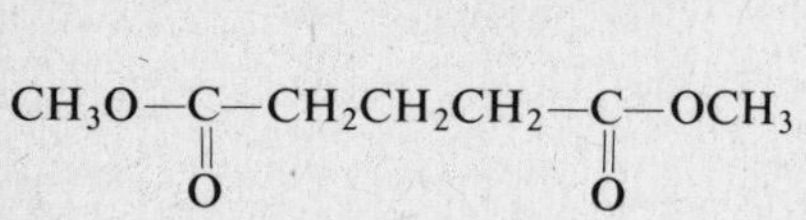

Methyl glutarate 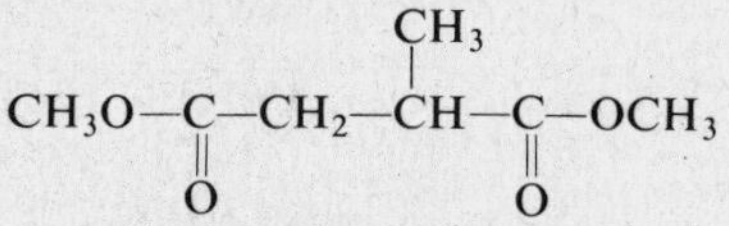   Methyl α-methylsuccinate 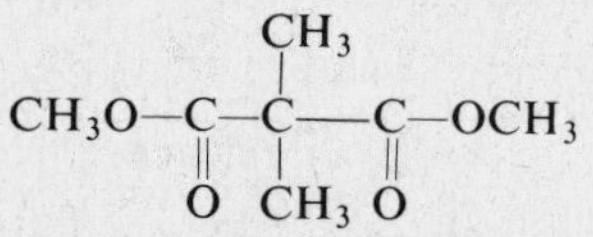   Methyl dimethylmalonate

2. (a) $n\text{-}C_3H_7COOH$
     n-Butyric acid

(b) $n\text{-}C_3H_7COOCH(CH_3)_2$
     Isopropyl *n*-butyrate

(c) 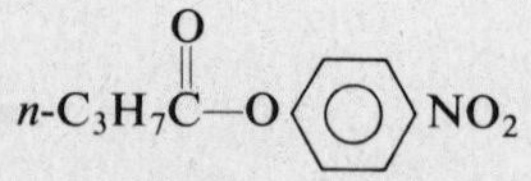 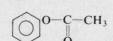
     p-Nitrophenyl *n*-butyrate

(d) $n\text{-}C_3H_7CONH_2$
     n-Butyramide

(e) 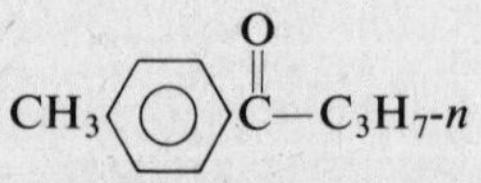 $CH_3\langle\bigcirc\rangle\overset{O}{\overset{\|}{C}}-C_3H_7\text{-}n$
     n-Propyl *p*-tolyl ketone
     (*p*-Methyl-*n*-butyrophenone)

(f) no reaction

(g) $n\text{-}C_3H_7COONa$
     Sodium *n*-butyrate

(h) $n\text{-}C_3H_7COOEt$ and AgCl
     Ethyl *n*-butyrate

(i) $n\text{-}C_3H_7-\overset{\overset{\displaystyle O}{\|}}{C}-NHCH_3$
N-Methyl-n-butyramide

$+$

$CH_3NH_3{}^+Cl$
Methylammonium
chloride

(j) $n\text{-}C_3H_7-\overset{\overset{\displaystyle O}{\|}}{C}-N(CH_3)_2$
N,N-Dimethyl-n-butyramide

$+$

$(CH_3)_2NH_2{}^+Cl^-$
Dimethylammonium
chloride

(k) no reaction

(l) $n\text{-}C_3H_7-\overset{\overset{\displaystyle O}{\|}}{C}-NHPh$
n-Butyranilide

$+$

$PhNH_3{}^+Cl^-$
Anilium chloride

(m) $Ph-\underset{\underset{\displaystyle O}{\|}}{C}-C_3H_7\text{-}n$
n-Butyrophenone
(n-Propyl phenyl ketone)

(n) $n\text{-}C_3H_7-\overset{\overset{\displaystyle Ph}{|}}{\underset{\underset{\displaystyle OH}{|}}{C}}-Ph$
Diphenyl-n-propylcarbinol
(1,1-Diphenyl-1-butanol)

3. In (b), (c), (e), and (l), an equimolar amount of acetic acid is formed along with the indicated product.

(a) $2CH_3COOH$
Acetic acid
(*Two moles*)

(b) $CH_3COOCH(CH_3)_2$
Isopropyl acetate

(c) 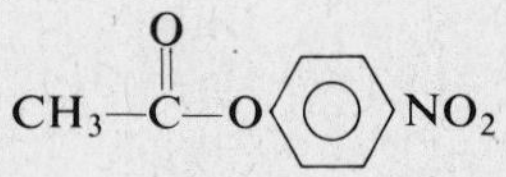 $CH_3-\overset{\overset{\displaystyle O}{\|}}{C}-O\langle\bigcirc\rangle NO_2$
p-Nitrophenyl acetate

(d) $CH_3CONH_2 + CH_3COO^-NH_4{}^+$
  Acetamide   Ammonium acetate

(e) 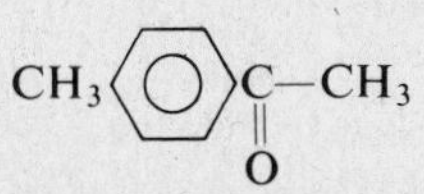 $CH_3\langle\bigcirc\rangle\underset{\underset{\displaystyle O}{\|}}{C}-CH_3$
Methyl p-tolyl ketone
(p-Methylacetophenone)

(f) no reaction

(g) $CH_3COO^-Na^+$
Sodium acetate

(h) $CH_3COOEt + CH_3COOAg$
  Ethyl acetate   Silver acetate

(i) $CH_3-\overset{\overset{\displaystyle O}{\|}}{C}-NHCH_3$
N-Methylacetamide

$+$

$CH_3COO^-{}^+H_3NCH_3$
Methylammonium acetate

(j) $CH_3-\overset{\overset{\displaystyle O}{\|}}{C}-N(CH_3)_2$
N,N-Dimethylacetamide

$+$

$CH_3COO^-{}^+H_2N(CH_3)_2$
Dimethylammonium acetate

(k) no reaction

(l) $CH_3CONHPh$
Acetanilide

4. (a) $Na^{+-}OOCCH_2CH_2COO^-Na^+$
Sodium succinate

(b) $NH_4^{+-}OOCCH_2CH_2CONH_2$
Ammonium succinamate

(c) $HOOC(CH_2)_2CONH_2$
Succinamic acid

(d)

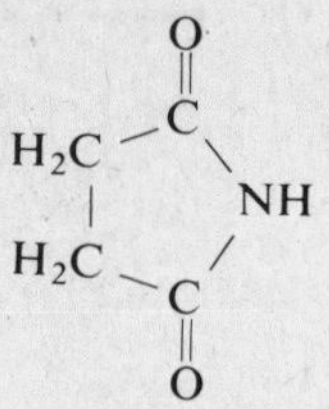

Succinimide

(e) $HOOC(CH_2)_2COOCH_2Ph$
Benzyl hydrogen succinate

(f) 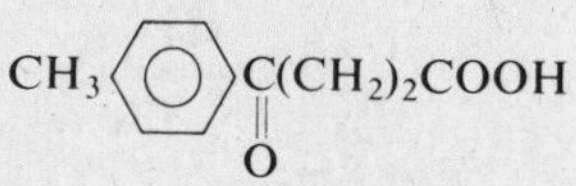 CH_3⟨◯⟩$C(CH_2)_2COOH$
        $\overset{\displaystyle O}{}$

β-(p-Toluyl)propionic acid

5. (a) $PhCH_2COOH + NH_4Cl$

(b) $PhCH_2COO^-Na^+ + NH_3$

6. (a) $PhCH_2COOH + NH_4Cl$

(b) $PhCH_2COO^-Na^+ + NH_3$

7. In each case an equimolar amount of methanol is formed along with the indicated product.

(a) $n\text{-PrCOOH}$
     n-Butyric acid

(b) $n\text{-PrCOO}^-K^+$
     Potassium n-butyrate

(c) $n\text{-PrCOOCH(CH}_3)_3$
     Isopropyl n-butyrate

(d) $n\text{-PrCOOCH}_2Ph$
     Benzyl n-butyrate

(e) $n\text{-PrCONH}_2$
     n-Butyramide

(f) $Ph_2C(OH)CH_2CH_2CH_3$
     Diphenyl-n-propylcarbinol

(g) $(iso\text{-Bu})_2C(OH)CH_2CH_2CH_3$
     Diisobutyl-n-propyl carbinol

(h) $n\text{-PrCH}_2OH$
     n-Butyl alcohol

8. (a) 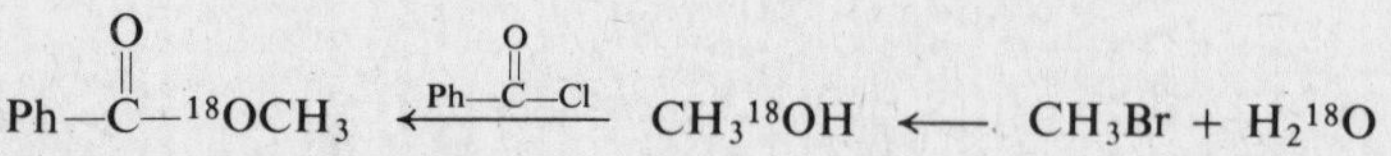 $Ph\overset{\displaystyle O}{-C}-^{18}OCH_3 \xleftarrow{Ph-\overset{O}{C}-Cl} CH_3{}^{18}OH \longleftarrow CH_3Br + H_2{}^{18}O$

(b) $Ph\overset{\displaystyle ^{18}O}{-C}-OCH_3 \xleftarrow{MeOH} Ph\overset{\displaystyle ^{18}O}{-C}-Cl \xleftarrow{SOCl_2} Ph\overset{\displaystyle ^{18}O}{-C}-^{18}OH \xleftarrow{H^+} PhC\equiv N + H_2{}^{18}O$

      $\xleftarrow{} PhCCl_3 + H_2{}^{18}O$ (one mole)

(c) $Ph\overset{\displaystyle ^{18}O}{-C}-^{18}OCH_3 \longleftarrow Ph\overset{\displaystyle ^{18}O}{-C}-Cl$ (from b) $+ CH_3{}^{18}OH$ (from a)

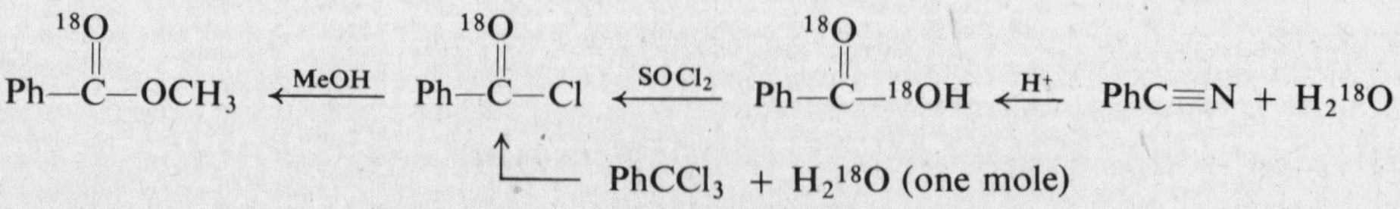

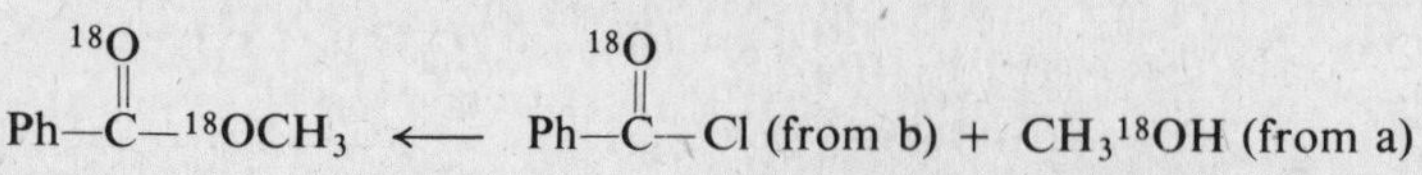

Hydrolysis of $PhCO^{18}OCH_3$ or $PhC^{18}O^{18}OCH_3$ gives $CH_3{}^{18}OH$; hydrolysis of $PhC^{18}OOCH_3$ gives CH_3OH.

9. (a) $Et{-}^{14}\overset{\|}{\underset{O}{C}}{-}Me \xleftarrow{Et_2Cd} Cl{-}^{14}\overset{\|}{\underset{O}{C}}{-}Me \xleftarrow{PCl_3} HOO^{14}C{-}Me \longleftarrow {}^{14}CO_2 + MeMgBr$

(b) $Et{-}\overset{\|}{\underset{O}{C}}{-}^{14}CH_3 \xleftarrow{EtCOCl} ({}^{14}CH_3)_2Cd \xleftarrow{CdCl_2} {}^{14}CH_3MgBr \xleftarrow{Mg} \xleftarrow{PBr_3} {}^{14}CH_3OH$

(c) $CH_3{-}^{14}CH_2{-}\overset{\|}{\underset{O}{C}}{-}CH_3 \xleftarrow{CH_3COCl} (CH_3{}^{14}CH_2)_2Cd \xleftarrow{CdCl_2} CH_3{}^{14}CH_2MgBr$

$CH_3{}^{14}CH_2MgBr \xleftarrow{Mg} \xleftarrow{PBr_3} CH_3{}^{14}CH_2OH \xleftarrow{LiAlH_4} CH_3{}^{14}COOH \xleftarrow{CH_3MgBr} {}^{14}CO_2$

(d) $^{14}CH_3CH_2{-}COCH_3 \xleftarrow{CH_3COCl} ({}^{14}CH_3CH_2)_2Cd \xleftarrow{as\ in\ (c)} {}^{14}CH_3CH_2OH$

$^{14}CH_3{-}CH_2OH \xleftarrow{HCHO} {}^{14}CH_3MgBr \longleftarrow as\ in\ (b)$

(e) $Ph^{14}CH_2CH_3 \xleftarrow[HCl]{Zn(Hg)} Ph^{14}\overset{\|}{\underset{O}{C}}{-}CH_3 \xleftarrow{Me_2Cd} Ph^{14}COCl \xleftarrow{PCl_5} Ph^{14}COOH \xleftarrow{PhMgBr} {}^{14}CO_2$

(f) $PhCH_2{}^{14}CH_3 \xleftarrow[HCl]{Zn(Hg)} Ph{-}\overset{\|}{\underset{O}{C}}{-}^{14}CH_3 \xleftarrow{PhCOCl} ({}^{14}CH_3)_2Cd \longleftarrow as\ in\ (b)$

(g) $Et{-}\overset{\|}{\underset{^{18}O}{C}}{-}Me \xleftarrow{Et_2Cd} Cl{-}\overset{\|}{\underset{^{18}O}{C}}{-}Me \xleftarrow{PCl_3} CH_3C^{18}O_2H \xleftarrow{CH_3C\equiv N} H_2{}^{18}O$

10. (a) An ester reacts with ammonia to give an amide and an alcohol.

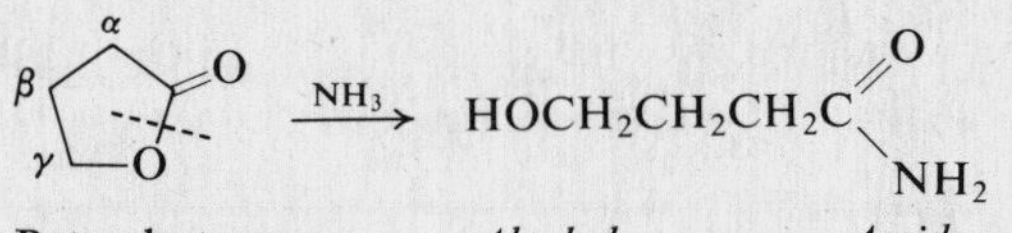

γ-Butyrolactone   Alcohol   Amide
Ester

(b) An ester is reduced by $LiAlH_4$ to two alcohols.

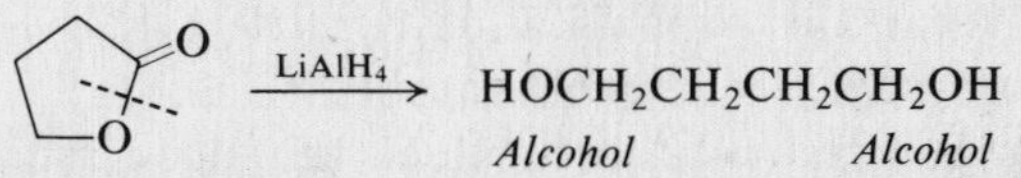

$\xrightarrow{LiAlH_4} HOCH_2CH_2CH_2CH_2OH$
Alcohol   Alcohol

(c) An ester undergoes transesterification.

$$\text{(lactone ring)} \xrightarrow{\text{EtOH, H}^+} \underset{\substack{\text{"New"}\\\text{alcohol}}}{\text{HOCH}_2\text{CH}_2\text{CH}_2\text{C}}\underset{\substack{\text{"New"}\\\text{ester}}}{\overset{\text{O}}{<}}_{\text{OEt}}$$

11. There are just three steps. A bond to chiral carbon cannot be broken in the first step,

$$\text{Ts}\!-\!\text{Cl} \quad \text{H}\!-\!\text{O}\!-\!\underset{\text{Me}}{\overset{\text{Et}}{\underset{|}{\overset{|}{\bigcirc}}}}\!-\!\text{H} \longrightarrow \text{Ts}\!-\!\text{O}\!-\!\underset{\text{Me}}{\overset{\text{Et}}{\underset{|}{\overset{|}{\bigcirc}}}}\!-\!\text{H} \xrightarrow{\text{inversion}} \text{H}\!-\!\underset{\text{Me}}{\overset{\text{Et}}{\underset{|}{\overset{|}{\bigcirc}}}}\!-\!\text{O}\!-\!\overset{\overset{\text{O}}{\|}}{\text{CPh}} \longrightarrow \text{H}\!-\!\underset{\text{Me}}{\overset{\text{Et}}{\underset{|}{\overset{|}{\bigcirc}}}}\!-\!\text{O}\!-\!\text{H}$$

$$\underset{(+)\text{-Alcohol}}{} \qquad \underset{\substack{(+)\text{-Tosylate}\\ {}^-\text{OOCPh}}}{} \qquad \underset{\substack{(-)\text{-Benzoate}\\ \text{HO}^-}}{} \qquad \underset{(-)\text{-Alcohol}}{}$$

because of the very nature of the reactants and products (see the answer to Problem 16.9, page 528), and is not broken in the third step (Sec. 16.9, page 528). Inversion *must*, then, take place in the attack by benzoate ion on *sec*-butyl tosylate, evidently in an S_N2 reaction.

12. Here, as in Problem 20.19, there is competition between a bimolecular reaction and a unimolecular reaction. With a high concentration of nucleophile (5N NaOH) there is (bimolecular) nucleophilic acyl substitution, with cleavage of the bond between oxygen and the acyl group:

$$\text{RCO}\!-\!\text{OR}'$$

The bond to the chiral carbon is not broken, and there is retention of configuration in the product.

With a low concentration of nucleophile (dil. NaOH), the bimolecular reaction is outpaced by the unimolecular reaction: an S_N1 reaction of the neutral ester, made possible by the stability of the benzylic-allylic cation generated. Both esters give the same hybrid

$$\left.\begin{array}{l}\underset{\substack{|\\\text{O}\!-\!\text{CR}\\ \|\\ \text{O}}}{\text{PhCH}\!-\!\text{CH}\!=\!\text{CH}\!-\!\text{CH}_3}\\[3em]\underset{\substack{|\\\text{O}\!-\!\text{CR}\\ \|\\ \text{O}}}{\text{PhCH}\!=\!\text{CH}\!-\!\text{CH}\!-\!\text{CH}_3}\end{array}\right\} \xrightarrow{-\text{RCOO}^-} \underset{\oplus}{\underbrace{\text{PhCH}\!=\!\!=\!\text{CH}\!=\!\!=\!\text{CH}}}\!-\!\text{CH}_3 \xrightarrow[\text{H}_2\text{O}]{\text{OH}^-} \underset{\substack{|\\\text{OH}}}{\text{PhCH}\!=\!\text{CH}\!-\!\text{CH}\!-\!\text{CH}_3}$$

cation and hence yield the same alcohol—the one with the double bond in conjugation with the ring. Configuration is lost in the achiral cation, and hence the alcohol is optically inactive.

13. The toluates and the phenylacetate can be distinguished by hydrolysis and determination of the m.p.'s of the resulting acids.

The two benzoates can be distinguished by saponification equivalent, or by m.p. of 3,5-dinitrobenzoate ester of the alcohol obtained by hydrolysis.

Benzyl acetate is the only one that gives a liquid, water-soluble acid upon hydrolysis.

14. (a) $NaHCO_3$(aq) gives CO_2 with acid.

(b) EtOH gives pleasant-smelling ester from RCOCl; RCOCl also gives CO_2 from $NaHCO_3$(aq).

(c) NaOH, heat; test for NH_3 by litmus.

(d) Br_2/CCl_4, or $KMnO_4$.

(e) NaOH, heat; test for NH_3 by litmus.   Also see slow dissolution of nitrile.

(f) $NaHCO_3$(aq), warm.

(g) CrO_3/H_2SO_4.

(h) NH_4^+ salt gives immediate evolution of NH_3 by action of cold NaOH(aq).

(i) Alcoholic $AgNO_3$.

15. (a)

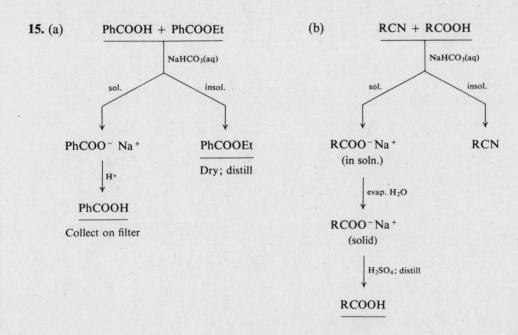

(c)

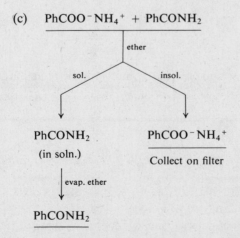

16. The THP ester is formed by acid-catalyzed addition of RCOOH to the double bond, made easy by the stability of the intermediate cation I, which is actually an oxonium ion:

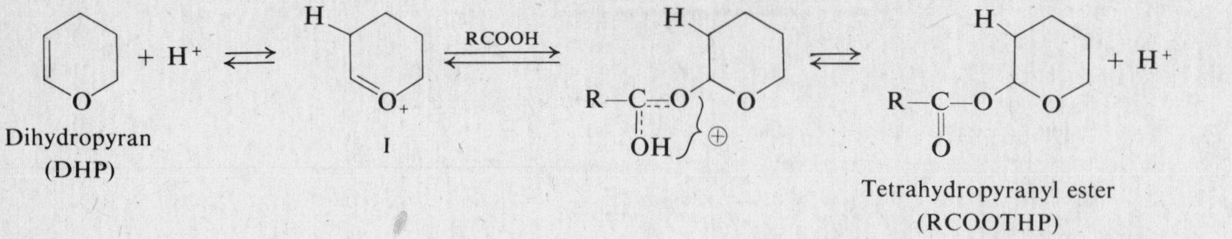

Dihydropyran
(DHP)

I

Tetrahydropyranyl ester
(RCOOTHP)

Acid-catalyzed hydrolysis of the THP ester is also easy, since it involves formation of the same cation I. A cyclic hemiacetal is formed and, like other hemiacetals, regenerates the aldehyde and alcohol, in this case parts of the same molecule. (Compare Problem 15, Chapter 19, page 649.)

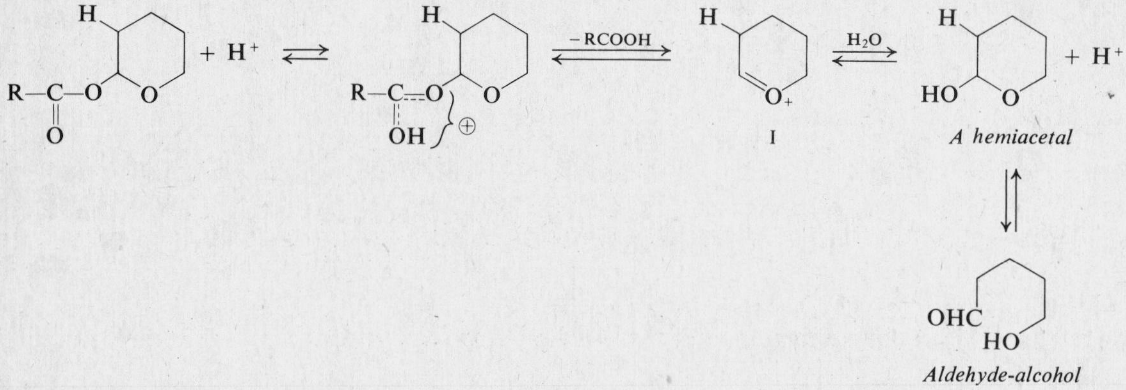

I

A hemiacetal

Aldehyde-alcohol

17.

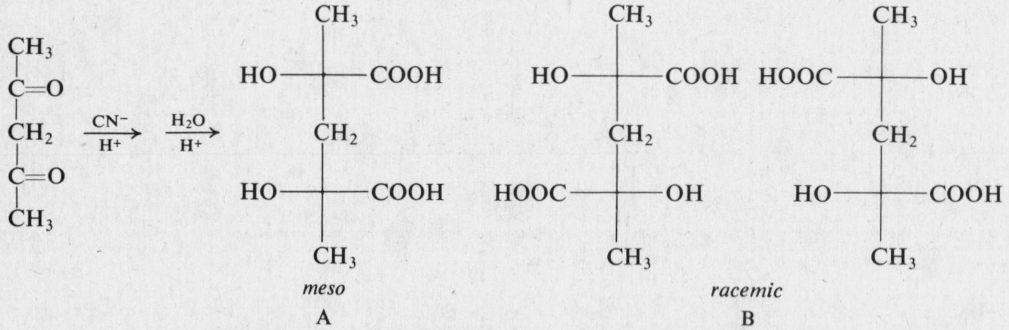

meso
A

racemic
B

B is the racemic modification. It gives monolactone II; but since the remaining —OH and —COOH are *cis*, II can react further to form the dilactone.

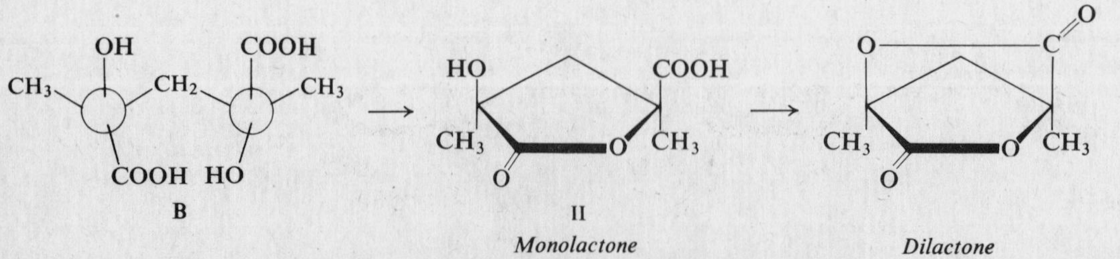

B

II
Monolactone

Dilactone

A is the *meso* compound. It gives monolactone III; here, the remaining —OH and —COOH are *trans*, and further reaction is not possible.

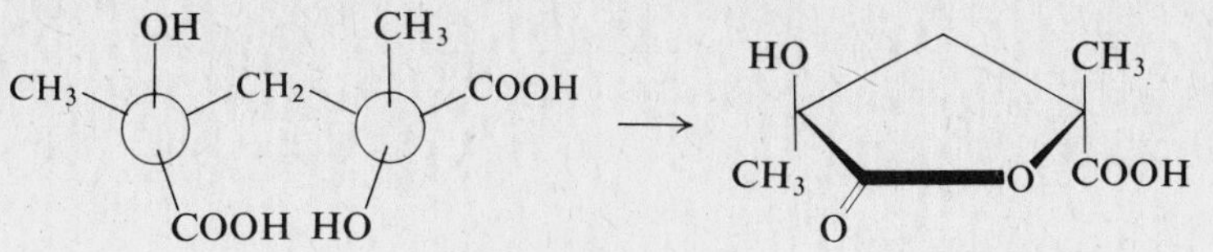

A

III
Monolactone

18. (a) $H_2N\!-\!\overset{\displaystyle O}{\underset{\displaystyle \|}{C}}\!-\!NH_2 \xrightarrow[\;H_2O\;]{\;NaOH\;} 2NH_3 + Na_2CO_3$

Di-amide   C

(b) $Cl\!-\!\underset{O}{C}\!-\!Cl \xrightarrow{\;EtOH\;} EtO\!-\!\underset{O}{C}\!-\!Cl \xrightarrow{\;NH_3\;} EtO\!-\!\underset{O}{C}\!-\!NH_2$

(Di)acid chloride   *Ester–acid chloride*   *Ester–amide*

D

(c) $PhBr \xrightarrow{\;Mg\;} PhMgBr \xrightarrow{\;\overset{CH_2-CH_2}{O}\;} \xrightarrow{\;H^+\;} PhCH_2CH_2OH \xrightarrow{\;PBr_3\;} PhCH_2CH_2Br$

E   F   G

$PhCH_2CH_2Br \xrightarrow{\;CN^-\;} PhCH_2CH_2C\!\equiv\!N \xrightarrow[\;H^+\;]{\;H_2O\;} PhCH_2CH_2COOH + NH_4^+$

G   H   I

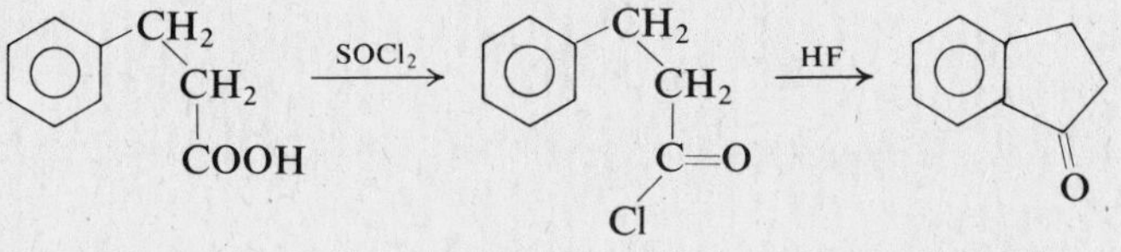

I   J   K
Friedel-Crafts acylation

1-Indanone
(See Problem 10a, page 649)

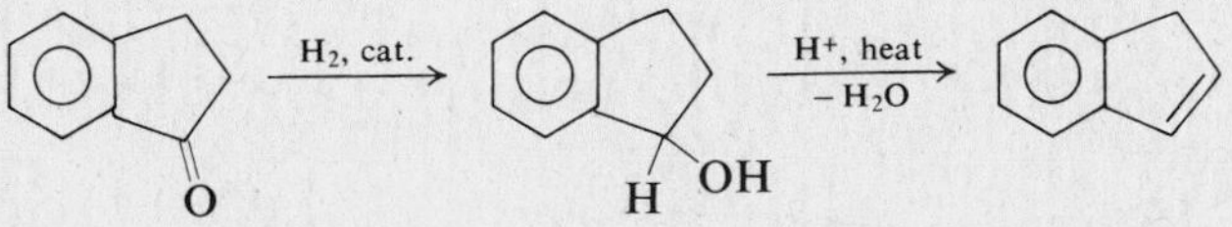

K   L   M
1-Indanol   Indene
(See answer to Problem 26, Chapter 12)

(d)

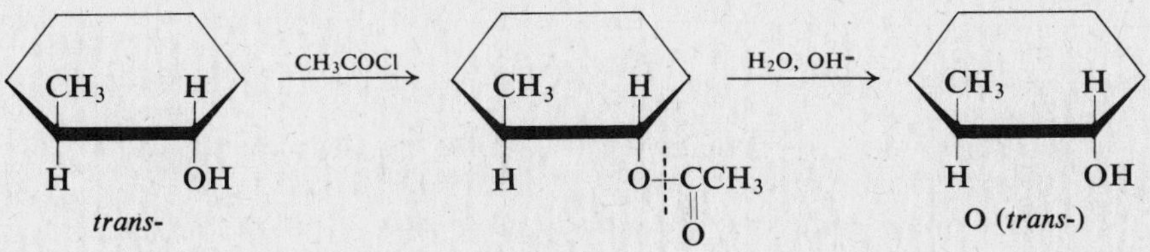

trans-   O (*trans-*)

N

19.

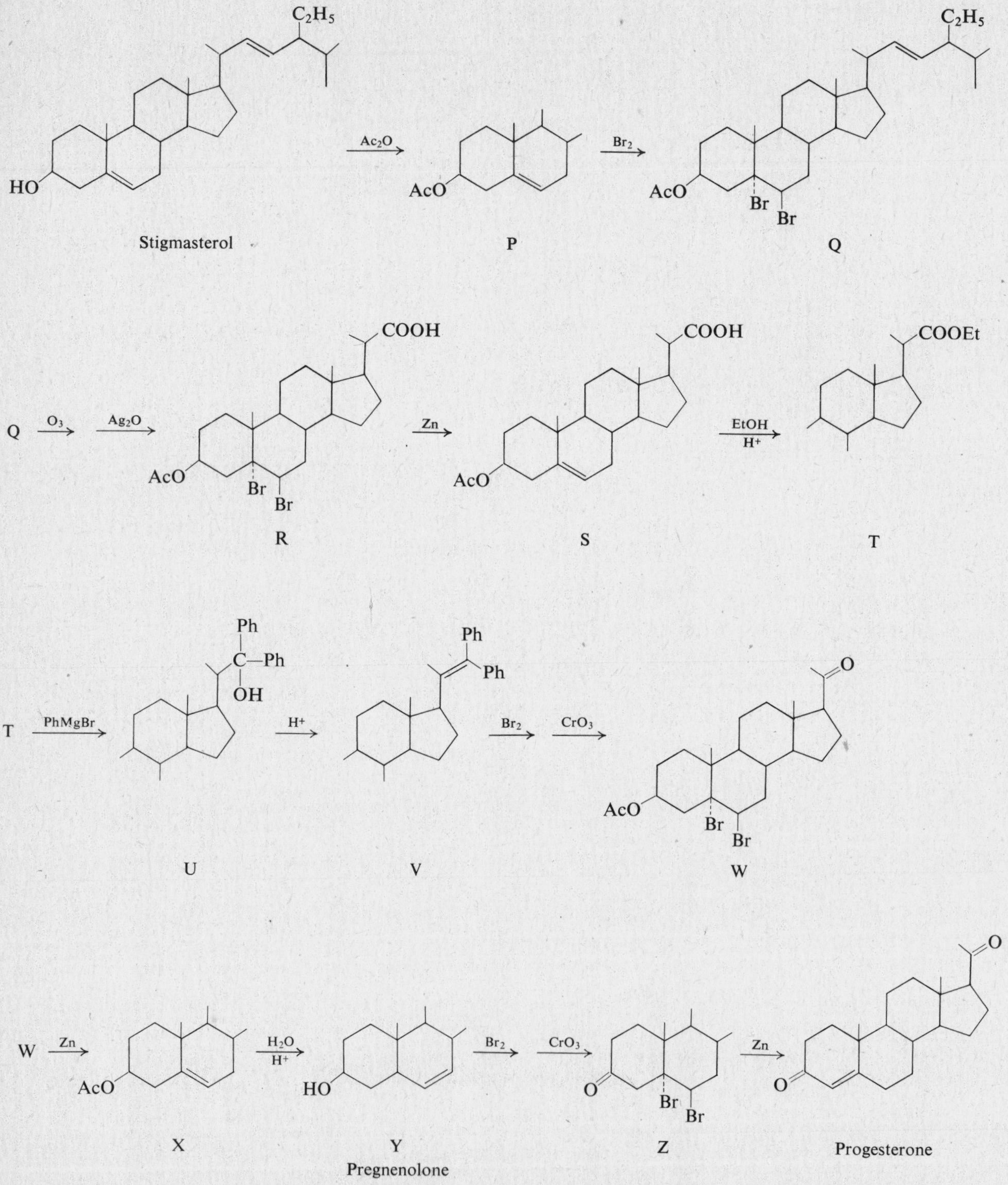

(b) The ultraviolet spectrum shows that progesterone has a double bond conjugated with the C–3 keto group. In the last step of the synthesis, the double bond shifts to yield the more stable product.

20. (a)

	NaHCO₃	Ac₂O	Tollens'	Schiff's	HIO₄
AA	not acid	two —OH	no —CHO	no —CHO	not glycol
BB	not acid	two —OH	no —CHO	no —CHO	1,2-glycol
CC	not acid	one —OH	no —CHO	no —CHO	not glycol
DD	not acid	no —OH	acetal	acetal	not glycol

$$CH_2CH_2CH_2 \qquad CH_3{-}CH{-}CH_2 \qquad CH_3OCH_2CH_2 \qquad H_2C{-}OCH_3$$
$$OH \quad\quad OH \qquad\quad OH \quad OH \qquad\qquad OH \qquad\qquad\quad OCH_3$$

AA BB CC DD

(b)

	NaHCO₃	Ac₂O	Tollens'	Schiff's	HIO₄
EE	not acid	one —OH	—CHO	—CHO	α-hydroxy aldehyde
FF	not acid	one —OH	α-hydroxy ketone	no —CHO	α-hydroxy ketone
GG	not acid	one —OH	—CHO	—CHO	not α-OH
HH	acid	no —OH	no —CHO	no —CHO	not glycol
II	Et ester	no —OH	—CHO	no —CHO (must be formate)	not glycol
JJ	not acid	no —OH	no —CHO	no —CHO	not glycol
KK	not acid	two —OH	no —CHO	no —CHO	1,2-glycol
LL	not acid	no —OH	acetal	acetal	1,2-glycol involved in acetal
MM	not acid	one —OH	no —CHO	no —CHO	1,2-glycol formed by action of H₂O

$$CH_3CHCHO \qquad CH_3{-}C{-}CH_2 \qquad CH_2CH_2CHO \qquad CH_3CH_2COOH \qquad HCOOC_2H_5$$
$$OH \qquad\qquad O \quad OH \qquad OH$$

EE FF GG HH II

$$CH_3COOCH_3 \qquad CH_2\!\!\begin{array}{c}CHOH\\ \\CHOH\end{array} \qquad H_2C\!\!\begin{array}{c}O{-}CH_2\\ \\O{-}CH_2\end{array} \qquad H_2C{-}CH{-}CH_2$$
$$O \quad OH$$

JJ KK LL MM

21. (a) NN and OO are substituted malonic acids; they lose CO₂ when heated ($C_9H_{14}O_4 -$ $CO_2 = C_8H_{14}O_2$). They are stereoisomers, and differ in the relationship of the —CH₃ groups, which can be *trans* or *cis*.

NN
trans
Racemic

PP
Racemic

One product

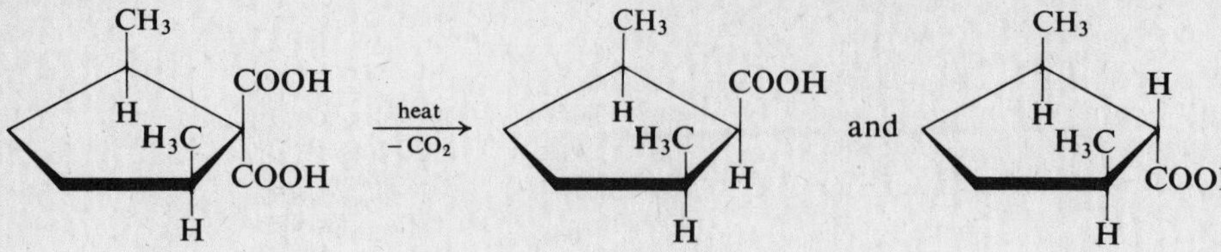

OO   QQ   RR   *Two products*

cis   Diastereomers

Achiral   *Achiral*   *Achiral*

meso

(b) Only the *trans*-isomer is resolvable; the *cis*-isomer is achiral and *meso*.

22. (a) Erythrose is an aldehyde and contains three —OH's. Cleavage by 3HIO$_4$ shows that it has the following structure, containing two chiral carbons:

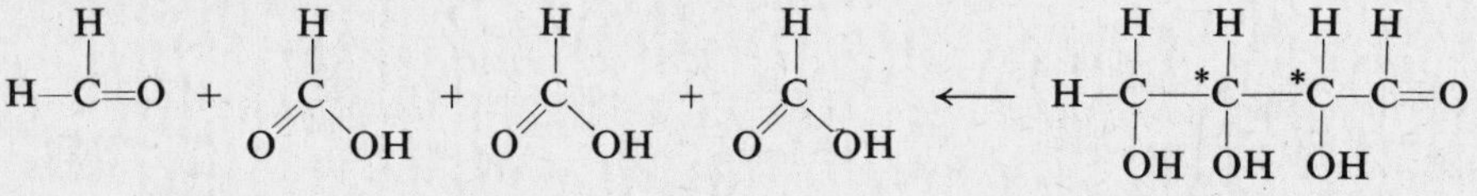

$$H-C=O \; + \; \underset{O}{\overset{H}{C}}OH \; + \; \underset{O}{\overset{H}{C}}OH \; + \; \underset{O}{\overset{H}{C}}OH \; \longleftarrow \; H-\overset{H}{\underset{OH}{C}}-\overset{H}{\underset{OH}{C}}^*-\overset{H}{\underset{OH}{C}}^*-\overset{H}{C}=O$$

(HIO$_4$)   (HIO$_4$)   (HIO$_4$)   Erythrose

Oxidation of (−)-erythrose yields a dicarboxylic acid, HOOC*CHOHCHOH*COOH. Since the acid is optically inactive, it must have the *meso* configuration, showing that (−)-erythrose is the 2R,3R- compound or its enantiomer.

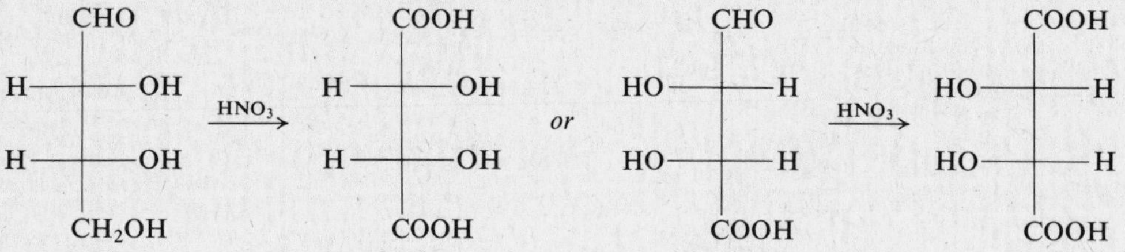

2R,3R-   *meso*-acid   2S,3S-   same *meso*-acid

(−)-Threose must be a diastereomer of erythrose, either the 2S,3R- or 2R,3S-isomer.

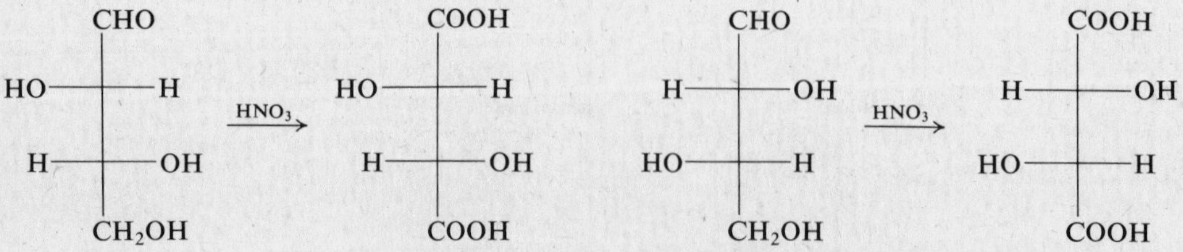

2S,3R-   Active acid   2R,3S-   Active acid

(b)

$$
\underset{\text{R-Glyceraldehyde}}{
\begin{array}{c}
\text{CHO} \\
\text{H}\!-\!\!-\!\!\overset{|}{\underset{|}{\text{C}}}\!\!-\!\!-\text{OH} \\
\text{CH}_2\text{OH}
\end{array}}
\xrightarrow[]{\text{CN}^-,\ \text{H}^+}
\begin{array}{c}
\text{CN} \\
\text{H}\!-\!\!-\!\text{OH} \\
\text{H}\!-\!\!-\!\text{OH} \\
\text{CH}_2\text{OH}
\end{array}
\quad + \quad
\begin{array}{c}
\text{CN} \\
\text{HO}\!-\!\!-\!\text{H} \\
\text{H}\!-\!\!-\!\text{OH} \\
\text{CH}_2\text{OH}
\end{array}
$$

$\downarrow\ \text{H}^+ \mid \text{H}_2\text{O}$ $\qquad$ $\downarrow\ \text{H}^+ \mid \text{H}_2\text{O}$

$$
\underset{(-)\text{-Erythrose}}{
\begin{array}{c}
\text{CHO} \\
\text{H}\!-\!\!-\!\text{OH} \\
\text{H}\!-\!\!-\!\text{OH} \\
\text{CH}_2\text{OH}
\end{array}}
\xrightarrow[]{\text{Br}_2,\ \text{H}_2\text{O}}
\begin{array}{c}
\text{COOH} \\
\text{H}\!-\!\!-\!\text{OH} \\
\text{H}\!-\!\!-\!\text{OH} \\
\text{CH}_2\text{OH}
\end{array}
\qquad
\begin{array}{c}
\text{COOH} \\
\text{HO}\!-\!\!-\!\text{H} \\
\text{H}\!-\!\!-\!\text{OH} \\
\text{CH}_2\text{OH}
\end{array}
\xleftarrow[]{\text{Br}_2,\ \text{H}_2\text{O}}
\underset{(-)\text{-Threose}}{
\begin{array}{c}
\text{CHO} \\
\text{HO}\!-\!\!-\!\text{H} \\
\text{H}\!-\!\!-\!\text{OH} \\
\text{CH}_2\text{OH}
\end{array}}
$$

(Compare with Problem 12, Chapter 19, page 649.)

The names of these compounds are the basis for the designations *erythro* and *threo* (see Problem 5, page 247) used to specify certain configurations of compounds containing two chiral carbons. The *erythro* isomer is the one that is convertible (in principle, at least) into a *meso* structure, whereas the *threo* isomer is convertible into a racemic modification. On page 481, for example, I and II are *erythro*, and III and IV are *threo*. On page 483, V is *erythro*, and VI is *threo*.

23. (a) $\text{CH}_3\text{C}\overset{\text{O}}{\underset{\text{OCH}_2\text{CH}_3}{\big<}}$ $\qquad$ (b) $\text{CH}_2{=}\underset{\text{CH}_3}{\text{C}}{-}\text{COOH}$ $\qquad$ (c) $\text{C}_6\text{H}_5\text{CH}_2\text{CONH}_2$

  Ethyl acetate $\qquad\qquad$ Methacrylic acid $\qquad\qquad$ Phenylacetamide

(See the labeled spectra on page 662 of this Study Guide.)

24. (a) $\text{HC}\overset{\text{O}}{\underset{\text{OCH}_2\text{CH}_2\text{CH}_3}{\big<}}$ $\quad$ (b) $\text{CH}_3\text{CH}_2\text{C}\overset{\text{O}}{\underset{\text{OCH}_3}{\big<}}$ $\quad$ (c) $\text{CH}_3\text{C}\overset{\text{O}}{\underset{\text{OCH}_2\text{CH}_3}{\big<}}$

  n-Propyl formate $\qquad\qquad$ Methyl propionate $\qquad\qquad$ Ethyl acetate

(See the labeled spectra on page 663 of this Study Guide.)

25. $\text{CH}_3\text{C}\overset{\text{O}}{\underset{\text{OCH}_2\text{C}_6\text{H}_5}{\big<}}$ $\qquad$ $\text{C}_6\text{H}_5\text{CH}_2\text{C}\overset{\text{O}}{\underset{\text{OCH}_3}{\big<}}$ $\qquad$ $\text{C}_6\text{H}_5\text{CH}_2\text{CH}_2\text{COOH}$

  Benzyl acetate $\qquad\qquad$ Methyl phenylacetate $\qquad\qquad$ Hydrocinnamic acid

  SS $\qquad\qquad\qquad\qquad\qquad$ TT $\qquad\qquad\qquad\qquad\qquad$ UU

(See the labeled spectra on pages 664 and 665 of this Study Guide.)

26. 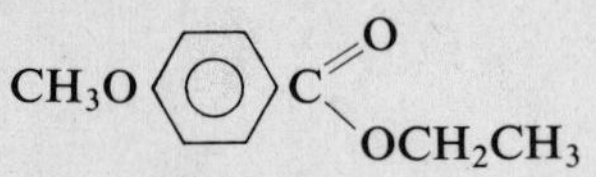 CH₃O⟨◯⟩C=O, OCH₂CH₃

Ethyl anisate

(See the labeled spectrum on page 666 of this Study Guide.)

27. 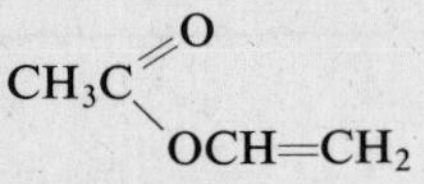 CH₃C=O, OCH=CH₂

Vinyl acetate
VV

(See the labeled spectra on page 666 of this Study Guide.)

28. (a) EtOOC(CH₂)₄COOEt   (b) Et—C—Ph   (c) CH₃—C—N—C—H

        Ethyl adipate

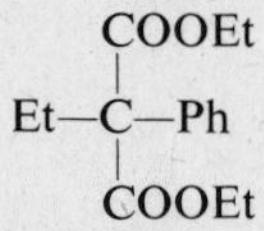

Ethyl
ethylphenylmalonate

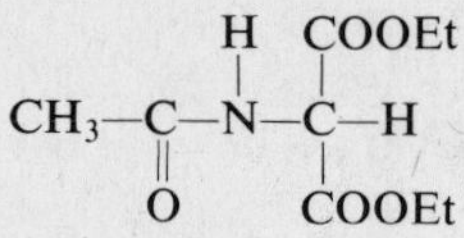

Ethyl
acetamidomalonate

(See the labeled spectra on page 667 of this Study Guide.)

Chapter 21

Carbanions I
Aldol and Claisen Condensations

21.1 III, in which the negative charge resides on oxygen, the atom that can best accommodate it.

21.2 Hydrogen on the carbon lying between the two carbonyl groups is the most acidic. Removal of a proton yields an anion that is highly stabilized by accommodation of the negative charge by two oxygens (rather than by only one).

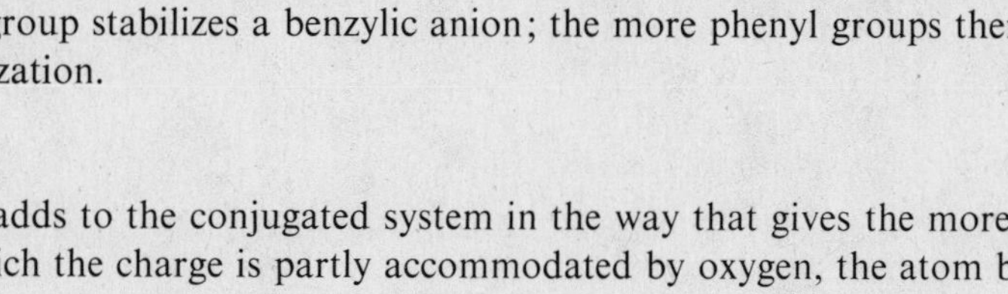

$$CH_3-\underset{O}{C}-CH_2-\underset{O}{C}-CH_3 \longrightarrow H^+ + CH_3-\underset{O}{C} \overset{\overset{H}{C}}{\underset{\ominus}{}} \underset{O}{C}-CH_3$$

21.3 Through resonance like that in free radicals (Sec. 12.14) and carbonium ions (Sec. 12.18), a phenyl group stabilizes a benzylic anion; the more phenyl groups there are, the greater the stabilization.

21.4 (a) CN⁻ adds to the conjugated system in the way that gives the more stable anion, the one in which the charge is partly accommodated by oxygen, the atom best able to do so.

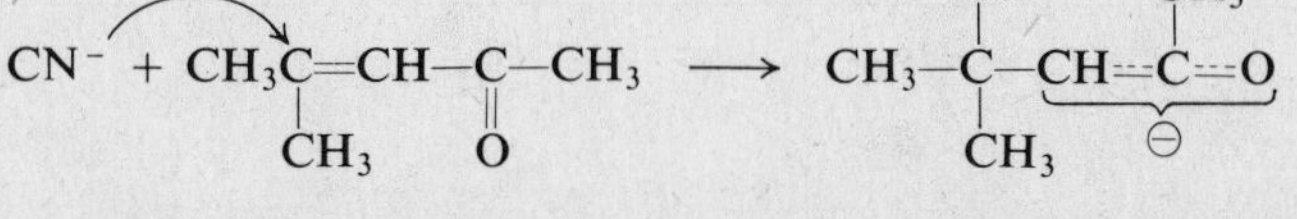

$$CN^- + CH_3\underset{CH_3}{C}=CH-\underset{O}{C}-CH_3 \longrightarrow CH_3-\underset{CH_3}{\overset{CN}{C}}-CH\!=\!\!\underset{\ominus}{\overset{CH_3}{C}}\!-\!O$$

(b) This anion accepts a proton at either the α-carbon or oxygen to yield either the keto or enol form of the product. In either case, the same product, chiefly keto, is finally obtained.

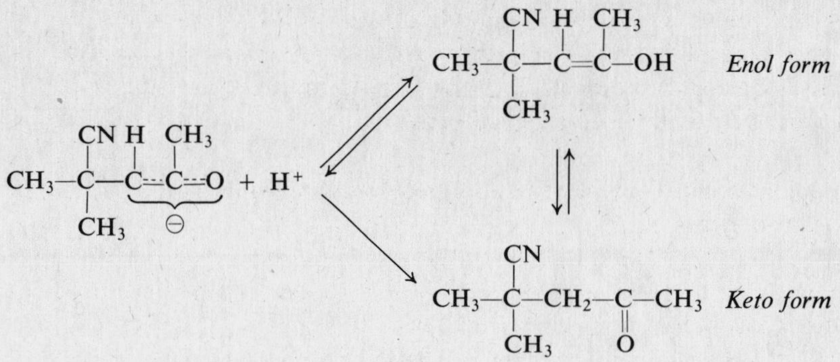

$$CH_3-\overset{\overset{\displaystyle CN}{|}}{\underset{\underset{\displaystyle CH_3}{|}}{C}}-\overset{\overset{\displaystyle H}{|}}{\underset{}{C}}=\overset{\overset{\displaystyle CH_3}{|}}{\underset{}{C}}-OH \quad \textit{Enol form}$$

$$CH_3-\overset{\overset{\displaystyle CN}{|}}{\underset{\underset{\displaystyle CH_3}{|}}{C}}-\overset{\overset{\displaystyle H}{|}}{\underset{\underset{\displaystyle \ominus}{}}{C}}=\overset{\overset{\displaystyle CH_3}{|}}{\underset{}{C}}=O \; + \; H^+$$

$$CH_3-\overset{\overset{\displaystyle CN}{|}}{\underset{\underset{\displaystyle CH_3}{|}}{C}}-CH_2-\overset{\underset{\displaystyle O}{\|}}{C}-CH_3 \quad \textit{Keto form}$$

21.5 (a) (1) $(+)$-Ph$-\overset{\underset{\displaystyle O}{\|}}{C}-\overset{\overset{\displaystyle CH_3}{|}}{C}H-C_2H_5$ + :B $\longrightarrow$ Ph$-\overset{\underset{\displaystyle O}{\|}}{C}=\overset{\overset{\displaystyle CH_3}{|}}{C}-C_2H_5$ + B:H *Slow*

        Optically active   *Achiral*

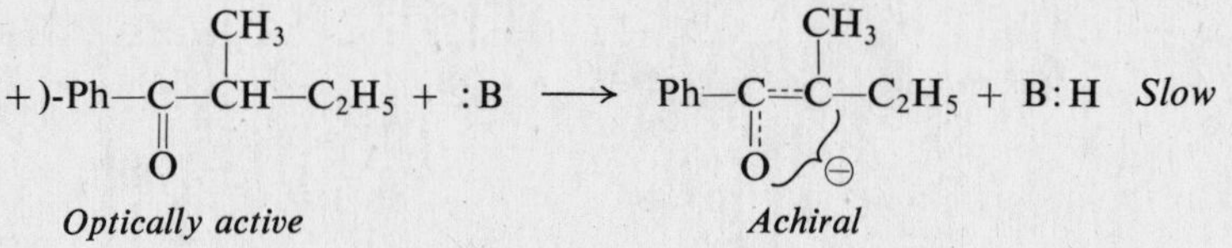

(2)   Ph$-\overset{\underset{\displaystyle O}{\|}}{C}=\overset{\overset{\displaystyle CH_3}{|}}{C}-C_2H_5$ + B:H $\longrightarrow$ $(\pm)$-Ph$-\overset{\underset{\displaystyle O}{\|}}{C}-\overset{\overset{\displaystyle CH_3}{|}}{C}H-C_2H_5$ + :B

                     Optically inactive:
                     racemic modification

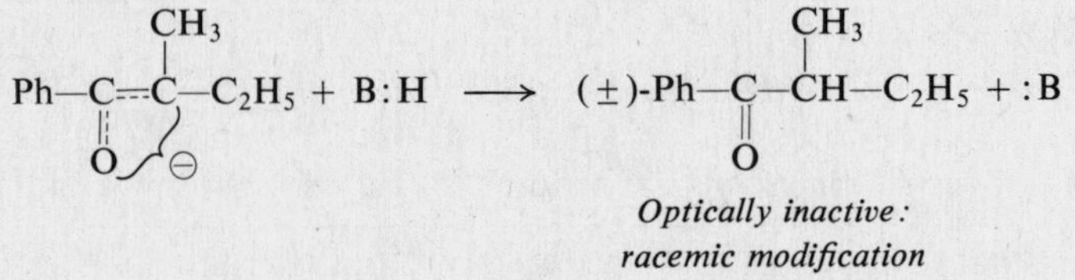

   The rate-determining step in racemization is the same as in bromination: formation of the carbanion. The carbanion is flat (and hence *achiral*) to permit accommodation of the negative charge by oxygen, which requires overlap of the *p*-orbital of negative

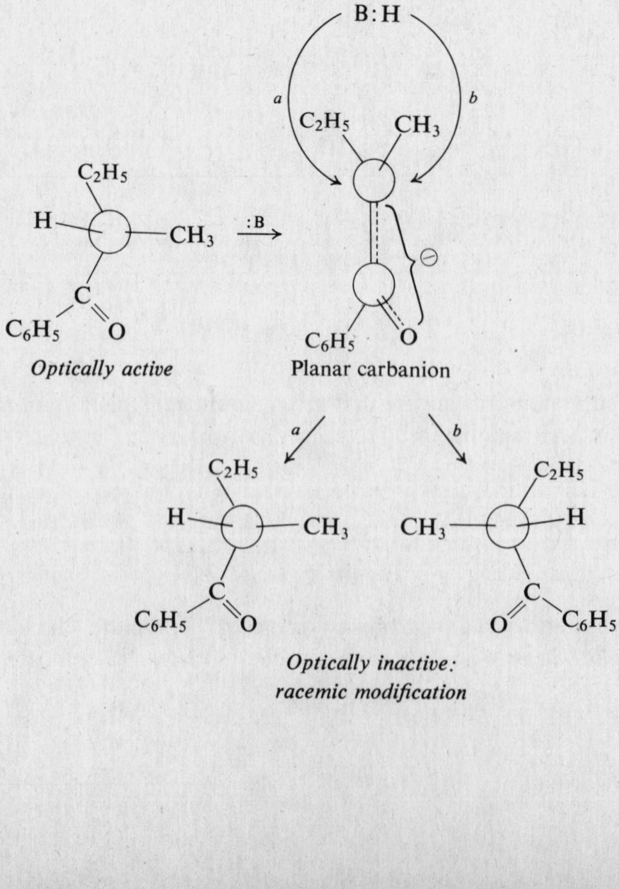

Optically active   Planar carbanion

Optically inactive:
racemic modification

carbon with the π cloud of the C=O group. When the carbanion takes back a proton from H:B the proton may become attached to either face of the carbanion and, depending upon which face, yields one or the other enantiomer. Since attachment to either face is equally likely, the enantiomers are formed in equal amounts.

(b) II can give a carbanion which, as in (a), loses configuration readily. III has no α-hydrogen, and cannot form a carbanion.

(c) The rate-determining step in hydrogen exchange is the same as in racemization, and hence in bromination: formation of the carbanion.

(1)
$$\text{Ph}-\overset{\displaystyle \underset{\displaystyle O}{\|}}{C}-\overset{\displaystyle \underset{\displaystyle H}{|}}{\overset{\displaystyle \overset{\displaystyle CH_3}{|}}{C}}-C_2H_5 + OD^- \longrightarrow \text{Ph}-\overset{\displaystyle O}{C}=\overset{\displaystyle \overset{\displaystyle CH_3}{|}}{C}-C_2H_5 \ominus + HOD \quad Slow$$

Unlabeled   Hydrogen lost
Optically active   Achiral

(2)
$$\text{Ph}-\overset{\displaystyle O}{C}=\overset{\displaystyle \overset{\displaystyle CH_3}{|}}{C}-C_2H_5 \ominus + D_2O \longrightarrow \text{Ph}-\overset{\displaystyle \underset{\displaystyle O}{\|}}{C}-\overset{\displaystyle \overset{\displaystyle CH_3}{|} \ \underset{\displaystyle D}{|}}{C}-C_2H_5$$

Labeled with D
Optically inactive

This is the final link in the chain of evidence showing that the loss of α-hydrogen is the rate-determining step in halogenation. When the carbanion regains a hydrogen ion, (i) this hydrogen ion, by statistical chance, is equally likely to attack either face of the carbanion and thus gives the racemic modification; and (ii) this hydrogen ion, also by statistical chance, is almost certain to be deuterium, since the acid (the solvent) is almost entirely D_2O. With the carbanion thus established as the intermediate in racemization and hydrogen exchange, and in view of the relationship between racemization and halogenation, we can quite reasonably conclude that the carbanion is the intermediate in halogenation, too.

21.6 (a) As in problem 21.5 (a), racemization takes place via an achiral carbanion, with accommodation of the negative charge by acyl oxygen.

$$\text{Ph}-\overset{\displaystyle \underset{\displaystyle OH}{|}}{C}H-\overset{\displaystyle \overset{\displaystyle O}{\|}}{C}-OEt \xrightarrow{\text{base}} \text{Ph}-\overset{\displaystyle \underset{\displaystyle OH}{|}}{C}=\overset{\displaystyle \ominus \ O}{C}-OEt$$

Achiral

(b) The doubly charged anion (carbanion and carboxylate ion) is more difficult to produce than a singly charged anion.

$$\text{Ph}-\overset{\displaystyle \underset{\displaystyle OH}{|}}{C}H-COO^- \xrightarrow{\text{base}} \left.\text{Ph}-\overset{\displaystyle \underset{\displaystyle OH}{|}}{C}=\overset{\displaystyle \overset{\displaystyle O}{\|}}{C}-O\right\} --$$

(c) Very slow (if it occurs at all), since there is no α-hydrogen available for removal to form carbanion.

21.7 We would expect the rate of racemization to be *twice* the rate of exchange, since the gain of one deuterium would bring about the loss of optical activity of two molecules. (See the answer to Problem 13, Chapter 13, page 488, in which the rate of racemization *is* twice the rate of isotopic exchange.)

21.8 (a) Both reactions go through the same slow step (2), on page 708, formation of the enol.
(b) Same as (a).

21.9 (a) HSO_4^- (b) H_2O or D_2O

21.10 (a)
$$CH_3CH_2\overset{\overset{\text{H}}{|}}{C}=O + [CH_3CHCHO]^- \longrightarrow CH_3CH_2\overset{}{C}H-\overset{\overset{\text{CH}_3}{|}}{\underset{\underset{\text{OH}}{|}}{C}}HCHO$$

(b)
$$CH_3\overset{\overset{\text{H}_3\text{C}}{|}}{C}=O + [CH_3COCH_2]^- \longrightarrow CH_3-\overset{\overset{\text{CH}_3}{|}}{\underset{\underset{\text{OH}}{|}}{C}}-CH_2COCH_3$$

(c)
$$Ph-\overset{\overset{\text{H}_3\text{C}}{|}}{C}=O + [PhCOCH_2]^- \longrightarrow Ph-\overset{\overset{\text{CH}_3}{|}}{\underset{\underset{\text{OH}}{|}}{C}}-CH_2COPh \xrightarrow{-H_2O} Ph-\overset{\overset{\text{CH}_3}{|}}{C}=CH-\overset{\overset{\text{Ph}}{|}}{C}=O$$

(d)

(e)
$$PhCH_2\overset{\overset{\text{H}}{|}}{C}=O + [PhCHCHO]^- \longrightarrow PhCH_2\overset{}{C}H-\overset{\overset{\text{Ph}}{|}}{\underset{\underset{\text{OH}}{|}}{C}}HCHO$$

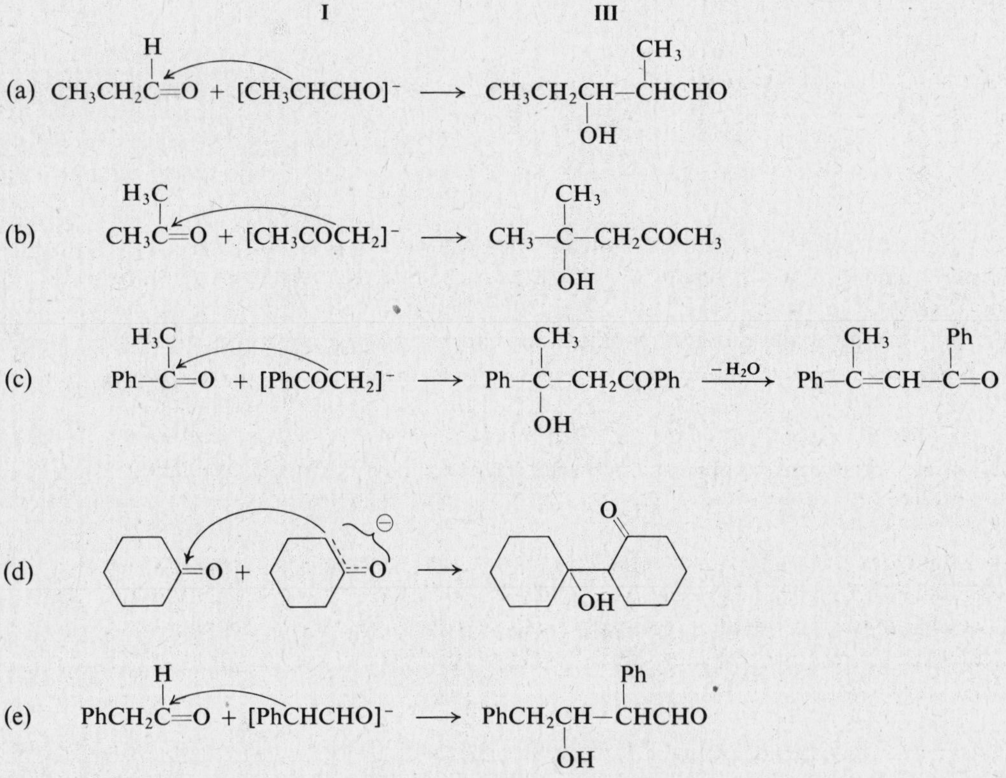

(For a review of the aldol condensation, see A. J. Nielsen and W. J. Houlihan, "The Aldol Condensation," O. R. XVI.)

21.11 It gives a mixture of aldol products, since either of two carbanions can form,

$$CH_3CH_2COCH_3 + :B \longrightarrow [CH_3CHCOCH_3]^- \quad or \quad [CH_3CH_2COCH_2]^-$$

and then add to the ketone.

21.12 $CH_3-\underset{\underset{O}{\|}}{C}-H + H^+ \rightleftarrows CH_3-\underset{\underset{\oplus OH}{\|}}{C}-H \rightleftarrows CH_2=\underset{\underset{OH}{|}}{C}-H + H^+$

$CH_3-\underset{\underset{\oplus OH}{|}}{C}-H + CH_2=\underset{\underset{OH}{|}}{C}-H \rightleftarrows CH_3-\underset{\underset{OH}{|}}{C}-\underset{\underset{\oplus OH}{|}}{CH_2-C}-H \rightleftarrows CH_3-\underset{\underset{OH}{|}}{\overset{\overset{H}{|}}{C}}-\underset{\underset{O}{\|}}{CH_2-C}-H + H^+$

Electrophile   Nucleophile   Protonated adduct   Aldol

21.13 (a) Step (2) is much faster than the reverse of step (1); as soon as a carbanion is formed (step 1), it reacts with a second CH_3CHO molecule (step 2) before it can react with water (reverse of step 1) to form an acetaldehyde molecule which, in D_2O, would contain deuterium.

More exactly,

$$rate_2 = k_2[\text{carbanion}][\text{aldehyde}]$$

$$rate_{-1} = k_{-1}[\text{carbanion}]$$

Evidently $k_2[\text{aldehyde}]$ is much bigger than k_{-1}, and $rate_2$ is much bigger than $rate_{-1}$.

(b) Second-order.

$$rate = k_1[\text{aldehyde}][\text{base}]$$

In the more exact terms of the fine print on pages 466–467,

$$rate = \cfrac{k_1[\text{aldehyde}][\text{base}]}{1 + \cfrac{k_{-1}}{k_2[\text{aldehyde}]}}$$

When [aldehyde] is high, $k_2[\text{aldehyde}]$ is much bigger than k_{-1}, and the term $k_{-1}/k_2[\text{aldehyde}]$ is very small relative to 1 and drops out.

(c) When [aldehyde] is low, $k_2[\text{aldehyde}]$ is no longer much bigger than k_{-1}, and $rate_2$ is no longer much bigger than $rate_{-1}$. As a result, a significant number of carbanions undergo the reverse of step (1) instead of undergoing step (2)—and pick up deuterium to form CH_2DCHO. When these labeled acetaldehyde molecules eventually undergo aldol condensation, they give labeled aldol.

(d) Step (2), carbonyl addition, is slower for a ketone, and is no longer much faster than the reverse of step (1).

21.14 Reverse of aldol condensation of acetone (*retrograde aldol* or *retro aldol*). According to the principle of microscopic reversibility (Problem 5.8, page 170), all steps in the aldol condensation are involved, in reverse order:

$$CH_3-\underset{\underset{OH}{|}}{\overset{\overset{CH_3}{|}}{C}}-\underset{\underset{O}{\|}}{CH_2-C}-CH_3 + OH^- \rightleftarrows CH_3-\underset{\underset{O_-}{|}}{\overset{\overset{CH_3}{|}}{C}}-\underset{\underset{O}{\|}}{CH_2-C}-CH_3 + H_2O$$

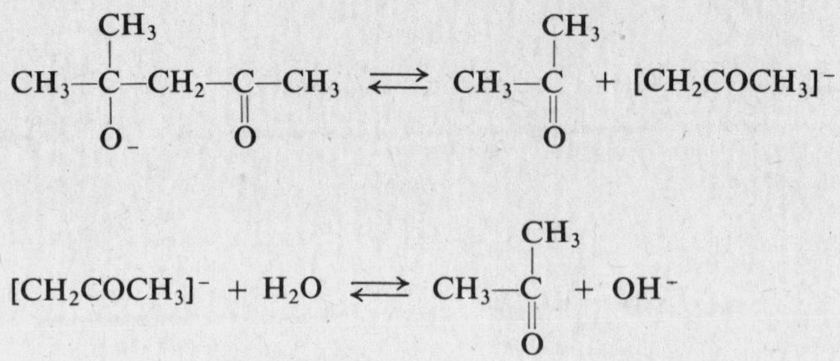

$$CH_3-\underset{\underset{O^-}{|}}{C}-CH_2-\underset{\underset{O}{\parallel}}{C}-CH_3 \rightleftarrows CH_3-\underset{\underset{O}{\parallel}}{\overset{\overset{CH_3}{|}}{C}} + [CH_2COCH_3]^-$$

$$[CH_2COCH_3]^- + H_2O \rightleftarrows CH_3-\underset{\underset{O}{\parallel}}{\overset{\overset{CH_3}{|}}{C}} + OH^-$$

21.15 $\left[-\overset{|}{\underset{}{C}}=\overset{|}{\underset{}{C}}-\overset{|}{\underset{}{C}}=O \quad -\overset{|}{\underset{}{C}}\cdots\overset{|}{\underset{}{C}}=\overset{|}{\underset{}{C}}\cdots\overset{}{\underset{}{O}} \right]$

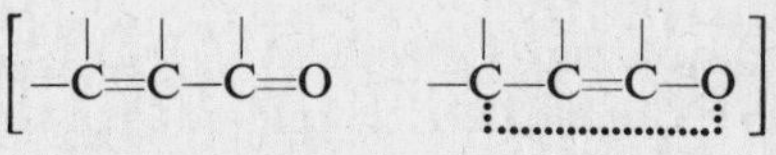

Overlap of π orbitals of carbon–carbon and carbon–oxygen double bonds, much as in conjugated dienes (see Figure 8.5, page 265) and in alkenylbenzenes.

21.16 (a) 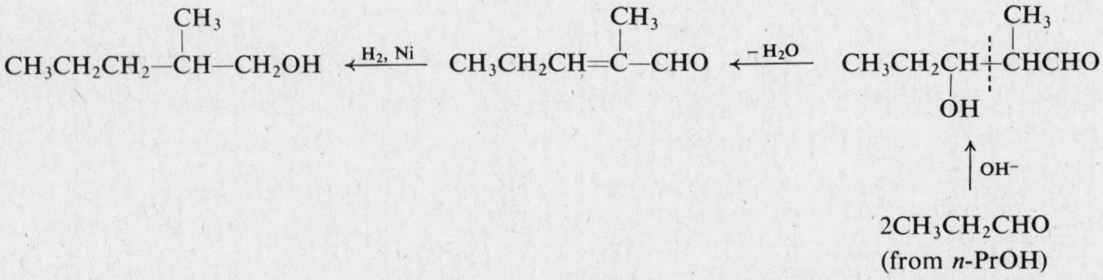 $CH_3CH_2CH_2-\overset{\overset{CH_3}{|}}{CH}-CH_2OH \xleftarrow{H_2,\ Ni} CH_3CH_2CH=\overset{\overset{CH_3}{|}}{C}-CHO \xleftarrow{-H_2O} CH_3CH_2\underset{\underset{OH}{|}}{CH}\!\vdots\!\overset{\overset{CH_3}{|}}{CH}CHO$

$$\uparrow OH^-$$

$$2CH_3CH_2CHO$$
$$(\text{from } n\text{-PrOH})$$

(b) $CH_3-\underset{\underset{OH}{|}}{CH}-CH_2-CH-CH_3 \xleftarrow{H_2,\ Ni} CH_3-\overset{\overset{CH_3}{|}}{C}=CH-\underset{\underset{O}{\parallel}}{C}-CH_3 \xleftarrow{-H_2O} CH_3\underset{\underset{OH}{|}}{C}\!\vdots\!CH_2\underset{\underset{O}{\parallel}}{C}CH_3$

$$\uparrow OH^-$$

$$2CH_3\underset{\underset{O}{\parallel}}{C}CH_3$$
$$(\text{from } i\text{-PrOH})$$

(c) $\xleftarrow{H_2,\ Ni}$ $\xleftarrow{-H_2O}$ $\xleftarrow{OH^-}$ 2
$$(\text{from cyclohexanol})$$

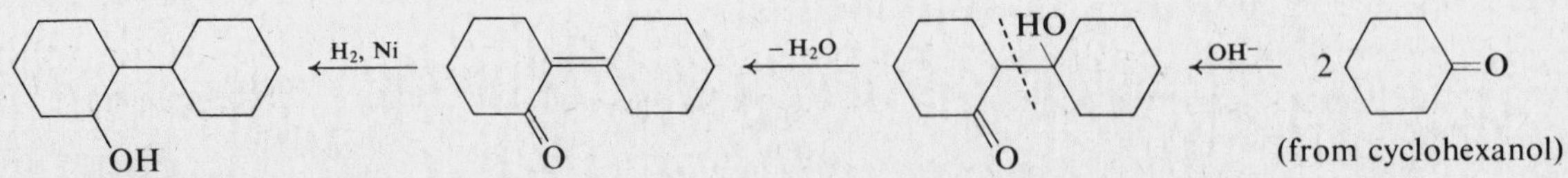

(d) 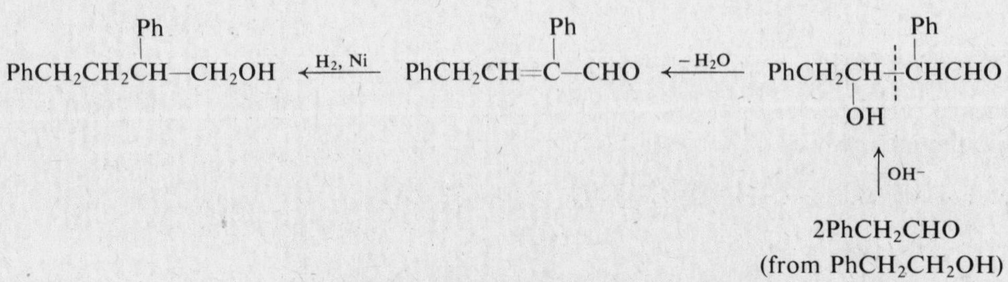 $PhCH_2CH_2\underset{}{CH}-CH_2OH \xleftarrow{H_2,\ Ni} PhCH_2CH=\overset{\overset{Ph}{|}}{C}-CHO \xleftarrow{-H_2O} PhCH_2\underset{\underset{OH}{|}}{CH}\!\vdots\!\overset{\overset{Ph}{|}}{CH}CHO$

$$\uparrow OH^-$$

$$2PhCH_2CHO$$
$$(\text{from } PhCH_2CH_2OH)$$

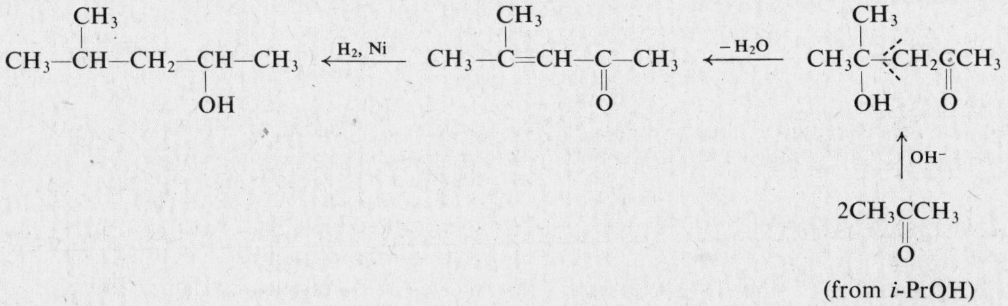

(e) $\underset{\underset{OH}{|}}{CH_3-C=CH-CH-Ph} \xleftarrow{NaBH_4} \underset{\underset{O}{||}}{CH_3-C=CH-C-Ph} \xleftarrow{-H_2O} \underset{\underset{OH}{|}\quad\underset{O}{||}}{CH_3C-CH_2CPh}$

$$\Big\uparrow OH^-$$

$$\underset{\underset{O}{||}}{2PhCCH_3}$$

$$\text{(from } PhCHOHCH_3\text{)}$$

21.17

$$\underset{\underset{OH}{|}}{EtCH_2CH-CHCH_2OH} \xleftarrow{redn.} \underset{\overset{|}{Et}}{\underset{\underset{OH}{|}}{EtCH_2CH-CHCHO}} \xleftarrow{OH^-} 2EtCH_2CHO \xleftarrow{oxidn.} n\text{-BuOH}$$
$$\overset{|}{Et}$$
$$\text{``6-12''}$$

$$\underset{\underset{OH}{|}\qquad\underset{OH}{|}}{CH_3-C-CH_2-CH-CH_3} \xleftarrow{redn.} \underset{\underset{OH}{|}\qquad\underset{O}{||}}{CH_3-C-CH_2-C-CH_3} \xleftarrow{OH^-} 2CH_3COCH_3$$
$$\overset{|}{CH_3}\qquad\qquad\qquad\overset{|}{CH_3}$$

21.18

$$2CH_3CHO \xrightarrow{OH^-} CH_3CH=CHCHO \xrightarrow{redn.} CH_3CH_2CH_2CH_2OH \xrightarrow{oxidn.} CH_3CH_2CH_2CHO$$
$$\text{Acetaldehyde}\qquad\quad\text{Crotonaldehyde}\qquad\qquad n\text{-Butyl alcohol}\qquad\qquad n\text{-Butyraldehyde}$$

$$CH_3CH_2CH_2CHO$$

$OH^- \mid CH_3CHO$	$OH^- \mid n\text{-PrCHO}$	$OH^- \mid CH_3CHO$

$$CH_3CH_2CH_2CH=CHCHO \qquad \underset{\overset{||}{C_2H_5}}{CH_3CH_2CH_2CH=C-CHO} \qquad \underset{\overset{||}{C_2H_5}}{CH_3CH=C-CHO}$$

$\downarrow$ redn.	$\downarrow$ redn.	$\downarrow$ redn.

$$CH_3CH_2CH_2CH_2CH_2CH_2OH \qquad \underset{\overset{|}{C_2H_5}}{CH_3CH_2CH_2CH_2-CH-CH_2OH} \qquad \underset{\overset{|}{C_2H_5}}{CH_3CH_2CHCH_2OH}$$
$$\text{1-Hexanol}\qquad\qquad\qquad\text{2-Ethyl-1-hexanol}\qquad\qquad\text{2-Ethyl-1-butanol}$$

21.19 (a) $\underset{\overset{|}{OH}}{PhCH_2CH_2CHCH_3} \xleftarrow{H_2,\ Ni} \underset{\overset{||}{O}}{PhCH=CHCCH_3} \xleftarrow[-H_2O]{OH^-} \begin{cases} PhCHO \text{ (from } PhCH_3\text{)} \\ \\ CH_3COCH_3 \text{ (from } i\text{-PrOH)} \end{cases}$

(b) $\underset{\overset{|}{OH}}{PhCH_2CH_2CHPh} \xleftarrow{H_2,\ Ni} \underset{\overset{||}{O}}{PhCH=CHCPh} \xleftarrow[-H_2O]{OH^-} \begin{cases} PhCHO \xleftarrow{} PhCH_3 \\ \\ CH_3CPh \text{ (page 623)} \\ \overset{|}{\underset{O}{||}} \end{cases}$

(c) $PhCH_2CH_2CH_2Ph$ $\xleftarrow{H_2, \text{ Ni}}$ $PhCH_2CH=CHPh$ $\xleftarrow{-H_2O}$ $PhCH_2CH_2\underset{\underset{OH}{|}}{C}HPh$ ← (b)

(d) $PhCH_2\overset{\overset{Ph}{|}}{C}HCH_2OH$ $\xleftarrow{H_2, \text{ Ni}}$ $PhCH=\overset{\overset{Ph}{|}}{C}CHO$ $\xleftarrow[-H_2O]{OH^-}$ ⎡ PhCHO (from $PhCH_3$)

PhCH$_2$CHO (from $PhCH_2COCl$)

(e) $PhCH=CHC\underset{\underset{O}{\|}}{C}H=CHPh$ $\xleftarrow[-2H_2O]{OH^-}$ ⎡ 2PhCHO (from $PhCH_3$)

CH_3COCH_3 (from i-PrOH)

21.20 (a) γ-Hydrogen will be acidic, since negative charge of anion is accommodated by electro-negative oxygen through the conjugated system.

$$-\overset{\overset{\gamma}{|}}{\underset{\underset{H}{|}}{C}}-\overset{\beta}{C}=\overset{\alpha}{C}-C=O \ + \ :B \ \rightleftarrows \ -\underset{\ominus}{\underbrace{C\cdots C\cdots C\cdots C}}=O \ + \ H:B$$

(b) $PhCH=CH-CH=CH-CHO$ $\xleftarrow[-H_2O]{OH^-}$ $PhCHO + CH_3CH=CHCHO$ $\xleftarrow[-H_2O]{OH^-}$ $2CH_3CHO$

21.21 CH_3COOEt $\xrightarrow{OEt^-}$ $[CH_2COOEt]^-$ $\xrightarrow{PhCHO}$ $PhCH-CH_2COOEt$ $\xrightarrow{+H^+}$ $PhCH-CH_2COOEt$

                                                              $\underset{O_-}{|}$ $\underset{OH}{|}$

$\downarrow -H_2O$

$PhCH=CHCOOEt$
Ethyl cinnamate

21.22 In each case (except d) base abstracts a proton that is *alpha* to one or more electron-withdrawing groups ($-NO_2$, $-C\equiv N$, $-C=O$, even $2,4-(NO_2)_2C_6H_3-$) to give one of the following anions:

(a) $[CH_2NO_2]^-$ (b) $[PhCHCN]^-$ (c) $[2,4-(O_2N)_2C_6H_3CH_2]^-$ (d) $HC\equiv C^-$

(e) $\left[CH_3\underset{\underset{O}{\|}}{C}-O-\underset{\underset{O}{\|}}{C}\cdots CH_2\right]^-$ (f) $[CH(COOEt)_2]^-$ (g) $[CH(CN)(COOEt)]^-$

The anion then adds to the aldehyde or ketone to give an aldol-like product, the alkoxy group is protonated, and water is lost. In (e), hydrolysis of an intermediate anhydride is needed as well.

(For a review of these reactions, see J. R. Johnson, "The Perkin Reaction and Related Reactions," O. R. I-8; G. Jones, "The Knoevenagel Condensation," O. R. XV-2.)

21.23 Elimination $\longrightarrow$ 1- and 2-butene.

(For a general discussion of the Wittig reaction, see S. Trippett, "The Wittig Reaction," Quart. Revs. (London), **17**, 406 (1963).)

21.24 In each case except (c), there are two combinations of reagents that, on paper, would give the desired product. In (a), (b), and (d), the better combination is shown; in each alternative, preparation of the ylide would be accompanied by much elimination. Compare Problem 21.23.

(a) $\text{CH}_3\text{CH}_2\text{CH}_2\text{CH}{=}\underset{\underset{\text{CH}_3}{|}}{\text{C}}\text{CH}_2\text{CH}_3 \longleftarrow \text{CH}_3\text{CH}_2\text{CH}_2\text{CH}{=}\text{PPh}_3 + \text{O}{=}\underset{\underset{\text{CH}_3}{|}}{\text{C}}\text{CH}_2\text{CH}_3$

                                                                            2-Butanone

(b) $\text{Ph}{-}\underset{\underset{\text{CH}_3}{|}}{\text{C}}{=}\text{CHCH}_2\text{Ph} \longleftarrow \text{Ph}{-}\underset{\underset{\text{CH}_3}{|}}{\text{C}}{=}\text{O} + \text{Ph}_3\text{P}{=}\text{CHCH}_2\text{Ph}$

(c) $\text{Ph}{-}\text{CH}{=}\text{CH}{-}\text{Ph} \longleftarrow \text{Ph}{-}\underset{\underset{\text{H}}{|}}{\text{C}}{=}\text{O} + \text{Ph}_3\text{P}{=}\text{CHPh}$

(d) $\bigpentagon{=}\text{CHCH}_3 \longleftarrow \bigpentagon{=}\text{O} + \text{Ph}_3\text{P}{=}\text{CHCH}_3$

(e) $\text{PhCH}{=}\text{CH}{-}\text{CH}{=}\text{CHPh} \longleftarrow \text{PhCH}{=}\text{CH}{-}\underset{\underset{\text{H}}{|}}{\text{C}}{=}\text{O} + \text{Ph}_3\text{P}{=}\text{CHPh}$

(f) $\text{CH}_2{=}\text{CHCH}{=}\underset{\underset{\text{CH}_3}{|}}{\text{C}}\text{COOMe} \longleftarrow \text{CH}_2{=}\text{CH}\underset{\underset{\text{H}}{|}}{\text{C}}{=}\text{O} + \text{Ph}_3\text{P}{=}\underset{\underset{\text{CH}_3}{|}}{\text{C}}\text{COOMe}$

21.25 (a) $\text{Ph}_3\text{P}{=}\text{CHC}_3\text{H}_7\text{-}n \xleftarrow[\text{base}]{\text{Ph}_3\text{P}} \text{BrCH}_2\text{C}_3\text{H}_7\text{-}n \xleftarrow{\text{PBr}_3} n\text{-BuOH}$

$\text{CH}_3\text{CH}_2\text{COCH}_3 \xleftarrow{\text{CrO}_3} sec\text{-BuOH}$

(b) $\text{PhCOCH}_3 \xleftarrow{\text{AlCl}_3}$ benzene + acetic anhydride

$\text{Ph}_3\text{P}{=}\text{CHCH}_2\text{Ph} \xleftarrow[\text{base}]{\text{Ph}_3\text{P}} \text{BrCH}_2\text{CH}_2\text{Ph} \xleftarrow{\text{PBr}_3} \text{HOCH}_2\text{CH}_2\text{Ph}$ (page 565)

(c) $\text{PhCHO} \xleftarrow{\text{OH}^-} \text{PhCHCl}_2 \xleftarrow{2\text{Cl}_2,\ \text{heat}} \text{PhCH}_3$

$\text{Ph}_3\text{P}{=}\text{CHPh} \xleftarrow[\text{base}]{\text{Ph}_3\text{P}} \text{ClCH}_2\text{Ph} \xleftarrow{\text{Cl}_2,\ \text{heat}} \text{PhCH}_3$

(d)

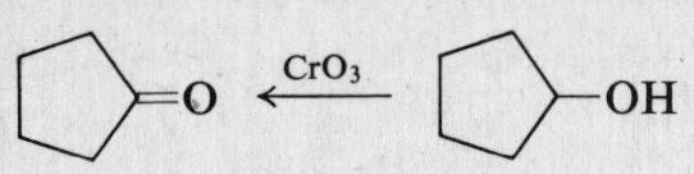

⬡=O $\xleftarrow{CrO_3}$ ⬡—OH

Ph_3P=$CHCH_3$ $\xleftarrow[base]{Ph_3P}$ $BrCH_2CH_3$ $\xleftarrow{PBr_3}$ EtOH

(e) $PhCH$=$CHCHO$ $\xleftarrow[-H_2O]{OH^-}$ $PhCHO$ (as in c) + CH_3CHO $\xleftarrow{K_2Cr_2O_7}$ EtOH

Ph_3P=$CHPh$ $\longleftarrow$ as in (c)

(f) CH_2=$CHCHO$ $\xleftarrow{heat}$ glycerol + $NaHSO_4$ (see Problem 27.5, p. 867)

Ph_3P=$\underset{CH_3}{C}COOMe$ $\xleftarrow[base]{Ph_3P}$ $Br\underset{CH_3}{C}HCOOMe$ $\xleftarrow{Br_2}{P}$ CH_3CH_2COOMe $\xleftarrow[H^+]{MeOH}$ CH_3CH_2COOH $\xleftarrow{KMnO_4}$ n-PrOH

21.26 $PhOCH_2Cl$ $\xrightarrow{Ph_3P}$ $PhOCH_2\overset{+}{P}Ph_3\ Cl^-$ $\xrightarrow{t\text{-}BuOK}$ $PhO\overset{H}{C}$=PPh_3

A

$PhO\overset{H}{C}$=PPh_3 + O=CCH_2CH_3 $\longrightarrow$ Ph_3PO + $PhO\overset{H}{C}$=$\underset{CH_3}{C}CH_2CH_3$

(with CH_3 below the central carbon of the ketone)

A   B

$PhO\overset{H}{C}$=$\underset{CH_3}{C}CH_2CH_3$ $\xrightarrow{H^+}$ $Ph\overset{+}{O}$=$\overset{H\ \ H}{C}$—$\underset{CH_3}{C}CH_2CH_3$ $\xrightarrow{H_2O}$ PhO—$\underset{HO\ \ CH_3}{C}$—CCH_2CH_3 (with H H above) $\longrightarrow$ O=$\overset{H}{C}$—$\underset{CH_3}{C}HCH_2CH_3$

B   Hemiacetal   C

A general rule: cleavage of vinyl ethers is particularly easy (compare Problem 15, Chapter 19, page 649).

The sequence is a general route to aldehydes.

21.27 (a) 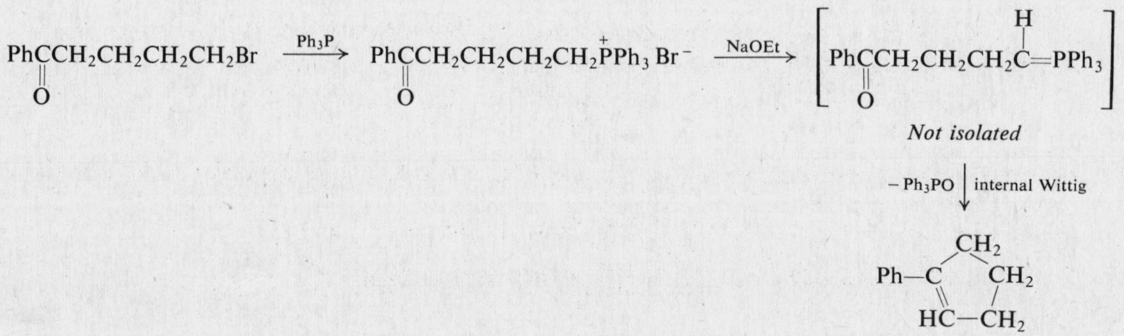 $Ph\underset{O}{C}CH_2CH_2CH_2CH_2Br$ $\xrightarrow{Ph_3P}$ $Ph\underset{O}{C}CH_2CH_2CH_2CH_2\overset{+}{P}Ph_3\ Br^-$ $\xrightarrow{NaOEt}$ $\left[Ph\underset{O}{C}CH_2CH_2CH_2\overset{H}{C}$=$PPh_3 \right]$

Not isolated

$-Ph_3PO$ $\downarrow$ internal Wittig

Ph—$C\overset{\displaystyle CH_2}{\underset{\displaystyle HC—CH_2}{\big\langle\ CH_2}}$

D

(b) $Ph_3P + Br(CH_2)_3Br + PPh_3 \longrightarrow Ph_3\overset{+}{P}(CH_2)_3\overset{+}{P}Ph_3 \xrightarrow{\text{base}}$

$Ph_3P=CCH_2C=PPh_3$

Br^- Br^-

E

double Wittig
$-2Ph_3PO$

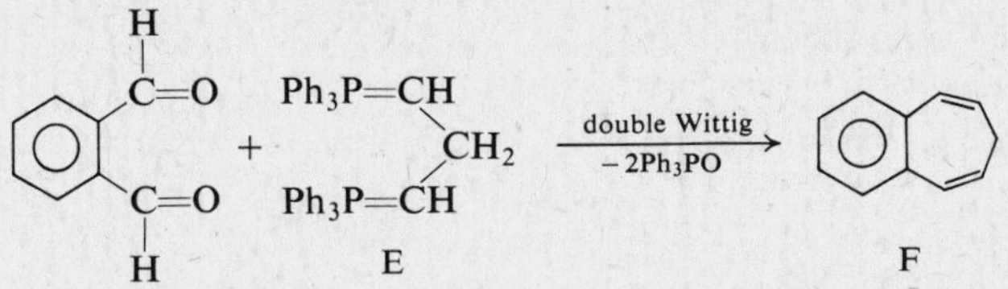

E   F

21.28

trans-2-Octene

$\xrightarrow{PhCO_2OH}$

and enantiomer
G

$\xrightarrow{Ph_2PLi}$

$\downarrow CH_3I$

cis-2-Octene

$+ Ph_2PO$
       $|$
      CH_3

$\xleftarrow{\textit{syn}\text{-elim.}}$

$\equiv$

and enantiomer
H

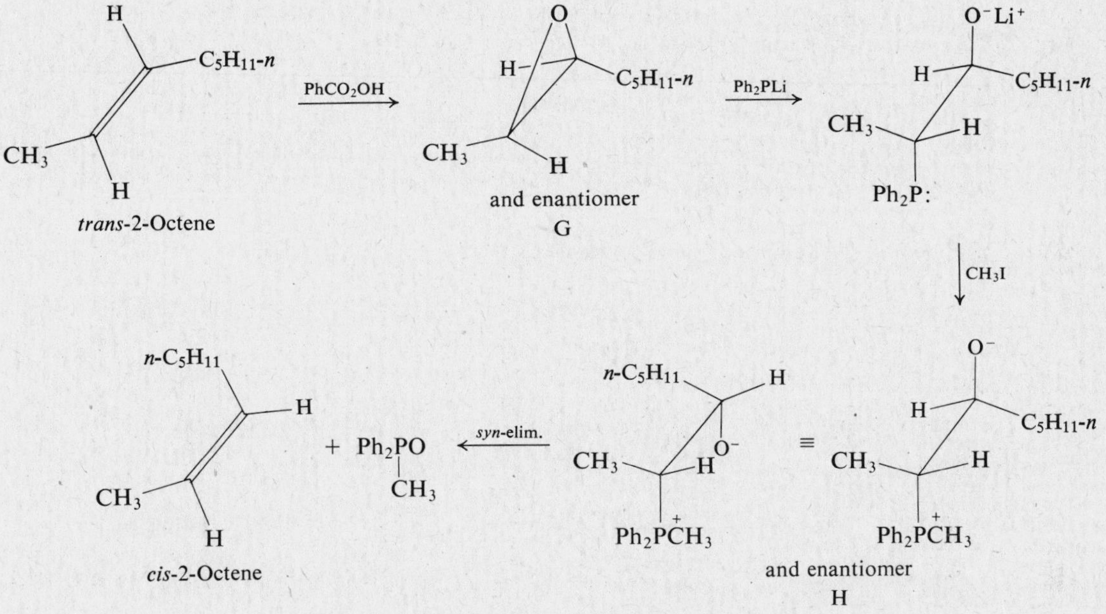

Here, a betaine is formed by nucleophilic attack on the epoxide by the basic and nucleophilic Ph_2P- group.

21.29 Equilibria (1) and (3) in the condensation (page 717) are reversed by EtOH. When the reaction is carried out in ether, the entire condensation proceeds further in the direction of product.

(For a review of the Claisen condensation, see C. R. Hauser and B. E. Hudson, Jr., "The Acetoacetic Ester Condensation and Certain Related Reactions," O. R. I–9.)

21.30 (a) *Dieckmann condensation:* intramolecular Claisen condensation leading to cyclization.

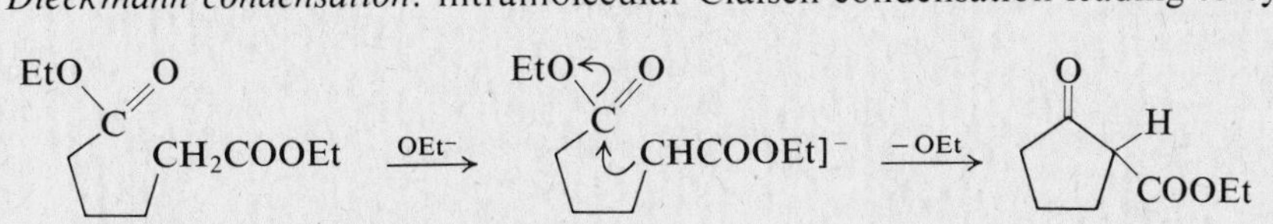

Ethyl adipate

$\xrightarrow{OEt^-}$

$\xrightarrow{-OEt}$

II

(b)

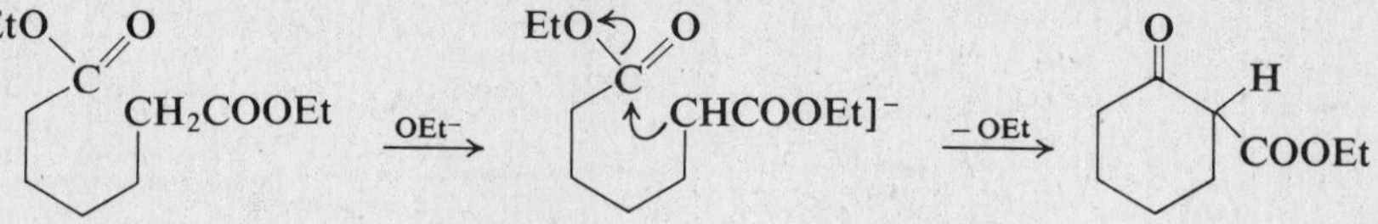

(c) No, because 3- and 4-membered rings are difficult to make.

The appearance of 12 carbons in the product shows that two molecules of ethyl succinate are involved; loss of $2C_2H_5OH$ suggests a double Claisen condensation.

$$
\begin{array}{ll}
C_{16}H_{28}O_8 & \text{2 ethyl succinate} \\
-C_{12}H_{16}O_6 & \text{product} \\
\hline
C_4H_{12}O_2 & \text{equivalent to } 2C_2H_5OH
\end{array}
$$

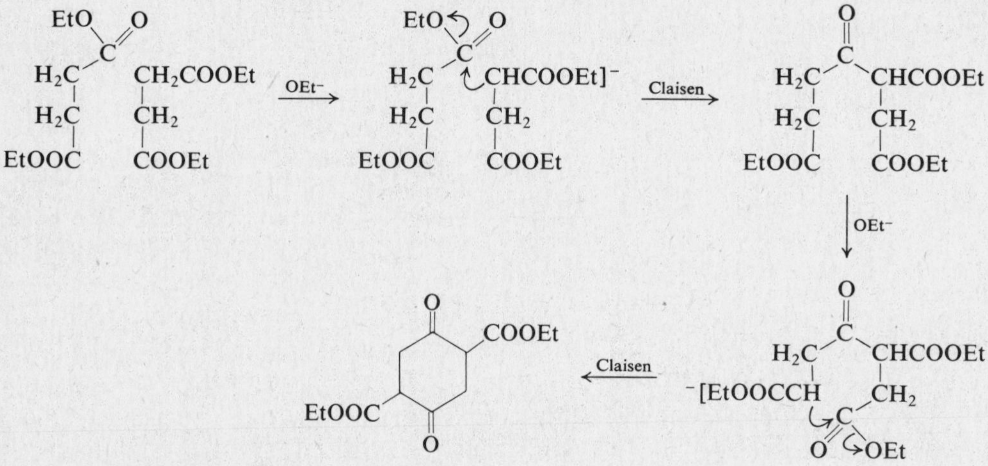

(For a review, see J. P. Schaefer and J. J. Bloomfield, "The Dieckmann Condensation," O. R. XV–1.)

21.31 PhCOOEt is mixed with NaOEt, and then CH_3COOEt is added slowly.
HCOOEt and NaOEt, then CH_3COOEt slowly.
Ethyl oxalate and NaOEt, then CH_3COOEt slowly.
Ethyl carbonate and NaOEt, then $PhCH_2COOEt$ slowly.

In every case, an ester without an α-H is mixed with the base, and then an ester carrying an α-H is added. The first ester cannot undergo self-condensation, and so is stable toward the base. When the second ester is added slowly, it is converted into an anion in the presence of a great deal of the first ester and very little of the second ester; for statistical reasons, then, it adds chiefly to the first ester to give a good yield of the crossed product.

21.32 (a) Replace anion I of page 717 by $^-CH_2COCH_3$, formed from acetone by the action of OEt^-.

(b) $CH_3CH_2C{-}OEt + {}^-CH_2COCH_3 \longrightarrow CH_3CH_2{-}C{-}CH_2{-}C{-}CH_3$

2,4-Hexanedione

(c) $PhC{-}OEt$ + $^-CH_2COPh$ $\longrightarrow$ $Ph{-}C{-}CH_2{-}C{-}Ph$

Dibenzoylmethane

(d)

$EtO{-}C{-}COOEt$ $\longrightarrow$

21.33 (a) $PhC{\vdots}CH{-}C{-}OEt$ $\xleftarrow{OEt^-}$ $PhCOOEt$ + $PhCH_2COOEt$

(b) $\xleftarrow{OEt^-}$ $EtOOC{-}COOEt$ + $EtOOCCH_2CH_2CH_2COOEt$

Ethyl oxalate   Ethyl glutarate

(c) $\xleftarrow{OEt^-}$ + CH_3COOEt

Ethyl phthalate

21.34 (a) $CH_3CH_2CH_2{\vdots}CH_2COOH$ $\xleftarrow{hydrol.}$ $\xleftarrow{H_2, Ni}$ $\xleftarrow{-H_2O}$ $CH_3CH_2CH{\vdots}CH_2COOEt$
$\qquad\qquad\qquad\qquad\qquad\qquad\qquad\qquad\qquad\qquad\qquad\qquad\qquad\qquad$ $\overset{|}{OH}$

$\qquad\qquad\qquad\qquad\qquad\qquad\qquad\qquad\qquad\qquad\qquad\qquad\qquad\qquad\qquad\qquad\quad$ $\uparrow$ Zn

$\qquad\qquad\qquad\qquad\qquad\qquad\qquad\qquad\qquad\qquad\qquad$ $BrCH_2COOEt$ + CH_3CH_2CHO

(b) $(CH_3)_2CHCH_2{\vdots}CH(CH_3)COOH$ $\xleftarrow{hydrol.}$ $\xleftarrow{H_2, Ni}$ $\xleftarrow{-H_2O}$ $(CH_3)_2CHCH{\vdots}CH(CH_3)COOEt$
$\qquad\qquad\qquad\qquad\qquad\qquad\qquad\qquad\qquad\qquad\qquad\qquad\qquad\qquad\qquad\qquad\qquad\qquad$ $\overset{|}{OH}$

$\qquad\qquad\qquad\qquad\qquad\qquad\qquad\qquad\qquad\qquad\qquad\qquad\qquad\qquad\qquad\qquad\qquad\qquad\quad$ $\uparrow$ Zn

$\qquad\qquad\qquad\qquad\qquad\qquad\qquad\qquad\qquad\qquad\qquad\quad$ $CH_3CHBrCOOEt$ + $(CH_3)_2CHCHO$

(c) $PhCH{=}CHCOOH$ $\xleftarrow{hydrol.}$ $\xleftarrow{-H_2O}$ $PhCH{\vdots}CH_2COOEt$ $\xleftarrow{Zn}$ $PhCHO$ + $BrCH_2COOEt$
$\qquad\qquad\qquad\qquad\qquad\qquad\qquad\qquad\qquad\qquad\quad$ $\overset{|}{OH}$

(d) $PhCH_2{\vdots}CH(CH_3)COOH$ $\xleftarrow{hydrol.}$ $\xleftarrow{H_2, Ni}$ $\xleftarrow{-H_2O}$ $PhCH{\vdots}CH(CH_3)COOEt$
$\qquad\qquad\qquad\qquad\qquad\qquad\qquad\qquad\qquad\qquad\qquad\qquad\qquad\qquad$ $\overset{|}{OH}$

$\qquad\qquad\qquad\qquad\qquad\qquad\qquad\qquad\qquad\qquad\qquad\qquad\qquad\qquad\quad$ $\uparrow$ Zn

$\qquad\qquad\qquad\qquad\qquad\qquad\qquad\qquad\qquad\qquad$ $CH_3CHBrCOOEt$ + $PhCHO$

In (a), (b), and (d), dehydration of the β-hydroxy ester initially formed is followed by ester hydrolysis and then catalytic hydrogenation.

(For a review, see R. L. Shriner, "The Reformatsky Reaction," O. R. I–1.)

21.35 (a) PhCH₂CH₂COOH $\xleftarrow{\text{hydrol.}}$ $\xleftarrow{\text{H}_2, \text{Ni}}$ $\xleftarrow{-\text{H}_2\text{O}}$ $\left[\begin{array}{c} \text{PhCHCH}_2\text{COOH} \\ | \\ \text{OH} \end{array} \right]$ $\xleftarrow{\text{Zn}}$ PhCHO + BrCH₂COOEt

(b) PhCH₂CH₂CHO $\xleftarrow{\text{LiAlH(OBu-}t)_3}$ PhCH₂CH₂COCl $\xleftarrow{\text{SOCl}_2}$ PhCH₂CH₂COOH ⟵ (a)

(c) PhCH₂CH₂CH₂CH₂COOH $\xleftarrow{\text{hydrol.}}$ $\xleftarrow{\text{H}_2, \text{Ni}}$ $\xleftarrow{-\text{H}_2\text{O}}$ PhCH₂CH₂CH—CH₂COOEt
$\qquad\qquad\qquad\qquad\qquad\qquad\qquad\qquad\qquad\qquad\qquad\qquad\qquad\quad$ |
$\qquad\qquad\qquad\qquad\qquad\qquad\qquad\qquad\qquad\qquad\qquad\qquad\qquad\quad$ OH
$\qquad\qquad\qquad\qquad\qquad\qquad\qquad\qquad\qquad\qquad\qquad\qquad\qquad\qquad\uparrow$

$\qquad\qquad\qquad\qquad\qquad$ (b) ⟶ PhCH₂CH₂CHO + BrZnCH₂COOEt

21.36 EtOOC—C—OEt + CH₃COOEt $\xrightarrow{\text{OEt}^-}$ EtOOC—C—CH₂COOEt
$\qquad\qquad\quad$ ‖ $\qquad\qquad\qquad\qquad\qquad\qquad\qquad\qquad\qquad\quad$ ‖
$\qquad\qquad\quad$ O $\qquad\qquad\qquad\qquad\qquad\qquad\qquad\qquad\qquad\quad$ O
$\qquad\qquad\qquad\qquad\qquad\qquad\qquad\qquad\qquad\qquad\qquad\qquad\quad$ A

COOEt $\qquad\qquad\qquad\qquad\qquad\qquad\qquad\qquad\qquad$ COOEt
| $\qquad\qquad\qquad\qquad\qquad\qquad\qquad\qquad\qquad\qquad\quad$ |
C=O $\qquad$ + BrZnCH₂COOEt ⟶ HO—C—CH₂COOEt
| $\qquad\qquad\qquad\qquad\qquad\qquad\qquad\qquad\qquad\qquad\qquad$ |
CH₂COOEt $\qquad\qquad\qquad\qquad\qquad\qquad\qquad\qquad$ CH₂COOEt
$\quad$ A $\qquad\qquad\qquad\qquad\qquad\qquad\qquad\qquad\qquad\qquad\quad$ B

$\qquad\qquad\qquad\qquad\qquad\qquad\qquad\qquad$ COOH
$\qquad\qquad\qquad\qquad\qquad\qquad\qquad\qquad\qquad$ |
B $\xrightarrow{\text{OH}^-}$ $\xrightarrow{\text{H}^+}$ HO—C—CH₂COOH + 3EtOH
$\qquad\qquad\qquad\qquad\qquad\qquad\qquad\qquad\qquad$ |
$\qquad\qquad\qquad\qquad\qquad\qquad\qquad$ CH₂COOH
$\qquad\qquad\qquad\qquad\qquad\qquad\quad$ C, citric acid

1. (a) PhCH₂C=O + [PhCHCHO]⁻ $\xrightarrow{\text{aldol cond.}}$ PhCH₂CH—CHCHO
$\qquad\qquad\quad$ | $\qquad\qquad\qquad\qquad\qquad\qquad\qquad\qquad\qquad\qquad\qquad$ |
$\qquad\qquad\quad$ H $\qquad\qquad\qquad\qquad\qquad\qquad\qquad\qquad\qquad\qquad$ OH
$\qquad\qquad\qquad\qquad\qquad\qquad\qquad\qquad\qquad\qquad\qquad\qquad\qquad\qquad\qquad\quad$ Ph

(b) As in (a), then PhCH₂CH—C—CHO $\xrightarrow[-\text{H}_2\text{O}]{\text{H}^+}$ PhCH₂CH=C—C=O
$\qquad\qquad\qquad\qquad\qquad\qquad\qquad$ | $\quad$ | $\qquad\qquad\qquad\qquad\qquad\qquad\qquad$ | $\quad$ |
$\qquad\qquad\qquad\qquad\qquad\qquad\quad$ OH $\,$ H $\qquad\qquad\qquad\qquad\qquad\qquad\qquad$ Ph $\,$ H

(c) Same as (a)

(d) $PhCH_2CHO$ $\xrightarrow{Br_2,\ CCl_4}$ $PhCHCHO$
$\qquad\qquad\qquad\qquad\qquad\quad |$
$\qquad\qquad\qquad\qquad\qquad\ Br$

(e) $PhCH_2\overset{\displaystyle H}{\underset{\displaystyle\parallel}{C}}{=}O + Ph_3P{=}CH_2$ $\xrightarrow{-\ Ph_3PO}$ $PhCH_2CH{=}CH_2$

2. (a) 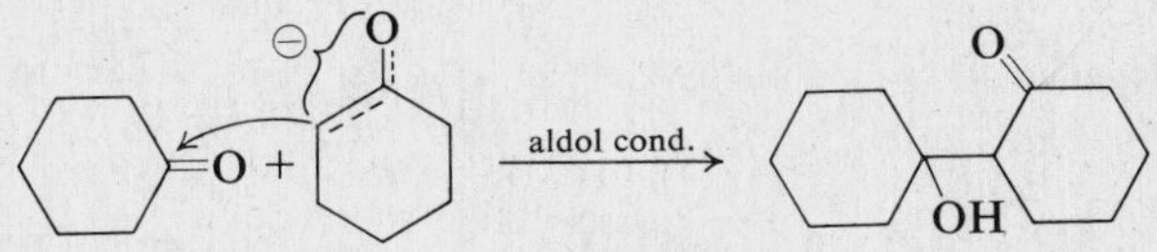 $\xrightarrow{\text{aldol cond.}}$

(b) As in (a), then $\xrightarrow[-H_2O]{H^+}$

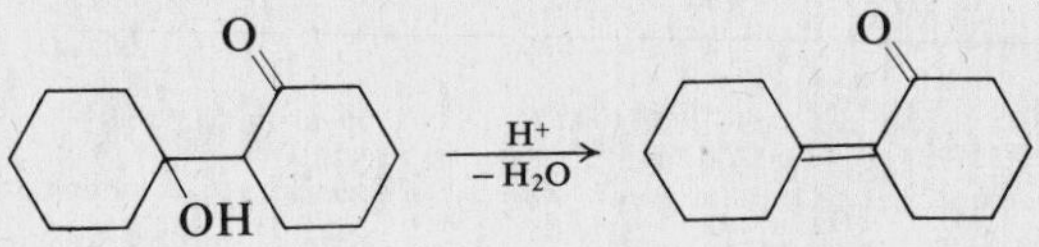

(c) Same as (a).

(d) 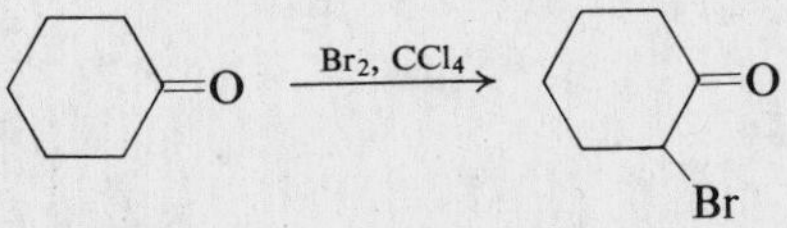 $\xrightarrow{Br_2,\ CCl_4}$

(e) 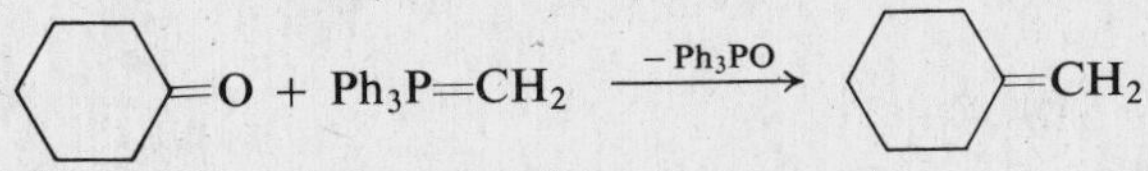 $O + Ph_3P{=}CH_2$ $\xrightarrow{-\ Ph_3PO}$ $={=}CH_2$ (see page 705)

3. (a) No reaction

(b) Cannizzaro reaction: $2PhCHO$ $\xrightarrow{OH^-}$ $PhCH_2OH + PhCOO^-Na^+$

(c) $Ph\overset{\displaystyle H}{\underset{\displaystyle\parallel}{C}}{=}O + CH_3{-}CHO$ $\xrightarrow{OH^-}$ $Ph\overset{\displaystyle H}{C}{=}\overset{\displaystyle H}{C}{-}\overset{\displaystyle H}{C}{=}O$
Cinnamaldehyde

(d) $Ph\overset{\displaystyle H}{\underset{\displaystyle\parallel}{C}}{=}O + \underset{\displaystyle CH_3}{CH_2}{-}CHO$ $\xrightarrow{OH^-}$ $Ph\overset{\displaystyle H}{C}{=}\underset{\displaystyle CH_3}{C}{-}\overset{\displaystyle H}{C}{=}O$

(e) $Ph\overset{\displaystyle H}{\underset{\displaystyle\parallel}{C}}{=}O + CH_3{-}\underset{\displaystyle\underset{\displaystyle O}{\parallel}}{C}{-}CH_3$ $\xrightarrow{OH^-}$ $Ph\overset{\displaystyle H}{C}{=}\overset{\displaystyle H}{C}{-}\underset{\displaystyle\underset{\displaystyle O}{\parallel}}{C}{-}CH_3$

Benzalacetone (*Benzal* is $PhCH{=}$)

(f)
$$\text{PhC}=\text{C}-\text{C}-\text{CH}_3 + \text{O}=\text{CPh} \xrightarrow{\text{OH}^-} \text{PhC}=\text{C}-\text{C}-\text{C}=\text{CPh}$$

Product (e)   Dibenzalacetone

(g)
$$\text{PhC}=\text{O} + \text{CH}_3-\text{C}-\text{Ph} \xrightarrow{\text{OH}^-} \text{PhC}=\text{C}-\text{C}-\text{Ph}$$

Benzalacetophenone
(Chalcone)

(h) Perkin reaction:
$$\text{PhC}=\text{O} + \text{CH}_3-\text{C} \underset{\text{CH}_3\text{C}}{\overset{\text{O}}{\Big<}}\text{O} \xrightarrow{\text{OAc}^-} \xrightarrow{\text{hydrolysis}} \text{PhC}=\text{C}-\text{C} \overset{\text{O}}{\underset{\text{OH}}{\Big<}}$$

Cinnamic acid

(i)
$$\text{PhC}=\text{O} + \text{CH}_3-\text{C} \overset{\text{O}}{\underset{\text{OEt}}{\Big<}} \xrightarrow{\text{OEt}^-} \text{PhC}=\text{C}-\text{C} \overset{\text{O}}{\underset{\text{OEt}}{\Big<}}$$

Ethyl cinnamate

(j)
$$\text{PhC}=\text{O} + \underset{\text{Ph}}{\text{CH}_2}-\text{C} \overset{\text{O}}{\underset{\text{OEt}}{\Big<}} \xrightarrow{\text{OEt}^-} \text{PhC}=\underset{\text{Ph}}{\text{C}}-\text{C} \overset{\text{O}}{\underset{\text{OEt}}{\Big<}}$$

(k) Crossed Cannizzaro: $\text{PhCHO} + \text{HCHO} \xrightarrow{\text{OH}^-} \text{PhCH}_2\text{OH} + \text{HCOO}^-\text{Na}^+$

(l)
$$\text{PhC}=\text{O} + \text{CH}_3\text{CH}=\text{CHCHO} \xrightarrow{\text{OH}^-} \text{PhCH}=\text{CH}-\text{CH}=\text{CH}-\text{CHO}$$

(Compare Problem 21.20, page 713.)

(m)
$$\text{PhC}=\text{O} + \text{Ph}_3\text{P}=\text{CHCH}=\text{CH}_2 \xrightarrow{-\text{Ph}_3\text{PO}} \text{PhCH}=\text{CH}-\text{CH}=\text{CH}_2$$

1-Phenyl-1,3-butadiene

(n)
$$\text{PhC}=\text{O} + \text{Ph}_3\text{P}=\text{C}-\text{OPh} \xrightarrow{-\text{Ph}_3\text{PO}} \text{PhC}=\text{C}-\text{OPh}$$

(o)
$$\text{PhC}=\text{C}-\text{OPh} \xrightarrow[\text{H}_2\text{O}]{\text{H}^+} \text{PhCH}_2\text{CHO}$$

(from n)   Phenylacetaldehyde

Vinyl ethers are easily cleaved; compare Problem 15, Chapter 19, page 649.

4. (a) $CH_3CH_2CH-CHCHO$ $\xleftarrow{OH^-}$ $2CH_3CH_2CHO$
                      | |
                     OH CH_3

(b) $CH_3CH_2CH_2CHCH_2OH$ $\xleftarrow{H_2,\ Ni}$ $CH_3CH_2CH=CCHO$ $\xleftarrow{-H_2O}$ product (a)
                | |
               CH_3 CH_3

(c) $CH_3CH_2CH=CCHO$ $\xleftarrow{-H_2O}$ product (a)
                 |
                CH_3

(d) $CH_3CH_2CH=CCH_2OH$ $\xleftarrow{NaBH_4}$ product (c)
                  |
                 CH_3

(e) $CH_3CH_2CH-CHCH_2OH$ $\xleftarrow{H_2,\ Ni}$ product (a)
                 | |
                OH CH_3

(f) $CH_3CH_2CH_2CHCOOH$ $\xleftarrow[heat]{KMnO_4}$ product (b)
                  |
                 CH_3

(g) $PhCH=CCHO$ $\xleftarrow[-H_2O]{OH^-}$ $PhCHO + CH_3CH_2CHO$
            |
         CH_3

(h) CH_3CD_2CHO $\longleftarrow$ $CH_3CH_2CHO + D_2O + OD^-$

Acidic α-hydrogen easily exchanged with deuterium.

(i) $CH_3CH_2CH^{18}O$ $\longleftarrow$ $CH_3CH_2CHO + H_2^{18}O + H^+$

Compare Problem 19.13, page 643.

                                H
                                |
(j) $CH_3CH_2CH=CHCH(CH_3)_2$ $\xleftarrow{Wittig}$ $CH_3CH_2C=O + Ph_3P=CHCH(CH_3)_2$ (from *iso*-BuBr)

5. (a) Need to remove 1C: try haloform reaction (Sec. 16.11).

$PhCOOH + CHI_3$ $\xleftarrow{H^+}$ $\xleftarrow{I_2,\ OH^-}$ $Ph-C-CH_3$
                                                ||
                                              O

         Ph
         |
(b) $CH_3-C=CH-C-Ph$ $\xleftarrow[-H_2O]{OH^-}$ $2PhCOCH_3$
                   ||
                   O

         Ph
         |
(c) $CH_3-CH-CH_2-CH-Ph$ $\xleftarrow{H_2,\ Ni}$ product (b)
                        |
                        OH

(d) 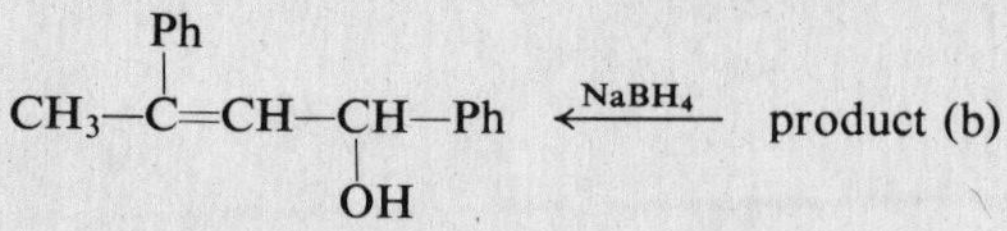　$\underset{\underset{OH}{|}}{\overset{\overset{Ph}{|}}{CH_3-C=CH-CH}}-Ph \xleftarrow{NaBH_4}$　product (b)

(e)　$PhCH=CHCOPh \xleftarrow[-H_2O]{OH^-} PhCHO + CH_3COPh$

(f) 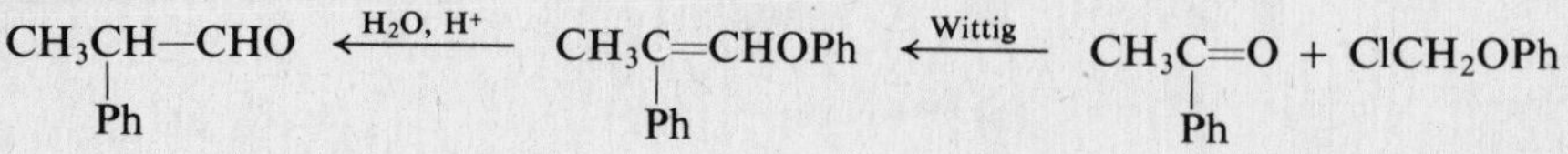　$\underset{\underset{Ph}{|}}{CH_3CH-CHO} \xleftarrow{H_2O,\ H^+} \underset{\underset{Ph}{|}}{CH_3C=CHOPh} \xleftarrow{Wittig} \underset{\underset{Ph}{|}}{CH_3C=O} + ClCH_2OPh$

6. These are all Claisen condensations, and give β-dicarbonyl products.

(a) 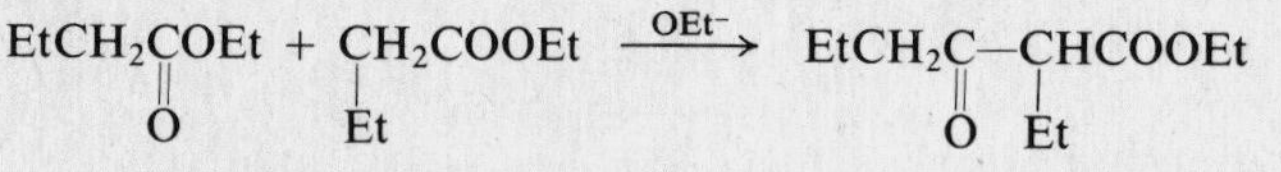　$\underset{\underset{O}{||}}{EtCH_2COEt} + \underset{\underset{Et}{|}}{CH_2COOEt} \xrightarrow{OEt^-} \underset{\underset{O}{||}\ \underset{Et}{|}}{EtCH_2C-CHCOOEt}$

(b) 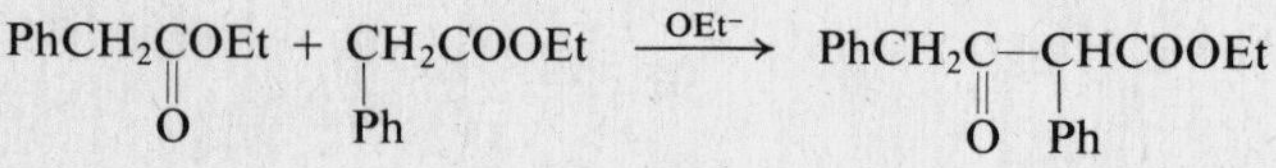　$\underset{\underset{O}{||}}{PhCH_2COEt} + \underset{\underset{Ph}{|}}{CH_2COOEt} \xrightarrow{OEt^-} \underset{\underset{O}{||}\ \underset{Ph}{|}}{PhCH_2C-CHCOOEt}$

(c) 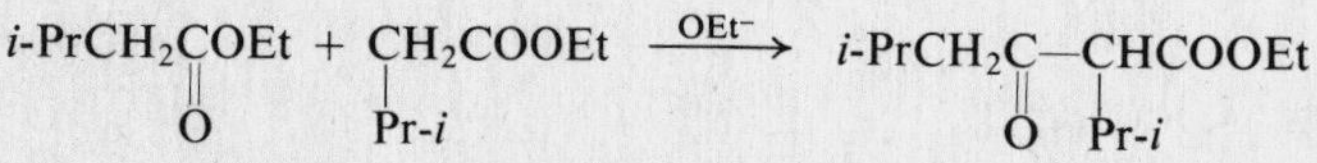　$\underset{\underset{O}{||}}{i\text{-}PrCH_2COEt} + \underset{\underset{Pr\text{-}i}{|}}{CH_2COOEt} \xrightarrow{OEt^-} \underset{\underset{O}{||}\ \underset{Pr\text{-}i}{|}}{i\text{-}PrCH_2C-CHCOOEt}$

(d) 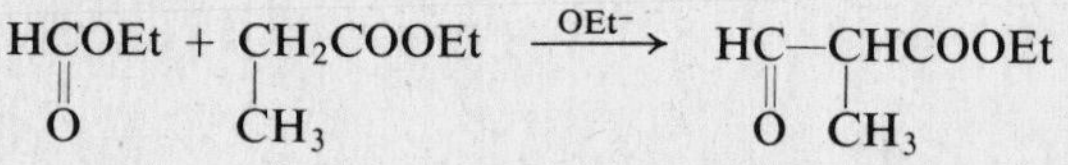　$\underset{\underset{O}{||}}{HCOEt} + \underset{\underset{CH_3}{|}}{CH_2COOEt} \xrightarrow{OEt^-} \underset{\underset{O}{||}\ \underset{CH_3}{|}}{HC-CHCOOEt}$

　　　　　　　　　　　　　(known only as the sodio derivative)

(e)　Ethyl oxalate is a diester, and can take part in two successive Claisen condensations:

$\underset{\underset{O}{||}}{EtOOC-COEt} + \underset{\underset{CH_2COOEt}{|}}{CH_2COOEt} \xrightarrow{OEt^-} \underset{\underset{O}{||}\ \underset{CH_2COOEt}{|}}{EtOOC-C-CHCOOEt}$

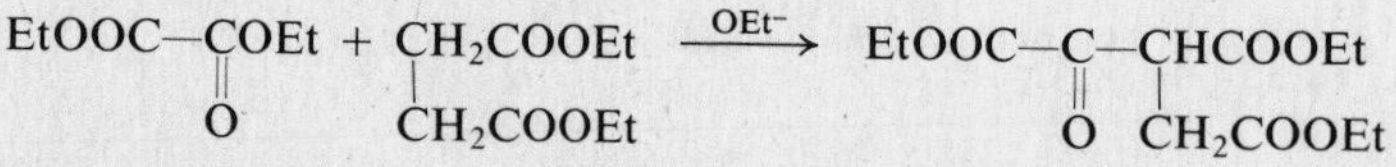

Ethyl oxalate　　Ethyl succinate

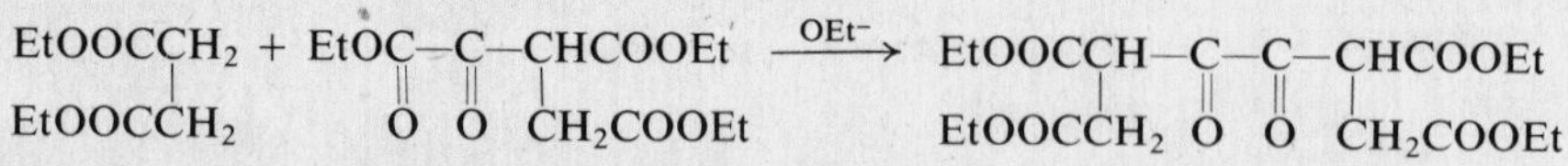

$\underset{\underset{EtOOCCH_2}{|}}{EtOOCCH_2} + \underset{\underset{O}{||}\ \underset{O}{||}\ \underset{CH_2COOEt}{|}}{EtOC-C-CHCOOEt} \xrightarrow{OEt^-} \underset{\underset{EtOOCCH_2}{|}\ \underset{O}{||}\ \underset{O}{||}\ \underset{CH_2COOEt}{|}}{EtOOCCH-C-C-CHCOOEt}$

Ethyl succinate　　Product of first step　　　　　　An α-diketone
Second mole

The initial product does not undergo an intramolecular Claisen condensation (Dieckmann condensation), evidently because of a reluctance to form a four-membered ring.

　　The crossed Claisen condensation of ethyl oxalate is the first step in a useful route to α-keto acids (see Problem 26.9, page 853) and thence to α-amino acids (See Problem 36.12, page 1140).

(f) 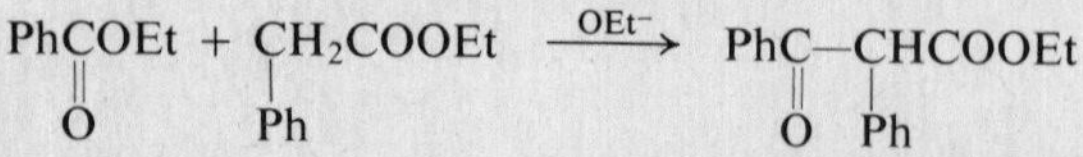　$\underset{\underset{O}{||}}{PhCOEt} + \underset{\underset{Ph}{|}}{CH_2COOEt} \xrightarrow{OEt^-} \underset{\underset{O}{||}\ \underset{Ph}{|}}{PhC-CHCOOEt}$

(g)   EtĊOEt + ⬡(C=O) $\xrightarrow{\text{OEt}^-}$ Et—C—⬡(C=O)
‖   ‖
O   O

(h) $PhCH_2COEt + CH_3CPh \xrightarrow{\text{OEt}^-} PhCH_2C—CH_2CPh$
         ‖   ‖   ‖   ‖
         O   O   O   O

(i) Ethyl carbonate, too, is a diester, and can take part in two successive Claisen condensations:

$$EtOCOEt + CH_3CPh \xrightarrow{\text{OEt}^-} EtOC—CH_2CPh$$
     ‖   ‖   ‖   ‖
     O   O   O   O

Ethyl carbonate

$$PhC—CH_3 + EtOC—CH_2CPh \xrightarrow{} PhCCH_2—C—CH_2CPh$$
     ‖   ‖   ‖   ‖   ‖   ‖
     O   O   O   O   O   O

Acetophenone   Product of   A triketone
Second mole   first step

    Ethyl carbonate is useful in synthetic work for the introduction of a carbethoxy group, —COOEt, as in the first step here. (See another example on page 719.)

7. (a) $CH_3C—CH_2COOEt$ $CH_3C—CHCOOEt$ $CH_3CH_2C—CHCOOEt$
         ‖   ‖ | ‖ |
         O   O CH_3 O CH_3

(b) No. It would give a poor yield of any one of them, contaminated by the other two.

8.

(a) $PhC—C—COEt \xleftarrow{\text{OEt}^-}$ ⎰ $\xrightarrow{a}$ $PhCOOEt + CH_3CH_2COOEt$
                                         ⎱ $\xrightarrow{b}$ $PhCOCH_2CH_3 + EtOCOEt$
                                                           ‖
                                                           O
                                                           Ethyl carbonate

Structure (a): PhC⦙C⦙COEt with CH₃ group above the central C (labeled a), H below the central C, and O double-bonds on the PhC and COEt carbons (labeled b).

(b) $PhCH_2C⦙C—COEt \xleftarrow{\text{OEt}^-} PhCH_2COOEt + PhCH_2COOEt$
             |
             Ph (above), H and O below

(c) $EtOOC—C⦙C—COEt \xleftarrow{\text{OEt}^-} EtOOC—COOEt + CH_3CH_2COOEt$
                  CH₃ (above), O H O (below)   Ethyl oxalate   *One mole*

(d) 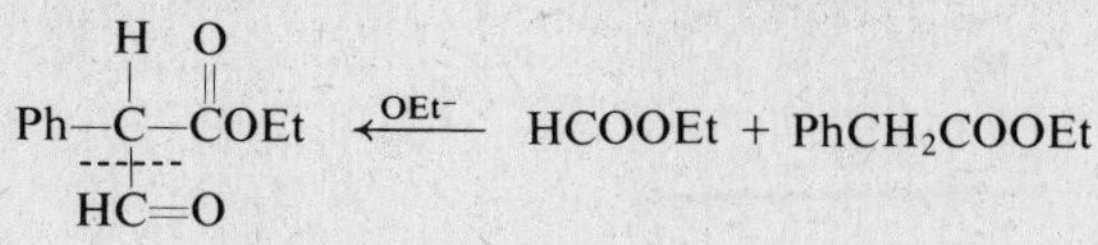 Ph—C—COEt $\xleftarrow{\text{OEt}^-}$ HCOOEt + PhCH$_2$COOEt

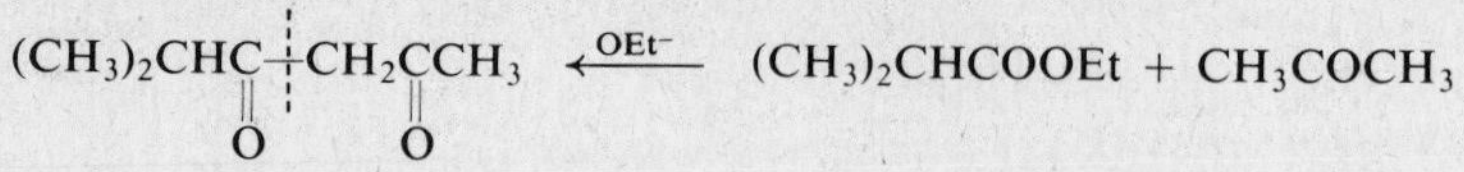

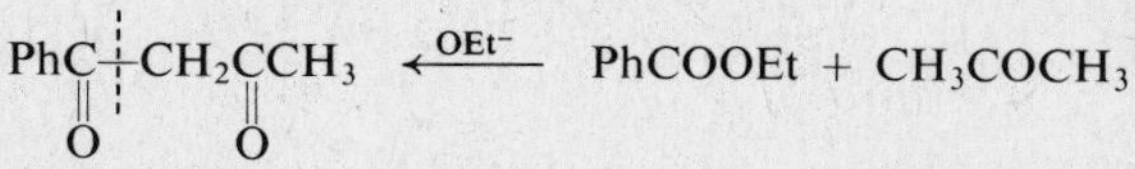

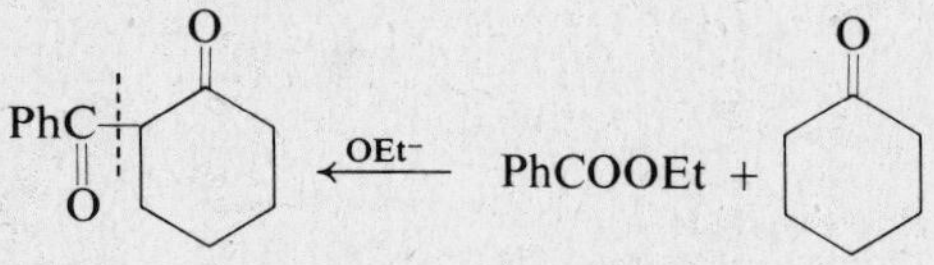

with structure showing Ph—C(H)(COEt with O)—HC=O

(e) (CH$_3$)$_2$CHC—CH$_2$CCH$_3$ $\xleftarrow{\text{OEt}^-}$ (CH$_3$)$_2$CHCOOEt + CH$_3$COCH$_3$

with O and O below

(f) PhC—CH$_2$CCH$_3$ $\xleftarrow{\text{OEt}^-}$ PhCOOEt + CH$_3$COCH$_3$

with O and O below

(g) PhC—(cyclohexanone ring) $\xleftarrow{\text{OEt}^-}$ PhCOOEt + (cyclohexanone)

with O below PhC and O on ring

(h) EtOC—CHCH$_2$COEt $\xleftarrow{\text{OEt}^-}$ HCOOEt + EtOOCCH$_2$CH$_2$COOEt

with O and O, and HC=O below

Ethyl formate   Ethyl succinate

One mole

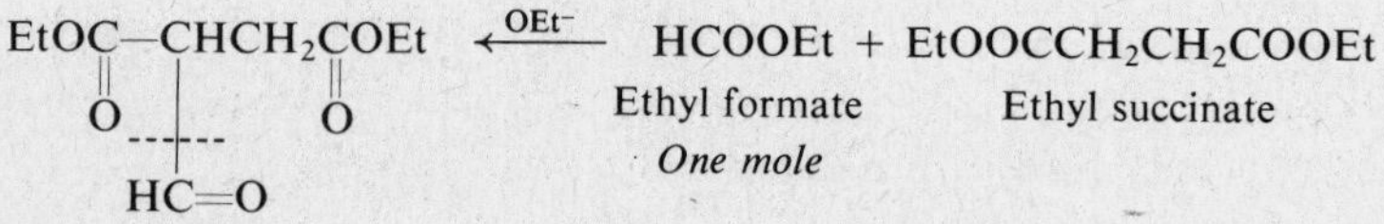

9. Remember that a *cis*-isomer can often be made by hydrogenation of a triple bond over Lindlar's or Brown's catalyst (Sec. 8.9).

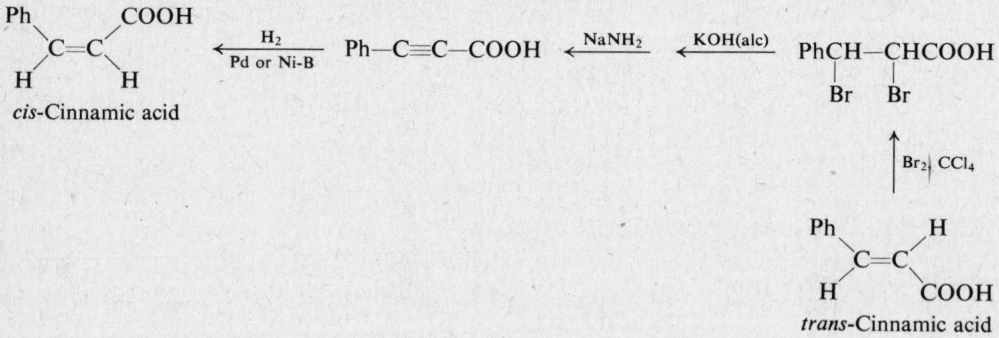

Ph/H C=C COOH/H (*cis*-Cinnamic acid) $\xleftarrow[\text{Pd or Ni-B}]{\text{H}_2}$ Ph—C≡C—COOH $\xleftarrow{\text{NaNH}_2}$ $\xleftarrow{\text{KOH(alc)}}$ PhCH—CHCOOH (with Br Br)

$\uparrow$ Br$_2$, CCl$_4$

Ph/H C=C H/COOH

trans-Cinnamic acid

10. (a) 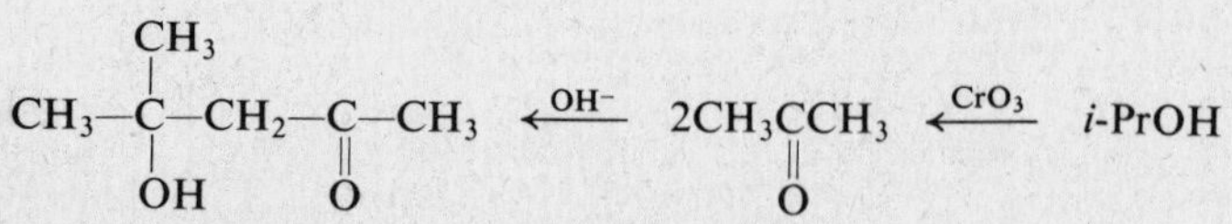 CH$_3$—C(CH$_3$)(OH)—CH$_2$—C(=O)—CH$_3$ $\xleftarrow{\text{OH}^-}$ 2CH$_3$CCH$_3$ (with O) $\xleftarrow{\text{CrO}_3}$ *i*-PrOH

(b) CH$_3$CH(CH$_3$)—CH$_2$CHCH$_3$ (with OH) $\xleftarrow{\text{H}_2, \text{Ni}}$ CH$_3$C(CH$_3$)=CHCCH$_3$ (with O) $\xleftarrow{-\text{H}_2\text{O}}$ CH$_3$CCH$_2$CCH$_3$ (with CH$_3$, OH and O) $\longleftarrow$ (a)

(c) CH$_3$CH=CHCHO $\xleftarrow{-\text{H}_2\text{O}}$ CH$_3$CH—CH$_2$CHO (with OH) $\xleftarrow{\text{OH}^-}$ 2CH$_3$CHO (from EtOH)

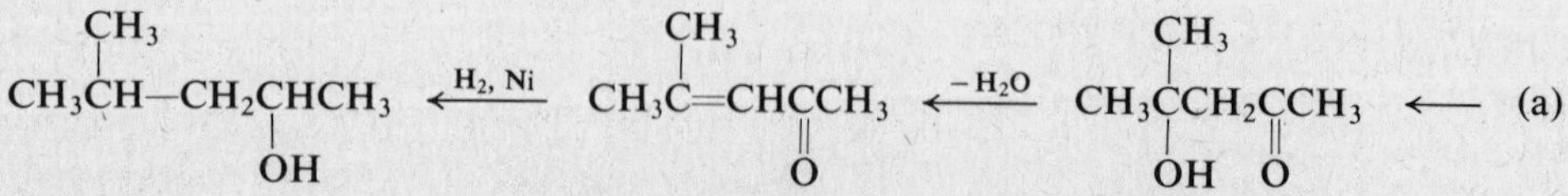

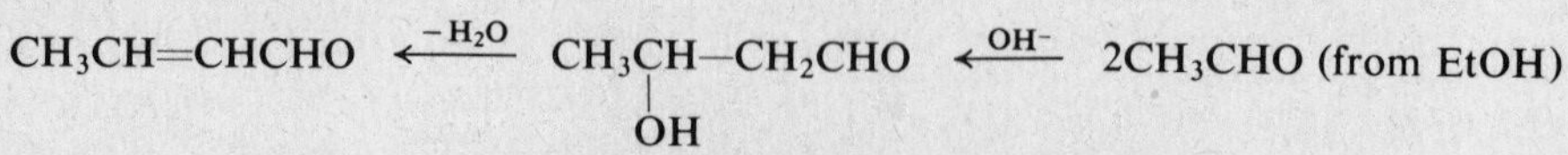

(d) $PhCH=CHCH_2OH$ $\xleftarrow{NaBH_4}$ $PhCH=CHCHO$ $\xleftarrow{-H_2O}$ $\left[\begin{array}{c} PhCH-CH_2CHO \\ | \\ OH \end{array}\right]$ $\longleftarrow$ $PhCHO$ + CH_3CHO
 (from (from
 $PhCH_3$) EtOH)

(e) $NO_2\langle\bigcirc\rangle CH=CHCHO$ $\xleftarrow{-H_2O}$ $\left[NO_2\langle\bigcirc\rangle\begin{array}{c} CH-CH_2CHO \\ | \\ OH \end{array}\right]$ $\xleftarrow{OH^-}$ $NO_2\langle\bigcirc\rangle CHO$ + CH_3CHO
 (page 622) (from EtOH)

(f) $\begin{array}{c} CH_3CH-CH_2CH_2 \\ | | \\ OH OH \end{array}$ $\xleftarrow{H_2,\ Ni}$ $\begin{array}{c} CH_3CH-CH_2CHO \\ | \\ OH \end{array}$ $\xleftarrow{OH^-}$ $2CH_3CHO$ (from EtOH)

(g) $CH_3\overset{\overset{\displaystyle CH_3}{|}}{C}=CHCOOH$ $\xleftarrow{OX^-}$ $CH_3\overset{\overset{\displaystyle CH_3}{|}}{C}=CH-\overset{\underset{\displaystyle O}{\|}}{C}-CH_3$ $\longleftarrow$ as in (b)

(h) $CH_3\overset{\overset{\displaystyle CH_3}{|}}{C}=CHCOOH$ $\xleftarrow{hydrol.}$ $\xleftarrow{-H_2O}$ $CH_3\overset{\overset{\displaystyle CH_3}{|}}{\underset{\underset{\displaystyle OH}{|}}{C}}-CH_2COOEt$ $\xleftarrow{Zn}$ CH_3COCH_3 + $BrCH_2COOEt$
 (from i-PrOH) (from EtOH, below)

$BrCH_2COOEt$ $\xleftarrow{EtOH,\ H^+}$ $BrCH_2COOH$ $\xleftarrow{Br_2,\ P}$ CH_3COOH
 (from EtOH)

or

$CH_3\overset{\overset{\displaystyle CH_3}{|}}{C}=CHCOOH$ $\xleftarrow[-H_2O]{H_2O,\ H^+}$ $CH_3-\overset{\overset{\displaystyle CH_3}{|}}{C}H-\overset{\underset{\underset{\displaystyle OH}{|}}{}}{C}H-CN$ $\xleftarrow{CN^-}$ $CH_3\overset{\overset{\displaystyle CH_3}{|}}{C}HCHO$
 (from i-BuOH)

(i) $CH_3CH_2\overset{\overset{\displaystyle CH_3}{|}}{\underset{\underset{\displaystyle OH}{|}}{C}}-C\equiv CH$ $\longleftarrow$ $CH_3CH_2\overset{\overset{\displaystyle CH_3}{|}}{C}=O$ + $NaC\equiv CH$
 (from sec-BuOH) (from acetylene)

(j) $PhCH=CHCH=CHCH=CH_2$ $\xleftarrow{-Ph_3PO}$ $\left\{\begin{array}{l} PhCH=CHCHO \\ \text{(page 713)} \\ \\ Ph_3P=CH-CH=CH_2 \xleftarrow[base]{Ph_3P} ClCH_2CH=CH_2 \end{array}\right.$

 $Cl_2,\ \uparrow heat$

 $CH_3CH=CH_2$
 (from i-PrOH)

(k) $PhCH=CHCH=CHCH=CHPh$ $\xleftarrow{-Ph_3PO}$ $\left\{\begin{array}{l} Ph_3P=CHPh \xleftarrow[base]{Ph_3P} ClCH_2Ph \\ \text{(from } PhCH_3) \\ \\ PhCH=CHCH=CHCHO \xleftarrow{OH^-} PhCH=CHCHO \\ \text{(page 713)} \\ + \\ CH_3CHO \\ \text{(from EtOH)} \end{array}\right.$

(l) $\underset{\text{C}_2\text{H}_5\text{C}=\overset{\text{H}_3\text{C}\quad\text{CH}_3}{\text{CCOOH}}}{}$ $\xleftarrow{\text{hydrol.}}$ $\xleftarrow{-\text{H}_2\text{O}}$ $\underset{\underset{\text{OH}}{|}}{\text{C}_2\text{H}_5\overset{\text{H}_3\text{C}\quad\text{CH}_3}{\underset{|}{\text{C}}}{-\text{CHCOOH}}}$ $\xleftarrow{\text{Zn}}$ $\underset{\text{(from }sec\text{-BuOH)}}{\text{C}_2\text{H}_5\overset{\text{CH}_3}{\underset{}{-}}\text{C}=\text{O}}$ + $\underset{\substack{\text{(from }n\text{-PrOH,}\\\text{below)}}}{\text{Br}\overset{\text{CH}_3}{\underset{}{}}\text{CHCOOEt}}$

$\underset{\text{Br}\overset{\text{CH}_3}{\underset{}{}}\text{CHCOOEt}}{}$ $\xleftarrow{\text{EtOH, H}^+}$ $\underset{\text{Br}\overset{\text{CH}_3}{\underset{}{}}\text{CHCOOH}}{}$ $\xleftarrow{\text{Br}_2,\text{ P}}$ $\underset{\text{(from }n\text{-PrOH)}}{\text{CH}_3\text{CH}_2\text{COOH}}$

(m) $\underset{\underset{\text{OH}}{|}}{\text{PhCH}_2\text{CHCH}_2\text{COOH}}$ $\xleftarrow{\text{Zn}}$ $\begin{cases}\text{BrCH}_2\text{COOEt (as in Problem 10h, above)}\\[2mm]\text{PhCH}_2\text{CHO} \xleftarrow{\text{K}_2\text{Cr}_2\text{O}_7} \text{PhCH}_2\text{CH}_2\text{OH (page 565)}\end{cases}$

(n) $\underset{\underset{\text{CH}_3}{|}}{n\text{-Bu}\overset{\text{CH}_3}{\underset{}{-}}\text{C}-\text{COOH}}$ $\xleftarrow{\text{H}_2,\text{ Ni}}$ $\underset{\underset{\text{CH}_3}{|}}{\text{C}_2\text{H}_5\text{CH}=\text{CH}\overset{\text{CH}_3}{\underset{}{-}}\text{C}-\text{COOH}}$ $\xleftarrow{-\text{H}_2\text{O}}$ $\xleftarrow{\text{hydrol.}}$ $\underset{\underset{\text{OH}\quad\text{CH}_3}{|\quad\quad|}}{\text{C}_2\text{H}_5\text{CH}_2\text{CH}\overset{\text{CH}_3}{\underset{}{-}}\text{C}-\text{COOEt}}$

$\underset{\underset{\text{CH}_3}{|}}{\text{Br}-\overset{\text{CH}_3}{\underset{}{}}\text{C}-\text{COOEt}}$ + $\underset{\text{(from }n\text{-BuOH)}}{\text{C}_2\text{H}_5\text{CH}_2\text{CHO}}$ $\quad\Big\uparrow\text{Zn}$

(below)

$\underset{\underset{\text{CH}_3}{|}}{\text{Br}-\overset{\text{CH}_3}{\underset{}{}}\text{C}-\text{COOEt}}$ $\xleftarrow{\text{EtOH, H}^+}$ $\xleftarrow{\text{Br}_2,\text{ P}}$ $\underset{\underset{\text{CH}_3}{|}}{\overset{\text{CH}_3}{\underset{}{}}\text{CH}-\text{COOH}}$ (from i-BuOH)

(o) $\xleftarrow{\text{HF}}$ $\xleftarrow{\text{SOCl}_2}$ $\xleftarrow{\text{H}_2,\text{ Ni}}$ $\text{CH}=\text{CHCOOH}$ (page 714)

Cinnamic acid

(p) $\equiv$ $\xleftarrow[\text{anti-addn.}]{\text{HCO}_2\text{OH}}$

trans-Cinnamic acid
(Problem 21.22e, page 714)

11. An aldol-like condensation leads to a γ-hydroxyacid, which subsequently forms the γ-lactone (cyclic ester).

$\xleftarrow{}$ $\underset{\underset{\text{OH}\quad\text{CH}_2\text{COOH}}{|\quad\quad\quad|}}{\overset{\gamma\quad\quad\beta}{\text{MeCH}-\text{CHCOOH}}}$ $\xleftarrow{}$ CH_3CHO + $\underset{\text{H}_2\text{CCOOH}}{\overset{\text{H}_2\text{CCOOH}}{\underset{}{|}}}$

A γ-lactone   A γ-hydroxyacid
A cyclic ester

12. $CH_3-\overset{\displaystyle O}{\underset{\displaystyle \|}{C}}-CH_2\!\!+\!\!CH\overset{CH_3}{\underset{CH_3}{\big<}}$ $\xleftarrow{H_2,\ Ni}$ $CH_3-\overset{\displaystyle O}{\underset{\displaystyle \|}{C}}-CH=C\overset{CH_3}{\underset{CH_3}{\big<}}$ $\xleftarrow[-H_2O]{base}$ $2CH_3COCH_3$

Mesityl oxide   Acetone
(page 711)

13. (a) Acylation can occur at either α-carbon of the ketone.

$$PhCOEt + CH_3-\overset{\displaystyle O}{\underset{\displaystyle \|}{C}}-CH_2CH_3 \longrightarrow$$

$$\begin{array}{l} \rightarrow PhC\!-\!CH_2CCH_2CH_3 \\ \qquad \| \qquad\quad \| \\ \qquad O \qquad\quad O \\[8pt] \rightarrow CH_3C\!-\!CHCH_3 \\ \qquad \|\quad\quad | \\ \qquad O\quad C\!=\!O \\ \qquad\qquad\quad | \\ \qquad\qquad\quad Ph \end{array}$$

(b) An iodoform test (Sec. 16.11) could distinguish between the possibilities.

14. (a) $CH_3-C\equiv C-COOEt$ $\xrightarrow[H^+,\ Hg^{++}]{H_2O}$ $\left[CH_3-\overset{\displaystyle}{\underset{\displaystyle OH}{C}}=CH-COOEt \right]$ $\longrightarrow$ $CH_3-\overset{\displaystyle O}{\underset{\displaystyle \|}{C}}-CH_2-COOEt$

(b) $CH_3-C\equiv C-COOEt$ $\xleftarrow[H^+]{EtOH}$ $CH_3-C\equiv C-COOH$ $\xleftarrow{H^+}$ $\xleftarrow{CO_2}$ $CH_3-C\equiv CMgBr$

$CH_3-C\equiv CMgBr$ $\xleftarrow{MeMgBr}$ $CH_3-C\equiv CH$ $\xleftarrow{CH_3Br}$ $Na^{+\ -}C\equiv CH$ $\xleftarrow{NaNH_2}$ $HC\equiv CH$
(See Problem 8.4, page 259.)

15. A crossed aldol converts HCHO into a *methylol* group, $-CH_2OH$.

$$O\!=\!\overset{H}{\underset{}{C}}\!-\!CH_3 + O\!=\!\overset{H}{\underset{}{C}}\!-\!H \xrightarrow{OH^-} O\!=\!\overset{H}{\underset{}{C}}\!-\!CH_2\!-\!CH_2OH$$

Two more of these conversions (there are two more α-hydrogens) gives a tri-methylol compound.

$$O\!=\!\overset{H}{\underset{}{C}}\!-\!CH_2\!-\!CH_2OH \xrightarrow{HCHO}{OH^-} O\!=\!\overset{H}{\underset{CH_2OH}{C}}\!-\!CH\!-\!CH_2OH \xrightarrow{HCHO}{OH^-} O\!=\!\overset{H}{\underset{CH_2OH}{C}}\!-\!\overset{CH_2OH}{\underset{}{C}}\!-\!CH_2OH$$

At this point, we have run out of α-hydrogens, but now the Cannizzaro reaction can take over.

$$HCHO + O\!=\!\overset{H\ \ CH_2OH}{\underset{CH_2OH}{C}\!-\!C}\!-\!CH_2OH \xrightarrow{base} HCOO^- + HOCH_2\!-\!\overset{CH_2OH}{\underset{CH_2OH}{C}}\!-\!CH_2OH$$

No α-hydrogens   Pentaerythritol

16. We exchange protium for deuterium conveniently and regiospecifically by treatment of the ketone with D_2O in the presence of base or acid (see Problem 21.5(c), page 707).

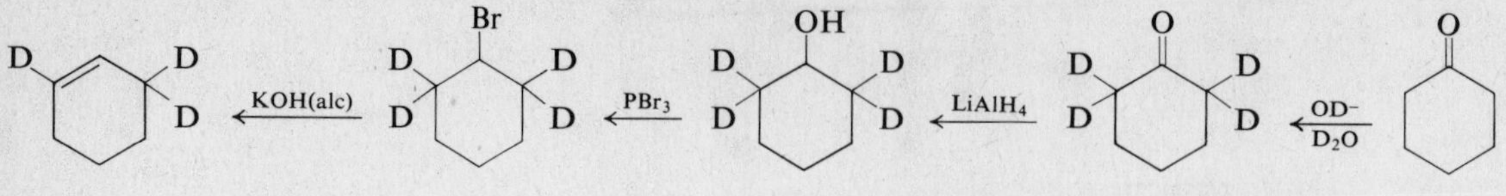

17. (a)

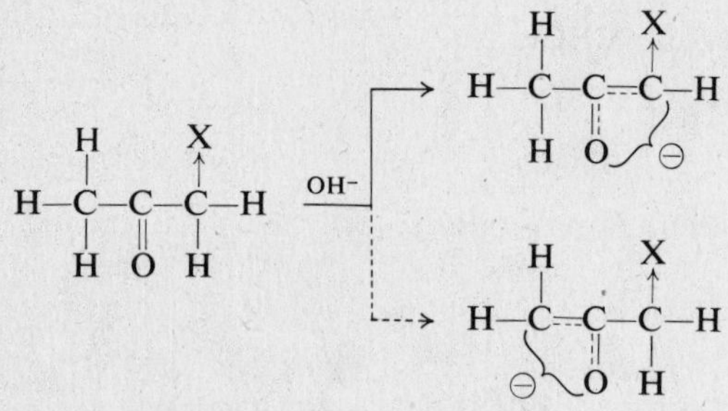

Electron withdrawal by halogen makes hydrogens on the carbon to which halogen has already become attached more acidic and hence more readily removed by base to give further substitution.

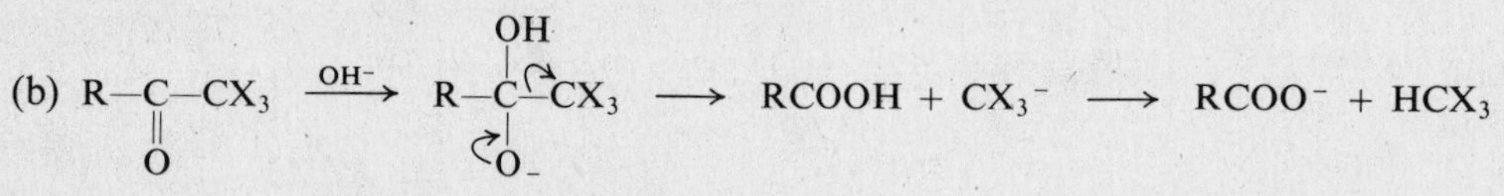

(b)

$$R-\underset{O}{\underset{\|}{C}}-CX_3 \xrightarrow{OH^-} R-\underset{O^-}{\overset{OH}{\underset{|}{C}}}CX_3 \longrightarrow RCOOH + CX_3^- \longrightarrow RCOO^- + HCX_3$$

Electron withdrawal by three halogens makes CX_3^- comparatively weakly basic (for a carbanion) and hence a good leaving group.

Thus both essential aspects of the haloform reaction—regiospecificity of halogenation, and cleavage—are controlled by the same factor: stabilization of a carbanion through electron withdrawal by halogen.

18. Recalling the acidity of γ-hydrogens of α,β-unsaturated aldehydes (Problem 21.20, page 713) and looking at the structure for citral (Problem 26, Chapter 19), we propose:

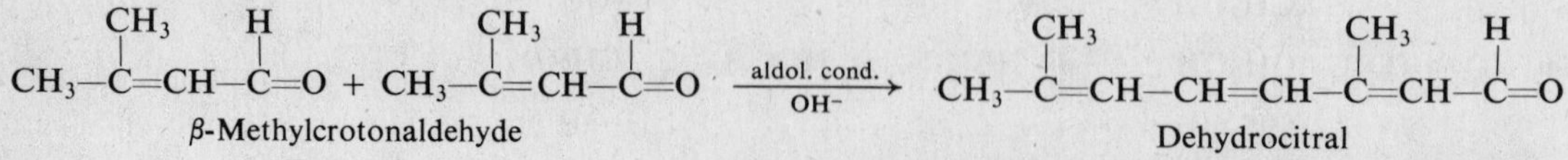

β-Methylcrotonaldehyde
Two moles

Dehydrocitral
(Compare citral, Problem 26, Chapter 19)

19. By Wittig reactions at two points in the molecule.

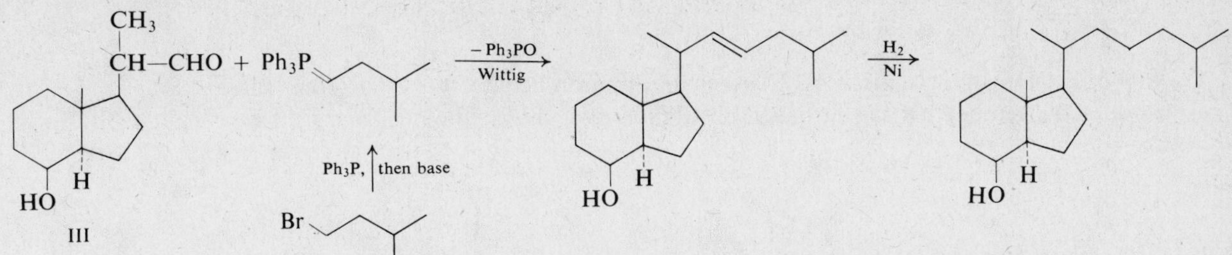
CH3
CH—CHO + Ph3P
– Ph3PO
Wittig
H2
Ni
Ph3P, then base
Br
H
HO
III
Isopentyl bromide

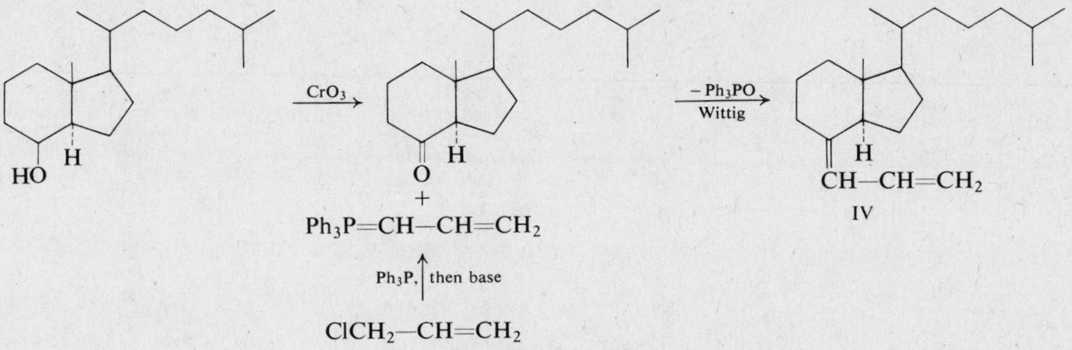
CrO3
– Ph3PO
Wittig
H
HO
O
+
Ph3P=CH—CH=CH2
Ph3P, then base
ClCH2—CH=CH2
CH—CH=CH2
IV

20. Acetoacetic ester is acidic enough to displace CH_4 from CH_3MgI.

$$CH_3COCH_2COOEt + CH_3MgI \longrightarrow \underset{\text{Methane}}{CH_4\uparrow} + CH_3COCHCOOEt\ ^{-}(MgI)^{+}$$

21. (a) Base-catalyzed tautomerization:

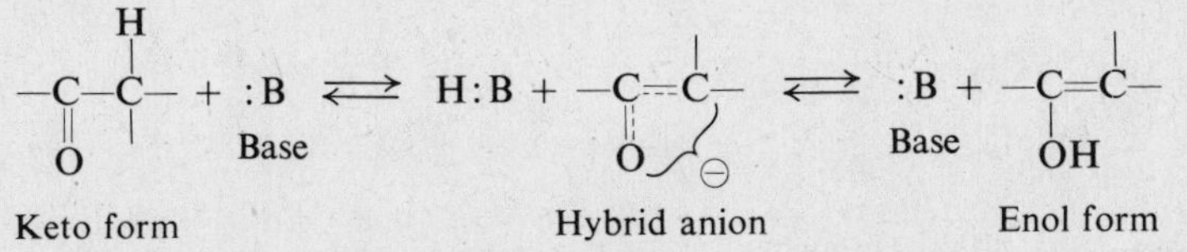
+ :B
Base
H:B +
:B +
Base
OH
Keto form
Hybrid anion
Enol form

Acid-catalyzed tautomerization:

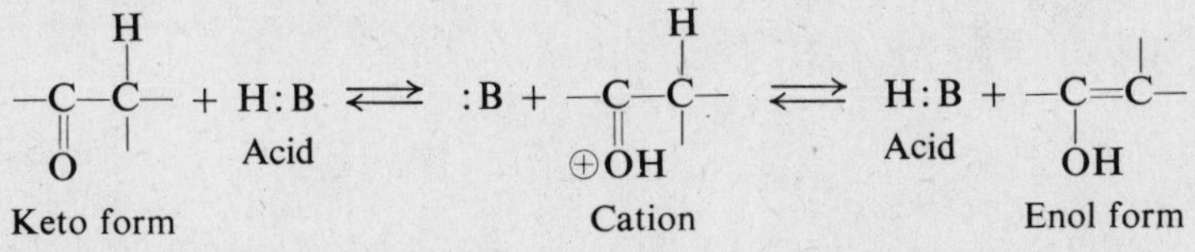
+ H:B
Acid
:B +
H:B +
Acid
⊕OH
OH
Keto form
Cation
Enol form

(b) and (c) The enol is stabilized by (i) conjugation of the remaining C=O with C=C, and (ii) intramolecular H-bonding:

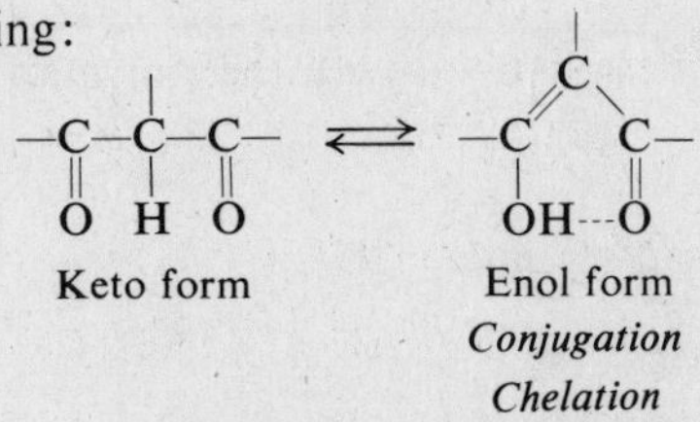
O H O
OH---O
Keto form
Enol form
Conjugation
Chelation

1,3- or β-Dicarbonyl compounds

22. (a) $C_{12}H_{20}O_3$ starting material
$\underline{-C_{10}H_{14}O_2}$ compound A
C_2H_6O means loss of EtOH

The loss of EtOH suggests an internal Claisen condensation between the ester group and an α-carbon of the ketone. Of the two such possibilities, we choose the one that gives a six-membered ring.

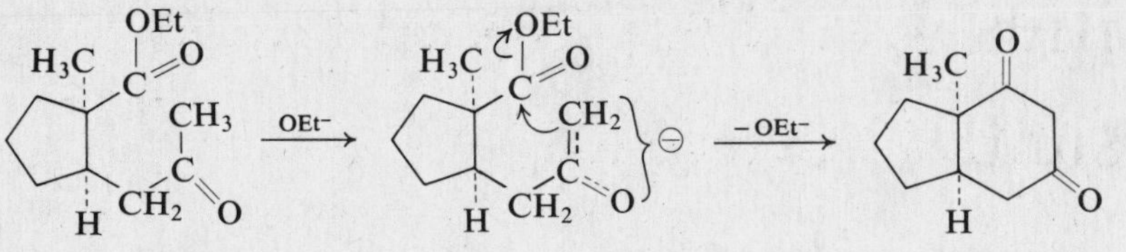

(b) C_4H_8O methyl ethyl ketone $C_{10}H_{18}O_5$
$\underline{+ C_6H_{10}O_4}$ ethyl oxalate $\underline{- C_6H_6O_3}$ compound B
$C_{10}H_{18}O_5$ $C_4H_{12}O_2$ means 2 EtOH

The loss of $2C_2H_5OH$ suggests a double Claisen condensation.

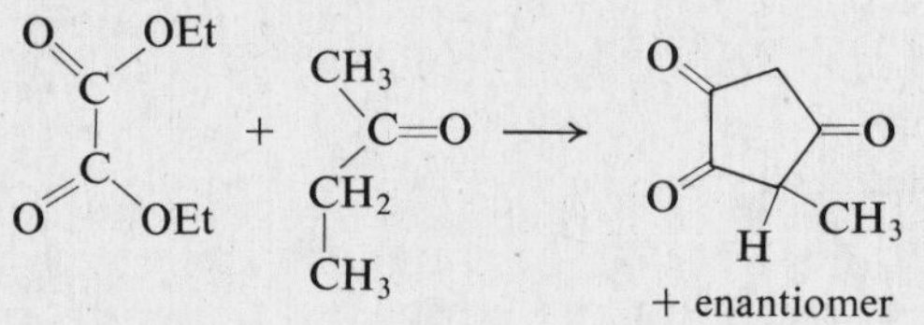

23. (a)   a, enol —CH₃   b, keto —CH₃   c, keto —CH₂—   d, enol —CH=   e, enol —OH

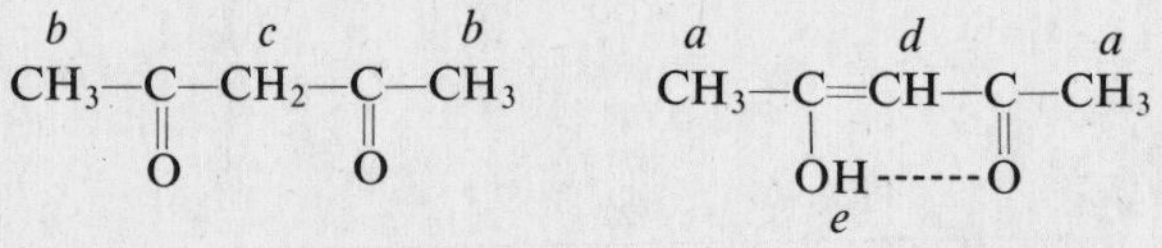

$$\overset{b}{C}H_3-\overset{c}{C}-CH_2-\overset{b}{C}-CH_3 \qquad \overset{a}{C}H_3-\overset{d}{C}=CH-\overset{a}{C}-CH_3$$

Ratios a:b and 2d:c are equal; the value of the ratios (5.6:1) shows that approximately 85% enol is present.

(b)   a, enol —CH₃   b, enol —CH=   c, C_6H_5—   d, enol —OH

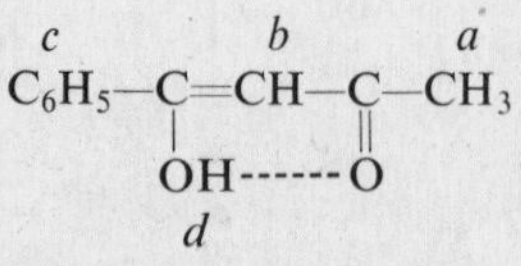

$$\overset{c}{C_6H_5}-\overset{}{C}=\overset{b}{C}H-\overset{a}{C}-CH_3$$

All enol. Conjugation of the carbon–carbon double bond with the ring stabilizes the enol to the exclusion of the keto form.

(See the labeled spectra on page 668 of this Study Guide.)

310

Chapter 22 | Amines I. Preparation and Physical Properties

22.1

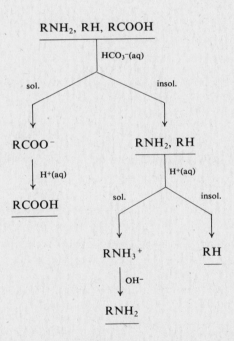

$$RNH_2, \; RH, \; RCOOH$$

HCO$_3^-$(aq)

sol.   insol.

RCOO$^-$ RNH$_2$, RH

H$^+$(aq)   H$^+$(aq)

RCOOH   sol.   insol.

RNH$_3{}^+$ RH

OH$^-$

RNH$_2$

The solid mixture can be treated directly, with each insoluble compound being collected on a filter. The liquids are easier to handle if they are dissolved in ether first; each separation is carried out in the separatory funnel, with the water-soluble salts being ether-insoluble, and the water-insoluble compounds being ether-soluble.

22.2 At room temperature, inversion about nitrogen is slow enough that nmr "sees" two kinds of ring protons: two protons *cis* and two protons *trans* to ethyl. At 120°, inversion is so fast that all four protons are seen in average positions: they are equivalent.

22.3 (a) IV exists as two pairs of enantiomers, each pair diastereomeric to the other pair. In one pair —CH$_3$ and —Cl are *cis*, in the other pair, *trans*.

cis   *trans*

Interconversion of diastereomers requires inversion about nitrogen, which—presumably because of the rigidity of the three-membered ring—is so slow at 25° that the diastereomers can be separated.

(b) Peroxidation of a carbon–nitrogen double bond, like that of a carbon–carbon double bond, produces a rigid three-membered ring containing nitrogen and oxygen.

$$Ph_2C=NCH_3 + R^*CO_2OH \longrightarrow Ph_2C{-}{-}NCH_3 + R^*COOH$$
$$\underset{O}{}$$

This exists in enantiomeric forms:

Enantiomers

The transition state for the epoxidation contains both the chiral center in the peroxy acid and the chiral center developing in the epoxide; diastereomeric configurations—of different energy—therefore exist. Since the peroxy acid is optically active, one diastereomer predominates in the transition state, and yields a predominance of one enantiomeric epoxide —and hence an optically active product. (This process is called *asymmetric induction*.) Interconversion of the enantiomeric epoxides requires inversion about nitrogen which, as in (a), is evidently very slow.

(S. J. Brois, "Aziridines. XII. Isolation of a Stable Nitrogen Pyramid," J. Am. Chem. Soc., **90**, 508 (1968).)

22.4 A carbanion, like ammonia, has a low, easily surmountable energy barrier between mirror-image pyramidal arrangements. Attachment of a fourth group to the carbanion can occur to either arrangement, and thus give the racemic modification.

22.5 (i) Hofmann degradation:

$$n\text{-}C_5H_{11}NH_2 \xleftarrow{\;OBr^-\;} n\text{-}C_5H_{11}C\underset{NH_2}{\overset{O}{\Vert}} \xleftarrow{\;NH_3\;} \xleftarrow{\;SOCl_2\;} n\text{-caproic acid}$$

(ii) Via nitrile:

$$n\text{-}C_5H_{11}NH_2 \xleftarrow{\text{H}_2,\text{ Ni}} n\text{-}C_4H_9C{\equiv}N \xleftarrow{\text{CN}^-} n\text{-BuBr} \xleftarrow{\text{PBr}_3} n\text{-BuOH}$$

(iii) Via halide:

$$n\text{-}C_5H_{11}NH_2 \xleftarrow{\text{NH}_3} n\text{-}C_5H_{11}Br \xleftarrow{\text{PBr}_3} n\text{-pentyl alcohol}$$

(iv) Reductive amination:

$$n\text{-}C_5H_{11}NH_2 \xleftarrow{\text{NH}_3,\text{ H}_2,\text{ Ni}} n\text{-}C_4H_9CHO \xleftarrow{\text{K}_2\text{Cr}_2\text{O}_7} n\text{-pentyl alcohol}$$

22.6 (a) $PhCH_2NH_2 \xleftarrow{\text{NH}_3} PhCH_2Cl \xleftarrow{\text{Cl}_2,\text{ heat}} PhCH_3$

(b) $CH_3\langle\bigcirc\rangle\underset{NH_2}{\overset{}{CHCH_3}} \xleftarrow{\text{NH}_3,\text{ H}_2,\text{ Ni}} CH_3\langle\bigcirc\rangle\underset{O}{\overset{}{CCH_3}} \xleftarrow{\text{Ac}_2\text{O, AlCl}_3} PhCH_3$

(c) $PhCH_2CH_2NH_2 \xleftarrow{\text{H}_2,\text{ Ni}} PhCH_2C{\equiv}N \xleftarrow{\text{CN}^-} PhCH_2Cl \text{ (as in a)}$

(d) $CH_3\langle\bigcirc\rangle NH_2 \xleftarrow{\text{Fe, H}^+} CH_3\langle\bigcirc\rangle NO_2 \xleftarrow{\text{HNO}_3,\text{ H}_2\text{SO}_4} PhCH_3$

(e) $PhNH_2 \xleftarrow{\text{OBr}^-} PhCONH_2 \xleftarrow{\text{NH}_3} \xleftarrow{\text{PCl}_5} PhCOOH \xleftarrow{\text{KMnO}_4} PhCH_3$

1. (a) 1° $CH_3CH_2CH_2CH_2{-}NH_2$ $CH_3CH_2\underset{NH_2}{\overset{}{CHCH_3}}$ $CH_3\overset{}{CHCH_2}{-}NH_2$ $CH_3{-}\overset{\overset{CH_3}{|}}{C}{-}CH_3$

                 n-Butylamine   *sec*-Butylamine   Isobutylamine   *tert*-Butylamine

2° $CH_3CH_2CH_2{-}\overset{\overset{H}{|}}{N}{-}CH_3$ $CH_3\overset{}{CH}{-}\overset{\overset{H}{|}}{N}{-}CH_3$ $CH_3CH_2{-}\overset{\overset{H}{|}}{N}{-}CH_2CH_3$

     Methyl-*n*-propylamine   Methylisopropylamine   Diethylamine

3° $CH_3CH_2{-}\overset{\overset{CH_3}{|}}{N}{-}CH_3$

    Ethyldimethylamine

(b) 1° $PhCH_2NH_2$, benzylamine; *o*-, *m*-, and *p*-$CH_3C_6H_4NH_2$ (the toluidines)

2° $Ph{-}\overset{\overset{H}{|}}{N}{-}CH_3$, N-methylaniline

3°   none

2. (a) $CH_3CH_2CHCH_3$ (b) ⬡$\begin{smallmatrix}CH_3\\NH_2\end{smallmatrix}$ (c) ⬡$NH_3{}^+Cl^-$ (d) $(C_2H_5)_2NH$
$\quad\quad\quad\;\; NH_2$

(e) H_2N⬡$COOH$ (f) ⬡CH_2NH_2 (g) $(CH_3)_2CHNH_3{}^+\,{}^-OOCPh$ (h) ⬡$\begin{smallmatrix}NH_2\\NH_2\end{smallmatrix}$

(i) ⬡$N(CH_3)_2$ (j) $HOCH_2CH_2NH_2$ (k) ⬡$CH_2CH_2NH_2$ (l) ⬡$-N(CH_3)_2$

(m) ⬡$-\overset{H}{N}-$⬡   (n) $2,4\text{-}(CH_3)_2C_6H_3NH_2$ (o) $(n\text{-}C_4H_9)_4N^+I^-$ (p) CH_3O⬡NH_2

3. (a) Excess NH_3 (b) $K_2Cr_2O_7$; NH_3, H_2, Ni
 (c) NH_3, H_2, Ni   (d) Fe, H^+, heat
 (e) H_2, Ni   (f) Br_2, OH^-
 (g) $K_2Cr_2O_7$; $SOCl_2$; NH_3; OBr^- (h) HBr, NaCN; H_2, Ni

Aniline: d (and, under very vigorous conditions, a)
Benzylamine: all (except first stage of h)

4. (a) $CH_3-\overset{\displaystyle CH_3}{\underset{\displaystyle NH_2}{CH}}-NH_2$ $\xleftarrow{NH_3,\ H_2,\ Ni}$ $CH_3-\overset{\displaystyle CH_3}{\underset{\displaystyle \|}{C}}=O$ $\xleftarrow{CrO_3}$ $CH_3CHOHCH_3$

(b) $n\text{-}BuCH_2NH_2$ $\xleftarrow{H_2,\ Ni}$ $n\text{-}Bu-C\equiv N$ $\xleftarrow{CN^-}$ $n\text{-}BuBr$ $\xleftarrow{HBr}$ $n\text{-}BuOH$

(c) $p\text{-}CH_3C_6H_4NH_2$ $\xleftarrow{Fe,\ H^+}$ $p\text{-}CH_3C_6H_4NO_2$ $\xleftarrow{HNO_3,\ H_2SO_4}$ $PhCH_3$

(d) $CH_3\overset{\displaystyle CH_3}{\underset{\displaystyle H}{\underset{|}{CH}}-\overset{|}{N}}-CH_2CH_3$ $\xleftarrow{H_2,\ Ni}$ $\Big[$ $EtNH_2$ $\xleftarrow{NH_3}$ $EtBr$ $\xleftarrow{HBr}$ $EtOH$
$\quad\quad\quad\quad\quad\quad\quad\quad\quad\quad\quad\quad\quad CH_3COCH_3$ $\xleftarrow{CrO_3}$ $CH_3CHOHCH_3$

(e) $PhCHCH_3$ $\xleftarrow{NH_3,\ H_2,\ Ni}$ $Ph-CCH_3$ $\xleftarrow{AlCl_3}$ $\Big[$ acetic anhydride (page 667)
$\quad\;\; NH_2 \quad\quad\quad\quad\quad\quad\quad\quad \overset{\|}{O} \quad\quad\quad\quad\quad\quad\quad\quad\quad C_6H_6$

(f) $PhCH_2CH_2NH_2$ $\xleftarrow{H_2,\ Ni}$ $PhCH_2-C\equiv N$ $\xleftarrow{CN^-}$ $PhCH_2Cl$ $\xleftarrow{Cl_2,\ heat}$ $PhCH_3$

(g) $m\text{-}ClC_6H_4NH_2$ $\xleftarrow{Fe,\ H^+}$ $m\text{-}ClC_6H_4NO_2$ $\xleftarrow[FeCl_3]{Cl_2}$ $PhNO_2$ $\xleftarrow{HNO_3,\ H_2SO_4}$ C_6H_6

(h) $p\text{-}H_2NC_6H_4COOH$ $\xleftarrow{Fe,\ H^+}$ $p\text{-}O_2NC_6H_4COOH$ (page 345)

(i) $n\text{-Bu}-\underset{\underset{\text{NH}_2}{|}}{\text{CH}}-\text{Et}$ $\xleftarrow{\text{NH}_3,\ \text{H}_2,\ \text{Ni}}$ $n\text{-Bu}-\underset{\underset{\text{O}}{\|}}{\text{C}}-\text{Et}$ $\longleftarrow$ $\begin{cases} \text{EtCOCl} \xleftarrow{\text{SOCl}_2} \text{EtCOOH (from } n\text{-PrOH)} \\[6pt] n\text{-Bu}_2\text{Cd} \xleftarrow{\text{CdCl}_2} n\text{-BuMgBr (from } n\text{-BuOH)} \end{cases}$

(j) $\text{PhNH}-\text{Et}$ $\longleftarrow$ $\begin{cases} \text{EtBr (from EtOH)} \\[6pt] \text{PhNH}_2 \xleftarrow{\text{Fe, H}^+} \text{PhNO}_2 \xleftarrow{\text{HNO}_3,\ \text{H}_2\text{SO}_4} \text{C}_6\text{H}_6 \end{cases}$

(k) 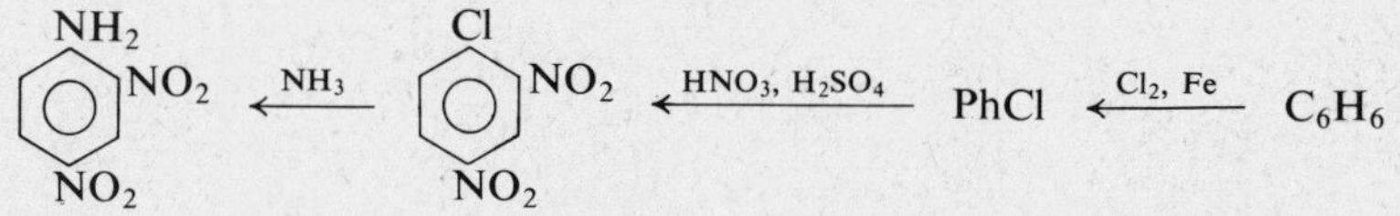 [benzene ring with NH$_2$, NO$_2$, NO$_2$ substituents] $\xleftarrow{\text{NH}_3}$ [benzene ring with Cl, NO$_2$, NO$_2$ substituents] $\xleftarrow{\text{HNO}_3,\ \text{H}_2\text{SO}_4}$ PhCl $\xleftarrow{\text{Cl}_2,\ \text{Fe}}$ C_6H_6

(l) $\text{PhCH}_2\underset{\underset{\text{NH}_2}{|}}{\text{CH}}\text{CH}_3$ $\xleftarrow{\text{NH}_3,\ \text{H}_2,\ \text{Ni}}$ $\text{PhCH}_2\underset{\underset{\text{O}}{\|}}{\text{C}}-\text{CH}_3$ $\longleftarrow$ $\begin{cases} \text{Me}_2\text{Cd} \xleftarrow{\text{CdCl}_2} \text{MeMgI (from MeOH)} \\[6pt] \text{PhCH}_2\text{COCl} \xleftarrow{\text{SOCl}_2} \text{PhCH}_2\text{COOH} \\[2pt] \hspace{5.5cm}\text{(page 587)} \end{cases}$

(m) 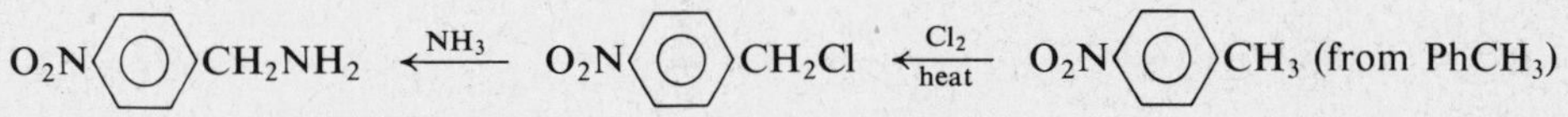 O_2N⟨O⟩CH_2NH_2 $\xleftarrow{\text{NH}_3}$ O_2N⟨O⟩CH_2Cl $\xleftarrow[\text{heat}]{\text{Cl}_2}$ O_2N⟨O⟩CH_3 (from PhCH$_3$)

(n) $\text{PhCH}-\underset{\underset{\text{NH}_2}{|}}{\text{CH}_2}$ $\underset{\underset{\text{OH}}{|}}{}$ $\xleftarrow{\text{H}_2,\ \text{Ni}}$ $\text{PhCH}-\underset{\underset{\text{OH}}{|}}{\text{C}}\!\equiv\!\text{N}$ $\xleftarrow[\text{H}^+]{\text{CN}^-}$ PhCHO $\xleftarrow{\text{H}_2\text{O, OH}^-}$ PhCHCl_2 (from PhCH$_3$)

5. (a) Same number of carbons:

$n\text{-C}_{15}\text{H}_{31}\text{CH}_2\text{NH}_2$ $\xleftarrow{\text{NH}_3}$ $n\text{-C}_{15}\text{H}_{31}\text{CH}_2\text{Br}$ $\xleftarrow{\text{HBr}}$ $n\text{-C}_{15}\text{H}_{31}\text{CH}_2\text{OH}$ $\xleftarrow{\text{LiAlH}_4}$ $n\text{-C}_{15}\text{H}_{31}\text{COOH}$

(b) One more carbon:

$n\text{-C}_{15}\text{H}_{31}\text{CH}_2\text{CH}_2\text{NH}_2$ $\xleftarrow{\text{H}_2,\ \text{Ni}}$ $n\text{-C}_{15}\text{H}_{31}\text{CH}_2\text{C}\!\equiv\!\text{N}$ $\xleftarrow{\text{CN}^-}$ $n\text{-C}_{15}\text{H}_{31}\text{CH}_2\text{Br}$ $\longleftarrow$ as in (a)

(c) One less carbon:

$n\text{-C}_{15}\text{H}_{31}\text{NH}_2$ $\xleftarrow{\text{OBr}^-}$ $n\text{-C}_{15}\text{H}_{31}\text{CONH}_2$ $\xleftarrow{\text{NH}_3}$ $n\text{-C}_{15}\text{H}_{31}\text{COCl}$ $\xleftarrow{\text{SOCl}_2}$ $n\text{-C}_{15}\text{H}_{31}\text{COOH}$

(d) Two C$_{16}$ units:

$n\text{-C}_{15}\text{H}_{31}-\underset{\underset{\text{NH}_2}{|}}{\text{CH}}-\text{CH}_2-\text{C}_{15}\text{H}_{31}\text{-}n$ $\xleftarrow{\text{NH}_3,\ \text{H}_2,\ \text{Ni}}$ $n\text{-C}_{15}\text{H}_{31}-\underset{\underset{\text{O}}{\|}}{\text{C}}-\text{CH}_2-\text{C}_{15}\text{H}_{31}\text{-}n$

$n\text{-C}_{15}\text{H}_{31}-\underset{\underset{\text{O}}{\|}}{\text{C}}-\text{CH}_2-\text{C}_{15}\text{H}_{31}\text{-}n$ $\longleftarrow$ $\begin{cases} n\text{-C}_{15}\text{H}_{31}\text{COCl} \longleftarrow \text{as in (c)} \\[8pt] (n\text{-C}_{15}\text{H}_{31}\text{CH}_2)_2\text{Cd} \longleftarrow n\text{-C}_{15}\text{H}_{31}\text{CH}_2\text{MgBr} \xleftarrow{\text{Mg}} n\text{-C}_{15}\text{H}_{31}\text{CH}_2\text{Br} \end{cases}$

$\uparrow$

(a)

6. (a) $BrCH_2CH_2Br \xrightarrow[\text{(+2CN, -2Br)}]{2CN^-} NCC_2H_4CN \xrightarrow[\text{(+8H)}]{\text{redn.}} H_2N(CH_2)_4NH_2$
Putrescine

(b) $Br(CH_2)_5Br \xrightarrow[\text{(+2NH}_2\text{, -2Br)}]{\text{excess NH}_3} H_2N(CH_2)_5NH_2$
Cadaverine

7. $H_2N(CH_2)_6NH_2 \xleftarrow{H_2,\ Ni} NC(CH_2)_4CN \xleftarrow{2CN^-} Cl(CH_2)_4Cl \xleftarrow{H_2,\ Ni} ClCH_2CH=CHCH_2Cl$

$\qquad\qquad\qquad\qquad\qquad\qquad\qquad\qquad\qquad\qquad\qquad\qquad\qquad\qquad$ 1,4-addn. $\uparrow$ Cl_2

$\qquad\qquad\qquad\qquad\qquad\qquad\qquad\qquad\qquad\qquad\qquad\qquad\qquad$ $CH_2=CH-CH=CH_2$

8. $H_2NCH_2CH_2COOH \xleftarrow{H^+} \xleftarrow{OBr^-} H_2NCCH_2CH_2COH \xleftarrow{NH_3}$
$\qquad\qquad\qquad\qquad\qquad\qquad\qquad\qquad\quad \| \qquad\quad \|$
$\qquad\qquad\qquad\qquad\qquad\qquad\qquad\qquad\ \ O \qquad\ \ O$

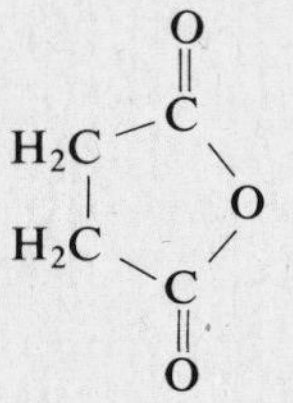

Succinic anhydride

9. (a) Pair of enantiomers: chiral carbon.

(b) One inactive compound: achiral.

(c) Pair of enantiomers: chiral nitrogen.

(d) Inactive *cis-trans* pair: compare 1,4-disubstituted cyclohexanes (Chapter 9).

(e) Pair of enantiomers: chiral molecule.   Planes of rings are perpendicular to each other. (Compare Problem 6(a), Chapter 9, page 315.)

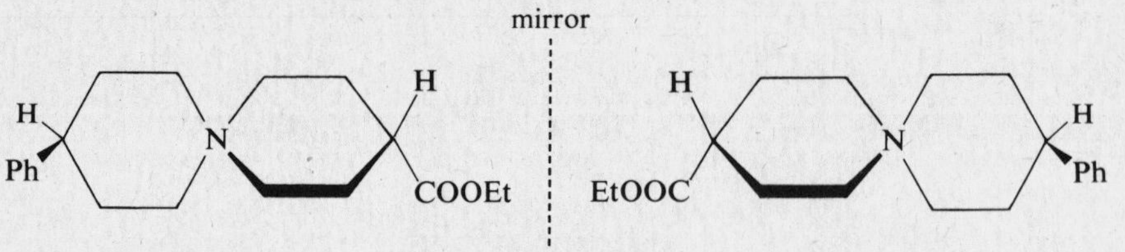

Not superimposable

(f) Pair of enantiomers: chiral nitrogen.

10. (a)

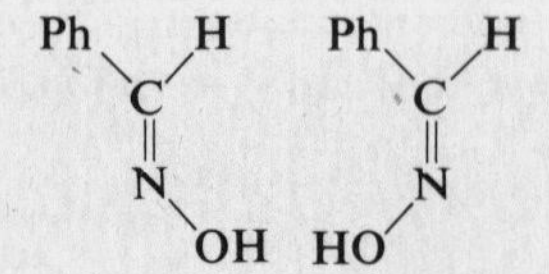

anti-   *syn-*

(b) The electronic configuration about C=N is similar to that about C=C (Secs. 5.2, 5.5, and 5.6). Where both sp^2 orbitals of C= hold substituents, one sp^2 orbital of N= contains an unshared pair. Hindered rotation about the carbon–nitrogen double bond permits stereoisomers to be isolated.

(c) Ph_2C=NOH: no (the two groups on carbon must be different, Sec. 5.6). The other two: yes.

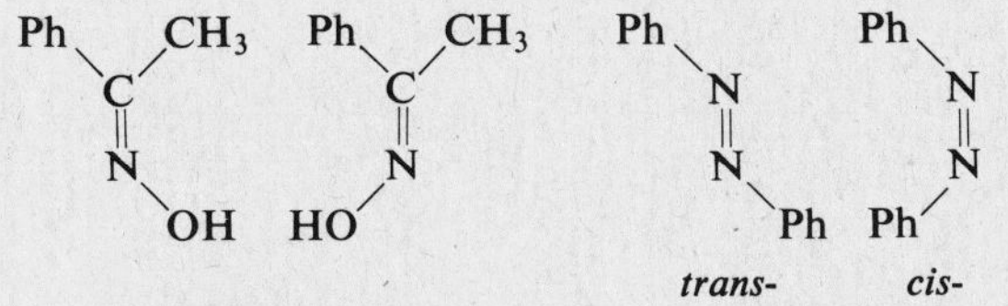

Acetophenoneoxime   Azobenzene

11. (a)

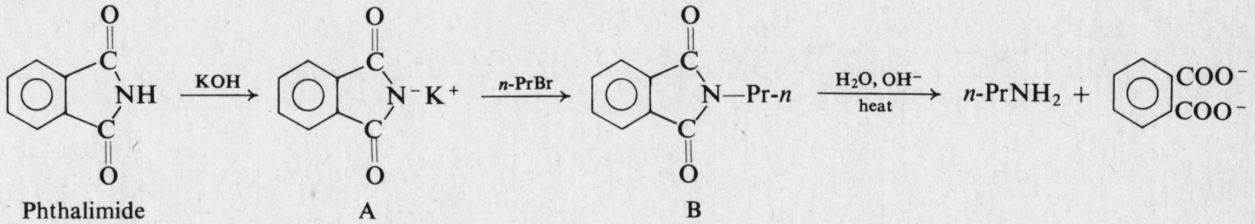

Phthalimide   A   B
Acidic   *Salt of imide*

(b) Because the N can be alkylated only once, the Gabriel synthesis produces *pure* primary amines, free from contaminating secondary or tertiary amines. The synthesis depends upon the acidity of the N—H bond of the imide (there are two acid-strengthening acyl groups present).

Chapter 23 | Amines II. Reactions

23.1 (a) A nitro group *ortho* or *para* to —NH₂ stabilizes the amine through structures like

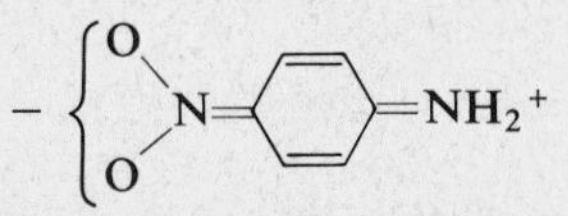

$$-\left\{\begin{array}{l}O\\O\end{array}\right.N-\bigcirc=NH_2{}^+$$

(b) No such structures are possible for *m*-isomer. (Try to draw them, just to prove it to yourself.)

23.2 $(CH_3)_3\overset{+}{N}:\overset{-}{B}F_3$

23.3 E2 elimination from 'onium ions (ammonium, sulfonium) proceeds via a carbanion-like transition state (Sec. 14.21). Orientation is Hofmann: there is preferential abstraction of a proton from the carbon that can best accommodate the partial negative charge, that is, preferential abstraction of the most acidic proton. For example:

> *Acidity of protons*   1° > 2° > 3° and —CH₂Cl > —CH₃

(a) Here, elimination can occur only in the 2-methyl-3-pentyl group. The question is: from which *branch* of that group is the proton lost?

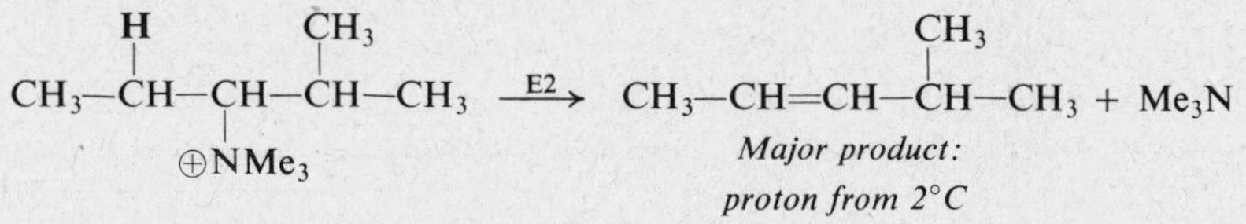

$$\underset{\oplus NMe_3}{CH_3-\underset{H}{CH}-CH-\underset{CH_3}{CH}-CH_3} \xrightarrow{E2} CH_3-CH=CH-\underset{CH_3}{CH}-CH_3 + Me_3N$$

Major product:
proton from 2°C

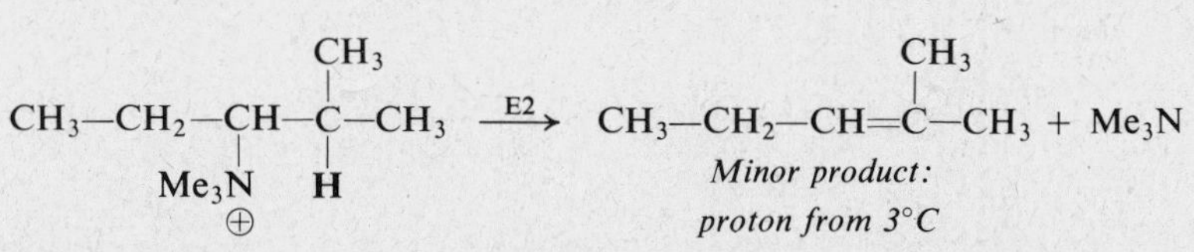

$$CH_3-CH_2-\underset{\underset{\oplus}{Me_3N}}{CH}-\underset{\underset{H}{}}{\overset{CH_3}{C}}-CH_3 \xrightarrow{E2} CH_3-CH_2-CH=\overset{CH_3}{C}-CH_3 + Me_3N$$

Minor product:
proton from 3°C

(b) Here, elimination can occur in either of two groups, an ethyl or a *n*-propyl.   The question is: from which group is the proton lost?

$$\underset{\substack{|\\\overset{\oplus}{N}\\|}}{CH_3-CH_2-CH_2-} \begin{matrix} CH_2CH_3 \\ \\ \\ CH_2-CH_2 \\ | \\ H \end{matrix} -CH_2-CH_2-CH_3 \xrightarrow{E2} CH_2{=}CH_2 + n\text{-}Pr_2NEt$$

Major product:
proton from 1°C

$$\underset{\substack{|\\H}}{CH_3-CH-CH_2-}\overset{\overset{\displaystyle CH_2CH_3}{|}}{\underset{\underset{\displaystyle CH_2-CH_3}{|}}{\overset{\oplus}{N}}}-CH_2-CH_2-CH_3 \xrightarrow{E2} CH_3CH{=}CH_2 + n\text{-}PrNEt_2$$

Minor product:
proton from 2°C

(c) As in (b), with the two groups being ethyl and 2-chloroethyl.

$$CH_3-CH_2-\overset{\overset{\displaystyle CH_3}{|}}{\underset{\underset{\displaystyle CH_3}{|}}{\overset{\oplus}{N}}}-CH_2-\underset{\underset{\displaystyle H}{|}}{CHCl} \xrightarrow{E2} CH_2{=}CHCl + EtNMe_2$$

Major product:
proton from —CH$_2$Cl

$$\underset{\underset{\displaystyle H}{|}}{CH_2}-CH_2-\overset{\overset{\displaystyle CH_3}{|}}{\underset{\underset{\displaystyle CH_3}{|}}{\overset{\oplus}{N}}}-CH_2-CH_2Cl \xrightarrow{E2} CH_2{=}CH_2 + ClCH_2CH_2NMe_2$$

Minor product:
proton from —CH$_3$

(d) As in (b), with the two groups being ethyl and *n*-propyl.

$$\underset{\underset{\displaystyle H}{|}}{CH_2}-CH_2-\overset{\overset{\displaystyle CH_3}{|}}{\underset{\underset{\displaystyle CH_3}{|}}{\overset{\oplus}{N}}}-CH_2-CH_2-CH_3 \xrightarrow{E2} CH_2{=}CH_2 + n\text{-}PrNMe_2$$

Major product:
proton from 1°C

$$CH_3-CH_2-\overset{\overset{\displaystyle CH_3}{|}}{\underset{\underset{\displaystyle CH_3}{|}}{\overset{\oplus}{N}}}-CH_2-\underset{\underset{\displaystyle H}{|}}{CH}-CH_3 \xrightarrow{E2} CH_2{=}CH-CH_3 + EtNMe_2$$

Minor product:
proton from 2°C

23.4 (a) Ethoxide ion is a strong base, and gives E2 elimination with the expected Hofmann orientation. Iodide ion and the solvent ethanol are weak bases, and E1 elimination is the principle reaction; the *tert*-pentyl cation is the one formed, and it loses a proton with the expected Saytzeff orientation.

(b) The ether is formed by the competing substitution reaction, which in this case is S$_N$1.

$$\underset{\underset{\displaystyle CH_3}{|}}{\overset{\overset{\displaystyle CH_3}{|}}{CH_3CH_2-C-SMe_2^{\oplus}}} \longrightarrow \underset{\underset{\displaystyle CH_3}{|}}{\overset{\overset{\displaystyle CH_3}{|}}{CH_3CH_2-C\oplus}}$$

$$\xrightarrow{-H^+} \underset{}{CH_3CH{=}\overset{\overset{\displaystyle CH_3}{|}}{C}-CH_3} \quad \textbf{E1}$$

$$\xrightarrow[-H^+]{EtOH} \underset{\underset{\displaystyle CH_3}{|}}{\overset{\overset{\displaystyle CH_3}{|}}{CH_3CH_2-C-OEt}} \quad \textbf{S}_N\textbf{1}$$

(c) Here, the substitution competing with elimination is S_N2, with preferential attack at the less hindered methyl group.

$$CH_3CH_2-\underset{\underset{CH_3}{|}}{\overset{\overset{CH_3}{|}}{C}}-\overset{\oplus}{\underset{\underset{CH_3}{|}}{S}}-CH_3 + OEt^- \longrightarrow CH_3-OEt + t\text{-PeSMe} \quad S_N2$$

Methyl ethyl
ether

23.5 (a) I $\longrightarrow CH_3CH_2CH_2CH_2CH_2NMe_2$

II $\longrightarrow CH_3CH_2\underset{\underset{CH_3}{|}}{CH}CH_2NMe_2$

(b) $CH_3CH_2CH_2CH_2CH_2Br + Me_2NH \longrightarrow CH_3CH_2CH_2CH_2CH_2NMe_2$

$CH_3CH_2\underset{\underset{CH_3}{|}}{CH}CH_2Br + Me_2N \longrightarrow CH_3CH_2\underset{\underset{CH_3}{|}}{CH}CH_2NMe_2$

23.6 I $\longrightarrow CH_2=CHCH_2CH=CH_2 \xrightarrow[\text{isom.}]{\text{heat}} CH_3CH=CHCH=CH_2$ *More stable conjugated diene*

II $\longrightarrow CH_2=CH-\underset{\underset{CH_3}{|}}{C}=CH_2$

23.7 (a)–(b) In the anion from ammonia, the negative charge is localized on nitrogen. In the anion from an amide, it is accommodated by nitrogen and one oxygen. In the anion from diacetamide, the charge is accommodated by nitrogen and *two* oxygens, as it is in the anion

$$\underset{\underset{CH_3}{|}}{\overset{\overset{CH_3}{|}}{N}}\left.\begin{array}{c} C=O \\ \\ C=O \end{array}\right\}\ominus \qquad CH_3-C\left.\begin{array}{c} \overset{O}{\diagup} \\ \diagdown \\ NH \end{array}\right\}\ominus \qquad NH_2^- \text{ or } RNH^-$$

Most stable   *Least stable*

from phthalimide (Problem 20.11, page 672) or the anion from a sulfonamide, and with much the same effect on acidity.

23.8 (i) *Nucleophilic attack at sulfonyl sulfur is more difficult than attack at acyl carbon.* Acyl carbon is trigonal; nucleophilic attack is unhindered, and involves formation of a tetrahedral intermediate. Sulfonyl sulfur is tetrahedral; nucleophilic attack is relatively hindered, and involves the temporary attachment of a fifth group.

$$R-C\underset{\diagdown W}{\overset{\diagup O}{}} + :Z \longrightarrow R-\underset{\underset{W}{|}}{\overset{\overset{O^-}{|}}{C}}-Z$$

**Acyl
nucleophilic
substitution**

Trigonal C   Tetrahedral C
*Attack relatively
unhindered*   *Stable octet*

$$Ar\text{—}\overset{\displaystyle O}{\underset{\displaystyle O}{S}}\text{—}W \;+\; :Z \longrightarrow \left[\,\overset{\displaystyle Ar}{\underset{\displaystyle O\quad O}{Z\text{—}S\text{—}W}}\,\right]^{-} \qquad \begin{array}{l}\textbf{Sulfonyl}\\ \textbf{nucleophilic}\\ \textbf{substitution}\end{array}$$

Tetrahedral S   Pentavalent S
Attack hindered   *Unstable decet*

(ii) *Nucleophilic attack at the alkyl carbon of a sulfonic ester is easier than attack at alkyl carbon of a carboxylic ester.* Displacement of the less basic sulfonate ion is easier than displacement of the carboxylate ion. Just as sulfonate separates with a pair of electrons from hydrogen more readily than does carboxylate (as shown by the relative acidities of the two kinds of acid), so sulfonate separates with a pair of electrons from an alkyl group more readily than does carboxylate.

$$Ar\text{—}\overset{\displaystyle O}{\underset{\displaystyle O}{S}}\text{—}O\text{—}R \qquad \longrightarrow \quad R\text{—}Z \;+\; ArSO_3^{-}$$
Z: *Weak base:*
good leaving group

$$R'\text{—}\overset{\displaystyle O}{C}\diagdown_{O\text{—}R} \qquad \longrightarrow \quad R\text{—}Z \;+\; R'COO^{-} \qquad \textit{Seldom happens}$$
Z: *Strong base:*
poor leaving group

23.9 A logical explanation would be the following. Although there is very little free aniline present, what there is is *very* much more reactive than the anilinium ion and gives the *para* product. (Activation by amino groups is enormous; in halogenation, where protonation of the amine is not important, dimethylaniline is 10^{19} times as reactive as benzene!)

However, from work of J. H. Ridd (University College, London), it appears that the anilinium ion itself undergoes considerable *para* substitution. He has proposed that, in comparison with $-N(CH_3)_3^{+}$, the positive charge on the $-NH_3^{+}$ is much dispersed by hydrogen bonding of its very acidic protons with the solvent.

23.10 An aromatic $-NH_2$ group ($K_a \sim 10^{-10}$) is considerably weaker than an aliphatic $-NH_2$ group ($K_a \sim 10^{-5}$), and cannot appreciably neutralize the $-COOH$ group ($K_a \sim 10^{-5}$). It *can*, of course, neutralize the strongly acidic $-SO_3H$ group, as in sulfanilic acid.

23.11 (a) *n*-Butyl cation.

(b)

$$CH_3CH_2CH_2CH_2NH_2 \xrightarrow[-N_2]{HONO} CH_3CH_2CH_2CH_2^{\oplus}$$

reaction	product	%
Cl^-	*n*-BuCl	5%
H_2O	*n*-BuOH	25%
$-H^+$	$CH_3CH_2CH{=}CH_2$	
$-H^+$	$CH_3CH{=}CHCH_3$	37%

rearr. ↓

$$CH_3CH_2\overset{\oplus}{C}HCH_3$$

| H_2O | *sec*-BuOH | 13% |
| Cl^- | *sec*-BuCl | 3% |

23.12 (a)

$$iso\text{-}BuNH_2 \xrightarrow[-N_2]{HONO} iso\text{-}Bu^+$$

$$iso\text{-}Bu^+ \xrightarrow{Cl^-} iso\text{-}BuCl$$
$$iso\text{-}Bu^+ \xrightarrow{H_2O} iso\text{-}BuOH$$
$$\xrightarrow{-H^+} (CH_3)_2C{=}CH_2$$

rearr. ↓

$$tert\text{-}Bu^+$$

$$tert\text{-}Bu^+ \xrightarrow{H_2O} tert\text{-}BuOH$$
$$tert\text{-}Bu^+ \xrightarrow{Cl^-} tert\text{-}BuCl$$

(b)

$$CH_3{-}\underset{\underset{CH_3}{|}}{\overset{\overset{CH_3}{|}}{C}}{-}CH_2NH_2 \xrightarrow[-N_2]{HONO} CH_3{-}\underset{\underset{CH_3}{|}}{\overset{\overset{CH_3}{|}}{C}}{-}CH_2{\oplus} \xrightarrow{rearr.} CH_3{-}\underset{\oplus}{\overset{\overset{CH_3}{|}}{C}}{-}CH_2{-}CH_3$$

1° cation   3° cation

$$\xrightarrow{H_2O} CH_3{-}\underset{\underset{OH}{|}}{\overset{\overset{CH_3}{|}}{C}}{-}CH_2CH_3$$

$$\xrightarrow{-H^+} CH_3{-}\overset{\overset{CH_3}{|}}{C}{=}CH{-}CH_3$$

$$\xrightarrow{-H^+} CH_2{=}\overset{\overset{CH_3}{|}}{C}{-}CH_2CH_3$$

23.13 (a) $ONO^- \underset{\rightleftharpoons}{\overset{H^+}{}} HONO \underset{\rightleftharpoons}{\overset{H^+}{}} H_2\overset{+}{O}{-}NO$

(b) $ArH + HONO \longrightarrow Ar\overset{\oplus}{\underset{NO}{\diagup}}\overset{H}{} + OH^-$

$Ar + H_2\overset{+}{O}{-}NO \longrightarrow Ar\overset{\oplus}{\underset{NO}{\diagup}}\overset{H}{} + H_2O$ *Less basic: better leaving group*

(c) $H_2\overset{+}{O}{-}NO + Cl^- \longrightarrow Cl{-}NO + H_2O$ S_N1-*like* or S_N2-*like*

(d) $ArH + Cl{-}NO \longrightarrow Ar\overset{\oplus}{\underset{NO}{\diagup}}\overset{H}{} + Cl^-$ *Weak base: good leaving group*

23.14 (a) Neither is likely. (i) The two groups, $-N(CH_3)_2$ and $-NHCH_3$, release electrons to the ring to about the same degree. (ii) Aside from a (probably modest) steric effect, nitrogen in the two groups should be about equally susceptible to attack.

(b) There is, of course, one way in which $-NHCH_3$ differs completely from $-N(CH_3)_2$: it contains a proton attached to nitrogen.

In both cases the initial electrophilic attack probably occurs at the same place, the place where unshared electrons are most available: at *nitrogen*, to give an unstable quaternary ammonium ion.

In the case of the secondary amine, this intermediate can lose a proton from nitrogen to give the N-nitroso product. This *N-nitrosation* is two-step electrophilic substitution, analogous to electrophilic aromatic substitution.

$$NO^+ + Ph-\underset{H}{\overset{CH_3}{N}}: \longrightarrow Ph-\underset{H}{\overset{CH_3}{\overset{\oplus}{N}}}-N=O \xrightarrow{-H} Ph-\overset{CH_3}{N}-N=O$$

N-nitroso compound

In the case of the tertiary amine, the intermediate quaternary ion has no proton to lose. Instead, it loses NO^+ and regenerates the amine, which eventually undergoes

$$Ph-\overset{\oplus}{\underset{CH_3}{N}}\overset{CH_3}{}-N=O \quad \text{\textit{Cannot lose a proton}}$$

$$NO^+ + Ph-\underset{CH_3}{\overset{CH_3}{N}}:$$

$$O=N\!\!-\!\!\langle\text{ring}\rangle\!\!=\!\!\overset{CH_3}{\overset{\oplus}{N}}\!\!-\!\!CH_3 \xrightarrow{-H^+} O=N\!\!-\!\!\langle\text{ring}\rangle\!\!-\!\!N\!\!\overset{CH_3}{\underset{CH_3}{}}$$

C-nitroso compound

substitution in the ring (*C-nitrosation*). C-nitrosation is thus a second choice to which the reaction turns when N-nitrosation is unsuccessful.

23.15 See answer to Problem 10 below for procedures in solving aromatic synthesis problems.

$$\underset{\text{\textit{m}-Nitrotoluene}}{\overset{CH_3}{\langle\;\rangle}NO_2} \xleftarrow{H_3PO_2} \overset{CH_3}{\underset{N_2^+}{\langle\;\rangle}NO_2} \xleftarrow{HONO} \overset{CH_3}{\underset{NH_2}{\langle\;\rangle}NO_2} \xleftarrow[H^+]{H_2O} \overset{CH_3}{\underset{NHCOCH_3}{\langle\;\rangle}NO_2} \xleftarrow[H_2SO_4]{HNO_3} \overset{CH_3}{\underset{NHCOCH_3}{\langle\;\rangle}} \quad \text{(page 771)}$$

$$\underset{\text{\textit{m}-Iodotoluene}}{\overset{CH_3}{\langle\;\rangle}I} \xleftarrow{KI} \overset{CH_3}{\underset{N_2^+}{\langle\;\rangle}} \xleftarrow{HONO} \overset{CH_3}{\underset{NO_2}{\langle\;\rangle}} \quad \text{(as above)}$$

$$\underset{\text{3,5-Dibromotoluene}}{Br\overset{CH_3}{\langle\;\rangle}Br} \xleftarrow{H_3PO_2} Br\overset{CH_3}{\underset{N_2^+}{\langle\;\rangle}}Br \xleftarrow{HONO} Br\overset{CH_3}{\underset{NH_2}{\langle\;\rangle}}Br \xleftarrow{2Br_2(aq)} \overset{CH_3}{\underset{NH_2}{\langle\;\rangle}} \quad \text{(page 770)}$$

$$\underset{\text{1,3,5-Tribromobenzene}}{Br\overset{Br}{\langle\;\rangle}Br} \xleftarrow{H_3PO_2} Br\overset{Br}{\underset{N_2^+}{\langle\;\rangle}}Br \xleftarrow{HONO} Br\overset{Br}{\underset{NH_2}{\langle\;\rangle}}Br \xleftarrow{3Br_2(aq)} \overset{Br}{\underset{NH_2}{\langle\;\rangle}} \quad \text{(page 733)}$$

$$\underset{\text{\textit{o}-Toluic acid}}{\overset{COOH}{\langle\;\rangle}CH_3} \xleftarrow[\text{heat}]{H_2O,\ H^+} \overset{CN}{\langle\;\rangle}CH_3 \xleftarrow{CuCN} \overset{N_2^+}{\langle\;\rangle}CH_3 \xleftarrow{HONO} \overset{NH_2}{\langle\;\rangle}CH_3 \quad \text{(page 770)}$$

$$\underset{\text{\textit{m}-Toluic acid}}{\overset{COOH}{\underset{CH_3}{\langle\;\rangle}}} \xleftarrow[\text{heat}]{H_2O,\ H^+} \overset{CN}{\underset{CH_3}{\langle\;\rangle}} \xleftarrow{CuCN} \overset{N_2^+}{\underset{CH_3}{\langle\;\rangle}} \xleftarrow{HONO} \overset{NH_2}{\underset{CH_3}{\langle\;\rangle}} \xleftarrow[H^+]{Fe} \overset{NO_2}{\underset{CH_3}{\langle\;\rangle}} \quad \text{(as above)}$$

COOH $\xleftarrow[\text{heat}]{\text{H}_2\text{O, H}^+}$ CN $\xleftarrow{\text{CuCN}}$ $\text{N}_2{}^+$ $\xleftarrow{\text{HONO}}$ NH_2 (page 770)

p-Toluic acid

OH, CH$_3$ $\xleftarrow[\text{warm}]{\text{H}_2\text{O, H}^+}$ $\text{N}_2{}^+$, CH$_3$ $\xleftarrow{\text{HONO}}$ NH_2, CH$_3$ (page 770)

o-Cresol

OH, CH$_3$ $\xleftarrow[\text{warm}]{\text{H}_2\text{O, H}^+}$ $\text{N}_2{}^+$, CH$_3$ $\xleftarrow{\text{HONO}}$ NH_2, CH$_3$ $\xleftarrow{\text{Fe, H}^+}$ NO_2, CH$_3$ (as above)

m-Cresol

OH, CH$_3$ $\xleftarrow[\text{warm}]{\text{H}_2\text{O, H}^+}$ $\text{N}_2{}^+$, CH$_3$ $\xleftarrow{\text{HONO}}$ NH_2, CH$_3$ (page 770)

p-Cresol

23.16   Br, Br $\xleftarrow{\text{CuBr}}$ $\text{N}_2{}^+$, Br $\xleftarrow{\text{HONO}}$ NH_2, Br $\xleftarrow{\text{Fe, H}^+}$ NO_2, Br   (page 771)

m-Dibromobenzene

I, Br $\xleftarrow{\text{KI}}$ $\text{N}_2{}^+$, Br   (as above)

m-Bromoiodobenzene

23.17   Br, CH$_3$, Br $\xleftarrow{\text{H}_3\text{PO}_2}$ Br, CH$_3$, Br, $\text{N}_2{}^+$ $\xleftarrow{\text{HONO}}$ Br, CH$_3$, Br, NH_2 $\xleftarrow[\text{H}^+]{\text{Fe}}$ Br, CH$_3$, Br, NO_2 $\xleftarrow[\text{Fe}]{2\text{Br}_2}$ CH$_3$, NO_2 (page 345)

2,6-Dibromo-
toluene

NO_2, Br, Br $\xleftarrow{\text{H}_3\text{PO}_2}$ NO_2, Br, Br, $\text{N}_2{}^+$ $\xleftarrow{\text{HONO}}$ NO_2, Br, Br, NH_2 $\xleftarrow[\text{Fe}]{2\text{Br}_2}$ NO_2, NH_2 (page 760)

3,5-Dibromo-
nitrobenzene

23.18 (a) Electron withdrawal by nitro groups makes diazonium ions more electrophilic.

(b) Less reactive due to electron release by —CH$_3$.

23.19 To prevent coupling of diazonium ion with unconsumed amine. In excess acid, most of the amine exists in the protonated form, which does not couple. The concentration of free amine is low, and its rate of coupling is slow.

23.20 (a) In an aromatic amine there are two electron-rich sites open to electrophilic attack: nitrogen and the ring. Like NO^+ (Problem 23.14), ArN_2^+ finds attack at nitrogen easier, and forms the intermediate I. In the case of a primary or secondary amine, I can lose a proton to yield II, a *diazoamino* compound, isomeric with the expected azo compound IV.

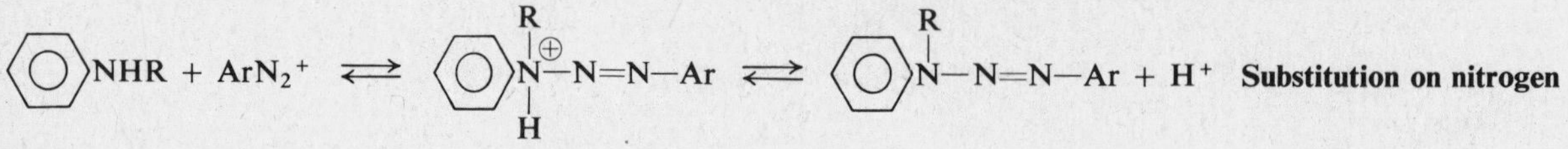

I   II

A diazoamino compound

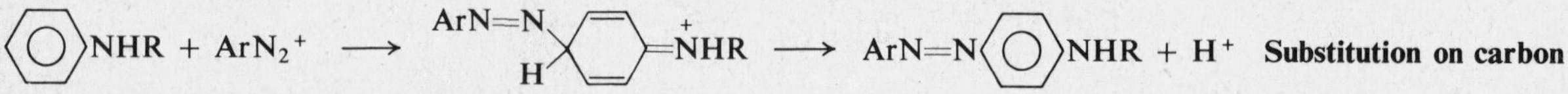

III   IV

An aminoazo compound

(b) Formation of II is easy, but reversible; electrophilic attack by H^+ regenerates the amine and ArN_2^+. These react again and again, and eventually with attack on the ring to form the *sigma* complex III and from it the azo compound IV. Once formed, IV persists; its formation is *not* reversible.

23.21 (a)

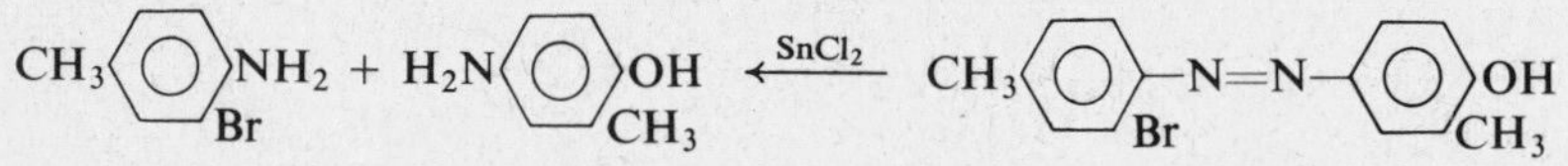

(b)

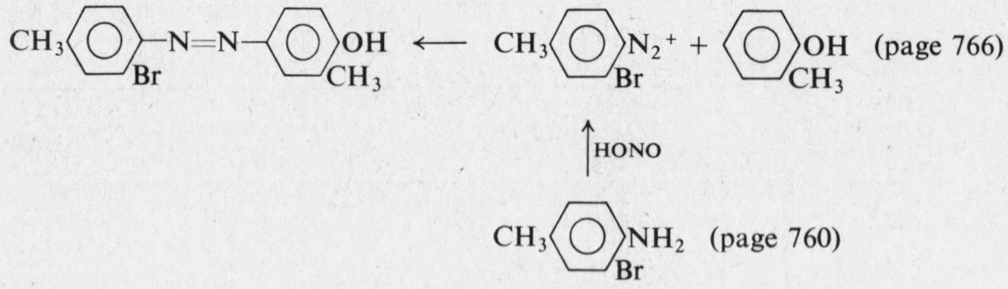

23.22

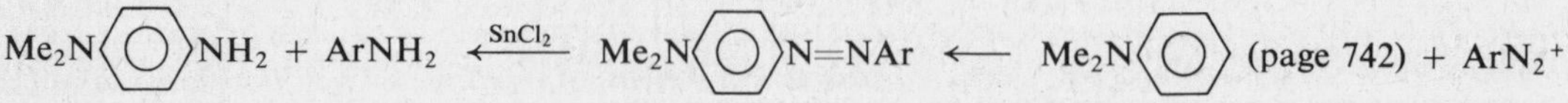

p-Amino-N,N-dimethylaniline

For ArN_2^+ one usually chooses $^-O_3SC_6H_4N_2^+$ (from sulfanilic acid, page 760); when sulfanilic acid is regenerated (as $ArNH_2$) in the reduction step, its solubility characteristics (see Sec. 23.8) differ enough from those of the desired amine to permit easy separation.

23.23 (a) Nucleophilic substitution.

$$Ph{-}\overset{\overset{\displaystyle O}{|}}{\underset{\underset{\displaystyle O}{|}}{S}}{-}\overset{\overset{\displaystyle CH_3}{|}}{\underset{\underset{\displaystyle CH_3}{|}}{\overset{+}{N}}}{-}CH_3 \quad :N(CH_3)_3 \longrightarrow Ph{-}\overset{\overset{\displaystyle O}{|}}{\underset{\underset{\displaystyle O}{|}}{S}}{-}\overset{\overset{\displaystyle CH_3}{|}}{\underset{\underset{\displaystyle CH_3}{|}}{N}} + CH_3\!:\!\overset{+}{N}(CH_3)_3$$

(b) A tertiary amine can react as in (a), to yield the derivative of a *secondary* amine. Incorrect conclusion: that the amine is secondary.

23.24 (a) Incorrect conclusion: that the amine is secondary.

(b) Filter (or separate) and acidify the aqueous solution. A precipitate will form if the amine was primary.

1. (a) $n\text{-}BuNH_3{}^+Cl^-$

(b) $n\text{-}BuNH_3{}^+HSO_4{}^-$

(c) $n\text{-}BuNH_3{}^+{}^-OOCCH_3$

(d) no reaction

(e) $CH_3\overset{\underset{\displaystyle O}{\|}}{C}{-}NH{-}Bu\text{-}n$ and product (c)

(f) $(CH_3)_2CH\overset{\underset{\displaystyle O}{\|}}{C}{-}NH{-}Bu\text{-}n$ and $n\text{-}BuNH_3{}^+{}^-OOCCH(CH_3)_2$

(g) $O_2N\langle\!\bigcirc\!\rangle\overset{\underset{\displaystyle O}{\|}}{C}{-}NH{-}Bu\text{-}n$ and $\langle\!\bigcirc\!\rangle NH^+Cl^-$

(h) $PhSO_2\overset{\overset{\displaystyle \ominus}{\cdot\cdot}}{N}{-}Bu\text{-}n\ Na^+$

(i) $n\text{-}BuNH_2{}^+Cl^-,\qquad n\text{-}Bu\overset{\overset{\displaystyle Et}{|}}{N}H^+Cl^-,\qquad n\text{-}Bu\overset{\overset{\displaystyle Et}{|}}{\underset{\underset{\displaystyle Et}{|}}{N}}{-}Et^+Cl^-$

(j) $n\text{-}Bu\overset{\overset{\displaystyle CH_2Ph}{|}}{N}H_2{}^+Cl^-$, etc.

(k) no reaction

(l) $n\text{-}BuNMe_3{}^+OH^-$

(m) $CH_3CH_2CH{=}CH_2 + Me_3N$

(n) $n\text{-}BuNHCH(CH_3)_2$

(o) see Problem 23.11, page 763

(p) $\overset{\underset{\displaystyle}{\bigcirc}}{}\ \begin{matrix}COOH\\ \overset{\underset{\displaystyle O}{\|}}{C}{-}NH{-}Bu\text{-}n\end{matrix}$

(q) $n\text{-}Bu\overset{\overset{\displaystyle H}{|}}{\underset{\underset{\displaystyle H}{|}}{\overset{\oplus}{N}}}{-}CH_2COO^-,\qquad$ N-(n-butyl)glycine

(r) $2,4,6\text{-}(O_2N)_3C_6H_2NH{-}Bu\text{-}n$

2. Aliphatic amines are more basic than aromatic amines (Sec. 23.3). Electron-releasing groups raise basicity; electron-withdrawing groups lower basicity (Sec. 23.4).

(a) $cyclo\text{-}C_6H_{11}NH_2 > NH_3 > C_6H_5NH_2$

(b) $CH_3CH_2NH_2$ > $HOCH_2CH_2CH_2NH_2$ > $HOCH_2CH_2NH_2$

(c) p-$CH_3OC_6H_4NH_2$ > $C_6H_5NH_2$ > p-$O_2NC_6H_4NH_2$

(d) m-$EtC_6H_4CH_2NH_2$ > $C_6H_5CH_2NH_2$ > m-$ClC_6H_4CH_2NH_2$

(e) p-ClC_6H_4NHMe > 2,4-$Cl_2C_6H_3NHMe$ > 2,4,6-$Cl_3C_6H_2NHMe$

3. In an aqueous solution of $Me_4N^+OH^-$ the base is OH^-, which is much stronger than Me_3N.

4. (a) All three form soluble ammonium salts.

(b) $PhNH_2$ $\xrightarrow{\text{HONO}}$ PhN_2^+ *Colorless solution*

$$PhN\overset{\displaystyle CH_3}{|}H \xrightarrow{\text{HONO}} PhN\overset{\displaystyle CH_3}{|}N=O$$ *Neutral, yellow compound*

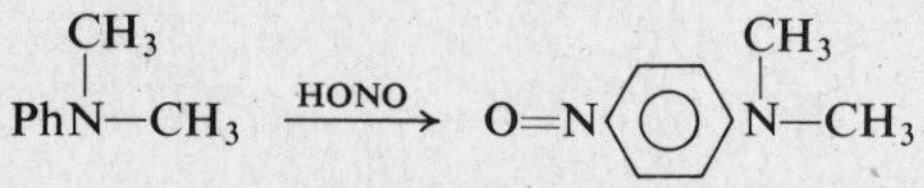

$$PhN\overset{\displaystyle CH_3}{|}CH_3 \xrightarrow{\text{HONO}} O=N\langle\bigcirc\rangle N\overset{\displaystyle CH_3}{|}CH_3$$ *Green solid*

(c) All three form quaternary ammonium salts.

(d) $PhNH_2$ $\longrightarrow$ $ArSO_2N\overset{\displaystyle H}{|}Ph$ **Acidic:** *soluble in aqueous base*

$PhN\overset{\displaystyle CH_3}{|}H$ $\longrightarrow$ $ArSO_2N\overset{\displaystyle CH_3}{|}Ph$ **Neutral:** *insoluble in acid or base*

$PhN\overset{\displaystyle CH_3}{|}CH_3$ $\longrightarrow$ no reaction **Remains basic:** *insoluble in base, soluble in acid*
                  If careful

(e) $PhNH_2$ $\longrightarrow$ $PhNHCOCH_3$ **Neutral:** *insoluble in dilute acid or base*

$PhN\overset{\displaystyle CH_3}{|}H$ $\longrightarrow$ $PhNCOCH_3\overset{\displaystyle CH_3}{|}$ **Neutral:** *insoluble in dilute acid or base*

$PhN\overset{\displaystyle CH_3}{|}CH_3$ $\longrightarrow$ no reaction **Remains basic:** *soluble in acid, insoluble in base*

(f) As in (e), with —COPh in place of —$COCH_3$.

(g) All three undergo very fast ring bromination to give tribromo products.

5. (a) All three form ammonium salts (same as 4a).

(b) $EtNH_2$ yields N_2; Et_2NH yields a neutral (yellow) N-nitroso compound; Et_3N yields the same N-nitroso compound, and cleavage products.

(c) All three form quaternary ammonium salts.

(d) $EtNH_2$ gives a solution, Et_2NH gives a solid, Et_3N does not react (as in 4d).

(e) Neutral amides from 1° and 2° amines, no reaction with 3° amine (as in 4e).

(f) Same as (e) (and as in 4f).

(g) We could not have known this, but 1° and 2° amines give N-bromoamines (compare page 888). No reaction with 3° amines.

6. (a) $p\text{-}CH_3C_6H_4N_2{}^+Cl^-$ (b) $p\text{-}O{=}NC_6H_4NEt_2$

(c) $n\text{-}PrOH$, $iso\text{-}PrOH$, $CH_3CH{=}CH_2$ (compare Problem 23.11, page 763)

(d) $p\text{-}{}^-O_3SC_6H_4N_2{}^+$ (e) $C_6H_5\overset{\displaystyle CH_3}{N}{-}N{=}O$

(f) 2-methyl-2-butene, *tert*-PeOH   (g) 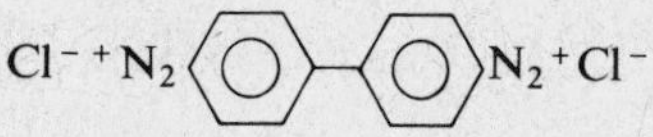 $Cl^-{}^+N_2$ ⟨○⟩⟨○⟩ $N_2{}^+Cl^-$

(h) $C_6H_5CH_2OH$

7. (a) 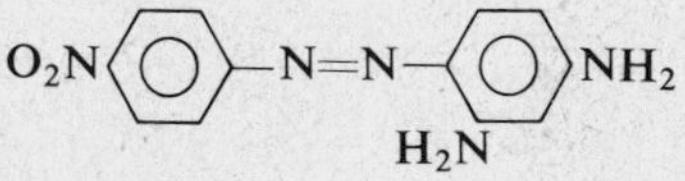 O_2N⟨○⟩$-N{=}N-$⟨○⟩NH_2 (with H_2N)   (b) O_2N⟨○⟩OH

(c) O_2N⟨○⟩Br (d) O_2N⟨○⟩$-N{=}N-$⟨○⟩ (with CH_3 and HO)

(e) O_2N⟨○⟩I (f) O_2N⟨○⟩Cl

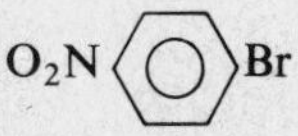

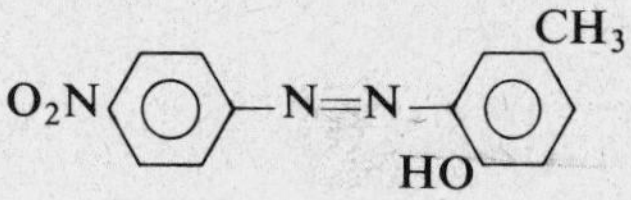

(g) O_2N⟨○⟩CN (h) O_2N⟨○⟩F

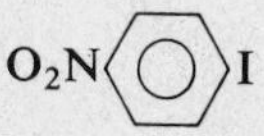

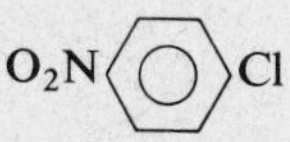

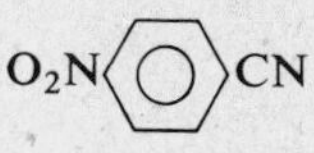

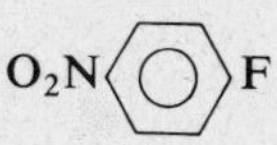

(i) O_2N⟨○⟩H

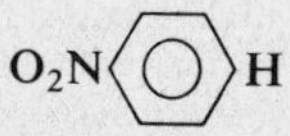

8. (a) H_3PO_2, H_2O.

(b) H_2O, H^+, heat, removing the cresol immediately by steam-distillation.

(c) CuCl. Heat solution of $ArN_2{}^+CuCl_2{}^-$ salt.

(d) CuBr.   Heat solution of $ArN_2{}^+CuBr_2{}^-$ salt.

(e) KI.

(f) HBF_4; isolate $ArN_2{}^+BF_4{}^-$; heat dry salt.

(g) CuCN.   Heat solution of $ArN_2{}^+Cu(CN)_2{}^-$ salt.

(h) 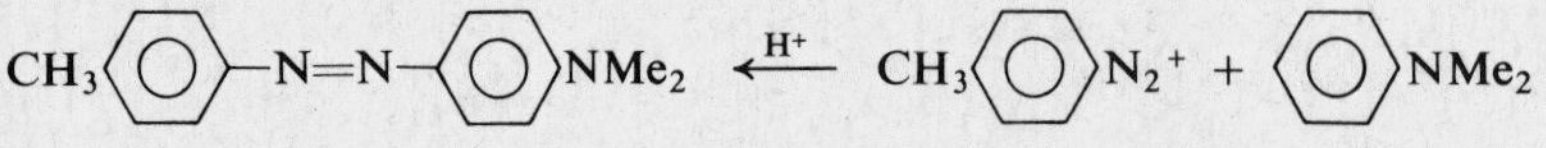 CH₃⟨◯⟩—N=N—⟨◯⟩NMe₂ $\xleftarrow{H^+}$ CH₃⟨◯⟩N₂⁺ + ⟨◯⟩NMe₂

(i) 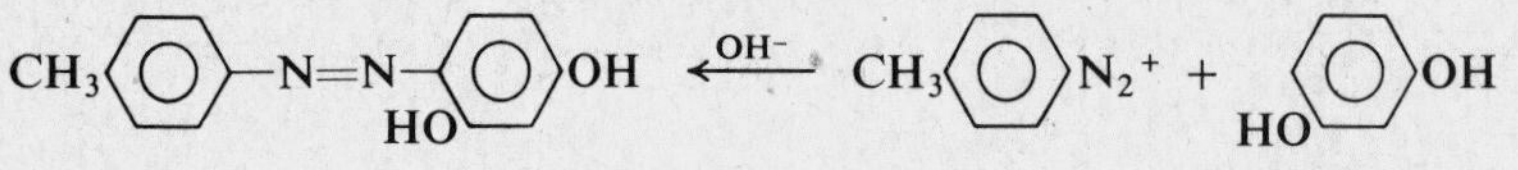 CH₃⟨◯⟩—N=N—⟨◯⟩OH $\xleftarrow{OH^-}$ CH₃⟨◯⟩N₂⁺ + HO⟨◯⟩OH
      HO
                                                          Resorcinol

9. (a) $n\text{-Pr}\overset{\text{O}}{\underset{\|}{\text{C}}}\!-\!\text{NHCH}_3$ + $\text{MeNH}_3{}^+\text{Cl}^-$ (b) $\text{PhN}\overset{\text{Me}}{|}\!-\!\overset{\text{O}}{\underset{\|}{\text{C}}}\!\text{CH}_3$ + CH_3COOH

   (c) $n\text{-Pr}_3\text{N}$ + $\text{CH}_3\text{CH}{=}\text{CH}_2$ (d) $\text{iso-BuC}\overset{\text{O}}{\underset{\|}{}}\!-\!\text{NEt}_2$ + $\text{Et}_2\text{NH}_2{}^+\text{Cl}^-$

   (e) MeOH + Me_3N (f) $\text{Me}_3\text{NH}^+\,{}^-\text{OOCCH}_3$

   (g) $\text{Me}_2\text{NH}_2{}^+\text{Cl}^-$ + CH_3COOH (h) PhNH_2 + $\text{PhCOO}^-\text{Na}^+$

   (i) MeOH + $\text{PhNH}{-}\overset{\text{O}}{\underset{\|}{\text{CH}}}$ (j) MeNHCNHMe (N,N′-dimethylurea) + $\text{MeNH}_3{}^+\text{Cl}^-$
                                                               $\overset{\|}{O}$

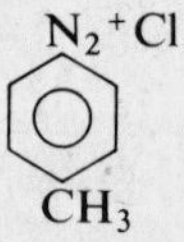

(k) 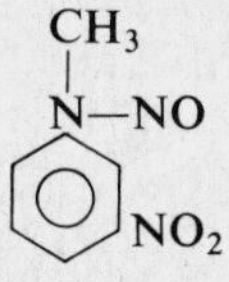 [ring with N—NO and CH₃ on N, NO₂ on ring]   (l) 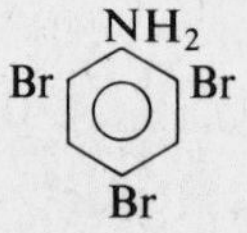 [ring with NH₂, Br, Br, Br]   (m) 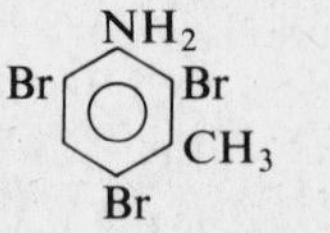 [ring with NH₂, Br, Br, CH₃, Br]   (n) 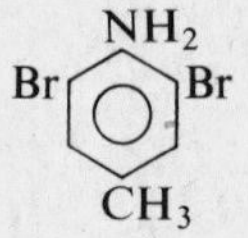 [ring with NH₂, Br, Br, CH₃]

(o) [ring with N₂⁺Cl and CH₃]   (p) 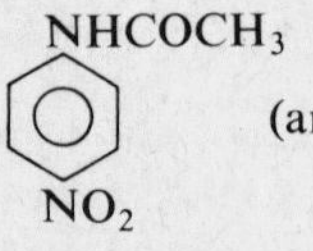 [ring with NHCOCH₃ and NO₂] (and o-isomer)   (q) 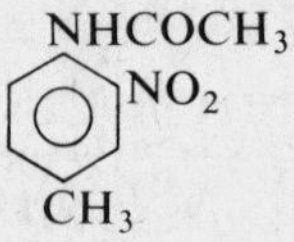 [ring with NHCOCH₃, NO₂, CH₃]

(r) 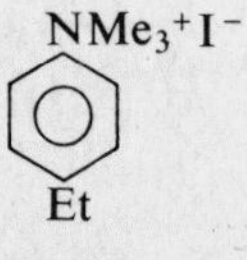 [ring with NMe₃⁺I⁻ and Et]   (s) 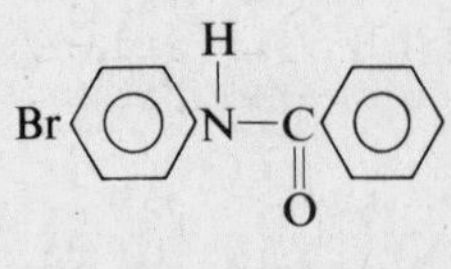 Br⟨◯⟩—N(H)—C(=O)⟨◯⟩ (and o-isomer)

Synthesis: A Systematic Approach

As the syntheses we plan become more and more complicated, we need to follow a systematic approach: something we have probably already been doing without fully realizing it. Let us outline formally the mental steps of such an approach. We shall use synthesis of aromatic compounds as our example here, but the approach is a general one and, with modifications, can be applied to synthesis of all kinds of complicated molecules.

We start, as usual, with the molecule we want—the target molecule—and work backwards. We ask ourselves these questions in this order:

Question (1). Can any of the substituents in the target molecule be introduced by direct electrophilic aromatic substitution, and with the proper orientation? If, not, then:

Question (2). Is there any substituent in the target molecule that can be formed by the oxidation or reduction of some related group? For example, $-COOH \longleftarrow -CH_3$ or $-NH_2 \longleftarrow -NO_2$. If not, then:

Question (3). Is there any substituent in the target molecule that can be introduced by displacement of some other group? For example, $-CN \longleftarrow -N_2^+$ or $-OH \longleftarrow -Cl$ (in activated rings). If not, then finally:

Question (4). Is there a hydrogen in the target molecule that got there by displacement of some substituent? For example, $-H \longleftarrow -N_2^+$ or $-H \longleftarrow -SO_3H$.

Whenever the answer to a question is "Yes": we take off that group (Question 1); or transform that group (Question 2) into its precursor ($-COOH \longleftarrow -CH_3$, say); or replace that group (Question 3) by its precursor(s) ($ArCN \longleftarrow ArN_2^+ \longleftarrow ArNH_2$, say); or (Question 4) replace that $-H$ (or perhaps $-D$ or $-T$, in labeled compounds) by the previous occupant of the site.

Having done the transformations implied by the "Yes" answer, we start the questions again with the *new* structure we have just generated. We repeat these steps as many times as we have to until we ultimately work back to a starting material like benzene or toluene.

We now have a set of backward steps called *transforms*. This uncovers the various needed reactions (nitration, reduction of a diazonium ion, Sandmeyer reaction, etc.), but not necessarily in the proper order. Next, then, we check our *anti*thetic procedure by arranging the separate *transforms* into a workable *syn*thetic sequence of forward steps (*reactions*).

In doing this, we should be aware of the frequent need for *control elements*: groups that we introduce to insure specificity (regiospecificity or stereospecificity) in a later step. Control elements may protect a group or lower its activating effect (acetylation of an amine, say); they may block a position or, conversely, increase reactivity at a particular spot. The need for these control elements may not be obvious as we work out the antithetic transforms, but will usually appear when we try to put everything together in a series of synthetic reactions.

(The particular way these questions are asked may have a familiar ring—and so it should. This approach is derived from the one being worked out—most notably by E. J. Corey of Harvard University—for the use of computers to design organic syntheses. See E. J. Corey and W. T. Wipke, Science, **166**, 178 (1969); E. J. Corey *et al.*, J. Am. Chem. Soc., **94**, 421, 431, 440, 460 (1972).)

10. (a)

$$\underset{NH_2}{\overset{CH_3}{\bigcirc}}Br$$

Let us apply our general procedure to the synthesis of this compound, and see how it works.

(i) Question (1): none of the groups can be introduced (with, of course, the correct orientation) by direct substitution, so we proceed to the next question.

Question (2): —NH_2 can be formed by reduction of —NO_2, so we write out this step in the synthesis.

$$\underset{NH_2}{\overset{CH_3}{\bigcirc}}Br \quad\xleftarrow{\text{Fe, H}^+}\quad \underset{NO_2}{\overset{CH_3}{\bigcirc}}Br$$

(ii) Having done this, we begin again, and ask our questions about the new (nitro) compound.

Question (1): —NO_2 could be put in by direct substitution. But so can —Br, and regiospecifically. So let us try a bromination step, working back to a simpler compound.

$$\underset{NO_2}{\overset{CH_3}{\bigcirc}}Br \quad\xleftarrow{\text{Br}_2,\text{ Fe}}\quad \underset{NO_2}{\overset{CH_3}{\bigcirc}}$$

(iii) Having done this, we begin again, and ask questions about this new compound.

Question (1): —NO_2 can be put in by nitration—and, incidentally, the *p*-isomer is easily separated from the *o*-isomer. Thus, we have worked back to toluene, an acceptable starting material.

$$\underset{NO_2}{\overset{CH_3}{\bigcirc}} \quad\xleftarrow{\text{HNO}_3,\text{ H}_2\text{SO}_4}\quad \overset{CH_3}{\bigcirc}$$

Putting all this together, we get:

$$\underset{NH_2}{\overset{CH_3}{\bigcirc}}Br \xleftarrow{\text{Fe, H}^+} \underset{NO_2}{\overset{CH_3}{\bigcirc}}Br \xleftarrow{\text{Br}_2,\text{ Fe}} \underset{NO_2}{\overset{CH_3}{\bigcirc}} \xleftarrow{\text{HNO}_3,\text{ H}_2\text{SO}_4} \overset{CH_3}{\bigcirc}$$

When we check this sequence of transforms, we find that it corresponds to a workable sequence of reactions, and constitutes an acceptable synthesis of the product we want.

(b)

$$\underset{NH_2}{\overset{CH_3}{\bigcirc}}Br$$

Applying our general procedure here, we arrive at the following sequence of transforms.

$$\underset{NH_2}{\overset{CH_3}{\bigcirc}}Br \xleftarrow[(\text{Br}_2)]{\text{Question 1}} \underset{NH_2}{\overset{CH_3}{\bigcirc}} \xleftarrow[(\text{reduction})]{\text{Question 2}} \underset{NO_2}{\overset{CH_3}{\bigcirc}} \xleftarrow[(\text{nitration})]{\text{Question 1}} \underset{\text{Toluene}}{\overset{CH_3}{\bigcirc}}$$

Now we check this to see if it corresponds to a workable synthetic route. We realize that, while we can indeed brominate *p*-toluidine, in doing this we would introduce two —Br. If, however, we acetylate the —NH$_2$ first, we slow down substitution enough to get mono-bromination; and, after this is done, we can easily remove the acetyl group.

We arrive, then, at the following acceptable synthesis.

$$\underset{\substack{NH_2}}{\overset{\substack{CH_3}}{\bigcirc}}\!Br \xleftarrow[H^+]{H_2O} \underset{\substack{NHCOCH_3}}{\overset{\substack{CH_3}}{\bigcirc}}\!Br \xleftarrow[Fe]{Br_2} \underset{\substack{NHCOCH_3}}{\overset{\substack{CH_3}}{\bigcirc}} \xleftarrow{Ac_2O} \underset{\substack{NH_2}}{\overset{\substack{CH_3}}{\bigcirc}} \quad \text{(page 770)}$$

We have added two steps, acetylation and subsequent hydrolysis, but by these steps we have gained control over the most critical part of the whole synthesis, introduction of bromine.

(c) $\underset{\substack{SO_2NHPh}}{\overset{\substack{NH_2}}{\bigcirc}}$

Here, our general procedure leads us to this sequence of transforms.

$$\underset{\substack{SO_2NHPh}}{\overset{\substack{NH_2}}{\bigcirc}} \xleftarrow[\text{(PhNH}_2)]{\text{Question 2}} \underset{\substack{SO_2Cl}}{\overset{\substack{NH_2}}{\bigcirc}} \xleftarrow[\text{(ClSO}_3\text{H)}]{\text{Question 2}} \overset{\substack{NH_2}}{\bigcirc} \xleftarrow{} PhNO_2 \xleftarrow{} C_6H_6$$

This we find to be defective as a synthetic route. We cannot have a free —NH$_2$ and —SO$_2$Cl in the same compound: they would react with each other. As in (b), we temporarily protect —NH$_2$ by acetylation. In the final step we remove the acetyl group by hydrolysis under conditions that do not also hydrolyze the less reactive sulfonamide group (see page 762).

$$\underset{\substack{SO_2NHPh}}{\overset{\substack{NH_2}}{\bigcirc}} \xleftarrow[H^+]{H_2O} \underset{\substack{SO_2NHPh}}{\overset{\substack{NHCOCH_3}}{\bigcirc}} \xleftarrow{} \underset{\text{(from C}_6\text{H}_6)}{PhNH_2 +} \underset{\substack{SO_2Cl}}{\overset{\substack{NHCOCH_3}}{\bigcirc}} \xleftarrow{ClSO_3H} \underset{\substack{}}{\overset{\substack{NHCOCH_3}}{\bigcirc}} \\ \text{(page 746)}$$

(d) $\underset{\substack{NH_2}}{\overset{\substack{NHCOCH_3}}{\bigcirc}} \xleftarrow{Fe, H^+} \underset{\substack{NO_2}}{\overset{\substack{NHCOCH_3}}{\bigcirc}} \quad \text{(page 760)}$

(e) $\underset{\substack{N=O}}{\overset{\substack{NEt_2}}{\bigcirc}} \xleftarrow{HONO} PhNEt_2 \xleftarrow{2EtI} PhNH_2 \quad \text{(page 733)}$

(f) $\underset{\substack{NH_2}}{\overset{\substack{COOH}}{\bigcirc}}\!NO_2 \xleftarrow[H^+]{H_2O} \underset{\substack{NHCOCH_3}}{\overset{\substack{COOH}}{\bigcirc}}\!NO_2 \xleftarrow[H_2SO_4]{HNO_3} \underset{\substack{NHCOCH_3}}{\overset{\substack{COOH}}{\bigcirc}} \xleftarrow[\text{warm}]{KMnO_4} \underset{\substack{NHCOCH_3}}{\overset{\substack{CH_3}}{\bigcirc}} \quad \text{(page 771)}$

(g) $\underset{\substack{CH(CH_3)_2}}{\overset{\substack{NH_2}}{Br\,\bigcirc\,Br}} \xleftarrow{2Br_2(aq)} \underset{\substack{CH(CH_3)_2}}{\overset{\substack{NH_2}}{\bigcirc}} \xleftarrow[H^+]{Fe} \underset{\substack{CH(CH_3)_2}}{\overset{\substack{NO_2}}{\bigcirc}} \xleftarrow[H_2SO_4]{HNO_3} \underset{\substack{CH(CH_3)_2}}{\overset{}{\bigcirc}} \quad \text{(from C}_6\text{H}_6)$

(h) $\underset{\substack{NH_2}}{\overset{\substack{CH_2NH_2}}{\bigcirc}} \xleftarrow{Fe \atop H^+} \underset{\substack{NO_2}}{\overset{\substack{CH_2NH_2}}{\bigcirc}} \xleftarrow{NH_3} \underset{\substack{NO_2}}{\overset{\substack{CH_2Cl}}{\bigcirc}} \xleftarrow[\text{heat}]{Cl_2} \underset{\substack{NO_2}}{\overset{\substack{CH_3}}{\bigcirc}} \xleftarrow[H_2SO_4]{HNO_3} C_6H_5CH_3$

(i) 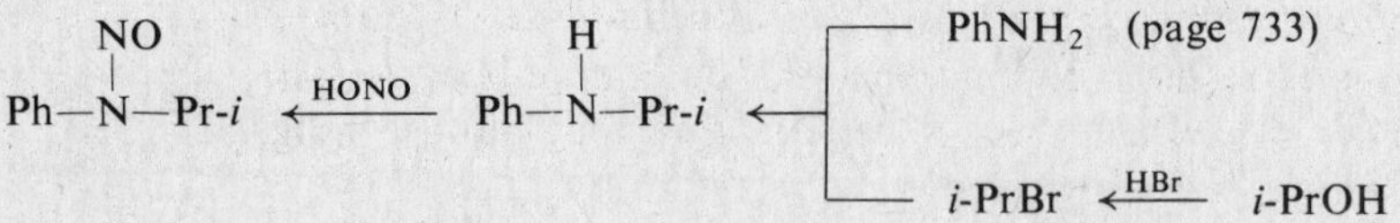 Ph—N—Pr-*i* $\xleftarrow{\text{HONO}}$ Ph—N—Pr-*i* $\longleftarrow$ PhNH₂ (page 733)

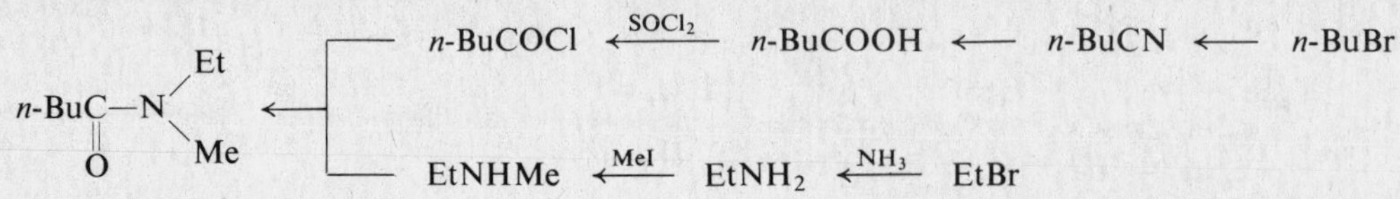

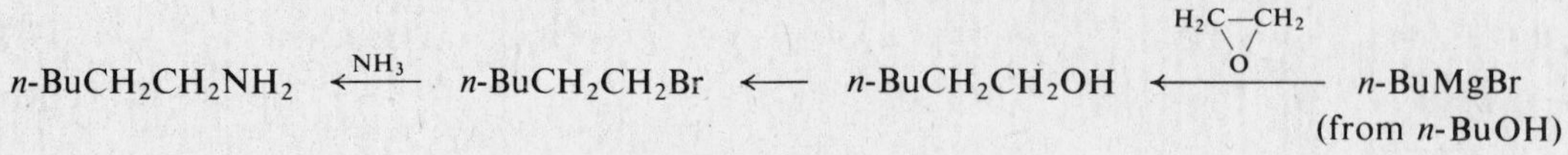

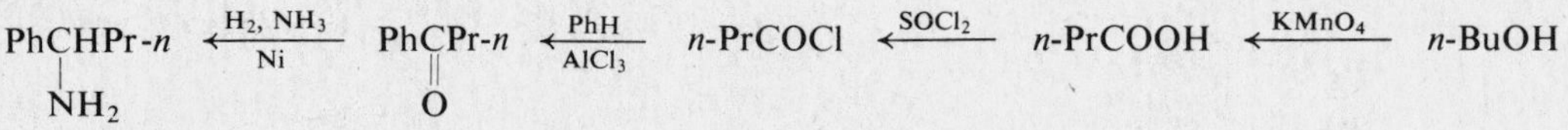

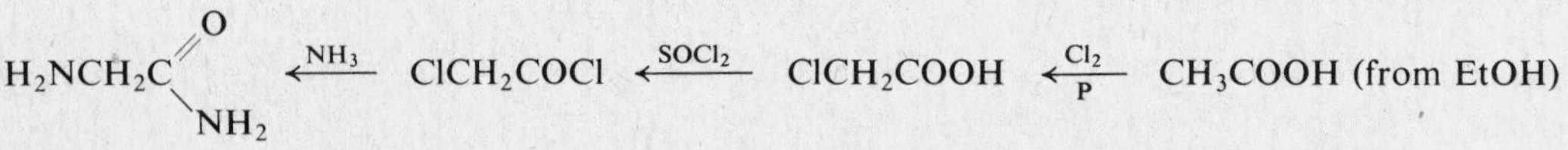

NO ... H

i-PrBr $\xleftarrow{\text{HBr}}$ *i*-PrOH

(j) *n*-BuC—N(Et)(Me) $\longleftarrow$ *n*-BuCOCl $\xleftarrow{\text{SOCl}_2}$ *n*-BuCOOH $\longleftarrow$ *n*-BuCN $\longleftarrow$ *n*-BuBr

EtNHMe $\xleftarrow{\text{MeI}}$ EtNH₂ $\xleftarrow{\text{NH}_3}$ EtBr

(k) *n*-BuCH₂CH₂NH₂ $\xleftarrow{\text{NH}_3}$ *n*-BuCH₂CH₂Br $\longleftarrow$ *n*-BuCH₂CH₂OH $\longleftarrow$ *n*-BuMgBr (from *n*-BuOH)

(l) PhCHPr-*n* (NH₂) $\xleftarrow[\text{Ni}]{\text{H}_2, \text{NH}_3}$ PhCPr-*n* (O) $\xleftarrow[\text{AlCl}_3]{\text{PhH}}$ *n*-PrCOCl $\xleftarrow{\text{SOCl}_2}$ *n*-PrCOOH $\xleftarrow{\text{KMnO}_4}$ *n*-BuOH

(m) H₂NCH₂C(O)(NH₂) $\xleftarrow{\text{NH}_3}$ ClCH₂COCl $\xleftarrow{\text{SOCl}_2}$ ClCH₂COOH $\xleftarrow[\text{P}]{\text{Cl}_2}$ CH₃COOH (from EtOH)

(n) PhCNHCH₂COOH (O) $\longleftarrow$ PhCOCl $\xleftarrow{\text{SOCl}_2}$ PhCOOH $\xleftarrow{\text{KMnO}_4}$ PhCH₃

H₂NCH₂COOH $\xleftarrow[\text{excess}]{\text{NH}_3}$ ClCH₂COOH (as in m, above)

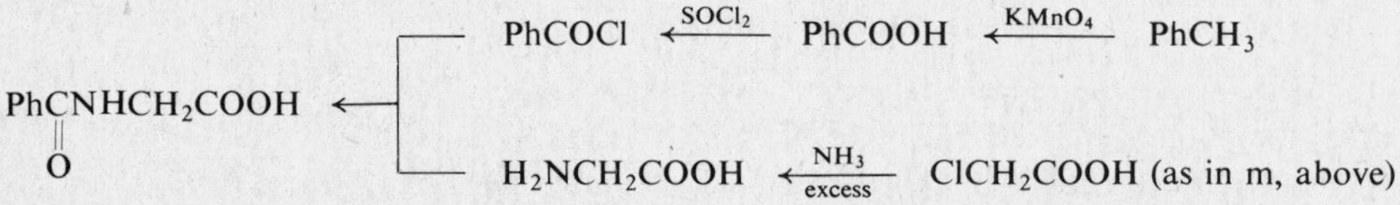

11. (a) Of all the six isomers, 2,3-dibromotoluene is the hardest to make. Use the antithetic approach outlined in the box above, and see how neatly the question-and-answer procedure gets you back to a readily available starting compound.

(i)

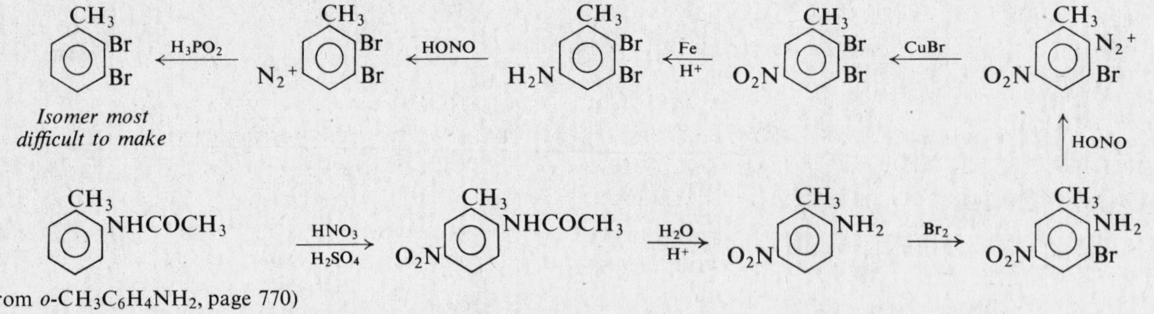

Isomer most difficult to make

(from *o*-CH₃C₆H₄NH₂, page 770)

About half-way back, you may face the following choice: which —Br should be introduced last, the one next to —CH₃, or the other one?

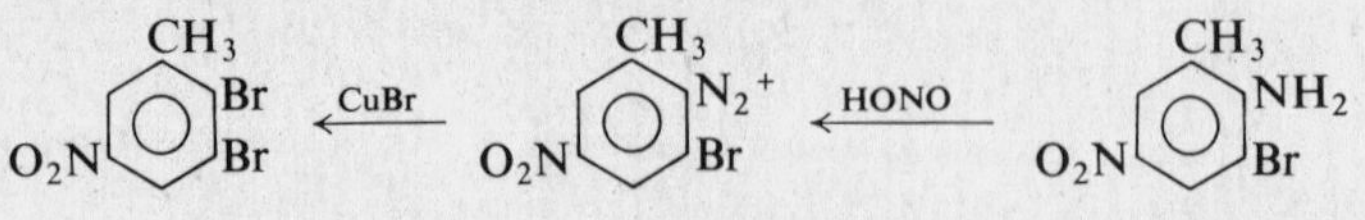

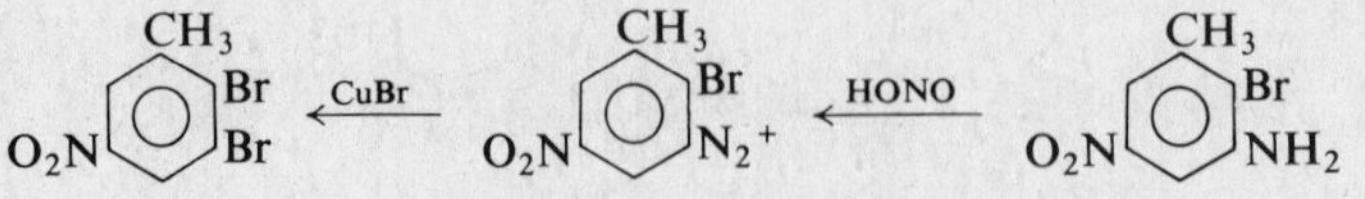

The answer lies in another question: which amine of the two shown above is the easier to make? The answer to this question is amine I, as shown in the complete answer outlined below, but you should try to work out a route to amine II just to prove to yourself which is easier to make.

A key role is played by —NO$_2$, which must be put on the ring temporarily to block one of the positions activated by —NH$_2$.

(ii) 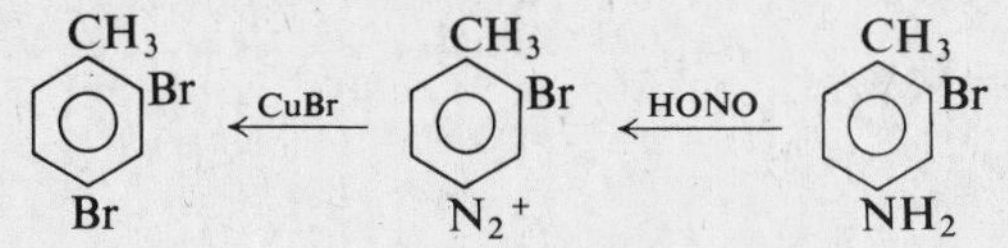   [structure: CH$_3$ ring with Br, Br] $\xleftarrow{\text{CuBr}}$ [CH$_3$ ring with Br, N$_2^+$] $\xleftarrow{\text{HONO}}$ [CH$_3$ ring with Br, NH$_2$] (as in Problem 10a)

(iii) 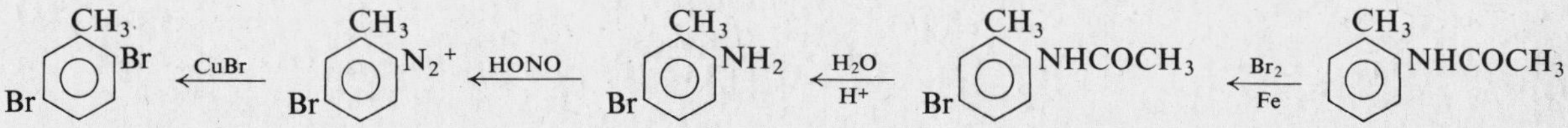 [CH$_3$ ring with Br, Br, Br] $\xleftarrow{\text{CuBr}}$ [CH$_3$ ring with Br, N$_2^+$] $\xleftarrow{\text{HONO}}$ [CH$_3$ ring with Br, NH$_2$] $\xleftarrow[\text{H}^+]{\text{H}_2\text{O}}$ [CH$_3$ ring with Br, NHCOCH$_3$] $\xleftarrow[\text{Fe}]{\text{Br}_2}$ [CH$_3$ ring with NHCOCH$_3$]

(from o-CH$_3$C$_6$H$_4$NH$_2$, page 770)

(iv) 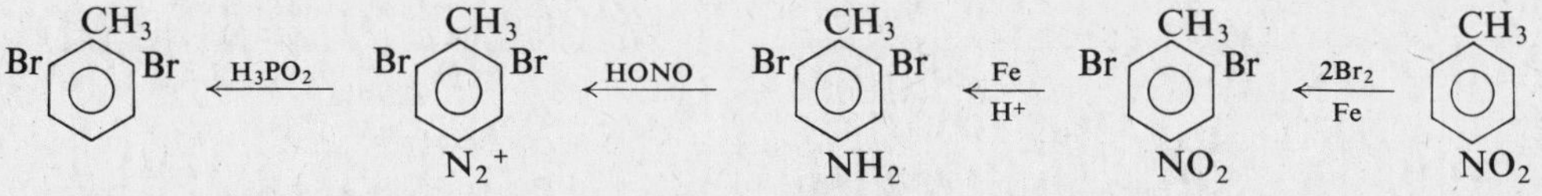 [CH$_3$ ring with Br, Br] $\xleftarrow{\text{H}_3\text{PO}_2}$ [CH$_3$ ring with Br, Br, N$_2^+$] $\xleftarrow{\text{HONO}}$ [CH$_3$ ring with Br, Br, NH$_2$] $\xleftarrow[\text{H}^+]{\text{Fe}}$ [CH$_3$ ring with Br, Br, NO$_2$] $\xleftarrow[\text{Fe}]{2\text{Br}_2}$ [CH$_3$ ring with NO$_2$] (page 345)

In making isomer iv we get all the way to Question 4 (see the box above) before we answer, "Yes." The sequence is —H ⟵ —N$_2^+$ ⟵ —NH$_2$ ⟵ —NO$_2$.

(v) 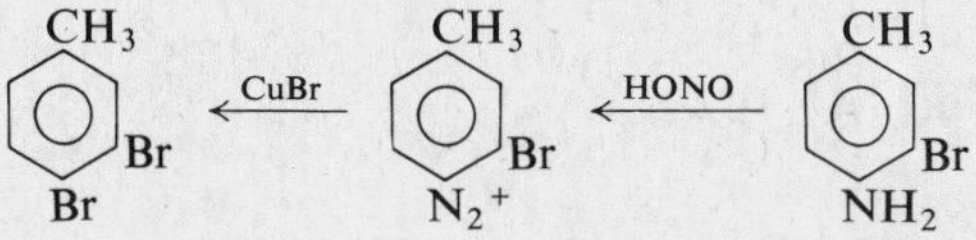 [CH$_3$ ring with Br, Br] $\xleftarrow{\text{CuBr}}$ [CH$_3$ ring with Br, N$_2^+$] $\xleftarrow{\text{HONO}}$ [CH$_3$ ring with Br, NH$_2$] (page 771)

(vi) 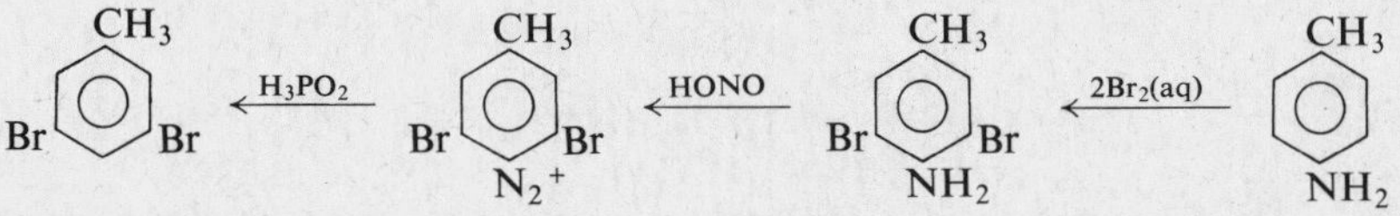 [CH$_3$ ring with Br, Br] $\xleftarrow{\text{H}_3\text{PO}_2}$ [CH$_3$ ring with Br, Br, N$_2^+$] $\xleftarrow{\text{HONO}}$ [CH$_3$ ring with Br, Br, NH$_2$] $\xleftarrow{2\text{Br}_2\text{(aq)}}$ [CH$_3$ ring with NH$_2$] (page 770)

In looking at isomer vi we may recognize that the two —Br's must have got onto the ring under the influence of a now-vanished group much more powerfully directing than —CH$_3$. The —NH$_2$ group is the logical choice, a powerful o,p-director easily removed by reductive deamination (diazotization and reduction).

(b) (i) 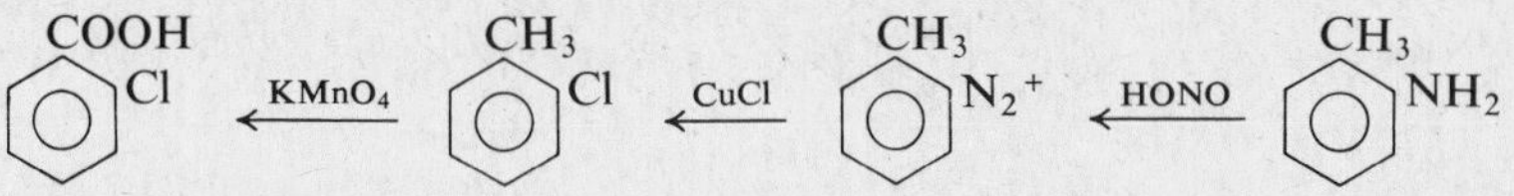 [COOH ring with Cl] $\xleftarrow{\text{KMnO}_4}$ [CH$_3$ ring with Cl] $\xleftarrow{\text{CuCl}}$ [CH$_3$ ring with N$_2^+$] $\xleftarrow{\text{HONO}}$ [CH$_3$ ring with NH$_2$] (page 770)

(ii) 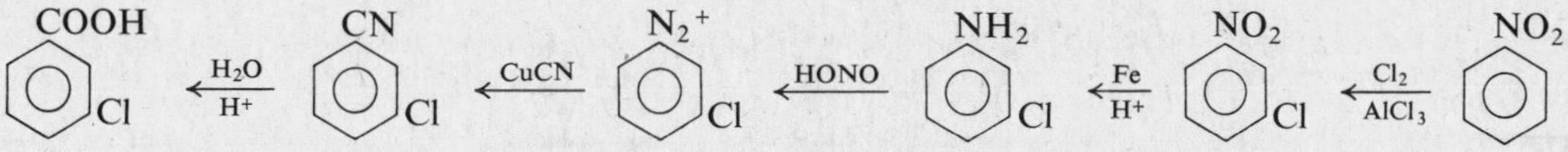 [COOH ring with Cl] $\xleftarrow[\text{H}^+]{\text{H}_2\text{O}}$ [CN ring with Cl] $\xleftarrow{\text{CuCN}}$ [N$_2^+$ ring with Cl] $\xleftarrow{\text{HONO}}$ [NH$_2$ ring with Cl] $\xleftarrow[\text{H}^+]{\text{Fe}}$ [NO$_2$ ring with Cl] $\xleftarrow[\text{AlCl}_3]{\text{Cl}_2}$ [NO$_2$ ring]

(Or: start with m-nitrotoluene, from Problem 23.15, page 771.)

(iii) 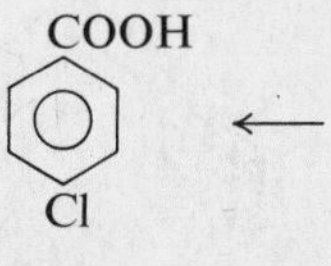 [COOH ring with Cl] ⟵ as for o-isomer, except start with p-nitrotoluene (page 345)

(c) (i)

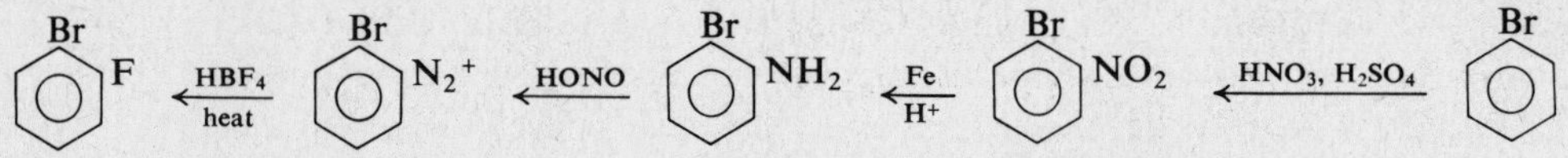

Br / F ←(HBF₄, heat)— Br / N₂⁺ ←(HONO)— Br / NH₂ ←(Fe/H⁺)— Br / NO₂ ←(HNO₃, H₂SO₄)— Br

(Separate from p-isomer)

(ii)

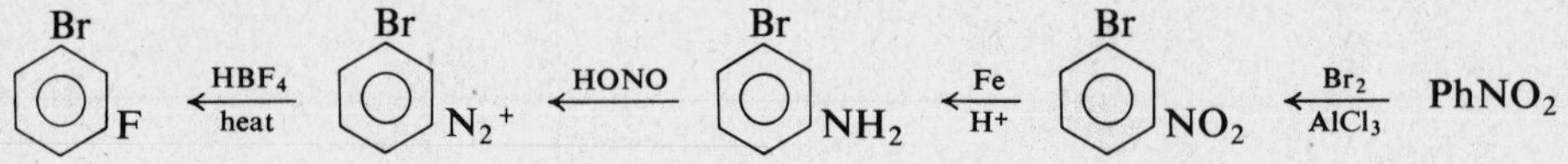

Br / F ←(HBF₄, heat)— Br / N₂⁺ ←(HONO)— Br / NH₂ ←(Fe/H⁺)— Br / NO₂ ←(Br₂, AlCl₃)— PhNO₂

(iii)

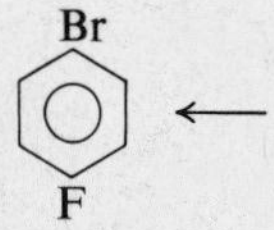

Br / F ← as for *o*-isomer, above, except start with *p*-bromonitrobenzene.

12. (a)

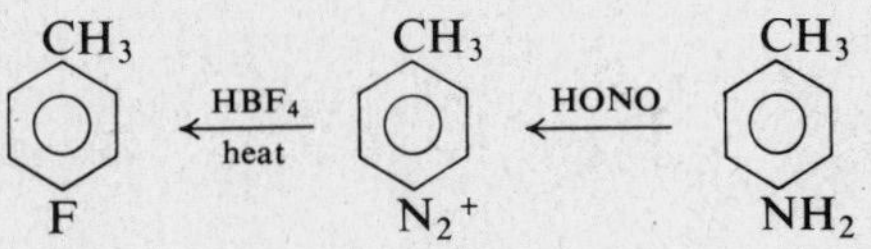

CH₃ / F ←(HBF₄, heat)— CH₃ / N₂⁺ ←(HONO)— CH₃ / NH₂   (page 770)

(b)

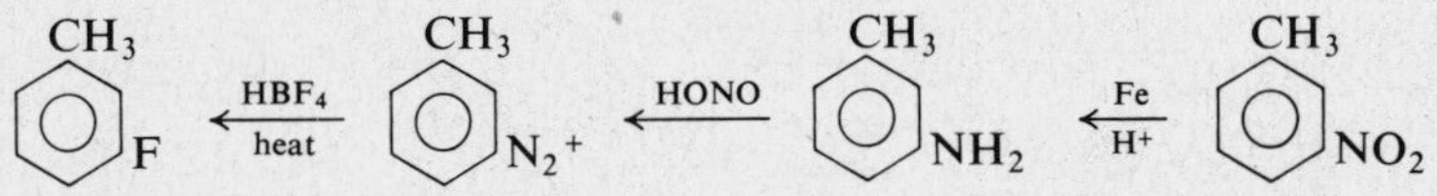

CH₃ / F ←(HBF₄, heat)— CH₃ / N₂⁺ ←(HONO)— CH₃ / NH₂ ←(Fe/H⁺)— CH₃ / NO₂   (as in Problem 23.15, page 771)

(c)

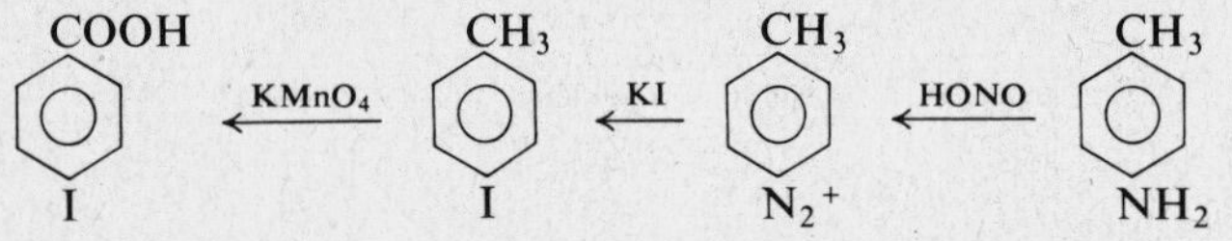

COOH / I ←(KMnO₄)— CH₃ / I ←(KI)— CH₃ / N₂⁺ ←(HONO)— CH₃ / NH₂   (page 770)

(d)

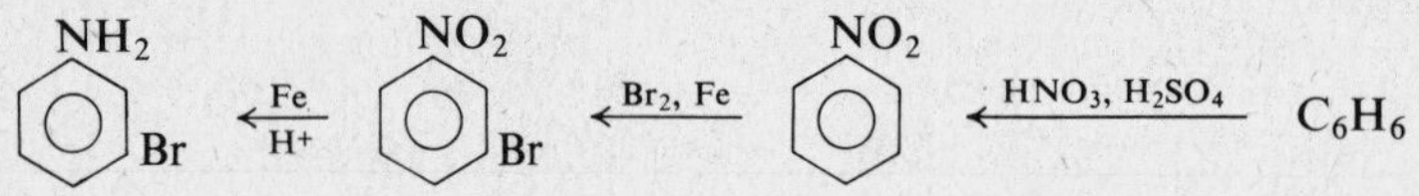

NH₂ / Br ←(Fe/H⁺)— NO₂ / Br ←(Br₂, Fe)— NO₂ ←(HNO₃, H₂SO₄)— C₆H₆

(e)

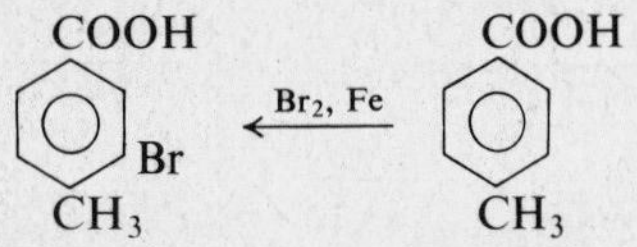

COOH / Br / CH₃ ←(Br₂, Fe)— COOH / CH₃   (page 769)

(f)

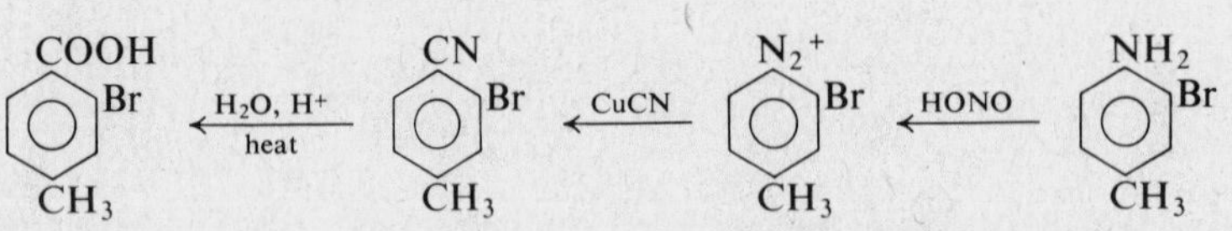

COOH / Br / CH₃ ←(H₂O, H⁺, heat)— CN / Br / CH₃ ←(CuCN)— N₂⁺ / Br / CH₃ ←(HONO)— NH₂ / Br / CH₃   (page 771)

(g)

OH / Et ←(H₂O, H⁺, warm)— N₂⁺ / Et ←(HONO)— NH₂ / Et ←(Fe/H⁺)— NO₂ / Et

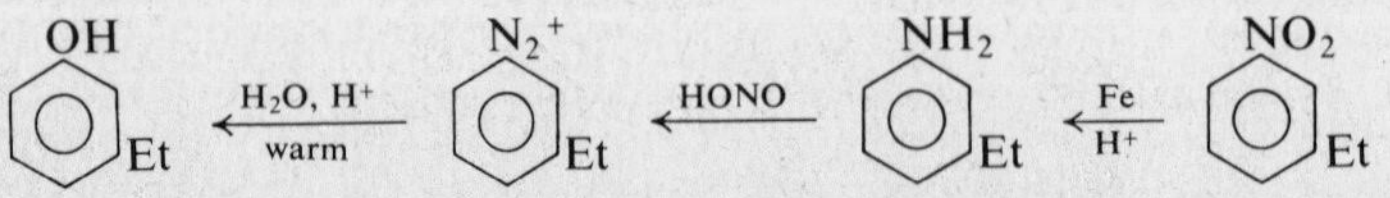

The required *m*-nitroethylbenzene is made in the same manner as *m*-nitrotoluene (in Problem 23.15), starting with ethylbenzene (page 395).

(h) Br—⟨NH₂/Br⟩ ←(Fe/H⁺)— Br—⟨NO₂/Br⟩ ← as in Problem 23.17

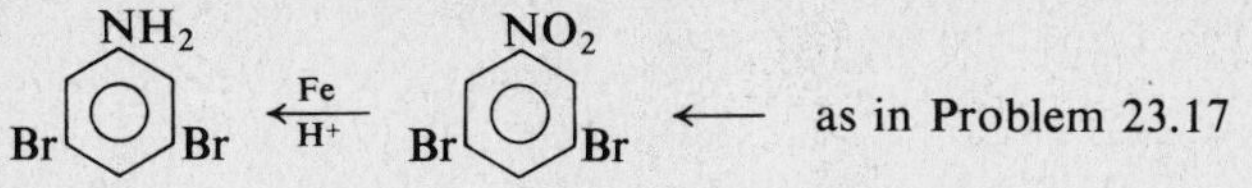

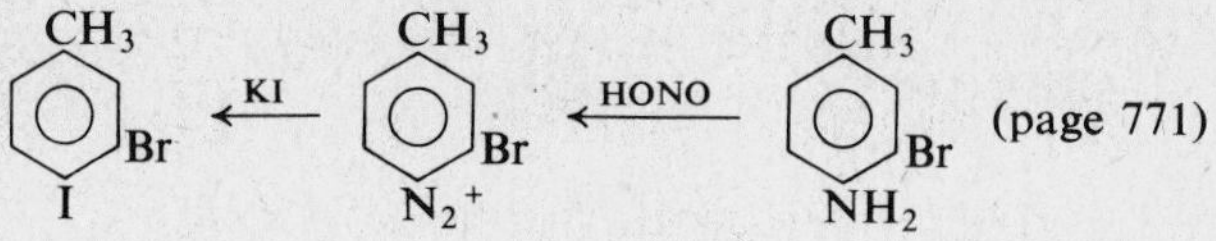

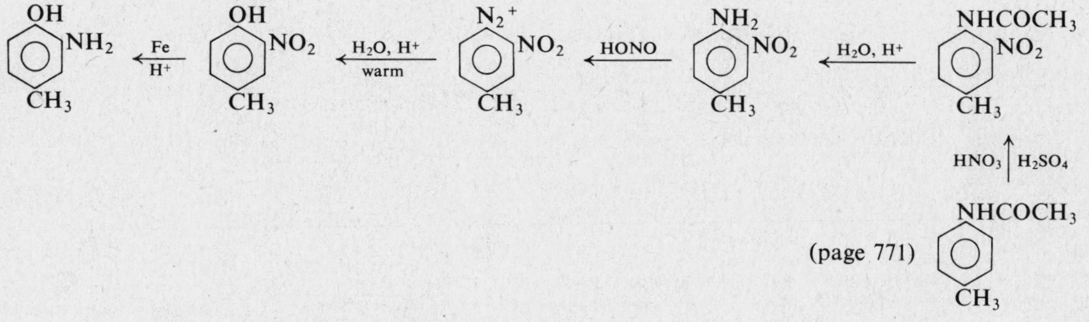

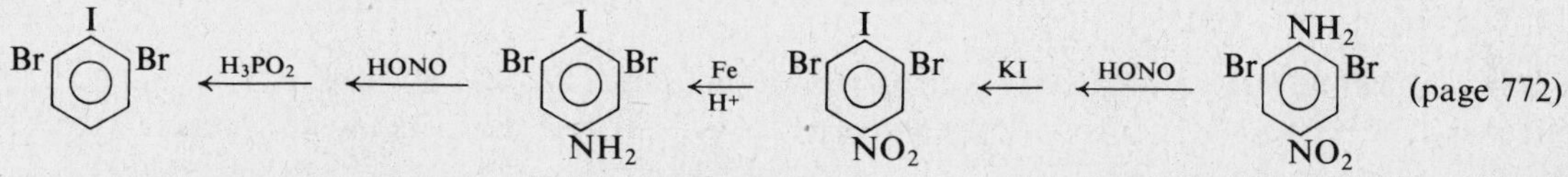

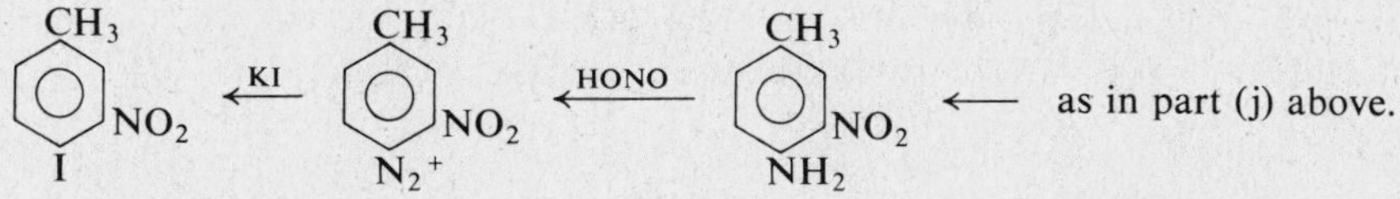

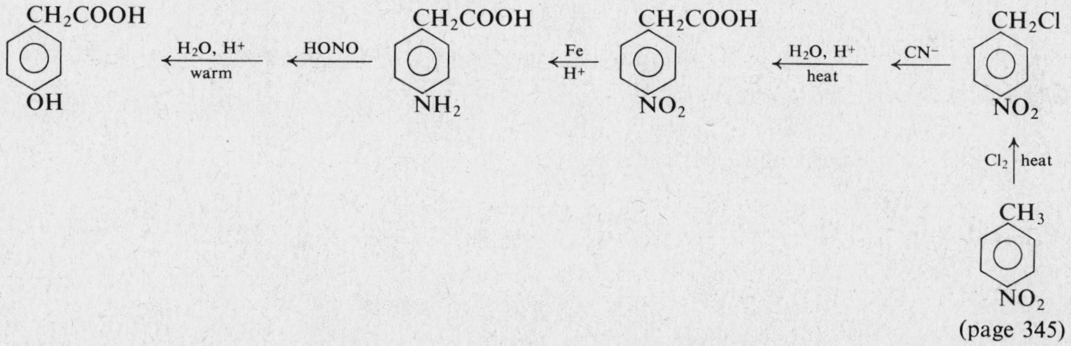

(i) ⟨CH₃/Br/I⟩ ←(KI)— ⟨CH₃/Br/N₂⁺⟩ ←(HONO)— ⟨CH₃/Br/NH₂⟩ (page 771)

(j) ⟨OH/NH₂/CH₃⟩ ←(Fe/H⁺)— ⟨OH/NO₂/CH₃⟩ ←(H₂O, H⁺ warm)— ⟨N₂⁺/NO₂/CH₃⟩ ←(HONO)— ⟨NH₂/NO₂/CH₃⟩ ←(H₂O, H⁺)— ⟨NHCOCH₃/NO₂/CH₃⟩

↑ HNO₃ | H₂SO₄

(page 771) ⟨NHCOCH₃/CH₃⟩

(k) Br—⟨I/Br⟩ ←(H₃PO₂)—(HONO)— Br—⟨I/Br/NH₂⟩ ←(Fe/H⁺)— Br—⟨I/Br/NO₂⟩ ←(KI)—(HONO)— Br—⟨NH₂/Br/NO₂⟩ (page 772)

(l) ⟨CH₃/NO₂/I⟩ ←(KI)— ⟨CH₃/NO₂/N₂⁺⟩ ←(HONO)— ⟨CH₃/NO₂/NH₂⟩ ← as in part (j) above.

(m) ⟨CH₂COOH/OH⟩ ←(H₂O, H⁺ warm)—(HONO)— ⟨CH₂COOH/NH₂⟩ ←(Fe/H⁺)— ⟨CH₂COOH/NO₂⟩ ←(H₂O, H⁺ heat)—(CN⁻)— ⟨CH₂Cl/NO₂⟩

↑ Cl₂ | heat

⟨CH₃/NO₂⟩
(page 345)

(n) ⟨CH₃/Br/Cl⟩ ←(CuCl)—(HONO)— ⟨CH₃/Br/NH₂⟩ ←(Fe/H⁺)— ⟨CH₃/Br/NO₂⟩ ←(Br₂/Fe)— ⟨CH₃/NO₂⟩ (page 345)

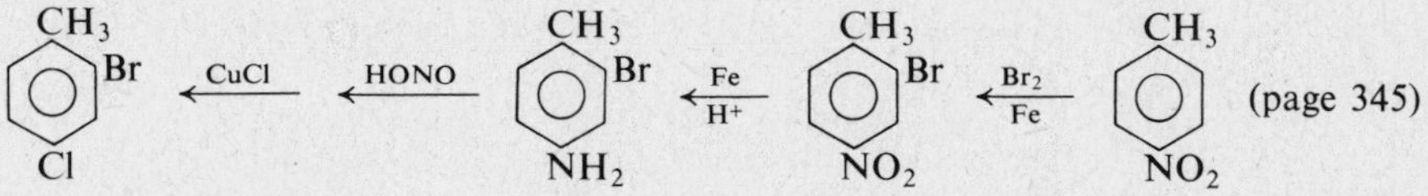

13. (a) HOOC(CH₂)₄COOH + H₂N(CH₂)₆NH₂ ⟶ salt

heat ↓ −H₂O

⁓C(CH₂)₄C—NH(CH₂)₆NH—C(CH₂)₄C—NH(CH₂)₆NH⁓
   ‖   ‖   ‖   ‖
   O   O   O   O

A polyamide

(b) Acidic hydrolysis of amide linkages.

14. A poor leaving group (OH^-) is converted into a good leaving group (OTs^-).

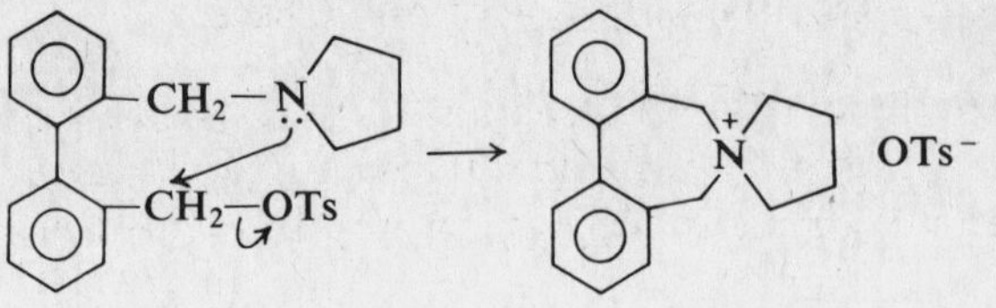

15. (a) Halide ion competes with water as the nucleophile:

Reaction of $Ph-N_2^+$ is S_N1-like:

(1) $Ph-N_2^+ \longrightarrow Ph^+ + N_2^+$ *Slow: rate-determining*

(2) $Ph^+ + :Z \longrightarrow Ph-Z$

The concentration and nature of the nucleophiles do not affect the *rate* of reaction (step 1), but they do affect *product composition*, controlled in step (2): the more halide ion present, the more halobenzene formed.

Electron-withdrawal by $-NO_2$ slows down formation of the aryl cation from $p\text{-}NO_2C_6H_4N_2^+$, and the reaction takes place by an S_N2-like substitution:

$$p\text{-}O_2NPh-N_2^+ \longrightarrow \left[p\text{-}O_2NPh \begin{array}{c} Z \\ N_2^{\delta+} \end{array} \right] \longrightarrow p\text{-}O_2NPh-Z + N_2$$

In this single-step reaction, the concentration and nature of the nucleophiles affect both rate and product composition.

(b) N_2 is an extraordinarily good leaving group.

16. (a) Hinsberg test   (b) HONO, then β-naphthol

(c) Hinsberg test   (d) Hinsberg test

(e) $3°$ salt $\xrightarrow{OH^-}$ odor or basic vapors of amine   (f) $AgNO_3$ ($\longrightarrow$ AgCl from $PhNH_3^+Cl^-$)

(g) CrO_3/H_2SO_4 (h) HCl(aq)

(i) $BaCl_2$ ($\longrightarrow$ $BaSO_4$ from sulfate)   (j) $AgNO_3$ ($\longrightarrow$ AgCl from $EtNH_3^+Cl$)

(k) HCl(aq) dissolves $PhNH_2$ (l) $BaCl_2$ ($\longrightarrow$ $BaSO_4$ from sulfate), or H_2O (dissolves salt), or NaOH (sets free $PhNH_2$)

17. Compare answer to Problem 22.1, above.

(a)–(d) Dissolve basic amine in aqueous acid. Neutral compound (alkane, ether, amide, dipolar ion) is unaffected and can be collected by distillation or filtration. Amine is then regenerated from the acidic solution of its salt by the action of strong base.

(e) Transform 2° amine into neutral amide by acetic anhydride, the 3° amine remaining unacetylated. Then proceed with separation of basic amine and neutral amide as in (c). Finally, hydrolyze the amide back to 2° amine by basic hydrolysis, and collect amine by distillation.

(f) Separate acid first by dissolving it in aqueous NaOH; then proceed to separate the in-soluble amine and hydrocarbon as in (a). Regenerate the acid from the alkaline solution of its salt by the action of strong acid.

(g)–(h) Proceed as in (c).

18. (a) HONO, then β-naphthol for aniline; then Hinsberg test (2° vs. 3°).

(b) HCl(aq): base vs. neutral amide.

(c) Hinsberg to distinguish 3° and 2° amines; only one of the 1° amines (o-toluidine) will give a diazotization and coupling test.

(d) Hot NaOH(aq) gives NH_3 from ethyl oxamate ($RCONH_2$).

(e) $HCONH_2$ is soluble in H_2O; basic 3° amine is soluble in HCl(aq), but neutral nitrile is insoluble.

(f) Only the basic amine is soluble in HCl(aq); hot NaOH(aq) on nitrile gives NH_3; $PhNO_2$ unreactive.

(g) Elemental analysis distinguishes tosyl chloride (S, Cl) and the two sulfonamides (S, N), which can be further distinguished by NaOH(aq) in manner of Hinsberg test.

   Elemental analysis distinguishes p-chloroaniline (Cl, N) and p-nitrobenzyl chloride (Cl, N) which can be further distinguished by $AgNO_3$ (side-chain Cl $\longrightarrow$ AgCl).

   Of the compounds remaining: only the 3° amine is colorless; o-nitroaniline (K_b 6 × 10^{-14}) is more soluble than 2,4-dinitroaniline (K_b about 10^{-19}) in HCl(aq); and 2,4-dinitro-aniline gives NH_3 by nucleophilic aromatic substitution (see Sec. 25.7) when warmed with NaOH(aq).

19. Test amine by Hinsberg procedure.
   If 3°, test for Cl.
   If 2°, make acetyl derivative, determine m.p.; if m.p. is 54–55°, make p-toluenesulfon-amide to distinguish between possibilities.
   If 1°, test for Cl.

20. (a)

	C_3H_9N	trimethylamine	$C_5H_{15}O_2N$	choline
+	C_2H_4O	ethylene oxide	− $C_5H_{13}ON$	
	$C_5H_{13}ON$		H_2O	has been added

$$H_2C\!-\!CH_2 + NMe_3 \longrightarrow {}^-OCH_2CH_2\overset{+}{N}Me_3 \xrightarrow{H_2O} HOCH_2CH_2NMe_3{}^+OH^-$$

$$\underset{O}{\qquad\qquad} \qquad\qquad (C_5H_{13}ON) \qquad\qquad\qquad \underset{(C_5H_{15}O_2N)}{\text{Choline}}$$

The quaternary ammonium hydroxide is completely ionized, and gives a strongly basic solution.

(b) The formula is increased by C_2H_2O, indicating replacement of H— by CH_3CO—.

$$CH_3COCH_2CH_2NMe_3{}^+OH^-$$
$$\underset{\text{O}}{\|}$$

Acetylcholine

21. (a) Paragraph 1: basic 1° aromatic amine.

Paragraph 2: hydrolysis of an acid derivative gives the salt of the acid and a water-soluble alcohol or amine.

Paragraph 3: pptn. of RCOOH. Subsequent dissolution indicates that the acid is amphoteric, with an amino group also present.

Chemical arithmetic:

$$
\begin{array}{ccc}
C_7H_7O_2N & C_6H_6N & \\
- \text{COOH} & - C_6H_4 & NH_2 \\
\hline
C_6H_6N & NH_2 & \textit{shows } C_6H_4 \\
& & \diagdown \\
& & COOH
\end{array}
$$

M.p. leaves no choice but *p*-aminobenzoic acid for A (Table 18.1, page 580).

Let us go back to the original Novocaine. Since it is an acid derivative, we can subtract the acyl group of A.

$$
\begin{array}{ll}
C_{13}H_{20}O_2N_2 & \text{Novocaine} \\
- H_2NC_6H_4CO & \text{acyl group} \\
\hline
C_6H_{14}ON &
\end{array}
$$

We have one nitrogen still unassigned; we cannot say yet whether it is in an amide or amino group, since in either case Novocaine would be acid-soluble by virtue of the —NH_2 group of A. So we must investigate the rest of the hydrolysis mixture, the ether solution of an amine (or possibly an aminoalcohol).

Paragraph 4: B is an amine. Reaction with acetic anhydride produces a material, C, which is still basic and therefore still an amine. B thus can only be a tertiary amine. (Tertiary amines do not form amides.) This means that Novocaine is an ester, not an amide. C must be an acetate ester of an alcohol.

$$
\begin{array}{ll}
C_8H_{17}O_2N & C \\
- C_2H_2O & \text{one ester group adds} \\
\hline
C_6H_{15}ON = B &
\end{array}
$$

Paragraph 5: $H_2C\!-\!CH_2 + NHEt_2 \longrightarrow HOCH_2CH_2NEt_2$
$$\underset{\text{O}}{\diagdown\!\diagup} \qquad\qquad\qquad\qquad\qquad\qquad B$$

Novocaine, therefore, is the *p*-aminobenzoic ester of alcohol B:

$$H_2N\langle\bigcirc\rangle C\!\!\nearrow^{O}_{\diagdown OCH_2CH_2N(C_2H_5)_2}$$

Novocaine

(b)

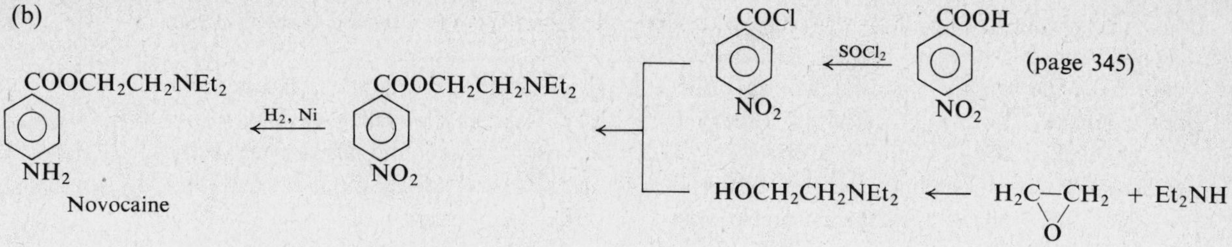
COOCH₂CH₂NEt₂
NH₂
Novocaine
H₂, Ni
COOCH₂CH₂NEt₂
NO₂
COCl
NO₂
SOCl₂
COOH
NO₂
(page 345)
HOCH₂CH₂NEt₂
H₂C—CH₂ + Et₂NH
O

22. D is a neutral compound. Its slow dissolution in hot NaOH points to its being an acid derivative. Formation of E, a base and thus an amine, points to D's being an amide. E itself is a 2° amine, giving G, $PhSO_2NRR'$, a neutral sulfonamide. The nitrogen-free acid F, m.p. 180°, is *p*-toluic acid (Table 18.1, page 580).

As usual, chemical arithmetic is helpful:

$C_{15}H_{15}ON$	cpd. D
$-\ CH_3C_6H_4CO$	acyl group of F
C_7H_8N	amino portion of amide

E begins to look like an aromatic amine,

$$\begin{array}{r} C_7H_8N \\ -\ C_6H_5 \\ \hline CH_3N \end{array}$$

and is undoubtedly N-methylaniline, $C_6H_5NHCH_3$. G, then, is $PhSO_2N(CH_3)Ph$.

Putting the pieces together gives us D:

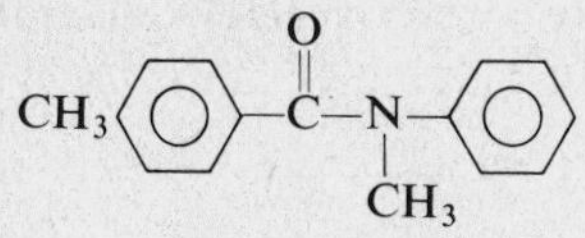
O
CH₃
C—N
CH₃

D

23.

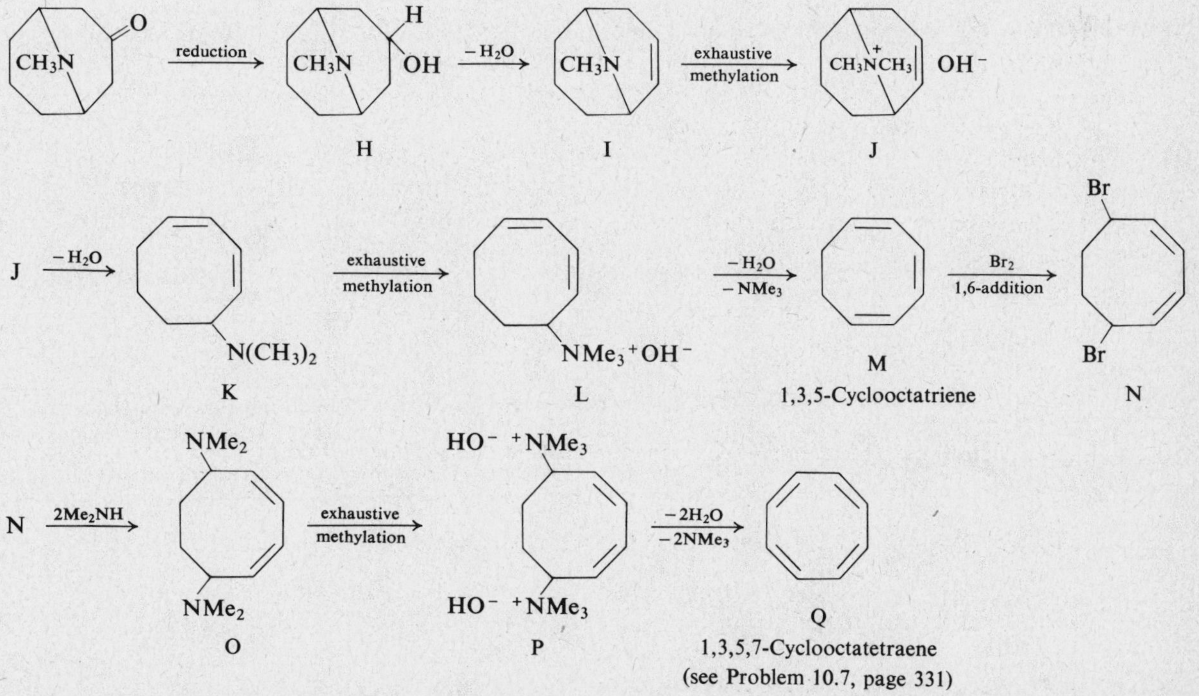
CH₃N
O
reduction
H
CH₃N
OH
H
−H₂O
CH₃N
I
exhaustive methylation
CH₃NCH₃
OH⁻
J
J
−H₂O
N(CH₃)₂
K
exhaustive methylation
NMe₃⁺OH⁻
L
−H₂O
−NMe₃
M
1,3,5-Cyclooctatriene
Br₂
1,6-addition
Br
Br
N
N
2Me₂NH
NMe₂
NMe₂
O
exhaustive methylation
HO⁻ ⁺NMe₃
HO⁻ ⁺NMe₃
P
−2H₂O
−2NMe₃
Q
1,3,5,7-Cyclooctatetraene
(see Problem 10.7, page 331)

This synthesis of cyclooctatetraene was carried out in 1911–1913 by the great German chemist and Nobel Prize winner Richard Willstätter.   As we saw earlier (Problem 10.7, page 331), the product was found to have the normal properties of an alkene, and hence is not aromatic.   Today, a beginning student finds this not surprising: ten π electrons does not fit Hückel's $4n + 2$ rule for aromaticity.   But fifty years ago this finding was one of the principal pieces of evidence that the benzene structure was very special—but in just what way was not realized until the Hückel rule in 1931.   (On various grounds, many chemists doubted that Willstätter had actually made cyclooctatetraene; the synthesis was laboriously repeated—and his findings verified—by other workers in 1948.   If Willstätter had observed the properties then *expected* of cyclooctatetraene, one wonders, would these doubts have been quite so strong?)

24. Pantothenic acid is a monocarboxylic acid: it yields a monosalt and a monoester.   It appears to be an amide as well: it contains non-basic nitrogen, and is hydrolyzed to β-aminopropionic acid and V.   Since V contains no nitrogen, β-aminopropionic acid must provide the nitrogen of the amide:

$$R-\underset{\underset{O}{\|}}{C}-NHCH_2CH_2COOH$$

Chemical arithmetic,

$$
\begin{array}{l}
C_9H_{17}O_5N \qquad\qquad \text{pantothenic acid}\\
- \; CONHCH_2CH_2COOH \quad \text{amide}\\
\hline
C_5H_{11}O_2 \quad \textit{indicates} \text{ that V is } (C_5H_{11}O_2)\text{—COONa}
\end{array}
$$

shows that V is saturated and open-chain, and contains oxygen in alcohol or ether linkages.

The synthetic route leads us to compound T.   This is a γ-hydroxy acid and spontaneously

Isobutyraldehyde   R   S   T

A γ-hydroxy acid

U

A γ-lactone

forms the γ-lactone U.   Treatment of U with base gives V, the sodium salt of T.

U   V

A lactone is an ester.   Esters react with ammonia to give amides (Sec. 20.19) and, as we would have expected, with amines to give substituted amides.   U reacts with the amino group of β-aminopropionic acid to give the substituted amide that is pantothenic acid.

$$\text{(lactone structure)} + NH_2CH_2CH_2COOH \longrightarrow HOCH_2-\underset{\underset{H_3C}{|}}{\overset{\overset{H_3C}{|}}{C}}-\underset{\underset{OH}{|}}{CH}-\overset{\overset{O}{||}}{C}-NHCH_2CH_2COOH$$

β-Aminopropionic acid

Pantothenic acid

The S-(+)-form is biologically active

25. Evidently, W is the hydrochloric acid salt (water-soluble, reacts with NaOH to lose Cl) of a 1° aromatic amine (diazotization and coupling with β-naphthol).   Assuming it is the salt of a monoamine, $ArNH_3{}^+Cl^-$, we can use the neutralization equivalent to determine the structure.

$$
\begin{array}{rl}
131 \pm 2 & ArNH_3{}^+Cl^- \\
-\quad 52.5 & NH_3{}^+Cl^- \\
\hline
78.5 \pm 2 & \textit{equivalent to } C_6H_5-
\end{array}
$$

W is anilinium chloride, $C_6H_5NH_3{}^+Cl^-$, and the liquid is aniline, $C_6H_5NH_2$.

26. (a) $CH_3CH_2CH_2CH_2NH_2$ (b) $HCONHCH_3$ (c) $m\text{-}CH_3OC_6H_4NH_2$

        n-Butylamine   N-Methylformamide   *m*-Anisidine

(See the labeled spectra on page 669 of this Study Guide.)

27. (a) $C_6H_5\underset{\underset{NH_2}{|}}{C}HCH_3$ (b) $C_6H_5CH_2CH_2NH_2$ (c) $p\text{-}CH_3C_6H_4NH_2$

                            β-Phenylethylamine   *p*-Toluidine

  α-Phenylethylamine

(See the labeled spectra on page 670 of this Study Guide.)

28. $p\text{-}CH_3CH_2OC_6H_4NH_2$ $C_6H_5CH_2NHCH_2CH_3$ $(CH_3)_2N\text{---}\langle\bigcirc\rangle\text{---}\overset{\overset{}{\underset{\underset{O}{||}}{C}}}{}\text{---}\langle\bigcirc\rangle\text{---}N(CH_3)_2$

      p-Phenetidine   Ethylbenzylamine   Michler's ketone

    (*p*-Ethoxyaniline)

        X   Y   Z

(See the labeled spectra on pages 671 and 672 of this Study Guide.)

Chapter 24
Phenols

24.1 *Inter*molecular H-bonding in the *m*- and *p*-isomers is diminished or eliminated by dilution with $CHCl_3$; *intra*molecular H-bonding in the *o*-isomer (page 789) is unaffected.

24.2 Intramolecular H-bonding:

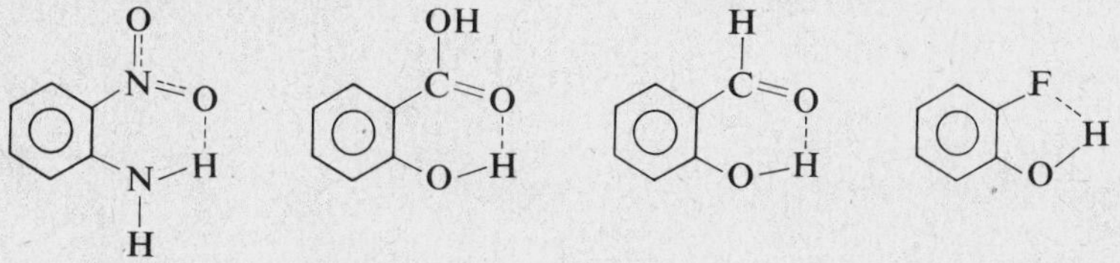

The geometry is wrong for the nitrile: the —C≡N group, with digonal (*sp*) carbon, is linear, putting nitrogen too far away from —OH. The —CH_3 group in *o*-cresol does not form H-bonds.

24.3 See Figure 24.7, page 347 of this Study Guide.

24.4 Friedel-Crafts alkylation: benzene, propylene (from cracking), HF.

24.5 (a) 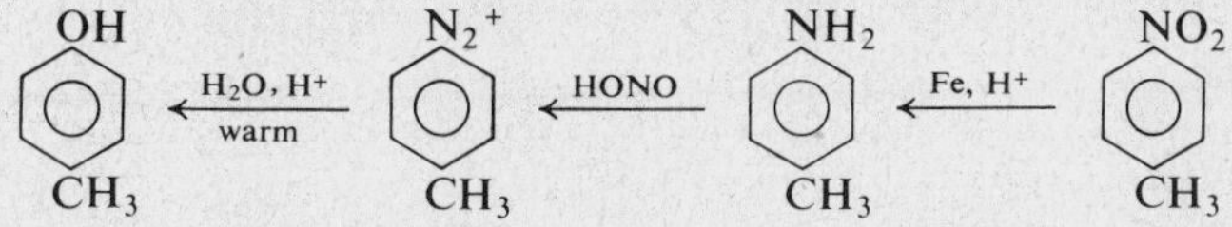 (page 345)

(b) 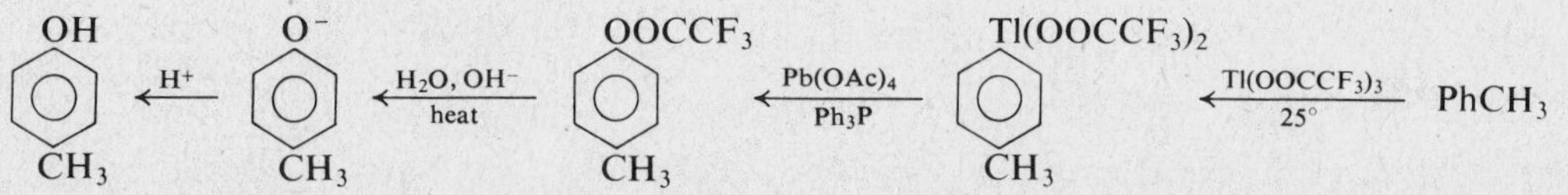

(c) 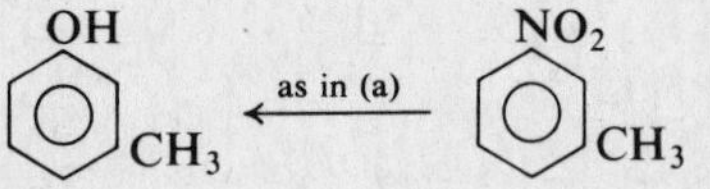 ⟵ (as in (a)) (Problem 23.15, page 771)

(d) 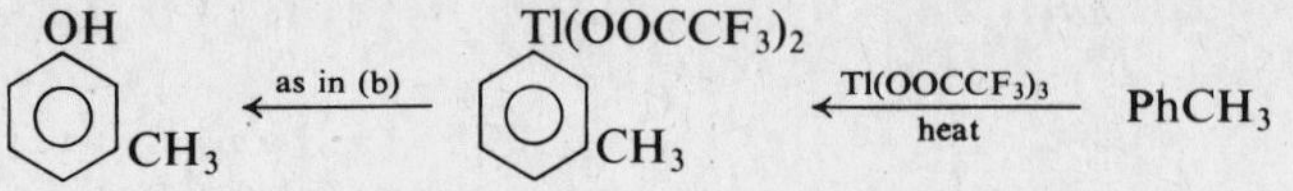 ⟵ (as in (b)) ⟵ $\frac{Tl(OOCCF_3)_3}{heat}$ PhCH₃

(See the reference given in the answer to Problem 9, Chapter 11; also E. C. Taylor, A. McKillop, *et al.*, "Thallium in Organic Synthesis. XV. Synthesis of Phenols and Aromatic Nitriles," J. Am. Chem. Soc., **92**, 3520 (1970).)

24.6 Due to acid-strengthening nitro groups, the acidity of these phenols is high enough (K_a's 10^{-4} and "very large") to permit reaction with weakly basic HCO_3^-.

24.7

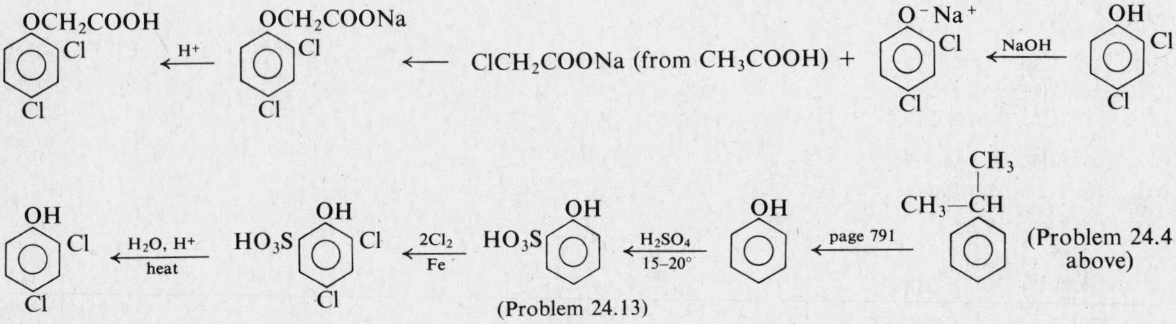

24.8

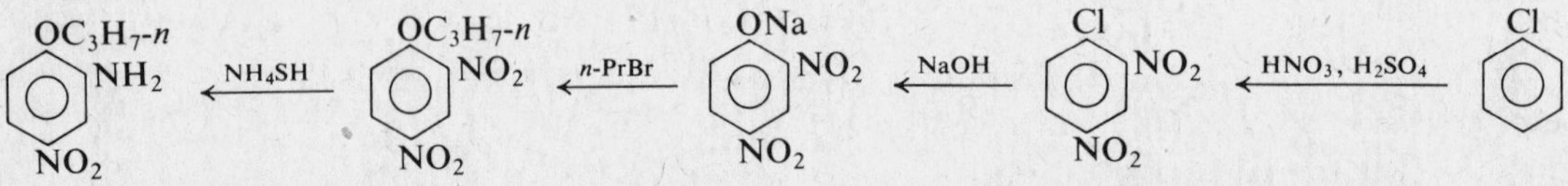

24.9 —O—C—R *activates*; —C—OR *deactivates*

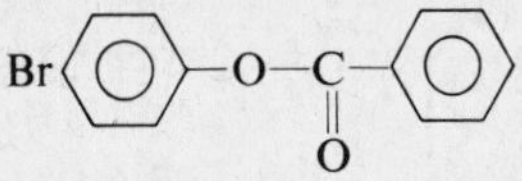

Compare Problem 6(c), page 369.

24.10 Volatility of the *o*-isomer is greater due to intramolecular H-bonding. (Compare Problem 24.2, page 790.)

24.11

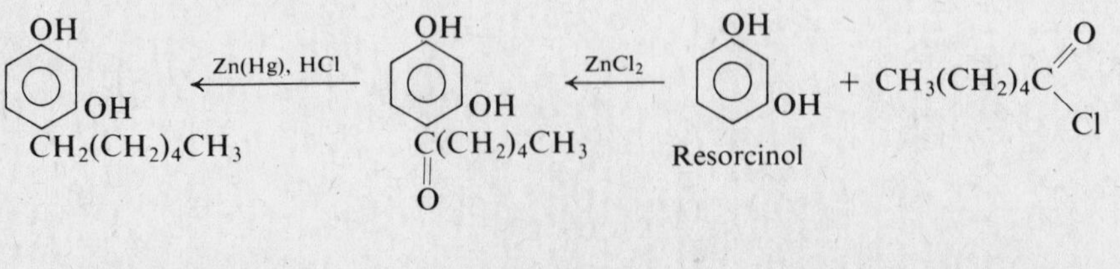

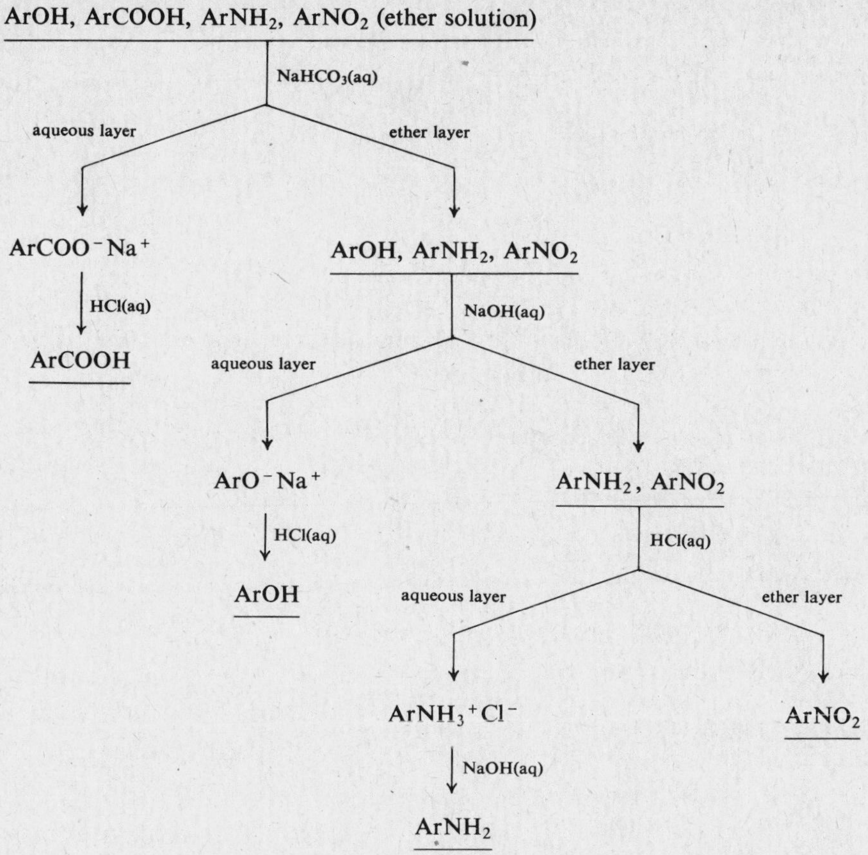
ArOH, ArCOOH, ArNH2, ArNO2 (ether solution)
NaHCO3(aq)
aqueous layer
ether layer
ArCOO⁻Na⁺
ArOH, ArNH2, ArNO2
HCl(aq)
NaOH(aq)
ArCOOH
aqueous layer
ether layer
ArO⁻Na⁺
ArNH2, ArNO2
HCl(aq)
HCl(aq)
ArOH
aqueous layer
ether layer
ArNH3⁺Cl⁻
ArNO2
NaOH(aq)
ArNH2

Figure 24.7. Separation of mixture of Problem 24.3.

24.12 (a) This is *nitrodesulfonation*: the —SO_3H is displaced by electrophilic reagents, in this case by $^{+}NO_2$.

$$ArSO_3^- + NO_2^+ \longrightarrow \overset{\oplus}{Ar}(SO_3^-)(NO_2) \longrightarrow ArNO_2 + SO_3$$

(b) There is less destructive oxidation by the nitrating agent.

24.13 Sulfonation is reversible. The *o*-isomer is formed more rapidly; the *p*-isomer is more stable. At 15–20°, there is rate control of product composition; at 100°, there is equilibrium control. (See Sec. 8.22.)

24.14

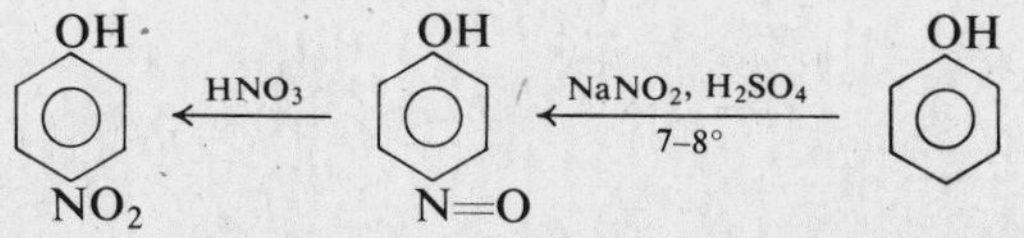
OH
NO2
HNO3
OH
N=O
NaNO2, H2SO4
7–8°
OH

24.15

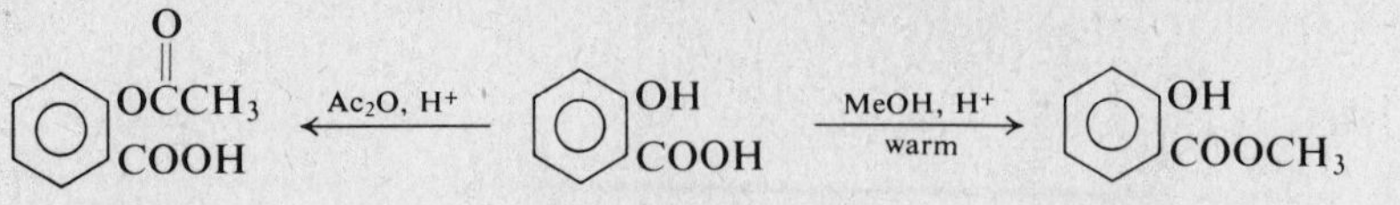

Aspirin ← Ac₂O, H⁺ — Salicylic acid (from phenol) → MeOH, H⁺ warm — Methyl salicylate

24.16 (a) Removal of the proton is easier from $CHCl_3$, which is more acidic than $CHCl_2$; it is impossible for CCl_4, which has no protons. Confirms step (1).

(b) Removal of the proton from $CHCl_3$ occurs to form an anion that can acquire D from D_2O to yield $CDCl_3$. Confirms step (1), and its reversibility.

(c) Step (2)—and with it step (1)—is reversed by excess Cl^-. Confirms step (2), and its reversibility.

(d) Step (2)—and with it step (1)—is reversed, this time by I^-. Confirms steps (1) and (2), and their reversibility.

(e) Nucleophilic Cl_3C^- adds to carbonyl group. Confirms step (1).

24.17 $ArONa$, $ClCH_2COONa$, warm; acidify. (Compare Problem 24.7, page 800.) Neutralization equivalents (Sec. 18.21) of the aryloxyacetic acids would be useful in identifying phenols.

1. (a) 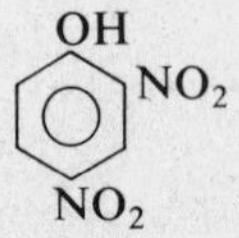 (b) 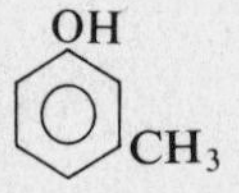 (c)  (d)

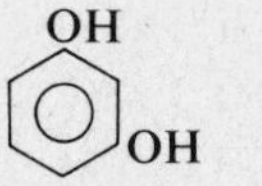

(e) 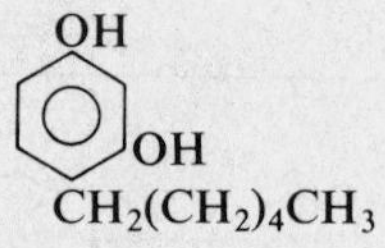 (f) 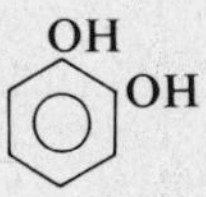 (g)

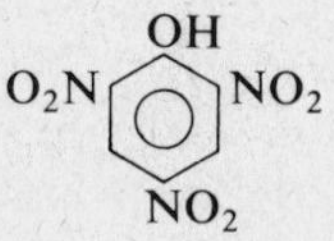

(h) 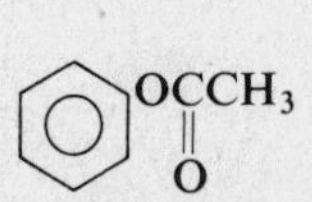 (i) 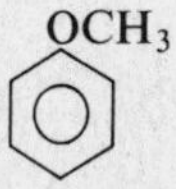 (j)  (k)

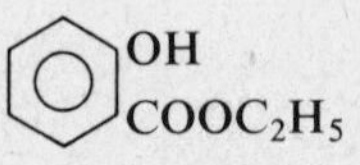

2. (a)

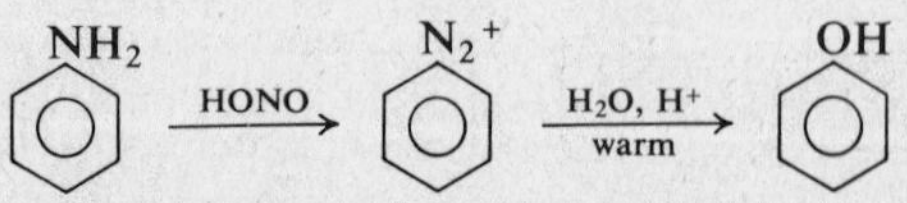

(b)

(c) [structure: Cl on benzene] $\xrightarrow[\text{heat, press.}]{\text{NaOH}}$ [structure: ONa on benzene] $\xrightarrow{\text{H}^+}$ [structure: OH on benzene]

(d) [structure: isopropylbenzene, C(CH$_3$)$_2$H] $\xrightarrow[\text{heat}]{\text{air}}$ [structure: cumene hydroperoxide, C(CH$_3$)$_2$OOH] $\xrightarrow{\text{H}^+}$ [structure: OH on benzene] $+ (CH_3)_2C=O$

3. (a) [structure: benzene with OH, OH] $\xleftarrow[\text{heat}]{\text{HBr}}$ [structure: benzene with OH, OCH$_3$]

(b) [structure: benzene with OH, OH] $\xleftarrow{\text{H}^+}$ [structure: benzene with ONa, ONa] $\xleftarrow[\text{strong heat}]{\text{NaOH}}$ [structure: benzene with ONa, SO$_3$Na] $\xleftarrow{\text{NaOH}}$ [structure: benzene with OH, SO$_3$H] $\xleftarrow[\text{low temp.}]{\text{H}_2\text{SO}_4}$ [structure: benzene with OH]

(c) [structure: benzene with OH, OH] $\xleftarrow{\text{H}^+}$ [structure: benzene with ONa, ONa] $\xleftarrow[\text{strong heat}]{\text{NaOH}}$ [structure: benzene with SO$_3$Na, SO$_3$Na] $\xleftarrow{\text{NaOH}}$ [structure: benzene with SO$_3$H, SO$_3$H] $\xleftarrow[\text{heat}]{\text{2SO}_3,\ \text{H}_2\text{SO}_4}$ [structure: benzene]

(d) [structure: benzene with OH, O$_2$N, NO$_2$, NO$_2$] $\xleftarrow{\text{HNO}_3,\ \text{H}_2\text{SO}_4}$ [structure: benzene with OH, NO$_2$, NO$_2$] $\xleftarrow{\text{H}^+}$ [structure: benzene with ONa, NO$_2$, NO$_2$] $\xleftarrow{\text{NaOH}}$ [structure: benzene with Cl, NO$_2$, NO$_2$] $\xleftarrow{\text{HNO}_3,\ \text{H}_2\text{SO}_4}$ [structure: benzene with Cl] (from C$_6$H$_6$)

(e) [structure: benzene with OCH$_3$, OCH$_3$] $\xleftarrow{\text{2Me}_2\text{SO}_4,\ \text{NaOH}}$ [structure: benzene with OH, OH]

4. (a) [structure: benzene with OH, CH$_3$] $\xleftarrow{\text{H}_2\text{O, H}^+}$ [structure: benzene with N$_2^+$, CH$_3$] $\xleftarrow{\text{HONO}}$ [structure: benzene with NH$_2$, CH$_3$] (page 770)

(b) [structure: benzene with OH, CH$_3$] $\xleftarrow{\text{as in (a)}}$ [structure: benzene with NH$_2$, CH$_3$] $\xleftarrow{\text{Fe, H}^+}$ [structure: benzene with NO$_2$, CH$_3$] (Problem 23.15, page 771)

(c) [structure: benzene with OH, CH$_3$] $\xleftarrow[\text{Ph}_3\text{P}]{\text{Pb(OAc)}_4}$ [structure: benzene with Tl(OOCCF$_3$)$_2$, CH$_3$] $\xleftarrow[\text{25}°]{\text{Tl(OOCCF}_3)_3}$ PhCH$_3$

(d) [structure: benzene with OH, I] $\xleftarrow[\text{Ph}_3\text{P}]{\text{Ph(OAc)}_4}$ [structure: benzene with Tl(OOCCF$_3$)$_2$, I] $\xleftarrow[\text{25}°]{\text{Tl(OOCCF}_3)_3}$ PhI (page 766)

(e) [structure: benzene with OH, Br] $\xleftarrow{\text{H}_2\text{O, H}^+}$ [structure: benzene with N$_2^+$, Br] $\xleftarrow{\text{HONO}}$ [structure: benzene with NH$_2$, Br] $\xleftarrow{\text{Fe, H}^+}$ [structure: benzene with NO$_2$, Br] $\xleftarrow{\text{Br}_2,\ \text{Fe}}$ [structure: benzene with NO$_2$] (from C$_6$H$_6$)

(f) [structure: benzene with OH, Br] $\xleftarrow{\text{H}_2\text{O, H}^+}$ [structure: benzene with N$_2^+$, Br] $\xleftarrow{\text{HONO}}$ [structure: benzene with NH$_2$, Br] $\xleftarrow[\text{H}^+]{\text{Fe}}$ [structure: benzene with NO$_2$, Br] $\xleftarrow{\text{HNO}_3,\ \text{H}_2\text{SO}_4}$ [structure: benzene with Br] (from C$_6$H$_6$)

(g) 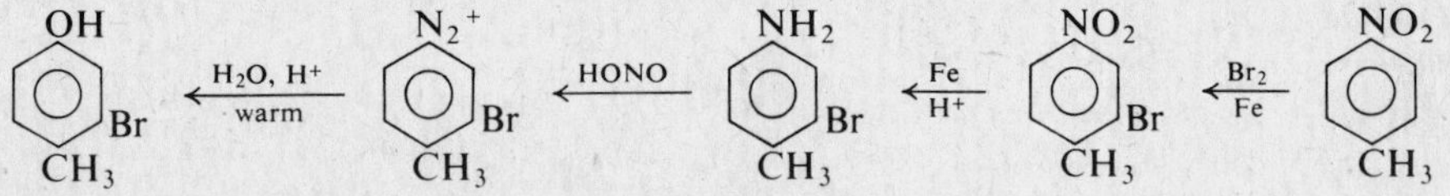 [OH, Br, CH₃] ←(H₂O, H⁺ / warm)— [N₂⁺, Br, CH₃] ←(HONO)— [NH₂, Br, CH₃] ←(Fe / H⁺)— [NO₂, Br, CH₃] ←(Br₂ / Fe)— [NO₂, CH₃] (from PhCH₃)

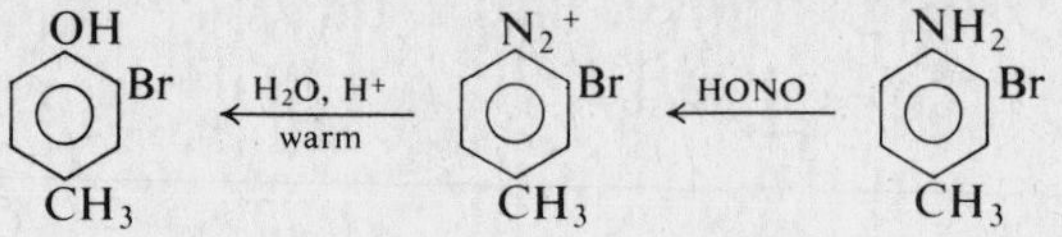

(h) [OH, Br, CH₃] ←(H₂O, H⁺ / warm)— [N₂⁺, Br, CH₃] ←(HONO)— [NH₂, Br, CH₃] (page 771)

(i) 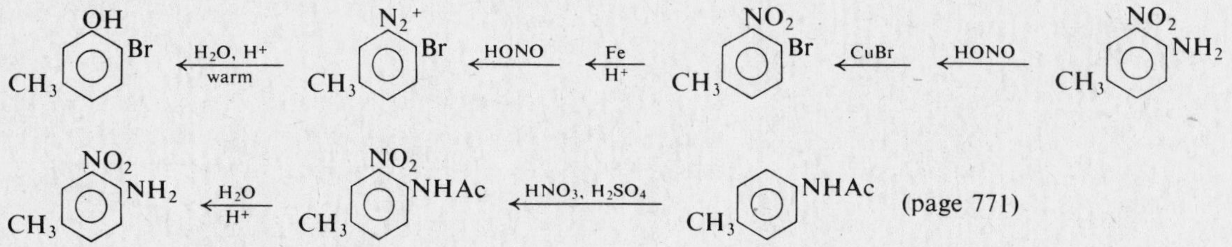 [CH₃, OH, Br] ←(H₂O, H⁺ / warm)— [CH₃, N₂⁺, Br] ←(HONO)— [CH₃, NO₂, Br] ←(Fe / H⁺)— ... ←(CuBr)— [CH₃, NO₂, NH₂] ←(HONO)—

[CH₃, NO₂, NH₂] ←(H₂O / H⁺)— [CH₃, NO₂, NHAc] ←(HNO₃, H₂SO₄)— [CH₃, NHAc] (page 771)

(j) 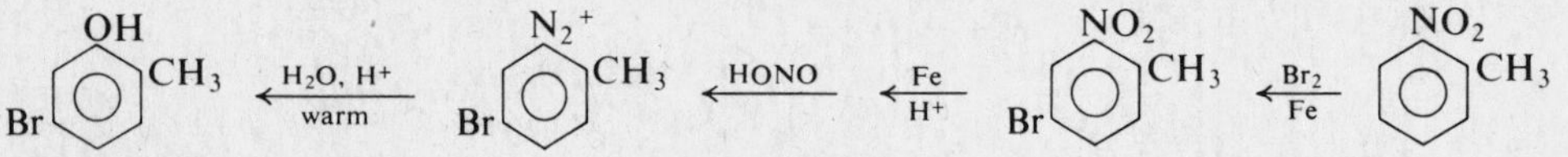 [OH, CH₃, Br] ←(H₂O, H⁺ / warm)— [N₂⁺, CH₃, Br] ←(HONO)— [NO₂, CH₃, Br] ←(Fe / H⁺)— ... ←(Br₂ / Fe)— [NO₂, CH₃] (from PhCH₃)

(k) 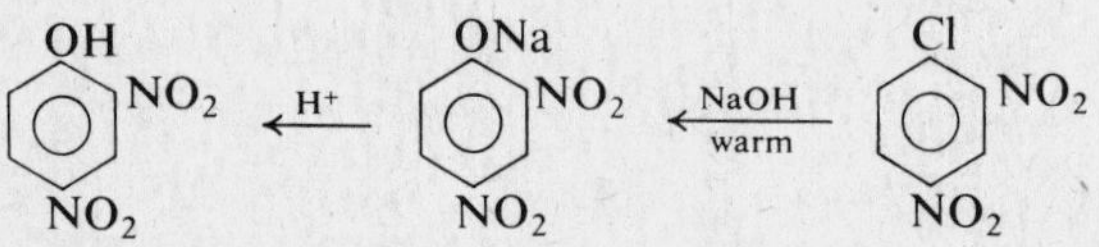 [OH, NO₂, NO₂] ←(H⁺)— [ONa, NO₂, NO₂] ←(NaOH / warm)— [Cl, NO₂, NO₂] (as in 3d, above)

(l) 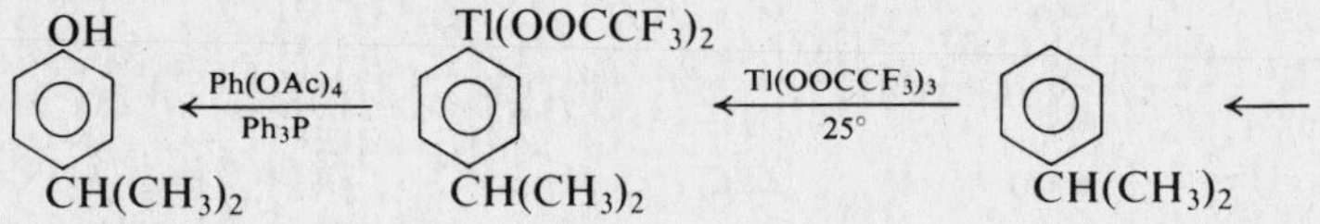 [OH, CH(CH₃)₂] ←(Ph(OAc)₄ / Ph₃P)— [Tl(OOCCF₃)₂, CH(CH₃)₂] ←(Tl(OOCCF₃)₃ / 25°)— [CH(CH₃)₂] ← as in Problem 24.4

(m) 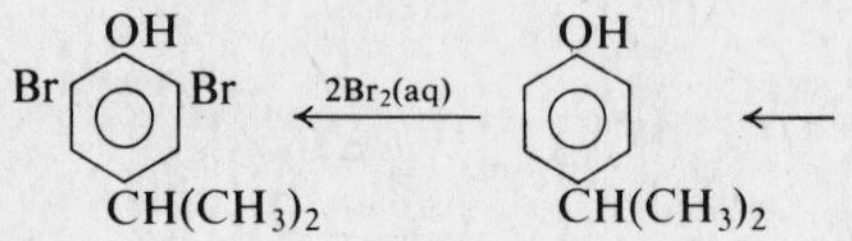 [Br, OH, Br, CH(CH₃)₂] ←(2Br₂(aq))— [OH, CH(CH₃)₂] ← as in (l)

(n) 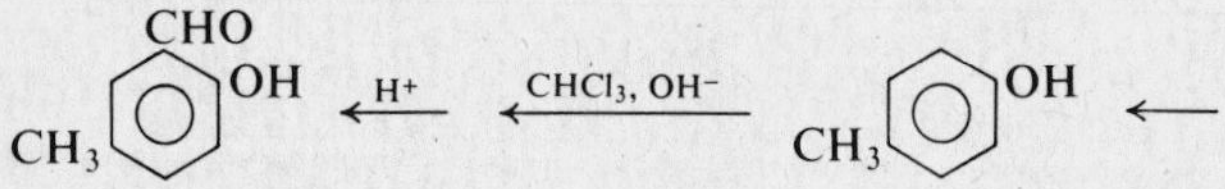 [CHO, OH, CH₃] ←(H⁺)— ←(CHCl₃, OH⁻)— [OH, CH₃] ← as in (c)

(o) 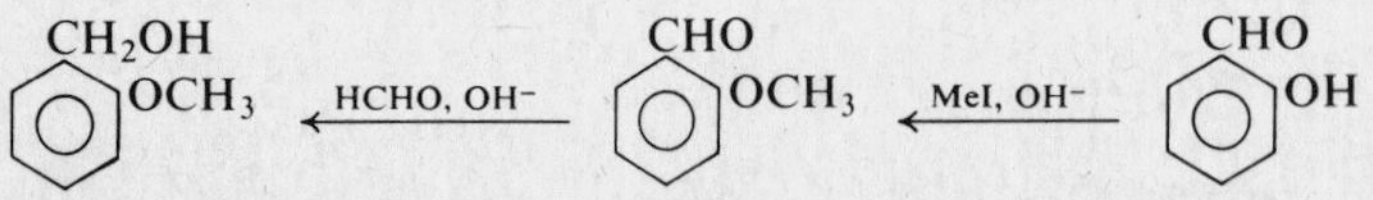 [CH₂OH, OCH₃] ←(HCHO, OH⁻)— [CHO, OCH₃] ←(MeI, OH⁻)— [CHO, OH] (page 804)

5. (a)  [ONa, CH₃] (b) no reaction (c) no reaction (d)  [OCH₃, CH₃]

(e) 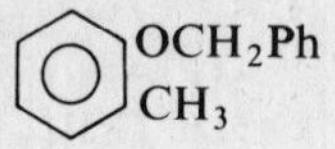 [OCH₂Ph, CH₃] (f) no reaction (g) 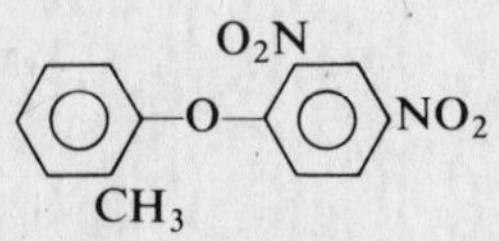 [O₂N, —O—, NO₂, CH₃] (h) 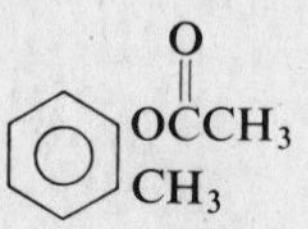 [OCCH₃(=O), CH₃] (see Sec. 18.16)

(i) 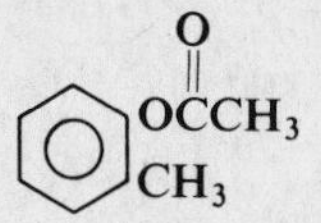 benzene ring with O–C(=O)CH$_3$ and CH$_3$ substituents

(j) 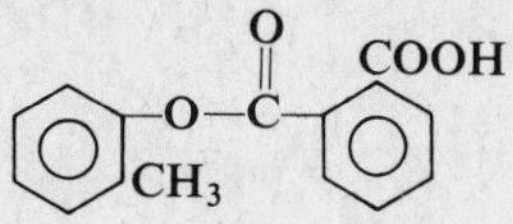 benzene ring with O–C(=O)– (attached to another benzene ring bearing COOH) and CH$_3$ substituents

(k) 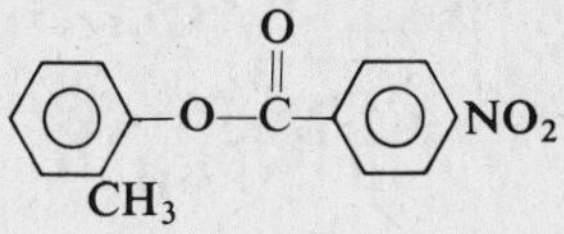 benzene ring with O–C(=O)– (attached to benzene ring bearing NO$_2$) and CH$_3$ substituents

(l) 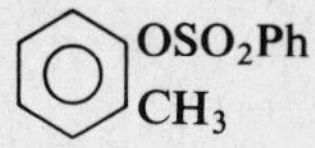 benzene ring with OSO$_2$Ph and CH$_3$ substituents

(m) 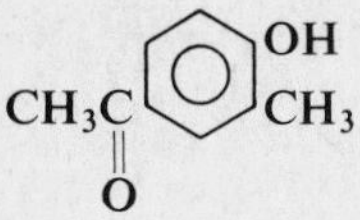 CH$_3$C(=O)– attached to benzene ring bearing OH and CH$_3$

(n) no reaction

(o) color produced

(p) 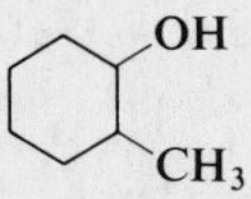 cyclohexane ring with OH and CH$_3$ substituents

(q) 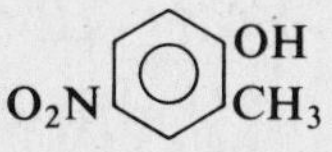 benzene ring with O$_2$N, OH, CH$_3$ substituents

(r) 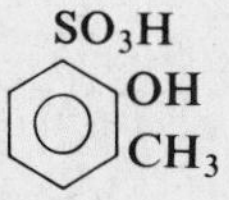 benzene ring with SO$_3$H, OH, CH$_3$ substituents

(s)  benzene ring with HO$_3$S, OH, CH$_3$ substituents

(t) 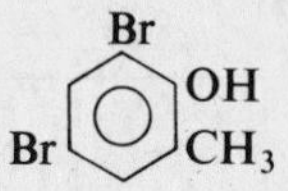 benzene ring with Br (top), Br, OH, CH$_3$ substituents

(u) 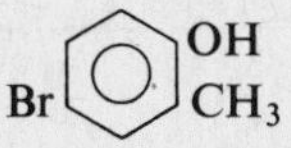 benzene ring with Br, OH, CH$_3$ substituents

(v) 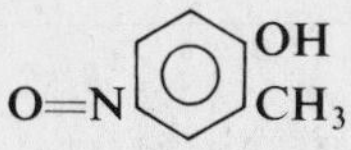 benzene ring with O=N, OH, CH$_3$ substituents

(w) 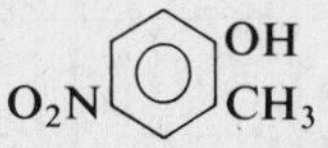 benzene ring with O$_2$N, OH, CH$_3$ substituents

(x) 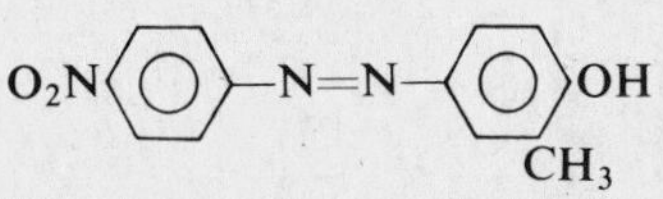 O$_2$N–(benzene ring)–N=N–(benzene ring bearing OH and CH$_3$)

(y) 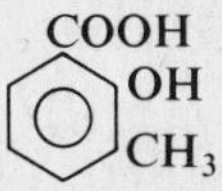 benzene ring with COOH, OH, CH$_3$ substituents

(z) 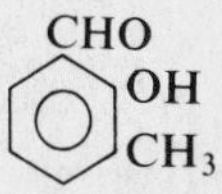 benzene ring with CHO, OH, CH$_3$ substituents

6. (c) PhOH + MeBr   (p) 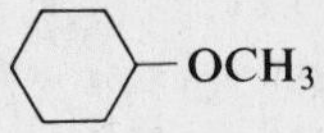 cyclohexane–OCH$_3$ (r) 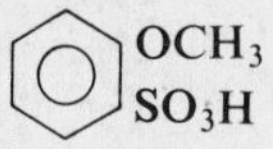 benzene ring with OCH$_3$, SO$_3$H and HO$_3$S–benzene ring–OCH$_3$

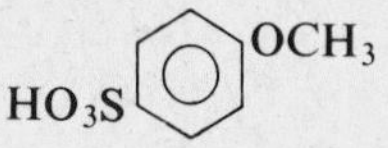

(s) same as (r)   (t)  benzene ring with OCH$_3$, Br and Br–benzene ring–OCH$_3$ (u) same as (t)

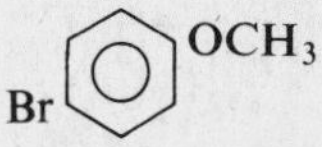

All others: no reaction

7. (c) PhCH$_2$Br   (h) PhCH$_2$OCCH$_3$ (C=O)   (i) same as (h)   (j) 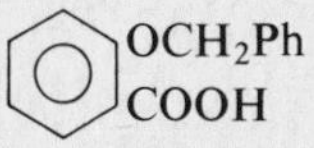 benzene ring with OCH$_2$Ph, COOH

(k) 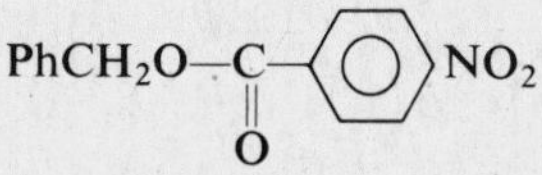 PhCH$_2$O–C(=O)–benzene ring–NO$_2$ (l) PhCH$_2$O–S(=O)$_2$–Ph   (n) PhCH$_2$Cl

All others: no reaction.

8. (a) PhSO$_3$H > PhCOOH > PhOH > PhCH$_2$OH

(b) H$_2$SO$_4$ > H$_2$CO$_3$ > PhOH > H$_2$O

(c) m-NO$_2$ > m-Br > unsubstd. > m-CH$_3$

(d) 2,4,6- > 2,4- > p-

9. (a) NaOH(aq)

(b) NaOH(aq) to detect phenol.   Then CrO_3/H_2SO_4 to distinguish alcohol from ether.

(c) $NaHCO_3$(aq) to detect the acid.   Then cold NaOH(aq) to distinguish phenol from ester.

(d) Dilute HCl.

(e) The two acids are soluble in $NaHCO_3$(aq), but only salicylic acid gives a color with $FeCl_3$.
    Of the remaining compounds, only ethyl salicylate (free phenolic —OH) is soluble in cold NaOH(aq).

(f) $NaHCO_3$(aq) to detect the acid (CO_2 evolved; dissolution).   Then NaOH(aq) dissolves the phenol.   Then dilute HCl dissolves the amine.   The neutral dinitrobenzene is unaffected.

10. (a)

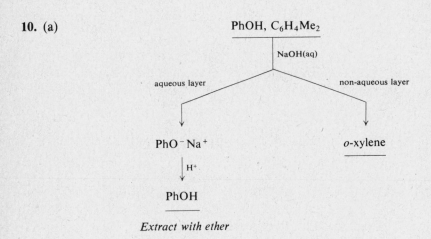

(c)

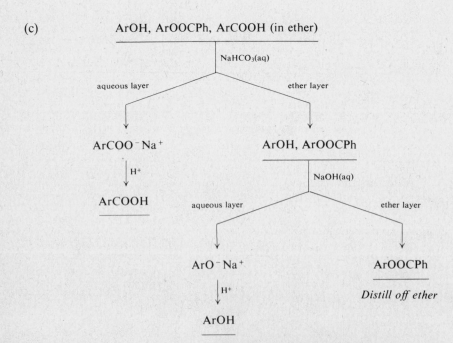

(d)

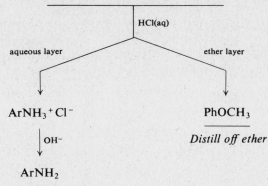

PhOMe, CH₃C₆H₄NH₂ (in ether)

$$PhOMe, \ CH_3C_6H_4NH_2 \ (in \ ether)$$

HCl(aq)

aqueous layer   ether layer

$ArNH_3^+Cl^-$ $PhOCH_3$

OH⁻   *Distill off ether*

$ArNH_2$

(f)

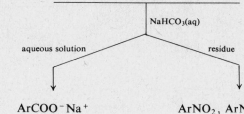

$ArNO_2$, $ArNH_2$, $ArCOOH$, $ArOH$

$NaHCO_3$(aq)

aqueous solution   residue

$ArCOO^-Na^+$ $ArNO_2$, $ArNH_2$, $ArOH$

H⁺   NaOH(aq)

$ArCOOH$ aqueous solution   residue

ArO^-Na^+ $ArNO_2$, $ArNH_2$

H⁺   HCl(aq)

$ArOH$ aqueous solution   residue

$ArNH_3^+Cl^-$ $ArNO_2$

OH⁻

$ArNH_2$

11. (a) 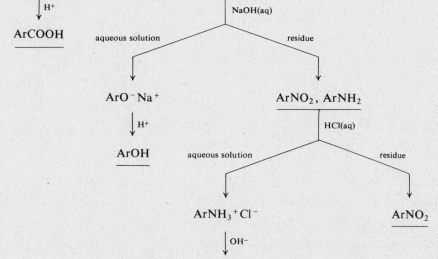 ⟵ Sn, H⁺ ⟵ as in 4(k) and 3(d), above

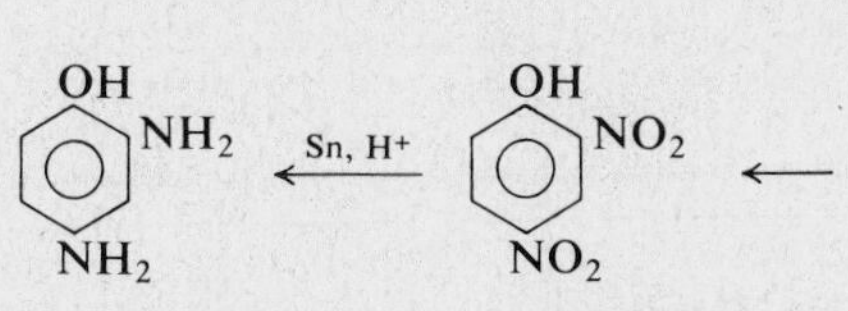

(b) ⟵ Fe, H⁺ ⟵ HNO₃, H₂SO₄ ⟵ 2Me₂SO₄, OH⁻

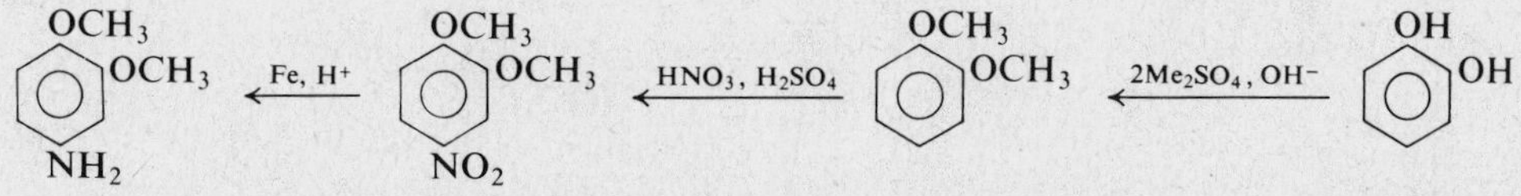

(c) [benzene ring with OH, NO₂, OH] $\xleftarrow[\text{heat}]{H_2O, H^+}$ [benzene ring HO₃S, OH, NO₂, OH, SO₃H] $\xleftarrow{HNO_3, H_2SO_4}$ [benzene ring HO₃S, OH, OH, SO₃H] $\xleftarrow{2H_2SO_4}$ [benzene ring OH, OH]

(d) [ring CH₃, OH, CH₃, CH₃] $\xleftarrow[\text{warm}]{H_2O, H^+}$ [ring CH₃, N_2^+, CH₃, CH₃] $\xleftarrow{HONO}$ $\xleftarrow{Fe, H^+}$ [ring CH₃, NO₂, CH₃, CH₃] $\xleftarrow{HNO_3, H_2SO_4}$ [ring CH₃, CH₃, CH₃] Mesitylene

(e) [ring OH, C(CH₃)₃] $\xleftarrow{\text{acid cat.}}$ [ring OH] + $(CH_3)_2C{=}CH_2$

(f) [ring OH, $CH_3{-}C{-}CH_2{-}C(CH_3)_3$, CH₃] $\xleftarrow{\text{acid cat.}}$ [ring OH] + $CH_3{-}\underset{CH_3}{C}{=}CH{-}\underset{CH_3}{\overset{CH_3}{C}}{-}CH_3$ (page 200)

(g) $BrCH_2CH_2OPh \longleftarrow BrCH_2CH_2Br + NaOPh \xleftarrow{NaOH(aq)} PhOH$

(h) $CH_2{=}CH{-}O{-}Ph \xleftarrow[\text{heat}]{KOH} BrCH_2CH_2OPh \longleftarrow$ (g)

(i) $CH_2{=}CH{-}O{-}Ph \xrightarrow{H^+} \underset{H}{CH_2}{-}CH{=}\overset{+}{O}{-}Ph \xrightarrow[-H^+]{H_2O} \left[\underset{OH}{CH_3}{-}CH{-}O{-}Ph\right] \longrightarrow CH_3CHO + PhOH$

A hemiacetal

(j) [ring O₂N, OCH₃, NO₂, CH₃, C(CH₃)₃] $\xleftarrow{HNO_3, H_2SO_4}$ [ring OCH₃, CH₃, C(CH₃)₃] $\xleftarrow{t\text{-BuCl, AlCl}_3}$ [ring OCH₃, CH₃] $\xleftarrow{Me_2SO_4, OH^-}$ [ring OH, CH₃]

(k) [ring CH₃, OH, OH] $\xleftarrow{H^+}$ $\xleftarrow[\text{heat}]{NaOH}$ [ring CH₃, SO₃H, SO₃H] $\xleftarrow{H_3PO_2}$ $\xleftarrow{HONO}$ [ring CH₃, SO₃H, NH₂, SO₃H] $\xleftarrow[\text{heat}]{2H_2SO_4}$ [ring CH₃, NH₂]

(page 770)

12. (a) [ring CH=CHCOOH, OH, OH] $\xleftarrow[\text{heat}]{HBr}$ [ring CH=CHCOOH, OCH₃, OH] $\xleftarrow[\substack{\text{Perkin reaction}\\\text{(Prob. 21.22e)}}]{Ac_2O, OAc^-}$ [ring CHO, OCH₃, OH]

Caffeic acid   Vanillin

(b)

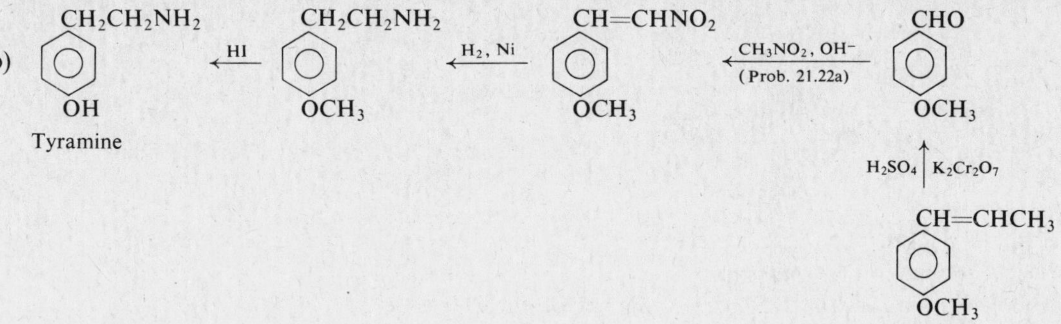

Tyramine

Anethole

(c)

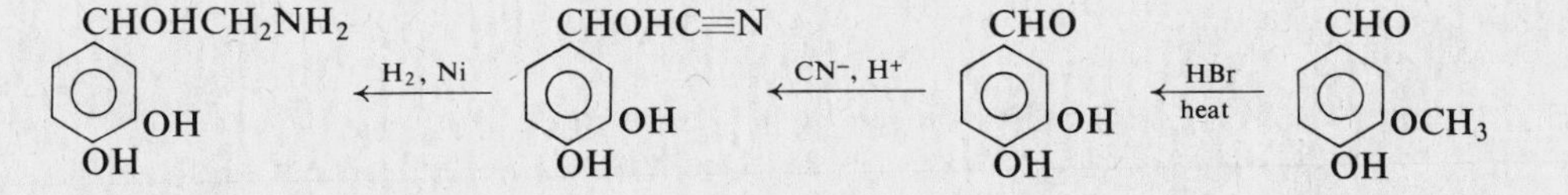

Noradrenaline

Vanillin

13. Phenoxide ion contains two sites of nucleophilic reactivity: the $—O:^-$, and the π electrons of the ring. It is an example of what Nathan Kornblum (Purdue University) has called an *ambident* anion (L.: *ambi*, both; *dens*, tooth).

(a) Nucleophilic aliphatic substitution on benzylic carbon, with $—O^-$ of phenoxide as the nucleophile:

$$PhCH_2Cl + {}^-OPh \longrightarrow PhCH_2OPh + Cl^-$$

(b) Nucleophilic aliphatic substitution, as before, but now the phenoxide ring competes with $—O^-$ as nucleophile:

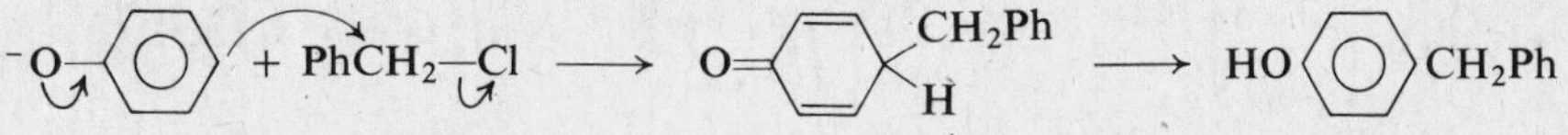

(From the standpoint of the ring, the reaction is electrophilic aromatic substitution.)

(c) In aprotic solvents, $—O^-$ is only weakly solvated and hence highly reactive; in water, solvation through H-bonding lowers the nucleophilic power of $—O^-$, and the ring begins to compete successfully with it.

(d) MeOH and EtOH are less effective than HOH at hydrogen bonding. The acidic protons of PhOH and CF_3CH_2OH hydrogen-bond strongly.

(N. Kornblum, P. J. Berrigan, and W. J. leNoble, "Solvation as a Factor in the Alkylation of Ambident Anions: the Importance of the Hydrogen Bonding Capacity of the Solvent," J. Am. Chem. Soc., **85**, 1141 (1963).)

14. Strong acid converts phloroglucinol (R = H) and its ethers (R = Me or Et) reversibly into protonated species, hybrids of this structure

R = H, Me, or Et

and corresponding structures with the charge on the other oxygens. These species are the familiar benzenonium ions (*sigma* complexes) that are intermediates in electrophilic aromatic substitution—this time with two protons at the site of attack (see Sec. 11.8; Problem 11.11, page 358; and Problem 3, Chapter 13, page 446).

The spectrum is due to protons a (δ 4.15) and b (δ 6.12), allylic and vinylic protons deshielded by the positive charge. On dilution with water, a proton is lost, and the original compounds are regenerated.

In D_2SO_4, all ring protons are gradually replaced by deuterons in electrophilic aromatic substitutions via similar benzenonium ions. From $1,3,5\text{-}C_6H_3(OCH_3)_3$ in D_2SO_4, we would expect to isolate $1,3,5\text{-}C_6D_3(OCH_3)_3$.

15.

Citral

(1) Protonation of aldehyde. (2) Electrophilic substitution, with protonated aldehyde as electrophile. (3) Protonation of —OH and loss of H_2O to form allylic cation. (4) Electrophilic addition of allylic cation to double bond to form 3° cation and generate second ring. (5) Combination of 3° cation with HO— of phenol to form protonated ether and generate third ring. (6) Loss of proton to yield I.

(E. C. Taylor, K. Lenard, and Y. Shvo, "Active Constituents of Hashish. Synthesis of *dl*-Δ^6-3,4-*trans*-Tetrahydrocannabinol," J. Am. Chem. Soc., **88**, 367 (1966).)

16. (a) $O_2N\text{—}\langle\bigcirc\rangle\text{—OH} \xrightarrow{OH^-} O_2N\text{—}\langle\bigcirc\rangle\text{—O}^- \xrightarrow{EtBr} O_2N\text{—}\langle\bigcirc\rangle\text{—OEt}$

A

$O_2N\text{—}\langle\bigcirc\rangle\text{—OEt} \xrightarrow{Sn, HCl} H_2N\text{—}\langle\bigcirc\rangle\text{—OEt} \xrightarrow{HONO} {}^+N_2\text{—}\langle\bigcirc\rangle\text{—OEt} \xrightarrow{PhOH} HO\text{—}\langle\bigcirc\rangle\text{—N}=\text{N—}\langle\bigcirc\rangle\text{—OEt}$

A   B   C

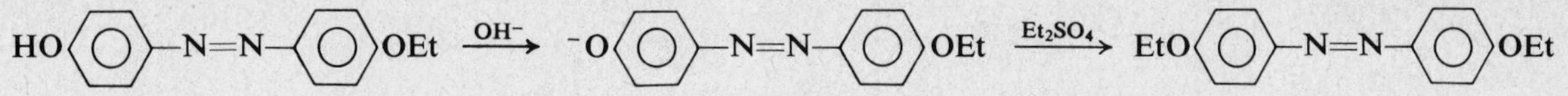

$$HO\!\!-\!\!\langle \rangle\!\!-\!\!N\!\!=\!\!N\!\!-\!\!\langle \rangle\!\!-\!\!OEt \xrightarrow{OH^-} \ ^-O\!\!-\!\!\langle \rangle\!\!-\!\!N\!\!=\!\!N\!\!-\!\!\langle \rangle\!\!-\!\!OEt \xrightarrow{Et_2SO_4} EtO\!\!-\!\!\langle \rangle\!\!-\!\!N\!\!=\!\!N\!\!-\!\!\langle \rangle\!\!-\!\!OEt$$

C   D

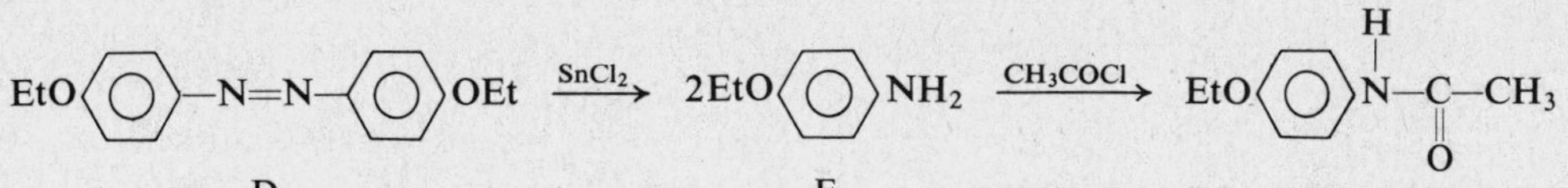

$$EtO\!\!-\!\!\langle \rangle\!\!-\!\!N\!\!=\!\!N\!\!-\!\!\langle \rangle\!\!-\!\!OEt \xrightarrow{SnCl_2} 2\,EtO\!\!-\!\!\langle \rangle\!\!-\!\!NH_2 \xrightarrow{CH_3COCl} EtO\!\!-\!\!\langle \rangle\!\!-\!\!\overset{H}{N}\!\!-\!\!\overset{O}{\underset{\|}{C}}\!\!-\!\!CH_3$$

D   E   Phenacetin

(b)

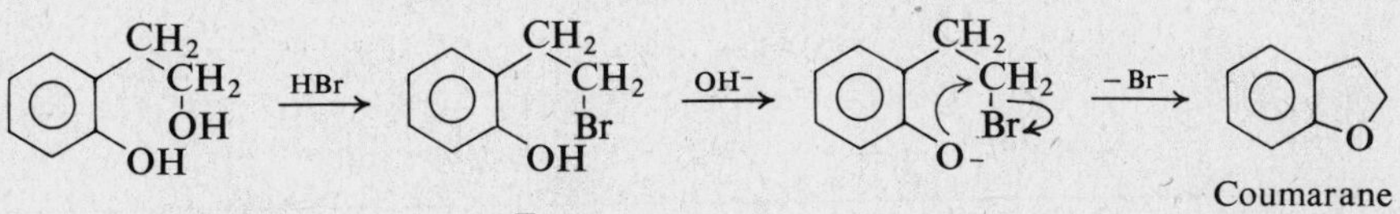

$$\xrightarrow{HBr} \qquad F \qquad \xrightarrow{OH^-} \qquad \xrightarrow{-Br^-}$$

Coumarane
No phenolic —OH;
NaOH-insoluble

(c)

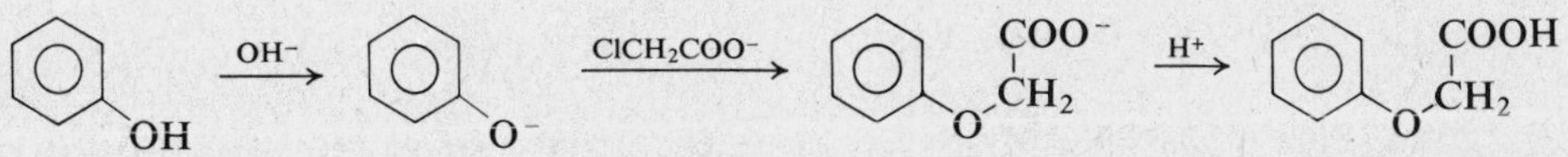

$$\xrightarrow{OH^-} \qquad \xrightarrow{ClCH_2COO^-} \qquad \xrightarrow{H^+}$$

G

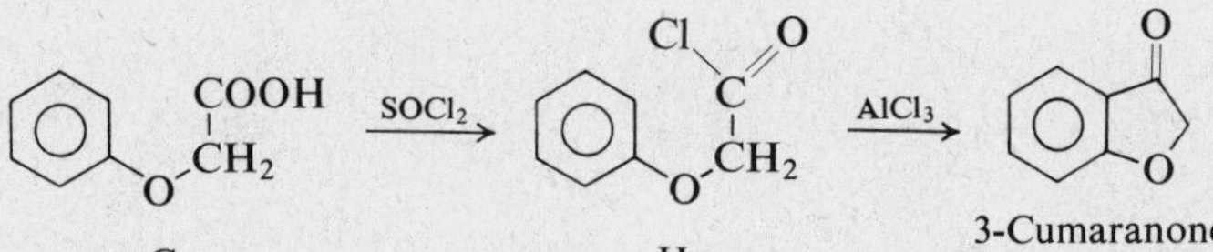

$$G \xrightarrow{SOCl_2} H \xrightarrow{AlCl_3}$$

G   H   3-Cumaranone

(d)

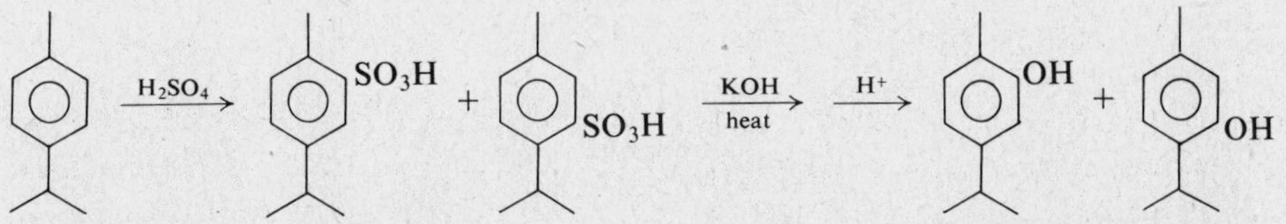

$$\xrightarrow{H_2SO_4} \quad SO_3H \ + \ SO_3H \xrightarrow[heat]{KOH} \xrightarrow{H^+} OH \ + \ OH$$

p-Cymene   I and J   Carvacrol and thymol
         Assignment uncertain   *Assignment uncertain*

$$I \xrightarrow[oxidn.]{HNO_3} \quad SO_3H \atop COOH \quad or \quad SO_3H \atop COOH$$

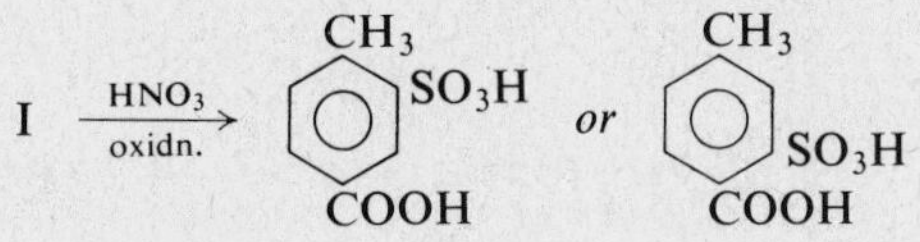

K
Assignment uncertain

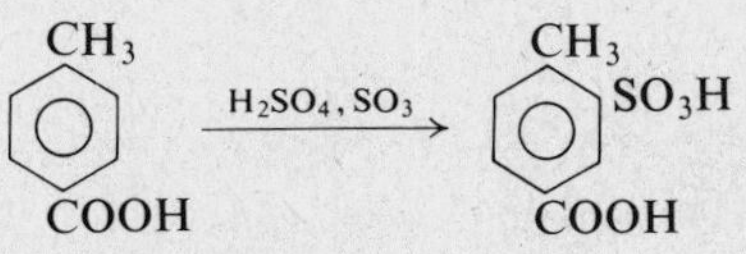

$$\xrightarrow{H_2SO_4,\,SO_3} \quad SO_3H \atop COOH$$

K

357

Therefore:

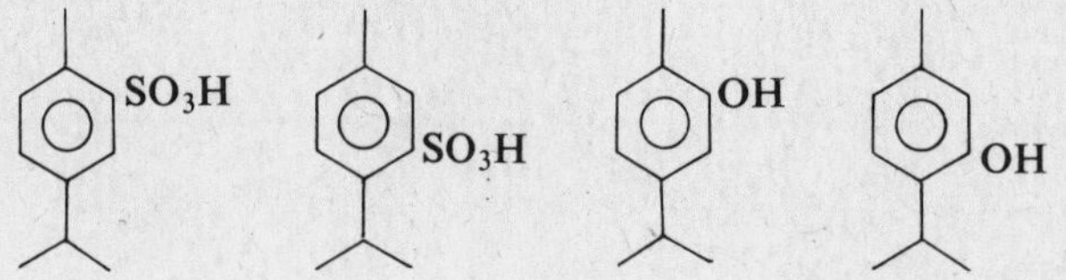

I   J   Carvacrol   Thymol

(e) CH_3O—⟨⟩—CH $\xrightarrow{HBr}$ CH_3O—⟨⟩—$CHBr$ $\xrightarrow{Mg}$ CH_3O—⟨⟩—CH—CH—⟨⟩—OCH_3

       $CHCH_3$ CH_2CH_3 Et   Et

Anethole   L   M

$\downarrow$ HBr, heat

HO—⟨⟩—CH—CH—⟨⟩—OH

Et   Et

Hexestrol

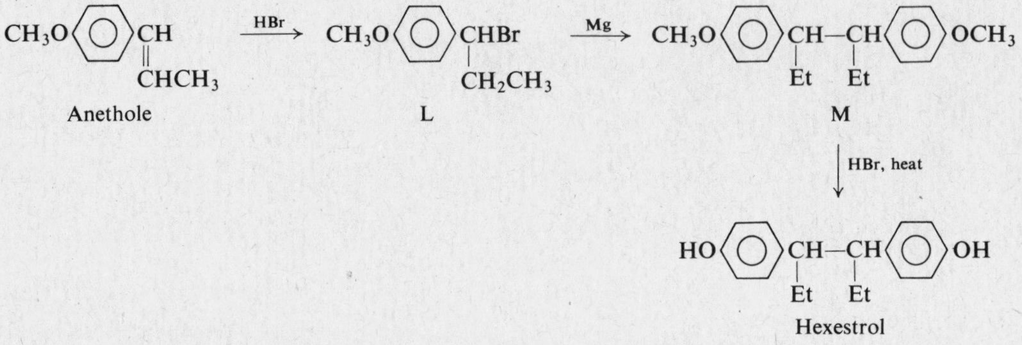

17.

$\xrightarrow[POCl_3]{ClCH_2COCl}$

N?   N?   N?

By ring acylation   *By ring alkylation*   *By esterification*

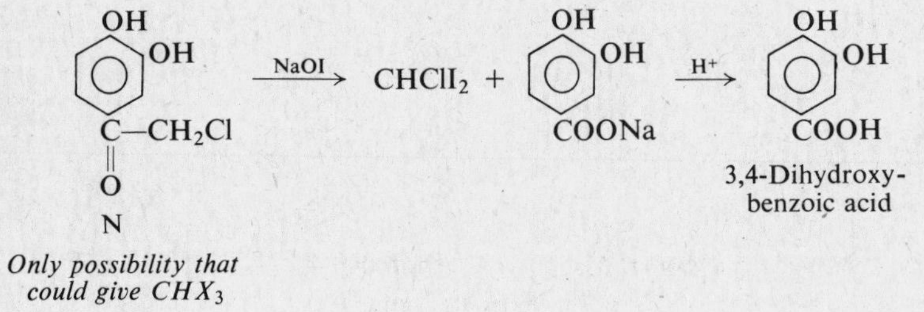

$\xrightarrow{NaOI}$ $CHClI_2$ +

$\xrightarrow{H^+}$

3,4-Dihydroxy-
benzoic acid

*Only possibility that
could give CHX₃*

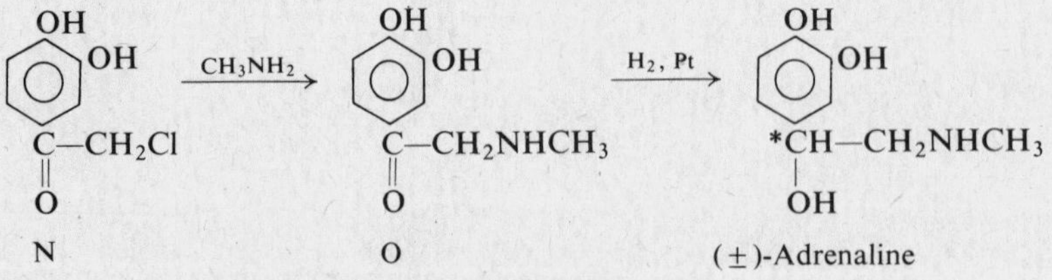

$\xrightarrow{CH_3NH_2}$

$\xrightarrow{H_2, Pt}$

N   O   (±)-Adrenaline

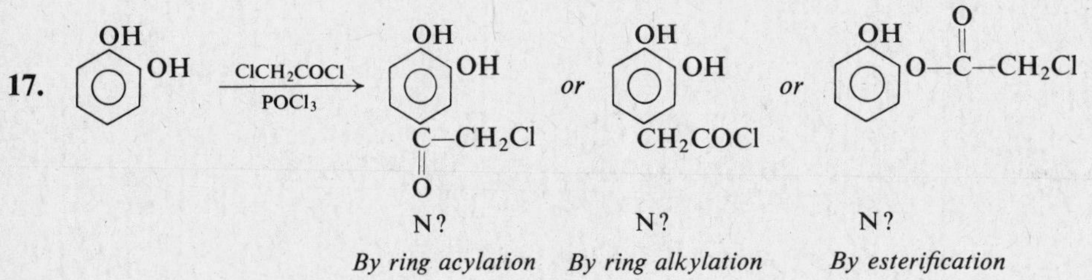

18. (a) Phellandral is an aldehyde (Tollens' test), oxidizable to an acid that contains one reducible unsaturation, undoubtedly a carbon–carbon double bond. With these facts in mind,
let us do some chemical arithmetic on the saturated acid.

$$C_{10}H_{18}O_2$$
$$\underline{-COOH}$$
$$C_9H_{17} \; means \; R$$

C_9H_{19} if R were open-chain
$\underline{-C_9H_{17}} \;\; R$
2H missing *means* 1 ring

We now know that phellandral contains one ring, one double bond, and a —CHO group. Synthesis by an unambiguous route gives us its structure.

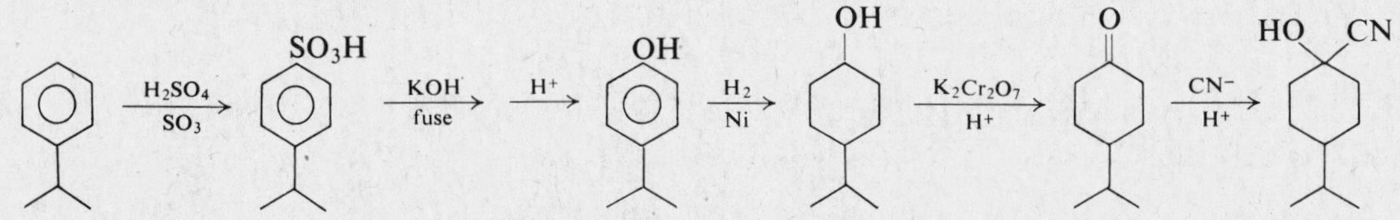

Isopropylbenzene   P   Q   R   S   T

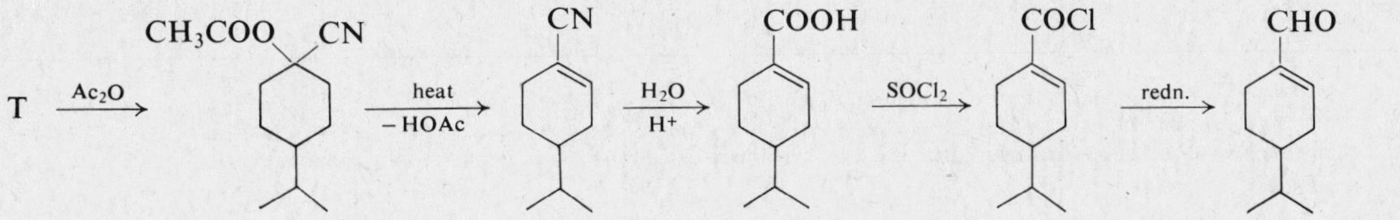

U   V   W   X   (±)-Phellandral

(b) Synthetic phellandral is the racemic modification:

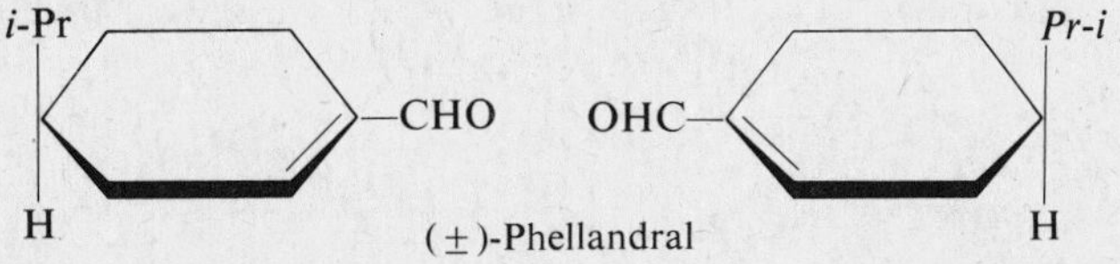

(±)-Phellandral

The chirality that persists into phellandral is generated when U is converted into V: the double bond can form in either direction from the carbon holding —CN, to give either of a pair of enantiomers. Elimination in either direction is equally likely, and the racemic modification is obtained. From this stage onward, there are no stereochemical changes.

Enantiomers
Formed in equal amounts

(c) Compare Problem 9.13(e) and (f), page 308.

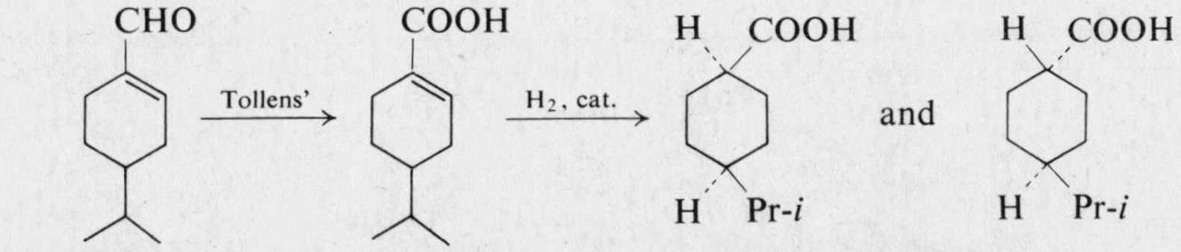

(−)-Phellandral   (−)-Phellandric acid   Dihydrophellandric acid
Diastereomers: each is achiral

19. Y is *m*-cresol, the only cresol that can give a tribromo compound.

$$\text{OH}\,\big|\,\text{C}_6\text{H}_3 \quad \xrightarrow{\text{Br}_2,\ \text{H}_2\text{O}} \quad \text{Br-substituted}$$

m-Cresol

20. (a) Z and AA are neutral: oxygen is —OH, —OR, or —C=O. They are unsaturated. Vigorous oxidation to anisic acid shows the presence (and retention) of an ether linkage, but loss of two carbons from a side chain. At this point, we can say that the following structural elements are present:

$$\text{CH}_3\text{O}-\big\langle\bigcirc\big\rangle-\text{C}- \ +\ \text{two more C's on side chain}$$

Chemical arithmetic

$$\begin{array}{ll} C_{10}H_{12}O & Z \text{ and } A \\ -\,C_7H_7O & CH_3OC_6H_4 \\ \hline C_3H_5 & \text{side chain} \end{array}$$

shows the side chain to be C_3H_5; the chemical tests have shown that this is unsaturated, not a cyclopropane ring.

Possible structures are, then:

cis- or *trans*-$CH_3O\langle\bigcirc\rangle CH{=}CHCH_3$ $CH_3O\langle\bigcirc\rangle CH_2CH{=}CH_2$ $CH_3O\langle\bigcirc\rangle C{=}CH_2$ with CH_3

(b) Hydrogenation to the same compound, $C_{10}H_{14}O$, shows that the carbon skeleton of the unsaturated side chain is the same in Z as in AA. This rules out the branched chain structure, and we are now down to three possibilities:

cis- or *trans*-$CH_3O\langle\bigcirc\rangle CH{=}CHCH_3$ $CH_3O\langle\bigcirc\rangle CH_2CH{=}CH_2$

(c) Ozonolysis; or isomerization of one into the other by strong heating with KOH.

(d) This synthesis involves the coupling of a Grignard reagent with an allyl halide (see Problem 16, Chapter 8, page 281).

$$CH_3O\langle\bigcirc\rangle Br \xrightarrow{Mg} CH_3O\langle\bigcirc\rangle MgBr + BrCH_2CH{=}CH_2 \longrightarrow CH_3O\langle\bigcirc\rangle CH_2CH{=}CH_2 + MgBr_2$$

p-Bromoanisole   Allyl bromide   *p*-Allylanisole

Z

(e) Heating with strong base converts the allylbenzene Z into the more stable AA, with the double bond conjugated with the ring. Of the two geometric isomers, we would expect to get the more stable *trans* isomer.

$$p\text{-}CH_3OC_6H_4CH_2CH{=}CH_2 \underset{+H^+}{\overset{-H^+}{\rightleftarrows}} p\text{-}CH_3OC_6H_4\overset{\ominus}{CH{\cdots}CH{\cdots}CH_2} \underset{-H^+}{\overset{+H^+}{\rightleftarrows}}$$

cis-*p*-Propenylanisole

$$+H^+ \updownarrow -H^+$$

trans-*p*-Propenylanisole

AA

Most stable isomer

(f) CH_3O⟨⟩$CH=CHCH_3$ $\xleftarrow{\text{Na, NH}_3}$ CH_3O⟨⟩$C\equiv CCH_3$ $\xleftarrow{\text{CH}_3\text{I}}$ $\xleftarrow{\text{NaNH}_2}$ CH_3O⟨⟩$C\equiv CH$

trans

CH_3O⟨⟩$C\equiv CH$ $\xleftarrow{\text{NaNH}_2}$ $\xleftarrow{\text{KOH(alc)}}$ CH_3O⟨⟩$\underset{\underset{Br\ \ \ Br}{|\ \ \ \ |}}{CH-CH_2}$ $\xleftarrow{\text{Br}_2}$ CH_3O⟨⟩$CH=CH_2$

CH_3O⟨⟩$CH=CH_2$ $\xleftarrow[\text{- H}_2\text{O}]{\text{H}^+}$ CH_3O⟨⟩$\underset{\underset{OH}{|}}{CHCH_3}$ $\xleftarrow{\text{CH}_3\text{CHO}}$ CH_3O⟨⟩$MgBr$ $\xleftarrow{\text{Mg}}$ CH_3O⟨⟩Br

21. BB is evidently a phenol; this accounts for (at least) one oxygen. It also appears to be an ester (accounting for the other two oxygens), slowly undergoing hydrolysis to an alcohol that gives a positive haloform test, that is, hydrolysis to $RCH(OH)CH_3$.

$$\begin{array}{lll}
C_{10}H_{12}O_3 & BB \\
\underline{-OH} & \text{phenolic} \\
C_{10}H_{11}O_2
\end{array}
\qquad
\begin{array}{lll}
C_{10}H_{11}O_2 & \\
\underline{-C_6H_4} & \text{aromatic ring} \\
C_4H_7O_2
\end{array}
\qquad
\begin{array}{lll}
C_4H_7O_2 & \\
\underline{-CO_2} & \text{ester} \\
C_3H_7
\end{array}$$

A residue of C_3H_7 shows that the alcohol is *iso*-PrOH.

$C_6H_4\begin{smallmatrix}\nearrow OH\\ \searrow COOPr\text{-}i\end{smallmatrix}$ $\xrightarrow[\text{heat}]{\text{OH}^-}$ *i*-PrOH $+ C_6H_4\begin{smallmatrix}\nearrow O^-\\ \searrow COO^-\end{smallmatrix}$ $\xrightarrow{\text{H}^+}$ $C_6H_4\begin{smallmatrix}\nearrow OH\\ \searrow COOH\end{smallmatrix}$

BB $\downarrow$ OI⁻   CC

CHI$_3$

The only hydroxybenzoic acid that is steam-volatile is the ortho-isomer, salicylic acid; CC must be this, and BB must be isopropyl salicylate.

22. Its solubility behavior shows that chavibetol is a phenol.

(a) Methylation with Me_2SO_4/NaOH introduces one methyl,

$$\begin{array}{ll}
C_{11}H_{14}O_2 & DD \\
\underline{-C_{10}H_{12}O_2} & \text{chavibetol} \\
CH_2
\end{array}$$

indicating that there is one phenolic group in chavibetol.

(b) Cleavage with hot HI to give CH_3I shows that chavibetol is a methyl ether, accounting for the second oxygen. Glancing ahead to the conversion into vanillin, we tentatively assume

$$\begin{array}{lll}
C_{10}H_{12}O_2 & \text{chavibetol} \\
\underline{-OH} & \text{phenolic} \\
C_{10}H_{11}O
\end{array}
\qquad
\begin{array}{lll}
C_{10}H_{11}O & \\
\underline{-OCH_3} & \text{ether} \\
C_9H_8
\end{array}
\qquad
\begin{array}{lll}
C_9H_8 & \\
\underline{-C_6H_3} & \text{trisubstituted aromatic} \\
C_3H_5
\end{array}$$

that, like vanillin, chavibetol is a trisubstituted aromatic, and that the ether oxygen is attached to the ring.

The residue of C_3H_5 from our chemical arithmetic corresponds to an unsaturated side chain (compare Problem 20, above).

(c) Treatment of chavibetol with strong, hot base brings about isomerization to EE; we can rationalize this conversion as:

$$\underset{\substack{\text{Chavibetol} \\ C_{10}H_{12}O_2}}{\underset{CH_3O}{\overset{HO}{\diagdown}} C_6H_3-CH_2-CH=CH_2} \xrightarrow[\text{heat}]{OH^-} \underset{\substack{\text{EE} \\ C_{10}H_{12}O_2}}{\underset{CH_3O}{\overset{HO}{\diagdown}} C_6H_3CH=CHCH_3}$$

The same isomerization undoubtedly occurs when DD is similarly treated.

$$\underset{\substack{\text{DD} \\ C_{11}H_{14}O_2}}{\underset{CH_3O}{\overset{CH_3O}{\diagdown}} C_6H_3-CH_2-CH=CH_2} \xrightarrow[\text{heat}]{OH^-} \underset{\substack{\text{FF} \\ C_{11}H_{14}O_2}}{\underset{CH_3O}{\overset{CH_3O}{\diagdown}} C_6H_3CH=CHCH_3}$$

Now we have to work out the orientation in the compounds. This problem is simplified by the ozonolysis.

$$FF \xrightarrow{O_3} \underset{OCH_3}{\overset{CHO}{\bigcirc}}OCH_3 \xleftarrow[OH^-]{Me_2SO_4} \underset{\substack{\text{Vanillin}}}{\underset{OH}{\overset{CHO}{\bigcirc}}OCH_3}$$

The result from FF is easy to interpret:

$$\underset{OCH_3}{\overset{CHO}{\bigcirc}}OCH_3 \xleftarrow{O_3} \underset{\substack{OCH_3 \\ FF}}{\overset{CH=CHCH_3}{\bigcirc}}OCH_3 \xleftarrow[\text{heat}]{OH^-} \underset{\substack{OCH_3 \\ DD}}{\overset{CH_2CH=CH_2}{\bigcirc}}OCH_3 \longleftarrow$$

$$\underset{OH}{\overset{CH_2CH=CH_2}{\bigcirc}}OCH_3$$

or   *One of these is chavibetol*

$$\underset{OCH_3}{\overset{CH_2CH=CH_2}{\bigcirc}}OH$$

Which of the two possibilities is chavibetol? We know that EE has essentially the same orientation of groups (1-carbon, 3-oxygen, 4-oxygen) as FF; hence the only possible isomer of vanillin that can be formed by cleavage of EE is the one in which the —OH and —OCH$_3$ groups are reversed.

$$EE \xrightarrow{O_3} \underset{\substack{OCH_3 \\ \text{Isomer of vanillin}}}{\overset{CHO}{\bigcirc}}OH$$

Finally, then, we arrive at the structure of chavibetol:

$$\underset{OCH_3}{\overset{CHO}{\bigcirc}}OH \xleftarrow{O_3} \underset{\substack{OCH_3 \\ EE}}{\overset{CH=CHCH_3}{\bigcirc}}OH \xleftarrow[\text{heat}]{OH^-} \underset{\substack{OCH_3 \\ \text{Chavibetol}}}{\overset{CH_2CH=CH_2}{\bigcirc}}OH$$

23. (a) Piperine is an amide of piperidine (Sec. 31.12) and an acid called piperic acid, RCOOH in which $R = C_{11}H_9O_2$.

Piperidine   ($R = C_{11}H_9O_2$)

(b) The m.w. of piperic acid ($C_{12}H_{10}O_4$) is 218; its N.E. is 215 ± 6. Clearly, piperic acid is a monocarboxylic acid.

From the low hydrogen content of R, it is apparent that it contains rings and/or unsaturated bonds. Bromination (without substitution) gives $C_{12}H_{10}O_4Br_4$; we conclude that there are probably two active double bonds present, requiring four of the twelve carbons.

$$
\begin{array}{ll}
C_{12}H_{10}O_4 & \text{piperic acid} \\
-\underline{C_4H_4} & \text{two double bonds} \\
C_8H_6O_4 &
\end{array}
\qquad
\begin{array}{ll}
C_8H_6O_4 & \\
-\underline{COOH} & \text{carboxylic acid} \\
C_7H_5O_2 &
\end{array}
$$

The residue points toward an aromatic ring (six C), leaving us with one as-yet-unaccounted-for carbon, and two oxygens. These cannot be in another —COOH, since piperic acid has only one such group (remember, m.w. = N.E.), which we have already taken into account.

Side-stepping this last problem for the moment, let us now take a look at the unsaturation. Oxidative cleavage of piperic acid gives a number of products, two of which account in a clear-cut way for all the carbons: tartaric acid, HOOCCHOHCHOHCOOH, and an acid, $C_8H_6O_4$, found to be different from any of the phthalic acids, $C_6H_4(COOH)_2$, and given the name of piperonylic acid by the original investigators.

We can relate piperic and piperonylic acid in the following way:

$$C_7H_5O_2\text{—COOH} + \underset{\underset{OH\quad OH}{|\quad\ \ |}}{HOOC\text{—CH—CH—COOH}} \xleftarrow{\text{oxidn.}} C_7H_5O_2\text{—CH}=\text{CH—CH}=\text{CH—COOH}$$

Piperonylic acid   Tartaric acid   Piperic acid
                                                         Partial

We can delay no longer our study of the "missing" carbon and two oxygens. Phenols and alcohols are impossible—oxidation would ruin them. A one-carbon carbonyl group would be an aldehyde; oxidized to —COOH it would lead to a phthalic acid instead of piperonylic acid. We are left with ether linkages as the answer for the oxygens. But how does one get two ether linkages from only *one* carbon?

Cleavage of the ether gives us our clue: formaldehyde and *two* phenolic groups are generated.

Furthermore, chemical arithmetic shows that only one H_2O is lost in the formation of these two ether linkages.

$$
\begin{array}{ll}
C_7H_6O_4 & \text{3,4-dihydroxybenzoic acid} \\
+\underline{HCHO} & \text{formaldehyde} \\
C_8H_8O_5 &
\end{array}
\qquad
\begin{array}{ll}
C_8H_8O_5 & \\
-\underline{C_8H_6O_4} & \text{piperonylic acid} \\
H_2O &
\end{array}
$$

The double ether is an acetal, and a cyclic one. Piperonylic acid can have only one structure:

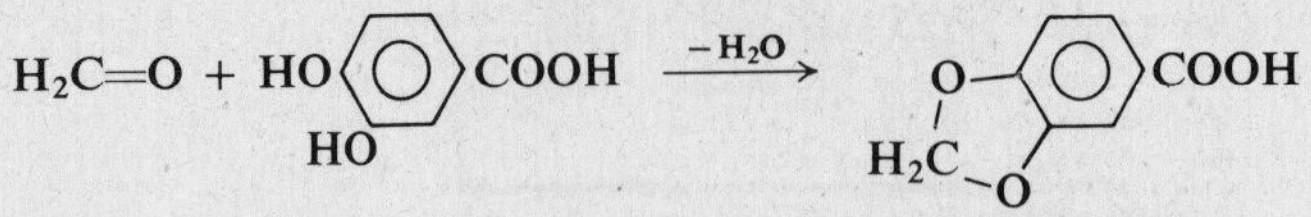

$$H_2C{=}O + HO{-}\langle\text{ring}\rangle{-}COOH \xrightarrow{-H_2O} \text{Piperonylic acid}$$

From this structure, the way is clear to piperic acid and piperine:

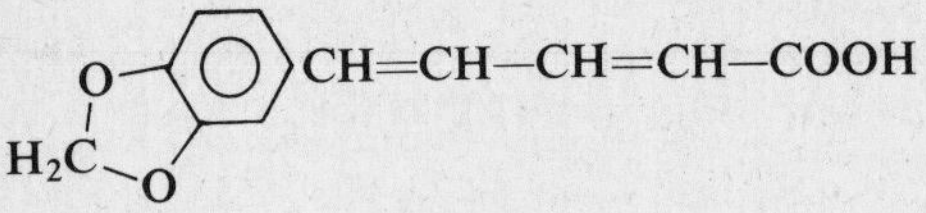

Piperic acid

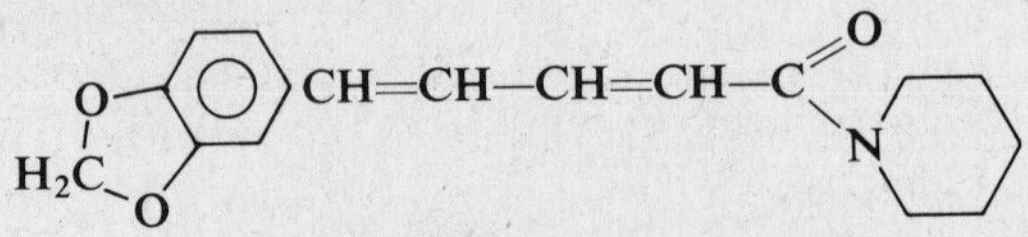

$$\langle\text{ring}\rangle CH{=}CH{-}CH{=}CH{-}COOH$$

Piperine

$$\langle\text{ring}\rangle CH{=}CH{-}CH{=}CH{-}C{\Big(}{=}O{\Big)}N\langle\text{ring}\rangle$$

(c) The synthetic sequence confirms our structure in every respect.

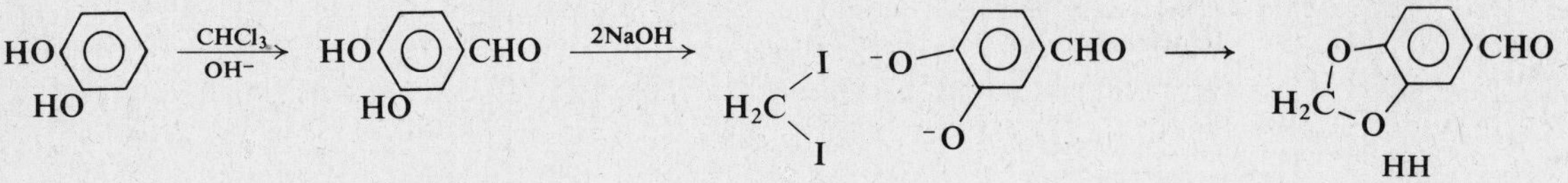

$$\text{Catechol} \xrightarrow[\text{OH}^-]{\text{CHCl}_3} \text{GG} \xrightarrow{\text{2NaOH}} \ \longrightarrow \ \text{HH}$$
$$\text{Piperonal}$$

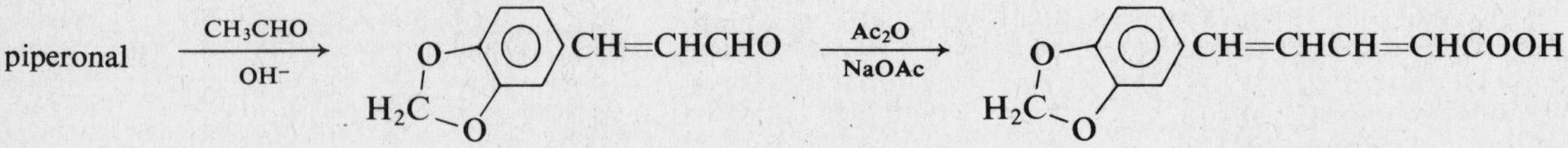

$$\text{piperonal} \xrightarrow[\text{OH}^-]{\text{CH}_3\text{CHO}} \text{II} \quad (CH{=}CHCHO) \xrightarrow[\text{NaOAc}]{\text{Ac}_2\text{O}} \text{Piperic acid} \quad (CH{=}CHCH{=}CHCOOH)$$

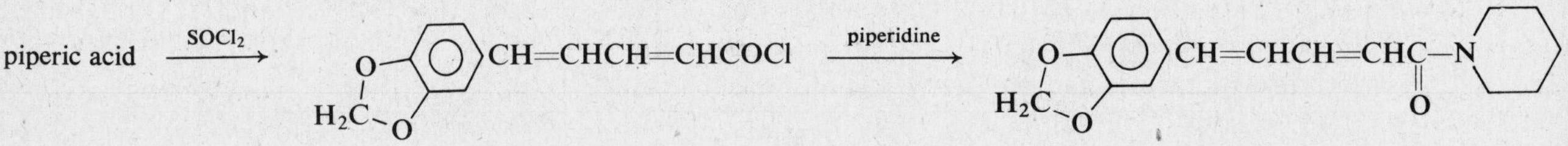

$$\text{piperic acid} \xrightarrow{\text{SOCl}_2} \text{JJ} \quad (CH{=}CHCH{=}CHCOCl) \xrightarrow{\text{piperidine}} \text{Piperine} \quad (CH{=}CHCH{=}CHC{-}N, {=}O)$$

24. Its solubility behavior shows that hordinene is a phenol and an amine. The Hinsberg test indicates that the amino grouping is tertiary; while the benzenesulfonyl chloride may react at the phenolic group, it leaves the amino group unchanged, and still basic.

Again, oxidative degradation gives much information but, because of the sensitivity of phenols, hordinene must be acetylated first.

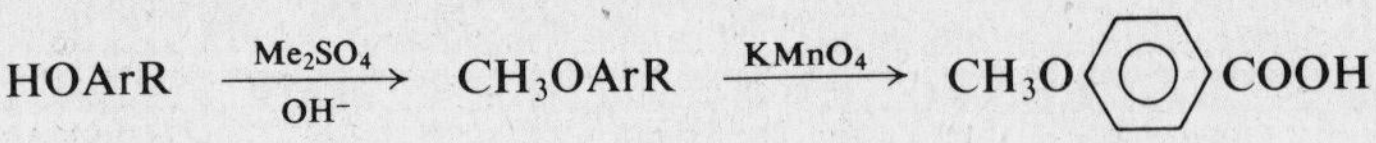

$$\underset{\text{Hordinene}}{\text{HOArR}} \xrightarrow[\text{OH}^-]{\text{Me}_2\text{SO}_4} \underset{\text{LL}}{\text{CH}_3\text{OArR}} \xrightarrow{\text{KMnO}_4} \underset{\text{Anisic acid}}{\text{CH}_3\text{O}\langle\text{ring}\rangle\text{COOH}}$$

Now we know that hordinene has a phenolic —OH *para* to a single side chain that carries the amino group. Since the amine is tertiary (requiring a minimum of three carbons), the side chain can contain no more than two carbons:

$$\begin{array}{ll}
\text{C}_{10}\text{H}_{15}\text{ON} & \text{hordinene} \\
-\text{HOC}_6\text{H}_4 & \\
\hline
\text{C}_4\text{H}_{10}\text{N} &
\end{array}
\qquad
\begin{array}{ll}
\text{C}_4\text{H}_{10}\text{N} & \\
-(\text{CH}_3)_2\text{N} & \text{smallest 3° amine} \\
\hline
\text{C}_2\text{H}_4 & \text{left for side-chain}
\end{array}$$

This is confirmed by the formation of *p*-methoxystyrene from LL with, presumably, elimination of Me$_2$NH.

$$\underset{p\text{-Methoxystyrene}}{\text{Me}_2\text{NH} + \text{CH}_3\text{O}\langle\text{ring}\rangle\text{CH}{=}\text{CH}_2} \xleftarrow{\text{heat}} \underset{\text{LL}}{\text{CH}_3\text{O}\langle\text{ring}\rangle\text{C}_2\text{H}_4\text{NMe}_2}$$

(a) So far, two structures are consistent with the data:

$$HO\langle O\rangle CH_2CH_2-\overset{\overset{\displaystyle CH_3}{|}}{N}-CH_3 \quad or \quad HO\langle O\rangle \overset{}{\underset{\underset{\displaystyle CH_3}{|}}{CH}}-\overset{\overset{\displaystyle CH_3}{|}}{N}-CH_3$$

(b) Either structure could be made by reductive amination (Sec. 22.11).

$$HO\langle O\rangle CH_2\overset{\overset{\displaystyle H}{|}}{C}=O \;+\; HN(CH_3)_2 \xrightarrow{H_2,\,Ni} HO\langle O\rangle CH_2CH_2-\overset{\overset{\displaystyle CH_3}{|}}{N}-CH_3$$

$$HO\langle O\rangle \underset{\underset{\displaystyle CH_3}{|}}{C}=O \;+\; HN(CH_3)_2 \xrightarrow{H_2,\,Ni} HO\langle O\rangle \underset{\underset{\displaystyle CH_3}{|}}{CH}-\overset{\overset{\displaystyle CH_3}{|}}{N}-CH_3$$

The first of these syntheses actually gives hordinene, and proves the structure to be

$$HO\langle O\rangle CH_2CH_2-\overset{\overset{\displaystyle CH_3}{|}}{N}-CH_3$$

Hordinene

25.

p-Toluic acid → MM → NN → OO → PP

$$PP \xrightarrow[heat]{base} QQ \xrightarrow[H^+]{EtOH} RR \xrightarrow{CH_3MgI} \xrightarrow{H_2O} \alpha\text{-Terpineol}$$

QQ (COOH)   RR (COOEt)   α-Terpineol

26. Coniferyl alcohol is soluble in NaOH but not in $NaHCO_3$, and hence must be a phenol. Treatment with benzoyl chloride (to give SS) increases the carbon number by 14, indicating that two benzoyl groups have been added, and hence that two —OH's have been esterified. Reaction with cold HBr to replace —OH by —Br shows that one of the —OH's is alcoholic. Cleavage by HI to give CH_3I reveals the presence of a methyl ether grouping.

Chemical arithmetic suggests the presence of an unsaturated side chain:

$C_{10}H_{12}O_3$	coniferyl alcohol	$C_{10}H_{11}O_2$		$C_{10}H_{10}O$		C_9H_7	
$-OH$	phenolic	$-OH$	alcoholic	$-OCH_3$	ether	$-C_6H_3$	aromatic ring
$C_{10}H_{11}O_2$		$C_{10}H_{10}O$		C_9H_7		C_3H_4	

In agreement, SS (with phenolic and alcoholic functions protected) was found to decolorize both $KMnO_4$ and Br_2/CCl_4: evidence for unsaturation. Structures possible at this point are:

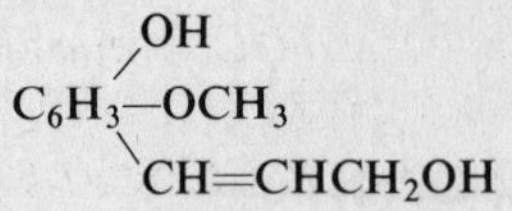

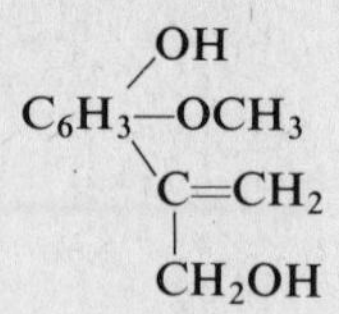

$$\begin{array}{cc}
\overset{\displaystyle OH}{\underset{\displaystyle CH=CHCH_2OH}{C_6H_3-OCH_3}} &
\overset{\displaystyle OH}{\underset{\displaystyle \overset{\displaystyle C=CH_2}{CH_2OH}}{C_6H_3-OCH_3}}
\end{array}$$

Finally, ozonolysis gives results that leave no doubts about the structures of coniferyl alcohol, SS, and TT.

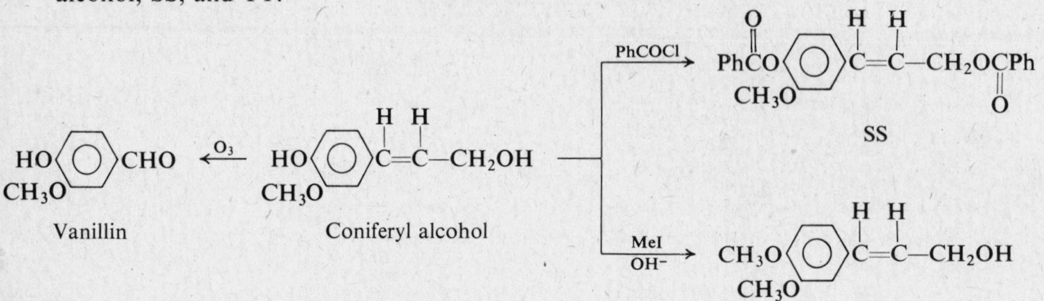

Vanillin   Coniferyl alcohol   SS

TT

27. (a) UU is an acetal (strictly, a ketal) and lactone. It is formed by intramolecular (nucleophilic) trapping of the benzononium ion that ordinarily leads to electrophilic aromatic substitution. (This seems to be the first such case to be discovered.)

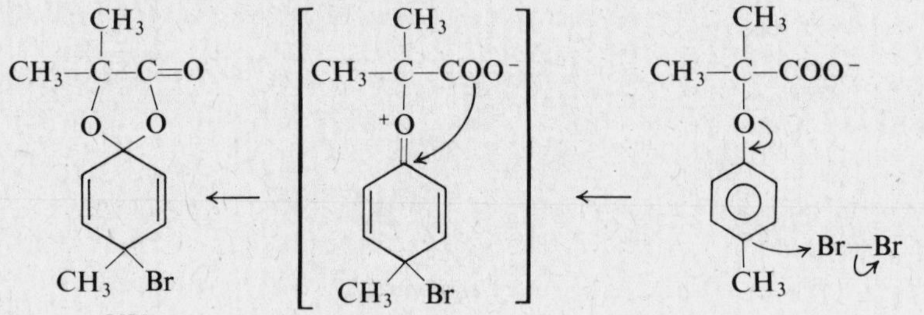

UU

A benzononium ion
(A *sigma* complex)

(b)

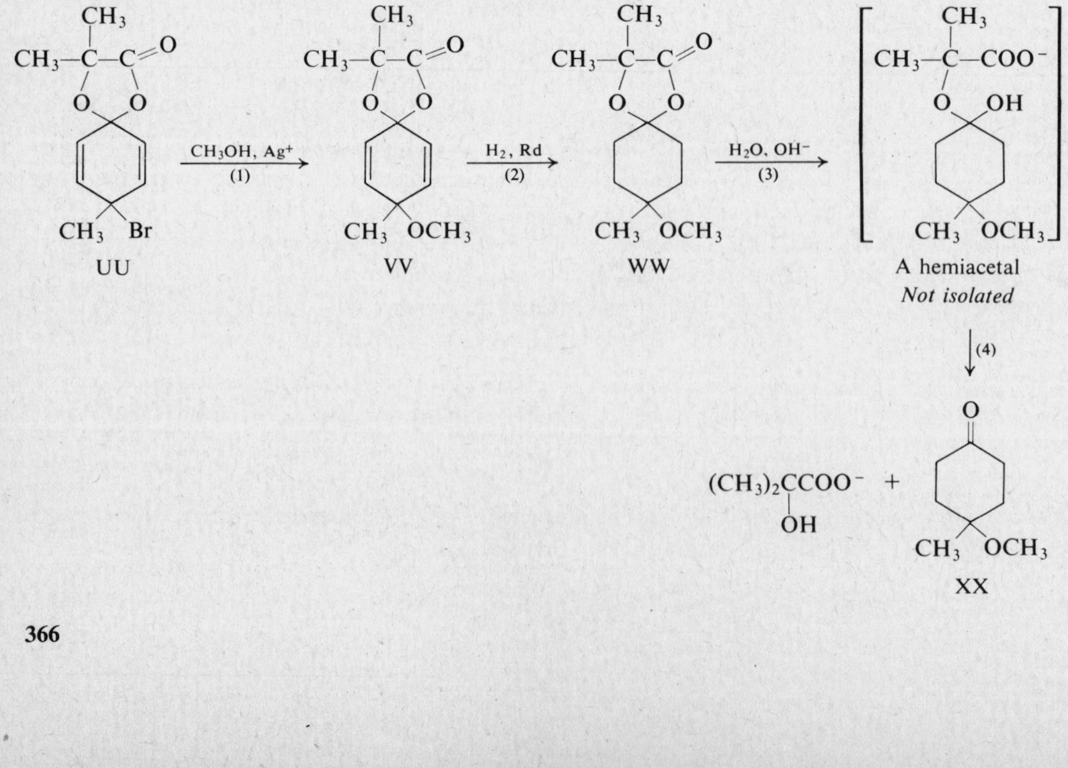

UU   VV   WW   A hemiacetal
Not isolated

$(CH_3)_2CCOO^-$ +

XX

(1) Nucleophilic substitution with (Lewis) acidic catalysis by Ag^+, which helps pull Br^- away to permit entry of $—OCH_3$. (2) Reduction of two double bonds. (3) Alkaline hydrolysis of a lactone (an ester) to yield an alcohol (a hemiacetal) and a carboxylate anion. (4) Like other hemiacetals, this one is unstable and yields the carbonyl compound and an alcohol (a hydroxy acid).

(c) This approach, with modifications, permits synthesis of a variety of non-benzenoid cyclic compounds from benzenoid starting materials. It is an alternative to reduction of aromatic rings by Li or Na in NH_3 (the Birch reduction, Problem 10, Chapter 30, page 998), which may affect other reducible functional groups in the molecule as well.

For example, cleavage of the ether linkage $(ROCH_3)$ in **XX**, and dehydration of the resulting alcohol would give rise to **VII**, and thus complete the transformation of a phenol into a polyfunctional aliphatic system. With a carbonyl group in the molecule—the most important functional group in organic chemistry—dozens of synthetic routes are open.

| VII | VIII | IX | X |

Groups other than $—OCH_3$ can be introduced in step (1): acetate, for example. The ketal–lactone can be cleaved before hydrogenation, to produce structures like **VIII**. The process can be extended to polycyclic aromatic systems (Chapter 30), with formation of compounds like **IX** and **X**.

(E. J. Corey, S. Barcza, and G. Klotmann, "A New Method for the Directed Conversion of the Phenoxy Grouping into a Variety of Cyclic Polyfunctional Systems," J. Am. Chem. Soc., **91**, 4782 (1969).)

28.

Piperonal (page 650)
AAA
(and HH in Problem 23, above)

Vanillin (page 792)
BBB

Eugenol (page 791)
CCC

Thymol (page 792)
DDD

Isoeugenol (page 791)
EEE

Safrole (page 792)
FFF

First, check your answers against those above. Whether you were right or wrong, try to fit the correct structures to the spectra. Next, turn to the labeled spectra on pages 673–675 of this Study Guide, and check your peak assignments against the ones shown there. Finally, turn to the analysis of these spectra outlined on page 676 of this Study Guide.

Chapter 25 | Aryl Halides
Nucleophilic Aromatic Substitution

25.1 (a) See Sec. 11.21.   (b) See Problem 11.13 (page 367) and its answer.

25.2 (a) 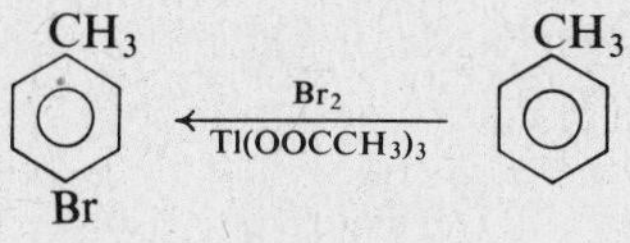
$$\text{(o-bromotoluene) } \xleftarrow[\text{Tl(OOCCH}_3)_3]{\text{Br}_2} \text{ (toluene)}$$

(b) 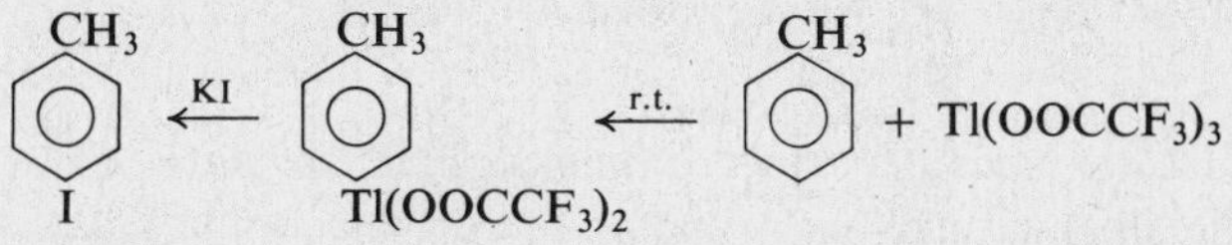
$$\text{(o-iodotoluene, I) } \xleftarrow{\text{KI}} \text{ (toluene-Tl(OOCCF}_3)_2) \xleftarrow{\text{r.t.}} \text{ (toluene) } + \text{Tl(OOCCF}_3)_3$$

(c) See page 771.

(d) 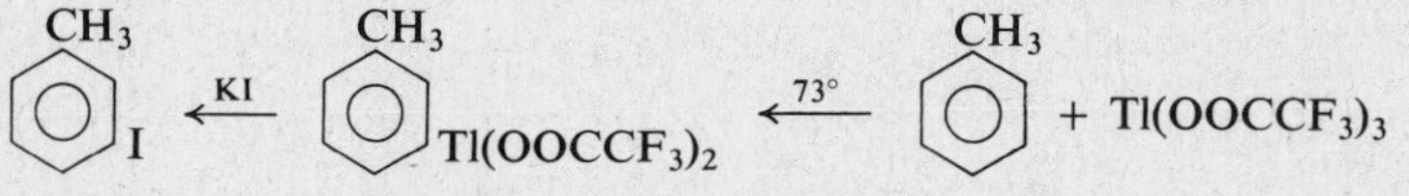
$$\text{(m-iodotoluene, I) } \xleftarrow{\text{KI}} \text{ (toluene-Tl(OOCCF}_3)_2) \xleftarrow{73^\circ} \text{ (toluene) } + \text{Tl(OOCCF}_3)_3$$

(e) 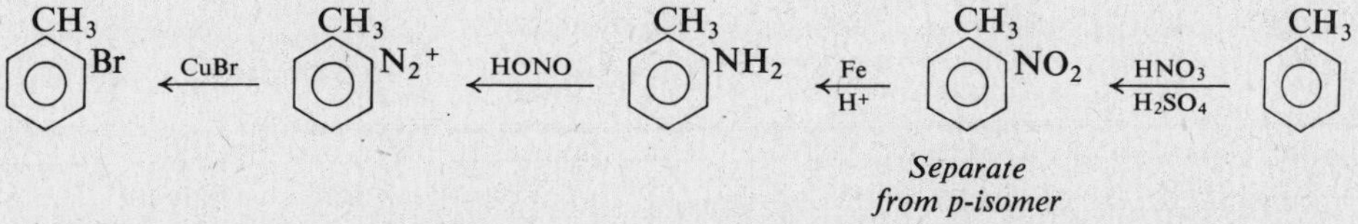
$$\text{(Br-toluene) } \xleftarrow{\text{CuBr}} \text{ (N}_2^+\text{-toluene) } \xleftarrow{\text{HONO}} \text{ (NH}_2\text{-toluene) } \xleftarrow[\text{H}^+]{\text{Fe}} \text{ (NO}_2\text{-toluene) } \xleftarrow[\text{H}_2\text{SO}_4]{\text{HNO}_3} \text{ (toluene)}$$

Separate from p-isomer

(See the reference given in the answer to Problem 9, Chapter 11; also A. McKillop, E. C. Taylor, *et al.*, "Thallium in Organic Synthesis. XXII. Electrophilic Aromatic Synthesis Using Thallium(III) Trifluoroacetate. A Simple Synthesis of Aromatic Iodides," J. Am. Chem. Soc., **93**, 4841 (1971); "XXIII. Electrophilic Aromatic Thallation. Kinetics and

Application to Orientation Control in the Synthesis of Aromatic Iodides," J. Am. Chem. Soc., **93**, 4845 (1971).)

25.3 There is a stronger Ar—O bond in phenols, due to partial double-bond character and/or sp^2 hybridization of aromatic carbon.

25.4 (a)

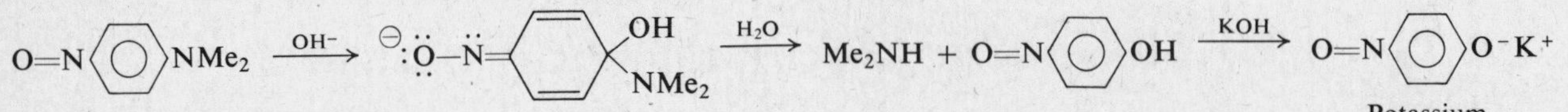

I

Potassium *p*-nitrosophenoxide

(b) Nucleophilic aromatic substitution.

(c) Electron withdrawal—or, more properly, electron acceptance—by the nitroso group, which stabilizes the transition state leading to intermediate I.

(d)

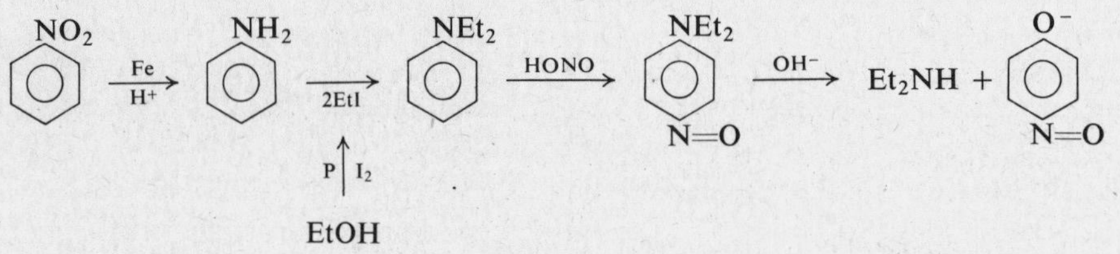

25.5 (a) Nucleophilic aromatic substitution of —OH for —OCH_3, with activation by the two —NO_2 groups *ortho* and *para* to —OCH_3.

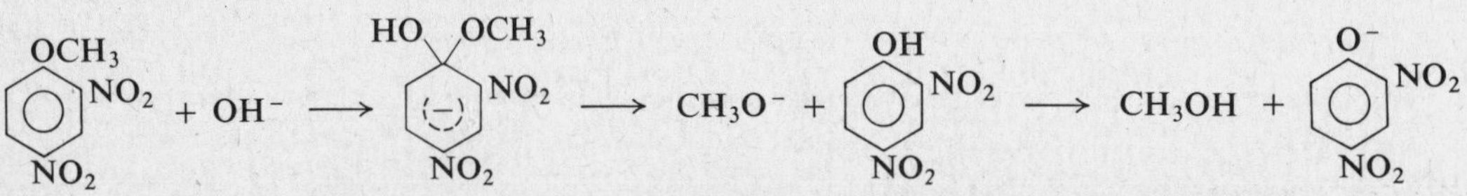

(b) The hydrolysis product, *p*-nitroaniline, is activated toward nucleophilic aromatic substitution by the *p*-nitro group. In the alkaline medium, some of the product suffers displacement of —NH_2 by —OH to give *p*-nitrophenol; the yield is lowered, and the phenol and its oxidation products make the product harder to purify.

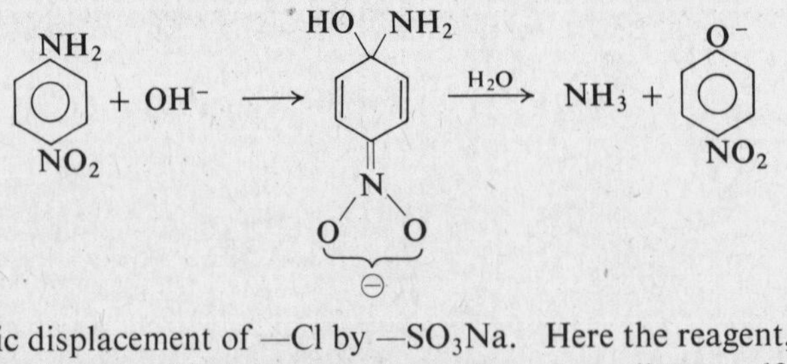

(c) Nucleophilic displacement of —Cl by —SO_3Na. Here the reagent, SO_3^{--}, is a nucleophile, with an unshared pair of electrons on sulfur. In ordinary sulfonation, the reagent, SO_3, is an electrophile, with electron-deficient sulfur.

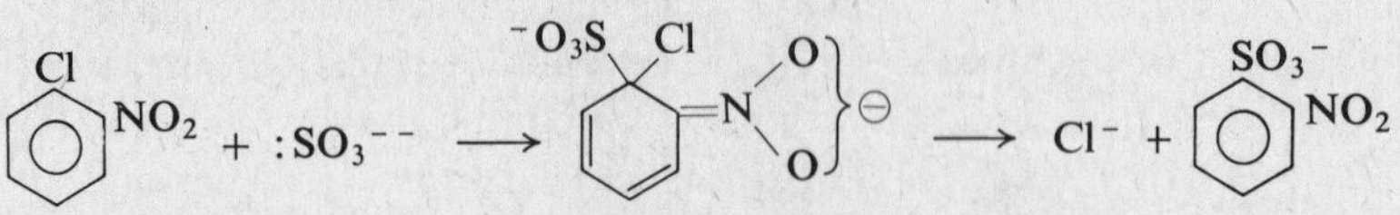

(d) It is not a general method, since it requires the ring to be activated by substituents like —NO_2 *ortho* and/or *para* to the point of attack. Furthermore, this kind of substitution cannot be used to displace H^-, but only weakly basic leaving groups like Cl^-. It could not be used to prepare benzenesulfonic acid.

(e) As in (c), there is nucleophilic aromatic substitution with SO_3^{--} as nucleophile and, this time, NO_2^- as leaving group. In the *o-* or *p-*isomer, each NO_2 activates the other, and a water-soluble benzenesulfonate salt is obtained:

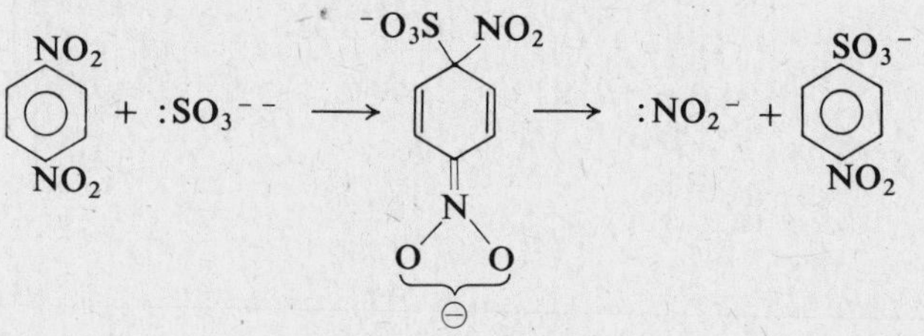

The *m*-isomer does not react, and remains insoluble.

25.6 The —NO can help disperse either kind of charge, negative or positive, developing in the transition state leading to the intermediate in aromatic substitution.

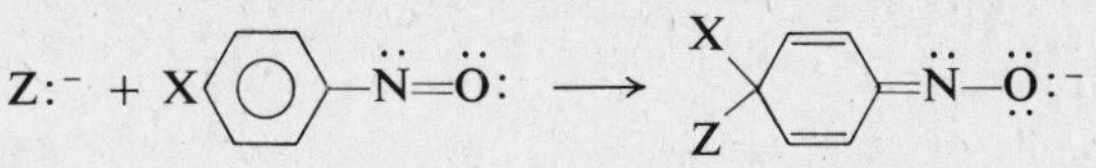

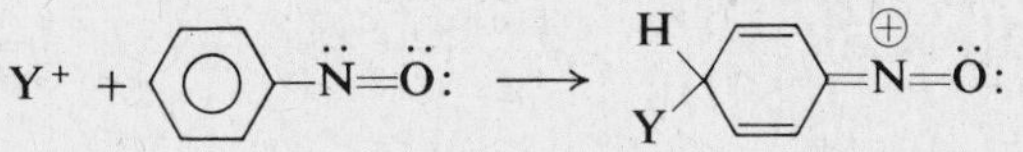

25.7 Chemical arithmetic indicates the addition of sodium ethoxide,

$$
\begin{array}{ll}
C_9H_{10}O_8N_3Na & \text{product} \\
- C_7H_5O_7N_3 & \text{starting material} \\
\hline
C_2H_5ONa & \text{equivalent to sodium ethoxide}
\end{array}
$$

or, in the second case, of sodium methoxide. The product I is the same from both reac-

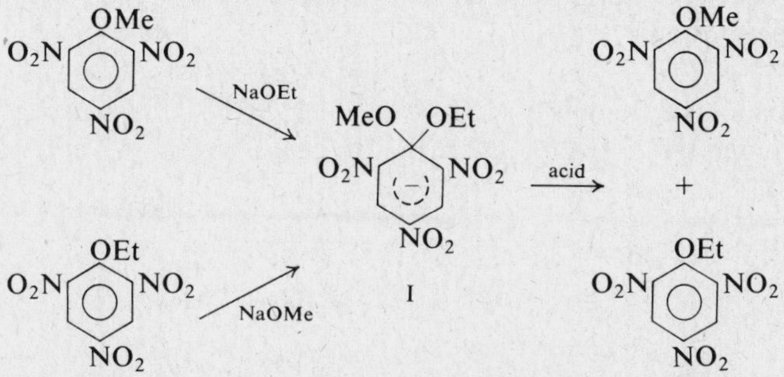

tions, and is a stable example of the intermediate in nucleophilic aromatic substitution by the bimolecular mechanism.

25.8 The phenyl anion, $C_6H_5^-$, is a stronger base than the $2\text{-F-}3\text{-CH}_3\text{OC}_6H_3^-$ anion, which contains two electron-withdrawing, base-weakening substituents, —F and —OCH$_3$. Conversely, *o*-fluoroanisole is a stronger acid than benzene.

25.9 The organolithium compound, like a Grignard reagent, reacts:

(a) with CO_2 to give a carboxylic acid:

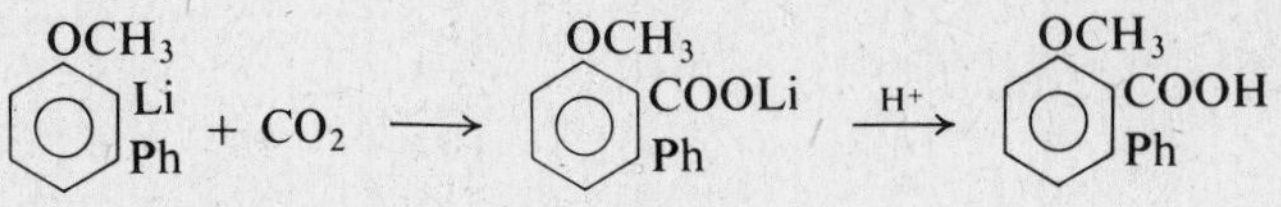

(b) with a ketone to give a tertiary alcohol:

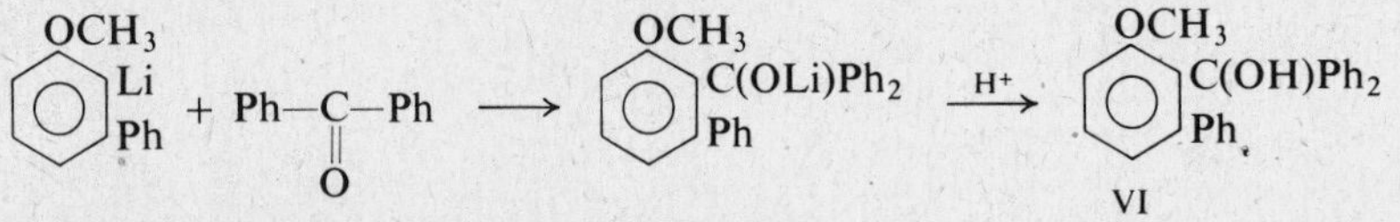

(c) Magnesium reacts at the —Br bond to give the Grignard reagent, which is analogous to the organolithium product of reaction (5), page 840, and reacts as in (6) to give the benzyne

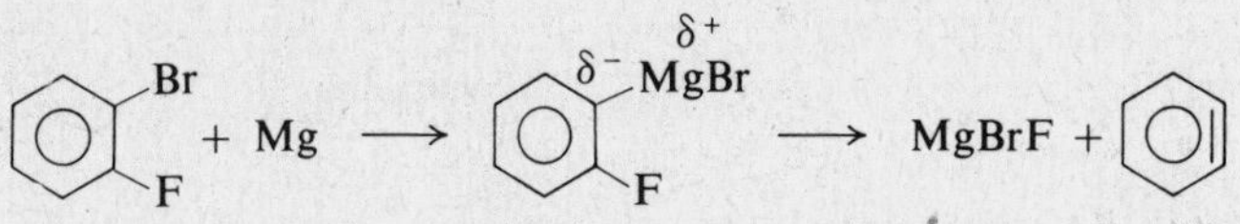

25.10 (a) AgNO$_3$ (b) AgNO$_3$

(c) Br$_2$/CCl$_4$, or KMnO$_4$ (d) iodoform test; or CrO$_3$/H$_2$SO$_4$

(e) iodoform test

25.11 (a) Oxidation to acids; then determination of their m.p.'s (but both are very high), or their neutralization equivalents (201 vs. 123).

(b) Ozonolysis and identification of fragments (CH$_3$CHO vs. HCHO), or isomerization of one isomer (the allylic compound) into the other (the propenyl isomer) by hot KOH (see Problem 20, Chapter 24).

1. (a) PhMgBr   (h) *o*- and *p*-BrC$_6$H$_4$NO$_2$ (*i*) *o*- and *p*-BrC$_6$H$_4$SO$_3$H   (j) *o*- and *p*-BrC$_6$H$_4$Cl
(m) *o*- and *p*-BrC$_6$H$_4$C$_2$H$_5$

      No reaction: b, c, d, e, f, g, k, l, n, o.

2. (a) *n*-BuMgBr   (b) *n*-BuOH

(c) $CH_3CH_2CH{=}CH_2$ (d) *n*-BuC≡CH

(e) *n*-Bu—O—Et   (f) *n*-BuNH$_2$

(g) *n*-BuCN   (l) Ph—Bu-*n* and Ph—Bu-*sec*

     No reaction: h, i, j, k, m, n, o.

3. (b) 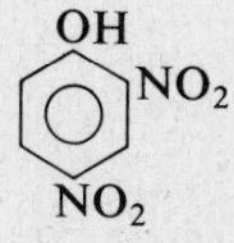   (e) 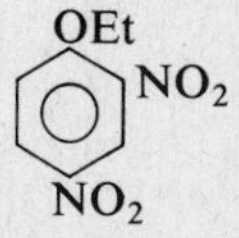   (f) 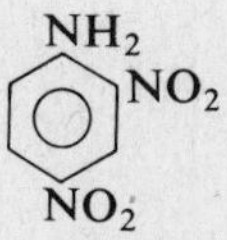   (g)

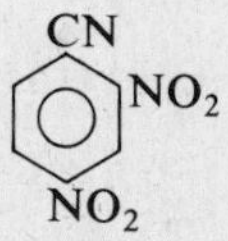

4. (a) Mg, anhydrous Et$_2$O; H$_2$O   (b) HNO$_3$, H$_2$SO$_4$, warm

(c) Cl$_2$, Fe   (d) SO$_3$, H$_2$SO$_4$

(e)

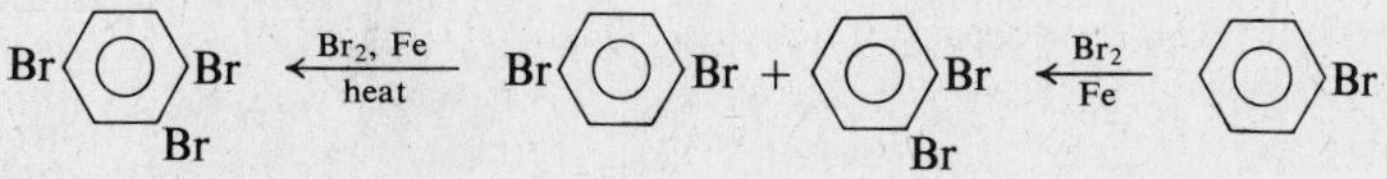

(f) CH$_3$Cl, AlCl$_3$

(g)

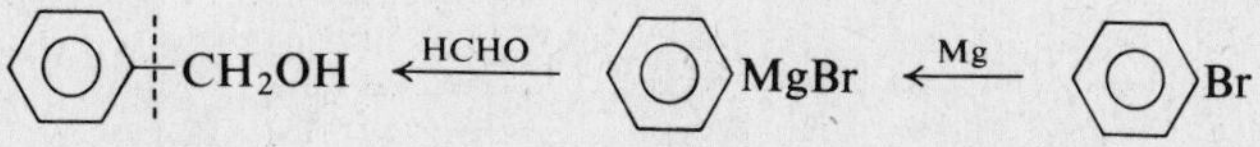

(h)

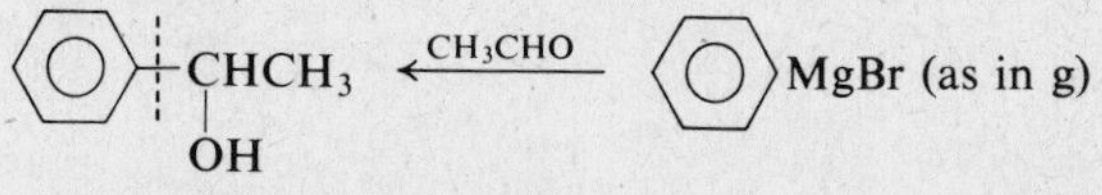

(i)

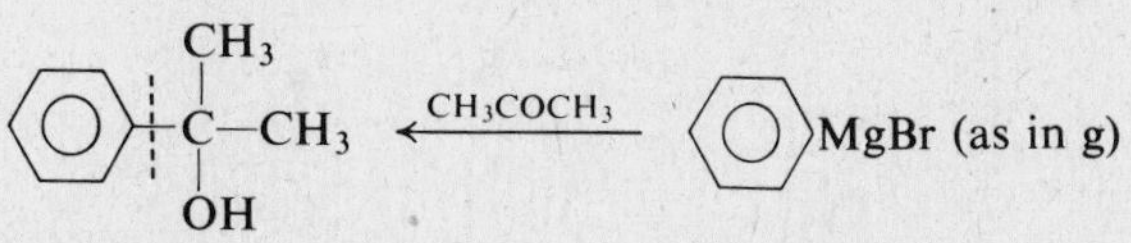

(j)

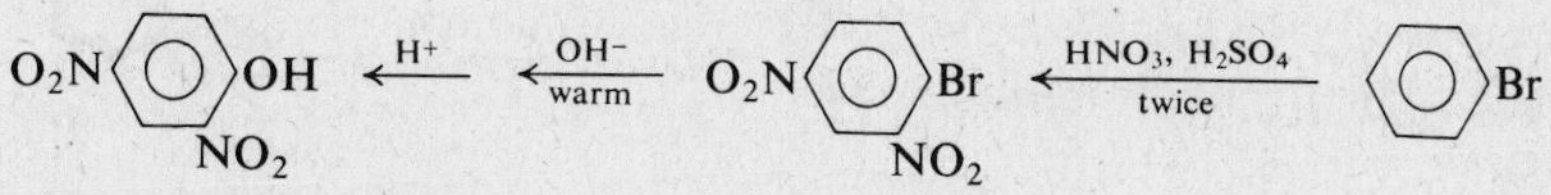

     (See Prob. 3b, above.)

(k)

     (Compare Prob. 16, Chapter 8, page 281.)

(l)

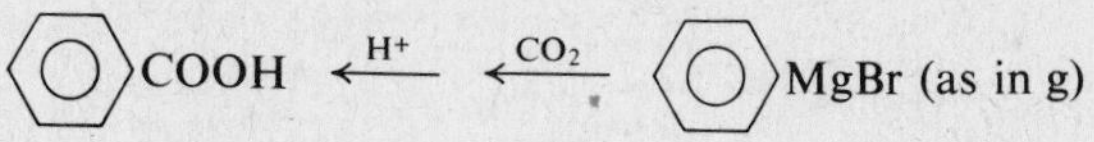

(m) NaNH$_2$/NH$_3$(liq.)

5. (a) C_6H_6 (b) C_6H_6 (c) C_6H_6 (d) $PhCH_2CH=CH_2$

(e) $PhCH_2OH$ (f) $PhCHOHCH_3$ (g) $PhCHOHPh$ (h) $p\text{-}CH_3C_6H_4CHOHPh$

(i) $Ph-\underset{\underset{OH}{|}}{\overset{\overset{CH_3}{|}}{C}}-CH_3$ (j) cyclohexane with OH and Ph   (k) cyclohexane with OH, Ph and two CH_3 (l) $Ph-\underset{\underset{OH}{|}}{\overset{\overset{Ph}{|}}{C}}-CH_3$

(m) Pn_3COH (n) $Ph-\underset{\underset{OH}{|}}{\overset{\overset{Ph}{|}}{C}}-\overset{\overset{CH_3}{|}}{C}H-C_2H_5$ (o) $C_6H_6 + HC\equiv CMgBr$

Optically active

Racemic modification: f, h, k. (Optically inactive.)
Optically active single compound: n.
All others: optically inactive single compounds.

6. (a) $2,4,6\text{-triNO}_2 > 2,4\text{-diNO}_2 > o\text{-NO}_2 > m\text{-NO}_2$, unsubstd.

(b) $PhCH_3 > PhH > PhCl > PhNO_2$

(c) $\underset{\underset{Br}{|}\text{ Allylic}}{CH_3CHCH=CH_2} > \underset{\underset{Br}{|}\text{ 1° alkyl}}{CH_2CH_2CH=CH_2} > \underset{\underset{Br}{|}\text{ Vinylic}}{CH_3CH_2CH=CH}$

(d) $PhMe > p\text{-}BrC_6H_4Me > PhBr > p\text{-}BrC_6H_4Br$

(e) $\underset{\text{Benzylic}}{PhCH_2Cl} > \underset{\text{1° alkyl}}{EtCl} > \underset{\text{Aryl}}{PhCl}$

(f) $\underset{\underset{Br}{|}\text{ Benzylic}}{PhCHCH_3} > \underset{\underset{Br}{|}\text{ 1° alkyl}}{PhCH_2CH_2} > \underset{\underset{Br}{|}\text{ Vinylic}}{PhCH=CH}$

7. Aqueous $NaHCO_3$, to minimize nucleophilic substitution by OH^- that would convert part of the product into 2,4-dinitrophenol (compare Problem 25.5b).

8. We must not forget the preference for the reaction of one halogen over another: (a) allyl over vinyl; (b) benzyl over aryl; (e) *para* over *meta*; (f) ArBr over ArCl; (i) benzylic over aryl.

(a) $\underset{\underset{Br\ \ OH}{|\ \ \ |}}{CH_2=C-CH_2}$ (b) $p\text{-}BrC_6H_4CH_2NH_2$ (c) $p\text{-}ClC_6H_4COOH$ (d) $m\text{-}BrC_6H_4\underset{\underset{Br\ \ Br}{|\ \ \ |}}{CH-CH_2}$

(e) [benzene ring with Cl, Cl, NO$_2$] $\xrightarrow{OCH_3^-}$ [benzene ring with OCH$_3$, Cl, NO$_2$]   (f) $p\text{-}ClC_6H_4MgBr$ (g) no reaction

(h) $p\text{-}BrC_6H_4CH_2Br$

(i) (C_6H_5 ring with $CH=CH_2$ and Cl)

(j) $p\text{-}BrC_6H_4CH_2Cl$

(k)

$$\text{C}_6\text{H}_4(\text{CF}_3)(\text{Br}) \xrightarrow{\text{NaNH}_2} \text{Aryne} \xrightarrow{\text{NH}_2^-} \underset{\text{(See Sec. 25.14)}}{\underset{\text{More stable anion}}{\text{C}_6\text{H}_4(\text{CF}_3)(\overset{:-}{\text{NH}_2})}} \xrightarrow{\text{NH}_3} \text{C}_6\text{H}_4(\text{CF}_3)(\text{NH}_2)$$

(l)

$$\text{C}_6\text{H}_4(\text{OCH}_3)(\text{Br}) \xrightarrow{\text{KNEt}_2} \text{Aryne} \xrightarrow{\text{Et}_2\text{N}^-} \underset{\text{(See Sec. 25.14)}}{\underset{\text{More stable anion}}{\text{C}_6\text{H}_4(\text{OCH}_3)(\overset{:-}{\text{NEt}_2})}} \xrightarrow{\text{Et}_2\text{NH}} \text{C}_6\text{H}_4(\text{OCH}_3)(\text{NEt}_2)$$

9. (a)

$$\text{C}_6\text{H}_4(\text{NO}_2)(\text{Cl}) \xleftarrow{\text{Cl}_2,\ \text{Fe}} \text{C}_6\text{H}_5\text{NO}_2 \xleftarrow{\text{HNO}_3,\ \text{H}_2\text{SO}_4} \text{C}_6\text{H}_6$$

(b)

$$o\text{-}\text{C}_6\text{H}_4(\text{NO}_2)(\text{Cl}) \xleftarrow{\text{HNO}_3,\ \text{H}_2\text{SO}_4} \text{C}_6\text{H}_5\text{Cl} \xleftarrow{\text{Cl}_2,\ \text{Fe}} \text{C}_6\text{H}_6$$

(c)

$$\text{C}_6\text{H}_4(\text{COOH})(\text{Br}) \xleftarrow[\text{heat}]{\text{Br}_2,\ \text{Fe}} \text{C}_6\text{H}_5\text{COOH} \xleftarrow{\text{KMnO}_4} \text{C}_6\text{H}_5\text{CH}_3$$

(d)

$$p\text{-}\text{C}_6\text{H}_4(\text{COOH})(\text{Br}) \xleftarrow{\text{KMnO}_4} p\text{-}\text{C}_6\text{H}_4(\text{CH}_3)(\text{Br}) \xleftarrow{\text{Br}_2,\ \text{Fe}} \text{C}_6\text{H}_5\text{CH}_3$$

(e)

$$\text{C}_6\text{H}_4(\text{CCl}_3)(\text{Cl}) \xleftarrow{\text{Cl}_2,\ \text{Fe}} \text{C}_6\text{H}_5\text{CCl}_3 \xleftarrow{3\text{Cl}_2,\ \text{heat}} \text{C}_6\text{H}_5\text{CH}_3$$

(f)

$$\text{C}_6\text{H}_3(\text{NO}_2)(\text{Br})(\text{Br}) \xleftarrow{\text{Br}_2,\ \text{Fe}} \text{C}_6\text{H}_4(\text{NO}_2)(\text{Br}) \xleftarrow{\text{HNO}_3,\ \text{H}_2\text{SO}_4} \text{C}_6\text{H}_5\text{Br} \xleftarrow{\text{Br}_2,\ \text{Fe}} \text{C}_6\text{H}_6$$

(g)

$$\text{C}_6\text{H}_3(\text{CHCl}_2)(\text{Br}) \xleftarrow{2\text{Cl}_2,\ \text{heat}} \text{C}_6\text{H}_4(\text{CH}_3)(\text{Br}) \xleftarrow{\text{Br}_2,\ \text{Fe}} \text{C}_6\text{H}_5\text{CH}_3$$

(h)

$$\text{C}_6\text{H}_2(\text{NH}_2)(\text{NO}_2)(\text{NO}_2) \xleftarrow[\text{warm}]{\text{NH}_3} \text{C}_6\text{H}_2(\text{Cl})(\text{NO}_2)(\text{NO}_2) \xleftarrow[\text{twice}]{\text{HNO}_3,\ \text{H}_2\text{SO}_4} \text{C}_6\text{H}_5\text{Cl} \xleftarrow{\text{Cl}_2,\ \text{Fe}} \text{C}_6\text{H}_6$$

(i)

$$\underset{Br}{\overset{CH=CH_2}{\bigcirc}} \xleftarrow{KOH(alc)} \underset{Br}{\overset{CHClCH_3}{\bigcirc}} \xleftarrow[heat]{Cl_2} \underset{Br}{\overset{CH_2CH_3}{\bigcirc}} \xleftarrow[Fe]{Br_2} \overset{Et}{\bigcirc} \quad \text{(page 395)}$$

(j)

$$\underset{Br}{\overset{COOH}{\bigcirc}}{Br} \xleftarrow{KMnO_4} \underset{Br}{\overset{CH_3}{\bigcirc}}{Br} \xleftarrow[twice]{Br_2, Fe} \overset{CH_3}{\bigcirc}$$

(k)

$$\underset{I}{\overset{CH_3}{\bigcirc}} \xleftarrow{KI} \underset{Tl(OOCCF_3)_2}{\overset{CH_3}{\bigcirc}} \xleftarrow[73°]{Tl(OOCCF_3)_3} \overset{CH_3}{\bigcirc} \quad \text{(Compare Prob. 25.2d)}$$

(l)

$$\underset{Br}{\overset{SO_3H}{\bigcirc}} \xleftarrow{SO_3, H_2SO_4} \underset{Br}{\bigcirc} \xleftarrow{Br_2, Fe} C_6H_6$$

(m)

$$\underset{Cl}{\overset{CH_2OH}{\bigcirc}} \xleftarrow{NaOH(aq)} \underset{Cl}{\overset{CH_2Cl}{\bigcirc}} \xleftarrow{Cl_2, heat} \underset{Cl}{\overset{CH_3}{\bigcirc}} \xleftarrow{Cl_2, Fe} \overset{CH_3}{\bigcirc}$$

(n)

$$\underset{CH_3}{\overset{H_3C\ CH_3\ CH}{\bigcirc}} \xleftarrow[Ni]{H_2} \underset{CH_3}{\overset{H_3C\ CH_2\ C}{\bigcirc}} \xleftarrow[heat]{H^+} \underset{CH_3}{\overset{OH\ CH_3-C-CH_3}{\bigcirc}} \xleftarrow{CH_3COCH_3} \underset{CH_3}{\overset{MgBr}{\bigcirc}} \xleftarrow{Mg} \underset{CH_3}{\overset{Br}{\bigcirc}} \quad \text{(as in d)}$$

10. See Sec. 31.10

11. In several of the steps, the reaction type depends upon the point of view.

Protonation of aldehyde.

$$Cl_3C-\overset{H}{\underset{}{C}}=O + H^+ \rightleftarrows Cl_3C-\overset{H}{\underset{}{C}}=OH^+$$

Electrophilic aromatic substitution; second step of acid-catalyzed nucleophilic addition.

$$Cl_3C-CHOH^+ + \bigcirc Cl \longrightarrow Cl_3C-\overset{H}{\underset{OH}{C}}-\bigcirc Cl + H^+$$

Protonation of alcohol.

$$Cl_3C-\overset{H}{\underset{OH}{C}}-\bigcirc Cl + H^+ \rightleftarrows Cl_3C-\overset{H}{\underset{\oplus OH_2}{C}}-\bigcirc Cl$$

Formation of carbonium ion: first step of S$_N$1 reaction of protonated alcohol.

$$Cl_3C-\underset{\oplus OH_2}{\overset{H}{C}}-\bigcirc Cl \longrightarrow Cl_3C-\underset{\oplus}{\overset{H}{C}}-\bigcirc Cl + H_2O$$

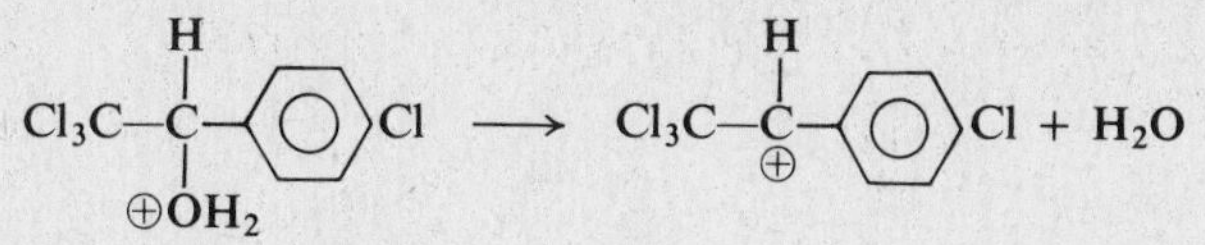

Electrophilic aromatic substitution: second step of S$_N$1 reaction of protonated alcohol.

$$Cl_3C-\underset{\oplus}{\overset{H}{C}}-\bigcirc Cl + \bigcirc Cl \longrightarrow Cl_3C-\overset{H}{C}-\bigcirc Cl + H^+$$

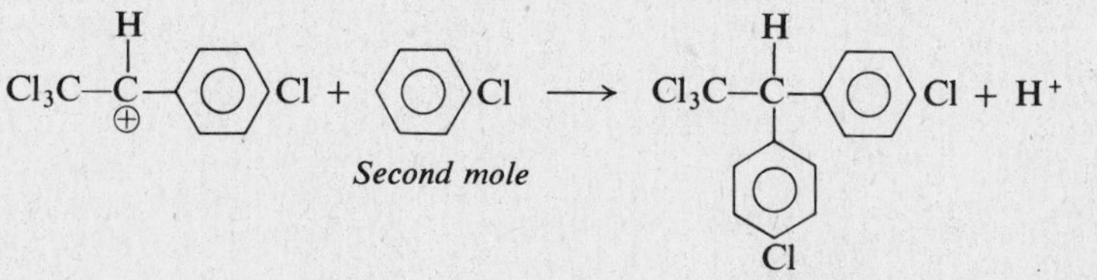

Second mole

DDT

12.

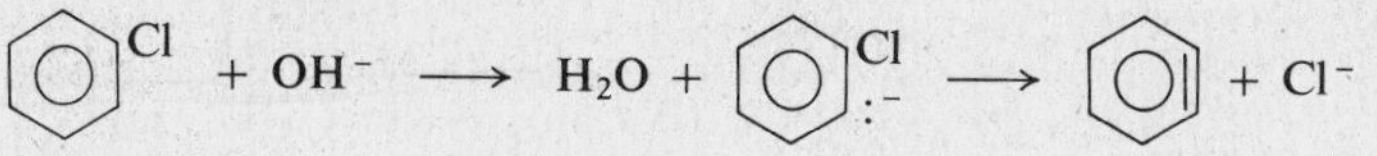

Chlorobenzene

Benzyne

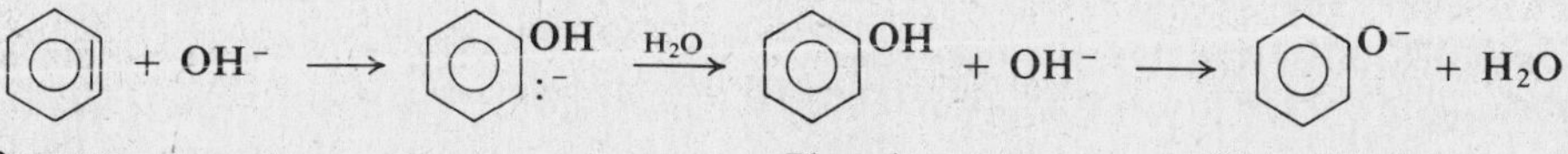

Benzyne

Phenol

Phenoxide ion

Phenoxide is an ambident anion, and can react at oxygen or at the ring; compare the answer to Problem 13, Chapter 24, page 808.

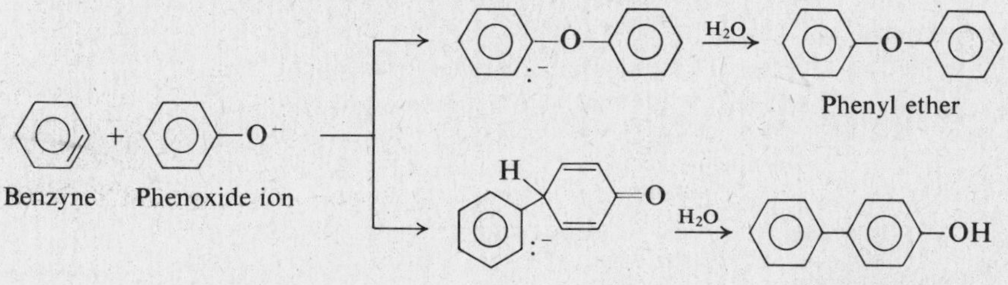

Benzyne   Phenoxide ion

Phenyl ether

Reaction at oxygen

p-Phenylphenol

Reaction at ring

13. Hydrogen exchange, we have seen, takes place via the carbanion (page 838). Its rate depends upon how fast the carbanion is formed. This, in turn, depends upon accommodation of the negative charge developing in the transition state leading to the formation of the carbanion. Through its inductive effect, fluorine helps to accommodate this negative charge. The inductive effect depends on the distance from the negative carbon, and is *much* stronger from the *ortho* position than from the *meta* or *para* positions. (See the reference given in the answer to Problem 23, below.)

14. The —N$_2^+$ group very powerfully activates the ring toward nucleophilic aromatic substitution. Chloride ion displaces much —Br before reduction occurs.

15. $ArX + R_2NH \underset{}{\overset{(1)}{\rightleftharpoons}} \ Ar\overset{\ominus}{\underset{\overset{+}{NHR_2}}{\diagup}}\diagdown X \ \overset{(2)}{\longrightarrow} \ X^- + Ar\overset{+}{N}HR_2$

$\longrightarrow ArNR_2$

X

$(3)\downarrow :B$

$Ar\overset{\ominus}{\underset{NR_2}{\diagup}}\diagdown X \ \overset{(4)}{\longrightarrow} \ X^- + ArNR_2$

XI

    Bunnett (page 478), who did this work, has given essentially the following interpretation. (There is additional evidence to help rule out certain alternative interpretations.)

    When X = F, intermediate X is formed reversibly (1). Some of X goes on to products (2) by loss of X$^-$, but some reverts to ArX by loss of weakly basic R$_2$NH; this reversal of (1) retards the over-all reaction. Via (3), base converts the intermediate X into XI, which rapidly goes on (4) to products by loss of X$^-$ rather than revert to ArX by loss of the strongly basic R$_2$N$^-$ anion.

    When X = Br, (2) is much faster because of weaker C—Br bond. X is formed irreversibly, all of it rapidly going on to product with or without action of base.

    (J. F. Bunnett and J. J. Randall, "Base Catalysis of the Reaction of N-Methylaniline with 2,4-Dinitrofluorobenzene. Proof of the Intermediate Complex Mechanism for Aromatic Nucleophilic Substitution," J. Am. Chem. Soc., **80**, 6020 (1958).)

16. (a) There is a primary isotope effect. Consider the answer to the preceding problem, with X = OPh or ^{18}OPh. Steps (2) and (3) are slower for the heavier isotope, and more of the intermediate reverts to the starting material via reverse of step (1).

(b) Piperidine speeds up (3) relative to the reverse of (1), so that a higher fraction of the intermediate goes on to product, regardless of whether OPh$^-$ or ^{18}OPh$^-$ is being lost.

17. In the aprotic solvent, azide ion is weakly solvated (no H-bonding) and hence is a stronger nucleophile.

18. (a) 28, N$_2$; 44, CO$_2$; 76, benzyne; 152, biphenylene, a dimer of benzyne which gradually forms.

       Benzyne   Biphenylene
      Two moles

(b)

o-Aminobenzoic acid
  (Anthranilic acid)

    (The dimerization of benzyne in part (a) is a [2 + 2] thermal cycloaddition, and is *symmetry-forbidden* (Sec. 29.9). Presumably, either the high energy of the particles involved here makes the difficult reaction possible, or reaction proceeds by a non-concerted, step-wise mechanism.)

19. Some chemical arithmetic first:

$$C_6H_5Cl \quad \text{chlorobenzene}$$
$$+C_{19}H_{15}K \quad Ph_3CK$$
$$\overline{C_{25}H_{20}ClK}$$

$$C_{25}H_{20}ClK \quad \text{reactants}$$
$$-C_{25}H_{20} \quad \text{products}$$
$$\overline{KCl \text{ is lost; suggests product is } Ph_4C}$$

On the assumption that the product is tetraphenylmethane, we can write the following series of reactions:

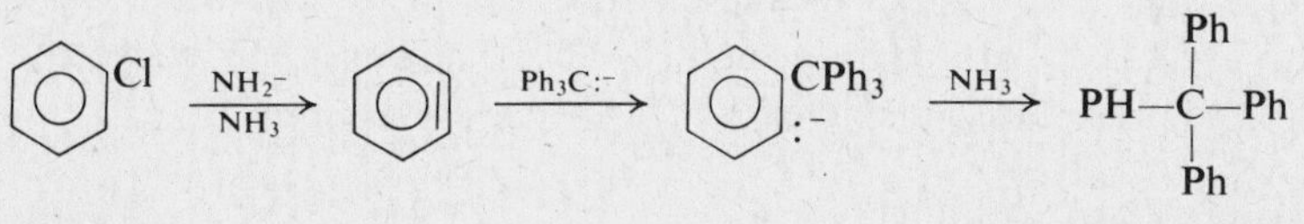

Chlorobenzene   Benzyne   Tetraphenylmethane

KNH_2 is needed to produce benzyne; the resonance-stabilized Ph_3C^- anion is too weakly basic to produce benzyne, and it is too weakly nucleophilic to bring about bimolecular displacement.

A benzylic carbanion, like a phenoxide ion (Problem 12, above) or an aromatic amine (Problem 23.14, page 765 and Problem 23.20, page 774), could be an ambident reagent (see the answer to Problem 13, Chapter 24, page 808). We should consider the possibility that—for steric reasons—benzyne might react not at side-chain carbon to give Ph_4Cl but at the ring to give XII, a structure reminiscent of the dimer of triphenylmethyl (I, page 393).

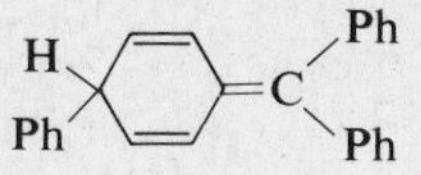

XII

20. (a) At 340° reaction proceeds, at least partly, via aryne XIII, which adds water in two different ways to give both *p*- and *m*-cresol.

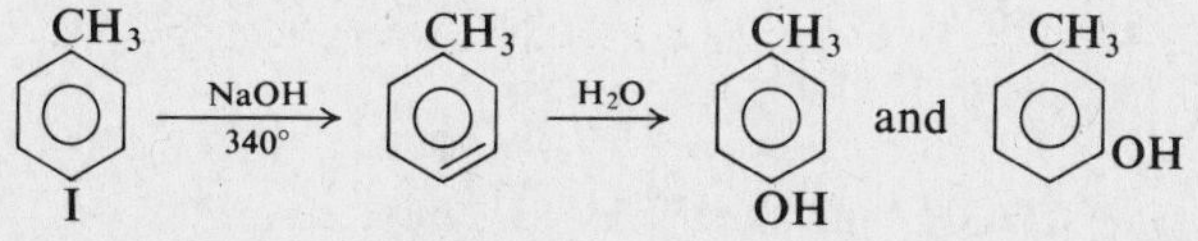

XIII

At 250°, reaction proceeds by bimolecular displacement to give only *p*-cresol.

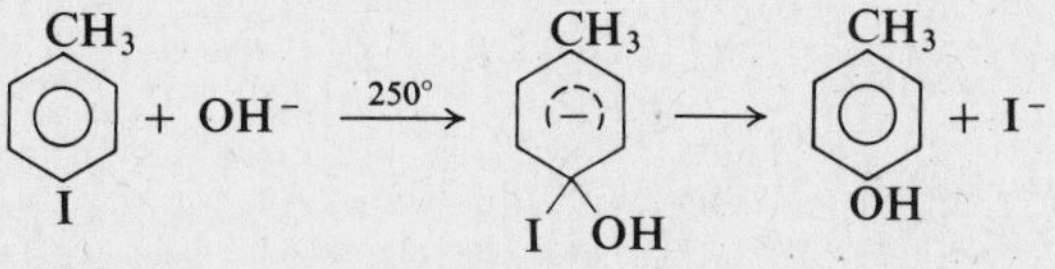

(b) Aryne XIV is formed (compare Problem 18, above) and adds *tert*-BuOH in the two possible ways.

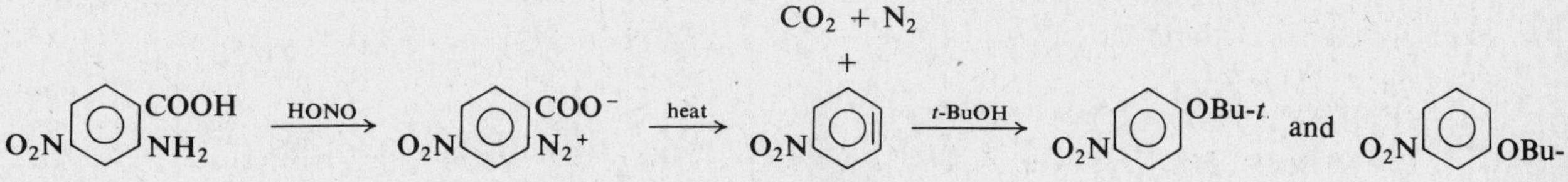

XIV

(c) Aryne XV is formed and adds not only NH_3 (as $^-NH_2$) but CH_3CN (as $^-CH_2CN$). Here the possible products are *ortho* and *meta* isomers and, as on page 839, the *meta* isomer

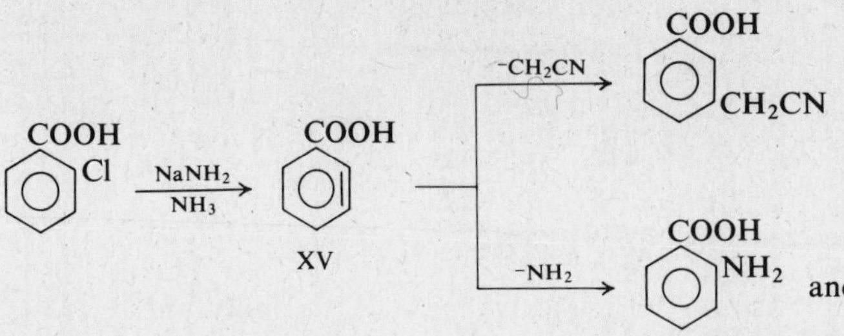

predominates: generation of the negative charge on the carbon next to the electron-withdrawing substituent is greatly preferred. (In parts (a) and (b) above, the choice is between *meta* and *para* isomers, and little preference for one over the other is shown.)

21. Both II and III give the same aryne, XVI, which by intramolecular reaction gives XVII.

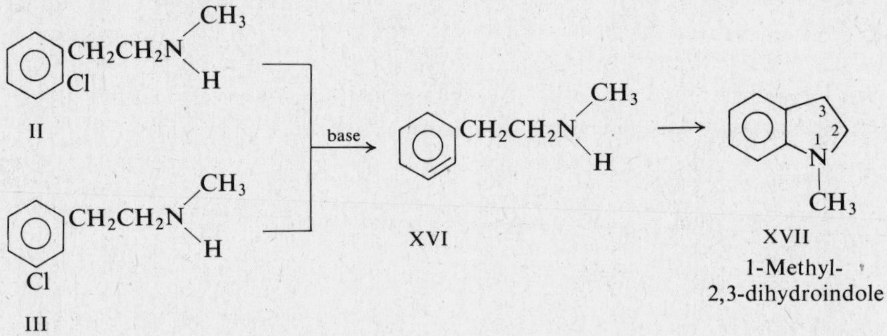

22. (a) Only $Br(CH_2)_5Br$ will react with $AgNO_3$.
Only p-$BrC_6H_4CH{=}CH_2$ will react with $KMnO_4$.

(b) Oxidize each to the acid, BrC_6H_4COOH, and determine the m.p.: o-, 148°; m-, 156°; p-, 254°.

(c) Only $PhCH_2Cl$ will react with $AgNO_3$.
o-$C_6H_4Cl_2$ will not react with hot $KMnO_4$.

(d) Only the alkyne decolorizes Br_2/CCl_4.
The alcohols react with CrO_3/H_2SO_4, and can be distinguished by elemental analysis (for Cl).
Of the remaining compounds, only ethylcyclohexane contains no Cl, only PhCl is soluble in fuming sulfuric acid.

(e) Aryl halide is negative toward $AgNO_3$.
Oxidize others to acids, and distinguish these by their m.p.'s (but both are very high), or by their neutralization equivalents (83 vs. 162).

23. (a) A benzyne mechanism will account for *some* of the facts.

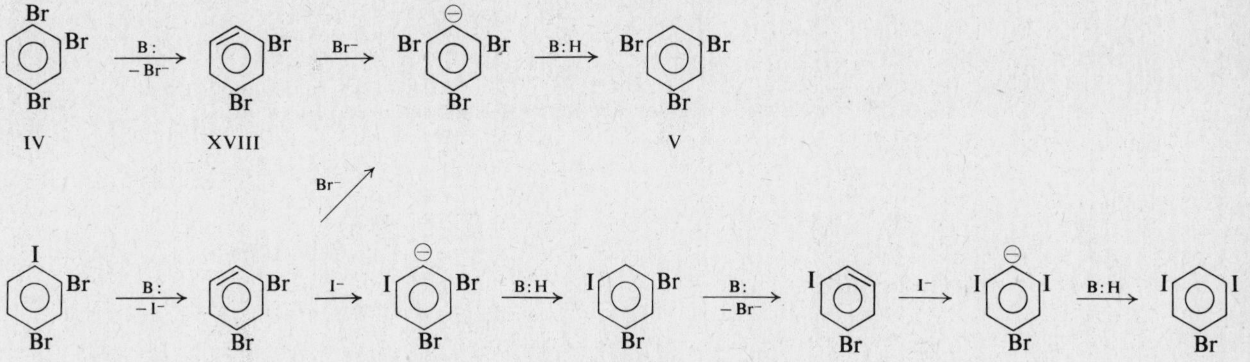

IV XVIII V VII VIII IX

(b) Benzynes are known to react with halide ion. If benzyne XVIII is an intermediate, as postulated in (a), it should react with the added iodide to give VIII—contrary to fact.

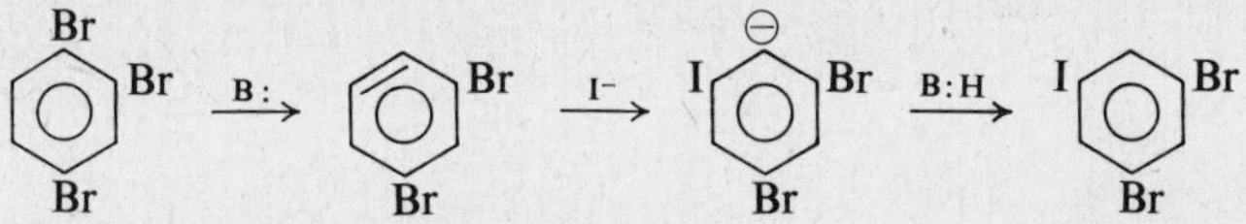

IV XVIII VIII

Is not obtained under these conditions

(c) The halogen dance is, basically, a two-step.

(i) Ar—H + B: $\rightleftarrows$ Ar$^{\ominus}$ + B:H

(ii) Ar$^{\ominus}$ + Ar′—X $\rightleftarrows$ Ar—X + Ar′$^{\ominus}$

Step (i) is the familiar abstraction of a proton by base to give a carbanion. Step (ii) is nucleophilic displacement *on halogen*, with an aryl carbanion as nucleophile, and another aryl carbanion as leaving group.

(ii) Ar$^{\ominus}$ + X—Ar′ $\longrightarrow$ $\left[\overset{\delta_-}{\text{Ar}}\text{---X---}\overset{\delta_-}{\text{Ar}'}\right]$ $\longrightarrow$ Ar—X + Ar′$^{\ominus}$

To account for all the facts, we need just these two steps plus one guiding rule: *the only carbanions stable enough to be involved are those with the negative charge ortho to halogen.* As exemplified by the data of Problem 13, above, the inductive effect of halogen is *much* stronger from the *ortho* position than from the more distant *meta* and *para* positions.

As suggested in the problem, let us start with V and the base, in the presence of VI, and see how IV can be formed. The key to this, we find, is the transformation of the tribromo carbanion XIX into the isomeric carbanion XX, with Br's distributed as in IV.

(1) 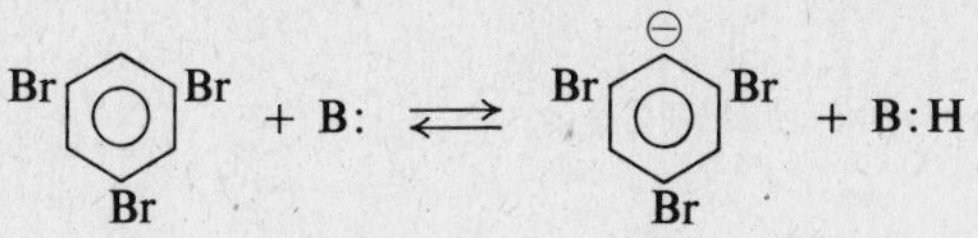

V XIX

(2) 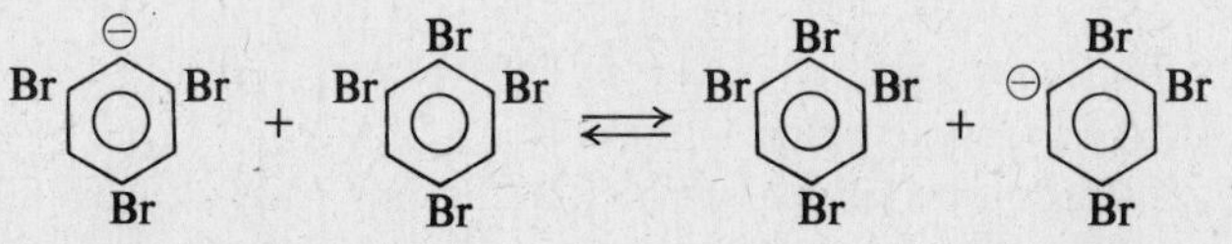

XIX VI VI XX

(3) 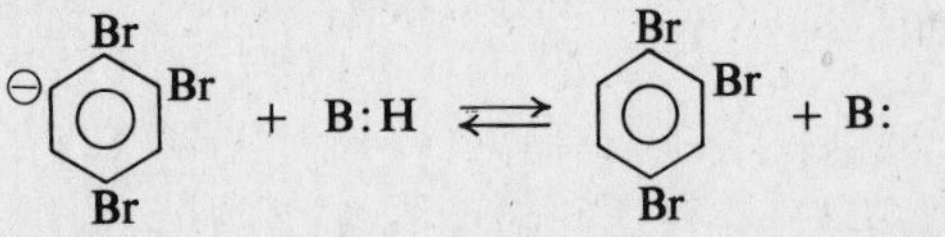 + B:H ⇌ + B:⁻

By the exact reverse of the above steps, we can account for the conversion of IV into V in the presence of VI.

(4) 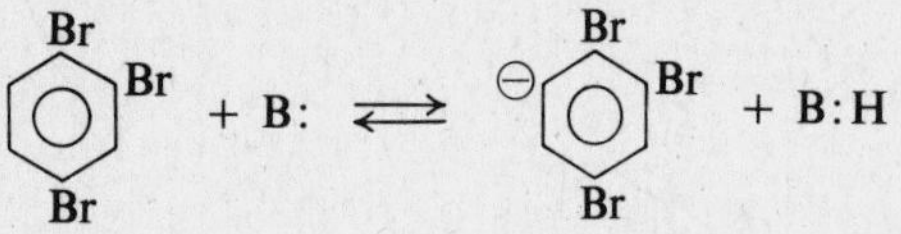 + B:⁻ ⇌ + B:H

(5) 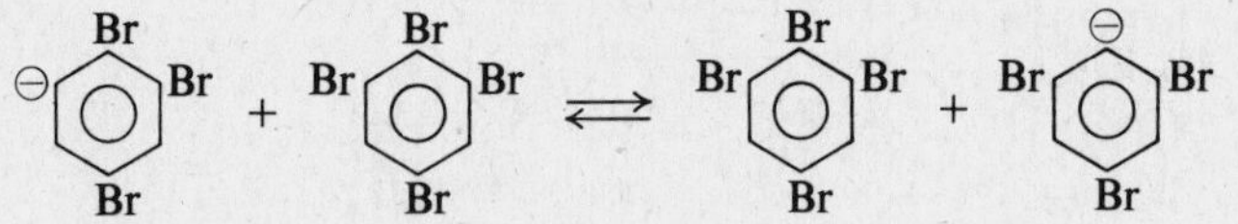 + ⇌ +

(6) 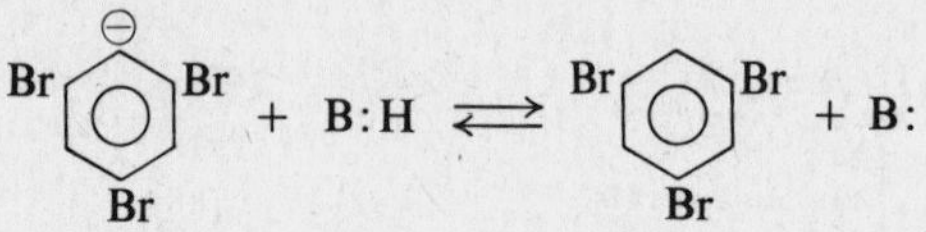 + B:H ⇌ + B:⁻

We begin to see the special role played by the tetrabromo compound VI: it contains Br's distributed *both* as in IV and as in V, and, by loss of one Br or another, can be converted into either IV or V.

Next, let us start with *only* IV and base, and see how all the products are formed.

(4) 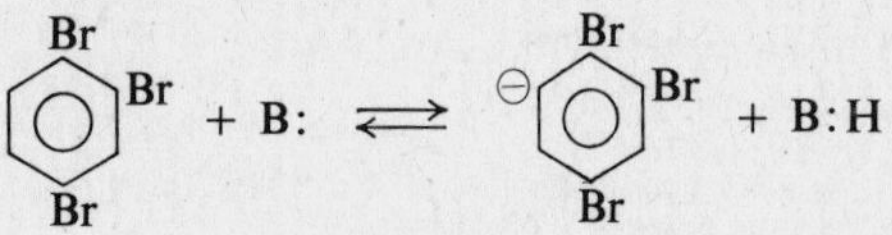 + B:⁻ ⇌ + B:H

(7)

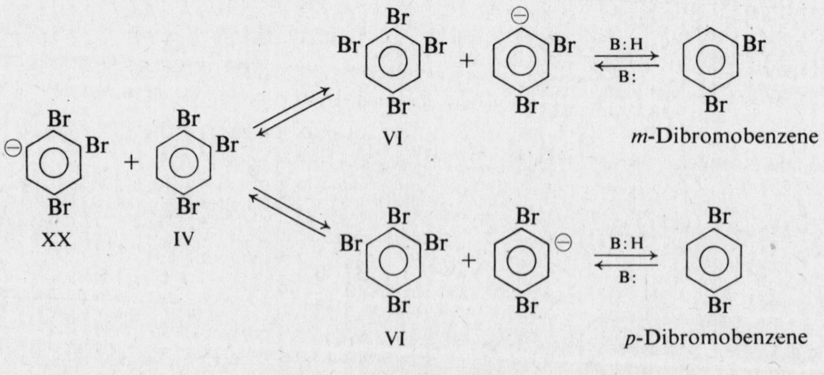

Sequence (4),(7) yields *m*- and *p*-dibromobenzenes and generates the tetrabromo compound VI. With VI now present, sequence (5),(6) can take place to yield V.

The reverse of (7),(4) converts *m*- and *p*-dibromobenzenes back into IV, but the Br that is gained need not be the same as the one originally lost from IV; the net result, after many cycles, is complete scrambling of a Br label in IV. (Write a set of equations to prove to yourself that this is so.)

The reactions of the dibromoiodo compound VII are analogous to (4) and (7). Tribromo and bromodiiodo compounds are obtained by a kind of "scrambling," this time not with isotopes but with the different halogens, Br and I. (Write equations to show this, too.)

Finally, the hardest part: why must VI be added to bring about isomerization of V but not the isomerization of IV? We write equations analogous to (4) and (7) starting with V instead of IV.

(1)

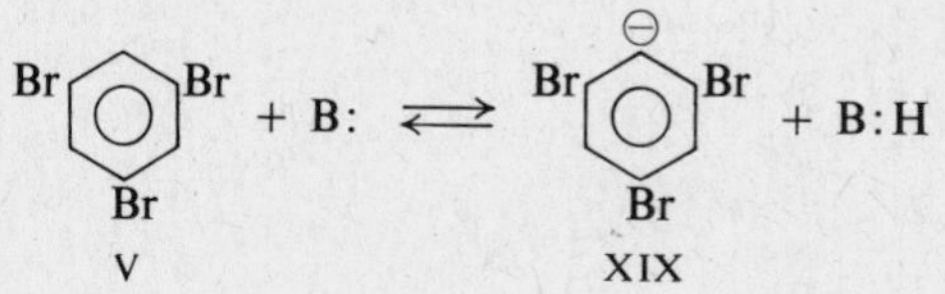

(8)

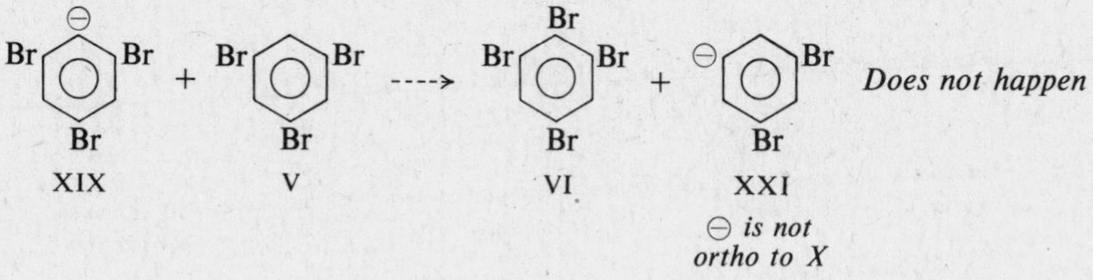

For V to generate the key tetrabromo compound VI, it must abstract Br from another molecule of V; but this it cannot do, because the reaction would require formation of carbanion XXI, in which the negative charge is not *ortho* to halogen.

Why is it *not* necessary to add V to bring about isomerization of IV? The answer, as Bunnett points out, "is already before our eyes": in sequence (4),(7) VI is *generated* from the reaction of base with IV alone.

(J. F. Bunnett, "The Base-Catalyzed Halogen Dance, and Other Reactions of Aryl Halides," Accounts Chem. Res., **5**, 139 (1972).)

Chapter 26

Carbanions II

Malonic Ester and Acetoacetic Ester Syntheses

26.1 Use the general procedure as given on pages 847–848.

	Acid wanted:	*Requires:*	*Alkylate malonic ester with:*
(a)	$CH_3CH_2CH_2$—CH_2COOH	R = *n*-Pr—	*n*-PrBr

$$\underset{\text{CH}_3}{\overset{|}{\text{CH}_3\text{CH}}}\text{—CH}_2\text{COOH} \qquad R = \textit{iso}\text{-Pr—} \qquad \textit{iso}\text{-PrBr}$$

$$CH_3CH_2\overset{\overset{\displaystyle CH_3}{|}}{-}CHCOOH \qquad \begin{array}{l} R = \text{Et—} \\ R' = \text{Me—} \end{array} \qquad \begin{array}{l} \text{EtBr} \\ \text{MeBr} \end{array}$$

One cannot make trisubstituted acetic acids by the malonic ester synthesis, because only *two* alkyl groups can be introduced.

(b) $(CH_3)_2CHCH_2$—$\underset{\underset{\text{NH}_2}{|}}{\text{CH}}COOH$ $\xleftarrow{\text{NH}_3}$ $(CH_3)_2CHCH_2$—$\underset{\underset{\text{Br}}{|}}{\text{CH}}COOH$ $\xleftarrow{\frac{\text{Br}_2}{\text{P}}}$ $(CH_3)_2CHCH_2$—CH_2COOH

Leucine

Malonic ester synthesis
with R = *iso*-Bu—

(Bromination and ammonolysis are best carried out on the free acid rather than on the ester.)

(c) $CH_3CH_2\overset{\overset{\displaystyle CH_3}{|}}{\text{CH}}$—$\underset{\underset{\text{NH}_2}{|}}{\text{CH}}COOH$ $\longleftarrow$ as in (b), from $CH_3CH_2\overset{\overset{\displaystyle CH_3}{|}}{\text{CH}}$—$CH_2COOH$

Isoleucine

Malonic ester synthesis
with R = *sec*-Bu—

385

26.2 (a) In the synthesis of adipic acid, the excess of sodiomalonic ester leads to displacement of both Br's from ethylene bromide.

$$(EtOOC)_2CH^- + BrCH_2CH_2Br + {}^-CH(COOEt)_2$$

$$\downarrow$$

$$(EtOOC)_2CH—CH_2CH_2—CH(COOEt)_2 \xrightarrow[\text{heat}]{OH^-} \xrightarrow[-CO_2]{H^+, \text{ heat}} HOOCCH_2—CH_2CH_2—CH_2COOH$$
Adipic acid

In the synthesis of the cyclic acid, the constant excess of organic halide ensures mono-substitution:

$$(EtOOC)_2CH^- + BrCH_2CH_2Br \longrightarrow (EtOOC)_2CH—CH_2CH_2Br + Br^-$$

Addition then of a second mole of ethoxide generates a carbanion which, by intramolecular nucleophilic attack, closes the ring.

$$(EtOOC)_2CH—CH_2CH_2Br \xrightarrow{OEt^-} (EtOOC)_2\overset{\ominus}{C}H—CH_2—CH_2—Br \longrightarrow (EtOOC)_2C\underset{CH_2}{\overset{CH_2}{\diagdown}}$$

with Br⁻ +

$$\downarrow OH^- \quad \text{heat}$$

$$\downarrow H^+, \text{ heat} \quad -CO_2$$

$$HOOC—HC\underset{CH_2}{\overset{CH_2}{\diagdown}}$$
Cyclopropanecarboxylic acid

(b) As for cyclopropanecarboxylic acid in (a), except use $BrCH_2CH_2CH_2CH_2Br$.

$$H_2C\overset{CH_2}{\diagup}\quad CH—COOH$$
$$H_2C—CH_2$$
Cyclopentanecarboxylic acid

26.3 (a) The Knoevenagel reaction is an aldol-like condensation.

$$CH_2(COOEt)_2 + PhCHO \xrightarrow{base} \left[\underset{OH}{PhCH—CH(COOEt)_2} \right] \xrightarrow{-H_2O} PhCH=C(COOEt)_2$$

(b) $PhCH=C(COOEt) \xrightarrow[\text{heat}]{OH^-, H_2O} PhCH=C(COO^-)_2 \xrightarrow{H^+} PhCH=C(COOH)_2 \xrightarrow[-CO_2]{\text{heat}} PhCH=CHCOOH$
Cinnamic acid

(c) By the Perkin condensation (Problem 21.22e, page 714).

26.4 (a) The Cope reaction is an aldol-like condensation (Problem 21.22g, page 714).

[cyclohexane]$=O + \underset{CN}{CH_2—COOEt} \xrightarrow{base} \left[\text{[cyclohexane]} \underset{CN}{\overset{OH}{\diagup} CH—COOEt} \right] \xrightarrow{-H_2O} \text{[cyclohexane]}=\underset{CN}{C—COOEt}$

(b) [cyclohexane]$=\underset{CN}{C—COOEt} \xrightarrow[\text{heat}]{H_2O, H^+} \text{[cyclohexane]}=\underset{COOH}{C—COOH} \xrightarrow{-CO_2} \text{[cyclohexane]}=CH—COOH$
Cyclohexylideneacetic acid

26.5. A is the last compound on page 873 with —H instead of the lowest —CH_3. It is formed as described in Sec. 27.7.

26.6 Nucleophilic substitution with competing elimination (Sec. 14.23). Yield of substitution product: 1° > 2° ≫3°. Aryl halides cannot be used (Sec. 25.4).

26.7 (a)

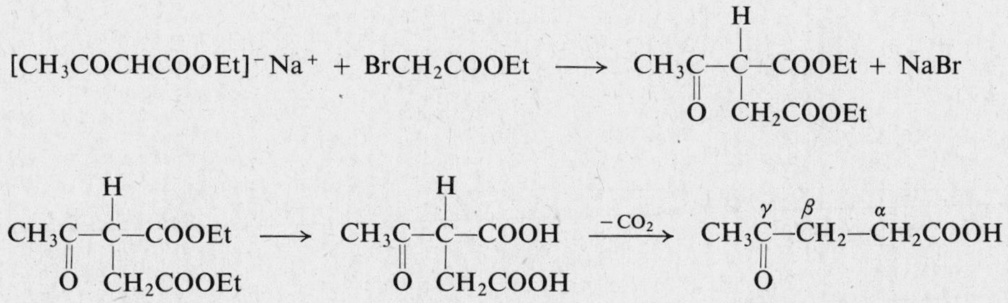

A γ-keto acid

We notice that, of the two carboxyl groups, it is the one to which the keto group is *beta* that is lost.

Use of the acidic bromoacetic acid would simply convert the sodio compound back into acetoacetic ester.

(b)

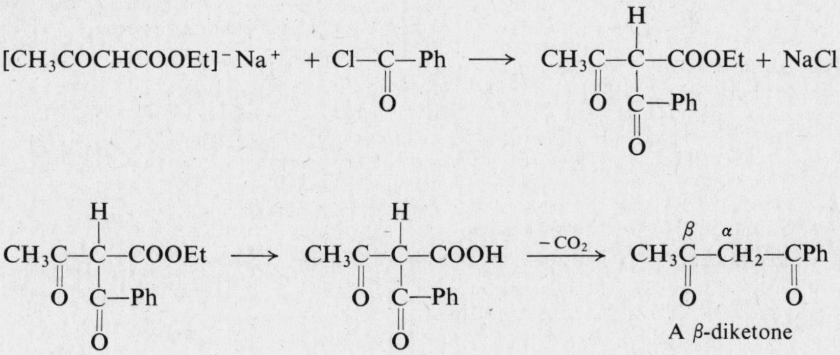

A β-diketone

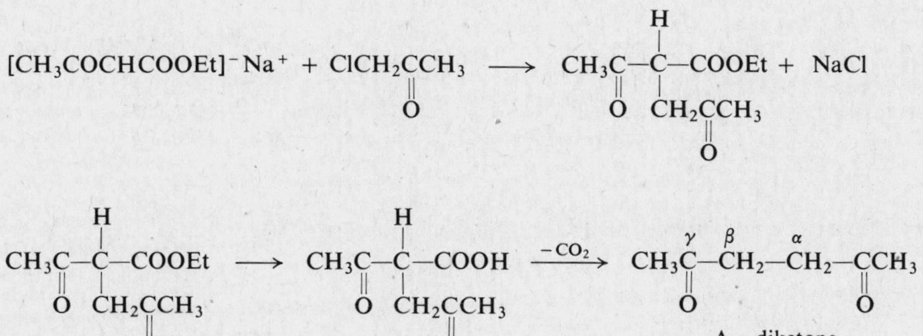

A γ-diketone

26.8 Follow the general procedure on page 851.

Ketone wanted:	*Requires:*	*Alkylate acetoacetic ester with:*

(a) $CH_3CH_2CH_2-CH_2\overset{\underset{\displaystyle \parallel}{O}}{C}CH_3$ R = *n*-Pr—   *n*-PrBr

(b) $CH_3\overset{\overset{\displaystyle CH_3}{|}}{C}H-CH_2\overset{\underset{\displaystyle \parallel}{O}}{C}CH_3$ R = *iso*-Pr—   *iso*-PrBr

(c) $CH_3CH_2-\overset{\overset{\displaystyle CH_3}{|}}{C}H\overset{\underset{\displaystyle \parallel}{O}}{C}CH_3$ R = Et—   EtBr
                              R′ = Me—   MeBr

(d) One cannot make trisubstituted acetones by the acetoacetic ester synthesis, because only *two* alkyl groups can be introduced.

(e) $CH_3\overset{\underset{\displaystyle \parallel}{O}}{C}-CH_2\overset{\underset{\displaystyle \parallel}{O}}{C}CH_3$ ⟵ acetoacetic ester synthesis with R = $CH_3\overset{\underset{\displaystyle \parallel}{O}}{C}$— (RX = CH_3COCl)

(f) $CH_3\overset{\underset{\displaystyle \parallel}{O}}{C}CH_2-CH_2\overset{\underset{\displaystyle \parallel}{O}}{C}CH_3$ ⟵ acetoacetic ester synthesis with R = $CH_3\overset{\underset{\displaystyle \parallel}{O}}{C}CH_2$— (RX = CH_3COCH_2Cl)

Or, more conveniently, proceed as follows:

$$2Na^{+-}[CH_3COCHCOOEt] + I_2 \longrightarrow 2NaI + CH_3\overset{\underset{\displaystyle \parallel}{O}}{C}\overset{\overset{\displaystyle EtOOC}{|}}{C}H-\overset{\overset{\displaystyle COOEt}{|}}{C}H\overset{\underset{\displaystyle \parallel}{O}}{C}CH_3 \xrightarrow{OH^-} \xrightarrow{H^+} \xrightarrow{-CO_2} CH_3\overset{\underset{\displaystyle \parallel}{O}}{C}CH_2-CH_2\overset{\underset{\displaystyle \parallel}{O}}{C}CH_3$$

(g) $Ph\overset{\underset{\displaystyle \parallel}{O}}{C}CH_2-CH_2\overset{\underset{\displaystyle \parallel}{O}}{C}CH_3$ ⟵ acetoacetic ester synthesis with R = $Ph\overset{\underset{\displaystyle \parallel}{O}}{C}CH_2$— (RX = $Ph\overset{\underset{\displaystyle \parallel}{O}}{C}CH_2Br$)

                                                   Phenacyl bromide

26.9 There is a crossed Claisen condensation to give A, followed by the hydrolysis and decarboxylation of a β-keto acid. (Here, as in Problem 26.7a, we see that, of two carboxyl groups, it is the one to which the keto group is *beta* that is lost.)

$$EtOOC-\overset{\underset{\displaystyle \parallel}{O}}{C}-OEt \quad CH_2COOEt \xrightarrow{-EtOH} EtOOC-\overset{\overset{\displaystyle CH_3}{|}}{\underset{\underset{\displaystyle \parallel}{O}}{C}}-CHCOOEt$$

                                                             A

$$A \xrightarrow[H_2O]{H^+} HOOC-\overset{\overset{\displaystyle CH_3}{|}}{\underset{\underset{\displaystyle \parallel}{O}}{C}}-CHCOOH \xrightarrow{-CO_2} HOOC-\overset{\underset{\displaystyle \parallel}{O}}{C}-CH_2CH_3$$

                                             An α-keto acid

26.10

(a) $\underset{\underset{\text{O}}{\|}}{\text{HOOCC}}-\text{CH}_2\text{CH}(\text{CH}_3)_2 \xleftarrow[\substack{-\text{CO}_2}]{\text{H}^+} \underset{\underset{\text{O}}{\|}}{\text{EtOOCC}}\overset{\overset{\text{COOEt}}{|}}{}\text{CHCH}(\text{CH}_3)_2 \xleftarrow{\text{OEt}^-} \underset{\underset{\text{O}}{\|}}{\text{EtOOC}}-\text{C}-\text{OEt} + \text{EtOOCCH}_2\text{CH}(\text{CH}_3)_2$

  α-Ketoisocaproic acid   Ethyl oxalate   Ethyl isovalerate

(b) $\underset{\underset{\text{O}}{\|}}{\text{HOOCC}}-\text{CH}_2\text{Ph} \xleftarrow[\substack{-\text{CO}_2}]{\text{H}^+} \underset{\underset{\text{O}}{\|}}{\text{HOOCC}}\overset{\overset{\text{COOEt}}{|}}{}\text{CHPh} \xleftarrow{\text{OEt}^-} \underset{\underset{\text{O}}{\|}}{\text{EtOOC}}-\text{C}-\text{OEt} + \text{EtOOCCH}_2\text{Ph}$

  α-Keto-β-phenyl-   Ethyl oxalate   Ethyl phenylacetate
  propionic acid

(c) $\underset{\underset{\text{O}}{\|}}{\text{HOOCC}}-\text{CH}_2\text{CH}_2\text{COOH} \xleftarrow[\substack{-\text{CO}_2}]{\text{H}^+} \underset{\underset{\text{O}}{\|}}{\text{EtOOCC}}\overset{\overset{\text{COOEt}}{|}}{}\text{CHCH}_2\text{COOEt}$

  α-Ketoglutaric acid $\uparrow$ OEt$^-$

$\underset{\underset{\text{O}}{\|}}{\text{EtOOC}}-\text{C}-\text{OEt} + \text{EtOOCCH}_2\text{CH}_2\text{COOEt}$

              Ethyl oxalate   Ethyl succinate

(d) $\underset{\underset{\text{NH}_2}{|}}{\text{HOOCCH}}-\text{CH}_2\text{CH}(\text{CH}_3)_2 \xleftarrow[\text{Ni}]{\text{NH}_3,\ \text{H}_2} \underset{\underset{\text{O}}{\|}}{\text{HOOCC}}-\text{CH}_2\text{CH}(\text{CH}_3)_2 \longleftarrow \text{ as in (a)}$

  Leucine

(e) $\underset{\underset{\text{NH}_2}{|}}{\text{HOOCCH}}-\text{CH}_2\text{CH}_2\text{COOH} \xleftarrow[\text{Ni}]{\text{NH}_3,\ \text{H}_2} \underset{\underset{\text{O}}{\|}}{\text{HOOCC}}-\text{CH}_2\text{CH}_2\text{COOH} \longleftarrow \text{ as in (c)}$

  Glutamic acid

26.11 (a) The charged end loses CO_2. The resulting anion (like I, page 853) is stabilized through accommodation of the negative charge by oxygen.

$$\text{HO}-\underset{\underset{\text{O}}{\|}}{\text{C}}-\text{CH}_2-\text{COO}^- \longrightarrow \text{HO}-\underset{\underset{\text{O}}{\|}}{\text{C}}=\text{CH}_2 + CO_2$$

(b) A doubly charged anion, $^-\text{OOC}-\text{CH}_2{}^-$, would be less stable, and the transition state leading to it would be reached with difficulty.

26.12 In this case decarboxylation gives $2,4,6\text{-}(NO_2)_3C_6H_2{}^-$, which is stabilized relative to most aryl anions by the three powerfully electron-withdrawing groups.

26.13 Intermediates that can react with Br_2 and I_2 are involved. Both the proposed intermediates in decarboxylation, the carbanion I (page 853) and the enol (page 854), are known to react with halogens. (Secs. 21.3 and 21.4.)

26.14 $HO-\overset{\underset{\|}{O}}{C}-CH_2-COOH \rightleftharpoons HO=\overset{\underset{\oplus}{C}}{C}\overset{\frown}{-}CH_2-COO^- \longrightarrow CO_2 + HO-\overset{\underset{\|}{OH}}{C}=CH_2$

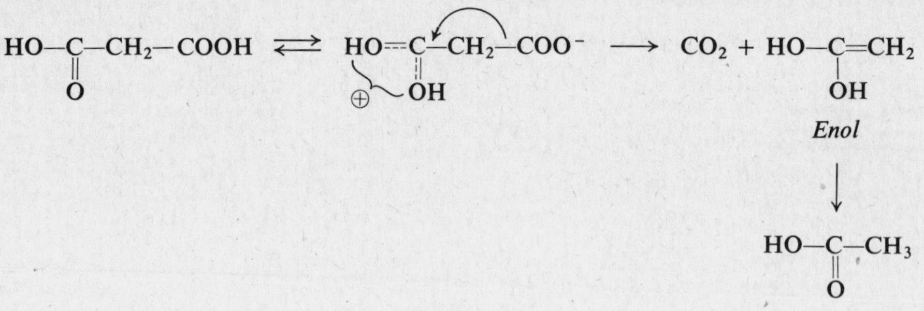

$\overset{}{\underset{\displaystyle Enol}{}}$

$\downarrow$

$HO-\overset{\underset{\|}{O}}{C}-CH_3$

26.15 $PhC\equiv\overset{\frown}{C}-CO_2^- \longleftarrow CO_2 + PhC\equiv C:^{\ominus}$

$\overset{H_2O}{\longrightarrow} PhC\equiv CH$

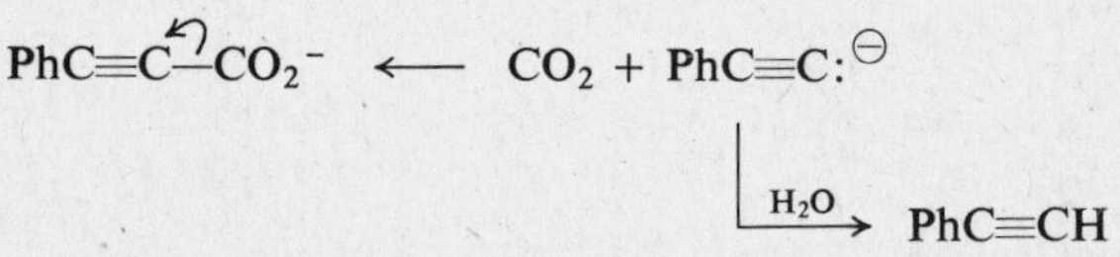

Decarboxylation gives $PhC\equiv C^-$, which, like anions from other terminal acetylenes, is relatively stable (Sec. 8.10).

26.16 Follow the general procedure on page 855.

	Acid wanted:	Requires:	Starting materials:
(a)	$CH_3CH_2-\underset{\underset{\textstyle H}{\|}}{CH}-COOH$	$R = H-$ $R' = CH_3CH_2-$	CH_3COOH $R'X = EtBr$
(b)	$CH_3-\underset{\underset{\textstyle CH_3}{\|}}{CH}-COOH$	$R = H-$ $R' = CH_3-$ $R'' = CH_3-$	CH_3COOH $R'X = MeBr$ $R''X = MeBr$
(c)	$CH_3-\underset{\underset{\textstyle CH_3}{\|}}{CH}-COOH$	$R = CH_3-$ $R' = CH_3-$	CH_3CH_2COOH $R'X = MeBr$
(d)	$PhCH_2-\underset{\underset{\textstyle H}{\|}}{CH}-COOH$	$R = H-$ $R' = PhCH_2-$	CH_3COOH $R'X = PhCH_2Cl$

(A. I. Meyers *et al.*, "Oxazolines. IX. Synthesis of Homologated Acetic Acids and Esters," J. Org. Chem., **39**, 2778 (1974).)

26.17 Like a Grignard reagent, the organolithium compound adds to the carbonyl group of the aldehyde to give a 2° alcohol.

(a) 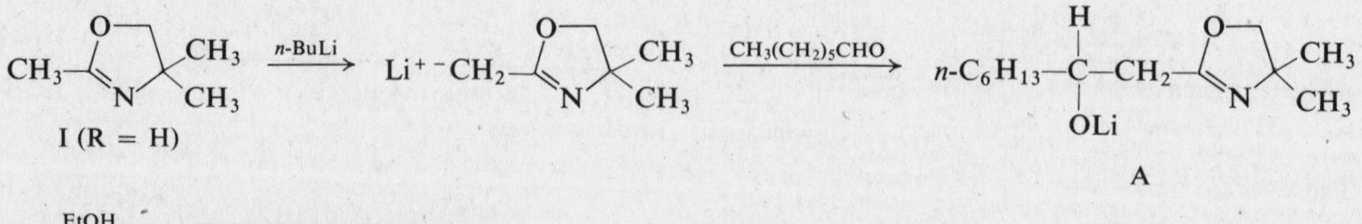 [oxazoline structure] $\xrightarrow{n\text{-BuLi}}$ [lithiated oxazoline] $\xrightarrow{CH_3(CH_2)_5CHO}$ [product A]

$I (R = H)$

A

$A \xrightarrow[H_2SO_4]{EtOH} n\text{-}C_6H_{13}\underset{\underset{\textstyle OH}{\|}}{CH}-CH_2COOEt$

B

A β-hydroxy ester

(b) As in (a), except use (n-Pr)$_2$C=O as the carbonyl compound.

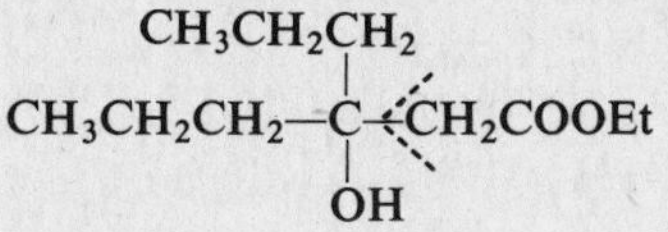

$$\begin{array}{c} CH_3CH_2CH_2 \\ | \\ CH_3CH_2CH_2-C{\dashleftarrow}CH_2COOEt \\ | \\ OH \end{array}$$

(c) As in (a), except use PhCHO and I (R = Et).

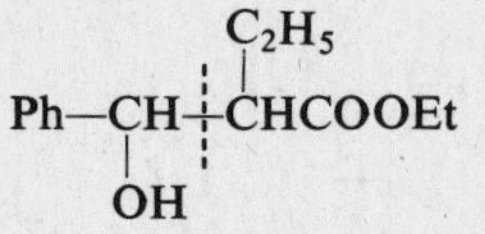

$$\begin{array}{c} C_2H_5 \\ | \\ Ph-CH{\vdots}CHCOOEt \\ | \\ OH \end{array}$$

26.18 The oxazoline ring protects the carboxyl group, so that Grignard reactions can be carried out on other parts of the molecule.

(a)

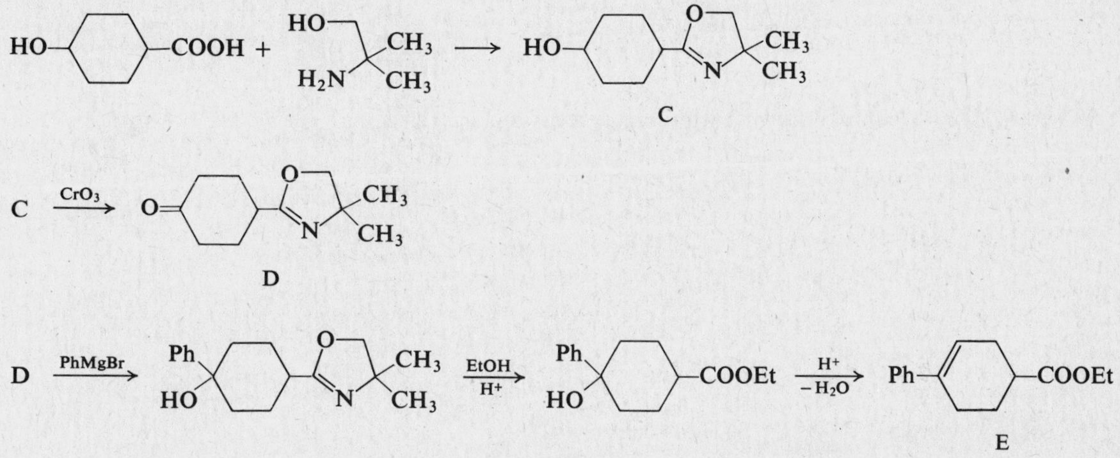

$$C \xrightarrow{CrO_3} D$$

$$D \xrightarrow{PhMgBr} \xrightarrow[H^+]{EtOH} \xrightarrow[-H_2O]{H^+} E$$

(b)

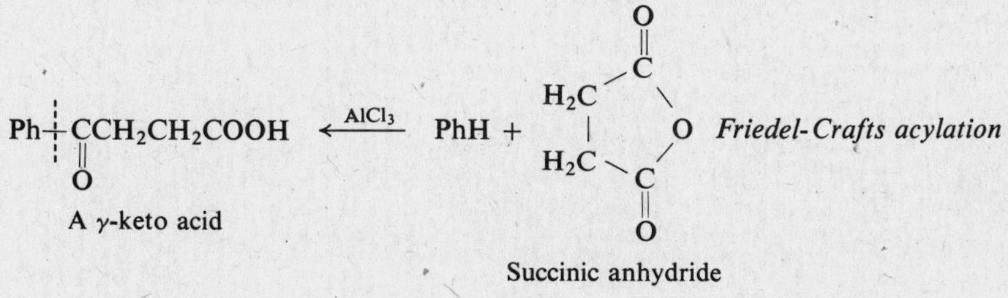

$$Ph{\dashleftarrow}CCH_2CH_2COOH \xleftarrow{AlCl_3} PhH + \text{Succinic anhydride} \quad \textit{Friedel-Crafts acylation}$$

$$\underset{O}{\overset{||}{}}$$

A γ-keto acid

(c)

$$\underset{Et}{\overset{Ph}{}}C=CHCH_2COOH \xleftarrow{-H_2O} Et-\underset{OH}{\overset{Ph}{C}}-CH_2CH_2COOH \xleftarrow{} Et-\underset{OH}{\overset{Ph}{C}}-CH_2CH_2\text{-oxazoline}$$

$$\uparrow EtMgBr$$

$$\text{as in (b)} \longrightarrow PhCCH_2CH_2COOH + \underset{H_2N}{\overset{HO}{}}C\underset{CH_3}{\overset{CH_3}{}} \longrightarrow PhCCH_2CH_2\text{-oxazoline}$$

$$\underset{O}{\overset{||}{}}$$

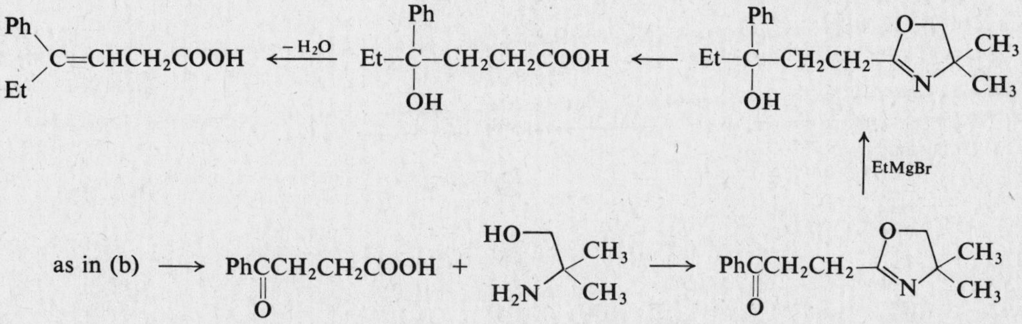

(d) 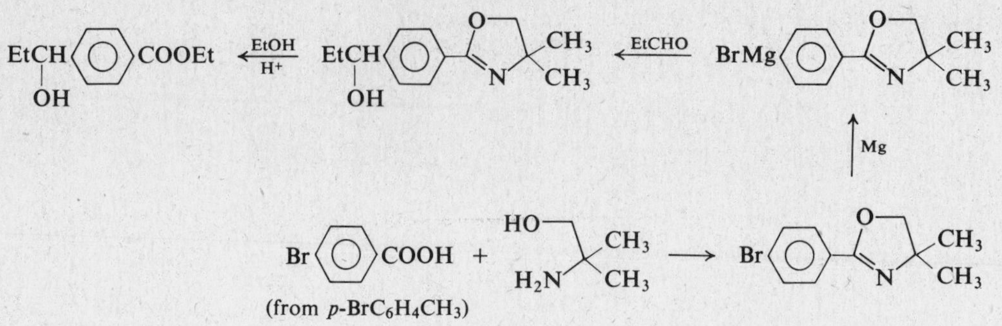 EtCH⟨O⟩COOEt ⟵EtOH/H⁺⟵ EtCH⟨O⟩[oxazoline with CH₃, CH₃] ⟵EtCHO⟵ BrMg⟨O⟩[oxazoline with CH₃, CH₃]
     |OH |OH

↑ Mg

Br⟨O⟩COOH + HO—C(CH₃)(CH₃)—NH₂ ⟶ Br⟨O⟩[oxazoline with CH₃, CH₃]

(from p-BrC₆H₄CH₃)

(e) 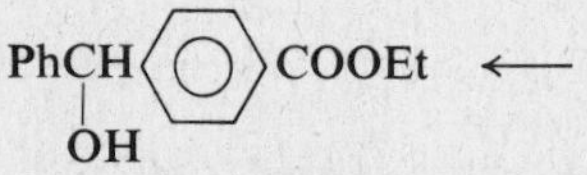 PhCH⟨O⟩COOEt ⟵ as in (d), except use PhCHO in place of EtCHO
     |OH

(A. I. Meyers *et al.*, "Oxazolines. XI. Synthesis of Functionalized Aromatic and Aliphatic Acids. A Useful Protecting Group for Carboxylic Acids against Grignard and Hydride Reagents," J. Org. Chem., **39**, 2787 (1974).)

26.19 An acid–base complex forms; within this, a proton is transferred via a cyclic transition state.

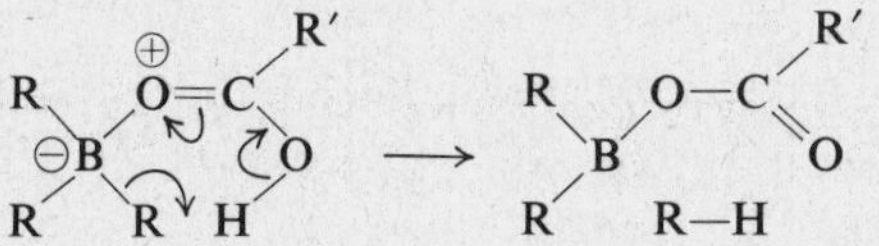

$$
\begin{array}{c}
\text{R}\!-\!\overset{\oplus}{\text{O}}\!=\!\text{C}\!-\!\text{R}' \\
\overset{\ominus}{\text{B}} \qquad \text{O} \\
\text{R} \quad \text{R}' \text{H}
\end{array}
\longrightarrow
\begin{array}{c}
\text{R} \quad \text{O}\!-\!\text{C}\!-\!\text{R}' \\
\text{B} \qquad \text{O} \\
\text{R} \quad \text{R}\!-\!\text{H}
\end{array}
$$

26.20. (a) 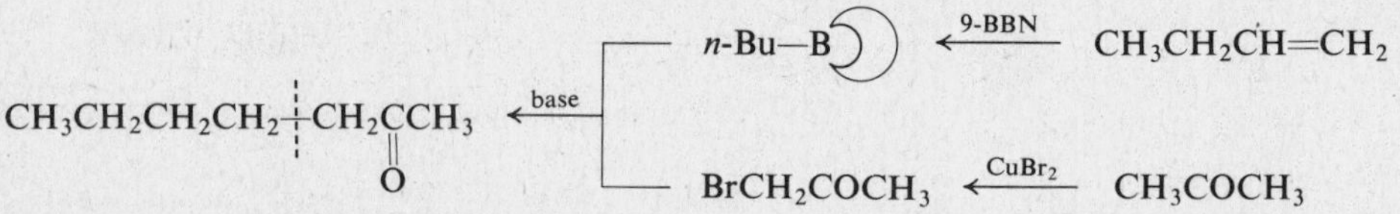 CH₃CH₂CH₂CH₂⫶CH₂CCH₃ ⟵base⟵ ⌐ *n*-Bu—B⟨⟩ ⟵9-BBN⟵ CH₃CH₂CH=CH₂
                        ‖ |
                        O   └ BrCH₂COCH₃ ⟵CuBr₂⟵ CH₃COCH₃

(b) CH₃CHCH₂⫶CH₂COOH ⟵H₂O/H⁺⟵ ⟵base⟵ ⌐ *iso*-Bu—B⟨⟩ ⟵9-BBN⟵ CH₃C(CH₃)=CH₂
     |CH₃ |
                                           └ BrCH₂COOEt (from CH₃COOH)

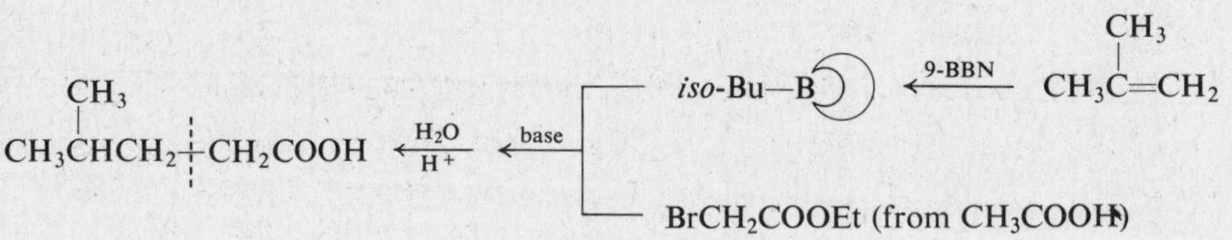

(c) 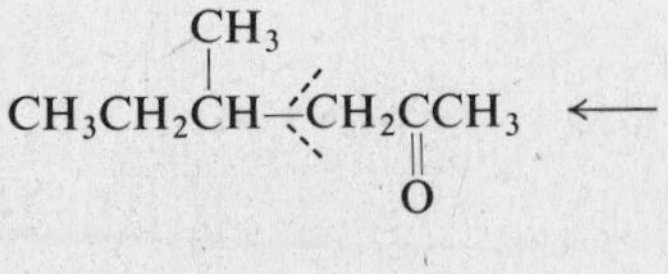 CH₃CH₂CH⫶CH₂CCH₃ ⟵ as in (a), except start with CH₃CH=CHCH₃
     |CH₃   ‖
             O

(d) 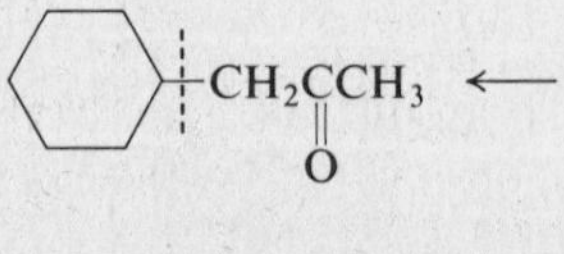 ⬡⫶CH₂CCH₃ ⟵ as in (a), except start with ⬡ (cyclohexene)
        ‖
        O

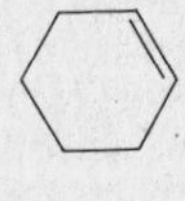

(e) 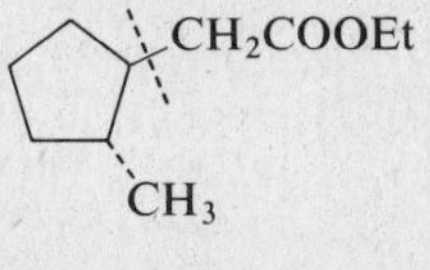 ⬠⫶CH₂COOEt ⟵ as in (b), except start with ⬠ (methylcyclopentene)
    |CH₃

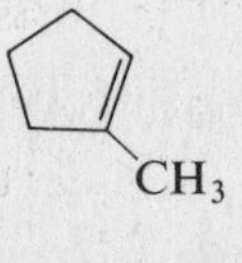

(f) 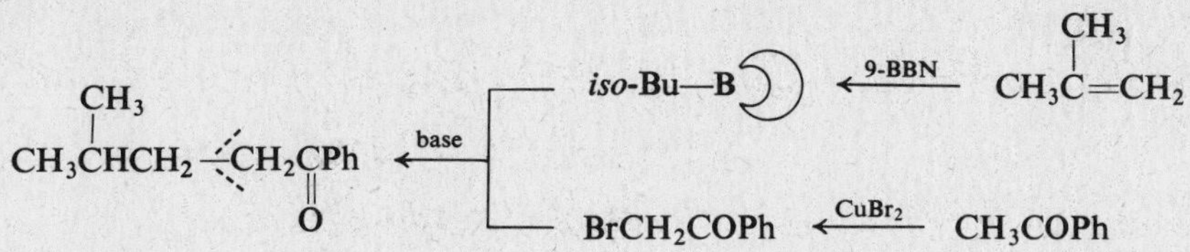 CH₃CHCH₂—CH₂CPh ←[base] iso-Bu—B◠ ←[9-BBN] CH₃C=CH₂
$\quad$ with CH₃ on the first carbon and O on the carbonyl; BrCH₂COPh ←[CuBr₂] CH₃COPh

(g) 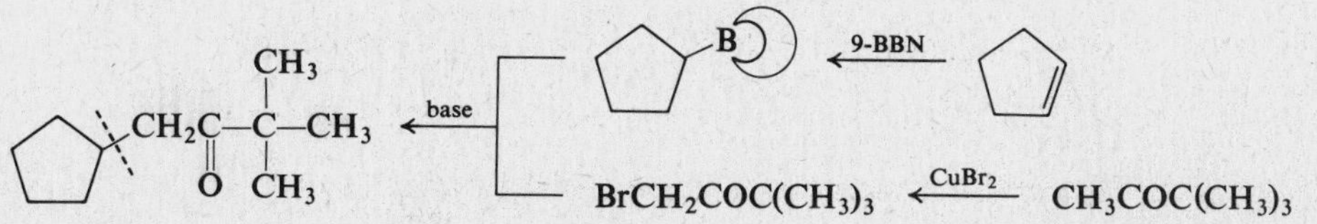 cyclopentyl—CH₂C—C—CH₃ ←[base] ◯—B◠ ←[9-BBN] cyclopentene
$\quad$ O and CH₃; BrCH₂COC(CH₃)₃ ←[CuBr₂] CH₃COC(CH₃)₃
$\quad\quad$ Methyl *tert*-butyl ketone

$\quad$ (H. C. Brown, *Boranes in Organic Chemistry*, Cornell University Press, Ithaca, N. Y.,
1972, pp. 372–391; or H. C. Brown, H. Nambu, and M. M. Rogic, J. Am. Chem. Soc.,
91, 6852, 6855 (1969).)

26.21 Follow the general procedure as illustrated on page 860.

To make:	Alkylate or acylate enamine of:	With:
(a) cyclohexanone + CH₂Ph	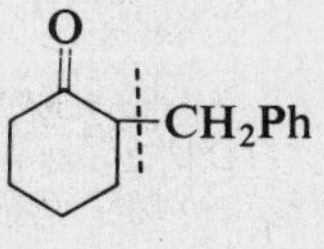 cyclohexanone	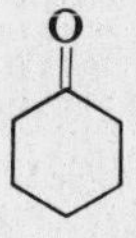 PhCH₂Br
(b) CH₂=CHCH₂—C(CH₃)₂—CHO	H—C(CH₃)₂—CHO	CH₂=CHCH₂Br
(c) cyclohexanone + CH(CH₃)COOEt	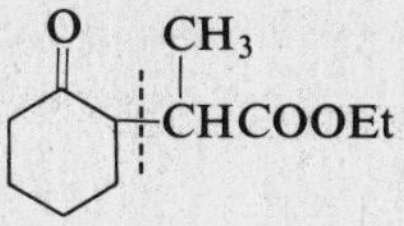 cyclohexanone	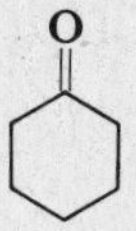 BrCH(CH₃)COOEt
(d) cyclohexanone + CH₂CCH₃ (O)	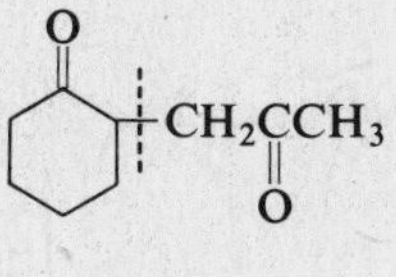 cyclohexanone	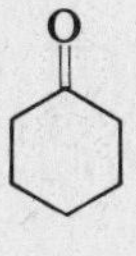 BrCH₂CCH₃ (O)
(e) cyclohexanone + (2,4-dinitrophenyl) 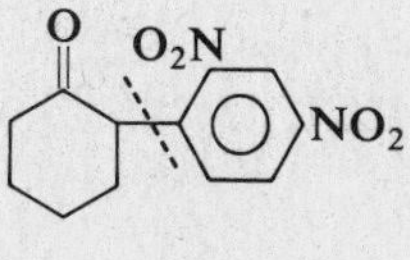 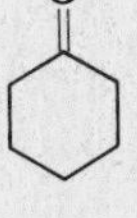 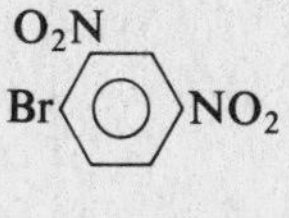	cyclohexanone	Br(2,4-dinitrophenyl)
(f) CH₃C(O)—C(CH₃)₂—CHO	H—C(CH₃)₂—CHO	CH₃—C(O)—Cl

$\quad$ (H. O. House, *Modern Synthetic Reactions*, W. A. Benjamin, New York, 1965,
pp. 197–201.)

26.22 (a) In an aldol-like condensation, nucleophilic carbon of the enamine attacks carbonyl carbon of the aldehyde.

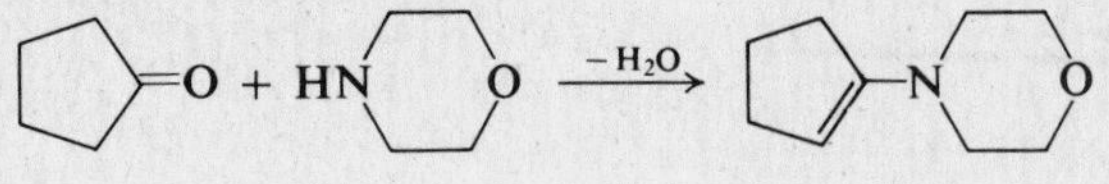

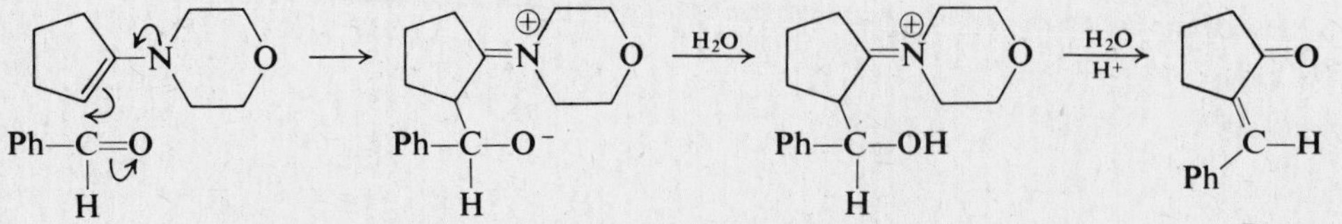

(b) Imine C is acidic enough to decompose the Grignard reagent and form the magnesium salt; this contains nucleophilic carbon, and is alkylated by benzyl chloride.

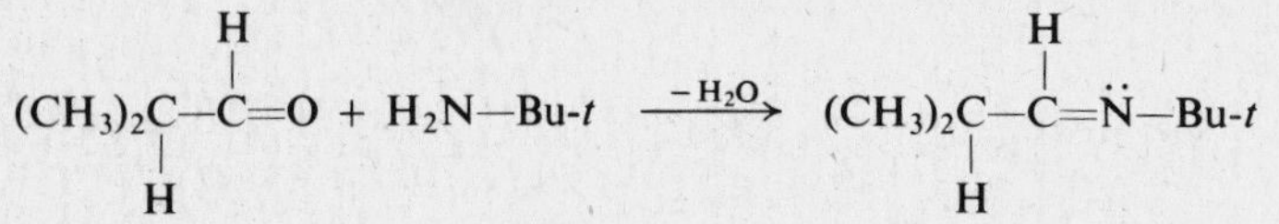

1. Follow the general procedure outlined on pages 847–848.

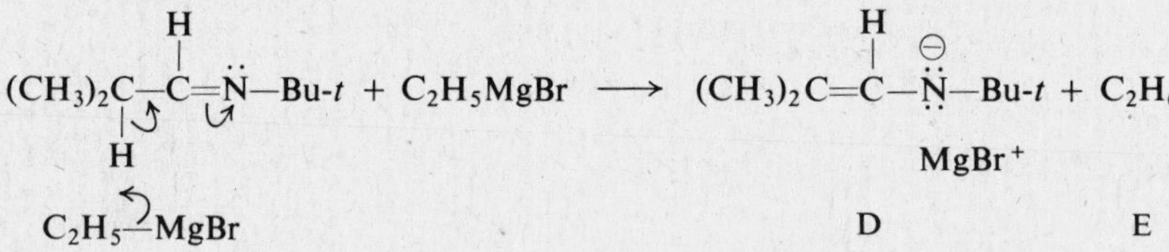

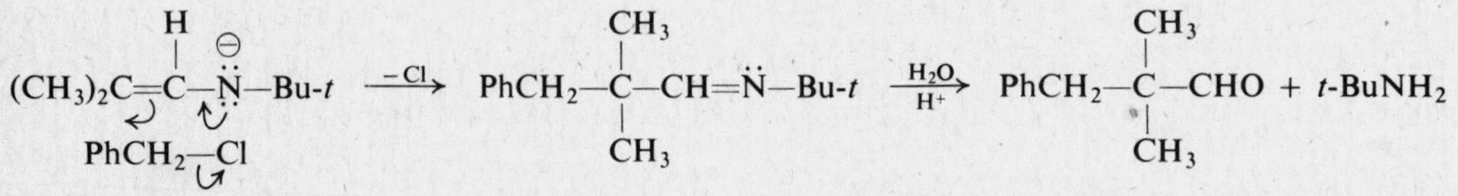

Acid wanted:	Requires:	Alkylate malonic ester with:
(a) $CH_3CH_2CH_2CH_2$—CH_2COOH	R = n-Bu—	n-BuBr
(b) CH_3—CHCOOH (with CH₃)	R = CH₃— R′ = CH₃—	CH_3Br CH_3Br
(c) CH_3CH—CH_2COOH (with CH₃)	R = iso-Pr—	iso-PrBr

$$\text{CH}_3$$

(d) CH$_3$CH—CHCOOH   R = iso-Pr—   iso-PrBr

　　　　　CH$_3$ R' = Me—   MeBr

(e) C$_2$H$_5$—CHCOOH   R = Et—   EtBr

　　　　　C$_2$H$_5$ R' = Et—   EtBr

(f) PhCH$_2$—CHCOOH   R = PhCH$_2$—   PhCH$_2$Br

　　　　　CH$_2$Ph   R' = PhCH$_2$—   PhCH$_2$Br

(g) HOOCCH—CHCOOH   R = CH$_3$CH—   CH$_3$CHBrCOOEt

　　　H$_3$C　　CH$_3$ 　　　　　COOEt

                                        R' = Me—   MeBr

Or, more conveniently, oxidize two moles of the sodiomethylmalonic ester with I$_2$. (Compare Problem 26.8f, page 852.)

$$2\text{Na}^{+-}[\text{CH}_3\text{C(COOEt)}_2] + \text{I}_2 \longrightarrow 2\text{NaI} + (\text{EtOOC})_2\text{C}-\text{C(COOEt)}_2 \xrightarrow{\text{OH}^-} \xrightarrow{\text{H}^+} \xrightarrow{-\text{CO}_2} \text{HOOCCH}-\text{CHCOOH}$$
$$\text{H}_3\text{C}\;\;\text{CH}_3 \qquad\qquad\qquad\qquad \text{H}_3\text{C}\;\;\;\text{CH}_3$$

(h) HOOCCH$_2$┊CH$_2$┊CH$_2$COOH

Add one mole of CH$_2$Br$_2$ to two moles of sodiomalonic ester; then hydrolyze and decarboxylate as usual. (See Problem 26.2, page 850.)

(i) ┊—CHCOOH
　　└ ─┊─

Add one mole of sodiomalonic ester to excess BrCH$_2$CH$_2$CH$_2$Br; then add one mole NaOEt; finally, hydrolyze and decarboxylate as usual. (See Problem 26.2, page 850.)

2. Follow the general procedure as outlined on page 851.

Ketone wanted:	Requires:	Alkylate acetoacetic ester with:
(a) CH$_3$—CH$_2$COCH$_3$	R = Me—	MeBr
(b) C$_2$H$_5$—CHCOCH$_3$　　　C$_2$H$_5$	R = Et—　　R' = Et—	EtBr　　EtBr
(c) CH$_3$CH$_2$CH$_2$—CHCOCH$_3$　　　　C$_2$H$_5$	R = n-Pr—　　R' = Et—	n-PrBr　　EtBr
(d) C$_2$H$_5$CHCH$_2$—CH$_2$COCH$_3$　CH$_3$	R = EtCHCH$_2$—　　　Me	EtCHCH$_2$Br (from active amyl alcohol)　Me
(e) (CH$_3$)$_2$CHCH$_2$CH$_2$—CHCOCH$_3$　　　　　CH$_3$	R = iso-Pe—　　R' = Me—	iso-PeBr　　MeBr

(f) **$HOOCCH(CH_3)—CH_2COCH_3$**   R = $HOOCCH(CH_3)—$ $CH_3CH(Br)COOEt$

(g) $HOOCCH_2—CH_2CH(OH)CH_3$ $\xleftarrow{H_2,\ Ni}$ **$HOOCCH_2—CH_2C(=O)CH_3$** $\longleftarrow$ a.a.e. synthesis with RX = $ClCH_2COOEt$

(h) $CH_3CH_2CH_2—CH(CH_3)CH(OH)CH_3$ $\xleftarrow{LiAlH_4}$ **$CH_3CH_2CH_2—CH(CH_3)C(=O)CH_3$** $\longleftarrow$ a.a.e synthesis with RX = *n*-PrBr and R′X = MeBr

(i) $CH_3CH_2CH(CH_3)CH_2—CH_2CH(CH_3)CH_3$ $\xleftarrow{H_2,\ Ni}$ $CH_3CH_2CH(CH_3)CH_2—CH{=}C(CH_3)CH_3$ (and 1-alkene) $\xleftarrow{KOH(alc)}$ $CH_3CH_2CH(CH_3)CH_2—CH_2C(CH_3)(Br)CH_3$

$\uparrow PBr_3$

(d) $\longrightarrow$ **$CH_3CH_2CH(CH_3)CH_2—CH_2C(=O)CH_3$** $\xrightarrow{MeMgBr}$ $CH_3CH_2CH(CH_3)CH_2—CH_2C(CH_3)(OH)CH_3$

(j) **$CH_3CH_2CH_2CH(CH_3)—CH_2COOH$** $\xleftarrow[\text{Prob. 11}]{\text{conc. }OH^-}$ a.a.e. alkylated by 2-PeBr

(k) **$CH_3CH(CH_3)—CH_2COOH$** $\xleftarrow[\text{Prob. 11}]{\text{conc. }OH^-}$ a.a.e. alkylated by *iso*-PrBr

(l) **$HOOCCH(CH_3)—CH_2COOH$** $\xleftarrow[\text{Prob. 11}]{\text{conc. }OH^-}$ a.a.e. alkylated by $CH_3CHBrCOOEt$

(m) $CH_3CH(OH)CH_2—CH_2CH(OH)CH_3$ $\xleftarrow{H_2,\ Ni}$ $CH_3C(=O)CH_2 \vdots CH_2C(=O)CH_3$ $\longleftarrow$ Problem 26.8f

3.

COOEt
OH⁻
COO⁻
−CO₂
2-Carbethoxy-
cyclopentanone
A β-keto ester
Cyclopentanone
CH₃
−CO₂⁻
COO⁻
CH₃
OH⁻
COOEt
CH₃
MeBr
OEt⁻
COOEt
2-Methyl-
cyclopentanone

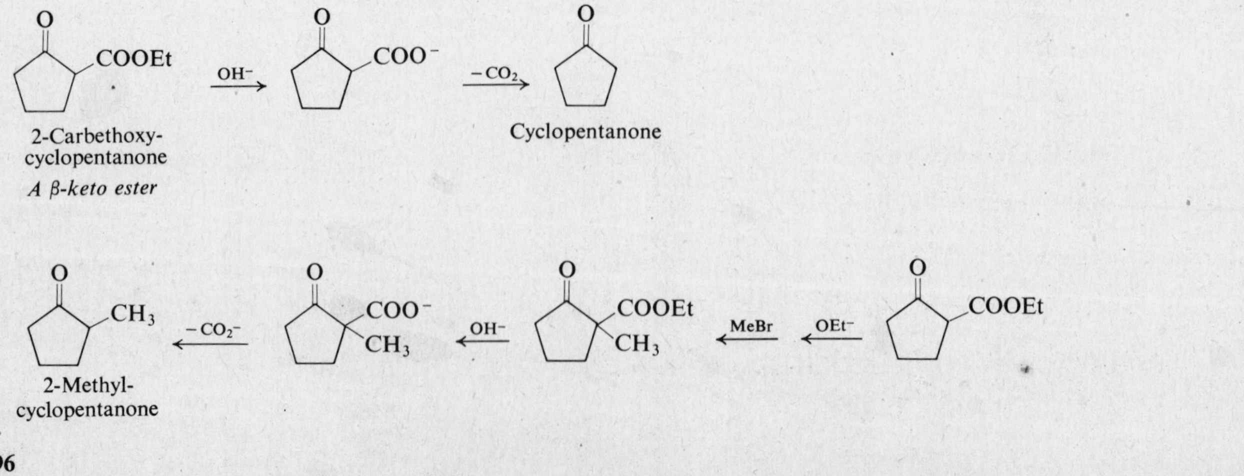

4. (a)

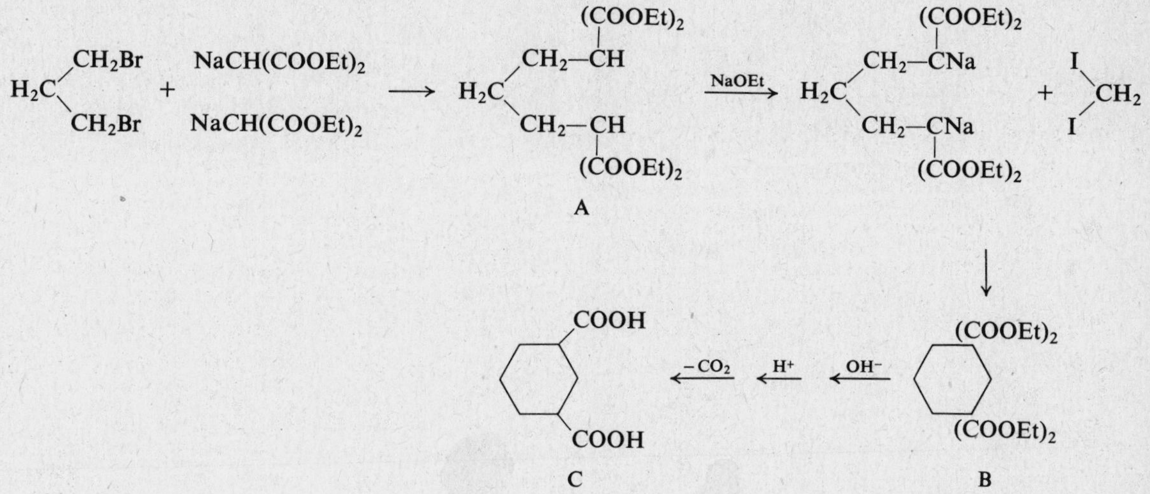

$$H_2C\begin{matrix}CH_2Br\\CH_2Br\end{matrix} + \begin{matrix}NaCH(COOEt)_2\\NaCH(COOEt)_2\end{matrix} \longrightarrow H_2C\begin{matrix}CH_2-CH(COOEt)_2\\CH_2-CH(COOEt)_2\end{matrix} \xrightarrow{NaOEt} H_2C\begin{matrix}CH_2-CNa(COOEt)_2\\CH_2-CNa(COOEt)_2\end{matrix} + \begin{matrix}I\\CH_2\\I\end{matrix}$$

A   B

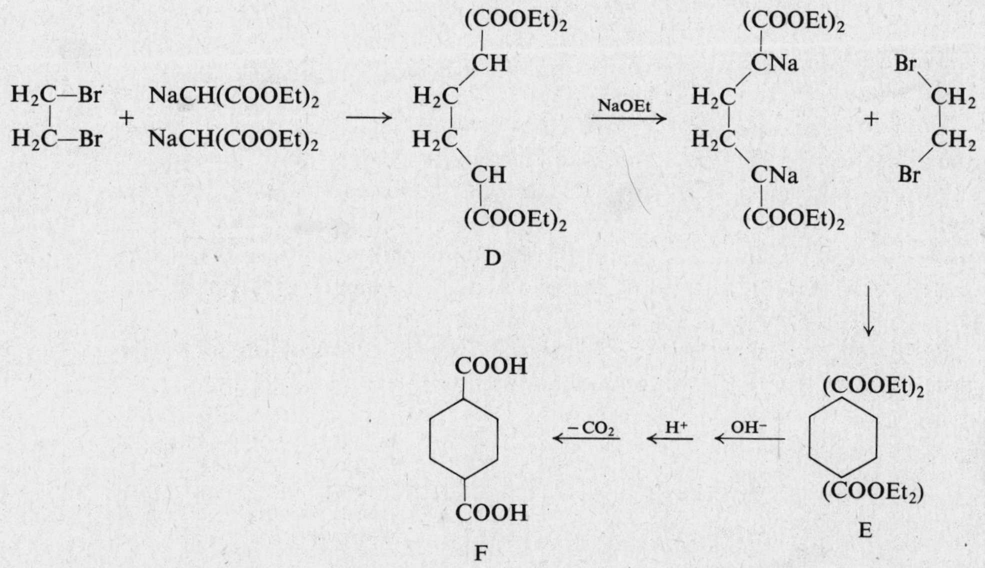

C $\xleftarrow{-CO_2}$ $\xleftarrow{H^+}$ $\xleftarrow{OH^-}$ B

Cyclohexane ring with COOH (top) and COOH (bottom) — C

Cyclohexane ring with (COOEt)$_2$ (top) and (COOEt)$_2$ (bottom) — B

(b)

$$\begin{matrix}H_2C-Br\\H_2C-Br\end{matrix} + \begin{matrix}NaCH(COOEt)_2\\NaCH(COOEt)_2\end{matrix} \longrightarrow \begin{matrix}(COOEt)_2\\CH\\H_2C\\H_2C\\CH\\(COOEt)_2\end{matrix} \xrightarrow{NaOEt} \begin{matrix}(COOEt)_2\\CNa\\H_2C\\H_2C\\CNa\\(COOEt)_2\end{matrix} + \begin{matrix}Br\\CH_2\\CH_2\\Br\end{matrix}$$

D

Cyclohexane ring with (COOEt)$_2$ (top) and (COOEt$_2$) (bottom) — E

F $\xleftarrow{-CO_2}$ $\xleftarrow{H^+}$ $\xleftarrow{OH^-}$ E

Cyclohexane ring with COOH (top) and COOH (bottom) — F

(c)

$$\begin{matrix}EtOOC\\HC^{\ominus}\\EtOOC\end{matrix} + \begin{matrix}COOEt\\{}^{\ominus}CH\\COOEt\end{matrix} + I_2 \longrightarrow \begin{matrix}EtOOC\;\;COOEt\\HC-CH\\EtOOC\;\;COOEt\end{matrix} + 2I^-$$

G

$$G \xrightarrow[heat]{OH^-} \xrightarrow{H^+} \begin{matrix}HOOC\;\;COOH\\HC-CH\\HOOC\;\;COOH\end{matrix} \xrightarrow[-2CO_2]{heat} HOOCCH_2-CH_2COOH$$

H

Succinic acid

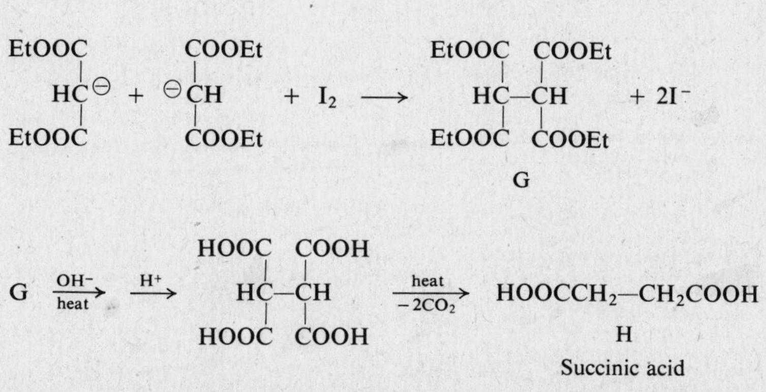

(d) D $\xrightarrow{2OEt^-}$ [structure with COOEt groups] $\xrightarrow{I_2}$ [structure I] $+ 2I^-$

I

I $\xrightarrow[heat]{OH^-}$ $\xrightarrow{H^+}$ [tetracarboxylic acid structure] $\xrightarrow[-2CO_2]{heat}$ [structure J]

J

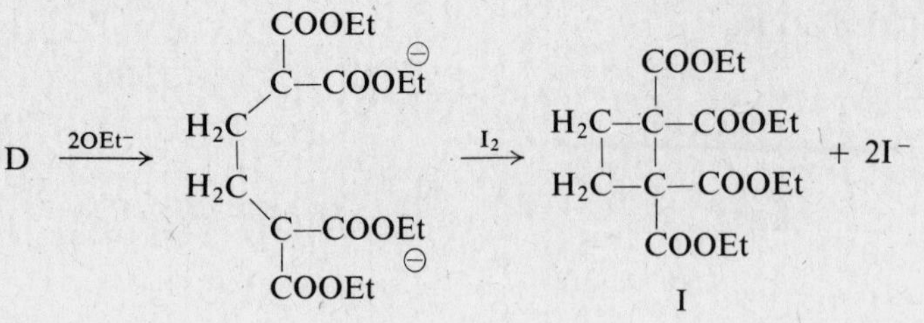

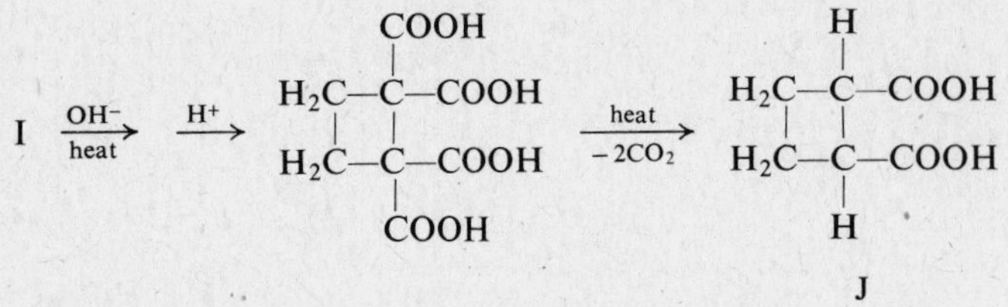

(e) [1,3-Cyclopentane structure]

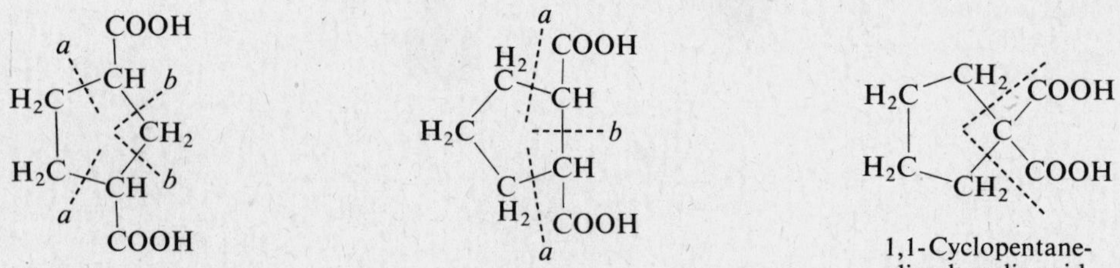

1,3-Cyclopentane-
dicarboxylic acid

[1,2-Cyclopentane structure]

1,2-Cyclopentane-
dicarboxylic acid

[1,1-Cyclopentane structure]

1,1-Cyclopentane-
dicarboxylic acid

Cyclopentane-1,3-dicarboxylic acid. Add one mole of $BrCH_2CH_2Br$ to an excess of sodiomalonic ester (as in Problem 26.2). Then add 2NaOEt, followed by CH_2I_2. Hydrolyze and decarboxylate.

Cyclopentane-1,2-dicarboxylic acid. As for the 1,3-compound, except use $BrCH_2CH_2$-CH_2Br instead of $BrCH_2CH_2Br$, and I_2 instead of CH_2I_2. (Compare Problem 4c, page . 861.)

Cyclopentane-1,1-dicarboxylic acid. As in Problem 1i (page 861), except use $BrCH_2CH_2$-CH_2CH_2Br instead of $BrCH_2CH_2CH_2Br$, and do not decarboxylate after hydrolysis.

5. $2CH_2{=}CHCH_2Br + Mg + BrCH_2CH{=}CH_2 \longrightarrow CH_2{=}CHCH_2{-}CH_2CH{=}CH_2 + MgBr_2$

K

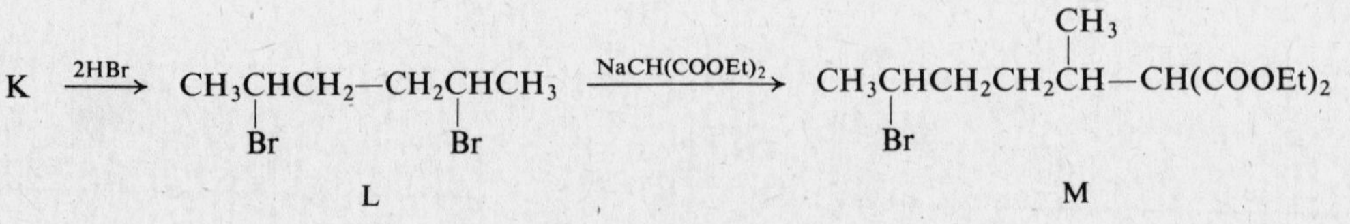

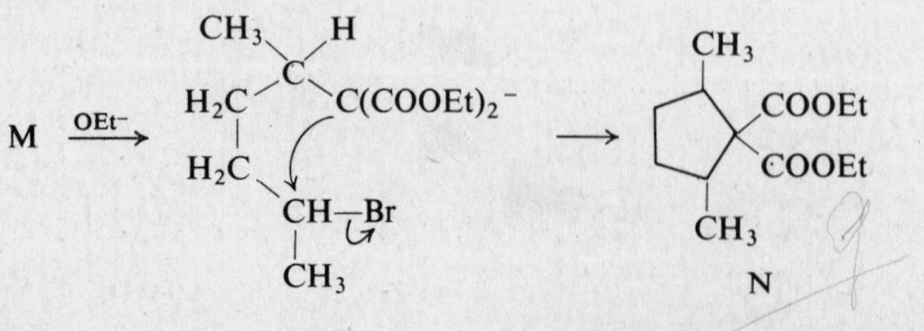

K $\xrightarrow{2HBr}$ $CH_3CHCH_2{-}CH_2CHCH_3$ $\xrightarrow{NaCH(COOEt)_2}$ $CH_3CHCH_2CH_2CH{-}CH(COOEt)_2$

Br   Br   Br

L   M

M $\xrightarrow{OEt^-}$ [cyclization structure] $\longrightarrow$ [structure N]

N

$$N \xrightarrow{OH^-} \xrightarrow{H^+} \text{[2,5-dimethylcyclopentane-1,1-dicarboxylic acid]} \xrightarrow[-CO_2]{heat} \text{[2,5-dimethylcyclopentanecarboxylic acid]}$$

6. (1)
$$Cl_3C\text{—}COO^- \xrightarrow{heat} CO_2 + Cl_3C:^-$$
Trichloroacetate

(2)
$$Cl_3C:^- \rightleftharpoons :CCl_2 + Cl^-$$
Dichloro-
carbene

(3)
$$\underset{\diagup}{\diagup}C{=}C\underset{\diagdown}{\diagdown} + :CCl_2 \longrightarrow \underset{\underset{Cl}{|}}{-}\overset{|}{C}\underset{\underset{Cl}{|}}{-}\overset{|}{C}-$$

Decarboxylation (1) of Cl_3CCOO^- is comparatively easy because of the stability of the carbanion being formed. (Compare Problem 26.12, page 853, and Problem 26.15, page 854.) Once Cl_3C^- is formed, (2) and (3) follow, as on page 311.

7. (a) Much as in Problem 26.2a, except that we use two moles of acetoacetic ester instead of malonic ester.

$$\underset{O}{\overset{\parallel}{\underset{}{}}}CH_3\overset{O}{\overset{\parallel}{C}}CH_2\,\vert\,CH_2CH_2\,\vert\,CH_2\overset{O}{\overset{\parallel}{C}}CH_3$$

$$\uparrow$$

$$CH_3\overset{O}{\overset{\parallel}{C}}\underset{COOEt}{\overset{|}{C}}H\text{—}CH_2CH_2\text{—}\underset{COOEt}{\overset{|}{C}}H\overset{O}{\overset{\parallel}{C}}CH_3 \xleftarrow{EtO^-} CH_3COCH_2 + BrCH_2CH_2Br + CH_2COCH_3$$
$$\underset{COOEt}{\qquad\qquad\qquad\qquad\qquad\qquad\qquad\qquad\qquad\quad}\quad\underset{COOEt}{}$$

(b) An intramolecular aldol condensation takes place. Of the two possibilities, the reaction takes place in the way that gives a five-membered ring, not a seven-membered ring.

[structures: 1-methyl-2-acetylcyclopentene $\xleftarrow{-H_2O}$ 1-methyl-1-hydroxy-2-acetylcyclopentane $\xleftarrow{EtO^-}$ open-chain diketone]

(c) As in (a), except use CH_2I_2 instead of CH_2BrCH_2Br.

$$CH_3\overset{O}{\overset{\parallel}{C}}CH_2\,\vert\,CH_2\,\vert\,CH_2\overset{O}{\overset{\parallel}{C}}CH_3 \longleftarrow CH_3\overset{O}{\overset{\parallel}{C}}\underset{COOEt}{\overset{|}{C}}H\text{—}CH_2\text{—}\underset{COOEt}{\overset{|}{C}}H\overset{O}{\overset{\parallel}{C}}CH_3 \xleftarrow{EtO^-} CH_3COCH_2 + CH_2I_2 + CH_2COCH_3$$
$$\underset{COOEt}{\qquad\qquad\qquad\qquad\qquad\qquad\qquad\qquad\qquad\qquad\qquad\qquad\quad}\quad\underset{COOEt}{}$$

(d) Probably an intramolecular aldol condensation, and in the way that gives a six-membered ring, not a four-membered ring.

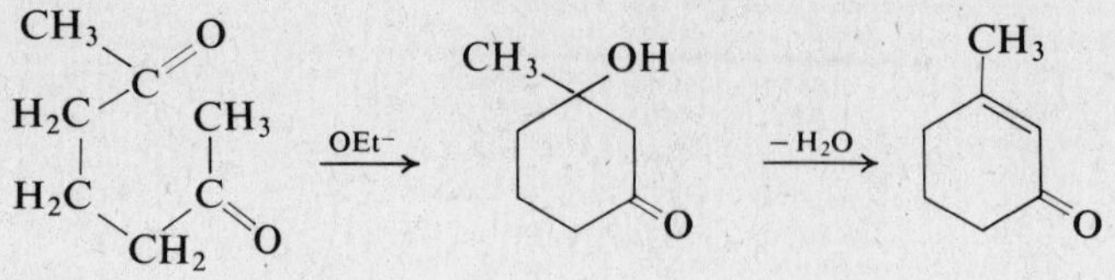

8. (a) In compound I, page 720, we see two —COOEt's *beta* to C=O and therefore susceptible to easy decarboxylation.

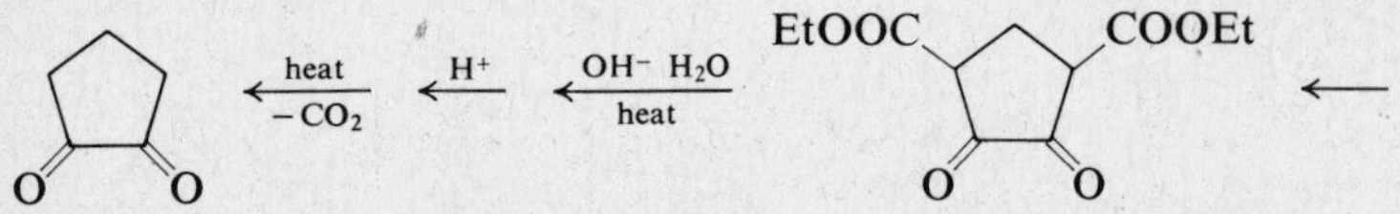

As in Problem 21.33b, page 770

(b) We follow the general procedure for preparation of α-keto acids, as illustrated in Problem 26.9 (page 853).

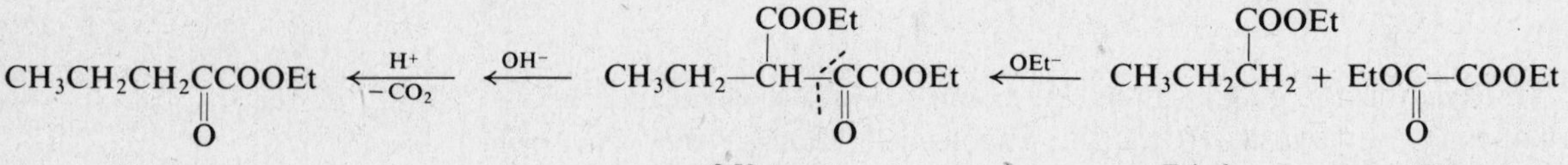

$$CH_3CH_2CH_2CCOOEt \xleftarrow[-CO_2]{H^+} \xleftarrow{OH^-} CH_3CH_2-CH-CCOOEt \xleftarrow{OEt^-} CH_3CH_2CH_2 + EtOC-COOEt$$

| Ketone | β-Keto ester | Ethyl *n*-butyrate | Ethyl oxalate |

9. These are all prepared by reaction of urea with the appropriate substituted malonic ester:

| Urea | Disubstituted malonic ester | Disubstituted barbituric acid |

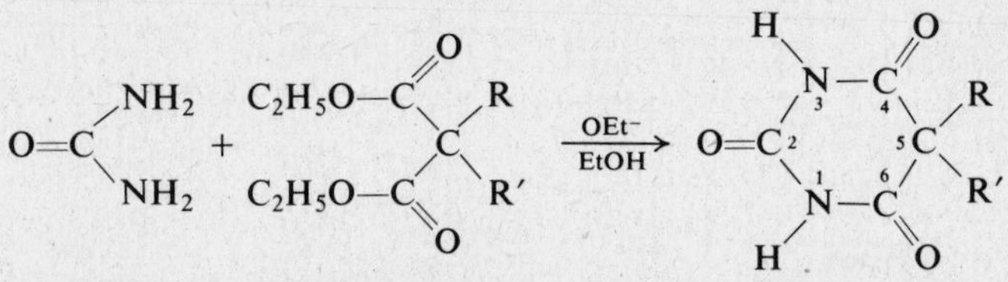

The malonic esters are prepared through successive alkylation with:

(a) C_2H_5Br, twice.

(b) $CH_3CH_2CH_2CHBrCH_3$; $CH_2=CHCH_2Br$.

(c) $(CH_3)_2CHCH_2CH_2Br$; C_2H_5Br.

10. (a) Removal of a C–5 proton from barbituric acid gives a resonance-stabilized aromatic anion (six π electrons: two from C–5 and two from each of the two nitrogens); Veronal has no C–5 protons, and cannot form such an anion.

(b) Veronal is an imide, with two acyl groups for each N—H group.

11. (a) This is a *retro*-Claisen condensation: a series of steps that are essentially the reverse of those by which acetoacetic ester is made. Here, the equilibria are shifted in the direction that produces resonance-stabilized, unreactive acetate ions. This reaction is called the *acid cleavage* of acetoacetic ester, in contrast to the more common *ketonic cleavage* used to produce ketones.

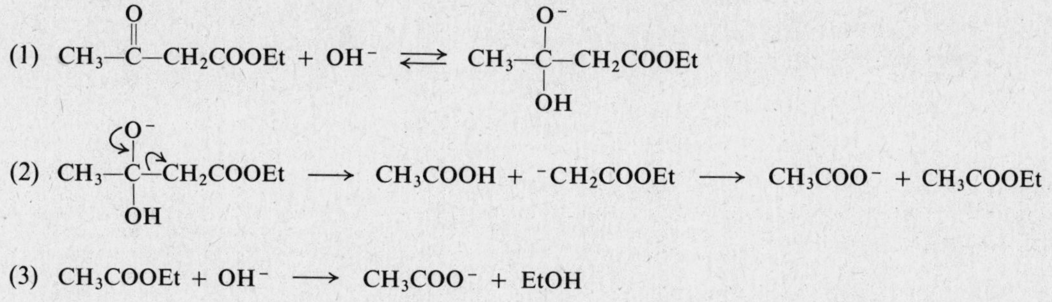

(1) CH₃—C—CH₂COOEt + OH⁻ ⇌ CH₃—C—CH₂COOEt
        ‖ |
        O   OH
                                            with O⁻ above C

(2) CH₃—C—CH₂COOEt ⟶ CH₃COOH + ⁻CH₂COOEt ⟶ CH₃COO⁻ + CH₃COOEt
        |
        OH

(3) CH₃COOEt + OH⁻ ⟶ CH₃COO⁻ + EtOH

(b) Alkylate acetoacetic ester as usual, then heat with concentrated base to cause *retro-*Claisen condensation (acid cleavage).

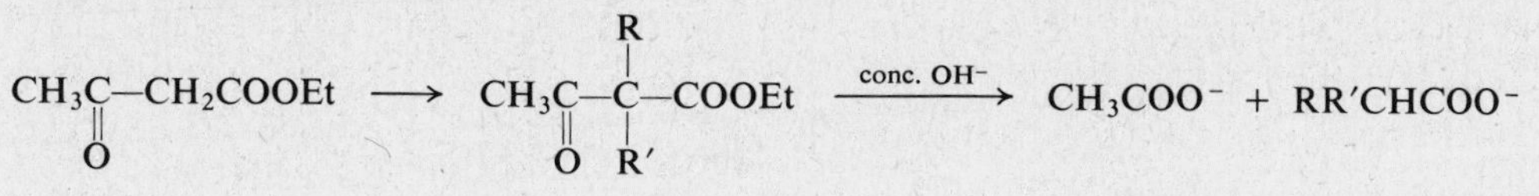

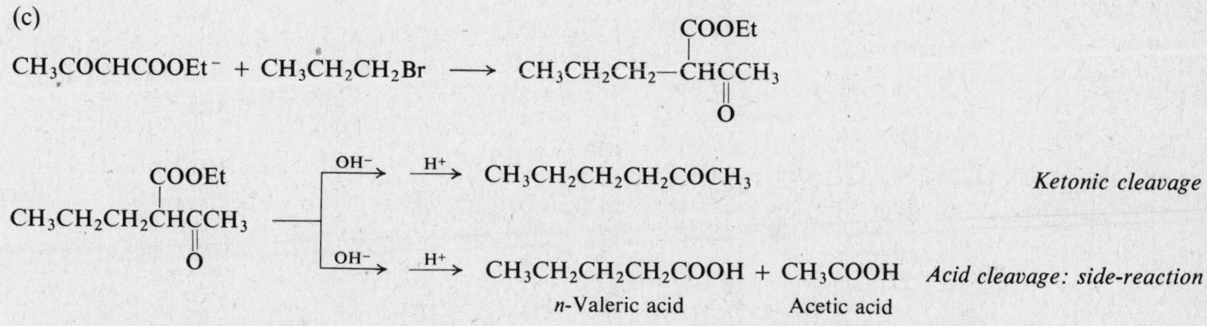

CH₃C—CH₂COOEt ⟶ CH₃C—C—COOEt $\xrightarrow{\text{conc. OH}^-}$ CH₃COO⁻ + RR′CHCOO⁻
      ‖   ‖ |
      O   O R′
                        (with R above central C)

(c)

CH₃COCHCOOEt⁻ + CH₃CH₂CH₂Br ⟶ CH₃CH₂CH₂—CHCCH₃
                                                        | ‖
                                                      (COOEt) O

(where COOEt is above CH and O below the C)

                                                                              Ketonic cleavage

CH₃CH₂CH₂CHCCH₃ $\xrightarrow{\text{OH}^-}$ $\xrightarrow{\text{H}^+}$ CH₃CH₂CH₂CH₂COCH₃
      | ‖
  (COOEt) O

$\xrightarrow{\text{OH}^-}$ $\xrightarrow{\text{H}^+}$ CH₃CH₂CH₂CH₂COOH + CH₃COOH   *Acid cleavage: side-reaction*
                                          n-Valeric acid   Acetic acid

The crude ketone will contain water, EtOH, acetic acid, and *n*-valeric acid. Remove the acids by washing with aqueous NaOH. Then distill, collecting water, EtOH, and the ketone. Remove the EtOH and some water by shaking with concentrated aqueous CaCl₂. Separate the ketone from the aqueous layer, dry over anhydrous MgSO₄, and distill.

12. (a) In Problem 21.32 (page 719), we saw that ketones undergo a crossed Claisen condensation with esters to give β-diketones. Like the ordinary Claisen condensation (Problem 11, above) this reaction, too, can be reversed, to give the ketone and (instead of the original ester) a carboxylate anion. To use the example in Problem 21.32:

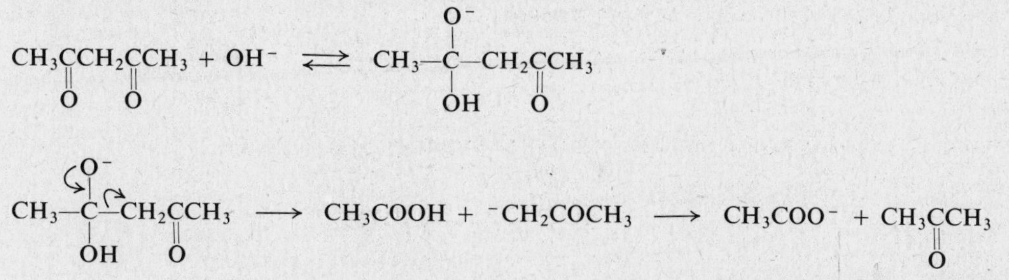

CH₃CCH₂CCH₃ + OH⁻ ⇌ CH₃—C—CH₂CCH₃
      ‖   ‖ | ‖
      O   O   OH   O
                              (with O⁻ above C)

CH₃—C—CH₂CCH₃ ⟶ CH₃COOH + ⁻CH₂COCH₃ ⟶ CH₃COO⁻ + CH₃CCH₃
        | ‖   ‖
        OH   O   O

If this *retro*-Claisen is applied to *cyclic* β-diketones, useful products can be obtained, with carboxylate and keto groups in the same molecule.

(b)

Ketone   Enamine   β-Diketone

$HOOC(CH_2)_5CPh$

N_2H_4 | base

$HOOC(CH_2)_6Ph$

(c) As in (b), except use $ClCO(CH_2)_7COOEt$ instead of PhCOCl. (Azaleic acid, HOOC-$(CH_2)_7COOH$, is available from the oxidative cleavage of oleic acid, from fats (page 579).)

13. $n\text{-}C_6H_{13}CHO + BrCH_2COOEt \xrightarrow{\text{Zn}} \xrightarrow{\text{H}_2\text{O}} n\text{-}C_6H_{13}CHCH_2COOEt$
                                                             |
                                                            OH
                                                             P

$n\text{-}C_6H_{13}CHCH_2COOEt \xrightarrow{\text{CrO}_3} n\text{-}C_6H_{13}CCH_2COOEt$
              | ‖
             OH   O
              P   Q
                                          A β-keto ester

$n\text{-}C_6H_{13}CCH_2COOEt \xrightarrow{\text{NaOEt}} \xrightarrow{\text{PhCH}_2\text{Cl}} n\text{-}C_6H_{13}C-CHCOOEt$
         ‖   ‖ |
         O   O CH_2Ph
         Q   R

$n\text{-}C_6H_{13}C-CHCOOEt \longrightarrow n\text{-}C_6H_{13}C-CHCOO^- \xrightarrow{-CO_2} CH_3(CH_2)_5CCH_2CH_2Ph$
         ‖ | ‖ | ‖
         O CH_2Ph   O CH_2Ph   O
         R   S
                                                                    1-Phenyl-3-nonanone

14.

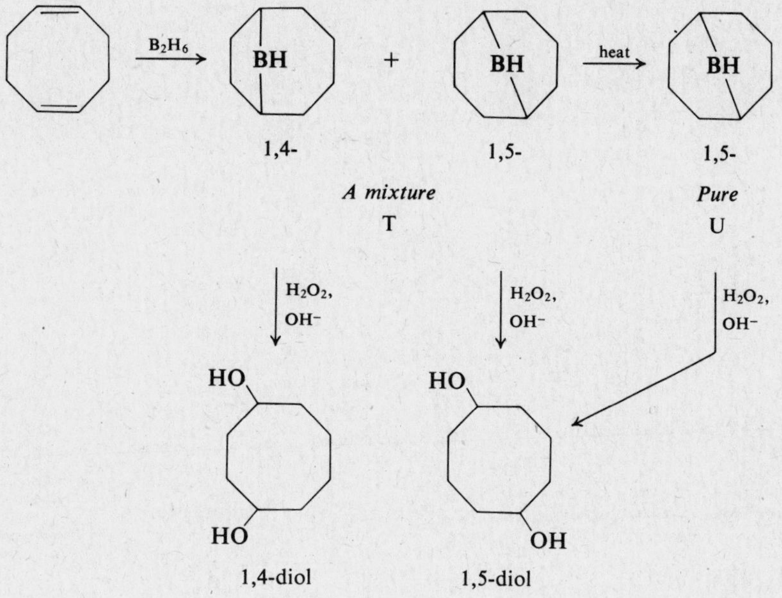

1,4-   1,5-   1,5-

A mixture   *Pure*

T   U

1,4-diol   1,5-diol

Hydroboration of the double bonds takes place in the two possible ways to give T, a mixture of 1,4- and 1,5-isomers. Heating causes equilibration, presumably through reversible hydroboration, to yield almost entirely the more stable 1,5-isomer, U, the one containing two six-membered rings. U is 9-BBN (Sec. 26.7).

The oxidation takes place with retention of configuration (Problem 15.12, page 509). Formation of the *cis*-diols shows, then, that bridging in both 1,4- and 1,5-boranes is *cis*, which is what we would expect on steric grounds.

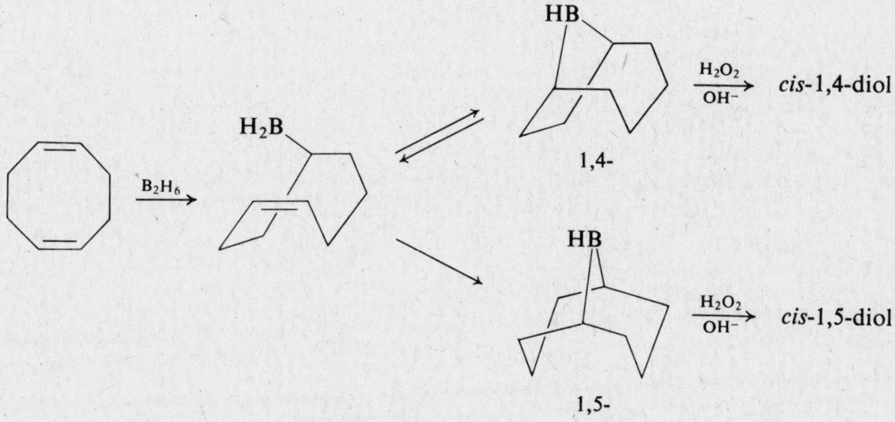

1,4-

cis-1,4-diol

1,5-

cis-1,5-diol

(E. F. Knight and H. C. Brown, "Cyclic Hydroboration of 1,5-Cyclooctadiene. A Simple Synthesis of 9-Borabicyclo[3.3.1]nonane, an Unusually Stable Dialkylborane," J. Am. Chem. Soc., **90**, 5280 (1968).)

15. (a) $ArCHO + OH^- \rightleftharpoons$ Ar—$\overset{\displaystyle H}{\underset{\displaystyle OH}{C}}$—$O^- + H_2O$

Ar—$\overset{\displaystyle H}{\underset{\displaystyle OH}{C}}$—$O^- + OH^- \rightleftharpoons$ Ar—$\overset{\displaystyle H}{\underset{\displaystyle O^-}{C}}$—$O^- + H_2O$

Ar—$\overset{\displaystyle H}{\underset{\displaystyle \underset{\displaystyle O^-}{|}}{C}}$—$O^-$ $\longrightarrow$ $Ar^- + HCOO^-$

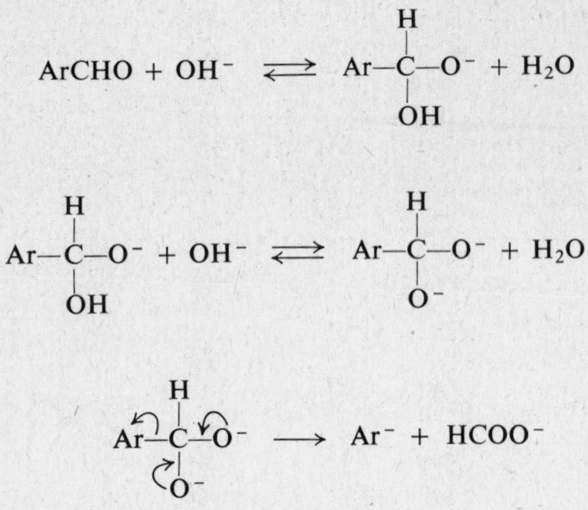

$Ar^- + H_2O \longrightarrow ArH + OH^-$

(b) The two o-Cl's stabilize Ar^- and the transition state leading to its formation.

(J. F. Bunnett, J. H. Miles, and K. V. Nahabedian, "Kinetics and Mechanism of the Alkali Cleavage of 2,6-Dihalobenzaldehydes," J. Am. Chem. Soc., **83**, 2512 (1961).)

16. Chemical arithmetic,

$$2C_4H_6O_2 - C_8H_{10}O_3 = H_2O$$

indicates combination of two moles of lactone with loss of one H_2O, and strongly suggests an aldol-like condensation followed by dehydration:

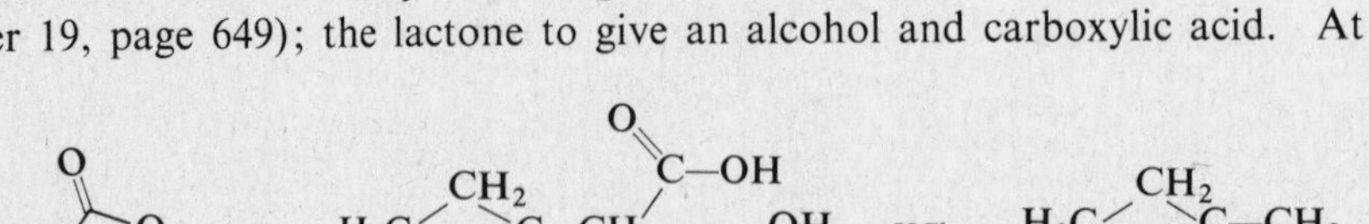

γ-Butyrolactone
Two moles

V

V is a lactone (an ester) and a vinyl ether. Both these functional groupings are cleaved by concentrated HCl: the vinyl ether to give a ketone and alcohol (see answer to Problem 15, Chapter 19, page 649); the lactone to give an alcohol and carboxylic acid. At this point

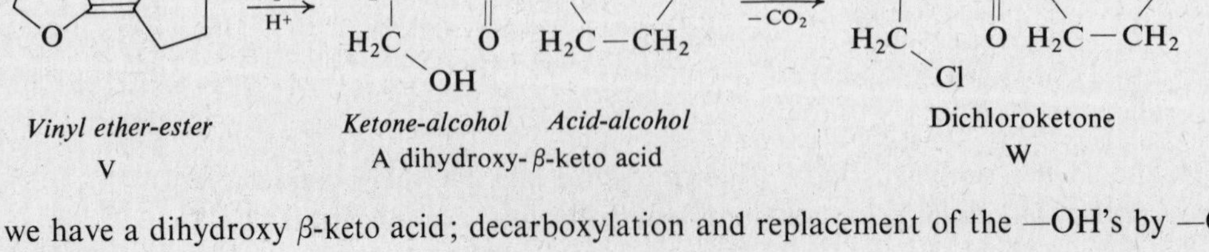

Vinyl ether-ester
V

Ketone-alcohol   Acid-alcohol
A dihydroxy-β-keto acid

Dichloroketone
W

we have a dihydroxy β-keto acid; decarboxylation and replacement of the —OH's by —Cl's give W.

In the presence of acid, the ketone W undergoes intramolecular alkylation at each of the α-carbons, to give dicyclopropyl ketone.

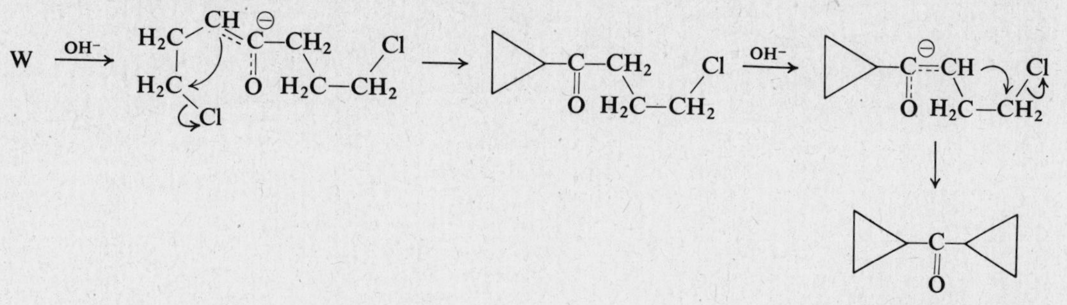

Dicyclopropyl ketone

17. (a) 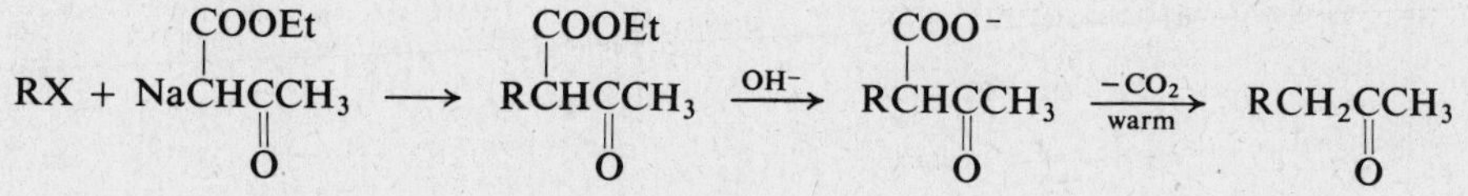

X Y

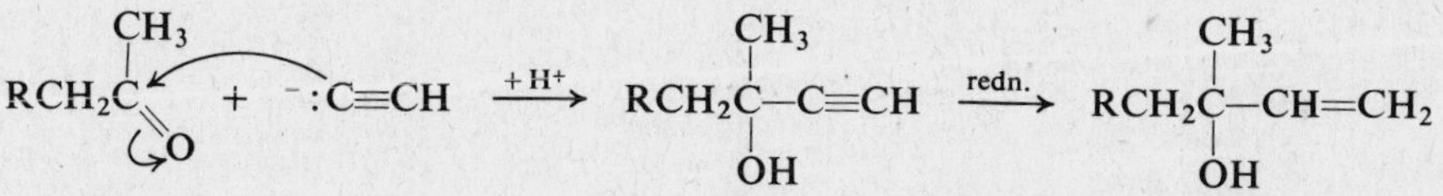

Y Z AA Nerolidol

(b) 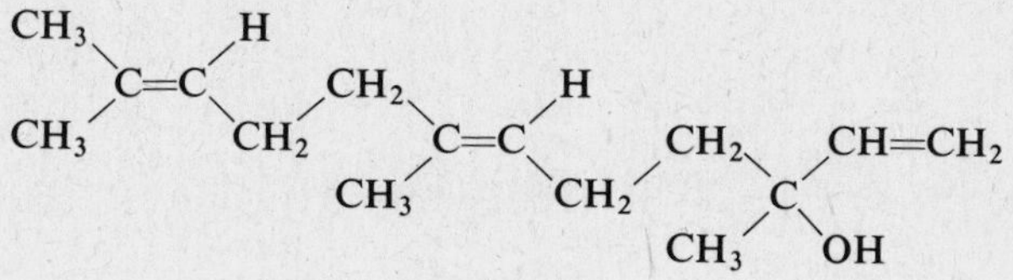

Nerolidol

18. (a) In the first step, C_2H_5OH is lost ($C_{12}H_{22}O_4 - C_{10}H_{16}O_3 = C_2H_5OH$) suggesting an intramolecular Claisen condensation (Dieckmann condensation, Problem 21.30, page 718). This could take place in either of two ways:

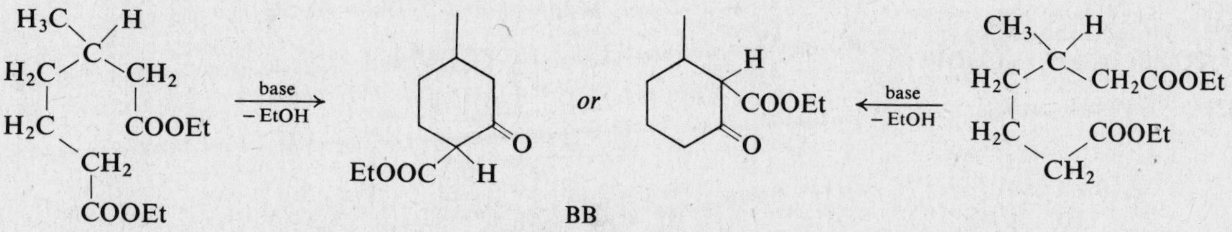

BB

Possible structures

Depending upon which is the correct structure for BB, menthone could have either of two structures:

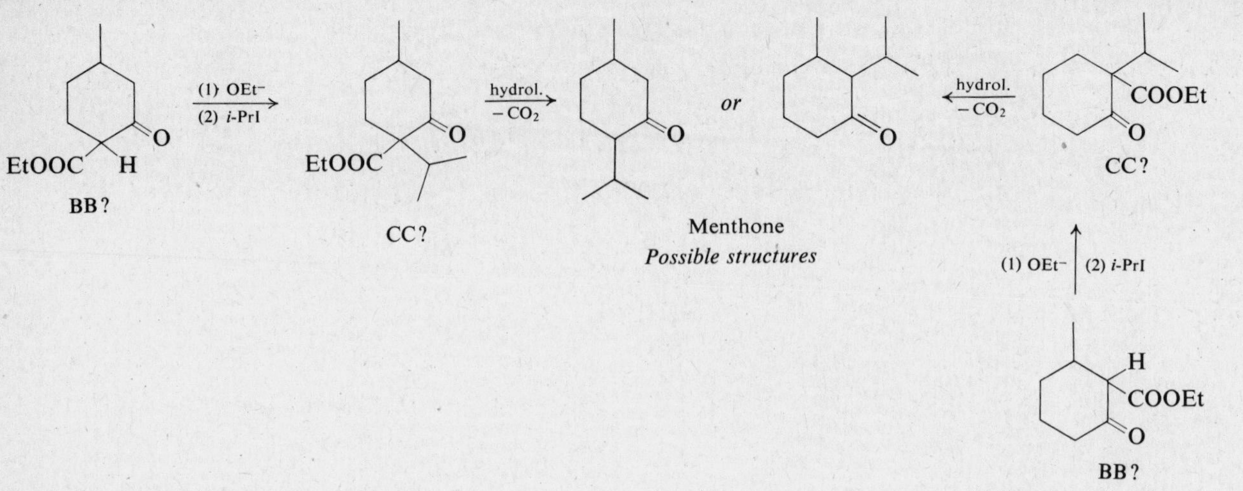

BB?   CC?   Menthone
Possible structures   CC?   BB?

(b) Of the two structures possible at this point, the more likely is the one that follows the isoprene rule.

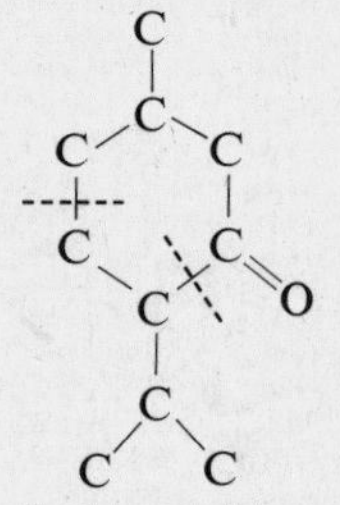

More likely: two isoprene units

(c) The reduction confirms the structure indicated in (b).

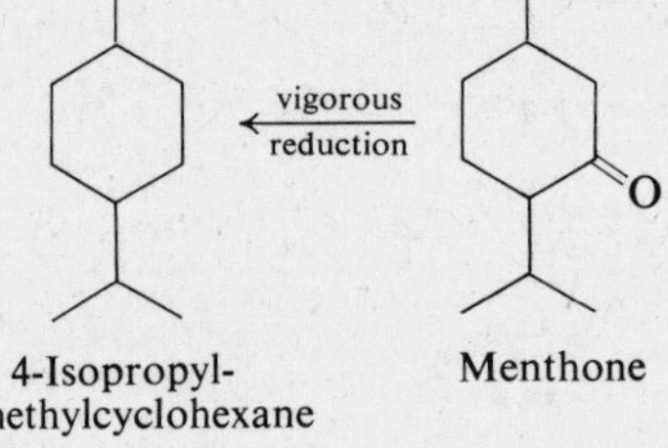

4-Isopropyl-
1-methylcyclohexane   Menthone

19.

$$[CH_3COCHCOOEt]^-Na^+ \xrightarrow{CH_3I} CH_3COCHCOOEt \xrightarrow{NaOEt} \xrightarrow{CH_3I} CH_3COCCOOEt$$

DD   EE

EE   FF

FF   GG   Camphoronic acid

406

20. $CH_3COCH^-Na^+$ $\xrightarrow{ClCH_2COOEt}$ $CH_3COCH{-}CH_2COOEt$ $\xrightarrow{CH_3MgI}$ $\xrightarrow{H_2O}$ $CH_3{-}\underset{\underset{H_3C}{|}}{\overset{\overset{OH}{|}}{C}}{-}\underset{\underset{COOEt}{|}}{CH}{-}CH_2COOEt$

         $\underset{COOEt}{|}$ $\underset{COOEt}{|}$

                                     HH   II

                                                       (Ketone reacts faster than ester)

II $\xrightarrow[heat]{OH^-}$ $\xrightarrow{H^+}$ $CH_3{-}\underset{\underset{H_3C}{|}}{\overset{\overset{OH}{|}}{C}}{-}\underset{\underset{COOH}{|}}{CH}{-}CH_2COOH$ $\xrightarrow{-H_2O}$ (γ-lactone structure: CH₃, CH₃ on ring, COOH, H)

                                       JJ   Terebic acid

                        A γ-hydroxy acid   *A γ-lactone*

$CH_3COCHCH_2COOEt$ $\xrightarrow{OEt^-}$ $CH_3CO\overset{\ominus}{C}CH_2COOEt$ $\xrightarrow{ClCH_2COOEt}$ $CH_3CO\underset{\underset{COOEt}{|}}{\overset{\overset{CH_2COOEt}{|}}{C}}CH_2COOEt$

  $\underset{COOEt}{|}$ $\underset{COOEt}{|}$

    HH   *Most acidic H removed*   KK

$CH_3CO\underset{\underset{COOEt}{|}}{\overset{\overset{CH_2COOEt}{|}}{C}}CH_2COOEt$ $\xrightarrow{OH^-}$ $\xrightarrow[-CO_2]{H^+}$ $CH_3CO\underset{\underset{H}{|}}{\overset{\overset{CH_2COOH}{|}}{C}}CH_2COOH$ $\xrightarrow{2EtOH,\ H^+}$ $CH_3CO\underset{\underset{H}{|}}{\overset{\overset{CH_2COOEt}{|}}{C}}CH_2COOEt$

     KK   LL   MM

A β-keto ester

MM $\xrightarrow{CH_3MgI}$ $\xrightarrow{H_2O}$ $CH_3{-}\underset{\underset{H_3C}{|}}{\overset{\overset{HO}{|}}{C}}{-}\underset{\underset{H}{|}}{\overset{\overset{CH_2COOEt}{|}}{C}}CH_2COOEt$ $\xrightarrow{OH^-}$ $\xrightarrow{H^+}$ $CH_3{-}\underset{\underset{H_3C}{|}}{\overset{\overset{HO}{|}}{C}}{-}\underset{\underset{H}{|}}{\overset{\overset{CH_2COOH}{|}}{C}}CH_2COOH$ $\xrightarrow{-H_2O}$ (γ-lactone structure: CH₃, CH₃ on ring, CH₂COOH, H)

                                  NN   OO   Terpenylic acid

                                                 A γ-hydroxy acid   *A γ-lactone*

21. $\underset{\underset{OPO_3H_2}{|}}{-\overset{\overset{-OOC}{|}}{C}-C-}$ $\longrightarrow$ $-C{=}C-$ $+\ CO_2\ +\ H_2PO_4^-$

    The dihydrogenphosphate ion, $H_2PO_4^-$, is a much better leaving group than the more strongly basic OH^-. Loss of CO_2 leads, not to the unstable carbanion, but—with $H_2PO_4^-$ carrying away the electrons—to a stable, neutral alkene.

    Carbanion chemistry—or something resembling it—runs throughout these biosyntheses. Acetate units are combined into acetoacetate by what amounts to a malonic ester synthesis (Sec. 37.6). In an aldol-like condensation, another acetate adds to acetoacetate to give 3-methyl-3-hydroxyglutarate, a six-carbon precursor of mevalonic acid. Here, we see the loss of CO_2 that finally gives the isoprenoid skeleton—and the way is clear to geraniol, farnesol, squalene, lanosterol, and cholesterol; or, in a different organism, to rubber. (See the reference given in the answer to Problem 27, Chapter 8, page 282.)

Chapter 27

α,β-Unsaturated Carbonyl Compounds
Conjugate Addition

27.1 (a) $CH_3CH_2CH_2COOH$

(b) $CH_3CH=CHCOO^- + EtOH$

(c) $PhCH=CHCOO^- + CHI_3$

(d) $CH_3CH=CHCH=NNHPh$

(e) $CH_3CH=CHCOO^- + Ag$

(f) $PhCHO + PhCOCHO$

(g) $CH_3CH_2CH_2CH_2OH$

(h) *meso*-$HOOCCHBrCHBrCOOH$

(i) racemic $HOOCCHOHCHOHCOOH$

27.2 A, $PhCH_2CH_2CHO$ B, $PhCH_2CH_2CH_2OH$ D, $PhCH=CHCH_2OH$

27.3 (a) $CH_3CH=CHCHO \xleftarrow[-H_2O]{H^+} CH_3CH-CHCHO \xleftarrow{OH^-} CH_3CHO \xleftarrow[H^+, Hg^{++}]{H_2O} HC\equiv CH$

OH (under $CH_3CH-CHCHO$)

(b) $PhCH=CHCHO \xleftarrow{-H_2O} \left[PhCH-CH_2CHO \right] \xleftarrow{OH^-} PhCHO + CH_3CHO$

OH (under $PhCH-CH_2CHO$)

(c) $PhCH=CHCOOH \xleftarrow[-H_2O]{OAc^-} PhCHO + (CH_3CO)_2O$ (Perkin condensation, Problem 21.22e)

(d) $(CH_3)_2CHCH=CHCOOH \xleftarrow{KOH(alc)} (CH_3)_2CHCH_2CHCOOH \longleftarrow$ as in Problem 26.1b

Br (under $CHCOOH$)

27.4 (a) $CH_2=CHCN$ $\xleftarrow[-H_2O]{H^+}$ CH_2CH_2CN $\xleftarrow{CN^-}$ CH_2-CH_2 $\xleftarrow{HOCl}$ $CH_2=CH_2$
  Acrylonitrile $\overset{|}{OH}$ $\overset{|}{OH}$ $\overset{|}{Cl}$

(b) $CH_2=CHCOOMe$ $\xleftarrow[-H_2O]{H^+}$ CH_2CH_2COOMe $\xleftarrow{MeOH}{H^+}$ CH_2CH_2CN $\longleftarrow$ as in (a)
  Methyl acrylate $\overset{|}{OH}$ $\overset{|}{OH}$

(c) $CH_2=\overset{\overset{\textstyle CH_3}{|}}{C}COOMe$ $\xleftarrow[-H_2O]{H^+}$ $CH_3\overset{\overset{\textstyle CH_3}{|}}{\underset{\underset{\textstyle OH}{|}}{C}}COOMe$ $\xleftarrow{MeOH}{H^+}$ $CH_3\overset{\overset{\textstyle CH_3}{|}}{\underset{\underset{\textstyle OH}{|}}{C}}CN$ $\xleftarrow{CN}{H^+}$ $CH_3\underset{\underset{\textstyle O}{\|}}{C}CH_3$
  Methyl methacrylate

(d) $\sim\!\!\sim CH_2\underset{\underset{\textstyle CN}{|}}{CH}\sim\!\!\sim$ $\sim\!\!\sim CH_2\underset{\underset{\textstyle COOMe}{|}}{CH}\sim\!\!\sim$ $\sim\!\!\sim CH_2\overset{\overset{\textstyle CH_3}{|}}{\underset{\underset{\textstyle COOMe}{|}}{C}}\sim\!\!\sim$
      Orlon   Acryloid   Lucite, Plexiglas

27.5 (a) $CH_2\underset{\underset{\textstyle OH}{|}}{CH}-\underset{\underset{\textstyle OH}{|}}{CH_2}$ (with OH OH OH) $\xrightarrow{-H_2O}$ $CH_2\underset{\underset{\textstyle OH}{|}}{CH}=\underset{\underset{\textstyle OH}{|}}{CH}$ $\xrightarrow{tautom.}$ $CH_2CH_2\underset{\underset{\textstyle OH}{|}}{CHO}$ $\xrightarrow{-H_2O}$ $CH_2=CHCHO$
  Glycerol   Acrolein

(b) $CH_2=CHCHO$ $\xrightarrow{Ag(NH_3)_2{}^+}$ $CH_2=CHCOOH$
  Acrolein   Acrylic acid

27.6 All are less stable than I.

27.7 IV is an amide, formed by nucleophilic addition and subsequent ring closure through amide formation.

$\begin{matrix} H & CH_2 \\ \diagdown C & \diagup \\ & C=O \\ CH_3 \diagdown N\!-\!N \diagup \\ H \quad\quad C_6H_5 \end{matrix}$ (IV) $\xleftarrow{-H_2O}$ $\begin{matrix} H & CH_2 \\ \diagdown C & \diagup \\ & COOH \\ CH_3 \diagdown N\!-\!N\!-\!H \\ H \quad\quad C_6H_5 \end{matrix}$ $\longleftarrow$ $CH_3CH=CHCOOH + C_6H_5NHNH_2$

27.8 Two successive nucleophilic additions.

$$CH_2=CHCN + NH_3 \longrightarrow CH_2CH_2CN \atop \underset{\textstyle NH_2}{|}$$

$$CH_2=CHCN + H_2NCH_2CH_2CN \longrightarrow H-N\!\!\begin{matrix} \\ \diagup CH_2CH_2CN \\ \diagdown CH_2CH_2CN \end{matrix}$$

27.9 Two successive nucleophilic additions.

$$CH_2{=}CHCOOEt + CH_3NH_2 \longrightarrow \underset{\underset{CH_3NH}{|}}{CH_2CH_2COOEt}$$

$$CH_2{=}CHCOOEt + \underset{\underset{CH_3NH}{|}}{CH_2CH_2COOEt} \longrightarrow CH_3{-}\overset{\overset{CH_2CH_2COOEt}{|}}{\underset{\underset{CH_2CH_2COOEt}{|}}{N}}$$

27.10 (a) $CH_3CH{=}CHCOOEt \longrightarrow \underset{\underset{\underset{COOEt}{|}}{H-C-COOEt}}{CH_3CHCH_2COOEt} \longrightarrow \longrightarrow \longrightarrow \underset{\underset{CH_2COOH}{|}}{CH_3CHCH_2COOH}$

$\underset{\underset{COOEt}{|}}{H_2C-COOEt}$ A   B

(b) $CH_2{=}CHCOOEt \longrightarrow \underset{\underset{\underset{COOEt}{|}}{H-C-COCH_3}}{CH_2CH_2COOEt} \xrightarrow{\text{ketonic cleavage}} \underset{\underset{CH_2COCH_3}{|}}{CH_2CH_2COOH}$

$\underset{\underset{COOEt}{|}}{H_2C-COCH_3}$ C   D

(c) $CH_2{=}CHCOCH_3 \longrightarrow \underset{\underset{\underset{COOEt}{|}}{H-C-COOEt}}{CH_2CH_2COCH_3}$

$\underset{\underset{COOEt}{|}}{H_2C-COOEt}$ E

(d) $PhCH{=}CHCOPh \longrightarrow \underset{\underset{CH_2COPh}{|}}{PhCHCH_2COPh}$

HCH_2COPh F

(e) $CH_2{=}CHCN \longrightarrow \underset{\underset{\underset{CH=CH_2}{|}}{H-C-CN}}{CH_2CH_2CN} \xrightarrow{H_2O,\ H^+} \underset{\underset{\underset{CH=CH_2}{|}}{H-C-COOH}}{CH_2CH_2COOH}$

$\underset{\underset{CH=CH_2}{|}}{H_2C-CN}$ G   H

(f) $EtOOCC{\equiv}CCOOEt \longrightarrow \underset{\underset{\underset{COOEt}{|}}{H-C-COCH_3}}{EtOOCC{=}CHCOOEt}$

$\underset{\underset{COOEt}{|}}{H_2C-COCH_3}$ I

(g) $\underset{\underset{\underset{COOEt}{|}}{H-C-COCH_3}}{EtOOCC{=}CHCOOEt} \xrightarrow{\text{strong OH}^-} \underset{\underset{CH_2COOH}{|}}{HOOCC{=}CHCOOH}$ *Cleavage to acids*

           I   J

27.11 (a)

$C_7H_{12}O_4$ malonic ester
$+ CH_2O$ formaldehyde

$C_8H_{14}O_5$

$C_8H_{14}O_5$
$- C_8H_{12}O_4$ product K

H_2O lost: suggests aldol condensation

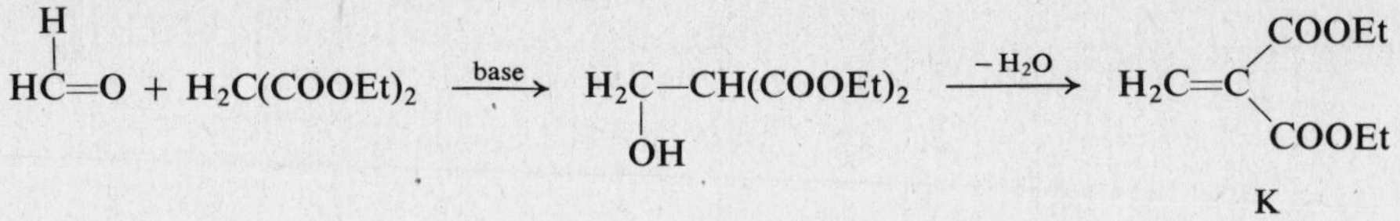

$HC{=}O + H_2C(COOEt)_2 \xrightarrow{\text{base}} H_2C{-}CH(COOEt)_2 \xrightarrow{-H_2O} H_2C{=}C\overset{\displaystyle COOEt}{\underset{\displaystyle COOEt}{}}$

K

(b)

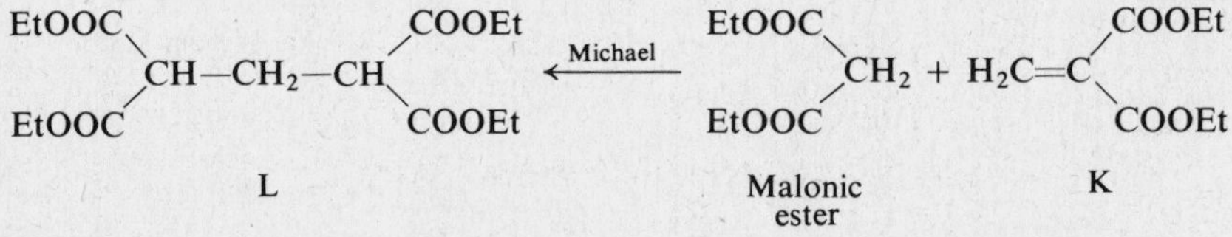

L $\xleftarrow{\text{Michael}}$ Malonic ester   K

(c)

L $\longrightarrow$ $\xrightarrow{-2CO_2}$ $HOOC(CH_2)_3COOH$

Glutaric acid

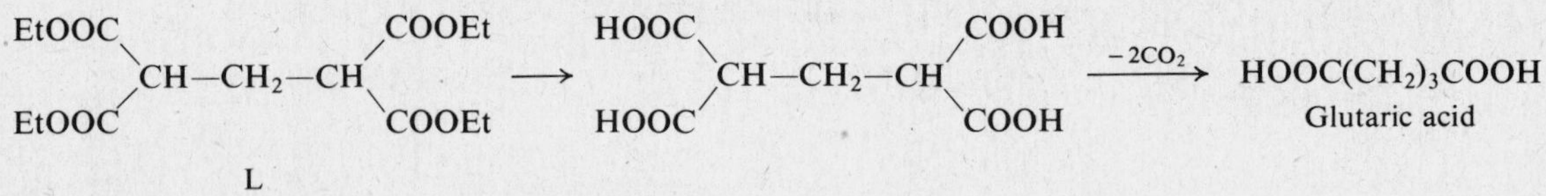

27.12

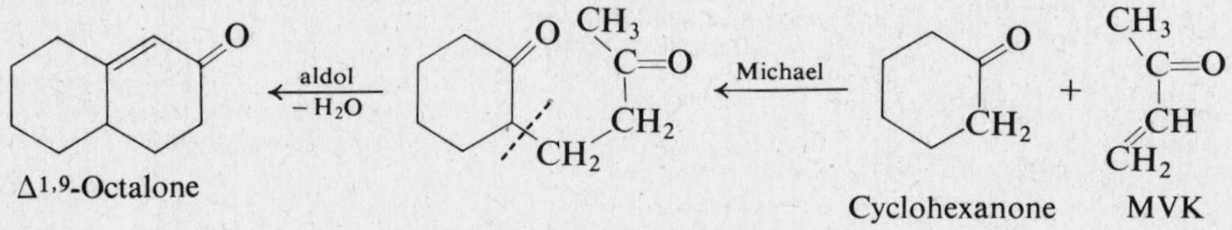

$\Delta^{1,9}$-Octalone $\xleftarrow[-H_2O]{\text{aldol}}$ $\xleftarrow{\text{Michael}}$ Cyclohexanone   +   MVK

27.13 (a)

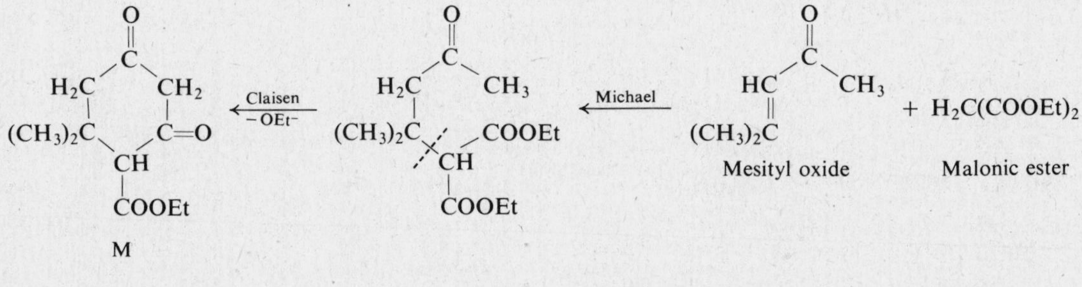

M $\xleftarrow[-OEt^-]{\text{Claisen}}$ $\xleftarrow{\text{Michael}}$ Mesityl oxide   + $H_2C(COOEt)_2$ Malonic ester

(b)

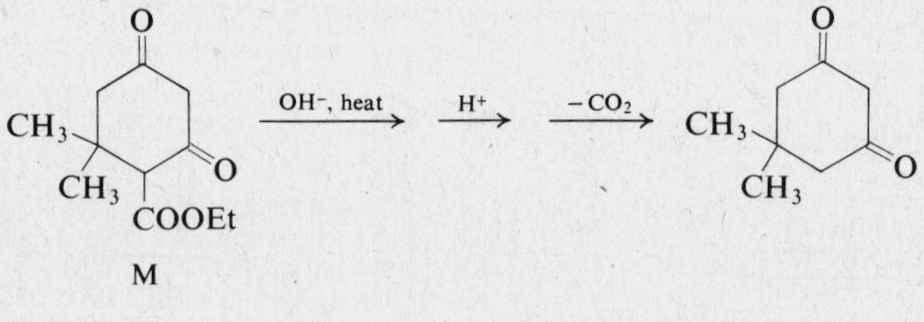

M $\xrightarrow{OH^-,\ heat}$ $\xrightarrow{H^+}$ $\xrightarrow{-CO_2}$

27.14 $PhCH{-}CH_2{-}C(...)Br$ $\xleftarrow{\text{Michael}}$ $PhCH{=}CH{-}C(...)Br$ +   1,3-Cyclopentadiene

Acidic (Sec. 10.10)

N

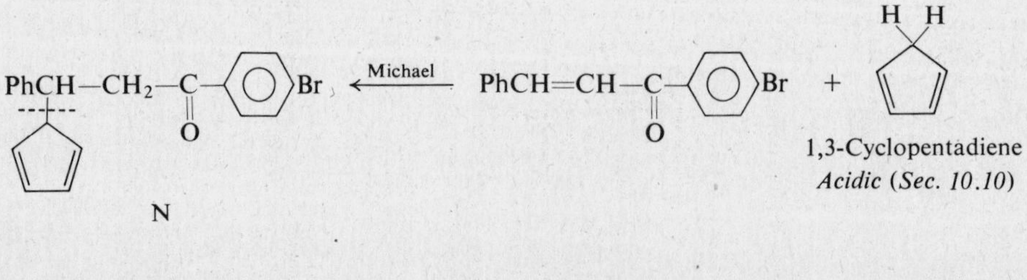

27.15

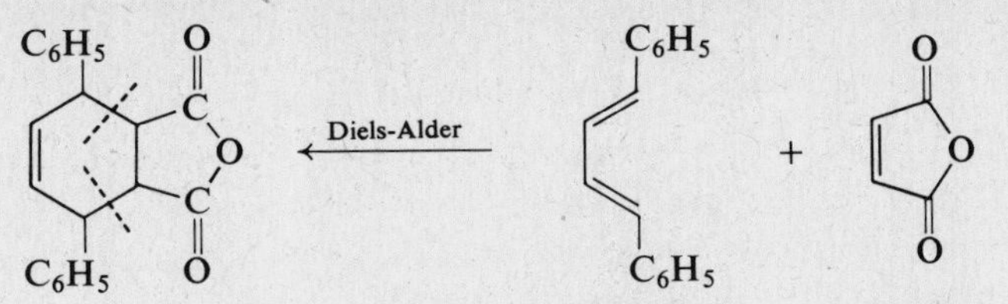

1,4-Diphenyl-
1,3-butadiene   Maleic
anhydride

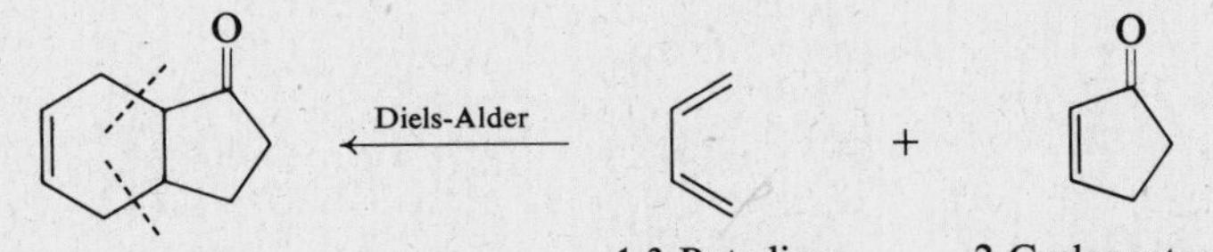

1,3-Butadiene   2-Cyclopentenone

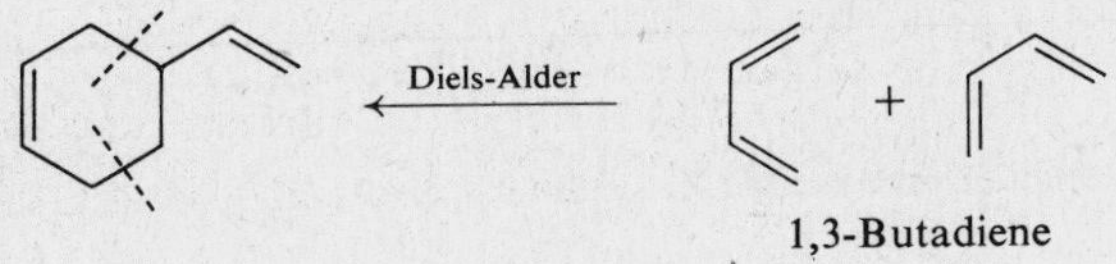

1,3-Butadiene
Two moles

27.16 (a)

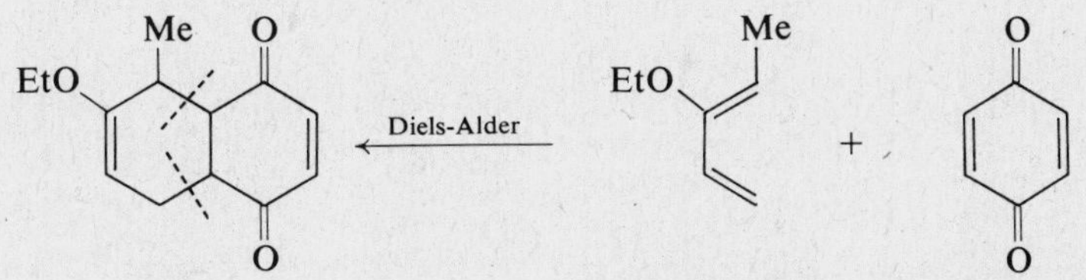

I   3-Ethoxy-
1,3-pentadiene   *p*-Benzoquinone

(b)

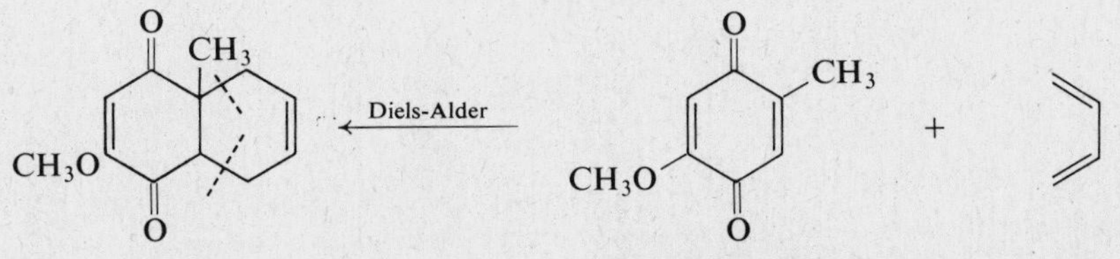

II   5-Methoxy-2-methyl-
1,4-benzoquinone   1,3-Butadiene

27.17

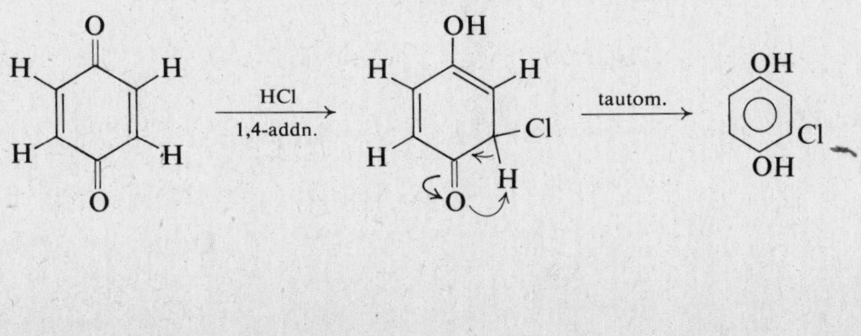

27.18 (a) Ease of oxidation.

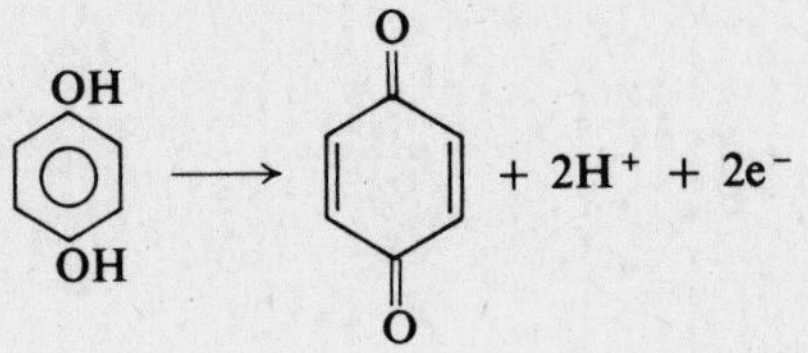

(b) Ease of reduction.

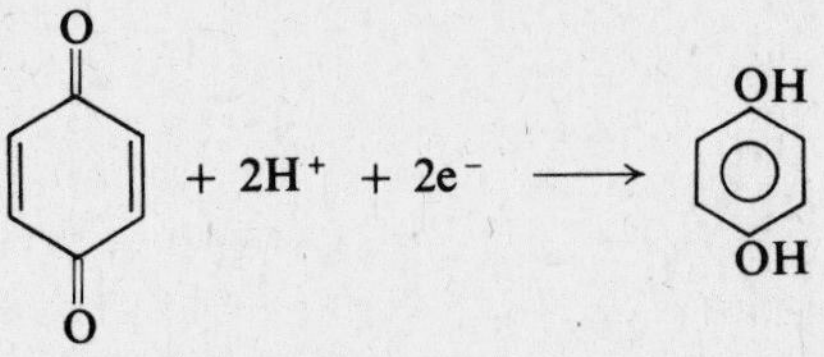

27.19

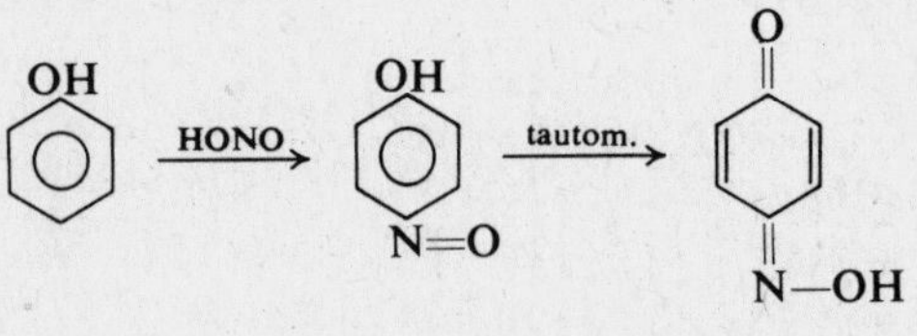

Phenol   *p*-Nitrosophenol   *p*-Benzoquinone
                                              monoxime

1.   Acrolein: $CH_2=CHCHO$ $\xleftarrow[-H_2O]{H^+}$ $\xleftarrow{OH^-}$ $HCHO + CH_3CHO$ (Or Problem 27.5, page 867.)

Crotonaldehyde: page 704.

Cinnamaldehyde: page 704.

Mesityl oxide: page 704.

Benzalacetone: page 704.

Dibenzalacetone: $PhCH=CHCCH=CHPh$ $\xleftarrow[-2H_2O]{aldol}$ $PhCHO + CH_3CCH_3 + OHCPh$
                                             $\overset{\|}{O}$ $\overset{\|}{O}$

Benzalacetophenone: page 704.

Dypnone: $PhC=CHCOPh$ $\xleftarrow[-H_2O]{aldol}$ $PhC=O + CH_3COPh$
                      $\underset{CH_3}{|}$ $\underset{CH_3}{|}$

Acrylic acid: as in Problem 27.4a.

Crotonic acid: *trans*-$CH_3CH=CHCOOH$ $\xleftarrow{Ag(NH_3)_2^+}$ *trans*-$CH_3CH=CHCHO$ (page 704)

Isocrotonic acid: *cis*-CH₃CH=CHCOOH $\xleftarrow{\text{H}_2,\text{ Pd}}$ CH₃C≡CCOOH $\xleftarrow{\text{CO}_2}$ CH₃C≡CMgBr $\xleftarrow{\text{MeMgBr}}$ CH₃C≡CH

Methacrylic acid: page 632.

Sorbic acid: CH₃CH=CHCH=CHCOOH $\xleftarrow{\text{Ag(NH}_3)_2^+}$ CH₃CH=CHCH=CHCHO

$$\text{OH}^- \uparrow -\text{H}_2\text{O}$$

CH₃CHO + CH₃CH=CHCHO (page 704)

Cinnamic acid: Problem 21.22e (Perkin condensation), page 714.

Maleic acid: *cis*-HOOCCH=CHCOOH $\xleftarrow{\text{H}_2\text{O}}$

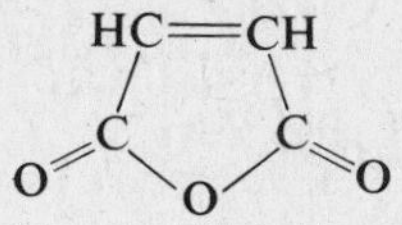

$\xleftarrow{250–300°}$ *cis*- and *trans*-HOOCCH=CHCOOH

(Problem 20.3, page 668)

$\uparrow$ KOH(alc)

HOOCCH₂CH₂COOH $\xrightarrow[\text{P}]{\text{Br}_2}$ HOOCCH₂CHCOOH | Br

(Problem 5a, page 610)

(Actually made via maleic anhydride by catalytic oxidation of benzene.)

Fumaric acid: *trans*-HOOCCH=CHCOOH $\xleftarrow[\text{light}]{\text{trace HBr}}$ *cis*-HOOCCH=CHCOOH
(See Problem 13, page 402)   Maleic acid
(above)

Maleic anhydride: 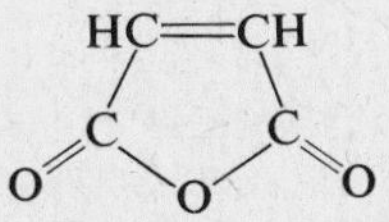 $\xleftarrow{\text{heat}}$ maleic or fumaric acid (see above)

Methyl acrylate: Problem 27.4b, page 867.

Methyl methacrylate: Problem 27.4c, page 867.

Ethyl cinnamate: PhCH=CHCOOEt $\xleftarrow{\text{OEt}^-}$ PhCHO + CH₃COOEt

Acrylonitrile: Problem 27.4a, page 867.

2. (a) PhCH₂CH₂CCH₃ + PhCH₂CH₂CHCH₃
     O   OH

(b) PhCH=CHCHCH₃
         OH

(c) PhCH=CHCOONa + CHI₃

(d) PhCHO + CH₃CCHO
         O

(e) PhCH—CH—CCH₃
     Br   Br   O

(f) PhCHCH₂CCH₃
   Cl   O

(g) PhCHCH₂CCH₃
   Br   O

(h) PhCHCH₂CCH₃
   OH   O

(i) PhCHCH₂CCH₃
  OCH₃   O

(j) PhCHCH₂CCH₃
   CN   O

(k) PhCHCH₂CCH₃
  CH₃—NH   O

(l) 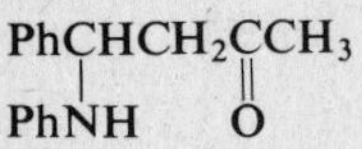 PhCHCH₂CCH₃
　　|　　　||
　　PhNH　　O

(m) 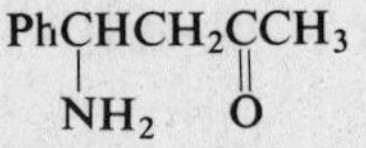 PhCHCH₂CCH₃
　　　|　　　||
　　　NH₂　　O

(n) 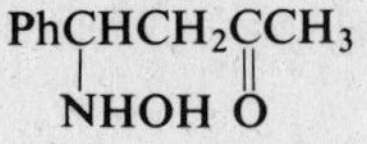 PhCHCH₂CCH₃
　　　|　　　||
　　　NHOH O

(o) PhCH=CHCCH=CHPh
　　　　　　||
　　　　　　O

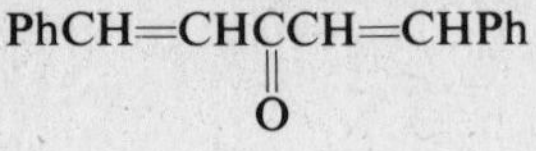

(p) 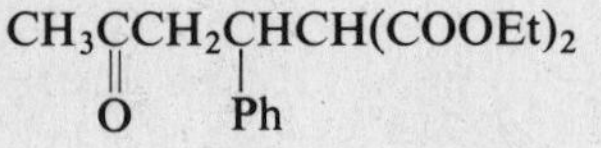 CH₃CCH₂CHCH(COOEt)₂
　　　||　　|
　　　O　　Ph

(q) 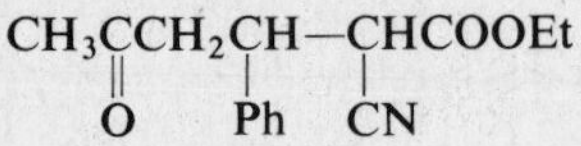 CH₃CCH₂CH—CHCOOEt
　　　||　　|　　|
　　　O　　Ph　CN

(r) 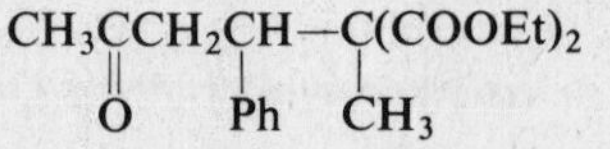 CH₃CCH₂CH—C(COOEt)₂
　　　||　　|　　|
　　　O　　Ph　CH₃

(s) 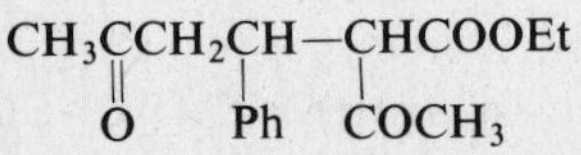 CH₃CCH₂CH—CHCOOEt
　　　||　　|　　|
　　　O　　Ph　COCH₃

(t) 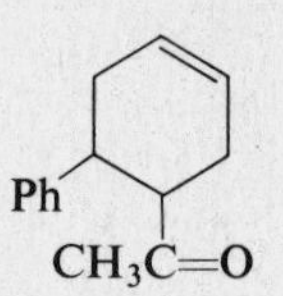
Ph
　CH₃C=O

(u) 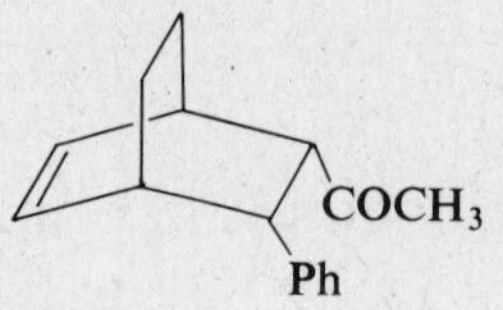
　　　　　COCH₃
　　　　　Ph

(v) 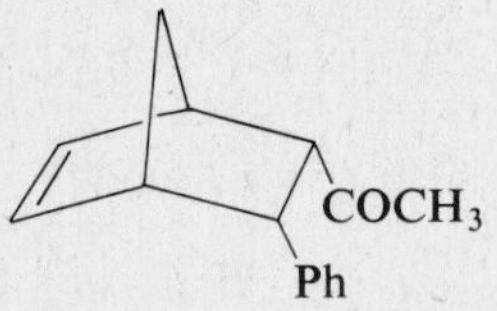
　　　COCH₃
　　Ph

3. The C—C bond formed in each Michael addition is indicated by the broken line. The
—COO— lost in subsequent decarboxylations (see Problem 4, below) is indicated by bold-
face type in parts (a)–(i).

(a) 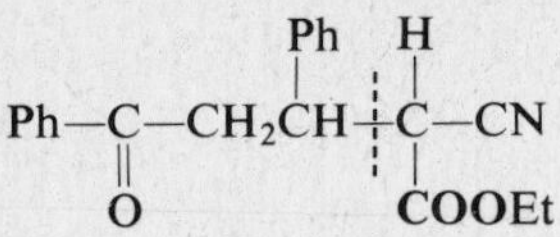
　　　　　　　Ph　H
　　　　　　　|　|
Ph—C—CH₂CH⫶C—CN
　||　　　　　|
　O　　　　COOEt

(b)
　　　　　　　Ph　H
　　　　　　　|　|
EtOOCCH₂CH⫶C—CN
　　　　　　　|
　　　　　　COOEt

(c)
　　　　　COOEt
　　　　　|
EtOOCCH₂CH⫶CH—COOEt
　　　　　|
　　　　COOEt

(d)
　　　　　COOEt
　　　　　|
EtOOCCH=C⫶CH—COOEt
　　　　　|
　　　　COOEt

(e)
　　　　H₃C　H
　　　　|　|
CH₃—C—CH₂C⫶C—COOEt
　||　　　|　|
　O　　H₃C COOEt

(f)
　　　　H₃C　　　O
　　　　|　　　　||
CH₃—C—CH₂C⫶CH—C—CH₃
　||　　　|
　O　　H₃C COOEt

(g)
　　　　H₃C　CH₃
　　　　|　|
EtOOCCH₂CH⫶C—COOEt
　　　　　　|
　　　　　COOEt

(h)
EtOOC—CH⫶CH₂⫶CH—COOEt
　　　　|　　　　　|
　　EtOOC　　　　COOEt

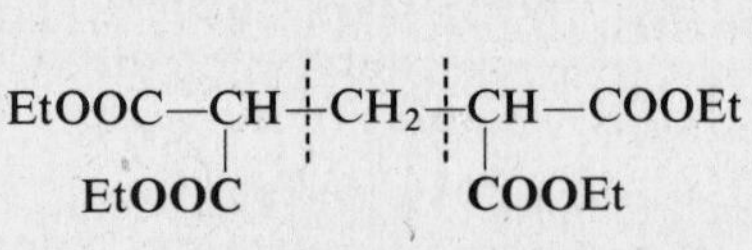

(i) 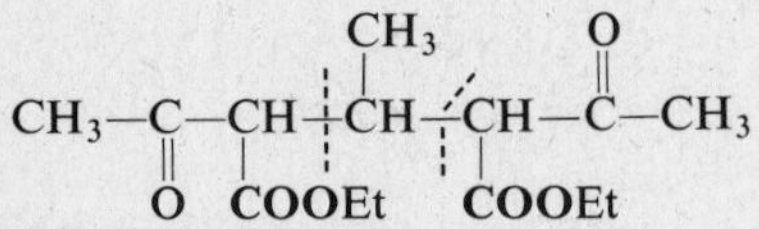
　　　　　　　CH₃　　O
　　　　　　　|　　　||
CH₃—C—CH⫶CH⫶CH—C—CH₃
　||　　|　　|
　O　COOEt COOEt

(j) 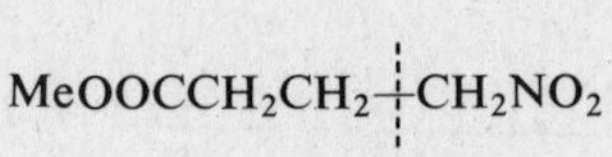 MeOOCCH₂CH₂⫶CH₂NO₂

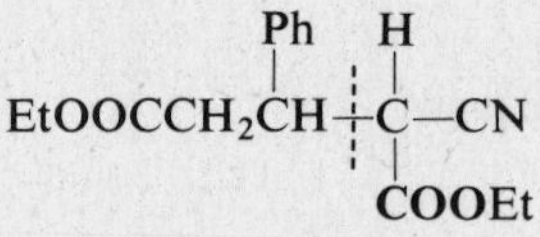

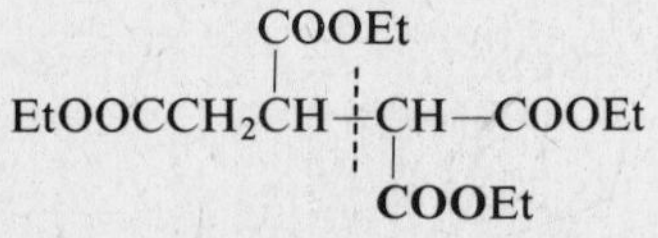

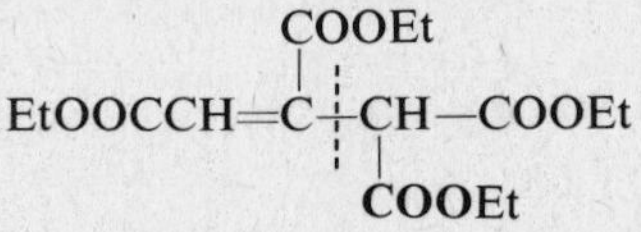

(k) 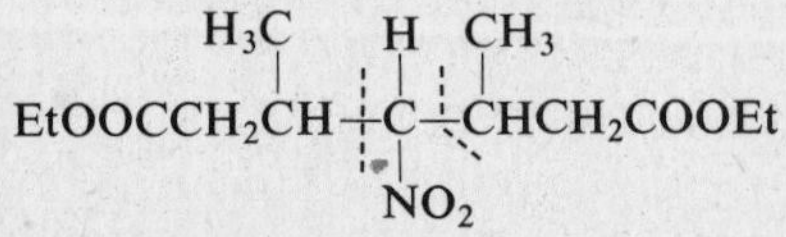
　　　　H₃C　H CH₃
　　　　|　| |
EtOOCCH₂CH⫶C⫶CHCH₂COOEt
　　　　　　|
　　　　　NO₂

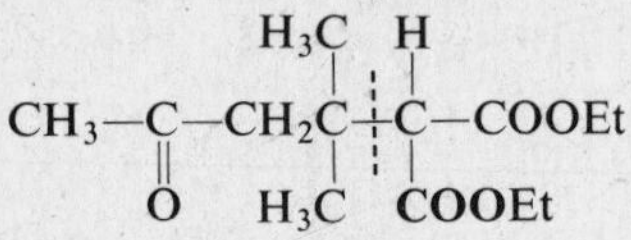

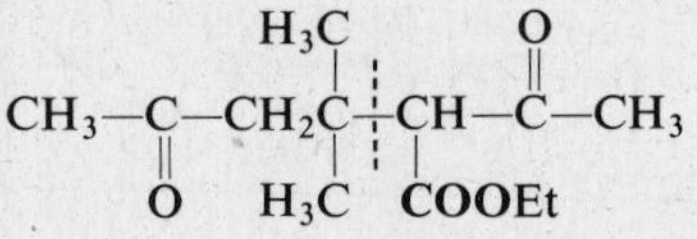

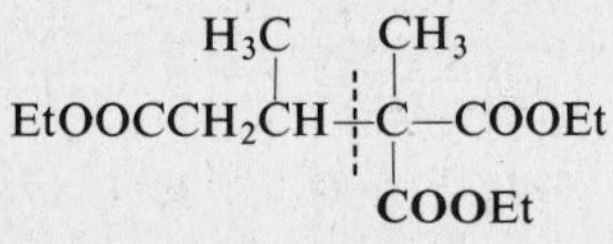

(l) 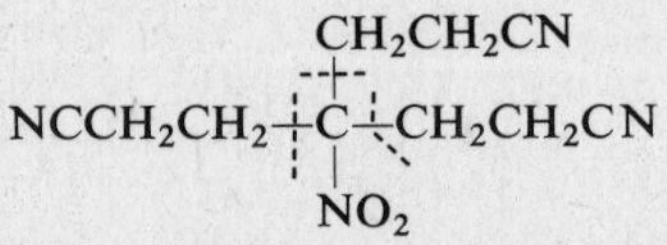
　　　　　　CH₂CH₂CN
　　　　　　⫶
NCCH₂CH₂⫶C⫶CH₂CH₂CN
　　　　　　|
　　　　　NO₂

(m) NCCH₂CH₂—CCl₃

4. (a) 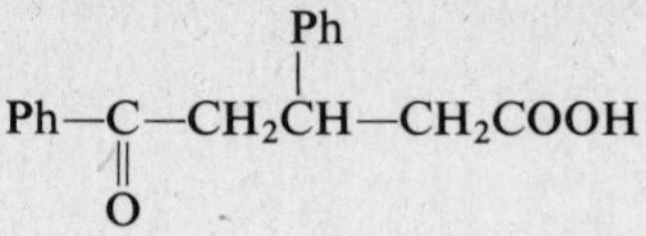 $Ph-\overset{\underset{\|}{O}}{C}-CH_2\overset{\underset{|}{Ph}}{CH}-CH_2COOH$

(b) $HOOCCH_2\overset{\underset{|}{Ph}}{CH}-CH_2COOH$

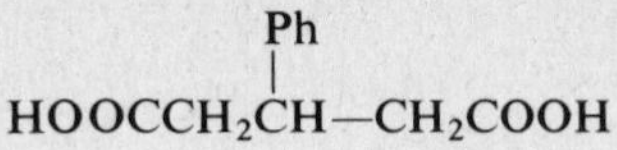

(c) $HOOCCH_2\overset{\underset{|}{COOH}}{CH}-CH_2COOH$

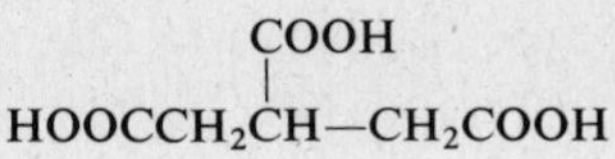

(d) 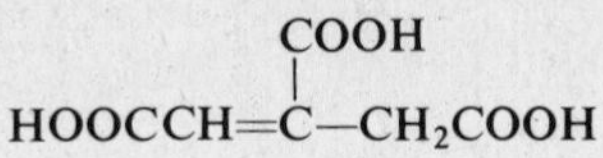 $HOOCCH=\overset{\underset{|}{COOH}}{C}-CH_2COOH$

(e) 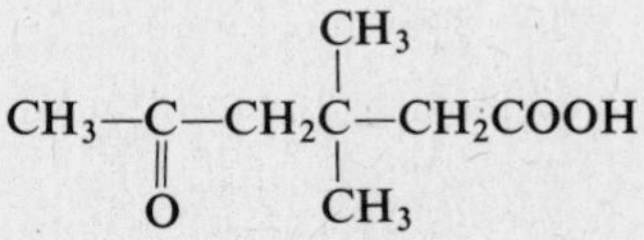 $CH_3-\overset{\underset{\|}{O}}{C}-CH_2\overset{\overset{CH_3}{|}}{\underset{\underset{|}{CH_3}}{C}}-CH_2COOH$

(f) 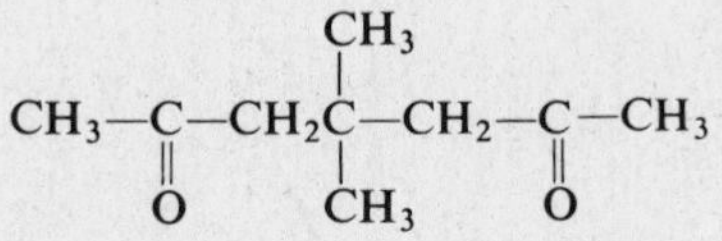 $CH_3-\overset{\underset{\|}{O}}{C}-CH_2\overset{\overset{CH_3}{|}}{\underset{\underset{|}{CH_3}}{C}}-CH_2-\overset{\underset{\|}{O}}{C}-CH_3$

(g) 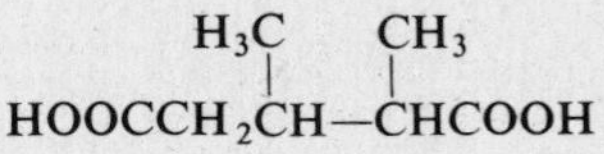 $HOOCCH_2\overset{\underset{|}{H_3C}}{CH}-\overset{\underset{|}{CH_3}}{CH}COOH$

(h) $HOOCCH_2-CH_2-CH_2COOH$

(i) 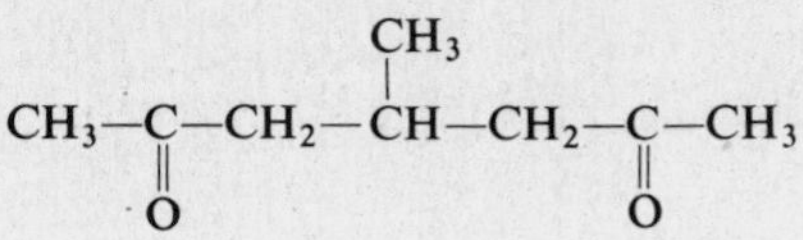 $CH_3-\overset{\underset{\|}{O}}{C}-CH_2-\overset{\overset{CH_3}{|}}{CH}-CH_2-\overset{\underset{\|}{O}}{C}-CH_3$

5.

$(EtOOC)_2CH_2 + PhCH=CHCCH=CHPh + H_2C(COOEt)_2 \xrightarrow[\text{Michael}]{\text{double}}$

where the central ketone is shown with $\overset{\|}{O}$

Product **A**: $EtOOC-\overset{\overset{H}{|}}{\underset{\underset{|}{EtOOC}}{C}}-\overset{\overset{Ph}{|}}{\underset{\underset{|}{H}}{C}}-CH_2\overset{\|}{C}CH_2-\overset{\overset{Ph}{|}}{\underset{\underset{|}{H}}{C}}-\overset{\overset{H}{|}}{\underset{\underset{|}{COOEt}}{C}}-COOEt$

$PhCH=CHCCH=CHPh + H_2C(COOEt)_2 \xrightarrow{\text{Michael}} PhCH=CHC-CH_2-\overset{\overset{Ph}{|}}{\underset{\underset{|}{H}}{C}}-\overset{\overset{H}{|}}{\underset{\underset{|}{COOEt}}{C}}-COOEt$

B

B $\xrightarrow{\text{internal Michael}}$ **C**

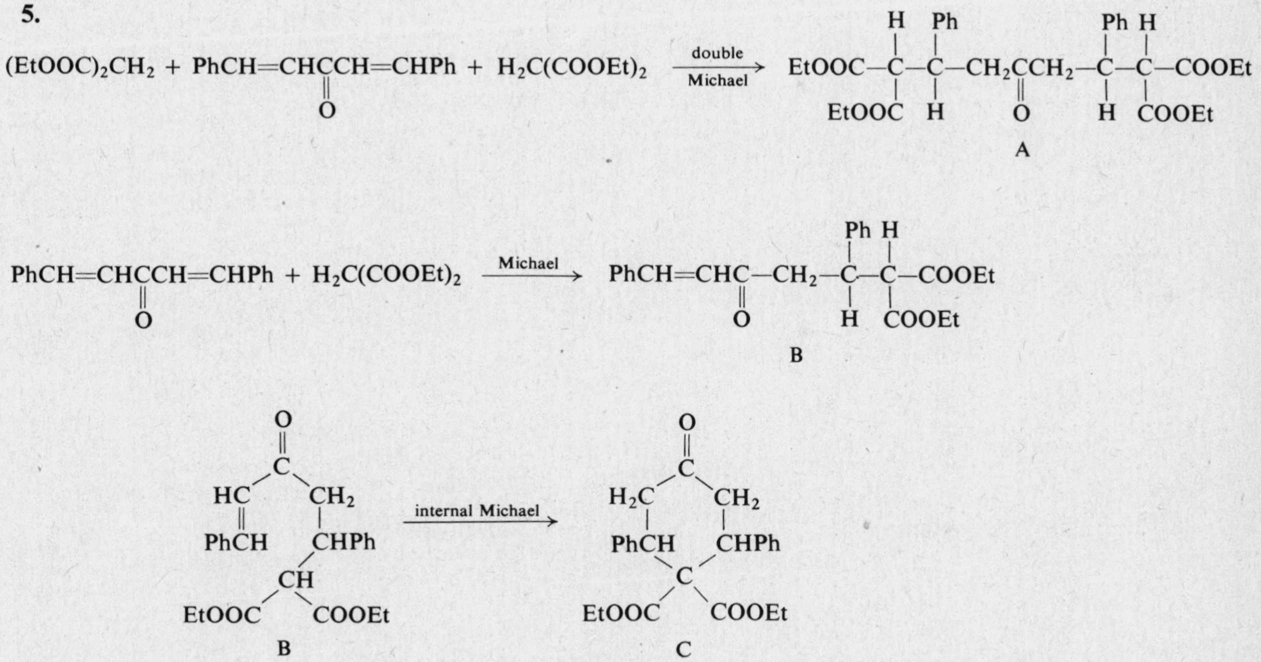

6. (a) 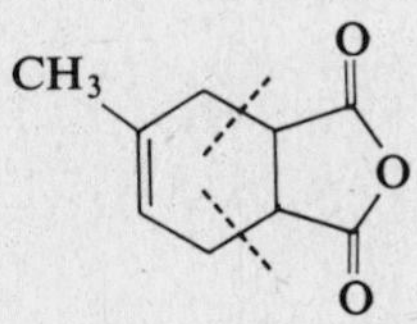 (b) 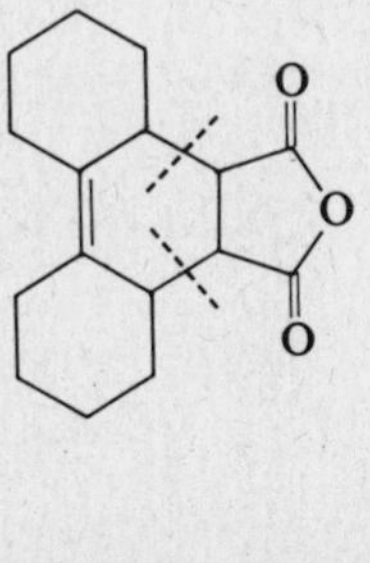 (c) 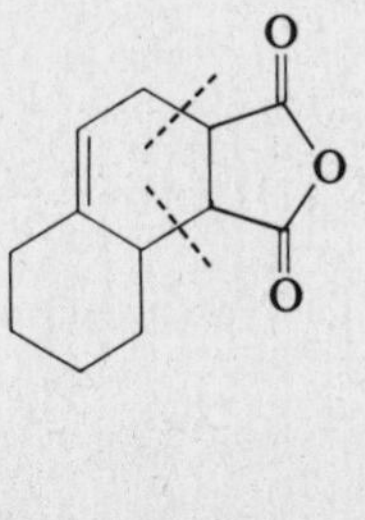 (d)

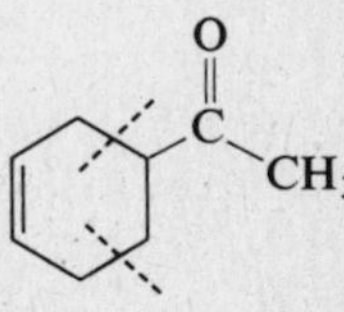

(e) 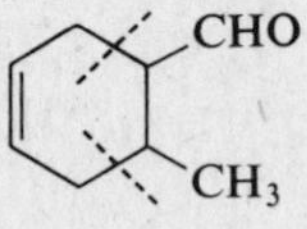　　　　(f) 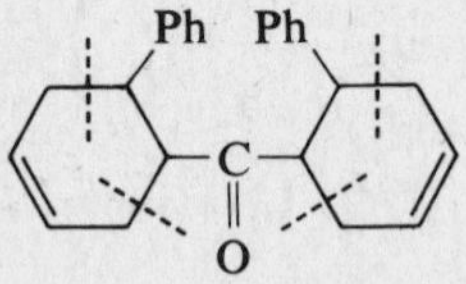　　　　(g) 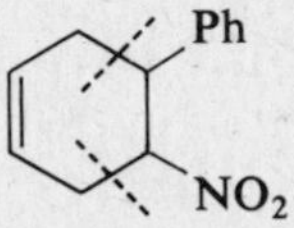　　　　(h)

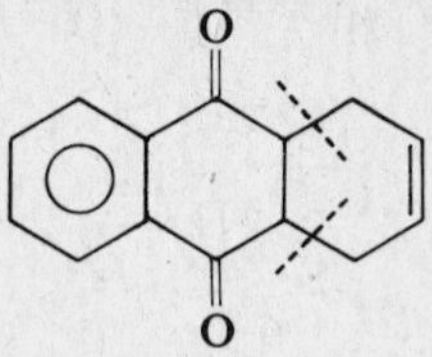

(i) 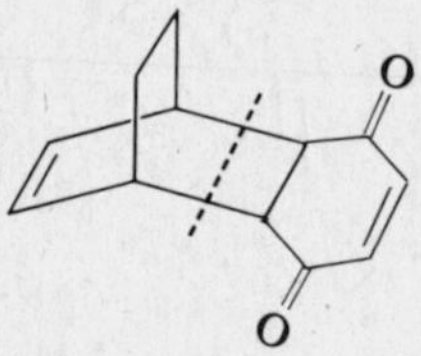　　　　(j) 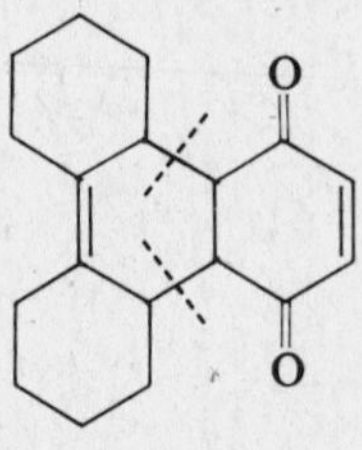　　　　(k)

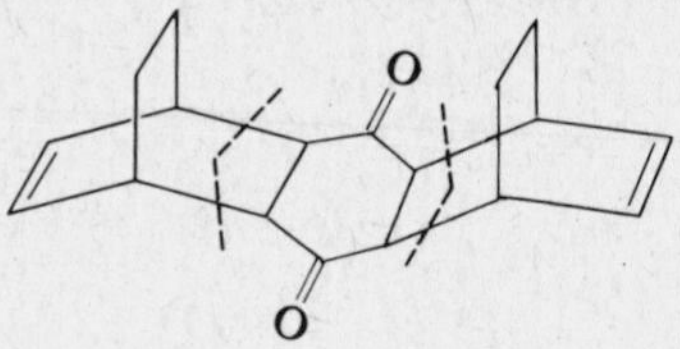

(l) 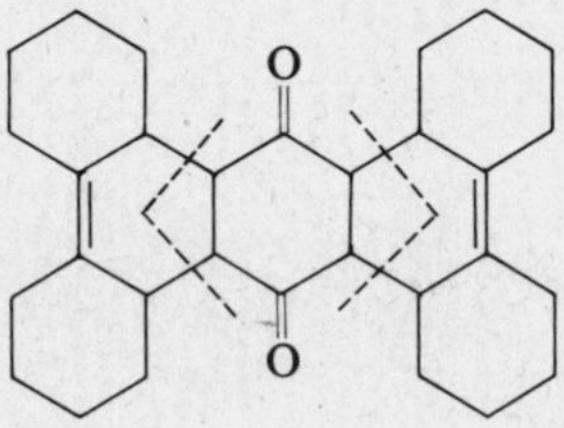　　　　(m) 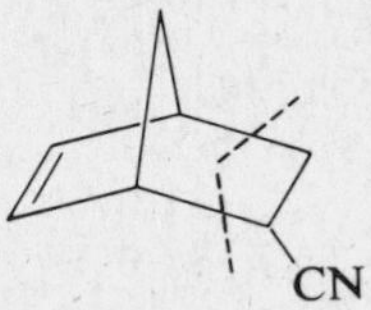　　　　(n)

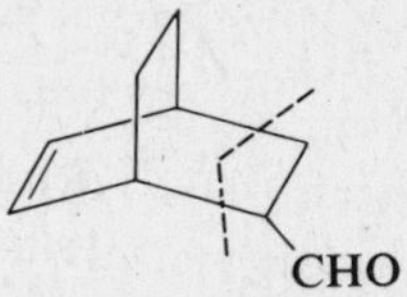

7. (a)

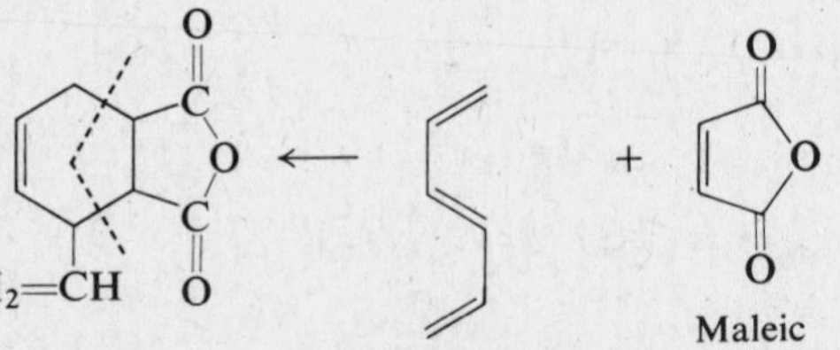

1,3,5-Hexatriene　　　Maleic anhydride

(b)

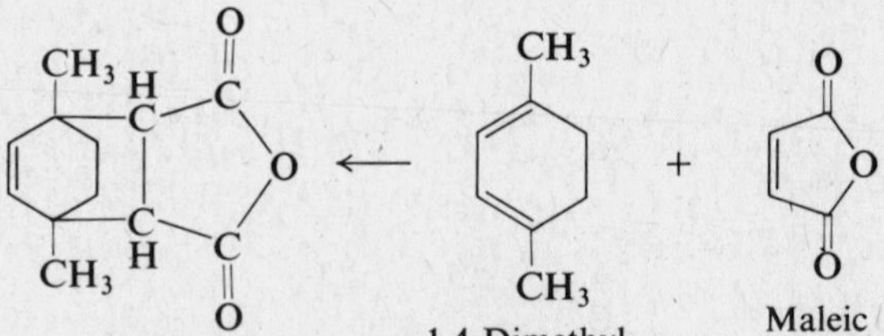

1,4-Dimethyl-
1,3-cyclohexadiene　　　Maleic anhydride

(c)

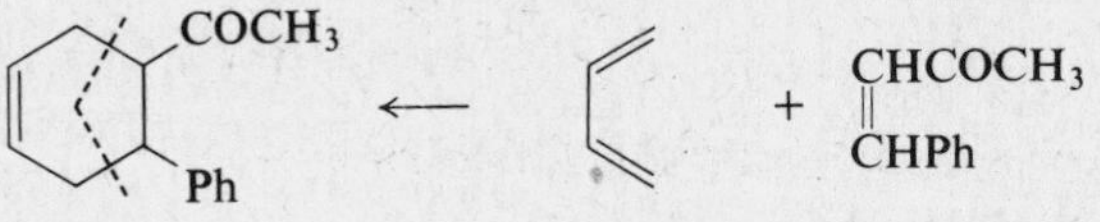

1,3-Butadiene　　　Benzalacetone

(d)

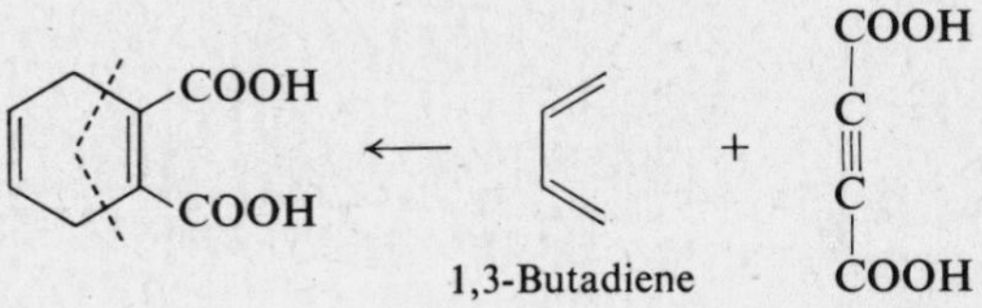

1,3-Butadiene　　　Acetylene-
dicarboxylic acid

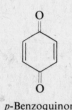

(e)

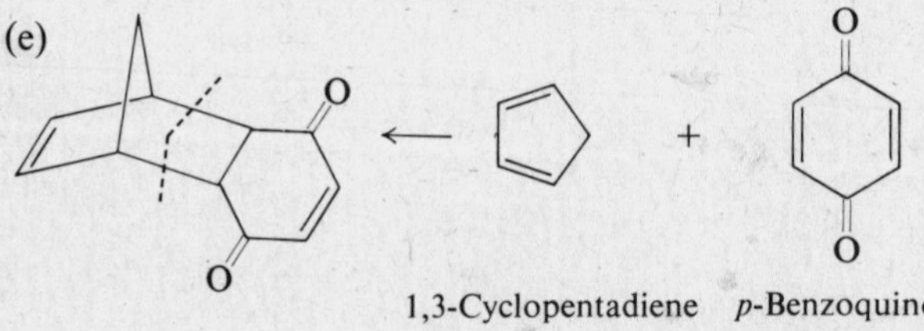

1,3-Cyclopentadiene　　p-Benzoquinone

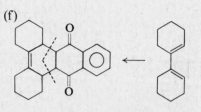

(f)

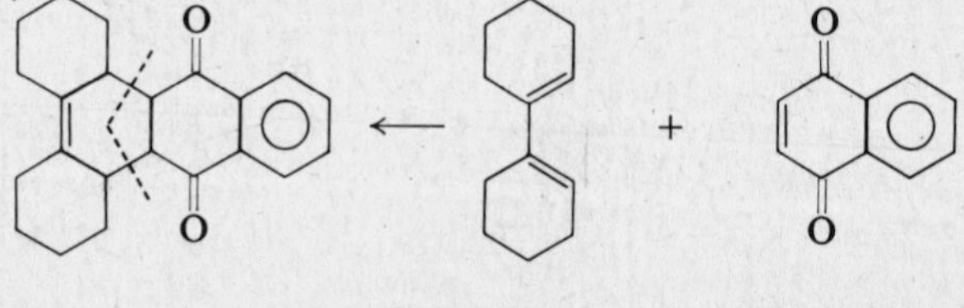

1,1'-Bicyclohexenyl
(I, page 880)　　　1,4-Naphthoquinone
(II, page 880)

(g)

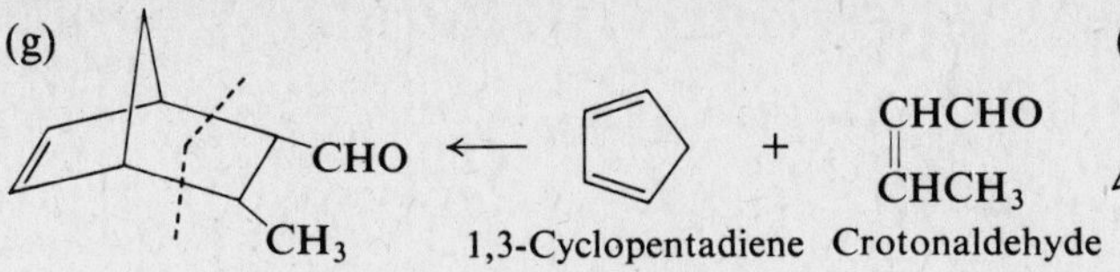

1,3-Cyclopentadiene   Crotonaldehyde

(h)

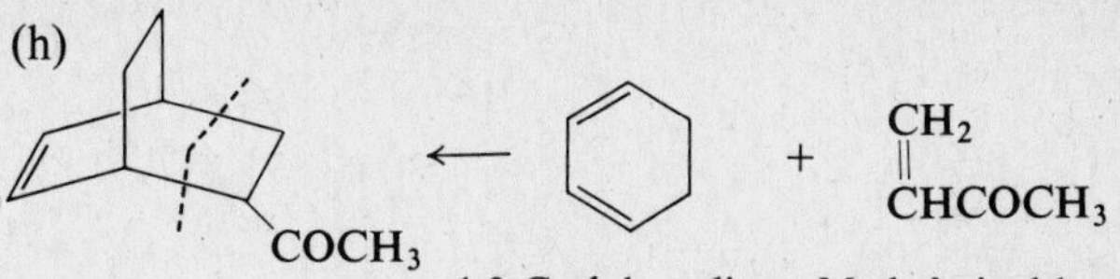

1,3-Cyclohexadiene   Methyl vinyl ketone
(MVK)

(i)

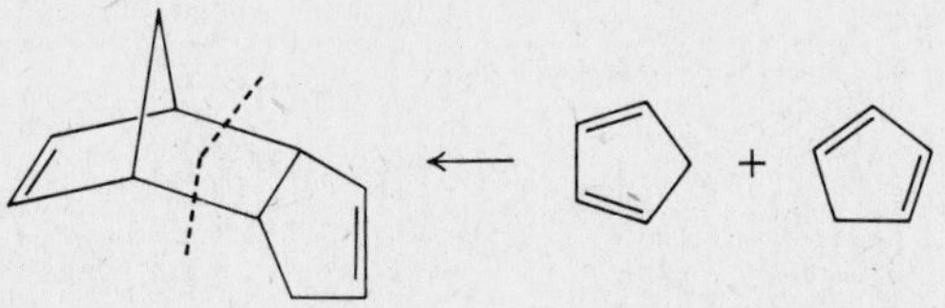

1,3-Cyclopentadiene
Two moles

8. If we work through this problem, neglecting for the moment stereoisomerism, we arrive at:

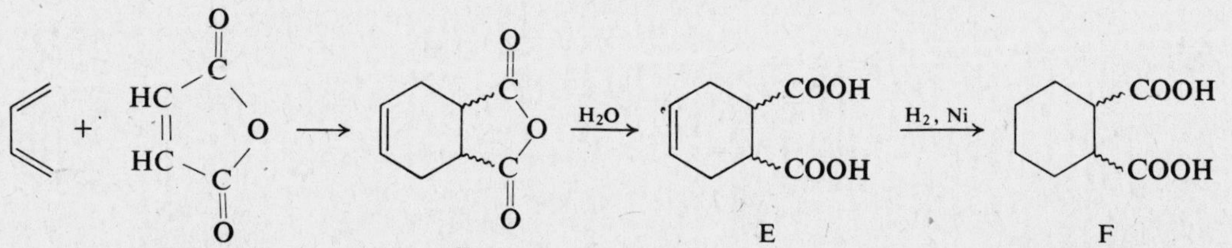

Maleic
anhydride

D

E

F
1,2-Cyclohexane-
dicarboxylic acid

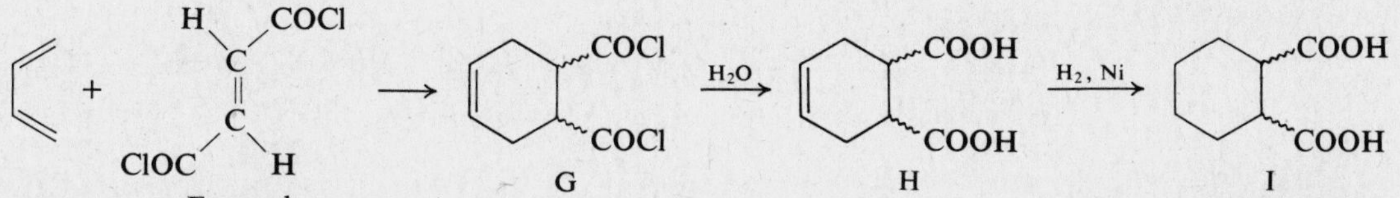

Fumaryl
chloride

G

H

I
1,2-Cyclohexane-
dicarboxylic acid

The final products G and I must be, then, the stereoisomeric 1,2-cyclohexanedicarboxylic acids:

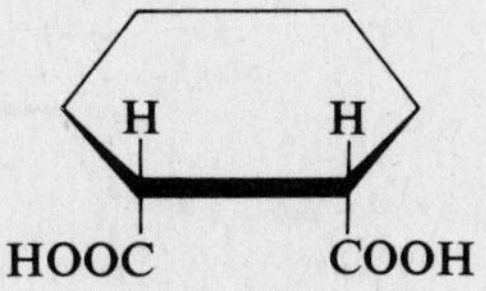

F
Meso: non-resolvable
cis-1,2-Cyclohexane-
dicarboxylic acid

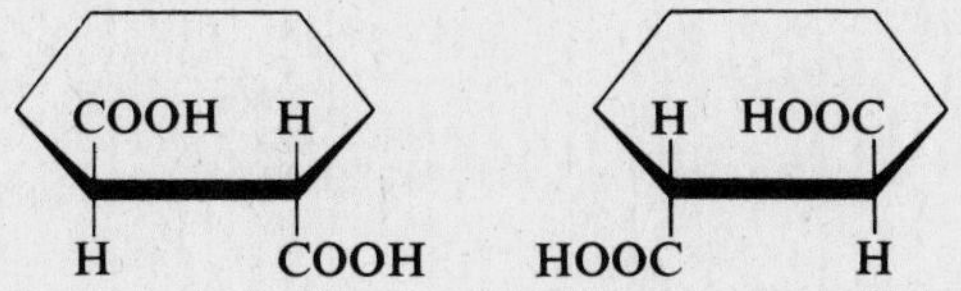

I
Racemic: resolvable
trans-1,2-Cyclohexane-
dicarboxylic acid

Working backwards from these, we see that **the Diels–Alder reaction involves *syn*-addition.**

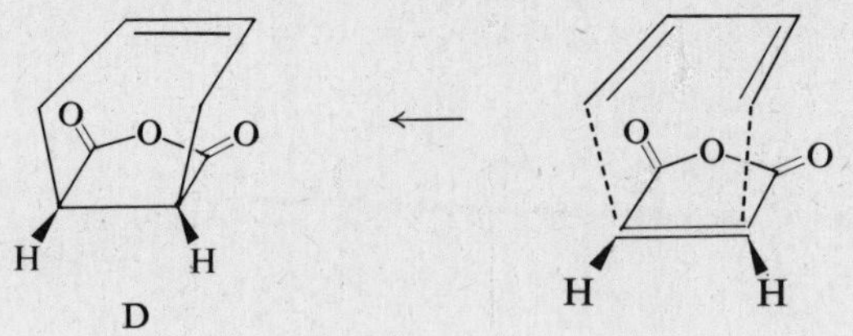

Syn-**addition**

D
cis-product

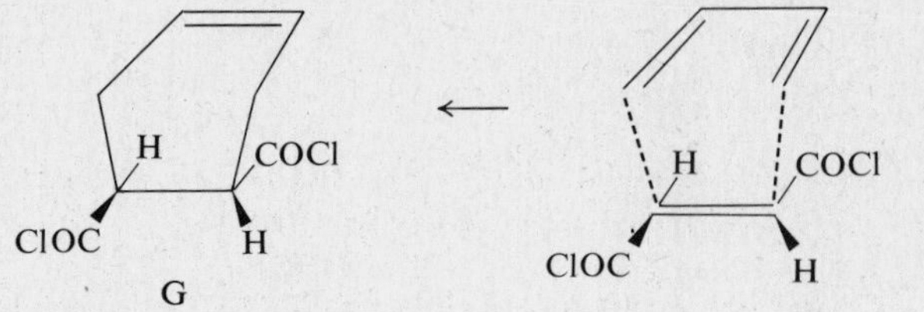

Syn-**addition**

G
trans-product

9. (a)

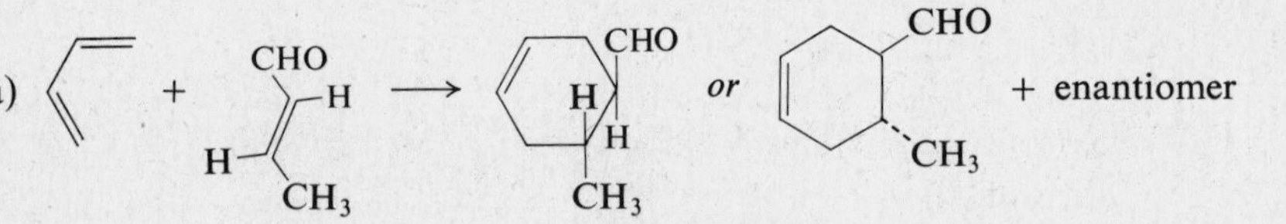

(b)

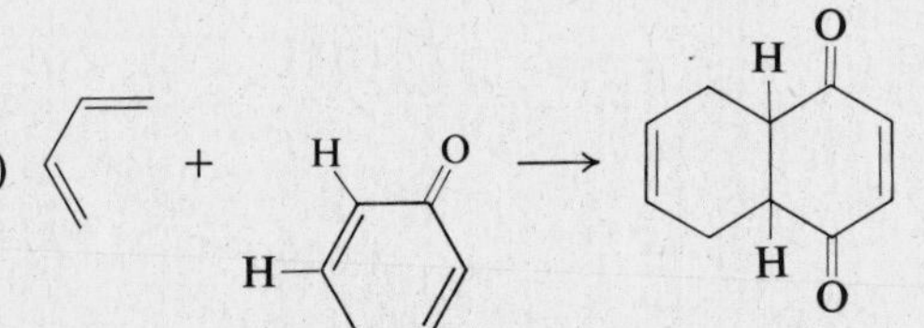

cis (*meso*)

(c)

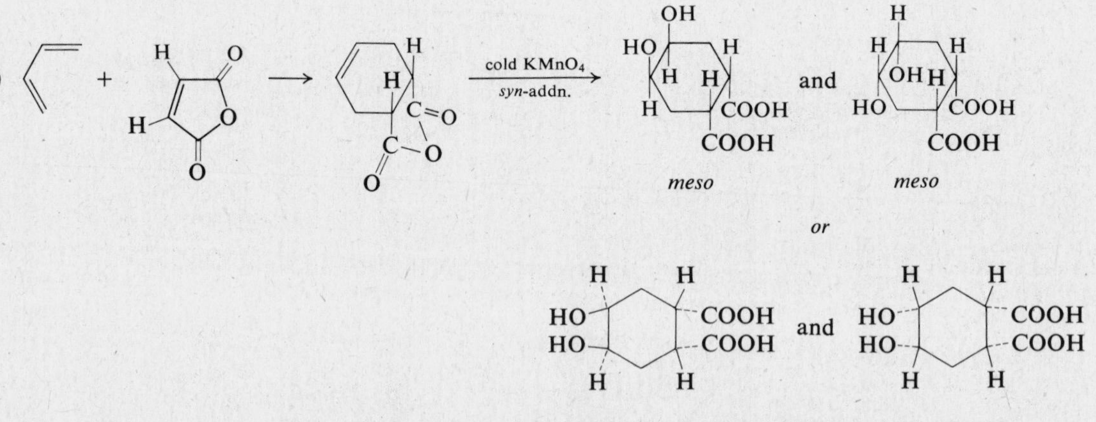

cold KMnO₄
syn-addn.

and

meso   *meso*

or

HO— —COOH   and   HO— —COOH
HO— —COOH   HO— —COOH

(d)

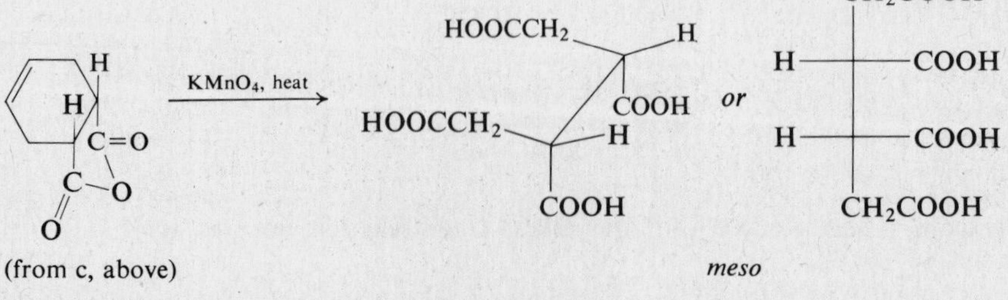

KMnO₄, heat

or

meso

(from c, above)

10. (a) There is a familiar 1,2-shift of methyl to give a more stable carbonium ion, followed by elimination of a proton.

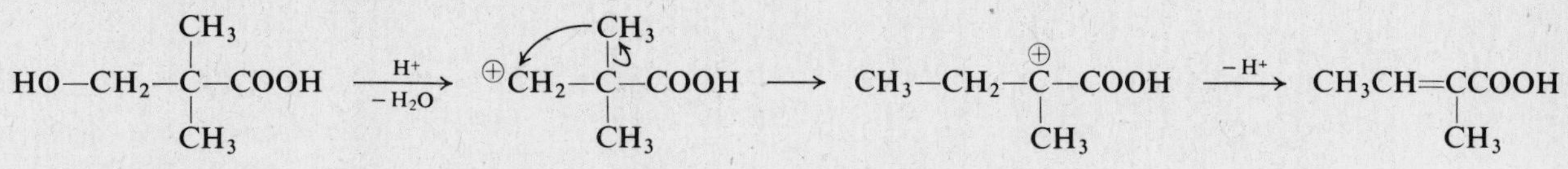

$$HO-CH_2-\underset{\underset{CH_3}{|}}{\overset{\overset{CH_3}{|}}{C}}-COOH \xrightarrow[-H_2O]{H^+} \overset{\oplus}{C}H_2-\underset{\underset{CH_3}{|}}{\overset{\overset{CH_3}{|}}{C}}-COOH \longrightarrow CH_3-CH_2-\underset{\underset{CH_3}{|}}{\overset{\oplus}{C}}-COOH \xrightarrow{-H^+} CH_3CH=\underset{\underset{CH_3}{|}}{C}COOH$$

(b) As we saw in Problem 21.20 (page 713), the γ-hydrogens of α,β-unsaturated carbonyl compounds are acidic; the carbanion formed by abstraction of such a proton is involved in a Claisen condensation.

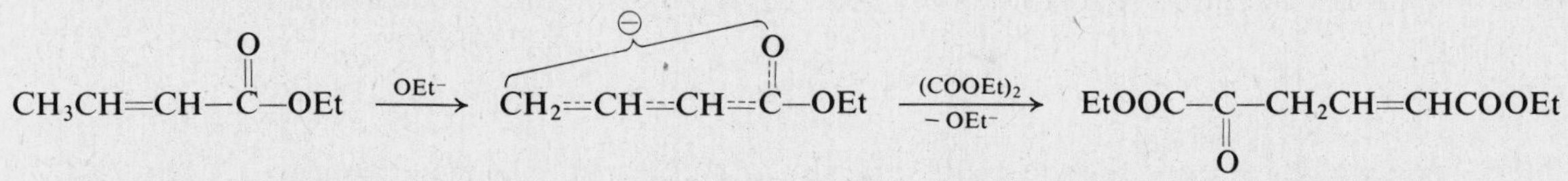

$$CH_3CH=CH-\overset{\overset{O}{||}}{C}-OEt \xrightarrow{OEt^-} CH_2=\!\!=\!\!CH=\!\!=\!\!CH=\!\!=\!\!\overset{\overset{O}{||}}{C}-OEt \xrightarrow[-OEt^-]{(COOEt)_2} EtOOC-\underset{\underset{O}{||}}{C}-CH_2CH=CHCOOEt$$

(c) The electron-withdrawing phosphonium group activates the double bond toward nucleophilic attack by the phenoxide ion; the addition product XII then undergoes an intramolecular Wittig reaction.

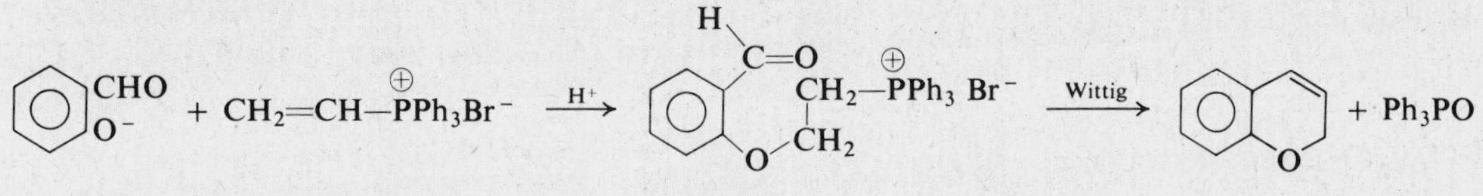

[reaction scheme leading to XII]

XII

(d) Carbon of the ylid has nucleophilic character and, in the first step of conjugate nucleophilic addition, attacks the α,β-unsaturated ester at C-3, to give anion XIII. PPh₃ is then displaced in an intramolecular nucleophilic substitution reaction.

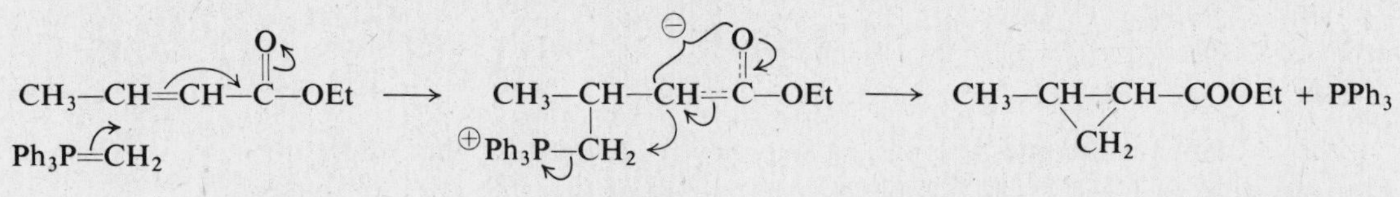

[reaction scheme]

XIII

(e) Benzyne is formed, and undergoes a Diels-Alder reaction with furan.

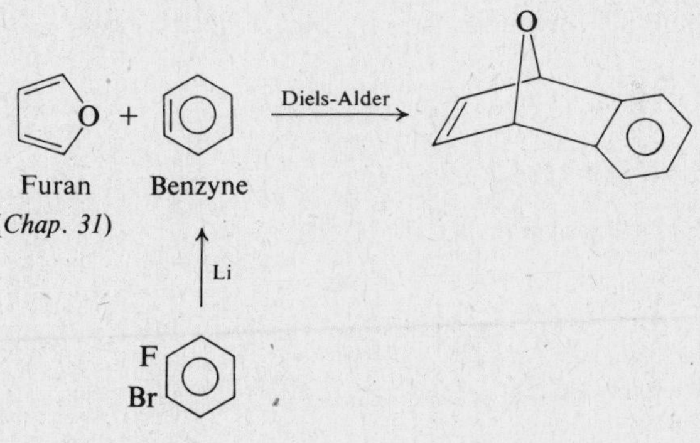

Furan Benzyne
(*Chap. 31*)

11. Nucleophilic addition of H₂O to an α,β-unsaturated aldehyde generates the aldol, XIV.

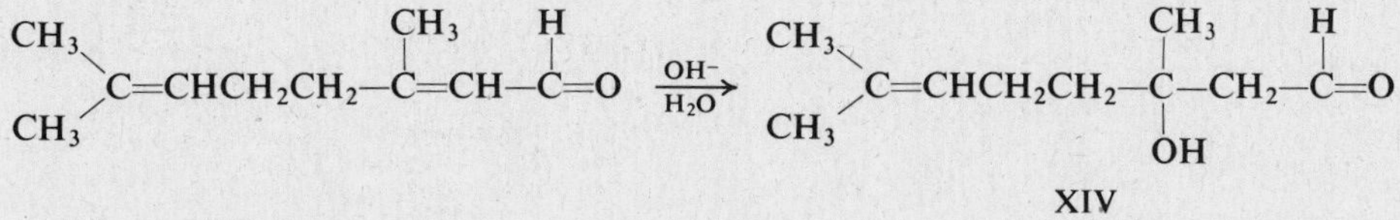

[reaction scheme]

XIV

This undergoes the base-catalyzed *retro*-aldol ("reverse aldol") reaction by reversal of steps (1)–(3) on pages 709–710.

XIV $\xrightarrow{OH^-}$ (CH₃)₂C=CHCH₂CH₂—C—CH₂—C=O ⟶ (CH₃)₂C=CHCH₂CH₂—C + ⁻CH₂CHO

$\downarrow$ H₂O

CH₃CHO

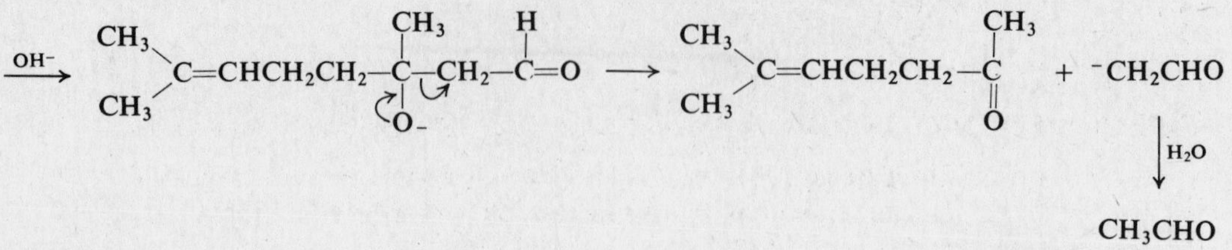

The equilibrium is displaced toward fragmentation by removal of one product (acetaldehyde) by distillation.

12. Like other nucleophilic reagents, the Grignard reagent can undergo conjugate addition to α,β-unsaturated carbonyl compounds.

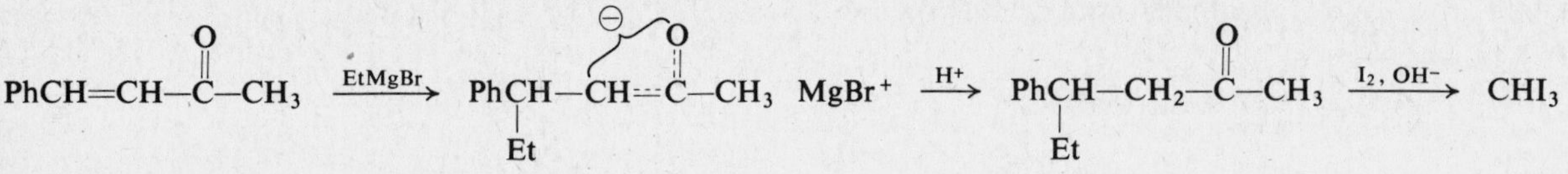

PhCH=CH—C—CH₃ $\xrightarrow{EtMgBr}$ PhCH—CH=C—CH₃ MgBr⁺ $\xrightarrow{H^+}$ PhCH—CH₂—C—CH₃ $\xrightarrow{I_2, OH^-}$ CHI₃

with Et groups

13.

F₃C,Ph C=C F,F + OEt⁻ ⟶ F₃C—C⁻—C(OEt)—F ⟶ (F₃C)(Ph)C=C(OEt)(F) + F⁻

ClF₂C,Ph C=C F,F + OEt⁻ ⟶ Cl—F₂C—C⁻—C(OEt)—F ⟶ (F₂C)(Ph)C—C(OEt)(F) + Cl⁻

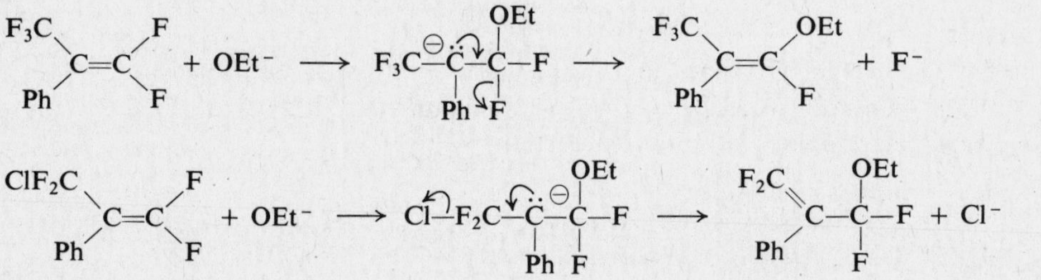

Formation of the carbanion is the rate-determining step, nearly identical for the two similar alkenes. Then the weakest C—X bond is broken, but the rate of this step does not affect the overall rate; that is, there is no "element effect" (see pages 478 and 834).

(H. F. Koch and A. J. Kielbanian, Jr., "Nucleophilic Reactions of Fluoroolefins. Evidence for a Carbanion Intermediate in Vinyl and Allyl Displacement Reactions," J. Am. Chem. Soc., **92**, 729 (1970).)

14. (a) 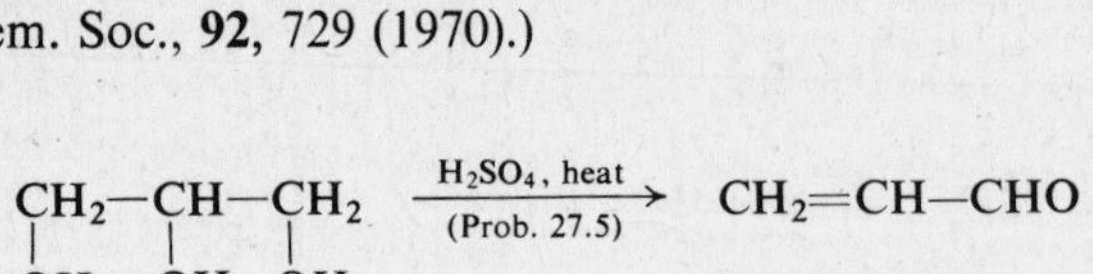 CH₂—CH—CH₂ (OH,OH,OH) $\xrightarrow[\text{(Prob. 27.5)}]{H_2SO_4, heat}$ CH₂=CH—CHO J, acrolein

CH₂=CH—CHO $\xrightarrow{HCl}$ [CH₂—CH₂—CHO, Cl] $\xrightarrow[H^+]{EtOH}$ CH₂—CH₂—C(H)(OEt)₂, Cl *Aldehyde protected by acetal formation*

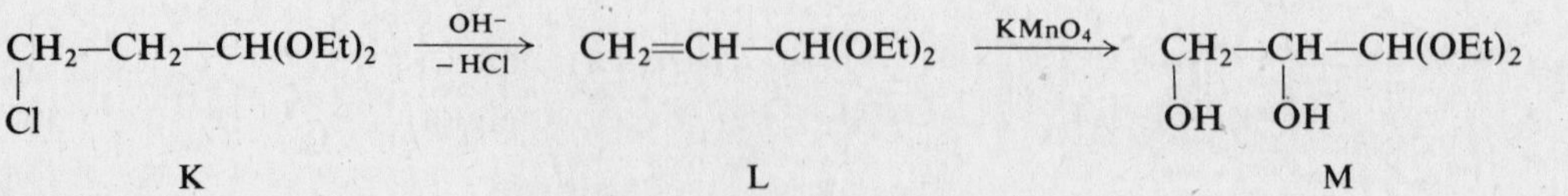

J   K

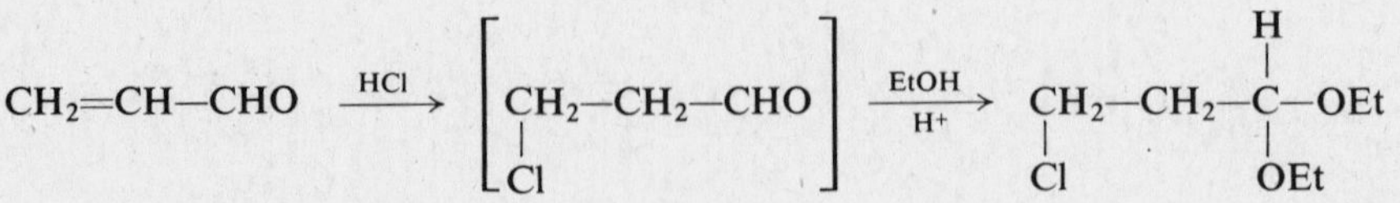

CH₂—CH₂—CH(OEt)₂, Cl $\xrightarrow[-HCl]{OH^-}$ CH₂=CH—CH(OEt)₂ $\xrightarrow{KMnO_4}$ CH₂—CH—CH(OEt)₂, OH OH

K   L   M

422

CH$_2$—CH—CH(OEt)$_2$ $\xrightarrow{H^+}$ CH$_2$—CH—CHO   *Cleavage of acetal*
| | | | *to aldehyde*
OH   OH   OH   OH

M   N, glyceraldehyde

(b) EtOOC—C≡C—COOEt + H$_2$C—COOEt $\xrightarrow{\text{Michael addn.}}$ EtOOC—C=CH—COOEt   (*Prob. 27.10f*)
                              | |
                           COOEt   HC—COOEt
                                                                             |
                                                                          COOEt
                                                                            O

O $\xrightarrow[\text{heat}]{OH^-}$ $^-$OOC—C=CH—COO$^-$ $\xrightarrow{H^+}$ $\xrightarrow[-CO_2]{\text{heat}}$ HOOC—C=CH—COOH
                        | |
                     H—C—COO$^-$ CH$_2$COOH
                        |
                     COO$^-$ P, aconitic acid

(c) EtOOC—CH=CH—COOEt + H$_2$C—COOEt $\xrightarrow{\text{Michael addn.}}$ EtOOC—CH—CH$_2$—COOEt
                              | |
                           COOEt   H—C—COOEt
                                                                             |
                                                                          COOEt
                                                                            Q

Q $\xrightarrow[\text{heat}]{OH^-}$ $^-$OOC—CH—CH$_2$—COO$^-$ $\xrightarrow{H^+}$ $\xrightarrow[-CO_2]{\text{heat}}$ HOOC—CH—CH$_2$—COOH
                        | |
                     H—C—COO$^-$ CH$_2$COOH
                        |
                     COO$^-$ R, tricarballylic acid

(d)

Ring forms by interaction of two bifunctional molecules

S, tetraphenylcyclopentadienone ("Tetracyclone")

S   Maleic anhydride   T   U, Tetraphenylphthalic anhydride

Aromatization through loss of CO + H$_2$

423

(e)

Ph—C(Ph)=C(Ph)—C(=O)—C(Ph)=C(Ph) (S) + HC≡CH →[Diels-Alder]→ (V) →[heat −CO]→ (W) Pentaphenylbenzene   *Aromatization through loss of CO*

(f) $CH_3-\overset{\overset{\displaystyle CH_3}{|}}{\underset{\underset{\displaystyle O}{||}}{C}}$ + BrMgC≡COEt → $CH_3-\overset{\overset{\displaystyle CH_3}{|}}{\underset{\underset{\displaystyle OMgBr}{|}}{C}}-C\equiv COEt$ →[H⁺]→ $CH_3-\overset{\overset{\displaystyle CH_3}{|}}{\underset{\underset{\displaystyle OH}{|}}{C}}-C\equiv COEt$ X

Pd/CaCO₃ | H₂

$CH_3-\overset{\overset{\displaystyle CH_3}{|}}{\underset{\underset{\displaystyle OH}{|}}{C}}-CH=CHOEt$ Y

$CH_3-\overset{\overset{\displaystyle CH_3}{|}}{\underset{\underset{\displaystyle OH}{|}}{C}}-CH=CHOEt$ →[H₂O / H⁺]→ $CH_3-\overset{\overset{\displaystyle CH_3}{|}}{\underset{\underset{\displaystyle OH}{|}}{C}}-CH_2-\overset{\overset{\displaystyle H}{|}}{C}=O$ *Easy cleavage of a vinyl ether (Problem 15, page 649)*

Y
A vinyl ether

$CH_3-\overset{\overset{\displaystyle CH_3}{|}}{\underset{\underset{\displaystyle OH}{|}}{C}}-CH_2-\overset{\overset{\displaystyle H}{|}}{C}=O$ →[H⁺ / −H₂O]→ $CH_3-\overset{\overset{\displaystyle CH_3}{|}}{C}=CH-\overset{\overset{\displaystyle H}{|}}{C}=O$

Z
β-Methylcrotonaldehyde
(*Has isoprene skeleton: useful in synthesis*)

(g) $CH_3-\overset{\overset{\displaystyle CH_3}{|}}{C}=CHCOOEt$ + $H_2\overset{\underset{\displaystyle COOEt}{|}}{C}-CN$ →[Michael addn.]→ $CH_3-\overset{\overset{\displaystyle CH_3}{|}}{\underset{\underset{\displaystyle \overset{H-C-CN}{\underset{\displaystyle COOEt}{|}}}{|}}{C}}-CH_2COOEt$ →[OH⁻ heat]→ $CH_3-\overset{\overset{\displaystyle CH_3}{|}}{\underset{\underset{\displaystyle \overset{H-C-COO^-}{\underset{\displaystyle COO^-}{|}}}{|}}{C}}-CH_2COO^-$

AA

H⁺

−CO₂ | heat

$CH_3-\overset{\overset{\displaystyle CH_3}{|}}{\underset{\underset{\displaystyle CH_2COOH}{|}}{C}}-CH_2COOH$

BB

(h)
$$CH_3-\underset{\underset{CH_3}{|}}{C}=CHCOCH_3 + H_2\underset{\underset{COOEt}{|}}{C}-COOEt \xrightarrow[\text{addn.}]{\text{Michael}} CH_3-\underset{\underset{H-\underset{\underset{COOEt}{|}}{C}-COOEt}{|}}{\overset{\overset{CH_3}{|}}{C}}-CH_2COCH_3$$

CC

$$CC \xrightarrow{OBr^-} CH_3-\underset{\underset{\underset{+}{COOEt}}{|}}{\overset{\overset{CH_3}{|}}{C}}-CH_2COO^- \xrightarrow[\text{heat}]{OH^-} CH_3-\underset{\underset{COO^-}{|}}{\overset{\overset{CH_3}{|}}{C}}-CH_2COO^- \xrightarrow{H^+} \xrightarrow[-CO_2]{\text{heat}} BB \text{ (see above)}$$

H–C–COOEt / H–C–COO⁻ labels, plus

CHBr₃

(i)
$$CH_3\overset{\overset{H}{|}}{C}=O + {}^-:C\equiv CCH_3 \longrightarrow CH_3-\underset{\underset{O^-}{|}}{\overset{\overset{H}{|}}{C}}-C\equiv CCH_3 \xrightarrow{H^+} CH_3-\underset{\underset{OH}{|}}{CH}-C\equiv CCH_3$$

DD

$$\downarrow K_2Cr_2O_7$$

$$CH_3-\underset{\underset{O}{\parallel}}{C}-C\equiv CCH_3$$

EE, 3-pentyn-2-one

(j)
$$CH_3-\underset{\underset{O}{\parallel}}{C}-C\equiv C-CH_3 \xrightarrow[H^+, Hg^{++}]{H_2O} CH_3-\underset{\underset{O}{\parallel}}{C}-\underset{\underset{H}{|}}{C}=\underset{\underset{OH}{|}}{C}-CH_3 \xrightarrow{\text{tautom.}} CH_3-\underset{\underset{O}{\parallel}}{C}-CH_2-\underset{\underset{O}{\parallel}}{C}-CH_3$$

Hydration of alkyne (Sec. 8.13)

EE (see above)   Enol   FF, acetylacetone (2,4-Pentanedione)

(k)
$$CH_3-\underset{\underset{\underset{O}{\parallel}}{}}{\overset{\overset{CH_3}{|}}{C}}=CH-\underset{\underset{O}{\parallel}}{C}-CH_3 \xrightarrow{OCl^-} CH_3-\overset{\overset{CH_3}{|}}{C}=CH-COO^- \xrightarrow{H^+} CH_3-\overset{\overset{CH_3}{|}}{C}=CH-COOH$$

+   GG

CHCl₃   (*Isoprene skeleton*)

(l)
$$\underset{\underset{Cl}{|}}{CH_2}-\overset{\overset{CH_3}{|}}{C}=CH_2 \xrightarrow{HOCl} \underset{\underset{Cl}{|}}{CH_2}-\underset{\underset{HO}{|}}{\overset{\overset{CH_3}{|}}{C}}-\underset{\underset{Cl}{|}}{CH_2} \xrightarrow{2CN^-} \underset{\underset{CN}{|}}{CH_2}-\underset{\underset{HO}{|}}{\overset{\overset{CH_3}{|}}{C}}-\underset{\underset{CN}{|}}{CH_2} \xrightarrow[\text{heat}]{H_2O, H^+} \underset{\underset{COOH}{|}}{HC}=\overset{\overset{CH_3}{|}}{C}-\underset{\underset{COOH}{|}}{CH_2}$$

HH   II   JJ

(m)

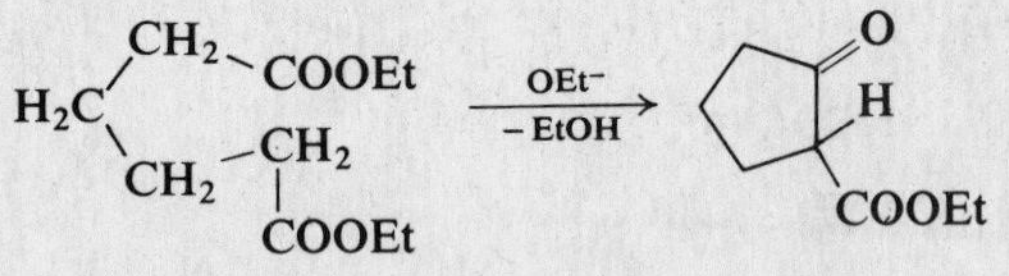

Intramolecular Claisen (Dieckmann) condensation
(Problem 21.30, page 718)

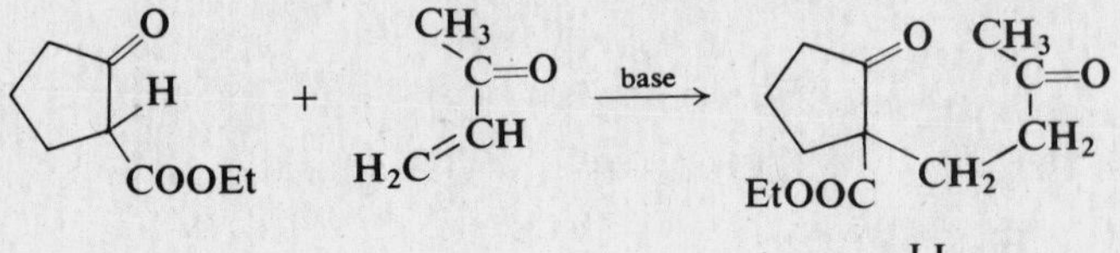

KK

KK + *Michael addition*

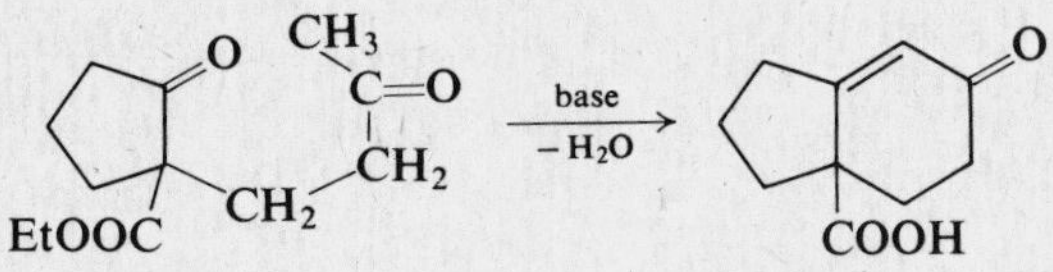

LL

LL *Intramolecular aldol condensation*

MM

(n) 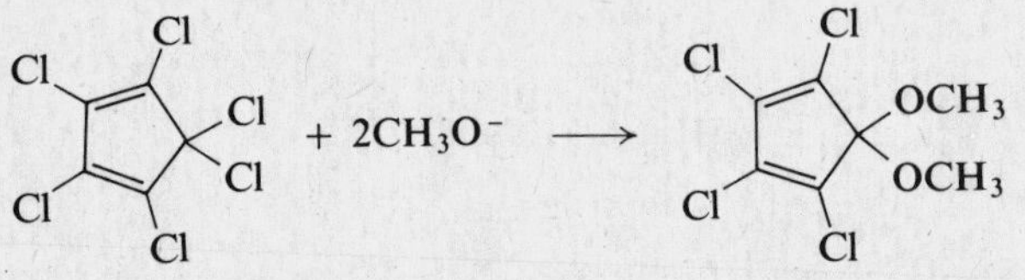 + 2CH₃O⁻ ⟶ *Allylic Cl's react; vinylic do not*

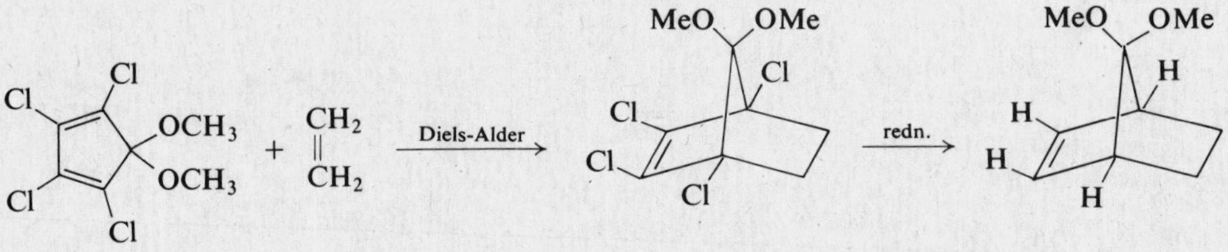

NN, a ketal

NN + Diels-Alder ⟶ OO redn. ⟶ PP

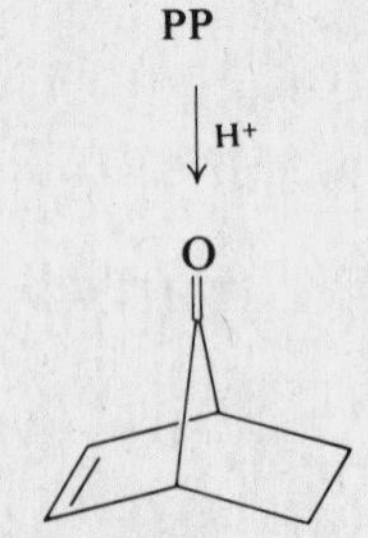

Cleavage of ketal unmasks ketone

QQ, 7-ketonorbornene

15. Spermine ←$_{Ni}^{H_2}$ NC—CH₂CH₂—N(CH₂)₄N—CH₂CH₂CN ←$_{addn.}^{nucleo.}$ 2CH₂=CHCN + NH₂(CH₂)₄NH₂

16. (a) HOOC—CH=CH—CH=CH—COOH $\xleftarrow{\text{KOH(alc)}}$ HOOCCHCH$_2$CH$_2$CHCOOH

                                                     Br   Br

                                                     2Br$_2$ $\Big\uparrow$ P

                                   HOOCCH$_2$CH$_2$CH$_2$CH$_2$COOH

                                               Adipic acid

(b) HC≡C—CHO ⟵ HC≡C—CH(OEt)$_2$ $\xleftarrow{\text{base}}$ $\underset{\text{Br Br}}{\overset{\text{H H}}{\text{HC—C—CH(OEt)}_2}}$ $\xleftarrow{\text{Br}_2}$ CH$_2$=CH—CH(OEt)$_2$

                                                                             (L of Prob. 14a, above)

                                                                             Protect —CHO

(c) $\underset{\text{O}}{\text{CH}_3\text{—C—CH=CH}_2}$ $\xleftarrow[-\text{H}_2\text{O}]{\text{H}^+}$ $\underset{\text{O} \quad \text{OH}}{\text{CH}_3\text{—C—CH}_2\!\!\mid\!\!\text{CH}_2}$ $\xleftarrow{\text{OH}^-}$ $\underset{\text{O}}{\text{CH}_3\text{—C—CH}_3}$ + O=CH$_2$

                                                    Aldol

(d) $\underset{\text{O}}{\text{CH}_3\text{—C—CH=CH}_2}$ $\xleftarrow{\text{tautom.}}$ $\underset{\text{OH}}{\text{CH}_2\text{=C—CH=CH}_2}$ $\xleftarrow[\text{H}^+]{\text{H}_2\text{O, Hg}^{++}}$ HC≡C—CH=CH$_2$

                                                                            Vinylacetylene

         Keto   *Enol*

(e) $\underset{\text{Ph}}{\text{HOOCCH}_2\text{—CHCH}_2\text{COOH}}$ $\xleftarrow{-\text{CO}_2}$ $\xleftarrow{\text{H}^+}$ $\xleftarrow{\text{OH}^-}$ $\underset{\text{EtOOC}}{\overset{\text{EtOOC}}{{>}}}\text{CH—}\underset{\text{Ph}}{\text{CHCH}_2\text{COOEt}}$

                                                            Michael $\Big\uparrow$ addn.

                                           $\underset{\text{EtOOC}}{\overset{\text{EtOOC}}{{>}}}\text{CH}_2$ + PhCH=CHCOOEt

                                                           Ethyl cinnamate

                                                           (Prob. 1, above)

Or via $\underset{\text{EtOOC}}{\overset{\text{EtOOC}}{{>}}}\text{CH}_2$ + PhCHO + H$_2$C$\underset{\text{COOEt}}{\overset{\text{COOEt}}{{<}}}$ (Knoevenagel followed by Michael)

(f) $\underset{\text{Ph}}{\text{HOOC—CH—CH}_2\text{COOH}}$ $\xleftarrow[\text{H}^+]{\text{H}_2\text{O}}$ $\underset{\text{Ph}}{\text{N≡C—CH—CH}_2\text{COOEt}}$ $\xleftarrow{\text{CN}^-}$ PhCH=CHCOOEt

                                                                           Ethyl cinnamate

                                                                           (Prob. 1, above)

(g) $\underset{\text{O}}{\text{CH}_3\text{CCH}_2}\text{—}\underset{}{\overset{\text{Ph}}{\text{CH}}}\text{—}\underset{\text{O}}{\text{CH}_2\text{CCH}_3}$ $\xleftarrow[-\text{CO}_2]{\text{H}^+}$ $\xleftarrow{\text{OH}^-}$ $\underset{\text{O COOEt}}{\text{CH}_3\text{C—CH}}\text{—}\underset{}{\overset{\text{Ph}}{\text{CH}}}\text{—}\underset{\text{O}}{\text{CH}_2\text{CCH}_3}$

                                                                 $\Big\uparrow$ Michael

PhCHO + CH$_3$COCH$_3$ $\xrightarrow[\text{OH}^-]{-\text{H}_2\text{O}}$ $\underset{\text{O}}{\text{PhCH=CHCCH}_3}$ + CH$_3$COCH$_2$COOEt

17. (a)

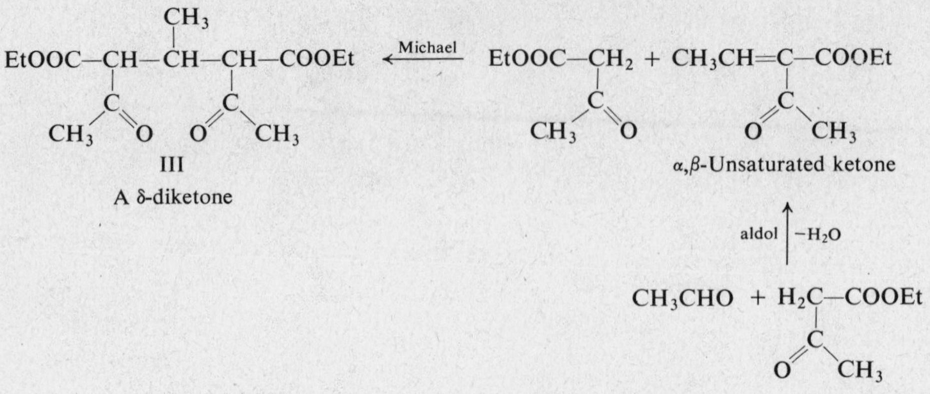

$$\text{EtOOC—CH—CH—CH—COOEt} \xleftarrow{\text{Michael}} \text{EtOOC—CH}_2 + \text{CH}_3\text{CH}=\text{C—COOEt}$$

III

A δ-diketone

α,β-Unsaturated ketone

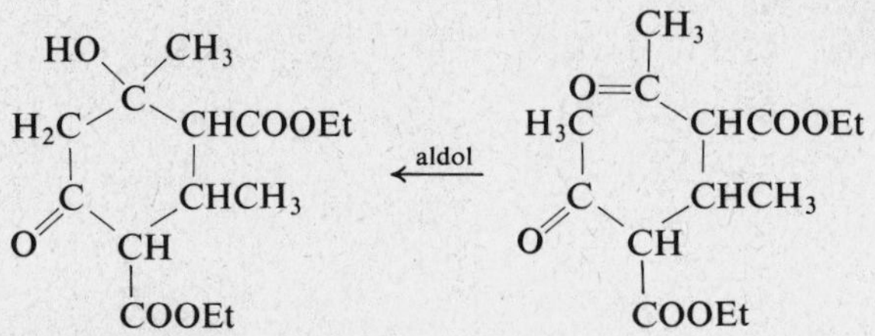

aldol |−H₂O

$$\text{CH}_3\text{CHO} + \text{H}_2\text{C—COOEt}$$

IV ← aldol ← III

A β-hydroxyketone

(b) IV is the correct structure.

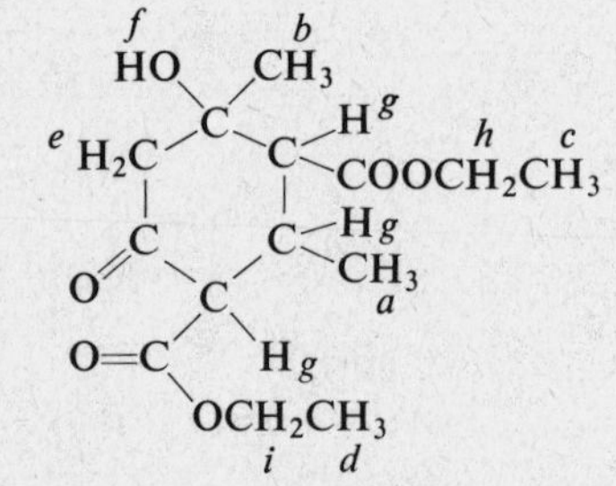

IV

The nmr spectrum is too complex for a molecule as symmetrical as III, from which we would expect only six signals.

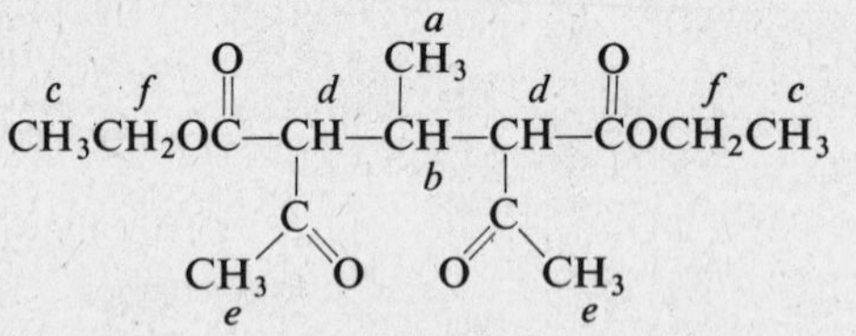

$$c \quad f \qquad \overset{a}{\text{CH}_3} \qquad d \quad \overset{O}{\parallel} \quad f \quad c$$

$$\text{CH}_3\text{CH}_2\text{OC—CH—CH—CH—COCH}_2\text{CH}_3$$

III

a doublet, 3H
b multiplet, 1H
c triplet, 6H
d doublet, 2H
e singlet, 6H
f quartet, 4H

(T. D. Binns and R. Brettle, "Anodic Oxidation. Part 1. The Electrolysis of Ethyl Sodioacetoacetate in Ethanolic Solutions," J. Chem. Soc. (C), 336 (1966).)

18.

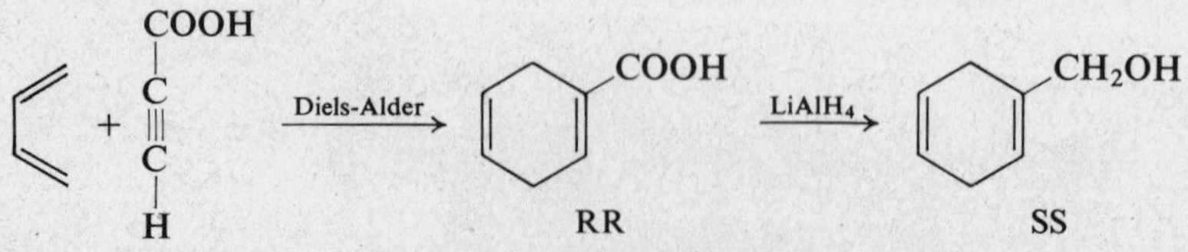

$$\xrightarrow{\text{Diels-Alder}} \text{RR} \xrightarrow{\text{LiAlH}_4} \text{SS}$$

$$\text{SS} + \text{Cl-C(=O)-OMe} \longrightarrow \text{TT} \xrightarrow[-\text{CO}_2, -\text{MeOH}]{\text{heat}} \text{Toluene} + \text{UU}$$

$$\text{(diene)} + \begin{matrix}\text{C(CN)}_2\\\text{C(CN)}_2\end{matrix} \xrightarrow{\text{Diels-Alder}} \text{VV}$$

The structure assigned to UU is indicated most strongly by its easy conversion into its isomer toluene; taken together with the fact that UU is *not* 1,3,5-cycloheptatriene, this conversion severely limits the number of possible structures. The ultraviolet spectrum (Sec. 13.5) indicates strong conjugation. (Actually, absorption is at precisely the wavelength predicted for this structure.) Conjugation is also shown by the occurrence of the Diels-Alder reaction. The infrared spectrum (Sec. 13.15) shows trigonal C—H stretch at 3020 cm^{-1}, conjugated C=C stretch at 1595, and the expected (see page 445) C—H out-of-plane bend, at 864 and 692.

Finally, the structure is consistent with the method of synthesis.

The conversion of TT into UU is worth looking at. TT is an allylic ester, and is believed to undergo an *allylic rearrangement* to XV before elimination. Such a rearrangement *could* proceed by ionization to an allylic cation and CH$_3$OCOO$^-$ followed by recombination at a different carbon of the hybrid cation; here, it seems more likely to involve concerted bond-breaking and bond-making as shown.

$$\text{TT} \longrightarrow \text{XV}$$

An allylic rearrangement

Formation of UU involves pyrolysis of an ester, a well-known and—as demonstrated here—mild method of elimination. From the stereochemistry of ester pyrolyses (*syn*), it seems likely that here, too, a concerted reaction is involved.

$$\text{XV} \longrightarrow \text{UU} + \left[\begin{matrix}\text{O}\\ \text{C-OCH}_3\\ \text{HO}\end{matrix}\right] \xrightarrow{\text{page 685}} \text{CH}_3\text{OH} + \text{CO}_2$$

The most interesting thing about all this is that UU is isolated at all. It is less stable by some 30 kcal than its aromatic isomer toluene, and we might expect it to rearrange practically instantaneously into the more stable compound. (In fact, such a prediction was made, in 1943.) Actually, the conversion, although easy, is far from instantaneous; it takes hours at room temperature. Evidently this highly exothermic reaction has a sizeable energy of activation, surprising for a simple transfer of hydrogen. The explanation for this did not emerge until the years following 1965, with the concept of *orbital symmetry* (Chapter 29). As we shall see (Problem 15(b), Chapter 29, page 965), the easy route from UU to toluene is "forbidden," and reaction is forced to follow a more difficult path.

(W. J. Bailey and R. A. Baylouny, "Cyclic Dienes. XXVI. 5-Methylene-1,3-cyclohexadiene, an Alicyclic Isomer of Toluene," J. Org. Chem., **27**, 3476 (1962).)

19. The enamine WW contains nucleophilic carbon, and undergoes a Michael addition to the α,β-unsaturated ester.

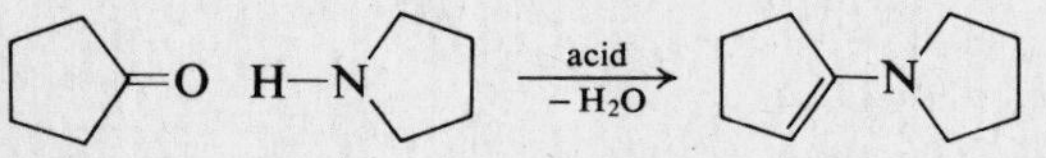

WW

An enamine (*Sec. 26.8*)

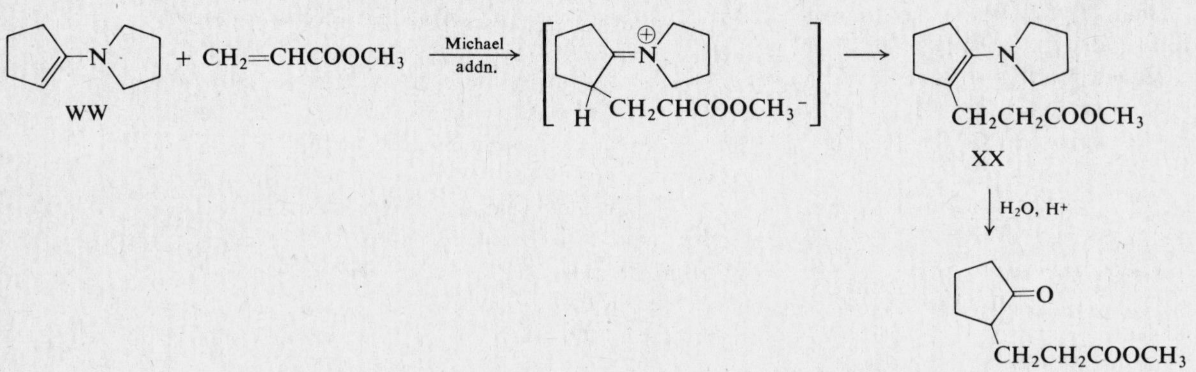

YY

20. (a)

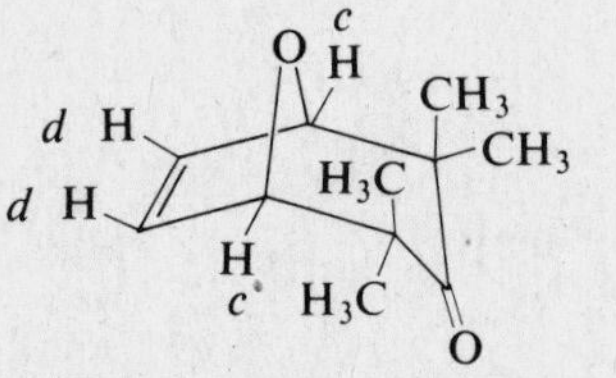

VII

CH_3's *a* and *b*

The four methyls are not equivalent: two are closer to the oxygen bridge than the other two.

(b)

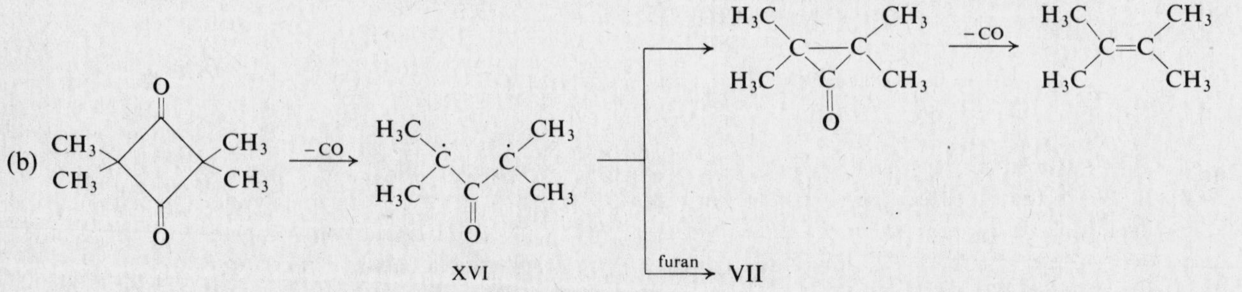

Loss of one CO gives XVI (which may be singlet or triplet, page 309). The terminal carbons of XVI each contribute an electron to form a bond and generate cyclopropanone, which loses a second CO to give the alkene. The intermediacy of XVI is shown by the fact that it can be trapped by furan to give the 1,4-addition product VII.

21. 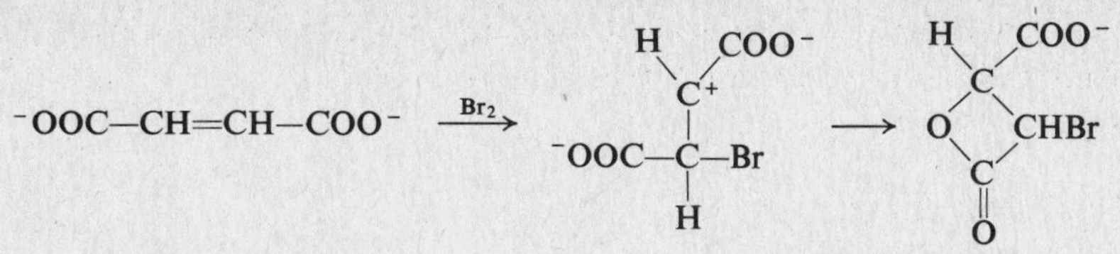 $^-$OOC—CH=CH—COO$^-$ $\xrightarrow{Br_2}$

VIII

This experiment gives evidence that the addition of bromine is two-step and electrophilic: the intermediate carbonium (or bromonium) ion is trapped—just as it was (Sec. 6.13) by Cl$^-$, Br$^-$, I$^-$, NO$_3{}^-$, etc.—by carboxylate ion within the same molecule.

(D. S. Tarbell and P. D. Bartlett, J. Am. Chem. Soc., **59**, 407 (1937).)

22. We have already seen (Problem 18, Chapter 25, page 844) that benzyne can be formed by loss of N$_2$ and CO$_2$ from an *ortho* diazonium carboxylate (I, page 844). We know (Sec. 10.10) that the cyclopentadienyl anion is aromatic. We consider it likely, therefore, that loss of N$_2$ and CO$_2$ from the salt of IX gives the aryne XVII related to this non-benzenoid aromatic compound.

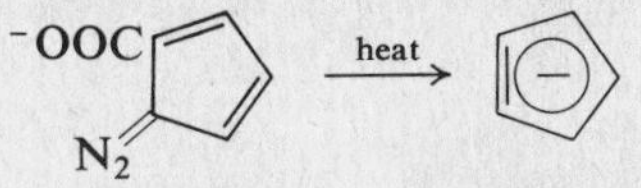

IX $\xrightarrow{heat}$ XVII

Dehydrocyclo-
pentadienyl anion
An aryne

Aryne XVII is trapped by X to give the Diels-Alder adduct XVIII which, by loss of CO, generates the new, central benzene ring of XI.

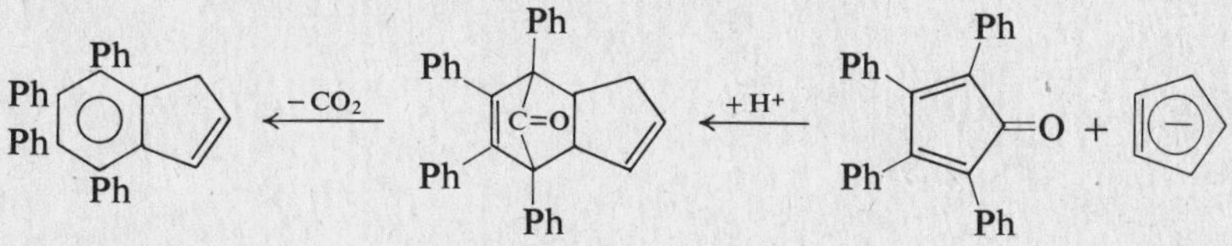

XI $\xleftarrow{-CO_2}$ XVIII $\xleftarrow{+H^+}$ X + XVII

These findings provide one more piece of evidence for the aromaticity of the cyclopentadienyl anion and for the validity of the Hückel $4n + 2$ rule (Secs. 10.10 and 29.6).

(J. C. Martin and D. R. Bloch, "The Dehydrocyclopentadienyl Anion. A New Aryne," J. Am. Chem. Soc., **93**, 451 (1971).)

Chapter 28

Rearrangements and Neighboring Group Effects
Nonclassical Ions

28.1 (a)

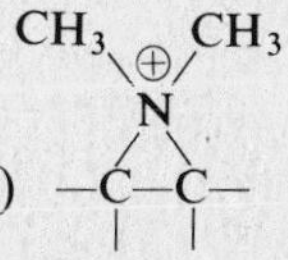

CH$_3$ CH$_3$
 N$^{\oplus}$
—C—C—

Ammonium ion

(b)

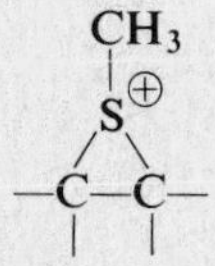

CH$_3$
S$^{\oplus}$
—C—C—

Sulfonium ion

(c)

H
O$^{\oplus}$
—C—C—

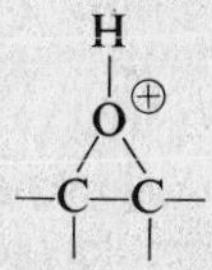

Protonated epoxide

(d)

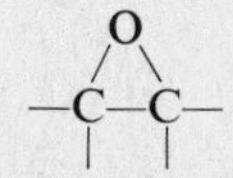

O
—C—C—

Epoxide

(e)

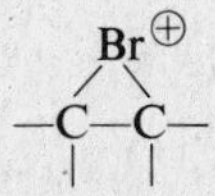

Br$^{\oplus}$
—C—C—

Bromonium ion

(f)

—C—C—

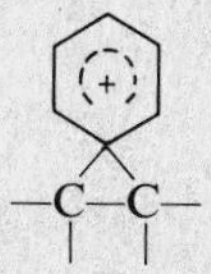

Benzenonium ion
(Phenonium ion)

(g)

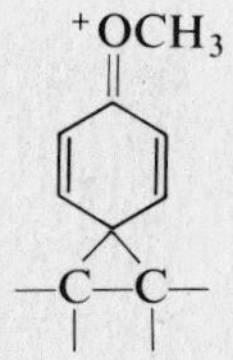

$^+$OCH$_3$

—C—C—

Oxonium ion

(h)

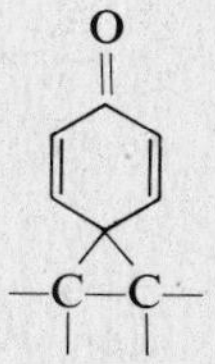

O

—C—C—

Ketone
(Dienone)

(i)

R   H
C+
CH
—C—C—

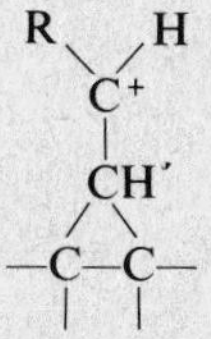

Cyclopropylcarbinyl
cation

28.2 The Curtius rearrangement involves a 1,2-shift to electron-deficient nitrogen, with N$_2$—that best of all leaving groups—being lost.

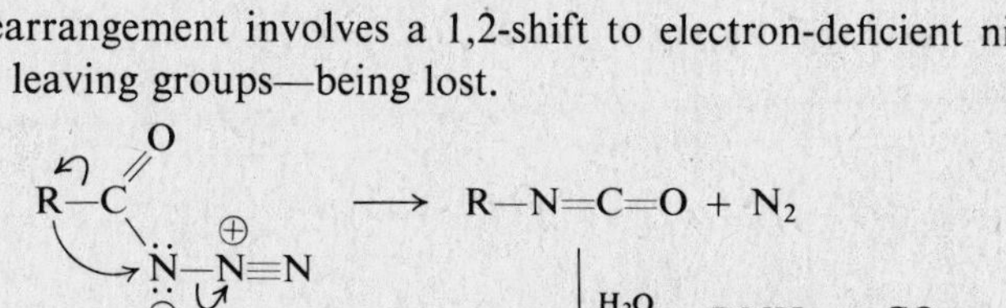

$$R-C \underset{\overset{\mid}{\underset{\ominus}{\ddot{N}}-N^{\oplus}\equiv N}}{\overset{O}{\parallel}} \longrightarrow R-N=C=O + N_2$$

$$\xrightarrow{\text{H}_2\text{O}} RNH_2 + CO_2$$

An acyl azide

(P. A. S. Smith, "The Curtius Rearrangement," O. R. III-8.)

28.3 The lactam IV can form only if III has the *cis* configuration.   Since II is also *cis*, the reaction must have taken place so that —NH$_2$ occupies the same position in III that —CONH$_2$

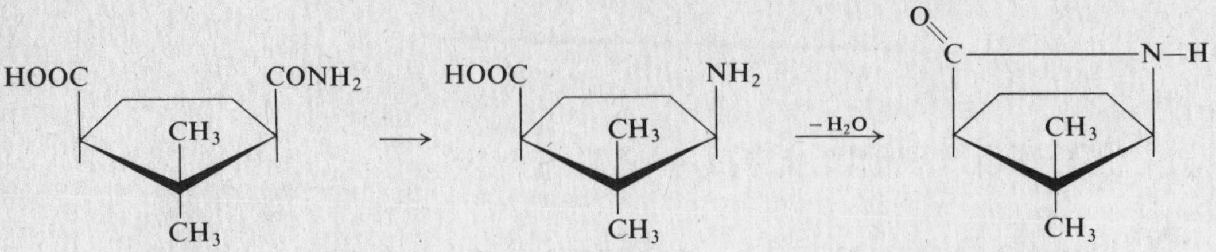

II	III	IV
Amide	Amine	Lactam
cis	*cis*	*cis*

occupied in II; that is, reaction must have taken place with retention of configuration at the migrating group.

28.4 The conclusion is exactly analogous to the one drawn on pages 892–893 for the Hofmann rearrangement.   The substituent effect (page 894) shows that migration occurs in the rate-determining step.   There are, then, two possibilities: that (2) is fast and reversible followed by a slow (3); or that (2) and (3) are concerted.   But the reverse of (2) is combination of the electron-deficient particle with water; if this happened, unrearranged hydroperoxide would contain oxygen-18—contrary to fact.   We are left with the concerted reaction (2,3).

(M. Bassey, C. A. Bunton, A. G. Davies, T. A. Lewis, and D. R. Llewellyn, "Organic Peroxides.   Part V.   Isotopic Tracer Studies on the Formation and Decomposition of Organic Peroxides," J. Chem. Soc., 2471 (1955).)

28.5 (a) *H migration*

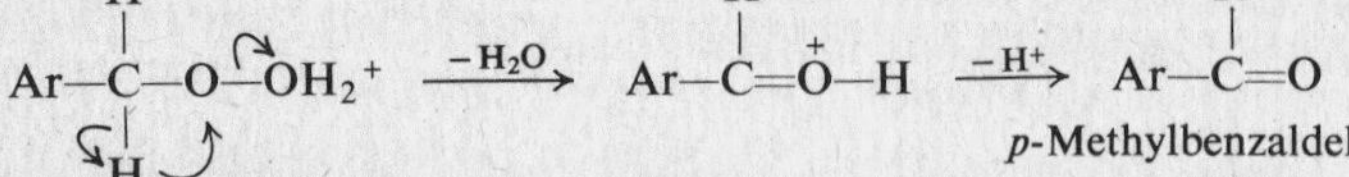

p-Methylbenzaldehyde
(68%)

(Ar = *p*-Tolyl)

Aryl migration

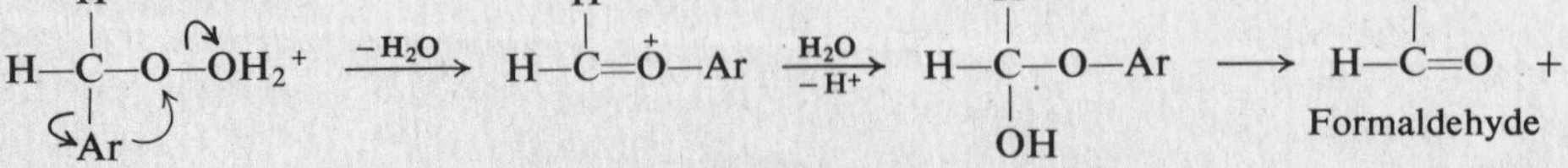

Hemiacetal

Formaldehyde   *p*-Cresol
(38%)

(b) H migrates faster than *p*-tolyl.

28.6

$$R-\overset{\overset{\text{H}}{|}}{\underset{\underset{\text{H}}{|}}{C}}-O-\overset{+}{O}H_2 \xrightarrow{-H_2O} R-\overset{\overset{\text{H}}{|}}{C}=\overset{+}{O}H \xrightarrow{-H^+} RCHO \qquad \textbf{H migration}$$

$$H-\overset{\overset{\text{H}}{|}}{\underset{\underset{\text{R}}{|}}{C}}-O-\overset{+}{O}H_2 \xdashrightarrow{-H_2O} H-\overset{\overset{\text{H}}{|}}{C}=\overset{+}{O}R \xrightarrow{H_2O} HCHO + ROH \qquad \textit{R migration}$$

$$R-\overset{\overset{\text{R}}{|}}{\underset{\underset{\text{H}}{|}}{C}}-O-\overset{+}{O}H_2 \xrightarrow{-H_2O} R-\overset{\overset{\text{R}}{|}}{C}=\overset{+}{O}H \xrightarrow{-H^+} R_2C=O \qquad \textbf{H migration}$$

$$R-\overset{\overset{\text{H}}{|}}{\underset{\underset{\text{R}}{|}}{C}}-O-\overset{+}{O}H_2 \xdashrightarrow{-H_2O} R-\overset{\overset{\text{H}}{|}}{C}=\overset{+}{O}R \xrightarrow{H_2O} RCHO + ROH \qquad \textit{R migration}$$

Since alcohols (HOR) are *not* obtained, H must migrate much faster than R.

28.7 Carbonium ions are formed in familiar ways,

$$R-NH_2 \xrightarrow{\text{HONO}} R-N_2^+ \longrightarrow R^+ + N_2 \quad (\textit{page 763})$$

and

$$R-X + Ag^+ \longrightarrow R^+ + AgX \quad (\textit{page 486})$$

and then undergo pinacol-like rearrangements.

(a)

$$Ph-\overset{\overset{\text{Ph}}{|}}{\underset{\underset{\text{HO}}{|}}{C}}-\overset{\overset{\text{Ph}}{|}}{\underset{\underset{\text{NH}_2}{|}}{C}}-CH_3 \xrightarrow[-N_2]{\text{HONO}} Ph-\overset{\overset{\text{Ph}}{|}}{\underset{\underset{\text{OH}}{|}}{C}}-\overset{\overset{\text{Ph}}{|}}{\underset{\oplus}{C}}-CH_3 \longrightarrow Ph-\overset{\overset{\text{Ph}}{|}}{\underset{\underset{\text{HO}^+}{|}}{C}}-\overset{\overset{}{|}}{\underset{\underset{\text{CH}_3}{|}}{C}}-Ph \xrightarrow{-H^+} Ph-\overset{}{\underset{\underset{\text{O}}{||}}{C}}-\overset{\overset{\text{Ph}}{|}}{\underset{\underset{\text{CH}_3}{|}}{C}}-Ph$$

(b)

$$CH_3-\overset{\overset{\text{Ph}}{|}}{\underset{\underset{\text{HO}}{|}}{C}}-\overset{}{\underset{\underset{\text{I}}{|}}{C}}H_2 \xrightarrow[-AgI]{Ag^+} CH_3-\overset{\overset{\text{Ph}}{|}}{\underset{\underset{\text{OH}}{|}}{C}}-CH_2^{\oplus} \longrightarrow CH_3-\overset{}{\underset{\underset{+\text{OH}}{||}}{C}}-CH_2Ph \xrightarrow{-H^+} CH_3\overset{}{\underset{\underset{\text{O}}{||}}{C}}-CH_2Ph$$

28.8 (a)

$$CH_3-\overset{}{\underset{\underset{\text{OH}}{|}}{C}}H-\overset{}{\underset{\underset{\text{OH}}{|}}{C}}H_2 \xrightarrow[-H_2O]{H^+} CH_3\overset{}{\underset{\oplus}{C}}H-\overset{\overset{\text{H}}{|}}{\underset{\underset{\text{OH}}{|}}{C}}-H \xrightarrow{-H^+} CH_3CH_2CHO$$

2° cation

(b)

$$CH_3-\overset{\overset{\text{H}_3\text{C}}{|}}{\underset{\underset{\text{OH}}{|}}{C}}-\overset{}{\underset{\underset{\text{OH}}{|}}{C}}H_2 \xrightarrow[-H_2O]{H^+} CH_3-\overset{\overset{\text{H}_3\text{C}}{|}}{\underset{\oplus}{C}}-\overset{\overset{\text{H}}{|}}{\underset{\underset{\text{OH}}{|}}{C}}-H \xrightarrow{-H^+} CH_3-\overset{\overset{\text{CH}_3}{|}}{\underset{}{C}}H-CHO$$

3° cation

(c) $Ph-CH-CH_2 \xrightarrow[-H_2O]{H^+} Ph-\overset{\oplus}{C}H-\overset{H}{\underset{OH}{C}}-H \xrightarrow{-H^+} PhCH_2CHO$
        | |
        OH   OH

Benzylic cation

(d) $Ph-\overset{Ph}{\underset{HO}{C}}-\overset{}{\underset{OH}{C}}H_2 \xrightarrow[-H_2O]{H^+} Ph-\overset{Ph}{\overset{\oplus}{C}}-\overset{H}{\underset{OH}{C}}-H \xrightarrow{-H^+} Ph_2CHCHO$

Benzylic cation

(e) $Ph-CH-CH-CH_3 \xrightarrow[-H_2O]{H^+} Ph-\overset{\oplus}{C}H-\overset{H}{\underset{OH}{C}}-CH_3 \xrightarrow{-H^+} PhCH_2CCH_3$
        | | ‖
        OH OH   O

Benzylic cation

(f) $Ph-\overset{Ph}{\underset{HO}{C}}-\overset{CH_3}{\underset{OH}{C}}-CH_3 \xrightarrow[-H_2O]{H^+} Ph-\overset{Ph}{\overset{\oplus}{C}}-\overset{CH_3}{\underset{OH}{C}}-CH_3 \xrightarrow{-H^+} Ph-\overset{Ph}{\underset{H_3C}{C}}-\overset{}{\underset{O}{C}}-CH_3$

Benzylic cation

(g) $Ph-\overset{Ph}{\underset{HO}{C}}-\overset{Ph}{\underset{OH}{C}}-CH_3 \xrightarrow[-H_2O]{H^+} Ph-\overset{Ph}{\overset{\oplus}{C}}-\overset{Ph}{\underset{OH}{C}}-CH_3 \xrightarrow{-H^+} Ph_3C-\overset{}{\underset{O}{C}}-CH_3$

Doubly benzylic cation

(h) $Me-\overset{Me}{\underset{HO}{C}}-\overset{Et}{\underset{OH}{C}}-Et \xrightarrow[-H_2O]{H^+}$

→ $Me-\overset{Me}{\overset{\oplus}{C}}-\overset{Et}{\underset{OH}{C}}-Et \xrightarrow{H^+} Et-\overset{Me}{\underset{Me}{C}}-\overset{}{\underset{O}{C}}-Et$

3° cation

→ $Me-\overset{Me}{\underset{OH}{C}}-\overset{Et}{\overset{\oplus}{C}}-Et \xrightarrow{H^+} Me-\overset{}{\underset{O}{C}}-\overset{Et}{\underset{Me}{C}}-Et$

3° cation

} *Mixture actually obtained*

(i) $An-\overset{An}{\underset{HO}{C}}-\overset{Ph}{\underset{OH}{C}}-Ph \xrightarrow[-H_2O]{H^+}$

(An = p-CH$_3$OC$_6$H$_4$)

→ $An-\overset{An}{\overset{\oplus}{C}}-\overset{Ph}{\underset{OH}{C}}-Ph \xrightarrow{H^+} An-\overset{An}{\underset{Ph}{C}}-\overset{}{\underset{O}{C}}-Ph$

Benzylic cation
More stable of the two

(Actually 72%)

→ $An-\overset{An}{\underset{OH}{C}}-\overset{Ph}{\overset{\oplus}{C}}-Ph \xrightarrow{H^+} An-\overset{}{\underset{O}{C}}-\overset{An}{\underset{Ph}{C}}-Ph$

Benzylic cation

(Actually 28%)

28.9 All three reactions give the same mixture because they all proceed via the same intermediate: an actual carbonium ion. This then reacts in either of two competing ways: by rearrangement to give pinacolone, or by combination with the solvent to give pinacol. Rearrangement is thus S_N1-like.

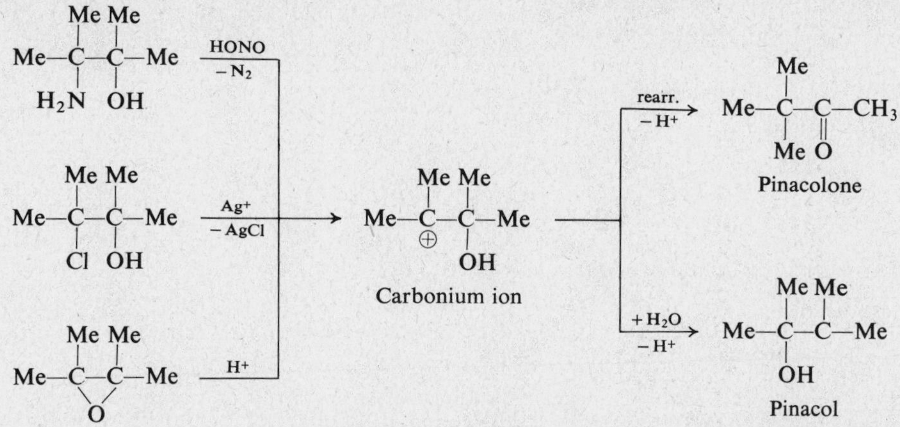

If rearrangement were S_N2-like, we would expect different proportions of products from the three reactions, since different leaving groups would be partially attached in the various transition states.

(For reviews of the pinacol rearrangement, see Y. Pocker, in P. de Mayo, *Molecular Rearrangements*, Part I, Wiley-Interscience, New York, 1963, pp. 15–25; C. J. Collins, "The Pinacol Rearrangement," Quart. Revs. (London), **14**, 357 (1960).)

28.10 Consider protonated pinacol. Competing with rearrangement is substitution of a molecule of solvent ($H_2^{18}O$) for the unlabeled $-OH_2^+$ group. Each of these processes could be either S_N1-like or S_N2-like. The evidence of Problem 28.9 indicates that both are S_N1-like. We conclude that carbonium ions are formed, which recombine with water two or

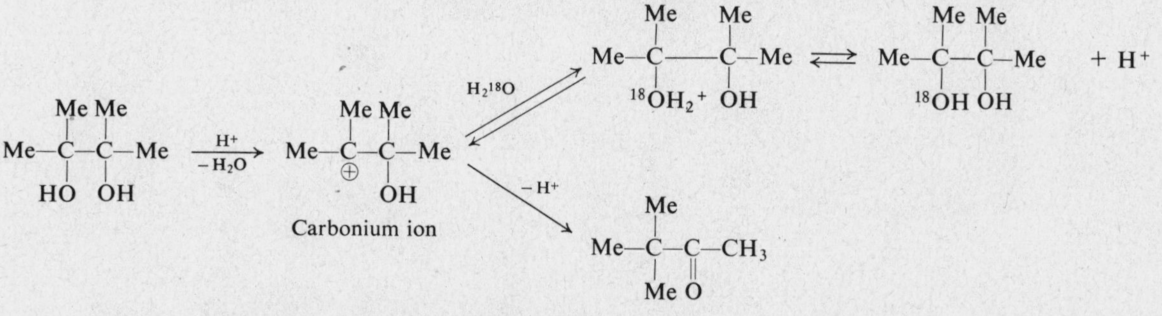

three times as fast as they undergo rearrangement.

(C. A. Bunton, T. Hadwick, D. R. Llewellyn, and Y. Pocker, "Tracer Studies in Alcohols. Part III. Intermediates in the Pinacol–Pinacone Rearrangement," J. Chem. Soc., 403 (1958).)

28.11 From the glycol we can obtain only the more stable carbonium ion; from the amino-alcohol we generate a carbonium ion in which the electron-deficient carbon must be the one that held the amino group.

(a) 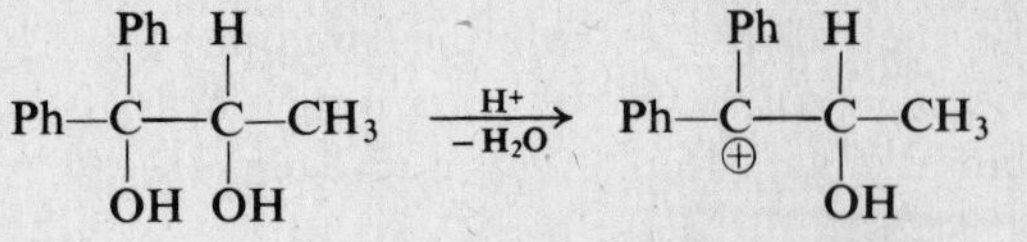 Ph—C(Ph)(H...)

$$
\text{(a)} \quad \underset{\substack{| \\ OH}}{\overset{\substack{Ph \\ |}}{Ph-C}}-\underset{\substack{| \\ OH}}{\overset{\substack{H \\ |}}{C}}-CH_3 \quad \xrightarrow[-H_2O]{H^+} \quad \underset{\substack{\oplus}}{\overset{\substack{Ph \\ |}}{Ph-C}}-\underset{\substack{| \\ OH}}{\overset{\substack{H \\ |}}{C}}-CH_3
$$

Benzylic cation
More stable cation

$$
\text{(b)} \quad \underset{\substack{| \\ OH}}{\overset{\substack{Ph \\ |}}{Ph-C}}-\underset{\substack{| \\ NH_2}}{\overset{\substack{H \\ |}}{C}}-CH_3 \quad \xrightarrow[-N_2]{HONO} \quad \underset{\substack{| \\ OH}}{\overset{\substack{Ph \\ |}}{Ph-C}}-\underset{\substack{\oplus}}{\overset{\substack{H \\ |}}{C}}-CH_3
$$

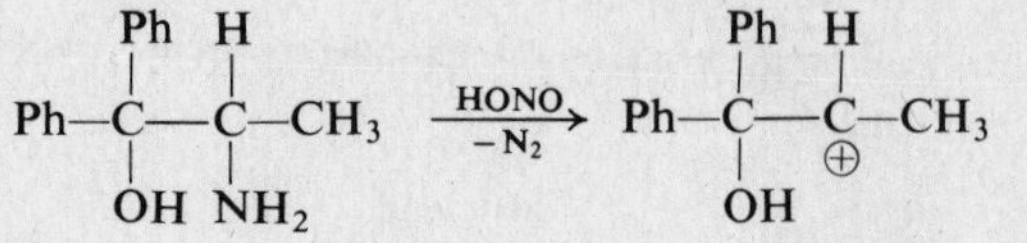

2° Alkyl cation
*Unobtainable from
glycol of part (a)*

28.12 Here there is no question as to which group will migrate: only Ar migrates. What is in question is whether migration of Ar takes place with inversion or retention. As we see, the favored reaction of each stereoisomer is the one that involves a *trans* transition state; that is, one with Ph— and CH₃— apart. For one diastereomer, this leads to predominant inversion; for the other, predominant retention.

VIII ⟶

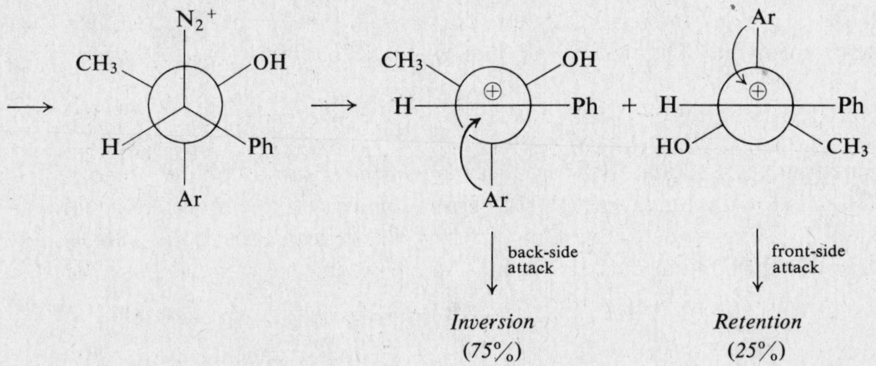

Inversion
(75%)

Retention
(25%)

IX ⟶

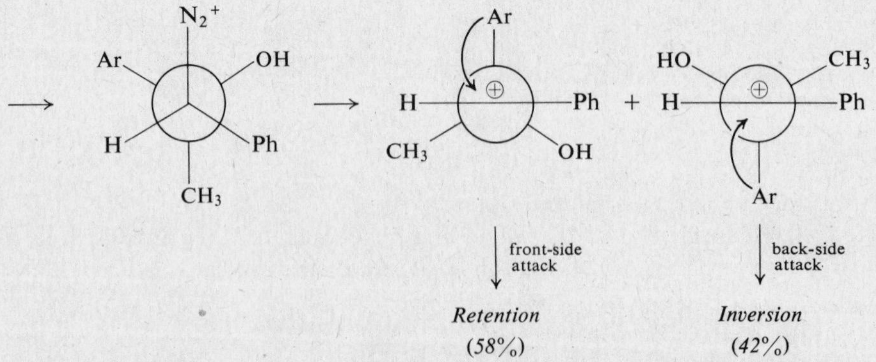

Retention
(58%)

Inversion
(42%)

(B. M. Benjamin and C. J. Collins, "Molecular Rearrangements. XVIII. The Deamination of *erythro-* and *threo-*1-Amino-1-phenyl-2-*p*-tolyl-2-propanol," J. Am. Chem. Soc., **83**, 3662 (1961). For a general discussion, see C. J. Collins, "Reactions of Primary Aliphatic Amines with Nitrous Acid," Accounts Chem. Res., **4**, 315 (1971).)

28.13

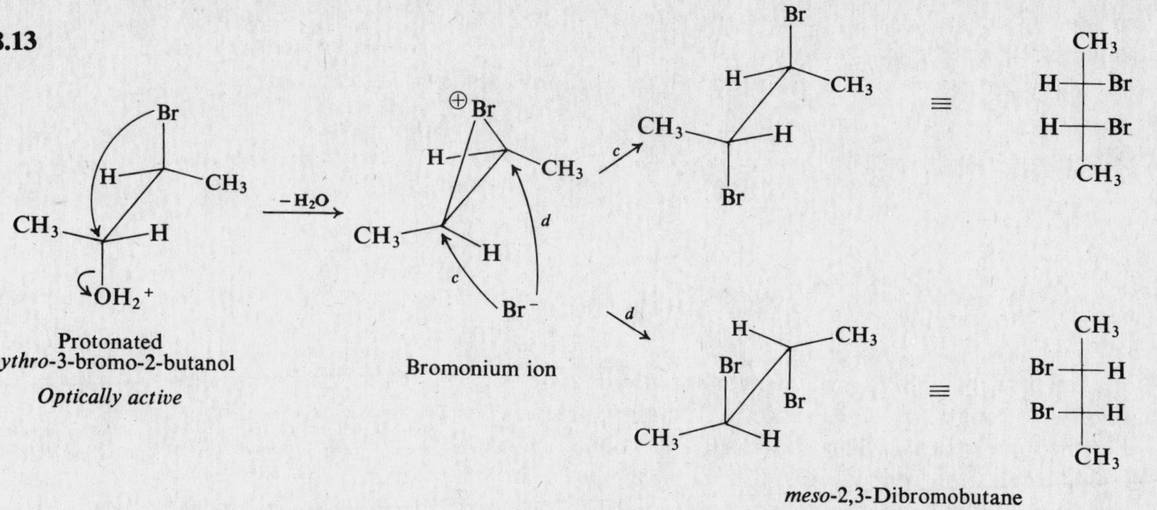

Protonated
erythro-3-bromo-2-butanol

Optically active

Bromonium ion

meso-2,3-Dibromobutane

Optically inactive

Attacks *c* and *d* both yield the same product: *meso*-2,3-dibromobutane.

(S. Winstein and H. J. Lucas, "Retention of Configuration in the Reaction of the 3-Bromo-2-butanols with Hydrogen Bromide," J. Am. Chem. Soc., **61**, 1576 (1939); "The Loss of Optical Activity in the Reaction of the Optically Active *erythro*- and *threo*-3-Bromo-2-butanols with Hydrogen Bromide," *ibid.*, 2845 (1939).)

28.14 There are two successive nucleophilic substitutions: first, intramolecular attack by —COO⁻ to give an α-lactone, and then attack on this lactone by hydroxide ion. Each proceeds with inversion, to give net retention.

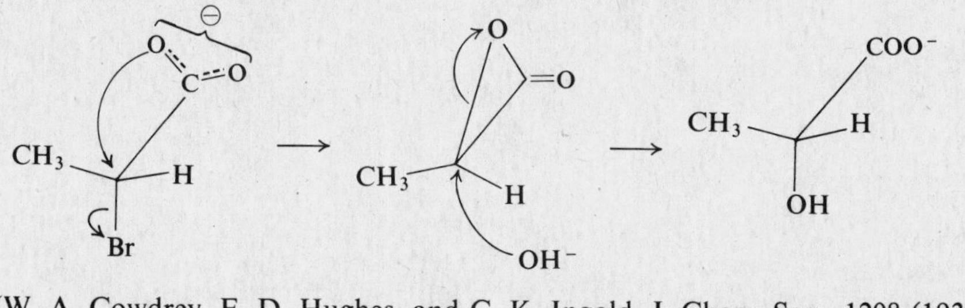

(W. A. Cowdrey, E. D. Hughes, and C. K. Ingold, J. Chem. Soc., 1208 (1937).)

28.15

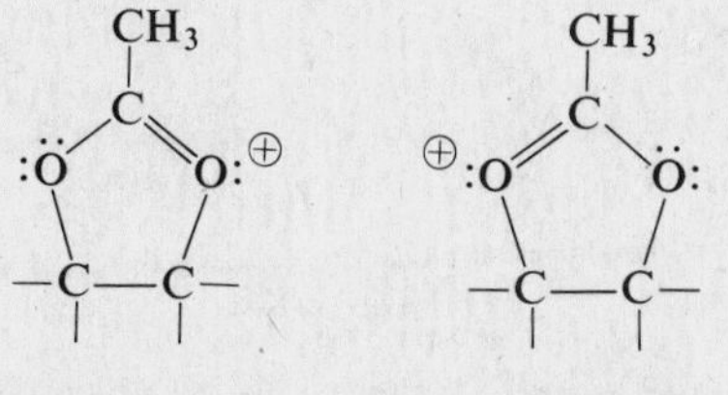

All atoms have complete octets

28.16 In solvolysis with S_N1 character (Sec. 14.17), halogen at the 2-position could exert two opposing effects. By its electron-withdrawing inductive effect, it could tend to slow down reaction by intensifying the positive charge developing on the carbon. Through anchimeric assistance, it could tend to speed up reaction. Only in the *trans* isomers can —X and —OBs take up the diaxial configuration required for anchimeric assistance.

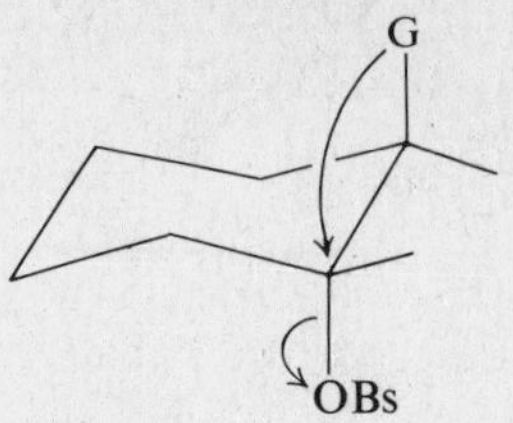

The data indicate no anchimeric assistance by *cis*-Cl or *trans*-Cl, or by *cis*-Br; there is only the strong deactivation expected from the inductive effect. The *trans*-Br compound is much more reactive than the *cis*-Br compound, indicating anchimeric assistance—although not strong enough to offset completely the inductive effect. Finally, *trans*-I provides powerful anchimeric assistance, more than offsetting any inductive effect, and giving a rate 5,000 times as fast as for the unsubstituted brosylate.

Clearly, the ability of halogens to give nucleophilic assistance falls in the order

$$I \gg Br > Cl$$

The —I group is the most nucleophilic: it is the least electronegative and hence the least reluctant to hold the positive charge that develops on it. (Another factor contributing to its nucleophilicity is its size: it is the "softest" and most easily deformable in the transition state; there are empty orbitals of not-too-high energy available for bonding.)

28.17 (a)

II and III are enantiomers
Racemic *threo*-3-phenyl-2-butyl acetate
Optically inactive

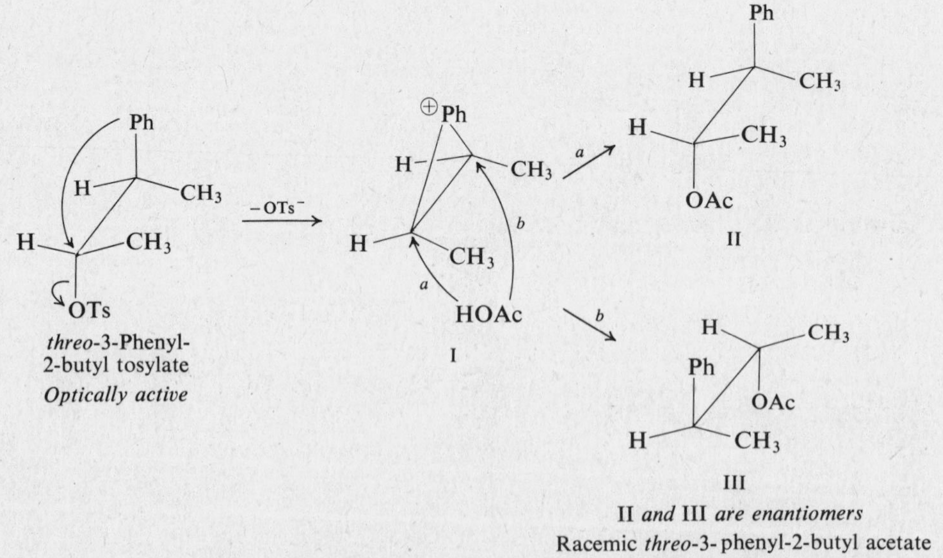

Bridged ion I is achiral, and must give optically inactive product: specifically, because attacks *a* and *b* give enantiomers, and in equal amounts.

(b)

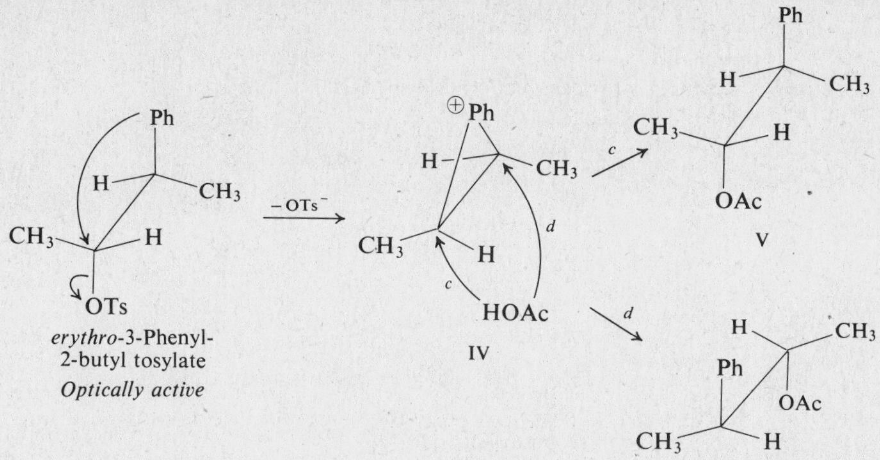

erythro-3-Phenyl-
2-butyl tosylate
Optically active

IV

V

VI

V *and* VI *are the same*
erythro-3-Phenyl-2-butyl acetate
Optically active

Bridged ion IV is chiral. Attack by either path *c* or path *d* gives the same, optically active compound.

(D. J. Cram, "Studies in Stereochemistry. V. Phenonium Sulfonate Ion-pairs as Intermediates in the Intramolecular Rearrangements and Solvolysis Reactions that Occur in the 3-Phenyl-2-butanol System," J. Am. Chem. Soc., **74**, 104 (1952).)

28.18

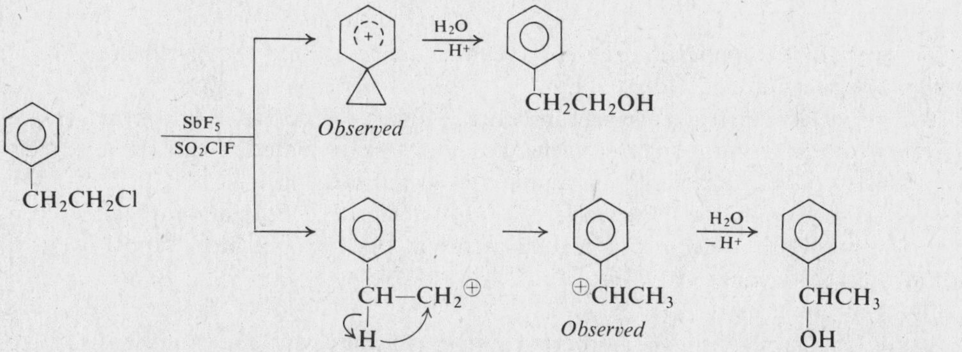

(G. A. Olah and R. D. Porter, "Stable Carbonium Ions. CXV. The Ethylene Phenonium and Ethylene-*p*-toluonium Ions," J. Am. Chem. Soc., **92**, 7627 (1970). For a brief history of the development of the carbonium ion theory, with emphasis on direct observation of long-lived carbonium ions, see G. A. Olah, "Stable, Long-lived Carbonium Ions," Chem. Eng. News, March 27, 1967, p. 76.)

28.19 (a) Starting with camphene hydrochloride, let us write equations analogous to (1) and (2) in Figure 28.9 on page 917. (See Figure 28.10, page 442 of this Study Guide.)

Chloride ion is lost from camphene hydrochloride with anchimeric assistance from C–6, to give bridged cation VI (a hybrid of structures I and II on page 915). Cation VI can undergo back-side attack at either C–1 (path *c*) or C–2 (path *d*). Unlike paths

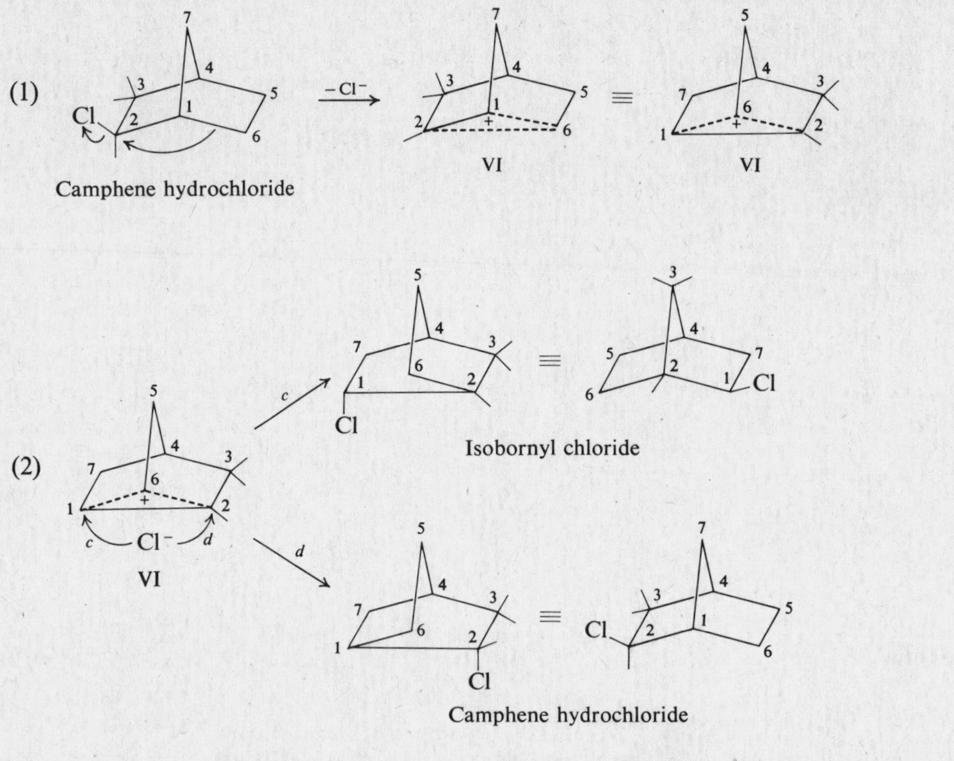

(1)

Camphene hydrochloride

(2)

Isobornyl chloride

Camphene hydrochloride

Figure 28.10. Rearrangement of camphene hydrochloride into isobornyl chloride: the nonclassical ion interpretation.

a and *b* for the norbornyl compound (page 917), paths *c* and *d* are *not* equally likely: C–2 carries a methyl group, and C–1 does not.

Attack at C–1 yields the rearrangement product, isobornyl chloride. Attack at C–2 (to the extent that it occurs) simply regenerates the starting material, camphene hydrochloride, which loses Cl⁻ again and has another try at rearrangement.

Attack *c* at C–1 gives only the *exo* chloride, isobornyl chloride. Formation of the *endo* diastereomer, bornyl chloride, would require attack at the opposite face ("top") of C–1, and this is partially bonded to C–6.

(b) Bridged cation III (page 917) in the norbornyl system is achiral, and the product is necessarily optically inactive; specifically, because attacks *a* and *b* yield enantiomers, and are equally likely.

The bridged cation here, VI, is *not* achiral: it is a less symmetrical structure than III because of the methyl at C–2 (as well as the methyls at C–3). It gives optically active products because paths *c* and *d* do not produce enantiomers—they give different compounds, each of which is optically active.

(For a series of papers on nonclassical ions, see P. D. Bartlett, *Nonclassical Ions*, W. A. Benjamin, New York, 1965. For the opposing point of view: H. C. Brown, "The Nonclassical Ion Problem," Chem. Eng. News, February 13, 1967, p. 86; or H. C. Brown, "The Norbornyl Cation—Classical or Non-classical?" Chemistry in Britain, **2**, 199 (May 1966). Finally G. A. Olah *et al.*, "Stable Carbonium Ions. C. The Structure of the Norbornyl Cation," J. Am. Chem. Soc., **92**, 4627 (1970).)

1. Successive hydrogen shifts occur to give rearrangement: first, from a 1° cation to a 2° cation; and then, among various 2° cations. As part of each cation is consumed by reaction with H_2O, there is less left for the next rearrangement.

(a) 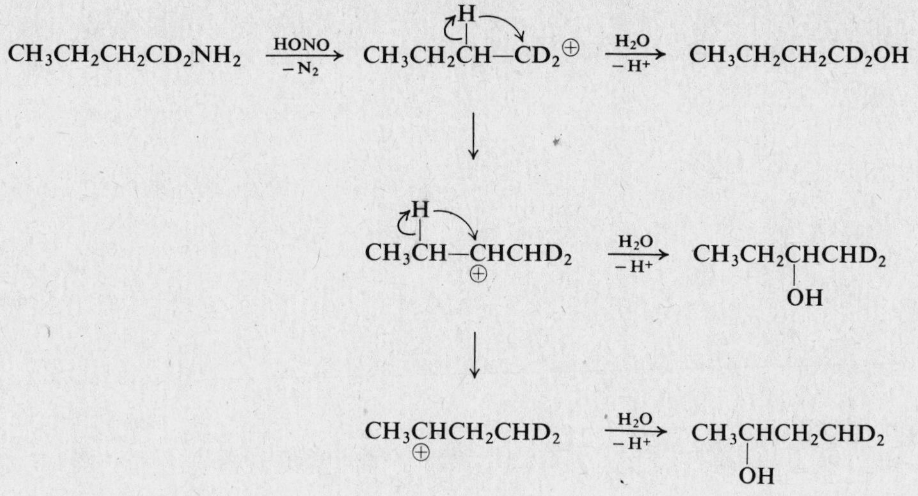 $CH_3CH_2CH_2CD_2NH_2 \xrightarrow[-N_2]{HONO} CH_3CH_2\overset{\curvearrowright H}{CH}-CD_2^{\oplus} \xrightarrow[-H^+]{H_2O} CH_3CH_2CH_2CD_2OH$

$$\downarrow$$

$CH_3\overset{\curvearrowright H}{CH}-\underset{\oplus}{C}HCHD_2 \xrightarrow[-H^+]{H_2O} CH_3CH_2\underset{\underset{OH}{|}}{C}HCHD_2$

$$\downarrow$$

$CH_3\underset{\oplus}{C}HCH_2CHD_2 \xrightarrow[-H^+]{H_2O} CH_3\underset{\underset{OH}{|}}{C}HCH_2CHD_2$

(b) 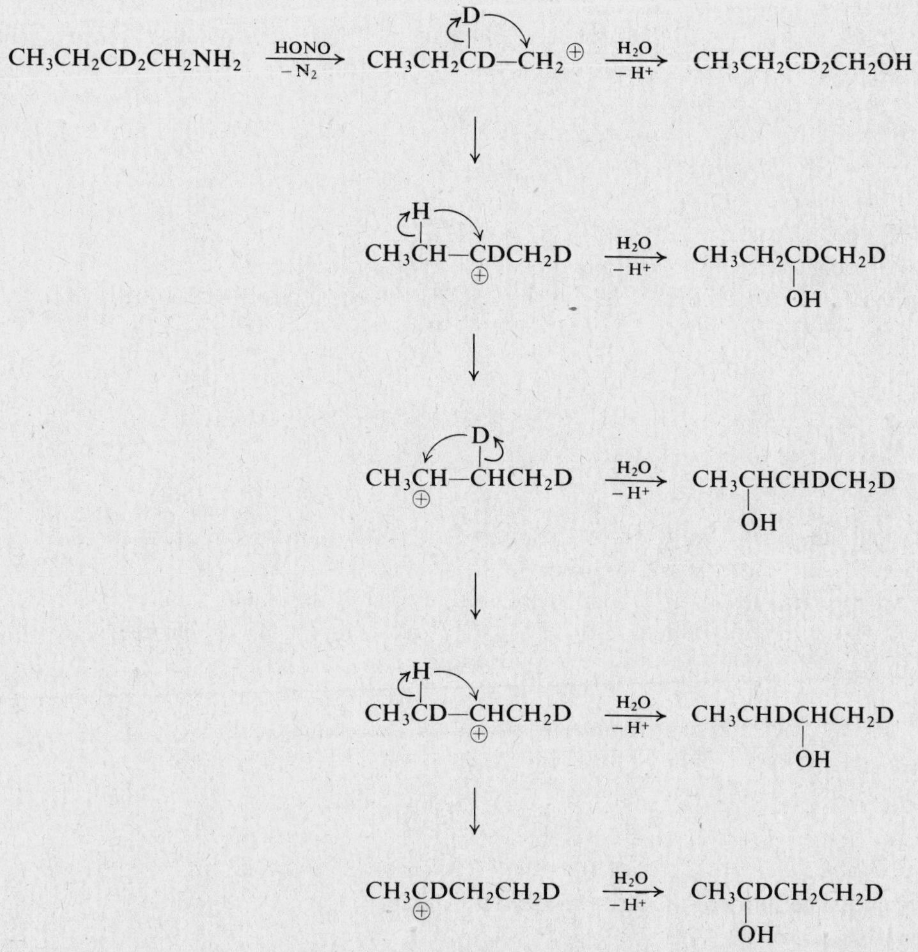 $CH_3CH_2CD_2CH_2NH_2 \xrightarrow[-N_2]{HONO} CH_3CH_2\overset{\curvearrowright D}{CD}-CH_2^{\oplus} \xrightarrow[-H^+]{H_2O} CH_3CH_2CD_2CH_2OH$

$$\downarrow$$

$CH_3\overset{\curvearrowright H}{CH}-\underset{\oplus}{C}DCH_2D \xrightarrow[-H^+]{H_2O} CH_3CH_2\underset{\underset{OH}{|}}{C}DCH_2D$

$$\downarrow$$

$CH_3\underset{\oplus}{C}H-\overset{\curvearrowleft D \curvearrowright}{C}HCH_2D \xrightarrow[-H^+]{H_2O} CH_3\underset{\underset{OH}{|}}{C}HCHDCH_2D$

$$\downarrow$$

$CH_3\underset{\oplus}{C}D-\overset{\curvearrowright H}{C}HCH_2D \xrightarrow[-H^+]{H_2O} CH_3\underset{\underset{OH}{|}}{C}HDCHCH_2D$

$$\downarrow$$

$CH_3\underset{\oplus}{C}DCH_2CH_2D \xrightarrow[-H^+]{H_2O} CH_3\underset{\underset{OH}{|}}{C}DCH_2CH_2D$

2. Migration of ring carbon occurs.

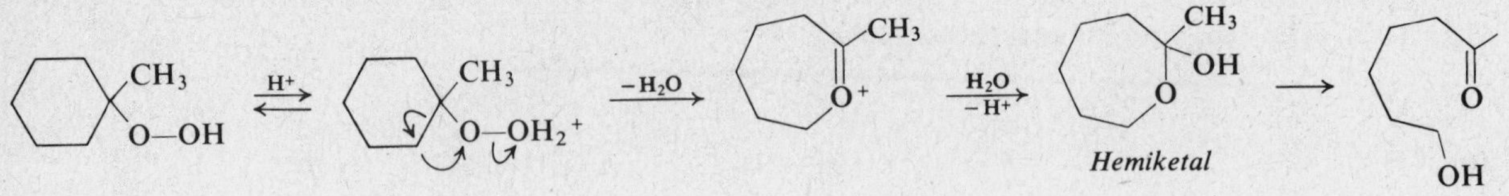

Hemiketal

or

$$HOCH_2CH_2CH_2CH_2CH_2CCH_3$$
$$\overset{\|}{O}$$

Alcohol–ketone
7-Hydroxy-2-heptanone

3. (a) Try iodoform and Tollens' tests: methyl benzyl ketone, positive iodoform; α-phenyl-propionaldehyde, positive Tollen's; ethyl phenyl ketone, negative iodoform and Tollens'.

(b) methyl benzyl ketone
   a singlet, 3H
   b singlet, 2H
   c (aromatic), 5H

α-phenylpropionaldehyde
   a doublet, 3H
   b quartet, 1H
      (probably split again
      into doublets)
   c (aromatic), 5H
   d doublet, 1H (CHO)

ethyl phenyl ketone
   a triplet, 3H
   b quartet, 2H
   c (aromatic) 5H

4. (a) The Lossen rearrangement involves migration to electron-deficient nitrogen, with $^-OOCR'$ playing the same part that Br^- does in the Hofmann rearrangement (Sec. 28.2) or N_2 in the Curtius rearrangement (Problem 28.2, page 889).

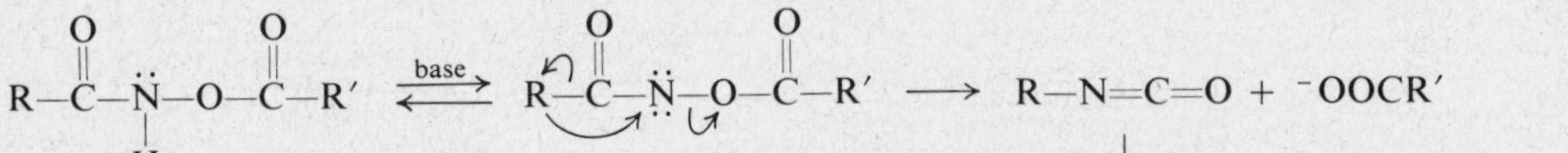

(b) Electron-releasing substituents in R speed up migration, as they do in the Hofmann rearrangement (Sec. 28.5). Electron-withdrawing substituents in R′ make $^-OOCR'$ less basic and hence a better leaving group. The overall rate evidently depends on both the rate of migration of R and the rate of departure of $^-OOCR'$: as it must if the two processes are concerted and occur in the same step.

(For a review of the Lossen and related rearrangements, see P. A. S. Smith in P. de Mayo, *Molecular Rearrangements*, Part 1, Wiley-Interscience, New York, 1963, pp. 528–558.)

5. (a) There are three possible shifts to the electron-deficient oxygen.
   (i) Migration of hydrogen:

444

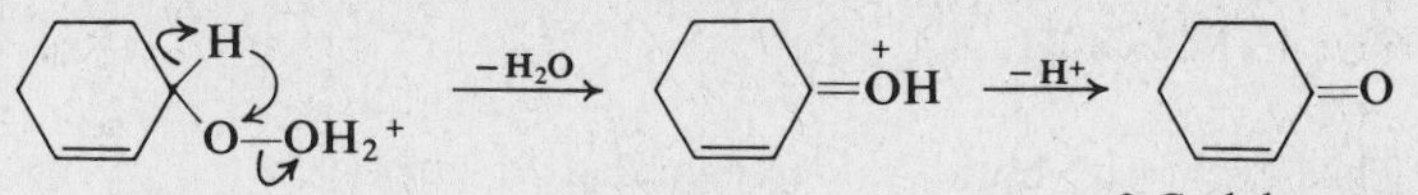

2-Cyclohexenone

(ii) Migration of an alkyl group, $-CH_2$:

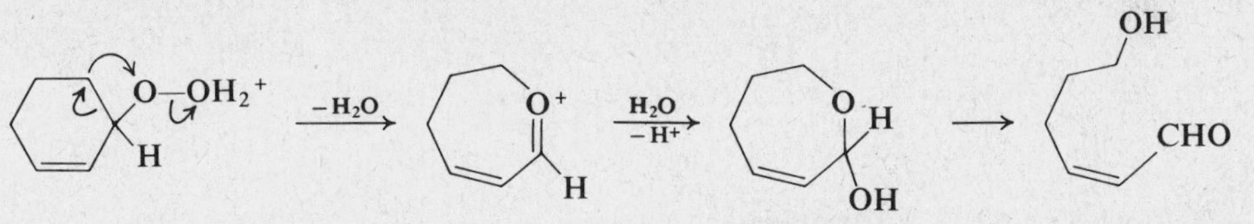

$$\xrightarrow{-H_2O} \quad \xrightarrow{H_2O \atop -H^+} \quad Hemiacetal$$

or

$$HOCH_2CH_2CH_2CH=CHCHO$$

Alcohol–aldehyde
6-Hydroxy-2-hexenal

(iii) Migration of a vinyl group, $-CH=CH$:

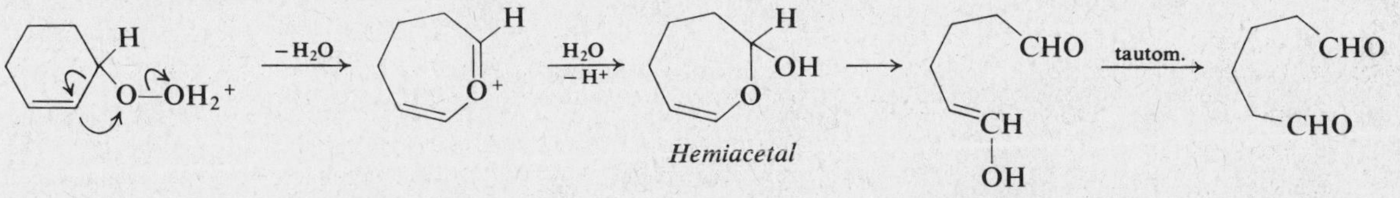

$$\xrightarrow{-H_2O} \quad \xrightarrow{H_2O \atop -H^+} \quad Hemiacetal \quad \xrightarrow{} \quad \xrightarrow{tautom.}$$

Alcohol–aldehyde
Enol

or

$$OHCCH_2CH_2CH_2CH_2CHO$$

Adipaldehyde
Keto

Of these only the last gives adipaldehyde. Furthermore, we realize that an intramolecular (acid-catalyzed) aldol condensation of adipaldehyde can give cyclopentene-1-carboxaldehyde. (In an acid-catalyzed aldol condensation, we remember (Problem 21.12, page 710), acid performs two functions: it catalyzes the enolization of one molecule (the nucleophile); and it protonates the other carbonyl group and makes it more electrophilic.)

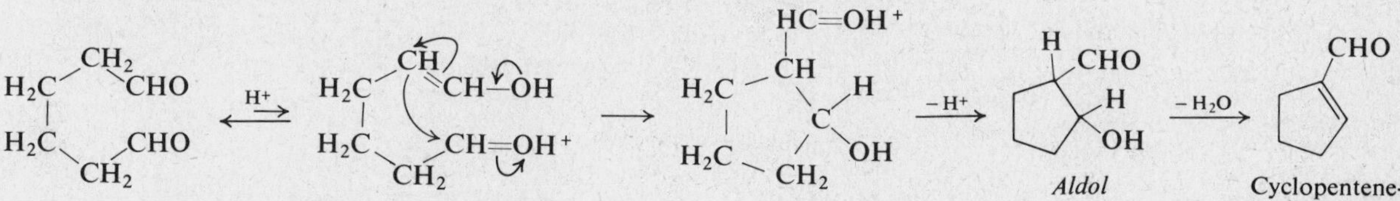

$$\xrightarrow{H^+} \quad \xrightarrow{} \quad \xrightarrow{-H^+} \quad \xrightarrow{-H_2O}$$

Enol–protonated aldehyde *Aldol* Cyclopentene-1-carboxaldehyde

(b) Evidently a vinyl group migrates faster than hydrogen or alkyl. We can see why this might be so: like aryl, vinyl has extra π electrons available for bonding. Migration of vinyl need not involve a transition state containing pentavalent carbon (IV, page 895), but instead proceeds via an epoxide intermediate.

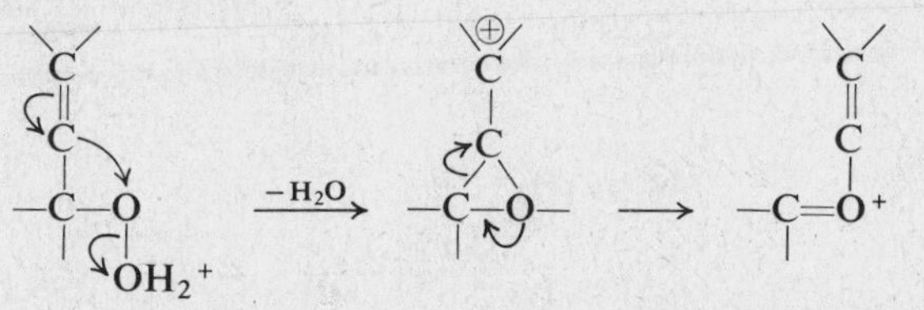

$$\xrightarrow{-H_2O} \quad \xrightarrow{}$$

6. (a) A, PhCONHPh; B, PhNH$_2$; C, PhCOOH.

(b) Clearly, a rearrangement has occurred: both Ph's are bonded to carbon in the oxime, and one is bonded to nitrogen in the product.

(c)

$$\underset{Oxime}{\underset{Ph}{\overset{Ph}{>}}C=N-OH} \;\rightleftharpoons\; \underset{Ph}{\overset{Ph}{>}}C=N \curvearrowright OH_2^{+} \;\longrightarrow\; Ph-\overset{\oplus}{C}=N-Ph \xrightarrow[-H^+]{H_2O} Ph-\underset{OH}{\overset{}{C}}=N-Ph \xrightarrow{taut.} Ph-\underset{\underset{Amide}{}}{\overset{O}{\overset{\|}{C}}}-\overset{H}{N}-Ph$$

N-Phenylbenzamide

(d) By conversion of —OH into a better (less basic) leaving group.

$$\underset{Ph}{\overset{Ph}{>}}C=N-OH \xrightarrow{PCl_5} \underset{Ph}{\overset{Ph}{>}}C=N \curvearrowright Cl \;\longrightarrow\; Ph-\overset{\oplus}{C}=N-Ph + Cl^-$$

$$\Big\downarrow H_2O \;\;-H^+$$

PhCONHPh

$$\underset{Ph}{\overset{Ph}{>}}C=N-OH \xrightarrow{TsCl} \underset{Ph}{\overset{Ph}{>}}C=N \curvearrowright OTs \;\longrightarrow\; Ph\overset{\oplus}{C}=N-Ph + OTs^-$$

$$\Big\downarrow H_2O \;\;-H^+$$

PhCONHPh

(Actually, the exact role of acids in the Beckmann rearrangement is not understood. In (c), we have shown sulfuric acid as protonating —OH, but it may instead convert it into —OSO$_3$H. PCl$_5$ may convert —OH into —OPCl$_4$. See the references given below.)

(e) acetone oxime $\rightleftharpoons$ $\underset{CH_3}{\overset{CH_3}{>}}C=N \curvearrowright OH_2^{+}$ $\longrightarrow$ $CH_3-\underset{}{\overset{O}{\overset{\|}{C}}}-\overset{H}{N}-CH_3$

N-Methylacetamide

(Z)-acetophenone oxime $\rightleftharpoons$ $\underset{Ph}{\overset{CH_3}{>}}C=N \curvearrowright OH_2^{+}$ $\longrightarrow$ $Ph-\underset{}{\overset{O}{\overset{\|}{C}}}-\overset{H}{N}-CH_3$

N-Methylbenzamide

(E)-acetophenone oxime $\rightleftharpoons$ $\underset{CH_3}{\overset{Ph}{>}}C=N \curvearrowright OH_2^{+}$ $\longrightarrow$ $CH_3-\underset{}{\overset{O}{\overset{\|}{C}}}-\overset{H}{N}-Ph$

N-Phenylacetamide

(Z)-p-nitrobenzophenone oxime $\rightleftharpoons$ $\underset{p\text{-}NO_2C_6H_4}{\overset{Ph}{>}}C=N \curvearrowright OH_2^{+}$ $\longrightarrow$ $p\text{-}NO_2C_6H_4-\underset{}{\overset{O}{\overset{\|}{C}}}-\overset{H}{N}-Ph$

N-Phenyl-p-nitrobenzamide

(E)-p-nitrobenzophenone oxime $\rightleftharpoons$ $\underset{Ph}{\overset{p\text{-}NO_2C_6H_4}{>}}C=N \curvearrowright OH_2^{+}$ $\longrightarrow$ $Ph-\underset{}{\overset{O}{\overset{\|}{C}}}-\overset{H}{N}-C_6H_4NO_2\text{-}p$

N-(p-Nitrophenyl)-benzamide

(Z)-methyl-n-propyl ketoxime $\xrightarrow{\text{H}^+}$

$$\underset{n\text{-Pr}}{\overset{CH_3}{\diagdown}}C=N\overset{\curvearrowright}{\underset{\curvearrowright OH_2^+}{}}$$

$\longrightarrow$

$$n\text{-Pr}-\overset{\overset{O}{\|}}{C}-\overset{\overset{H}{|}}{N}-CH_3$$
N-Methyl-n-butyramide

(E)-methyl-n-propyl ketoxime $\xrightarrow{\text{H}^+}$

$$\underset{CH_3}{\overset{n\text{-Pr}}{\diagdown}}C=N\overset{\curvearrowright}{\underset{\curvearrowright OH_2^+}{}}$$

$\longrightarrow$

$$CH_3-\overset{\overset{O}{\|}}{C}-N-Pr\text{-}n$$
N-n-Propylacetamide

(f) Hydrolyze amides and identify the resulting amines and acids.

Suitably substituted oximes exist as geometric isomers (Problem 10, Chapter 22, page 744), and the product of rearrangement depends upon which isomer is involved: as shown in (e), the group *trans* (*anti*) to the leaving group is the one that migrates.

(For a review of the Beckmann rearrangement, see: P. B. D. de la Mare in P. de Mayo, *Molecular Rearrangements*, Part 1, Wiley-Interscience, New York, 1963, pp. 483–507; or W. Z. Heldt and L. G. Donaruma, O. R. XI-1.)

7. Hofmann rearrangement with R = NH_2.

$$\underset{\overset{\|}{O}}{H_2N-C-NH_2} \xrightarrow{\text{OBr}^-} \underset{\overset{\|}{O}}{H_2N-C-NHBr} \xrightarrow{\text{OH}^-} \underset{\overset{\|}{O}}{H_2N-\overset{\curvearrowright}{C}-\overset{\ominus}{N}-Br} \xrightarrow{-Br^-} O=C=N-NH_2$$
Urea

$$\downarrow \text{OH}^-$$

$$H_2N-NH_2 + CO_3{}^{--}$$
Hydrazine

$$\downarrow \text{oxid.}$$

$$N_2$$

8. (a) First, a hydroperoxide is formed, probably by a mechanism like this:

$$Ar_3C-OH \underset{\xleftarrow{\text{}}}{\overset{\text{H}^+}{\rightleftharpoons}} Ar_3C-OH_2{}^+ \xrightarrow{-H_2O} Ar_3C^{\oplus} \xrightarrow{H_2O_2} Ar_3C-\overset{\overset{H}{|\oplus}}{O}-OH \xrightarrow{-H^+} Ar_3C-O-OH$$
A hydroperoxide

This then undergoes rearrangement as discussed in Sec. 28.6:

$$\underset{\textit{Hydroperoxide}}{Ar_3C-O-OH} \underset{\xleftarrow{\text{}}}{\overset{\text{H}^+}{\rightleftharpoons}} Ar_2\overset{\overset{\curvearrowright Ar}{|}}{C}-\underset{\curvearrowright}{O}-OH_2{}^+ \longrightarrow Ar_2C=\overset{+}{O}-Ar \xrightarrow[-H^+]{H_2O} \underset{\textit{Hemiketal}}{Ar_2\overset{\overset{OH}{|}}{C}-OAr} \longrightarrow \underset{\textit{Ketone}}{Ar_2C=O} + \underset{\textit{Phenol}}{ArOH}$$

(b) The relative migratory aptitude (Sec. 28.7) of the aryl groups involved is

$$p\text{-}CH_3OC_6H_4- > Ph-\cdot > p\text{-}ClC_6H_4-$$

The expected products are:

$$\underset{\overset{\|}{O}}{C_6H_5-C-C_6H_5} + p\text{-}CH_3OC_6H_4OH \quad \text{and} \quad \underset{\overset{\|}{O}}{p\text{-}ClC_6H_4-C-C_6H_5} + C_6H_5OH$$

9. (a) Either of two 3° cations can form, followed by migration either of —CH$_2$ of the ring, or of C$_2$H$_5$.

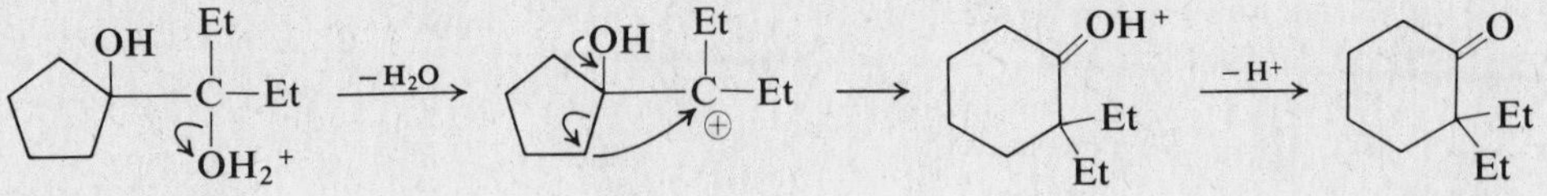

Protonated I　　　　　　　3° cation　　　　　　　　　　　　　　　　　II
(R = Et)　　　　　　　　　　　　　　　　　　　　　　　　　　　　　　　(R = Et)

Protonated I　　　　　　　3° cation　　　　　　　　　　　　　　　　　III
(R = Et)　　　　　　　　　　　　　　　　　　　　　　　　　　　　　　　(R = Et)

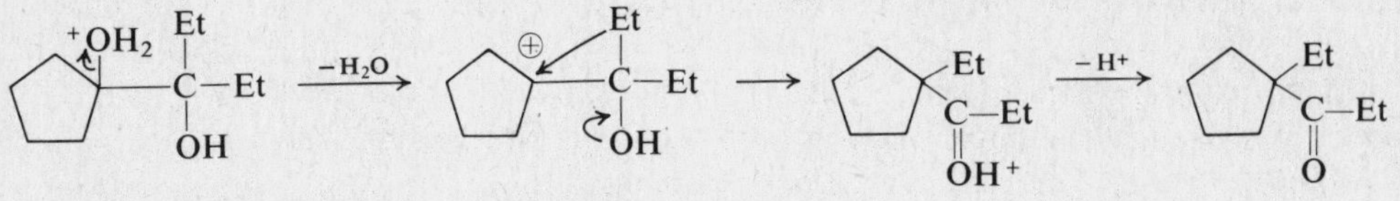

(b) Only the more stable, benzylic cation forms, so that only —CH$_2$ of the ring can migrate.

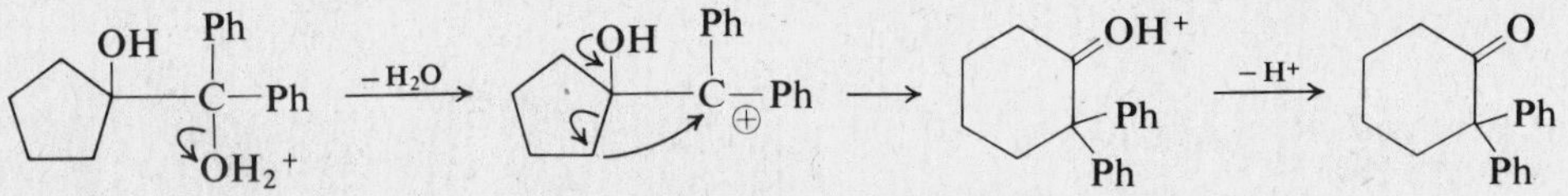

Protonated I　　　　　Benzylic cation　　　　　　　　　　　　　　　　II
(R = Ph)　　　　　　　　　　　　　　　　　　　　　　　　　　　　　　　(R = Ph)

(c) Either 3° cation can form, followed by migration of CH$_3$ or —CH$_2$ of the ring.

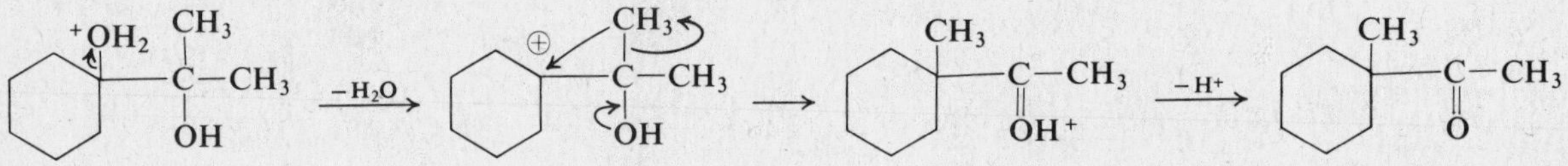

3° cation

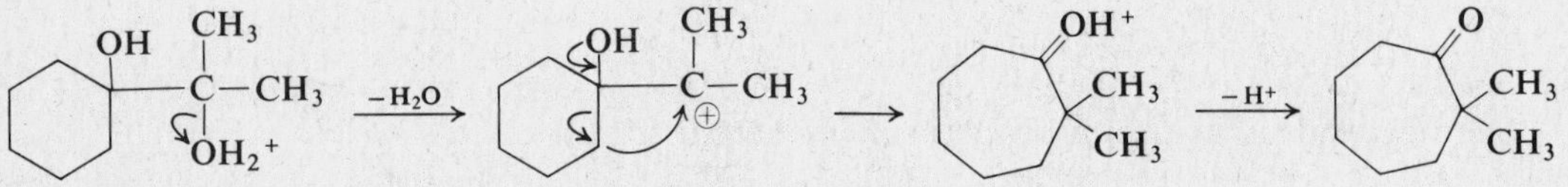

3° cation

(d) Loss of either —OH gives the same cation; we predict preferred migration of Ph.

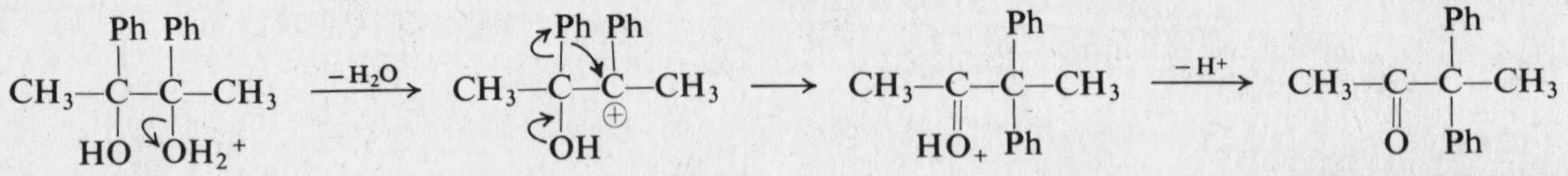

3,3-Diphenyl-2-butanone
Positive iodoform test
Predicted product

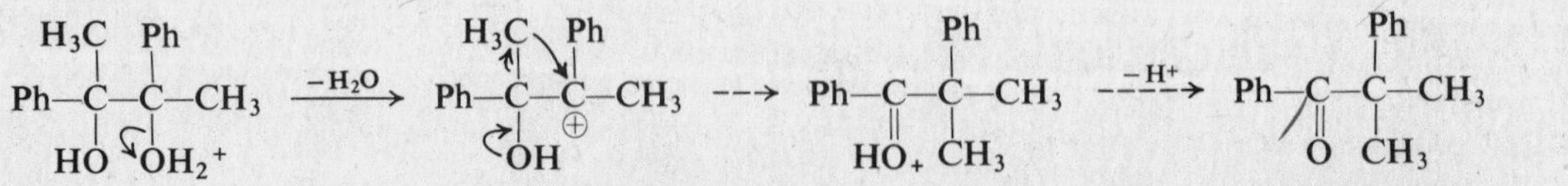

1,2-Diphenyl-
2-methylpropanone
Negative iodoform test

448

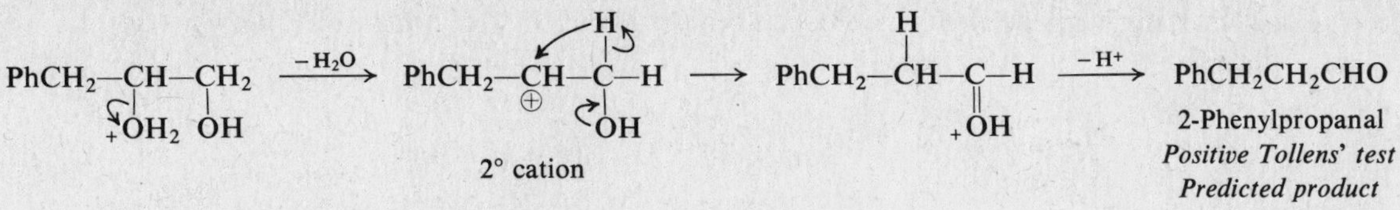

$$PhCH_2\!-\!CH\!-\!CH_2 \xrightarrow{-H_2O} PhCH_2\!-\!CH\!-\!C\!-\!H \longrightarrow PhCH_2\!-\!CH\!-\!C\!-\!H \xrightarrow{-H^+} PhCH_2CH_2CHO$$

2° cation

2-Phenylpropanal
Positive Tollens' test
Predicted product

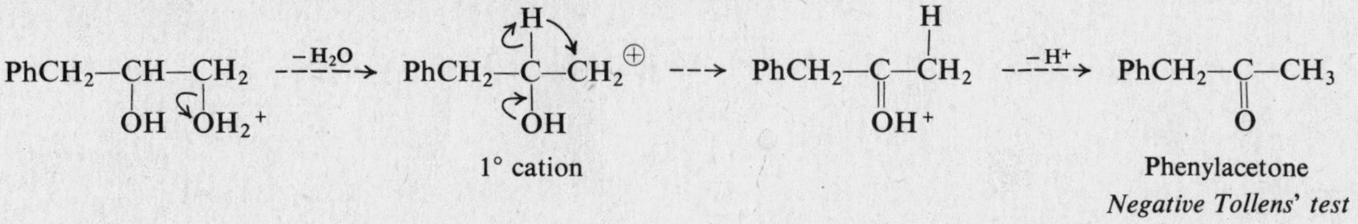

$$PhCH_2\!-\!CH\!-\!CH_2 \xrightarrow{-H_2O} PhCH_2\!-\!C\!-\!CH_2 \dashrightarrow PhCH_2\!-\!C\!-\!CH_2 \xrightarrow{-H^+} PhCH_2\!-\!C\!-\!CH_3$$

1° cation

Phenylacetone
Negative Tollens' test

10. 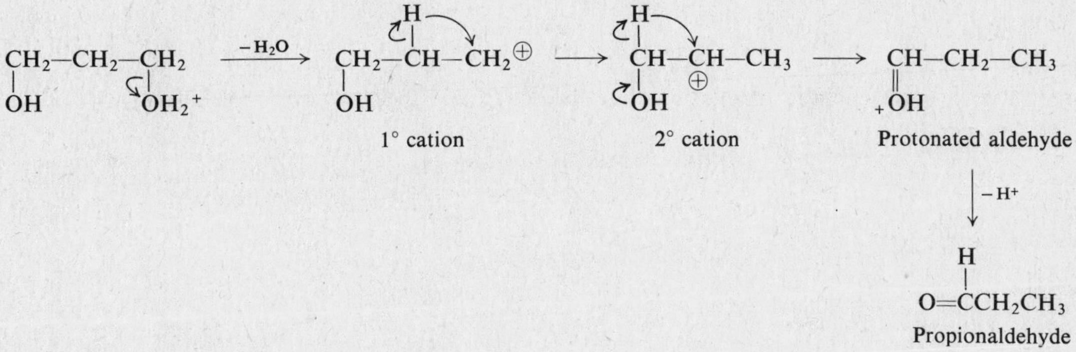
$$CH_2\!-\!CH_2\!-\!CH_2 \xrightarrow{-H_2O} CH_2\!-\!CH\!-\!CH_2 \longrightarrow CH\!-\!CH\!-\!CH_3 \longrightarrow CH\!-\!CH_2\!-\!CH_3$$

1° cation   2° cation   Protonated aldehyde

$$\downarrow {-H^+}$$

$$O\!=\!CCH_2CH_3$$
Propionaldehyde

11. (a) 1,2-shift of R from boron to oxygen.

(1) $R_3B + OOH^- \longrightarrow R_3B\!-\!O\!-\!OH$

(2) $R_2B\!-\!O\!-\!OH \longrightarrow R_2B\!-\!O\!-\!R + OH^-$

then (1), (2), (1), (2) for other R's

(b) Retention of configuration in R. This is what is observed for other 1,2-shifts (Sec. 28.4).

12. (a) Both *cis-* and *trans-*alcohols give the same cyclic bromonium ion upon loss of water from an initial oxonium ion: the *trans-*alcohol probably by an S_N2-like reaction (reaction (1), page 906); the *cis-*alcohol necessarily via an open carbonium ion. Back-side attack by Br^- on the bridged ion gives the *trans-*dibromide.

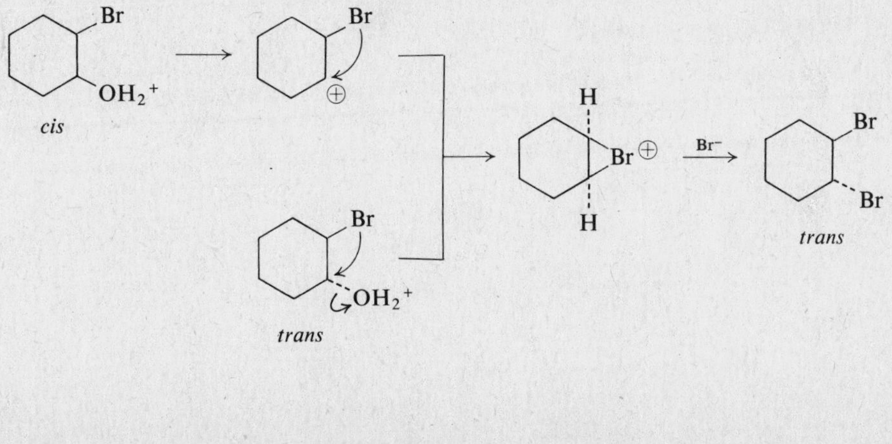

(b) In each case there is nucleophilic attack by OH⁻ at the least hindered position: at —CH₂Br in IV to give VI directly;

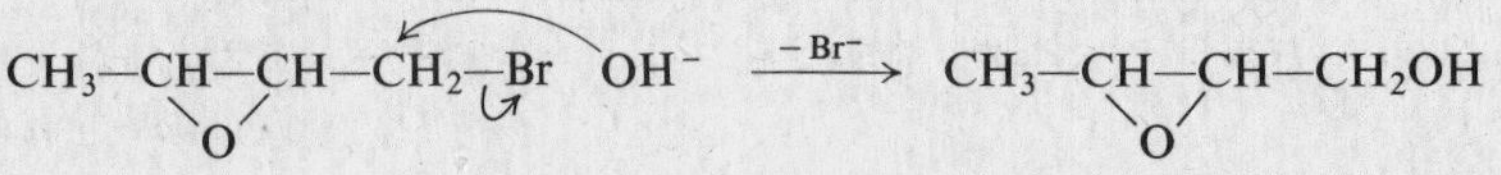

$$CH_3-CH-CH-CH_2-Br \quad OH^- \xrightarrow{-Br^-} CH_3-CH-CH-CH_2OH$$

IV   VI

at C–1 of V to form XXVI, which reacts further intramolecularly (page 563) to yield VI.

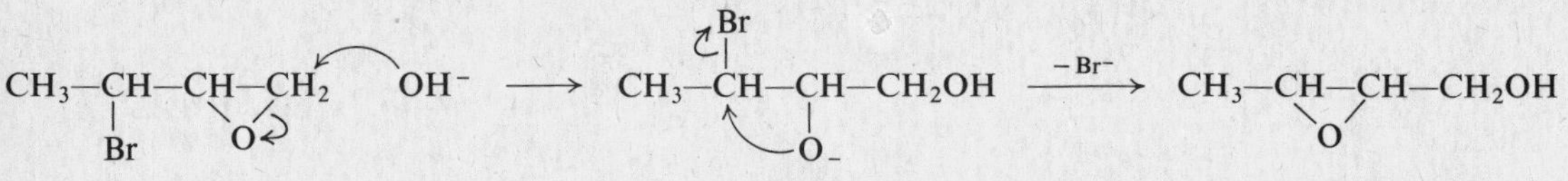

$$CH_3-CH-CH-CH_2 \quad OH^- \longrightarrow \overset{Br}{CH_3-CH-CH-CH_2OH} \xrightarrow{-Br^-} CH_3-CH-CH-CH_2OH$$

V   XXVI   VI

(c) Three successive nucleophilic attacks: one intermolecular and then two intramolecular.

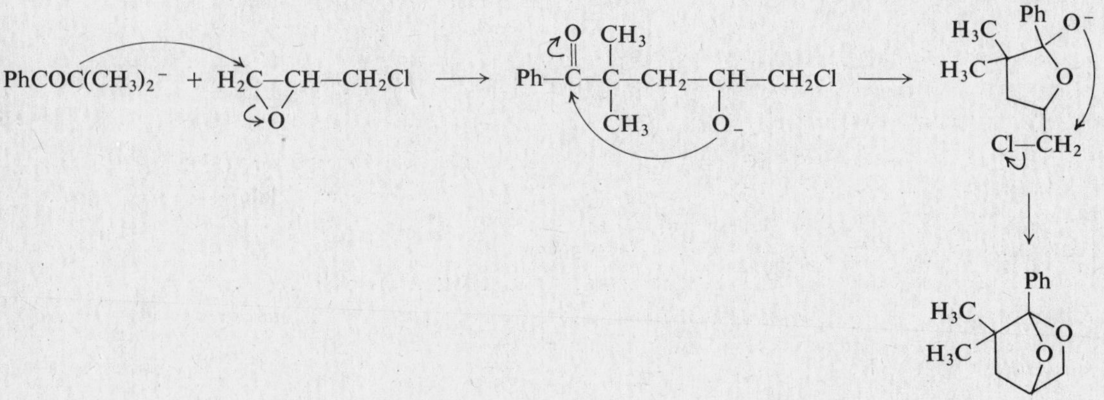

$$PhCOC(CH_3)_2^- + H_2C-CH-CH_2Cl \longrightarrow Ph-C-C-CH_2-CH-CH_2Cl \longrightarrow$$

(d) Total solvolysis is the sum of two reactions: one *solvent-assisted* and the other *aryl-assisted* (page 914). For this primary substrate the solvent-assisted reaction is probably

Aryl assisted

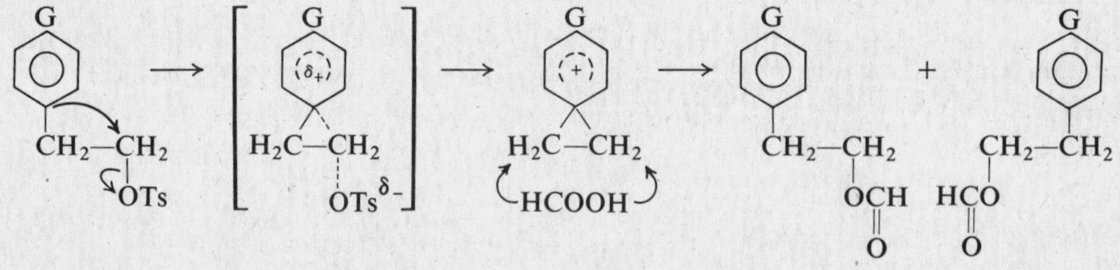

Solvent-assisted

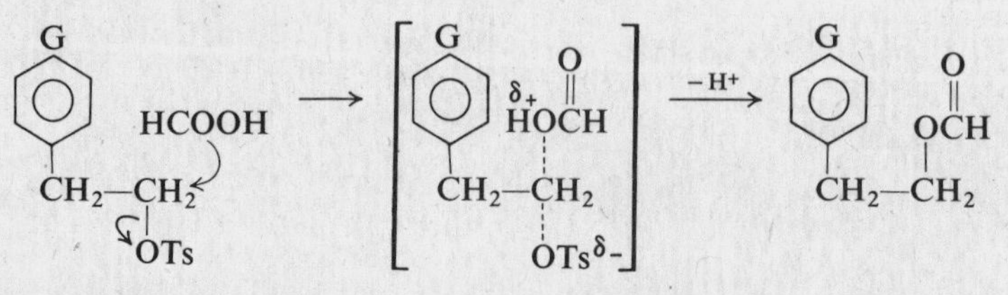

essentially S_N2. The aryl-assisted reaction gives the bridged cation, which is then attacked by solvent.

As the electron-releasing tendency of G increases along the series $-H$, $-OCH_3$, $-O^-$, so does the nucleophilic power of aryl. As aryl becomes more nucleophilic, two things

$$CH_2{-}CH_2{-}OTs \longrightarrow \text{Benzenonium ion}$$

$$\overset{OCH_3}{CH_2{-}CH_2{-}OTs} \longrightarrow \overset{+OCH_3}{\text{Oxonium ion}}$$

$$\overset{O^-}{CH_2{-}CH_2{-}OTs} \longrightarrow \overset{O}{\text{Ketone (Dienone)}}$$

happen: the fraction of reaction that is aryl-assisted rises; and the strength of that assistance increases. The results are the enormous increases in rate that are observed.

The facts agree with this interpretation. The solvent-assisted reaction should give no scrambling of the label in $PhCH_2{}^{14}CH_2OTs$; the aryl-assisted reaction should give scrambling. The solvent-assisted reaction should give inversion of configuration: *threo*-PhCHD-CHDOTs, say, should give *erythro*-acetate. The aryl-assisted reaction should give net retention as on page 912: *threo*-PhCHDCHDOTs should give *threo*-acetate. The *amount* of retention should exactly equal the *amount* of scrambling. (Put differently, the amount of retention should be exactly twice the amount of "rearrangement" of the label.) All this was found to be true. (J. L. Coke *et al.*, "Carbonium Ions. II. Mechanism of Acetolysis of 2-Phenylethyl Tosylate," J. Am. Chem. Soc., **91**, 1154 (1969); R. J. Jablonski and E. I. Snyder, "Stereochemistry of Solvolysis of 2-Phenylethyl Sulfonate Esters. The Phenonium Ion–Equilibrating Classical Ion Problem," *ibid.*, 4445 (1969).)

The above facts show that there are two competing paths and that one involves a symmetrical intermediate. But the symmetrical intermediate need not be a bridged ion, and it need not be formed with anchimeric assistance; a pair of rapidly equilibrating open carbonium ions would lead to the same results. By the approach of page 914, however, Schleyer showed that the effects of various G's on the rate were greater than those expected (from Hammett constants, Sec. 18.11) for formation of open cations, and that the amount of "extra" speed matches the scrambling and stereochemical data, and hence is due to anchimeric assistance.

(P. von R. Schleyer *et al.*, "Participation by Neighboring Aryl Groups. V. Determination of Assisted and Non-assisted Rates in Primary Systems. Rate–Product Correlations," J. Am. Chem. Soc., **91**, 7508 (1969). The "white paper" that marked *détente* between Schleyer and H. C. Brown (page 914) is: H. C. Brown, C. J. Kim, C. J. Lancelot, and P. von R. Schleyer "Product-Rate Correlation in Acetolysis of *threo*-3-Aryl-2-butyl Brosylates. Supporting Evidence for the Existence of Two Discrete Pathways," J. Am. Chem. Soc., **92**, 5244 (1970).)

(e) As in the base-catalyzed cleavage of epoxides (Sec. 17.15), attack by OH^- is at the least hindered position.

$$CH_3{-}CH{-}CH_2{-}Cl \;(:NEt_2) \longrightarrow CH_3{-}CH{-}CH_2 \;\overset{Et\;\;Et}{\underset{\oplus}{N}} + OH^- \longrightarrow CH_3{-}CH{-}CH_2OH \;(NEt_2)$$

(f) Formation of the bridged sulfonium ion may be either S_N1-like (two steps) or S_N2-like (one step).

Cleavage involves attack by the weak nucleophile Cl^-, and takes place either by an S_N1 reaction (two steps) or, like acid-catalyzed cleavage of epoxides (Sec. 17.15), by a single step that has considerable S_N1-character; in either case, considerable positive charge develops in the transition state, and reaction occurs at the carbon that can best accommodate that positive charge, the secondary carbon.

13. The results suggest the operation of conformational factors, and so we draw what we would expect to be the most stable conformations of the diazonium ions: **XXVII***a* from the *erythro* amine, and **XXVIII***a* from the *threo* amine. We find that in **XXVII***a* phenyl can undergo

	XXVII*a*	XXVIII*a*
	Erythro	*Threo*

back-side migration via a *trans* transition state (methyls apart), and that in **XXVIII***a* methyl can undergo back-side migration via a *trans* transition state (phenyl and methyl apart).

Analogous conformations can be drawn for the brosylates (Sec. 28.12); yet there, only phenyl migrates. How are we to account for the difference? Loss of brosylate occurs with anchimeric help from the neighboring group; phenyl is much more effective at this than methyl (Sec. 28.7) and is the only group migrating. That is to say, migration is S_N2-like; of the competing nucleophiles, phenyl and methyl, phenyl is much more powerful. De-amination, on the other hand, involves the best of all leaving groups, N_2. Reaction needs—and gets—little or no anchimeric assistance from the migrating group. That is to say, reaction is essentially S_N1-like; N_2 is lost, and the group that then migrates is chiefly the one in the best position to do so.

Let us think further about this reaction. Even if migration of phenyl does not provide anchimeric assistance, we would expect it to give the symmetrical bridged cation, which then reacts as in reaction (2) on page 913.

Whether one starts with optically active reactants or not, the bridged cation I (see answer to Problem 28.17) is achiral, and will give racemic *threo* acetate. Bridged cation IV, on the other hand, is chiral and will give optically active *erythro* acetate.

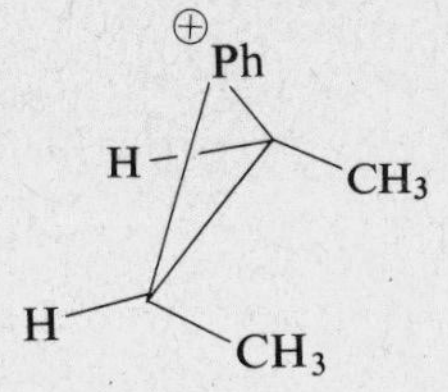

$$\text{I} \xrightarrow{\text{HOAc}} \textit{racemic threo-3-phenyl-2-butyl acetate}$$

I
Achiral

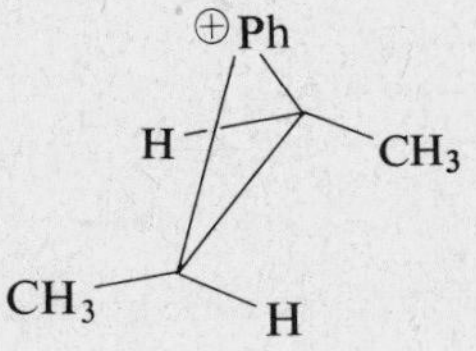

Optically inactive

$$\text{IV} \xrightarrow{\text{HOAc}} \textit{erythro-3-phenyl-2-butyl acetate}$$

IV
Chiral

Optically active

With this in mind, and in light of Collins' work on pinacolic deamination, we can propose this more detailed hypothesis:

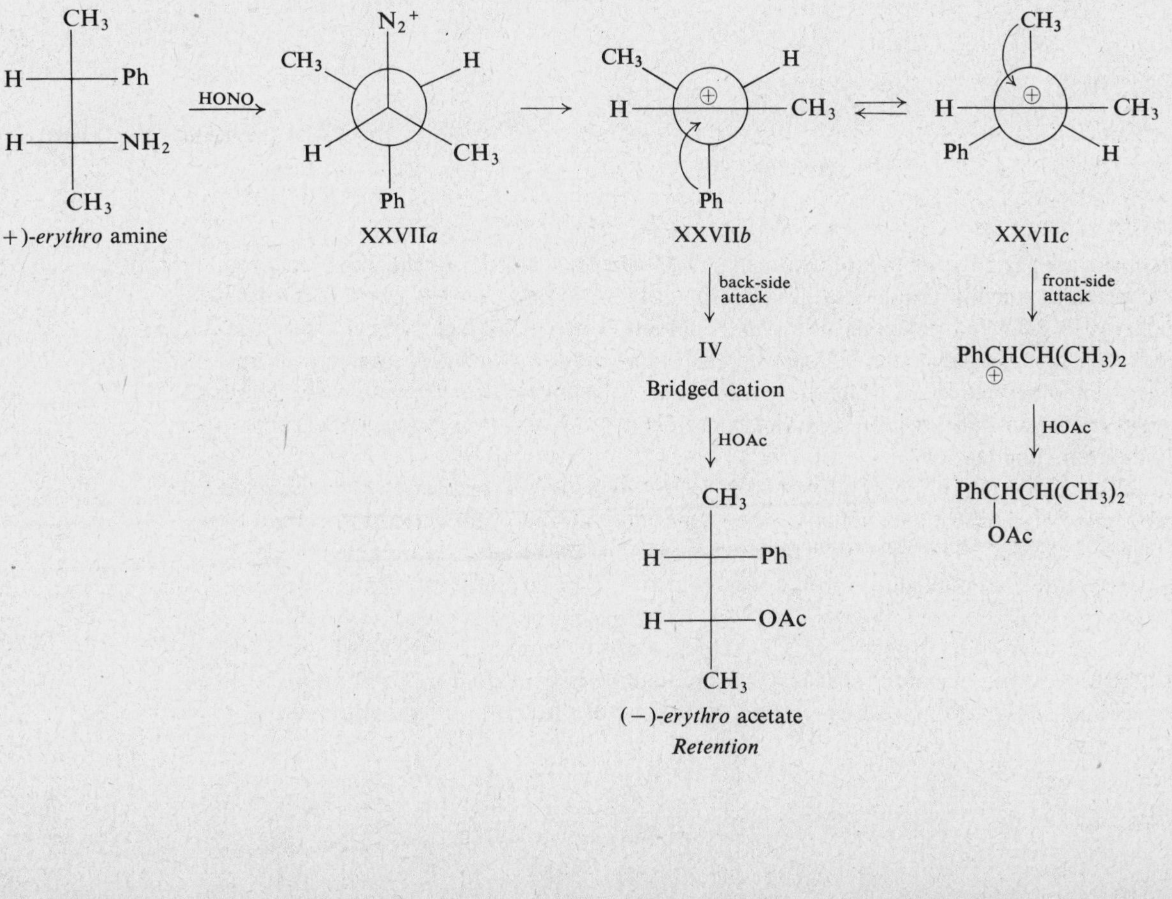

(+)-*erythro* amine $\xrightarrow{\text{HONO}}$ XXVIIa $\longrightarrow$ XXVIIb $\rightleftharpoons$ XXVIIc

XXVIIb — back-side attack → IV (Bridged cation) $\xrightarrow{\text{HOAc}}$ (−)-*erythro* acetate (*Retention*)

XXVIIc — front-side attack → $\underset{\oplus}{\text{PhCHCH(CH}_3)_2}$ $\xrightarrow{\text{HOAc}}$ $\text{PhCHCH(CH}_3)_2$ with OAc

453

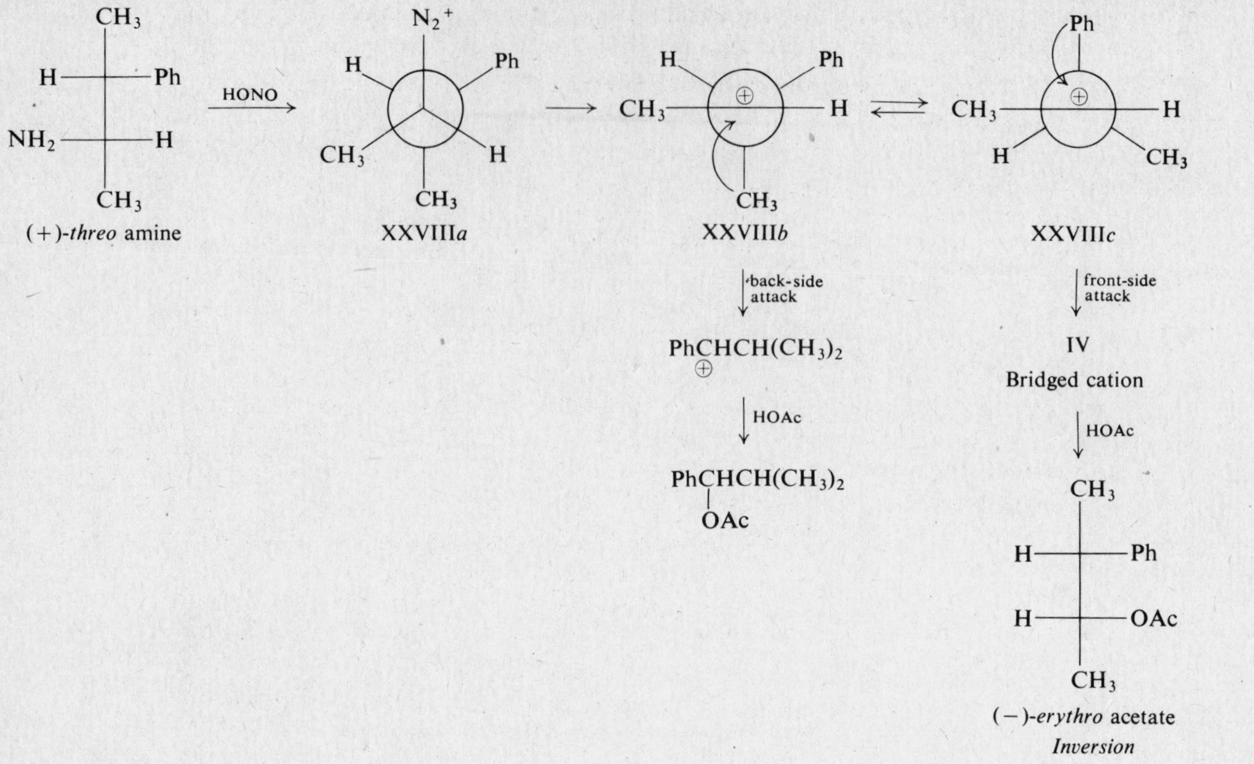

$$CH_3$$
$$H——Ph$$
$$NH_2——H$$
$$CH_3$$
$(+)$-*threo* amine

→ HONO →

XXVIIIa

→

XXVIIIb

⇌

XXVIIIc

back-side attack

PhCHCH(CH_3)_2
⊕

↓ HOAc

PhCHCH(CH_3)_2
OAc

front-side attack

IV
Bridged cation

↓ HOAc

$$CH_3$$
$$H——Ph$$
$$H——OAc$$
$$CH_3$$
$(-)$-*erythro* acetate
Inversion

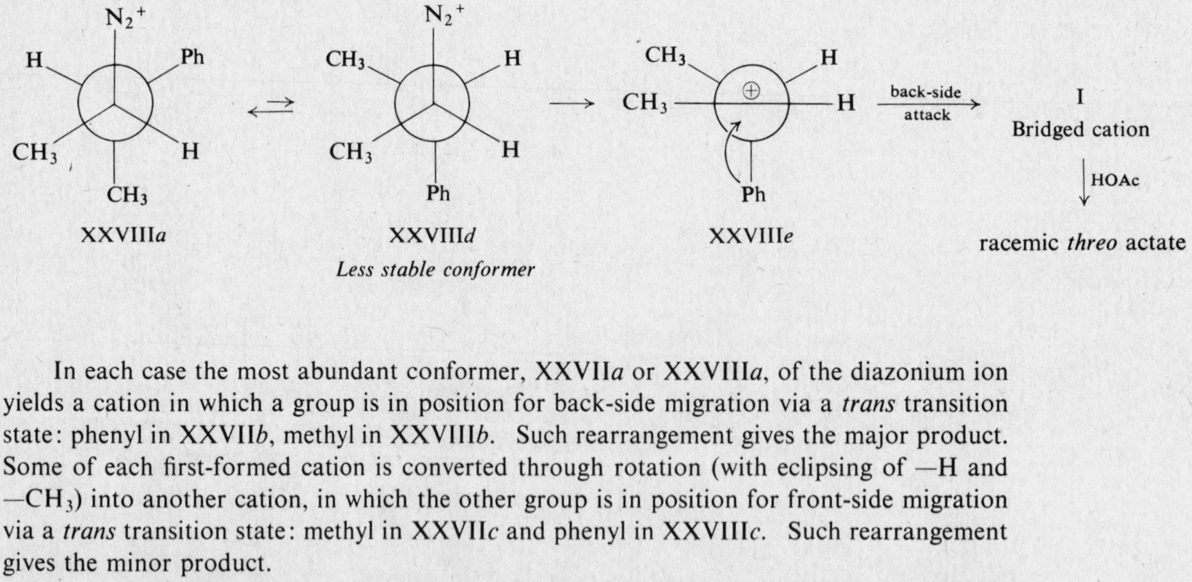

XXVIIIa   ⇌   XXVIIId   →   XXVIIIe   →   I
                Less stable conformer   back-side attack   Bridged cation

↓ HOAc

racemic *threo* acetate

In each case the most abundant conformer, XXVIIa or XXVIIIa, of the diazonium ion yields a cation in which a group is in position for back-side migration via a *trans* transition state: phenyl in XXVIIb, methyl in XXVIIIb. Such rearrangement gives the major product. Some of each first-formed cation is converted through rotation (with eclipsing of —H and —CH₃) into another cation, in which the other group is in position for front-side migration via a *trans* transition state: methyl in XXVIIc and phenyl in XXVIIIc. Such rearrangement gives the minor product.

Now, back-side migration of phenyl in XXVIIb gives bridged cation IV, which yields optically active *erythro* acetate. That is, the $(+)$-*erythro* amine is converted into optically active *erythro* acetate. This is what Cram (who did this work) actually obtained; indeed, it was this retention of configuration that was used to measure the extent of phenyl migration.

Front-side migration of phenyl in XXVIIIc also gives bridged cation IV, which yields optically active *erythro* acetate. That is, $(+)$-threo amine is converted into optically active *erythro* acetate. This, too, was observed by Cram; most of the *erythro* acetate obtained was optically active.

454

(Cram's work was done before Collins', and Cram did not consider the possibility of front-side migration; he attributed optically active *erythro* acetate to simple inversion without phenyl migration. This interpretation may, of course, be the right one. If our interpretation is correct, however, the estimate of phenyl migration must be increased to include the optically active *erythro* product; phenyl migration would then be about equal to methyl migration, still far lower than the 8:1 ratio for the *erythro* amine.)

The *threo* amine also yields racemic *threo* acetate, and this we attribute to bridged cation I, formed by back-side migration of phenyl in a cation (XXVIII*e*) derived from the less stable conformer XXVIII*d*.

The important thing in all this is that methyl migration is much more important in the reaction of one diastereomer than in the reaction of the other, and this must almost certainly be a conformational effect. Analysis of the effect indicates a preference for back-side migration over front-side migration. This, in turn, suggests that migration within the initial cation XXVII*b* or XXVIII*b* is fast compared with rotation to cation XXVII*c* or XXVIII*c*; in addition, the leaving group, N_2 (and the counter-ion), may still shield the front side. Conceivably, migration may take place while $-N_2{}^+$ is still weakly bonded.

Besides phenyl and methyl migration Cram observed, as we might expect, considerable H migration.

(D. J. Cram and J. E. McCarty, "Studies in Stereochemistry. XXVII. Conformational Control of the Migrating Group in the Deamination of 3-Phenyl-2-butylamine," J. Am. Chem. Soc., **79**, 2866 (1957).)

14.

(Y. Pocker and B. P. Ronald, "Kinetics and Mechanism of *vic*-Diol Dehydration. I. The Origin of Epoxide Intermediates in Certain Pinacolic Rearrangements," J. Am. Chem. Soc., **92**, 3385 (1970).)

15. There is competition between reaction without aryl migration, in which the label is undisturbed,

$$ArCH_2\overset{*}{C}H_2-N_2{}^+ \xrightarrow[-H^+]{H_2O} ArCH_2\overset{*}{C}H_2-OH + N_2$$

$$\downarrow \text{oxidn.}$$

$$ArCOOH$$

and with aryl migration, in which the label is scrambled.

$$ArCOOH \qquad HOO\overset{*}{C}Ar$$
$$(50\%) \qquad\qquad (50\%)$$

Since the bridged cation gives product with *half* the label in each position, the fraction of reaction that involves this intermediate is:

$$p\text{-}NO_2C_6H_4 \ 16\%, \quad C_6H_5 \ 54\%, \quad p\text{-}CH_3OC_6H_4 \ 90\%.$$

As expected (see the answer to Problem 12d, above), electron-release in Ar favors migration.

(J. D. Roberts and C. M. Regan, "Rearrangements in the Reactions of 2-(4-substituted)-phenylethylamines-1-C[14]," J. Am. Chem. Soc., **75**, 2069 (1953).)

The absence of the product of hydrogen migration, $ArCHOHCH_3$, led Roberts and Regan to suggest that the open cation $ArCH_2CH_2{}^+$ may not be an intermediate, but that, instead, aryl migration is synchronous with loss of N_2. More recent stereochemical work supports this proposal. It was found that 52% of $PhCHDCHDNH_2$ reacts with retention of configuration, in close agreement with the value of 54% above. This strongly suggests complete stereospecificity in each process: inversion in the non-scrambling reaction, and retention in the scrambling reaction. The most likely interpretation is that competition is between direct S_N2 displacement by solvent and S_N2-like migration, that is, aryl-assisted displacement of $-N_2{}^+$. (E. I. Snyder, "Stereochemistry of Deamination of 2-Phenylethyl-amine," J. Am. Chem. Soc., **91**, 5118 (1969).)

These results are analogous to what is observed for solvolysis of the corresponding tosylates, $ArCH_2CH_2OTs$, as described in the answer to Problem 12d, above. For these primary compounds, both $-OTs$ and $-N_2{}^+$ are displaced with aryl assistance. In contrast, for the secondary (3-phenyl-2-butyl) compounds, deamination takes a different course than solvolysis of the tosylates (Problem 13, above). Where a more stable open carbonium ion can be formed, only the tosylates react with aryl assistance.

16. (a) In the rate-determining step, $-OAc$ is lost to give an open carbonium ion; this cation rapidly rearranges by phenyl migration to give statistical equilibration of each label before the cation reacts with acetic acid to give product: the chain-C* is equally divided between the two possible positions, and the ring-C* is equally divided among the three possible positions (Figure 28.11, page 457 of this Study Guide).

The rate (k_1) at which chain-C* approaches a 50:50 distribution, the rate (k_2) at which ring-C* approaches 33:67 distribution, and the rate at which $-OAc$ is replaced by $-OAc*$ are all equal, since each is equal to the rate at which $-OAc$ is lost.

If, instead, a bridged cation (formed in either an S_N1-like or S_N2-like process) were the intermediate, each loss of $-OAc$ would lead to a 50:50 mixture of ring-C* products. That

is, the rate that would equal k_1 and k_3 would be the rate at which ring-C* approaches 50:50 distribution, not 33:67 distribution. Although ring-C* might ultimately reach the statistical 33:67 distribution, the rate at which it approaches this (k_2) would be *less than* k_1—contrary to fact.

Study of reaction (3) was necessary to rule out a possibility that might not have occurred to us, but which was never the less a real possibility: equilibration via *internal return*. Under

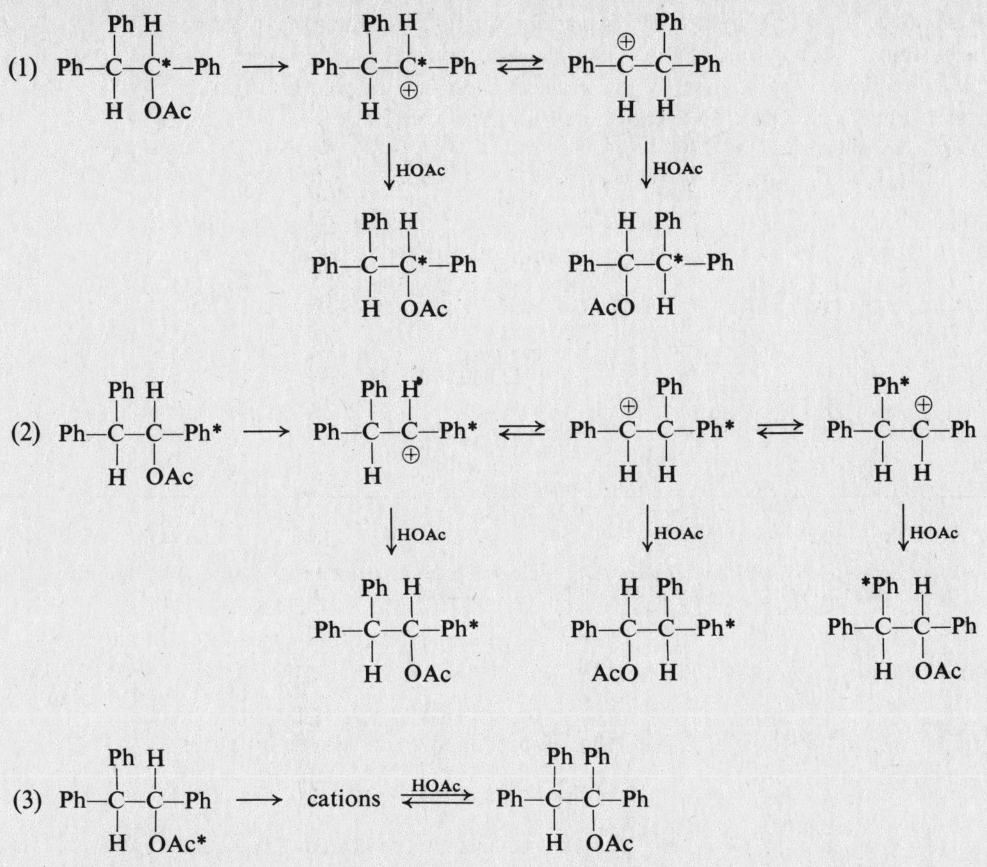

(1)   Ph—C(Ph)(H)—C*(H)(OAc)—Ph ⟶ Ph—C(Ph)(H)—C*⊕(H)—Ph ⇌ Ph—C⊕(Ph)—C(H)(H)—Ph

     ↓ HOAc   ↓ HOAc

     Ph—C(Ph)(H)—C*(H)(OAc)—Ph   Ph—C(H)(Ph)—C*(AcO)(H)—Ph

(2)   Ph—C(Ph)(H)—C(H)(OAc)—Ph* ⟶ Ph—C(Ph)(H)—C⊕(H)—Ph* ⇌ Ph—C⊕(Ph)—C(H)(H)—Ph* ⇌ Ph*—C⊕(Ph)—C(H)(H)—Ph

     ↓ HOAc   ↓ HOAc   ↓ HOAc

     Ph—C(Ph)(H)—C(H)(OAc)—Ph*   Ph—C(H)(Ph)—C(AcO)(H)—Ph*   Ph—C(*Ph)(H)—C(H)(OAc)—Ph

(3)   Ph—C(Ph)(H)—C(H)(OAc*)—Ph ⟶ cations ⇌(HOAc) Ph—C(Ph)(H)—C(Ph)(OAc)—Ph

Figure 28.11. Reaction of triply labeled 1,1,2-triphenylethyl acetate in acetic acid (Problem 16).

certain circumstances, solvolysis gives ion pairs (page 474) which recombine (undergo internal return) many times before the cation finally reacts with an outside nucleophile. If that were to happen here, equilibration of chain-C* and ring-C*—via bridged or open cations —could take place before reaction with external (labeled) acetic acid. In that case, k_1 and k_2 would be equal even if bridged cations were involved. *But k_1 and k_2 would be greater than k_3*—contrary to fact.

Note: When we rule out bridged cations as intermediates here, we are ruling out bridged cations as *the intermediates that combine with acetic acid.* The rapid rearrangement of phenyl takes place via a species in which—at some instant—phenyl is equally bonded to the two carbons. This species may well correspond to a (perhaps shallow) minimum in potential energy, and hence be an intermediate compound; but it does not last long enough to react with acetic acid.

(b) In this system we are dealing with relatively stable benzylic cations, which have less need of stabilization by bridging than, say, the 2° cations in the 3-phenyl-2-butyl system.

(For a detailed description of this elegant work, see W. A. Bonner and C. J. Collins, "Molecular Rearrangements. IV. Triple-labeling Experiments on the Isotope Position Isomerization of 1,2,2-Triphenylethyl Acetate," J. Am. Chem. Soc., **77**, 99 (1955).)

17. (a) Formation of VIII could arise from straight-forward substitution. But the abnormally fast rate makes us curious, and isolation of the cyclopropyl compound IX gives us the clue. Reaction occurs with anchimeric assistance by the π electrons of the double bond, to give the cyclopropylcarbinyl cation XXIX. This cation can undergo elimination to yield IX or substitution, with ring opening, to yield VIII.

VII XXIX

XXIX IX

XXIX VIII

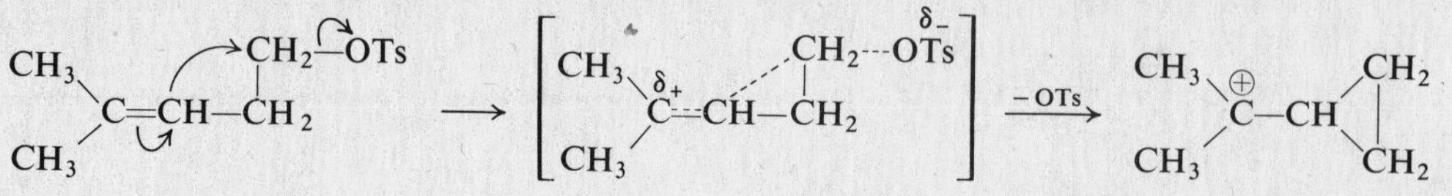

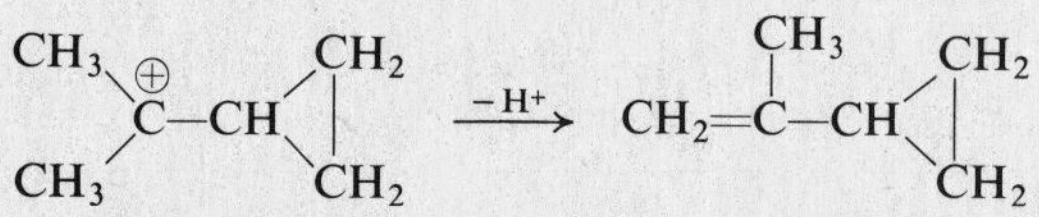

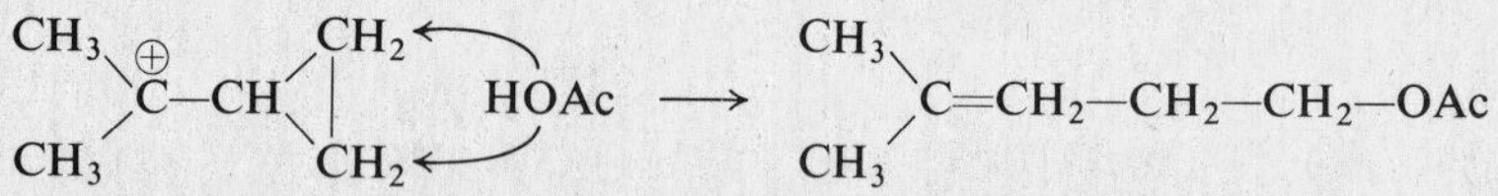

This interpretation is confirmed by the behavior of the labeled compound VII*a*, which yields the mixture of isomers, VIII*a* and VIIIb, that is predicted from attack by paths *a* and *b* on the cation XXIXa.

VII*a* XXIX*a* VIII*a* VIII*b*

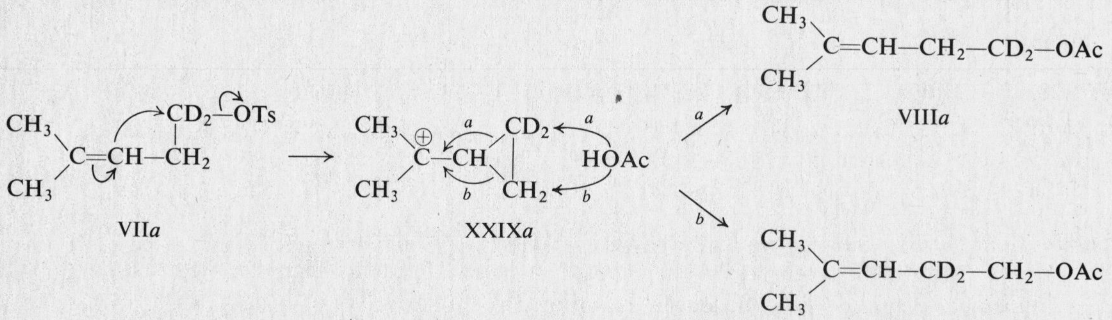

(J. B. Rogan, "The Acetolysis of 4-Methyl-3-Penten-1-yl *p*-Toluenesulfonate," J. Org. Chem., **27**, 3910 (1962). The labeling experiment was not actually carried out on this reaction, but complete scrambling was obtained in the closely related substitution of chloride for chloride (H. Hart *et al.*, "Alkylation of Phenol with a Homoallylic Halide," J. Am. Chem. Soc., **85**, 3269 (1963)).)

(b) The π electrons of the double bond give anchimeric assistance to expulsion of ONs^-. The product is the norbornyl cation which—bridged or not—reacts as usual (Sec. 28.13) to yield the *exo*-norbornyl product.

X Norbornyl cation XII

exo-Norbornyl acetate

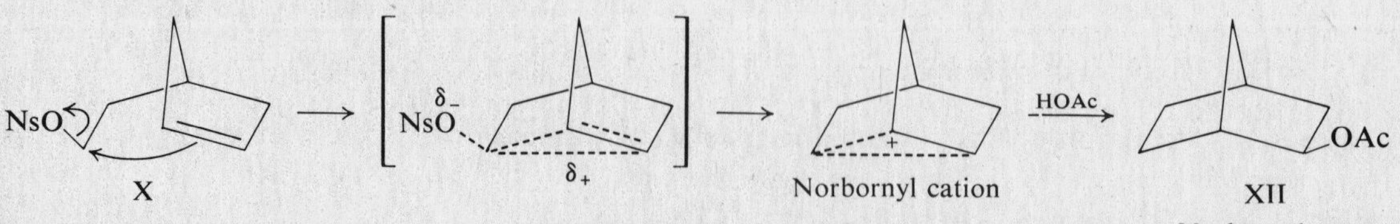

(c) Through back-side attack at C–7, the π electrons of the double bond help to expel OTs$^-$. The product is the bridged cation XXX. This reacts with HOAc, with a second inversion, to give acetate XIV of the same configuration as the starting tosylate.

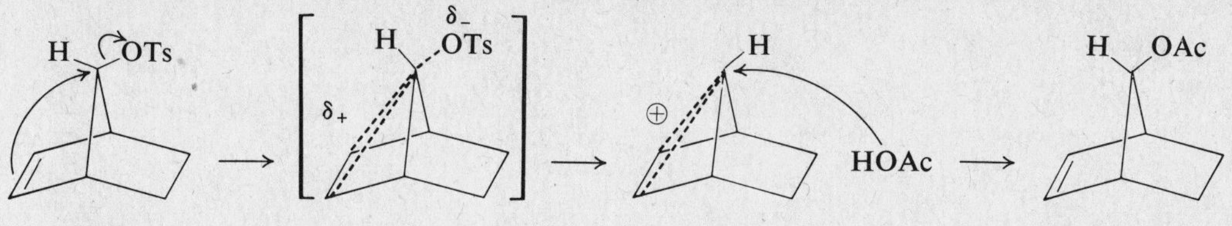

In the presence of NaBH$_4$, XXX suffers attack by H:$^-$ at either of two places: at C–7 to give XV, or at C–2 (or C–3) to give XVI.

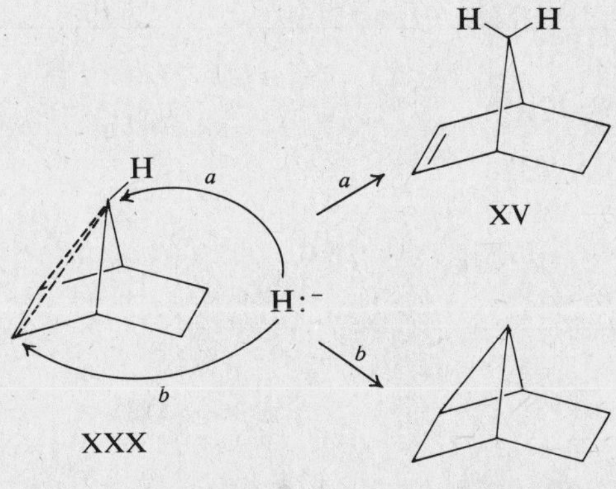

(The historic discovery of this enormous *homoallylic* assistance to ionization was reported by S. Winstein, M. Shatavsky, C. Norton, and R. B. Woodward, "7-Norbornenyl and 7-Norbornyl Cations," J. Am. Chem. Soc., **77**, 4183 (1955).)

18. (a)–(b) Starting with XVII, let us write equations analogous to those in Figure 28.9 (page 917) for *exo*-norbornyl chloride. These are shown in Figure 28.12 (page 460 of this Study Guide).

Brosylate ion is lost (1) with anchimeric assistance from C–6, to give bridged cation XXXI. Cation XXXI undergoes back-side attack at either C–2 (path *c*) to give XVIII or C–1 (path *d*) to give XIX.

In the norbornyl reaction, the intermediate cation (III, page 917) was symmetrical: attacks at the two positions gave enantiomers, and in equal amounts, that is, gave a racemic product.

Here, the intermediate cation XXXI is *not* symmetrical: there is a one-carbon bridge (C–3) on one side, and a two-carbon bridge (C–7 and C–8) on the other. Attacks by paths *c* and *d* are not equally likely, and do not give enantiomers: they give different compounds, each of which is optically active.

Path *c* gives XVIII, of the same configuration as the starting material XVII. The overall retention is the result of two successive inversions: first in the attack by C–6, and then in the attack by HOAc.

Path *d* gives XIX but not XX. Formation of XX would require attack by HOAc on the face of C–1 that is bonded to C–6.

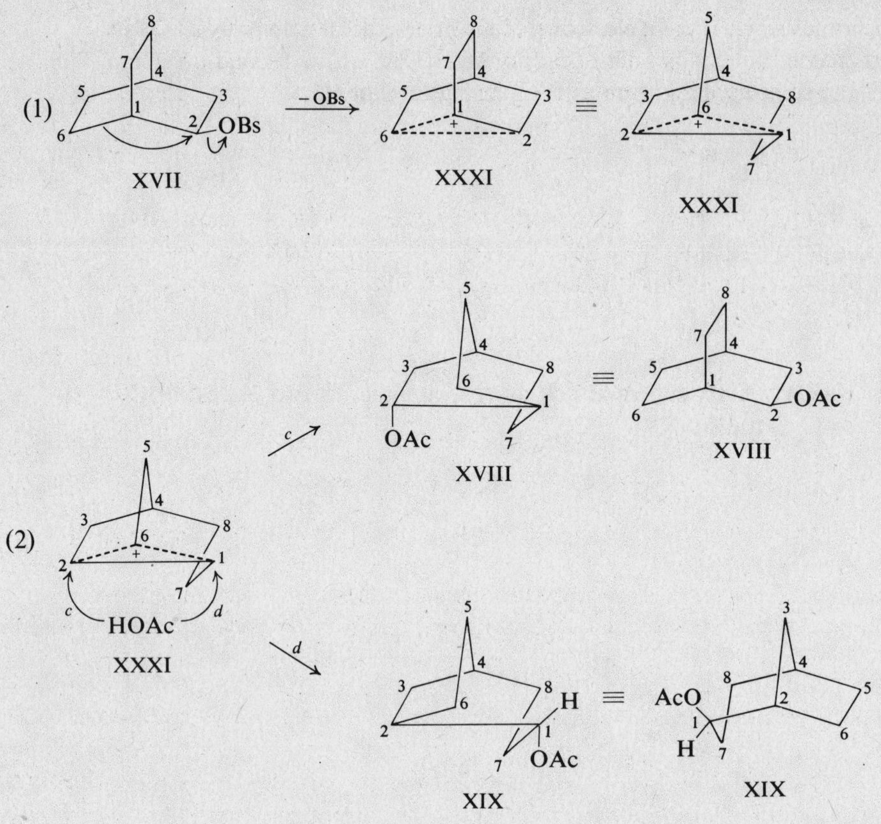

Figure 28.12. Acetolysis of optically active XVII (Problem 18).

(c) Brosylate ion is lost from XXI with anchimeric assistance from the π electrons of the double bond, to give bridged cation XXXII. Acetic acid attacks this symmetrical cation,

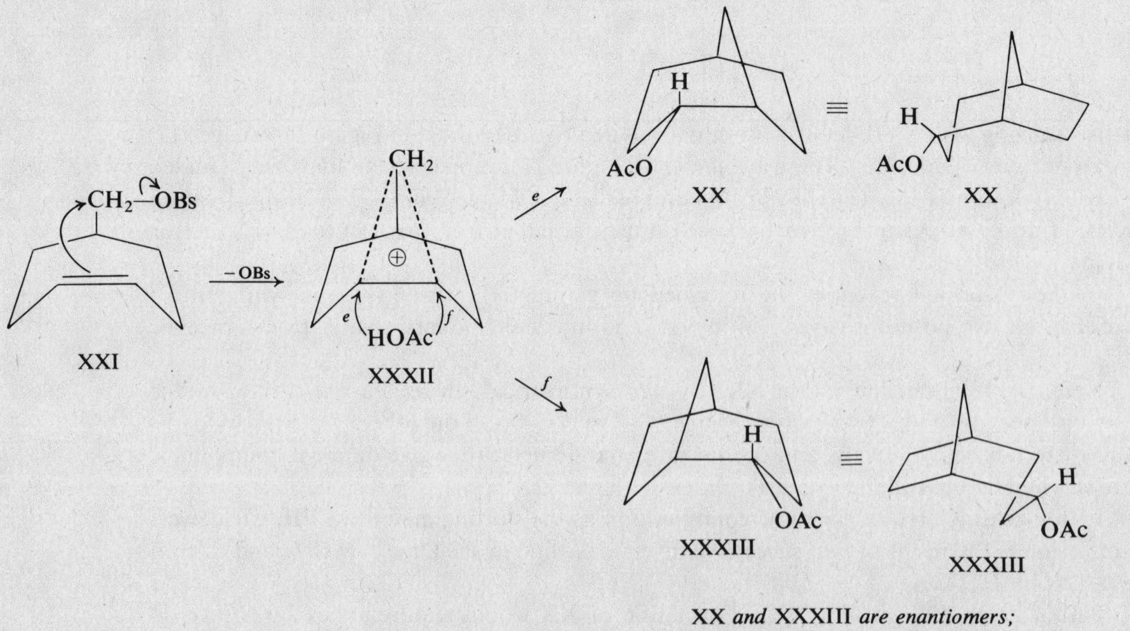

XX *and* XXXIII *are enantiomers;*
formed in equal amounts

by either path *e* or path *f*, to give either of a pair of enantiomers: XX and XXXIII. Paths *e* and *f* are equally likely, and the racemic modification is obtained.

XIX is not formed, since this would require attack on the side of the carbon bonded to the (originally) side-chain —CH_2.

19. (a) XXII is a lactone and hence an ester. Like any ester, it reacts with OCH_3^- to undergo nucleophilic acyl substitution: transesterification (Sec. 20.20). The product is a new ester and a new alkoxide (or, in this case, a phenoxide); these are part of the same molecule, XXXIV.

XXII XXXIV XXXIV XXIII

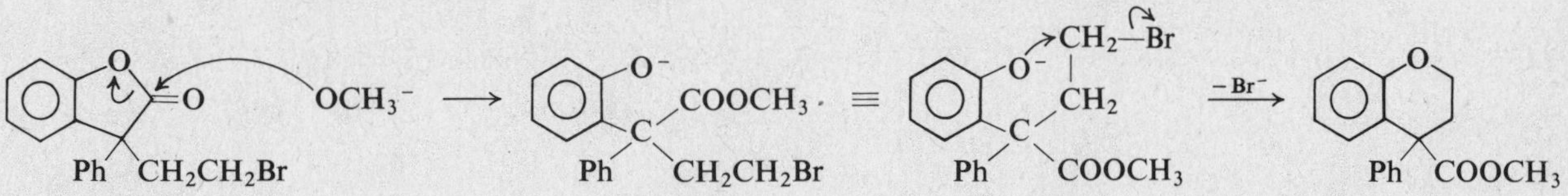

XXXIV undergoes intramolecular nucleophilic substitution—a Williamson synthesis, actually—to give the cyclic ether XXIII.

An analogous reaction with R_2NH gives XXXV, which is both an amide and a phenoxide. As before, there is an internal Williamson reaction to yield the ether, XXIV.

XXII XXXV XXXV XXIV

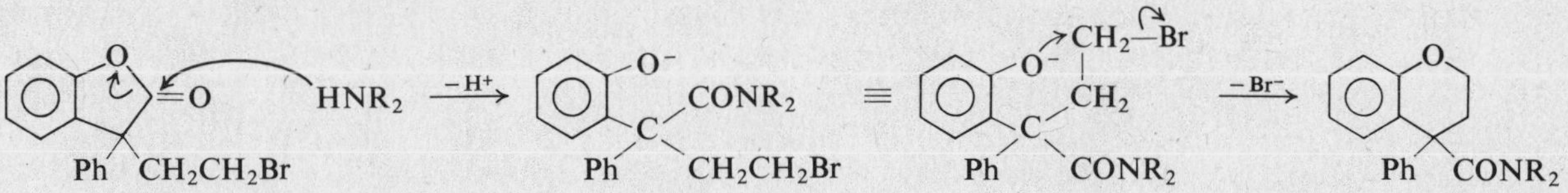

(b) Nucleophilic acyl substitution, it has been proposed (Sec. 20.4), involves attachment of the nucleophile to acyl carbon to give a tetrahedral intermediate which then, in a second step, ejects the leaving group. We have seen (Sec. 20.17) evidence of isotopic exchange which supports that proposal. In the reaction of XXII with HNR_2, for example, the intermediate would be XXXVI. Expulsion of phenoxide would yield the amide XXXV.

XXII XXXVI XXXV

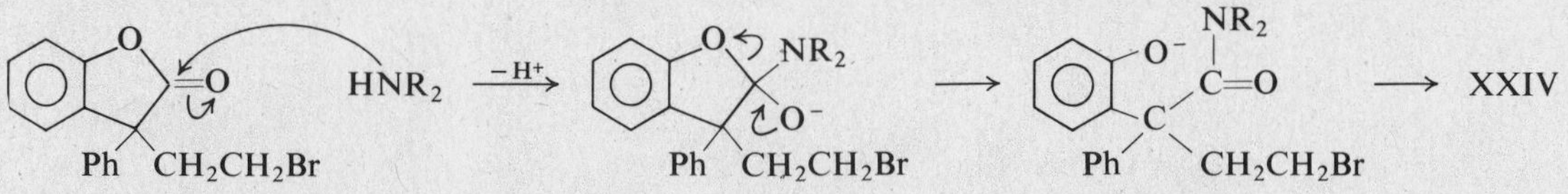

But in intermediate XXXVI the acyl group is temporarily converted into —O^-. This group is capable of nucleophilic attack on the carbon carrying —Br to yield XXV. Compound XXV is thus the product of internal trapping of the tetrahedral intermediate, and its isolation is strong evidence for the existence of this intermediate.

XXV

XXXVI

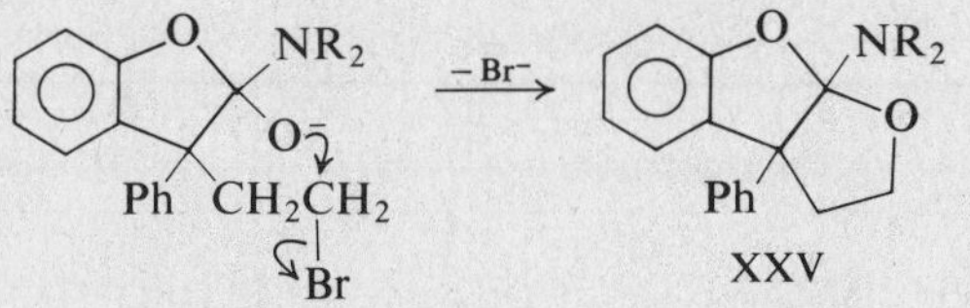

Molecular Orbitals.
Orbital Symmetry

29.1 The first-formed monocation is slowly converted into the—evidently more stable—dication. The stability of this latter species, despite its double positive charge, is consistent with aromaticity; it contains 2 π electrons, $(4n + 2, n = 0)$.

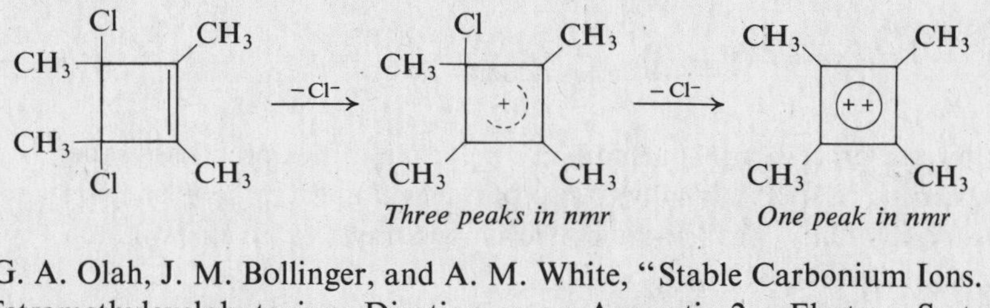

Three peaks in nmr   *One peak in nmr*

(G. A. Olah, J. M. Bollinger, and A. M. White, "Stable Carbonium Ions. LXXXIX. The Tetramethylcyclobutenium Dication as an Aromatic 2 π-Electron System," J. Am. Chem. Soc., **91**, 3667 (1969).)

29.2 (a) The facts indicate that compound I has a dipolar structure; it is basic because acceptance of a proton eliminates the separation of charge and thus stabilizes the molecule.

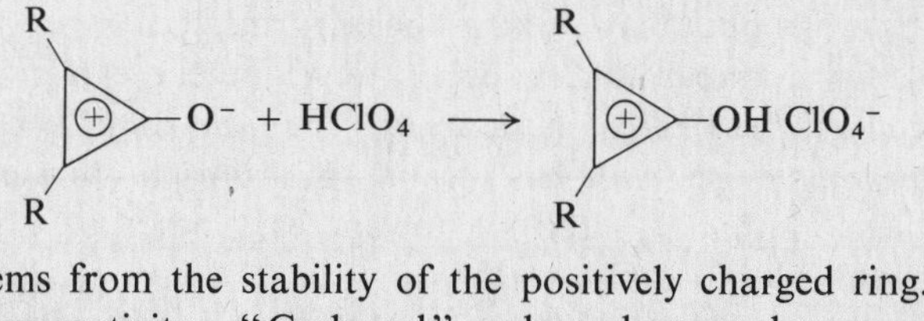

All this stems from the stability of the positively charged ring. This, in turn, we attribute to its aromaticity. "Carbonyl" carbon shares only one pair of electrons with oxygen; the empty p orbital of this carbon contributes no π electrons, and the total for the cyclic system is the Hückel number 2 $(4n + 2$, with $n = 0)$.

(b) The product is the aromatic 1,2,3-triphenylcyclopropenyl cation (see the answer to Problem 10.6, page 330).

(c) Decarboxylation of trichloroacetate is easy because of the relative stability of the Cl_3C^- anion being formed.

$$Cl_3CCOO^- \xrightarrow{\text{heat}} CO_2 + Cl_3C:^- \xrightarrow{-Cl^-} :CCl_2$$

(For a general discussion of aromaticity, see R. Breslow, "Aromatic Character," Chem. Eng. News, June 28, 1965, p. 90.)

29.3 (a) The ring closure involves four double bonds of the polyene, and hence 8 π electrons; this corresponds to $4n$ ($n = 2$). Table 29.1 predicts that for such a system a thermal electrocyclic reaction will take place with conrotatory motion.

I or III $\longrightarrow$ *trans* product

II $\longrightarrow$ *cis* product

(b) Photochemical ring closure in this same system will take place with disrotatory motion.

I or III $\longrightarrow$ *cis* product

II $\longrightarrow$ *trans* product

Let us look at the reverse process, ring-opening. In principle, if not practically, the *trans*-dimethylcyclooctatriene is related thermally to *two* polyenes, I and III, *both* by symmetry-allowed conrotatory motion. The *cis*-dimethylcyclooctatriene is related photochemically to *two* polyenes, I and III.

There will always be two possible conrotatory modes,

and   **Conrotatory**

and two possible disrotatory modes.

and   **Disrotatory**

These may lead to the same product or to different products, depending on the particular case. Where two products are possible, both may or may not be formed.

In Figure 29.12 (page 940), for example, the *cis* dimethylcyclobutene can give by conrotatory motion only one product, the *cis,trans* hexadiene. The *trans* dimethylcyclobutene could give either the *trans,trans* diene (with H's turning toward each other in the transition state) or the *cis,cis* diene (with CH$_3$'s turning toward each other). Only the *trans,trans* diene is obtained, undoubtedly for steric reasons.

In the reverse reaction, ring closure (Figure 29.15, page 943), the *cis,trans* hexadiene can give only the *cis* dimethylcyclobutene.   The *trans,trans* hexadiene yields one enantiomer of the dimethylcyclobutene by one conrotatory mode, and the other enantiomer by the other mode; here the two modes are equally likely, and the racemic product is obtained.

We shall encounter other reactions where this factor enters in, and must carefully consider each case individually.

29.4 (a) ψ_1 (Figure 29.7, page 933). The allyl cation has 2 π electrons.

(b) $4n + 2$ thermal: expect disrotatory motion.

(c) In the allyl anion, there are 4 π electrons. $4n$ thermal: expect conrotatory motion.

(d) In the pentadienyl cation, there are 4 π electrons. $4n$ thermal: expect conrotatory motion.

29.5 (a) In the polyene (the tropylidene) three double bonds are involved, or 6 π electrons. $4n + 2$ thermal: disrotatory ring-opening.

This interconversion takes place so fast that only at $-120°$ does the nmr spectrum of the equilibrium mixture show completely separate signals for the two compounds. (E. Ciganek, "The Direct Observation of a Norcaradiene–Cycloheptatriene Equilibrium," J. Am. Chem. Soc., **87**, 1149 (1965).)

We notice that, of the two possible disrotatory modes, only one can occur: the angular H's must turn *away from* each other as the ring opens. The alternative mode would generate two *trans* double bonds, geometrically impossible in a ring this size. Photochemical ring-opening cannot occur, since this would require conrotatory motion, with the formation of one *trans* double bond in the seven-membered ring, also geometrically impossible.

(b) In the polyene (the reactant) three double bonds are involved, or 6 π electrons. $4n + 2$ thermal: disrotatory ring-closing.

(c) Three double bonds or 6 π electrons ($4n + 2$) are involved throughout.

First step. Thermal disrotatory ring-closing.

Second step. Photochemical conrotatory ring-opening.

Third step. Thermal disrotatory ring-closing.

(d) In the first reaction, only two double bonds of the polyene are involved; in the second reaction, all three double bonds are involved.

First step. 4 π electrons ($4n$): thermal conrotatory ring-opening.

Second step. 6 π electrons ($4n + 2$): thermal disrotatory ring closure.

(e)

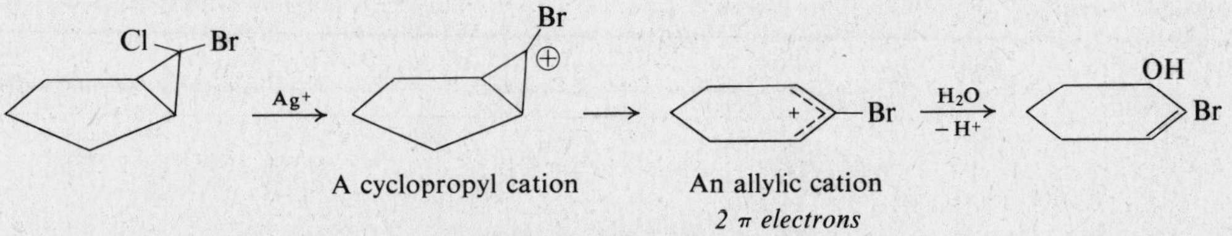

A cyclopropyl cation   An allylic cation
                               2 π electrons

This is an example of the conversion of a cyclopropyl cation into an allylic cation (Problem 29.4, page 946): 2 π electrons ($4n + 2$), thermal disrotatory ring-opening. The allylic cation then combines with water.

Actually, ring-opening is stereospecific in a further way, indicating that ring-opening is concerted with loss of halide ion. Calculations lead to another rule, by which one can predict *which* of the two possible disrotatory modes will occur: the substituents on the same side of the cyclopropyl ring as the leaving group will rotate inward (toward each other), not outward. In the present case, the substituents on one side of the cyclopropyl ring are carbons of the five-membered ring: these must rotate toward each other since they are tied together in a smallish (5 $\longrightarrow$ 6) ring. As a result, the leaving group is the *endo* halogen, as shown.

(For a discussion of this point, see one of the following: R. B. Woodward and R. Hoffmann, *The Conservation of Orbital Symmetry*, Academic Press, New York, 1970, pp. 46–48; C. H. DePuy, "The Chemistry of Cyclopropanols," Accounts Chem. Res., **1**, 33 (1968).)

(f) The protonated ketone is equivalent to the pentadienyl cation VIII (4 π electrons, 4*n*), which undergoes thermal conrotatory ring-closing to the cyclopentenyl cation IX. (See Problem 29.4d, page 946.)

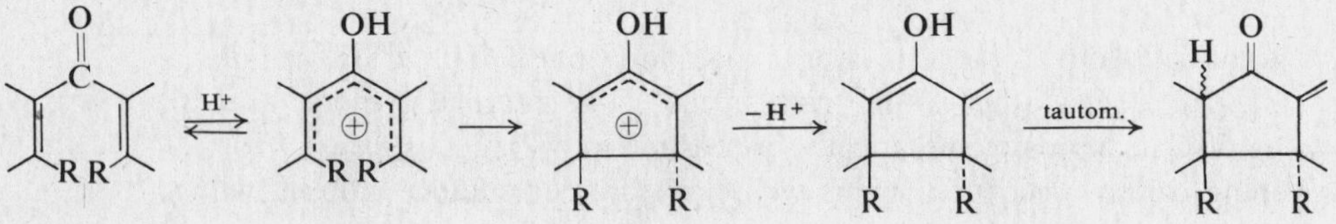

29.6 Interconversion takes place via the cyclobutene. Two double bonds of the polyene are involved, or 4 π electrons. 4*n* thermal: conrotatory ring-closing and ring-opening.

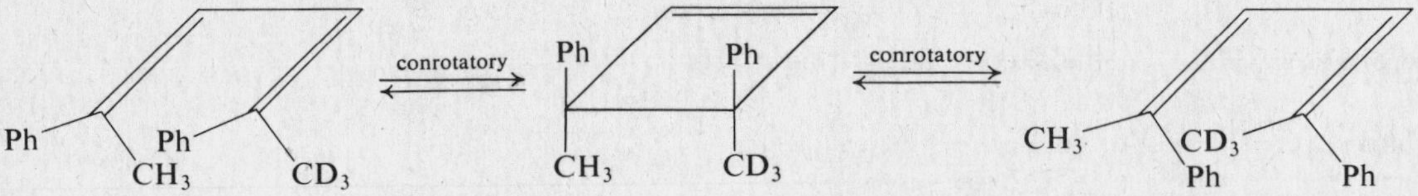

Here we see the two possible conrotatory modes of ring-opening (see the answer to Problem 29.3, page 946), one giving IV and the other V.

29.7 (a) This is a [4 + 2] thermal cycloaddition, the familiar Diels-Alder reaction. The symmetry-allowed reaction is *supra,supra*, which is also geometrically easy. Putting the diene in the necessary *s-cis* configuration, we arrive at our product:

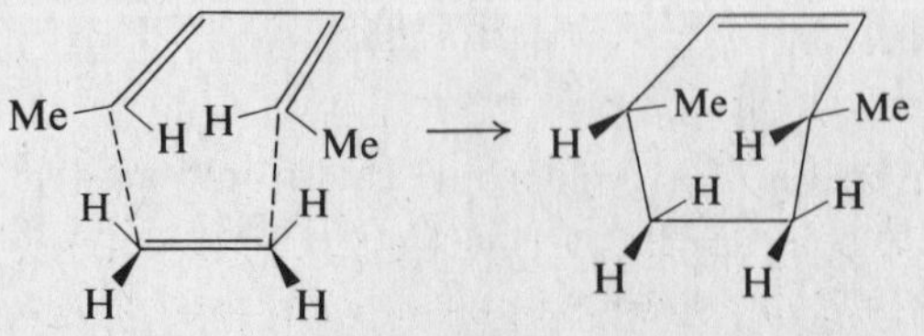

We predict, then, formation of *cis*-3,6-dimethylcyclohexene.

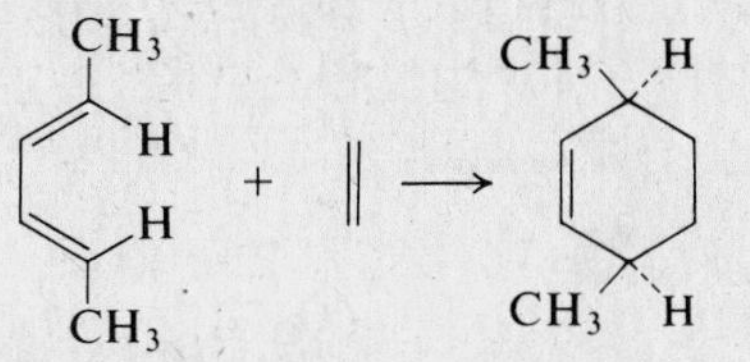

trans,trans-   Ethylene   *cis*-3,6-Dimethyl-
2,4-Hexadiene   cyclohexene

(b) Another [4 + 2] thermal cycloaddition: *supra,supra*. With the diene in the *s-cis* configuration, we bring up the dienophile. With regard to the dienophile, reaction leads

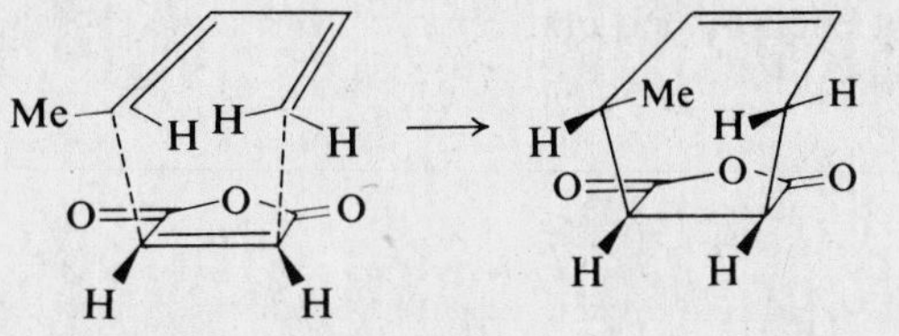

to *syn* addition: in the product, the two angular H's are *cis* to each other. The reaction is also *endo*, with the anhydride bridge near the developing double bond (beneath it, in the representation above): in the product, the methyl group is *cis* to the anhydride bridge.

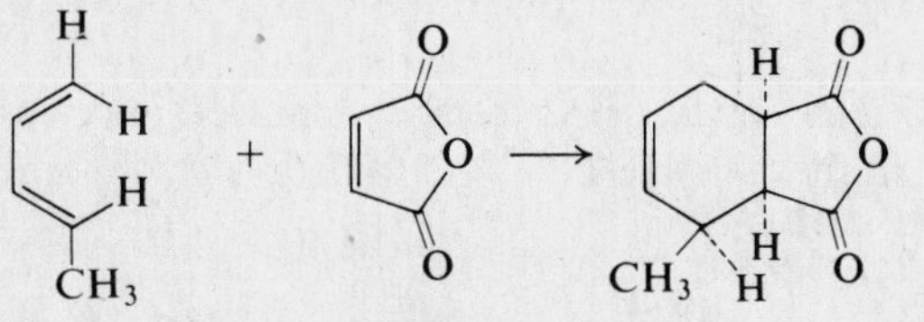

trans-1,3-Pentadiene   Maleic
                          anhydride

(c) As in part (b), [4 + 2] thermal *supra,supra* cycloaddition. Once more we remember: *s-cis*, *syn*, and *endo*.

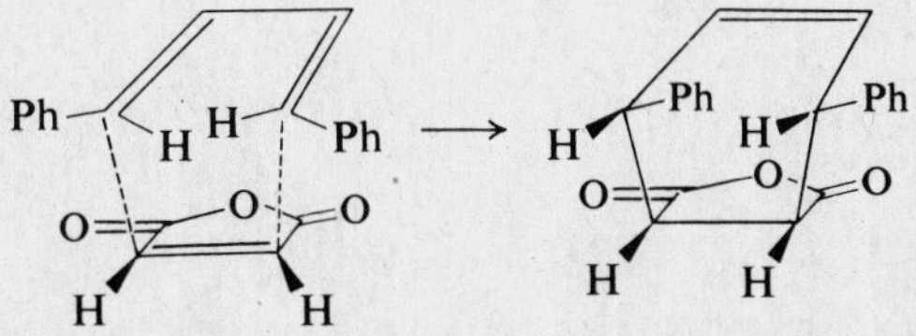

In the product the Ph's are *cis* to each other (*syn* addition) and *cis* to the anhydride bridge (*endo* reaction).

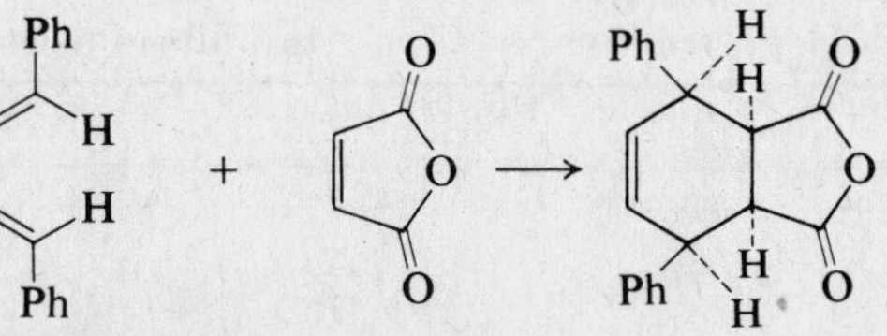

trans,trans-1,4-Diphenyl-   Maleic
1,3-butadiene   anhydride

(d) [2 + 2] photochemical cycloaddition: *supra,supra*. This amounts to *syn*-addition to each alkene, so that the two methyls of each alkene unit remain *cis* to each other. But there

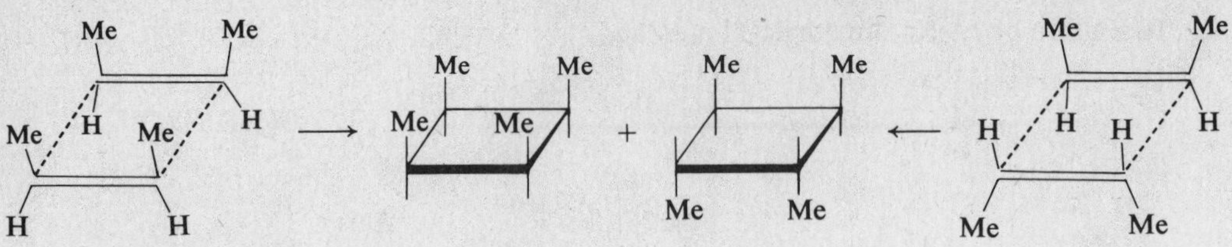

cis-2-Butene
Two moles

B

A

cis-2-Butene
Two moles

are two ways in which the alkene units can come together, and two products; one set of methyls can be *cis* or *trans* to the other set.

(e) As in (d). Here the methyls of each alkene unit remain *trans* to each other. Again there are two combinations, and two products.

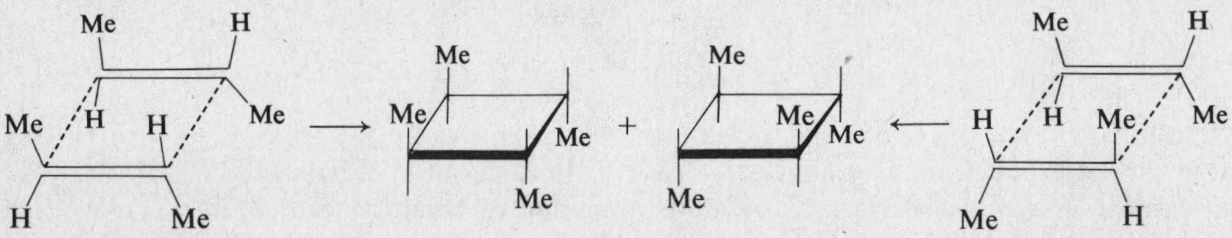

trans-2-Butene
Two moles

A

C

trans-2-Butene
Two moles

(f) *cis*-Alkene can add to *cis*-alkene as in (d), *trans* to *trans* as in (e), and, besides, *cis* to *trans*. There are four products, all the possible 1,2,3,4-tetramethylcyclobutanes.

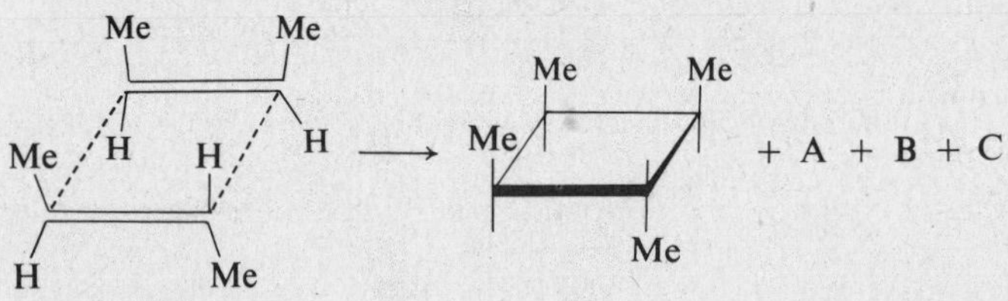

cis-2-Butene
and
trans-2-Butene

D

29.8 (a) Dicyclopentadiene is formed by a Diels-Alder reaction in which one molecule of cyclopentadiene acts as diene, and the other molecule acts as dienophile. Regeneration of cyclopentadiene is a *retro*-Diels-Alder reaction, with the equilibrium being shifted to replace the more volatile component, which is being removed by distillation.

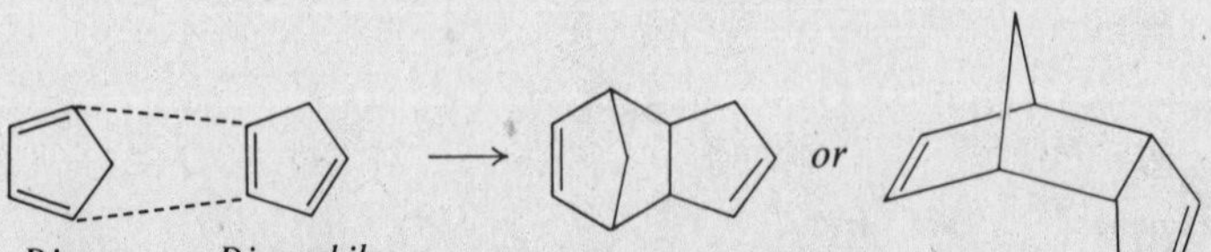

Diene   *Dienophile*
Cyclopentadiene
Two moles

I
Dicyclopentadiene

(b) Reaction is *endo*, with the ring of the dienophile close to the developing double bond in the diene moiety.

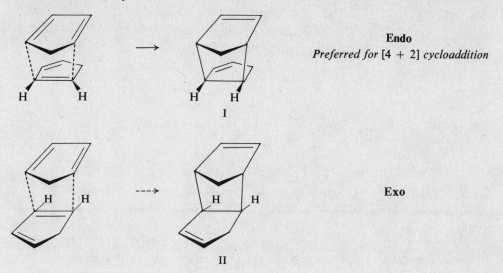

Endo
Preferred for [4 + 2] cycloaddition

I

Exo

II

29.9 (a) Only two double bonds of the triene are involved. Reaction is thus [4 + 2] cyclo-addition, which is *supra,supra* and easy, instead of [6 + 2] cycloaddition, which would have to be *supra,antara* and hence geometrically difficult.

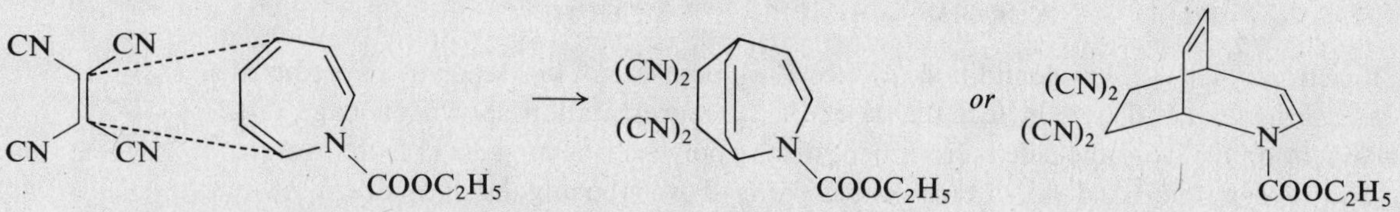

(b) An intramolecular [2 + 2] photochemical cycloaddition. The *supra,supra* stereo-chemistry gives the all-*cis* configuration to the cyclobutane ring.

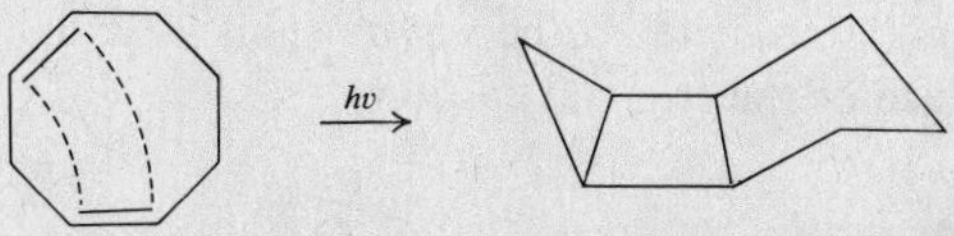

(c) Reaction is a [6 + 4] thermal cycloaddition: predicted to be *supra,supra*—and hence geometrically easy—on the basis that $i + j$ equals 10 and hence is $4n + 2$.

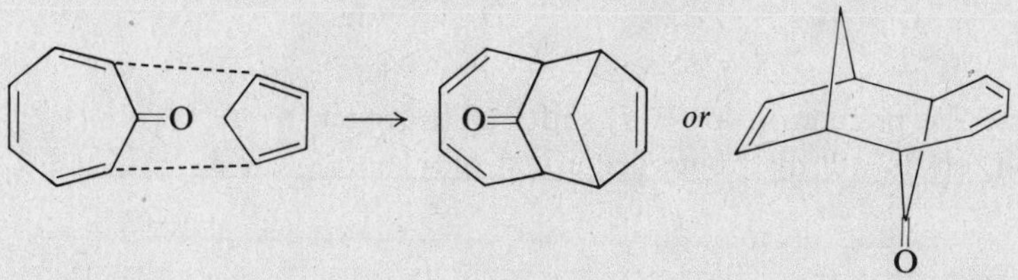

This was the first example of a [6 + 4] cycloaddition to be reported. (R. C. Cook-son, B. V. Drake, J. Hudec, and A. Morrison, "The Adduct of Tropone and Cyclopenta-diene: A New Type of Cyclic Reaction," Chem. Commun., 15 (1966).)

A closer look shows us that, in contrast to [4 + 2] cycloaddition, this reaction takes place in the *exo* sense.

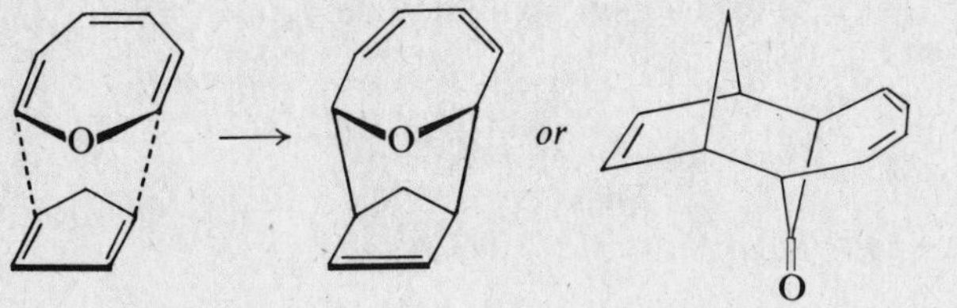

Exo
Preferred for [6 + 4] *cycloaddition*

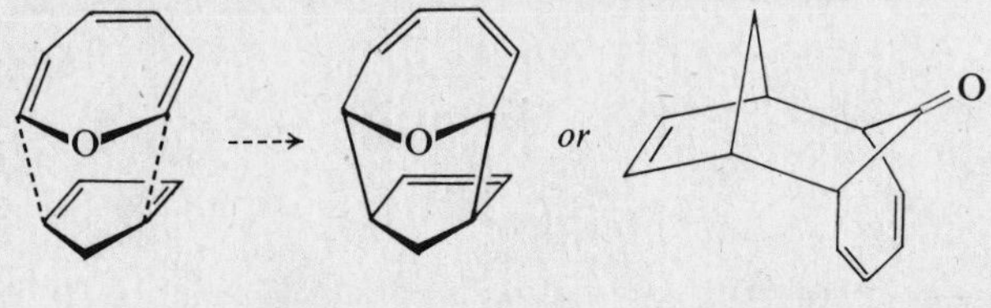

Endo

We shall see later (Problem 13, page 965) that both the *exo* preference in [6 + 4] cycloaddition and the *endo* preference in [4 + 2] cycloaddition are accounted for by orbital symmetry theory.

(d) Reaction is [8 + 2] thermal cycloaddition. As in (c), $i + j$ equals 10, and reaction was predicted to be *supra,supra* and hence geometrically easy—as it has turned out to be.

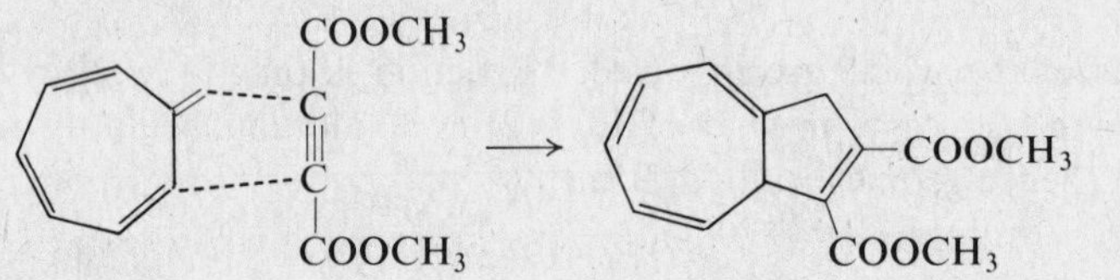

(e) Reaction is [14 + 2] cycloaddition. Since $i + j$ equals 16 ($4n$), reaction is predicted to be *supra,antara*. This is in fact the case: in the representation shown on page 954, attachment on the left-hand side is from beneath the polyene (the angular H is up, toward us), and on the right-hand side attachment is from above (the angular H is down, away from us).

29.10 (a) The reaction is a [1,5]-H shift, and is predicted to be *supra*. This is actually the case, with transfer taking place to either face of the trigonal carbon.

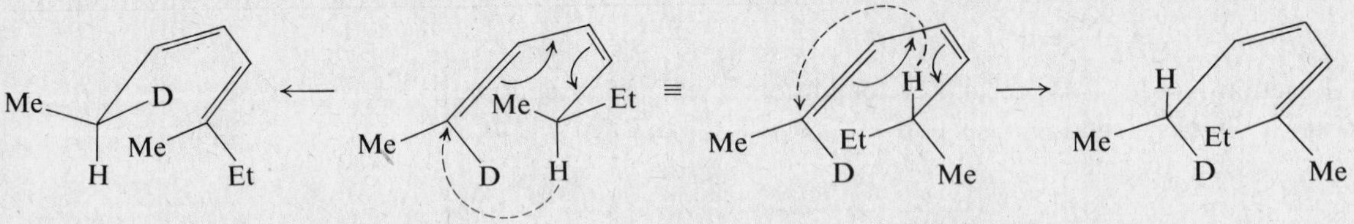

(b) These results illustrate the preference for [1,5] shifts of hydrogen over [1,3] or [1,7] shifts. A series of [1,5]-D shifts will distribute deuterium only among the 3, 4, 7, and 8 positions, as is observed.

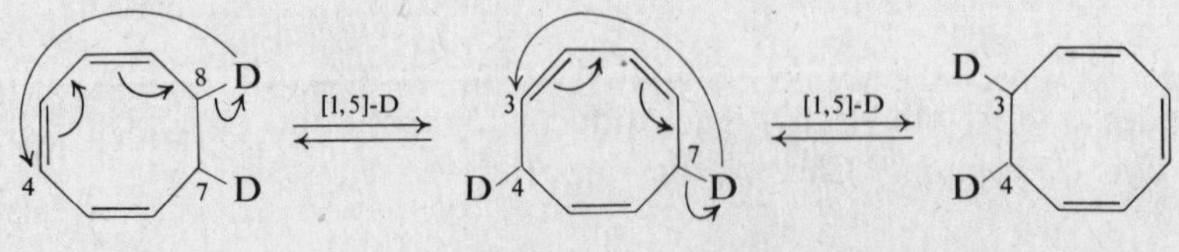

A series of [1,3]-D shifts (or of [1,7]-D shifts), on the other hand, would scramble the label among all positions—contrary to fact.

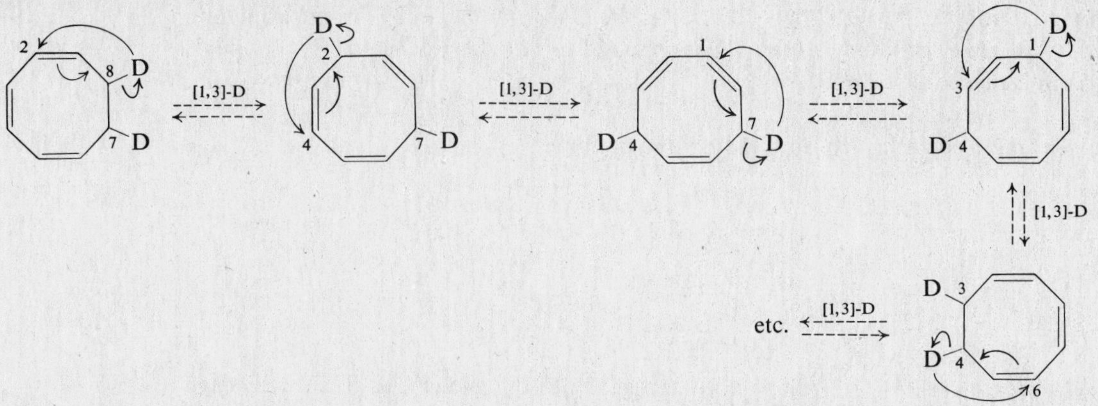

(c) A [1,3]-C shift (*supra*) occurs with *inversion* in the migrating group, as predicted.   (If,

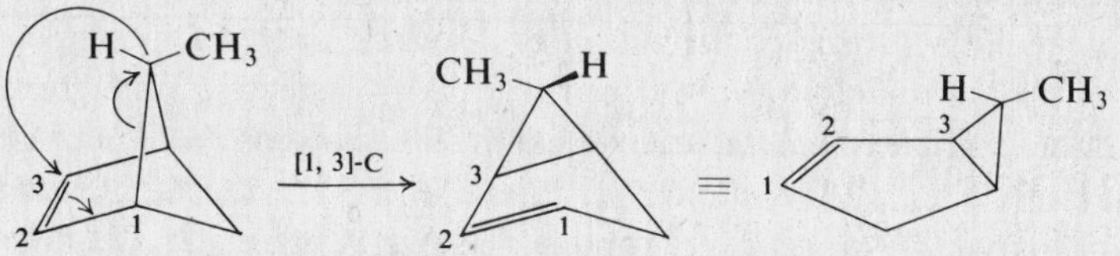

instead, the front leg had simply *stepped* from C–1 to C–3 with retention, the —CH$_3$ would have swung into the *endo* position, over the ring—contrary to fact.)

1. (a) In chemical properties tropolone resembles phenols.   It is not adequately represented by structure I.

(b) To be a phenol, tropolone must have an aromatic ring.   Aromaticity is also indicated by other properties: flatness, bond lengths, heat of combustion.   The dipole moment suggests a dipolar structure.

   Tropolone is a hybrid of seven dipolar structures like these:

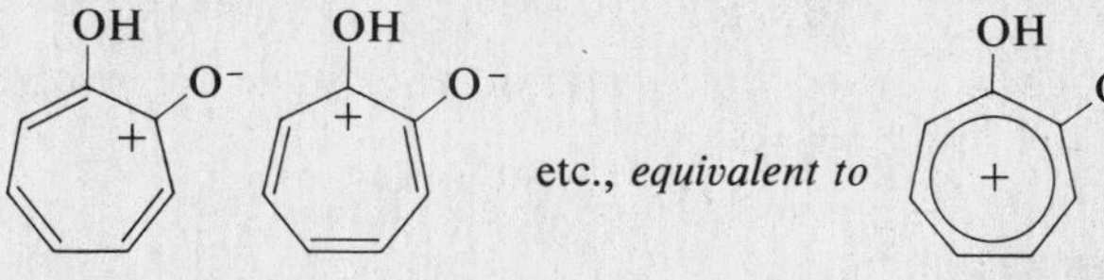

etc., *equivalent to*

Tropolone
Six π electrons
Aromatic

The ring is that of the cycloheptatrienyl cation, which has 6 π electrons and is aromatic (see Secs. 10.10 and 29.6).

(c) The dipole moment of 5-bromotropolone is smaller than that of tropolone, showing that the strong dipole of tropolone is in the opposite direction from that of C—Br, that is, *toward* "carbonyl" oxygen.   This is consistent with the dipolar structure.   (The dipole moment for

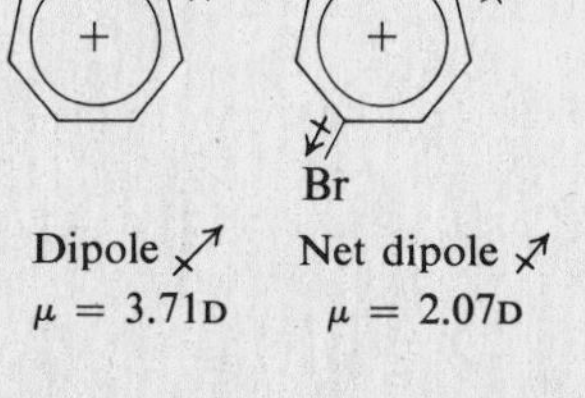

Dipole
$\mu = 3.71$D

Net dipole
$\mu = 2.07$D

bromobenzene (Table 25.2, page 825) is 1.71 D.   If the C—Br moment of 5-bromotropolone were roughly the same, we would expect the compound to have $\mu = 3.71 - 1.71 = 2.00$ D, rather close to observed value of 2.07 D.)

(d) There is intramolecular hydrogen bonding in tropolone.

Tropolone

Hückel proposed his $4n + 2$ rule in 1931 (page 936).   In 1945, M. J. S. Dewar (page 267) postulated tropolone as the unit responsible for the unusual—that is, aromatic—properties of several natural products.   Tropolone itself was synthesized in 1951 by W. von E. Doering and L. H. Knox, and it was this confirmation of Dewar's proposal—coupled with the recognition in 1952 of the aromaticity of ferrocene (page 329)—that set off the great wave of experimental work on non-benzenoid aromaticity that is still going on.

(W. von E. Doering and L. H. Knox, "Tropolone," J. Am. Chem. Soc., **73**, 828 (1951); H. J. Dauben, Jr., and H. J. Ringold, "Synthesis of Tropone," J. Am. Chem. Soc., **73**, 876 (1951); W. von E. Doering and F. L. Detert, "Cycloheptatrienylium Oxides," J. Am. Chem. Soc., **73**, 876 (1951).   For a review, see D. Ginsburg, ed., *Non-Benzenoid Aromatic Compounds*, Interscience, New York, 1959.)

2. So that you can keep track of carbons in the following transformations, the numbers used in the starting material are retained throughout (even though the numbers thus assigned in a product may not conform to accepted usage).   In many cases, only one of two (equivalent) possibilities is shown: for example, in the first product in (a), the *trans* double bond is shown between C–2 and C–3 and not between C–8 and C–9.

(a) **First step.**   The cyclobutene ring opens.   Two double bonds in the polyene are involved, or 4 π electrons. $4n$ thermal: conrotatory ring-opening.

Second step.   A [1,5]-H shift, thermal: *supra.*

(b) **First step.**   The cyclobutene ring opens.   Two double bonds of the polyene are involved, or 4 π electrons. $4n$ thermal: conrotatory ring-opening.

Second step.   Three double bonds of the polyene are involved, or 6 π electrons. $4n + 2$ thermal: disrotatory ring-closing.

(c) For some molecules there is closure of a cyclobutene ring. Two double bonds of the polyene are involved, or 4 π electrons. $4n$ photochemical: disrotatory ring-closing.

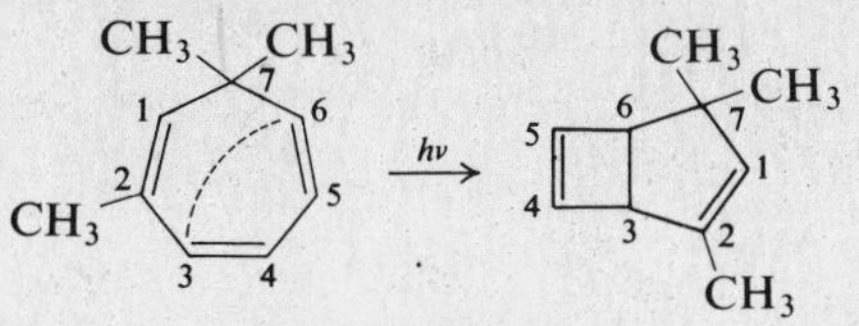

The rest of the molecules undergo the following.

First step. A [1,7]-C shift, photochemical: *supra.*

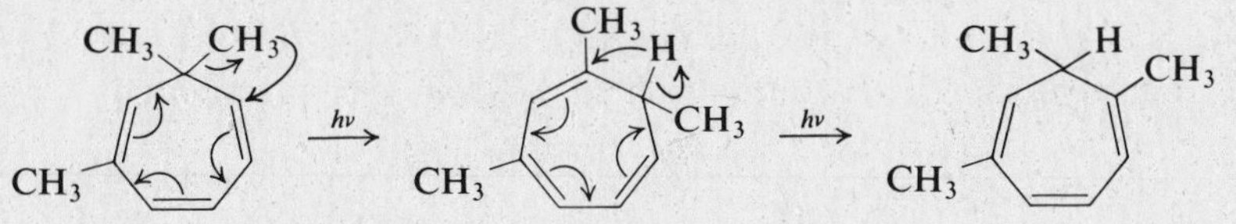

Second step. A [1,7]-H shift, photochemical: *supra.*

(d) **First step.** An intramolecular [4 + 4] cycloaddition, $i + j$ equals 8 ($4n$), photochemical: *supra,supra.*

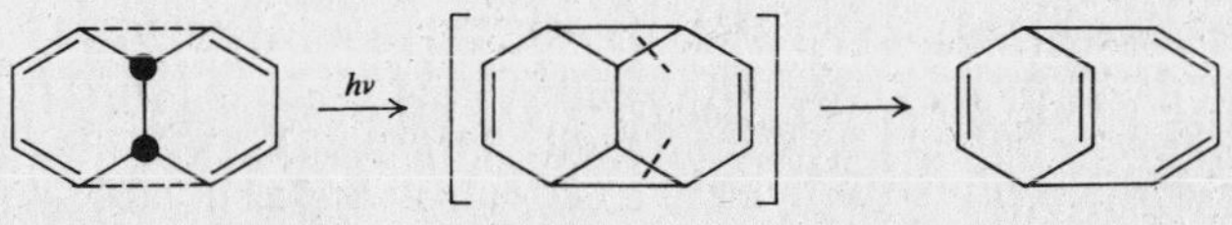

Second step. A [4 + 2] cycloreversion, $i + j$ equals 6 ($4n + 2$), présumably thermal: *supra,supra.*

(e) An allylic cation (2 π electrons) undergoes a [4 + 2] cycloaddition (thermal, *supra,supra*) to give cation XIII, which then loses a proton from either of two positions.

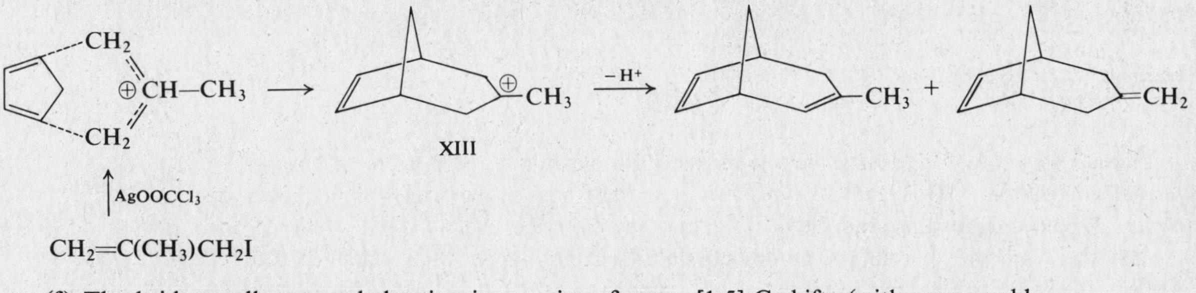

(f) The bridge walks around the ring in a series of *supra* [1,5]-C shifts (with, presumably, retention at the migrating carbon).

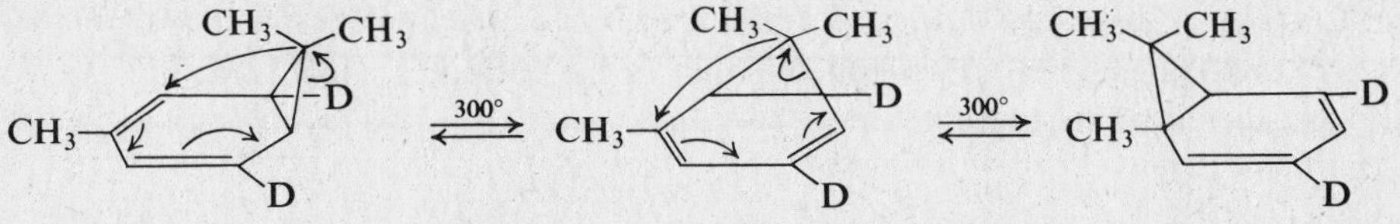

3. (a) **First step.** Ring closure involving four double bonds of the polyene, or 8 π electrons ($4n$): thermal, conrotatory. One R swings up, the other R down, so that they are *trans* in A.

A

trans-7,8-Dialkyl-*cis,cis,cis*-
cycloocta-1,3,5-triene

Second step.   Ring closure involving three double bonds of the polyene, or 6 π electrons ($4n + 2$): thermal, disrotatory.   The R's retain, of course, their *trans* relationship;
the disrotatory motion gives *cis* fusion of the rings.

(b) First step.   There is only one possible [1,5]-H shift (which, as we have seen, is preferred
over a [1,3]-H shift).

B
1-Isopropenyl-
2,3,3-trimethylcyclobutene

2,4,5-Trimethyl-3-methylene-1,4-hexadiene

Second step.   There is only one possible electrocyclic ring opening.   It involves two
double bonds of the polyene, or 4 π electrons ($4n$); being thermal, it is presumably conrotatory.

(c) First step.   The cyclopropane ring opens.   Four double bonds in the polyene (D) are
involved, or 8 π electrons. $4n$ thermal: conrotatory.

D

Second step.   A ring closure involving three double bonds, or 6 π electrons. $4n + 2$
thermal: disrotatory.   This leads to *trans* fusion of the rings: as we draw our formulas, one
angular H comes toward us, and the other goes away from us.

But there are *two* disrotatory modes possible (see the answer to Problem 29.3).   The
angular H on C–1 can come toward us or go away from us; that is, it can be on the same
side of the ring as Me, or on the same side as Et.   (The angular H on C–6, of course, does
the opposite in each case.)   The two modes give different (diastereomeric) products.

(d) First step.   Closure of cyclobutene ring.   Two double bonds of the polyene are involved,
or 4 π electrons. $4n$ photochemical: disrotatory.

	E		F		G
cis,cis-Cyclo-	*cis*-Bicyclo[5.2.0]-		*cis, trans*-Cyclo-		*trans*-Bicyclo[5.2.0]-
nona-1,3-diene	nona-8-ene		nona-1,3-diene		nona-8-ene

Second step. As in first step, but thermal: conrotatory.

Third step. Closure of cyclobutene ring. As in the first step, $4n$ photochemical: disrotatory.

(e) The H shifts are *supra* for geometric reasons. They are [1,5] if thermal, [1,7] if photochemical.

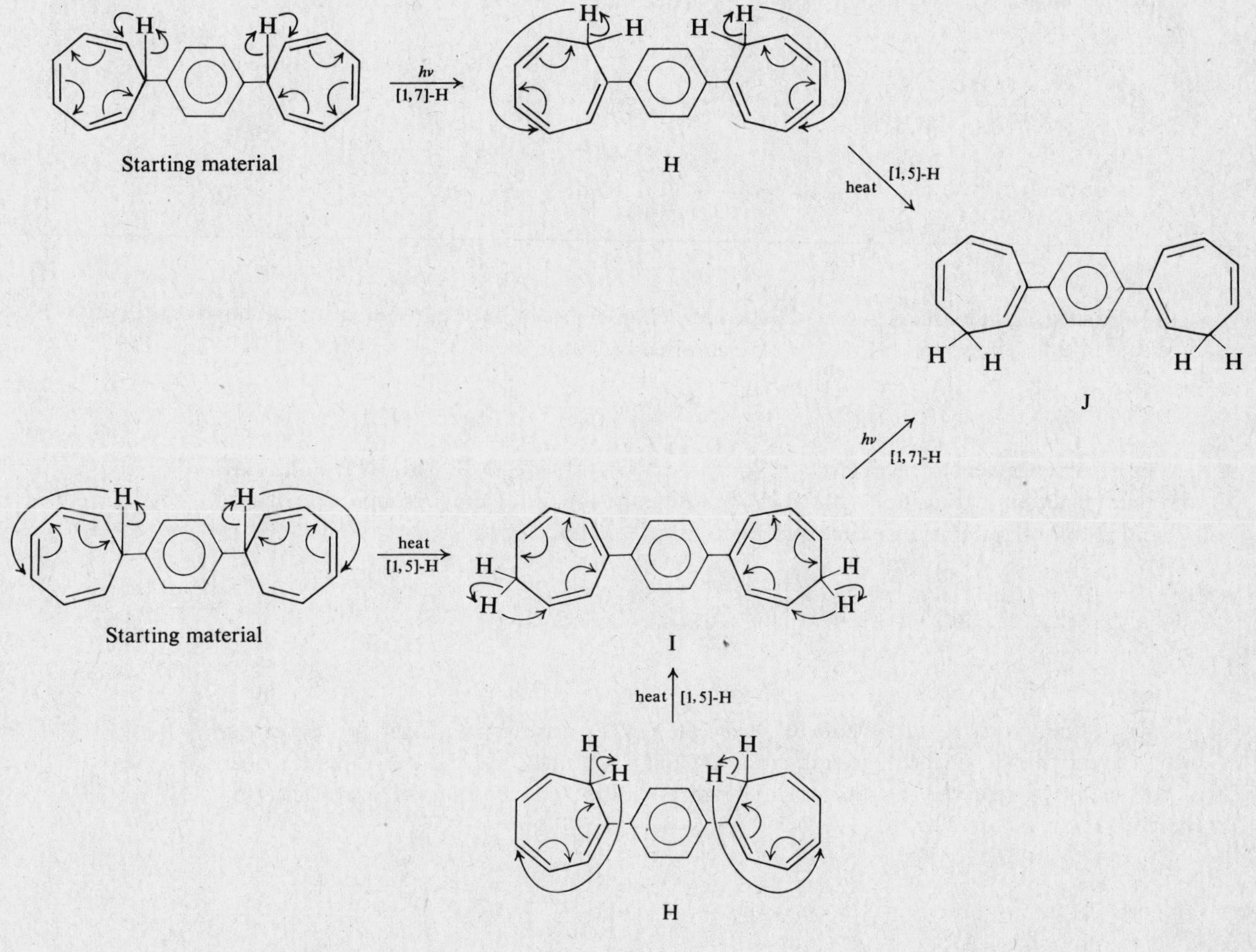

4. Two double bonds of the polyene are involved, or 4 π electrons. $4n$ thermal: conrotatory. Conrotatory motion must convert a *cis*-3,4-disubstituted cyclobutene into a *cis,trans* diene. This is easy for the first reactant, and is what actually happens. The second reactant, however, yields a seven-membered ring, which is too small to accommodate a *trans* double bond. The second reaction takes the disrotatory path, probably by a high-energy non-concerted (stepwise) mechanism.

5. **First step.** Closure of cyclobutane ring. Three double bonds of the polyene are involved or 6 π electrons. $4n + 2$ thermal: disrotatory to give K with *cis* fusion.

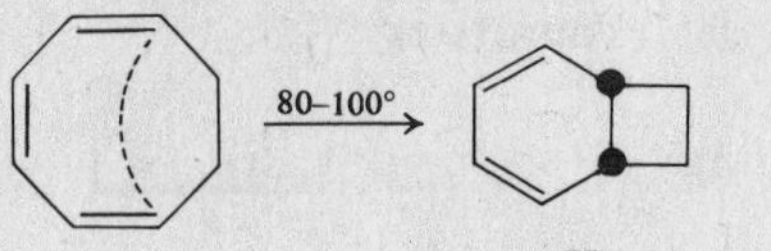

cis,cis,cis-Cyclo-
octa-1,3,5-triene

cis-Bicyclo[4.2.0]-
octa-2,4-diene

Second step. K undergoes a [4 + 2] cycloaddition (Diels-Alder reaction) to give L: *supra,supra* and *endo*.

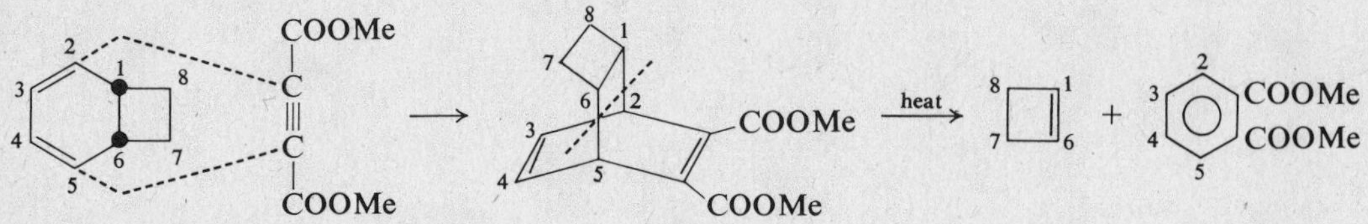

Cyclobutene

Dimethyl
phthalate

Third step. L undergoes a [4 + 2] cycloreversion (*retro*-Diels-Alder reaction) but with breaking of different bonds than those formed in the second step.

6. (a) The rearrangement of a carbonium ion, we have said (Sec. 5.22), involves a 1,2-shift, and takes place via a transition state, XIV, in which the migrating alkyl group (or hydrogen) is bonded to both migration source and migration terminus.

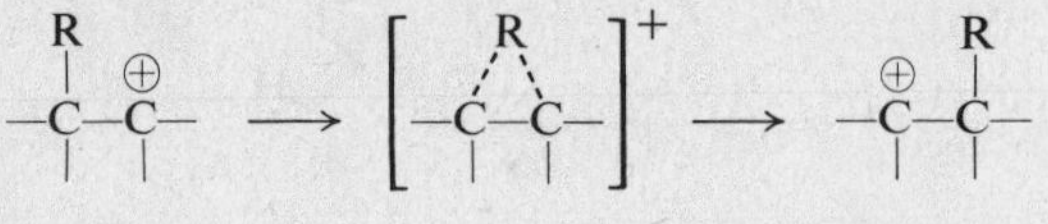

XIV

Using the approach of Sec. 29.10, we treat XIV as arising from overlap between an orbital of a free radical (the migrating group R) and a π framework. The π framework here is not an allylic free radical, but a vinyl radical-cation: ethylene minus one π electron. HOMO is π in Figure 29.5 (page 931).

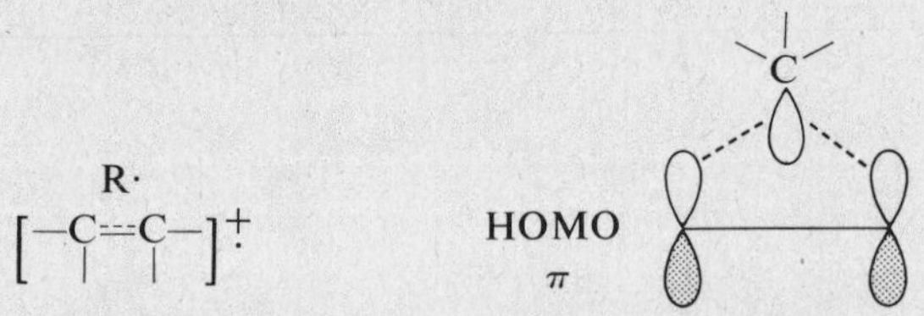

XIV

HOMO
π

1,2-shift in carbonium ion

Suprafacial
Retention in migrating group

On this basis, we see, the symmetry-allowed process is suprafacial and hence feasible geometrically. It should proceed with retention in the migrating group, the stereochemistry actually observed for 1,2-shifts to electron-deficient atoms (Sec. 28.4).

Rearrangement of carbonium ions by [1,2] suprafacial sigmatropic shifts proceeds with retention of configuration in the migrating group.

(b) By the approach of part (a), we consider the π framework to be a diene radical-cation: a diene minus one π electron. HOMO is ψ_2 in Figure 29.6 (page 932). For a [1,4] suprafacial sigmatropic shift in an allylic cation, we predict *inversion of configuration* in the migrating group.

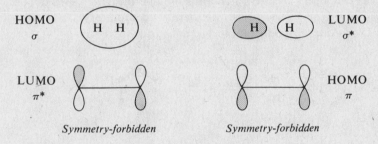

$[C{=}C{-}C{=}C]^{+}$
R·

HOMO
ψ_2

1,4-shift in allylic cation

Suprafacial
Inversion in migrating group

Although such [1,4] shifts are not common, they do occur, and have been found to proceed with the predicted inversion of configuration. (See, for example, H. Hart, T. R. Rodgers, and J. Griffiths, "Stereochemistry of the Rapid Equilibration of Protonated Bicyclo-[3.1.0]hexenones," J. Am. Chem. Soc., **91**, 754 (1969).)

7. For geometric reasons, such a reaction would have to be *supra,supra*, and that is symmetry-forbidden. For H_2, HOMO is σ and LUMO is σ^* (Figure 29.4, page 930). For ethylene, LUMO is π^* and HOMO is π (Figure 29.5, page 931).

HOMO
σ

LUMO
σ^*

LUMO
π^*

HOMO
π

Symmetry-forbidden   *Symmetry-forbidden*

8. An intramolecular Diels-Alder reaction converts II into the intermediate XV. This undergoes either of two (almost) equivalent *retro*-Diels-Alder reactions: one, to regenerate II, and the other—by breaking of bonds as shown—to form III.

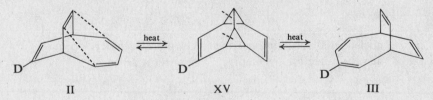

II   XV   III

9. (a) The diazonium salt IV forms benzyne (Problem 18, Chapter 25, page 844),

$\xrightarrow[-N_2, -CO_2]{\text{heat}}$

IV   Benzyne

which undergoes a [4 + 2] *supra,supra* cycloaddition with the diene to give V. (See the answer to Problem 10e, Chapter 27, page 881.)

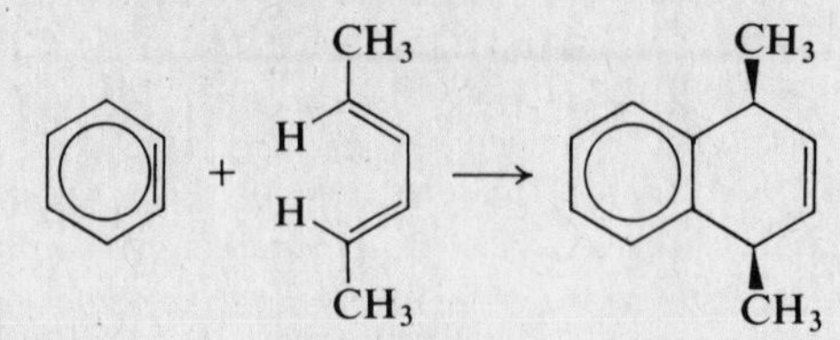

Benzyne   *trans,trans-*   *cis*-V
                2,4-Hexadiene

(b) A [2 + 2] thermal cycloaddition is symmetry-forbidden. Reaction is non-stereospecific because it is not concerted: it probably takes place stepwise via diradicals. For example:

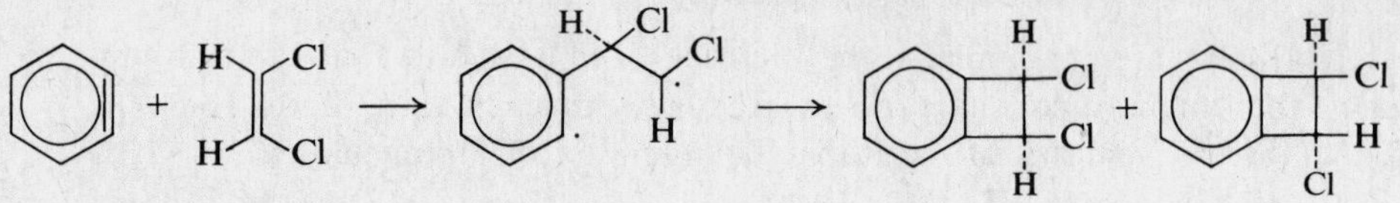

Benzyne *cis*-1,2-Dichloro-   *A diradical*   *cis*-VI   *trans*-VI
             ethene

(M. Jones, Jr., and R. H. Levin, "The Stereochemistry of the 2 + 2 and 2 + 4 Cyclo-additions of Benzyne," J. Am. Chem. Soc., **91**, 6411 (1969).)

10. In each case the cyclobutane ring opens to give a non-aromatic intermediate. Two double bonds of the polyene are involved, or 4 π electrons. 4n thermal: conrotatory. This intermediate then undergoes a Diels-Alder reaction with maleic anhydride.

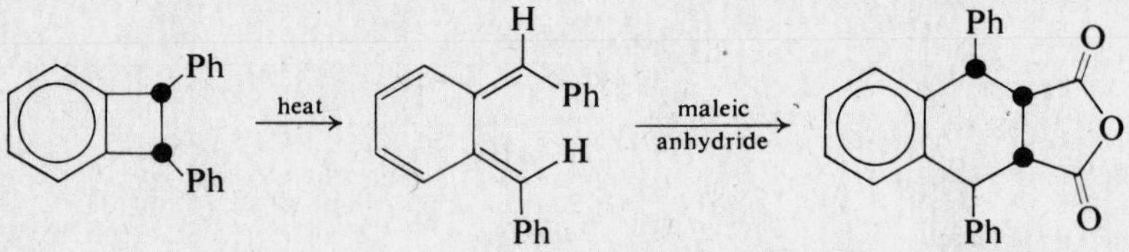

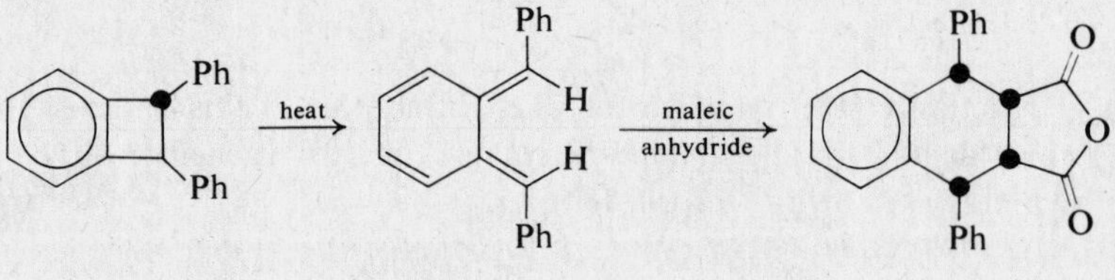

11. (a) There are two successive nucleophilic substitutions: the nucleophile is the cyclopenta-dienyl anion in the first, and the substituted cyclopentadienyl anion in the second.

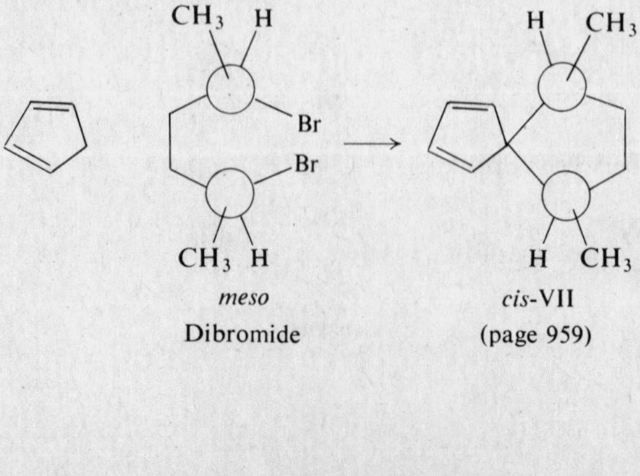

                                    meso   *cis*-VII
                                 Dibromide   (page 959)

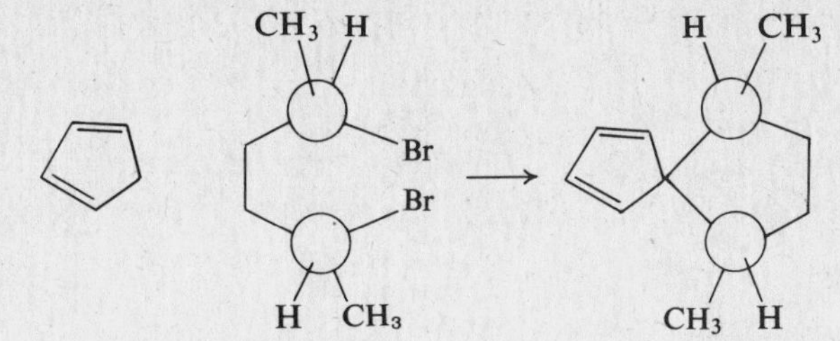

racemic
Dibromide

trans-VII

(b) *cis*-VII contains four non-equivalent olefinic protons. *trans*-VII contains two pairs of equivalent protons. (Remember, the planes of the two rings are perpendicular to each other. *Use models.*)

(H. Kloosterziel *et al.*, "Stereospecificity and Stereochemistry of a Thermal Sigmatropic [1,5]-Shift of an Alkyl Group," Chem. Commun., 1168 (1970).)

12. (a) M and N are position isomers, both from *syn exo* addition of B_2D_6. Similarly, O and P are position isomers

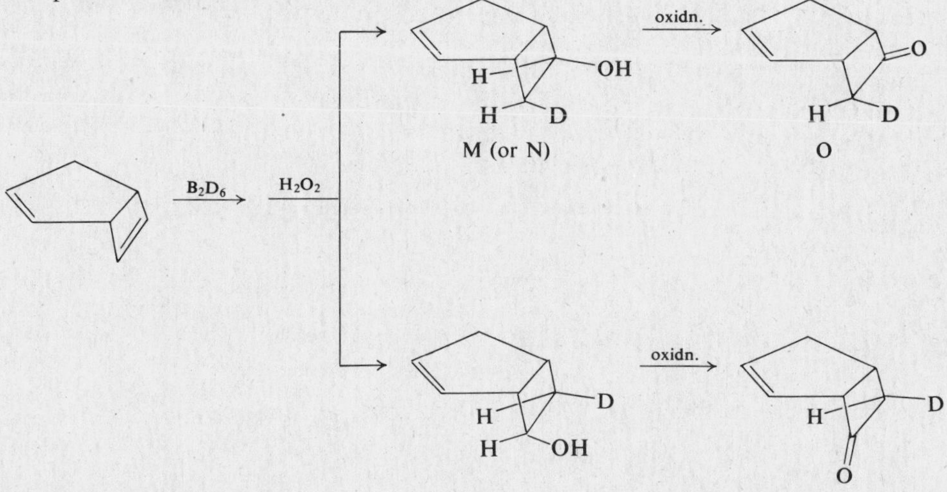

(We know that hydroboration-oxidation is *syn* (Sec. 15.10). We have not learned, however, that hydroboration is stereospecifically *exo* in reactions like these—although this is the case. Let us consider the possibility that, instead of being position isomers, M and N were stereoisomers resulting from *exo* and *endo* hydroboration. In that case, O and P would also be stereoisomers; but they would differ only as to whether —H or —D were *exo* or *endo*, and would hardly have been separable.)

Reduction of O converts the carbonyl group into a new chiral center, and gives both possible configurations about that center (Figure 29.27, page 480 of this Study Guide).

(b) V undergoes a *retro*-Diels-Alder reaction (Figure 29.28, page 480 of this Study Guide).

(J. A. Berson and G. L. Nelson, "Inversion of Configuration in the Migrating Group of a Thermal 1,3-Sigmatropic Rearrangement," J. Am. Chem. Soc., **89**, 5503 (1967); J. A. Berson, "The Stereochemistry of Sigmatropic Rearrangements. Test of the Predictive Power of Orbital Symmetry Rules," Accounts Chem. Res., **1**, 152 (1968).)

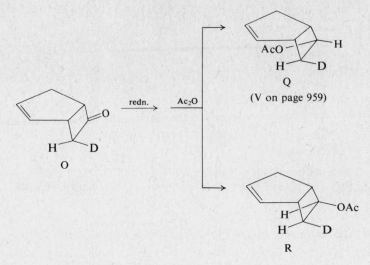

Figure 29.27.   Formation of Q and R (Problem 12).

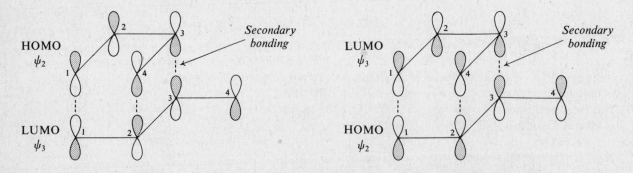

Figure 29.28.   *Retro*-Diels-Alder reaction of VI (Problem 12).

13. (a) Let us make a diagram similar to Figure 29.20 (page 950) for the *endo* dimerization of butadiene.

We see, of course, the overlap between lobes on C–1 and C–4 of the diene and lobes on C–1 and C–2 of the ene. In addition, we see that a lobe on C–3 of the diene is brought close to a lobe *of the same phase* on C–3 of the ene—carbons that are not even bonded to each other in the product. Weak (temporary) bonding between these atoms helps to stabilize the transition state. Such secondary bonding could not occur if reaction were *exo*, since the atoms concerned would be far apart.

(b) Let us make a diagram similar to the one in (a) for the *endo* [6 + 4] cycloaddition of a triene with a diene.

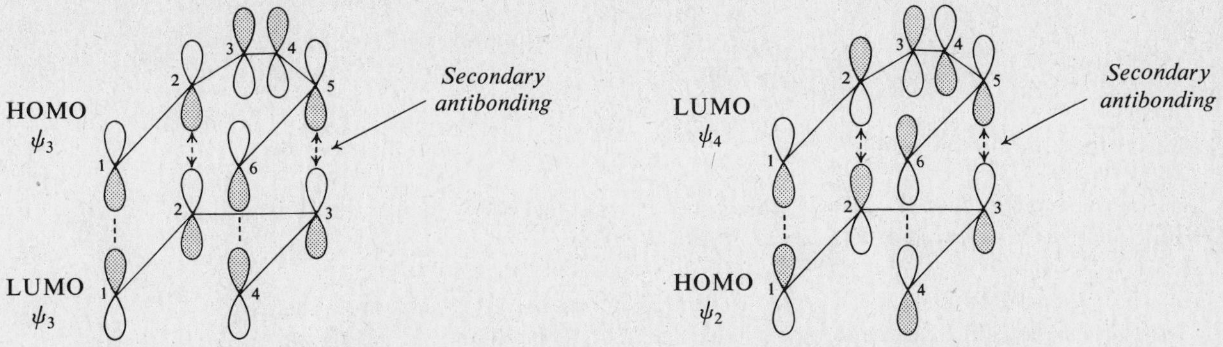

We see the overlap that leads to bond formation: between lobes on C–1 and C–6 of the triene and C–1 and C–4 of the diene. But juxtaposed lobes on C–2 and C–5 of the triene and C–2 and C–3 of the diene are *of opposite phase*; instead of giving secondary bonding in the transition state, such interactions would give antibonding. As a result, [6 + 4] cycloadditions take place in the *exo* manner.

This was predicted by Woodward and Hoffmann in 1965, and when, shortly afterward, the first [6 + 4] cycloaddition was recognized, the prediction was found to be correct (see Problem 29.9c, page 953).

14.

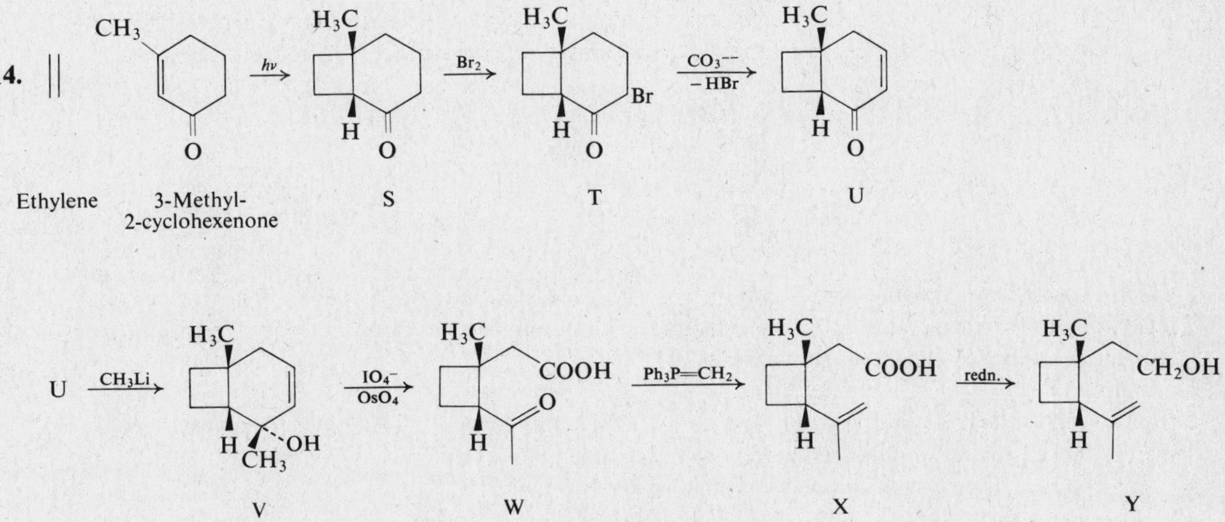

The first step is a [2 + 2] photochemical *supra,supra* cycloaddition. This is the key step in the synthesis. It generates the needed cyclobutane ring, and with the proper stereochemical relationship between —CH₃ and —H. At the same time, it provides the first of the carbonyl groups on whose properties depend most of the straight-forward chemistry that follows: *alpha*-bromination to give T; addition of CH_3Li to give V, which opens the way to W, with a new carbonyl group; the Wittig reaction to give X with the terminal olefinic group. Several steps deserve a closer look.

Conversion of U into V. CH_3Li attacks U from the most open side (by path a), away from the fold in the molecule.

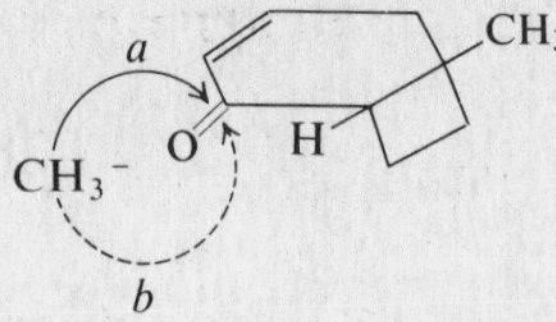

Conversion of V into W. The double bond of V is presumably hydroxylated to give a triol, which is then cleaved at two places to give W.

(b) The molecular formula for Z indicates intramolecular solvomercuration. This is geometrically possible only for the isomer in which the double bond and $—CH_2OH$ are *cis*.

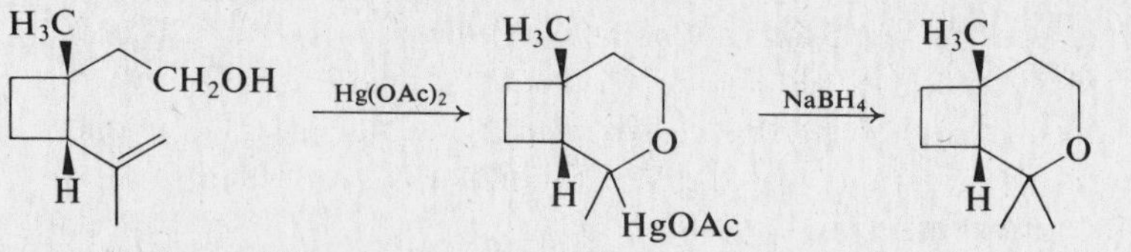

(R. Zurflüh, L. L. Dunham, V. L. Spain, and J. B. Siddall, "Synthetic Studies on Insect Hormones. IX. Stereoselective Total Synthesis of a Racemic Boll Weevil Pheromone," J. Am. Chem. Soc., **92**, 425 (1970).)

15. (a) Conversion of VII into benzene requires opening of a cyclobutene ring, and involves 4π electrons of the "polyene" (benzene). The symmetry-allowed process is conrotatory,

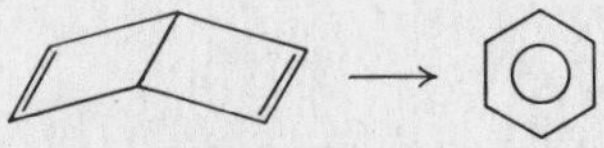

VII

Bicyclo[2,2,0]-
hexa-2,5-diene

"Dewar benzene"

Benzene

which would yield the impossibly strained *cis,cis,trans*-cyclohexa-1,3,5-triene. Reaction must go with forbidden (that is, difficult) disrotatory motion—or, perhaps, by a mechanism that is not concerted at all, but step-wise.

(b) Conversion of VIII into toluene requires a [1,3]-H shift (or a [1,7]-H shift). The symmetry-allowed *antara* shift is geometrically impossible. Reaction must go either by the

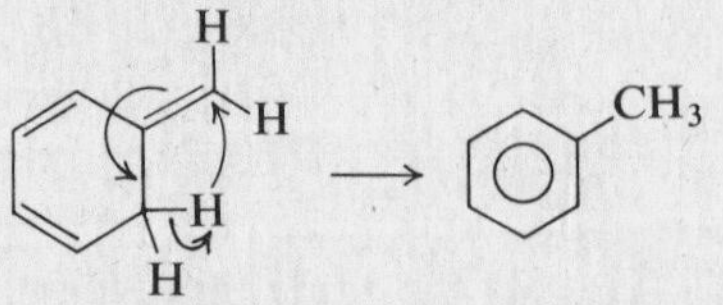

VIII

5-Methylene-
1,3-cyclohexadiene

Toluene

forbidden *supra* path, or by a non-concerted mechanism, also difficult.

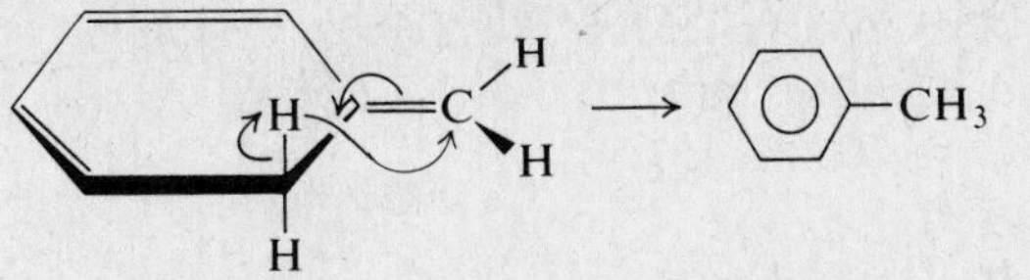

Antara
Symmetry-allowed,
geometrically impossible

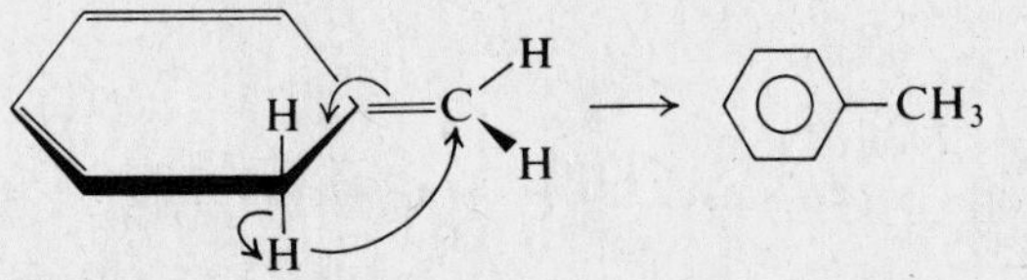

Supra
Geometrically easy,
symmetry-forbidden

16. (a) **First step.** An electrocyclic opening of ring B. In the polyene three double bonds are involved, or 6 π electrons. $4n + 2$ photochemical: conrotatory opening.

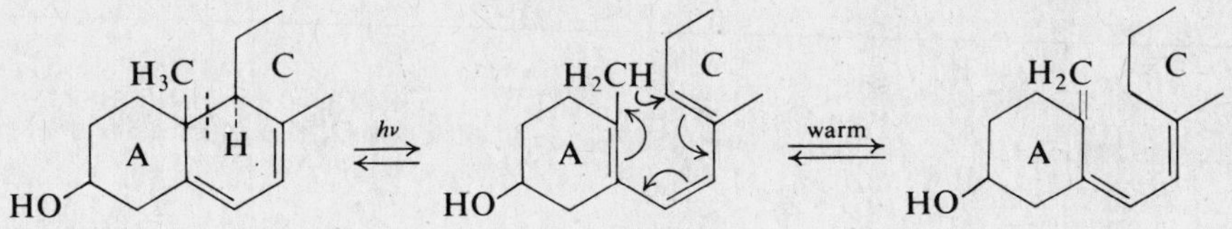

7-Dehydrocholesterol   Pre-cholecalciferol   Cholecalciferol

(A *thermal* ring-opening cannot take place: the symmetry-allowed disrotatory motion would generate a *trans* double bond in either ring A or C, geometrically impossible in a six-membered ring.)

Second step. A [1,7]-H shift which, since it is thermal, is presumably *antara*.

(b) These have structures analogous to those of pre-cholecalciferol and cholecalciferol, but with a nine-carbon unsaturated side chain:

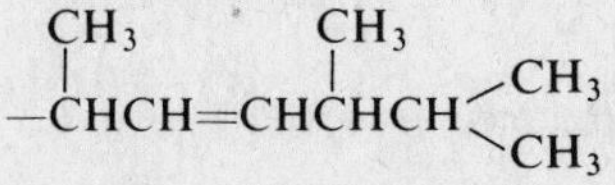

$$
\begin{array}{cc}
\text{CH}_3 & \text{CH}_3 \\
| & | \\
\end{array}
$$
$$-\text{CHCH}=\text{CHCHCH}\overset{\text{CH}_3}{\underset{\text{CH}_3}{\diagdown}}$$

(c) Electrocyclic closure of ring B, involving three double bonds of the polyene, or 6 π electrons. $4n + 2$ thermal: disrotatory.

As discussed in the answer to Problem 3c, above, there are two possible disrotatory modes,

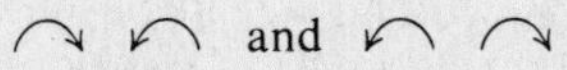

and

and here, as in that earlier problem, they yield different stereoisomers: IX and X. In ergosterol, the C–1 methyl is β and the C–9 H is α (see page 514). In IX and X, both substituents are α in one and both are β in the other.

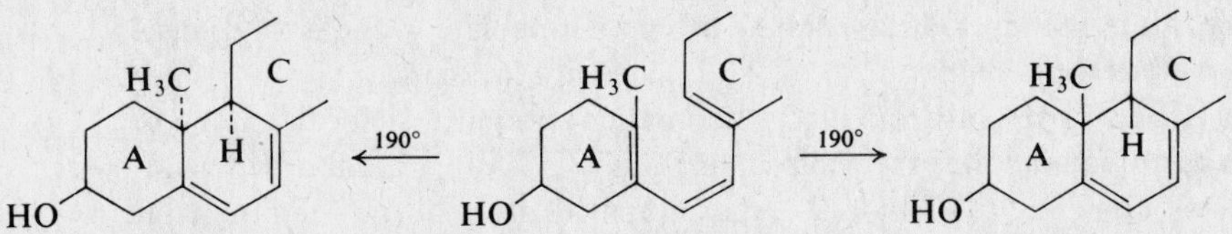

IX (or X)   Pre-ergocalciferol   X (or IX)

(d) Electrocyclic opening of ring B, involving 6 π electrons. $4n + 2$ photochemical: conrotatory, as in the first step of part (b).

Consider the ring-closing process. There are *two* conrotatory modes,

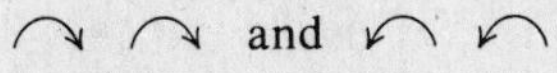

and

and in this case they yield different stereoisomers. Pre-ergocalciferol is thus related by conrotatory motion to two compounds: ergosterol and XI. In ergosterol the C-1 methyl is β, and the C-9 H is α. In XI the C-1 methyl is α, and the C-9 H is β.

17. (a) *trans*-XII undergoes ring-opening to give a cyclodecapentaene. This involves three double bonds of the polyene, or 6 π electrons. $4n + 2$ photochemical: conrotatory ring-opening.

There are two conrotatory modes (see the answer to Problem 29.3, above) to give two possible products: in XVIa the three new double bonds are all *cis*; in XVIb they are *trans,cis,-trans*.

XVI is evidently thermally unstable, and at room temperature undergoes a ring closure involving three double bonds, or 6 π electrons. $4n + 2$ thermal: disrotatory motion to give *cis*-XII. XVI is thus related to both stereoisomers of XII: to one photochemically, and to the other, thermally.

At $-190°$ XVI is stable, and is not converted into *cis*-XII. It can, however, be reduced to cyclodecane.

When *trans*-XII is photolyzed at $-190°$, XVI is formed. Warming to room temperature converts it into *cis*-XII, which remains unchanged when the temperature is lowered again.

(b) Cyclodecapentaene is a cyclic, completely conjugated polyene with five double bonds, or 10 π electrons; that is, it is [10]*annulene* (Problem 9, Chapter 10, page 336). But 10 is a Hückel number ($4n + 2$, with $n = 2$), and—*geometry permitting*—should give rise to aromaticity. (As we saw in Problem 10.7, page 331, the cyclooctatetraenyl dianion, with 10 π electrons, *is* aromatic.)

The question is, *does* the geometry permit aromaticity? And the answer seems clearly to be: *no*. There is little doubt that cyclodecapentaene is the intermediate in these reactions; there is equally little doubt that it is highly unstable and hence not aromatic.

The next question is: *why* is it not aromatic? Aromaticity requires delocalization of π electrons, and this in turn requires that the molecule be flat. If XVIa were flat, it would be a regular decagon, with bond angles of 144°, 24° more than the normal trigonal angle of 120°. The angle strain accompanying this deviation is evidently more than the molecule can afford to accept for aromaticity. (The aromatic cyclooctatetraenyl dianion has bond angles of 135°, 15° bigger than normal trigonal angles: this much strain the molecule evidently *can* accept—along with a double negative charge.)

XVIb could be flat with no C—C—C angle strain, but here another problem arises: the hydrogens on C-1 and C-6 would point toward each other and be hopelessly crowded.

(Not until the [18]annulene (page 935) is the ring big enough to accommodate the inward-pointing hydrogens.)

The isolation of a stable compound to establish aromaticity can be difficult, but is straight-forward.   To interpret *failure to isolate* a supposed intermediate as evidence of non-aromaticity is much trickier.   Besides isolation of the reduction product, cyclodecane, the strength of the argument here depends on the design of the experiments, and rests ultimately on the strength of the orbital symmetry rules.

(E. E. van Tamelen, T. L. Burkoth, and R. H. Greeley, "The *trans*-9,10-Dihydronaphthalene–Cyclodecapentaene Valence Bond Isomer System," J. Am. Chem. Soc., **93**, 6120 (1971).)

Chapter 30 | Polynuclear Aromatic Compounds

30.1

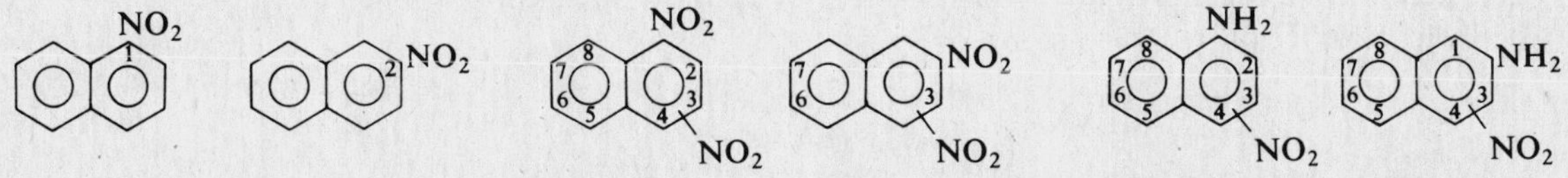

![Naphthalene with NO₂ at position 1] ![Naphthalene with NO₂ at position 2]

Mononitro: two

![Naphthalene numbered 8,7,6,5,2,3,4 with NO₂ at 1 and NO₂ at 4]
Seven

![Naphthalene with NO₂ at 1 and NO₂ at 3]
Three

Dinitro: ten

![Naphthalene with NH₂ and NO₂]
Seven

![Naphthalene with NH₂ and NO₂]
Seven

Nitronaphthylamines: fourteen

30.2
$$\text{ArH} \xrightarrow[\text{Crafts}]{\text{Friedel-}} \underset{\substack{\parallel \\ \text{O}}}{\text{ArCCH}_3} \xrightarrow{\text{OI}^-} \text{CHI}_3 + \text{ArCOO}^- \xrightarrow{\text{H}^+} \text{ArCOOH}$$

α or β Iodoform α or β

30.3 (a)

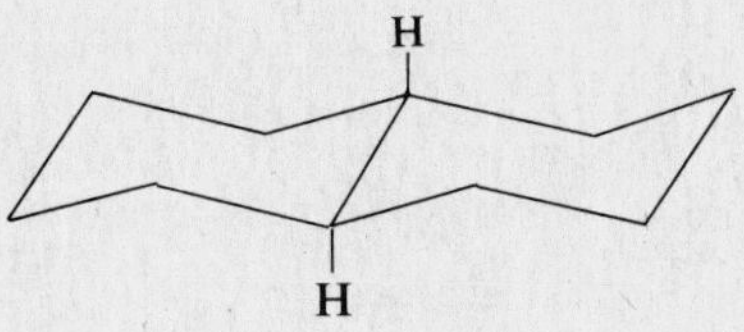

trans-Decalin 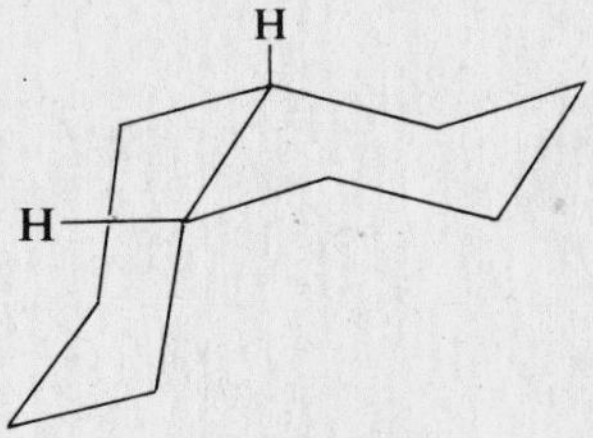   *cis*-Decalin

(b)

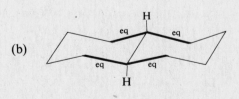

trans: 2 equatorial bonds

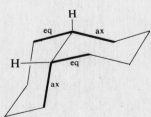

cis: 1 equatorial bond, 1 axial bond

487

trans-Decalin should be the more stable: both large groups (the other ring) on each ring are equatorial.

(c) Hydrogenation is reversible: *syn*-hydrogenation to give *cis*-decalin is rate-controlled; *anti*-hydrogenation to give *trans*-decalin is equilibrium-controlled. This is comparable to rapid 1,2-addition vs. slower formation of the more stable 1,4-addition product from conjugated dienes; conversion of initially formed *o*- and *p*-xylenes into more stable *m*-xylene in Friedel-Crafts methylation of toluene; conversion of *o*-phenolsulfonic acid into the more stable *p*-isomer; conversion of 1-naphthalenesulfonic acid into the more stable 2-naphthalenesulfonic acid.

30.4

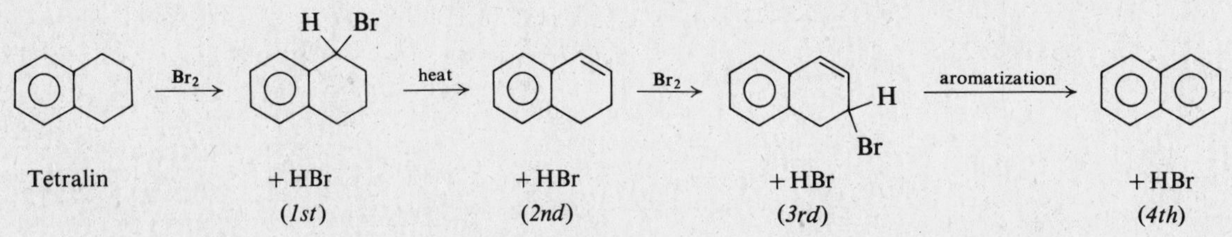

Each step yields a mole of HBr in, respectively: (1) benzylic substitution, (2) elimination, (3) allylic substitution, (4) elimination.

30.5

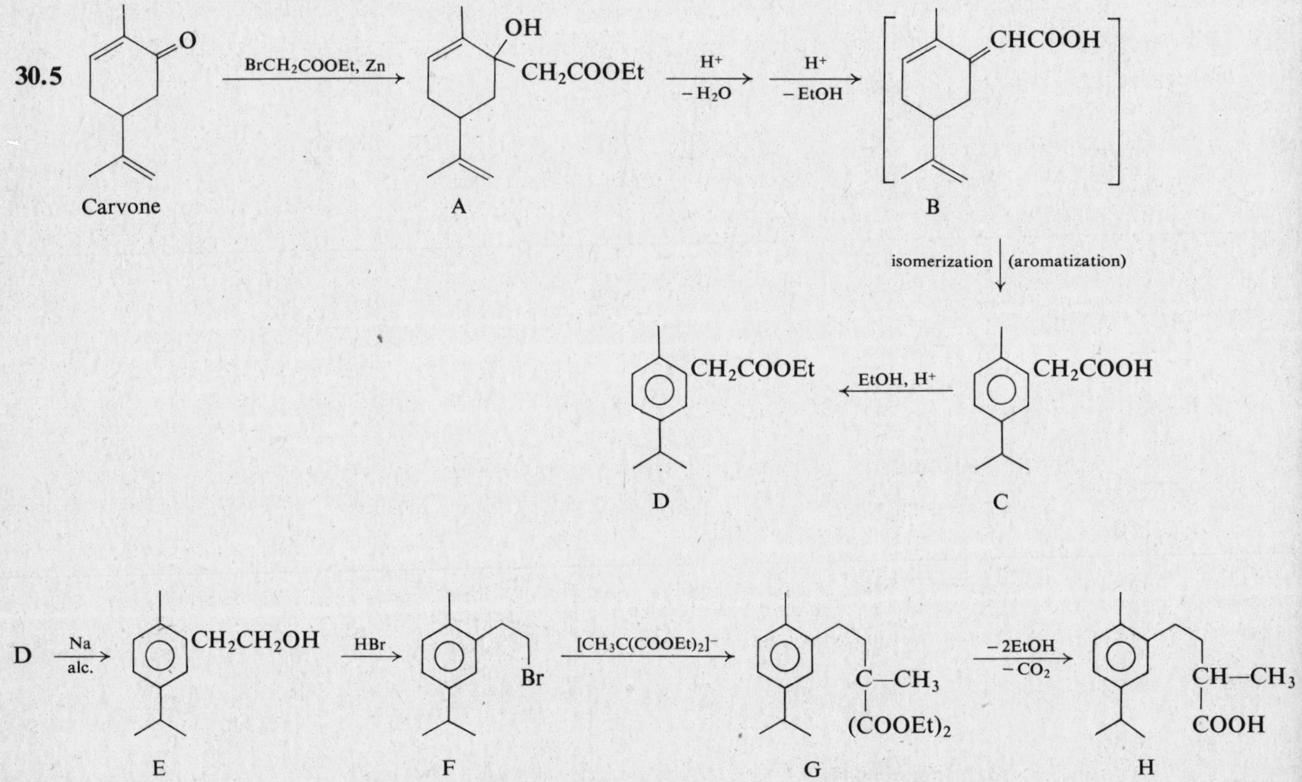

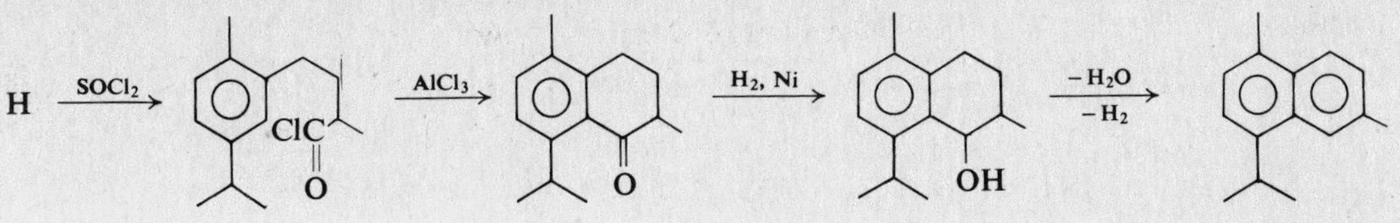

$$H \xrightarrow{\text{SOCl}_2} I \xrightarrow{\text{AlCl}_3} J \xrightarrow{\text{H}_2, \text{Ni}} K \xrightarrow[\,-\text{H}_2\,]{-\text{H}_2\text{O}} L$$

L
Cadalene
4-Isopropyl-1,6-dimethylnaphthalene

(b) Cadinene has the same carbon skeleton as cadalene; its structure follows the isoprene rule.

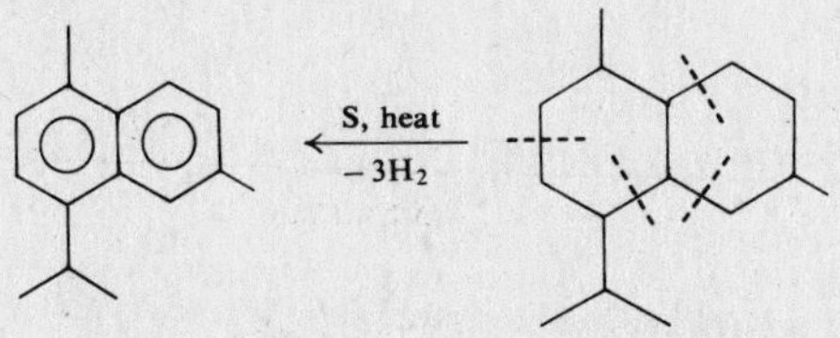

Cadalene $\xleftarrow[\,-3\text{H}_2\,]{\text{S, heat}}$ Cadinene
Probable skeleton

30.6 (a) $\xleftarrow{\text{Fe, H}^+}$

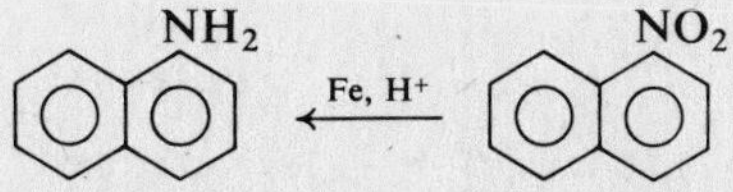

(b) $\xleftarrow{\text{KI}}$ $\xleftarrow{\text{HONO}}$ $\longleftarrow$ as in (a)

(c) $\xleftarrow{\text{CuCN}}$ $\longleftarrow$ as in (b)

(d) $\xleftarrow[\text{heat}]{\text{H}_2\text{O, H}^+}$ $\longleftarrow$ as in (c)

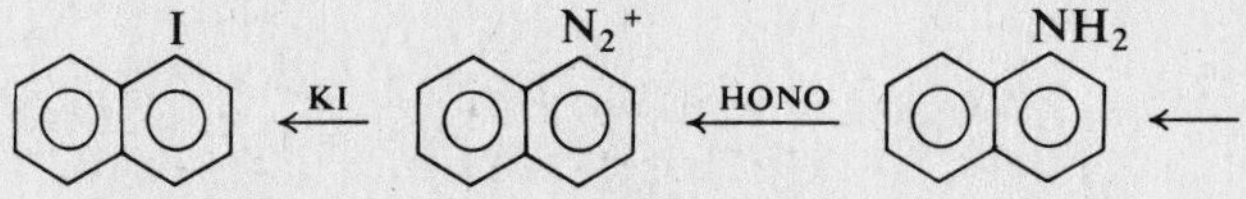

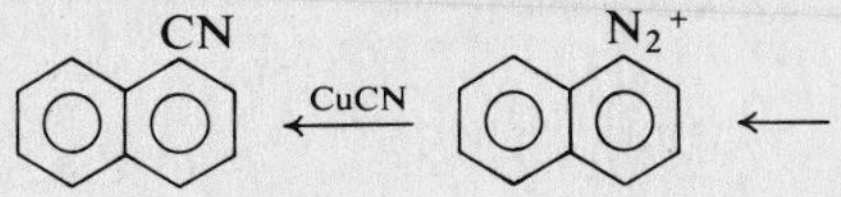

(e) $\xleftarrow{\text{SOCl}_2}$ $\longleftarrow$ as in (d)

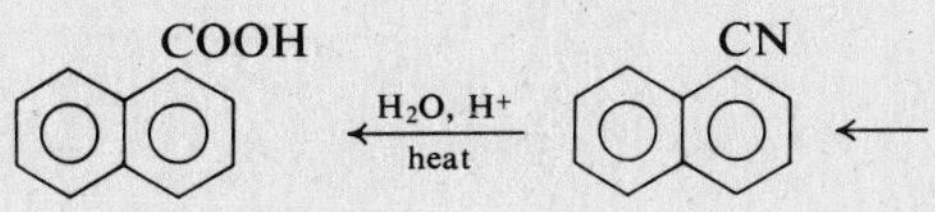

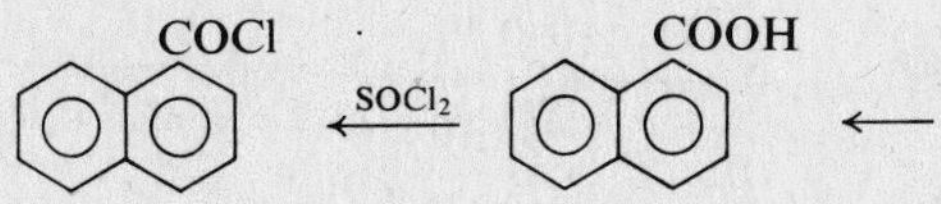

(f) $\xleftarrow{\text{Et}_2\text{Cd}}$ $\longleftarrow$ as in (e)

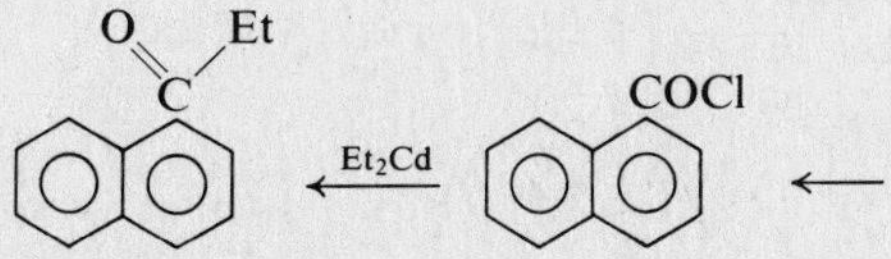

(g) $\xleftarrow{2\text{H}_2, \text{Ni}}$ $\longleftarrow$ as in (c)

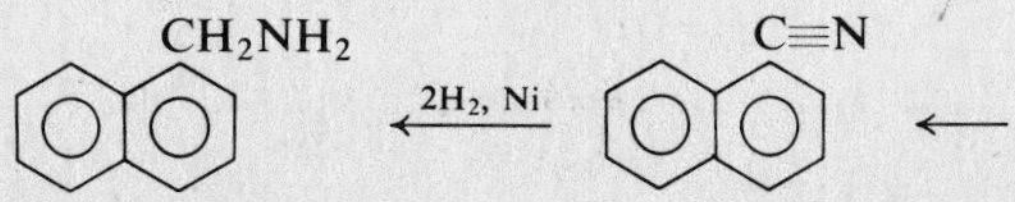

(h) 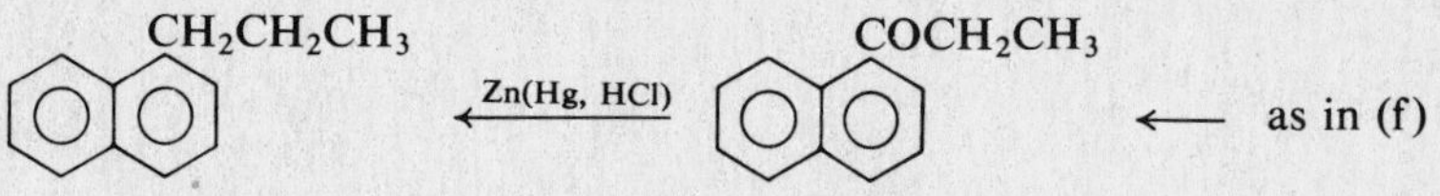 $CH_2CH_2CH_3$ (naphthalene) $\xleftarrow{\text{Zn(Hg, HCl)}}$ $COCH_2CH_3$ (naphthalene) $\longleftarrow$ as in (f)

(i) 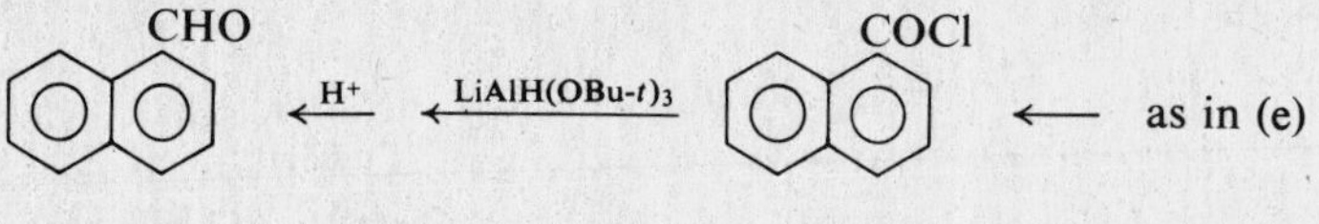   CHO (naphthalene) $\xleftarrow{H^+}$ $\xleftarrow{\text{LiAlH(OBu-}t)_3}$ COCl (naphthalene) $\longleftarrow$ as in (e)

(j) 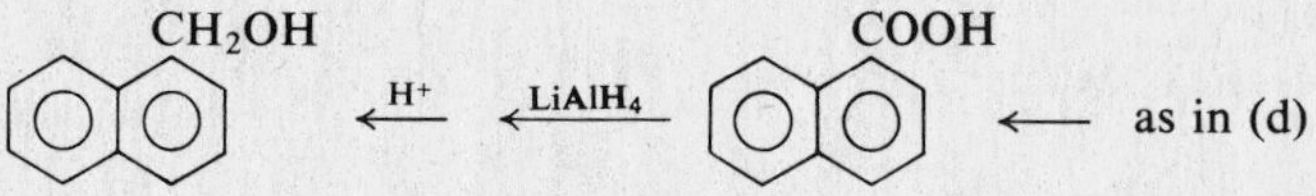 CH_2OH (naphthalene) $\xleftarrow{H^+}$ $\xleftarrow{\text{LiAlH}_4}$ COOH (naphthalene) $\longleftarrow$ as in (d)

(k) 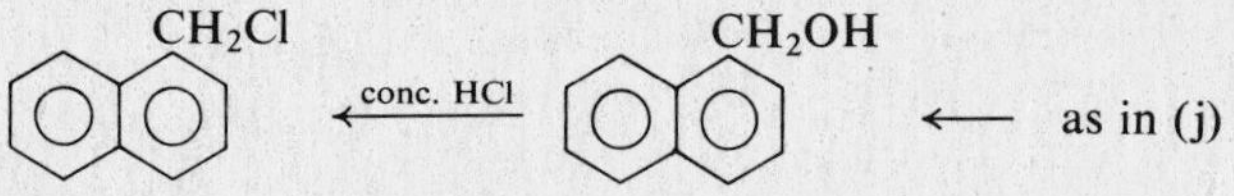 CH_2Cl (naphthalene) $\xleftarrow{\text{conc. HCl}}$ CH_2OH (naphthalene) $\longleftarrow$ as in (j)

(l)  CH_2COOH (naphthalene) $\xleftarrow[\text{heat}]{H_2O, H^+}$ CH_2CN (naphthalene) $\xleftarrow{CN^-}$ CH_2Cl (naphthalene) $\longleftarrow$ as in (k)

(m) 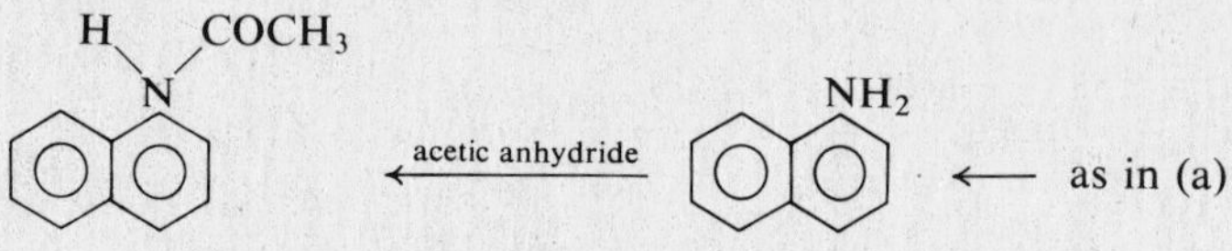 $\underset{N}{\overset{H \quad COCH_3}{}}$ (naphthalene) $\xleftarrow{\text{acetic anhydride}}$ NH_2 (naphthalene) $\longleftarrow$ as in (a)

30.7 (a) 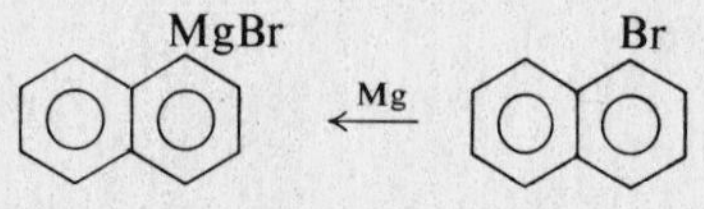   MgBr (naphthalene) $\xleftarrow{\text{Mg}}$ Br (naphthalene)

(b) 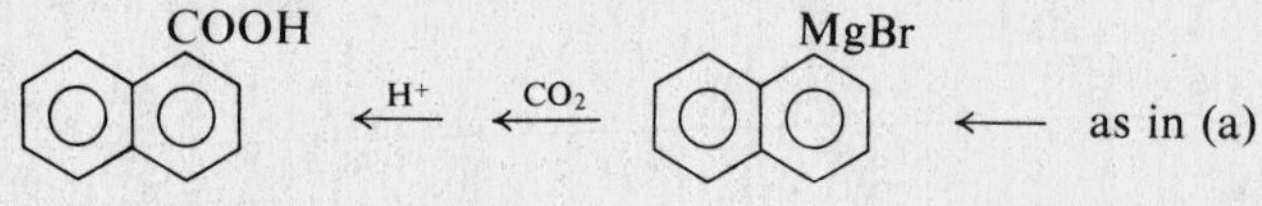   COOH (naphthalene) $\xleftarrow{H^+}$ $\xleftarrow{CO_2}$ MgBr (naphthalene) $\longleftarrow$ as in (a)

(c) 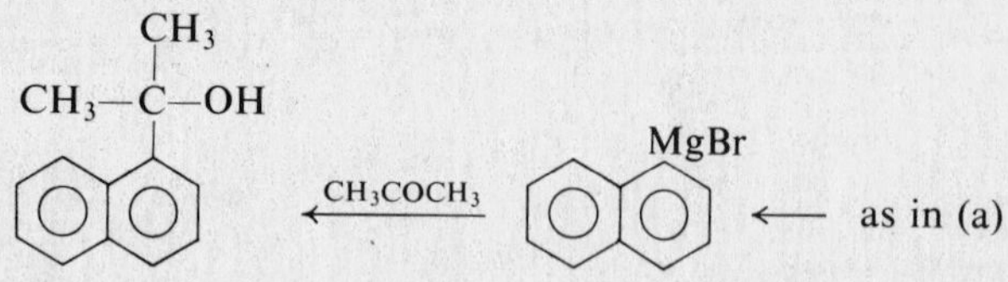 $CH_3-\overset{CH_3}{\underset{|}{C}}-OH$ (naphthalene) $\xleftarrow{CH_3COCH_3}$ MgBr (naphthalene) $\longleftarrow$ as in (a)

(d) 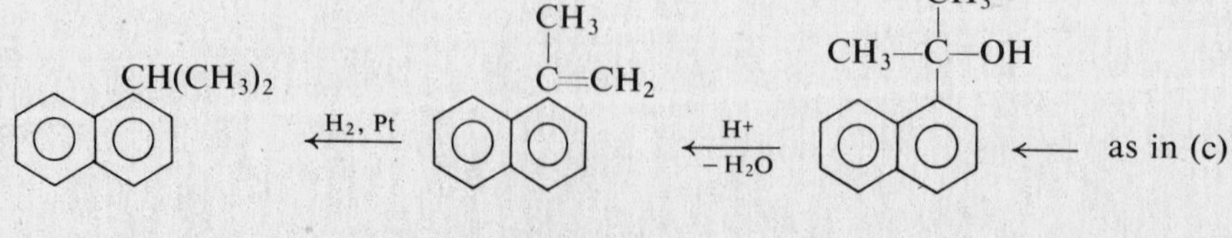 $CH(CH_3)_2$ (naphthalene) $\xleftarrow{H_2, Pt}$ $\overset{CH_3}{\underset{}{C}}=CH_2$ (naphthalene) $\xleftarrow[-H_2O]{H^+}$ $CH_3-\overset{CH_3}{\underset{|}{C}}-OH$ (naphthalene) $\longleftarrow$ as in (c)

(e) 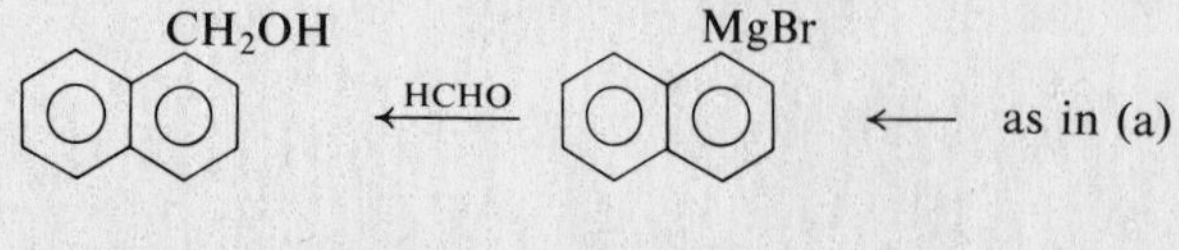 CH_2OH (naphthalene) $\xleftarrow{HCHO}$ MgBr (naphthalene) $\longleftarrow$ as in (a)

(f) 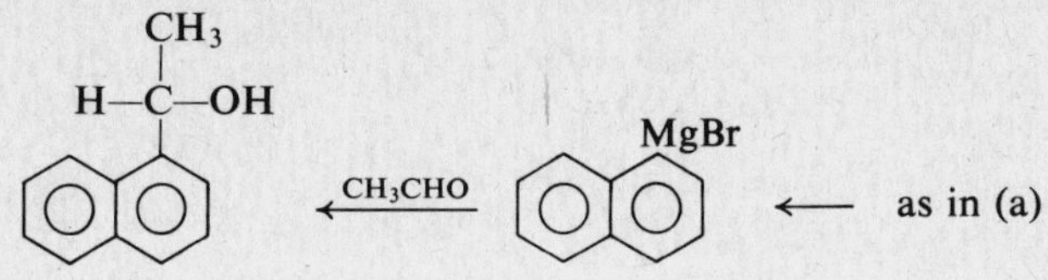   CH$_3$CHO $\longleftarrow$ MgBr $\longleftarrow$ as in (a)

(g) 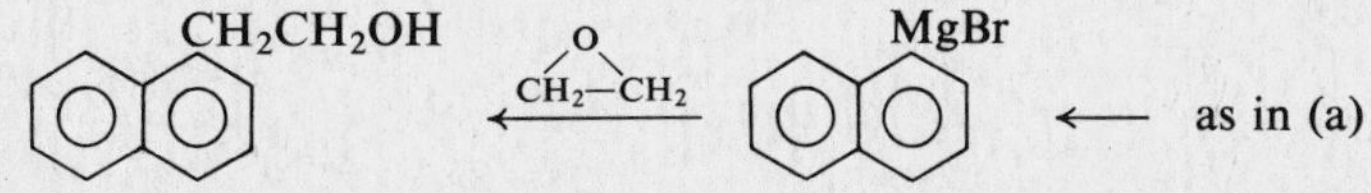   CH$_2$CH$_2$OH   CH$_2$—CH$_2$ $\longleftarrow$ MgBr $\longleftarrow$ as in (a)

30.8 (a) 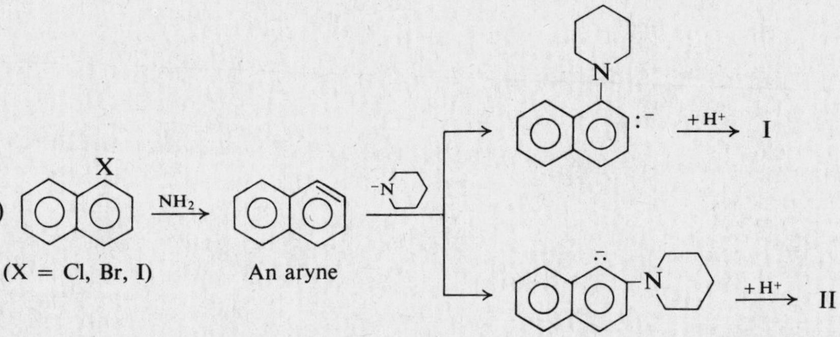   (X = Cl, Br, I) $\xrightarrow{NH_2}$ An aryne $\xrightarrow{-N}$ $\xrightarrow{+H^+}$ I $\xrightarrow{+H^+}$ II

The three different halides give the same product ratio because they all form the same intermediate: the aryne. II is the favored product because it is formed via the more stable carbanion: the one in which the negative charge is closer to the electron-withdrawing substituent—the other ring.

(b) By direct bimolecular displacement of —F by the amine. The 2-CH$_3$ makes formation of an aryne impossible.

(c) Both bimolecular displacement and elimination-addition occur. Fluoronaphthalene is the least reactive toward aryne formation (Sec. 25.14), but the most reactive toward bimolecular displacement (Sec. 25.12).

30.9   COOH $\xrightarrow{oxidn.}$ HOOC   COOH   HOOC

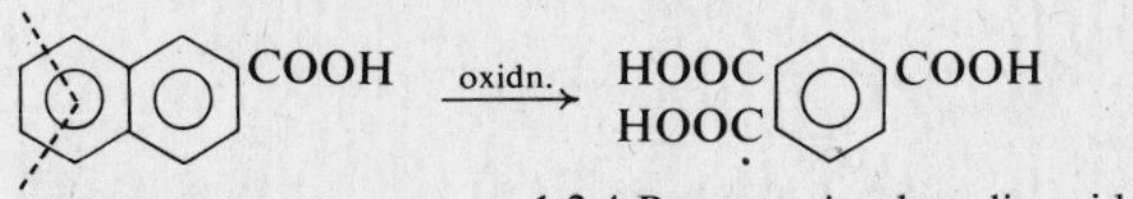

2-Naphthoic acid   1,2,4-Benzenetricarboxylic acid
                        M.p. 225–235° dec.

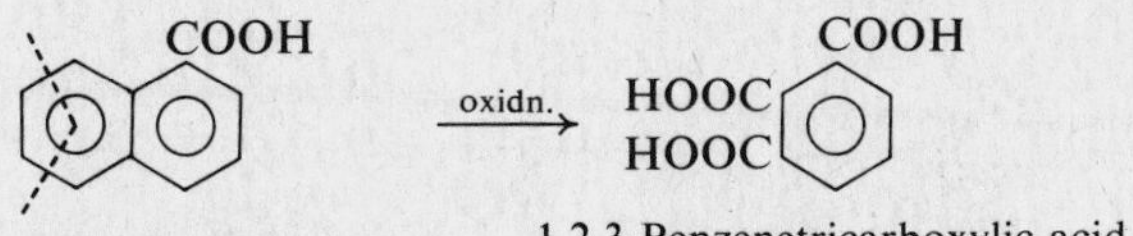

   COOH $\xrightarrow{oxidn.}$ COOH   HOOC   HOOC

1-Napththoic acid   1,2,3-Benzenetricarboxylic acid
                        M.p. 190–197° dec.

30.10 (a)    Et $\xleftarrow[H^+]{Zn(Hg)}$ COCH$_3$ $\xleftarrow[PhNO_2, AlCl_3]{CH_3COCl}$

(b) 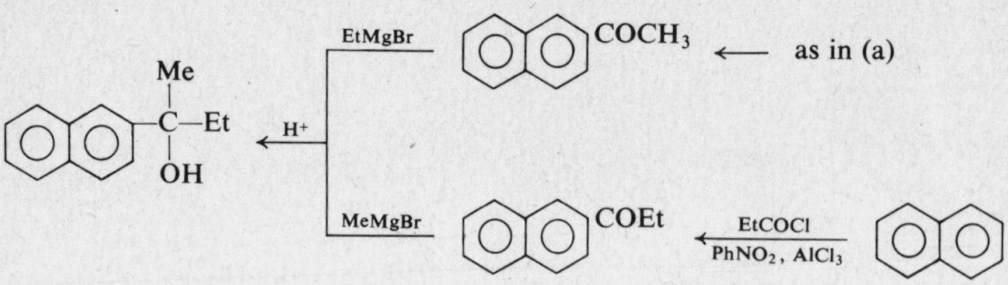 reaction scheme:

$\xrightarrow{\text{EtMgBr}}$ (naphthalene)COCH₃ $\longleftarrow$ as in (a)

(b) (naphthalene)C(Me)(Et)OH $\xleftarrow{\text{H}^+}$

$\xrightarrow{\text{MeMgBr}}$ (naphthalene)COEt $\xleftarrow[\text{PhNO}_2,\ \text{AlCl}_3]{\text{EtCOCl}}$ (naphthalene)

(c) (naphthalene)C(Me)(Et)H $\xleftarrow[\text{Pt}]{\text{H}_2}$ (naphthalene)C(Me)=CHCH₃ $\xleftarrow[-\text{H}_2\text{O}]{\text{H}^+}$ product (b)

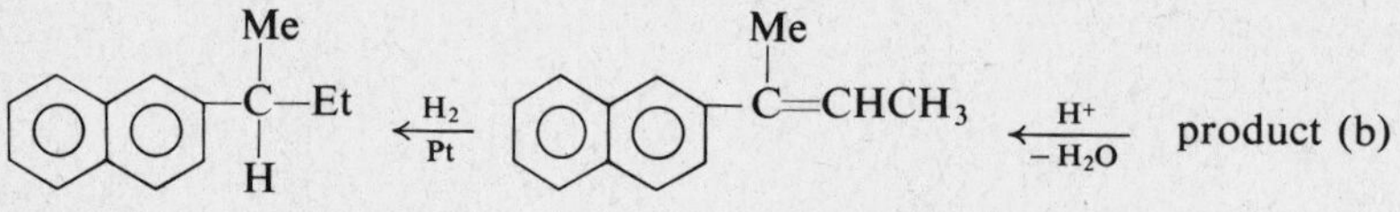

(d) 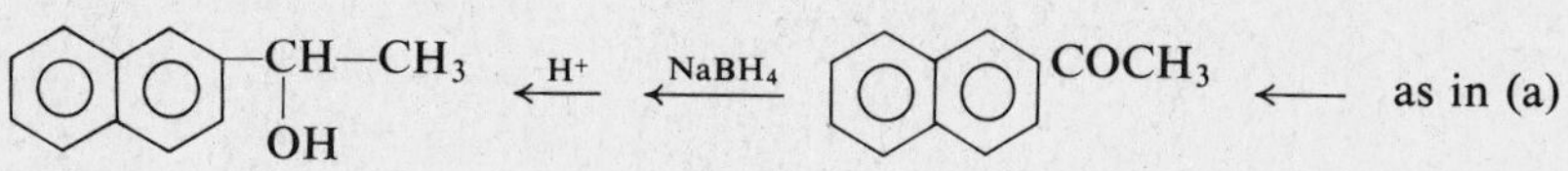 (naphthalene)CH(CH₃)OH $\xleftarrow{\text{H}^+}$ $\xleftarrow{\text{NaBH}_4}$ (naphthalene)COCH₃ $\longleftarrow$ as in (a)

(e) 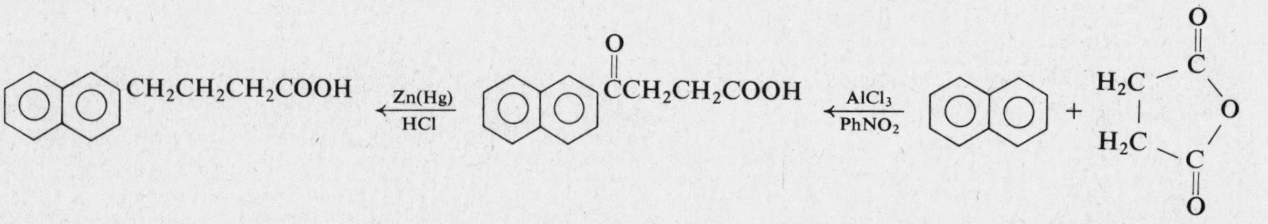 (naphthalene)CH₂CH₂CH₂COOH $\xleftarrow[\text{HCl}]{\text{Zn(Hg)}}$ (naphthalene)COCH₂CH₂COOH $\xleftarrow[\text{PhNO}_2]{\text{AlCl}_3}$ (naphthalene) + Succinic anhydride

Succinic anhydride

(f) 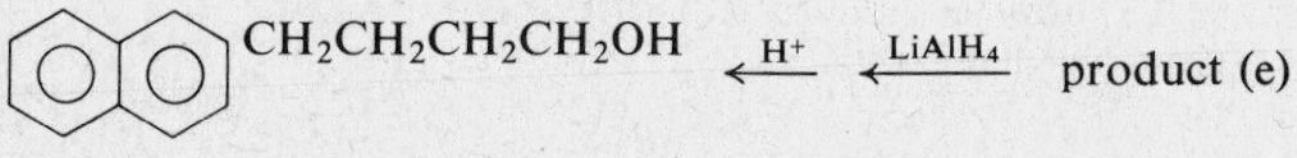 (naphthalene)CH₂CH₂CH₂CH₂OH $\xleftarrow{\text{H}^+}$ $\xleftarrow{\text{LiAlH}_4}$ product (e)

(g) 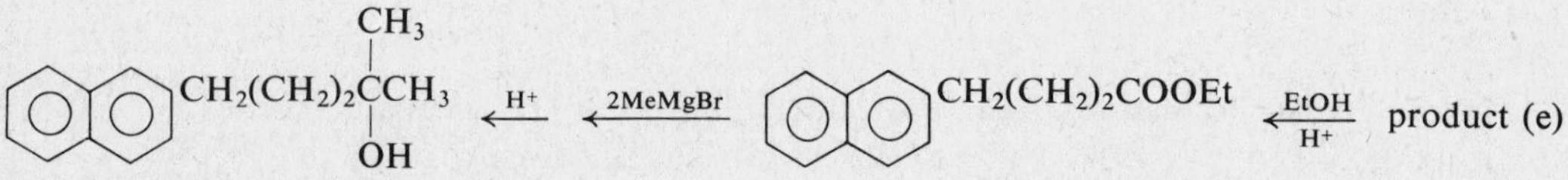 (naphthalene)CH₂(CH₂)₂C(CH₃)₂OH $\xleftarrow{\text{H}^+}$ $\xleftarrow{\text{2MeMgBr}}$ (naphthalene)CH₂(CH₂)₂COOEt $\xleftarrow[\text{H}^+]{\text{EtOH}}$ product (e)

(h) (naphthalene)CH₂(CH₂)₂CH(CH₃)CH₃ $\xleftarrow[\text{Ni}]{\text{H}_2}$ $\xleftarrow[-\text{H}_2\text{O}]{\text{H}^+}$ product (g)

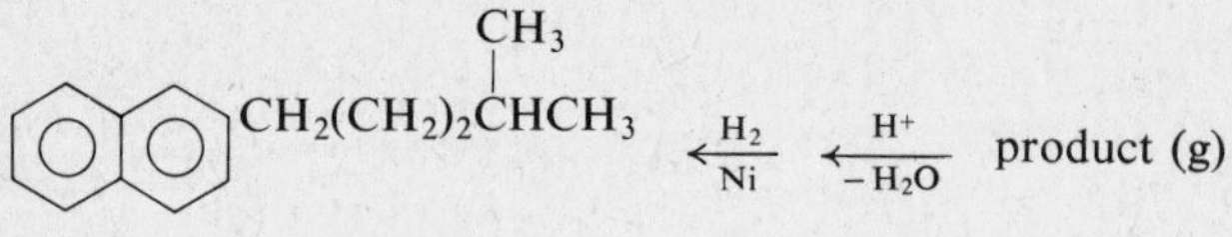

(i) 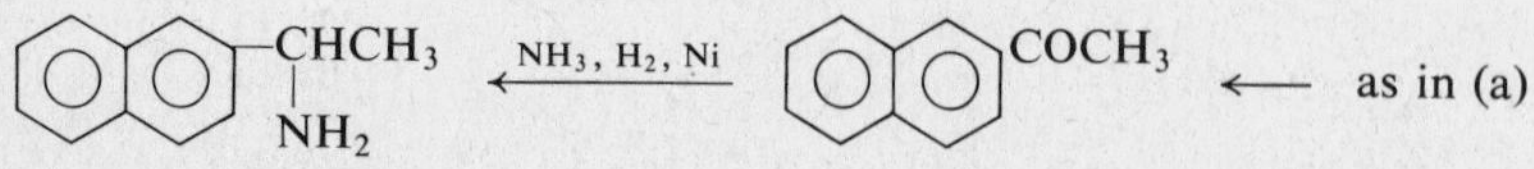 (naphthalene)CH(CH₃)NH₂ $\xleftarrow{\text{NH}_3,\ \text{H}_2,\ \text{Ni}}$ (naphthalene)COCH₃ $\longleftarrow$ as in (a)

(j) 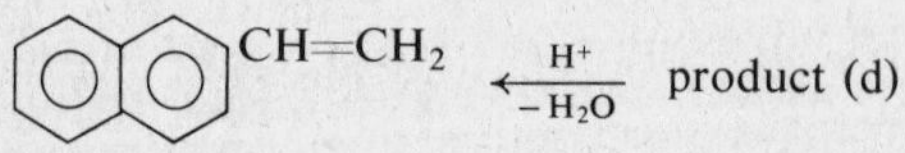 (naphthalene)CH=CH₂ $\xleftarrow[-\text{H}_2\text{O}]{\text{H}^+}$ product (d)

30.11 (a) Sulfonation as in Sec. 11.9; desulfonation as in Sec. 11.12.

(b) See Figure 30.4, page 493 of this Study Guide.

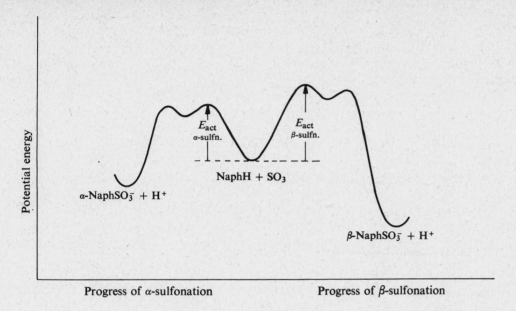

Figure 30.4. Potential energy changes during course of reaction: α- vs. β-sulfonation of naphthalene (Problem 30.11).

30.12 (a) Br ← CuBr — N₂⁺ ← HONO — NH₂ (page 982)

(b) F ← HBF₄/heat — N₂⁺ ← as in (a)

(c) CN ← CuCN — N₂⁺ ← as in (a)

(d) COOH ← H₂O, H⁺/heat — product (c)

or COOH ← H⁺ ← CO₂ — MgBr ← Mg — product (a)

or COOH ← OI⁻ — COCH₃ (page 979) (Compare Problem 30.2.)

(e) CHO ← H⁺ ← LiAlH(OBu-t)₃ — COCl ← SOCl₂ — product (d)

(f) C=C—COOH (with H H on the two carbons) ← Ac₂O, NaOAc/Prob. 21.22e — product (e) (Or via Reformatsky.)

493

30.13 Replace CuBr in Problem 30.12a by NO_2^-, catalyst.

30.14 (a) 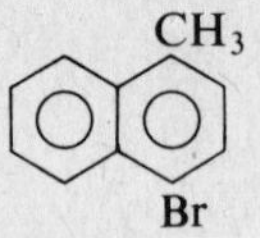 *Alpha* orientation in activated ring

(b) 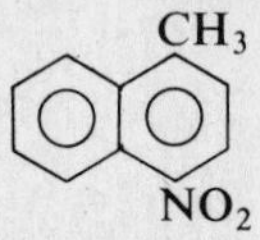 *Alpha* orientation in activated ring

(c) 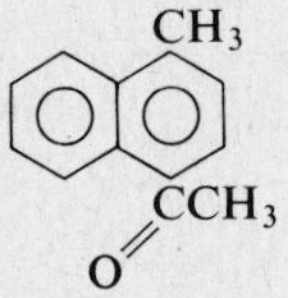 *Alpha* orientation in activated ring

(d) 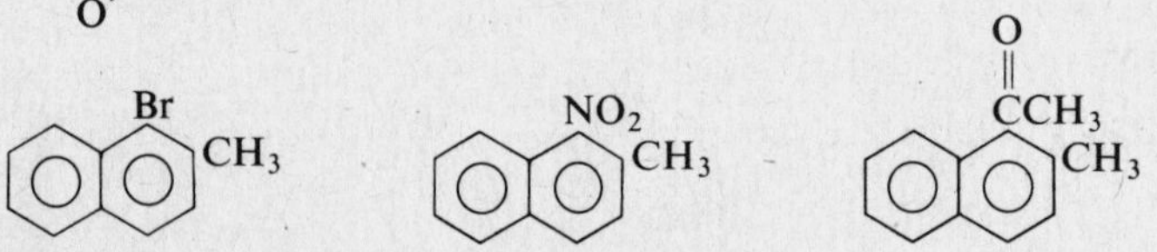 *Alpha* (position 1, not 4) orientation in activated ring

(e) 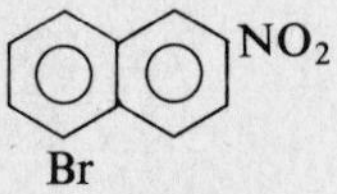 and 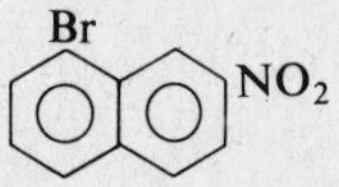 *Alpha* orientation in other ring

(f) 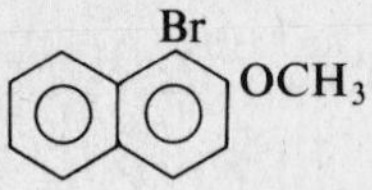 *Alpha* (position 1, not 4) orientation in activated ring

30.15 (a) 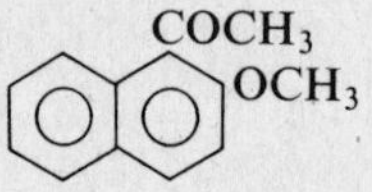 *Alpha* (position 1, not 4) orientation in activated ring

(b) 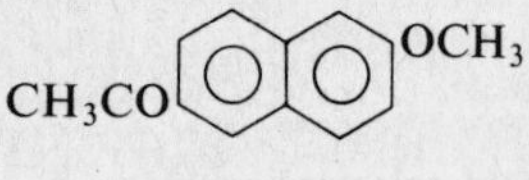 Bulky complex seeks least hindered position: *beta* substitution in other ring

(c) 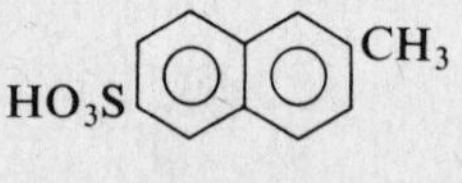 High-temperature sulfonation (equilibrium-control) gives *beta* substitution at least hindered position

(d) 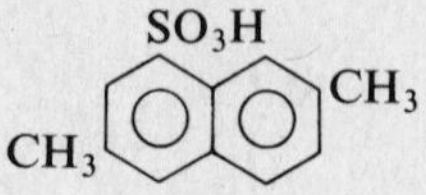 Low-temperature sulfonation (rate-control) gives *alpha* substitution at least hindered position

(e) 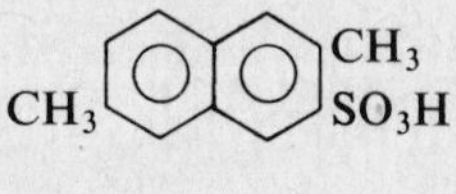 High-temperature sulfonation gives *beta* substitution

(f) 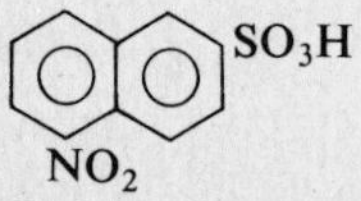 and 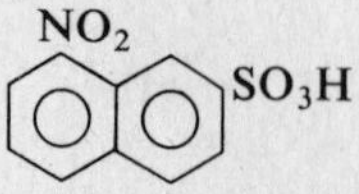 *Alpha* orientation in other ring

30.16 (a) Compare with the preparation of *p*-nitroaniline (page 760).

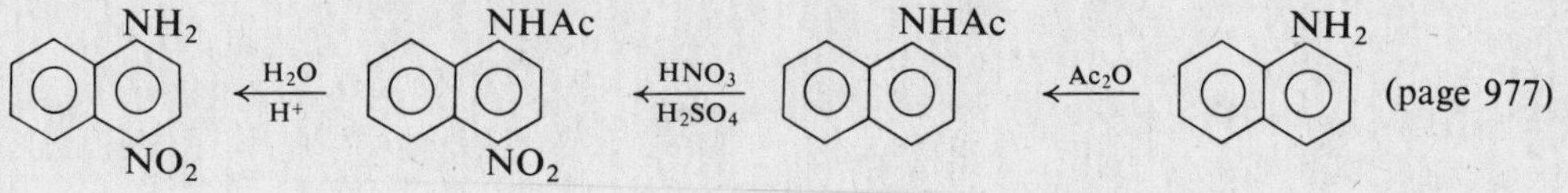

(b)
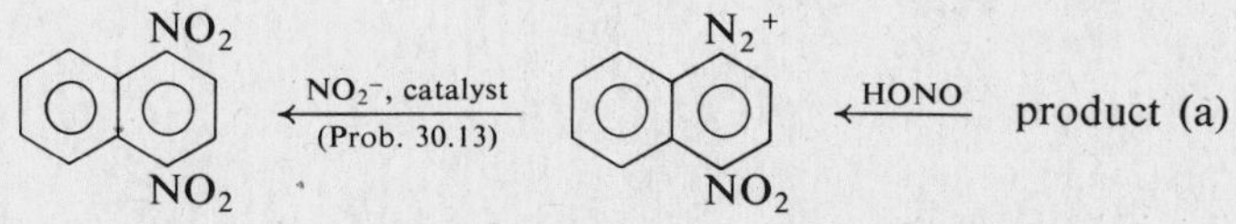

(c)
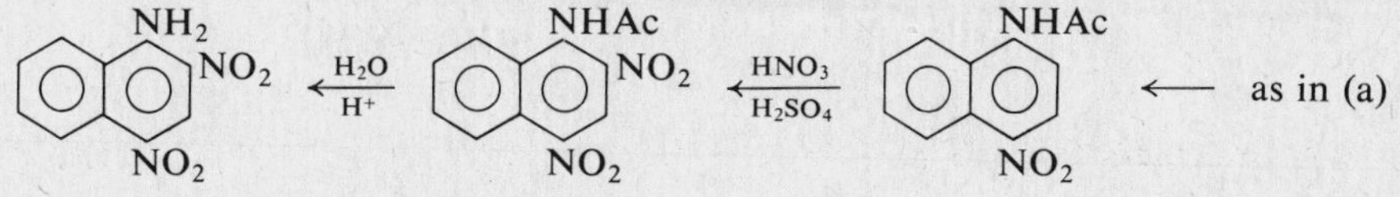

(d)
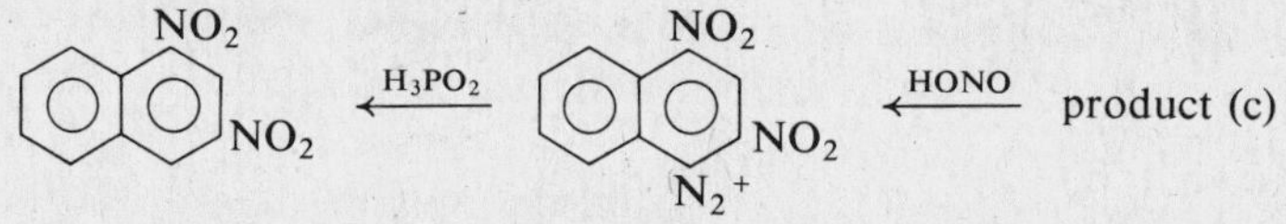

(e)
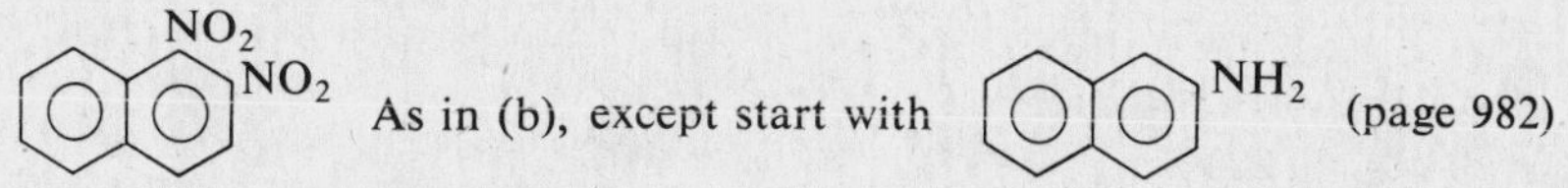

(f) Compare with the preparation of sulfanilic acid (Sec. 23.8).

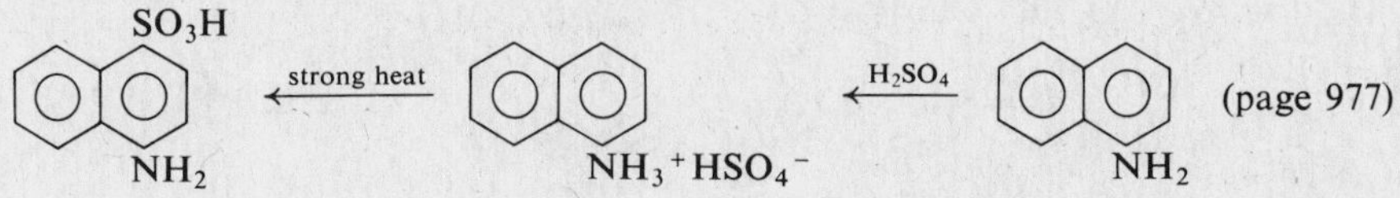

(g)
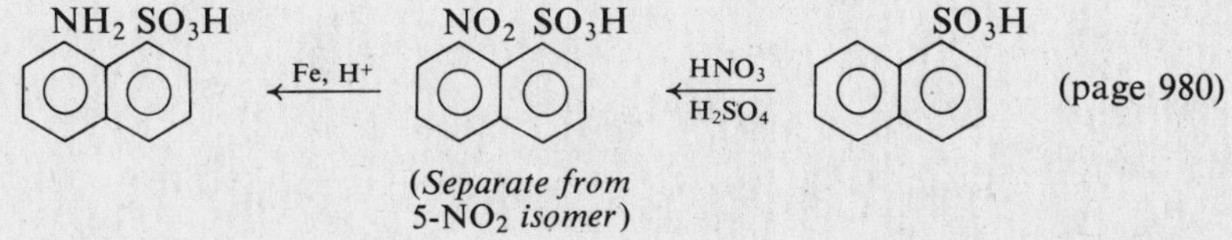

(h)
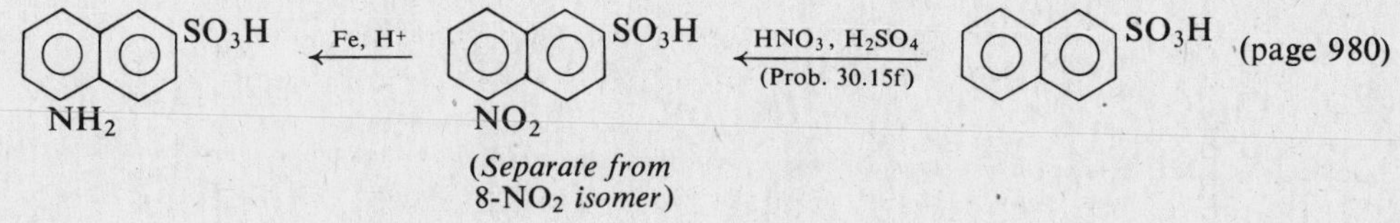

(i)
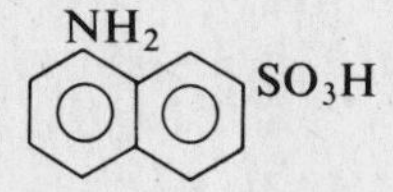

As in (h), except use 8-nitro isomer.

30.17 A deactivating acyl group is transformed into an activating alkyl group.

30.18 (a) Start Haworth sequence (page 987) with toluene (from benzene), and acylate initially *para* to $—CH_3$.

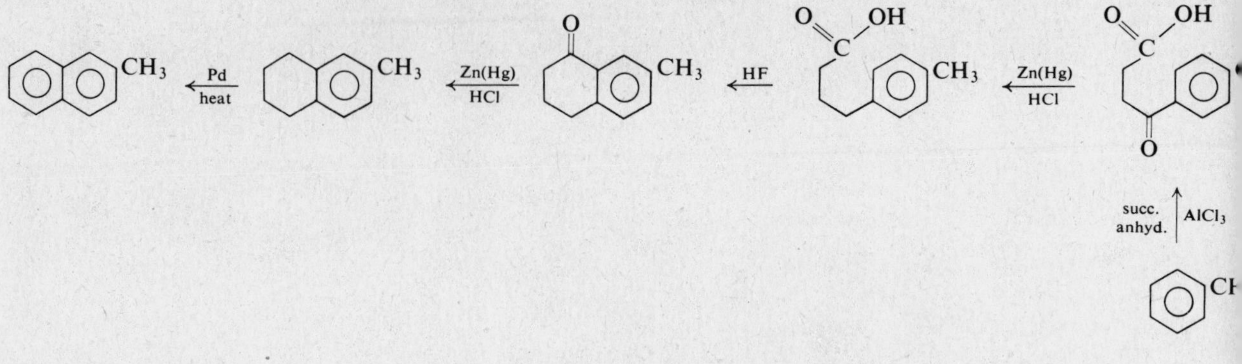

(b)

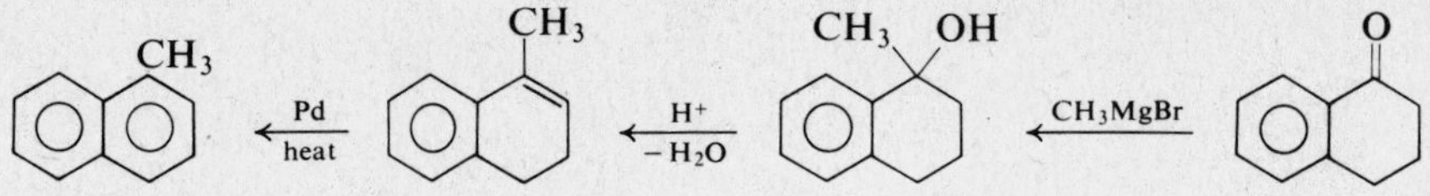

(page 987)

α-Tetralone

(c) Use *p*-xylene in Haworth sequence (page 987).

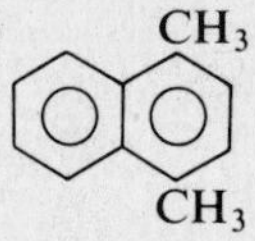

(d)

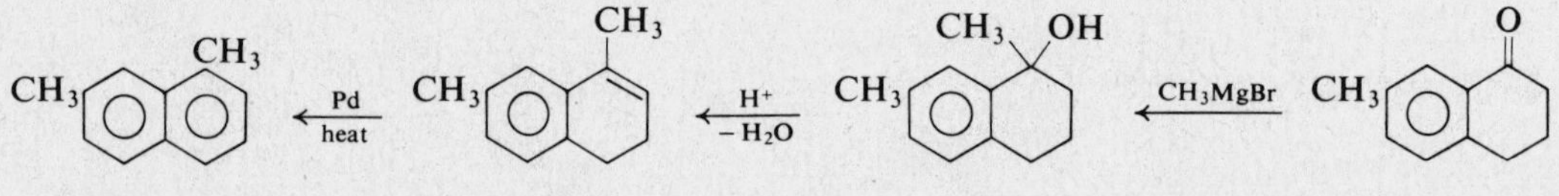

(as in a)

(e)

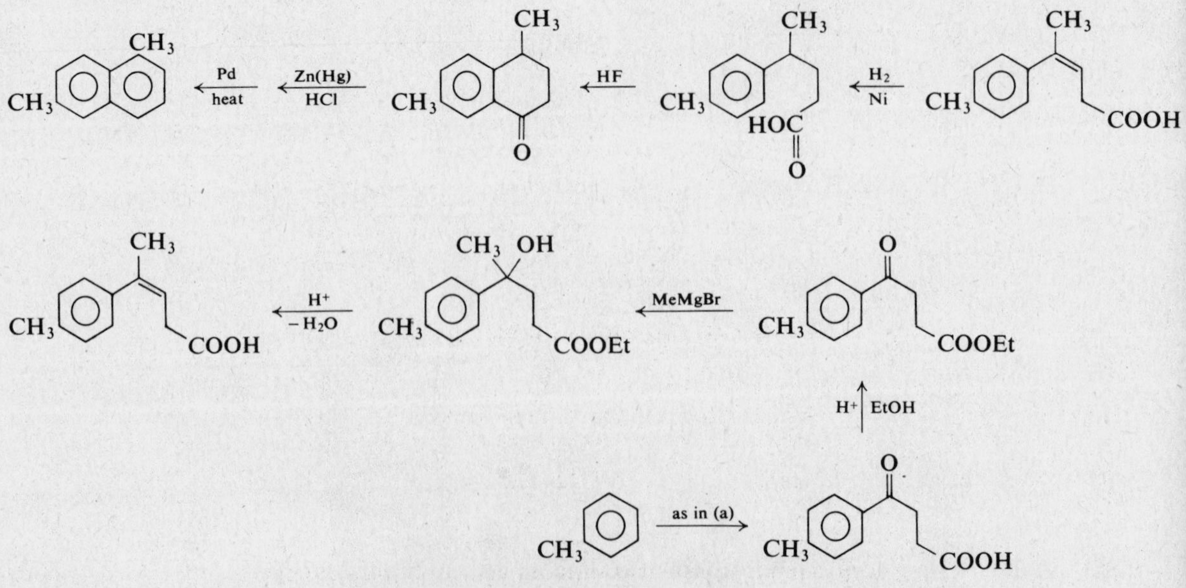

(f) 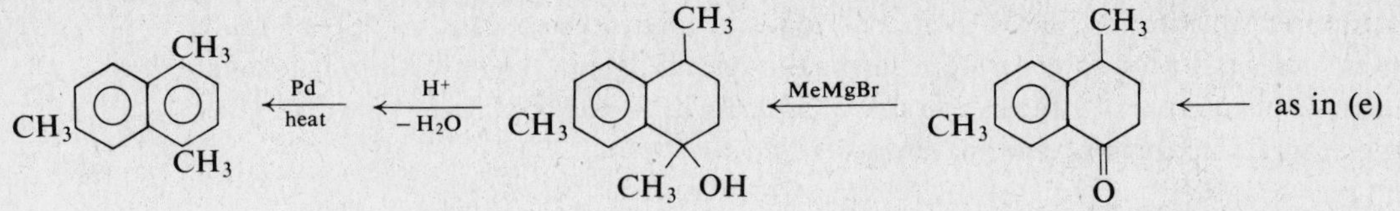
$$CH_3 \xleftarrow[heat]{Pd} \xleftarrow[-H_2O]{H^+} \xleftarrow{MeMgBr} \xleftarrow{\text{as in (e)}}$$

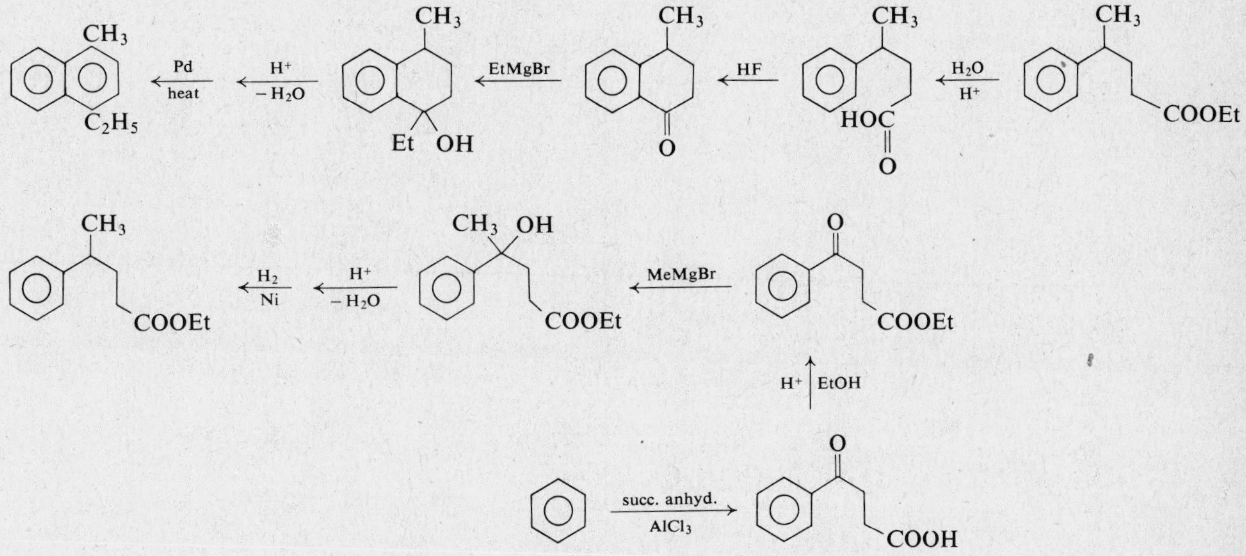

(g)
$$C_2H_5 \xleftarrow[heat]{Pd} \xleftarrow[-H_2O]{H^+} \xleftarrow{EtMgBr} \xleftarrow{HF} \xleftarrow[H^+]{H_2O}$$

$$\xleftarrow[Ni]{H_2} \xleftarrow[-H_2O]{H^+} \xleftarrow{MeMgBr}$$

$$H^+ \mid EtOH$$

$$\xrightarrow[AlCl_3]{succ.\ anhyd.}$$

(h)
$$\xleftarrow[heat]{Pd} \xleftarrow[H_2O]{H^+} \xleftarrow{EtMgBr} \xleftarrow{HF} \xleftarrow[HCl]{Zn(Hg)} \xleftarrow[AlCl_3]{succ.\ anhyd.}$$

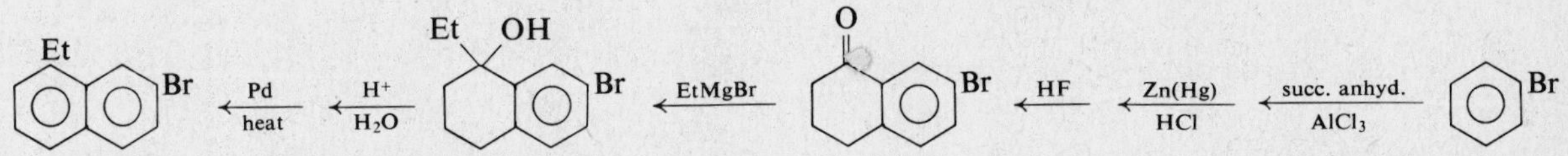

(i) 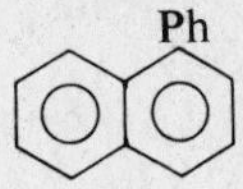
As in (b) above, except use PhMgBr in place of MeMgBr.

30.19 See Figure 30.3, page 995.

30.20

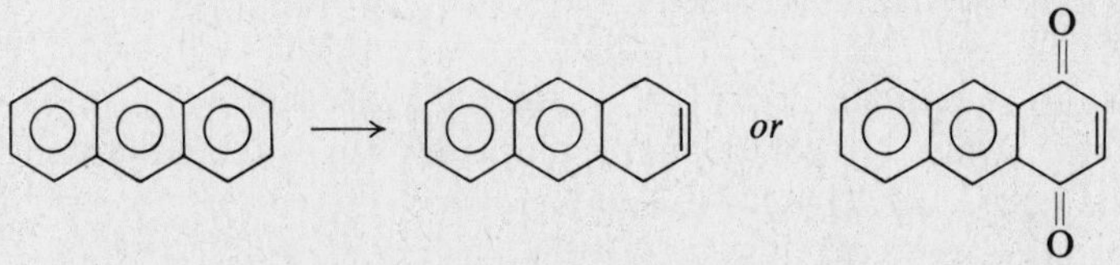

$$\longrightarrow \quad or$$

Anthracene (84 kcal) − Naphthalene (61 kcal) = 23-kcal sacrifice

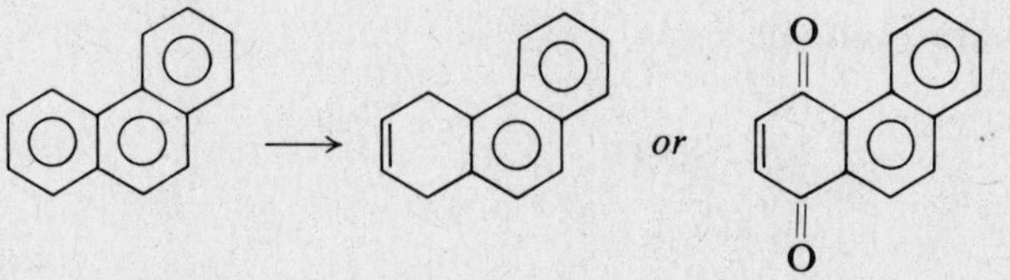

$$\longrightarrow \quad or$$

Phenanthrene (92 kcal) − Naphthalene (61 kcal) = 31-kcal sacrifice

30.21 In electrophilic substitution in the benzene ring, the intermediate cation (benzenonium ion) always loses a proton to generate an aromatic system again. Here, with only a small sacrifice in resonance energy, the cation can combine with a nucleophile to give a product with two separate benzene rings remaining.

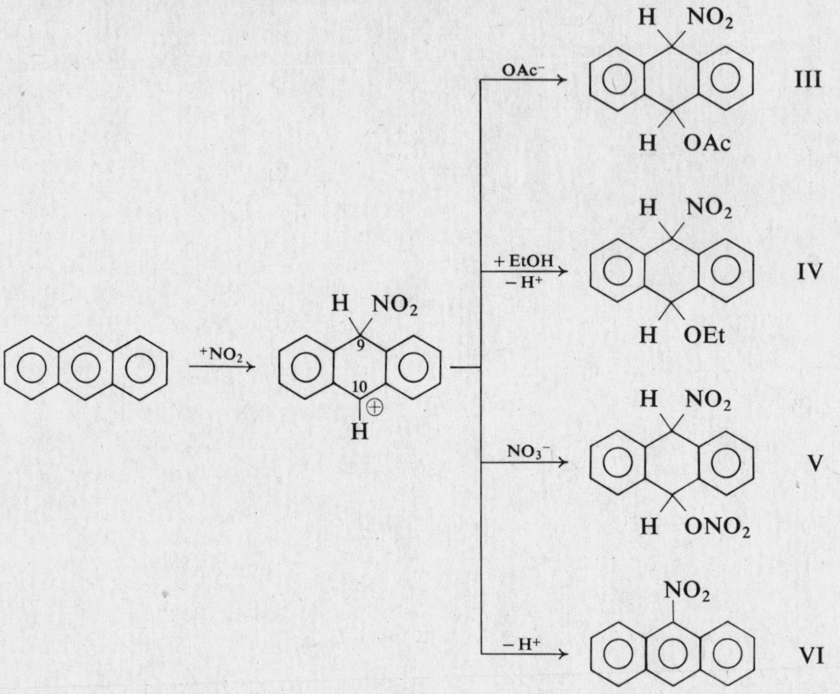

30.22 (a) The most stable *di*hydroanthracene is (Sec. 30.17) the 9,10 compound. The most stable *tetra*hydroanthracene is, however,

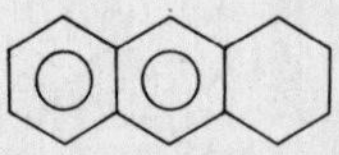

1,2,3,4-Tetrahydroanthracene

in which the aromaticity of only one ring is destroyed. Catalytic hydrogenation is reversible, and even if reaction goes via the 9,10 compound, it ultimately yields the most stable tetrahydro product.

(b) Sulfonation is reversible. Although sulfonation—like bromination and nitration—is probably fastest at the 9-position, so is desulfonation. Reaction is equilibrium-controlled, and yields the more stable 1-sulfonic acid.

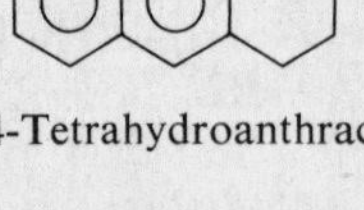

Anthracene-1-sulfonic acid

30.23 (a)

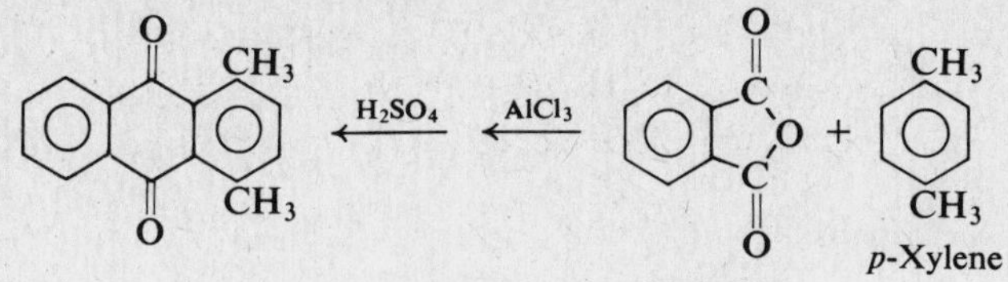

Phthalic
anhydride

p-Xylene

(b)

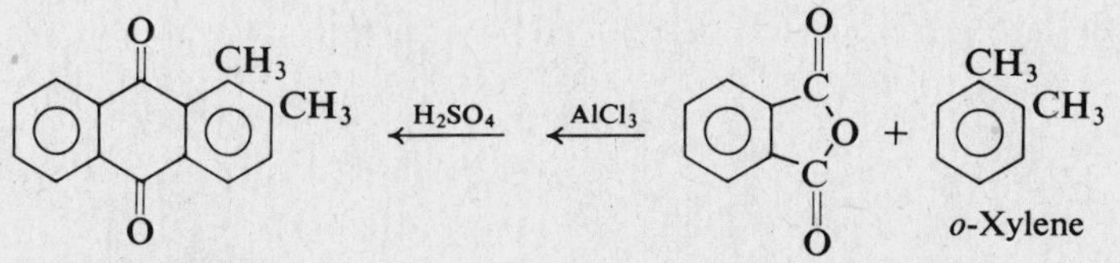

20% yield
Separated from
2,3-product

o-Xylene

(c)

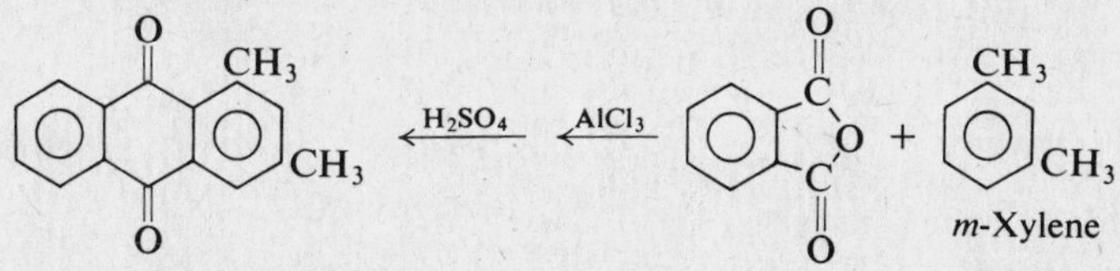

m-Xylene

(d)

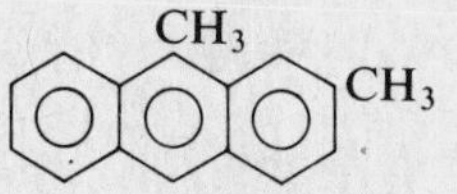

As under (c) on page 993, except start with toluene
instead of benzene; RMgX = MeMgBr

(We would not, instead, make 2-methylanthraquinone, and then allow this to react
with MeMgBr: anthraquinones do not show very much carbonyl reactivity, and in any
case, so far as we know, addition might be chiefly to the 10-position.)

(e)

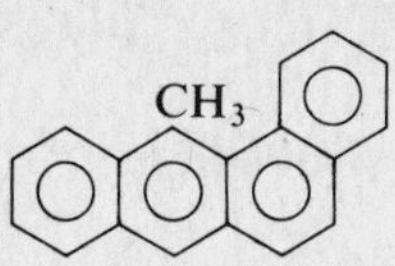

As in (d), except start with β-acylation of naphthalene

30.24 The acyl group nearer the electron-withdrawing nitro group is the more reactive. (See,
for example, Problem 20.14, page 680.)

(a)

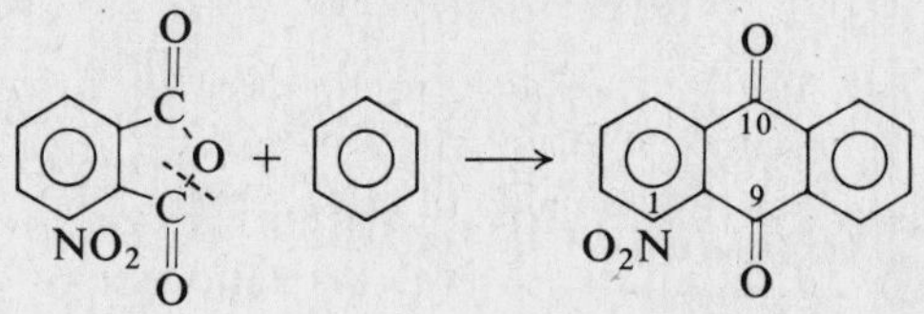

1-Nitro-9,10-anthraquinone

(b)

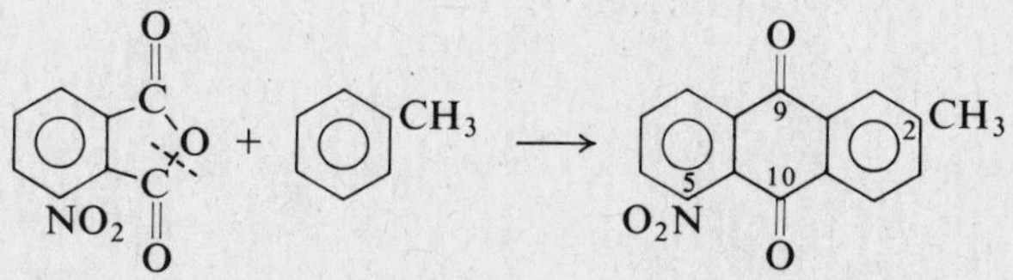

+ some 8-NO$_2$ isomer

5-Nitro-2-methyl-
9,10-anthraquinone

30.25   (a) As in Problem 30.18(a), except use 1-methylnaphthalene in place of toluene.

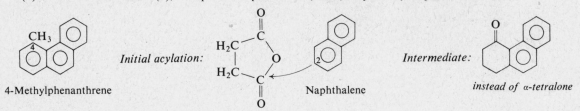

9-Methylphenanthrene   *Initial acylation:*   1-Methylnaphthalene

(b) As in Problem 30.18(b), except use naphthalene (with β-acylation) in place of benzene.

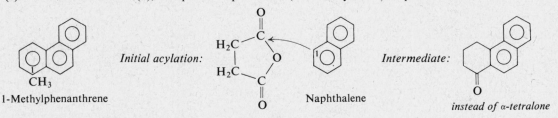

4-Methylphenanthrene   *Initial acylation:*   Naphthalene   *Intermediate:*   *instead of α-tetralone*

(c) As in Problem 30.18(b), except use naphthalene (with α-acylation) in place of benzene.

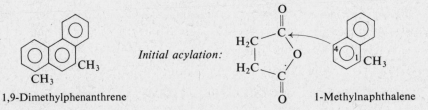

1-Methylphenanthrene   *Initial acylation:*   Naphthalene   *Intermediate:*   *instead of α-tetralone*

(d) As in Problem 30.18(d), except use 1-methylnaphthalene in place of toluene.

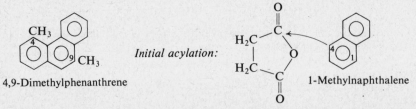

1,9-Dimethylphenanthrene   *Initial acylation:*   1-Methylnaphthalene

(e) As in Problem 30.18(e), except use 1-methylnaphthalene in place of toluene.

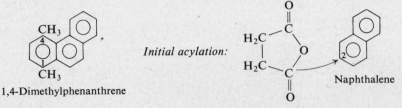

4,9-Dimethylphenanthrene   *Initial acylation:*   1-Methylnaphthalene

(f) As in Problem 30.18(g), except use naphthalene (with β-acylation) in place of benzene, and a second molecule of MeMgBr in place of EtMgBr in the final alkylation step.

1,4-Dimethylphenanthrene   *Initial acylation:*   Naphthalene

(α-Acylation would appear to be a valid alternative; the reported synthesis actually used the β-product.)

(g) As in Problem 30.25(f), except use 1-methylnaphthalene (with α-acylation) in place of naphthalene.

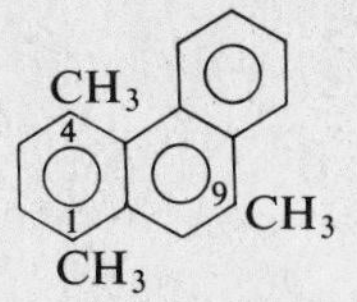

1,4,9-Trimethylphenanthrene

Initial acylation:

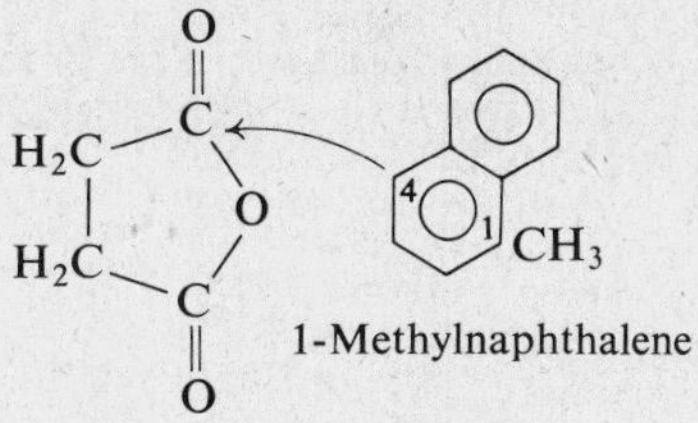

1-Methylnaphthalene

(h) Problem 30.15(b), page 984, states that acylation of 2-methoxynaphthalene can occur at the 6-position (β-acylation). In this case, the Haworth sequence can begin with succinoylation of 2-methoxynaphthalene at the 6-position, and proceed in standard fashion (Figure 30.3, page 995) to the target structure.

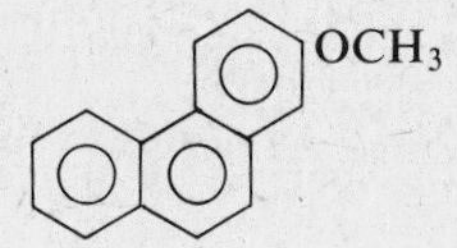

2-Methoxyphenanthrene

30.26

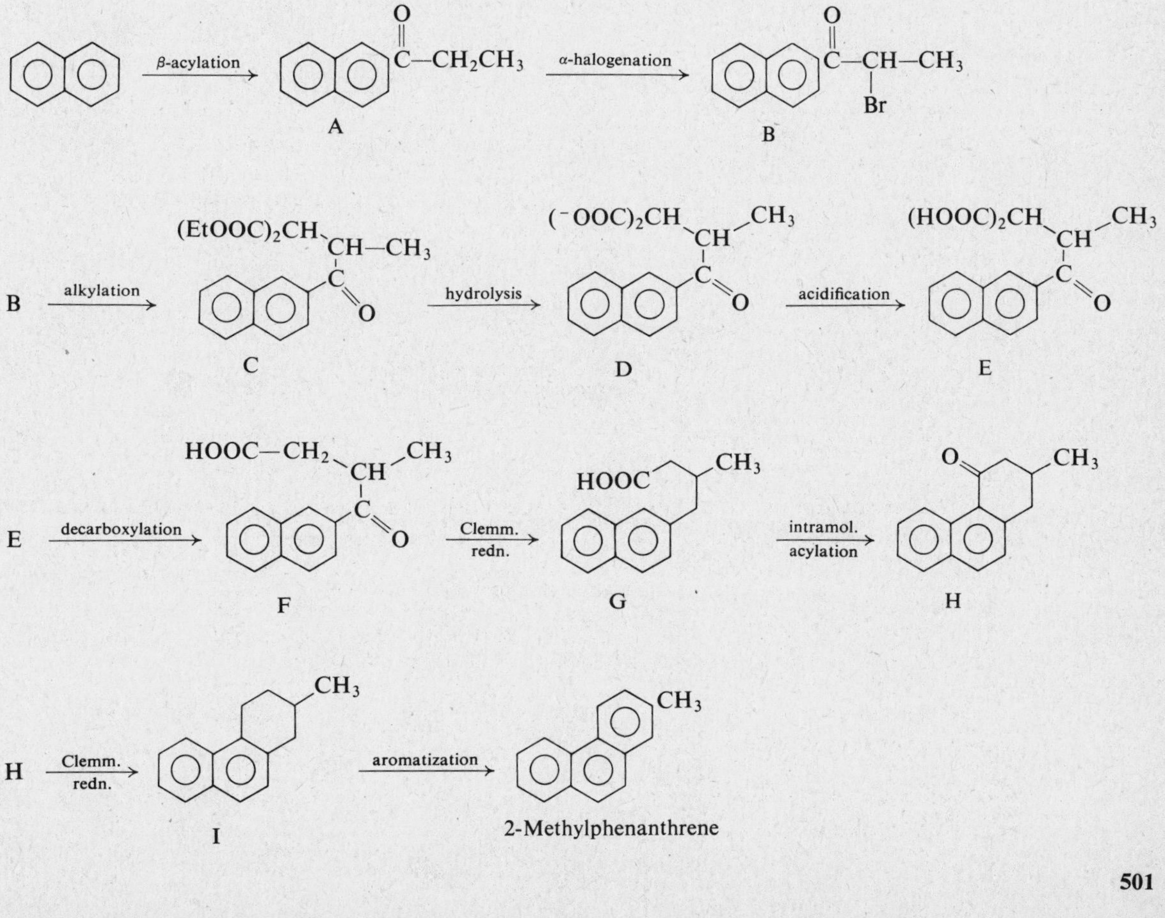

2-Methylphenanthrene

30.27

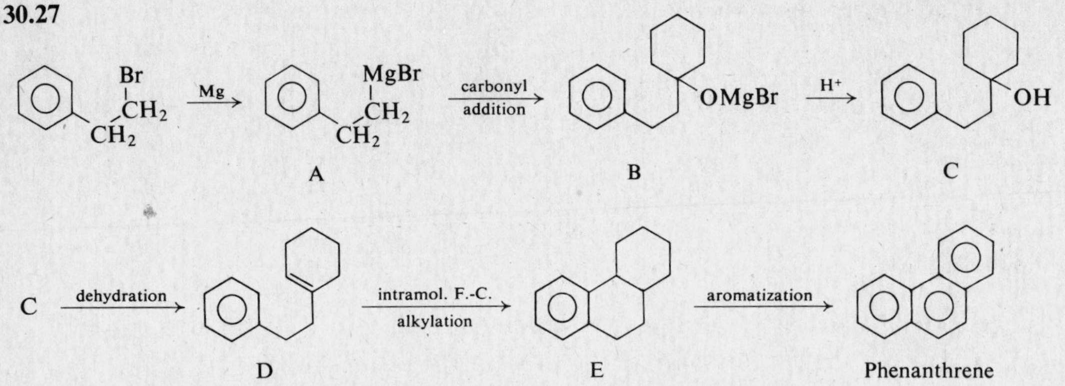

The needed β-phenylethyl bromide can be made as follows:

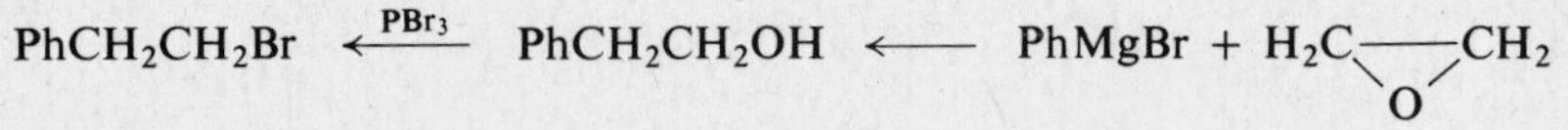

$$PhCH_2CH_2Br \xleftarrow{\ PBr_3\ } PhCH_2CH_2OH \longleftarrow PhMgBr + H_2C\overset{\displaystyle\diagup\!\!\diagdown}{\underset{O}{}}CH_2$$

30.28

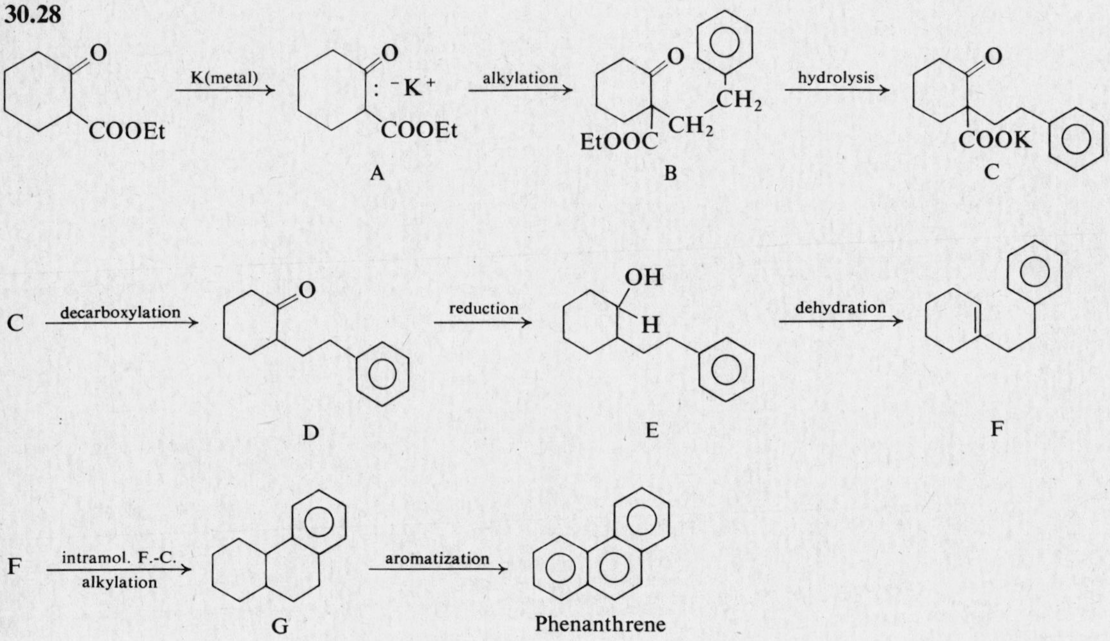

30.29

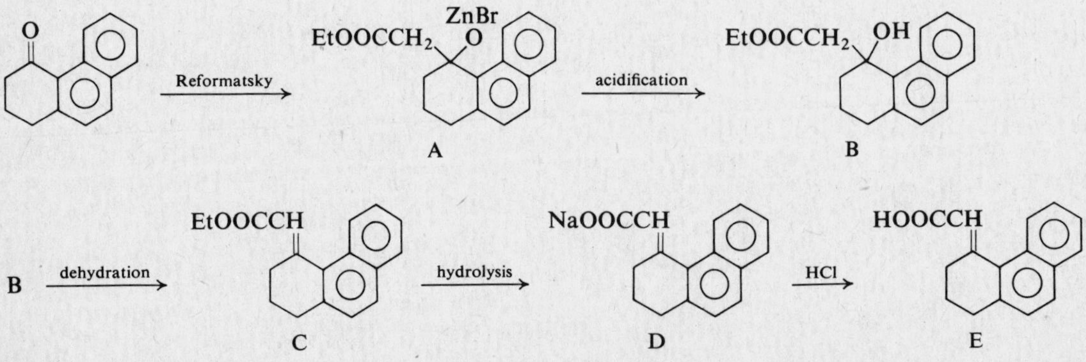

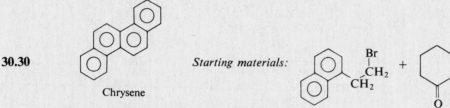

$$E \xrightarrow[\text{acylation}]{\text{intramol.}} F \xrightarrow[\text{redn.}]{\text{Clemm.}} G \xrightarrow{\text{aromatization}} \text{Pyrene}$$

The starting material is one of the intermediates in Problem 30.25(b).

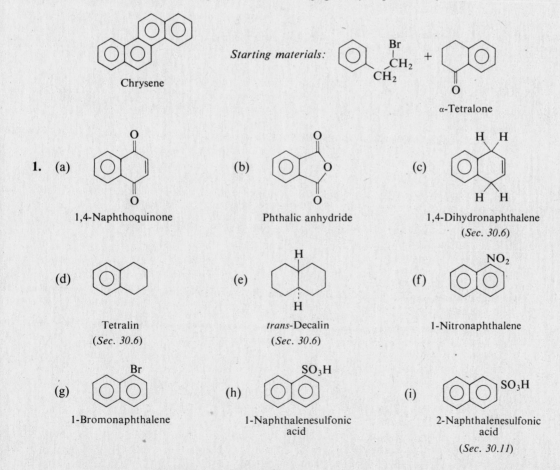

30.30   Chrysene   *Starting materials:*

The starting halide is made as follows:

$$\xleftarrow{\text{PBr}_3} \xleftarrow{\hspace{1cm}} \text{Prob. 30.7(g)}$$

30.31 As in Problem 30.27, except use α-tetralone (page 987) in place of cyclohexanone.

Chrysene   *Starting materials:*   α-Tetralone

1. (a)   1,4-Naphthoquinone

(b)   Phthalic anhydride

(c)   1,4-Dihydronaphthalene
        (*Sec. 30.6*)

(d)   Tetralin
       (*Sec. 30.6*)

(e)   *trans*-Decalin
       (*Sec. 30.6*)

(f)   1-Nitronaphthalene

(g)   1-Bromonaphthalene

(h)   1-Naphthalenesulfonic
       acid

(i)   2-Naphthalenesulfonic
       acid
       (*Sec. 30.11*)

(j) 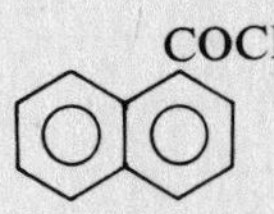 Methyl α-naphthyl ketone

(k) 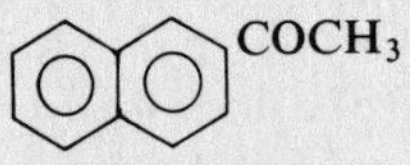 Methyl β-naphthyl ketone
(Sec. 30.10)

(l) 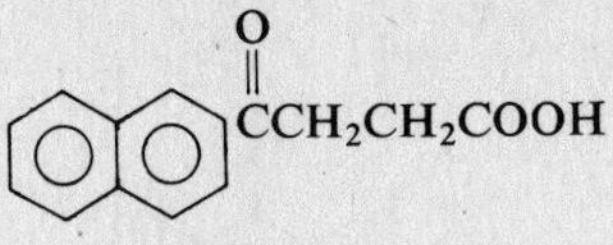 (Sec. 30.19)

2. (a) 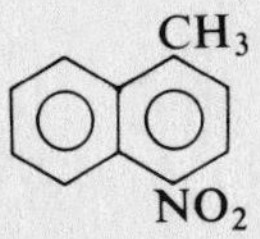 4-Nitro-1-methylnaphthalene

(b) 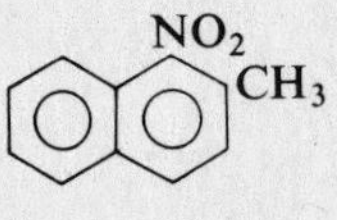 1-Nitro-2-methylnaphthalene

(c) 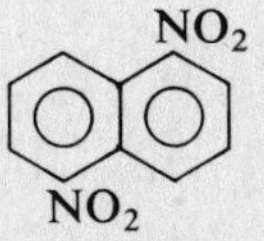 1,5-Dinitro-naphthalene   and 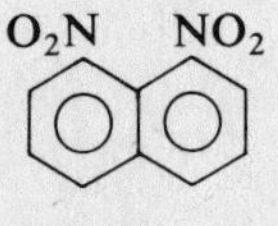   1,8-Dinitro-naphthalene

(d) 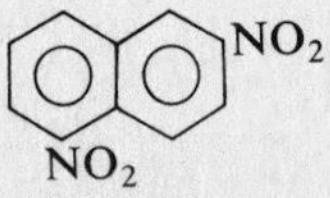 1,6-Dinitro-naphthalene   and 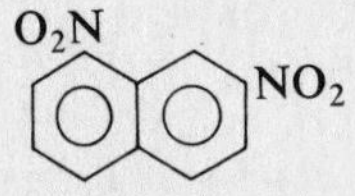   1,7-Dinitro-naphthalene

(e) 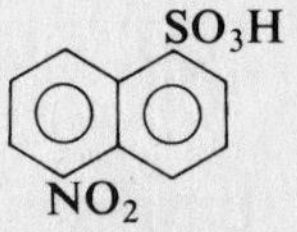 5-Nitronaphthalene-1-sulfonic acid   and 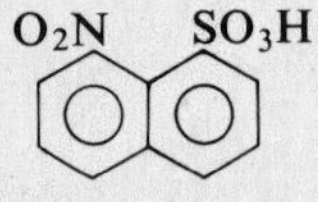   8-Nitronaphthalene-1-sulfonic acid

(f) 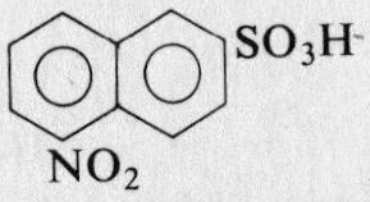 5-Nitronaphthalene-2-sulfonic acid   and    8-Nitroanaphthalene-2-sulfonic acid

(g) 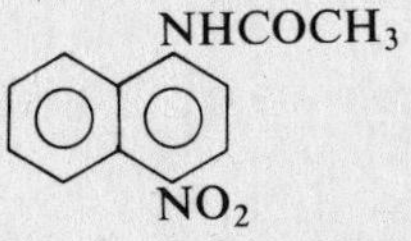 4-Nitro-1-acetamidonaphthalene

(h) 1-Nitro-2-acetamidonaphthalene

(i) 4-Nitro-1-naphthol

(j) 1-Nitro-2-naphthol

(k) 9-Nitroanthracene

3.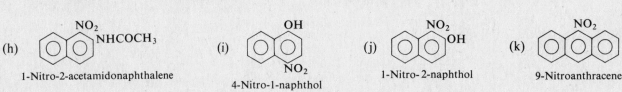

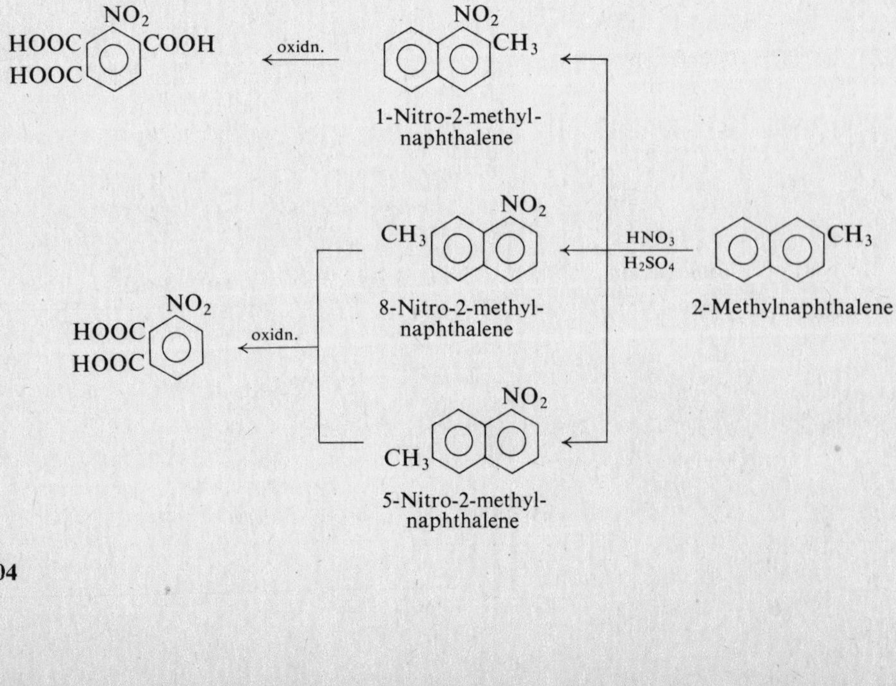

HOOC, NO2, COOH (from oxidn. of) 1-Nitro-2-methyl-naphthalene

8-Nitro-2-methyl-naphthalene

HOOC, NO2, HOOC (from oxidn. of) 8-Nitro-2-methyl-naphthalene

5-Nitro-2-methyl-naphthalene

2-Methylnaphthalene + HNO3/H2SO4

4. (a) 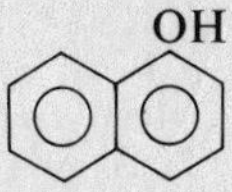 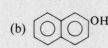 OH
(*Sec. 30.12*)

(b) OH
(*Sec. 30.12*)

(c) 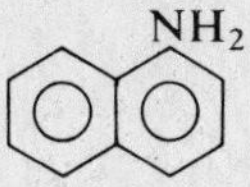 NH_2
(*Sec. 30.8*)

(d) 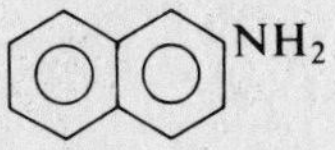 NH_2
(*Bucherer reaction,
Sec. 30.12*)

(e) 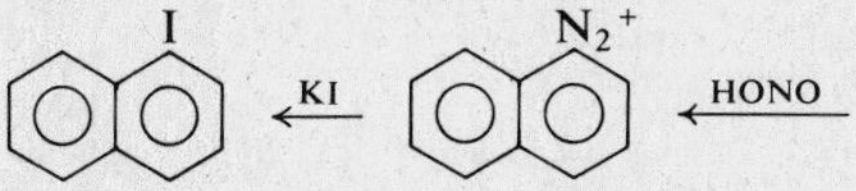 I $\xleftarrow{KI}$ $N_2{}^+$ $\xleftarrow{HONO}$ product (c)

(f) 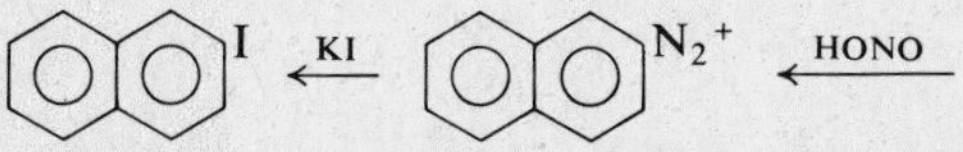 I $\xleftarrow{KI}$ $N_2{}^+$ $\xleftarrow{HONO}$ product (d)

(g) 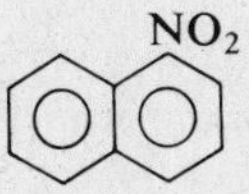 NO_2 (page 971)

(h) 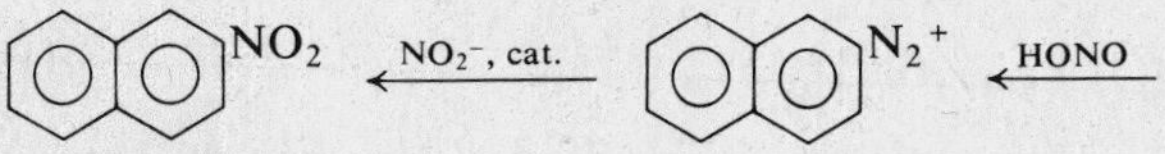 NO_2 $\xleftarrow{NO_2{}^-,\ cat.}$ $N_2{}^+$ $\xleftarrow{HONO}$ product (d)

*Not available by
direct nitration*

(i) 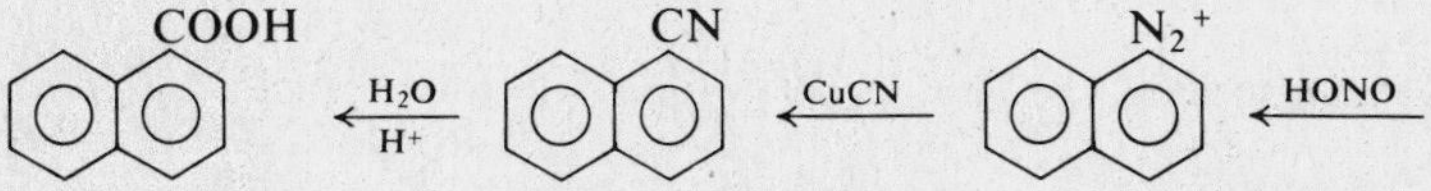 COOH $\xleftarrow[\ H^+\ ]{H_2O}$ CN $\xleftarrow{CuCN}$ $N_2{}^+$ $\xleftarrow{HONO}$ product (c)

or 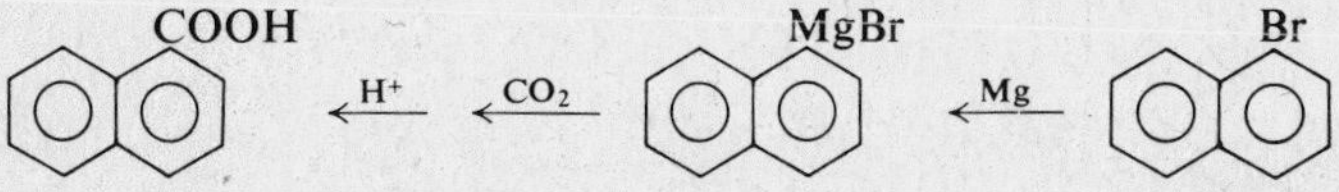 COOH $\xleftarrow{H^+}$ $\xleftarrow{CO_2}$ MgBr $\xleftarrow{Mg}$ Br (page 972)

or 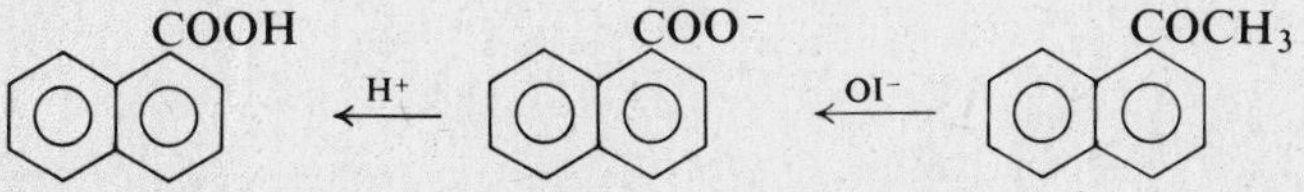 COOH $\xleftarrow{H^+}$ COO^- $\xleftarrow{OI^-}$ $COCH_3$ (page 979)

(j) 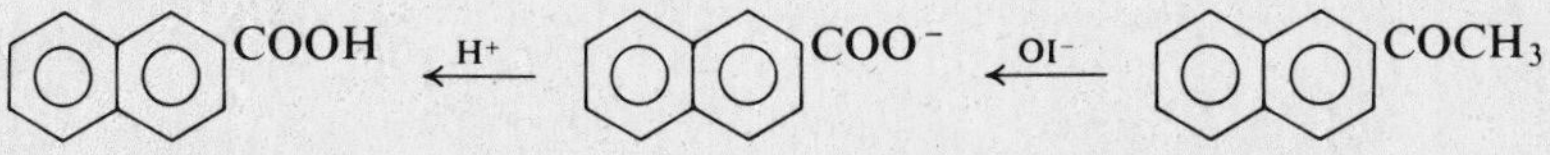 COOH $\xleftarrow{H^+}$ COO^- $\xleftarrow{OI^-}$ $COCH_3$ (page 979)

(k) 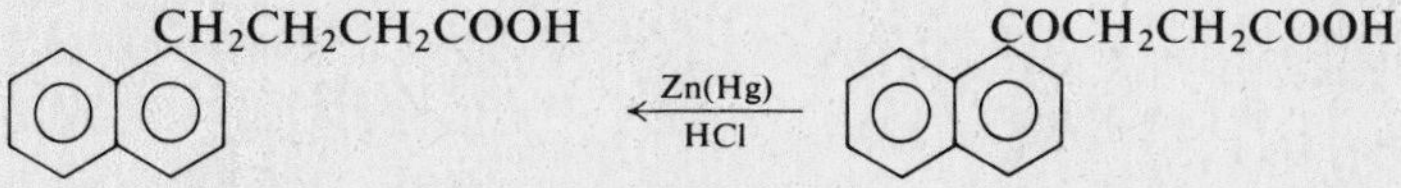 $CH_2CH_2CH_2COOH$ $\xleftarrow[HCl]{Zn(Hg)}$ $COCH_2CH_2COOH$ (page 980)

(l) 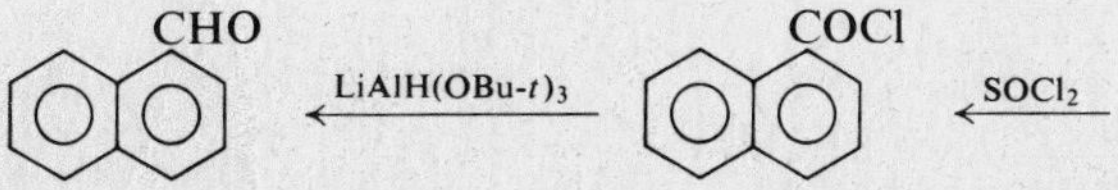 CHO $\xleftarrow{LiAlH(OBu\text{-}t)_3}$ COCl $\xleftarrow{SOCl_2}$ product (i)

(m) 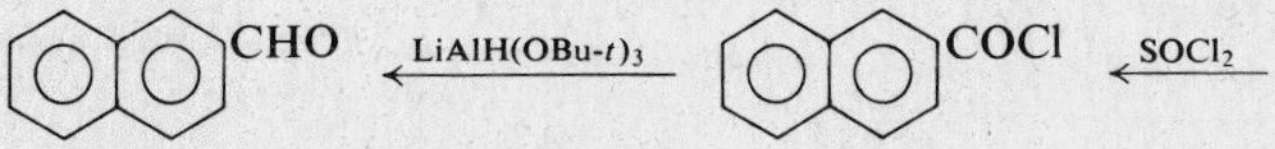 CHO $\xleftarrow{LiAlH(OBu\text{-}t)_3}$ COCl $\xleftarrow{SOCl_2}$ product (j)

(n) 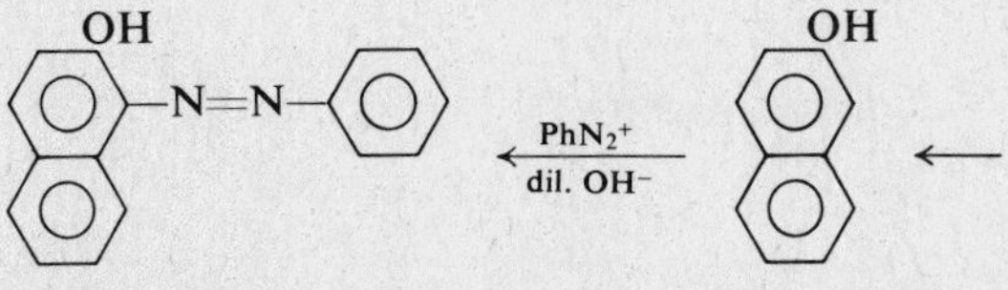 OH —N=N— $\xleftarrow[dil.\ OH^-]{PhN_2{}^+}$ OH $\xleftarrow{}$ as in (b)

(*Sec. 30.13*)

(o) 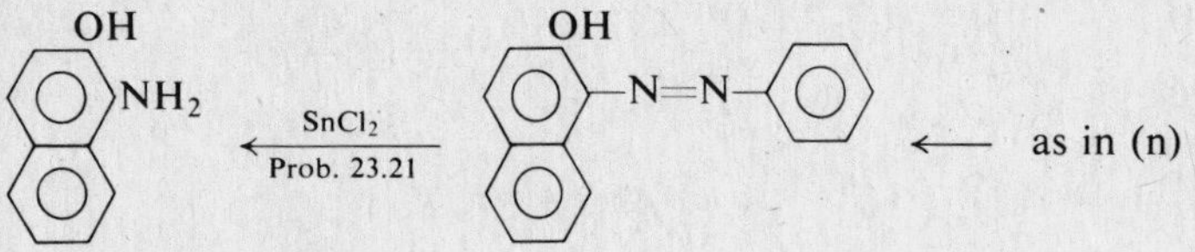
$\underset{\text{Prob. 23.21}}{\overset{\text{SnCl}_2}{\longleftarrow}}$
$\longleftarrow$ as in (n)

(p) 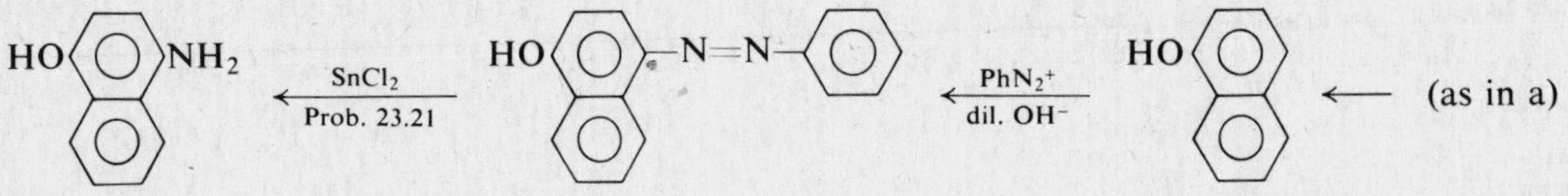
$\underset{\text{Prob. 23.21}}{\overset{\text{SnCl}_2}{\longleftarrow}}$
$\underset{\text{dil. OH}^-}{\overset{\text{PhN}_2^+}{\longleftarrow}}$
$\longleftarrow$ (as in a)

(*Sec. 30.13*)

(q) 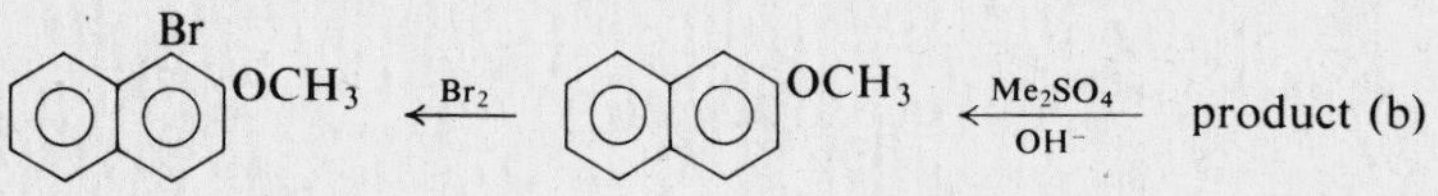
$\overset{\text{Br}_2}{\longleftarrow}$
$\underset{\text{OH}^-}{\overset{\text{Me}_2\text{SO}_4}{\longleftarrow}}$ product (b)

(*Prob. 30.14f*)

(r) 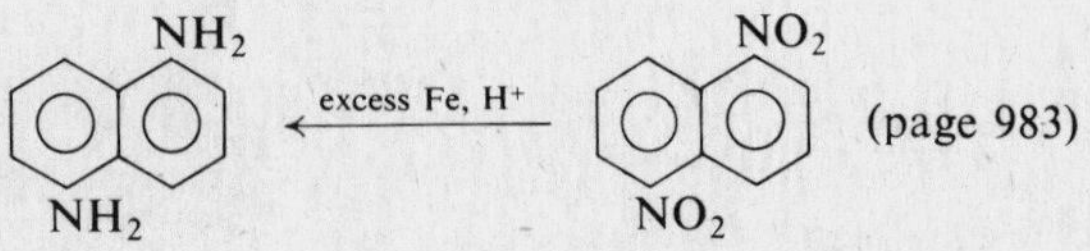
$\overset{\text{excess Fe, H}^+}{\longleftarrow}$ (page 983)

(s) 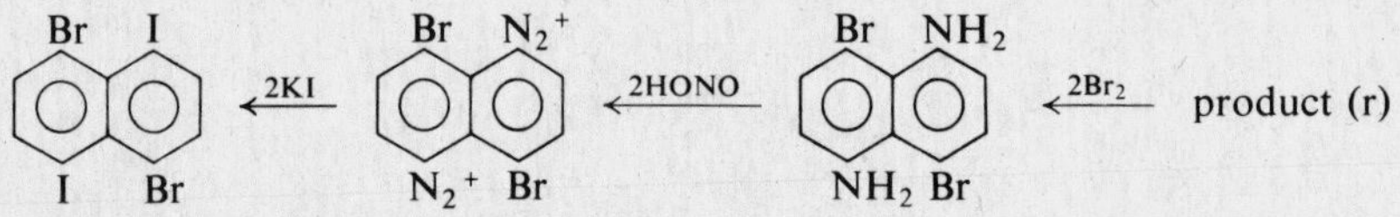
$\overset{\text{2KI}}{\longleftarrow}$
$\overset{\text{2HONO}}{\longleftarrow}$
$\overset{\text{2Br}_2}{\longleftarrow}$ product (r)

(t) 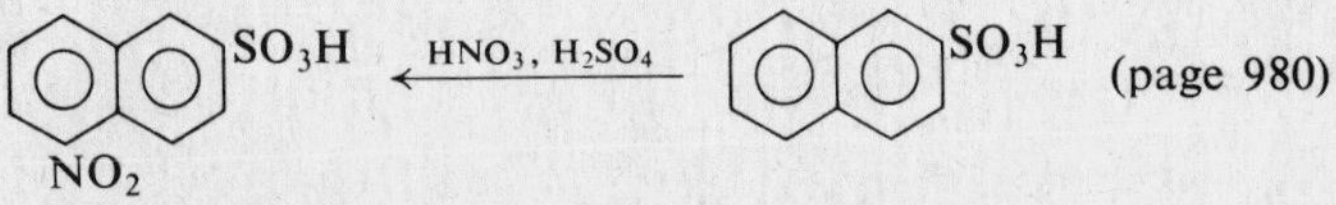
$\overset{\text{HNO}_3,\ \text{H}_2\text{SO}_4}{\longleftarrow}$ (page 980)

(*Prob. 30.15f*)

(u) 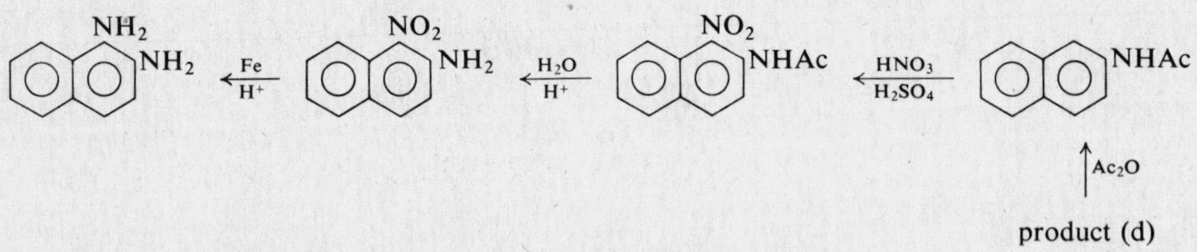
$\overset{\text{Fe}}{\underset{\text{H}^+}{\longleftarrow}}$
$\overset{\text{H}_2\text{O}}{\underset{\text{H}^+}{\longleftarrow}}$
$\underset{\text{H}_2\text{SO}_4}{\overset{\text{HNO}_3}{\longleftarrow}}$
$\uparrow \text{Ac}_2\text{O}$

product (d)

(v) 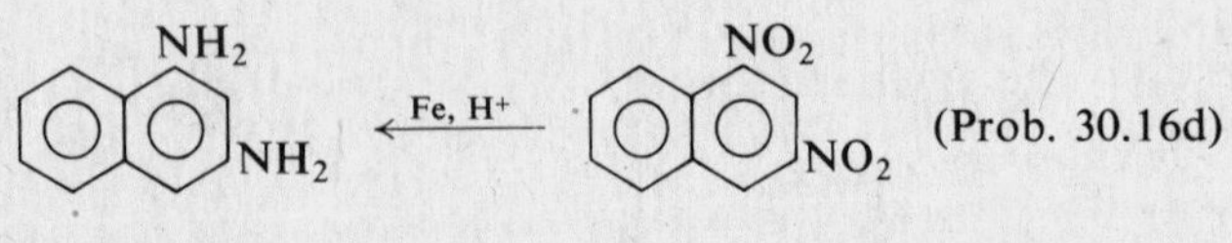
$\overset{\text{Fe, H}^+}{\longleftarrow}$ (Prob. 30.16d)

(w) 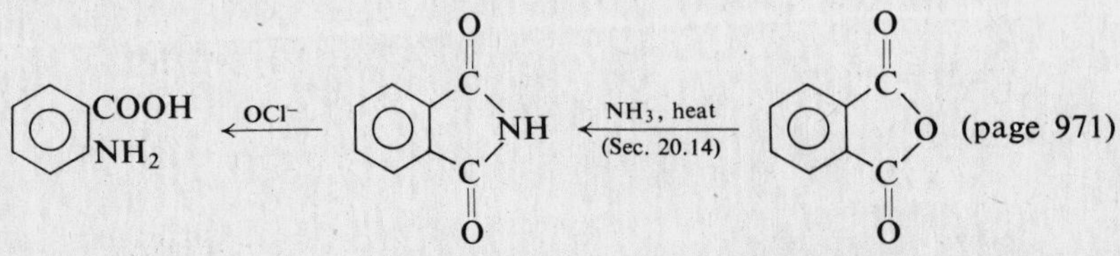
$\overset{\text{OCl}^-}{\longleftarrow}$
$\underset{\text{(Sec. 20.14)}}{\overset{\text{NH}_3,\ \text{heat}}{\longleftarrow}}$ (page 971)

Phthalimide   Phthalic anhydride

(x) 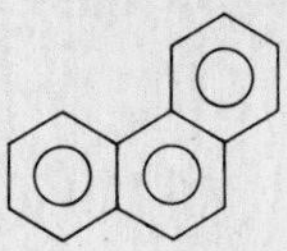 As on page 995.

(y) 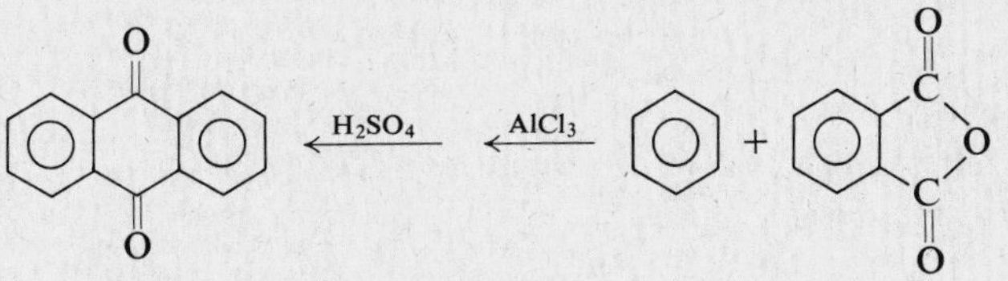 $\xleftarrow{\text{H}_2\text{SO}_4}$ $\xleftarrow{\text{AlCl}_3}$ + (page 971)

(z) 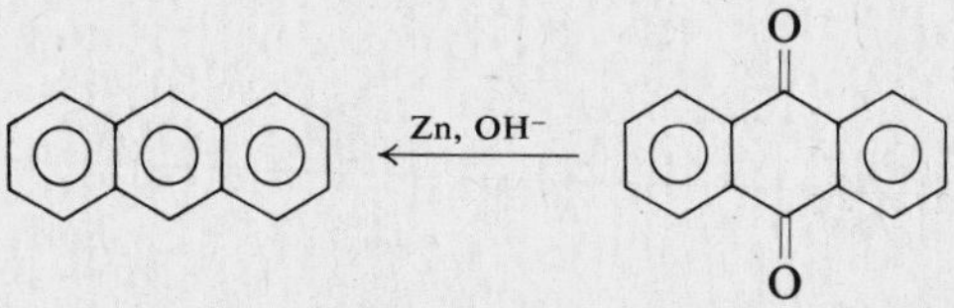 $\xleftarrow{\text{Zn, OH}^-}$ (from y)

5.

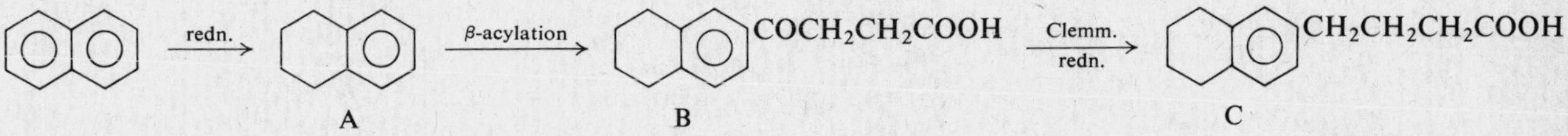

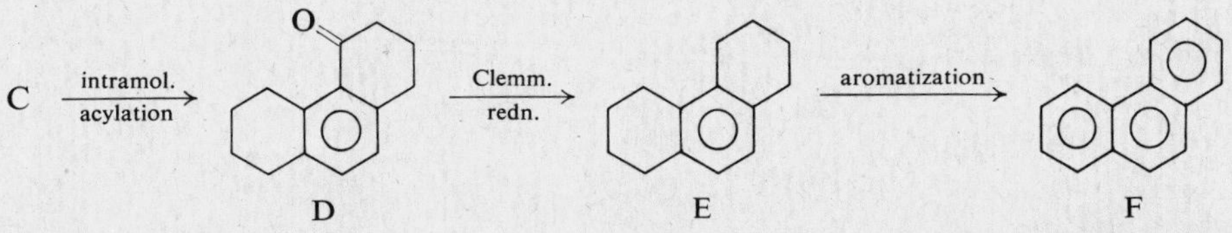

$\xrightarrow{\text{redn.}}$ $\xrightarrow{\beta\text{-acylation}}$ COCH₂CH₂COOH $\xrightarrow[\text{redn.}]{\text{Clemm.}}$ CH₂CH₂CH₂COOH

A   B   C

C $\xrightarrow[\text{acylation}]{\text{intramol.}}$ $\xrightarrow[\text{redn.}]{\text{Clemm.}}$ $\xrightarrow{\text{aromatization}}$

D   E   F
                                       Phenanthrene

Acylation of A is actually *beta*, evidently for steric reasons. (Whatever the orientation, however, phenanthrene would be the final product.)

6. (a) As in Problem 30.18(f), except: start with anisole in place of benzene, and use PhMgBr in place of one of the moles of MeMgBr.

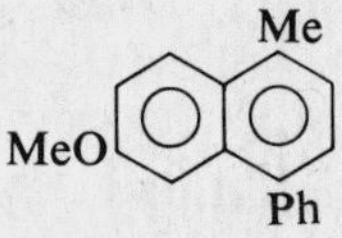

(b) 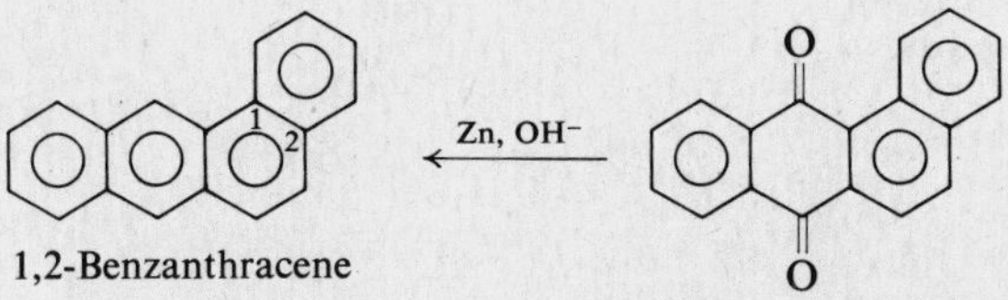 $\xleftarrow{\text{Zn, OH}^-}$ *Initial acylation:*

1,2-Benzanthracene   1,2-Benz-9,10-anthra-   Naphthalene
                                   quinone (page 993)

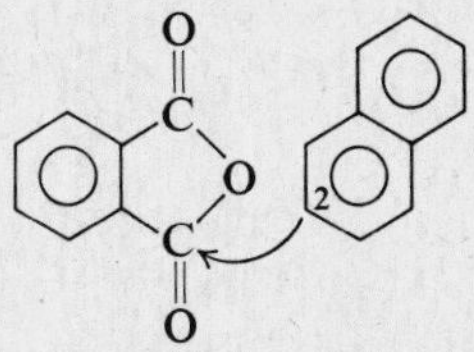

(c)

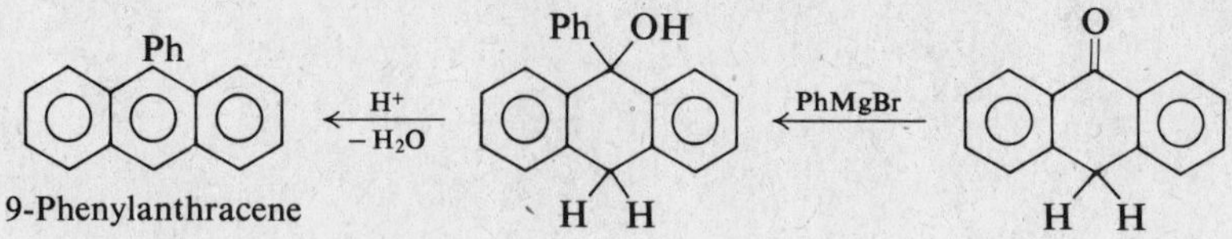

9-Phenylanthracene $\xleftarrow[-H_2O]{H^+}$ (Ph, OH compound, H H) $\xleftarrow{PhMgBr}$ Anthrone (page 993)

(d) As in Problem 30.25(c), except use PhMgBr in place of MeMgBr in the reaction with the second ketone encountered in the Haworth sequence.

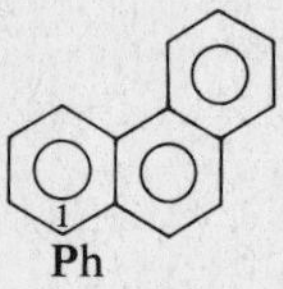

1-Phenylphenanthrene

(e) As in Problem 6(d) above, except use 1-phenylnaphthalene (made earlier in Problem 30.18(i)) in place of naphthalene.

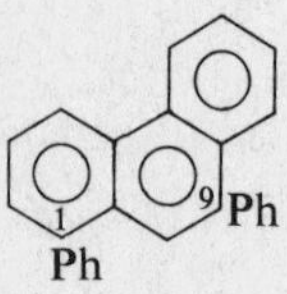

1,9-Diphenyl-phenanthrene

7.

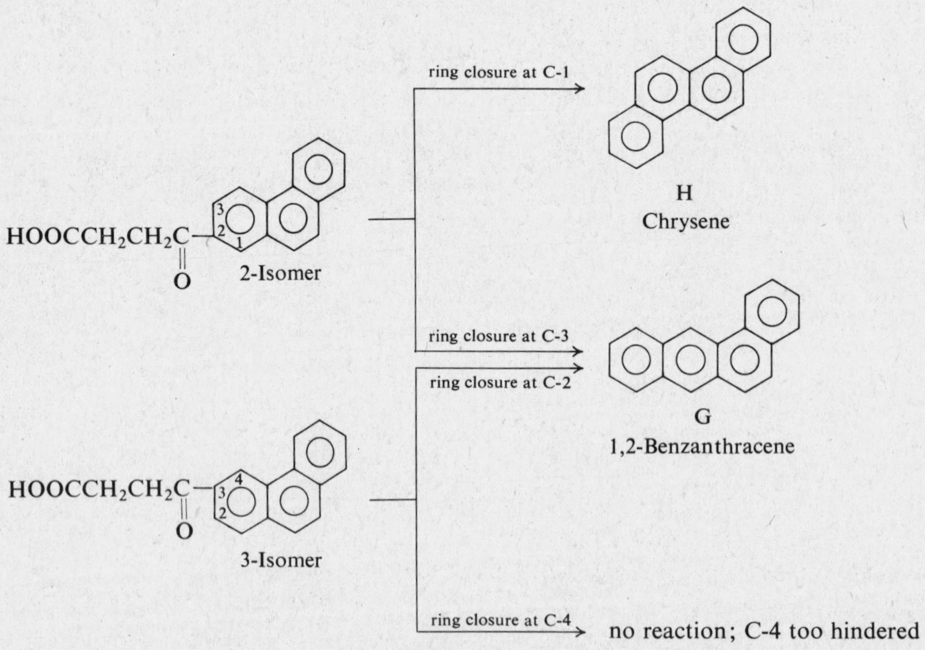

$HOOCCH_2CH_2C(=O)-$ 2-Isomer

ring closure at C-1 → Chrysene H

ring closure at C-3 →
ring closure at C-2 → 1,2-Benzanthracene G

$HOOCCH_2CH_2C(=O)-$ 3-Isomer

ring closure at C-4 → no reaction; C-4 too hindered

8. The final product has all the characteristics of a phenol: it is soluble in NaOH(aq), and insoluble in $NaHCO_3$(aq); it couples with $PhN_2^+Cl^-$ to give a colored azo compound. The formula indicates a naphthol. A reasonable reaction sequence leads only to 1-naphthol.

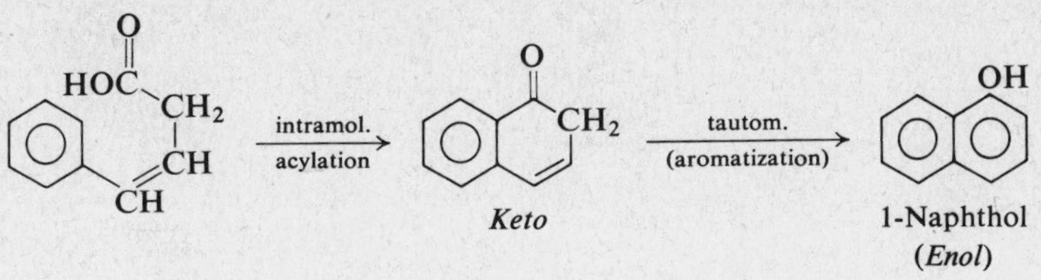

Keto

1-Naphthol
(Enol)

9. Through addition to the 9,10-positions—which requires only a small sacrifice of aromatic stabilization—anthracene acts as the diene in Diels-Alder reactions (Secs. 27.8 and 29.9). Reaction is *supra,supra* and results in *syn* addition to each component. (See Figure 30.5, page 510 of this Study Guide.)

10. (a)

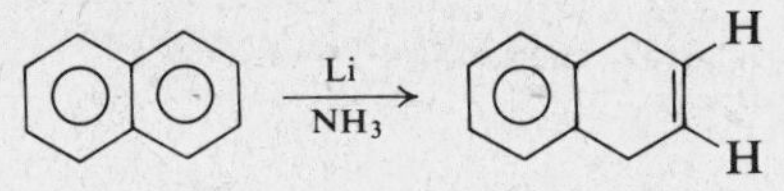

(b)–(d)

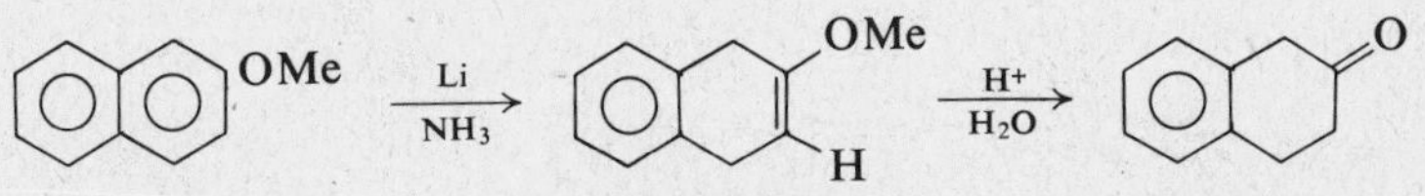

Enol ether
A vinylic ether
(*Prob. 15, page 649*)

β-Tetralone

Here and in the following problem, we catch a glimpse of the *Birch reduction*, whose enormous value lies in the synthesis *from aromatic starting materials* of a wide variety of non-aromatic cyclic compounds. In part (b) above, for example, a carbonyl group is generated, and this opens the way to a host of other compounds. (Compare the newer approach illustrated in Problem 27, Chapter 24, page 811.)

(A. J. Birch and H. Smith, "Reduction by Metal–Amine Solutions: Applications in Synthesis and Determination of Structure," Quart. Revs. (London), **12**, 17 (1958).)

11. (a)–(b)

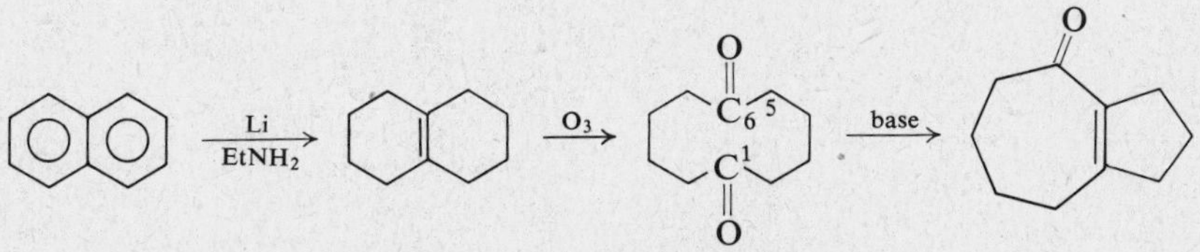

1,6-Cyclodecanedione

The diketone undergoes an intramolecular aldol condensation (addition, say, of C–5 to C–1) to yield a bicyclic ketone with a five-membered ring and a seven-membered ring.

(c)

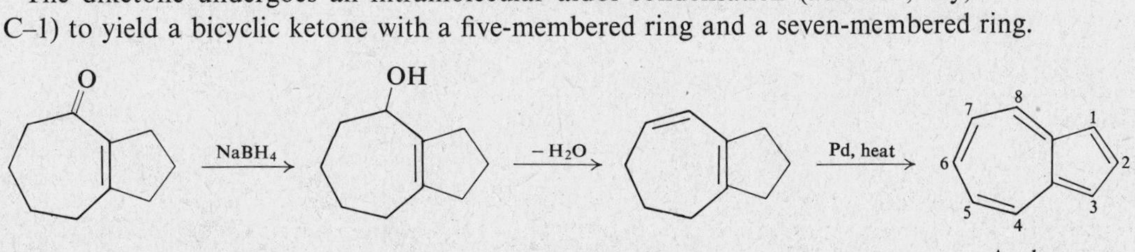

Azulene

509

(a)–(c)

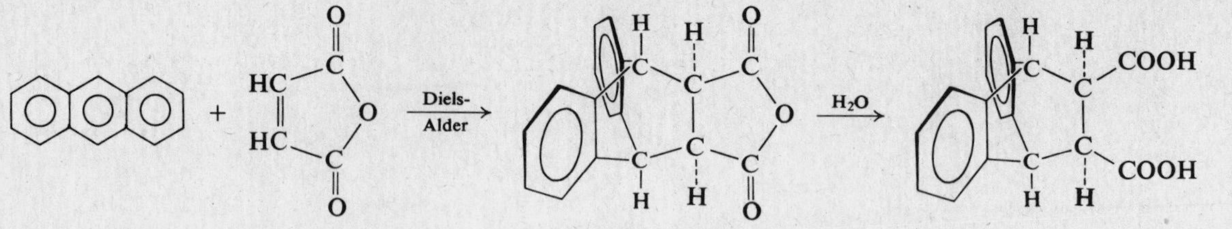

Maleic anhydride
(*cis*)

I
Achiral; meso

J
Achiral; meso

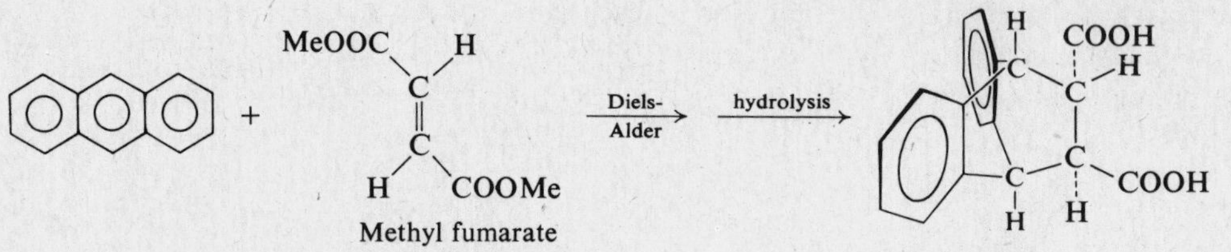

MeOOC   C   H

H   C   COOMe

Methyl fumarate
(*trans*)

K
Chiral; racemic

(d)

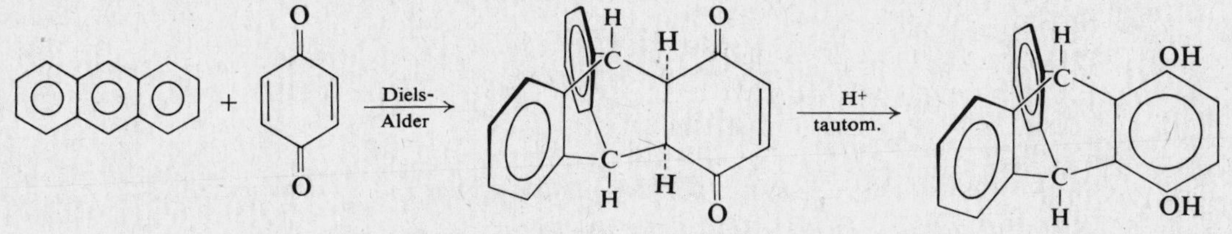

L
Diels-Alder adduct

M
A hydroquinone

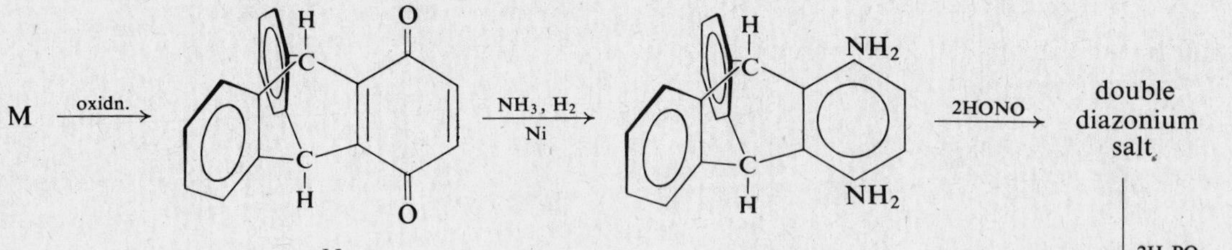

M

N
A quinone

O
A diamine

double
diazonium
salt

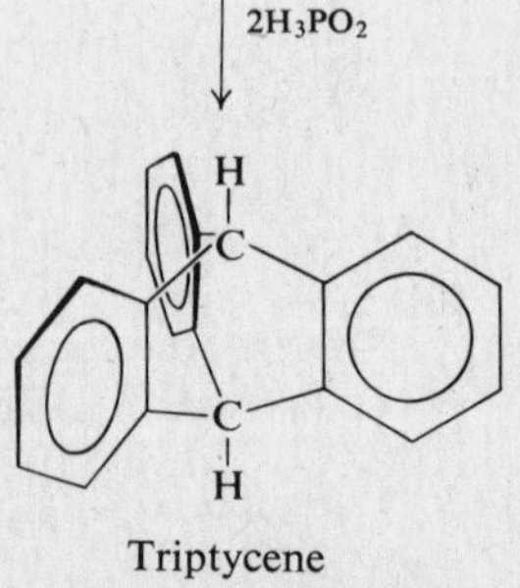

Triptycene
(Gr.: *triptychos*, consisting
of three plates)

Figure 30.5. Reactions of Problem 9.

12. (a) Azulene is clearly aromatic. It has a structure in which the seven-membered ring contains a positive charge, and the five-membered ring contains a negative charge. This provides

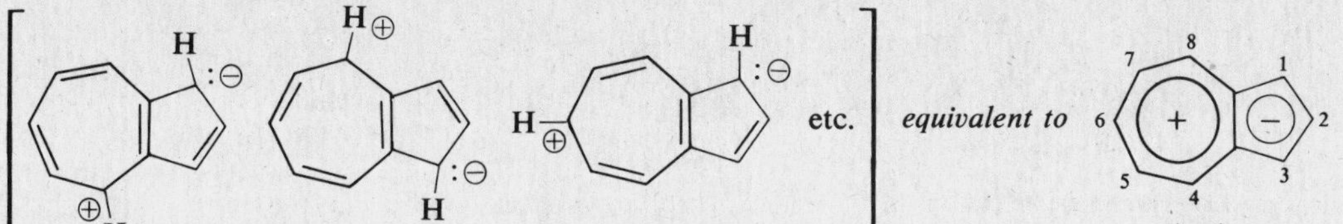

six π electrons—the aromatic sextet—for each ring. Just as naphthalene consists of two benzene rings fused, so azulene consists of a tropylium cation and a cyclopentadienyl anion fused.

(b) The dipole moment of azulene as we have represented it should be from the 7-ring toward the 5-ring. It should be augmented by the C—Cl dipole, as in fact it is.

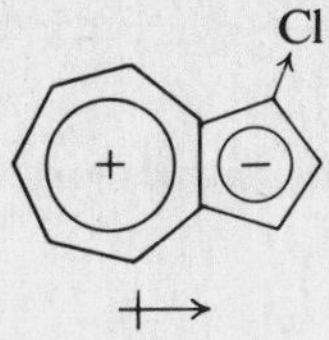

Dipole

13. (a)

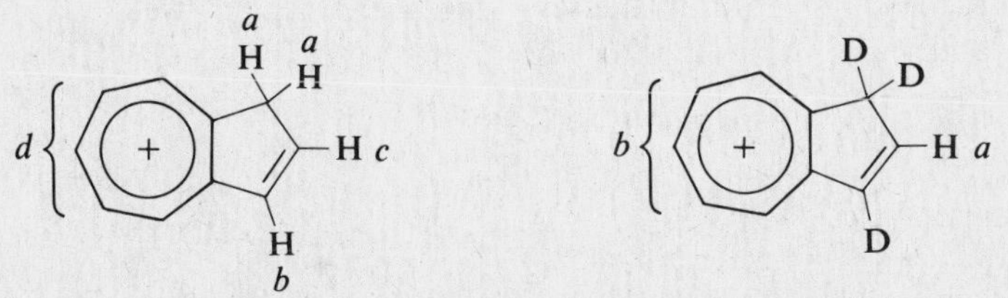

We note that signals *a* and *b* observed in CF_3COOD correspond to signals *c* and *d* observed in CF_3COOH. Deshielding due to the positive charge shifts all signals downfield from their usual positions.

(b) The first step of electrophilic aromatic substitution occurs, to yield the cation comparable to the benzenonium ion formed from benzene derivatives. (Compare Problem 11.11, page 358, and Problem 14, Chap. 24, page 808.)

Attack occurs where it does, at C–1 in the five-membered ring, because this gives the most stable cation: the only one in which the aromaticity of one of the rings—the seven-membered—is preserved. (Draw structures for the product of attack at C–2 in the five-membered ring or at any position in the seven-membered ring.)

(J. Schultze and F. A. Long, "Hydrogen Exchange of Azulenes. I. Structure of the Conjugate Acids of Azulenes," J. Am. Chem. Soc., **86**, 322 (1964).)

(c)

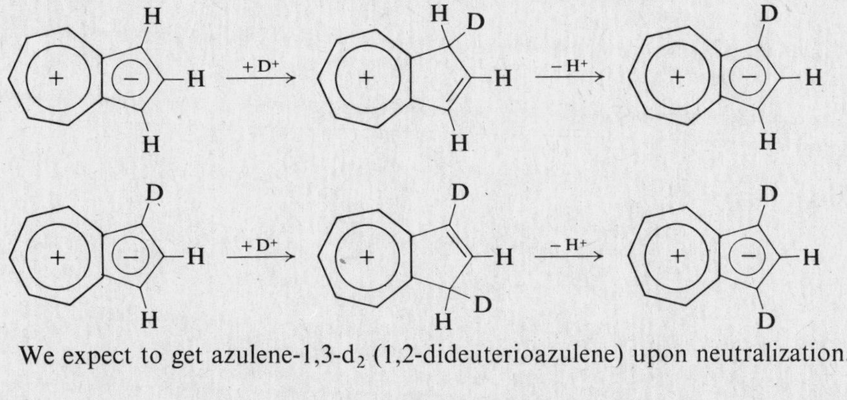

We expect to get azulene-1,3-d_2 (1,2-dideuterioazulene) upon neutralization.

(d) Electrophilic aromatic substitution should take place—and does—at C–1, since this proceeds via the most stable cation. (See the answer to part (b), above.)

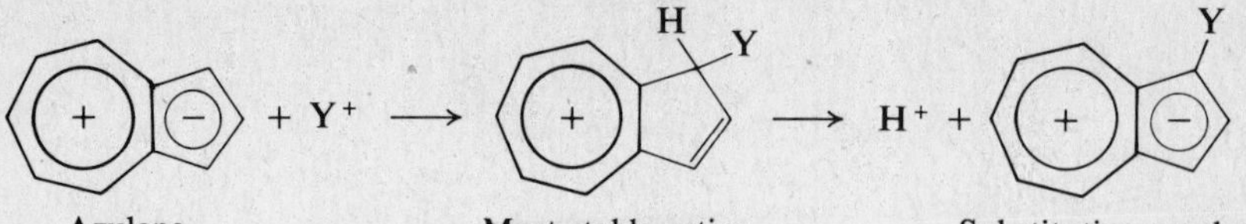

Azulene — Most stable cation — Substitution product

Electrophilic substitution

14. These are nucleophilic aromatic substitution reactions (Sec. 25.8), and should proceed via the most stable carbanion: the one resulting from attack at C–4 in the seven-membered ring.

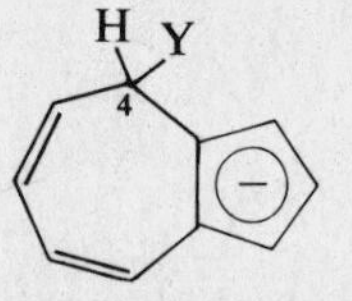

Nucleophilic substitution

Most stable carbanion

Only attack at C–4 or C–6 preserves the aromaticity of one of the rings—the five-membered—and attack at C–4 gives diene conjugation in the seven-membered ring as well. (Draw structures for the product of attack at C–6 or any other position.)

15.

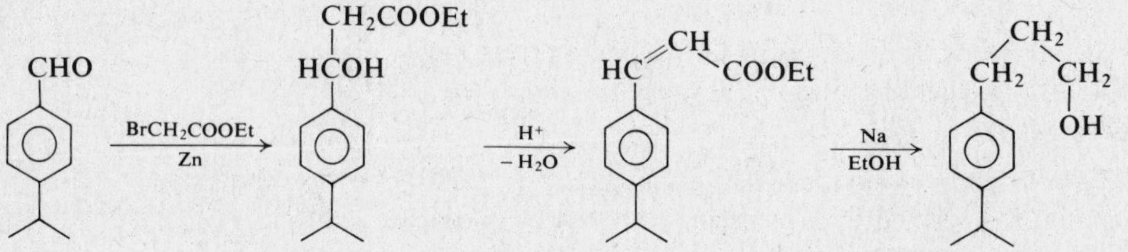

P Q R

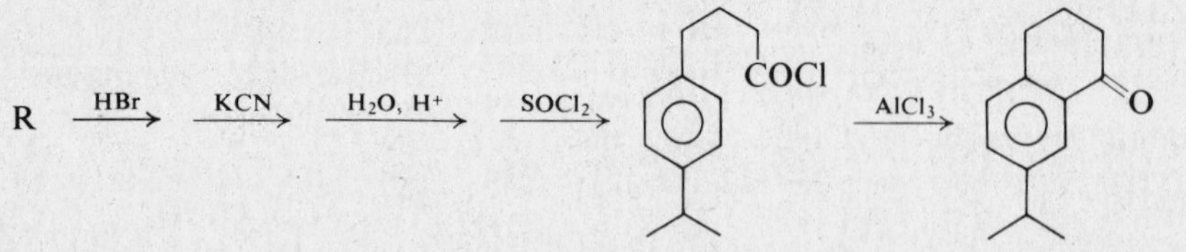

S T

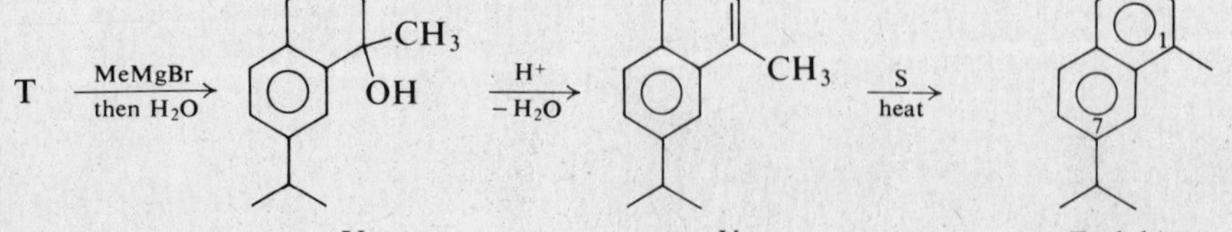

U V Eudalene
1-Methyl-7-isopropylnaphthalene

16. (a)

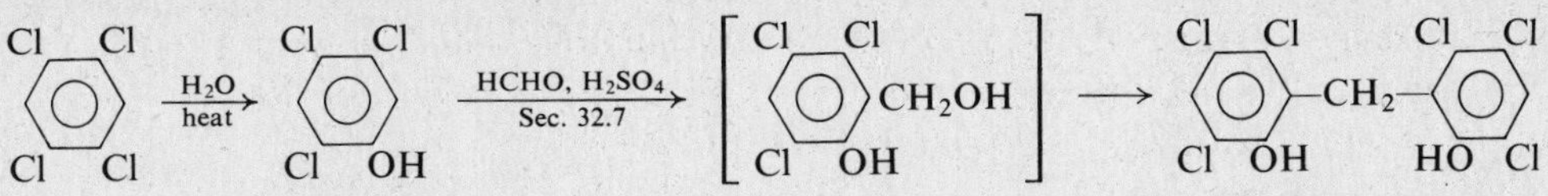

W

2,4,5-Trichlorophenol

X

2,4,5-Trichloro-
6-(hydroxymethyl)phenol

Y

2,2′,3,3′,5,5′-Hexachloro-
6,6′-dihydroxydiphenylmethane
"Hexachlorophene"

(b)

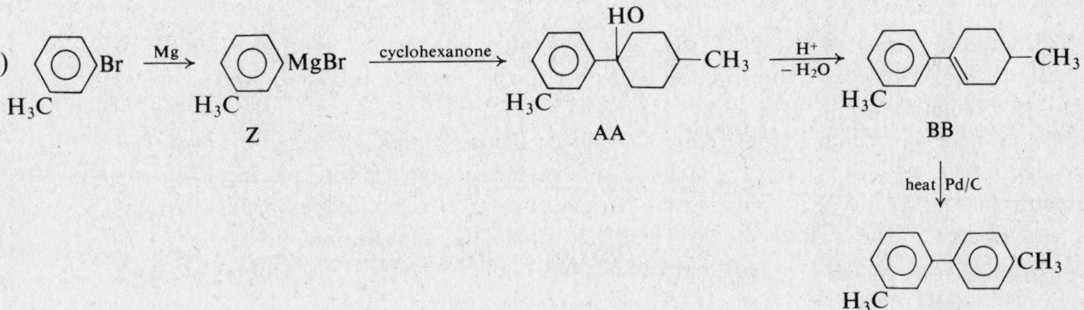

Z

AA

BB

heat | Pd/C

CC

3,4′-Dimethylbiphenyl

(c) $PhCOOEt + 2PhMgBr \longrightarrow Ph_3COH \xrightarrow{HBr} Ph_3CBr \xrightarrow{Ag}$

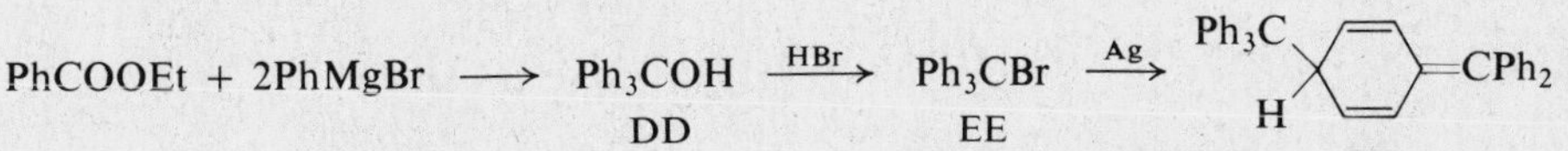

DD

EE

FF

(I, page 393)

(d) $Ph_3COH + PhNH_2 \xrightarrow{H^+} Ph_3C\langle\bigcirc\rangle NH_2 \xrightarrow{HONO} \xrightarrow{H_3PO_2} Ph_3C\langle\bigcirc\rangle$

GG

HH

Tetraphenylmethane

(e)

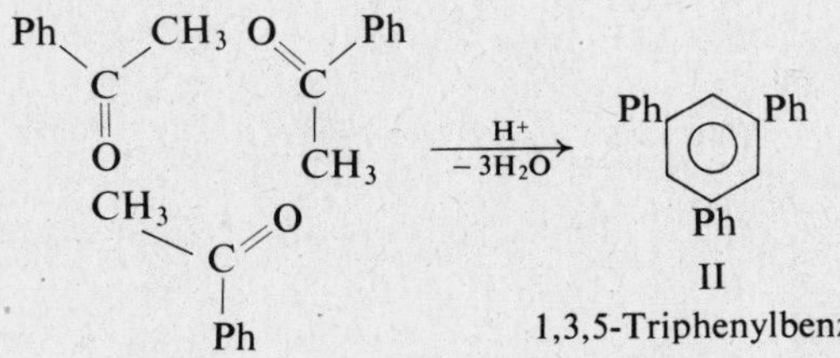

$\xrightarrow[-3H_2O]{H^+}$ *Triple aldol condensation*

II

1,3,5-Triphenylbenzene

17.

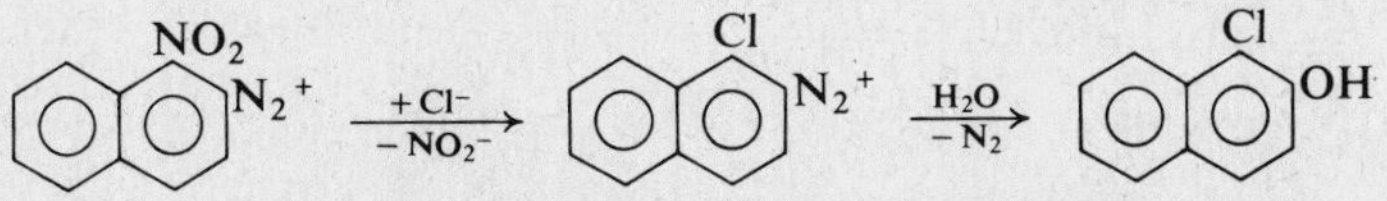

The —N_2^+ group (positively charged, electron-withdrawing) powerfully activates the molecule toward nucleophilic aromatic substitution. Chloride ion displaces —NO_2 in some molecules before —N_2^+ is converted into —OH. (Compare Problem 14, Chapter 25, page 843.)

18. (a)

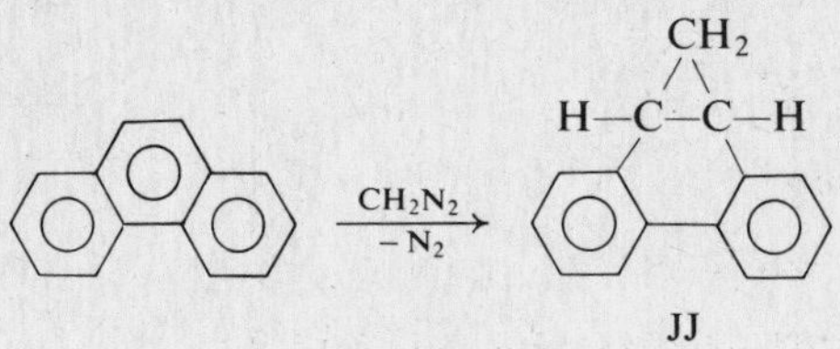

$$\text{phenanthrene} \xrightarrow[-N_2]{CH_2N_2} JJ$$

Addition of methylene across the 9,10-bond demands only a small sacrifice of resonance stabilization (Sec. 30.17).

(b) Methylene regenerated from JJ inserts itself into the carbon–hydrogen bonds of *n*-pentane (page 311). The driving force is the restoration of the resonance-stabilized aromatic system of phenanthrene.

We see here something we have not yet studied: the *randomness* of insertion of singlet methylene into every carbon–hydrogen bond of a hydrocarbon. (Compare the 34:17:49 ratio with the 4:2:6 ratio of hydrogens on C–2, C–3, and C–1 of *n*-pentane.) This randomness is attributed to the high energy of the singlet methylene initially generated; every collision, whether at a primary or secondary bond, has enough energy for reaction. (It is as though reaction were carried out at a very high temperature (Sec. 3.23).)

(c) Three insertion products and one addition product:

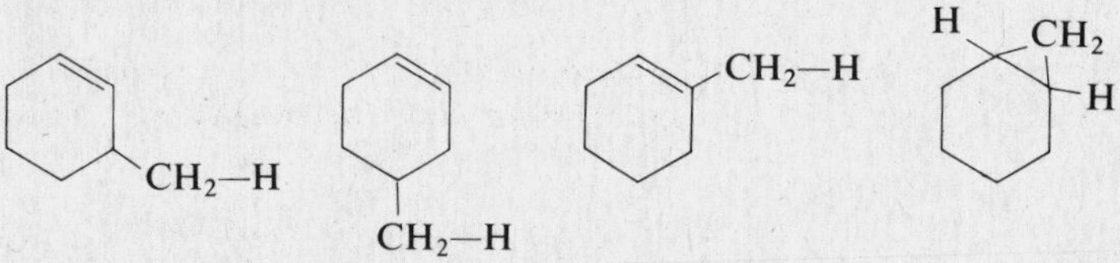

(d) There will be a number of insertion products, but our interest here—with diastereomeric alkenes as reactants—obviously lies in the addition reaction. Since reaction necessarily takes place in the liquid phase (consider the likely boiling point of a molecule the size of JJ), we expect singlet methylene to be involved, and hence predict stereospecific (*syn*) addition (Sec. 9.15).

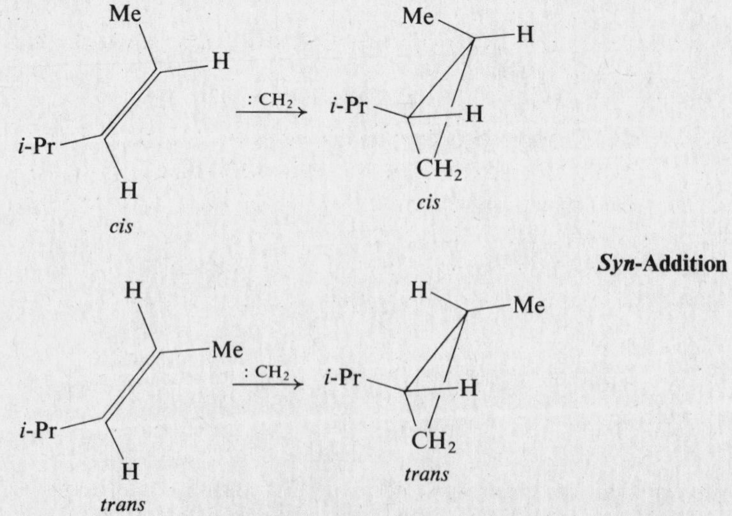

Syn-**Addition**

(D. B. Richardson *et al.*, "Generation of Methylene by Photolysis of Hydrocarbons," J. Am. Chem. Soc., **87**, 2763 (1965).)

19.

$$\text{Dihydropentalene} \xrightarrow{2n\text{-BuLi}} 2n\text{-BuH} + 2\text{Li}^+ \quad \text{KK}$$

KK

Lithium pentalenide

Like cyclopentadiene—and for the same reason—dihydropentalene is acidic, and decomposes n-BuLi with formation of the salt KK. The doubly charged anion KK is aromatic, with each ring containing six π electrons. It corresponds to two cyclopentadienyl anions fused.

Nmr signal a is 4H, split by one proton; signal b is 2H, split by two protons.

$$b \overset{a \quad a}{\underset{a \quad a}{\langle - | - \rangle}} b$$

(T. J. Katz and M. Rosenberger, "The Pentadienyl Dianion," J. Am. Chem. Soc., **84**, 865 (1962).)

The parent hydrocarbon is *pentalene*: KK without the negative charges. It is unknown (although the existence of 1-methylpentalene at low temperatures has been detected), and is presumably much less stable than the dianion—as we would expect it to be.

20. (a) 1-Chloronaphthalene and 2-chloronaphthalene yield the same mixture of products because they both form the same intermediate aryne, 1,2-dehydronaphthalene. This is the only aryne 1-chloronaphthalene can form. It is the *preferred* aryne from 2-chloronaph-

products
↑

| 1-Chloro-naphthalene | | *Only carbanion possible* | | 1,2-Dehydro-naphthalene | | *More stable carbanion* | | 2-Chloro-naphthalene |

An aryne

thalene, since it is formed via the more stable carbanion: the one with the negative charge next to the electron-withdrawing substituent—the other ring.

(b) In the case of the fluoro compound, direct bimolecular displacement (which can yield only the 1-product) accompanies elimination-addition (which yields both 1- and 2-products). *Fluoro*naphthalene is the least reactive toward aryne formation (Sec. 25.14), and (often) the most reactive toward bimolecular displacement (Sec. 25.12).

The effect of changing the concentration of piperidine (B:H) is understandable. As we saw on page 838, when X = F, $k_{-1} \gg k_2$, so that the initial carbanion is formed rapidly

(1) $\quad$ X $\quad$ + B:$^-$ $\underset{k_{-1}}{\overset{k_1}{\rightleftharpoons}}$ X $^{:-}$ + B:H

(2) $\quad$ X $^{:-}$ $\xrightarrow{k_2}$ + X$^-$

for X = F, $\quad k_{-1} \gg k_2$

and reversibly (1), and occasionally loses halide ion (2) to give the aryne. The rate of aryne formation depends, then, on the equilibrium concentration of the carbanion; this concentration is lowered as the piperidine concentration is raised and shifts the equilibrium to the left. As aryne formation becomes slower, bimolecular displacement becomes relatively more important.

21. Looking over the problem first, we find a useful clue to keep in mind as we work our way through: the final product, UU, is not only aromatic, but its nmr spectrum shows unusual, far-upfield peaks like those we saw for protons *inside* the ring of [18]annulene (Problem 5, Chapter 13, page 447; also see page 935).

$C_8H_8OBr_2$ | LL $C_{10}H_8ON_2$ | MM $C_{12}H_{14}O_5$ | NN $C_{10}H_{14}O_3$

OO $C_{10}H_{12}OBr_2$ | PP $C_{20}H_{24}O_2$ | QQ $C_{18}H_{18}O_2$

RR $C_{18}H_{16}O_2Na_2$ | SS $C_{18}H_{14}O_2$

TT $C_{18}H_{18}$

UU, $C_{18}H_{16}$
trans-15,16-Dimethyldihydropyrene
Eighteen π electrons
Aromatic

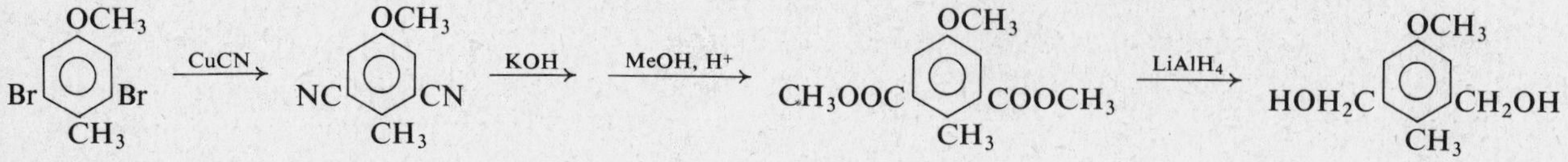

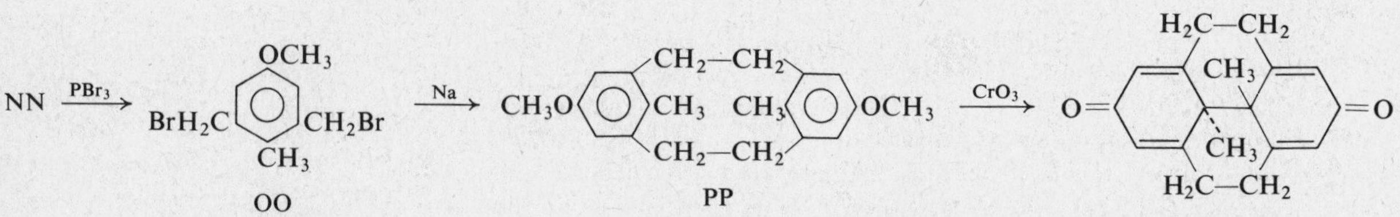

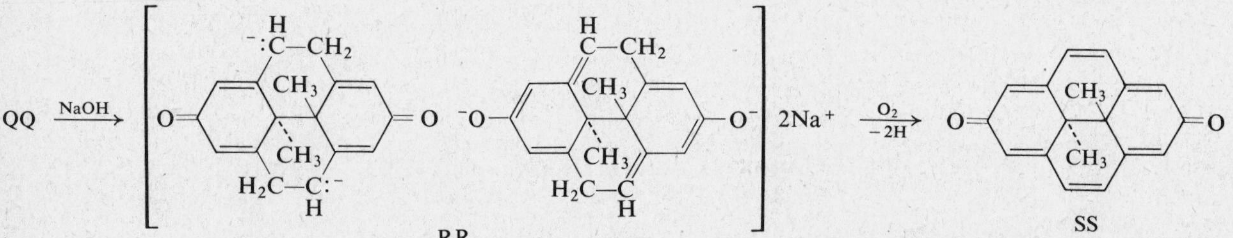

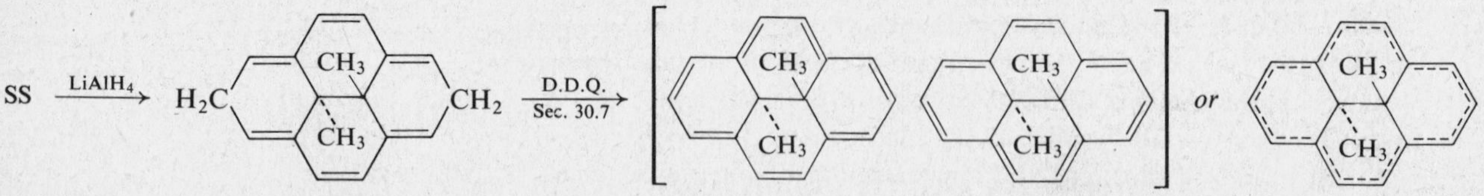

We have previously encountered aromatic compounds with the Hückel numbers of 2, 6, 10, and 18 π electrons. Here, we find one with the Hückel number of 14 ($n = 3$).

Aromaticity requires overlap of *p* orbitals, and this in turn requires planarity. For a ring of 10 carbons, this poses a special problem (see the answer to Problem 17, Chapter 29, page 966): an all-*cis* polyene would be a regular decagon, with angles much bigger than the normal 120° of trigonal carbon—too much bigger, evidently, to allow the molecule to be flat, and aromatic. The alternative flat polyene (*cis,trans,cis*) would contain certain hydrogens turned inward, and there evidently is not room enough for them. The result is that [10]annulene is a non-planar non-aromatic compound. (For an aromatic compound with 10 π electrons, we must look to an *eight*-membered ring: the cyclooctatetraenyl dianion of Problem 10.7, page 331.)

The ring of [18]annulene, on the other hand, *is* big enough to accommodate inward-pointing hydrogens, and the compound is aromatic.

The [14]annulene has been prepared; its instability and nmr spectrum show it to be essentially non-aromatic. Here again, there is not enough room inside the ring for hydrogen atoms. With UU, this difficulty has been cleverly overcome; the ring is forced into planarity by bridging between carbons on opposite sides of the ring. At the same time, it is the first compound to have substituents (methyl groups) inside the π cavity. Their far-upfield absorption is exactly what is predicted for these methyl protons (see Figure 13.4, page 419), and provides powerful evidence for the aromaticity of UU and the validity of the Hückel $4n + 2$ rule.

(V. Boekelheide and J. B. Phillips, "Aromatic Molecules Bearing Substituents within the Cavity of the π-Electron Cloud. Synthesis of *trans*-15,16-Dimethyldihydropyrene," J. Am. Chem. Soc., **89**, 1695 (1967); V. Boekelheide and T. A. Hylton, "General Method for the Synthesis of *trans*-15,16-Dialkyldihydropyrene," *ibid.*, **92**, 3669 (1970).

For a general discussion, see F. Sondheimer, "The Annulenes," Accounts Chem. Res., **3**, 81 (1972).)

Heterocyclic Compounds

31.1

$$CH_3COCH_2COOEt \xrightarrow{NaOEt} [CH_3COCHCOOEt]^- \xrightarrow{I_2}$$

A

$$\underset{\text{A double }\beta\text{-keto ester}}{\underset{B}{\overset{EtOOC \quad COOEt}{CH_3COCH-CHCOCH_3}}} \xrightarrow[\text{H}_2\text{O}]{\text{H}^+}$$

$$\overset{HOOC \quad COOH}{CH_3COCH-CHCOCH_3}$$

$$\downarrow -2CO_2$$

$$CH_3CCH_2CH_2CCH_3$$
$$\overset{\parallel}{O} \qquad \overset{\parallel}{O}$$

2,5-Hexanedione

31.2
Ph⟨O⟩Ph $\xleftarrow{P_2O_5}$ $\underset{O \quad O}{\overset{H_2C-CH_2}{PhC \qquad CPh}}$ $\xleftarrow[\text{Prob. 31.1}]{\text{as in}}$ PhCOCH$_2$COOEt $\xleftarrow{OEt^-}$ PhCOOEt + CH$_3$COOEt

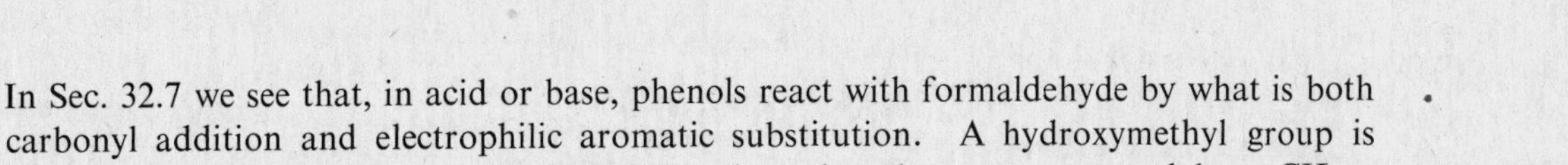

2,5-Diphenylfuran   A 1,4-diketone

31.3 Ring-opening involves electrophilic attack by H$^+$ (see the answer to Problem 4(b), below), which is slowed down by the electron-withdrawing —COOH group.

31.4 In Sec. 32.7 we see that, in acid or base, phenols react with formaldehyde by what is both carbonyl addition and electrophilic aromatic substitution. A hydroxymethyl group is introduced into the ring, and, on further reaction, rings become connected by —CH$_2$— linkages.

Like phenol, furan contains an activated ring, and it undergoes an analogous reaction.

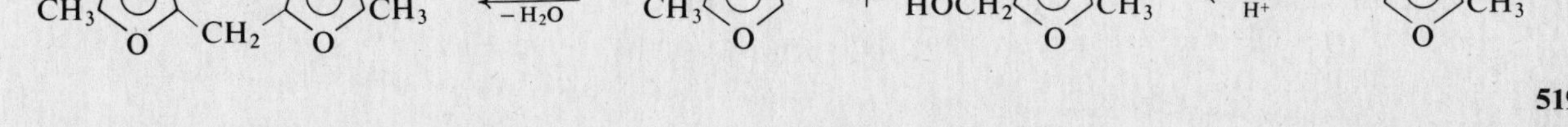

31.5 Like an aryl aldehyde, furfural has no α-hydrogens, and undergoes the Cannizzaro reaction (Sec. 19.16).

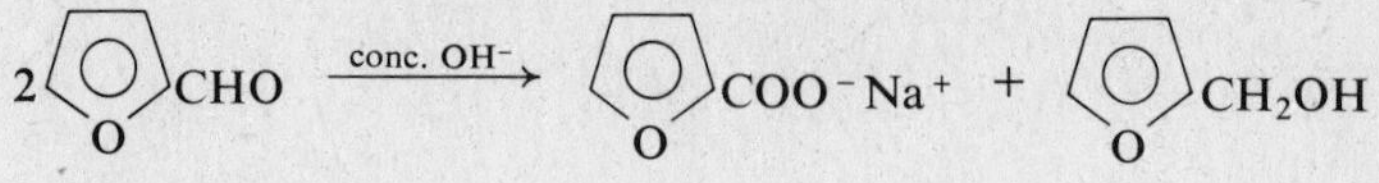

Furfural   Sodium furoate   Furfuryl alcohol

31.6

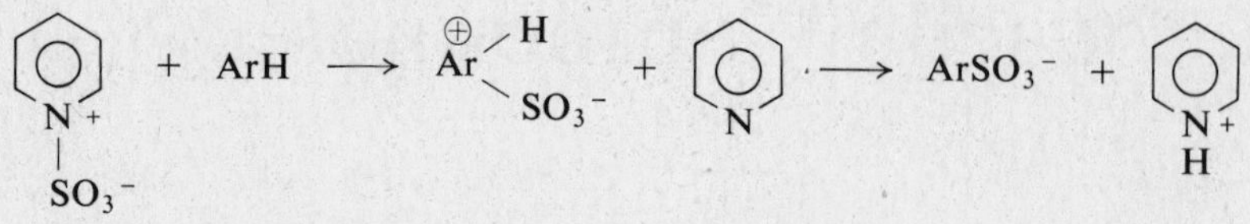

31.7 (a) Overlap of sextets of pyrrole and benzene rings, much as in naphthalene (Sec. 30.3).

(b) Attack at the 3-position gives the most stable cation: aromaticity of the six-membered ring is preserved, and the positive charge is accommodated by nitrogen. (See Secs. 30.9 and 31.4.)

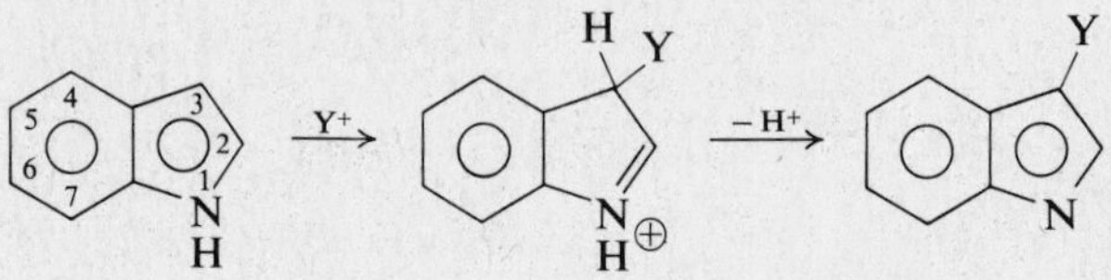

31.8

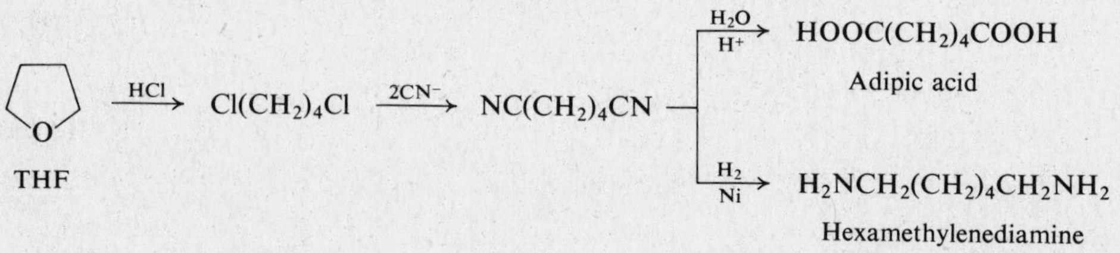

HOOC(CH₂)₄COOH

Adipic acid

H₂NCH₂(CH₂)₄CH₂NH₂

Hexamethylenediamine

31.9 (a)   (b) no reaction   (c)   (d)   (e)

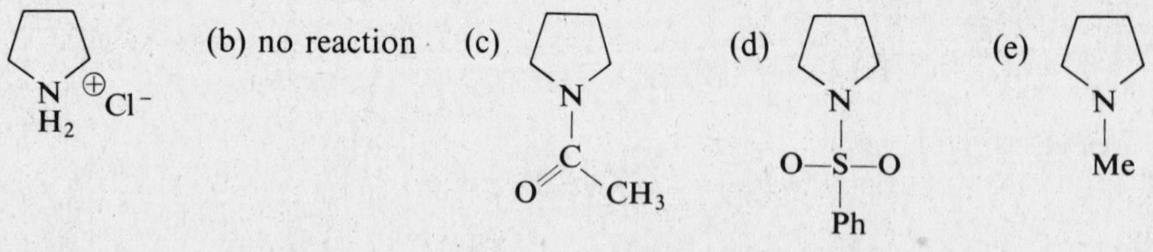

(f)

→ Me₂NCH₂CH₂CH=CH₂

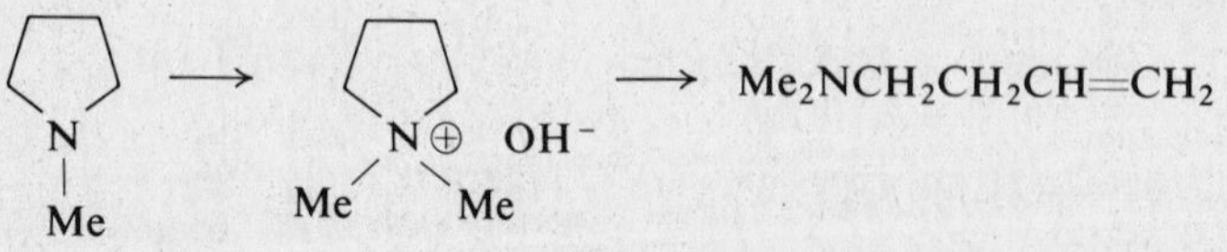

Exhaustive methylation (Sec. 23.5)

31.10 Hygrine is basic, a 3° amine. It contains a carbonyl group, and is a methyl ketone (Sec. 16.11). Vigorous oxidation removes one more carbon (and two more hydrogens) than the haloform reaction: a tentative conclusion is that there is a —CH_2COCH_3 group attached to a resistant nucleus.

$$RCH_2COCH_3 \xrightarrow{\text{oxidn.}} RCOOH$$

Hygrine   Hygrinic acid

Synthesis of hygrinic acid gives us the structures.

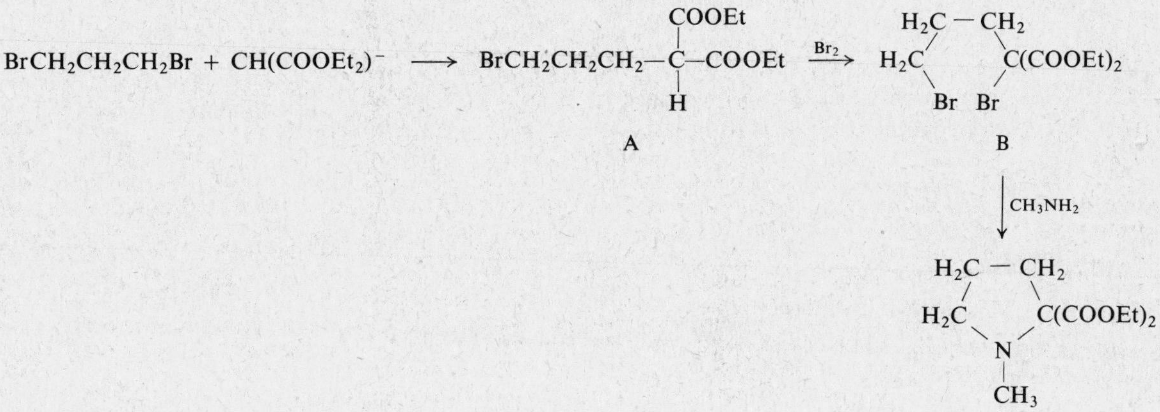

$$BrCH_2CH_2CH_2Br + CH(COOEt_2)^- \longrightarrow BrCH_2CH_2CH_2-\underset{\underset{H}{|}}{\overset{\overset{COOEt}{|}}{C}}-COOEt \xrightarrow{Br_2}$$

A

B

$\downarrow$ CH_3NH_2

C

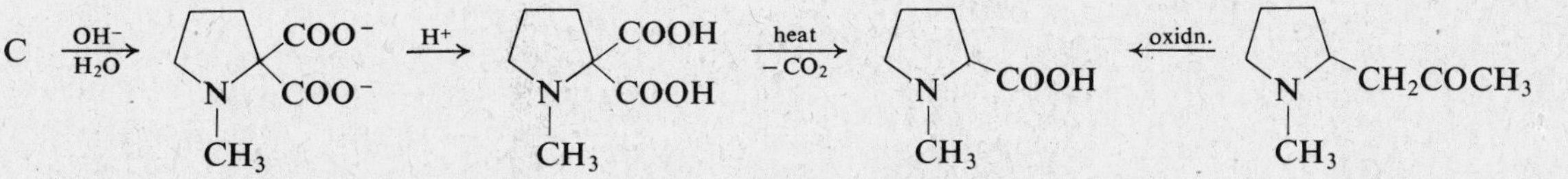

   D   E   Hygrinic acid   Hygrine

31.11 The orientation ("para") and reactivity are controlled by the activating —NH_2 group (as in aniline).

31.12

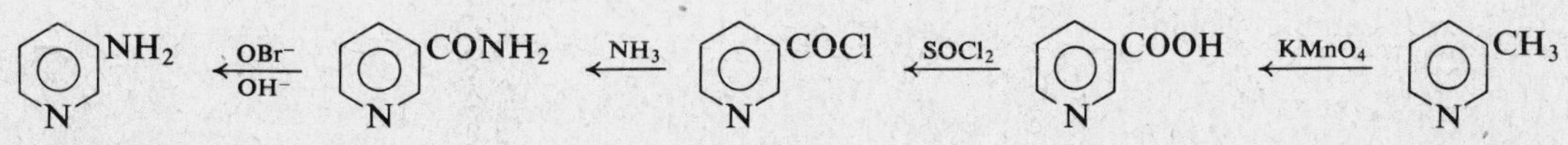

3-Aminopyridine β-Picoline

31.13 Basicity falls off with increasing s-character of the orbital holding the unshared pair.

amine > imine > nitrile
 (sp^3)   (sp^2)   (sp)

31.14 Pyridine is the base needed for the dehydrohalogenation. It does not cleave the ester as alcoholic KOH would.

31.15 Electrophilic substitution proceeds via intermediate I in which the positive charge is carried by nitrogen, and all atoms have complete octets.

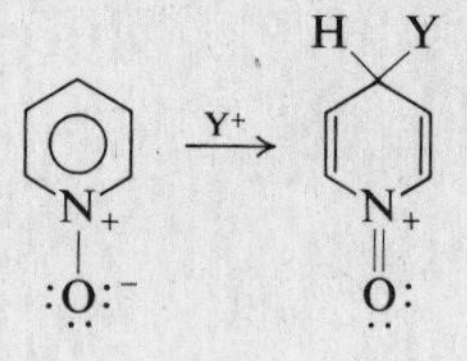

Electrophilic substitution

I

31.16 Nucleophilic substitution proceeds via intermediate II in which the negative charge is carried by the electronegative element oxygen.

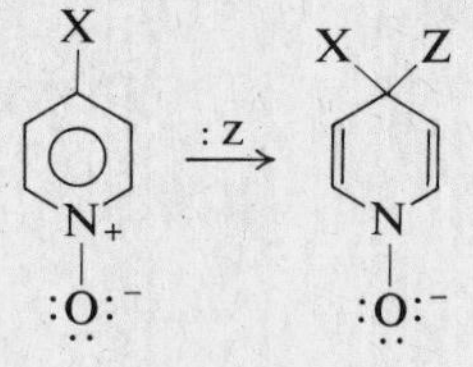

Nucleophilic substitution

II

31.17

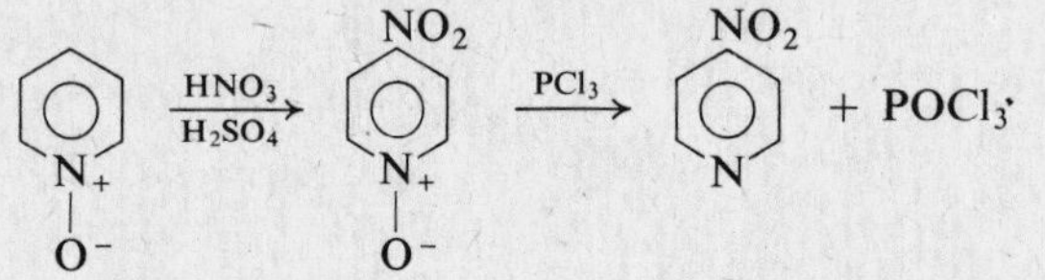

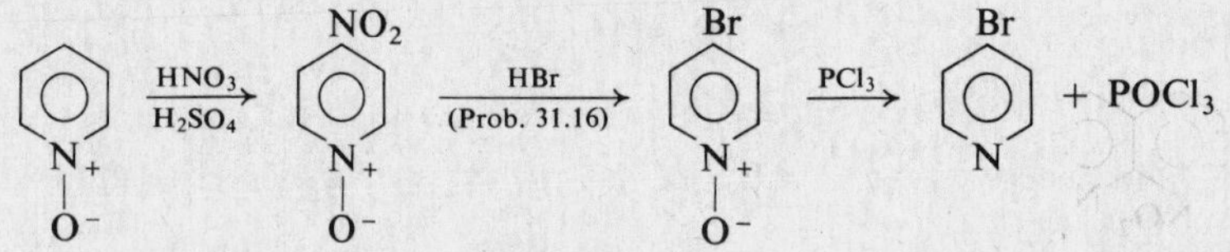

31.18 Piperidine, a 2° amine, would itself be acylated.

31.19 (a) The nitrogen-containing ring is deactivated, as in pyridine; electrophilic substitution thus occurs at an α-position in the less deactivated ring. (Compare Secs. 30.9 and 30.13.)

(b) The benzenoid ring is more easily attacked by oxidizing agents than the deactivated nitrogen-containing ring.

(c) Nucleophilic substitution takes place in the (activated) pyridine-like ring (compare Sec. 31.10).

31.20

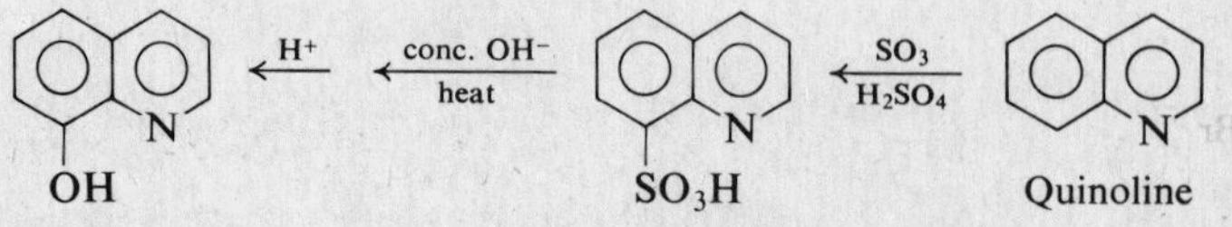

OH
8-Quinolinol

SO₃H
*Separate from
5-isomer*

Quinoline

31.21

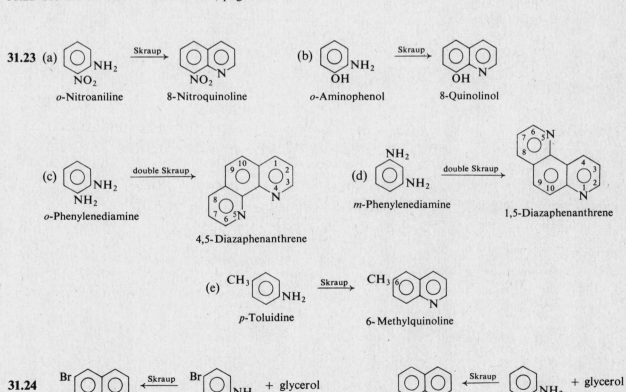

31.22 See the answer to Problem 27.5, page 867.

31.23 (a) *o*-Nitroaniline → Skraup → 8-Nitroquinoline   (b) *o*-Aminophenol → Skraup → 8-Quinolinol

(c) *o*-Phenylenediamine → double Skraup → 4,5-Diazaphenanthrene

(d) *m*-Phenylenediamine → double Skraup → 1,5-Diazaphenanthrene

(e) *p*-Toluidine → Skraup → 6-Methylquinoline

31.24 6-Bromoquinoline ← Skraup ← *p*-Bromoaniline + glycerol   8-Methylquinoline ← Skraup ← *o*-Toluidine + glycerol

31.25 (a)

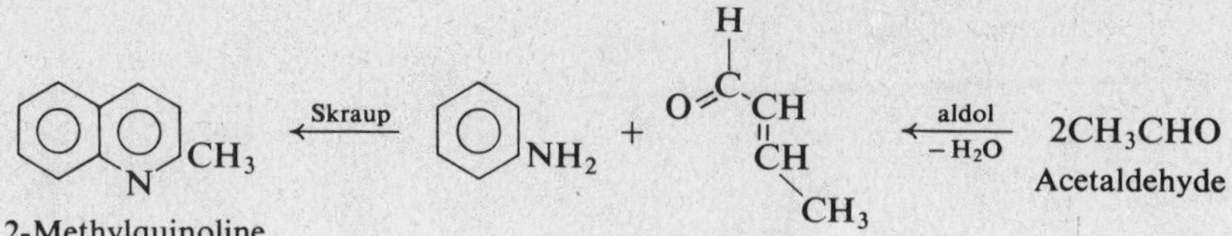

2-Methylquinoline

(*As in Problem 31.21*)

Acetaldehyde

(b)

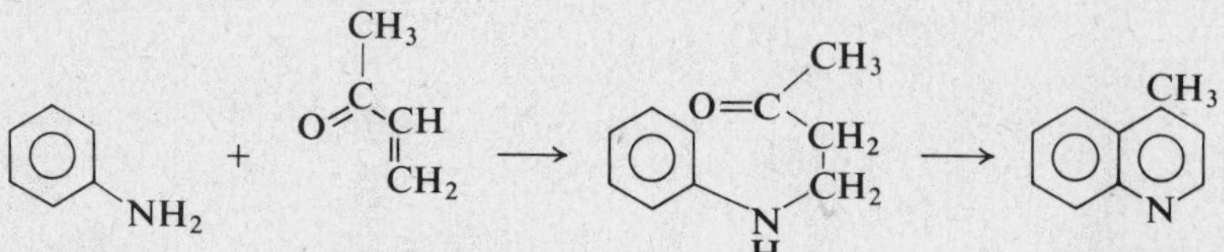

Aniline   MVK

4-Methyl-
quinoline

(c)   PhCHO + CH₃COCOOH

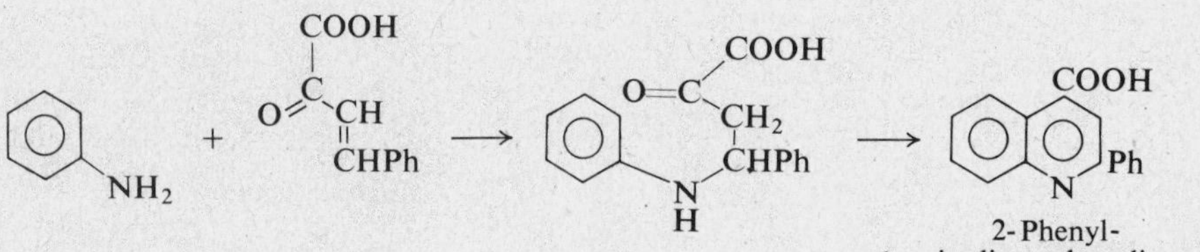

2-Phenyl-
4-quinolinecarboxylic acid

31.26

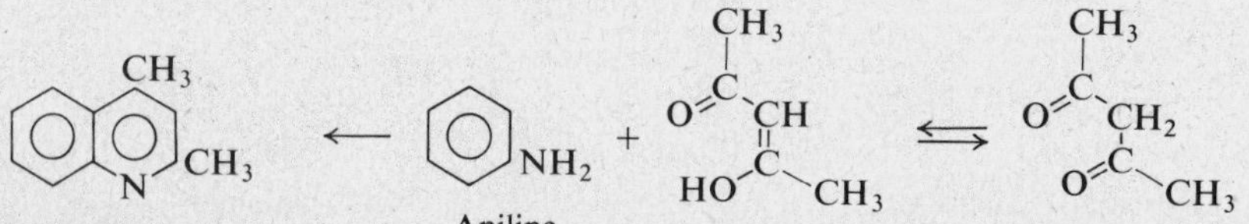

2,4-Dimethylquinoline   Aniline

Acetylacetone

Enol   *Keto*

The enol is the α,β-unsaturated ketone that undergoes conjugate addition of the aniline.

31.27 (a) The nitrogen-containing (pyridine-like) ring is deactivated, and electrophilic substitution takes place at an α-position of the other ring (Secs. 30.9 and 30.13).

(b) Nucleophilic substitution takes place in the activated (pyridine-like) ring, and at the position that allows the other ring to preserve its aromaticity and the negative charge to be accommodated by nitrogen.

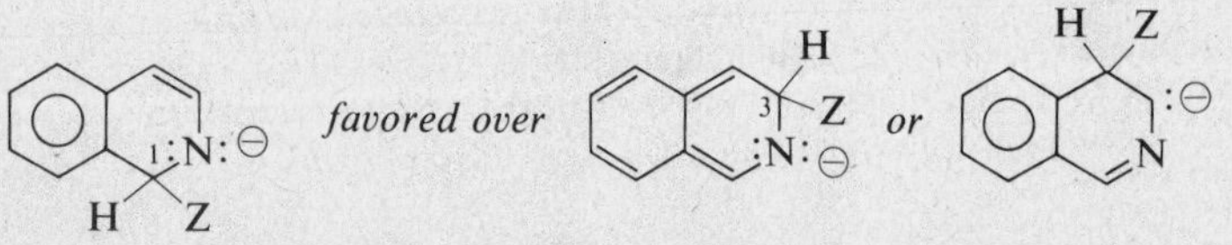

favored over   *or*   **Nucleophilic substitution**

(Z = NH₂⁻ or R⁻)

(c) Reaction involves an aldol-like addition to benzaldehyde of a carbanion derived from 1-methylquinoline.

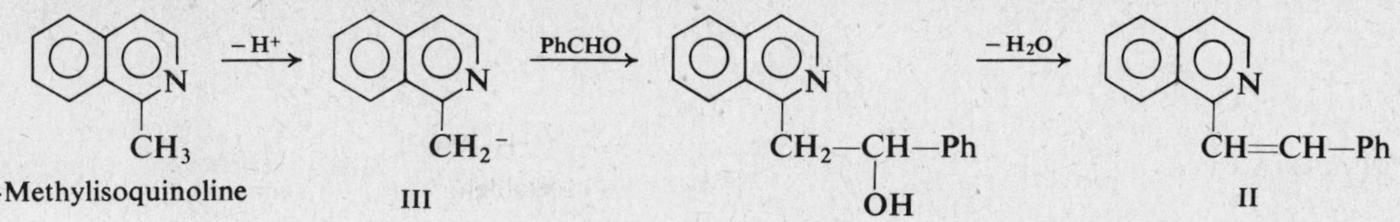

1-Methylisoquinoline   III   II

1-Methylquinoline yields a carbanion, III, in which the aromaticity of the other ring is preserved. 2-Methylquinoline can yield only the less stable carbanion, IV.

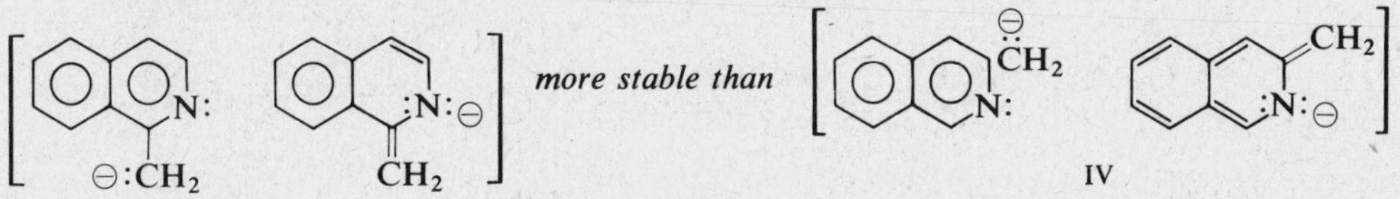

more stable than

III   IV

31.28 Electrophilic aromatic substitution or acid-catalyzed carbonyl addition, depending upon one's viewpoint. (See equation (3), page 1019.)

31.29 $PhCH_2CH_2NHCOCH_3 \xleftarrow{Ac_2O} PhCH_2CH_2NH_2 \xleftarrow[Ni]{H_2} PhCH_2CN \xleftarrow{CN^-} PhCH_2Cl \xleftarrow[heat]{Cl_2} PhCH_3$

1. (a) 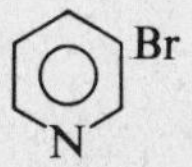 3-Bromopyridine

(b) 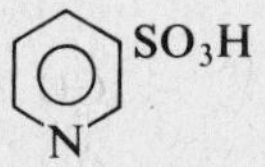 3-Pyridinesulfonic acid

(c) no reaction

(d) 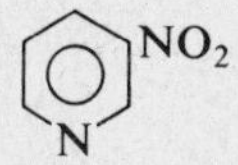 3-Nitropyridine

(e) 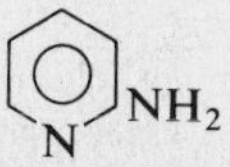 2-Aminopyridine (*Sec. 31.10*)

(f) 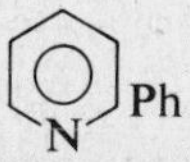 2-Phenylpyridine (*Sec. 31.10*)

(g) 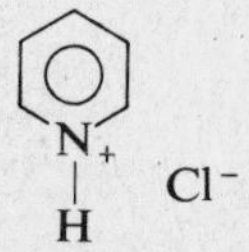 Pyridinium chloride

(h) no reaction

(i) no reaction

(j) no reaction

(k) 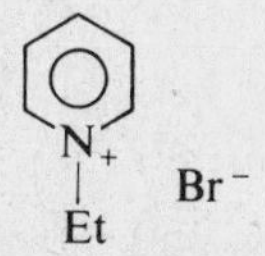 N-Ethylpyridinium bromide

(l) 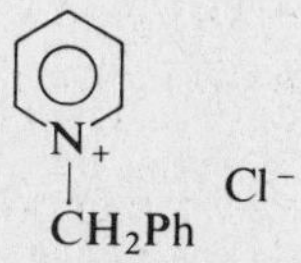 N-Benzylpyridinium chloride

(m) 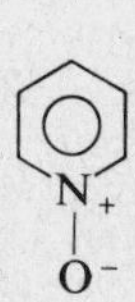 Pyridine N-oxide

(n) 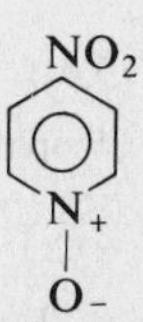 4-Nitropyridine N-oxide (*Problem 31.15*)

(o) 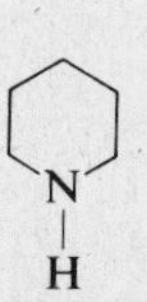 Piperidine (*Sec. 31.12*)

2. (a)  2-Thiophenesulfonic acid — SO$_3$H

(b) 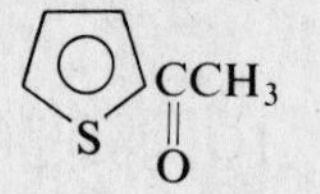 2-Acetylthiophene — CCH$_3$, O

(c) 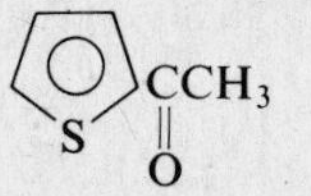 2-Acetylthiophene — CCH$_3$, O

(d) 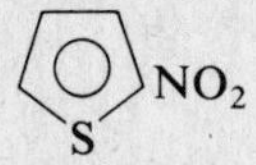 2-Nitrothiophene — NO$_2$

(e) 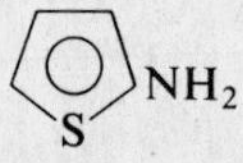 2-Aminothiophene — NH$_2$

(f)  2-Bromothiophene — Br

(g)  2-Thiophenecarboxylic acid — COOH

(h) 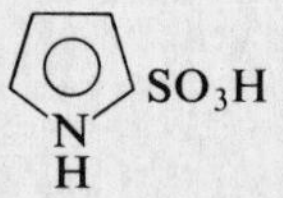 2-Pyrrolesulfonic acid — SO$_3$H

(i) 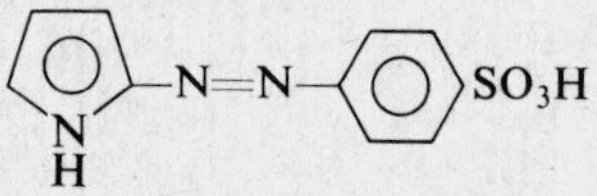 p-(2-Pyrrylazo)benzenesulfonic acid — N=N—SO$_3$H

(j) 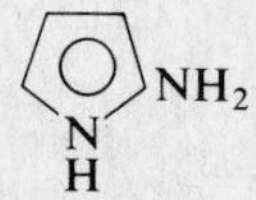 2-Aminopyrrole — NH$_2$

(k) 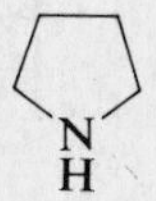 Pyrrolidine

(l) 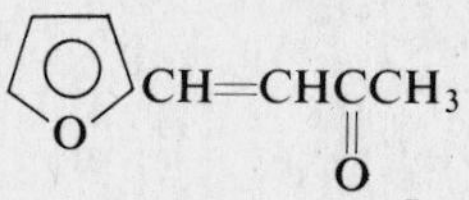 Furfurylideneacetone — CH=CHCCH$_3$, O

(m) 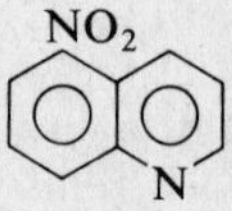 5- and 8-Nitroquinolines — NO$_2$ and NO$_2$

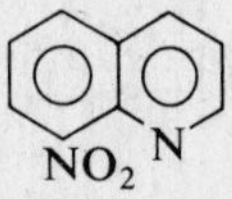

(n) 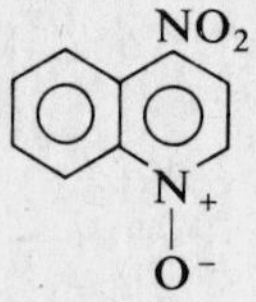 4-Nitroquinoline N-oxide — NO$_2$, N$^+$, O$^-$

(o) 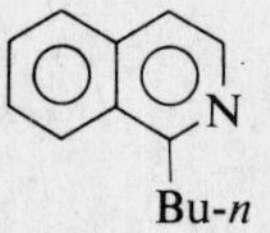 1-(n-Butyl)isoquinoline (*Problem 31.27b*) — Bu-n

3. (a)

Pyrrole $\xrightarrow{+2H}$ Δ^3-Pyrroline   or Δ^2-Pyrroline

(b) 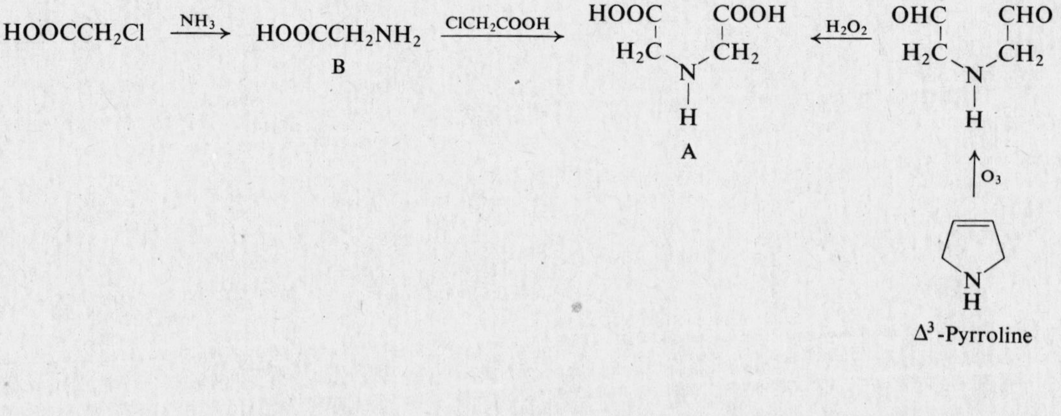 HOOCCH$_2$Cl $\xrightarrow{NH_3}$ HOOCCH$_2$NH$_2$ $\xrightarrow{ClCH_2COOH}$

B

$$\text{HOOC—CH—COOH} \quad \xleftarrow{H_2O_2} \quad \text{OHC—CH—CHO}$$
$$\text{H}_2\text{C} \quad \text{CH}_2 \qquad \qquad \text{H}_2\text{C} \quad \text{CH}_2$$
$$\text{N} \qquad \qquad \qquad \text{N}$$
$$\text{H} \qquad \qquad \qquad \text{H}$$
$$\text{A}$$

$\uparrow O_3$

Δ^3-Pyrroline

4. (a) 2CHI$_3$ + HOOCCH$_2$CH$_2$COOH $\xleftarrow{NaOI}$ CH$_3$CCH$_2$CH$_2$CCH$_3$
                  Succinic acid   O   O
                                                              C
                                                       2,5-Hexanedione

(b)

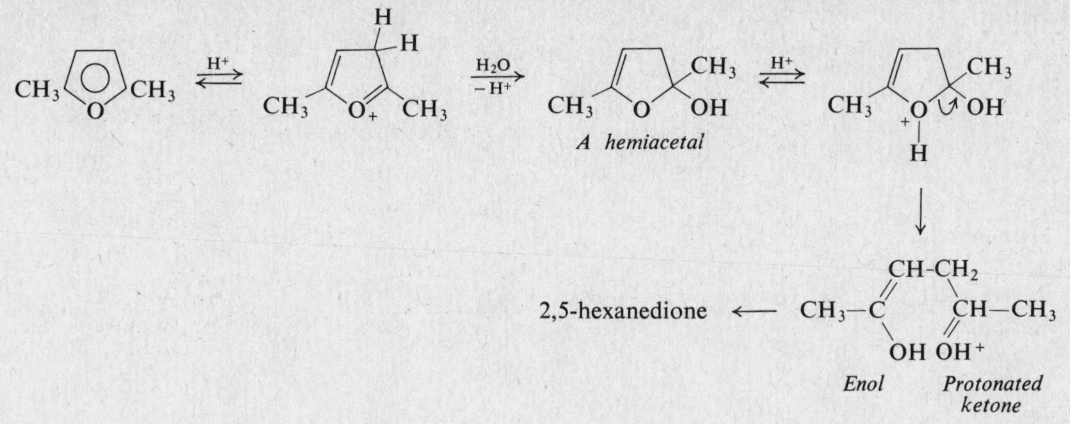

A hemiacetal

2,5-hexanedione ←── CH_3-C ... $CH-CH_3$

Enol   *Protonated ketone*

5. First, some arithmetic:

$(C_4H_5N)_4$ 4 pyrrole $(C_5H_7ON)_4$
$+(HCHO)_4$ 4 HCHO $-(C_5H_5N)_4$ product
$(C_5H_7ON)_4$ $4H_2O$ lost

Like phenol (or furan, Problem 31.4, page 1009), pyrrole contains an activated ring, and we suspect that a reaction analogous to that described in Sec. 32.7 has taken place here:

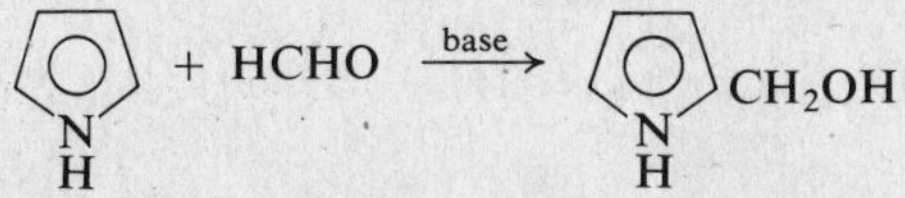

Four of these units are joined together, with loss of $4H_2O$. The structure must therefore be *cyclic*: four units, four new linkages.

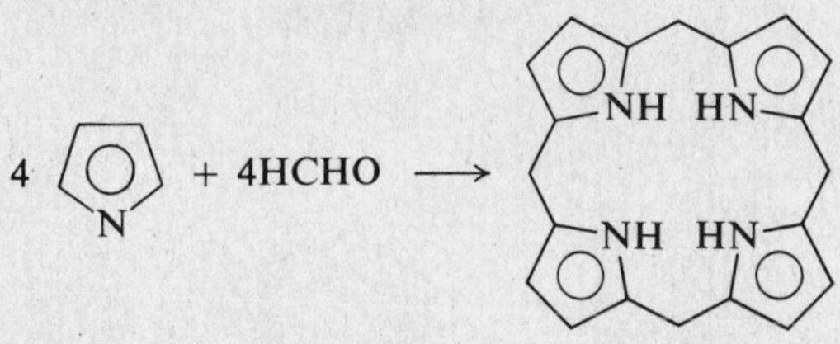

Compare ring structure of chlorophyll (*p. 1004*) *and* heme (*p. 1152*)

Porphin

6.

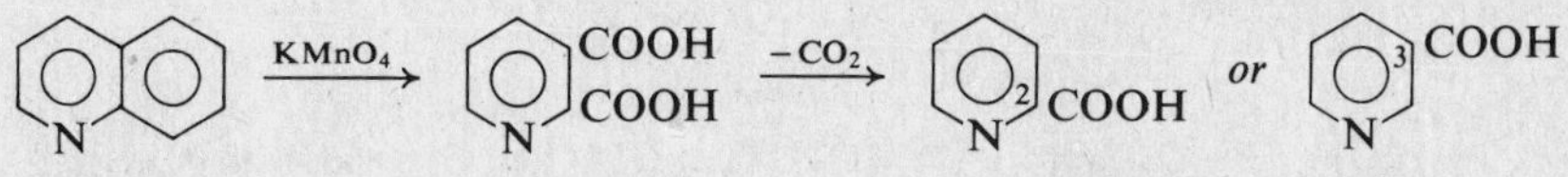

Quinoline

E *must be one of these*

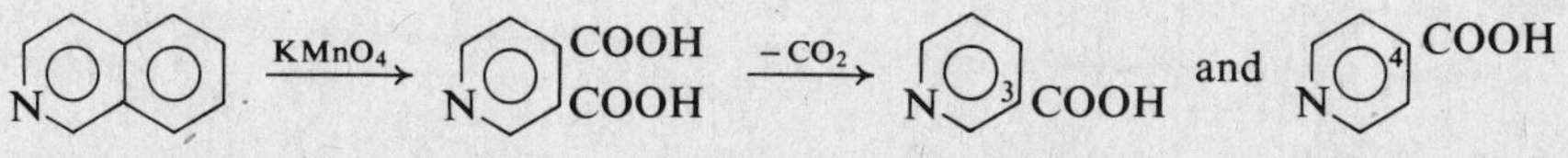

Isoquinoline

This must be E   *This must be* F

The only acid left must be D:

D

7. (a)

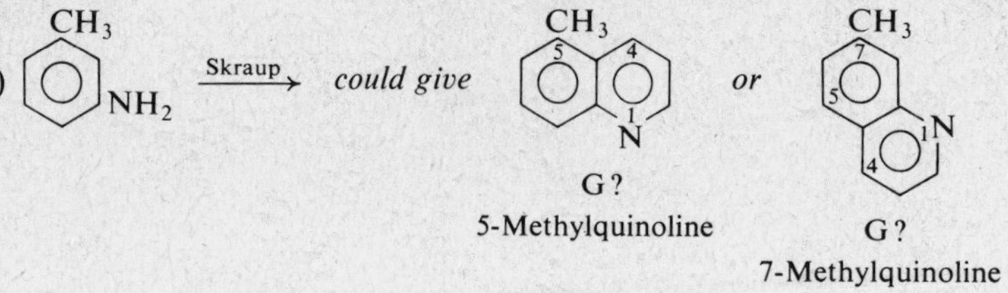

CH_3—benzene—NH_2 $\xrightarrow{\text{Skraup}}$ *could give*   5-Methylquinoline   **G?**   *or*   7-Methylquinoline   **G?**

(b)

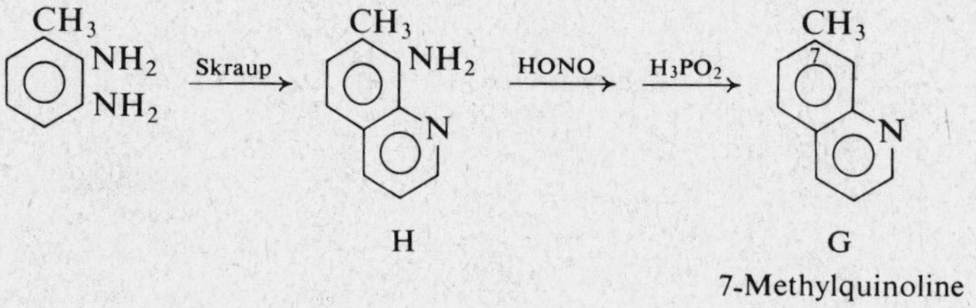

CH_3-diaminobenzene $\xrightarrow{\text{Skraup}}$ **H** $\xrightarrow{\text{HONO}}$ $\xrightarrow{H_3PO_2}$ **G**

7-Methylquinoline

8. (a)

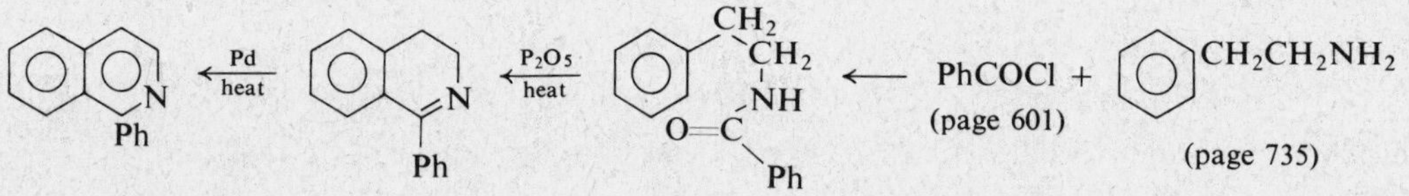

isoquinoline(Ph) $\xleftarrow[\text{heat}]{\text{Pd}}$ $\xleftarrow[\text{heat}]{P_2O_5}$ $\xleftarrow{}$ PhCOCl + $C_6H_5CH_2CH_2NH_2$

(page 601)   (page 735)

(b)

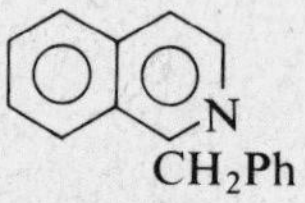

isoquinoline(CH_2Ph)   As in (a), except use $PhCH_2COCl$ (from $PhCH_2COOH$, page 587)

(c)

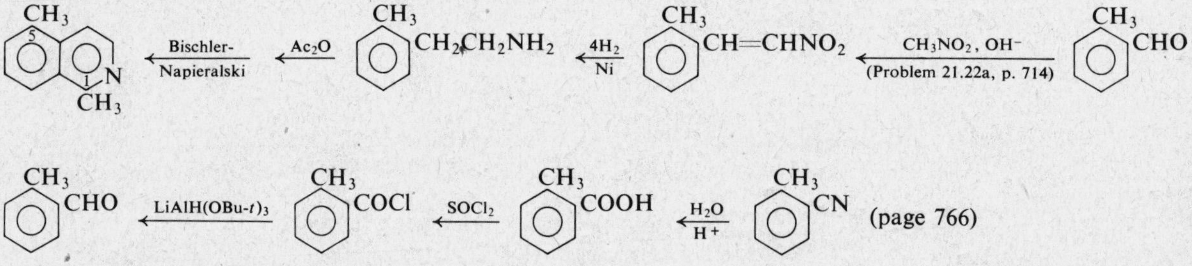

$\xleftarrow[\text{Napieralski}]{\text{Bischler-}}$ $\xleftarrow{Ac_2O}$ $CH_2CH_2NH_2$ $\xleftarrow[\text{Ni}]{4H_2}$ $CH=CHNO_2$ $\xleftarrow[\text{(Problem 21.22a, p. 714)}]{CH_3NO_2,\ OH^-}$ CHO

CHO $\xleftarrow{LiAlH(OBu\text{-}t)_3}$ $COCl$ $\xleftarrow{SOCl_2}$ $COOH$ $\xleftarrow[\text{H}^+]{H_2O}$ CN (page 766)

(d)

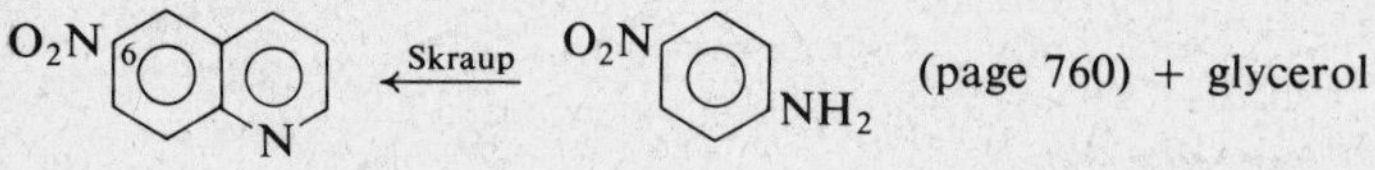

O_2N-quinoline $\xleftarrow{\text{Skraup}}$ O_2N-aniline   (page 760) + glycerol

(e)

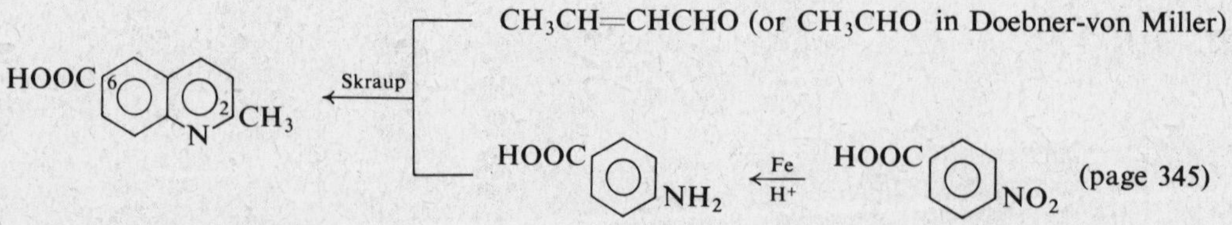

$CH_3CH=CHCHO$ (or CH_3CHO in Doebner-von Miller)

HOOC-quinoline-CH_3 $\xleftarrow{\text{Skraup}}$ HOOC-aniline $\xleftarrow[\text{H}^+]{\text{Fe}}$ HOOC-nitrobenzene   (page 345)

(f)

$\xleftarrow[\text{Skraup}]{\text{double}}$ H_2N-benzene-NH_2 $\xleftarrow[\text{H}^+]{\text{Fe}}$ O_2N-benzene-NH_2 (page 760)

p-Phenylenediamine

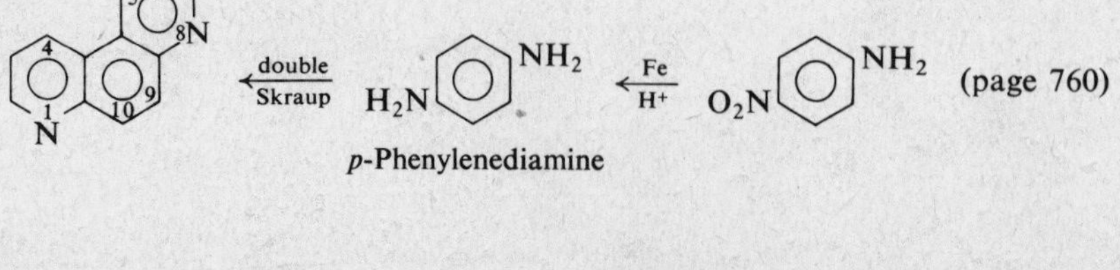

9. (a)

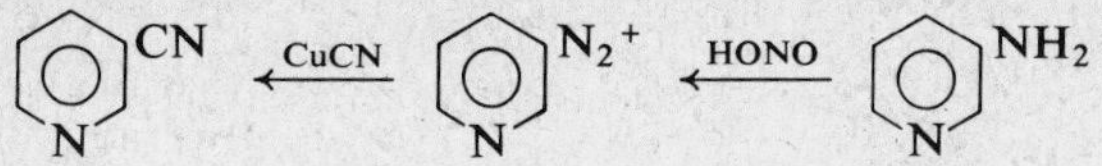

(Problem 31.12)

(b)

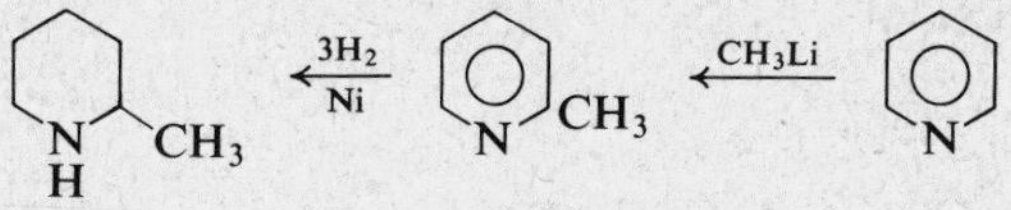

(c)

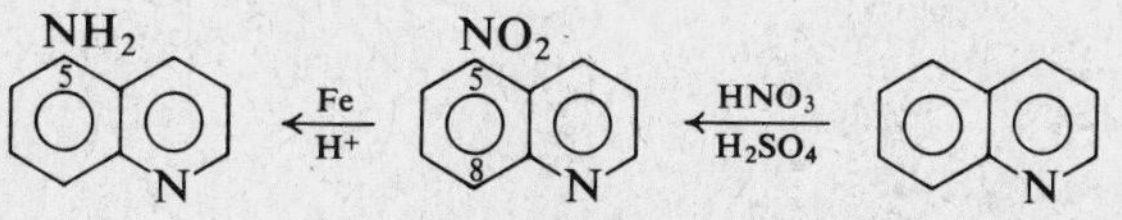

(*Separate from 8-isomer*)

(d)

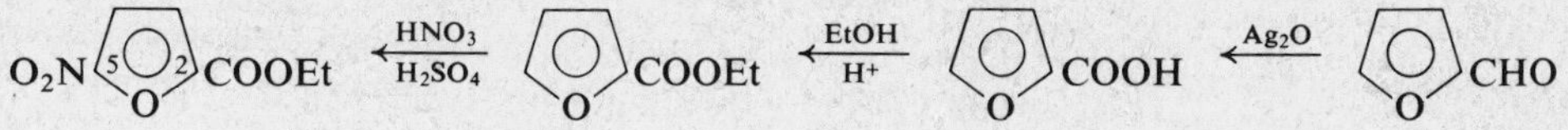

(e)

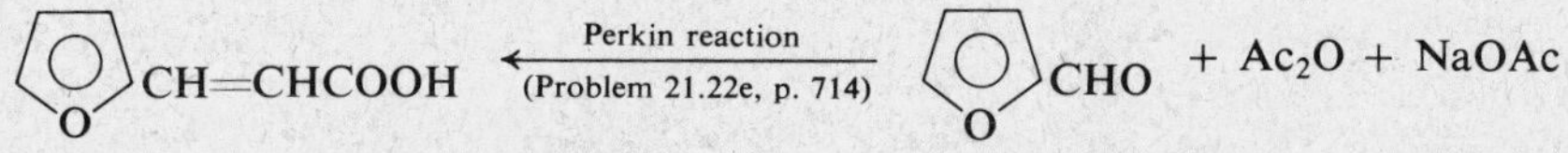

(f)

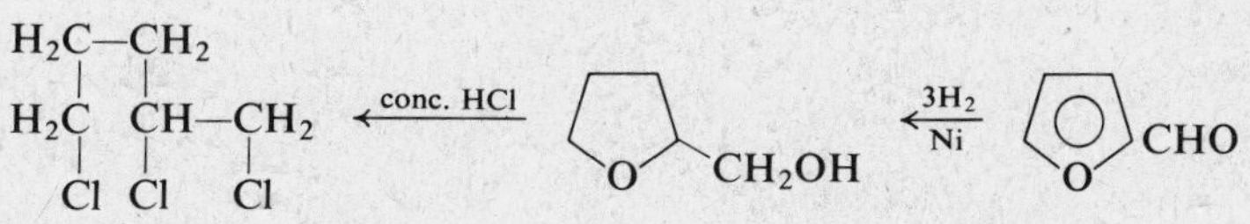

(g)

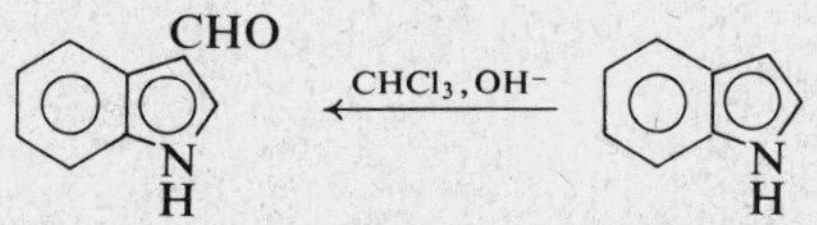

10. The parent ring systems—for consultation *after* you have worked this problem—are given on page 535 of this Study Guide.

(a) $C_7H_{12}O_4$ ethyl malonate $C_8H_{16}O_5N_2$
       + CH_4ON_2 urea $- C_4H_4O_3N_2$ **I**
       $C_8H_{16}O_5N_2$ $C_4H_{12}O_2$ *or* $2C_2H_5OH$ *lost:* $-\overset{\displaystyle O}{\underset{}{C}}-OEt + H_2N- \longrightarrow -\overset{\displaystyle O}{\underset{}{C}}-\overset{\displaystyle H}{\underset{}{N}}-$

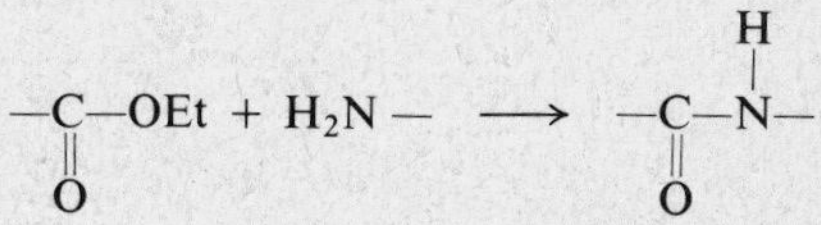

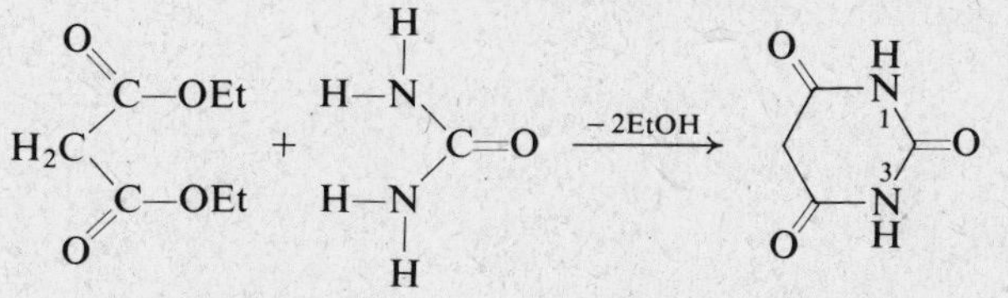

A pyrimidine
(*A 1,3-diazine*)

(b) $C_6H_{10}O_2$ 2,5-hexanedione $C_6H_{14}O_2N_2$
 $+ N_2H_4$ hydrazine $- C_6H_{10}N_2$ **J**
 ─────── $C_6H_{14}O_2N_2$ H_4O_2 *or* $2H_2O$ *lost:* $C=O + H_2N- \longrightarrow C=N-$

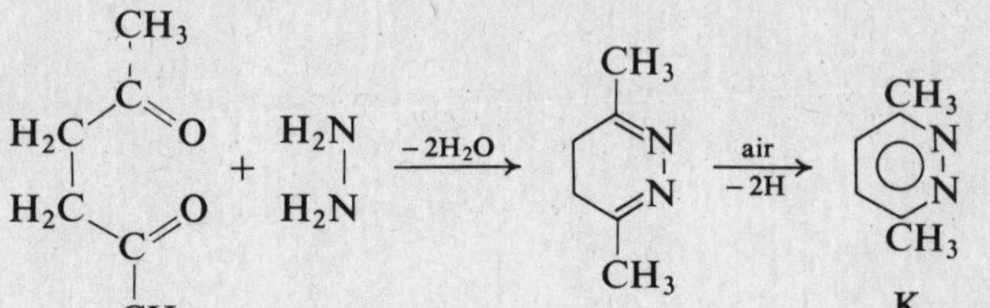

J

K

A pyridazine
(A 1,2- diazine)

(c) $C_5H_8O_2$ 2,5-pentanedione $C_5H_{12}O_2N_2$
 $+ N_2H_4$ hydrazine $- C_5H_8N_2$ **L**
 ─────── $C_5H_{12}O_2N_2$ H_4O_2 *or* $2H_2O$ *lost:* $C=O + H_2N- \longrightarrow C=N-$

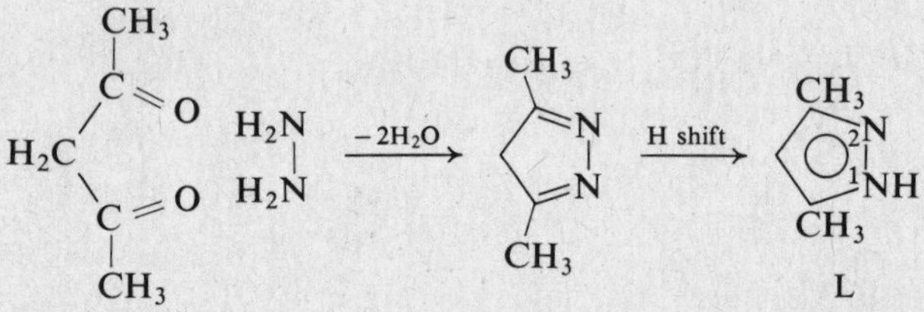

L

A pyrazole
(A 1,2-diazole)
(Compare pyrazole, p. 1002)

(d) $C_4H_6O_2$ butanedione $C_{10}H_{14}O_2N_2$
 $+ C_6H_8N_2$ o-$C_6H_4(NH_2)_2$ $- C_{10}H_{10}N_2$ **M**
 ─────── $C_{10}H_{14}O_2N_2$ H_4O_2 *or* $2H_2O$ *lost:* $C=O + H_2N- \longrightarrow C=N-$

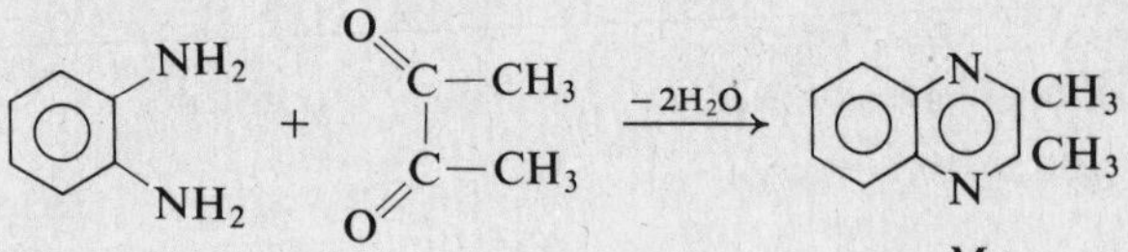

M

A quinoxaline

(e) $C_2H_6O_2$ ethylene glycol $C_3H_6O_3Cl_2$
 $+ COCl_2$ phosgene $- C_3H_4O_3$
 ─────── $C_3H_6O_3Cl_2$ H_2Cl_2 *or* $2HCl$ *lost:* $-OH + Cl-\underset{O}{C}- \longrightarrow -O-\underset{O}{C}-$

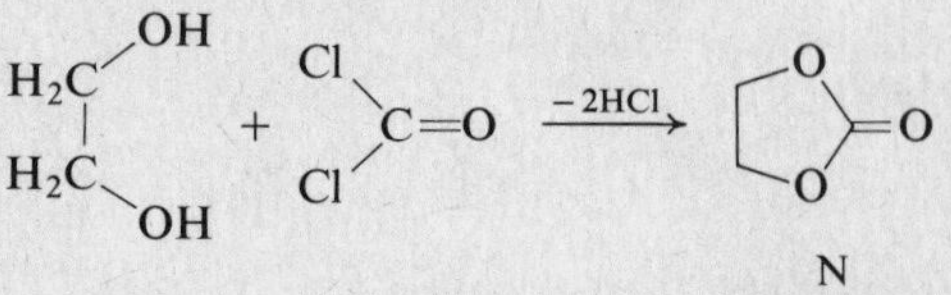

N

A 1,3-dioxolane
derivative

(f) $C_7H_7O_2N$ o-$NH_2C_6H_4COOH$ $C_9H_{10}O_4NCl$ $C_9H_9O_4N$ O
  $+C_2H_3O_2Cl$ chloroacetic acid $-C_9H_9O_4N$ $-C_8H_7ON$ P
  ───────   ──────────────   ───────────────────
  $C_9H_{10}O_4NCl$ HCl *lost:* $-NH_2 + Cl-R \longrightarrow -N-R$ CH_2O_3 *or* $CO_2 + H_2O$ *lost*
                                                              H

O

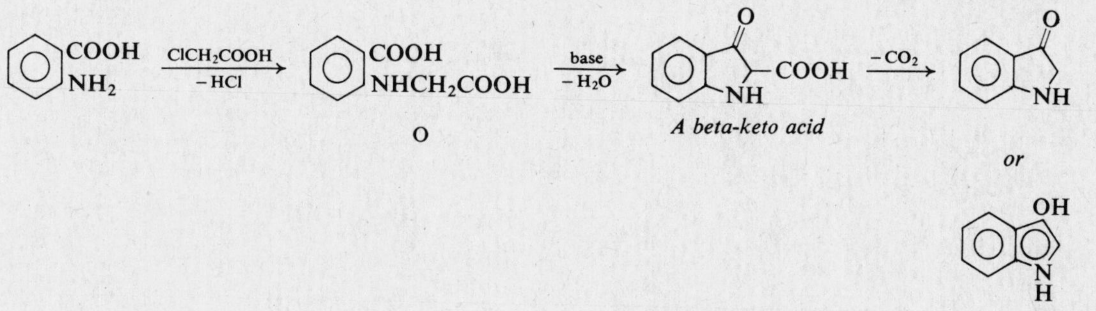

A beta-keto acid

or

P
3-Hydroxyindole

The ring closure is like a Dieckmann reaction (Problem 21.30), and is the same reaction that is involved in adipic acid $\longrightarrow$ cyclopentanone, page 667.

(g) $C_6H_{14}O_2N_2$ $2NH_2CH_2COCH_3$ $C_6H_{10}N_2$ Q
  $-C_6H_{10}N_2$ Q $-C_6H_8N_2$ R
  ─────────   ──────────
  H_4O_2 *or* $2H_2O$ *lost:* $\diagup C=O + H_2N- \longrightarrow \diagup C=N- $ 2H *lost:* oxidation

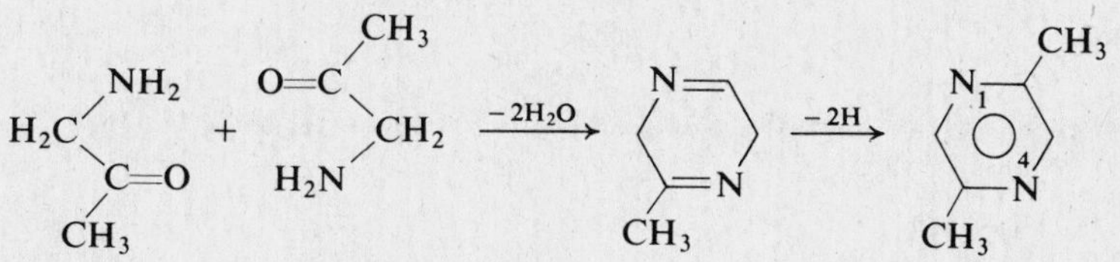

R
A pyrazine
(A 1,4-diazine)

(h) $C_2H_8N_2$ ethylenediamine $C_7H_{18}O_3N_2$
  $+ C_5H_{10}O_3$ ethyl carbonate $-C_3H_6ON_2$ S
  ──────────   ─────────────────
  $C_7H_{18}O_3N_2$ $C_4H_{12}O_2$ *or* 2EtOH *lost:* $-NH_2 + EtO-C- \longrightarrow -NH-C-$
                                                              ‖   ‖
                                                              O   O

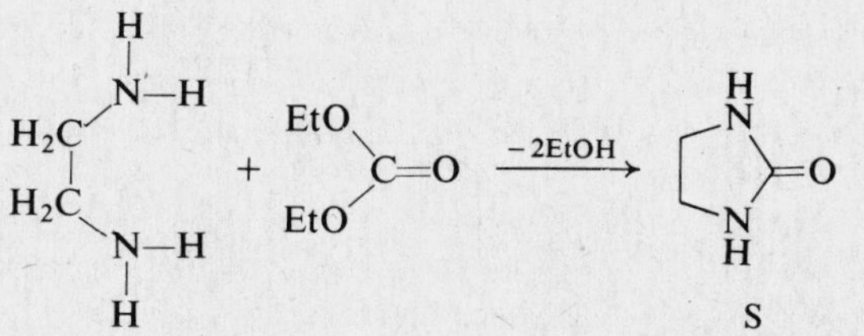

S
*A derivative of
imidazole*

(i) $C_6H_8N_2$ $o\text{-}C_6H_4(NH_2)_2$ $C_8H_{12}O_2N_2$
      $+C_2H_4O_2$ HOAc $-C_8H_8N_2$ T
      $\overline{C_8H_{12}O_2N_2}$ $\overline{H_4O_2 \text{ or } 2H_2O \text{ lost}}$

and $-NH_2 + O{=}C\diagdown \longrightarrow -N{=}C\diagdown$

$-NH_2 + HO{-}\underset{O}{C}{\diagdown} \longrightarrow -NH{-}\underset{O}{C}{\diagdown}$

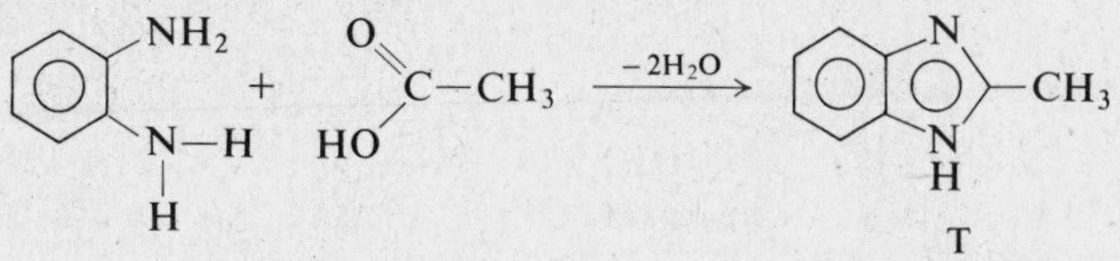

A benzimidazole

(j) $C_9H_{11}O_2N$ $o\text{-}NH_2C_6H_4COOEt$ $C_{16}H_{23}O_6N$
      $+C_7H_{12}O_4$ malonic ester $-C_{14}H_{17}O_5N$
      $\overline{C_{16}H_{23}O_6N}$ $\overline{C_2H_6O \text{ or } EtOH \text{ lost:}}$ $-NH_2 + EtO{-}\underset{O}{C}{-} \longrightarrow -NH{-}\underset{O}{C}{-}$

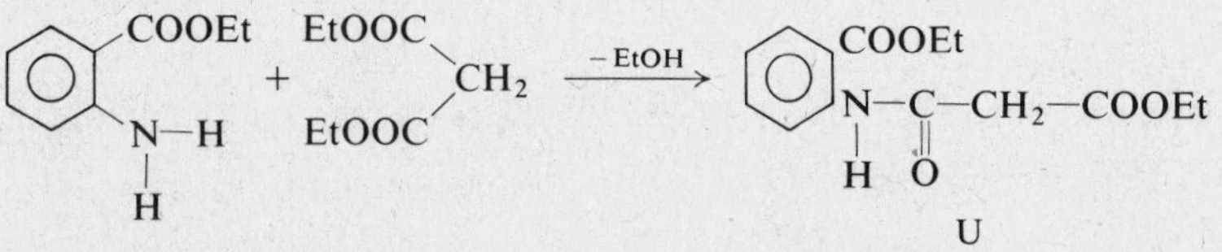

Reaction at $-NH_2$

U

Insoluble in acid

$C_{14}H_{17}O_5N$ U
$\underline{-C_{12}H_{11}O_4N}$ V
$C_2H_6O \text{ or } EtOH \text{ lost:}$ $-\underset{O}{C}{-}OEt + H{-}\underset{H}{\overset{|}{C}}{-}\underset{O}{C}{-} \longrightarrow -\underset{O}{C}{-}\underset{H}{\overset{|}{C}}{-}\underset{O}{C}{-}$

This step appears to be an intramolecular Claisen condensation (Secs. 21.11 and 21.12).

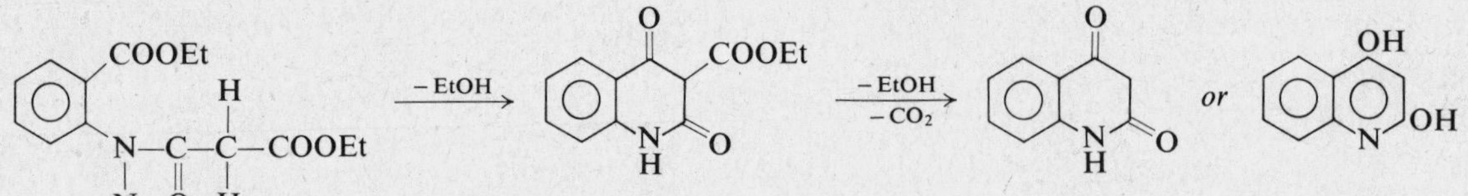

A β-keto ester   V   Keto   Enol
                                                   W
                                        2,4-Dihydroxyquinoline

(k)

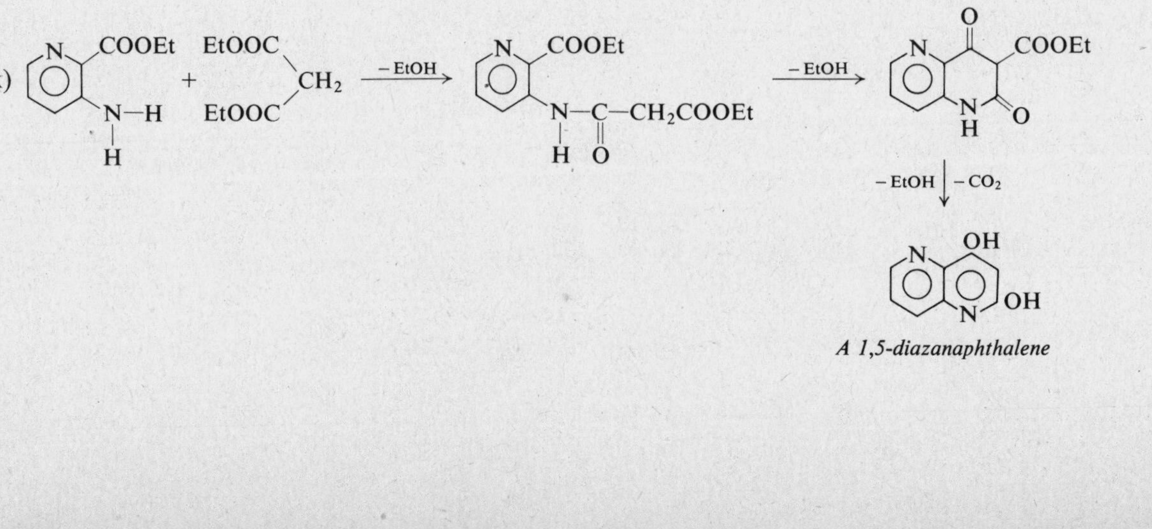

A 1,5-diazanaphthalene

(l) $C_{15}H_{12}O$ PhCOCH=CHPh $C_{16}H_{13}ON$ X $C_{17}H_{19}O_3N$

    $+ HCN$ $+ CH_4O$ $- C_{17}H_{16}O_3$

      $C_{16}H_{13}ON$ = product X $+ H_2O$ NH_3 *lost:* hydrolysis of —CN

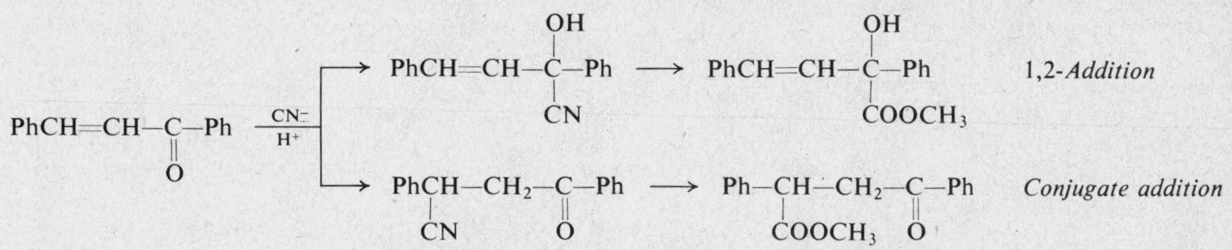

          Simple addition $C_{17}H_{19}O_3N$

X *is one of these*   Y *is one of these*

    $C_{17}H_{16}O_3$ Y $C_{23}H_{24}O_3N_2$

    $+ C_6H_8N_2$ $PhNHNH_2$ $- C_{22}H_{18}ON_2$

     $C_{23}H_{24}O_3N_2$ $CH_6O_2 = CH_3OH$ *and* H_2O *lost*

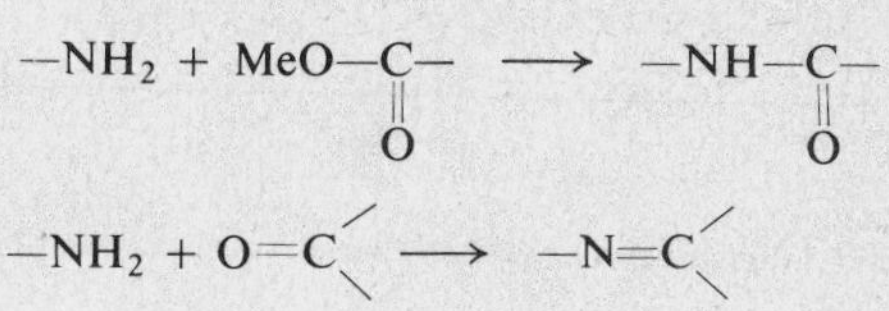

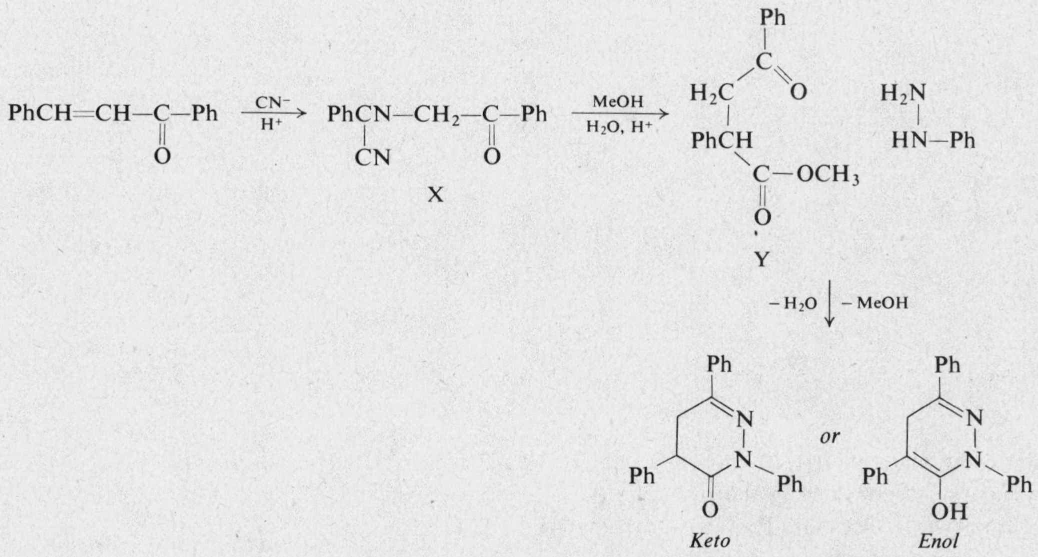

$-H_2O \downarrow -MeOH$

       Keto   *Enol*

            Z

    6-Hydroxy-1,3,5-triphenyl-
    1,4-dihydro-1,2-diazine

(m) $C_3H_4O_2$ acrylic acid $C_3H_8O_2N_2$ AA

    $+ N_2H_4$ hydrazine $- C_3H_6ON_2$ BB

     $C_3H_8O_2N_2$ = AA H_2O *lost:* —C—OH + H_2N— $\longrightarrow$ —C—NH—

      Simple addition   O   O

A pyrazolidione

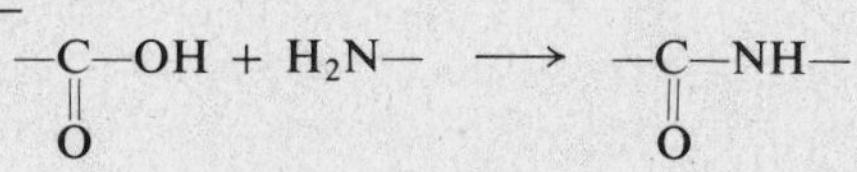

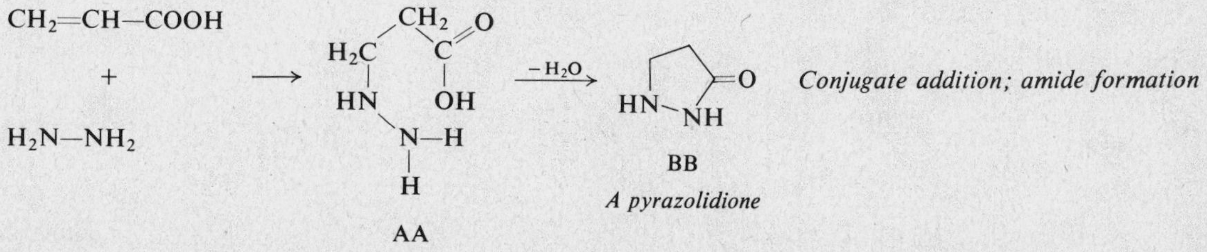

(n)

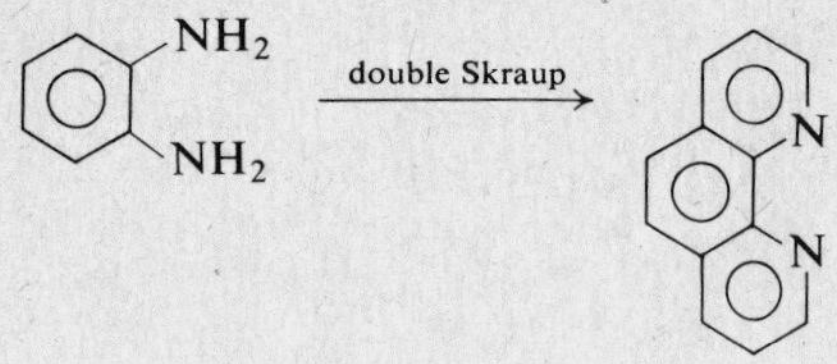

CC

4,5-Diazaphenanthrene
(*Problem 31.23c*)

(o)

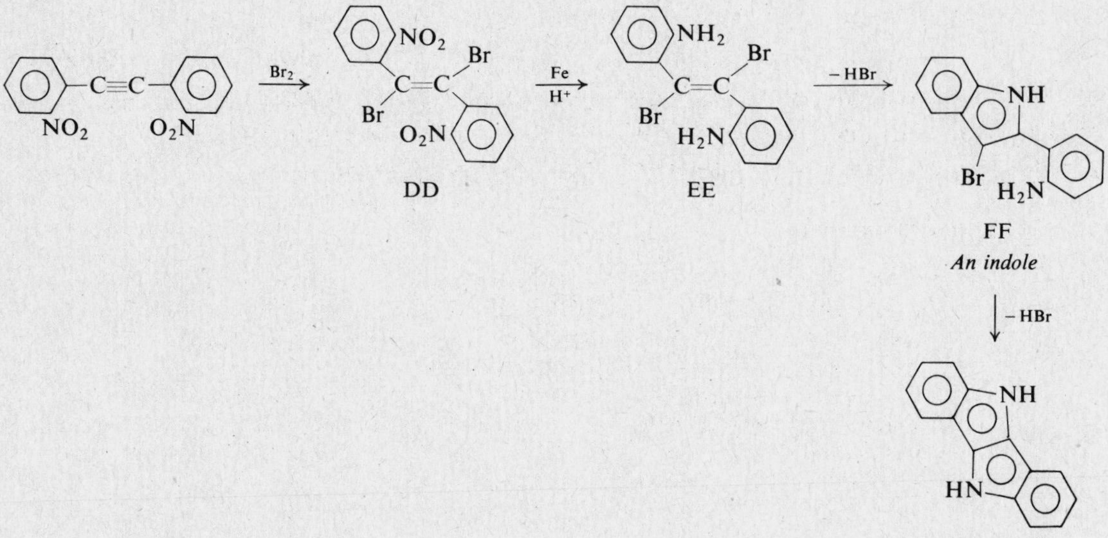

DD   EE   FF
An indole

GG

*Two indole units
fused 2,3 to 3′,2′*

(p) The starting material has 10 carbons, and so has product HH. The reaction is thus intramolecular, and considering the reactants (halide and phenyllithium), we suspect it of involving an aryne. (Compare Problem 21, page 844.)

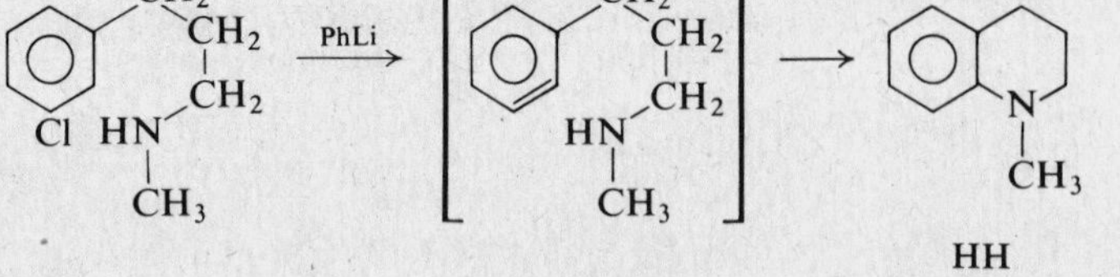

An aryne   HH
                      A tetrahydroquinoline

(q)

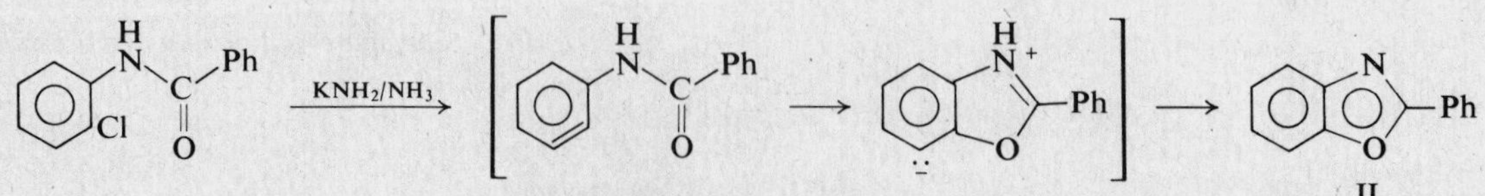

An aryne   II

A benzoxazole

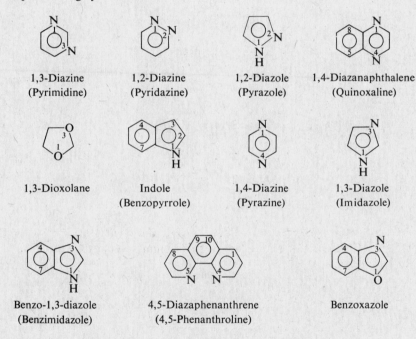

(r)

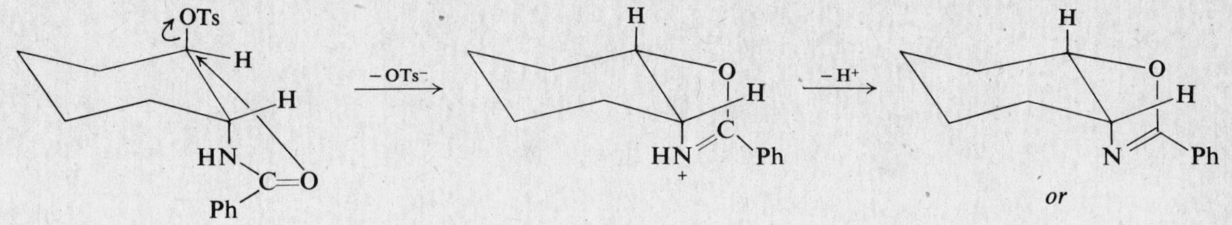

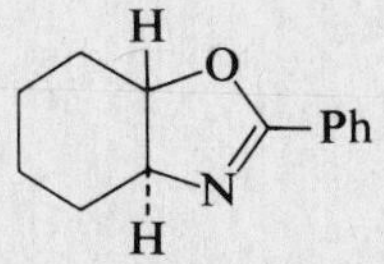

JJ
An oxazoline

(s) *cis*-I cannot get into the diaxial (*anti*) conformation that is needed for intramolecular displacement of tosylate anion. Presumably it reacts by (S_N1-like) slow loss of OTs⁻ to give a cation, which then reacts intramolecularly. (See Sec. 28.11.)

The parent ring systems are:

1,3-Diazine
(Pyrimidine)

1,2-Diazine
(Pyridazine)

1,2-Diazole
(Pyrazole)

1,4-Diazanaphthalene
(Quinoxaline)

1,3-Dioxolane

Indole
(Benzopyrrole)

1,4-Diazine
(Pyrazine)

1,3-Diazole
(Imidazole)

Benzo-1,3-diazole
(Benzimidazole)

4,5-Diazaphenanthrene
(4,5-Phenanthroline)

Benzoxazole

For definitive nomenclature rules, see The Ring Index.

11.

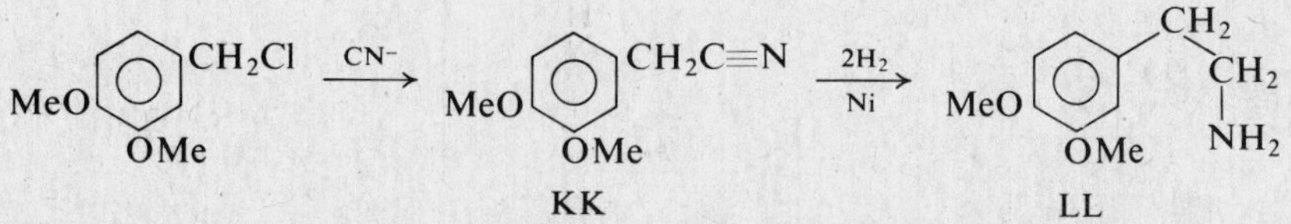

KK

LL
An amine

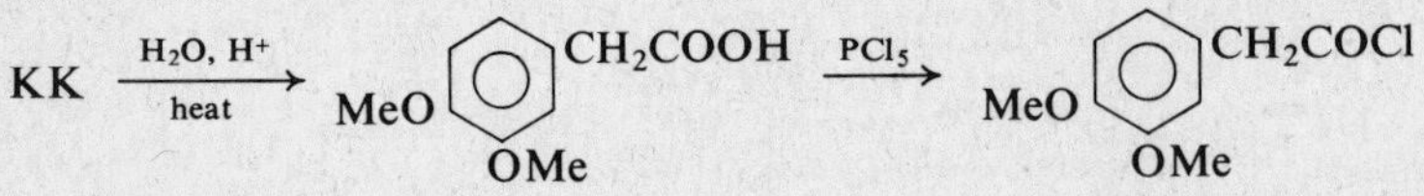

$$KK \xrightarrow[\text{heat}]{H_2O, H^+} \quad MeO-\underset{OMe}{\bigodot}-CH_2COOH \xrightarrow{PCl_5} \quad MeO-\underset{OMe}{\bigodot}-CH_2COCl$$

MM

NN

An acid chloride

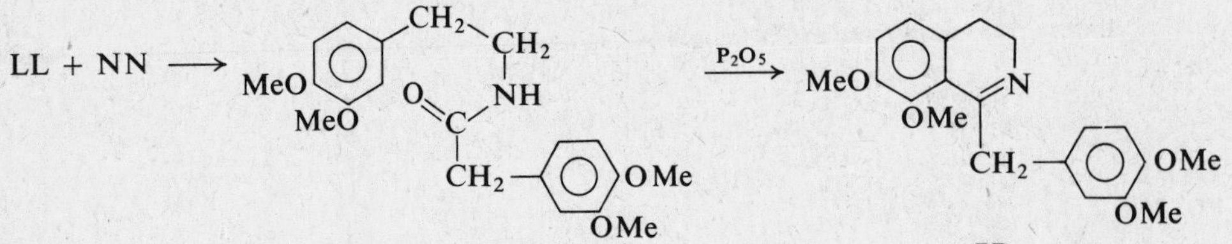

$$LL + NN \longrightarrow$$

MeO — ... — CH₂—CH₂ / NH / O=C / CH₂ — ... — OMe, OMe

$$\xrightarrow{P_2O_5}$$

MeO — ... — N / OMe / CH₂ — ... — OMe, OMe

PP

A dihydroisoquinoline

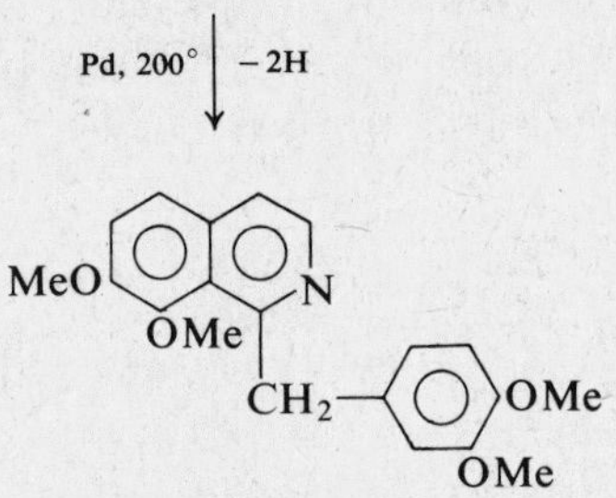

Pd, 200° | −2H

MeO — ... — N / OMe / CH₂ — ... — OMe, OMe

Papaverine

An opium alkaloid

12. $CH_2-CH_2 + Et_2NH \longrightarrow Et_2NCH_2CH_2OH \xrightarrow{SOCl_2} Et_2NCH_2CH_2Cl$

(epoxide with O)

QQ

RR

$$RR + [CH_3COCHCOOEt]^- \longrightarrow CH_3-\underset{O}{C}-\underset{CH_2CH_2NEt_2}{CH}-COOEt \xrightarrow[\text{warm}]{H_2O, H^+} CH_3CCH_2CH_2CH_2NEt_2$$

SS

TT (with O)

$$H_2 \bigg| Ni$$

$$CH_3CHCH_2CH_2CH_2NEt_2 \xleftarrow{HBr} CH_3CHCH_2CH_2CH_2NEt_2$$

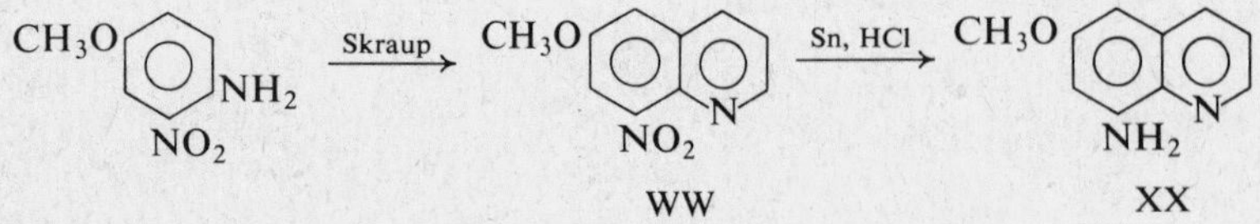

Br (below first)

OH (below second)

VV

UU

$CH_3O-\underset{NO_2}{\bigodot}-NH_2 \xrightarrow{Skraup} CH_3O-\underset{NO_2}{\bigodot\bigodot} \xrightarrow{Sn, HCl} CH_3O-\underset{NH_2}{\bigodot\bigodot}$

WW

XX

$$VV + XX \longrightarrow$$

CH₃O — ... (quinoline ring)

H—N

CH / CH₃ — CH₂ — CH₂ — CH₂ — NEt₂

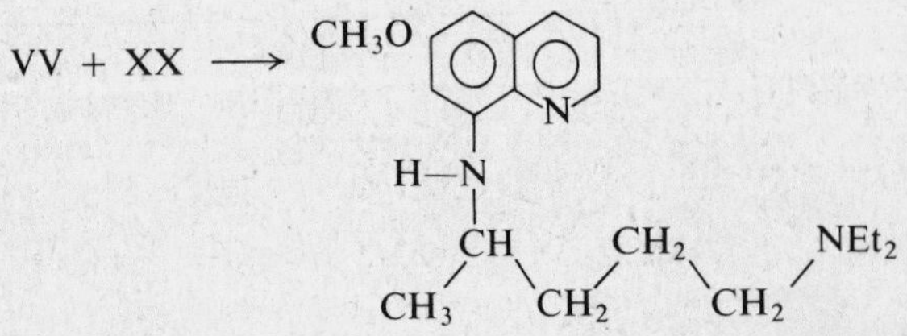

Plasmochin

13.

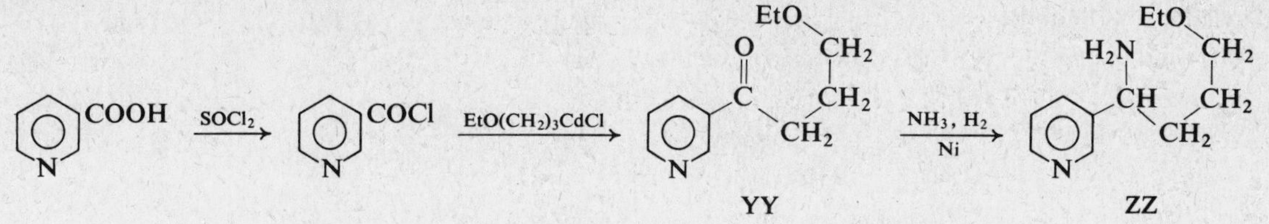

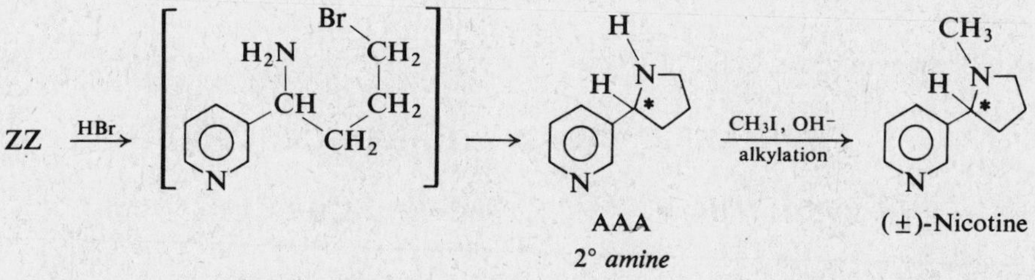

$$\text{(pyridine)COOH} \xrightarrow{SOCl_2} \text{(pyridine)COCl} \xrightarrow{EtO(CH_2)_3CdCl} \text{YY} \xrightarrow[Ni]{NH_3, H_2} \text{ZZ}$$

$$\text{ZZ} \xrightarrow{HBr} \left[\quad \right] \longrightarrow \text{AAA} \atop \text{2° amine} \xrightarrow[\text{alkylation}]{CH_3I, OH^-} (\pm)\text{-Nicotine}$$

BBB and CCC are diastereomeric salts.

14. (a)

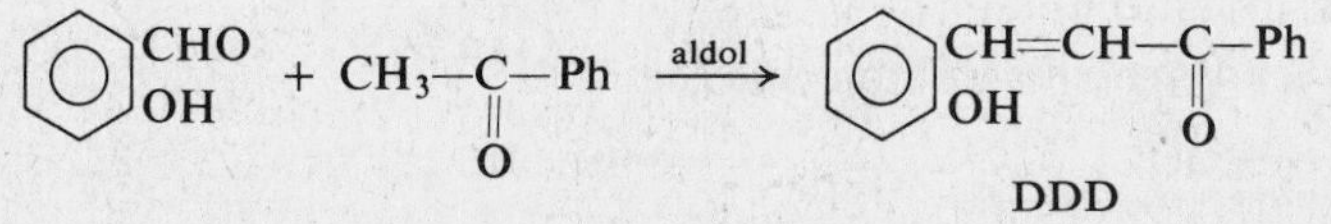

$$\text{(phenol)} {CHO \atop OH} + CH_3 - \underset{O}{\overset{}{C}} - Ph \xrightarrow{aldol} \text{DDD}$$

(b)

Flavylium chloride ← $-H_2O$ ← (intermediate, Cl^-) ← HCl ← DDD

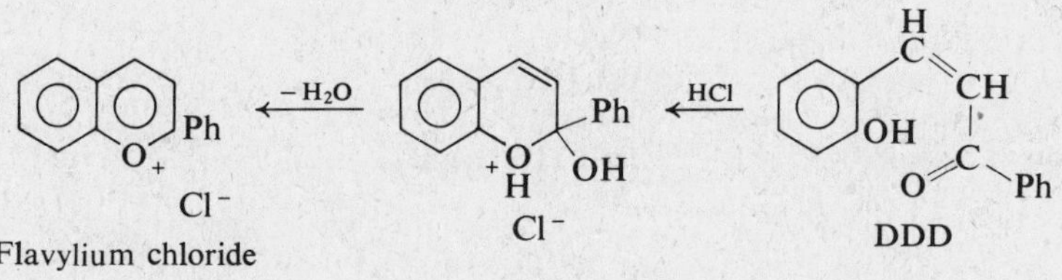

(c) Oxygen contributes a pair of electrons to complete the aromatic sextet, thus giving a system analogous to that of naphthalene.

15. Tropinic acid, $C_8H_{13}O_4N$, has a m.w. of 187. Since its N.E. is 94 ± 1 ($= $ m.w./2), it must be a diacid. It is a 3° amine, contains no easily oxidizable functions, and is saturated. Chemical arithmetic,

$$
\begin{array}{ll}
C_6H_{15}N & \text{sat'd open chain amine} \\
+C_2O_4 & \text{two —COO— groups} \\
\hline
C_8H_{15}O_4N &
\end{array}
\qquad
\begin{array}{l}
C_8H_{15}O_4N \\
-C_8H_{13}O_4N \quad \text{tropinic acid} \\
\hline
\text{2H missing: } \textit{means} \text{ one ring}
\end{array}
$$

indicates that tropinic acid contains *one ring*.

Let us turn to the exhaustive methylation data:

$$
\begin{array}{l}
C_9H_{16}O_4NI \quad \text{EEE} \\
-C_8H_{13}O_4N \quad \text{tropinic acid} \\
\hline
CH_3I \textit{ means} \text{ one Me introduced}
\end{array}
$$

$$\text{EEE} \xrightarrow{-H_2O} \text{FFF } \textit{with no loss of N}$$

Tropinic acid must be a 3° amine. The N must be part of the ring, since otherwise it would have been lost as $MeNR_2$.

A repetition of the procedure does split out Me_3N, and leaves an unsaturated molecule,

HHH, which undoubtedly contains the two —COOH's.   Only one carbon is lost ($C_8 \longrightarrow C_7$), clearly indicating a —N—CH$_3$ unit in tropinic acid.

$$C_7H_{12}O_4 \quad \text{sat'd open-chain diacid}$$
$$-C_7H_8O_4 \quad \text{HHH}$$

$$\overline{\text{4H missing: \textit{means} two double bonds}}$$

HHH is a doubly unsaturated diacid; hydrogenation to heptanedioic acid shows that the carbon chain is unbranched.

HHH thus can be one of several acids,

$$HOOC—CH=CH—CH=CH—CH_2—COOH \quad HOOC—CH=CH—CH_2—CH=CH—COOH$$

$$HOOC—CH=C=CH—CH_2—CH_2—COOH \quad HOOC—CH_2—CH=C=CH—CH_2—COOH$$

although the allenic compounds are unlikely, and we are discounting the likelihood of an acetylenic structure.   (From what size ring could an alkyne structure, or an allenic structure, arise through exhaustive methylation?)

(a) At this point, depending upon which structure for HHH and which point of cleavage of the ring we postulate, possible structures for tropinic acid include these:

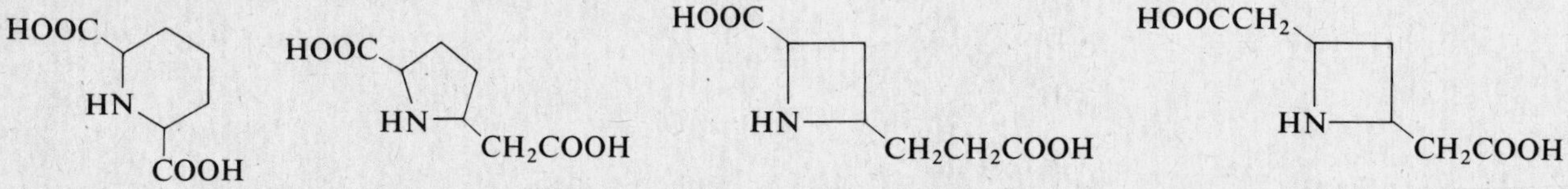

(Starting with each of these, give possible structures for FFF and HHH.)

(b)

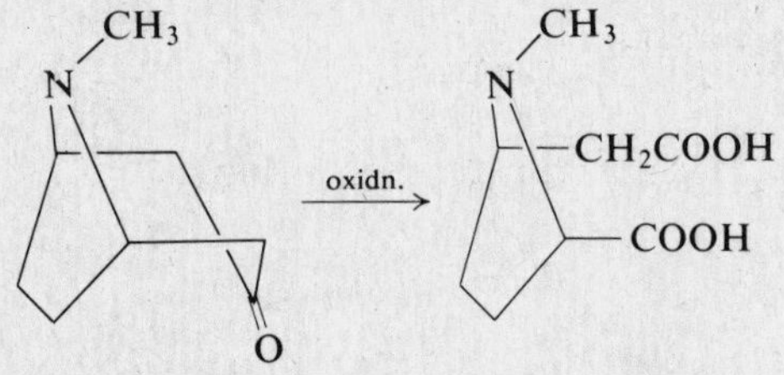

Tropinone   Tropinic acid

The various other structures are:

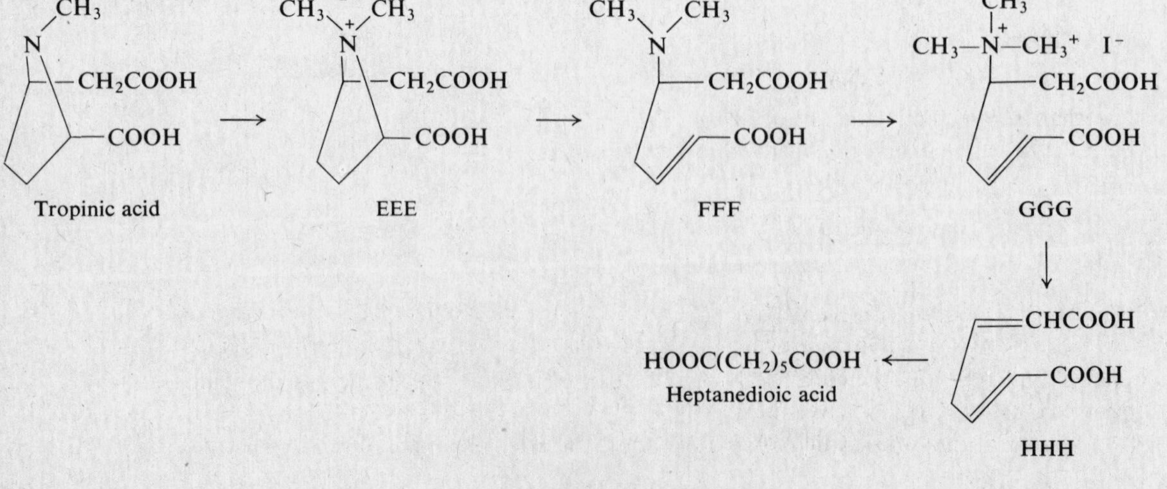

Tropinic acid   EEE   FFF   GGG

HOOC(CH$_2$)$_5$COOH $\longleftarrow$
Heptanedioic acid

HHH

16.

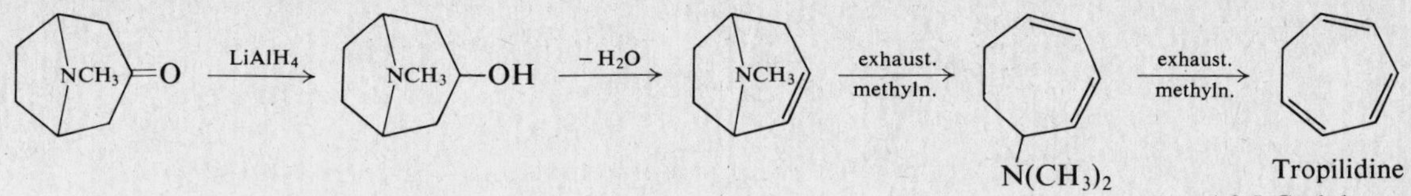

Tropilidine
1,3,5-Cycloheptatriene

17.

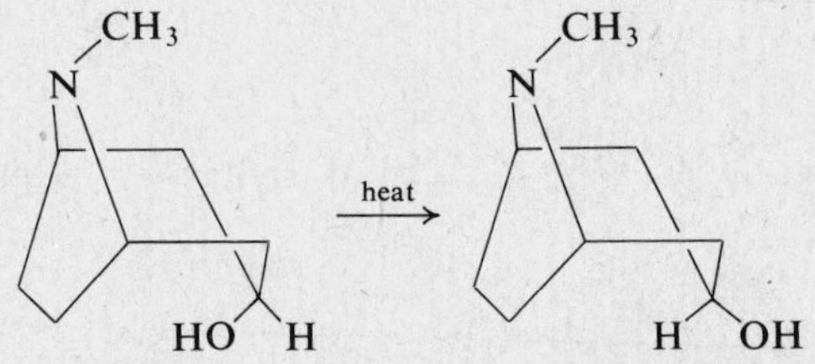

Tropine   Pseudotropine

Pseudotropine is the more stable, and hence is probably the one with the equatorial —OH.

18. (a)

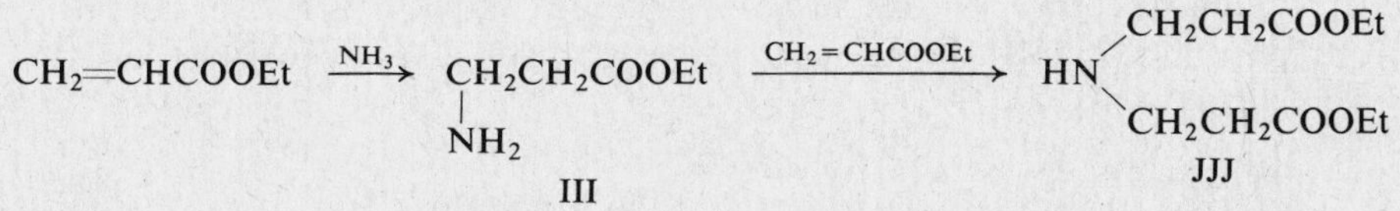

III   JJJ

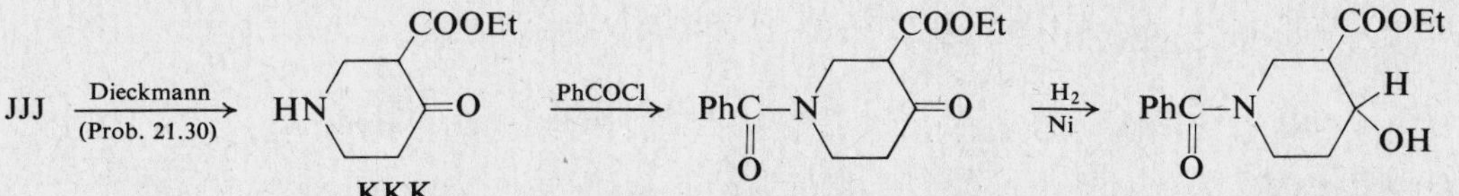

JJJ $\xrightarrow[\text{(Prob. 21.30)}]{\text{Dieckmann}}$

KKK   LLL   MMM

MMM $\xrightarrow[\text{H}_2\text{O}]{\text{H}^+}$

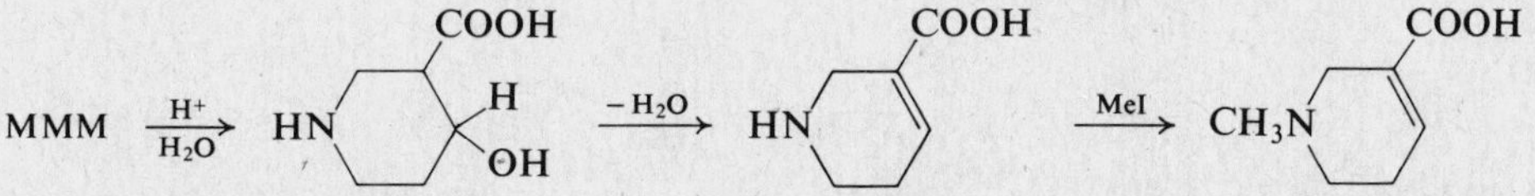

NNN
Guvacine   Arecaidine

(b) Guvacine $\xrightarrow{-4\text{H}}$

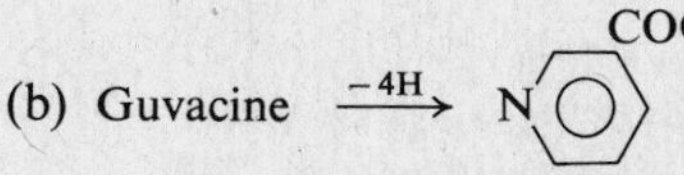

Nicotinic acid

19. Phenol (Sec. 32.7), furan (Problem 31.4, page 1009), and pyrrole (Problem 5, page 1022) contain rings reactive enough to react with formaldehyde: first, with the introduction of —CH$_2$OH; and then, with the connecting of rings by —CH$_2$— linkages.

Here we have another reactive aromatic ring, and here we have, if not formaldehyde, another carbonyl compound. Chemical arithmetic,

$$C_8H_8S_2 \quad \text{2 thiophene} \qquad\qquad C_{14}H_{20}OS_2$$
$$\underline{+ C_6H_{12}O \quad \text{3-hexanone}} \qquad \underline{- C_{14}H_{18}S_2} \quad OOO$$
$$C_{14}H_{20}OS_2 \qquad\qquad\qquad H_2O \text{ missing}$$

indicates that two moles of thiophene combine with one of ketone and lose one H_2O.   We postulate, therefore:

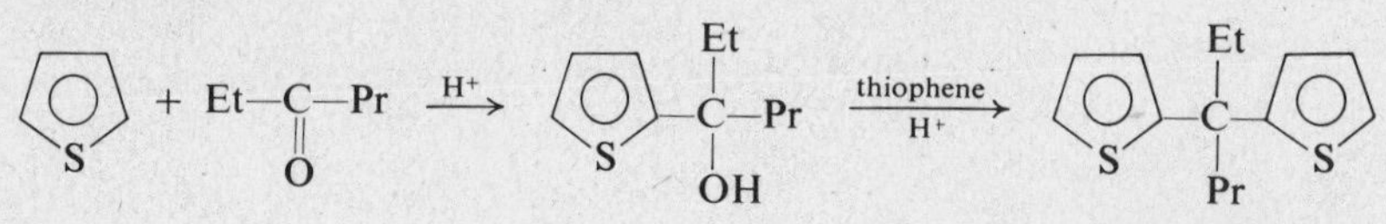

Next, Friedel-Crafts acylation at the usual 2-position, followed by Wolff-Kishner reduction (Sec. 19.10):

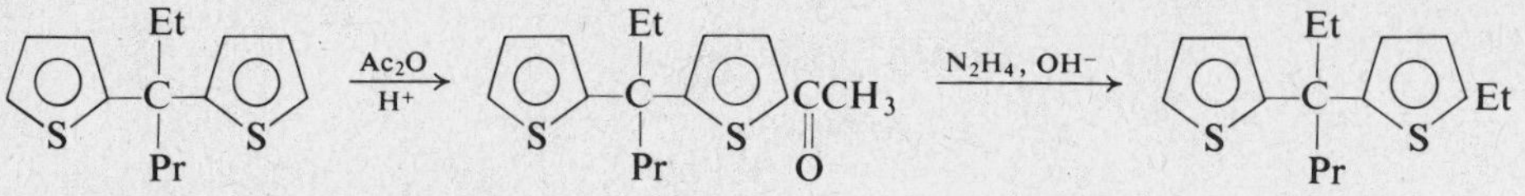

We have not previously encountered an amide as an acylating agent (although, of course, an amide acylates water, in hydrolysis), but the hint plus an atom count leaves us with only a decision about the position of formylation, and we pick the only 2-position left open. Oxidation with Tollens' reagent gives SSS; the acidic handle permits resolution (see Problem 20.9, page 670).

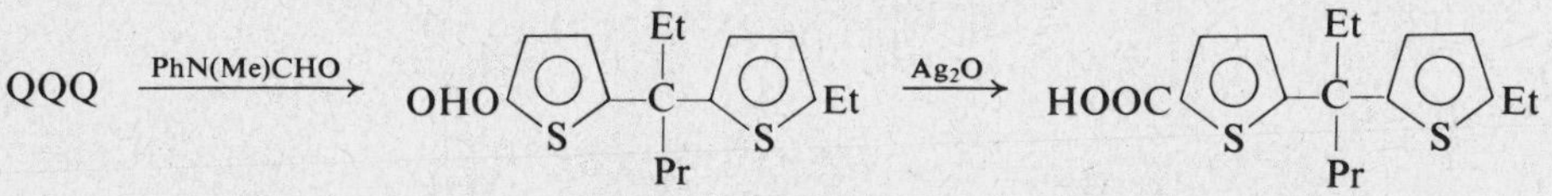

Resolved into enantiomers

Following decarboxylation there is, finally, hydrogenation, which removes sulfur, adds hydrogen, and clearly opens the rings.   We are left with a saturated open-chain alkane (C_nH_{2n+2}).

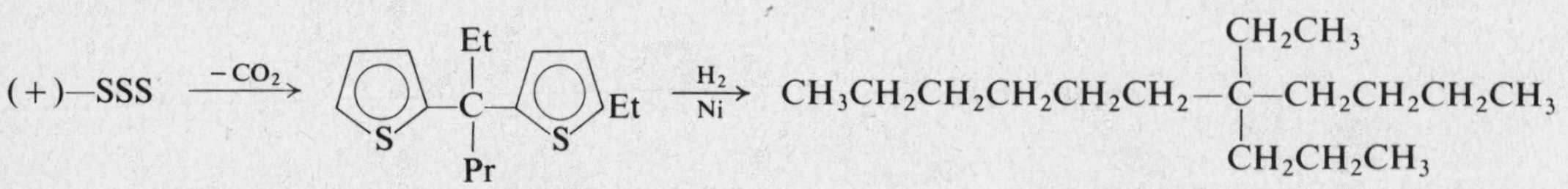

UUU
Ethyl-*n*-propyl-*n*-butyl-*n*-hexylmethane
Optically inactive

UUU is a single enantiomer of a chiral compound and, by all rights, should be optically active.   But, as we learned earlier (Sec. 4.13), chirality does not always lead to *measurable* optical activity.   This particular compound, it has been estimated, should have only the undetectable specific rotation of 0.00001°.

(H. Wynberg *et al.*, "The Optical Activity of Butylethylhexylpropylmethane," J. Am. Chem. Soc., **87**, 2635 (1965).)

In the last step of this synthesis, the heterocyclic rings are opened and even the heteroatom, sulfur, is removed.   Yet the presence of these rings was vital to each preceding step, providing the regiospecificity necessary for the building of the molecule that was wanted. (For an overview of the potential of the heterocyclic field, see A. R. Katritzky, "Heterocycles," Chem. Eng. News, April 13, 1970, p. 80.)

540

20. (a) "Pyrrole" nitrogen—the ring N with the —H attached—contributes two π electrons, and the other atoms (including the "pyridine" N) contribute one each, to give the aromatic sextet (Sec. 31.2).

(b)

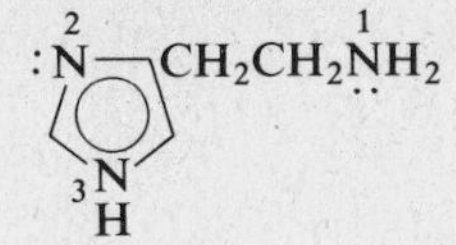

Histamine

$$\text{aliphatic} -NH_2 > \text{"pyridine" N} > \text{"pyrrole" N}$$

The unshared pair is in, respectively, an sp^3 orbital, an sp^2 orbital, and the π cloud. (See Secs. 31.2 and 31.11.)

21. (a)

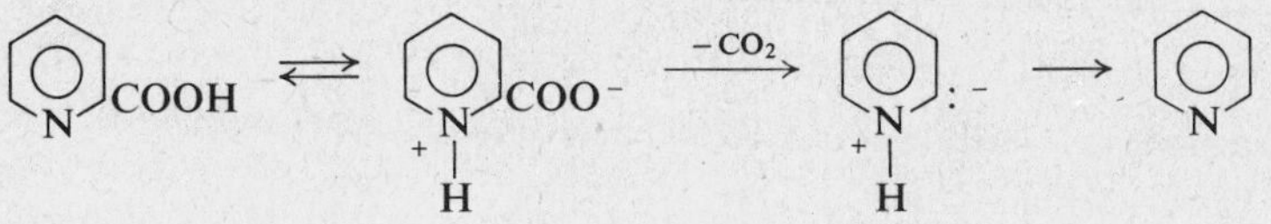

The dipolar ion (VI) loses CO_2 to form a carbanion (VII) stabilized by the positive charge on nitrogen. Either acid or base decreases the concentration of dipolar ion, and slows down reaction. The N-methyl derivative necessarily exists entirely as dipolar ion, and hence reacts faster.

Like other carbanions, VII can add to a carbonyl group, and in the presence of the ketone is (partially) trapped.

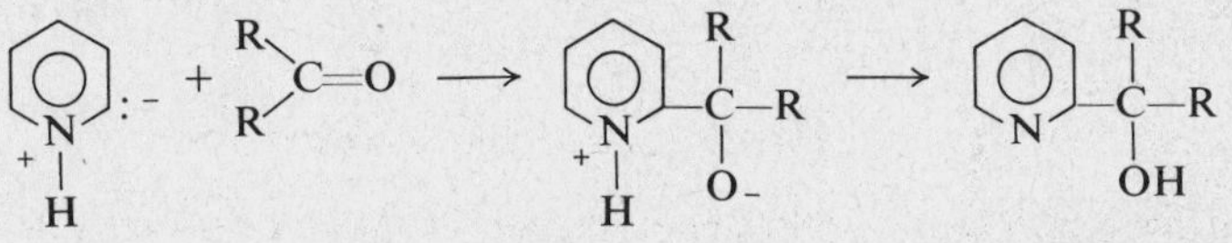

(b) Carbanion VII and its isomers are stabilized by the inductive effect of the positive charge on nitrogen, which becomes weaker with distance. The 2- and 4-pyridinacetic acids give benzylic-like carbanions (VIII) stabilized by resonance involving structures like IX. Such resonance is not possible for the "*meta*" 3-pyridinacetic acid.

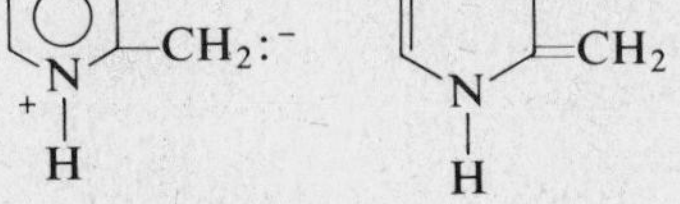

VIII   IX

Chapter 32

Macromolecules. Polymers and Polymerization

32.1 (a) Amide (see Sec. 32.7). $HOOC(CH_2)_4COOH + H_2N(CH_2)_6NH_2$
    Adipic acid   Hexamethylenediamine

(b) Amide. $HOOC(CH_2)_5NH_2$ (Monomer is actually caprolactam, Problem 12, p. 1051.)
    ε-Aminocaproic acid

(c) Ether. $H_2C—CH_2$, ethylene oxide
         \\O/

(d) Chloroalkene. $CH_2=C—CH=CH_2$, chloroprene
                    |
                    Cl

(e) Chloroalkane. $CH_2=C—Cl$, vinylidene chloride
                    |
                    Cl

32.2 (a) Amide. $^+H_3NCHRCOO^-$, amino acid

(b) Ester. H_3PO_4 +
     Phosphoric
     acid

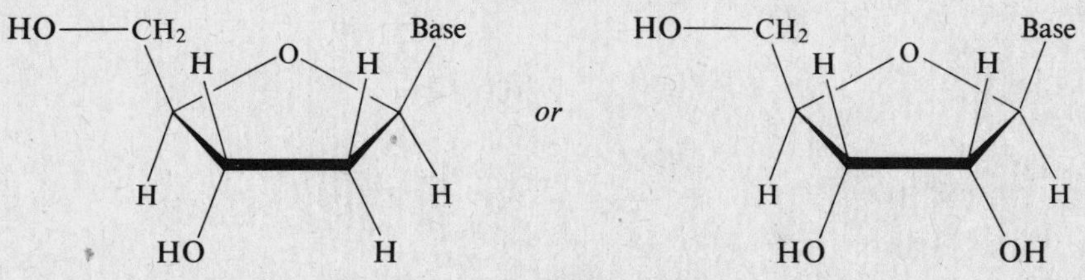

A sugar

(c) Acetal.

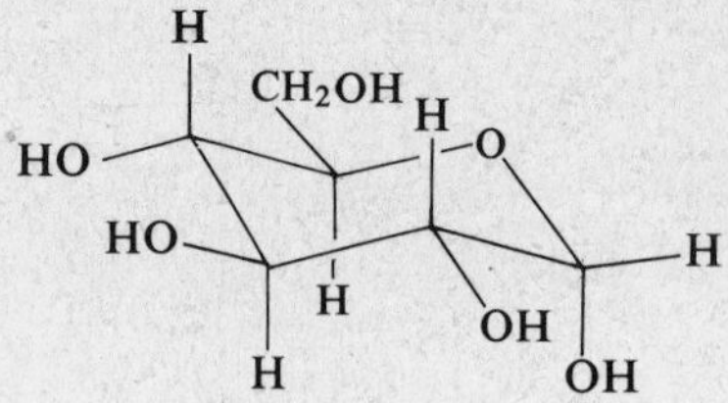

α-D-Glucopyranose

(d) Acetal.

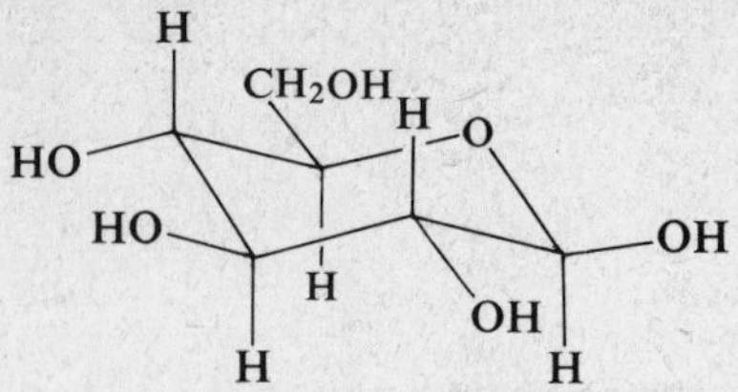

β-D-Glucopyranose

32.3 Rad· + CH_2=CH—CH=CH_2 ⟶ Rad—CH_2—CH⸺CH⸺CH_2

I

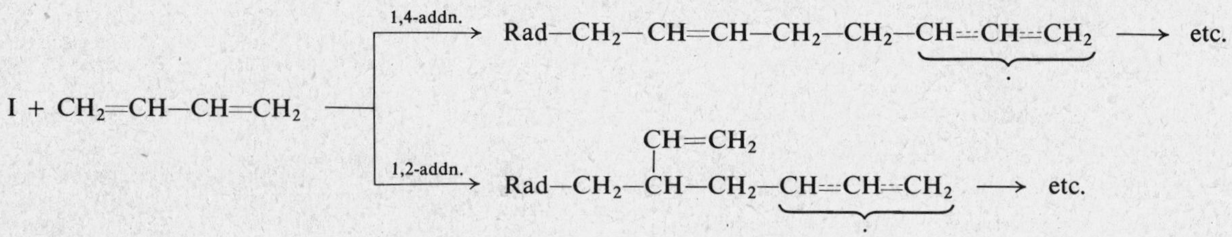

I + CH_2=CH—CH=CH_2 →

1,4-addn. → Rad—CH_2—CH=CH—CH_2—CH_2—CH⸺CH⸺CH_2 ⟶ etc.

1,2-addn. → Rad—CH_2—CH—CH_2—CH⸺CH⸺CH_2 ⟶ etc.

with CH=CH_2 substituent

32.4 Combination.

32.5 The effectiveness of a chain-transfer agent depends upon the ease of abstraction of an atom from it; this depends on, among other things, the stability of the radical being formed.

(a) H: aryl, 1°, 1° benzylic, 2° benzylic, 3° benzylic.

(b) H: 2°, allylic.

(c) X: C—Br is weaker than C—Cl.

32.6 (a) The chain-transfer agent can be another molecule of polymer. For example:

⁓CH_2—CH· + ⁓CH_2—CH—CH_2—CH⁓ ⟶ ⁓CH_2—CH_2 + ⁓CH_2—Ċ—CH_2—CH⁓

with G substituents

Growing radical   Polymer

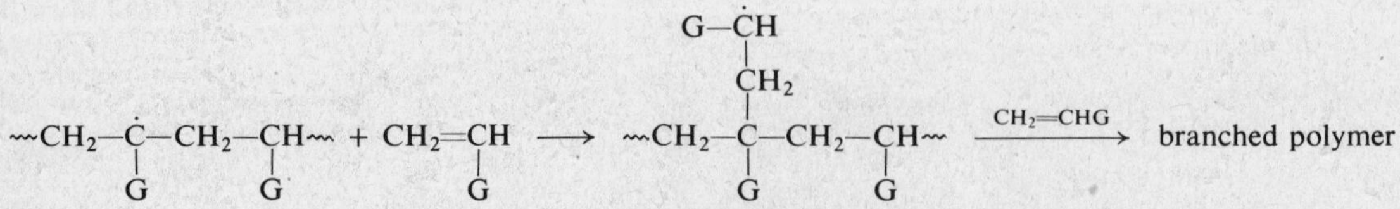

⁓CH_2—Ċ—CH_2—CH⁓ + CH_2=CH ⟶ ⁓CH_2—C—CH_2—CH⁓ $\xrightarrow{CH_2=CHG}$ branched polymer

with G substituents and G—CH—CH_2 branch

Monomer   Branched growing radical

544

(b) The chain-transfer agent can be the growing polymer molecule itself: the growing end abstracts ("bites") hydrogen from a position four or five carbons back along the chain. For example:

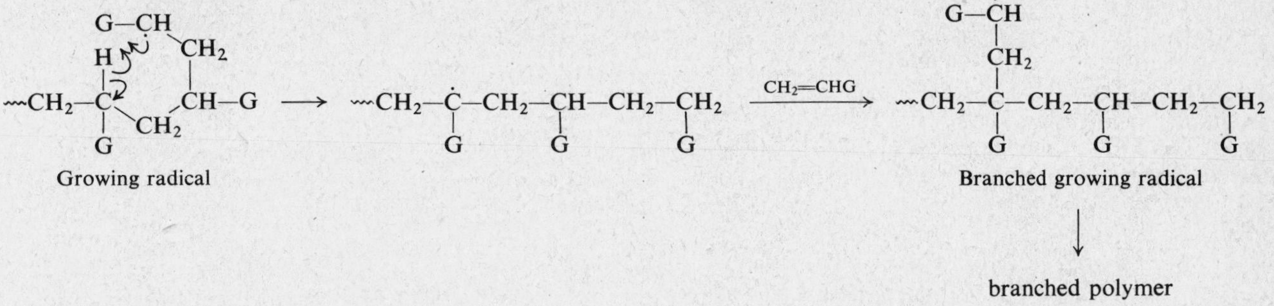

Growing radical   Branched growing radical

↓

branched polymer

32.7 (a) In acrylonitrile, the —CN group is electron-withdrawing. In butadiene, the —CH=CH$_2$ group releases electrons through resonance.

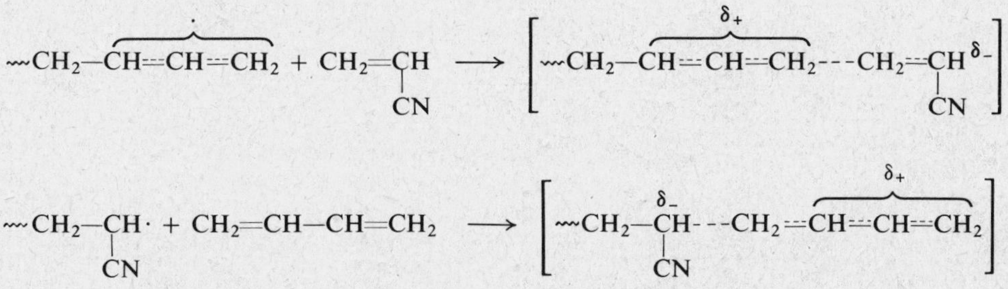

(b) Through delocalization of the odd electron in the transition state, butadiene is more reactive toward *any* radical than acrylonitrile is.

32.8 (a) The polybutadiene still contains double bonds and hence easily abstracted allylic hydrogens. The allylic free radicals thus formed add to styrene to start a reaction sequence that grafts polystyrene onto the polybutadiene chain. The process resembles the branching in Problem 32.6(a).

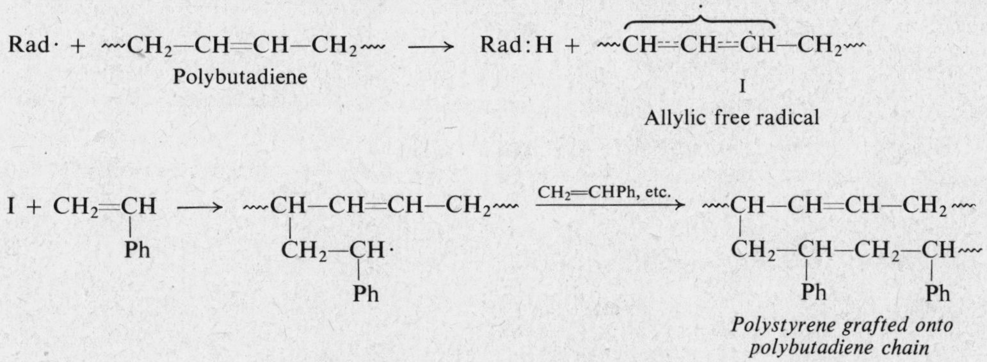

Polybutadiene   I

Allylic free radical

Polystyrene grafted onto
polybutadiene chain

(b) Similar to part (a), with hydrogen being abstracted from the carbon carrying —Cl.

PhCOO· + ⌇CH₂—CH—CH₂—CH⌇ ⟶ PhCOOH + ⌇CH₂—Ċ—CH₂—CH⌇
                  | | | |
                  Cl   Cl   Cl   Cl

II

$$\text{II} + CH_2=C\underset{COOMe}{\overset{Me}{|}} \longrightarrow \text{⌇CH}_2-\underset{\underset{Cl}{|}}{C}-CH_2-\underset{\underset{Cl}{|}}{CH}\text{⌇} \xrightarrow{CH_2=C(Me)COOMe,\ etc.}$$

(with Me—Ċ·, COOMe, CH₂ group shown on the intermediate)

Product:

    COOMe COOMe
       | |
Me—C—CH₂—C⌇
       | |
      CH₂   Me
⌇CH₂—C—CH₂—CH⌇
       | |
      Cl   Cl

*Poly(methyl methacrylate) grafted
onto poly(vinyl chloride) chain*

32.9 (a) Chain-transfer is involved.

$$CH_3O(CH_2CH_2O)_nCH_2CH_2O^- + CH_3OH \longrightarrow CH_3O(CH_2CH_2O)_nCH_2CH_2OH + CH_3O^-$$

(b) $CH_3OCH_2CH_2OH$, 2-methoxyethanol.

32.10 (a) ⌇CH₂—CH:⁻ + H₂O ⟶ ⌇CH₂—CH₂ + OH⁻
                        | |
                        Ph   Ph

(b) ⌇CH₂—CH:⁻ + CO₂ ⟶ ⌇CH₂—CH—COO⁻ $\xrightarrow{H_2O}$ ⌇CH₂CHCOOH + OH⁻
                 | | |
                 Ph   Ph   Ph

(c) ⌇CH₂—CH:⁻ + H₂C—CH₂ ⟶ ⌇CH₂CHCH₂CH₂O⁻ $\xrightarrow{H_2O}$ ⌇CH₂CHCH₂CH₂OH + OH⁻
                 | \ / | |
                 Ph   O   Ph   Ph

(d) ⌇CH₂—CHCH₂CH₂O⁻ + nH₂C—CH₂ ⟶ ⌇CH₂CH(CH₂CH₂O)ₙCH₂CH₂O⁻
                 | \ / |
                 Ph   O   Ph
     (from part c)   ↓ H₂O

                    ⌇CH₂CH(CH₂CH₂O)ₙCH₂CH₂OH + OH⁻
                        |
                        Ph

32.11 (a) Primary —OH is esterified more rapidly than secondary —OH.

(phthalic anhydride) + CH₂—CH—CH₂ $\xrightarrow{-H_2O}$ ⌇OCH₂CHCH₂O—C(=O)—(C₆H₄)—C(=O)—OCH₂CHCH₂O⌇
                         | | | | |
                         OH   OH   OH   OH   OH

(b) Cross-linking results from esterification of the secondary —OH groups.

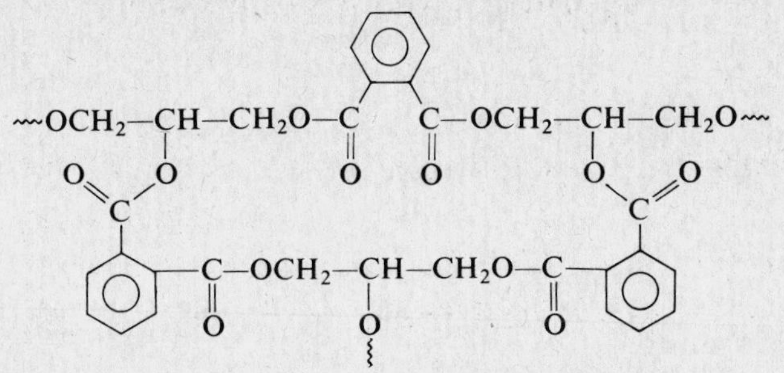

32.12 HOCH$_2$CH$_2$OH + O=C=N ⟶ N=C=O + HOCH$_2$CH$_2$OH
Diol   *Diol*

Diisocyanate

Polyurethane

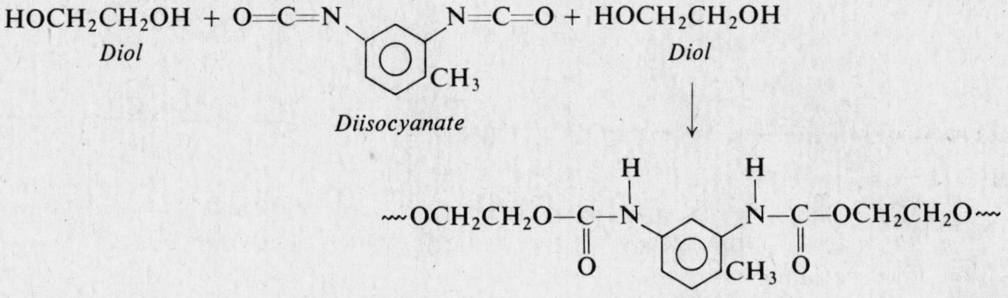

32.13 (a) Hydrolysis of an amide: hot aqueous acid or base.

(b) Hydrolysis of an ester: hot aqueous acid or base.

(c) Hydrolysis of an acetal: aqueous acid.

(d) Hydrolysis of an acetal: aqueous acid.

(Or, in each case, an enzyme-catalyzed hydrolysis.)

32.14 (a) Transesterification.

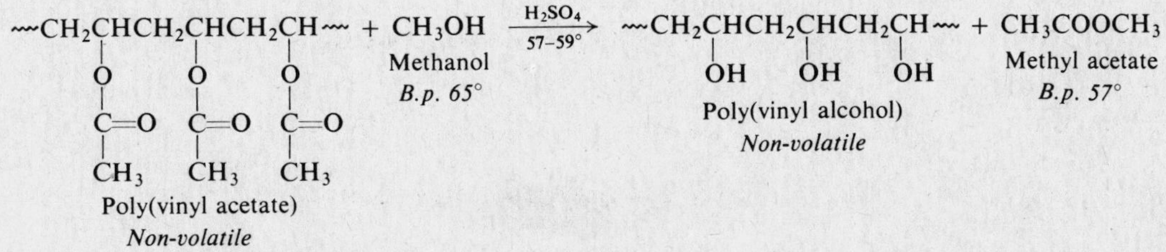

The hypothetical monomer of the new polymer is vinyl alcohol, which exists in the keto form, acetaldehyde.

(b) Formation of a cyclic acetal.

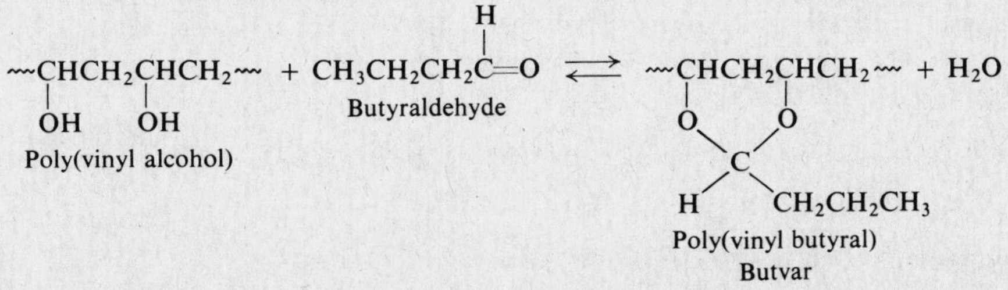

32.15

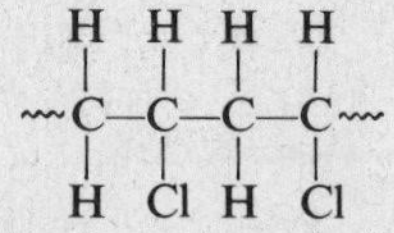

H H H H
~~~C─C─C─C~~~
H Cl H Cl

Poly(vinyl chloride)

H Cl H Cl
~~~C─C─C─C~~~
H Cl H Cl

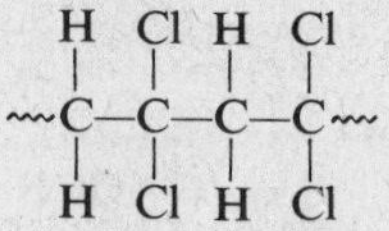

Poly(vinylidene chloride)
Saran

Poly(vinyl chloride) can show the same stereoisomerism as polypropylene (Figure 32.1, page 1041, with —Cl in place of —CH$_3$). Formed by a free-radical process, it is atactic; the molecules fit together poorly.

Poly(vinylidene chloride) has two identical substituents on each carbon, and the chains fit together well.

32.16 (a) The chains are irregularly substituted, and fit together poorly; the intermolecular forces are weak. (The diene provides double bonds, and hence allylic hydrogen, in the polymer; vulcanization causes cross-linking as with rubber, page 276.)

(b) Abstraction of —H from the polymer generates free radicals, which combine. The cross-links here are carbon–carbon bonds.

1. The more stable particle is formed in each step, with the same orientation.

2. There is acid-catalyzed polymerization of the alkene that is easily formed from a 2° or 3° alcohol.

3. Each carbon of polyisobutylene carries two identical substituents (two —H's or two —CH$_3$'s); this symmetry prevents the existence of stereoisomeric polymers. On the other hand, 1-butene should form exactly the same kind of isomeric polymers as propylene (Figure 32.1, page 1041, with —C$_2$H$_5$ in place of —CH$_3$).

4. Nucleophilic carbonyl addition; anionic chain-reaction polymerization.

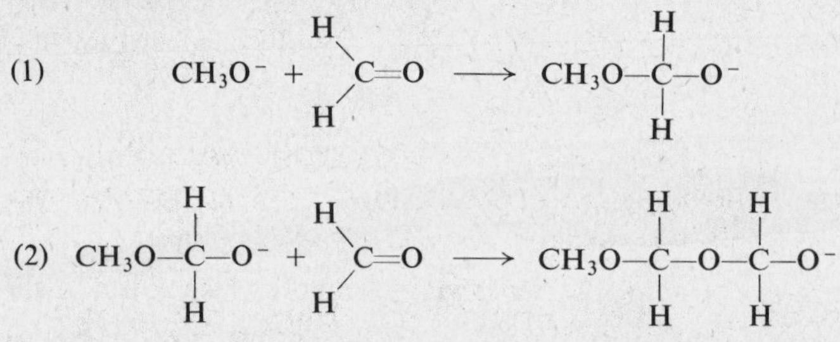

(1)   CH$_3$O$^-$ + $\overset{H}{\underset{H}{C}}$=O ⟶ CH$_3$O─$\overset{H}{\underset{H}{C}}$─O$^-$

(2) CH$_3$O─$\overset{H}{\underset{H}{C}}$─O$^-$ + $\overset{H}{\underset{H}{C}}$=O ⟶ CH$_3$O─$\overset{H}{\underset{H}{C}}$─O─$\overset{H}{\underset{H}{C}}$─O$^-$

then steps like (2), *to give* CH$_3$O(CH$_2$—O)$_n$CH$_2$OH

5. Polyurethanes contain ester and amide linkages. Hydrolysis—with water in sealed containers at 200°—gives diols, diamines, and CO$_2$. For example:

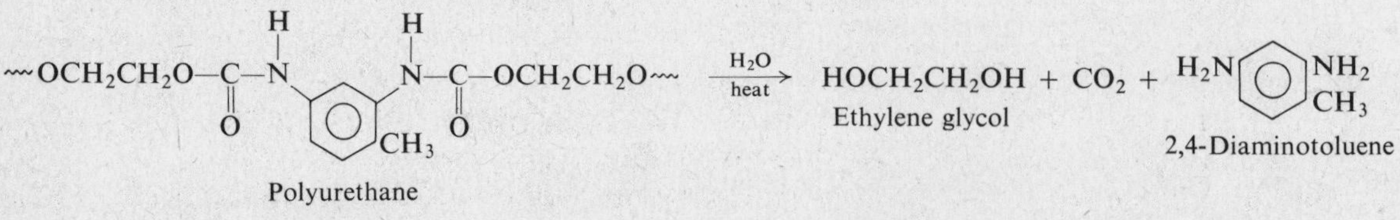

~~~OCH$_2$CH$_2$O─$\overset{}{\underset{O}{C}}$─$\overset{H}{\underset{}{N}}$⬡$\underset{CH_3}{}$$\overset{H}{\underset{}{N}}$─$\overset{}{\underset{O}{C}}$─OCH$_2CH_2$O~~~ $\xrightarrow[\text{heat}]{\text{H}_2\text{O}}$ HOCH$_2$CH$_2$OH + CO$_2$ + H$_2$N⬡$\underset{CH_3}{}$NH$_2$

Polyurethane   Ethylene glycol   2,4-Diaminotoluene

6. Reactivity toward addition of $BrCCl_3$ (Sec. 8.24) depends on how fast this step occurs:

$$RCH=CH_2 + \cdot CCl_3 \longrightarrow R\overset{\cdot}{C}HCH_2CCl_3$$

$PhCH=CH_2$ is more reactive than 1-octene because an incipient benzylic free radical is more stable than an incipient 2° free radical.

The rest of the series we attribute to a polar effect. The $\cdot CCl_3$ radical is electrophilic, and reactivity toward it is decreased by the electron-withdrawing inductive effect of the substituents.

We should notice the difference between $PhCH=CH_2$ and $PhCH_2CH=CH_2$. In $PhCH=CH_2$ the phenyl group can help stabilize the incipient free radical through resonance, and also through resonance probably helps to disperse a partial positive charge in the transition state. In $PhCH_2CH=CH_2$ the phenyl group is once removed from the reaction site, and can exert only its electron-withdrawing inductive effect.

7. Let us examine the mechanism for this reaction given on page 205.

(1) $\qquad\qquad\qquad\qquad$ peroxide $\longrightarrow$ Rad$\cdot$

(2) $\qquad\qquad\qquad$ Rad$\cdot$ + Cl:CCl_3 $\longrightarrow$ Rad:Cl + $\cdot CCl_3$

(3) $\qquad\qquad\qquad$ $\cdot CCl_3$ + $RCH=CH_2$ $\longrightarrow$ $R\overset{\cdot}{C}H-CH_2-CCl_3$

(4) $\qquad$ $R\overset{\cdot}{C}H-CH_2-CCl_3$ + Cl:CCl_3 $\longrightarrow$ $RCH-CH_2-CCl_3$ + $\cdot CCl_3$
$\qquad\qquad\qquad\qquad\qquad\qquad\qquad\qquad\qquad\qquad$ |
$\qquad\qquad\qquad\qquad\qquad\qquad\qquad\qquad\qquad\qquad$ Cl

then (3), (4), (3), (4), etc.

(a) The free radicals produced in step (3) are shown as abstracting an atom from CCl_4 in step (4). They can, instead, *add to the alkene* (5) to form a new, bigger radical which then attacks (6) CCl_4 to yield the 2:1 adduct.

(5) $\qquad$ $RCH=CH_2$ + $R\overset{\cdot}{C}H-CH_2-CCl_3$ $\longrightarrow$ $R\overset{\cdot}{C}H-CH_2-\overset{\overset{\textstyle R}{|}}{C}H-CH_2-CCl_3$

(6) $R\overset{\cdot}{C}H-CH_2-\overset{\overset{\textstyle R}{|}}{C}H-CH_2-CCl_3$ + Cl:CCl_4 $\longrightarrow$ $RCH-CH_2-\overset{\overset{\textstyle R}{|}}{C}H-CH_2-CCl_3$ + $\cdot CCl_3$
$\qquad\qquad\qquad\qquad\qquad\qquad\qquad\qquad\qquad\qquad\qquad\qquad\qquad$ |
$\qquad\qquad\qquad\qquad\qquad\qquad\qquad\qquad\qquad\qquad\qquad\qquad\qquad$ Cl

(b) Two reagents are competing for the organic free radical: CX_4 in reaction (4), and $RCH=CH_2$ in reaction (5). CBr_4 is more reactive than CCl_4—it contains weaker C–X bonds—and, other things being equal, abstraction (4) is favored.

$$R\overset{\cdot}{C}HCH_2CX_3 \begin{cases} \xrightarrow[(4)]{CX_4} RCHCH_2CX_3 \quad\text{(X below)}\quad \textbf{Abstraction} \\ \\ \xrightarrow[(5)]{RCH=CH_2} R\overset{\cdot}{C}HCH_2\overset{\overset{\textstyle R}{|}}{C}HCH_2CX_3 \quad \textbf{Addition} \end{cases}$$

(c) Again there is competition. Styrene is a more reactive alkene than 1-octene and, other things being equal, addition (5) is favored—in this case, to such an extent that only polymerization (with occasional chain-transfer) is observed. (There is an additional factor. The intermediate benzylic radical formed from styrene is relatively unreactive and therefore selective (Secs. 2.23 and 3.28). In choosing between abstraction and the easier addition, it tends to choose addition.)

In these examples of free-radical addition we see a type intermediate between the simple 1:1 addition at one end of the reaction spectrum, and straight-forward polymerization at the other end.

8. (a) An anionic chain-reaction polymerization to give I, a polymer with terminal —OH groups; then a step-reaction polymerization with the diisocyanate to give the block co-polymer.

$$CH_2\!-\!CH_2 + OH^- \longrightarrow HOCH_2CH_2(OCH_2CH_2)_nOCH_2CH_2OH$$
$$\underset{O}{} \quad \underset{\textit{Limited amount}}{} \qquad\qquad\qquad\qquad I$$

$$I + OCN\!\!\left\langle\!\!\bigcirc\!\!\right\rangle\!\!CH_3 \longrightarrow polymer$$
$$\underset{NCO}{}$$

(b) A step-reaction polymerization (esterification) to give the unsaturated linear polyester II; then chain-reaction vinyl copolymerization with styrene to provide cross-links.

$$HOCH_2CH_2OH + \quad CH\!\!=\!\!CH \longrightarrow \sim\!\!OCH_2CH_2O\!-\!CO\!-\!CH\!\!=\!\!CH\!-\!CO\!-\!OCH_2CH_2O\!\sim$$
$$\underset{OC}{}\quad\underset{CO}{} \qquad\qquad\qquad\qquad II$$
$$\underset{O}{}$$

$$II + C_6H_5CH\!\!=\!\!CH_2 \xrightarrow[\text{initiator}]{\text{free-radical}} polymer$$

(c) An anionic chain-reaction polymerization. (Compare the reaction of ethylene oxide with aniline, Problem 17.19(c), page 567.)

$$CH_3CH\!\!-\!\!CH_2 + H_2NCH_2CH_2NH_2 \longrightarrow polymer$$
$$\underset{O}{}\quad \underset{\textit{Limited amount}}{}$$

(d) An anionic chain-reaction polymerization to give living polystyrene, which is then killed with ethylene oxide to give a block copolymer.

$$PhCH\!\!=\!\!CH_2 \xrightarrow[\text{naphthalene}]{\text{Na}} ^-CHCH_2(CHCH_2)_n(CH_2CH)_nCH_2CH^-$$
$$\underset{Ph}{}\quad\underset{Ph}{}\qquad\underset{Ph}{}\quad\underset{Ph}{}$$
$$III$$
$$\textit{Living polymer}$$

$$III + CH_2\!\!-\!\!CH_2 \longrightarrow polymer$$
$$\underset{O}{}$$

(e) A free-radical chain-reaction vinyl copolymerization to give the saturated linear polymer IV. Abstraction of hydrogen from IV by free radicals from benzoyl peroxide generates free-radical sites on IV; at these places, poly(methyl methacrylate) branches grow by free-radical chain-reaction polymerization.

$$CH_2=CH + CH_2=CH \xrightarrow[\text{initiator}]{\text{free-radical}} \sim\sim CH_2CHCH_2CHCH_2CHCH_2CHCH_2CHCH_2CH\sim\sim$$

with Cl, OAc substituents on left reactants, and OAc, Cl, OAc, Cl, Cl, OAc substituents on the product chain.

IV

$$IV + CH_2=\underset{\underset{COOCH_3}{|}}{\overset{\overset{CH_3}{|}}{C}} \xrightarrow[\text{peroxide}]{\text{benzoyl}} \text{polymer}$$

9. The product is a polyester, formed by an anionic chain-reaction polymerization; each step involves nucleophilic substitution at the acyl group of the cyclic ester.

$$\text{Base:} \quad \underset{O}{\overset{CH_2-CH_2}{C-O}} \longrightarrow \text{Base}-\underset{O}{C}-CH_2CH_2-O^-$$

$$\text{Base}-\underset{O}{C}-CH_2CH_2-O^- \quad \underset{O}{\overset{CH_2-CH_2}{C-O}} \longrightarrow \text{Base}-\underset{O}{C}-CH_2CH_2-O-\underset{O}{C}-CH_2CH_2-O^-, \ \textit{etc.}$$

10. $NH_2CH_2CH\sim\sim CH_2CH^- + NH_3 \longrightarrow NH_2CH_2CH\sim\sim CH_2CH_2$

with Ph, Ph substituents on the left and Ph, Ph substituents on the right.

11. Cleavage by HIO_4 indicates occasional vicinal diol groupings, $-CH(OH)CH(OH)-$, and hence vicinal diacetate groups, $-CH(OAc)CH(OAc)-$. These show that *some* head-to-head polymerization has taken place along with the predominant head-to-tail orientation.

12. (a) $\sim\sim\underset{\underset{O}{||}}{N}(CH_2)_5\overset{\overset{H}{|}}{C}-\underset{\underset{O}{||}}{N}(CH_2)_5\overset{\overset{H}{|}}{C}\sim\sim$

Nylon 6

(b) $\text{Base:} \quad \overset{C-N}{\underset{O \quad H}{}} \longrightarrow \text{Base}-\underset{O}{C}(CH_2)_5-NH^-$

Caprolactam

$\text{Base}-\underset{O}{C}(CH_2)_5-NH^- \quad \overset{C-N}{\underset{O \quad H}{}} \longrightarrow \text{Base}-\underset{O}{C}(CH_2)_5-\overset{\overset{H}{|}}{N}-\underset{O}{C}(CH_2)_5NH^-, \ \textit{etc.}$

Caprolactam

The reaction is anionic chain-reaction polymerization, involving nucleophilic substitution at the acyl group of the cyclic amide. The base could be OH^- itself or the anion formed by abstraction of the —NH proton from a molecule of lactam.

13.

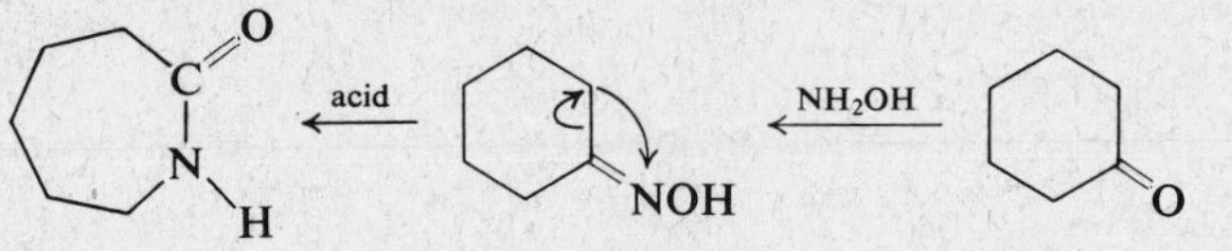

Caprolactam Cyclohexanonoxime Cyclohexanone

14. The *para* isomer gives straight, symmetrical chains that fit together well.

15. These compounds are ionic—or essentially so—due to the stability of benzylic anions.

16. (a)–(b) Diastereomers: A is *meso*, B is racemic.
(c) A resembles the isotactic polymer; B resembles the syndiotactic polymer.

17. First step. Excess ethylene glycol ensures terminal —OH's in C.

$$HOCH_2CH_2O{-}\underset{\displaystyle O}{\underset{\|}{C}}(CH_2)_4\underset{\displaystyle O}{\underset{\|}{C}}{-}O{\sim\sim}CH_2CH_2OH$$

C

Second step. Excess diisocyanate ensures terminal —NCO's in D;

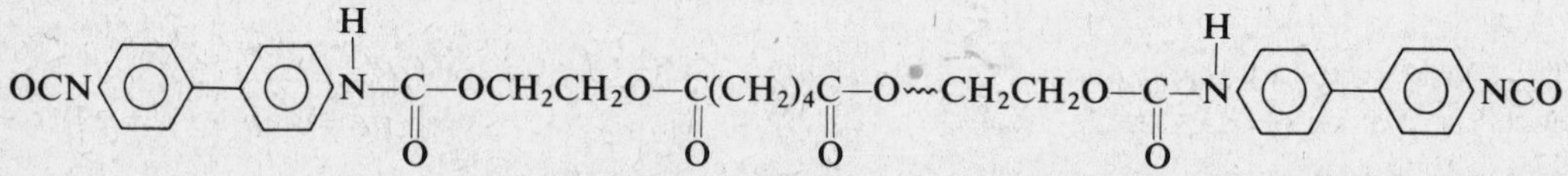

D

it also provides cross-linking through reaction at urethane (amide) —NH.

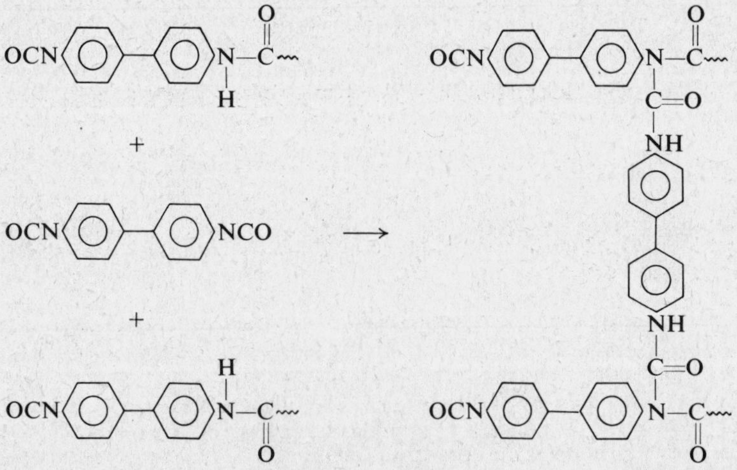

Third step. Water reacts with some of the —NCO groups in D to release CO_2 and provide $—NH_2$ groups.

D
$\downarrow$ H₂O

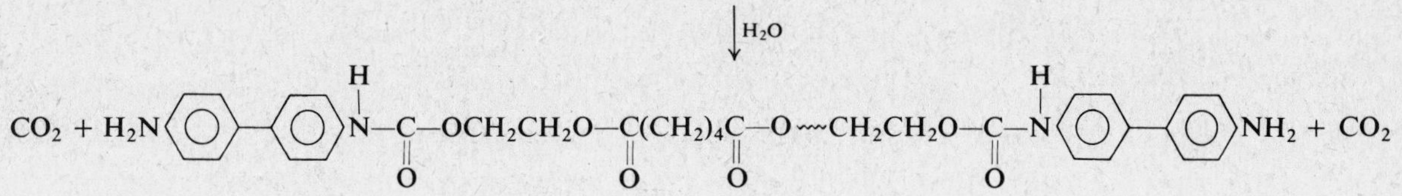

$CO_2 + H_2N$⟨O⟩—⟨O⟩—N(H)—C(=O)—OCH_2CH_2O—C(=O)(CH₂)₄C(=O)—O—CH_2CH_2O—C(=O)—N(H)⟨O⟩—⟨O⟩$NH_2 + CO_2$

E

The —NH₂ groups react with other free —NCO groups to give ureas (amides) with further cross-linking. The CO₂ is dispersed in the polymer to give a foam.

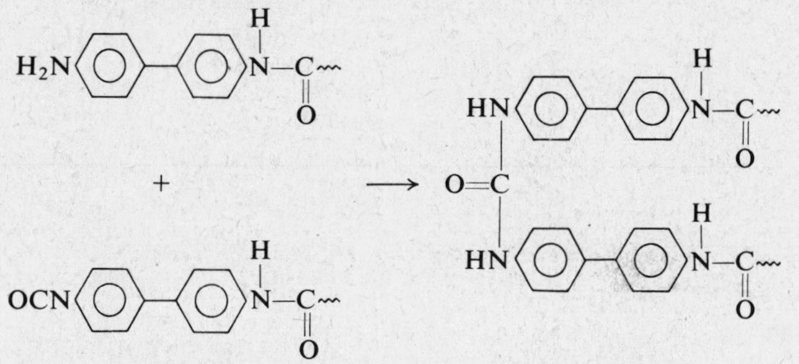

H₂N⟨O⟩—⟨O⟩N(H)—C(=O)⌇

+

OCN⟨O⟩—⟨O⟩N(H)—C(=O)⌇

⟶

HN⟨O⟩—⟨O⟩N(H)—C(=O)⌇
|
O=C
|
HN⟨O⟩—⟨O⟩N(H)—C(=O)⌇

18. The monomer (which can lose a hydrogen atom to give an allylic free radical) serves as a chain-transfer agent, and limits the size of the polymer. Abstraction of deuterium is more difficult, and the labeled monomer is a poorer chain-transfer agent. (Here again we see a free radical undergoing the competing reactions, addition and abstraction—this time at two sites in the *same* molecule.)

19. Oxygen abstracts hydrogen atoms to form allylic free radicals, $R\cdot$. These combine with oxygen to produce —O—O— cross-links; the gain in weight is due to the gain of oxygen. The mechanism seems to be:

$$R\cdot + O_2 \longrightarrow RCOO\cdot$$

$$2RCOO\cdot \longrightarrow ROOR + O_2$$

(Oxygen serves the same purpose here that sulfur does in the vulcanization of rubber, Sec. 8.25.)

20. (a) The two reactive groups of the epichlorohydrin are (i) C—Cl, which reacts with phenoxide in a Williamson synthesis, and (ii) the epoxide group, which reacts with phenoxide as in Sec. 17.13. The epoxide group is the less reactive; the excess of epichlorohydrin assures unreacted epoxide groups at both ends. The cement is:

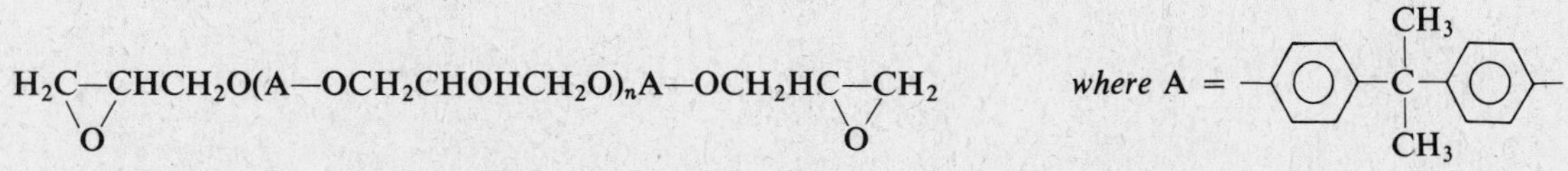

H_2C—$CHCH_2O(A$—$OCH_2CHOHCH_2O)_nA$—OCH_2HC—CH_2 *where* A = —⟨O⟩—C(CH₃)(CH₃)—⟨O⟩—

(b) During hardening, the amine reacts—at three sites in the molecule—with the epoxide groups in the cement, to form cross-links and thus generate a space-network polymer.

$$\sim\sim NHCH_2CH_2NCH_2CH_2NHCH_2CHCH_2O(A-OCH_2CHOHCH_2O)_nA-OCH_2CHCH_2NH\sim\sim$$

$$\underset{OH}{|} \qquad\qquad\qquad\qquad\qquad\qquad\qquad\qquad\qquad \underset{OH}{|}$$

(c) Use excess phenol, acetone, and acid or base (compare page 1043).

21. In F the methylene protons ($-CH_2-$) are equivalent, and give one signal, a singlet (2H per monomer unit). F must therefore be *syndiotactic*.

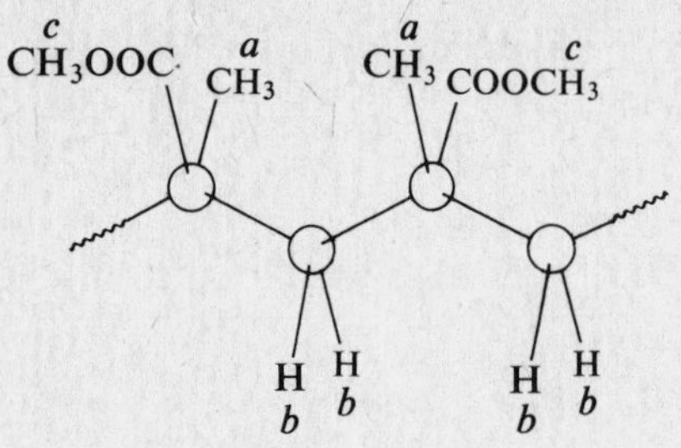

F
Syndiotactic

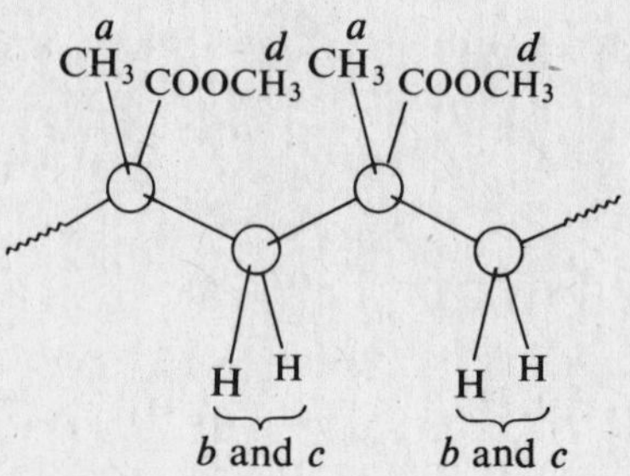

G
Isotactic

In G the methylene protons are not equivalent, and give two 1H signals, each split into a doublet by the other. G must therefore be *isotactic*. The methylene protons here are diastereotopic (Sec. 13.7); in the representation we have shown, for example, one H is near two $-CH_3$'s and the other H is near two $-COOCH_3$'s.

Chapter 33 | Fats

33.1 Decarboxylation. This finding supports the theory that the hydrocarbons in oil shales were formed by such algae.

$$^{14}CH_3(CH_2)_{15}CH_2COOH \longrightarrow {}^{14}CH_3(CH_2)_{15}CH_3 + CO_2$$

Stearic acid-18-^{14}C   *n*-Heptadecane-1-^{14}C

(J. Han, H. W.-S. Chau, and M. Calvin, "A Biosynthesis of Alkanes in *Nostoc muscorum*," J. Am. Chem. Soc., **91**, 5156 (1969).)

33.2 (a)

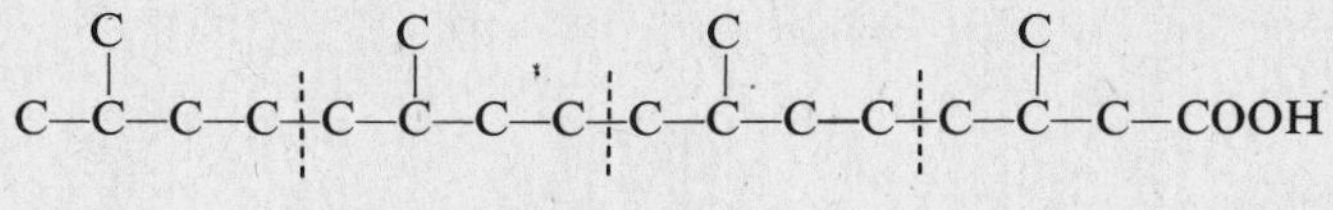

$$C-C-C-C \mathbin{\vert} C-C-C-C \mathbin{\vert} C-C-C-C \mathbin{\vert} C-C-C-COOH$$

with C substituents above as shown

Isoprene units

(b) The configurations at C–7 and C–11 are identical in phytol and the carboxylic acid. In phytol, C–3 is not a chiral center and, significantly, *both* configurations at this center were found in the carboxylic acid. These facts strongly indicate that these acids came from chlorophyll and hence from green plants.

(R. E. Cox, J. R. Maxwell, G. Eglinston, and C. T. Pillinger, "The Geological Fate of Chlorophyll: the Absolute Stereochemistries of a Series of Acyclic Isoprenoid Acids in a 50 Million Year Old Lacustrine Sediment," Chem. Commun., 1639 (1970).)

Actually, acetate is the building block here, too, since the isoprene units come from isopentenyl pyrophosphate (Sec. 8.26), which comes ultimately from acetate. (See Problem 21, Chapter 26, page 864 and its answer.)

33.3 Tung oil contains a high proportion of eleostearic acid (Table 33.1, page 1057). Abstraction of hydrogen atoms (see the answer to Problem 19, page 1052) gives allylic free radicals with delocalization over three double bonds.

33.4 Alkoxide is a poor leaving group.

33.5 Unsaturation preserves the semiliquidity of the membranes in the colder part of the body.

1.

$$CH_3(CH_2)_{22}COOH \xleftarrow[Ni]{H_2} CH_3(CH_2)_7CH{=}CH(CH_2)_{13}COOH \xrightarrow{\text{oxidn.}} CH_3(CH_2)_7COOH + HOOC(CH_2)_{13}COOH$$

Tetracosanoic   cis- or trans-15-Tetracosenoic acid   Nonanoic acid   Pentadecanedioic acid

acid   Nervonic acid   N.E. = m.w. = 158   N.E. = m.w./2 = 136

                              (*Actually trans*)

2. Since the acid units (and the alcohol unit, glycerol) remain the same, the only change can be in the *distribution* of acyl groups among the glyceride molecules. Something like this, say:

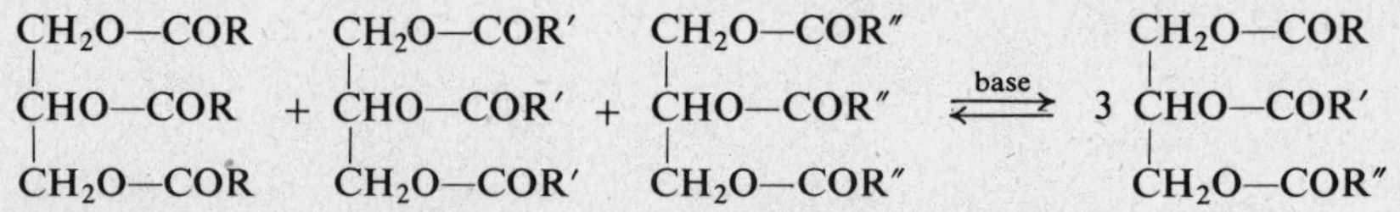

$$
\begin{array}{c}
CH_2O{-}COR \\
| \\
CHO{-}COR \\
| \\
CH_2O{-}COR
\end{array}
+
\begin{array}{c}
CH_2O{-}COR' \\
| \\
CHO{-}COR' \\
| \\
CH_2O{-}COR'
\end{array}
+
\begin{array}{c}
CH_2O{-}COR'' \\
| \\
CHO{-}COR'' \\
| \\
CH_2O{-}COR''
\end{array}
\underset{\text{base}}{\rightleftharpoons}
3
\begin{array}{c}
CH_2O{-}COR \\
| \\
CHO{-}COR' \\
| \\
CH_2O{-}COR''
\end{array}
$$

This would involve transesterification (Sec. 20.20), for which sodium methoxide would be an excellent catalyst:

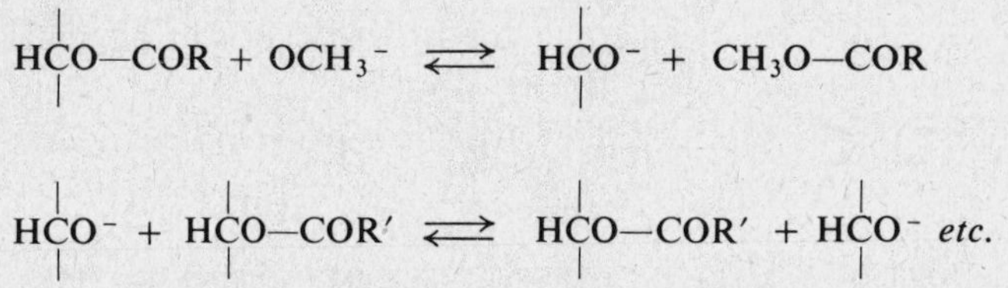

$$HCO{-}COR + OCH_3^- \rightleftharpoons HCO^- + CH_3O{-}COR$$

$$HCO^- + HCO{-}COR' \rightleftharpoons HCO{-}COR' + HCO^- \ \textit{etc.}$$

3.

$$R{-}H + O_2 \longrightarrow H{-}O{-}O\cdot + R\cdot$$

$$R\cdot + O_2 \longrightarrow R{-}O{-}O\cdot$$

$$R{-}O{-}O\cdot + R{-}H \longrightarrow R{-}O{-}O{-}H + R\cdot \ \textit{etc.}$$

In the case of methyl oleate:

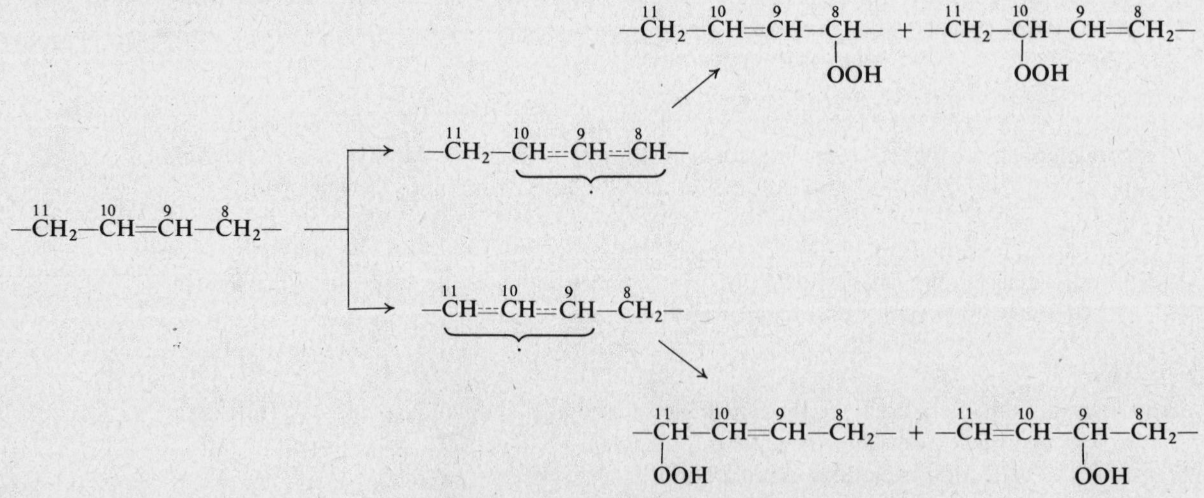

4. The weakly basic 2,4-dinitrophenoxide ion is a good leaving group.

5. $n\text{-}C_{15}H_{31}COO^- + n\text{-}C_{15}H_{31}CH_2OH \xleftarrow[\;H_2O\;]{OH^-}$

$$n\text{-}C_{15}H_{31}\overset{\displaystyle O}{\overset{\displaystyle \|}{C}}\text{—}O\text{—}CH_2C_{15}H_{31}\text{-}n$$

B

1-Hexadecanol

Spermaceti

n-Hexadecyl hexadecanoate

N.E. = 480

$\Big\downarrow H^+$

$n\text{-}C_{15}H_{31}COOH$

A

Hexadecanoic acid

(Palmitic acid)

M.p. 63°;

N.E. = m.w. = 256

6. (a) Cleavage of the monoanion as a dipolar ion I (or with simultaneous transfer of a proton) is easiest because of: (i) protonation of the alkoxy group, which makes it a better leaving group; and (ii) the double negative charge on the other oxygens, which helps in the transfer of electrons from P to O.

$$R\text{—}O\text{—}PO_3H^- \;\rightleftarrows\; R\text{—}\overset{+}{O}\text{—}PO_3^{--} \xrightarrow{H_2O}\; ROH + H_2PO_4^-$$

$$\underset{\text{H}}{|}$$

I

(b) Only a fraction of the ester exists as I. As the acidity is raised above the optimum point, this fraction decreases, and I is replaced as the reacting species by II, and then by III. Mole-

$$R\text{—}\overset{+}{O}\text{—}PO_3H^- \qquad\qquad R\text{—}\overset{+}{O}\text{—}PO_3H_2$$

$$\underset{\text{H}}{|} \qquad\qquad\qquad\quad \underset{\text{H}}{|}$$

II   III

cule for molecule, each of these is probably less reactive than I (since they lack the double negative charge); but as the acidity is raised higher and higher, the fraction of ester existing as III continues to climb, and the rate continues to increase.

(C. A. Bunton, "Hydrolysis of Monosubstituted Orthophosphate Esters," J. Chem. Educ., **45**, 21 (1968).)

7. $n\text{-}C_6H_{13}Cl \xrightarrow{NaC\equiv CH} n\text{-}C_6H_{13}C\equiv CH \xrightarrow{Na, NH_3} n\text{-}C_6H_{13}C\equiv C^-Na^+ \xrightarrow{I(CH_2)_9Cl} n\text{-}C_6H_{13}C\equiv C(CH_2)_9Cl$

C   D

$D \xrightarrow{KCN} n\text{-}C_6H_{13}C\equiv C(CH_2)_9CN \xrightarrow[\text{heat}]{OH^-} \xrightarrow{H^+} n\text{-}C_6H_{13}C\equiv C(CH_2)_9COOH$

E   F

$Pd \Big\downarrow H_2$

$$\begin{array}{ccc} n\text{-}C_6H_{13} & & (CH_2)_9COOH \\ & C=C & \\ H & & H \end{array}$$

Vaccenic acid

(*cis*)

8. $n\text{-}C_{13}H_{27}CH_2Br + Na[CH(COOEt)_2] \longrightarrow n\text{-}C_{13}H_{27}CH_2CH(COOEt)_2 \xrightarrow{\text{1KOH}} n\text{-}C_{13}H_{27}CH_2CH\begin{smallmatrix}\text{COOH}\\\\\text{COOEt}\end{smallmatrix}$

                                                   G   **H**

$\downarrow$ DHP

$n\text{-}C_{13}H_{27}CH_2CH\begin{smallmatrix}\text{COOTHP}\\\\\text{COOEt}\end{smallmatrix}$

                        I

$cis\text{-}n\text{-}C_6H_{13}CH{=}CH(CH_2)_7COOH \xrightarrow{\text{SOCl}_2} cis\text{-}n\text{-}C_6H_{13}CH{=}CH(CH_2)_7COCl$

                                                                        J

$I \xrightarrow{\text{Na}} \longrightarrow J \longrightarrow cis\text{-}n\text{-}C_{13}H_{27}CH_2{-}\underset{\underset{\text{THPOOC}}{\big|}}{\overset{\overset{\text{COOEt}}{\big|}}{C}}{-}\underset{\underset{\text{O}}{\big\|}}{C}(CH_2)_7CH{=}CHC_6H_{13}\text{-}n$

                                                          K

$\downarrow$ dil. H$^+$

$cis\text{-}n\text{-}C_{13}H_{27}CH_2{-}\underset{\underset{\text{H}}{\big|}}{\overset{\overset{\text{COOEt}}{\big|}}{C}}{-}\underset{\underset{\text{O}}{\big\|}}{C}(CH_2)_7CH{=}CHC_6H_{13}\text{-}n + CO_2$

                                                               L

$\downarrow$ NaBH$_4$

$cis\text{-}n\text{-}C_{13}H_{27}CH_2{-}\underset{\underset{\text{H}}{\big|}}{\overset{\overset{\text{COOEt}}{\big|}}{C}}{-}\underset{\underset{\text{OH}}{\big|}}{C}H(CH_2)_7CH{=}CHC_6H_{13}\text{-}n$

                                                              M

heat $\downarrow$ OH$^-$

$\downarrow$ H$^+$

$cis\text{-}n\text{-}C_{13}H_{27}CH_2{-}\underset{}{\overset{\overset{\text{COOH}}{\big|}}{CH}}{-}CHOH(CH_2)_7CH{=}CHC_6H_{13}\text{-}n$

                      ($\pm$)-Corynomycolenic acid

9. $n\text{-}C_8H_{17}\text{—}\underset{\underset{OH}{|}}{CH}\text{—}CH_3 \xrightarrow{PBr_3} n\text{-}C_8H_{17}\text{—}\underset{\underset{Br}{|}}{CH}\text{—}CH_3 + Na[CH(COOEt)_2] \longrightarrow \underset{\underset{CH_3}{|}}{\overset{\overset{n\text{-}C_8H_{17}}{|}}{CH}}\text{—}CH(COOEt)_2$

$\qquad\qquad\qquad\qquad\qquad\qquad\qquad\qquad\qquad\qquad\qquad N$

$\underset{\underset{CH_3}{|}}{\overset{\overset{n\text{-}C_8H_{17}}{|}}{CH}}CH(COOEt)_2 \xrightarrow{\underset{heat}{OH^-}} \xrightarrow{H^+} \xrightarrow{heat} \underset{\underset{CH_3}{|}}{\overset{\overset{n\text{-}C_8H_{17}}{|}}{CH}}CH_2COOH \xrightarrow{SOCl_2} \underset{\underset{CH_3}{|}}{\overset{\overset{n\text{-}C_8H_{17}}{|}}{CH}}CH_2COCl$

$\qquad\qquad\qquad\qquad\qquad\qquad\qquad\qquad\qquad\qquad O \qquad\qquad\qquad\qquad\qquad\qquad P$

$P \xrightarrow{EtOH} \underset{\underset{CH_3}{|}}{\overset{\overset{n\text{-}C_8H_{17}}{|}}{CH}}CH_2COOEt \xrightarrow{LiAlH_4} \underset{\underset{CH_3}{|}}{\overset{\overset{n\text{-}C_8H_{17}}{|}}{CH}}CH_2CH_2OH \xrightarrow{PBr_3} \underset{\underset{CH_3}{|}}{\overset{\overset{n\text{-}C_8H_{17}}{|}}{CH}}CH_2CH_2Br$

$\qquad\qquad\qquad Q \qquad\qquad\qquad\qquad\qquad R \qquad\qquad\qquad\qquad\qquad S$

$S \xrightarrow{Mg} \underset{\underset{CH_3}{|}}{\overset{\overset{n\text{-}C_8H_{17}}{|}}{CH}}CH_2CH_2MgBr \xrightarrow{CdCl_2} \left[\underset{\underset{CH_3}{|}}{\overset{\overset{n\text{-}C_8H_{17}}{|}}{CH}}CH_2CH_2\right]_2 Cd$

$\phantom{S \xrightarrow{Mg}}\qquad\qquad\qquad\qquad\qquad\qquad\qquad\qquad\qquad EtOOC(CH_2)_5COCl$

$\underset{\underset{CH_3}{|}}{\overset{\overset{n\text{-}C_8H_{17}}{|}}{CH}}CH_2CH_2\text{—}\underset{\underset{O}{\|}}{C}\text{—}(CH_2)_5COOEt$

$ T$

$T \xrightarrow{\underset{HCl}{Zn(Hg)}} \underset{\underset{CH_3}{|}}{\overset{\overset{n\text{-}C_8H_{17}}{|}}{CH}}CH_2CH_2CH_2\text{—}(CH_2)_5COOEt \xrightarrow{\underset{heat}{OH^-}} \xrightarrow{H^+} \underset{\underset{CH_3}{|}}{\overset{\overset{n\text{-}C_8H_{17}}{|}}{CH}}CH_2CH_2CH_2(CH_2)_5COOH$

$\qquad\qquad\qquad\qquad\qquad\qquad\qquad U \qquad\qquad\qquad\qquad\qquad\qquad\qquad\qquad$ Tuberculostearic acid

 10-Methyloctadecanoic acid

10. $C_{26}H_{53}COOH$ sat'd open-chain

$\underline{-C_{26}H_{51}COOH}$

 2H missing *means* one double bond (or one ring)

$\underset{\substack{| \\ C_{27}\text{-Phthienoic acid}}}{-C}\overset{\overset{CH_3}{|}}{\parallel}C\text{—}COOH \xrightarrow{O_3} \xrightarrow{\underset{H_2O}{Zn}} O=\overset{\overset{CH_3}{|}}{C}\text{—}COOH$

$\underset{\substack{| \\ C_{27}\text{-Phthienoic acid} \\ (C_{27}H_{52}O_2)}}{-C}\overset{\overset{H\ \ CH_3}{|\ \ \ |}}{\parallel}C\text{—}COOH \xrightarrow{KMnO_4} -COOH$

$ V$

$ (C_{24}H_{48}O_2)$

$$\underset{\substack{\text{V}\\(C_{24}H_{48}O_2)}}{\overset{\displaystyle|}{\underset{\displaystyle H}{C}}-COOH} \xrightarrow[H^+]{MeOH} \overset{\displaystyle|}{\underset{\displaystyle H}{C}}-COOMe \xrightarrow{2PhMgBr} \xrightarrow{H_2O} \underset{\substack{\text{W}\\(C_{36}H_{58}O)}}{\overset{\displaystyle|}{\underset{\displaystyle H}{C}}-\overset{\displaystyle Ph}{\underset{\displaystyle OH}{C}}-Ph} \xrightarrow[-H_2O]{H^+} \underset{\substack{X\\(C_{36}H_{56})}}{\overset{\displaystyle|}{C}=\overset{\displaystyle Ph}{C}-Ph}$$

$$\downarrow CrO_3$$

$$\overset{\displaystyle|}{C}=O \ + \ O=\overset{\displaystyle Ph}{C}-Ph$$

Y, a ketone
$(C_{23}H_{46}O)$

$$\underset{Y}{C_{21}H_{43}-\overset{\displaystyle CH_3}{C}=O} \xrightarrow[NaOH]{I_2} C_{21}H_{43}COONa \ + \ CHI_3$$

A methyl ketone

At this point, then:

$$\underset{Y}{C_{21}H_{43}-\overset{\displaystyle CH_3}{C}=O} \qquad \underset{V}{C_{21}H_{43}-\overset{\displaystyle CH_3}{\underset{\displaystyle H}{C}}-COOH} \qquad C_{21}H_{43}-\overset{\displaystyle CH_3}{\underset{\displaystyle H}{C}}-CH=\overset{\displaystyle CH_3}{C}-COOH$$

C_{27}-Phthienoic acid

Next:

$$\underset{\substack{V\\(C_{24}H_{48}O_2)}}{\overset{\displaystyle|}{\underset{\displaystyle H}{C}}-\overset{\displaystyle CH_3}{\underset{\displaystyle H}{C}}-COOH} \xrightarrow[P]{Br_2} \underset{Z}{\overset{\displaystyle|}{\underset{\displaystyle H}{C}}-\overset{\displaystyle CH_3}{\underset{\displaystyle Br}{C}}-COOH} \xrightarrow[-HBr]{KOH(alc)} \underset{\substack{AA\\(C_{24}H_{46}O_2)}}{\overset{\displaystyle|}{C}=\overset{\displaystyle CH_3}{C}-COOH}$$

$$\underset{\substack{AA\\(C_{24}H_{46}O_2)}}{CH_3(CH_2)_{17}CH \dashdot CH=\overset{\displaystyle CH_3}{C}-COOH} \xrightarrow{KMnO_4} \underset{\substack{Identified\\(C_{20}H_{40}O)}}{CH_3(CH_2)_{17}\overset{\displaystyle CH_3}{C}=O} + \text{other products}$$

Hence:

$$\underset{Y}{CH_3(CH_2)_{17}\overset{\displaystyle CH_3}{CH}CH_2\overset{\displaystyle CH_3}{C}=O} \qquad \underset{V}{CH_3(CH_2)_{17}\overset{\displaystyle CH_3}{CH}CH_2\overset{\displaystyle CH_3}{CH}COOH} \qquad CH_3(CH_2)_{17}\overset{\displaystyle CH_3}{CH}CH_2\overset{\displaystyle CH_3}{CH}CH=\overset{\displaystyle CH_3}{C}-COOH$$

C_{27}-Phthienoic acid

11.
$$\overset{a}{C}H_3\overset{b}{(CH_2)}_{15}\overset{c}{C}H_2\overset{d}{C}OOH$$

$$\overset{a}{C}H_3\overset{c}{(CH_2)}_{14}\overset{b}{\underset{\overset{|}{\underset{d}{H}}}{\overset{CH_3}{\overset{|}{C}}}}\overset{e}{C}OOH$$

CC   DD

Octadecanoic acid   2-Methylheptadecanoic acid

Stearic acid

12. $C_2H_5-\overset{CH_3}{\overset{|}{C}}=O + [(MeO)_2P(O)CHCOOMe]^-Na^+ \longrightarrow$

$$\left[\begin{array}{c} CH_3 \\ | \\ C_2H_5-C-CHCOOMe \\ | \quad\quad | \\ Na^+{}^-O \quad P(O)(OMe)_2 \end{array} \right] \longrightarrow$$

EE

$$\underset{C_2H_5}{\overset{CH_3}{>}}C=C\underset{COOMe}{\overset{H}{<}}$$

FF

FF $\xrightarrow{\text{LiAlH}_4}$ $\underset{C_2H_5}{\overset{CH_3}{>}}C=C\underset{CH_2OH}{\overset{H}{<}}$ GG $\xrightarrow{\text{PBr}_3}$ $\underset{C_2H_5}{\overset{CH_3}{>}}C=C\underset{CH_2Br}{\overset{H}{<}}$ HH

$\downarrow$ $[C_2H_5COCHCOOEt]^-Na^+$

$$\underset{C_2H_5}{\overset{CH_3}{>}}C=C\underset{CH_2CHCOC_2H_5}{\overset{H\quad\overset{COOEt}{\overset{|}{C}}}{<}}$$

II

II $\xrightarrow[\text{H}_2\text{O}]{\text{OH}^-}$ $\xrightarrow{\text{H}^+}$ $\xrightarrow{\text{heat}}$ $\underset{C_2H_5}{\overset{CH_3}{>}}C=C\underset{CH_2CH_2COC_2H_5}{\overset{H}{<}}$

JJ

$\downarrow$ $[(MeO)_2P(O)CHCOOMe]^-Na^+$

$$\left[\begin{array}{c} \underset{C_2H_5}{\overset{CH_3}{>}}C=C\underset{CH_2CH_2C-CHCOOMe}{\overset{H\quad\quad\overset{C_2H_5}{\overset{|}{C}}}{<}} \\ \qquad\qquad Na^+{}^-O \quad P(O)(OMe)_2 \end{array} \right]$$

KK

$\downarrow$

$$\underset{C_2H_5}{\overset{CH_3}{>}}C=C\underset{CH_2-CH_2}{\overset{H}{<}}\quad\underset{H}{\overset{C_2H_5}{>}}C=C\underset{H}{\overset{COOMe}{<}}$$

LL

LL $\xrightarrow{\text{LiAlH}_4}$ $\underset{C_2H_5}{\overset{CH_3}{>}}C=C\underset{CH_2-CH_2}{\overset{H}{<}}\underset{H}{\overset{C_2H_5}{>}}C=C\underset{H}{\overset{CH_2OH}{<}}$ MM $\xrightarrow{\text{PBr}_3}$ $\underset{C_2H_5}{\overset{CH_3}{>}}C=C\underset{CH_2-CH_2}{\overset{H}{<}}\underset{H}{\overset{C_2H_5}{>}}C=C\underset{H}{\overset{CH_2Br}{<}}$ NN

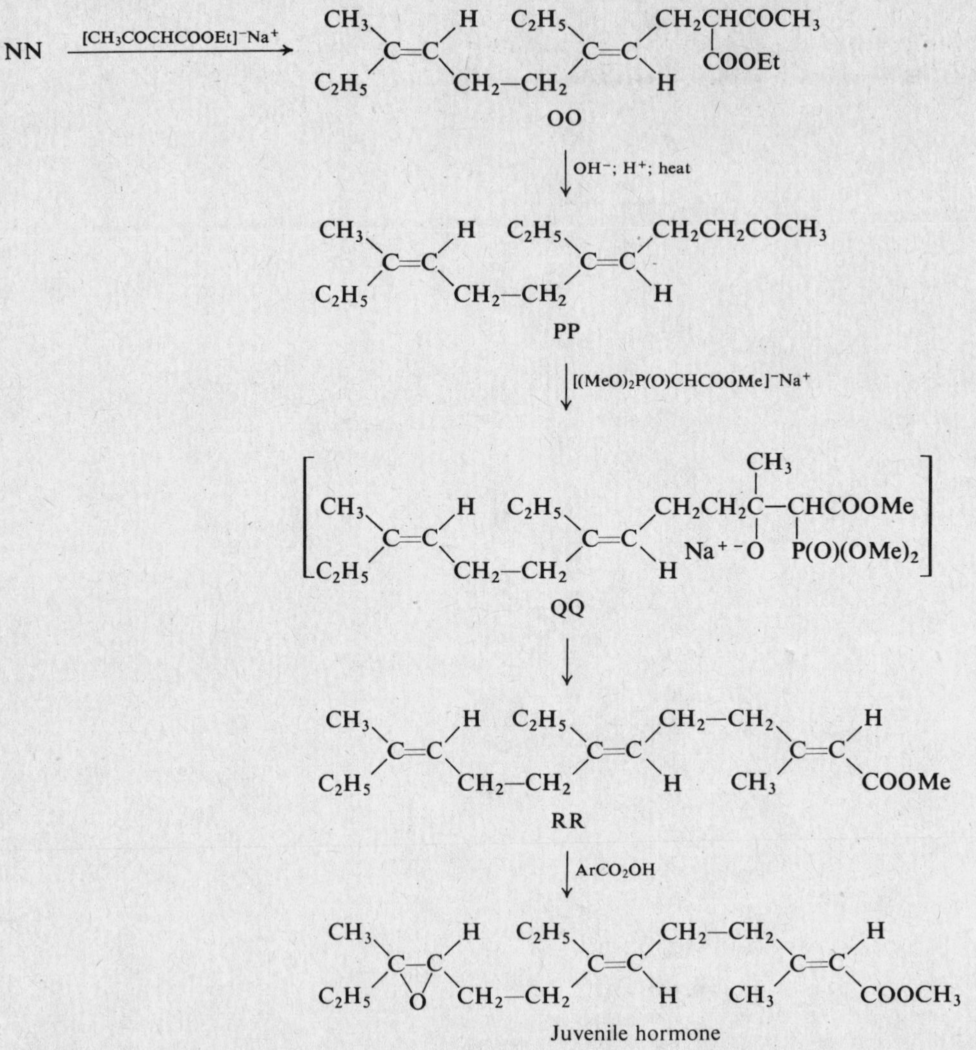

$$ NN \xrightarrow{[CH_3COCHCOOEt]^- Na^+} $$

CH₃ structures...

$$\text{NN} \xrightarrow{[CH_3COCHCOOEt]^-Na^+}$$

OO

OH⁻; H⁺; heat

PP

[(MeO)₂P(O)CHCOOMe]⁻Na⁺

QQ

RR

ArCO₂OH

Juvenile hormone

(K. H. Dahm, B. M. Trost, and H. Röller, "The Juvenile Hormone. V.   Synthesis of the Racemic Juvenile Hormone," J. Am. Chem. Soc., **89**, 5292 (1967); B. M. Trost, "The Juvenile Hormone of *Hyalophora cecropia*," Accounts Chem. Res., **3**, 120 (1970).)

In the preparation of FF, LL, and RR, we encounter a modification of the Wittig reaction (Sec. 21.10), in which a *phosphonate* is used in place of the usual ylide.   A base

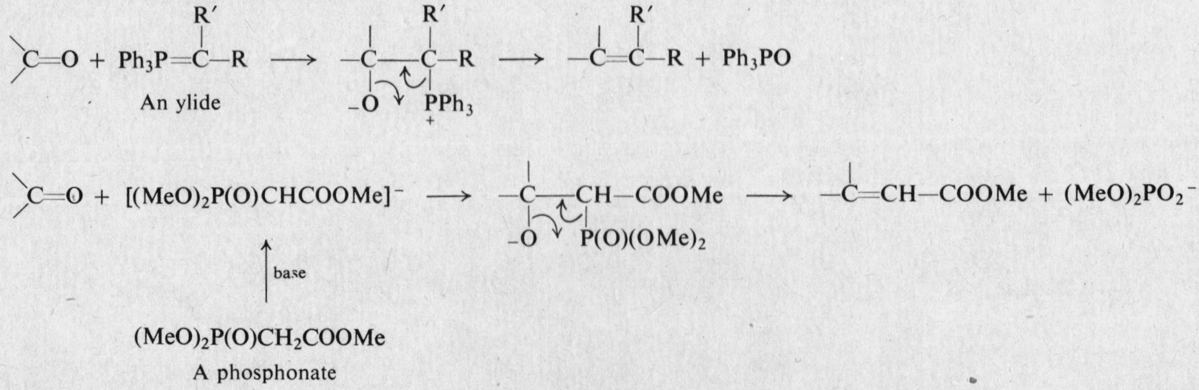

$$\begin{aligned} \text{C=O} + Ph_3P=C-R' \longrightarrow \quad \underset{-O \;\; +PPh_3}{-C-C-R'} \longrightarrow -C=C-R' + Ph_3PO \end{aligned}$$

An ylide

$$\text{C=O} + [(MeO)_2P(O)CHCOOMe]^- \longrightarrow \underset{-O \;\; P(O)(OMe)_2}{-C-CH-COOMe} \longrightarrow -C=CH-COOMe + (MeO)_2PO_2^-$$

base

$(MeO)_2P(O)CH_2COOMe$

A phosphonate

562

(NaH, in this particular example) converts the phosphonate into an anion which adds, in a Reformatsky-like reaction, to the carbonyl group. Elimination of $(MeO)_2PO_2^-$, the anion of dimethyl phosphate, occurs readily to generate the double bond. (S. Trippett, "The Wittig Reaction," Quart. Rev. (London), **17**, 406 (1963), especially pp. 431–434.)

Chapter 34

Carbohydrates I. Monosaccharides

34.1 (a) In each case, you would observe a change in the molecular formula which, interpreted in light of the reagent used and supported by the properties of the product, would tell you what had happened. For example, in the second reaction,

$$C_6H_{12}O_6 \xrightarrow{Br_2(aq)} C_6H_{12}O_7$$

addition of one O with no loss of C or H indicates oxidation of an aldehyde to a (mono)-carboxylic acid:

$$RCHO \longrightarrow RCOOH$$

This would be supported by the acidic properties of the product, and the fact that N.E. = m.w.

(b) Starting at the top, each reaction shows that glucose contains:

(i) $-\underset{\underset{O}{\|}}{C}-$ (ii) —CHO   (iii) —CHO and —CH$_2$OH   (iv) five —OH

(v) C—C—C—C—C—C   (vi) C—C—C—C—C—CHO

34.2 Formulas I–VIII, page 1082.

34.3 (a) Three chiral centers:

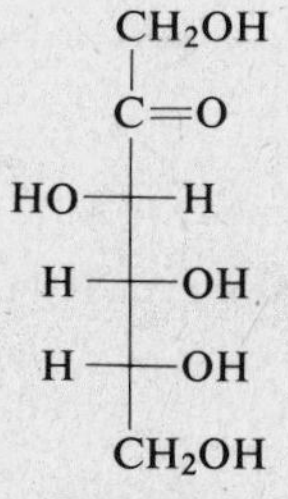

$$
\begin{array}{c}
\text{CH}_2\text{OH} \\
| \\
\text{C}=\text{O} \\
\text{HO}\!-\!\!|\!-\!\text{H} \\
\text{H}\!-\!\!|\!-\!\text{OH} \\
\text{H}\!-\!\!|\!-\!\text{OH} \\
\text{CH}_2\text{OH}
\end{array}
$$

Fructose

(b) There should be eight 2-ketohexoses ($2^3 = 8$): four pairs of enantiomers.

(c)

1	CH_2OH	CH_2OH	CH_2OH	CH_2OH
2	$C{=}O$	$C{=}O$	$C{=}O$	$C{=}O$
3	H——OH	HO——H	H——OH	HO——H
4	H——OH	H——OH	HO——H	HO——H
5	H——OH	H——OH	H——OH	H——OH
6	CH_2OH	CH_2OH	CH_2OH	CH_2OH
	Psicose	Fructose	Sorbose	Tagatose

34.4 Remembering our approach to Problem 16.11 (page 538), we write:

$$CH_2{\vdots}CH{\vdots}CH{\vdots}CH{\vdots}CH{\vdots}CHO \xrightarrow{5HIO_4} CH_2{-}OH + 4\ HO{-}\overset{\displaystyle H}{\underset{\displaystyle OH}{C}}{-}OH + HO{-}\overset{\displaystyle H}{C}{=}O$$

with OH groups on the first five carbons.

$$\Big\downarrow {-}H_2O \qquad\qquad \Big\downarrow {-}H_2O \qquad\qquad \Big\downarrow$$

$$HCHO \qquad\qquad 4HCOOH \qquad\qquad HCOOH$$

34.5 Remembering our approach to Problem 16.12 (page 538), we write:

HCHO   HCOOH   HCOOH   HCOOH   OHC—COOH

$$\uparrow \qquad\qquad \uparrow \qquad\qquad \uparrow \qquad\qquad \uparrow \qquad\qquad \uparrow$$

$$H_2\overset{\displaystyle}{\underset{\displaystyle OH}{C}}{-}OH \quad HO{-}\overset{\displaystyle H}{\underset{\displaystyle OH}{C}}{-}OH \quad HO{-}\overset{\displaystyle H}{\underset{\displaystyle OH}{C}}{-}OH \quad HO{-}\overset{\displaystyle H}{\underset{\displaystyle OH}{C}}{-}OH \quad HO{-}\overset{\displaystyle H}{\underset{\displaystyle OH}{C}}{-}COOH$$

$$\uparrow 4HIO_4$$

$$CH_2{\vdots}CH{\vdots}CH{\vdots}CH{\vdots}CH{-}COOH$$

with OH groups below each carbon.

A
Gluconic acid

$$HCHO + HCOOH + HCOOH + HCOOH + HCOOH + HCHO \xleftarrow{5HIO_4} CH_2{\vdots}CH{\vdots}CH{\vdots}CH{\vdots}CH{\vdots}CH_2$$

with OH groups below each carbon.

B
Glucitol

$$HOOC{-}CHO + HCOOH + HCOOH + OHC{-}COOH \xleftarrow{3HIO_4} HOOC{-}CH{\vdots}CH{\vdots}CH{\vdots}CH{-}COOH$$

with OH groups below each carbon.

C
Glucaric acid

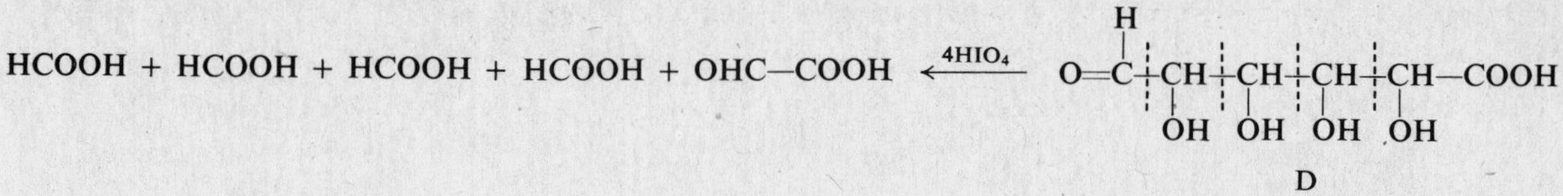

$$\text{HCOOH} + \text{HCOOH} + \text{HCOOH} + \text{HCOOH} + \text{OHC-COOH} \xleftarrow{\text{4HIO}_4}$$

O=C—CH┊CH┊CH┊CH—COOH
  |
  H   OH OH OH OH

D
Glucuronic acid

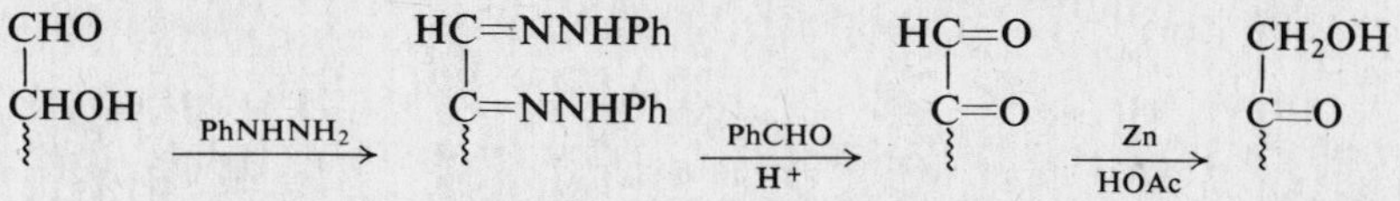

34.6

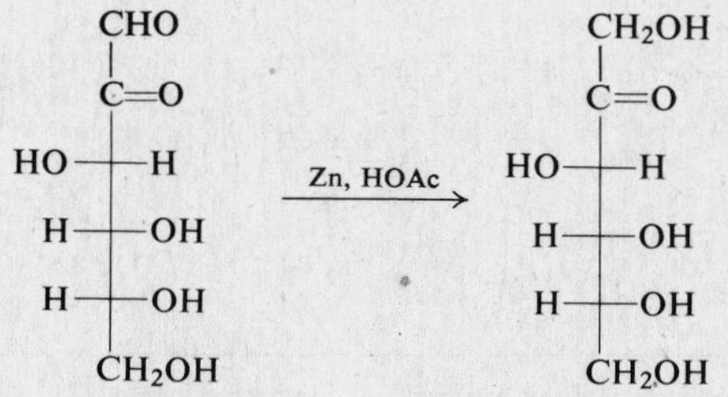

CHO	CH₂OH
C=O	C=O
HO—H	HO—H
H—OH	H—OH
H—OH	H—OH
CH₂OH	CH₂OH
Glucosone	Fructose

$$\xrightarrow{\text{Zn, HOAc}}$$

CHO
CHOH
~
Aldose

$$\xrightarrow{\text{PhNHNH}_2}$$

HC=NNHPh
C=NNHPh
~
Osazone

$$\xrightarrow[\text{H}^+]{\text{PhCHO}}$$

HC=O
C=O
~
Osone

$$\xrightarrow[\text{HOAc}]{\text{Zn}}$$

CH₂OH
C=O
~
2-Ketose

34.7 All three sugars have identical configurations at C–3, C–4, and C–5.

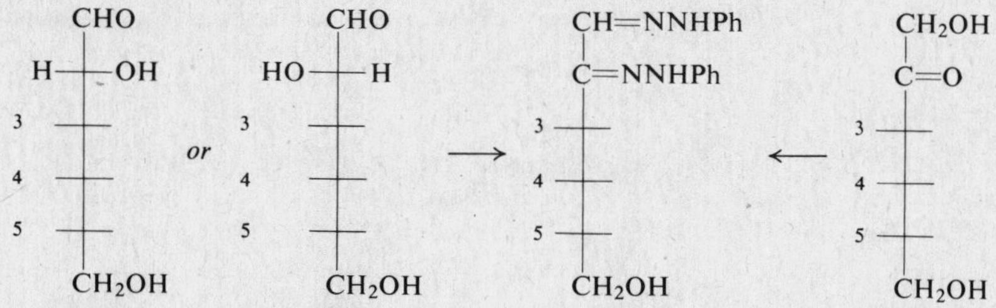

CHO	CHO	CH=NNHPh	CH₂OH
H—OH	HO—H	C=NNHPh	C=O
3	3	3	3
4	4	4	4
5	5	5	5
CH₂OH	CH₂OH	CH₂OH	CH₂OH
Glucose or mannose	Mannose or glucose	Osazone	Fructose

or $\longrightarrow$ $\longleftarrow$

34.8 Addition of the lactone to NaBH₄—even to a limited quantity—means a temporary excess of reducing agent, which reduces the lactone all the way to the alcohol. For example:

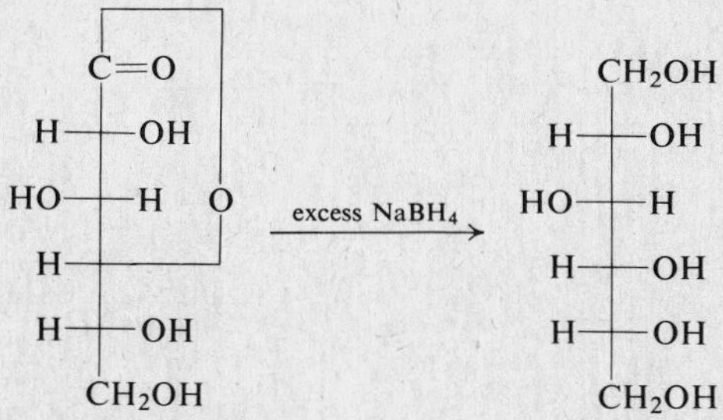

C=O	CH₂OH
H—OH	H—OH
HO—H O	HO—H
H—	H—OH
H—OH	H—OH
CH₂OH	CH₂OH
Gluconolactone	Glucitol

$$\xrightarrow{\text{excess NaBH}_4}$$

34.9

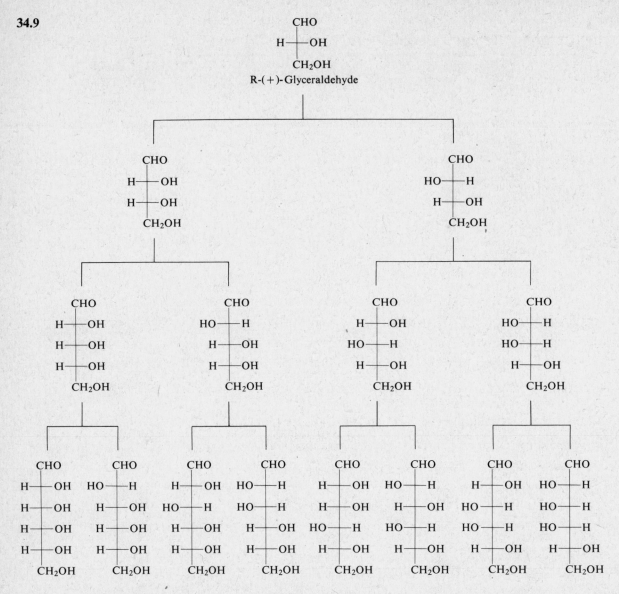

The lowest chiral center has the —OH on the right, since this configuration is undisturbed in building up from R-(+)-glyceraldehyde.

34.10 (a)

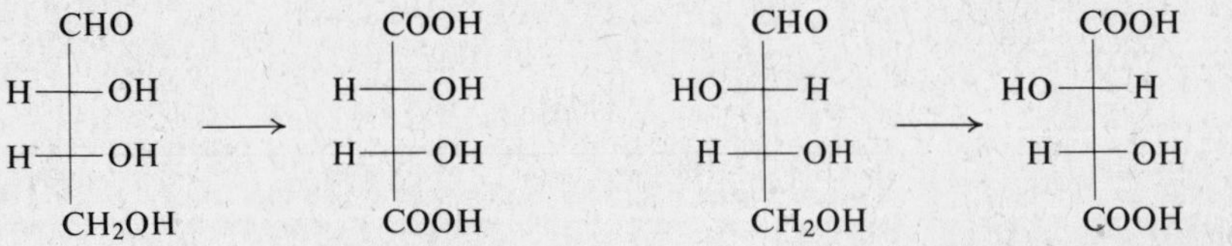

(R,S)-Tartaric acid
(Mesotartaric acid)

Inactive

(S,S)-Tartaric acid

Active

(b) Simply carry out the oxidation, and see whether the product is optically active or inactive.

34.11 (a) Because they give the same osazone, (+)-galactose and (+)-talose must differ only in the configuration about C–2; they are a pair of epimers.   Thus, they can be I and II, or VII and VIII, but not VI.

```
      CHO            CHO                 CHO            CHO
  H──┼──OH      HO──┼──H            H──┼──OH      HO──┼──H
  H──┼──OH       H──┼──OH          HO──┼──H      HO──┼──H
  H──┼──OH       H──┼──OH          HO──┼──H      HO──┼──H
  H──┼──OH       H──┼──OH           H──┼──OH      H──┼──OH
     CH₂OH          CH₂OH              CH₂OH          CH₂OH
       I              II                VII            VIII
```

Of these, I and II would be degraded to ribose:

```
                      CHO
                  H──┼──OH
     I or II ──→   H──┼──OH
                  H──┼──OH
                     CH₂OH
                    Ribose
```

VII and VIII would be degraded to lyxose, and hence must be the structures of galactose and talose.

```
                      CHO
                 HO──┼──H
    VII or VIII ──→ HO──┼──H
                  H──┼──OH
                     CH₂OH
                    Lyxose
```

Since (−)-lyxose belongs to the same family as (+)-glucose—the R-(+)-glyceraldehyde family—so must (+)-galactose and (+)-talose.

This leaves only the problem of deciding which configuration, VII or VIII, represents galactose and which represents talose.   On oxidation with nitric acid, VII would give an optically inactive (*meso*) aldaric acid, and VIII would give an optically active aldaric acid.

```
      CHO                 COOH              CHO                 COOH
  H──┼──OH            H──┼──OH          HO──┼──H            HO──┼──H
 HO──┼──H    HNO₃    HO──┼──H          HO──┼──H    HNO₃    HO──┼──H
 HO──┼──H    ──→     HO──┼──H          HO──┼──H    ──→     HO──┼──H
  H──┼──OH            H──┼──OH           H──┼──OH            H──┼──OH
     CH₂OH               COOH              CH₂OH               COOH
      VII               Meso               VIII              Active
  (+)-Galactose       Inactive         (+)-Talose
```

(+)-Galactose must therefore be VII and (+)-talose must be VIII.

(b) (+)-Allose and (+)-altrose are epimers, differing only in configuration about C–2, the new chiral center generated in the Kiliani-Fischer synthesis. Since (−)-ribose belongs

$$
\begin{array}{ccccc}
& \text{CHO} & & \text{CHO} \\
& \text{H}\!-\!\text{OH} & & \text{HO}\!-\!\text{H} \\
\text{CHO} & \text{H}\!-\!\text{OH} & & \text{H}\!-\!\text{OH} \\
\text{H}\!-\!\text{OH} & \text{H}\!-\!\text{OH} & \text{and} & \text{H}\!-\!\text{OH} \\
\text{H}\!-\!\text{OH} \longrightarrow & \text{H}\!-\!\text{OH} & & \text{H}\!-\!\text{OH} \\
\text{H}\!-\!\text{OH} & \text{CH}_2\text{OH} & & \text{CH}_2\text{OH} \\
\text{CH}_2\text{OH} & & & \\
(-)\text{-Ribose} & \text{I} & & \text{II}
\end{array}
$$

to the same family as (+)-glucose, so do (+)-allose and (+)-altrose. They must be I and II, the only enantiomeric pair left.

Aldose I would give an optically inactive (*meso*) aldaric acid on oxidation, and an optically inactive alditol on reduction. The corresponding products from II would be optically active. (+)-Allose must therefore be I, and (+)-altrose must be II.

$$
\begin{array}{cccc}
\text{CHO} & \text{CH}_2\text{OH} & \text{CHO} & \text{CH}_2\text{OH} \\
\text{H}\!-\!\text{OH} & \text{H}\!-\!\text{OH} & \text{HO}\!-\!\text{H} & \text{HO}\!-\!\text{H} \\
\text{H}\!-\!\text{OH} \xrightarrow{\text{redn.}} & \text{H}\!-\!\text{OH} & \text{H}\!-\!\text{OH} \xrightarrow{\text{redn.}} & \text{H}\!-\!\text{OH} \\
\text{H}\!-\!\text{OH} & \text{H}\!-\!\text{OH} & \text{H}\!-\!\text{OH} & \text{H}\!-\!\text{OH} \\
\text{H}\!-\!\text{OH} & \text{H}\!-\!\text{OH} & \text{H}\!-\!\text{OH} & \text{H}\!-\!\text{OH} \\
\text{CH}_2\text{OH} & \text{CH}_2\text{OH} & \text{CH}_2\text{OH} & \text{CH}_2\text{OH} \\
\text{I} & \textit{Meso} & \text{II} & \textbf{Active} \\
(+)\text{-Allose} & \textbf{Inactive} & (+)\text{-Altrose} &
\end{array}
$$

(c) This leaves only structure VI for idose. Since (−)-idose and (−)-gulose give the same osazone, they are epimers. (−)-Gulose (structure V, page 1085) belongs to the same

$$
\begin{array}{ccc}
\text{CHO} & \text{HC}\!=\!\text{NNHPh} & \text{CHO} \\
\text{HO}\!-\!\text{H} & \text{C}\!=\!\text{NNHPh} & \text{H}\!-\!\text{OH} \\
\text{H}\!-\!\text{OH} & \text{H}\!-\!\text{OH} & \text{H}\!-\!\text{OH} \\
\text{HO}\!-\!\text{H} \longrightarrow & \text{HO}\!-\!\text{H} \longleftarrow & \text{HO}\!-\!\text{H} \\
\text{H}\!-\!\text{OH} & \text{H}\!-\!\text{OH} & \text{H}\!-\!\text{OH} \\
\text{CH}_2\text{OH} & \text{CH}_2\text{OH} & \text{CH}_2\text{OH} \\
\text{VI} & & \\
(-)\text{-Idose} & \text{Osazone} & (-)\text{-Gulose}
\end{array}
$$

family as (+)-glucose, and so, therefore, does (−)-idose.

34.12 See Figure 34.15, page 571 of this Study Guide.

Derived in this systematic way, the aldoses fall into the order given by: **All Altruists Gladly Make Gum In Gallon Tanks.**

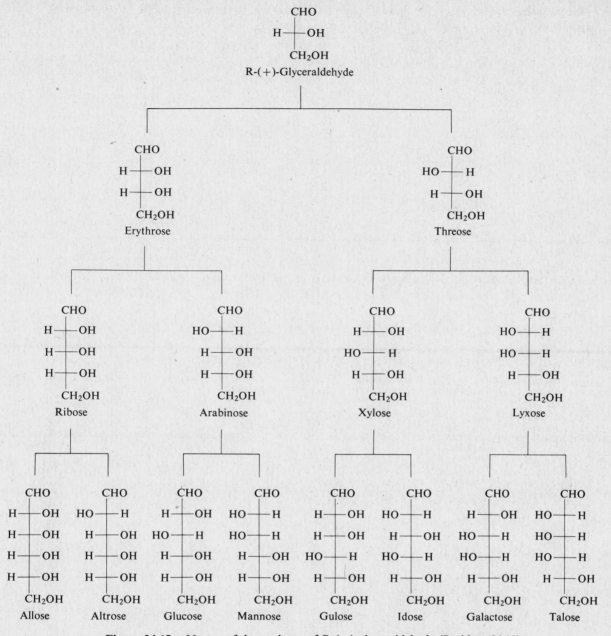

Figure 34.15. Names of descendants of R-(+)-glyceraldehyde (Problem 34.12).

34.13 Since (−)-fructose gives the same osazone as (+)-glucose, it must have this configuration:

34.14 These are the enantiomers of III and IV (page 1085), and of the fructose structure in the answer to Problem 34.13 (above).

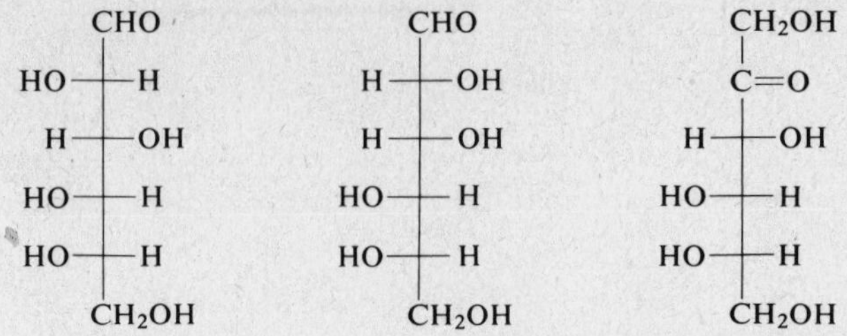

CHO	CHO	CH_2OH
HO——H	H——OH	C=O
H——OH	H——OH	H——OH
HO——H	HO——H	HO——H
HO——H	HO——H	HO——H
CH_2OH	CH_2OH	CH_2OH
(−)-Glucose	(−)-Mannose	(+)-Fructose

34.15

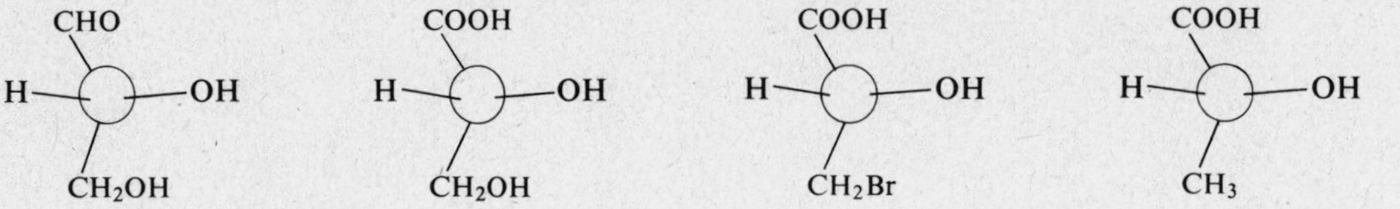

CHO — H—C—OH — CH_2OH — R

COOH — H—C—OH — CH_2OH — R

COOH — H—C—OH — CH_2Br — S

COOH — H—C—OH — CH_3 — R

34.16

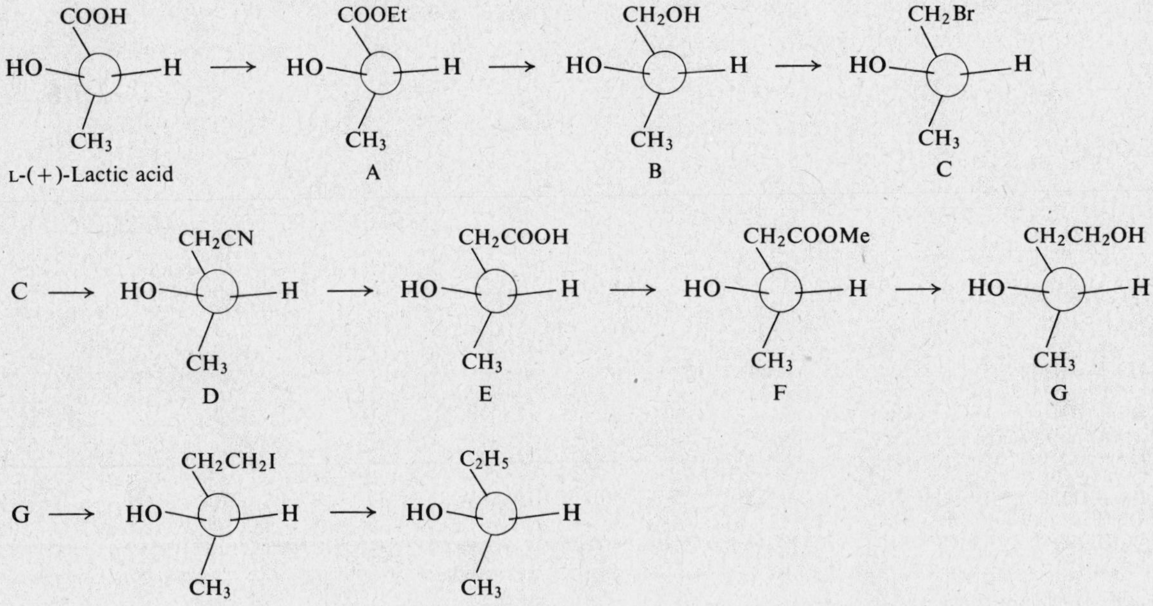

COOH — HO—C—H — CH_3 — L-(+)-Lactic acid → COOEt — HO—C—H — CH_3 — A → CH_2OH — HO—C—H — CH_3 — B → CH_2Br — HO—C—H — CH_3 — C

C → CH_2CN — HO—C—H — CH_3 — D → CH_2COOH — HO—C—H — CH_3 — E → CH_2COOMe — HO—C—H — CH_3 — F → CH_2CH_2OH — HO—C—H — CH_3 — G

G → CH_2CH_2I — HO—C—H — CH_3 — H → C_2H_5 — HO—C—H — CH_3 — S-(+)-2-Butanol

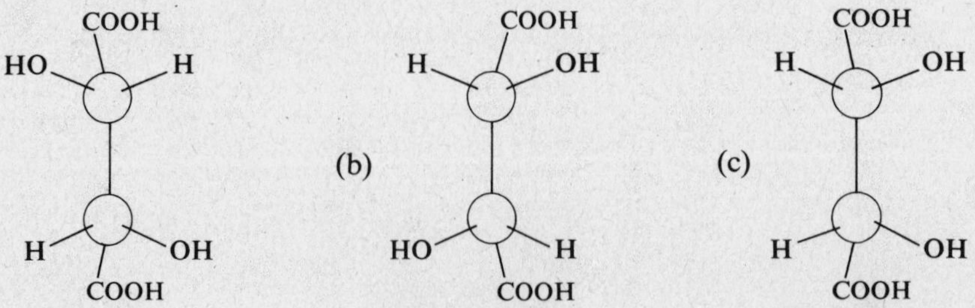

34.17 (a)

COOH — HO—C—H — H—C—OH — COOH — (−)-Tartaric acid (S,S)

(b)

COOH — H—C—OH — HO—C—H — COOH — (+)-Tartaric acid (R,R)

(c)

COOH — H—C—OH — H—C—OH — COOH — Mesotartaric acid (R,S)

34.18 (a) This ratio is determined in the initial step: formation of the cyanohydrins. A chiral center is already present in the reactants; the transition states, like the products, are diastereomeric and of different stabilities.

(b)

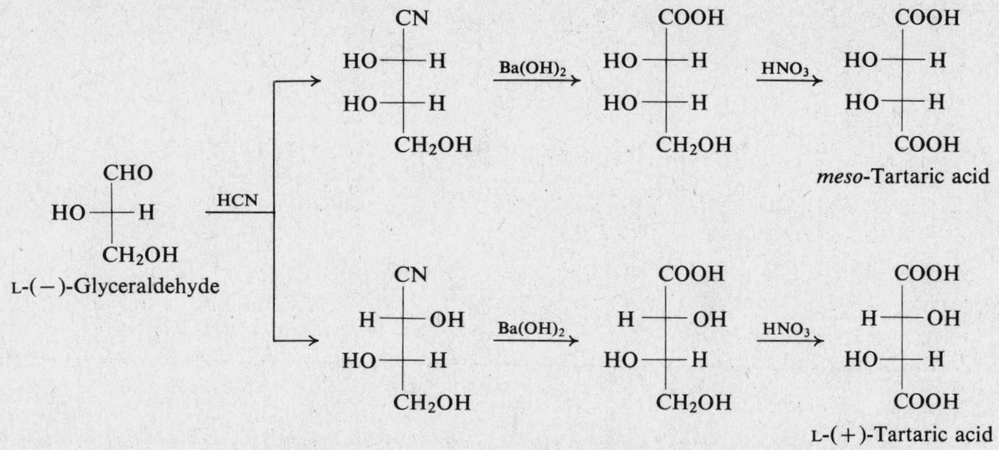

L-(−)-Glyceraldehyde

CHO
HO—H
CH₂OH

$\xrightarrow{\text{HCN}}$

CN
HO—H
HO—H
CH₂OH

$\xrightarrow{\text{Ba(OH)}_2}$

COOH
HO—H
HO—H
CH₂OH

$\xrightarrow{\text{HNO}_3}$

COOH
HO—H
HO—H
COOH

meso-Tartaric acid

CN
H—OH
HO—H
CH₂OH

$\xrightarrow{\text{Ba(OH)}_2}$

COOH
H—OH
HO—H
CH₂OH

$\xrightarrow{\text{HNO}_3}$

COOH
H—OH
HO—H
COOH

L-(+)-Tartaric acid

The ratio *meso*:active will again be 1:3. In cyanohydrin formation, the two transition states are enantiomers of the transition states in the reaction of D-(+)-glyceraldehyde. The energy difference between transition states is the same as in the other case, and so, also, is the ratio between the products.

(c) Inactive, because the enantiomeric cyanohydrins will be formed in equal amounts. The isomer favored in the L-series will be exactly balanced by its enantiomer, favored in the D-series.

34.19

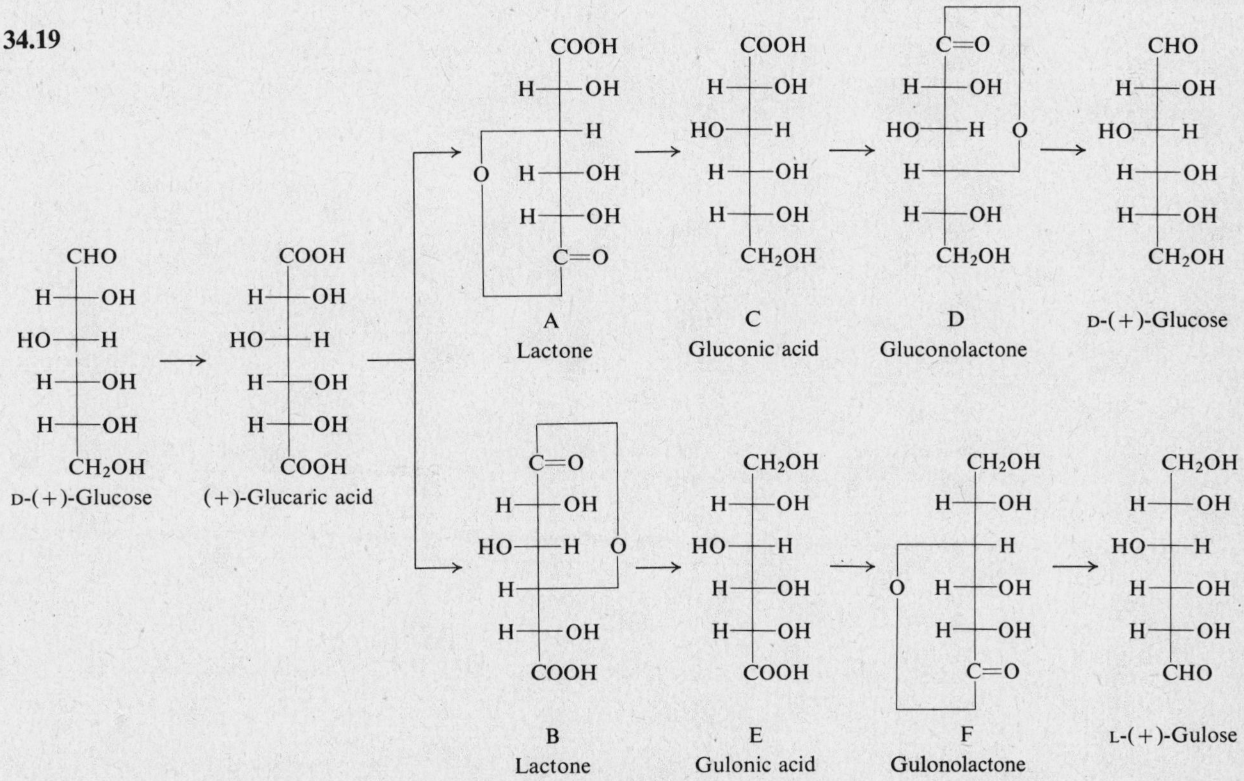

D-(+)-Glucose

CHO
H—OH
HO—H
H—OH
H—OH
CH₂OH

→

(+)-Glucaric acid

COOH
H—OH
HO—H
H—OH
H—OH
COOH

→

A — Lactone

COOH
H—OH
H
H—OH
H—OH
C=O

→

C — Gluconic acid

COOH
H—OH
HO—H
H—OH
H—OH
CH₂OH

→

D — Gluconolactone

C=O
H—OH
HO—H
H
H—OH
CH₂OH

→

D-(+)-Glucose

CHO
H—OH
HO—H
H—OH
H—OH
CH₂OH

B — Lactone

C=O
H—OH
HO—H
H
H—OH
COOH

→

E — Gulonic acid

CH₂OH
H—OH
HO—H
H—OH
H—OH
COOH

→

F — Gulonolactone

CH₂OH
H—OH
H
H—OH
H—OH
C=O

→

L-(+)-Gulose

CH₂OH
H—OH
HO—H
H—OH
H—OH
CHO

(+)-Gulose is a member of the L-family since, with —CHO at the top, the —OH on the bottom chiral carbon is on the left. It is the enantiomer of V on page 1085.

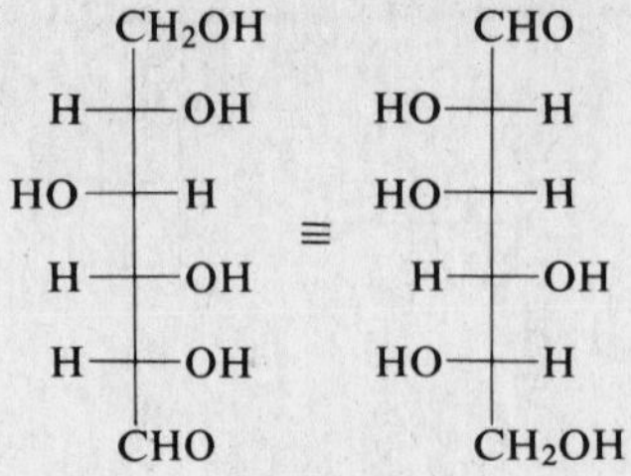
CH2OH
CHO
H—OH
HO—H
HO—H
HO—H
H—OH
H—OH
H—OH
HO—H
≡
CHO
CH2OH

L-(+)-Gulose

34.20 (a)

$$N_\alpha \times 112° + N_\beta \times 19° = 52.7°$$

and

$$N_\alpha + N_\beta = 1$$

where N is the mole fraction of each anomer.

Solving, we find

$$N_\alpha = 36.2\% \qquad N_\beta = 63.8\%$$

(b) The β-form predominates because in it the anomeric —OH group is equatorial, not axial as it is in the α-form.

34.21

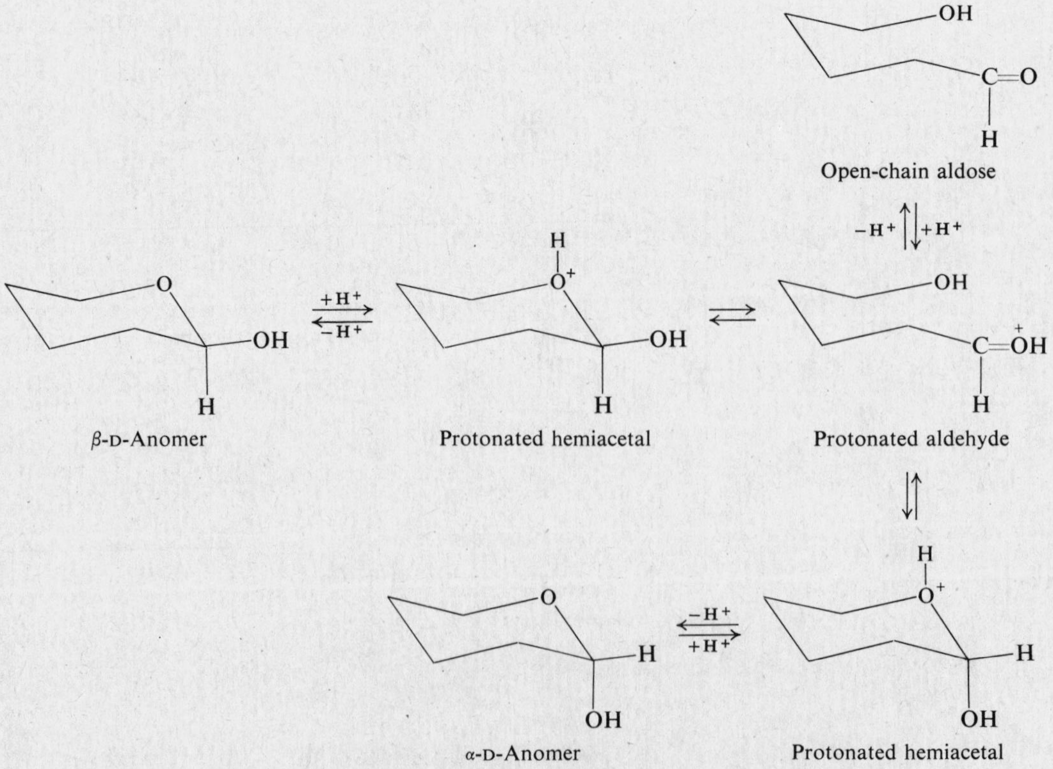
OH
C=O
H
Open-chain aldose
−H⁺
+H⁺
O
OH
H
β-D-Anomer
+H⁺
−H⁺
H
O⁺
OH
H
Protonated hemiacetal
OH
C=OH⁺
H
Protonated aldehyde
O
H
OH
α-D-Anomer
−H⁺
+H⁺
H
O⁺
H
OH
Protonated hemiacetal

34.22 (+)-Glucose $\xrightarrow{Ac_2O}$

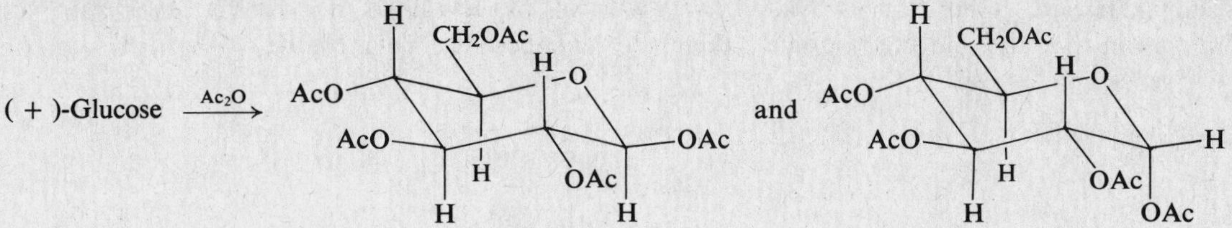

β-Pentaacetate   and   α-Pentaacetate

Acetylation of D-glucose does not give, as we might naively have assumed, an open-chain aldehyde with acetate groups at C–2 through C–6. Instead, the products are cyclic; acetylation occurs, not at C–5, but at C–1, and to give α- and β-forms.

34.23 (a)

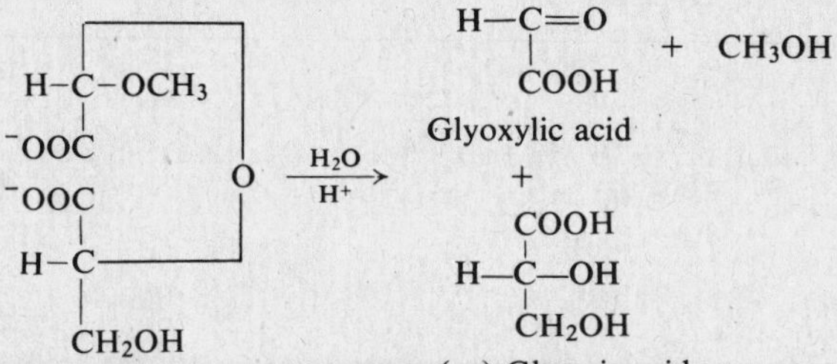

Acetal $\xrightarrow[H^+]{H_2O}$

H–C=O / COOH   Glyoxylic acid   + CH$_3$OH

+

COOH / H–C–OH / CH$_2$OH   D-(−)-Glyceric acid

These products are common to *all* methyl D-aldohexopyranosides.

(b) (+)-Glucose gave the same (−)-glyceric acid as that obtained by oxidation of (+)-glyceraldehyde.

34.24

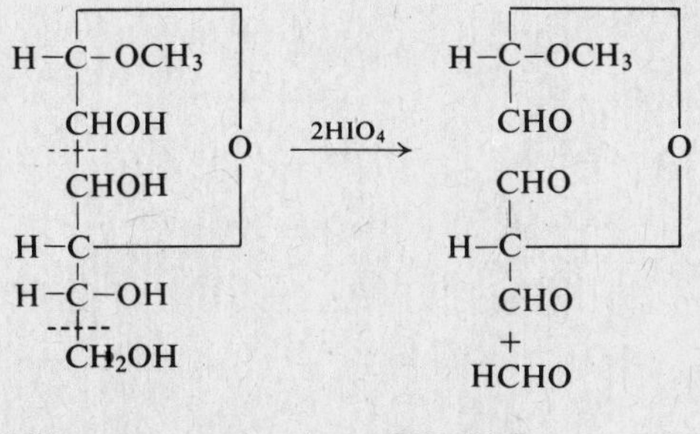

$\xrightarrow{2HIO_4}$

+ HCHO

The easily detected difference: production of HCHO.

34.25 (a)

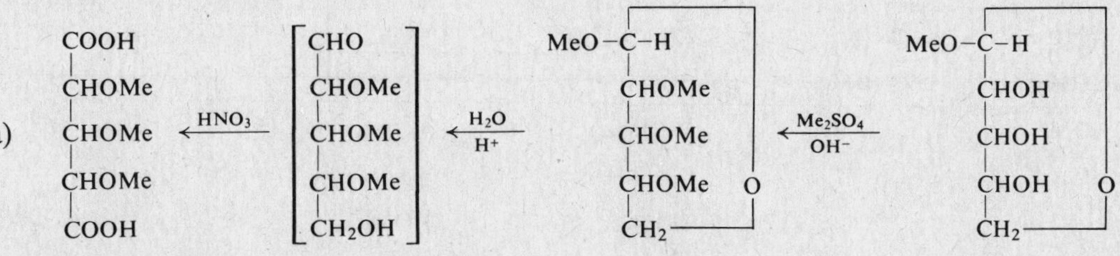

Trimethoxyglutaric acid $\xleftarrow{HNO_3}$ 2,3,4-Tri-O-methylpentose $\xleftarrow[H^+]{H_2O}$ Methyl 2,3,4-tri-O-methylpentoside $\xleftarrow[OH^-]{Me_2SO_4}$ Methyl pentoside

The —OH on C–5 was not methylated because it was not available: it must have been tied up in the ring; and the ring must, therefore, have been six-membered (C–1 through C–5, plus an oxygen).

(b)

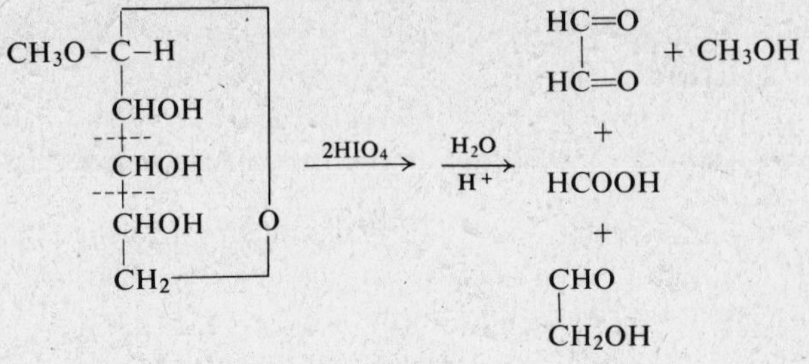

The —OH on C–6 can be oxidized because it was not methylated: it must have been tied up in the ring; and the ring must, therefore, have been six-membered (C–2 through C–6, plus one oxygen).

34.26 (a)

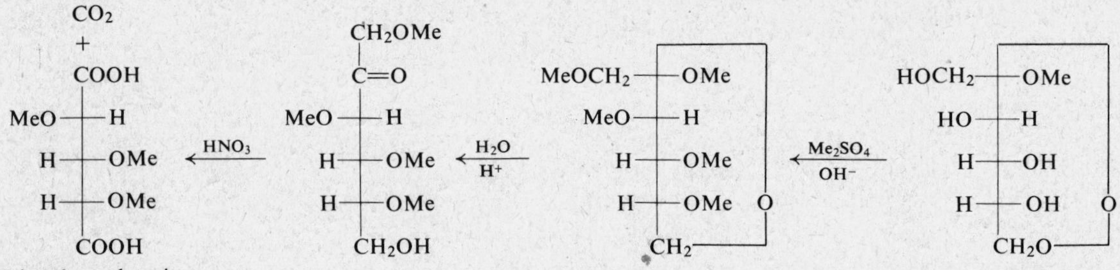

Trimethoxyglutaric acid

Methyl α-D-fructoside

(b)

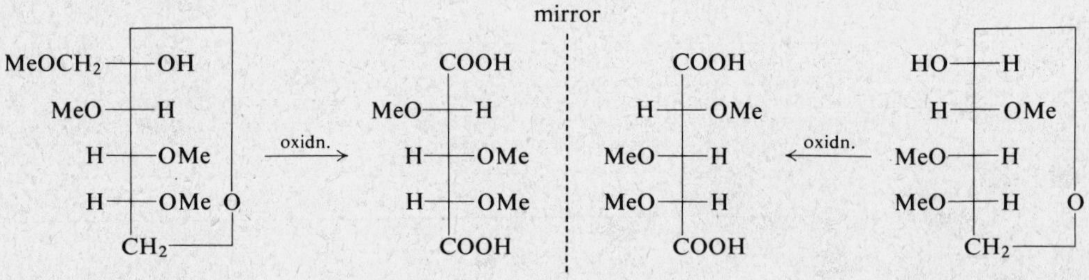

1,3,4,5-Tetra-O-methyl-D-fructopyranose

Enantiomers

2,3,4-Tri-O-methyl-L-arabinopyranose

mirror

34.27 (a)

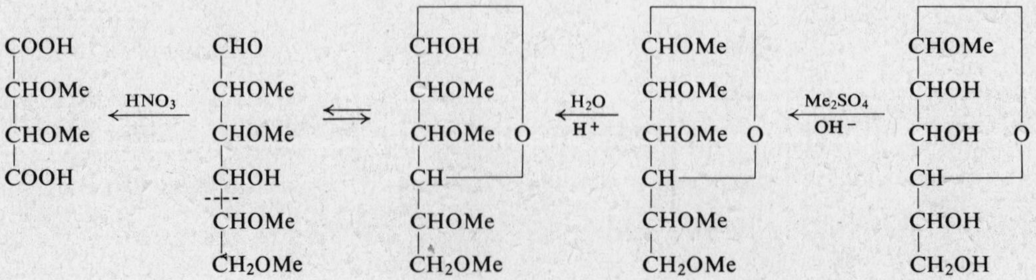

γ-Glucoside

The formation of a dimethoxysuccinic acid (without accompanying glutaric acid) indicates a "free" —OH on C-4. This means a five-membered ring in the γ-glucoside.

(b)

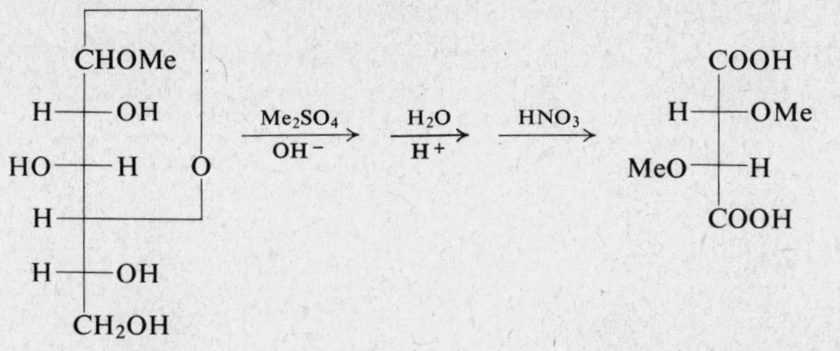

Methyl γ-glucoside   L-Dimethoxysuccinic acid
                                                          Optically active

(c) Following the argument of (a), the γ-fructoside also contains a five-membered ring, the dimethoxysuccinic acid coming from C-2, C-3, C-4, and C-5.

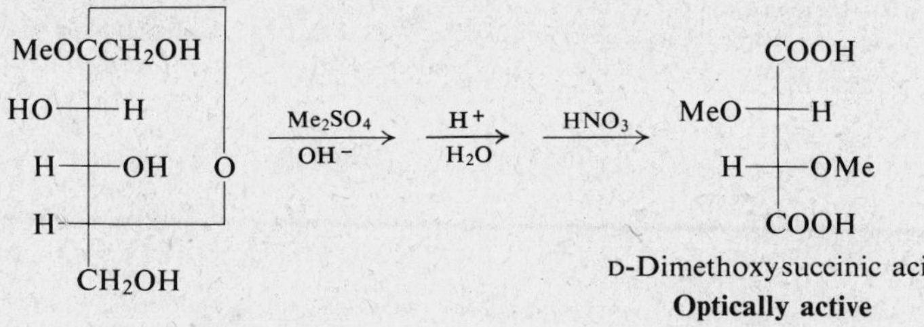

D-Dimethoxysuccinic acid
Optically active

This acid, too, is optically active; it is the enantiomer of the acid from the γ-glucoside.

34.28 In each case, the prediction given below has been confirmed by experiment.

(a)

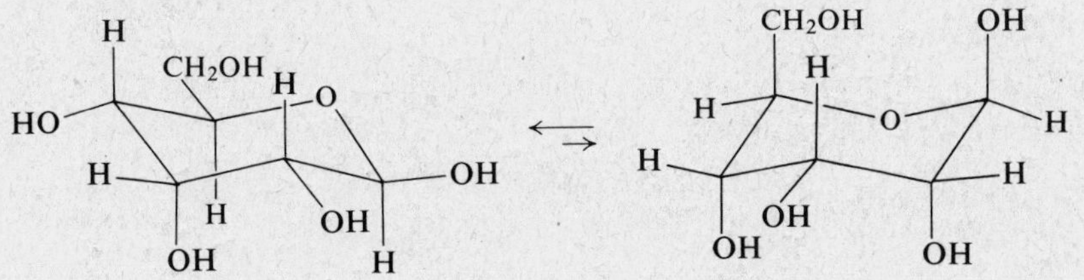

More stable

β-D-Allopyranose

(b)

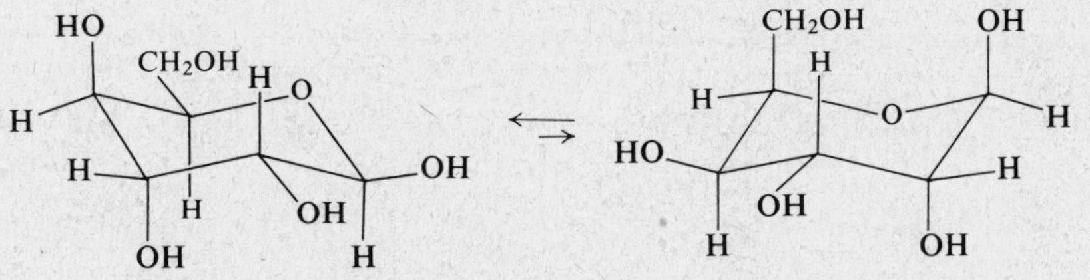

More stable

β-D-Gulopyranose

(c)

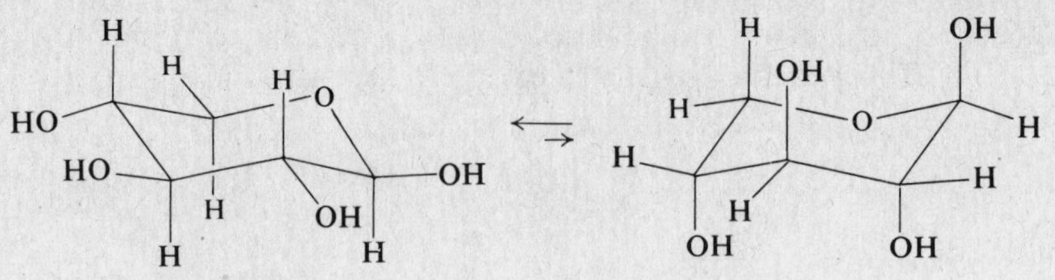

β-D-Xylopyranose

(d)

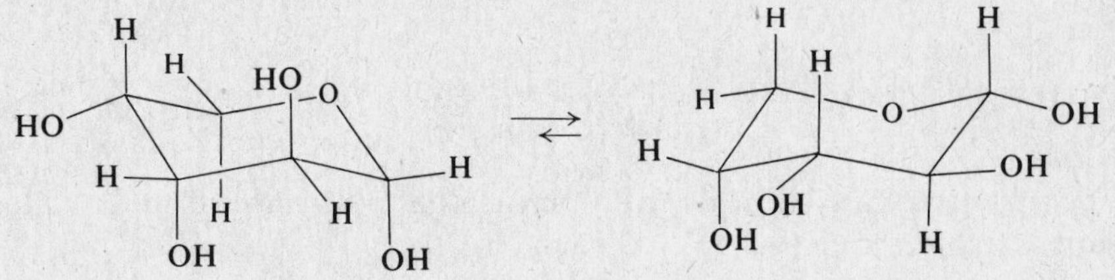

α-D-Arabinopyranose

(e)

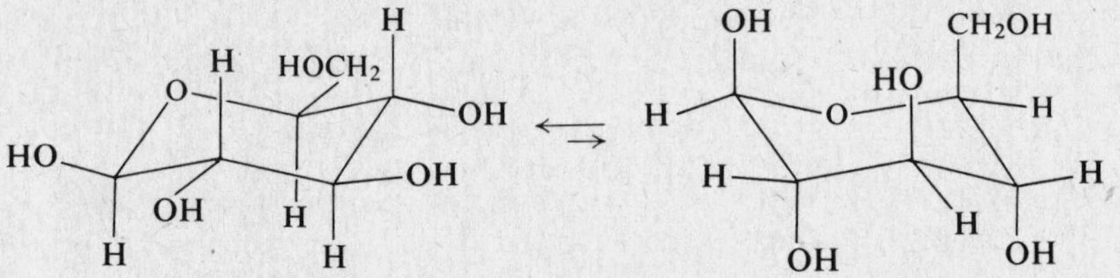

β-L-Glucopyranose

(f)

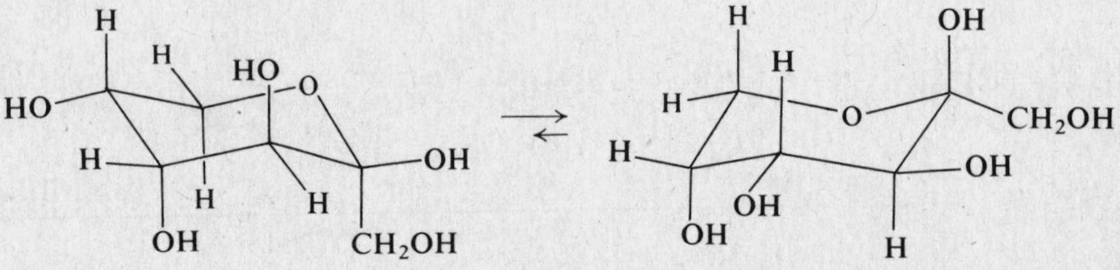

β-D-Fructopyranose

1. (a)

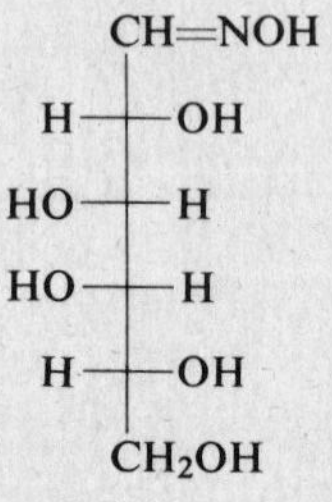

```
CH=NOH
H ——— OH
HO ——— H
HO ——— H
H ——— OH
CH₂OH
```

D-Galactosoxime

(b)

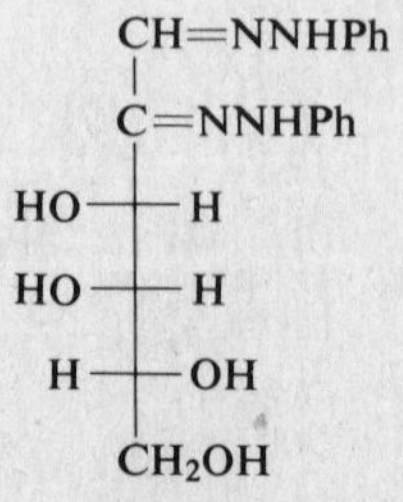

```
CH=NNHPh
C=NNHPh
HO ——— H
HO ——— H
H ——— OH
CH₂OH
```

D-Galactosazone

(c)

```
COOH
H ——— OH
HO ——— H
HO ——— H
H ——— OH
CH₂OH
```

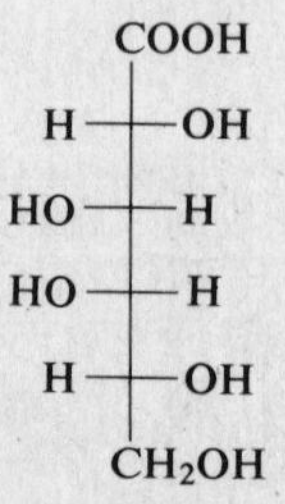

D-Galactonic acid
(or lactone)

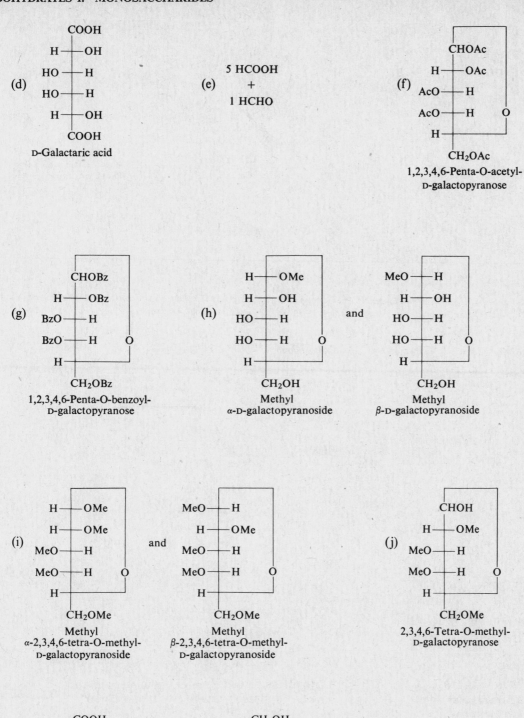

(d)

```
        COOH
   H ——— OH
  HO ——— H
  HO ——— H
   H ——— OH
        COOH
```
D-Galactaric acid

(e)

5 HCOOH
+
1 HCHO

(f)

```
      ┌─────────┐
      CHOAc
   H ——— OAc
  AcO ——— H
  AcO ——— H ─── O
   H
      CH₂OAc
```
1,2,3,4,6-Penta-O-acetyl-
D-galactopyranose

(g)

```
      ┌─────────┐
      CHOBz
   H ——— OBz
  BzO ——— H
  BzO ——— H ─── O
   H
      CH₂OBz
```
1,2,3,4,6-Penta-O-benzoyl-
D-galactopyranose

(h)

```
      ┌─────────┐
   H ——— OMe
   H ——— OH
  HO ——— H
  HO ——— H ─── O
   H
      CH₂OH
```
Methyl
α-D-galactopyranoside

and

```
      ┌─────────┐
  MeO ——— H
   H ——— OH
  HO ——— H
  HO ——— H ─── O
   H
      CH₂OH
```
Methyl
β-D-galactopyranoside

(i)

```
      ┌─────────┐
   H ——— OMe
   H ——— OMe
  MeO ——— H
  MeO ——— H ─── O
   H
      CH₂OMe
```
Methyl
α-2,3,4,6-tetra-O-methyl-
D-galactopyranoside

and

```
      ┌─────────┐
  MeO ——— H
   H ——— OMe
  MeO ——— H
  MeO ——— H ─── O
   H
      CH₂OMe
```
Methyl
β-2,3,4,6-tetra-O-methyl-
D-galactopyranoside

(j)

```
      ┌─────────┐
      CHOH
   H ——— OMe
  MeO ——— H
  MeO ——— H ─── O
   H
      CH₂OMe
```
2,3,4,6-Tetra-O-methyl-
D-galactopyranose

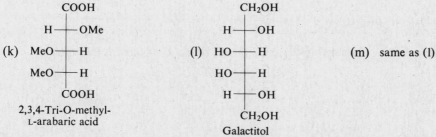

(k)

```
        COOH
   H ——— OMe
  MeO ——— H
  MeO ——— H
        COOH
```
2,3,4-Tri-O-methyl-
L-arabaric acid

(l)

```
        CH₂OH
   H ——— OH
  HO ——— H
  HO ——— H
   H ——— OH
        CH₂OH
```
Galactitol

(m)   same as (l)

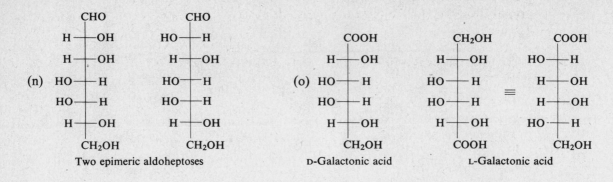

(n) Two epimeric aldoheptoses   (o) D-Galactonic acid   ≡   L-Galactonic acid

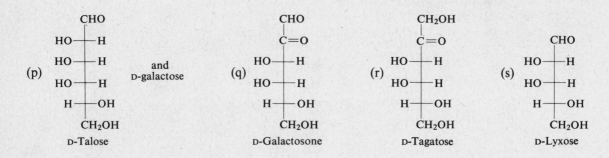

(p) and D-galactose   D-Talose   (q) D-Galactosone   (r) D-Tagatose   (s) D-Lyxose

(t) No reaction

(u) Product on page 1099
from α-anomer;
β-anomer
gives opposite
configuration
at C–1.

(v)

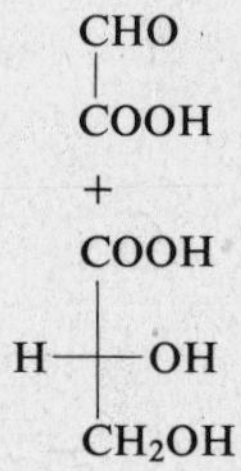

$$\begin{array}{c} CHO \\ | \\ COOH \end{array}$$

+

$$\begin{array}{c} COOH \\ | \\ H \!-\!\!\!-\! OH \\ | \\ CH_2OH \end{array}$$

D-Glyceric acid

2. (a) MeOH, HCl.

(b) Product (a), Me_2SO_4, OH^-.

(c) Product (b), dil. HCl.

(d) See Figure 34.4 (page 1081). The starting material shown there is D-glucose, and the product is D-mannose.

(e) See answer to Problem 34.19, above.

(f) See Figure 34.3 (page 1080). The starting material shown there is D-glucose, and the product is D-arabinose.

(g) Product (f) treated as at top of page 1087.

(h) $NaBH_4$; then Ac_2O, H^+.

(i) $3PhNHNH_2$; PhCHO, H^+ (osone formation); Zn, HOAc.   (Compare the answer to Problem 34.6, above.)

(j) Kiliani-Fischer synthesis (Sec. 34.8): CN^-, H^+; H_2O, H^+, warm; separate the two lactones; Na(Hg), CO_2.

3. (a) See answer to Problem 34.3(c).

(b) D-Psicose: gives the same osazone as D-allose or D-altrose.
D-Sorbose: gives the same osazone as D-gulose or D-idose.
D-Tagatose: gives the same osazone as D-galactose or D-talose.

4. (a)

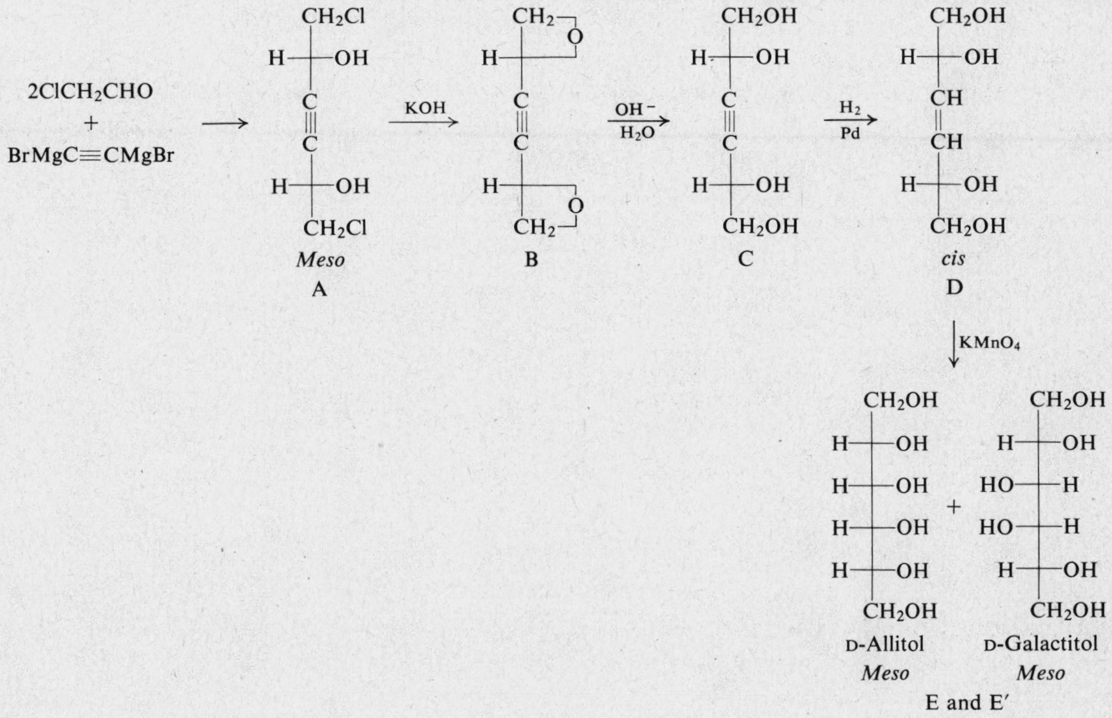

$$2ClCH_2CHO + BrMgC{\equiv}CMgBr \longrightarrow$$

Meso
A

$\xrightarrow{KOH}$ B

$\xrightarrow[H_2O]{OH^-}$ C

$\xrightarrow[Pd]{H_2}$ cis D

$\downarrow KMnO_4$

D-Allitol *Meso*   D-Galactitol *Meso*

E and E′

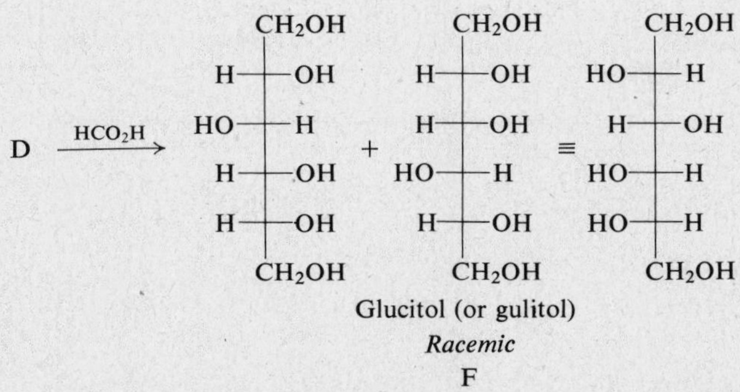

D $\xrightarrow{HCO_2H}$ Glucitol (or gulitol) *Racemic*   F

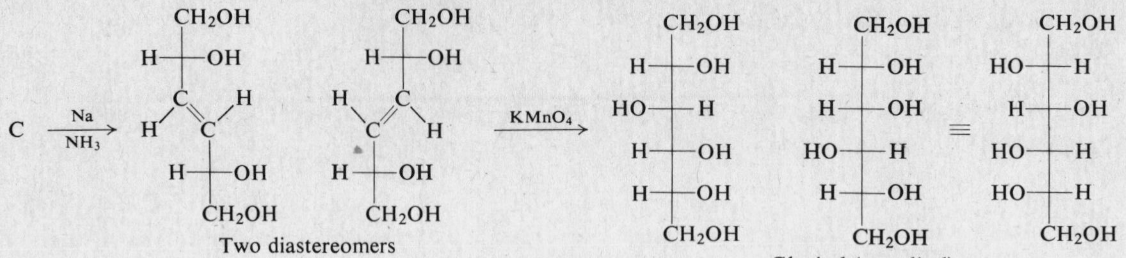

$$C \xrightarrow[\text{NH}_3]{\text{Na}}$$

Two diastereomers

G

$$\xrightarrow{\text{KMnO}_4}$$

Glucitol (or gulitol)

Racemic

H

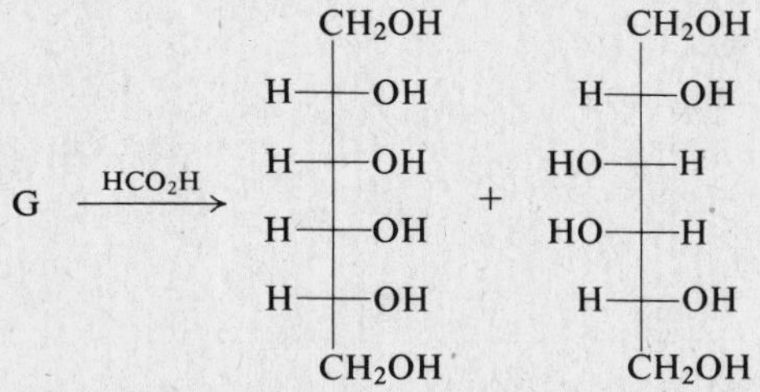

$$G \xrightarrow{\text{HCO}_2\text{H}}$$

+

D-Allitol

Meso

D-Galactitol

Meso

I and I′

(b)

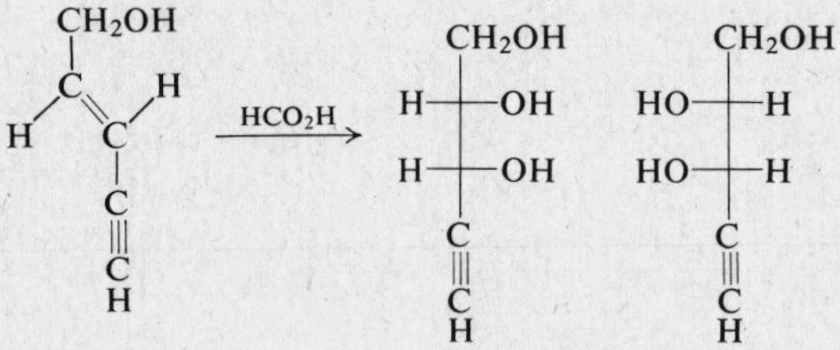

$$\xrightarrow{\text{HCO}_2\text{H}}$$

Racemic

J

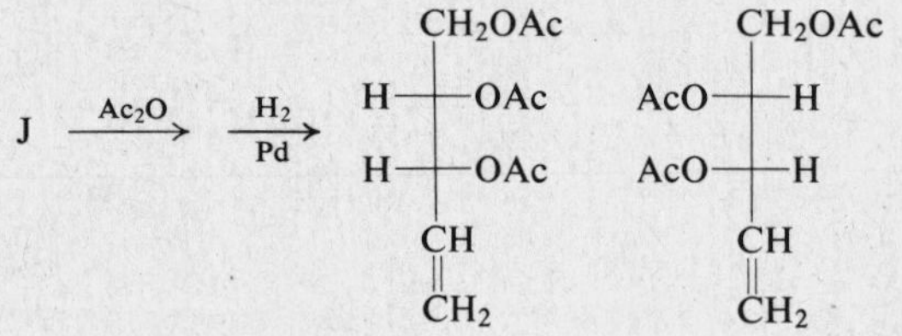

$$J \xrightarrow{\text{Ac}_2\text{O}} \xrightarrow[\text{Pd}]{\text{H}_2}$$

Racemic

K

$$\downarrow \text{HOBr}$$

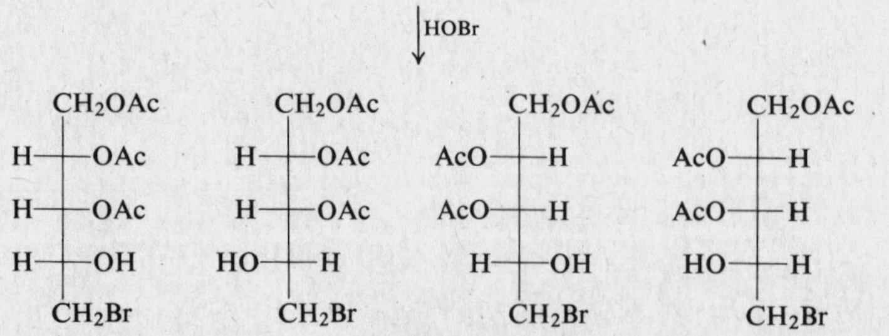

Enantiomers

Enantiomers

L and M

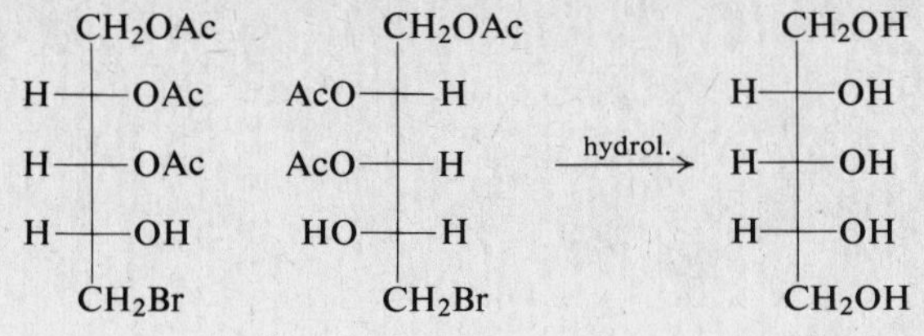

CH₂OAc | CH₂OAc   CH₂OH

L $\xrightarrow{\text{hydrol.}}$

Meso
N
Ribitol

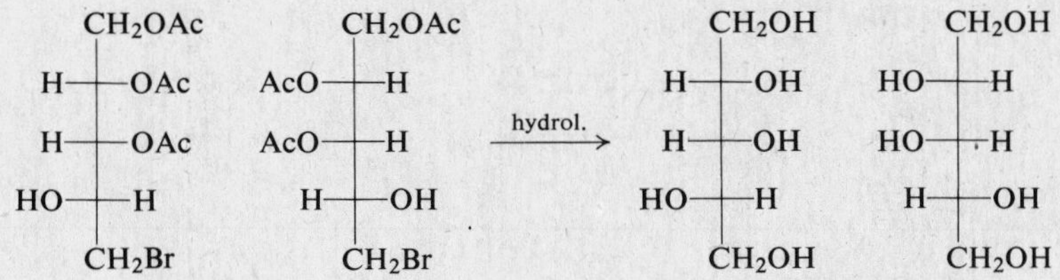

M $\xrightarrow{\text{hydrol.}}$ *Racemic*
O
Arabitol
(Lyxitol)

(c) There are two possible routes to each.

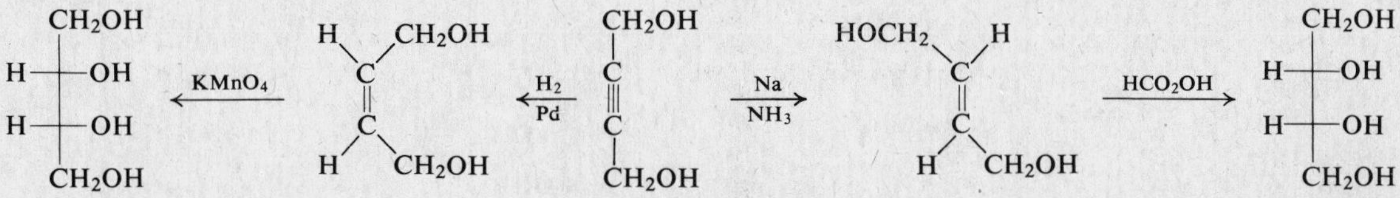

Erythritol $\xleftarrow{\text{KMnO}_4}$ *cis* $\xleftarrow[\text{Pd}]{\text{H}_2}$ 2-Butyn-1,4-diol $\xrightarrow[\text{NH}_3]{\text{Na}}$ *trans* $\xrightarrow{\text{HCO}_2\text{OH}}$ Erythritol

Starting material

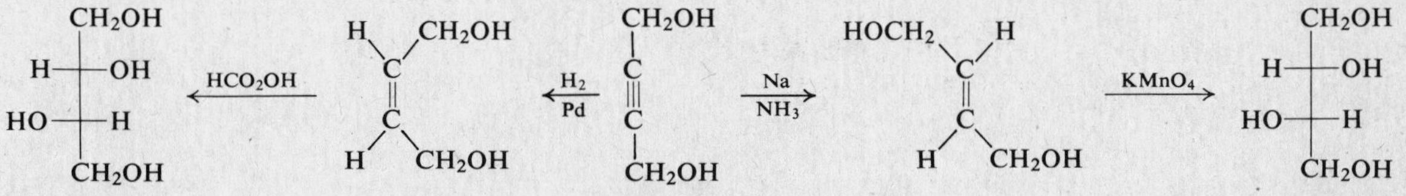

+ enantiomer
DL-Threitol $\xleftarrow{\text{HCO}_2\text{OH}}$ *cis* $\xleftarrow[\text{Pd}]{\text{H}_2}$ 2-Butyn-1,4-diol $\xrightarrow[\text{NH}_3]{\text{Na}}$ *trans* $\xrightarrow{\text{KMnO}_4}$ + enantiomer
DL-Threitol

Starting material

(d) Nucleophilic carbonyl addition.

5. (a)

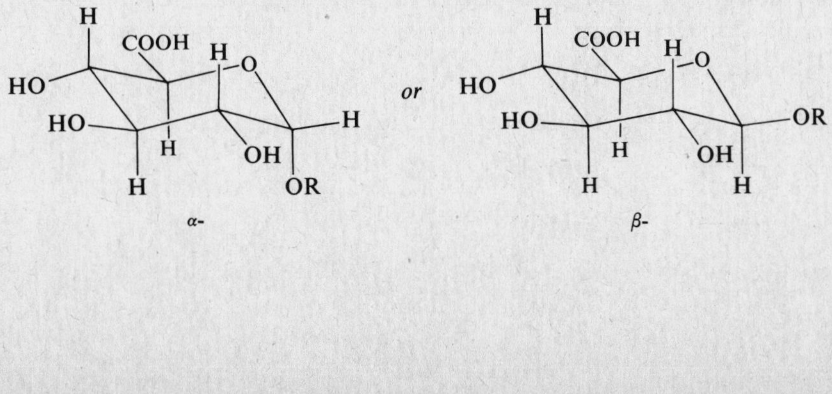

α-   *or*   β-

(b)

```
      CHO                    CHO
   H——OH               HO——H
  HO——H                HO——H
  HO——H                 H——OH
   H——OH                H——OH
     COOH                  COOH
 D-Galacturonic acid   D-Mannuronic acid
```

(c)

```
   CH2OH                          COOH
    C=O                            C=O                    COOH              COOH
  HO——H          ———→         HO——H         ———→      H——OH            HO——H
   H——OH                        H——OH                  HO——H       and   HO——H
   H——OH                        H——OH                   H——OH             H——OH
     CH2OH                         CH2OH                 H——OH             H——OH
  D-Fructose                  2-Keto-D-gluconic            CH2OH             CH2OH
                                   acid               D-Gluconic acid    D-Mannonic acid

                                  CH2OH                 CH2OH             CH2OH
                                   C=O               H——OH            HO——H
                             HO——H          ———→     HO——H       and   HO——H
                               H——OH                  H——OH             H——OH
                               H——OH                  H——OH             H——OH
                                 COOH                   COOH              COOH
                             5-Keto-L-gulonic       L-Gulonic acid    D-Mannonic acid
                                   acid
```

(d)

```
      CHO                       COOH
       C=O          Br2(aq)      C=O
   HO——H          ————→      HO——H
    H——OH                      H——OH
    H——OH                      H——OH
      CH2OH                      CH2OH
   D-Glucosone             2-Keto-D-gluconic acid
```

6. The rate-determining step involves OH^- before reaction with Cu^{++}: probably abstraction of a proton leading to the formation of the enediol (page 1076). (Compare Section 21.3 and the answer to Problem 21, Chapter 21, page 724.)

```
      O                        O                    OH
      ‖          OH⁻            ‖          H⁺          |
      C—H      ————→    ⊖{    C—H      ————→       C—H      ——Cu⁺⁺——→   oxidation products
      |         -H⁺            ‖                    ‖
  H—C—OH                      C—OH                 C—OH

    Aldose                  Carbanion              Enediol
```

7. (a)

$$
\begin{array}{c}
\text{H–C–OMe} \\
\text{CHO} \\
\text{CHO} \quad \text{O} \\
\text{H–C} \\
\text{CH}_2\text{OH}
\end{array}
\quad \xleftarrow{\text{HIO}_4} \quad
\begin{array}{c}
\text{H–C–OMe} \\
\text{CHOH} \\
\text{CHOH} \quad \text{O} \\
\text{H–C} \\
\text{CH}_2\text{OH} \\
\mathbf{Q} \\
\textit{Five-membered ring}
\end{array}
$$

(b) The configuration is known for C–1 and C–4.

(c)

$$
\begin{array}{c}
\text{COOH} \\
\text{MeO}\!-\!\!-\!\text{H} \\
\text{H}\!-\!\!-\!\text{OMe} \\
\text{COOH}
\end{array}
\xleftarrow{\text{HNO}_3}
\begin{array}{c}
\text{CHO} \\
\text{MeO}\!-\!\!-\!\text{H} \\
\text{H}\!-\!\!-\!\text{OMe} \\
\text{H}\!-\!\!-\!\text{OH} \\
\text{CH}_2\text{OMe}
\end{array}
\xleftarrow[\text{H}^+]{\text{H}_2\text{O}}
\begin{array}{c}
\text{H–C–OMe} \\
\text{MeO}\!-\!\!-\!\text{H} \\
\text{H}\!-\!\!-\!\text{OMe} \quad \text{O} \\
\text{H} \\
\text{CH}_2\text{OMe}
\end{array}
$$

Di-O-methyl ether of
D-(−)-tartaric acid

$$\text{OH}^- \Big| \text{Me}_2\text{SO}_4$$

$$
\begin{array}{c}
\text{H–C–OMe} \\
\text{HO}\!-\!\!-\!\text{H} \\
\text{H}\!-\!\!-\!\text{OH} \quad \text{O} \\
\text{H} \\
\text{CH}_2\text{OH}
\end{array}
\quad or \quad
\text{Methyl } \alpha\text{-D-arabinofuranoside}
$$

8. Salicin is a non-reducing sugar and evidently a β-glucoside (hydrolyzed by emulsin). From

$$
\text{D-glucose} + \text{ROH} \quad \xleftarrow[\text{emulsin}]{\text{H}_2\text{O}} \quad
\begin{array}{c}
\text{RO}\!-\!\!-\!\text{H} \\
\text{H}\!-\!\!-\!\text{OH} \\
\text{HO}\!-\!\!-\!\text{H} \\
\text{H}\!-\!\!-\!\text{OH} \quad \text{O} \\
\text{H} \\
\text{CH}_2\text{OH}
\end{array}
$$

Saligenin
$(C_7H_8O_2)$

Salicin

(Pyranose ring assumed)

its formula $(C_7H_8O_2)$ saligenin, ROH, appears to be aromatic.

The HNO_3 treatment evidently does not affect the glucose moiety—subsequent hydrolysis gives D-glucose unchanged—and hence it must be a group in R of saligenin that is oxidized. The isolation of salicylaldehyde shows that this can only involve oxidation of —CH$_2$OH to —CHO. We now know the structure of a saligenin and hence of salicin.

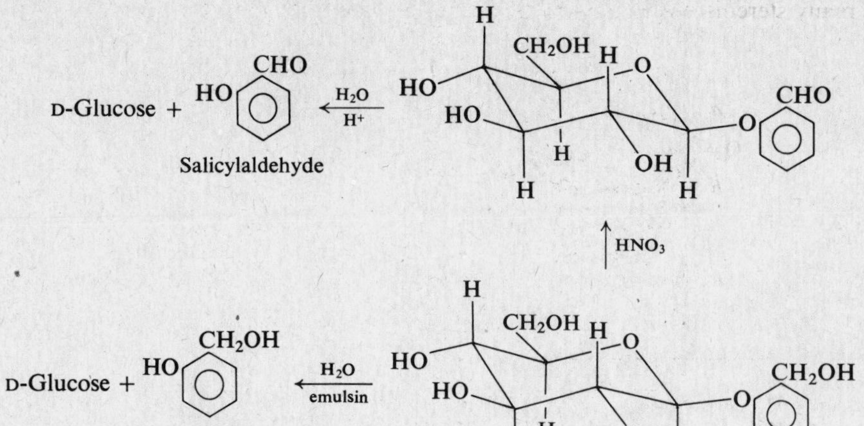

D-Glucose + Salicylaldehyde $\xleftarrow[\text{H}^+]{\text{H}_2\text{O}}$

D-Glucose + Saligenin
o-(Hydroxymethyl)phenol $\xleftarrow[\text{emulsin}]{\text{H}_2\text{O}}$

Salicin
o-(Hydroxymethyl)phenyl
β-D-glucopyranoside

Methylation and hydrolysis confirms the assumption that salicin is a pyranoside, not a furanoside. (The fifth methyl of pentamethylsalicin is in —CH$_2$OCH$_3$ attached to the aromatic ring.)

9. Let us examine R and S first. Chemical arithmetic,

$$\text{C}_6\text{H}_{14}\text{O}_6 \quad \text{sat'd open-chain}$$
$$-\ \text{C}_6\text{H}_{12}\text{O}_6 \quad \text{R or S}$$
$$\text{2H}\ \ \text{missing } \textit{means}\text{ one ring}$$

indicates—in the absence of unsaturation—the presence of one ring. The acetylation data indicate six —OH's in R and S, and the HIO$_4$ results narrow the possibilities down to stereo-isomers of 1,2,3,4,5,6-hexahydroxycyclohexane (*inositol*).

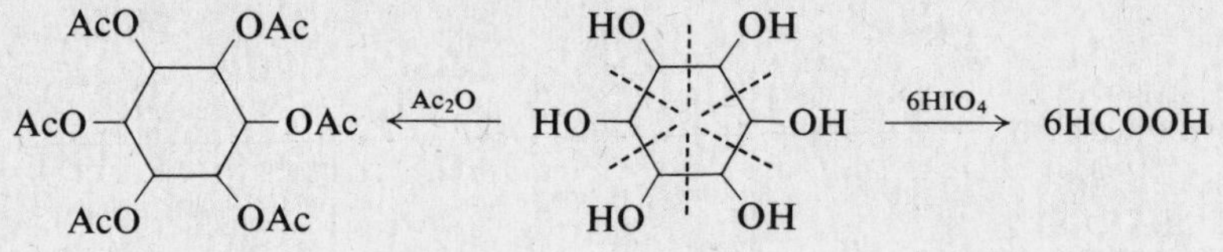

Hexaacetates
C$_{18}$H$_{24}$O$_{12}$

R and S
Inositols
C$_6$H$_{12}$O$_6$

→ 6HCOOH

Bio-inonose is, then, a pentahydroxycyclohexanone. It reduces Benedict's solution because it is an α-hydroxyketone (Sec. 34.6). It does not reduce bromine water because it is not an aldehyde or hemiacetal.

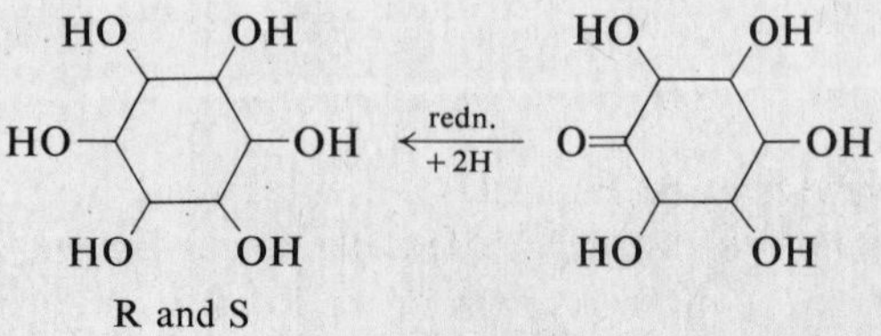

R and S
Inositols

Bio-inonose (*partial structure*)
Positive Benedict's

But which of the many stereoisomeric possibilities is bio-inonose? Vigorous oxidation gives the answer.

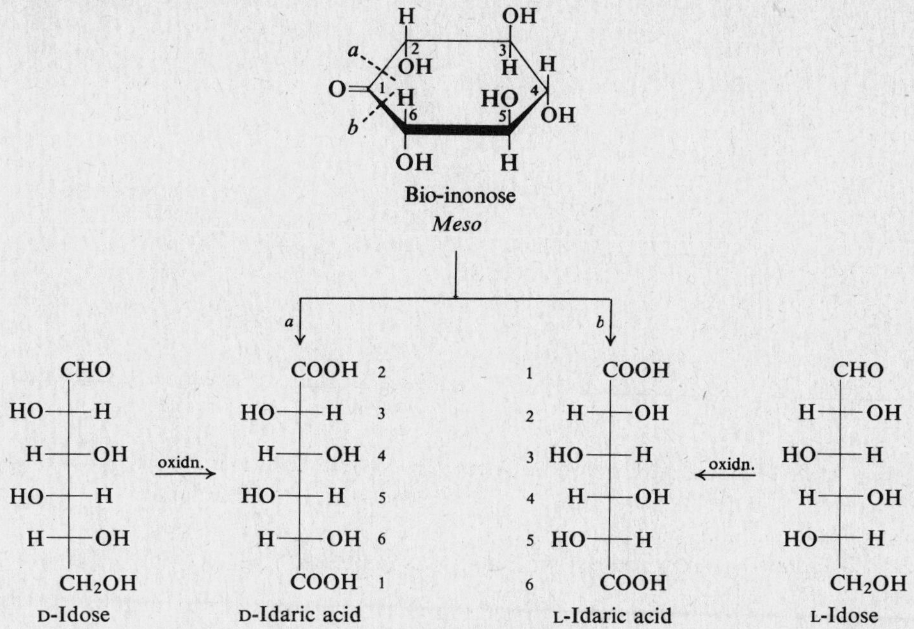
Bio-inonose
Meso
oxidn.
oxidn.
D-Idose
D-Idaric acid
L-Idaric acid
L-Idose

Reduction of bio-inonose gives both possible configurations about the new chiral center, and the diastereomers R and S are formed.

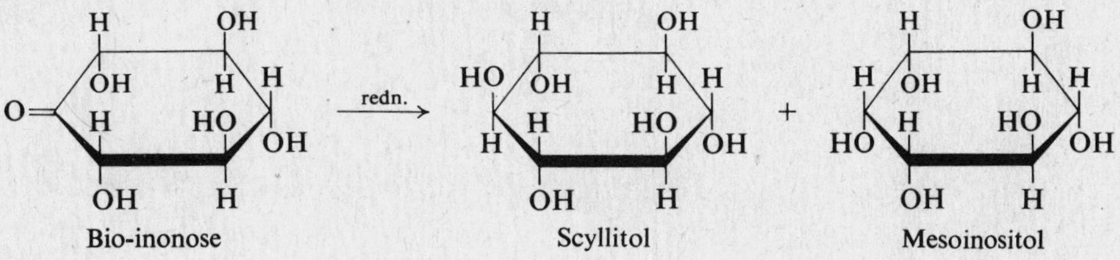
redn.
Bio-inonose
Scyllitol
Meso
Mesoinositol
Meso

R and S

It is not hard to figure out the more stable configuration of each compound.

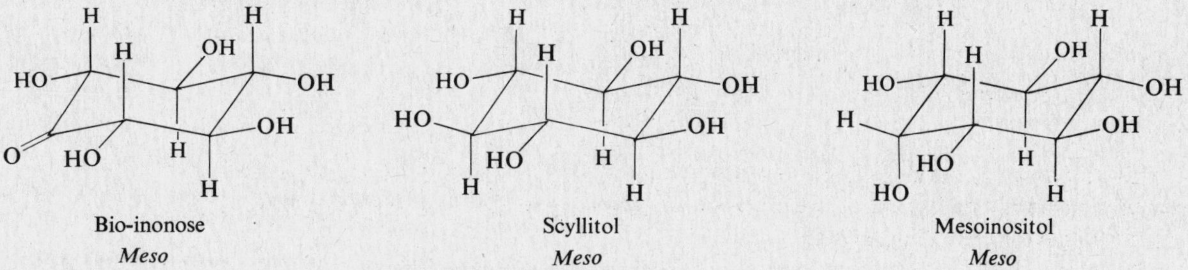
Bio-inonose
Meso
Scyllitol
Meso
Mesoinositol
Meso

R and S

10. (a) $\overset{6}{CH_2OH}-\overset{5}{CHOH}-\overset{4}{CHOH}-\overset{3}{CHOH}-\overset{2}{CHOH}+\overset{1}{CHO} \xrightarrow{\text{Ruff}}$

Glucose

$\overset{6}{CH_2OH}-\overset{5}{CHOH}-\overset{4}{CHOH}-\overset{3}{CHOH}-\overset{2}{CHO} + CO_2$

Arabinose C-1

$$\overset{6}{CH_2OH}-\overset{5}{CHOH}-\overset{4}{CHOH}-\overset{3}{CHOH}\vdots\overset{2}{CHO} \xrightarrow{\text{Ruff}} \begin{matrix} CO_2 \\ \textbf{C-2} \end{matrix}$$

$$\underset{\text{Arabinose}}{}$$

$$\overset{6}{CH_2OH}\vdots\overset{5}{CHOH}\vdots\overset{4}{CHOH}\vdots\overset{3}{CHOH}\vdots\overset{2}{CHOH}\vdots\overset{1}{CHO} \xrightarrow{\text{HIO}_4} \begin{matrix} HCHO + 5HCOOH \\ \textbf{C-6} \end{matrix}$$

$$\underset{\text{Glucose}}{}$$

Glucose $\xrightarrow[\text{HCl}]{\text{CH}_3\text{OH}}$ $\overset{6}{CH_2OH}-\overset{5}{CH}-\overset{4}{CHOH}\vdots\overset{3}{CHOH}\vdots\overset{2}{CHOH}-\overset{\overset{OCH_3}{|}}{CH}$ $\xrightarrow{\text{HIO}_4}$ $\begin{matrix} HCOOH \\ \textbf{C-3} \end{matrix}$

$$\underset{O}{\underbrace{\hspace{4cm}}}$$

(See page 1099)

$$\begin{matrix} CO_2 \\ \textbf{C-4} \\ + \end{matrix}$$

$$\overset{6}{CH_3}\vdots\overset{5}{CHO} \xrightarrow{\text{NaOI}} \begin{matrix} CHI_3 + HCOOH \\ \textbf{C-6} \quad\quad \textbf{C-5} \end{matrix}$$

$$\uparrow \text{KMnO}_4$$

$$\overset{6}{CH_3}\overset{5}{CHOH}\vdots\overset{4}{COOH}$$

$$\overset{6}{CH_2OH}\overset{5}{CHOH}\overset{4}{CHOH}\vdots\overset{3}{CHOH}\overset{2}{CHOH}\overset{1}{CHO} \xrightarrow{\textit{L. casei}}$$

$$\underset{\text{Glucose}}{}$$

$$\overset{3}{HOOC}\vdots\overset{2}{CHOH}\overset{1}{CH_3}$$

$$\downarrow \text{KMnO}_4$$

$$\overset{2}{OHC}\vdots\overset{1}{CH_3} \xrightarrow{\text{NaOI}} \begin{matrix} HCOOH + CHI_3 \\ \textbf{C-2} \quad\quad \textbf{C-1} \end{matrix}$$

$$\begin{matrix} + \\ CO_2 \\ \textbf{C-3} \end{matrix}$$

(b) $\overset{5}{CH_2OH}\vdots\overset{4}{CHOH}\vdots\overset{3}{CHOH}\vdots\overset{2}{\underset{\overset{||}{O}}{C}}-\overset{1}{CH_2OH}$ $\xrightarrow{\text{HIO}_4}$ $\begin{matrix} HCHO + 2HCOOH + HOOC-\overset{1}{CH_2OH} \\ \textbf{C-5} \quad\quad \textbf{C-4 and C-3} \quad\quad \textbf{C-2 and C-1} \end{matrix}$

$$\underset{\text{Ribulose}}{}$$

ribulose $\xrightarrow[\text{Pt}]{\text{H}_2}$ $\overset{5}{CH_2OH}\vdots\overset{4}{CHOH}\vdots\overset{3}{CHOH}\vdots\overset{2}{CHOH}\vdots\overset{1}{CH_2OH}$ $\xrightarrow{\text{HIO}_4}$ $\begin{matrix} 2HCHO \quad + \quad 3HCOOH \\ \textbf{C-5 and C-1} \quad \textbf{C-4, C-3, and C-2} \end{matrix}$

ribulose $\xrightarrow{\text{PhNHNH}_2}$ $\overset{5}{CH_2OH}\vdots\overset{4}{CHOH}\vdots\overset{3}{CHOH}-\overset{2}{\underset{\overset{||}{NNHPh}}{C}}-\overset{1}{CH}=NNPh$ $\xrightarrow{\text{HIO}_4}$

$$HCHO + HCOOH + OHC-\underset{\overset{||}{NNHPh}}{C}-CH=NNPh$$

$$\textbf{C-5} \quad\quad\quad \textbf{C-4} \quad\quad\quad \textbf{C-3, C-2, and C-1}$$

A problem in algebra: can you determine the activity of *each individual position* of glucose and of ribulose?

11. (a) Sugar T is $C_5H_{10}O_5$ and is oxidized by $Br_2(aq)$: it is an aldopentose. It is levorotatory and gives an optically inactive aldaric acid: it must be either D-(−)-ribose (IX, page 1086) or L-(−)-xylose (enantiomer of XI, page 1086).

$$
\begin{array}{ccc}
\begin{array}{c}
\text{CHO} \\
\text{H}{-}\text{OH} \\
\text{H}{-}\text{OH} \\
\text{H}{-}\text{OH} \\
\text{CH}_2\text{OH} \\
\text{D-(−)-Ribose} \\
\text{T?}
\end{array}
&
\xrightarrow{\text{HNO}_3}
&
\begin{array}{c}
\text{COOH} \\
\text{H}{-}\text{OH} \\
\text{H}{-}\text{OH} \\
\text{H}{-}\text{OH} \\
\text{COOH} \\
\textit{Meso} \text{ aldaric acid} \\
C_5H_8O_7 \\
\textbf{Inactive}
\end{array}
\qquad
\begin{array}{c}
\text{CHO} \\
\text{HO}{-}\text{H} \\
\text{H}{-}\text{OH} \\
\text{HO}{-}\text{H} \\
\text{CH}_2\text{OH} \\
\text{L-(−)-Xylose} \\
\text{T?}
\end{array}
\xrightarrow{\text{HNO}_3}
\begin{array}{c}
\text{COOH} \\
\text{HO}{-}\text{H} \\
\text{H}{-}\text{OH} \\
\text{HO}{-}\text{H} \\
\text{COOH} \\
\textit{Meso} \text{ aldaric acid} \\
C_5H_8O_7 \\
\textbf{Inactive}
\end{array}
\end{array}
$$

The epimer of D-(−)-ribose is D-(−)-arabinose (X, page 1084); the epimer of L-(−)-xylose is L-(+)-lyxose (enantiomer of XII, page 1086). The epimer of T is (−)-U. From its sign of rotation alone, we conclude that (−)-U must be D-(−)-arabinose. This is confirmed by its degradation and oxidation of the resulting tetrose to inactive (*meso*) tartaric acid. (Similar treatment of L-(+)-lyxose would have given active L-(+)-tartaric acid.)

$$
\begin{array}{c}
\text{COOH} \\
\text{H}{-}\text{OH} \\
\text{H}{-}\text{OH} \\
\text{COOH} \\
\text{Mesotartaric} \\
\text{acid}
\end{array}
\xleftarrow{\text{HNO}_3}
\begin{array}{c}
\text{CHO} \\
\text{H}{-}\text{OH} \\
\text{H}{-}\text{OH} \\
\text{CH}_2\text{OH}
\end{array}
\xleftarrow{\text{degradn.}}
\begin{array}{c}
\text{CHO} \\
\text{HO}{-}\text{H} \\
\text{H}{-}\text{OH} \\
\text{H}{-}\text{OH} \\
\text{CH}_2\text{OH} \\
\text{D-(−)-Arabinose} \\
\text{U}
\end{array}
\qquad
\begin{array}{c}
\text{CHO} \\
\text{H}{-}\text{OH} \\
\text{H}{-}\text{OH} \\
\text{H}{-}\text{OH} \\
\text{CH}_2\text{OH} \\
\text{D-(−)-Ribose} \\
\text{T}
\end{array}
$$

T, then, is D-(−)-ribose.

(b) Of the possible phosphates of D-(−)-ribose, only the 3-phosphate can give the *meso* compound V.

$$
\begin{array}{c}
\text{CHO} \\
\text{H}{-}\text{OH} \\
\text{H}{-}\text{OPO}_3\text{H}_2 \\
\text{H}{-}\text{OH} \\
\text{CH}_2\text{OH} \\
\text{T phosphate}
\end{array}
\xrightarrow[\text{Pt}]{\text{H}_2}
\begin{array}{c}
\text{CH}_2\text{OH} \\
\text{H}{-}\text{OH} \\
\text{H}{-}\text{OPO}_3\text{H}_2 \\
\text{H}{-}\text{OH} \\
\text{CH}_2\text{OH} \\
\text{V} \\
\textit{Meso} \\
\textbf{Inactive}
\end{array}
\xrightarrow{\text{H}_2\text{O}}
\begin{array}{c}
\text{CH}_2\text{OH} \\
\text{H}{-}\text{OH} \\
\text{H}{-}\text{OH} \\
\text{H}{-}\text{OH} \\
\text{CH}_2\text{OH} \\
\text{W} \\
\textit{Meso} \\
\textbf{Inactive}
\end{array}
\xrightarrow{\text{Ac}_2\text{O}}
\begin{array}{c}
\text{CH}_2\text{OAc} \\
\text{H}{-}\text{OAc} \\
\text{H}{-}\text{OAc} \\
\text{H}{-}\text{OAc} \\
\text{CH}_2\text{OAc} \\
\text{X} \\
\textit{Meso} \\
\textbf{Inactive}
\end{array}
$$

(c) Phosphate is located, then, at the 3-position of ribose. At which point is adenosine attached? Adenylic acid does not reduce Tollens' reagent or Benedict's solution. *Conclusion:* that C–1 of ribose is tied up as in a glycoside. If the nitrogen of adenosine is joined to a carbon atom in T, it must be to C–1.

Finally, what is the ring size of the ribose unit? This is revealed by the usual sequence of methylation, hydrolysis, and vigorous oxidation.

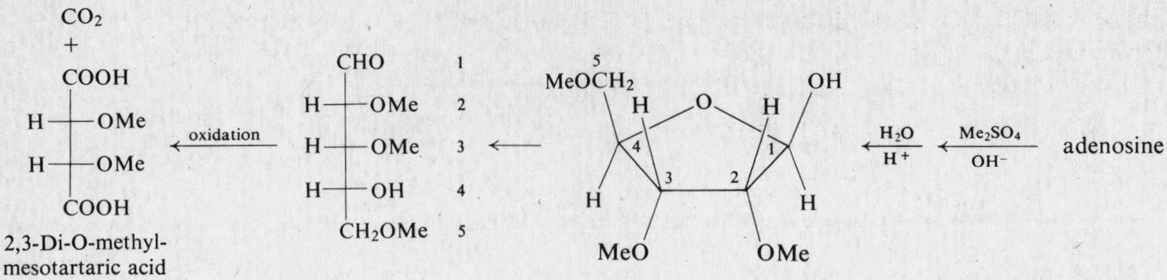

$$CO_2$$
$$+$$

2,3-Di-O-methyl-mesotartaric acid

$$\begin{array}{c} COOH \\ H{\rule{1cm}{0.4pt}}OMe \\ H{\rule{1cm}{0.4pt}}OMe \\ COOH \end{array}$$

$\xleftarrow{\text{oxidation}}$

$$\begin{array}{cc} CHO & 1 \\ H{\rule{1cm}{0.4pt}}OMe & 2 \\ H{\rule{1cm}{0.4pt}}OMe & 3 \\ H{\rule{1cm}{0.4pt}}OH & 4 \\ CH_2OMe & 5 \end{array}$$

$\longleftarrow$

Y

$C_8H_{16}O_5$

$\xleftarrow[H^+]{H_2O} \xleftarrow[OH^-]{Me_2SO_4}$ adenosine

Taking the evidence of synthesis, we arrive at the structure of adenylic acid (compare Figure 37.5, page 1179).

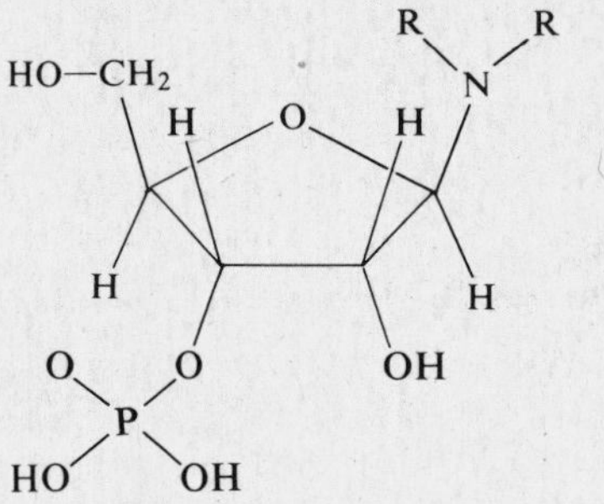

Adenylic acid

12. (a) Chemical arithmetic,

$$\begin{array}{ll} C_6H_{12}O_6 & \text{glucose} \\ + \; C_3H_6O & \text{1 acetone} \\ \hline C_9H_{18}O_7 & \end{array}$$

$$\begin{array}{l} C_9H_{18}O_7 \\ - \; C_9H_{16}O_6 \quad \text{AA} \\ \hline H_2O \; \text{lost} \end{array}$$

$$\begin{array}{ll} C_6H_{12}O_6 & \text{glucose} \\ + \; C_6H_{12}O_2 & \text{2 acetone} \\ \hline C_{12}H_{24}O_8 & \end{array}$$

$$\begin{array}{l} C_{12}H_{24}O_8 \\ - \; C_{12}H_{20}O_6 \quad \text{Z} \\ \hline 2H_2O \; \text{lost} \end{array}$$

shows that glucose combines with one (AA) or two (Z) molecules of acetone to split out, respectively, one or two molecules of water. The products are resistant to alkali and hydrolyzed by acid, properties we expect of acetals. Each acetone reacts with two —OH's of glucose to give a cyclic ketal, called an O-*isopropylidene derivative*:

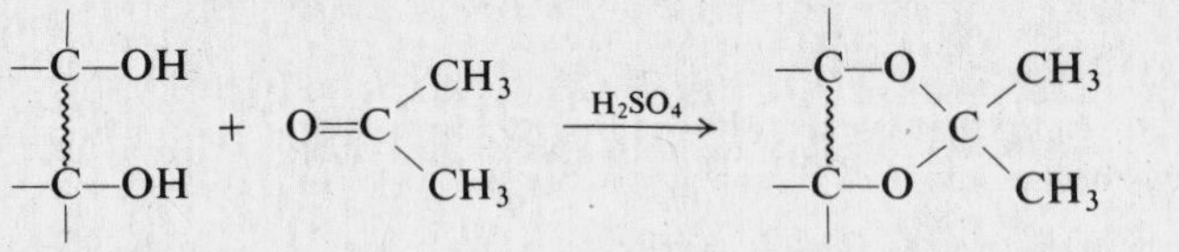

$$\begin{array}{c} -\overset{|}{\underset{|}{C}}-OH \\ -\overset{|}{\underset{|}{C}}-OH \end{array} + O{=}C\begin{array}{c} CH_3 \\ CH_3 \end{array} \xrightarrow{H_2SO_4} \begin{array}{c} -\overset{|}{C}-O \\ -\overset{|}{C}-O \end{array}C\begin{array}{c} CH_3 \\ CH_3 \end{array}$$

Glucose   Acetone   O-Isopropylideneglucose

A cyclic ketal

Hydrolysis of Z gives AA as an intermediate. Since aqueous acid is an unfavorable medium for re-formation of acetals, it seems highly likely that the ketal bridge AA is the same as one of the ketal bridges in Z.

Now, to which pairs of carbons in glucose are the O-isopropylidene groups attached?

(b) Neither Z nor AA is a reducing sugar. *Conclusion:* in both compounds an O-isopropylidene group is attached to C–1.

(c) Z is methylated at one position. Acid cleaves the ketals, but not the ether: the methoxyl group remains. This O-methyl ether gives an osazone. Evidently methylation did not occur at either C–1 or C–2. *Conclusion:* in Z, an O-isopropylidene group is attached to C–1 and C–2.

CH=NNHPh   CHOH   1 CH—O
C=NNHPh   CHOH   O   2 CH—O $\,$C(CH$_3$)$_2$ O
     $\xleftarrow{\text{PhNHNH}_2}$ $\xleftarrow[\text{H}^+]{\text{H}_2\text{O}}$ BB $\xleftarrow[\text{OH}^-]{\text{Me}_2\text{SO}_4}$

'DD   CC   One O-isopropylidene grouping in Z

(d) The evidence is as in (c). *Conclusion:* in AA, the only O-isopropylidene group is attached to C–1 and C–2.

CH=NNHPh   CHOH   1 CH—O
C=NNHPh   CHOH   O   2 CH—O $\,$C(CH$_3$)$_2$ O
     $\xleftarrow{\text{PhNHNH}_2}$ $\xleftarrow[\text{H}^+]{\text{H}_2\text{O}}$ EE $\xleftarrow[\text{OH}^-]{\text{Me}_2\text{SO}_4}$

GG   FF   The O-isopropylidene grouping in AA

(e) Methylation shows that C–4 was tied up in the acetal ring of FF, and hence of EE and AA, too. *Conclusion:* these compounds contain furanose rings. We now know the structure of AA.

CHOH   CHOH   CH—O   CH—O
CHOMe   CHOH   CH—O $\,$C(CH$_3$)$_2$ CH—O $\,$C(CH$_3$)$_2$
CHOMe   O $\xleftarrow[\text{OH}^-]{\text{Me}_2\text{SO}_4}$ CHOMe   O $\xleftarrow[\text{H}^+]{\text{H}_2\text{O}}$ CH—OMe   O $\xleftarrow[\text{OH}^-]{\text{Me}_2\text{SO}_4}$ CH—OH   O
CHO—   CH—   CH—   CH—
CHOMe   CHOMe   CH—OMe   CH—OH
CH$_2$OMe   CH$_2$OMe   CH$_2$—OMe   CH$_2$—OH

2,3,5,6-Tetra-O-methyl-   FF   EE   AA
D-glucofuranose   1,2-O-Isopropylidene- D-glucofuranose

(f) HH is a six-carbon dicarboxylic acid. One carbon is in the O-methyl group, leaving five carbons for the chain; the carboxyl groups are C–1 and C–5 of the O-methyl-D-glucose, CC. *Conclusion:* C$_5$-OH is not methylated, and hence must have been tied up by an O-isopropylidene group.

In CC, methoxyl is not at C–1 or C–2 (O-isopropylidene bridge), not at C–4 (furanose ring), and, now, evidently not at C–5, leaving only C–3 or C–6.

(g) Formation of the δ-lactone II shows that C–5 of this compound, *and hence C-4 of CC* carries a free —OH. *Conclusion:* the methoxyl group of CC is at C–3.

This leaves only C–5 and C–6 for the second O-isopropylidene bridge in Z. We now know the structure of Z.

$$
\begin{array}{cccc}
\text{C}{=}\text{O} & \text{CN} & \text{CHO} & \text{CH—O} \\
1\ \text{CHOH} & 1\ \text{CHOH} & 1\ \text{CHOH} & \text{CH—O}\!\!>\!\!\text{C(CH}_3)_2 \\
2\ \text{CHOH} & 2\ \text{CHOH} & 2\ \text{CHOH} & \text{CH—OMe} \\
3\ \text{CHOMe} & 3\ \text{CHOMe} & 3\ \text{CHOMe} & \text{CH} \\
4\ \text{CH} & 4\ \text{CHOH} & 4\ \text{CHOH} & \text{CH—O} \\
5\ \text{CHOH} & 5\ \text{CHOH} & 5\ \text{CHOH} & \text{CH}_2\text{—O}\!\!>\!\!\text{C(CH}_3)_2 \\
6\ \text{CH}_2\text{OH} & 6\ \text{CH}_2\text{OH} & 6\ \text{CH}_2\text{OH} &
\end{array}
$$

and epimer   CC   BB   Z

II   (As open-chain)   1,2:5,6-Di-O-isopropylidene-

A δ-lactone   D-glucofuranose

13. Carbon–carbon bonds are formed; with the reactants aldehydes and ketones, and the reagent base, we think of the aldol condensation. Either D-glyceraldehyde (an aldose) or dihydroxyacetone (a ketose) gives the same products; Sec. 34.6 reminds us that base can cause isomerization—among other things, between an aldose and a ketose.

We postulate an equilibrium between the two three-carbon reagents:

$$
\begin{array}{ccccc}
\text{CHO} & & \text{CH—OH} & & \text{CH}_2\text{OH} \\
\text{H—C—OH} & \underset{\text{base}}{\rightleftharpoons} & \text{CH—OH} & \underset{\text{base}}{\rightleftharpoons} & \text{C}{=}\text{O} \\
\text{CH}_2\text{OH} & & \text{CH}_2\text{OH} & & \text{CH}_2\text{OH}
\end{array}
$$

D-Glyceraldehyde   Enediol   Dihydroxyacetone

An aldose   *A ketose*

Aldol and cross-aldol condensations yield the final products.

$$
\begin{array}{ccc}
\text{CH}_2\text{OH} & \text{CH}_2\text{OH} & \text{CH}_2\text{OH} \\
\text{C}{=}\text{O} & \text{C}{=}\text{O} & \text{C}{=}\text{O} \\
\text{CH}_2\text{OH} & \text{HO—C—H} & \text{H—C—OH} \\
& \text{H—C—OH} & \text{HO—C—H} \\
\text{H} & \text{H—C—OH} & \text{H—C—OH} \\
\text{C}{=}\text{O} & \text{CH}_2\text{OH} & \text{CH}_2\text{OH} \\
\text{H—C—OH} & \text{D-Fructose} & \text{D-Sorbose} \\
\text{CH}_2\text{OH} & &
\end{array}
$$

$$
\begin{array}{ccc}
\text{CH}_2\text{OH} & \text{CH}_2\text{OH} & \text{CH}_2\text{OH} \\
\text{C}{=}\text{O} & \text{C}{=}\text{O} & \text{C}{=}\text{O} \\
\text{CH}_2\text{OH} & \text{H—C—OH} & \text{HO—C—H} \\
& \text{HOCH}_2\text{—C—OH} & \text{HO—C—CH}_2\text{OH} \\
\text{HOCH}_2\text{—C}{=}\text{O} & \text{CH}_2\text{OH} & \text{CH}_2\text{OH} \\
\text{CH}_2\text{OH} & \text{DL-Dendroketose} &
\end{array}
$$

14. Hydrolysis is S_N1-like, with the separation of an especially stable cation: an oxonium ion, not a carbonium ion.

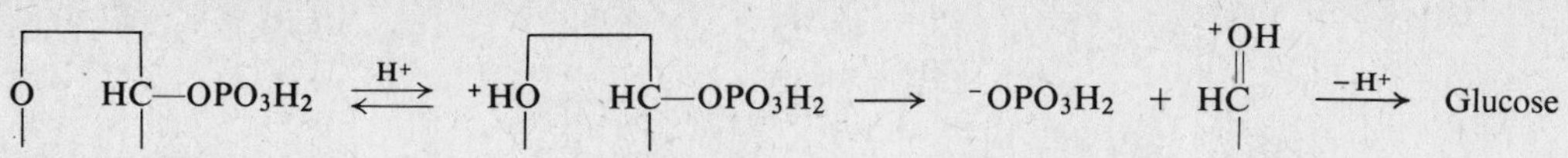

Glucose-1-phosphate $\xrightarrow{H^+}$ Protonated acetal $\longrightarrow$ Phosphate + Protonated aldehyde $\xrightarrow{-H^+}$ Glucose

Oxonium ion

15. (a) The downfield signal is due to the proton on C–1, which is deshielded by *two* oxygens.

(b) Our interpretation (based on Chapter 13) would be: KK has an equatorial —H on C–1 (signal farther downfield); JJ has *anti* (axial,axial) protons on C–1 and C–2 (larger *J* value). Only in the β-anomer are the protons on C–1 and C–2 *trans*, and capable of being axial,axial. We are led to these structures:

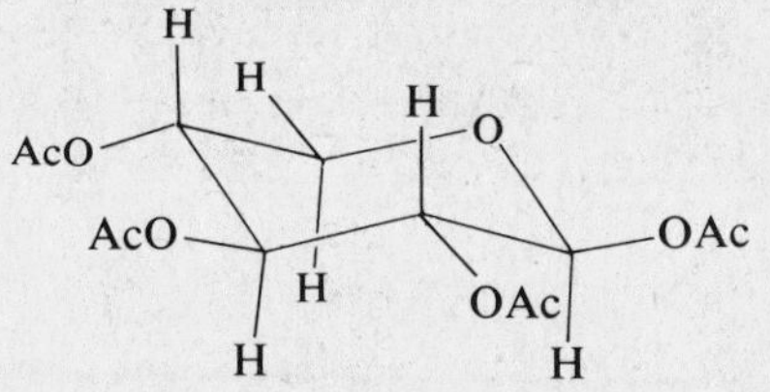

JJ
β-Anomer
Axial —H on C-1
Anti protons on C-1, C-2

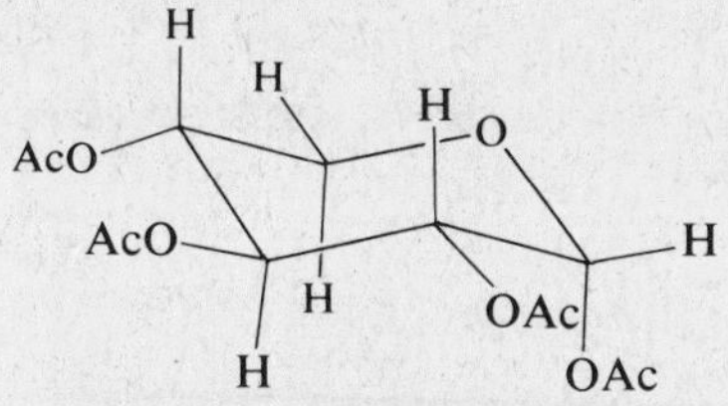

KK
α-Anomer
Equatorial —H on C-1
Gauche protons on C-1, C-2

D-Tetra-O-acetylxylopyranose

Each conformation, we find, is the one we would expect to be the more stable for the particular anomer. In the β-anomer, the anomeric effect (page 1106) is weaker than the steric effect: four equatorial —OAc's. (In the α-anomer, the anomeric effect simply reinforces the steric effect.)

(c) Our interpretation: MM has an equatorial —H on C–1 (signal farther downfield); LL has *anti* (axial,axial) protons on C–1 and C–2 (larger *J* value). Again, only in the β-anomer can the protons on C–1 and C–2 be axial,axial.

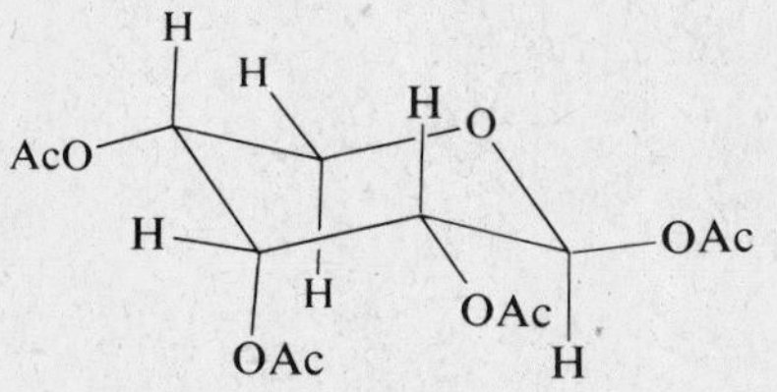

LL
β-Anomer
Axial —H on C-1
Anti protons on C-1, C-2

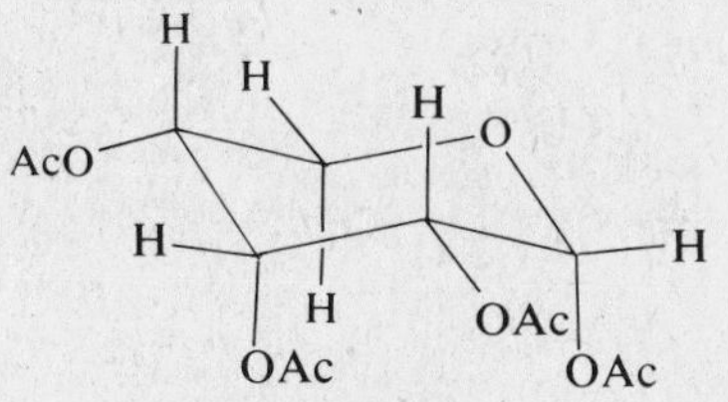

MM
α-Anomer
Anomeric effect
(Sec. 34.20)

D-Tetra-O-acetylribopyranose

In the β-anomer, the anomeric effect is weaker than the steric effect: three out of four —OAc's equatorial. In the α-anomer, where there are two equatorial —OAc's in either conformation, the anomeric effect is the deciding factor.

(d) This time, let us begin by drawing what we would expect to be the more stable conformation of each compound.

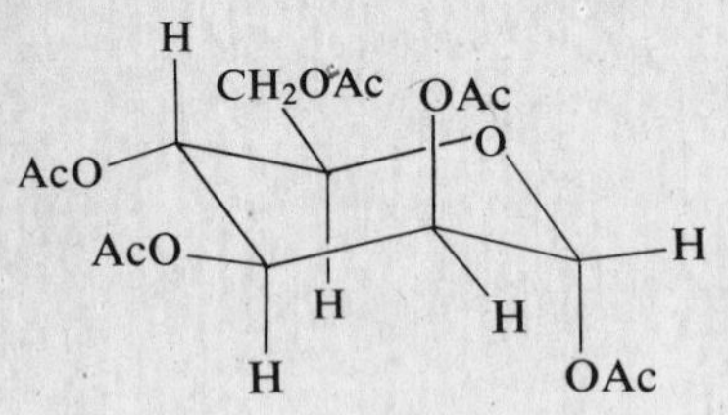

NN
α-Anomer
Equatorial —H on C-1
Gauche protons on C-1, C-2

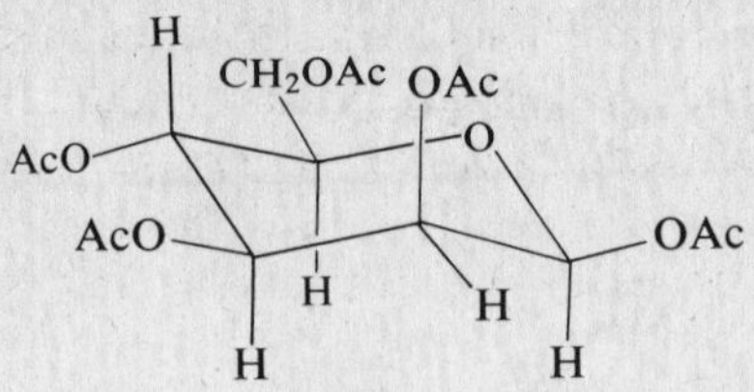

OO
β-Anomer
Axial —H on C-1
Gauche protons on C-1, C-2

D-Penta-O-acetylmannopyranose

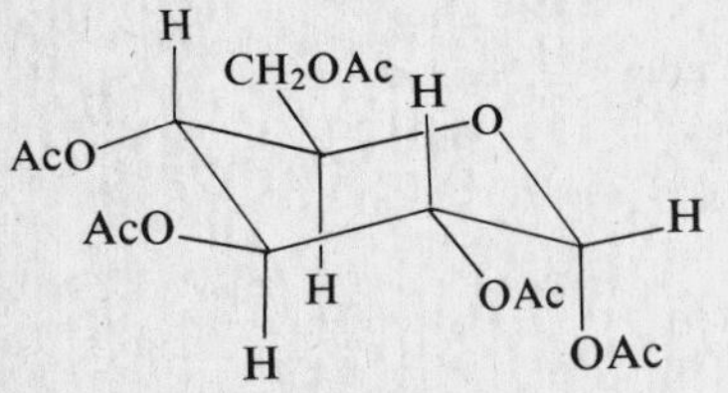

QQ
α-Anomer
Equatorial —H on C-1
Gauche protons on C-1, C-2

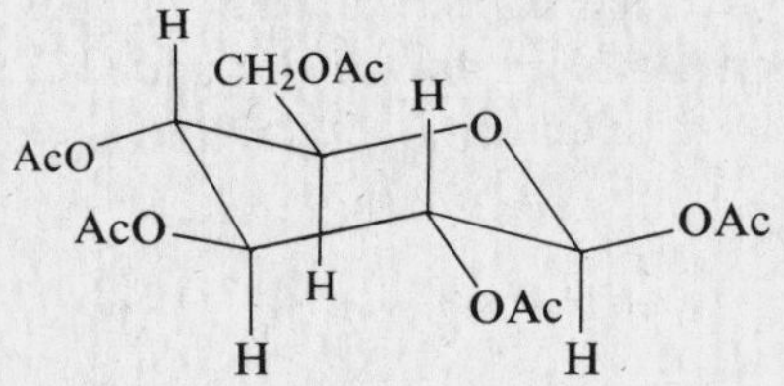

PP
β-Anomer
Axial —H on C-1
Anti protons on C-1, C-2

D-Penta-O-acetylglucopyranose

We can readily assign an α-structure (equatorial —H) to NN (δ 5.97) and QQ (δ 5.99), and a β-structure (axial —H) to OO (δ 5.68) and PP (δ 5.54). Only one of the four anomers, the β-glucose derivative, has *anti* protons; only PP has a large value for *J*, the coupling constant. Therefore PP is the β-glucose derivative; its anomer, QQ, must be the α-glucose derivative. Of necessity, then, NN is the α-mannose derivative, and OO is the β-mannose derivative.

16. (a)

(i)

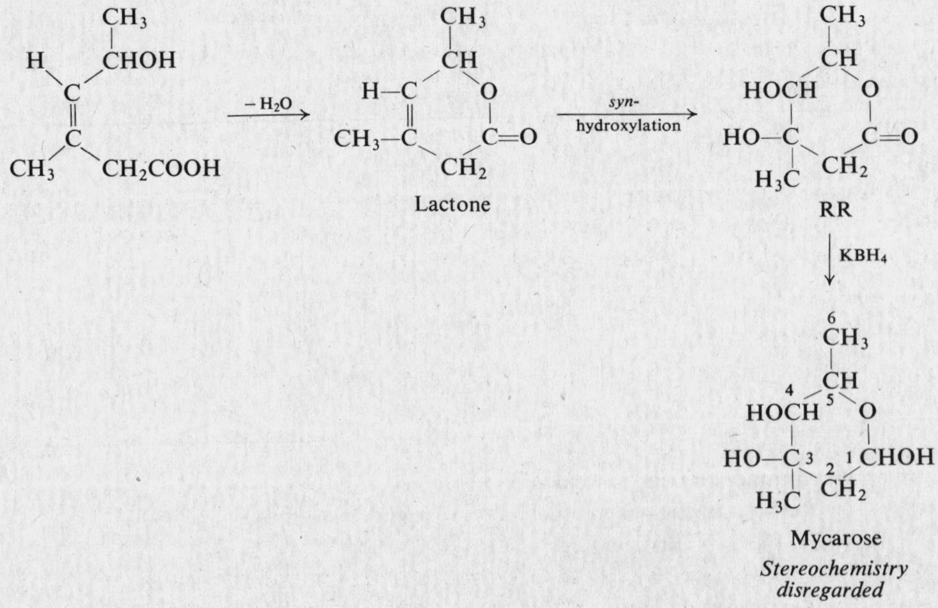
$-H_2O$
syn-
hydroxylation
Lactone
RR
KBH_4
Mycarose
Stereochemistry disregarded

(Only the *Z*-isomer of the starting material can form a lactone.)

(b) *syn*-Hydroxylation (i) of the double bond in the cyclic compound gives a *cis*-diol; hence in RR and mycarose C_3–OH and C_4–OH are *cis*.   The large coupling constant (ii) shows that C_4–H and C_5–H are *anti* (axial,axial) and hence *trans* to each other.

(c) The absolute configuration about C–5 is determined exactly as was done for D-glucose (Sec. 34.17).

(iii)

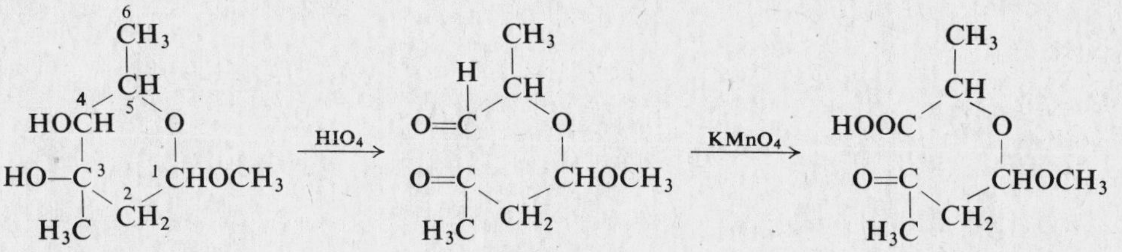

$$\text{Methyl mycaroside} \xrightarrow{HIO_4} \text{SS} \xrightarrow{KMnO_4} \text{TT} \xrightarrow[\;]{H^+ \,\big|\, H_2O} \text{L-Lactic acid}$$

Methyl mycaroside
(*Based on i, above*)

SS

TT

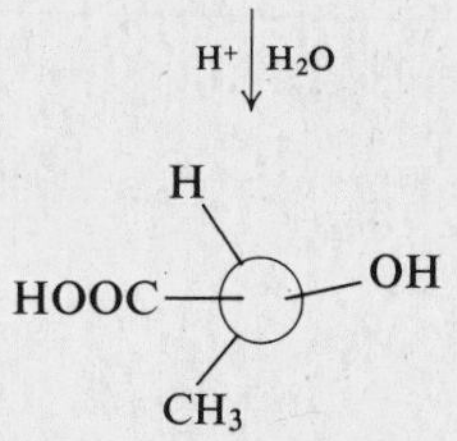

L-Lactic acid
((S)-Lactic acid)

(d) Taking the information gathered so far, we can draw:

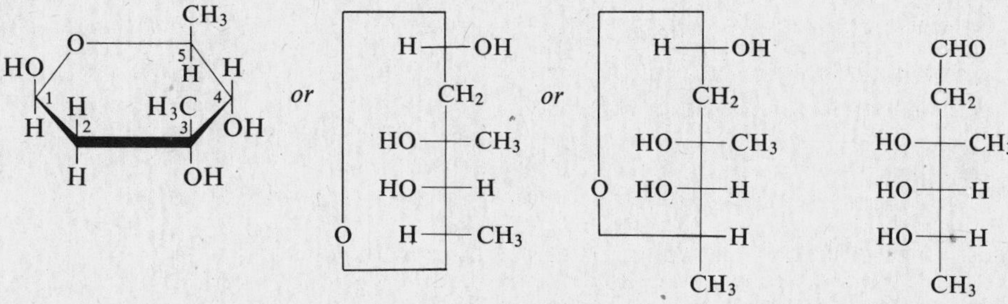

L-(−)-Mycarose
(β-anomer)

Open-chain

Since —OH on the lowest chiral carbon is on the left, (−)-mycarose belongs to the L-family.

The size of the coupling constant ($J = 9.5$–9.7 Hz) between C_4–H and C_5–H indicates that these protons are axial,axial.   Starting from there, we arrive at the following preferred conformation:

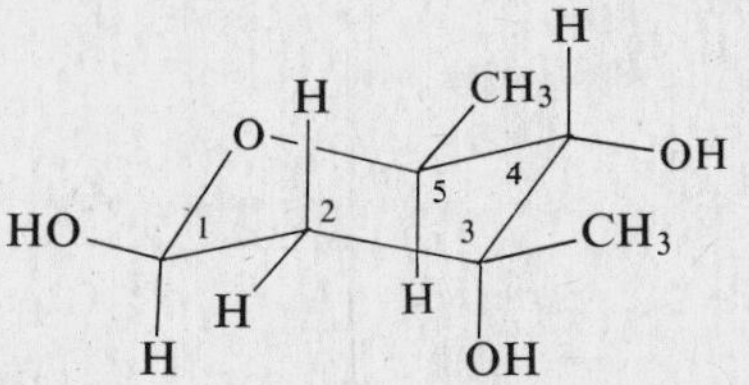

L-(−)-Mycarose

Disregarding C–1 for the moment, we see that in this conformation three of four large groups are equatorial; this is the conformation we would have predicted to be the more stable.

(e) The downfield peak is, of course, due to C_1–H. In the α-anomer C_1–H is *gauche* to both H's on C–2. The signal is split into a triplet, with a low *J* value, as observed.

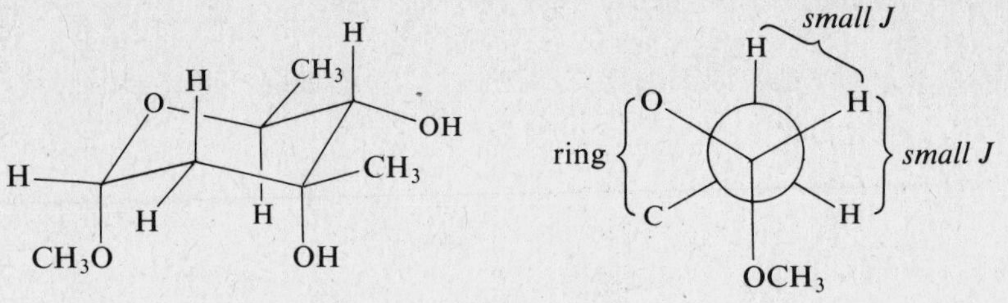

α-Anomer

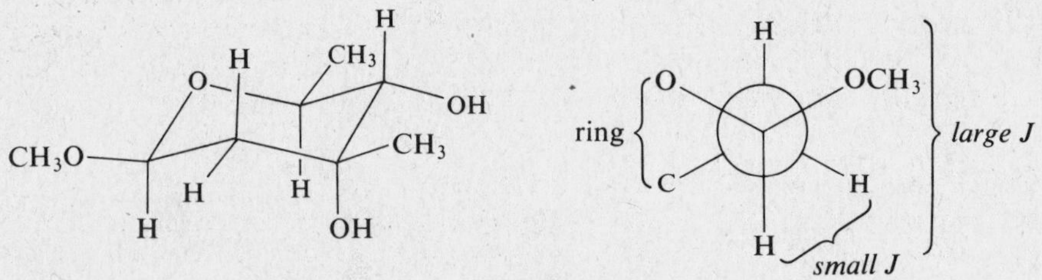

β-Anomer

In the β-anomer, C_1–H is *gauche* to one C_2–H and *anti* to the other C_2–H. We expect two signals for C_1–H, each a doublet. There will be two *J* values: one small, comparable to that (2.4 Hz) for the α-anomer; and the other larger.

(f) Here we see exactly the kind of nmr spectrum just predicted. Evidently free mycarose exists preferentially as the β-anomer, as shown in (d) above.

17. (a) The anomeric effect (page 1106) stabilizes the α-anomer relative to the β-anomer.

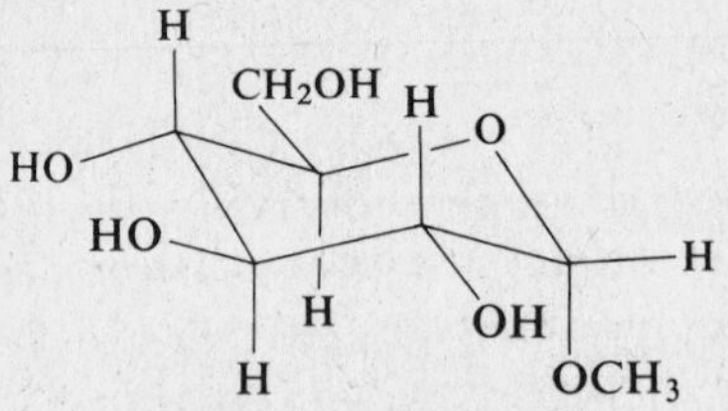

Methyl-α-D-Glucoside
Favored anomer

(*Note:* Here we are not just discussing which conformation of a particular compound is more stable. We are discussing which *compound*—the α- or β-anomer—is the more stable. Our interpretation is this: that the α-anomer is more stable than the β-anomer because the preferred conformation of the α-anomer is more stable than the preferred conformation of the β-anomer.)

(b) This is evidently an anomeric-like effect. Repulsion between the dipoles associated with the ring oxygens and the C—Cl bonds favors the diaxial conformation.

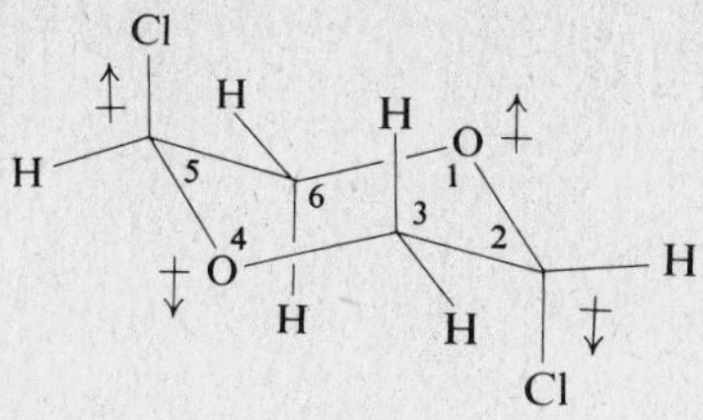

trans-2,5-Dichloro-1,4-dioxane
Favored conformation

18. (a)

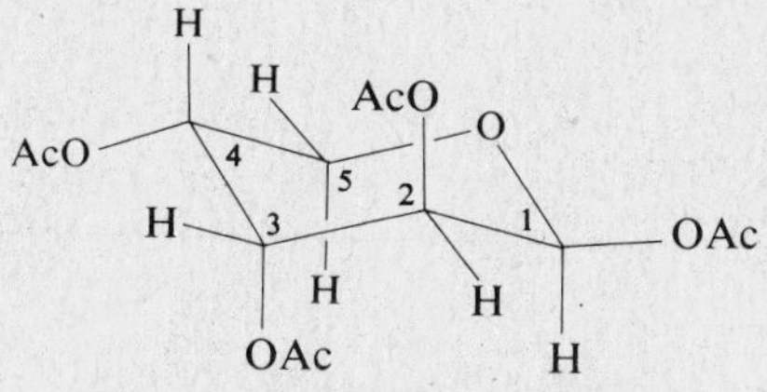

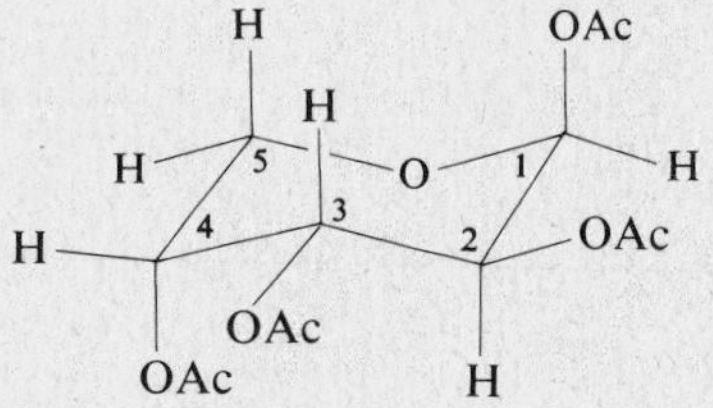

Tetra-O-acetyl-β-
L-arabinopyranose

I

Two —OAc's axial
Two —OAc's equatorial

II

Two —OAc's axial
Two —OAc's equatorial
Favored by anomeric effect

Neither is favored on steric grounds, but the anomeric effect (page 1106) would favor conformation II, with axial C_1–OAc.

(b) This tells nothing: in either conformation, I or II, two —OAc's are equatorial, and two are axial.

(c) We are now considering only the protons in —OAc's on C_2, C_3, and C_4. The $e:a$

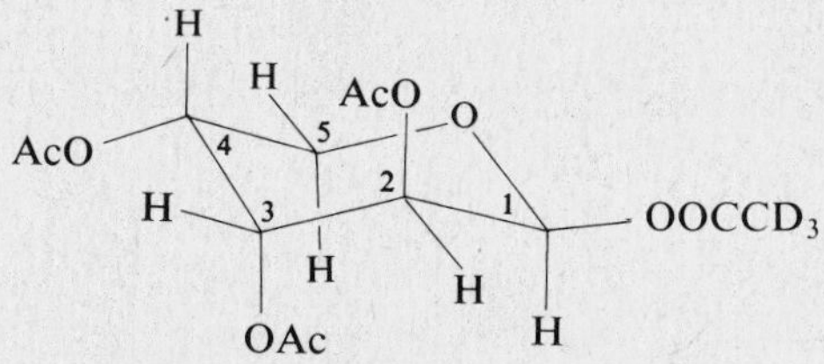

I-*d*
$e:a = 0.5:1$

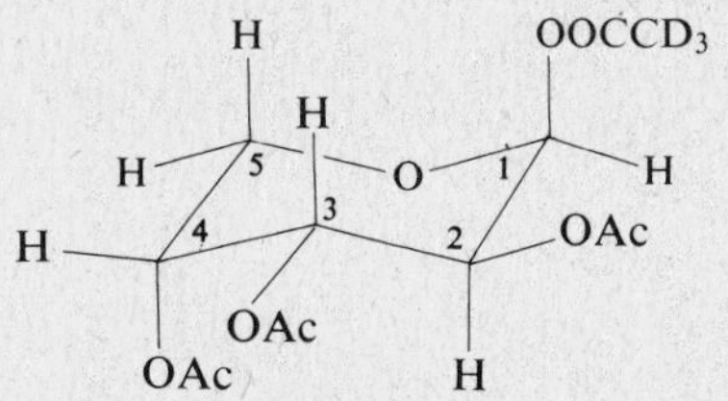

II-*d*
$e:a = 2:1$

ratio would be 0.5:1 for conformation I-*d*, 2:1 for conformation II-*d*, and 1:1 for a 50:50 mixture of I-*d* and II-*d*. The observed $e:a$ ratio of 1.46:1.00 shows that the predominant conformation (78%) is II-*d*, with C_1–OAc axial. The anomeric effect is evidently the deciding factor.

$$\frac{e\text{-OAc}}{\text{total OAc}} = \frac{1}{3}N_I + \frac{2}{3}N_{II} = \frac{1.46}{1.46 + 1.00}$$

$$\frac{1}{3}(1 - N_{II}) + \frac{2}{3}N_{II} = \frac{1.46}{2.46}$$

$$N_{II} = 0.78$$

$$N_I = 0.22$$

Chapter 35

Carbohydrates II. Disaccharides and Polysaccharides

35.1

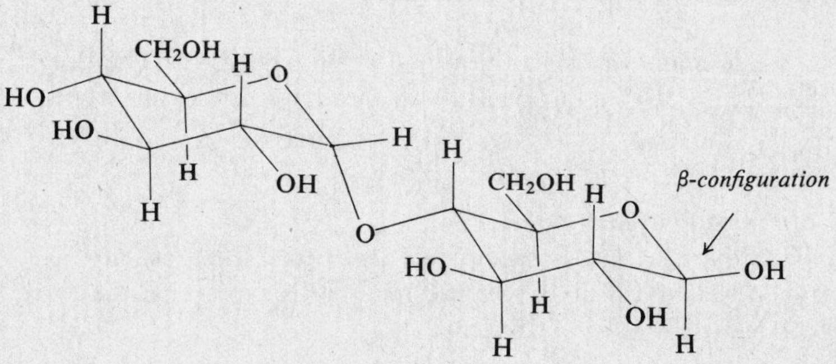

β-Maltose

α- and β-maltose differ in configuration only at C–1 of the reducible glucose unit. (If the two glucose moieties were *joined* by a β-linkage, the compound would not be maltose, but cellobiose.)

35.2

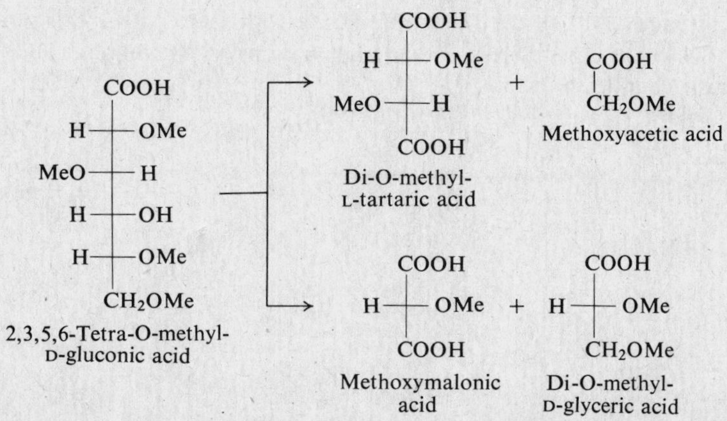

2,3,5,6-Tetra-O-methyl-
D-gluconic acid

Di-O-methyl-
L-tartaric acid

Methoxyacetic acid

Methoxymalonic
acid

Di-O-methyl-
D-glyceric acid

35.3

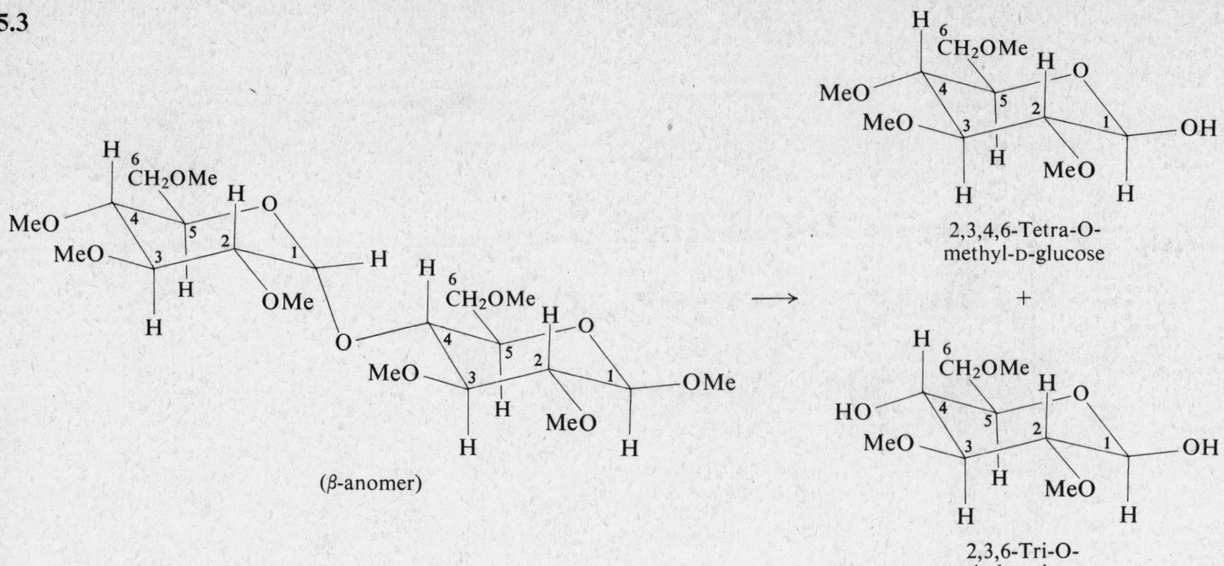

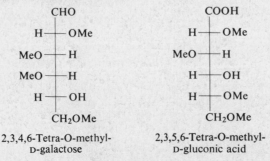

2,3,4,6-Tetra-O-
methyl-D-glucose

+

2,3,6-Tri-O-
methyl-D-glucose

(β-anomer)

Besides, of course, knowing that maltose is a glucosyl glucose, we would know that the non-reducing glucose unit contains a pyranose ring. But we would be lacking knowledge about the reducing unit: its ring size, and the point of attachment of the glycosidic linkage. It could be either a 4-O-substituted-D-glucopyranose *or* a 5-O-substituted-D-glucofuranose.

Oxidation before methylation and hydrolysis opens the hemiacetal ring. The only unmethylated position is then the point of attachment; with that known, we can deduce the ring size.

35.4 See Figure 35.4, page 602 of this Study Guide.

After loss of two carbons, the C_3–CHO has no —OH *alpha* to it (that is, no free —OH on C_4), so that osazone formation is no longer possible. This shows that the two rings are joined at C–4 of the reducing moiety of maltose.

35.5 There is less chance of breaking the glycosidic linkage, with formation of monosaccharides.

35.6 Everything as in Figure 35.1 (page 1114), except that there is a β-glycosidic linkage in the first three formulas. The final products are identical to those from (+)-maltose.

35.7 (a) Similar to Figure 35.1 (page 1114), except that one starts with lactose: there is a β-glycosidic linkage, and C–4 in the non-reducing moiety has the opposite configuration. The final methylated monosaccharides are:

```
        CHO                      COOH
  H ——— OMe                H ——— OMe
 MeO ——— H               MeO ——— H
 MeO ——— H                 H ——— OH
  H ——— OH                 H ——— OMe
      CH2OMe                  CH2OMe
```

2,3,4,6-Tetra-O-methyl-   2,3,5,6-Tetra-O-methyl-
D-galactose   D-gluconic acid

(b)

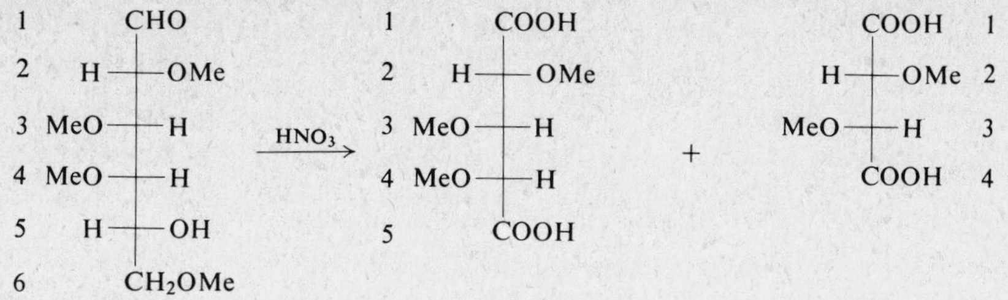

1	CHO
2	H—OMe
3	MeO—H
4	MeO—H
5	H—OH
6	CH₂OMe

2,3,4,6-Tetra-O-methyl-
D-galactose

$\xrightarrow{\text{HNO}_3}$

1	COOH
2	H—OMe
3	MeO—H
4	MeO—H
5	COOH

2,3,4-Tri-O-methyl-
L-arabinaric
(or L-lyxaric) acid

+

COOH	1
H—OMe	2
MeO—H	3
COOH	4

Di-O-methyl-
L-tartaric acid

1	COOH
2	H—OMe
3	MeO—H
4	H—OH
5	H—OMe
6	CH₂OMe

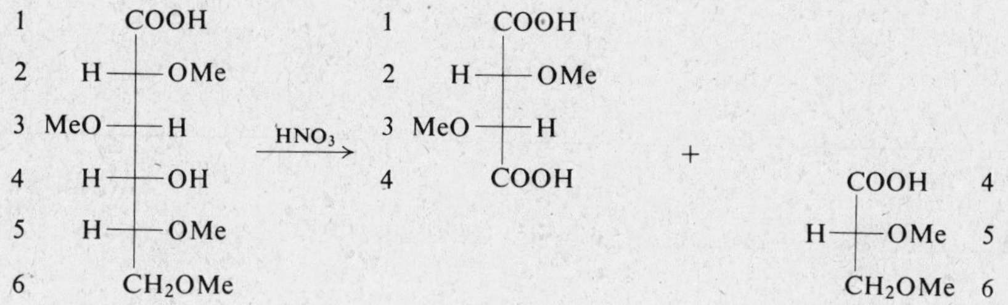

2,3,5,6-Tetra-O-methyl-
D-gluconic acid

$\xrightarrow{\text{HNO}_3}$

1	COOH
2	H—OMe
3	MeO—H
4	COOH

Di-O-methyl-
L-tartaric acid

+

COOH	4
H—OMe	5
CH₂OMe	6

Di-O-methyl-
D-glyceric acid

35.8

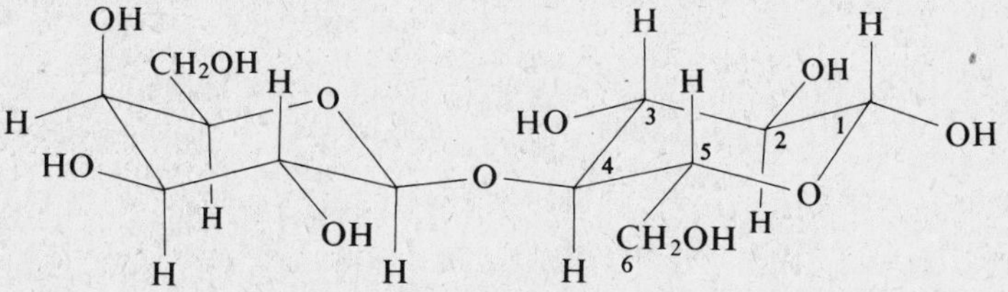

Lactose
(β-anomer)

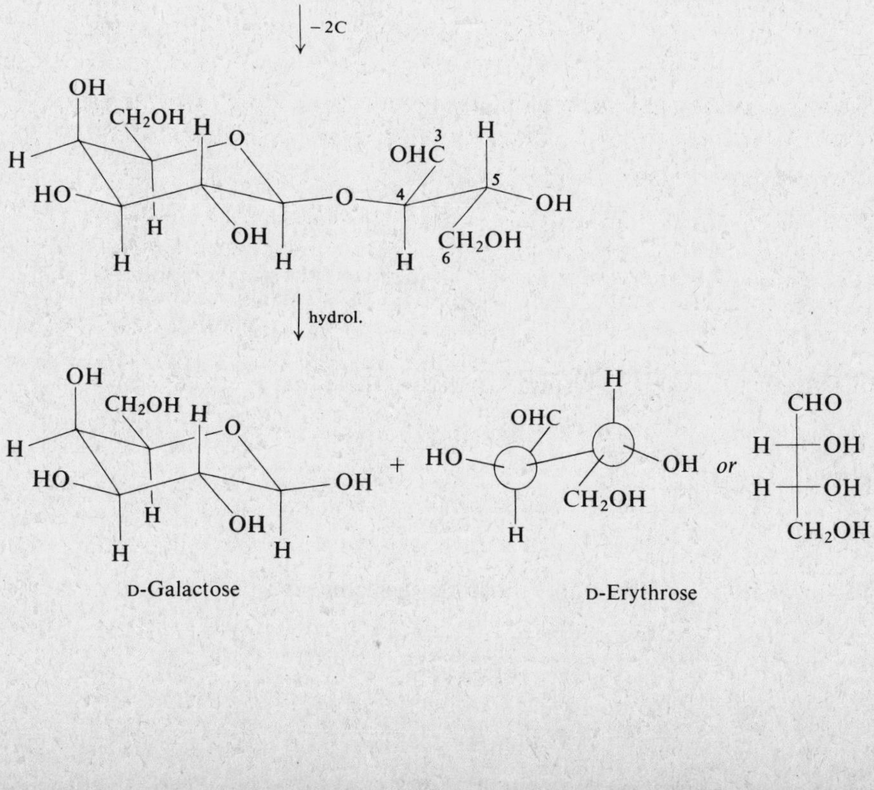

↓ −2C

↓ hydrol.

D-Galactose   +   D-Erythrose

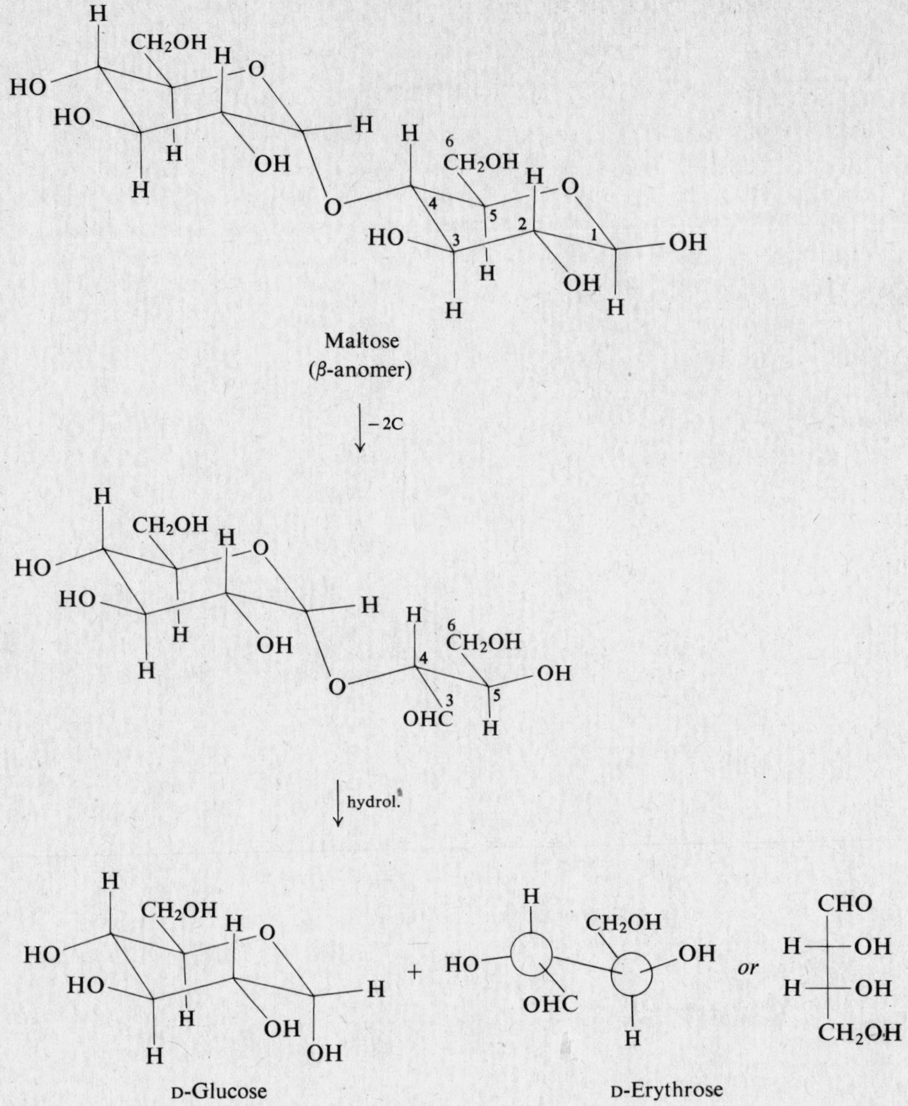

Maltose
(β-anomer)

↓ −2C

↓ hydrol.

D-Glucose   +   D-Erythrose   or

Figure 35.4. Reactions of Problem 35.4.

35.9 $(-92.4° + 52.7°)/2 = -19.9°$

Divided by 2 since g/cc is based on the combined weight of the *two* (isomeric) compounds.

35.10 There would be two glycosidic linkages, and hence loss of two moles of water:

$$C_6H_{12}O_6 + C_6H_{12}O_6 - 2H_2O = C_{12}H_{20}O_{10}$$

It would be non-reducing.

35.11 The glucose unit probably has the α-configuration, and hence sucrose is an α-glucoside.

35.12 (a) The methylated monosaccharides from sucrose are:

```
        CHO                         CH2OMe
   H —— OMe                      C=O
  MeO —— H                   MeO —— H
   H —— OMe                     H —— OMe
   H —— OH                      H —— OH
        CH2OMe                      CH2OMe
```

2,3,4,6-Tetra-O-methyl-   1,3,4,6-Tetra-O-methyl-
D-glucose   D-fructose

(b)
```
1       CHO              1       COOH                   COOH  1
2  H —— OMe              2  H —— OMe               H —— OMe  2
3 MeO —— H    HNO3→      3 MeO —— H        +      MeO —— H   3
4  H —— OMe              4  H —— OMe               COOH  4
5  H —— OH               5       COOH
6       CH2OMe
```

2,3,4,6-Tetra-O-methyl-   2,3,4-Tri-O-methyl-   Di-O-methyl-
D-glucose   xylaric acid   L-tartaric acid

```
1       CH2OMe
2       C=O                         COOH  2
3  MeO —— H    HNO3→          MeO —— H   3
4   H —— OMe                   H —— OMe  4
5   H —— OH                    COOH  5
6       CH2OMe
```

1,3,4,6-Tetra-O-methyl-   Di-O-methyl-
D-fructose   D-tartaric acid

35.13

One break	⟶	2000 avg. length	1/4000 or 0.025% links broken
Three breaks	⟶	1000 avg. length	3/4000 or 0.075% links broken
Nine breaks	⟶	400 avg. length	9/4000 or 0.225% links broken

35.14 (a) The largest groups attached to each ring are the other glucose units of the chain. One or the other of these would have to be axial in a chair conformation.

(b) In a twist-boat conformation, these (as well as the —CH₂OH at C–5) would be in quasi-equatorial positions: at, say, the positions marked 1, 4, and 5 below. (Compare Sec. 9.11.)

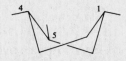

35.15 (a) As the structural formula given on page 1124 shows, we get 3 molecules of HCOOH (and one molecule of HCHO) per molecule of amylose:

> 1HCOOH from the terminal non-reducing unit;
> no small cleavage products from non-terminal units;
> 2HCOOH + 1HCHO from the terminal reducing unit.

(b)

$$\text{moles HCOOH}/3 = \text{moles amylose}$$

$$\text{wt. amylose/moles amylose} = \text{m.w. amylose}$$

$$\text{m.w. amylose/wt. (of 162) per glucose unit} = \text{glucose units per molecule of amylose.}$$

(c) 980 glucose units per molecule. (What is the approximate m.w. of this sample of amylose?)

35.16

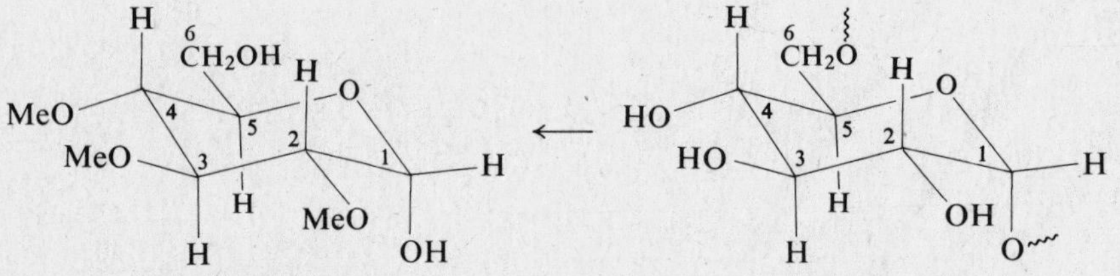

2,3,4-Tri-O-methyl-
D-glucose
Chief product

*Attachment at
C-1 and C-6*

Chain-forming unit

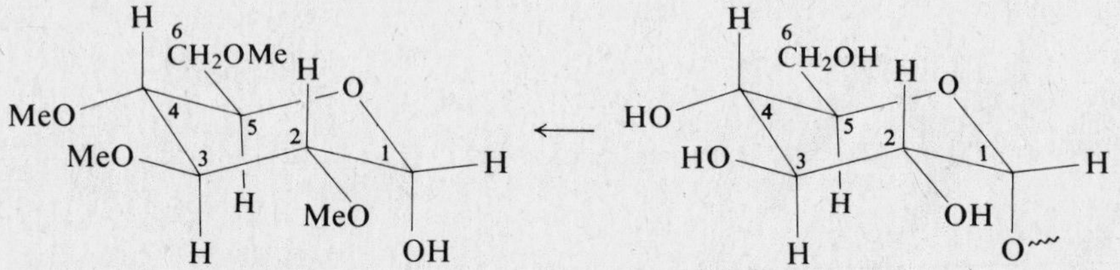

2,3,4,6-Tetra-O-methyl-
D-glucose

Attachment at C-1

Chain-terminating unit

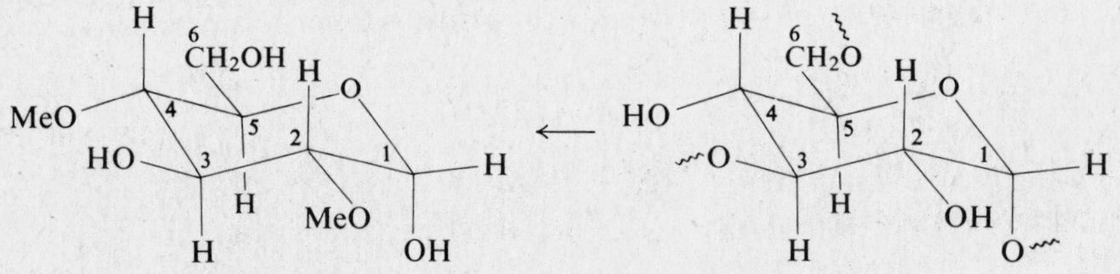

2,4-Di-O-methyl-
D-glucose

*Attachment at
C-1, C-3, C-6*

Chain-linking unit

Dextran is a poly-α-D-glucopyranoside.

> Chain-forming unit: attachment at C–1 and C–6.
> Chain-terminating unit: attachment at C–1.
> Chain-linking unit: attachment at C–1, C–3, and C–6.

Thus, much of the dextran is made up of simple α-glucoside chains with units joined to each other by linkages between C–1 and C–6; but every once in a while two of these chains are joined together by a linkage involving C–3.   Schematically, something like this:

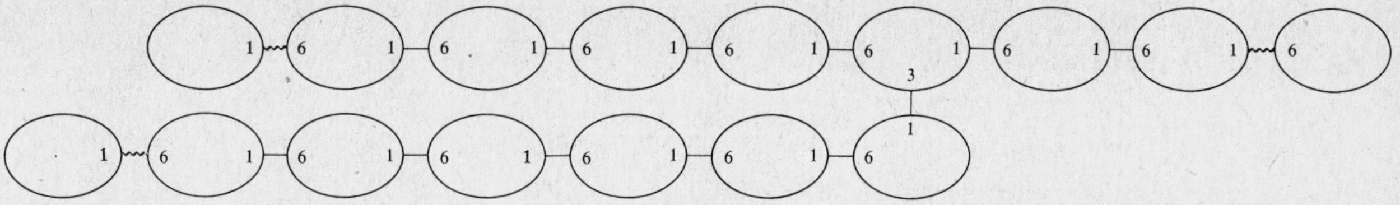

A dextran

35.17

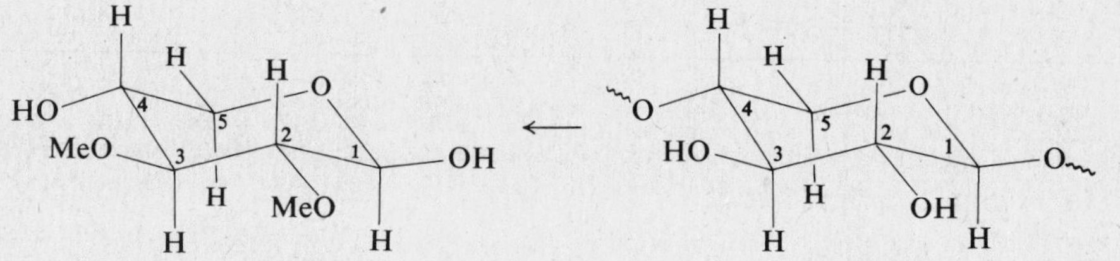

2,3-Di-O-methyl-
D-xylose

Attachment at
C-1 and C-4

Chain-forming unit

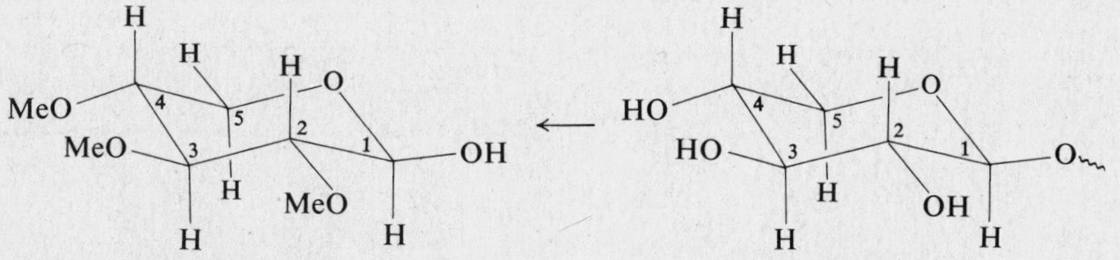

2,3,4-Tri-O-methyl-
D-xylose

Attachment at C-1

Chain-terminating unit

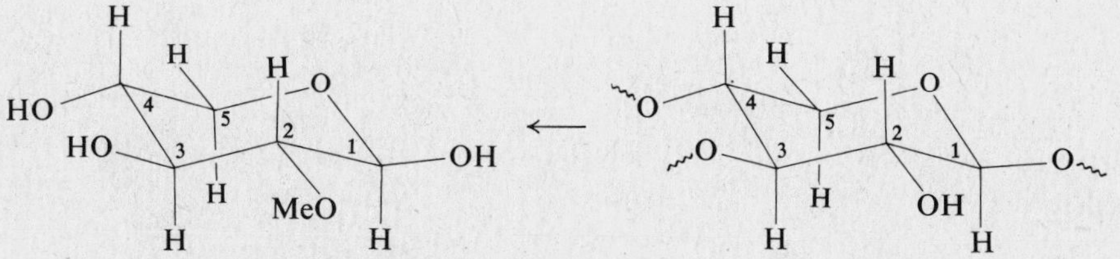

2-O-Methyl-D-xylose

Attachment at
C-1, C-3, and C-4

Chain-linking unit

The xylan is a poly-β-D-xylopyranoside.

Chain-forming unit: attachment at C–1 and C–4.
Chain-terminating unit: attachment at C–1.
Chain-linking unit: attachment at C–1, C–3, and C–4.

Much of the xylan is made up of simple β-xyloside chains, with units joined to each other by linkages between C–1 and C–4; every once in a while two chains are joined together by a linkage involving C–3.

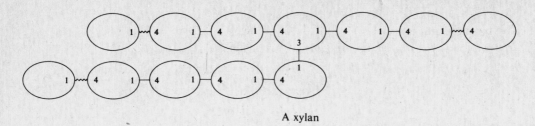

A xylan

1.

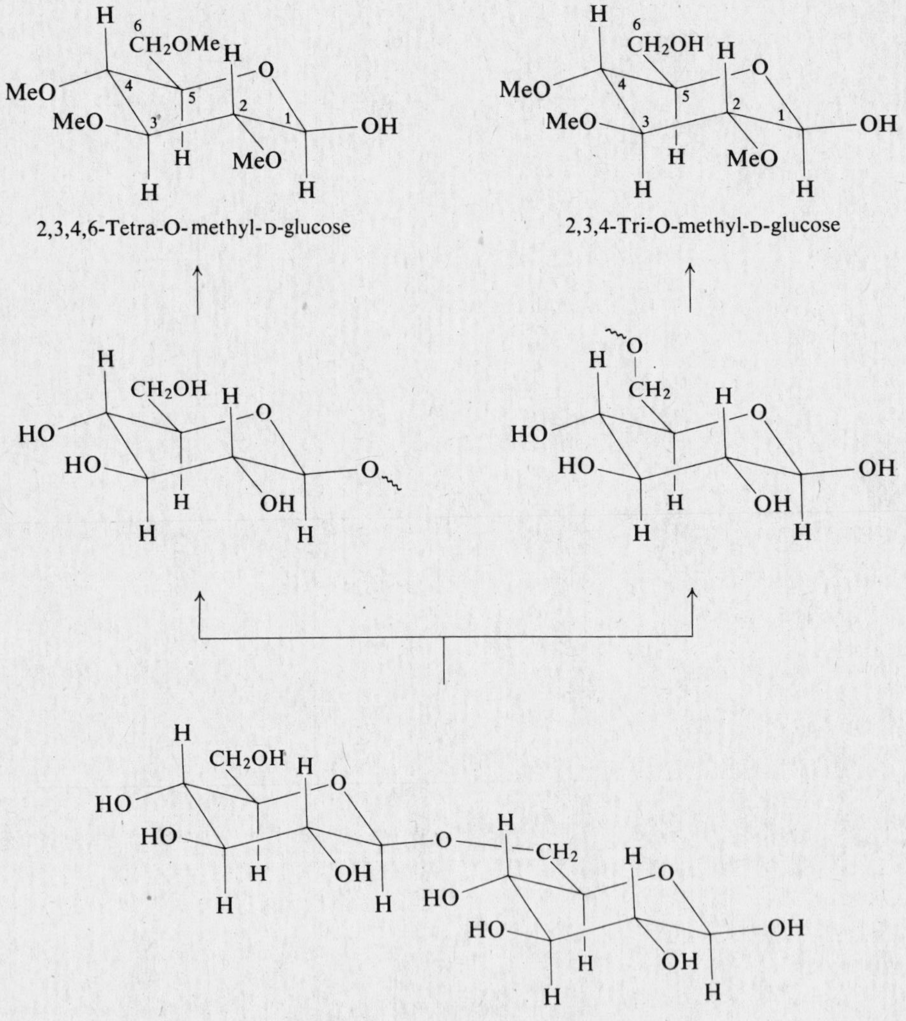

2,3,4,6-Tetra-O-methyl-D-glucose

2,3,4-Tri-O-methyl-D-glucose

6-O-(β-D-Glucopyranosyl)-D-glucopyranose
(β-anomer)
(+)-Gentiobiose

2. (a) Trehalose is a non-reducing sugar: the glucose units must be joined by a C_1—C_1 linkage. It is hydrolyzed by maltase only: the glycosidic linkage is *alpha* to each ring. The C_5-OH is not methylated: the rings are pyranose.

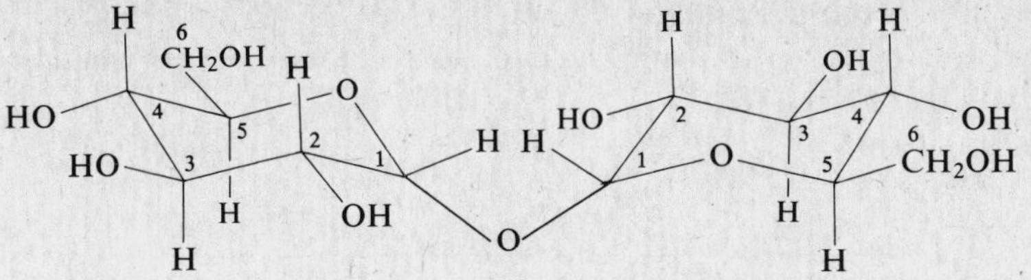

α-D-Glucopyranosyl-α-D-glucopyranoside
Trehalose

(b) Isotrehalose is hydrolyzed by either emulsin or maltose: the glycoside linkage is *alpha* to one ring, and *beta* to the other. In neotrehalose, the linkage is *beta* to each ring.

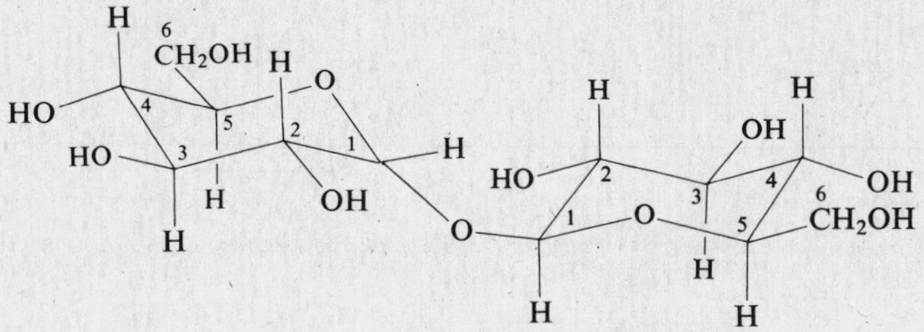

α-D-Glucopyranosyl-β-D-glucopyranoside
Isotrehalose

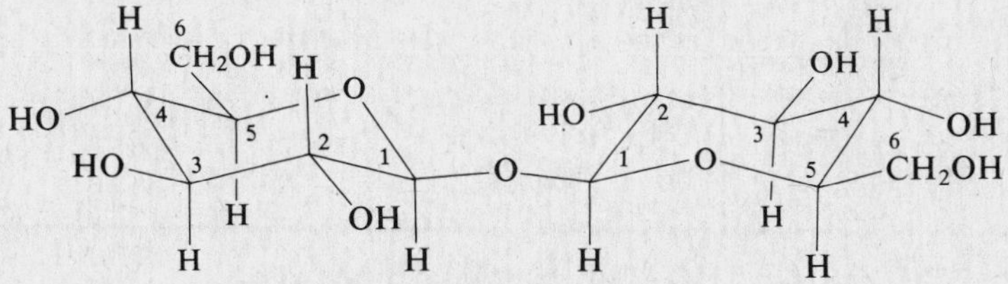

β-D-Glucopyranosyl-β-D-glucopyranoside
Neotrehalose

3.

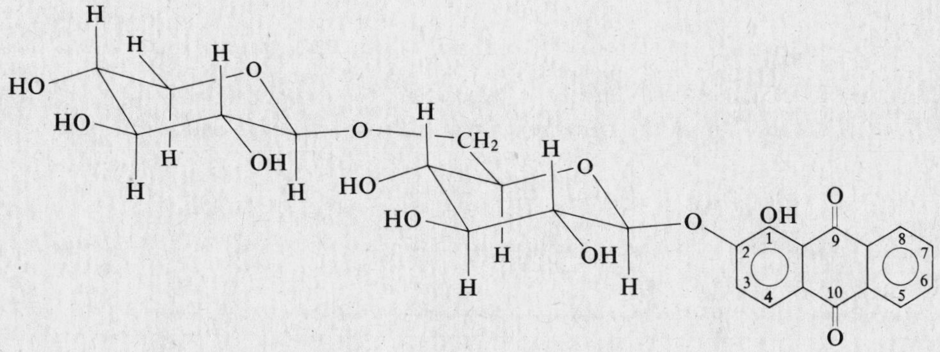

Ruberythric acid
2-(1-Hydroxy-9,10-anthraquinonyl)-β-6-O-(β-D-xylopyranosyl)-D-glucopyranoside

From the data provided two questions remain unanswered. (*i*) Which —OH of alizarin is involved? Methylate ruberythric acid, hydrolyze, and see which —OH is unmethylated. (Glycoside linkages are cleaved under conditions where phenolic ether linkage is preserved.)

(*ii*) Are the glycoside linkages (between alizarin and C–1 of the glucose unit, and between C–6 of the glucose unit and C–1 of the xylose unit) *alpha* or *beta*? Study the enzymatic hydrolysis of ruberythric acid and of primeverose.

4.

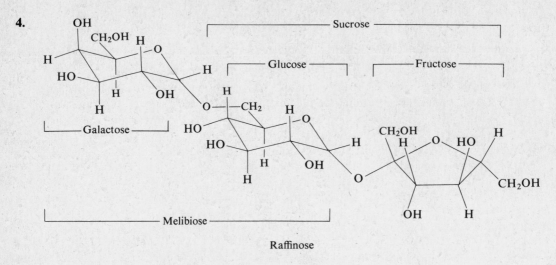

Raffinose

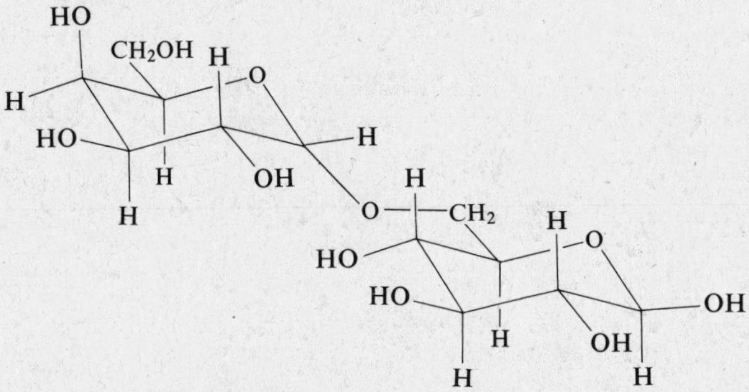

6-O-(α-D-Galactopyranosyl)-D-glucopyranose
(β-anomer)
Melibiose

5. (a) See Figure 35.5, page 609 of this Study Guide. In melezitose, the α-D-glucopyranosyl unit (maltase test) is attached at C–3 of the fructose unit of sucrose (no methylation of C$_3$–OH of fructose).

(b)

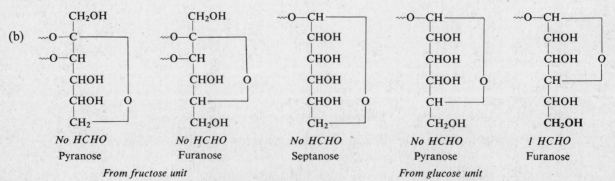

| *No HCHO* | *No HCHO* | *No HCHO* | *No HCHO* | *1 HCHO* |
| Pyranose | Furanose | Septanose | Pyranose | Furanose |

From fructose unit   *From glucose unit*

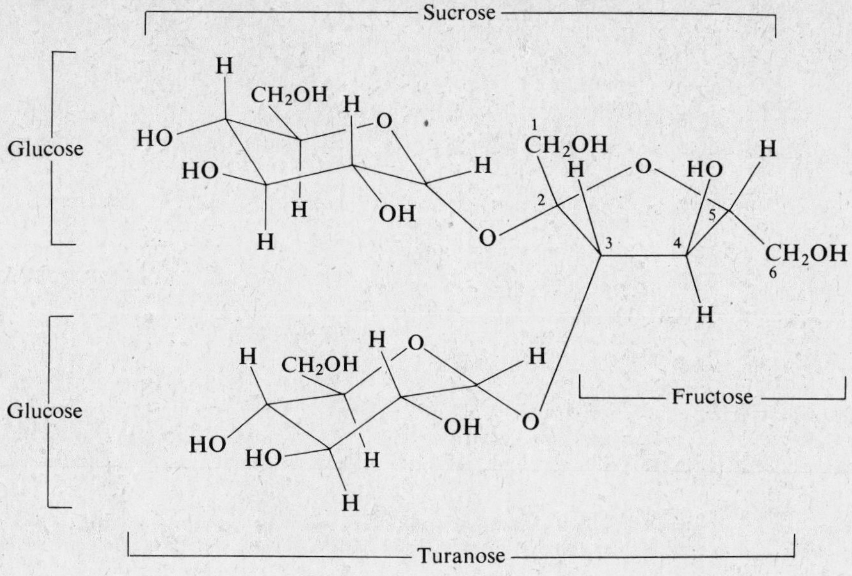

Figure 35.5. Melezitose and turanose (Problem 5a).

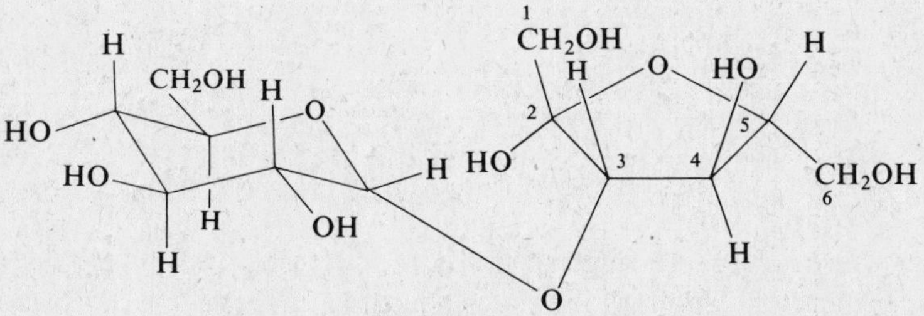

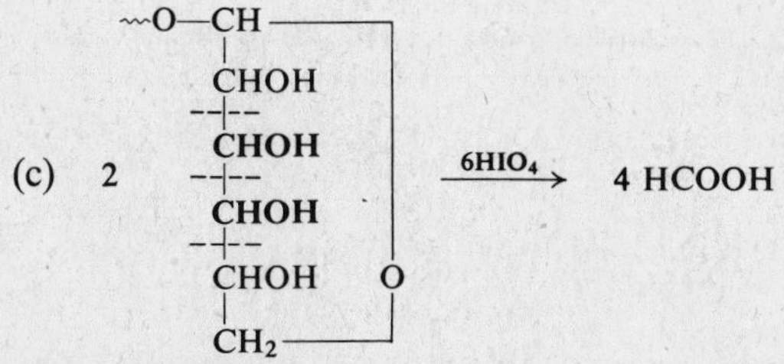

(d)

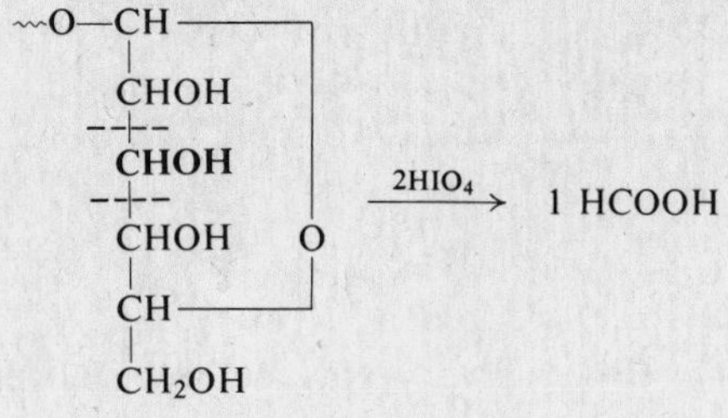
2HIO₄
1 HCOOH

Pyranose

$$3HIO_4 \text{ (septanose)} + 2HIO_4 \text{ (pyranose)} = 5HIO_4$$

$$2HCOOH \text{ (septanose)} + 1HIO_4 \text{ (pyranose)} = 3HCOOH$$

(e) $2HIO_4$ (pyranose) × 2 = $4HIO_4$

1HCOOH (pyranose) × 2 = 2HCOOH

(f) Both glucose rings must be pyranose, since actually only $4HIO_4$ was consumed and 2HCOOH was formed.

(g)

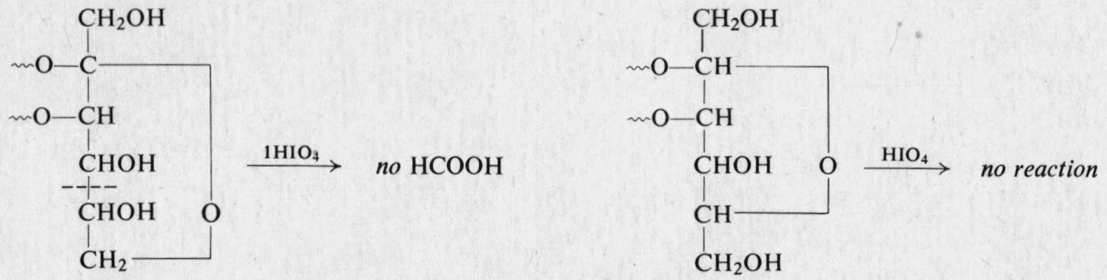
1HIO₄
no HCOOH
HIO₄
no reaction

Pyranose Furanose

(h) The fructose unit must have a furanose ring: all the HIO_4 was used up by the glucose units ($4HIO_4$), and hence the fructose ring must have consumed no HIO_4.

(i) Perfectly consistent, as examination of the structure given in (a) will show.

6. (a)

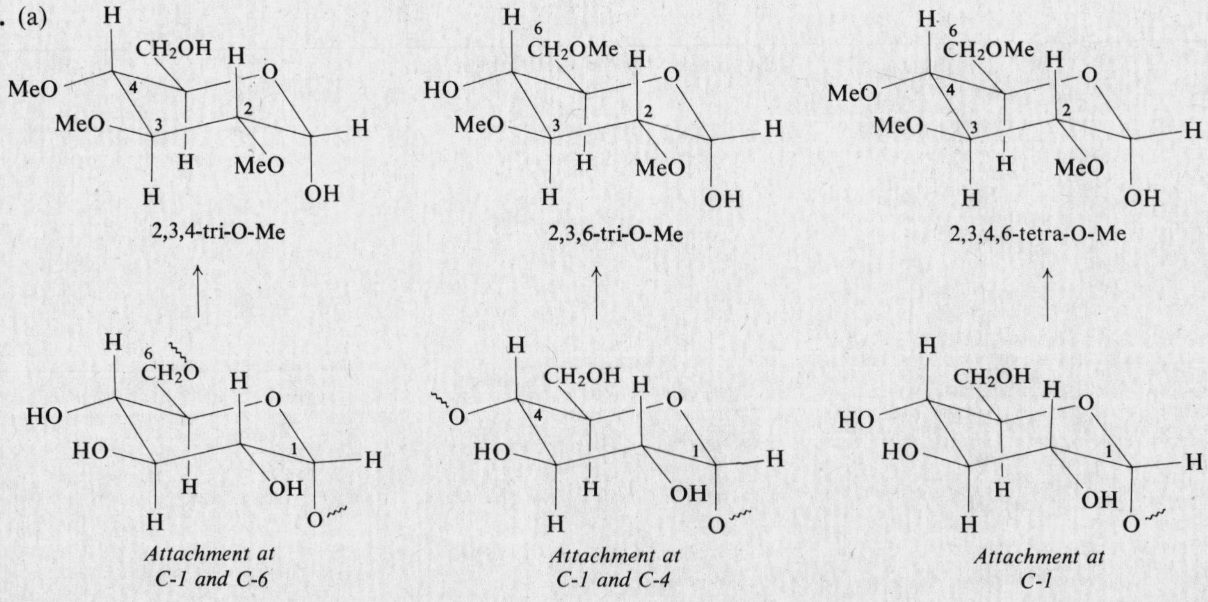
2,3,4-tri-O-Me
2,3,6-tri-O-Me
2,3,4,6-tetra-O-Me
Attachment at C-1 and C-6
Attachment at C-1 and C-4
Attachment at C-1

These three units could be arranged in two different ways:

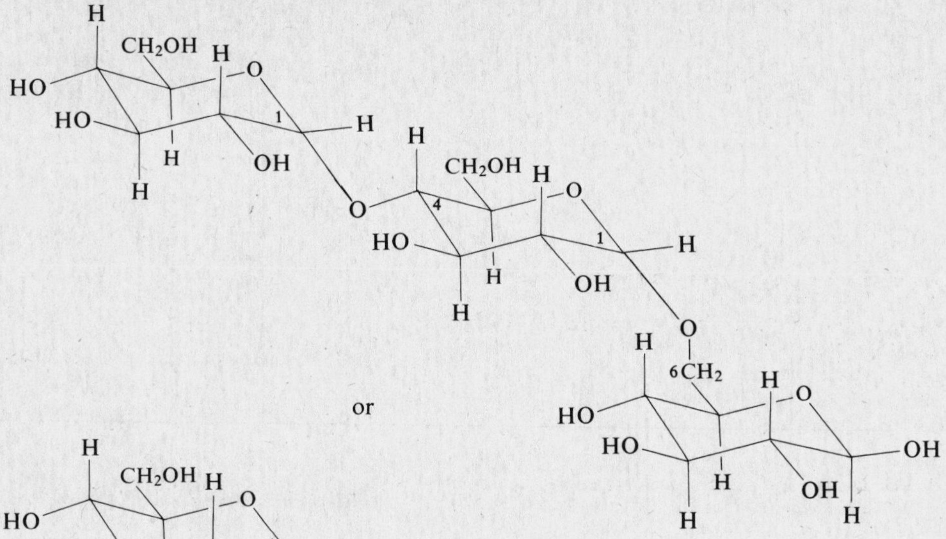

or

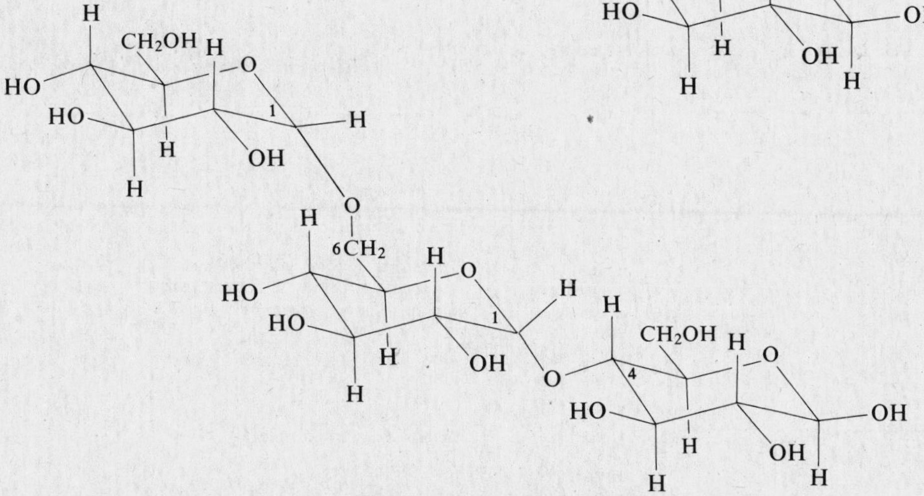

Actual structure

(*Chair conformations assumed*)
Panose

(b) Panose must have the second of these structures, since the maltose unit must include the oxidizable (and reducible) glucose unit.

Isomaltose is like maltose except that linkage is through C–6 instead of C–4.

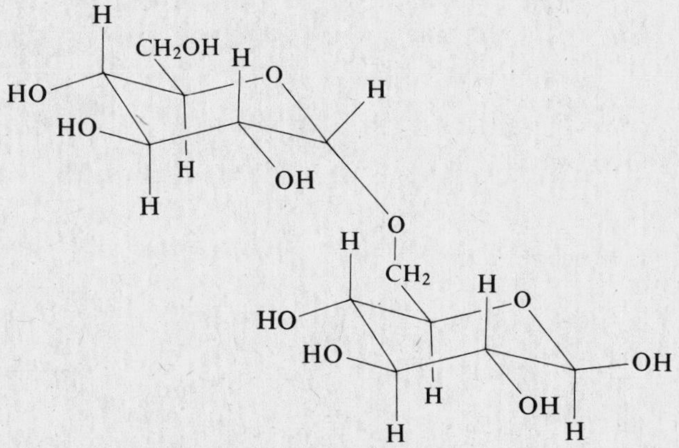

6-O-(α-D-Glucopyranosyl)-D-glucopyranose
Isomaltose

611

(c) This confirms the branched structure of amylopectin (page 1125).

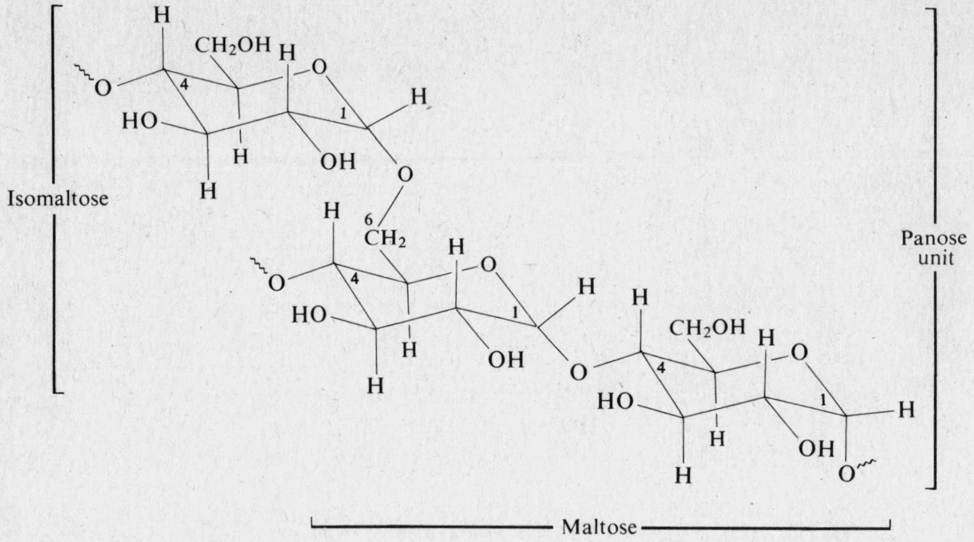

Amylopectin

7.

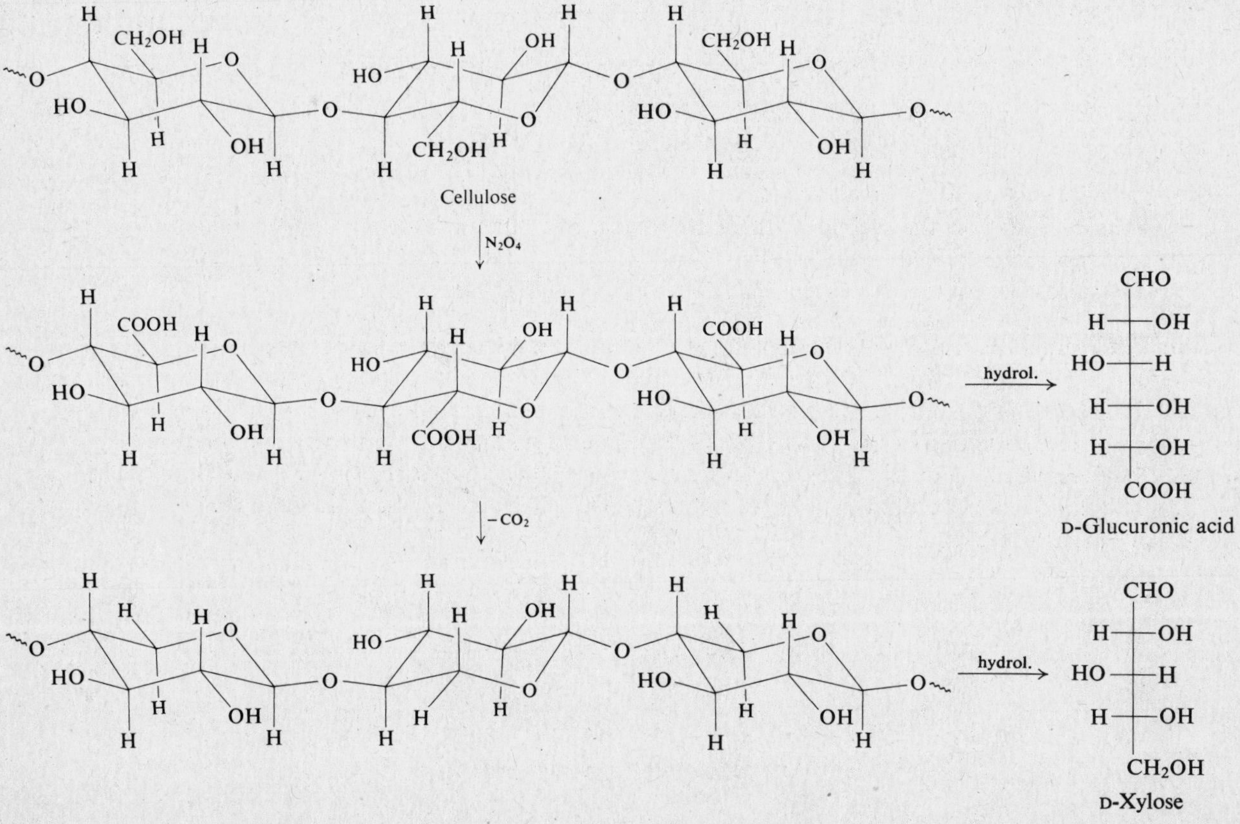

Cellulose

D-Glucuronic acid

D-Xylose

8. (a) The araban is a poly-L-arabinofuranoside.

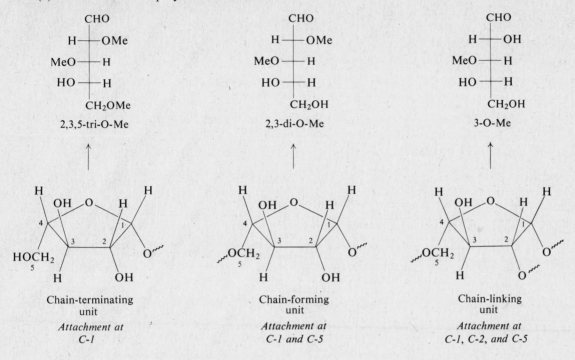

2,3,5-tri-O-Me	2,3-di-O-Me	3-O-Me

Chain-terminating unit	Chain-forming unit	Chain-linking unit
Attachment at C-1	*Attachment at C-1 and C-5*	*Attachment at C-1, C-2, and C-5*

It evidently has many two-unit branches:

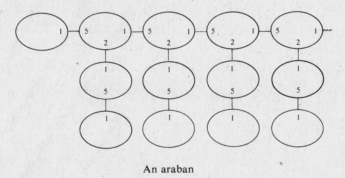

An araban

(b) The mannan is a poly-D-mannopyranoside. (See Figure 35.6, page 614 of this Study Guide.)

It has one-unit branches at (on the average) every other chain-forming unit. If we assume a maximum of regularity in its biosynthesis, we might arrive at something like this: three different units alternate regularly along the chain; those linked through C_1 and C_6 generally (but not quite always—witness the small amount of the 2,3,4-tri-O-methyl compound) branch at C_2.

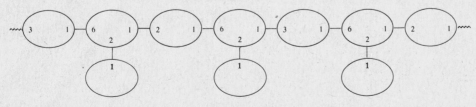

A mannan

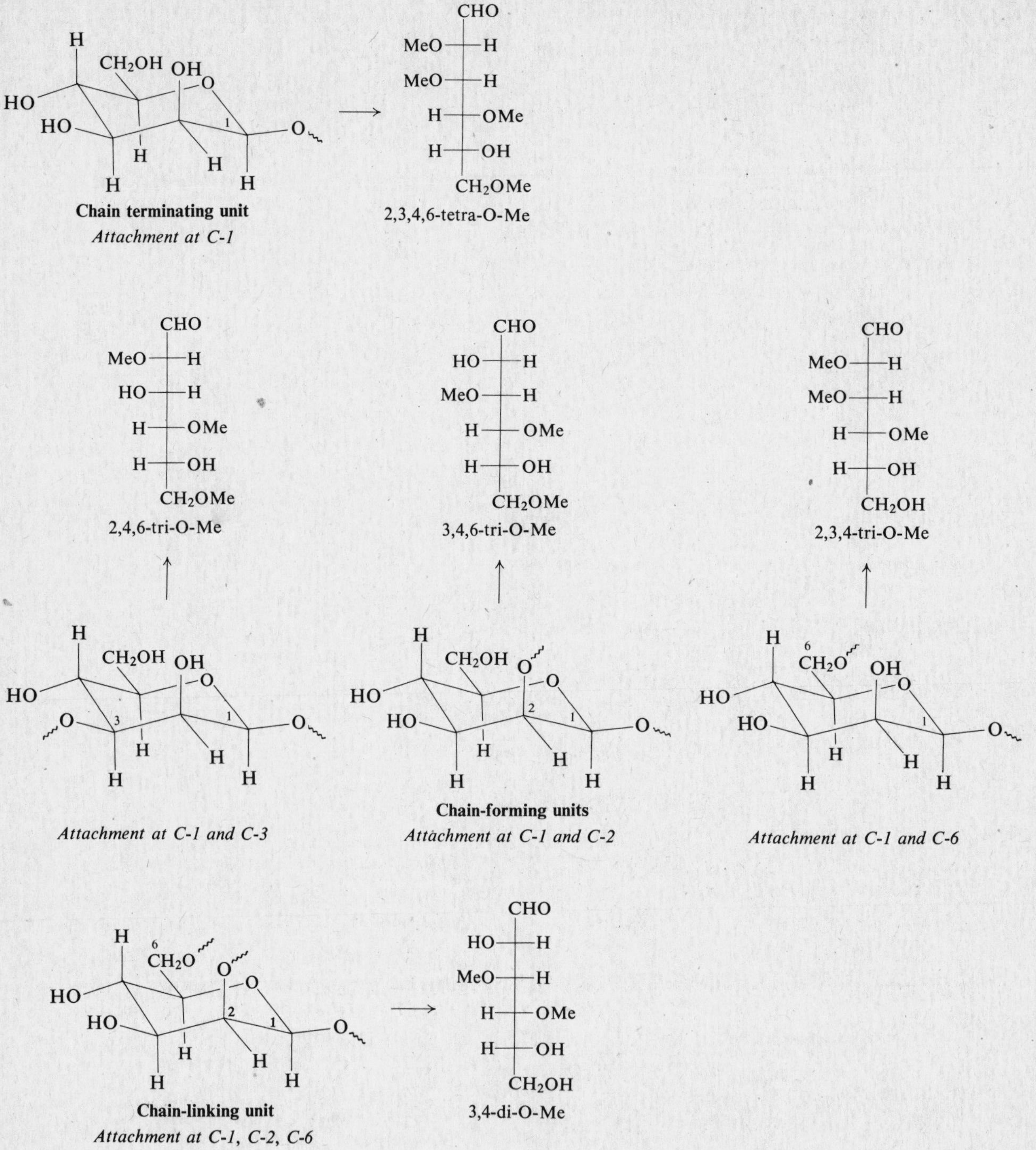

Chain terminating unit
Attachment at C-1

2,3,4,6-tetra-O-Me

2,4,6-tri-O-Me

3,4,6-tri-O-Me

2,3,4-tri-O-Me

Attachment at C-1 and C-3

Chain-forming units
Attachment at C-1 and C-2

Attachment at C-1 and C-6

Chain-linking unit
Attachment at C-1, C-2, C-6

3,4-di-O-Me

Figure 35.6. A mannan: methylation, hydrolysis (Problem 8b).

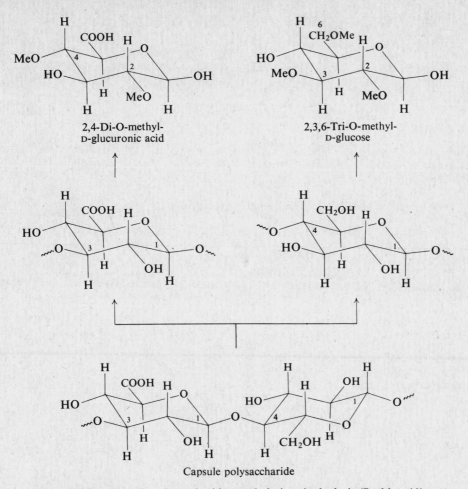

2,4-Di-O-methyl-
D-glucuronic acid

2,3,6-Tri-O-methyl-
D-glucose

Capsule polysaccharide

Figure 35.7. Capsule polysaccharide: methylation, hydrolysis (Problem 11).

9. $HOOC(CH_2)_4COOH \xleftarrow[H^+]{H_2O} N{\equiv}C(CH_2)_4C{\equiv}N \xleftarrow{CN^-} Cl(CH_2)_4Cl$

                F   E   D

    Adipic acid $C_6H_8N_2$ $C_4H_8Cl_2$

$Cl(CH_2)_4Cl \xleftarrow{HCl} \text{(ring, O)} \xleftarrow[cat.]{H_2} \text{(ring, O)} \xleftarrow[-CO_2]{heat} \text{(ring, O)}{-}COOH \xleftarrow{KMnO_4} \text{(ring, O)}{-}CHO$

   D   C   B   A

           C_4H_8O C_4H_4O $C_5H_4O_3$ $C_5H_4O_2$

    Tetrahydrofuran   Furan   Furoic acid   Furfural

       THF   (Sec. 31.2)

All this does not reveal the point of attachment of —CHO to the furan ring. Since furfural is formed here from a straight-chain pentose unit, it would seem likely that attachment is at the 2-position. (It actually *is*.) We can picture something like this happening:

$$\text{An ether} \xleftarrow{-H_2O} [\text{A di(enol)}] \xleftarrow{-H_2O} \text{A pentose}$$

An ether   A di(enol)   A pentose

615

10. (a) Alginic acid is a poly(β-D-mannuronic acid), with units attached at C–1 and C–4.

CHO
MeO——H
MeO——H
H——OH
H——OH
COOH

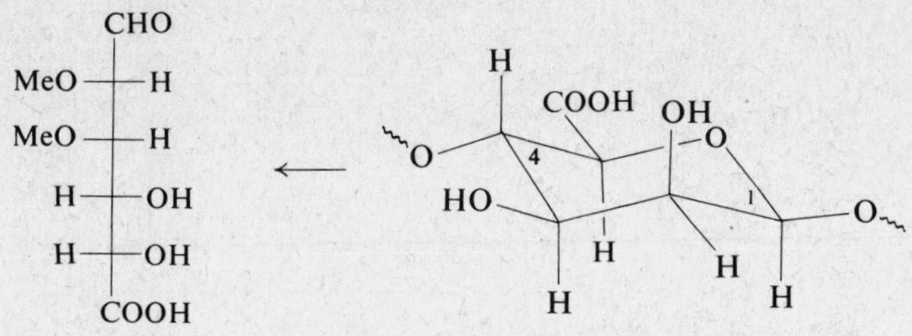

2,3-Di-O-methyl-
D-mannuronic acid

←

Attachment at C-1 and C-4

(b) Pectic acid is a poly(α-D-galacturonic acid), with units attached at C–1 and C–4.

CHO
H——OMe
MeO——H
HO——H
H——OH
COOH

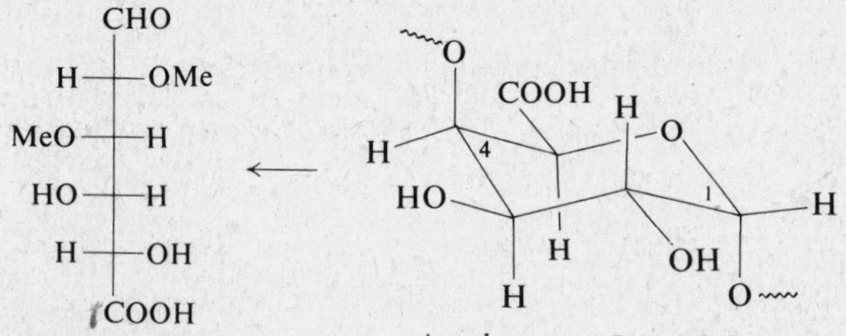

2,3-Di-O-methyl-
D-galacturonic acid

←

Attachment at C-1 and C-4

(c) Agar is a polygalactopyranoside.

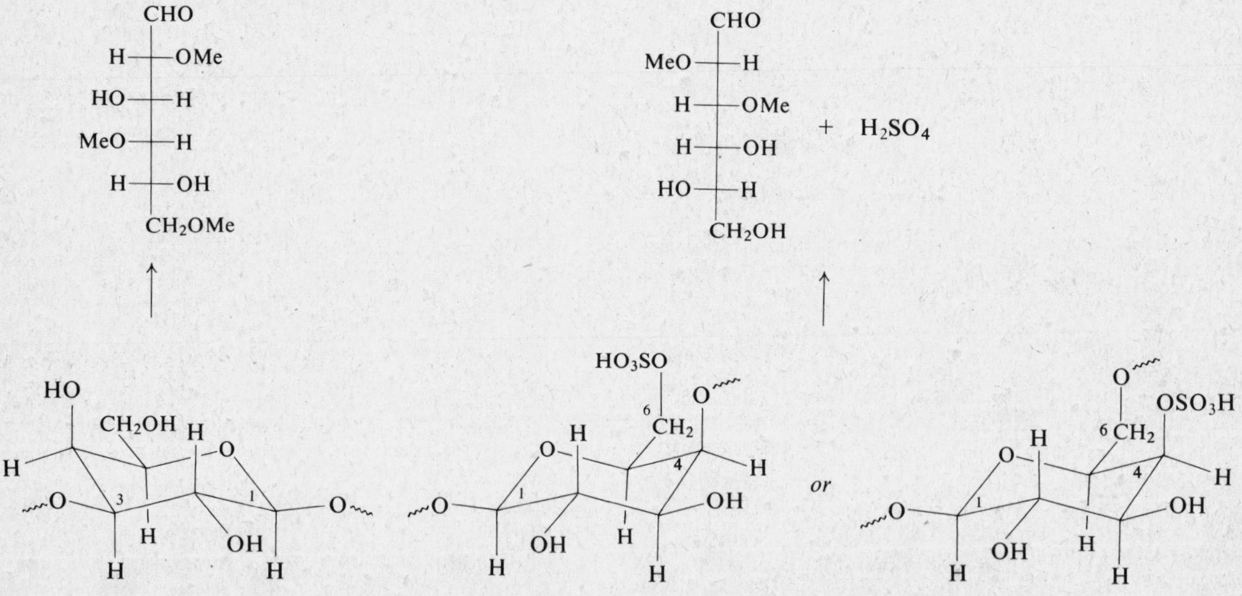

CHO
H——OMe
HO——H
MeO——H
H——OH
CH₂OMe

CHO
MeO——H
H——OMe
H——OH
HO——H
CH₂OH

+ H₂SO₄

Most of the units are D-galactose, attached at C–1 and C–3. About every tenth unit is L-galactose esterified by sulfuric acid. This unit is either attached through C–1 and C–4 with sulfate at C–6, or attached through C–1 and C–6 with sulfate at C–4. (Or, conceivably, it has a furanose ring, with attachment at C–6 and sulfate at C–5, or vice-versa.)

11. This capsule polysaccharide is a poly(cellobiuronic acid) with pyranose rings, and chain-forming attachments at C–1 and C–4 of the D-glucose units and at C–1 and C–3 of the D-glucuronic acid units. (See Figure 35.7, page 615 of this Study Guide.)

12.

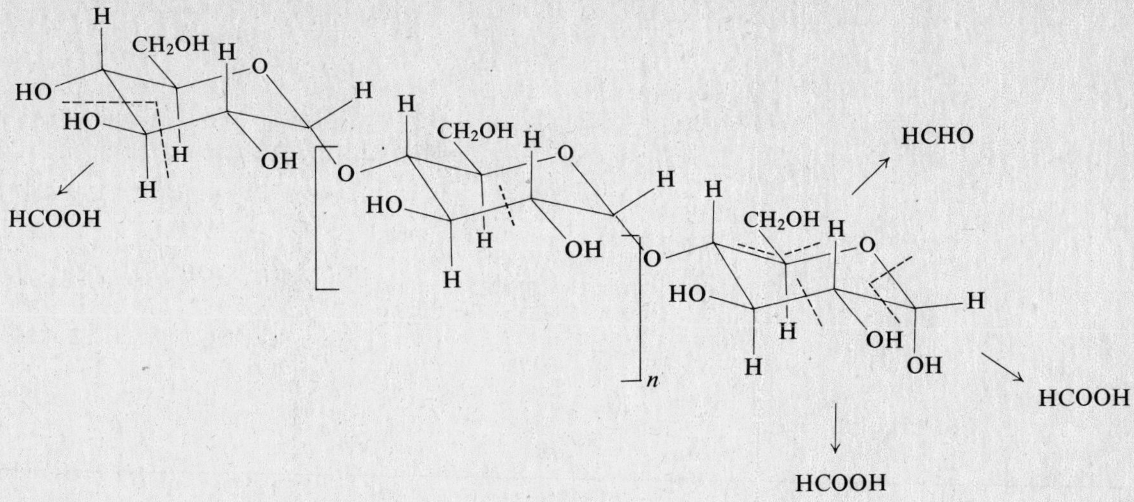

Amylose

HIO₄

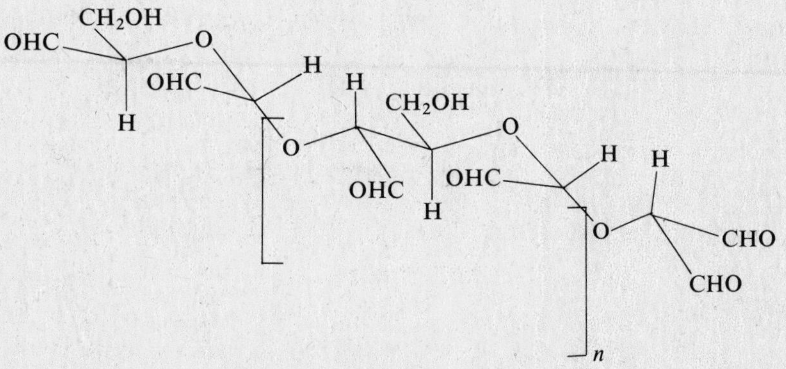

G

Br₂(aq)

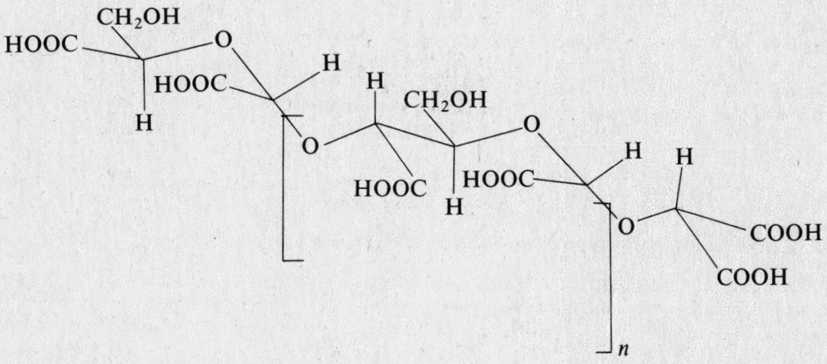

H

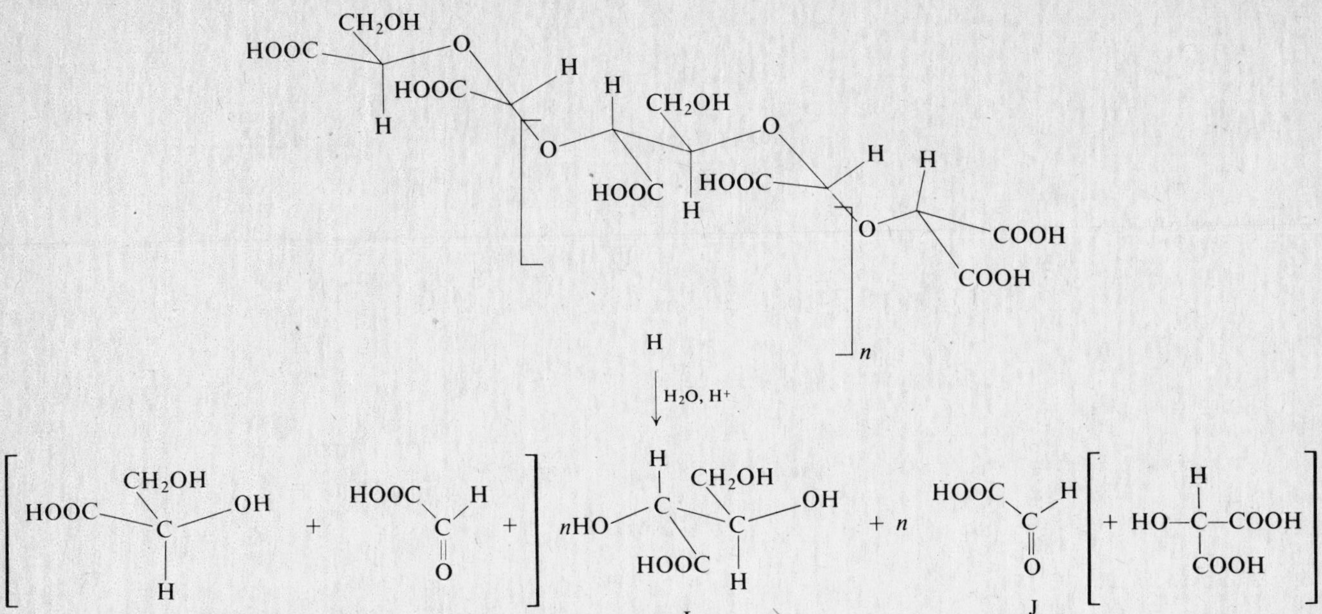

I

C4H8O5

J

C2H2O3

Negligible product   Negligible product

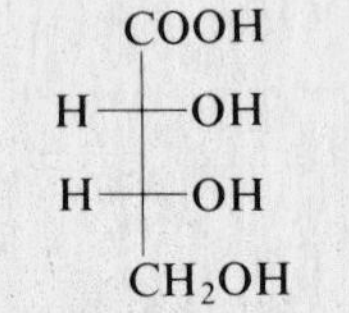

D-Erythronic acid

I

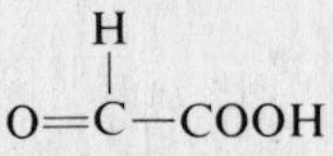

Glyoxylic acid

J

13. (a)

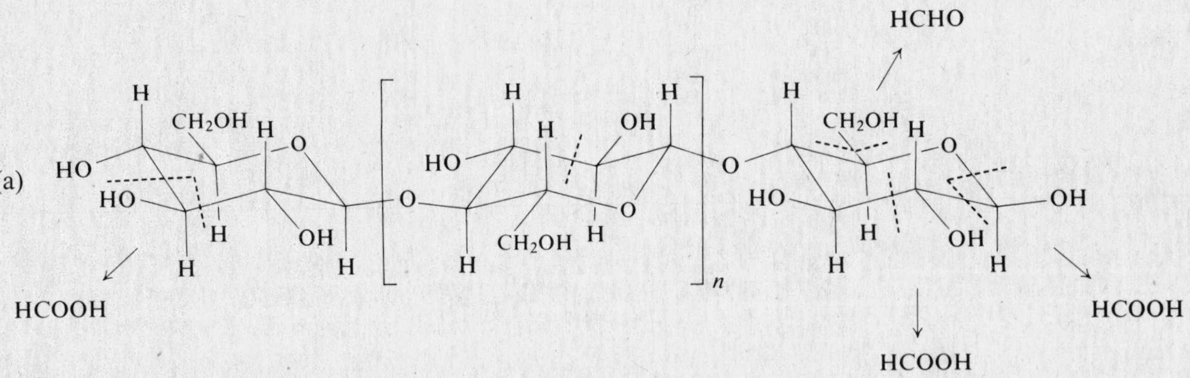

Cellulose

There would be formed 3HCOOH (and 1HCHO) per molecule of cellulose: 2HCOOH from the reducing terminal unit, and 1HCOOH from the non-reducing terminal unit.

(b) See Problem 35.15(b), substituting cellulose for amylose.

(c)

$$0.0027/3 = 0.0009 \text{ mmole cellulose}$$

$$203 \text{ mg}/0.0009 \text{ mmol} = \text{m.w. cellulose}$$

$$\text{m.w.}/162 = \frac{203/0.0009}{162} = 1390 \text{ glucose units per molecule}$$

Amino Acids and Proteins

36.1 $-NH_2 > -COO^-$

$K_b \sim 10^{-4}$ $K_b = \dfrac{K_w}{K_a} = \dfrac{10^{-14}}{\sim 10^{-3}} = \sim 10^{-11}$

Proton goes to $-NH_2$: $H_2NCHRCOO^- + H^+ \longrightarrow {}^+H_3NCHRCOO^-$

36.2 $-COOH > -NH_3{}^+$

$K_a \sim 10^{-3}$ $K_a = \dfrac{K_w}{K_b} = \dfrac{10^{-14}}{\sim 10^{-4}} = \sim 10^{-10}$

Proton leaves $-COOH$: ${}^+H_3NCHRCOOH \longrightarrow H^+ + {}^+H_3NCHRCOO^-$

36.3 K_b for aromatic amines is quite low, about 10^{-10}. (The $-COO^-$ group, with K_b 10^{-9}, is actually a slightly stronger base than aromatic $-NH_2$.) As a result, $-NH_2$ is not neutralized by $-COOH$. But $-SO_3H$ is strongly acidic and can neutralize an aromatic $-NH_2$ group.

36.4 (a) Esterify amino acids in strongly acidic solutions where the major component, ${}^+H_3NCHRCOOH$, contains $-COOH$.

(b) Acylate amino acids in strongly alkaline solutions where the major component, $H_2NCH_2COO^-$, contains $-NH_2$.

36.5 (a) On the acid side. The ionization of the extra $-COOH$ ($K_a \sim 10^{-5}$) has to be suppressed by a considerable excess of H^+.

$$HOOC\text{\~}CH\text{—}COO^- \rightleftarrows H^+ + {}^-OOC\text{\~}CH\text{—}COO^-$$

$$\underset{NH_3{}^+}{|} \qquad\qquad\qquad\qquad \underset{NH_3{}^+}{|}$$

Neutral species

(b) On the basic side. The ionization of the extra —NH_2 ($K_b \sim 10^{-4}$) to give OH^- has to be suppressed by a considerable excess of OH^-.

$$^+H_3N\text{\~}CH\text{—}COO^- + H_2O \rightleftarrows OH^- + {}^+H_3N\text{\~}CH\text{—}COO^-$$

$$\underset{NH_2}{|} \qquad\qquad\qquad\qquad\qquad \underset{NH_3{}^+}{|}$$

Neutral species

(c) More acidic and more basic than for glycine.

36.6 In alkali, sulfanilic acid, $p\text{-}{}^+H_3NC_6H_4SO_3{}^-$, gives up a proton to OH^- to form a water-soluble sulfonate, $p\text{-}H_2NC_6H_4SO_3{}^-$. But sulfonic acids are very strong acids, and sulfonate anions are correspondingly weak bases; in acid solution —$SO_3{}^-$ does not accept a proton, and sulfanilic acid remains unchanged.

36.7 Use differential migration at different pH's; or use differences in solubility at different pH's; or precipitate acidic amino acids as salts of certain bases, and basic amino acids as salts of certain acids.

36.8

COO⁻	COO⁻	COO⁻	COO⁻	CHO	CHO
H——NH₃⁺	⁺H₃N——H	⁺H₃N——H	H——NH₃⁺	HO——H	H——OH
H——OH	HO——H	H——OH	HO——H	H——OH	HO——H
CH₃	CH₃	CH₃	CH₃	CH₂OH	CH₂OH
		L-Threonine		Threose	
		(L *based on C-2*)		(Sec. 34.12)	

36.9 CyS—SCy: one *meso*, 1 pair of enantiomers. Hylys: 2 pairs of enantiomers.
Hypro: 2 pairs of enantiomers. Ileu: 2 pairs of enantiomers.

36.10 Direct ammonolysis (as on page 1139).

$$\underset{\underset{Glycine}{\underset{|}{{}^+NH_3}}}{CH_2COO^-} \longleftarrow \underset{Acetic\ acid}{CH_3COOH}$$

$$\underset{\underset{Alanine}{\underset{|}{{}^+NH_3}}}{CH_3CHCOO^-} \longleftarrow \underset{Propionic\ acid}{CH_3CH_2COOH}$$

$$\underset{\underset{Valine}{\underset{|}{{}^+NH_3}}}{(CH_3)_2CHCHCOO^-} \longleftarrow \underset{Isovaleric\ acid}{(CH_3)_2CHCH_2COOH}$$

(CH₃)₂CHCH₂CHCOO⁻ ⟵ (CH₃)₂CHCH₂CH₂COOH
$\quad\quad\quad$ |
$\quad\quad\quad$ ⁺NH₃ $\quad\quad\quad\quad\quad\quad\quad$ Isocaproic acid
$\quad\quad\quad$ Leucine

HOOCCH₂CHCOO⁻ ⟵ HOOCCH₂CH₂COOH
$\quad\quad$ |
$\quad\quad$ ⁺NH₃ $\quad\quad\quad\quad\quad\quad\quad$ Succinic acid
$\quad$ Aspartic acid

Gabriel synthesis (as on page 1139).

CH₂COO⁻ ⟵ ClCH₂COOEt
|
⁺NH₃ $\quad\quad\quad\quad$ Ethyl chloracetate
$\quad$ Glycine

(CH₃)₂CHCH₂CHCOO⁻ ⟵ (CH₃)₂CHCH₂CHCOOEt
$\quad\quad\quad$ |
$\quad\quad\quad$ ⁺NH₃ $\quad\quad\quad\quad\quad\quad\quad\quad$ Cl
$\quad\quad\quad$ Leucine $\quad\quad\quad\quad$ Ethyl α-chloroisocaproic acid

Malonic ester synthesis (as on page 1139).

(CH₃)₂CH—CHCOO⁻ ⟵ (CH₃)₂CHBr
$\quad\quad\quad$ |
$\quad\quad\quad$ ⁺NH₃ $\quad\quad\quad\quad$ Isopropyl bromide
$\quad\quad$ Valine

$\quad\quad\quad$ CH₃ $\quad\quad\quad\quad\quad\quad\quad$ CH₃
$\quad\quad\quad$ | $\quad\quad\quad\quad\quad\quad\quad$ |
CH₃CH₂CH—CHCOO⁻ ⟵ CH₃CH₂CHBr
$\quad\quad\quad\quad$ |
$\quad\quad\quad\quad$ ⁺NH₃ $\quad\quad\quad\quad$ sec-Butyl bromide
$\quad\quad$ Isoleucine

Phthalimidomalonic ester method (as on page 1140).

HOCH₂—CHCOO⁻ ⟵ HCHO, OH⁻ (aldol condensation)
$\quad\quad\quad$ |
$\quad\quad\quad$ ⁺NH₃
$\quad\quad$ Serine

HOOCCH₂CH₂—CHCOO⁻ ⟵ EtOOCCH₂CH₂Br
$\quad\quad\quad\quad\quad$ |
$\quad\quad\quad\quad\quad$ ⁺NH₃ $\quad\quad\quad$ Ethyl β-bromopropionate
$\quad$ Glutamic acid

HOOCCH₂—CHCOO⁻ ⟵ EtOOCCH₂Br
$\quad\quad\quad\quad$ |
$\quad\quad\quad\quad$ ⁺NH₃ $\quad\quad\quad$ Ethyl bromoacetate
$\quad$ Aspartic acid

36.11 The Strecker synthesis converts an aldehyde into an α-amino nitrile:

$$CH_3\underset{\underset{^+NH_3}{|}}{CH}COO^- \xleftarrow{H_2O,\ H^+} CH_3\underset{\underset{NH_2}{|}}{CH}{-}CN \xleftarrow[NH_4^+]{CN^-} CH_3CHO \qquad CH_2\underset{\underset{^+NH_3}{|}}{}COO^- \longleftarrow HCHO\ (\text{from MeOH})$$

$$\text{Alanine} \qquad\qquad\qquad\qquad\qquad\qquad\qquad\qquad\qquad \text{Glycine}$$

$$(CH_3)_2CHCH_2\underset{\underset{^+NH_3}{|}}{CH}COO^- \longleftarrow (CH_3)_2CHCH_2CHO \longleftarrow (CH_3)_2CHCH_2CH_2OH$$

$$\text{Leucine} \qquad\qquad\qquad\qquad\qquad\qquad\qquad\qquad \text{Isopentyl alcohol}$$

$$CH_3CH_2CH(CH_3)\underset{\underset{^+NH_3}{|}}{CH}COO^- \longleftarrow CH_3CH_2CH(CH_3)CHO \longleftarrow CH_3CH_2CH(CH_3)CH_2OH$$

$$\text{Isoleucine} \qquad\qquad\qquad\qquad\qquad\qquad\qquad\qquad \text{2-Methyl-1-butanol}$$

$$(CH_3)_2\underset{\underset{^+NH_3}{|}}{CH}CHCOO^- \longleftarrow (CH_3)_2CHCHO \longleftarrow (CH_3)_2CHCH_2OH$$

$$\text{Valine} \qquad\qquad\qquad\qquad\qquad\qquad \text{Isobutyl alcohol}$$

$$HOCH_2\underset{\underset{^+NH_3}{|}}{CH}COO^- \xleftarrow{\text{acid}} EtOCH_2\underset{\underset{^+NH_3}{|}}{CH}COO^- \xleftarrow[H^+]{H_2O} EtOCH_2\underset{\underset{NH_2}{|}}{CH}CN \xleftarrow[NH_4^+]{CN^-} EtOCH_2CHO$$

$$\text{Serine}$$

36.12 (a) $(CH_3)_2CHCH_2COOEt + EtOOC{-}COOEt \xrightarrow{OEt^-} (CH_3)_2CHCH{-}\underset{\underset{O}{\|}}{C}{-}COOEt$ (with COOEt above)

A

$A \xrightarrow[\text{heat}]{H^+} (CH_3)_2CHCH{-}\underset{\underset{O}{\|}}{C}{-}COOH$ (with COOH above) $\xrightarrow{-CO_2} (CH_3)_2CHCH_2{-}\underset{\underset{O}{\|}}{C}{-}COOH \xrightarrow[Pd]{NH_3,\ H_2} (CH_3)_2CHCH_2{-}\underset{\underset{^+NH_3}{|}}{CH}{-}COO^-$

B   Leucine

(b) $CH_3\underset{\underset{^+NH_3}{|}}{CH}COO^- \longleftarrow CH_3COOEt$ in place of ethyl isovalerate

Alanine   Ethyl acetate

$HOOCCH_2CH_2\underset{\underset{^+NH_3}{|}}{CH}COO^- \longleftarrow EtOOCCH_2CH_2COOEt$ in place of ethyl isovalerate

Glutamic acid   Ethyl succinate

36.13 (a) $H_2NCH_2COO^-Na^+$ (b) $Cl^-\ ^+H_3NCH_2COOH$

(c) $Ph\underset{\underset{O}{\|}}{C}{-}NHCH_2COOH$ (d) $CH_3\underset{\underset{O}{\|}}{C}{-}NHCH_2COOH$

(e) $HOCH_2COOH + N_2$ (f) $HSO_4^-\ ^+H_3NCH_2COOEt$

(g) $PhCH_2O\underset{\underset{O}{\|}}{C}{-}NHCH_2COOH$ (Cbz.Gly)

36.14 (a) PhCNHCH$_2$COCl
$\quad\quad\quad\quad$‖
$\quad\quad\quad\quad$O

(b) PhCNHCH$_2$CONH$_2$
$\quad\quad\quad$‖
$\quad\quad\quad$O

(c) PhCNHCH$_2$CNHCHCOOH
$\quad\quad\quad$‖$\quad\quad$‖$\quad\quad$CH$_3$
$\quad\quad\quad$O$\quad\quad$O
$\quad\quad\quad\quad\quad$(Bz.Gly.Ala)

(d) PhCNHCH$_2$COOEt
$\quad\quad\quad$‖
$\quad\quad\quad$O

(e) 3,5-Dibromotyrosine

HO〈○〉CH$_2$CHCOO$^-$ with Br at 3,5 positions, $^+$NH$_3$

3,5-Dibromotyrosine

(f) $^-$OOCCH$_2$CHCOO$^-$
$\quad\quad\quad\quad\quad$|
$\quad\quad\quad\quad\quad$$^+NH_3$

(g) N,N-Dimethylproline

〈ring〉—COO$^-$ with $^+$N, CH$_3$ CH$_3$

N,N-Dimethylproline

(h) MeO〈○〉CH$_2$CHCOO$^-$
$\quad\quad\quad\quad\quad\quad\quad$|
$\quad\quad\quad\quad\quad\quad\quad$$^+NH_3$

(i) Na$^+$ $^-$OOCCH$_2$CH$_2$CHCOO$^-$
$\quad\quad\quad\quad\quad\quad\quad\quad\quad$|
$\quad\quad\quad\quad\quad\quad\quad\quad\quad$$^+NH_3$
$\quad\quad\quad$Monosodium glutamate

(j) EtOOCCH$_2$CH$_2$CHCOOEt
$\quad\quad\quad\quad\quad\quad\quad\quad$|
$\quad\quad\quad\quad\quad\quad\quad\quad$$^+NH_3HSO_4$$^-$

36.15 (a) 0.001 mole or 22.4 cc N$_2$ from monoamine (Leu).

(b) 0.002 mole or 44.8 cc N$_2$ from diamine (Lys).

(c) No N$_2$ from 2° amine.

36.16 $\quad\quad\quad\quad\quad 2.01 \times \dfrac{748}{760} \times \dfrac{273}{293} = $ ml N$_2$ at S.T.P.

$$\frac{\text{ml N}_2}{22.4 \text{ ml/mmol}} = \text{millimoles N}_2 = \text{millimoles —NH}_2 = \text{millimoles amino acid if one —NH}_2$$

$$\text{if monoamino: } 0.0821 \text{ mmole amino acid}$$

$$\frac{9.36 \text{ mg}}{0.0821 \text{ mmole}} = \text{mol. wt.} = 114 = \text{minimum mol. wt.}$$

The amino acid could be valine, C$_5$H$_{11}$O$_2$N, m.w. 117. It could *not* be proline (m.w. 115), since proline has no —NH$_2$ present.

36.17 (a)

—C=Ö: with N—H ⟷ —C=Ö:$^{\ominus}$ with $^{\oplus}$N—H

(b) Overlap of *p*-orbital of N with π-orbital of $\rangle$C=O.

36.18 (a)

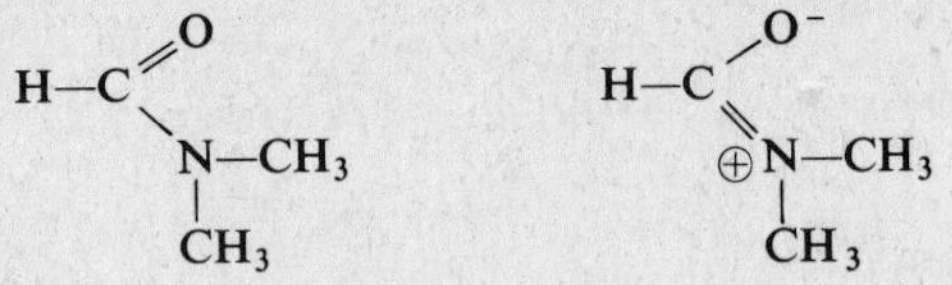

Partial double-bond character of the carbon–nitrogen bond causes hindered rotation, resulting in diastereotopic —CH_3 groups that produce separate nmr signals. These signals coalesce at higher temperatures as the —CH_3's become equivalent through rapid rotation.

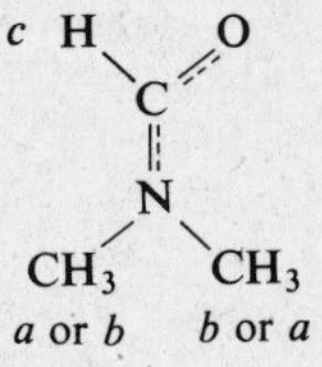

(b) This confirms the partial double-bond character of the carbon–nitrogen bond in the peptide linkage.

36.19

Ala: $\dfrac{0.89 \text{ g}}{89 \text{ g/mole}} = 0.01$ mole in 100 g salmine

Arg: $\dfrac{86.04 \text{ g}}{174 \text{ g/mole}} = 0.50$ mole

Gly: $\dfrac{3.01 \text{ g}}{75 \text{ g/mole}} = 0.04$ mole

Ileu: $\dfrac{1.28 \text{ g}}{131 \text{ g/mole}} = 0.01$ mole

Pro: $\dfrac{6.90 \text{ g}}{115 \text{ g/mole}} = 0.06$ mole

Ser: $\dfrac{7.29 \text{ g}}{105 \text{ g/mole}} = 0.07$ mole

Val: $\dfrac{3.68 \text{ g}}{117 \text{ g/mole}} = 0.03$ mole

The empirical formula is thus $AlaArg_{50}Gly_4IleuPro_6Ser_7Val_3$. The weights of the amino acids add up to more than 100 g because water is taken up in the hydrolysis of the peptide links.

36.20 Since there is 0.01 mole Ala in 100 g of salmine (Problem 36.19, above), there must be 1 mole Ala in 10,000 g of salmine. This means that there is only one Ala per molecule, and the molecular formula is the same as the empirical formula.

36.21 If we assume only one Tyr per molecule,

$$0.29\% \text{ Tyr} = \frac{\text{m.w. Tyr}}{\text{m.w. protein}} \times 100 = \frac{204}{\text{m.w.}} \times 100$$

$$\text{m.w.} = 70{,}300$$

This is the *minimum* m.w.; if there are two Tyr per molecule, the m.w. is $2 \times 70{,}300$, etc.

36.22 (a) If we assume only one Fe per molecule,

$$0.335\% \text{ Fe} = \frac{\text{at.wt. Fe}}{\text{m.w. protein}} \times 100 = \frac{55.8}{\text{m.w.}} \times 100$$

$$\text{m.w.} = 16{,}700$$

This is the *minimum* m.w.

(b) $$\frac{67{,}000}{16{,}700} = 4\text{Fe per molecule}$$

36.23

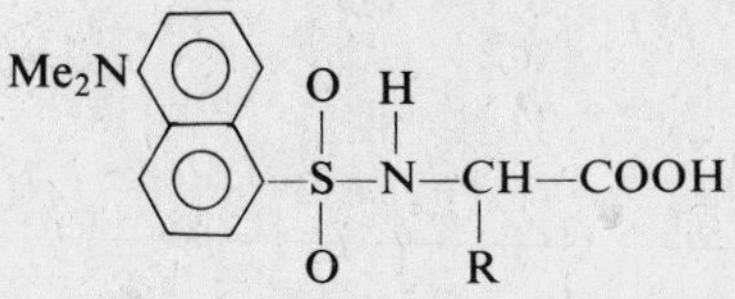

A sulfonamide:
not easily hydrolyzed

Dansyl derivative

36.24 (a)

Val.Asp
Phe.Val
　　　Asp.Glu
　　　　　Glu.His
_____
Phe.Val.Asp.Glu.His

(b)

　　　　　CySH.Gly.Ser
His.Leu.CySH
　　　　　　　Ser.His.Leu
_____
His.Leu.CySH.Gly.Ser.His.Leu

(c)

　　　　Val.CySH.Gly
Tyr.Leu.Val
　　　　　　Gly.Glu.Arg
　　　　　　Glu.Arg.Gly
　　　　　　　　Gly.Phe.Phe
_____
Tyr.Leu.Val.CySH.Gly.Glu.Arg.Gly.Phe.Phe

(One must be careful of placement of Gly units.)

36.25 (a) Cbz.Gly.Ala + $SOCl_2$; then Phe; then H_2, Pd.

(b) $PhCH_2OCOCl$ + Ala; then $SOCl_2$; then Gly; finally H_2, Pd.

36.26

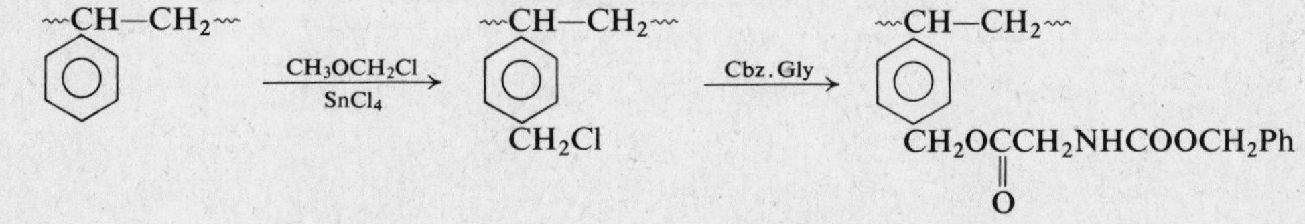

| Polystyrene | *Friedel-Crafts alkylation* | A | *Nucleophilic substitution* | B, a carbamate |

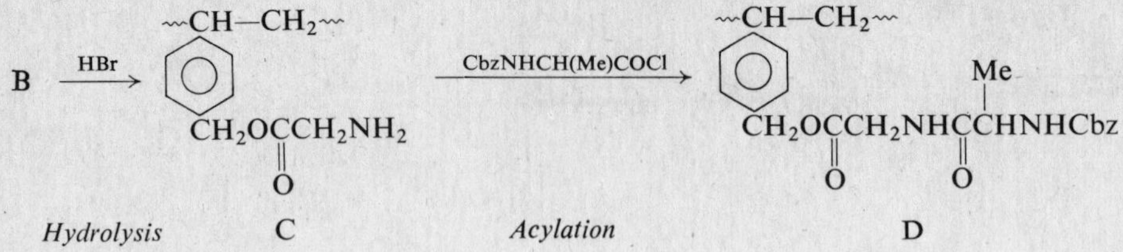

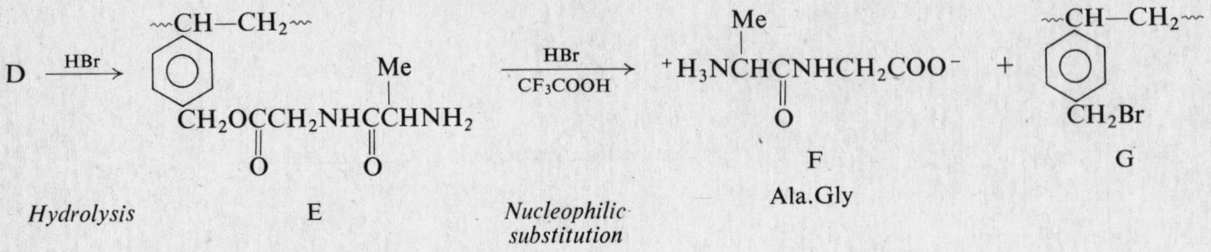

B $\xrightarrow{HBr}$ [ring structure] $\xrightarrow{CbzNHCH(Me)COCl}$ [ring structure]

~CH—CH₂~

CH₂OCCH₂NH₂
‖
O

Hydrolysis   C   *Acylation*   D

~CH—CH₂~

CH₂OCCH₂NHCCHNHCbz
‖   ‖
O   O   Me

D $\xrightarrow{HBr}$ [ring structure] $\xrightarrow[CF_3COOH]{HBr}$ $^+H_3NCHCNHCH_2COO^-$ + [ring structure]

~CH—CH₂~

CH₂OCCH₂NHCCHNH₂
‖   ‖
O   O   Me

Hydrolysis   E   *Nucleophilic substitution*

Me
|
$^+H_3NCHCNHCH_2COO^-$
‖
O
F
Ala.Gly

~CH—CH₂~
CH₂Br
G

(John F. Henahan, "R. Bruce Merrifield.   Designer of Protein-Making Machine,"
Chem. Eng. News, August 2, 1971, pp. 22–25.)

1. (a) 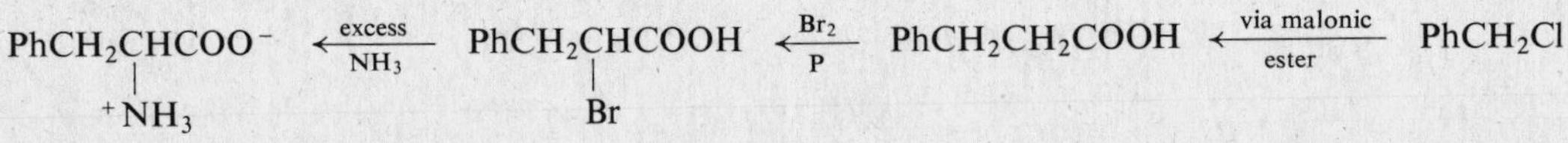 $PhCH_2CHCOO^-$ $\xleftarrow[NH_3]{excess}$ $PhCH_2CHCOOH$ $\xleftarrow[P]{Br_2}$ $PhCH_2CH_2COOH$ $\xleftarrow[ester]{via\ malonic}$ $PhCH_2Cl$

$^+NH_3$; Br

(b) 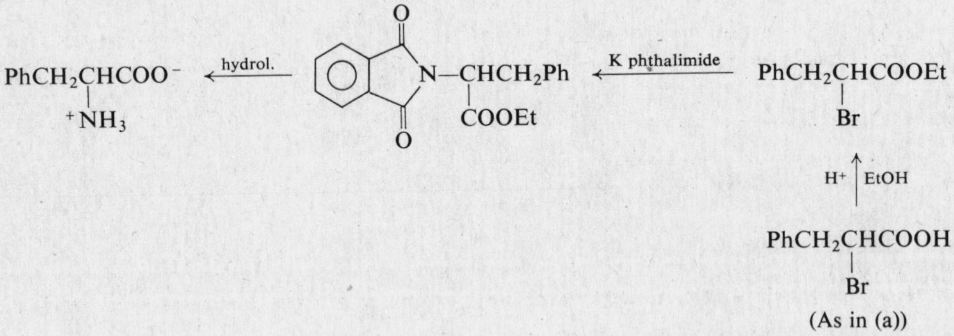 $PhCH_2CHCOO^-$ $\xleftarrow{hydrol.}$ [phthalimide structure] N—CHCH₂Ph $\xleftarrow{K\ phthalimide}$ $PhCH_2CHCOOEt$

$^+NH_3$; COOEt ; Br

$\uparrow$ H⁺ | EtOH

$PhCH_2CHCOOH$
|
Br
(As in (a))

(c) 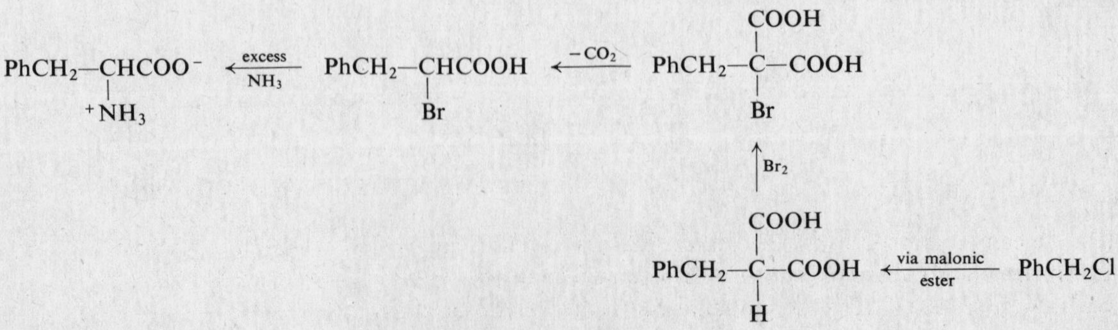 $PhCH_2—CHCOO^-$ $\xleftarrow[NH_3]{excess}$ $PhCH_2—CHCOOH$ $\xleftarrow{-CO_2}$ $PhCH_2—C—COOH$

$^+NH_3$; Br ;

COOH
|
PhCH₂—C—COOH
|
Br

$\uparrow$ Br₂

COOH
|
PhCH₂—C—COOH $\xleftarrow[ester]{via\ malonic}$ $PhCH_2Cl$
|
H

626

(d) PhCH₂—CHCOO⁻ $\xleftarrow{\text{hydrol.}}$ [phthalimide-N-CH(COOH)-CH₂Ph] $\xleftarrow[-CO_2]{H^+,\ heat}$ [phthalimide-N-CH(COOEt)₂] $\xleftarrow{PhCH_2Cl}$

(d) PhCH₂—CHCOO⁻ with ⁺NH₃

Phthalimidomalonic ester

$\uparrow$ K phthalimide

COOEt
BrCH
COOEt

Bromomalonic ester

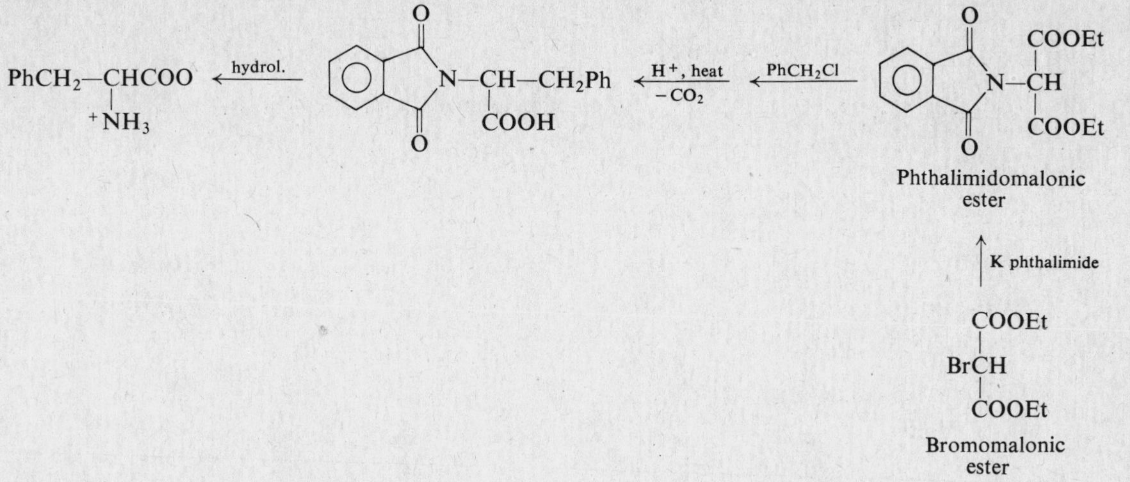

(e) PhCH₂CH—COO⁻ $\xleftarrow[H^+]{H_2O}$ PhCH₂CH—CN $\xleftarrow[NH_4^+]{CN^-}$ PhCH₂CHO $\xleftarrow{LiAlH(OBu\text{-}t)_3}$ PhCH₂COCl

with ⁺NH₃ ... NH₂

$\uparrow$ SOCl₂

PhCH₂COOH
(page 587)

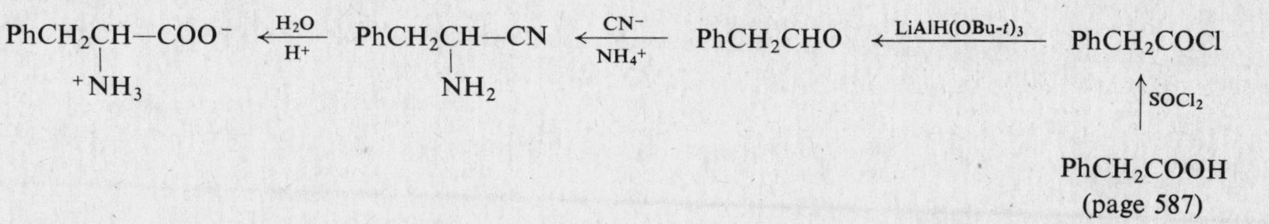

(f) PhCH₂CHCOO⁻ $\xleftarrow[Pd]{NH_3,\ H_2}$ PhCH₂CCOOH $\xleftarrow[-CO_2]{H^+,\ heat}$ PhCHCCOOEt $\xleftarrow[NaOEt]{(COOEt)_2}$ PhCH₂COOEt

with ⁺NH₃ ... O ... O (with COOEt)

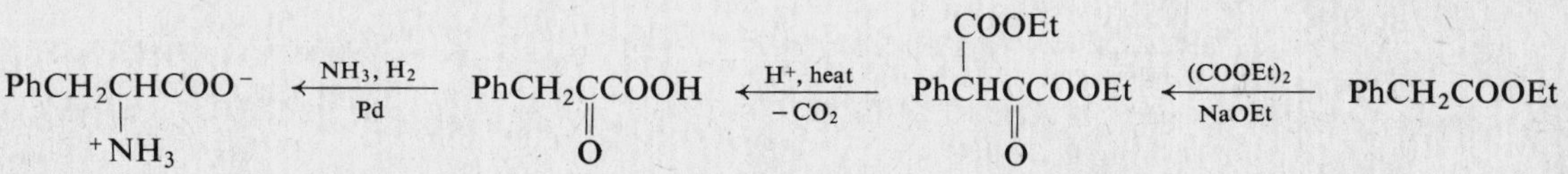

2. (a)

phthalimide-N⁻K⁺ + BrCH(COOEt)₂ → A $\xrightarrow[NaOEt]{Br(CH_2)_3Br}$ B

A

B

B $\xrightarrow{KOAc}$ C $\xrightarrow[heat]{OH^-}$ $\xrightarrow[-CO_2]{H^+}$ ⁺H₃N—C—CH₂CH₂CH₂OH (COO⁻, H) D

C

D

D $\xrightarrow{HCl}$ [⁺H₃N—C—CH₂CH₂CH₂Cl (COO⁻, H)] → Proline

E

Proline

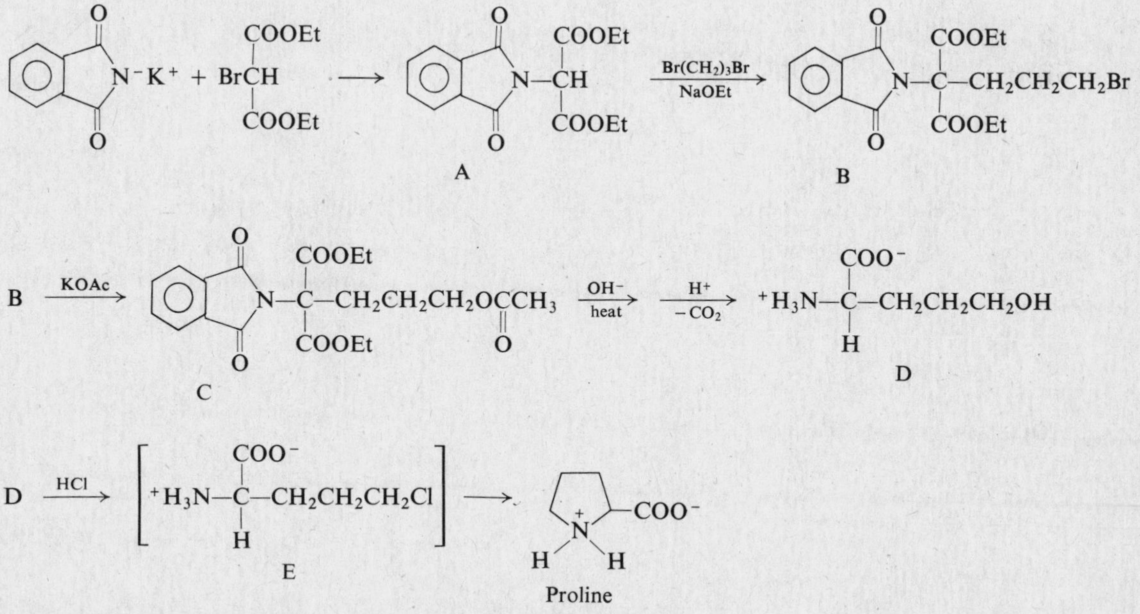

(b) We can see a good example of the "backwards" procedure of working out a synthesis in the structure of lysine; here the various synthetic units are easily recognized, once we commit ourselves to using the phthalimidomalonic ester procedure.

NH_3 ----→ ^+H_3N⊢$CH_2CH_2CH_2CH_2$⊣$CHCOO^-$

$BrCH_2CH_2CH_2CH_2Br$ ⇠   ⇢ $\underset{NH_2}{}$ ⇠ phthalimidomalonic ester

Lysine

$^+H_3N(CH_2)_4\underset{NH_2}{CHCOO^-}$ $\xleftarrow{\text{base}}$ [phthalimide]–$\overset{COOEt}{\underset{COOEt}{C}}$–$(CH_2)_4$—$NH_2$ $\xleftarrow[NH_3]{\text{excess}}$ [phthalimide]–$\overset{COOEt}{\underset{COOEt}{C}}$–$(CH_2)_4Br$

Lysine

$\xuparrow{Br(CH_2)_4Br}$

[phthalimide]N—$\overset{COOEt}{\underset{COOEt}{C}}$:$^-Na^+$

(from phthalimodomalonic ester)

3. (a)

$CH_3CON\underset{COOEt}{\overset{COOEt}{H\dot{C}H}}$ $\xrightarrow[\text{Michael}]{CH_2=CHCHO}$ $CH_3CON\underset{COOEt}{\overset{COOEt}{H\dot{C}}}$—$CH_2CH_2CHO$ $\xrightarrow[H^+]{CN^-}$ $CH_3CON\underset{COOEt}{\overset{COOEt}{H\dot{C}}}CH_2CH_2\underset{OH}{CHCN}$

                                                          F   G

G $\xrightarrow[\text{heat}]{\text{acid}}$ $CH_3CON\underset{COOEt}{\overset{COOEt}{H\dot{C}}}CH_2CH=CHCN$ $\xrightarrow[\text{cat.}]{H_2}$ $CH_3CON\underset{COOEt}{\overset{COOEt}{H\dot{C}}}CH_2CH_2CH_2CH_2NH_2$

                                        H   I

I $\xrightarrow{Ac_2O}$ $CH_3CON\underset{COOEt}{\overset{COOEt}{H\dot{C}}}(CH_2)_4NHAc$ $\xrightarrow[\text{heat}]{OH^-}$ $\xrightarrow{H^+}$ $\xrightarrow{-CO_2}$ $H_2N\underset{COO^-}{CH}(CH_2)_4NH_3^+$

                                        J   (±)-Lysine

(b)

$\underset{COOEt}{\overset{COOEt}{H\dot{C}H}}$ $\xrightarrow[\text{Michael}]{CH_2=CHCN}$ $\underset{COOEt}{\overset{COOEt}{H\dot{C}}}$—$CH_2CH_2CN$ $\xrightarrow[\text{cat.}]{H_2}$ $\left[\underset{COOEt}{\overset{COOEt}{H\dot{C}}}-CH_2CH_2CH_2NH_2\right]$ $\xrightarrow{-\text{EtOH}}$ [piperidinone structure] EtOOC—[ring]—NH, O

                      K   L   M

                                                                                                            An amide

M $\xrightarrow[\alpha\text{-substn.}]{SO_2Cl_2}$ [Cl-substituted piperidinone] EtOOC—[ring]—NH, O $\xrightarrow[-\text{EtOH}, -CO_2]{\text{HCl, heat}}$ $^+H_3NCH_2CH_2CH_2\underset{Cl}{CHCOO^-}$ $\xrightarrow[-\text{HCl}]{\text{base}}$ [pyrrolidine ring]$\overset{}{\underset{H\ \ \ H}{N^+}}COO^-$

                      N   O   (±)-Proline

628

(c) 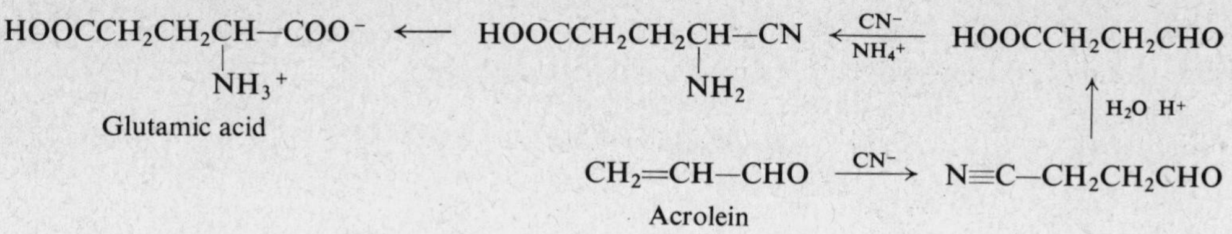 $HOOCCH_2CH_2CH—COO^-$ $\longleftarrow$ $HOOCCH_2CH_2CH—CN$ $\xleftarrow[NH_4^+]{CN^-}$ $HOOCCH_2CH_2CHO$

$\overset{|}{NH_3^+}$ $\overset{|}{NH_2}$

Glutamic acid

$\uparrow$ H_2O H^+

$CH_2=CH—CHO$ $\xrightarrow{CN^-}$ $N\equiv C—CH_2CH_2CHO$

Acrolein

4.

(a)

$$\xrightarrow[-2H_2O]{heat}$$

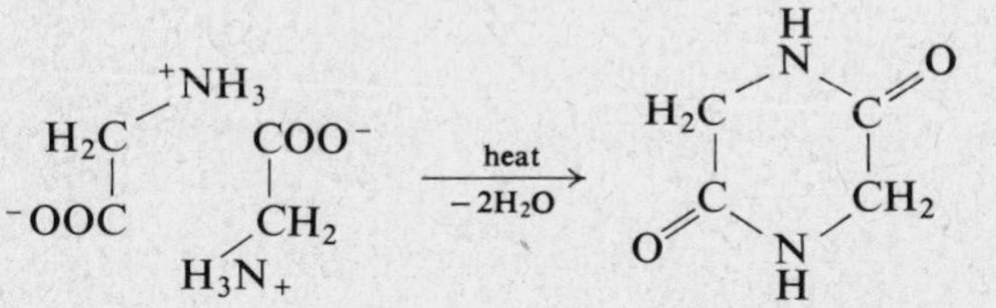

Diketopiperazine
A cyclic diamide

(b) $CH_3CHCH_2COO^-$ $\xrightarrow[-NH_3]{heat}$ $CH_3CH=CHCOOH$

$\overset{|}{^+NH_3}$ α,β-Unsaturated acid

(c)

$$\xrightarrow[-H_2O]{heat}$$

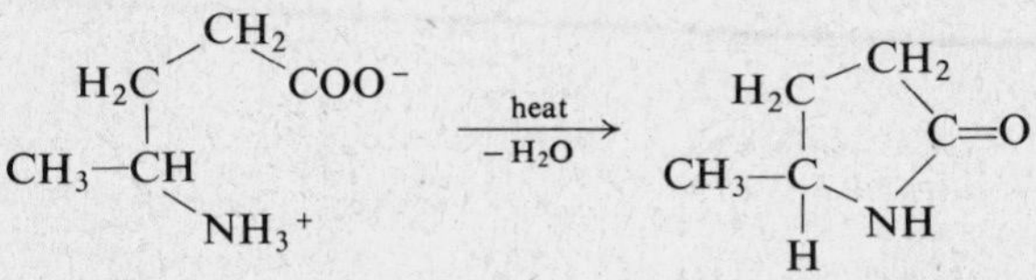

A γ-lactam
A cyclic amide

(d)

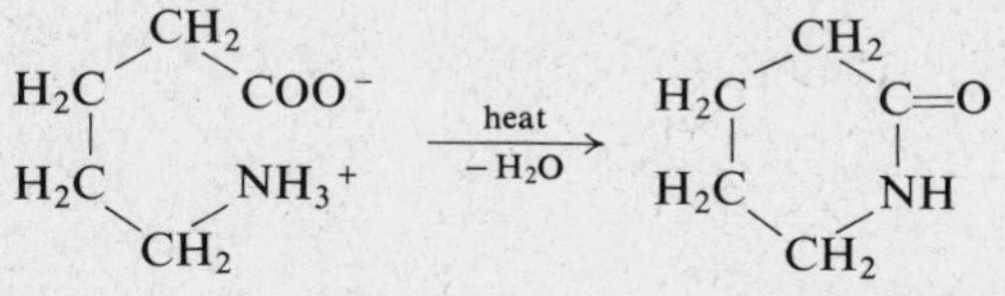

$$\xrightarrow[-H_2O]{heat}$$

A δ-lactam
A cyclic amide

5. (a)

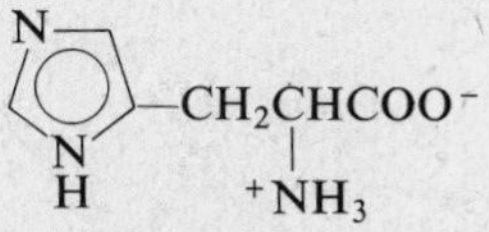

Histidine

The proton goes to the most basic nitrogen:

$$\text{aliphatic} —NH_2 > \text{"pyridine" N} > \text{"pyrole" NH}$$

(See Problem 20(b), page 1026.)

(b) $^+H_3N(CH_2)_4CHCOO^-$ or $H_2N(CH_2)_4CHCOO^-$

$\overset{|}{NH_2}$ $\overset{|}{^+NH_3}$

Actual structure

Lysine

The —COO$^-$ group is electron-withdrawing and hence base-weakening. The more distant ε-amino group is affected less, is more basic, and gets the proton.

(c)   HOOCCH$_2$CHCOO$^-$ or $^-$OOCCHCHCOOH
                | |
              $^+$NH$_3$ $^+$NH$_3$

Actual structure
Aspartic acid

The inductive effect of —NH$_2$ is electron-withdrawing and hence acid-strengthening. The nearer —COOH is affected more, is more acidic, and loses the proton.

(d)   H$_2$N—C—N(CH$_2$)$_3$CHCOO$^-$ $^+$H$_3$N—C—N(CH$_2$)$_3$CHCOO$^-$
               H || | H | |
             $^+$NH$_2$ NH$_2$ NH   NH$_2$

Actual structure
Argenine

     H$_2$N—C—N(CH$_2$)$_3$CHCOO$^-$ H$_2$N—C—N(CH$_2$)$_3$CHCOO$^-$
          H $^+$| | H | |
        HN H   NH$_2$ NH $^+$NH$_3$

The =NH nitrogen is the most basic, since the acceptance of a proton there gives a resonance-stabilized cation (compare the guanidium ion, Problem 20.24, page 686).

(e)   HO⟨◯⟩CH$_2$CHCOO$^-$ or $^-$O⟨◯⟩CH$_2$CHCOOH
                     | |
                   $^+$NH$_3$ $^+$NH$_3$

Actual structure
Tyrosine

The —COOH group (K_a about 10^{-3}) is much more acidic than phenolic —OH (K_a about 10^{-10}), and loses the proton.

6. (a) Chemical arithmetic,

C$_2$H$_3$O$_2$Cl	ClCH$_2$COOH	C$_5$H$_{12}$O$_2$NCl		C$_2$H$_5$O$_2$N	glycine	C$_5$H$_{14}$O$_2$NI$_3$
+ C$_3$H$_9$N	Me$_3$N	− C$_5$H$_{11}$O$_2$N	betaine	+ C$_3$H$_9$I$_3$	3MeI	− C$_5$H$_{11}$O$_2$N	betaine
C$_5$H$_{12}$O$_2$NCl		HCl lost		C$_5$H$_{14}$O$_2$NI$_3$		3HI lost

leads us to the same product for both reactions:

$$\text{Me}_3\text{N} + \text{ClCH}_2\text{COOH} \xrightarrow{-\text{HCl}} \text{Me}_3\overset{+}{\text{N}}\text{CH}_2\text{COO}^-$$
Betaine

$$3\text{MeI} + {}^+\text{H}_3\text{NCH}_2\text{COO}^- \xrightarrow{-3\text{HI}} \text{Me}_3\overset{+}{\text{N}}\text{CH}_2\text{COO}^-$$
Betaine

The properties are those expected of a dipolar compound: a crystalline, water-soluble high-melting solid. Reaction with acid involves the —COO$^-$ group.

$$Me_3\overset{+}{N}CH_2COO^- + H^+ + Cl^- \longrightarrow Me_3\overset{+}{N}CH_2COOH$$

Betaine

$$Cl^-$$

(b)

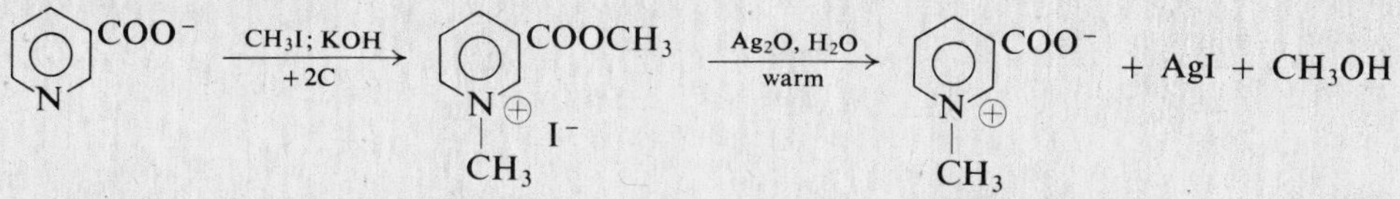

Anion of
nicotinic acid

P
*Methyl ester and
4° ammonium salt*

Trigonelline
Dipolar ion

$$+ AgI + CH_3OH$$

7. As the water content of the solvent is lowered, the hydrophobic parts no longer hide themselves. This breaks up soap micelles, and changes the characteristic shape—and with it characteristic properties—of a globular protein.

8.

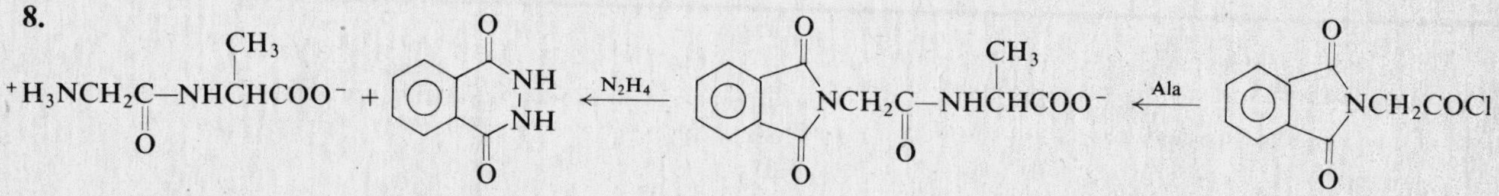

Gly.Ala

Phthalhydrazide

I

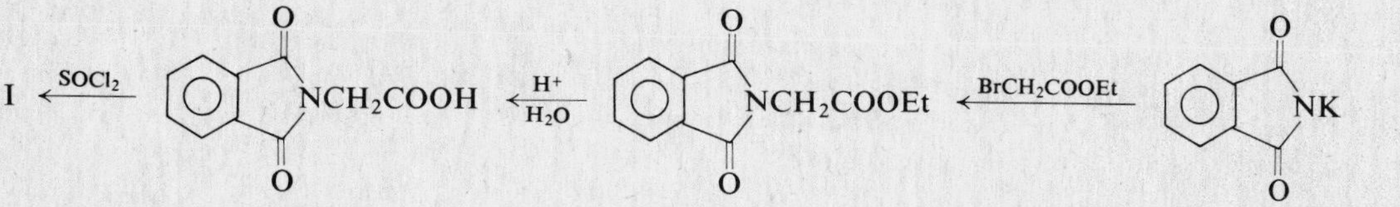

Potassium phthalimide

For Ala.Gly: start with $BrCHCOOEt$ (with CH_3), and then use $^+H_3NCH_2COO^-$

9. If one Fe atom per molecule:

$$0.43\% \text{ Fe} = \frac{55.8}{\text{m.w.}} \times 100$$

$$\text{m.w.} = 13,000$$

If one S atom per molecule:

$$1.48\% \text{ S} = \frac{32.1}{\text{m.w.}} \times 100$$

$$\text{m.w.} = 2,170$$

The minimum m.w. is 13,000: 1Fe and 6S.

10. (a)

$$\frac{1.31 \text{ mg } NH_3}{17 \text{ mg/mmole}} = 0.077 \text{ mmole } NH_3$$

$$\frac{100 \text{ mg protein}}{42,020 \text{ mg/mmole}} = 0.00238 \text{ mmole protein}$$

$$\frac{0.077 \text{ mmole } NH_3}{0.00238 \text{ mmole protein}} = \text{approx. } 32 \; -CONH_2 \text{ per molecule}$$

(b)

$$0.00238 \text{ mmole protein requires } \frac{17 \text{ mg } H_2O}{18 \text{ mg/mmole}} = 0.944 \text{ mmole } H_2O$$

$$\frac{0.944 \text{ mmole } H_2O}{0.00238 \text{ mmole protein}} = \text{approx. } 397 \text{ amide linkages (peptide } + \; 32 \; -CONH_2)$$

(c) $397 - 32 = 365$ peptide links $+ \, 4$ N-terminal groups $= 369$ amino acid residues

11. (a)

Leu (m.w.)	131
Orn	132
Phe	165
Pro	115
Val	117
	———
	660

A m.w. of 1300 means that there are two of each residue, and the molecular formula is $Leu_2Orn_2Phe_2Pro_2Val_2$.

(b) The DNP derivative is simply the one expected from the side-chain of ornithine. Hence there is no C-terminal *or* N-terminal group: the polypeptide must be *cyclic*.

(c)

```
Leu.Phe
    Phe.Pro
    Phe.Pro.Val
            Val.Orn.Leu
                Orn.Leu
            Val.Orn
        Pro.Val.Orn
```
_____

Leu.Phe.Pro.Val.Orn.Leu

From the molecular formula, we see that Orn, Phe, Pro, Val are still to be accounted for. These must be Phe.Pro.Val.Orn. The Orn combines with Leu at the other end to complete the ring:

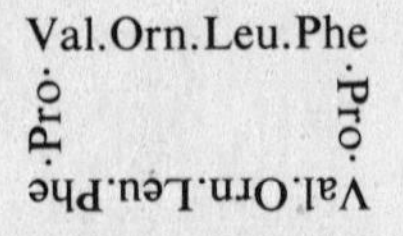

Val.Orn.Leu.Phe

Gramicidin S

Cyclic decapeptide

12. (a) In chain B, six amino acids occur only once: Pro, Arg, Lys, Asp, Ser, Thr. Let us look first for tripeptides and dipeptides that contain these amino acids.

Thr Pro Lys	**Arg**	**Ser**	**Asp**
Thr.Pro		Ser.His.Leu	Phe.Val.Asp
Pro.Lys.Ala	Arg.Gly		Val.Asp.Glu
	Gly.Glu.Arg		

Thr.Pro.Lys.Ala	Gly.Glu.Arg.Gly		Phe.Val.Asp.Glu

Four amino acids occur only twice: Ala, CySH, His, Tyr.   Let us look for these next.

His

Ser.His.Leu   Glu.His.Leu   His.Leu.CySH   (Two of these overlap)

CySH

His.Leu.CySH   Leu.Val.CySH

Val.CySH.Gly

Leu.Val.CySH.Gly

Tyr	**Ala**	
Tyr.Leu.Val	Val.Glu.Ala	Thr.Pro.Lys.Ala (derived above)

One Tyr is missing.

Gly occurs three times, and we have located all three:

Gly.Glu.Arg.Gly   Leu.CySH.Gly   Leu.Val.CySH.Gly

Two of these three peptides must overlap (four Gly appear here), but which two is uncertain at this point.

Val occurs three times:

Leu.Val.CySH.Gly   Leu.Val.Glu   Phe.Val.Asp.Glu   Tyr.Leu.Val

Val.Glu.Ala

Leu.Val.Glu.Ala

There is overlapping of Tyr.Leu.Val with one of the first two peptides.

Glu occurs three times:

Glu.His.Leu	Gly.Glu.Arg	Leu.Val.Glu
Val.Asp.Glu	Gly.Glu.Arg.Gly	Val.Glu.Ala

Val.Asp.Glu.His.Leu   Gly.Glu.Arg.Gly   Leu.Val.Glu.Ala

Phe.Val.Asp.Glu

Phe.Val.Asp.Glu.His.Leu

There is no overlapping.

Phe occurs three times; two of these are missing from the tripeptides.

Now for **Leu**, which occurs four times.   We find seven tripeptides containing Leu; there must be much overlapping.

Glu.His.Leu   His.Leu.CySH   Leu.CySH.Gly   Leu.Val.CySH

Leu.Val.Glu   Ser.His.Leu   Tyr.Leu.Val

Certainly Leu.CySH.Gly, Leu.Val.CySH, and Leu.Val.Glu have no overlapping, and the fourth Leu must be found in Glu.His.Leu or Ser.His.Leu.   This means, then, that we can propose the following combinations.

<pre>
     Leu.CySH.Gly   Leu.Val.CySH   or   Leu.Val.Glu
     His.Leu.CySH   Tyr.Leu.Val   Tyr.Leu.Val
    ─────────────   ──────────────   ──────────────
     His.Leu.CySH.Gly   Tyr.Leu.Val.CySH or Tyr.Leu.Val.Glu
</pre>

Phe.Val.Asp.Glu.His.Leu   (derived above, under Glu)

or

Ser.His.Leu

We have located three Leu; one is still missing.

At this point we have derived the following:

One hexapeptide: Phe.Val.Asp.Glu.His.Leu

Five tetrapeptides: Thr.Pro.Lys.Ala   Gly.Glu.Arg.Gly   Leu.Val.CySH.Gly

Leu.Val.Glu.Ala   His.Leu.CySH.Gly

Two tripeptides: Ser.His.Leu   Tyr.Leu.Val

Missing: 2Phe, 1Tyr, 1Leu

(b)
<pre>
 Tyr.Leu.Val.CySH   Ser.His.Leu.Val   Glu.His.Leu
   Leu.Val.CySH.Gly   His.Leu.Val.Glu   His.Leu.CySH.Gly
 ──────────────────   ───────────────   ────────────────
 Tyr.Leu.Val.CySH.Gly Ser.His.Leu.Val.Glu   Glu.His.Leu.CySH.Gly
                          Leu.Val.Glu.Ala   Phe.Val.Asp.Glu.His.Leu
                        ─────────────────   ────────────────────
              Ser.His.Leu.Val.Glu.Ala Phe.Val.Asp.Glu.His.Leu.CySH.Gly
</pre>

This takes care of most of the overlapping except for Gly.

Now we have the following pieces put together:

One octapeptide: Phe.Val.Asp.Glu.His.Leu.CySH.Gly

One hexapeptide: Ser.His.Leu.Val.Glu.Ala

One pentapeptide: Tyr.Leu.Val.CySH.Gly

Two tetrapeptides: Thr.Pro.Lys.Ala   Gly.Glu.Arg.Gly

Missing: 2Phe, 1Tyr, 1Leu

(c)
<pre>
                                        Val.Glu.Ala.Leu
                                      Ser.His.Leu.Val.Glu.Ala
                                      ──────────────────────
                                      Ser.His.Leu.Val.Glu.Ala.Leu
                         His.Leu.CySH.Gly.Ser.His.Leu
                         ────────────────────────────
                         His.Leu.CySH.Gly.Ser.His.Leu.Val.Glu.Ala.Leu
          Phe.Val.Asp.Glu.His.Leu.CySH.Gly
          ──────────────────────────────────
 Phe.Val.Asp.Glu.His.Leu.CySH.Gly.Ser.His.Leu.Val.Glu.Ala.Leu
</pre>

This takes care of 15 residues out of 30.   Ten more are found in

Tyr.Leu.Val.CySH.Gly.Glu.Arg.Gly.Phe.Phe

and the remaining five in

Tyr.Thr.Pro.Lys.Ala

Since Phe.Val is the N-terminal end (DNP data) and Ala is the C-terminal end (carboxy-peptidase), the three pieces can be put together in only one way, and the structure of the B chain of beef insulin must be:

Phe.Val.Asp.Glu.His.Leu.CySH.Gly.Ser.His.Leu.Val.Glu.Ala.Leu.Tyr.Leu.Val.CySH.Gly.Glu.Arg.Gly.⌐
⌊Phe.Phe.Tyr.Thr.Pro.Lys.Ala

B-chain

(d) In chain A, three amino acids appear once: Gly, Ala, Ileu. They are the first ones to look for among the peptides.

Gly Ileu	Ala
Gly.Ileu.Val.Glu.Glu	CySH.CySH.Ala

Five amino acids appear twice: Val, Leu, Asp, Ser, Tyr.

Val

Ser.Val.CySH   Gly.Ileu.Val.Glu.Glu
(N-terminal
residue)

Leu

Glu.Leu.Glu   Leu.Tyr.Glu
Ser.Leu.Tyr
_____
Ser.Leu.Tyr.Glu

Tyr

Leu.Tyr.Glu   Tyr.CySH
Ser.Leu.Tyr   Glu.Asp.Tyr
_____   _____
Ser.Leu.Tyr.Glu   Glu.Asp.Tyr.CySH

Asp

Glu.Asp.Tyr   CySH.Asp
(C-terminal
residue)

Ser

Ser.Leu.Tyr.Glu   Ser.Val.CySH

Two amino acids appear four times: **Glu, CySH**.

Gly.Ileu.Val.Glu.Glu
(N-terminal   Glu.Glu.CySH
residue)   Glu.CySH.CySH
_____
Gly.Ileu.Val.Glu.Glu.CySH.CySH
              CySH.CySH.Ala
_____
Gly.Ileu.Val.Glu.Glu.CySH.CySH.Ala

Glu.Leu.Glu
Leu.Tyr.Glu
_____
Leu.Tyr.Glu.Leu.Glu
Ser.Leu.Tyr.Glu
_____
Ser.Leu.Tyr.Glu.Leu.Glu
              Glu.Asp.Tyr.CySH
_____
Ser.Leu.Tyr.Glu.Leu.Glu.Asp.Tyr.CySH

Two other fragments are found, one of which is the C-terminal residue.

Ser.Val.CySH   CySH.Asp
                 (C-terminal
                 residue)

At this point, there are no amino acids missing, and we have put together the following pieces:

One nonapeptide: Ser.Leu.Tyr.Glu.Leu.Glu.Asp.Tyr.CySH

One octapeptide: Gly.Ileu.Val.Glu.Glu.CySH.CySH.Ala
                     (N-terminal)

One tripeptide: Ser.Val.CySH

One dipeptide: CySH.Asp
                   (C-terminal)

We still do not know which of the non-terminal fragments comes first, since the C-terminal CySH of either could overlap the N-terminal CySH of the dipeptide.

(e) The only way to get Ser.Val.CySH and Ser.Leu together is to have a center portion that runs . . . Ser.Val.CySH.Ser.Leu.Tyr. . . .   The complete A chain of beef insulin is thus:

Gly.Ileu.Val.Glu.Glu.CySH.CySH.Ala.Ser.Val.CySH.Ser.Leu.Tyr.Glu.Leu.Glu.Asp.Tyr.CySH.Asp

A-chain

(f) **Chain A:**

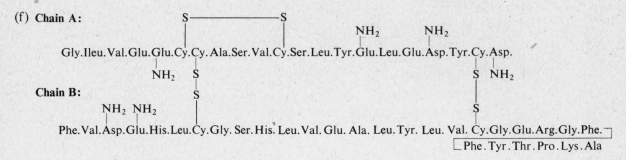

Beef insulin

(g) The other could have been

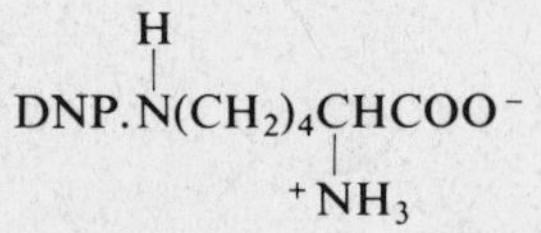

$$\text{DNP.N(CH}_2)_4\overset{\overset{\displaystyle H}{|}}{\underset{\underset{\displaystyle +NH_3}{|}}{C}}\text{HCOO}^-$$

from the ε-amino group of Lys. But if Lys had been N-terminal, we would have obtained a double DNP derivative of it, and no DNP.Phe—contrary to fact.

Chapter 37

Biochemical Processes
Molecular Biology

1. CO_2 becomes the —COOH of malonyl–CoA in reaction (1), page 1176; this is the carbon lost in reaction (4).

2. Consider the hydrolysis shown in equations (1) and (2) on page 1168, with chymotrypsin playing the part of both B: and H—B. This is shown as proceeding via the tetrahedral intermediate I (compare Sec. 20.17).

(1)

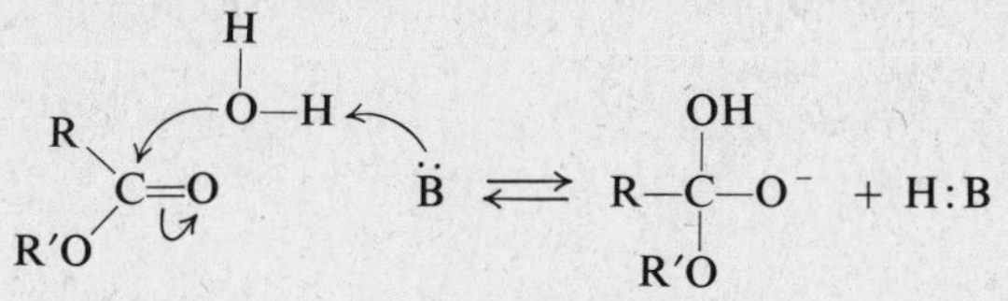

(2)

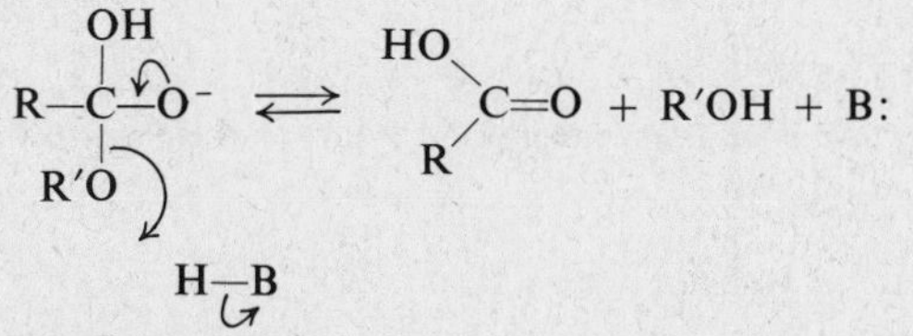

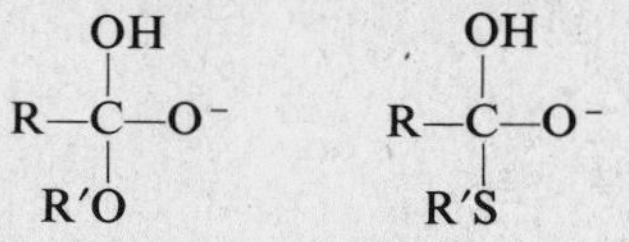

The facts of this problem lead us to two conclusions: (i) reaction does indeed proceed via the tetrahedral intermediate I; and (ii) step (1) is slow and rate-determining, and step (2) is fast. (That is, k_2 is much greater than k_{-1}.) Under these conditions formation of intermediate I from the acetate and intermediate II from the thioacetate is the same; the fact that the subsequent step (2) is much faster for II than for I has no effect on the overall rates. Furthermore, each molecule of I rapidly yields product instead of undergoing the proton-transfer shown on page 679 and reverting to (labeled) starting material.

If, instead, reaction (1) were fast and reversible, we would expect two things: first, reaction of the thioester would be faster, since a greater fraction of II than I would go on to product in a given period of time; and second, oxygen exchange would take place as described in Sec. 20.17.

(A single-step displacement involving no tetrahedral intermediate at all would account for the lack of oxygen exchange, but *not* the equality of rate for the two esters; the bond to OR or SR would be breaking in the (only) transition state, and the thioester would react faster.)

3. (a)

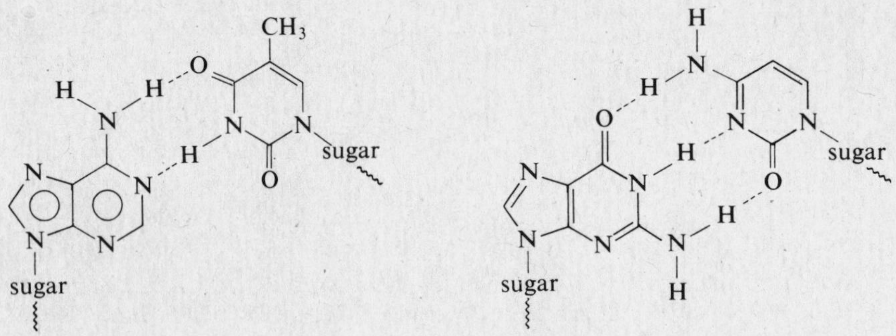

Adenine   Thymine   Guanine   Cytosine

(b) Three hydrogen bonds are stronger than two.

4. (a) An aldol-like condensation between the ester and the keto group of oxaloacetate.

(b) An aldol-like condensation between the ester and the keto group of acetoacetyl–CoA. Then reduction of an ester to a primary alcohol by hydride transfer.

5. (a)

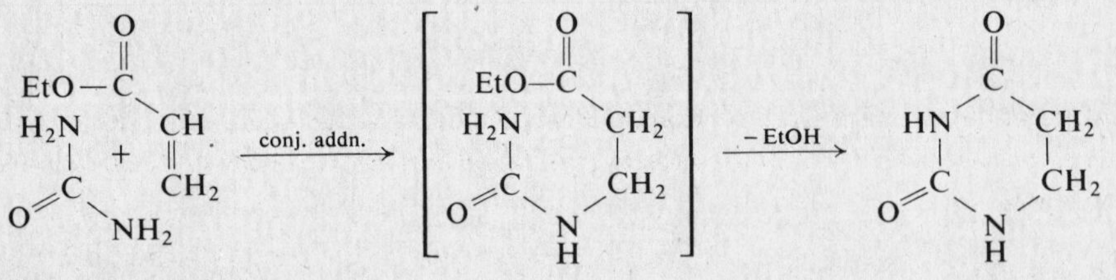

Urea   Ethyl acrylate   A   B

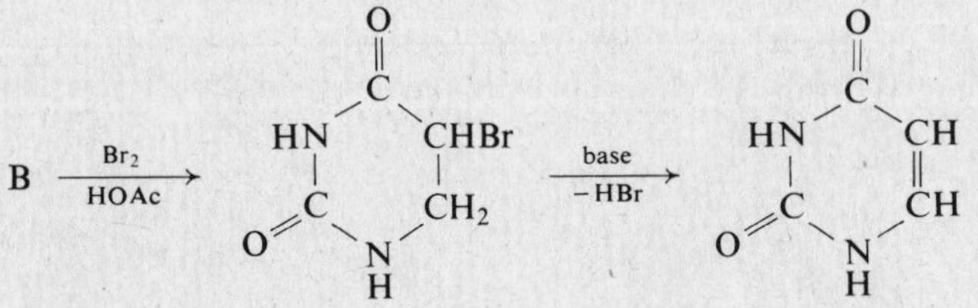

C   Uracil

(b)

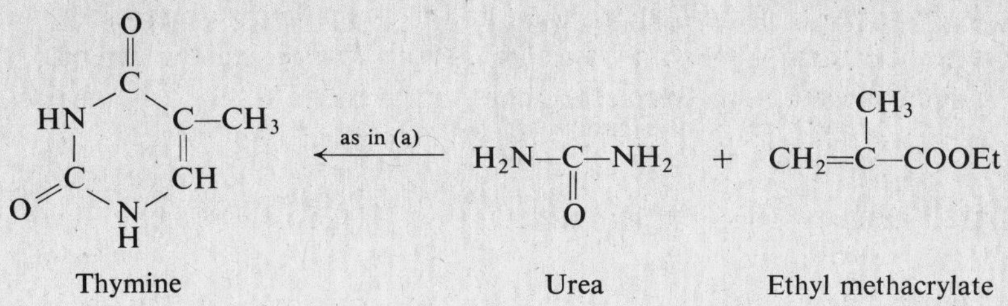

Thymine   as in (a) ←   Urea   Ethyl methacrylate

$$H_2N-C-NH_2 \quad + \quad CH_2=C-COOEt$$

(c)

Uracil $\xrightarrow{POCl_3}$ D $\xrightarrow{NH_3}$ E $\xrightarrow{OCH_3^-}$ G $\xrightarrow{H^+}$ F   Cytosine

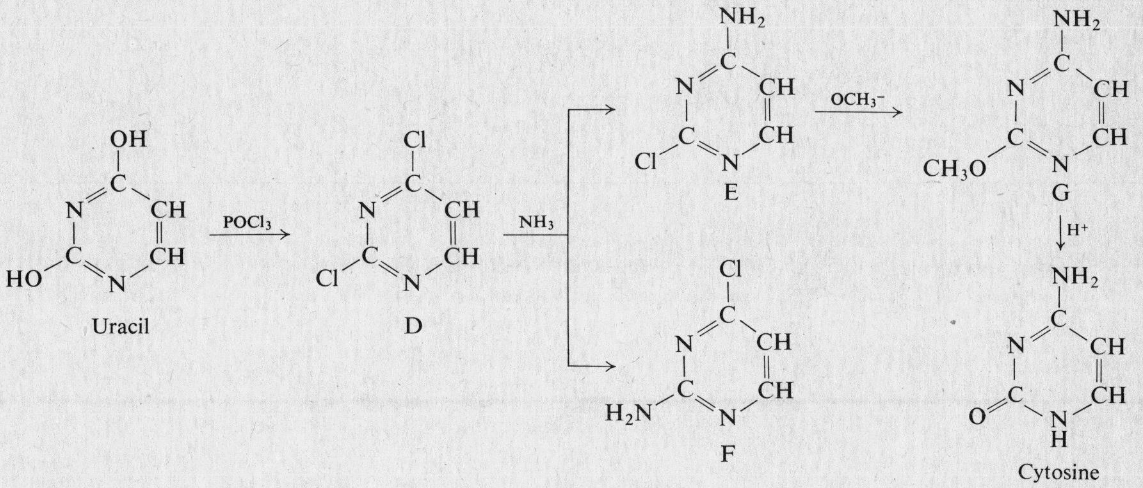

(d)

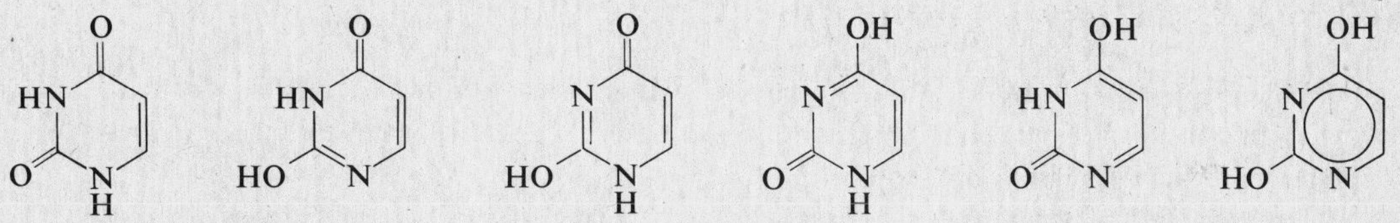

6. Biological oxidation of fatty acids removes two carbons at a time, starting at the carboxyl end: "*beta*-oxidation."

7. A *retro* (reverse) aldol condensation, similar to that in Problem 21.14 (page 711). Let us write equations neglecting the ionization state of the phosphate, and considering the enzyme to act as both base and acid.

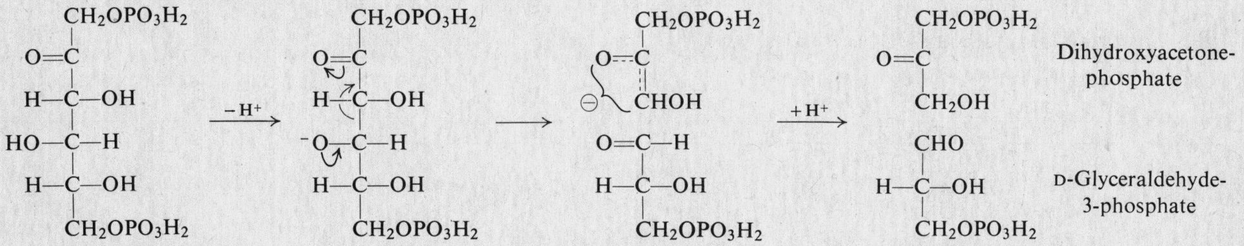

$$\xrightarrow{-H^+} \qquad \xrightarrow{\qquad} \qquad \xrightarrow{+H^+}$$

Dihydroxyacetone-phosphate

D-Glyceraldehyde-3-phosphate

We have actually encountered almost exactly the reverse of this reaction in the (base-catalyzed) formation of D-fructose from either D-glyceraldehyde or dihydroxyacetone (Problem 13, Chapter 34, page 1109).

8. There is *direct* transfer of a hydride ion from C–1 of ethanol to C–4 of the pyridine ring of NAD$^+$ (see page 1153) to give reduced NAD (NADH or NADD). There are now, we see, two hydrogens on C–4: the one originally there, and the one transferred.

$$\text{NAD}^+ \ (\text{pyridine-CONH}_2) + CH_3CH_2OH \xrightarrow[\text{enzyme}]{D_2O} CH_3CHO + \text{NADH} \ (H,H\text{-}CONH_2) + H^+$$

$$\text{NAD}^+ \ (\text{pyridine-CONH}_2) + CH_3CD_2OH \xrightarrow[\text{enzyme}]{H_2O} CH_3CDO + \text{NADD} \ (H,D\text{-}CONH_2) + H^+$$

(b) Of the two hydrogens on C–4 of reduced NAD, *only* the one received from ethanol in part (a) is transferred back to the aldehyde. How are we to account for this remarkable

$$H^+ + \text{NADD} \ (H,D\text{-}CONH_2) + CH_3CHO \xrightarrow[\text{enzyme}]{H_2O} \underset{(Evidently)}{CH_3CHDOH} + \text{NAD}^+ \ (\text{pyridine-CONH}_2)$$

NADD *From part (a)*   NAD$^+$ *Contains no D*

specificity? How can the molecule "remember" which hydrogen was transferred to it? Clearly, it must keep this particular hydrogen in a different "drawer" than the other hydrogen. How can this be, if both are on C–4?

Let us examine the structure of reduced NAD. If one of the two hydrogens on C–4 is deuterium, C–4 is a chiral center; because of the chirality of the rest of the molecule, this new center creates the possibility of two diastereomers, III and IV.

III (H,D-CONH$_2$)   IV (D,H-CONH$_2$)

Diastereomers
NADD

(In the terminology of Sec. 13.7, the two protons on C–4 of ordinary NADH are *diastereotopic protons*.)

The specificity of hydride transfer is *stereospecificity*. Hydride is transferred to only one face of NAD—the "back" face, let us assume, to give diastereomer III. In the reverse

$$\text{NAD}^+ \ (\text{pyridine-CONH}_2) + CH_3CD_2OH \xrightarrow[\text{enzyme}]{} CH_3CDO + \text{NADD} \ (H,D\text{-}CONH_2) + H^+$$

NAD$^+$ NADD

reaction then—catalyzed by the same enzyme and passing through the same transition state—hydride is transferred *from* the back face.

(c) Enzymatic oxidation by D-glucose (a different enzyme from the one in part (b)) is evidently completely stereospecific, too, but with the *opposite* stereospecificity. If hydride is transferred only from the back face in part (b), it is transferred only from the *front* face in part (c).

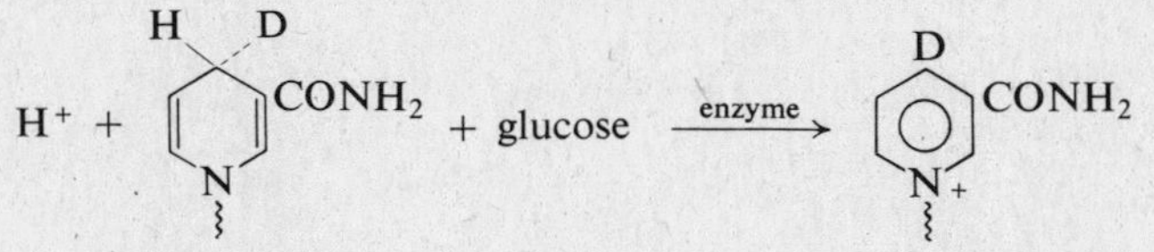

$$H^+ + \quad \text{NADD} \quad + \text{ glucose} \xrightarrow{\text{enzyme}} \quad \text{NAD}^+$$

NADD
From part (a)

NAD$^+$
Retains D

(d) Enzymatic oxidation by acetaldehyde is, of course, still stereospecific, removing hydride from (say) the back face of NADD. The 44% of deuterium left in the NAD$^+$ must, therefore, have been attached to the front face of the NADD; that is, the NADD was evidently a mixture of 56% III and 44% IV. Not surprisingly, then, chemical reduction of NAD$^+$ is (nearly) non-stereospecific.

(e) Clearly, hydride transfer is highly specific with regard to the acetaldehyde/ethanol, too. Of the two hydrogens in ethanol, *only* the one originally received from reduced NAD is transferred back to NAD$^+$.

Let us examine the structure of ethanol. If one of the hydrogens on C–1 is deuterium, ethanol is chiral, and can exist as a pair of enantiomers.

(In the terminology of Sec. 13.7, the C–1 protons of CH$_3$CH$_2$OH are *enantiotopic protons*.)

Here, too, the specificity is *stereospecificity*. From NADD, D is transferred to only one face of CH$_3$CHO, to give only one enantiomer, X, of CH$_3$CHDOH. From NADH, H is transferred to only one face—*the same face*—of CH$_3$CDO, to give the other enantiomer, Y.

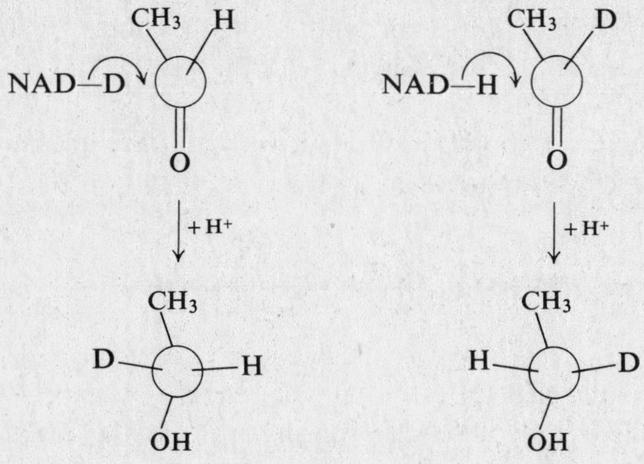

Enantiomers

On reoxidation, NAD$^+$ takes hydride only from the same preferred location in the ethanol molecule. In enantiomer X the hydrogen at that location is D; in enantiomer Y the hydrogen at that location is H.

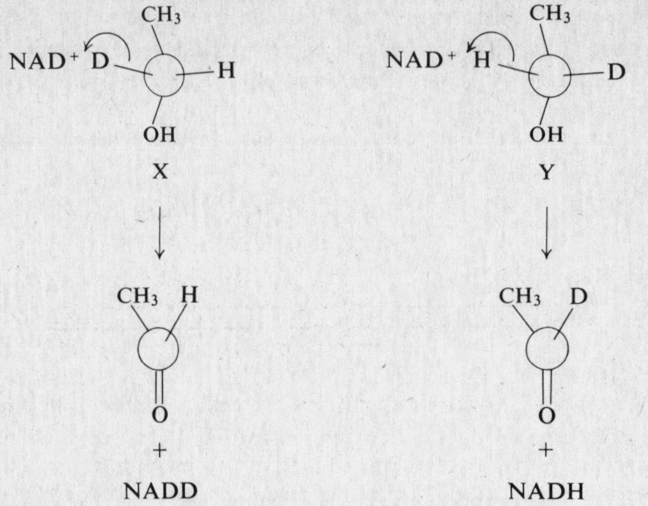

We know that in general racemic modifications are obtained in chemical reactions, and that in general enantiomers undergo identical reactions—but *not* necessarily if one reagent is optically active. The transition states for the reactions (either oxidation or reduction) at the two faces of C–1 are diastereomeric, and hence of different stabilities.

One further point. There can be no doubt that the same stereospecificity exists whether the compounds contain deuterium or not: one particular enantiotopic H of CH_3CH_2OH or one particular diastereotopic H of NADH is transferred in all these reactions. The deuterium label is used simply to *reveal* what is going on.

Most of this elegant work was done by Frank H. Westheimer and Birget Vennesland at the University of Chicago. The formation of different enantiomers in (e) was demonstrated unequivocally without measurement of optical rotation—there was not enough material available! It was not until four years later that the optical activity of the CH_3CHDOH was measured directly. The enantiomers have the configuration shown above: X is the R-(+)-isomer, and Y is the S-(−)-isomer.

For a general discussion, see H. R. Levy, P. Talalay, and B. Vennesland, "The Steric Course of Enzymatic Reaction at Meso Carbon Atoms: Application of Hydrogen Isotopes," *Progress in Stereochemistry*, Vol. 3, Butterworths, Washington, 1962, pp. 299–344. For the work of part (e), see F. A. Loewus, F. H. Westheimer, and B. Vennesland, "Enzymatic Synthesis of Enantiomorphs of Ethanol-1-d," J. Am. Chem. Soc., **75**, 5018 (1953).

It seems fitting to end this book, not with an answer, but with another question. In the work of part (e), a further experiment was carried out. Labeled ethanol Y was treated in the following way,

$$CH_3CHDOH \xrightarrow[\text{pyridine}]{\text{TsCl}} CH_3CHDOTs \xrightarrow{\text{OH}^-} CH_3CHDOH$$
$$\quad\quad\text{Y} \quad\quad\quad\quad\quad\quad\quad\quad\quad\quad\quad\quad\quad\quad\quad\text{Z}$$

and the product was oxidized with (unlabeled) NAD$^+$. Which would you expect to get, NADH or NADD? What is Z? Where can you check your answer to this question?

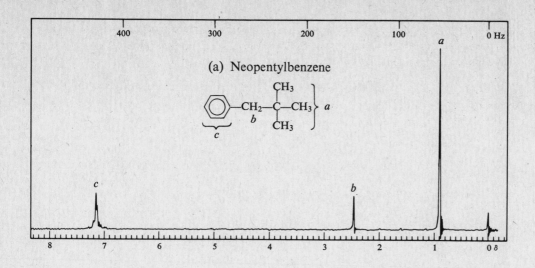

(a) Neopentylbenzene

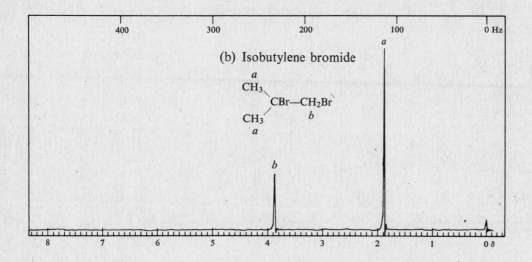

(b) Isobutylene bromide

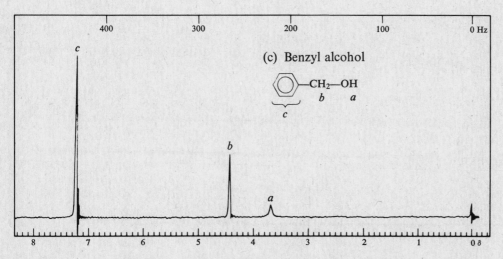

(c) Benzyl alcohol

Figure 13.7. Nmr spectra for Problem 13.9, p. 425.

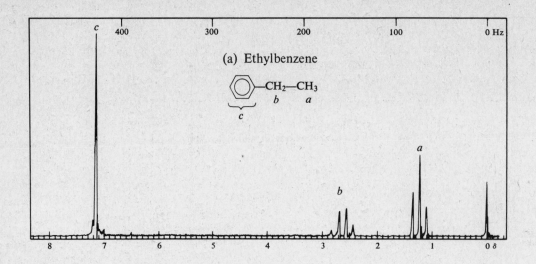

(a) Ethylbenzene

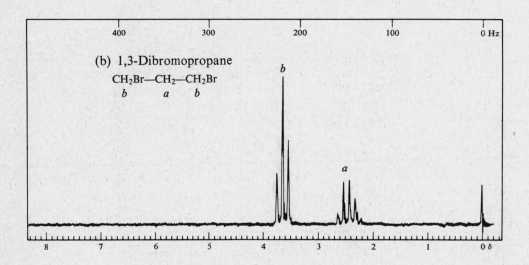

(b) 1,3-Dibromopropane

CH_2Br—CH_2—CH_2Br
   b   a   b

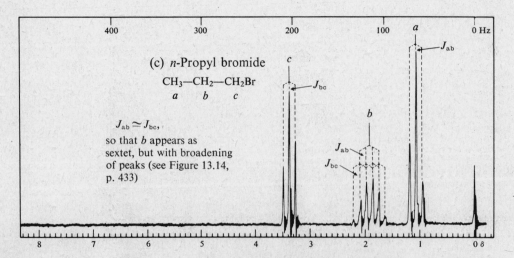

(c) n-Propyl bromide

CH_3—CH_2—CH_2Br
  a   b   c

$J_{ab} \simeq J_{bc}$,

so that b appears as
sextet, but with broadening
of peaks (see Figure 13.14,
p. 433)

Figure 13.16. Nmr spectra for Problem 13.12, p. 435.

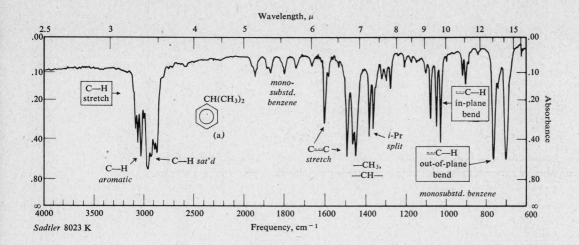

Sadtler 8023 K

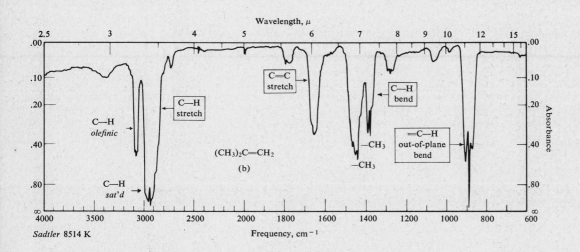

Sadtler 8514 K

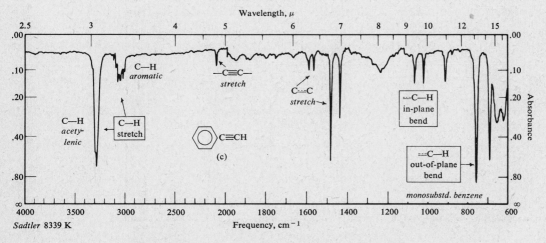

Sadtler 8339 K

Figure 13.19. Infrared spectra for Problem 10, p. 447.

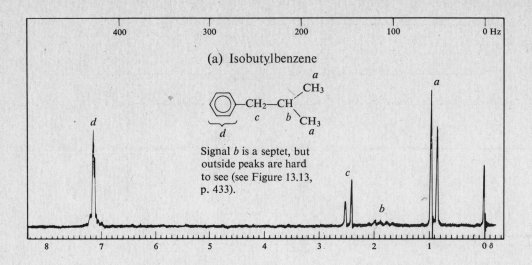

(a) Isobutylbenzene

Signal *b* is a septet, but outside peaks are hard to see (see Figure 13.13, p. 433).

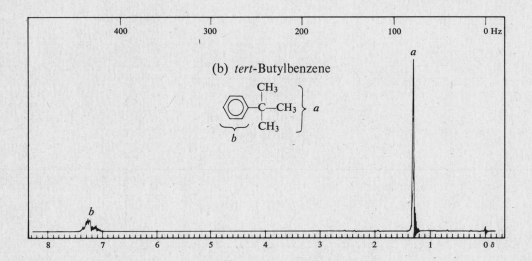

(b) *tert*-Butylbenzene

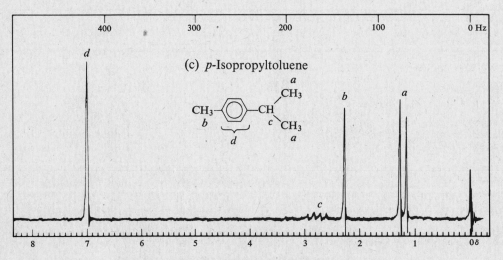

(c) *p*-Isopropyltoluene

Figure 13.20. Nmr spectra for Problem 11, p. 447.

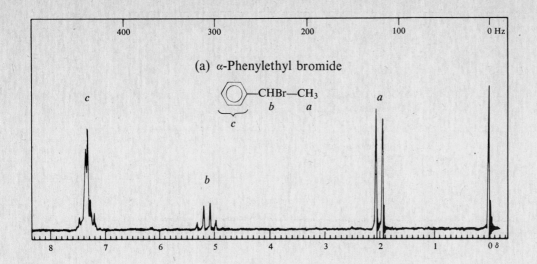

(a) α-Phenylethyl bromide

⬡—CHBr—CH₃
b   a

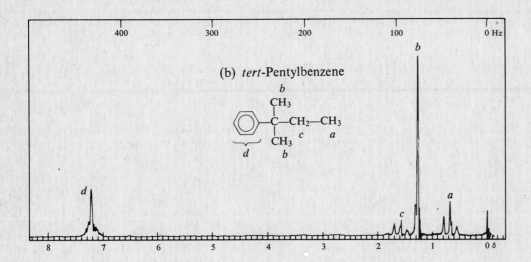

(b) *tert*-Pentylbenzene

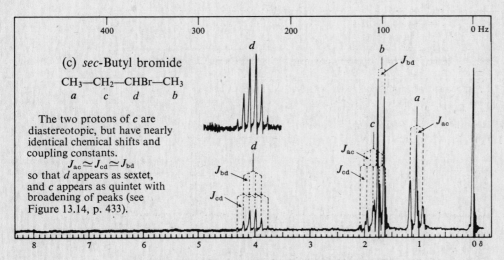

(c) *sec*-Butyl bromide

CH_3—CH_2—CHBr—CH_3
a   c   d   b

The two protons of *c* are diastereotopic, but have nearly identical chemical shifts and coupling constants.

$J_{ac} \simeq J_{cd} \simeq J_{bd}$

so that *d* appears as sextet, and *c* appears as quintet with broadening of peaks (see Figure 13.14, p. 433).

Figure 13.21. Nmr spectra for Problem 12. p. 447.

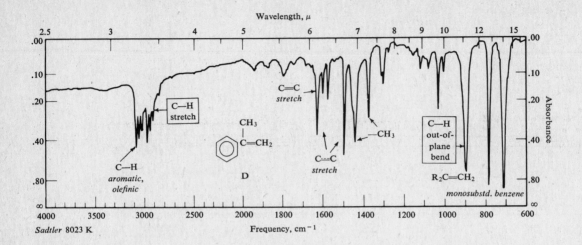

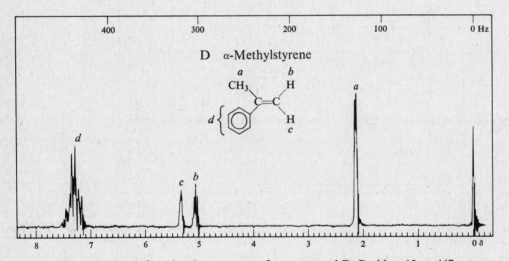

Figure 13.22. Infrared and nmr spectra for compound D, Problem 13, p. 447.

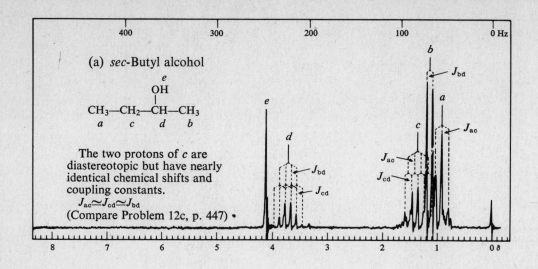

(a) *sec*-Butyl alcohol

The two protons of *c* are diastereotopic but have nearly identical chemical shifts and coupling constants.

$J_{ac} \simeq J_{cd} \simeq J_{bd}$

(Compare Problem 12c, p. 447) •

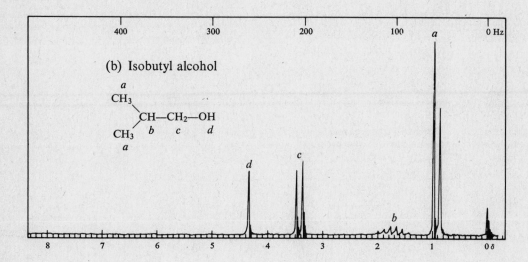

(b) Isobutyl alcohol

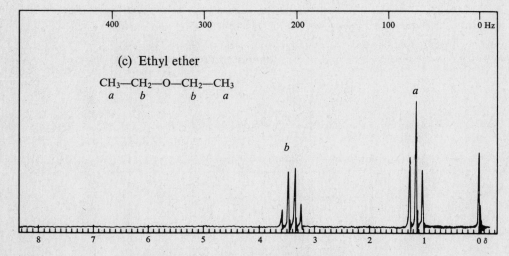

(c) Ethyl ether

Figure 16.2. Nmr spectra for Problem 22, p. 546.

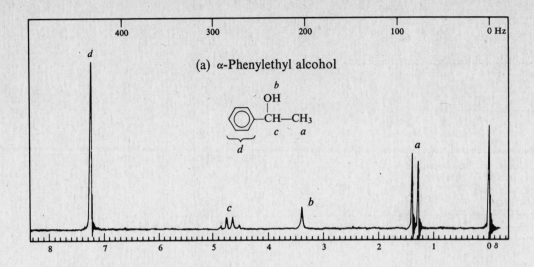

(a) α-Phenylethyl alcohol

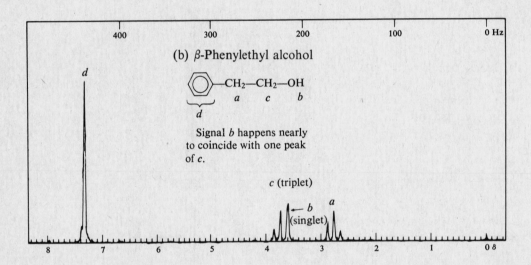

(b) β-Phenylethyl alcohol

Signal *b* happens nearly to coincide with one peak of *c*.

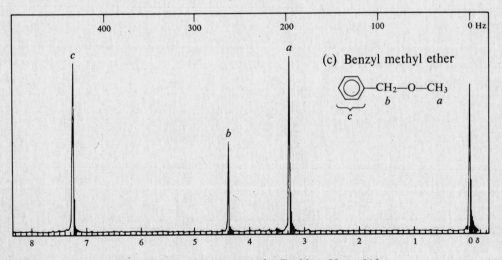

(c) Benzyl methyl ether

Figure 16.3. Nmr spectra for Problem 23, p. 546.

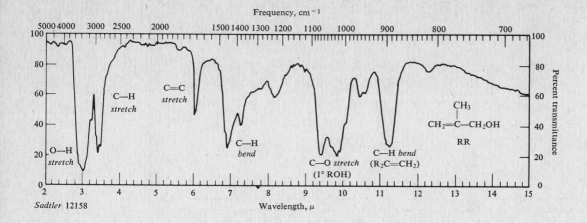

Sadtler 12158

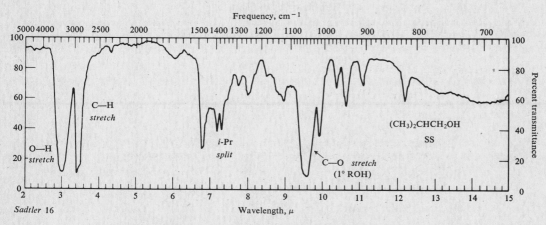

Sadtler 16

Figure 16.4. Infrared spectra for Problem 24, p. 546.

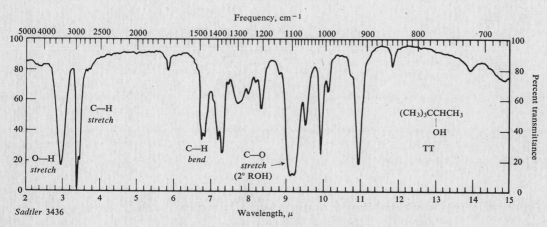

Sadtler 3436

Figure 16.5. Infrared spectrum for Problem 25, p. 546.

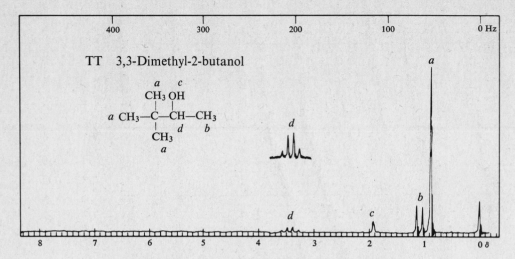

Figure 16.6. Nmr spectrum for Problem 25, p. 546.

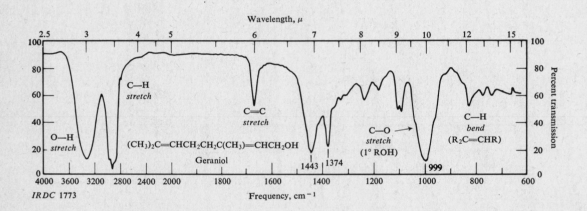

IRDC 1773

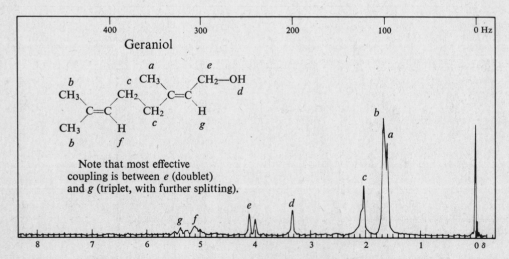

Figure 16.7. Infrared and nmr spectra for Problem 26, p. 546.

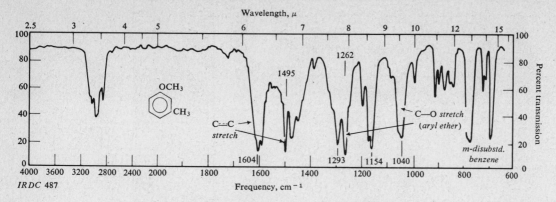

Wavelength, μ

1495

1262

C≡C
stretch

C—O stretch
(aryl ether)

1604

1293 1154 1040

m-disubstd.
benzene

IRDC 487

Frequency, cm⁻¹

Figure 17.2. Infrared spectra for Problem 16, p. 574.

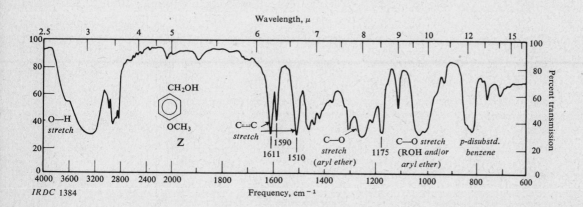

Wavelength, μ

CH₂OH

OCH₃

Z

O—H
stretch

C≡C
stretch

1590

C—O
stretch
(aryl ether)

C—O stretch
(ROH *and/or*
aryl ether)

p-disubstd.
benzene

1611 1510

1175

IRDC 1384

Frequency, cm⁻¹

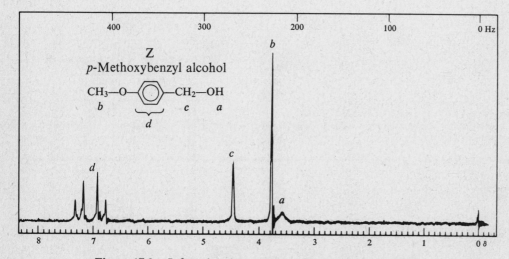

Z
p-Methoxybenzyl alcohol

CH₃—O—⬡—CH₂—OH
b   *c*   *a*
     d

b

c

a

d

Figure 17.3. Infrared and nmr spectra for Problem 17, p. 574.

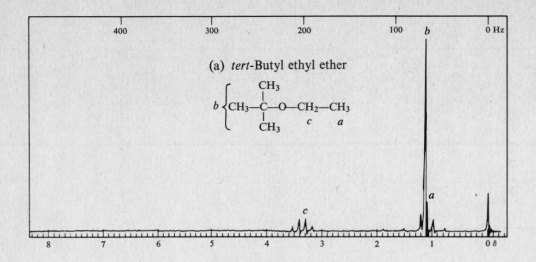

(a) *tert*-Butyl ethyl ether

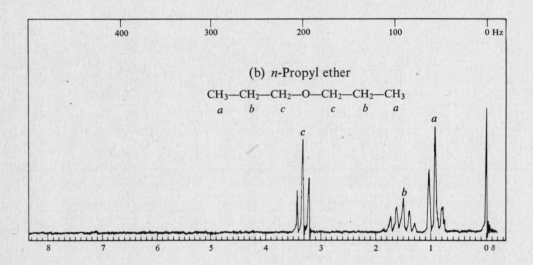

(b) *n*-Propyl ether

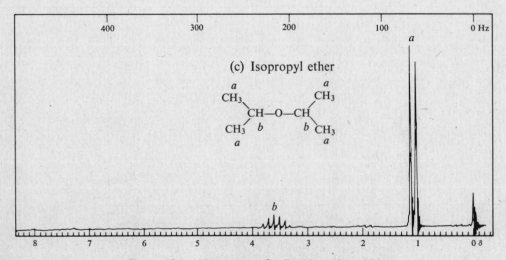

(c) Isopropyl ether

Figure 17.4. Nmr spectra for Problem 18, p. 574.

654

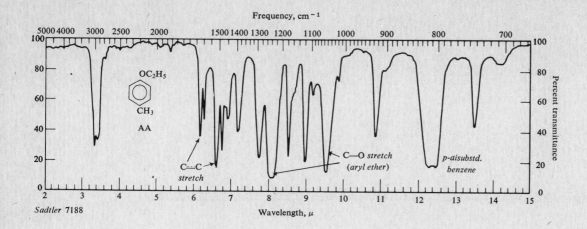

Frequency, cm⁻¹

OC₂H₅ / CH₃

AA

C═C stretch

C—O stretch (aryl ether)

p-aisubstd. benzene

Sadtler 7188

Wavelength, μ

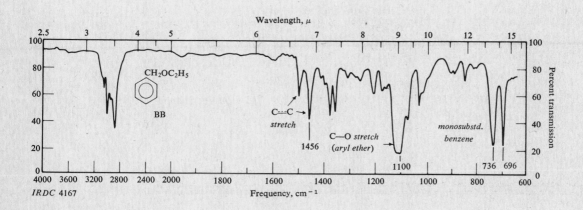

Wavelength, μ

CH₂OC₂H₅

BB

C═C stretch

1456

C—O stretch (aryl ether)

1100

monosubstd. benzene

736 696

IRDC 4167

Frequency, cm⁻¹

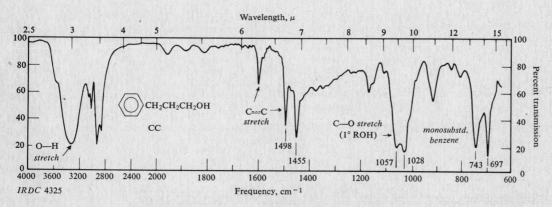

Wavelength, μ

CH₂CH₂CH₂OH

CC

O—H stretch

C═C stretch

1498

1455

C—O stretch (1° ROH)

1057 1028

monosubstd. benzene

743 697

IRDC 4325

Frequency, cm⁻¹

Figure 17.5. Infrared spectra for Problem 19, p. 574.

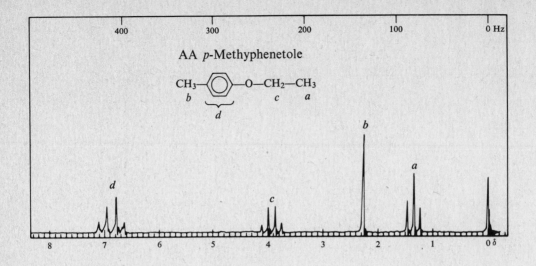

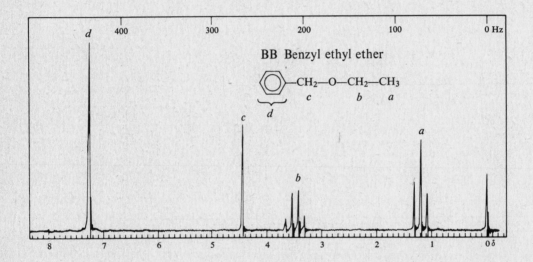

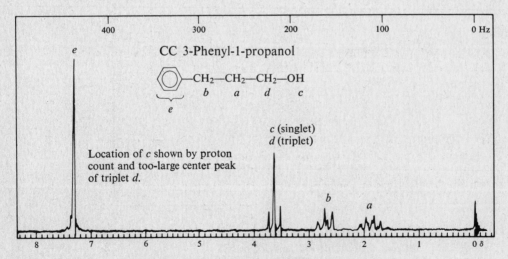

Figure 17.6. Nmr spectra for Problem 19, p. 574.

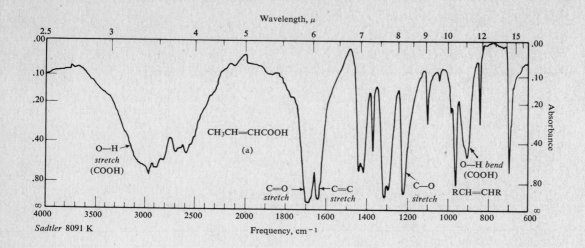

O—H
stretch
(COOH)

CH₃CH=CHCOOH
(a)

C=O
stretch

C=C
stretch

C—O
stretch

O—H bend
(COOH)

RCH=CHR

Sadtler 8091 K

Wavelength, μ

Frequency, cm⁻¹

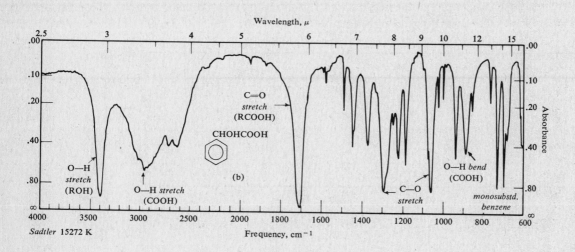

O—H
stretch
(ROH)

O—H stretch
(COOH)

C=O
stretch
(RCOOH)

CHOHCOOH

(b)

C—O
stretch

O—H bend
(COOH)

*monosubstd.
benzene*

Sadtler 15272 K

Wavelength, μ

Frequency, cm⁻¹

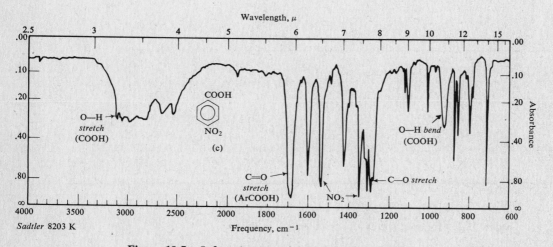

O—H
stretch
(COOH)

COOH

NO₂

(c)

C=O
stretch
(ArCOOH)

NO₂

C—O stretch

O—H bend
(COOH)

Sadtler 8203 K

Wavelength, μ

Frequency, cm⁻¹

Figure 18.5. Infrared spectra for Problem 30, p. 615.

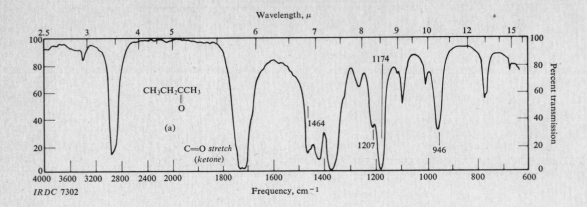

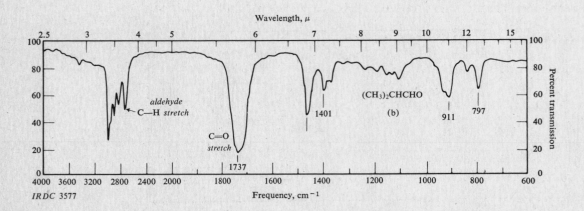

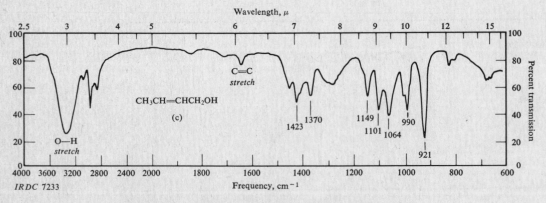

Figure 19.2. Infrared spectra for Problem 28, p. 653.

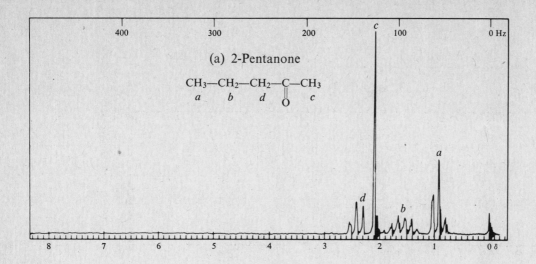

(a) 2-Pentanone

CH₃—CH₂—CH₂—C—CH₃
 a   b   d   ‖   c
                    O

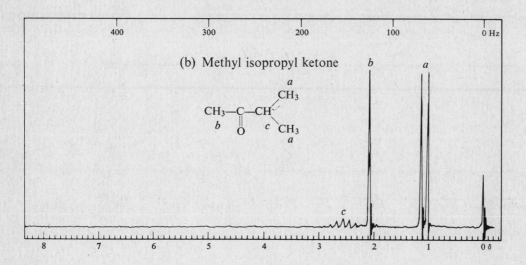

(b) Methyl isopropyl ketone

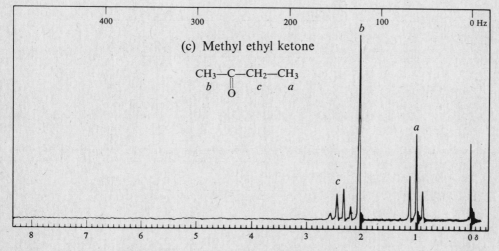

(c) Methyl ethyl ketone

CH₃—C—CH₂—CH₃
 b   ‖   c   a
     O

Figure 19.3. Nmr spectra for Problem 29, p. 653.

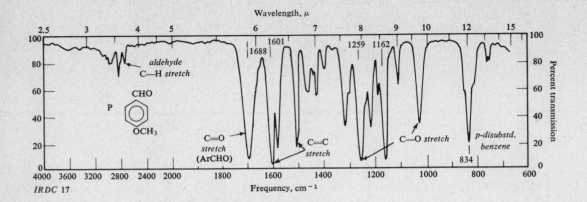

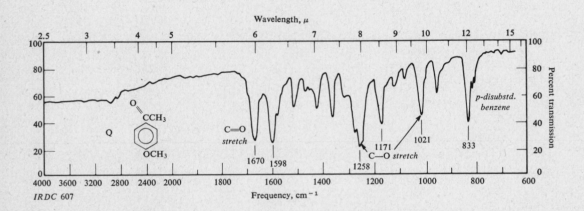

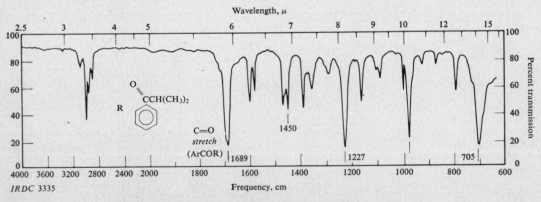

Figure 19.4. Infrared spectra for Problem 30, p. 653.

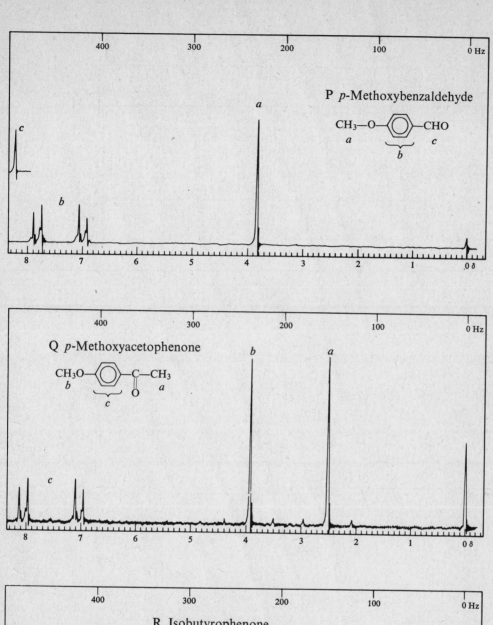

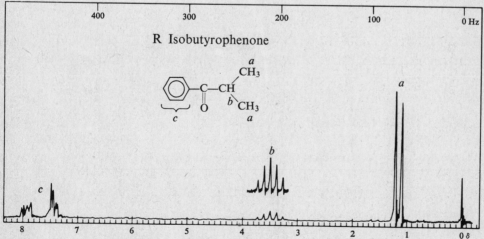

Figure 19.5. Nmr spectra for Problem 30, p. 653.

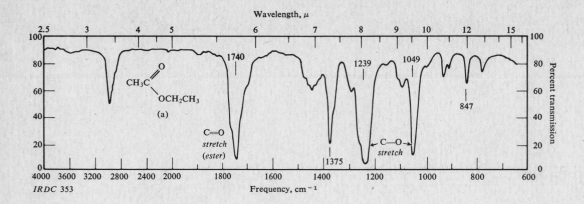

IR DC 353

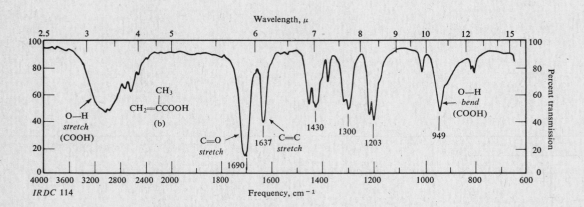

IR DC 114

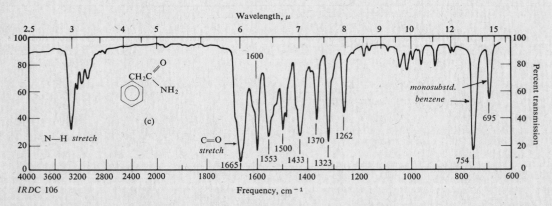

IR DC 106

Figure 20.2. Infrared spectra for Problem 23, p. 694.

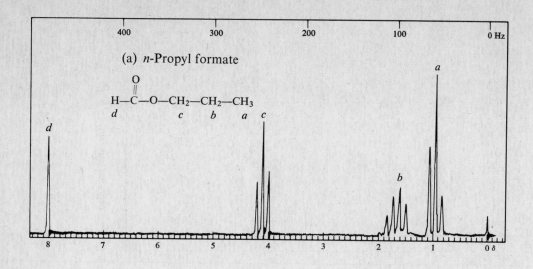

(a) *n*-Propyl formate

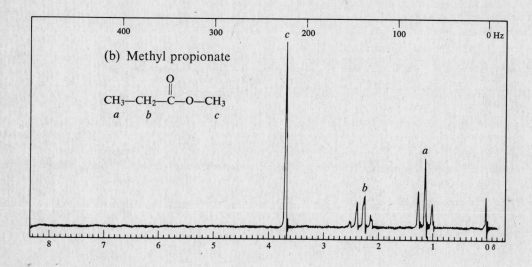

(b) Methyl propionate

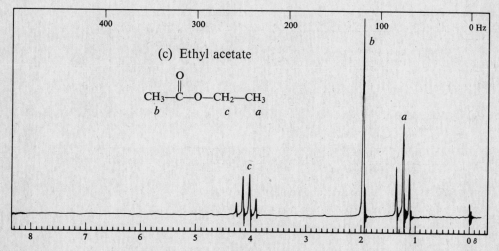

(c) Ethyl acetate

Figure 20.3. Nmr spectra for Problem 24, p. 694.

663

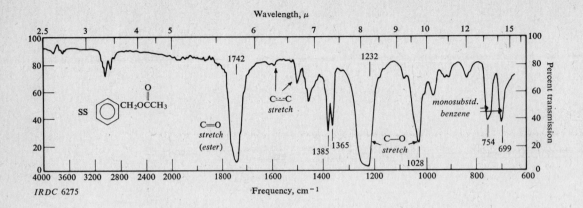

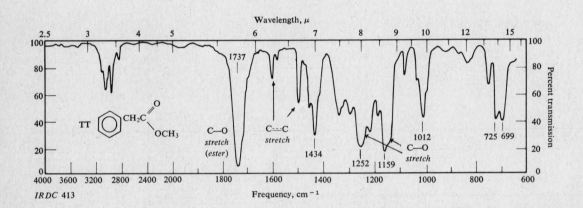

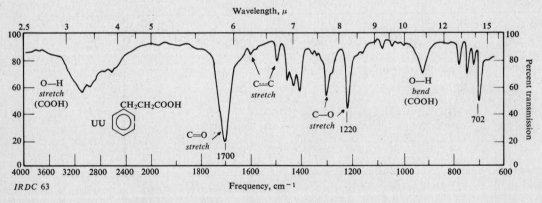

Figure 20.4. Infrared spectra for Problem 25, p. 694.

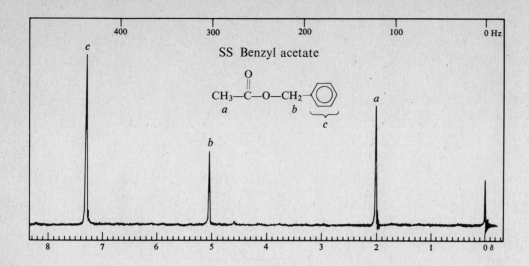

SS Benzyl acetate

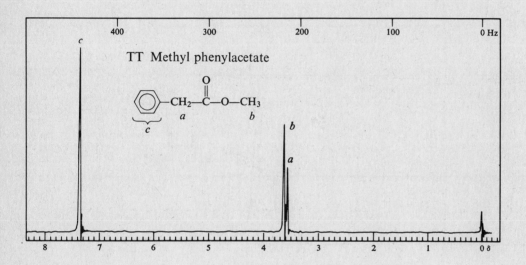

TT Methyl phenylacetate

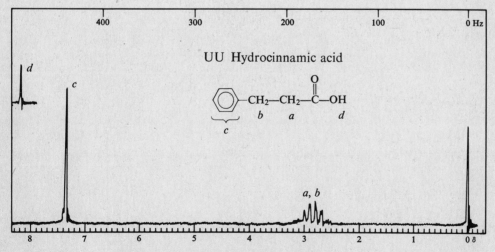

UU Hydrocinnamic acid

Figure 20.5. Nmr spectra for Problem 25, p. 694.

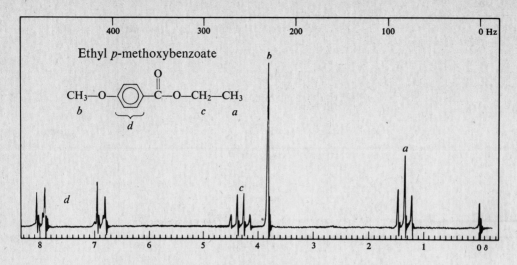

Figure 20.6. Nmr spectrum for Problem 26, p. 694.

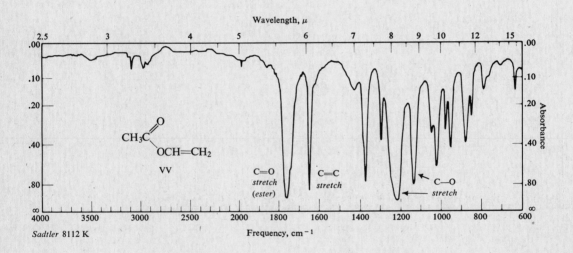

Sadtler 8112 K

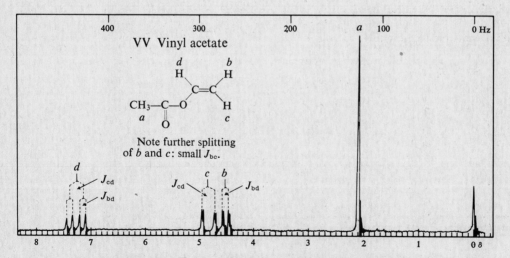

Figure 20.7. Infrared and nmr spectra for Problem 27, p. 694.

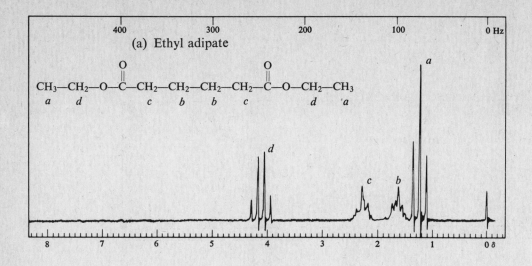

(a) Ethyl adipate

$$CH_3{-}CH_2{-}O{-}C{-}CH_2{-}CH_2{-}CH_2{-}CH_2{-}C{-}O{-}CH_2{-}CH_3$$
a   *d*   *c*   *b*   *b*   *c*   *d*   *a*

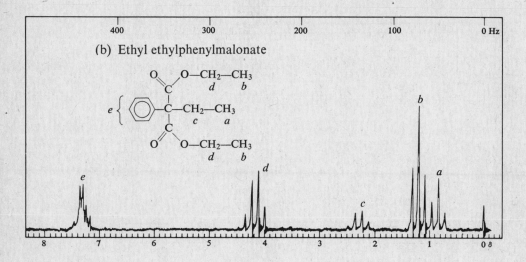

(b) Ethyl ethylphenylmalonate

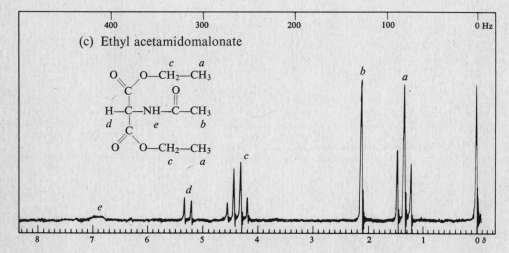

(c) Ethyl acetamidomalonate

Figure 20.8. Nmr spectra for Problem 28, p. 694.

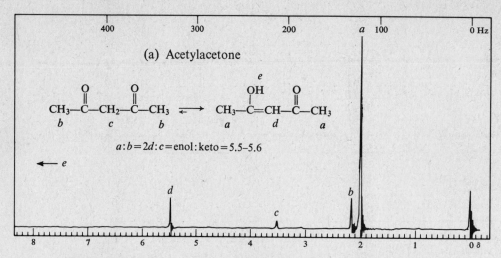

Figure 21.1(a). Nmr spectrum of acetylacetone (Problem 23, p. 725).

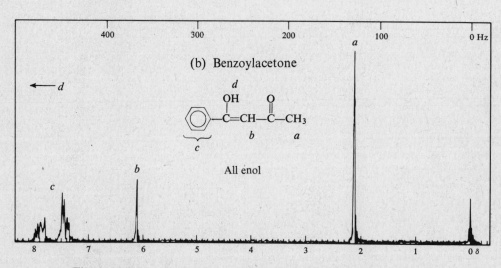

Figure 21.1(b). Nmr spectrum of benzoylacetone (Problem 23, p. 725).

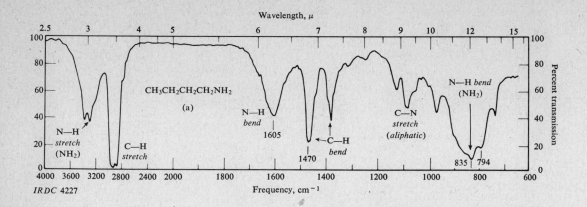

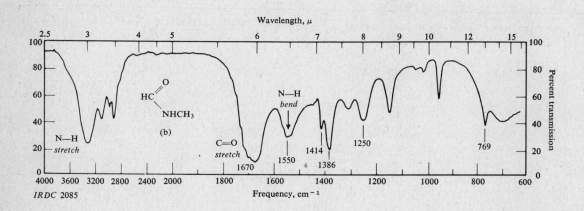

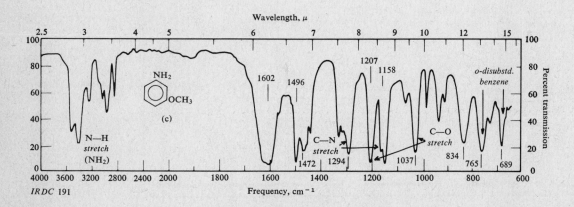

Figure 23.3. Infrared spectra for Problem 26, p. 782.

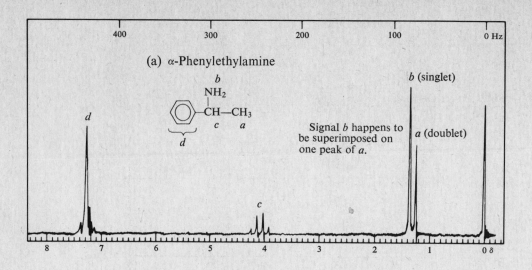

(a) α-Phenylethylamine

b (singlet)

a (doublet)

Signal *b* happens to be superimposed on one peak of *a*.

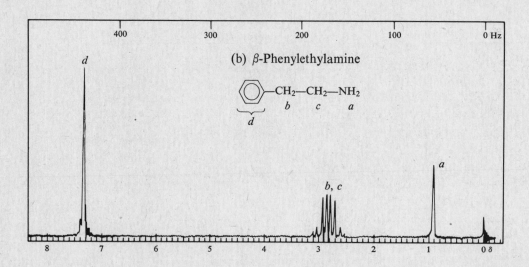

(b) β-Phenylethylamine

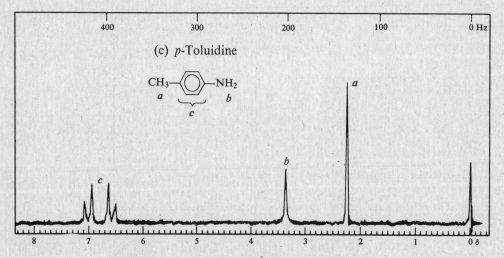

(c) *p*-Toluidine

Figure 23.4. Nmr spectra for Problem 27, p. 782.

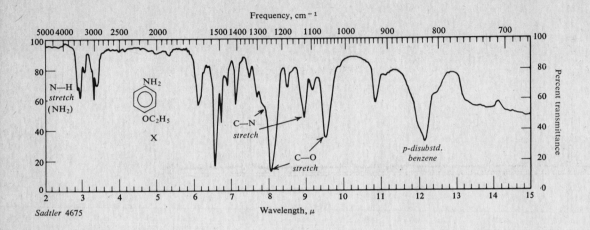

Sadtler 4675

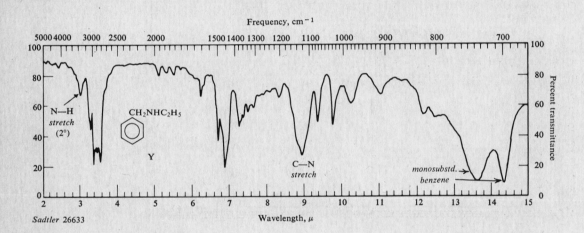

Sadtler 26633

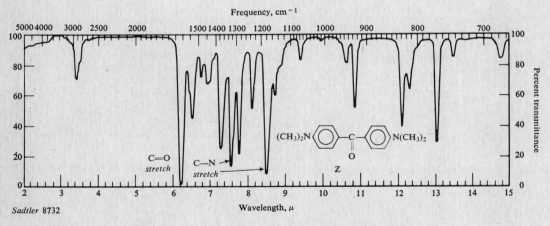

Sadtler 8732

Figure 23.5. Infrared spectra for Problem 28, p. 782.

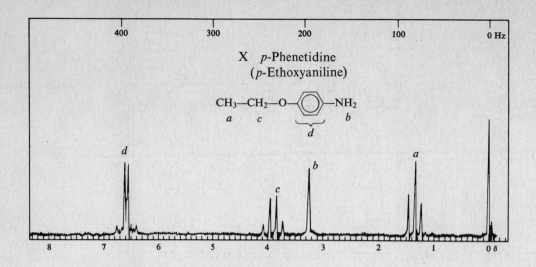

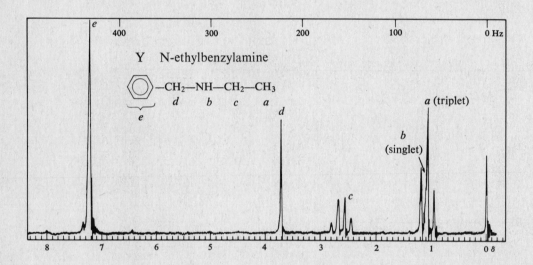

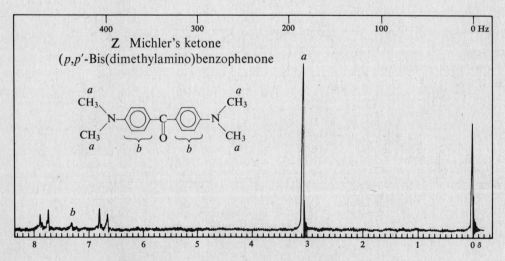

Figure 23.6. Nmr spectra for Problem 28, p. 782.

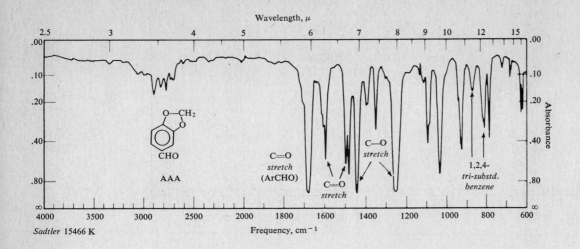

Wavelength, μ

O—H *stretch*

C=O *stretch* (ArCHO)

C=O *stretch*

C—O *stretch*

1,2,4-*tri-substd. benzene*

AAA

Sadtler 15466 K

Frequency, cm⁻¹

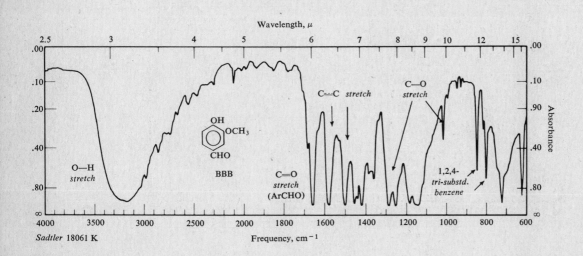

Wavelength, μ

O—H *stretch*

BBB

C=O *stretch* (ArCHO)

C=C *stretch*

C—O *stretch*

1,2,4-*tri-substd. benzene*

Sadtler 18061 K

Frequency, cm⁻¹

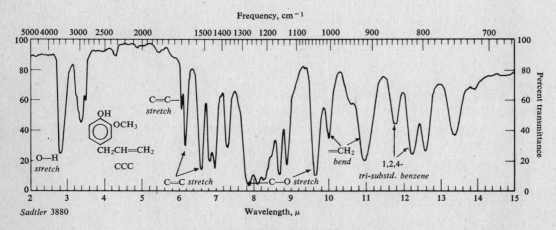

Frequency, cm⁻¹

O—H *stretch*

C=C *stretch*

C=C *stretch*

C—O *stretch*

=CH₂ *bend*

1,2,4-*tri-substd. benzene*

CCC

Sadtler 3880

Wavelength, μ

Figure 24.3. Infrared spectra for Problem 28, p. 811.

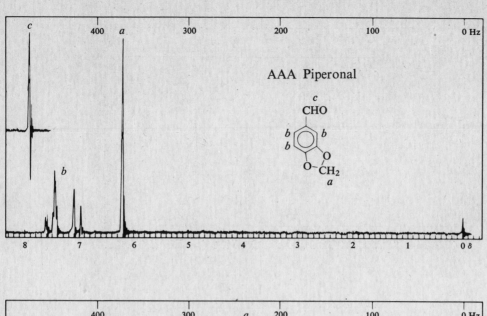

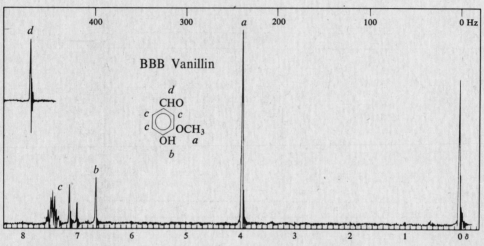

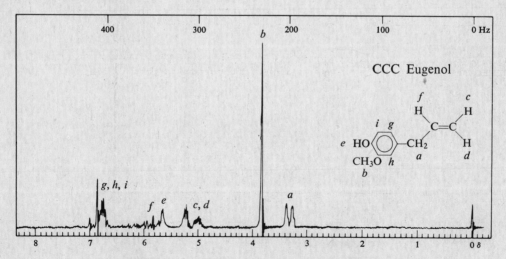

Figure 24.4. Nmr spectra for Problem 28, p. 811.

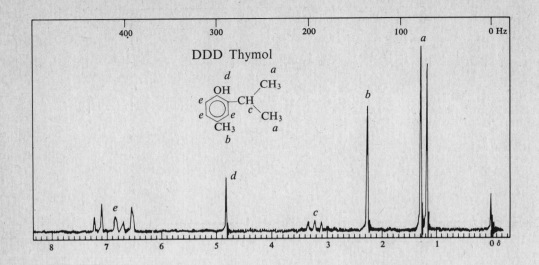

DDD Thymol

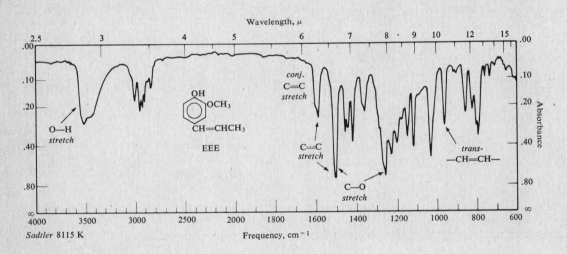

Sadtler 8115 K

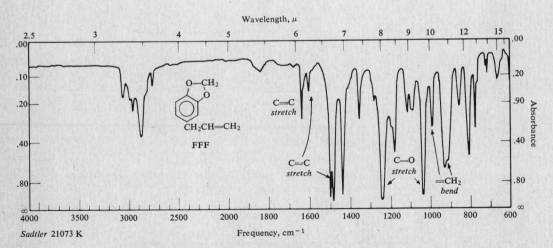

Sadtler 21073 K

Figure 24.6. Infrared spectra for Problem 28, p. 811.

28. (Chapter 28, page 811.)

AAA and BBB show the C=O stretching band at 1700 cm^{-1}, and must be piperonal and vanillin, the only carbonyl compounds of the set; this is confirmed by the far downfield —CHO proton absorption in their nmr spectra. Of the two, BBB shows O—H stretching at 3200 cm^{-1}, and hence is vanillin; AAA is piperonal. These assignments are confirmed by the nmr spectra: proton counting (relative to the —CHO) reveals —OCH$_3$ (plus —OH) in contrast to —OCH$_2$O—; the two oxygens of —OCH$_2$O— cause a much stronger downfield shift than the single oxygen of —OCH$_3$.

Of the remaining, CCC and EEE show O—H stretching, and must belong to the group of unassigned phenols: eugenol, isoeugenol, and thymol. Of the two, CCC shows the C=C stretching at 1650 cm^{-1} expected of an unconjugated C=C, and hence is eugenol; this is confirmed by the C—H out-of-plane bending bands at about 915 and 1000 cm^{-1}, characteristic of a terminal =CH$_2$. Compound EEE is, then, either isoeugenol or thymol. In contrast to thymol, isoeugenol has an unsaturated side chain, but C=C stretching in the conjugated system might well be hidden by the aromatic stretching band at 1600 cm^{-1}.

The nmr spectrum of DDD shows that it can be, of all seven possibilities, only thymol. The large doublet at δ 1.25 is too far upfield to be due to any allylic, vinylic, or alkoxy protons, and can be due only to the six β-protons of the isopropyl side chain of thymol. Inspection shows the entire spectrum to fit this structure neatly.

With thymol eliminated, EEE must now be isoeugenol. The shape of the 1600 cm^{-1} band hints at a hidden C=C absorption, and we see a band (about 965 cm^{-1}) where we would expect C—H bending for a *trans* —CH=CH—.

Finally, FFF shows no O—H band, and so must be, of the remaining possibilities, either safrole or anethole. The C=C stretching band at 1650 cm^{-1} indicates that the side chain is unconjugated, and hence that FFF is safrole; this is confirmed by the C—H bending bands (about 920 and 1000 cm^{-1}) expected of a terminal =CH$_2$.